AF782464

BIOGEOCHEMICAL DYNAMICS AT MAJOR RIVER-COASTAL INTERFACES

Linkages with Global Change

This volume provides a state-of-the-art summary of biogeochemical dynamics at major river-coastal interfaces for advanced students and researchers. River systems play an important role (via the carbon cycle) in the natural self-regulation of Earth's surface conditions by serving as a major sink for anthropogenic CO_2. Approximately 90 percent of global carbon burial occurs in ocean margins, with the majority of this thought to be buried in large delta-front estuaries (LDEs). This book provides information on how humans have altered carbon cycling, sediment dynamics, CO_2 budgets, wetland dynamics, and nutrients and trace element cycling at the land-margin interface. Many of the globally important LDEs are discussed across a range of latitudes, elevations, and climates in the drainage basin, coastal oceanographic setting, and nature and degree of human alteration. It is this breadth of examination that provides the reader with a comprehensive understanding of the overarching controls on major river biogeochemistry.

Thomas S. Bianchi is a Professor in the Department of Geological Sciences at the University of Florida, Gainesville, where he holds the Jon and Beverly Thompson Endowed Chair of Geological Sciences. His general areas of expertise are organic geochemistry, biogeochemical dynamics of aquatic food chains, carbon cycling in estuarine and coastal ecosystems, and biochemical markers of colloidal and particulate organic carbon. He has published more than 130 articles in refereed journals and 4 books, including *Biogeochemistry of Gulf Mexico Estuaries* (1999, lead co-editor with Pennock and Twilley), *Biogeochemistry of Estuaries* (2007), *Hypoxia in the Northern Gulf of Mexico* (2010, co-author with Dale et al.), and *Chemical Biomarkers in Aquatic Ecosystems* (2011, lead co-author with Canuel). In 2012, he was elected as a Fellow of the American Association for Advancement of Science.

Mead A. Allison is the Director of Physical Processes and Sediment Systems at The Water Institute of the Gulf in Baton Rouge, Louisiana, and a Professor of Earth and Environmental Sciences at Tulane University, New Orleans, Louisiana. His general areas of expertise are sedimentology of the continental margin, particle-reactive radioisotopes, seafloor mapping, geomorphic impact of cyclonic storms, and the impact of human alteration of coastal environments. He has worked in riverine, coastal, estuarine, and shelf systems around the world, with particular emphasis on the continental margins of the Mississippi-Atchafalaya, Amazon, and Ganges-Brahmaputra rivers. He has published more than 80 articles in refereed journals and is the primary author (with DeGaetano and Pasachoff) of the high-school-level textbook *Earth Sciences* (2008).

Wei-Jun Cai is a Professor in the School of Marine Science and Policy at the University of Delaware, Newark. Prior to this position, he was a Professor at the University of Georgia. He studies air-sea CO_2 exchange, carbon cycling, and ocean acidification in coastal waters and marine sediments and develops sensors for carbon cycle research. He has worked in coastal systems around the world, including the U.S. southeastern rivers and shelf, the Mississippi River plume and northern Gulf of Mexico shelf system, the South and East China Seas, and the Arctic Ocean. He has published more than 80 articles in refereed journals. Cai is currently Associate Editor for the journal *Marine Chemistry*. He has served on many national committees and is currently a member of the U.S. Carbon Cycle Science Steering Group.

To our families for their unending support and patience through the years.

"No man ever steps in the same river twice, for it's not the same river and he's not the same man."

– Heraclitus

BIOGEOCHEMICAL DYNAMICS AT MAJOR RIVER-COASTAL INTERFACES

Linkages with Global Change

Edited by

THOMAS S. BIANCHI

University of Florida, Gainesville

MEAD A. ALLISON

The Water Institute of the Gulf, Baton Rouge, and Tulane University, New Orleans

WEI-JUN CAI

University of Delaware

CAMBRIDGE
UNIVERSITY PRESS

CAMBRIDGE
UNIVERSITY PRESS

32 Avenue of the Americas, New York, NY 10013-2473, USA

Cambridge University Press is part of the University of Cambridge.

It furthers the University's mission by disseminating knowledge in the pursuit of education, learning, and research at the highest international levels of excellence.

www.cambridge.org
Information on this title: www.cambridge.org/9781107022577

© Cambridge University Press 2014

This publication is in copyright. Subject to statutory exception and to the provisions of relevant collective licensing agreements, no reproduction of any part may take place without the written permission of Cambridge University Press.

First published 2014

Printed in the United States of America

A catalog record for this publication is available from the British Library.

Library of Congress Cataloging in Publication Data
Biogeochemical dynamics at major river-coastal interfaces : linkages with global change / [edited by]
Thomas S. Bianchi, University of Florida, Gainesville, Mead A. Allison, The Water Institute of the Gulf, Baton Rouge, Wei-Jun Cai, University of Delaware.
pages cm
Includes index.
ISBN 978-1-107-02257-7 (hardback)
1. Biogeochemical cycles. 2. Estuarine ecology. I. Bianchi, Thomas S. II. Allison, Mead A. (Mead Ashton) III. Cai, Wei-Jun, 1960–
QH344.B525 2014
577′.14–dc23 2013013368

ISBN 978-1-107-02257-7 Hardback

Cambridge University Press has no responsibility for the persistence or accuracy of URLs for external or third-party Internet Web sites referred to in this publication and does not guarantee that any content on such Web sites is, or will remain, accurate or appropriate.

Contents

List of Contributors

M. A. Allison The Water Institute of the Gulf and Tulane University

D. M. Alongi Australian Institute of Marine Science

N. Bi College of Marine Geosciences, Ocean University of China

T. S. Bianchi Department of Geological Sciences, University of Florida

G. Billen CNRS/UPMC, UMR7619 Sisyphe

N. Blair Civil and Environmental Engineering and Earth and Planetary Sciences, Northwestern University

D. Bombar Ocean Sciences Department, University of California Santa Cruz

A. Borges University of Liège, Chemical Oceanography Unit, Institut de Physique

S. Bouillon Department of Analytical and Environmental Chemistry, Vrije Universiteit Brussel

W. P. Broussard III Institute for Coastal Ecology and Engineering, University of Louisiana Lafayette

W.-J. Cai The University of Delaware/School of Marine Science and Policy

J. Callens CNRS/UPMC, UMR7619 Sisyphe

S. Chakraborty School for Marine Science and Technology, University of Massachusetts, Dartmouth

C. T. Arthur Chen National Sun Yat-sen (Zhongshan) University, Marine Geology and Chemistry

N. Chen Department of Natural Sciences, University of Maryland Eastern Shore

D. R. Corbett Department of Geological Sciences, Institute of Coastal Science and Policy, East Carolina University

M. Dai State Key Laboratory of Marine Environmental Science, Xiamen University

J. W. Day, Jr. Department of Oceanography and Coastal Sciences, Louisiana State University

J. W. Dippner Leibniz Institute for Baltic Sea Research Warnemünde

S. Duan Earth System Science Interdisciplinary Center, University of Maryland

C. Duarte UWA Oceans Institute, The University of Western Australia (M470)

T. I. Eglinton Geological Institute, Department of Earth Sciences, ETH Zurich

G. Erkens Deltares, Utrecht, The Netherlands

C. France-Lanord CRPG UPR 2300 CNRS/INSU, Université de Lorraine

J. Gaillardet Institut de Physique du Globe de Paris, Sorbonne Paris Cité, Université Paris Diderot

V. Galy Woods Hole Oceanographic Institution

J. Gan Department of Mathematics and Division of Environment, Hong Kong University of Science and Technology

J. Garnier CNRS/UPMC, UMR7619 Sisyphe

M. Goñi College of Oceanic and Atmospheric Sciences, Oregon State University

S. L. Goodbred, Jr. Vanderbitt University, Nashville, TN

K. Gundersen Department of Marine Science, University of Southern Mississippi, Stennis Space Center

L. Guo School of Freshwater Sciences, University of Wisconsin

D. Nhu Hai Institute of Marine Geology and Geophysics, Vietnam Academy of Science and Technology

A. Han State Key Laboratory of Marine Environmental Science, Xiamen University

P. J. Harrison Atmosphere, Marine and Coastal Environment Program, Hong Kong University of Science and Technology

C. Hein Woods Hole Oceanographic Institution

P. J. Hernes Department of Land, Air and Water Resources, University of California

R. D. Hetland Department of Oceanography, Texas A&M University

R. M. Holmes Woods Hole Research Center

T. J. Hsu Center for Applied Coastal Research, Civil and Environmental Engineering, University of Delaware

G. Hunsinger Department of Geology and Geophysics, Yale University

A. Kolker Louisiana Universities Marine Consortium (LUMCON)

S. A. Kuehl Virginia Institute of Marine Science, Greta Point, VA

H. S. Kung Atmosphere, Marine and Coastal Environment Program, Hong Kong University of Science and Technology

Z. Lai School of Marine Sciences, Sun Yat-sen (Zhongshan) University

N. Ngoc Lam Institute of Marine Geology and Geophysics, Vietnam Academy of Science and Technology

E. L. Leithold Department of Marine, Earth, and Atmospheric Sciences, North Carolina State University

P. Liu Department of Marine, Earth, and Atmospheric Sciences, North Carolina State University

S. E. Lohrenz School for Marine Science and Technology, University of Massachusetts, Dartmouth

N. Loick-Wilde School of Biology, Georgia Institute of Technology

R. Macdonald Institute of Ocean Sciences

B. A. McKee Department of Marine Sciences, University of North Carolina

E. Meselhe Department of Civil Engineering, University of Louisiana Lafayette

H. Middelkoop Department of Physical Geography, Utrecht University

S. Mitra Geological Sciences, East Carolina University

W. Moufaddal National Institute of Oceanography and Fisheries (NIOF), Egypt

M. C. Murrell US EPA Gulf Ecology Division

C. A. Nittrouer School of Oceanography, Department of Earth and Space Sciences, and Quaternary Research Center, University of Washington

A. S. Ogston School of Oceanography, University of Washington

P. Passy CNRS/UPMC, UMR7619 Sisyphe

M. van der Perk Department of Physical Geography, Utrecht University

A. Ramanathan School of Environmental Science, Jawaharlal Nehru University

P. A. Raymond Yale School of Forestry and Environmental Studies

A. I. Robertson School of Plant Biology, The University of Western Australia

B. E. Rosenheim Department of Earth and Environmental Sciences, Tulane University

G. P. Shaffer Department of Biological Sciences, Southeastern Louisiana University

A. M. Shiller Department of Marine Science, University of Southern Mississippi, Stennis Space Center

M. Silvestre CNRS/UPMC, UMR7619 Sisyphe

R. G. M. Spencer Woods Hole Research Center

R. G. Striegl US Geological Survey

A. Stubbins Skidaway Institute of Oceanography, University of Georgia

S. E. Tank Department of Geography, York University

V. Thieu CNRS/UPMC, UMR7619 Sisyphe

J. M. Visser Institute for Coastal Ecology and Engineering, School of Geosciences, University of Louisiana Lafayette

M. Voss Leibniz Institute for Baltic Sea Research

J. P. Walsh Department of Geological Sciences, Institute of Coastal Science and Policy, East Carolina University

H. Wang College of Marine Geosciences, Ocean University of China

W. R. Woerner Department of Geosciences, Stony Brook University

Y. Wu State Key Laboratory of Estuarine and Coastal Research, East China Normal University

J. Xu Atmosphere, Marine and Coastal Environment Program, Hong Kong University of Science and Technology

Z. Yang College of Marine Geosciences, Ocean University of China

K. Yin School of Marine Sciences, Sun Yat-sen (Zhongshan) University

Z. Yin State Key Laboratory of Marine Environmental Science, Xiamen University

G. L. Zhang College of Chemistry and Chemical Engineering, Ocean University of China

J. Zhang State Key Laboratory of Estuarine and Coastal Research, East China Normal University

Z. Y. Zhu State Key Laboratory of Estuarine and Coastal Research, East China Normal University

A. R. Zimmerman Department of Geological Sciences, University of Florida

Preface

Approximately 87% of Earth's land surface is connected to the ocean by rivers. Over the past 60 years, increases in the human population have had severe, globally significant effects on large-river systems through enhanced fertilizer usage, damming, deforestation, and many other land-use changes. Many countries in the world are experiencing potable and agricultural water shortages, particularly in Asia, which contains 30% ($13{,}500 \times 10^9$ m^3 yr^{-1}) of the world's ($42{,}700 \times 10^9$ m^3 yr^{-1}) renewable water resources. The world's 25 largest rivers drain approximately half of the continental surface and transport approximately 50% of the freshwater and 40% of the particulate materials entering the ocean. Moreover, it has been estimated that 80% of the total organic carbon preserved in marine sediments occurs in "terrigenous-deltaic" or large delta-front estuaries (LDEs).

Organic carbon (OC) burial in marine sediments is only second to silicate weathering and subsequent carbonate precipitation as sink of atmospheric CO_2, and much of this occurs in deltaic regions of the world. Thus rivers play a vital role, delivering a large flux of OC from the continents to the oceans. Recent work has also documented global decreases in water and/or sediment discharge to the coastal ocean in numerous LDEs such as the Mississippi, Nile, Indus, Changjiang, and Huanghe systems. Although humans have increased riverine sediment transport within the continents through soil erosion by an estimated 2.3 ± 0.6 Pg yr^{-1}, the actual amount reaching the ocean has decreased by 1.4 ± 0.3 Pg yr^{-1}, mainly as a result of dams and reservoirs. These reductions play an important role in deltaic coastal retreat, where a large fraction of the human population lives, and which, due to their low elevation, are highly susceptible to rising sea levels. Consequently, there has been increased interest in understanding how the flux of materials from rivers to the ocean have been altered, including global community programs such as the International Geosphere Biosphere Programme (IGBP) and its major project, Land Ocean Interaction in the Coastal Zone (LOICZ). A number of investigations have demonstrated relationships between fisheries' yields and the high nutrient loads and freshwater inputs associated with LDEs. For example, a great fraction of the harvested secondary production in the Gulf of Mexico "fertile crescent" is derived from estuarine ecosystems, including areas on the shallow shelf influenced by the Mississippi-Atchafalaya river plumes, as has been found for other fisheries linked with plumes from rivers such as the Nile, Mekong, and the Changjiang, just to mention a few.

The coastal ocean is a dynamic region where the rivers, estuaries, ocean, land, and atmosphere interact. Although relatively small in area, this region, having 30% of the total net oceanic productivity, supports as much as 90% of the global fish catch. In particular, LDEs have historically played an important role in the advance of human civilizations (via trade, transportation, and food resources). This relationship between humans and rivers began some 5,000 years ago with the demands of hydraulic power in Mesopotamia, as well as in the Nile, Huanghe, and Indus valleys. It has been estimated that approximately 61% of the world population lives along the coastal boundary. By 2025, an estimated 75% of the world's population is expected to live in the coastal zone, with many of the remaining 25% living near major rivers. One of the most challenging issues concerning large river fluxes is to better understand the presumably major changes that they have undergone over the Anthropocene as a result of land-use changes (agriculture and urbanization) and river basin and delta alterations, and the resultant impact of these changes on the land-ocean material transfer term, both quantitatively and qualitatively.

Our main objectives in this book are to provide the reader with a comprehensive overview of what is known about the biogeochemical processes of the major LDEs around the world – the natural and anthropogenic factors that control and regulate them, in the western and eastern hemispheres. Thus we have divided this book into the following four sections: Section I. Introduction; Section II. Water and Sediment Dynamics from Source to Sink; Section III. Eastern Hemisphere Systems; and Section IV. Western Hemisphere Systems. In Section I we provide the reader with an overview of how we define these regions, why they are important to the global carbon cycle, and how climate change may be impacting these systems as they are changing rapidly in the Anthropocene. In Section II, the chapters primarily address the physical processes that determine how fluvial inputs of water, sediment, carbon, and nutrients are modified by tidal modulations, ocean wave incursion, wind-driven currents, estuarine circulation, and passage through wetlands and shallow water bodies and are then distributed along and across the upper continental margin. These chapters address how these processes control the fate of nutrients, sediments, and plume waters, including hydrodynamic sorting processes that control the fate of particulate and dissolved organic carbon sources (e.g., terrestrial, marine, and black). Section III, which focuses on LDEs in the Eastern Hemisphere, provides a comprehensive view of the biogeochemical dynamics of major rivers that drain the Himalayas, such as the Changjiang, Huanghe, Pearl, Ganges-Bramaputra, and Mekong, and empty in a range of coastal settings from tropical to temperate. These systems are being rapidly urbanized, owing to the highest rates of population growth in the world, and also are experiencing the most rapid rates of dam building; these factors are discussed in the context of comparing natural versus anthropogenic factors and how these processes are likely to be modulated by climate change. Finally, Section IV, focused on LDEs in the Western Hemisphere, is the largest section of the book, because there they have been the focus of considerably more biogeochemical research. Once again, the chapters provide the reader with an excellent overview of what is known about the biogeochemical dynamics of LDEs such as the Nile, Mississippi-Atchafalaya, Yukon, Congo, Reine, Seine, and many Arctic systems. We believe this is the most comprehensive presentation to date of how large-river systems are being altered globally due to human and climate change and also provides important information on the role of these systems in understanding the global carbon cycle.

Our special thanks go to the authors of each chapter, who supported our overall goal of providing the most comprehensive view of the biogeochemical cycling of the large-river LDEs across the globe. We would also like to thank the editors we worked with at Cambridge University Press, Amanda O'Connor, and Adrian Pereira at Aptara Inc., for their invaluable guidance through this effort.

Section I

Introduction

1

An introduction to the biogeochemistry of river-coastal systems

T. S. Bianchi, M. A. Allison, and W.-J. Cai

1. Introduction

The coastal zone is a dynamic region where the rivers, estuaries, ocean, land, and the atmosphere interact (Walsh 1988; Wollast 1998; Liu et al. 2009; Cai 2011; Canuel et al. 2012). Although the coastal zone, as defined by the recent Land-Ocean Interactions in the Coastal Zone (LOICZ) project, comprises less than 20% of the Earth's surface, it contains greater than 40% of the human population, borders 75% of the megacities in the world, and yields 90% of the global fisheries (Crossland et al. 2005). Rivers provide connectivity between terrestrial and ocean environments, the two largest global sinks for atmospheric CO_2; however, this connectivity often is greatly modified in the coastal zone. The world's 25 largest rivers drain nearly half of the Earth's land surface and account for approximately 40% of the fluvial sediments and 50% of the freshwater entering the ocean (Milliman and Meade 1983; Meade 1996; Vorosmarty and Peterson 2000). However, studies of small and medium-sized rivers form the basis of most of our knowledge about how rivers function (Milliman and Farnsworth 2011).

A relatively comprehensive understanding about the flux, characteristics, and processing of carbon exists for smaller rivers, relative to the world's largest rivers, which are often in remote areas that are logistically challenging to study. In most cases, large rivers have been examined in an "expedition" mode. Our knowledge of carbon in these rivers is based on only a few field studies or relies on data collected at the last hydrological station of a river above the reach of tides, which is often hundreds of kilometers upstream from the ocean interface. Nevertheless, comparisons that have been made between small "flashy" rivers (e.g., Eel River [United States]) and larger systems such as the Amazon have shown the relative importance of storage and transit time on the composition of particulate organic carbon (POC) delivered to the continental margin (Blair et al. 2004). For example, in the Eel River system, there is "mass wasting" in the watershed, which delivers bedrock and vascular vegetation (as indicated by the $\delta^{13}C$-depleted values) from soils – with minimal transformation during transport to the coast (Blair et al. 2003). Conversely, the Amazon, which has a much larger watershed, with extensive storage and processing time of organic carbon (OC) in lowland soils, allows for enough time for the OC signature to be altered before being deposited and buried on the continental margin. In the present, riverine delivery of organic matter (OM) to continental margins is the dominant means by which terrigenous production is preserved, influencing global biogeochemical cycles and the ocean's ability to sequester atmospheric CO_2. The main premise of this introductory chapter is that

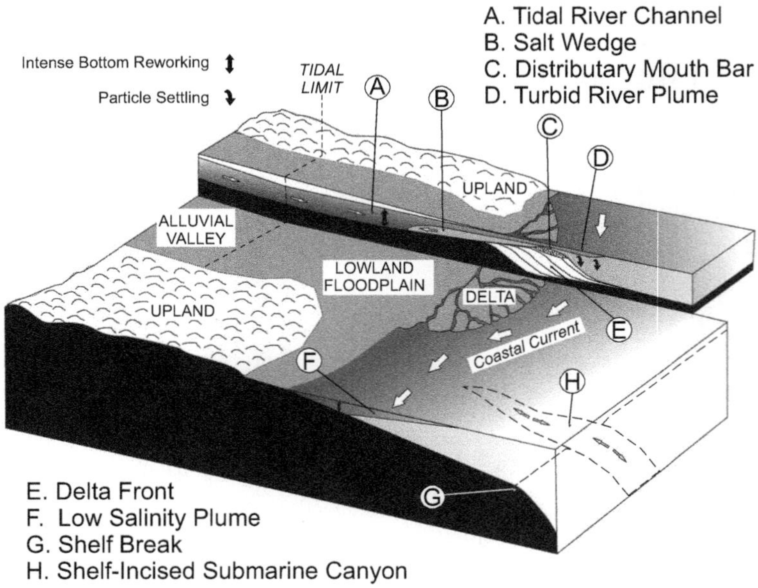

Figure 1.1. There are regional geomorphogical domains and associated sedimentary deposits with river-ocean mixing zone that have significant effects on biogeochemical processes in the coastal zone (modified from Bianchi and Allison, 2009).

the material flux from rivers has a profound impact on the biogeochemistry of the world's ocean and, through this river-ocean interaction, plays an important role in global biogeochemical cycles of elements such as carbon and nitrogen and in sediment delivery to the global ocean.

River systems and their directly affected coastal margin areas can be divided into three strongly interactive domains: (1) the drainage basin, extending from upland streams to the tidally influenced lower river; (2) the river-ocean mixing zone, extending from the tidally influenced freshwater portion of the river to the river mouth; and (3) the margin, which extends from the inner shelf, adjacent to the river mouth, to the shelf-slope break/canyon (Fig. 1.1). The river-ocean mixing zone, however, often extends to the shelf or further in the form of a low-salinity plume. The importance of large river source-to-sink systems to global OC burial (Hedges and Keil 1995) is evidenced by the sheer magnitude of material fluxes to the margin in these systems (Milliman and Farnsworth 2011). Our conceptual framework here is built on the idea that strong interactions and feedbacks between physical, geological, and biogeochemical processes control the net change in OC within the coastal zone (Fig. 1.1). For example, the position and intensity of the frontal zone between the freshwater outflow and marine water (controlled by the interaction of discharge, channel morphology, and tidal forcing) greatly influences the dominant sediment processes (e.g., flocculation, sorting, sinking, deposition, and resuspension) for a given time and space scale. These physical processes thus determine the microenvironment of particles, which in turn influences the rates and mechanisms of chemical alterations within the land-water interface. Understanding these complex interactions (and feedbacks) between processes is a critical first step toward building a predictive capability for estimating changes in OM inputs to the ocean under changing global conditions – such as climate changes that result in an increase/decrease in precipitation and runoff, or changes in the magnitude and/or phasing of annual river hydrographs due to natural or anthropogenic causes (Arora and Boer 2001). By better understanding the processing of OM within the river-ocean mixing zone, we will be more fully prepared to predict how the global carbon cycle, including CO_2 fluxes between various carbon reservoirs, will be affected by river responses to global change.

2. Riverine inputs of particulate and dissolved organic matter to the coastal ocean

Approximately 87% of Earth's land surface is connected to the ocean by rivers (Ludwig and Probst 1998). The annual flux of carbon from rivers to the ocean is approximately 1 Pg C (1 Pg $= 10^{15}$ g, or = 1 gigaton), which is of the same order as the global net ocean-atmosphere exchange of CO_2 (Mackenzie et al. 2004; Mackenzie and Lerman 2006 and references therein; Takahashi et al. 2009). Annually, rivers transport an estimated 20 Pg of fluvial sediments to the coastal margin (Meybeck 1982; Meade 1996). Associated with this sediment loading is an estimated 0.21 Pg of POC (Hedges and Keil 1995). Global estimates of the riverine flux of dissolved organic carbon (DOC) to the oceans range from about 0.25 to 0.36 Pg C y^{-1} (Meybeck 1981; Degens et al. 1991; Aitkenhead and McDowell 2000). Thus a terrestrial total DOC (TDOC) flux is about 0.5 Pg C y^{-1}, although a few others have suggested a much higher flux (see Cai 2011). In modern marine environments, riverine delivery of OM to continental margins is the dominant means by which terrigenous OM production is preserved (through burial), influencing global biogeochemical cycles and the ocean's ability to sequester atmospheric CO_2. However, there remains considerable uncertainty in our ability to adequately quantify carbon exchange from land to the coastal ocean and in our understanding of the processes influencing the fate of terrigenous carbon in coastal and deep sea sediments (Berner 1982; Sarmiento and Sundquist 1992; Hedges and Keil 1995; Schlünz and Schneider 2000; Galy et al. 2007, 2008; Hilton et al. 2010; Cai 2011; Galy and Eglinton 2011; Hilton et al. 2011).

2.1. The conundrum of missing OC

Berner (1989) suggested that most POC burial occurs in river deltas at a rate of 0.114 Pg C yr^{-1}. In contrast, little POC is buried in open ocean sediments (~0.006 Pg C yr^{-1}). Hedges and Keil (1995), however, suggested that POC burial in non-deltaic shelf sediments is equally as important as that in deltaic sediments (each ~0.070 Pg C yr^{-1}). Burdige (2007) further estimated an additional POC burial of 0.070 Pg C yr^{-1} in the 200- to 1,000-m depth range. In summary, among the 0.5 Pg C yr^{-1} of terrestrial POC export, no more than 25% to <50% (0.12 to 0.22 Pg C yr^{-1}) is found in marine sediments (also, isotopic signal of the POC suggests that they are only partially terrestrial). Therefore, one must conclude that most terrestrial POC is respired in sediments and water column via microbial decomposition and that the respiration occurs mostly in ocean margins, although we know little about the detailed pathway of terrestrial POC decomposition in ocean margins.

Interestingly, another conundrum in chemical oceanography over the past few decades has been that although the amount of DOC discharged by rivers can account for the renewal of DOC in the global ocean (every ~4,000–6,000 yr; Williams and Gordon 1970), riverine DOC, which is widely believed to be composed of TDOC, is recognized as accounting for only a small fraction of oceanic DOC (Meyers-Schulte and Hedges 1986; Hedges et al. 1997; Opsahl and Benner 1997). Interestingly, there is two times as much C that is delivered to inland waters (1.9 Pg C yr^{-1}) than is delivered to the ocean (0.9 Pg C yr^{-1}) – suggesting that in these systems, C is consumed more actively than previously thought. Moreover, recent estimates of CO_2 efflux from streams and rivers suggest that TDOC is not as recalcitrant as previously thought (Cole et al. 2007 and references therein).

One possibility is the "priming" of OM degradation, a process discovered by Lohnis (1926), who revealed that rates of soil humus mineralization were enhanced by the addition of fresh organic residues. Although the importance of photodegradation and bacterial consumption of TDOC in freshwaters have been widely investigated, the role of priming processes has been totally ignored. Therefore, we propose that a focus on priming will transform our understanding of carbon cycling in inland and coastal waters. Past work has shown that during transport through the freshwater-marine continuum, TDOM is continually exposed to solar irradiance. Here, photoreactions convert chromophoric dissolved OM (DOM) to dissolved inorganic carbon (DIC) in a globally significant component of the carbon cycle (e.g., Kieber et al. 1990; Mopper et al. 1991; Miller and Moran 1997; Xie et al. 2004). Recently, it was observed that at the mouth of the Congo River, which delivers 12.4 Tg DOC yr^{-1} to the ocean (5% of the global DOC export by rivers), much of TDOC is altered by photoreactions to produce carboxylic-rich alicyclic molecules (CRAM) (Stubbins et al. 2010) – highly recalcitrant and ubiquitous components of oceanic DOM (Hertkorn et al. 2006). Other work has shown that bacterial consumption of modified lignin derivatives can be responsible for as much as 30% of the carbon photooxidation in surface waters of the Mississippi River plume (Hernes and Benner 2003). Recently, Bianchi (2011) suggested that priming is likely important in carbon transformations in many aquatic ecosystems, because recent work has shown much higher TDOC consumption in inland and coastal waters than reported in the past (Cole et al. 2007). This change in perception has occurred due to more extensive and accurate measurements of CO_2 fluxes and TDOC in inland waters (e.g., rivers lakes), confirming that TDOC, from soils and plant litter, is being consumed more efficiently than previously reported. This greater apparent stability may in part be the result of priming, which to date has not been investigated. Although priming studies have been conducted extensively in soil systems, research in aquatic systems has lagged considerably. Although mention of priming, or in some cases "cometabolism" effects, can be found in the aquatic literature, the process has largely been supported by superficial or equivocal evidence. We posit here that this process is more important than previously recognized in understanding coastal OM processing.

3. Transport processes and controls on OM preservation

Controls on the abundance of terrestrial and algal-derived POC in river/coastal systems include losses from in situ bacterial and photochemical breakdown, flocculation/sinking, and deposition followed by diagenesis in bed sediments (Amon and Benner 1996; Opsahl and Benner 1998; McKee and Baskaran 1999; Guo and Santschi 2000; Mannino and Harvey 2000; Mitra et al. 2000; Benner and Opsahl 2001; Duan and Bianchi 2006). Losses of terrestrial POC in river-dominated continental margins such as the Fly (New Guinea), Huang He or Yellow River (China), Columbia (United States), and Mississippi indicate globally significant losses of fluvial POM in these highly dynamic regions.

Another possible control on the loss of terrestrially derived OM in rivers is through the mixing of algal and terrestrial OM, which is believed to promote oxidation of the more refractory terrestrial material through the action of cometabolism or co-oxidation (Canfield 1994). OC in the lower Mississippi River consists of POC and DOC derived from both terrestrial and phytoplankton sources (Benner and Opsahl 2001; Bianchi et al. 2002, 2004; Duan and Bianchi 2006). The relative importance of these different sources is generally controlled by discharge and suspended particulate matter (SPM), whereby more phytoplankton occur during low-flow stages, when there is more light availability – nutrient limitation is generally not a controlling factor here (Duan and Bianchi 2006). The higher the abundance of phytoplankton-derived POC, the greater the contribution of phytoplankton-derived DOC will be during these low-flow periods (Bianchi et al. 2004). The influence of marine-derived phytoplankton also varies proportionally with the intensity of the frontal zone exchange processes – primarily as river discharge changes seasonally.

Coagulation processes are affected by factors such as ionic strength, particle size and composition, and the concentration of DOM – all of which show sharp concentration gradients in coastal zones where the salt wedge migrates between the lower river and the shelf. The abundance of bacterioplankton and phytoplankton can also affect rates of flocculation by producing polymers (e.g., mucopolysaccharides), to which particles can adhere (van Loosdrecht et al. 1990). These flocs can also selectively scavenge components of the DOM pool as they move through the water column, thereby altering the composition of DOM before it is released to the coastal ocean. Although the complex interaction of salinity, organic content, suspended particle concentration, and turbulence that determines floc size remains unresolved (Geyer et al. 2004), recent results suggest that flocs may form or reform rapidly in the tidal freshwater zone of rivers and have settling rates that could allow the riverine layer to be the source of the ephemeral mud deposits below the salt wedge (Galler and Allison 2008). Observations by Eisma (1996), Milligan and Hill (1998), and Hill et al. (2001) in coastal environments suggests that floc size is independent of turbulence at low to moderate energies and then decreases abruptly at higher turbulences.

The importance of large river source-to-sink systems to global OC burial (Hedges and Keil 1995) is evidenced by the sheer magnitude of material fluxes to the margin in these systems. Approximately 80% of the total OC preserved in marine sediments occurs in "terrigenous-deltaic" regions near river mouths (Romankevich 1984; Berner 1989). Bianchi and Allison (2009) posited that large river delta-front estuaries (LDE) are both "drivers" and "recorders" of natural and anthropogenic environmental change. However, despite the importance of these environments, there remains a fundamental lack of understanding about OC dynamics operating within these regions and how this influences the magnitude and character of OC exported offshore. This lack of understanding is primarily due to the high heterogeneity in the sources of OC, which include (1) primary production by phytoplankton, and (2) discharge of terrestrially derived OC (OC_{terr}) by rivers, as well as the complex architecture of margin environments influencing pathways and transport rates along and cross-margin.

Typically, when sediment and OC is transported across the shelf break to the slope, there is only modest early diagenesis taking place in the first few tens of centimeters below the sediment-water interface and have oxygen penetration depths of several centimeters that tend to increase with increasing water depth (e.g., Reimers et al. 1992; Cai and Sayles 1996). This reflects increased oxygen exposure time and extent of organic matter degradation (e.g., Hartnett et al. 1998; Hedges et al. 1999).

Notable exceptions are sediments found beneath highly productive overlying waters that have a high rain rate of POC, resulting in rapid and extensive diagenesis near the sediment-water interface, where formation of high concentrations of authigenic minerals is commonly observed (e.g., Rowe and Howarth 1985; Hedges and Keil 1995; Arthur et al. 1998).

There is now clear evidence from measurements of diagenetic processes that relatively rapid down-slope transport of OM, often associated with submarine canyons, can lead to significantly greater benthic metabolism and associated diagenesis than can be explained by POC input from overlying waters (e.g., Jahnke et al. 1990; DeMaster et al. 1994; Jahnke and Jahnke 2000). Studies such as those by Silverberg et al. (2000) and Arnosti and Holmer (2003) have also demonstrated that the intensity of diagenetic activity is often not well predicted by sediment total organic carbon (TOC) content and characteristics in continental margin sediments. These earlier studies provide sound examples of the importance of measuring diagenetic processes in association with the study of OM transport and deposition in slope environments. It has been shown that labile OM, likely produced by in situ diatom production in the Mississippi River plume, is rapidly transported to the Mississippi Canyon (Bianchi et al. 2006; Waterson and Canuel 2008; Sampere et al. 2008, 2011). Preliminary results suggest that productive LDEs, such as the Mississippi, may be important conduits for transporting fixed carbon from highly productive plume waters on the shelf to deeper benthic communities. Observations of an active benthic community within the Mississippi Canyon provide further support for this view (Bianchi et al. 2006). The pioneering work of Aller and his associates (e.g., Aller et al. 1996; Aller 1998; Aller and Blair 2006; Blair and Aller 2012) has shown the importance of mobile muds as an "incinerator" of OC and, in the proposed diagenetic studies, mobile muds are also inherently linked with the transport of OC offshore. Complementing this body of work, Arzayus and Canuel (2004) showed that degradation of more refractory OM was facilitated in mobile, physically mixed estuarine sediments.

Canyons are deep incisions of the continental shelf and slope. Despite their prevalence on both active and passive continental margins (Harris and Whiteway 2011), they remain understudied regions with respect to OC dynamics. One reason for this paucity of data is their complex wall and floor terrain, which results in high spatial variability and makes detailed OC studies based on only a few sediment cores difficult. Despite a rarity of in-depth OC studies, canyons are known to be (1) characterized by high faunal biomass, (2) important conduits for delivery and burial of OC in the oceans, and (3) "fast-track corridors" for sediment and carbon transported from land to the deep sea (Weaver et al. 2000; Liu and Lin 2004; Kiriakoulakis et al. 2011). The proximity and orientation of river mouth–canyon systems relative to alongshore sediment transport pathways lead to differences in the capture of riverine sediment and OC between large river source-to-sink systems (McKee et al. 2004; Walsh and Nittrouer 2009).

The OC and sediment connectivity between modern coastal systems and the deep sea has been the subject of considerable attention in recent years. For example, a primary goal of the National Science Foundation MARGINS Source-to-Sink program was to develop a quantitative understanding of sediment dispersal systems, including OC export from river mouth and shelf regions. Many large river margins are believed to export a relatively small volume of river-derived particulate matter seaward of the shelf break either due to (1) their location on wide, passive continental margins where deltaic sedimentation is confined to the inner shelf, or (2) because they are located where along-shelf–dominant coastal currents are present (McKee et al. 2004). This belief has led to the prevailing view that sediments reach the continental slope through hemipelagic means (Shanmugam and Moiola 1985; Stow et al. 1985). However, it has been increasingly recognized that some large river margins (e.g., Sepik, Congo, Ganges-Brahmaputra, Eel, Rhone) are characterized by direct export of large volumes of sediment and OC to the lower continental margin due to either a (1) narrow (active continental margin) shelf (Kineke et al. 2000; Mullenbach and Nittrouer 2006); (2) progradation of the deltaic clinoform in the late Holocene near to the shelf edge (Coleman et al. 1998); (3) density-driven, cold shelf water (winter) advection (Puig and Palanques 1998); (4) landward incision of the associated submarine canyon so that it intercepts along-shelf transport pathways (Johnson et al. 2001; Michels et al. 2003); or (5) large-scale sediment remobilization from hurricane activity (Sampere et al. 2008, 2011).

4. CO_2 fluxes in coastal waters

Globally, estuaries are a major source of CO_2 to the atmosphere. In general, CO_2 degassing flux is more intensive in lower latitude estuaries than in high latitudes. Earlier estimates suggested that estuarine CO_2 degassing flux was as high as 0.5 Pg C yr^{-1}, equivalent to the annual riverine TOC flux (Borges 2005). More recently, it is generally agreed that the likely flux is about 0.25 Pg C yr^{-1}, even though both spatial and temporary resolutions of the available field measurements are low and thus uncertainty of the global flux synthesis is still high (Cai 2011 and references therein). In addition, the exact sources supporting this CO_2 degassing flux are unknown but include microbial decomposition of riverine terrestrial materials, inputs of high CO_2 soil and groundwater in the river basin, and inputs of both CO_2 and OC (and the subsequent respiration) from the coastal wetlands. Cai (2011) postulated that the latter is the major source supporting CO_2 degassing in low- to middle-latitude estuaries.

Beyond estuaries in the continental shelves, CO_2 is generally supersaturated with respect to the atmosphere in proximal areas (i.e., inshore and near-shore areas and the inner shelf, generally to about 20 m) but undersaturated in distal areas (middle and outer shelves to about 200 m) (Cai et al. 2006). In addition, although shelves located in middle and high latitudes have lower surface water pCO_2, those in tropical and subtropical areas have higher pCO_2. Our knowledge of CO_2 distribution and flux in the continental shelves has improved rapidly in the past decade, although further improvement is still needed (Borges 2005; Cai et al. 2006; Chen and Borges 2009). The current consensus is that net global continental shelf sea-air CO_2 flux is an uptake of 0.25 Tg C yr^{-1} from the atmosphere (Cai 2011). Synthesis also revealed a striking latitudinal contrast in shelf sea-air CO_2 flux. Present-day shelves located between 30° and 90° are sinks of atmospheric CO_2 with a total uptake flux of 0.35 Pg C yr^{-1}, whereas those located from 0 to 30° are sources of CO_2 to the atmosphere with a total release flux of 0.10 Pg C yr^{-1} (Cai et al. 2006).

5. Possible links between coastal water CO_2 flux and riverine loadings

Rivers bring both inorganic nutrients and OC to the coastal zone and thus can influence metabolic state, carbon cycle, and air-sea CO_2 flux differently. In principle, terrestrial TOC loading drives the oceanic systems toward net heterotrophy, as it is eventually decomposed to CO_2. However, dissolved inorganic nitrogen (DIN) flux should lead coastal waters to net autotrophy because the use of DIN in the ocean by phytoplankton leads to net synthesis of OC and net removal of CO_2. Globally, gross terrestrial "heterotrophic loading" greatly exceeds that of the "autotrophic loading" (Table 1–1, Cai 2011). However, in a few mid-latitude large rivers near population centers, for example, the Mississippi River, such loading ratios are more toward net autotrophy because of human-induced nutrient loadings (Cai and Lohrenz 2010). Most importantly, the relative terrestrial loadings have been altered in the past, in particular, recently from Holocene to Anthropocene. Historically, TOC loading had been increased over many centuries as humans expanded their occupation of the earth surface, but has been decreased substantially in the past century as humans built dams and reservoirs (Syvitski et al. 2003, 2005). Most noticeably, inorganic nutrient loadings, which have been increased greatly in many rivers, can potentially alter the trophic state of the affected coastal zones. However, the net role of this increased nutrient loading is unclear, as eventually most nitrogen nutrient is denitrified in coastal waters (Seitzinger et al. 2006).

It is also noted that rivers deliver freshwater, carbon, and sediments to the ocean unevenly (Borges 2005; Cai et al. 2006; Cai 2011). About two thirds of the terrestrial OC is supplied to lower latitude coastal oceans (Ludwig et al. 1996). In addition, it appears that OC is better preserved in cold high-latitude river delta sediments than in warm low-latitude marine sediments (de Haas et al. 2002). For example, terrestrial OC burial is ~50% in the Mackenzie River delta but less than 20% in the Amazon (de Haas et al. 2002; Showers and Angle 1986) and the Mississippi (Cai and Lohrenz 2010) systems. It is thus not unreasonable to speculate that lower latitude coastal oceans would release more CO_2. If 75% of the OC delivered to low-latitude coastal oceans is decomposed there, then this alone would generate a CO_2 flux of 0.25 Pg C yr^{-1} (note that two thirds of the global OC is delivered to the low-latitude margins; e.g., 0.5 Pg C yr^{-1} × 2/3 × 75%; see also Keil et al. [1997] for a similar estimate). Therefore, the observed air-sea CO_2 flux pattern in coastal estuaries and shelves are consistent with the nature of river inputs. However, with climate change resulting in greater flooding in some coastal regions, for example, coastal river plume regions that are typically net sinks for CO_2 (see Cai 2011; Cai et al. Chapter 7 of this book), recent work has shown that a large flooding event, such the flood of 2011 into the northern Gulf of Mexico, can be changed for a short period of time into a net CO_2 source with high loading and bacterial consumption of TDOC (Bianchi et al. 2013).

6. Directions for future research

In the coming decades, predicted changes in global climate are likely to have a dramatic impact on sediment and OC delivery of large rivers to the global ocean. Alteration to the magnitude and character of materials arriving from the drainage basin as a result of changing precipitation patterns, denudation rates, and land use are likely to even more strongly override biogeochemical processes in the coastal zone. Although existing studies outlined in the following chapters suggest that these changes to the OC budget of the oceans and atmosphere will be profound, a new generation of process studies will be necessary to quantify these impacts. By necessity, these studies will have to be coupled with a more comprehensive monitoring of key parameters in the lower rivers, coastal zone, and margin far-range areas.

One extreme example of the two central issues discussed in this book – land-ocean connectivity and impact of climate changes – can be found in Arctic rivers. First, although the Arctic Ocean has only 1% of global ocean volume, it receives about 10% of the global freshwater input and associated suspended and dissolved materials, and thus the impact of terrestrial flux on the ocean is expected to be intensive. Second, the Arctic Ocean and adjacent land areas are experiencing arguably the most intensive climate warming in the earth system. Projections of anthropogenic (greenhouse gas emission) global warming by 2100 suggest that the largest increases (3–6°C) will take place in the highest latitudes, particularly in the Arctic (Arctic Climate Impact Assessment, 2005, Intergovernmental Panel on Climate Change, 2007). These reports also suggest this change will be coupled with an increase in precipitation, resulting in a warmer and wetter Arctic in both winter and summer, most notably over land. Evidence of this shift began in the latter half of the twentieth century (approximately the time span of the instrumented record), with increased melting of permafrost and glacial ice, decreasing summer sea ice extent, increasing surface air temperatures, increased coastal erosion, and changing ocean circulation (Guo and Macdonald 2006; McGuire et al. 2009). For example, it was shown that coastal erosion and rivers bring to the East Siberian Arctic shelf an integrated signal of terrestrial OM released from thawing permafrost (Anderson et al. 2011).

In the lower and middle-latitude coastal zones, changes in carbon pools and key biogeochemical processes associated with sea level rise and loss of coastal wetlands are expected to be a more important and focused research area in the coming years. It has been known that estuaries are a strong source of CO_2 to the atmosphere. Although the exact CO_2 degassing flux is uncertain at approximately 0.2 to 0.5 Pg C yr^{-1} (Chen and Borges 2009), this flux is in the same order of magnitude of riverine terrestrial OC export. Thus it presents an important question regarding the fate of terrestrial carbon – whether terrestrial OC is largely respired in estuaries or, as the community largely believes, in ocean margins (Hedges and Kiel 1995). On the basis of inorganic carbon flux studies in coastal marsh-surrounded estuaries in Georgia, United States, Cai (2011) suggested that estuarine CO_2 degassing is supported by lateral transport for CO_2 and OC from coastal wetlands, whereas riverine POC and DOC largely bypass estuarine interface and contribute to ocean margin respiration. The relative importance of various carbon sources and sinks and the exact routes or processes of carbon transport and transformation are far from clear in the river-coast-ocean continuum and require much intensive research in the context of global climate changes and human impacts.

References

Arctic Climate Impact Assessment. 2005. Arctic Climate Impact Assessment. New York: Cambridge University Press. http://www.acia.uaf.edu.

Aitkenhead, J.A., and W.H. McDowell. 2000. Soil C:N ratio as a predictor of annual riverine DOC flux at local and global scales. *Global Biogeochem. Cycles* 14: 127–138.

Aller, R.C. 1998. Mobile deltaic and continental shelf muds as suboxic, fluidized bed reactors. *Mar. Chem.* 61: 143–155.

Aller, R.C., and N.E. Blair. 2006. Carbon remineralization in the Amazon – Guianas mobile mudbelt: a sedimentary incinerator. *Cont. Shelf Res.* 26: 2241–2259.

Aller, R.C., N.E. Blair, Q. Xia, and P.D. Rude. 1996. Remineralization rates, recycling and storage of carbon in Amazon shelf sediments. *Cont. Shelf Res.* 16: 753–786.

Amon, R.M.W., and R. Benner. 1996. Photochemical and microbial consumption of dissolved organic carbon and dissolved oxygen in the Amazon River system. *Geochim. Cosmochim. Acta* 60: 1783–1792.

Anderson, L.G., G. Björk, S. Jutterström, I. Pipko, N. Shakhova, I. Semiletov, and I. Wåhlström. 2011. East Siberian Sea, an Arctic region of very high biogeochemical activity. *Biogeosciences* 8: 1745–1754.

Arnosti, C., and M. Holmer. 2003. Carbon cycling in a continental margin sediment: contrasts between organic matter characteristics and remineralization rates and pathways. *Estuar. Coast. Shelf Sci.* 58: 197–208.

Arora, V.K., and G.J. Boer. 2001. Effect of simulated climate change on the hydrology of major river basins. *J. Geophys. Res.* 106: 3335–3348.

Arthur, M.A., G.B. Coltharp, and D.L. Brown. 1998. Effects of best management practices on forest stream water quality in eastern Kentucky. *J. Am. Water Res. Assoc.* 34: 481–495.

Arzayus, K.M., and E.A. Canuel. 2004. Organic matter degradation in sediments of the York River estuary: Effects of biological vs. physical mixing. *Geochim. Cosmochim. Acta* 69: 455–463.

Benner, R., and S. Opsahl. 2001. Molecular indicators of the sources and transformations of dissolved organic matter in the Mississippi River plume. *Org. Geochem.* 32: 597–611.

Berner, R.A. 1982. Burial of organic carbon and pyrite sulfur in the modern ocean: its geochemical and environmental significance. *Am. J. Sci.* 282: 451–473.

Berner, R.A. 1989. Biogeochemical cycles of carbon and sulfur and their effect on atmospheric oxygen over Phanerozoic time. *Global Planet. Change* 75: 97–122.

Bianchi, T.S. 2011. The role of terrestrially derived organic carbon in the coastal ocean: A changing paradigm and the priming effect. *Proc. Natl. Acad. Sci.* 108: 19473–19481.

Bianchi, T.S., and M.A. Allison. 2009. Large-river delta-front estuaries as natural "recorders" of global environmental change. *Proc. Natl. Acad. Sci.* 106: 8085–8092.

Bianchi, T.S., T. Filley, K. Dria, and P. Hatcher. 2004. Temporal variability in sources of dissolved organic carbon in the lower Mississippi River. *Geochim. Cosmochim. Acta* 68: 959–967.

Bianchi, T.S., F. Garcia-Tigreros, S. Yvon-Lewis, M. Shields, H.J. Mills, D. Butman, C. Osburn, P. Raymond, C. Shank, S. F. DiMarco, N. Walker, B. Reese, R. Mullins, A. Quigg, G.R. Aiken, and E.L. Grossman. 2013. Enhanced transfer of terrestrially-derived carbon to the atmosphere in a flooding event. *Geophys. Res. Lett.* 40: 1–7.

Bianchi, T.S., S. Mitra, and B.A. McKee. 2002. Sources of terrestrially-derived organic carbon in lower Mississippi River and Louisiana shelf sediments: implications for differential sedimentation and transport at the coastal margin. *Mar. Chem.* 77: 211–223.

Bianchi, T.S., T. Sampere, M. Allison, E.A. Canuel, B.A. McKee, S. Wakeham, and B. Waterson. 2006. Rapid export of organic matter to the Mississippi Canyon. *Eos* 87: 565, 572–573.

Blair, N.E., and R.C. Aller. 2012. The fate of terrestrial organic carbon in the marine environment. *Ann. Rev. Mar. Sci.* 4: 401–423.

Blair, N.E., E.I. Leithold, and R.C. Aller. 2004. From bedrock to burial: the evolution of particulate organic carbon across coupled watershed-continental margin systems. *Mar. Chem.* 92: 141–156.

Blair, N.E., E.I. Leithold, S.T. Ford, K.A. Peeler, J.C. Holmes, and D.W. Perkey. 2003. The persistence of memory: the fate of ancient sedimentary organic carbon in a modern sedimentary system. *Geochim. Cosmochim. Acta* 67: 63–73.

Borges, A.V. 2005. Do we have enough pieces of the jigsaw to integrate CO_2 fluxes in the coastal ocean? *Estuaries* 28: 3–27.

Beusen, A.H.W., A.L.M. Dekkers, A.F. Bouwman, W. Ludwig, and J. Harrison. 2005. Estimation of global river transport of sediments and associated particulate C, N, and P, Global Biogeochem. *Cycles* 19: GB4S05.

Burdige, D.J. 2007. Preservation of organic matter in marine sediments: controls, mechanisms, and an imbalance in sediment organic carbon budgets? *Chem. Rev.* 107: 467–485.

Cai, W.-J. 2011. Estuarine and coastal ocean carbon paradox: CO_2 sinks or sites of terrestrial carbon incineration? *Ann. Rev. Mar. Sci.* 3: 123–145.

Cai, W.-J., M. Dai, and Y. Wang. 2006. Air-sea exchange of carbon dioxide in ocean margins: a province-based synthesis. *Geophys. Res. Lett.* 33: L12603.

Cai, W.-J., and S. Lohrenz. 2010. Carbon, nitrogen, and phosphorus fluxes from the Mississippi River and the transformation and fate of biological elements in the river plume and the adjacent margin, In: Liu, K.-K., Atkinson, L., Quiñones, R., Talaue-McManus, L. (eds.), Carbon and Nutrient Fluxes in Continental Margins: A Global Synthesis. Berlin, Germany: Springer-Verlag.

Cai, W.J., and F.L. Sayles. 1996. Oxygen penetration depths and fluxes in marine sediments. *Mar. Chem.* 52: 123–131.

Canfield, D.E. 1994. Factors influencing organic carbon preservation in marine sediments. *Chem. Geol.* 114: 315–329.

Canuel, E.A., S.S. Cammer, H.A. McIntosh, and C.R. Pondell. 2012. Climate change impacts on the organic carbon cycle at the land-ocean interface. *Ann. Rev. Earth Planet. Sci.* 40: 685–711.

Chen, C.-T.A., and A.V. Borges. 2009. Reconciling opposing views on carbon cycling in the coastal ocean: continental shelves as sinks and near-shore ecosystems as sources of atmospheric CO_2. *Deep-Sea Res. Part II Top. Stud. Oceanogr.* 56: 578–590.

Cole, J.J., Y.T. Prairie, N.F. Caraco, W.H. McDowell, L.J. Tranvik, R.G. Striegl, C.M. Duarte, P. Kortelainen, J.A. Downing, J.J. Middleburg, and J. Melakck. 2007. Plumbing the global carbon cycle: integrating inland waters into the terrestrial carbon budget. *Ecosystems* 10: 171–184.

Coleman, D.C., D.A. Crossley Jr., M.H. Beare, and P.E. Hendrix. 1988. Interactions of organisms at root/soil and litter/soil interfaces in terrestrial ecosystems. *Agr. Ecosyst. Environ.* 24: 117–134.

Crossland, C.J., H.H. Kremer, H.J. Lindeboom, J.I. Marshall Crossland, and M.D. Le Tissier (eds.). 2005. Coastal Fluxes in the Anthropocene. Berlin, Germany: Springer-Verglag.

Degens, E.T., S. Kempe and J.E. Richey (eds.). 1991. Biogeochemistry of Major Rivers. New York: Wiley and Sons.

de Haas, H., T.C. Evan Weering, and H. de Stieger, H. 2002. Organic carbon in shelf seas: sinks or sources, processes and products. *Cont. Shelf Res.* 22: 691–717.

DeMaster, D.J., R.H. Pope, L.A. Levin, and N.E. Blair. 1994. Biological mixing intensity and rates of organic carbon accumulation in North Carolina slope sediments. *Deep-Sea Res.* 41: 735–753.

Duan, S.W., and T.S. Bianchi. 2006. Seasonal changes in the abundance and composition of plant pigments in particulate organic carbon in the lower Mississippi and Pearl Rivers (USA). *Estuar. Coasts* 29: 427–442.

Eisma, D. 1996. Flocculation and deflocculation of suspended matter in estuaries. *Nether. J. Sea Res.* 20: 183–199.

Galler, J.J. and M.A. Allison. 2008. Estuarine controls on fine-grained sediment storage in the lower Mississippi and Atchafalaya Rivers. *Geol. Soc. Am. Bull.* 120: 386–398.

Galy, V., and T.I. Eglinton. 2011. Protracted storage of biospheric carbon in the Ganges-Brahmaputra basin. *Nat. Geosci.* 4: 843–847.

Galy, V., J. Bouchez, and C. France-Lanord. 2007. Determination of total organic carbon content and $\delta^{13}C$ in carbonate rich detrital sediments. *Geostand. Geoanal. Res.* 31: 199–207.

Galy, V., C. France-Lanord, and B. Lartiges. 2008. Loading and fate of particulate organic carbon from the Himalaya to the Ganga-Brahmaputra delta. *Geochim. Cosmochim. Acta* 72: 1767–1787.

Geyer, W.R, P.S. Hill, and G.C. Kineke. 2004. The transport, transformation and dispersal of sediment by buoyant coastal flows. *Cont. Shelf Res.* 24: 927–949.

Guo, L., and R.W. Macdonald. 2006. Sources and transport of terrigenous organic matter in the upper Yukon River: evidence from isotope (d^{13}C, D^{14}C, and d^{15}N) composition of dissolved, colloidal, and particulate phases. Global Biogeochem. *Cycles* 20: GB2011.

Guo, L., and P.H. Santschi. 2000. Sedimentary sources of old high molecular weight dissolved organic carbon from the ocean margin benthic nepheloid layer. *Geochim. Cosmochim. Acta* 64: 651–660.

Harris, P.T. and T. Whiteway. 2011. Global distribution of large submarine canyons: geomorphic differences between active and passive continental margins. *Mar. Geol.* 285: 69–86.

Hartnett, H.E., R.G. Keil, J.I. Hedges and A.H. Devol. 1998. Influence of oxygen exposure time on organic carbon preservation in continental margin sediments. *Nature* 391: 572–574.

Hedges, J.I., and R. Keil. 1995. Sedimentary organic matter preservation; an assessment and speculative synthesis. *Mar. Chem.* 49: 81–115.

Hedges, J.I., R. Keil, and R. Benner. 1997. What happens to terrestrially-derived organic matter in the ocean? *Org. Geochem.* 27: 195–212.

Hedges J.I., F.S. Hu, A.H. Devol, H.E. Hartnett, E. Tsamakis, and R.G. Keil. 1999. Sedimentary organic matter preservation: a test for selective degradation under toxic conditions. *Am. J. Sci.* 299: 529–555.

Hernes P.J., and R. Benner. 2003. Photochemical and microbial degradation of dissolved lignin phenols: implications for the fate of terrigenous organic matter in marine systems. *J. Geophys. Res. Oceans* 108(C9).

Hertkorn, N., R. Benner, M. Frommberger, P. Schmitt-Kopplin, M. Witt, K. Kaiser, A. Kettrup, and J.I. Hedges. 2006. Characterization of a major refractory component of marine dissolved organic matter. *Geochim. Cosmochim. Acta* 70: 2990–3010.

Hill, P.S., G. Voulgaris, and J.H. Trowbridge. 2001. Controls on floc size in a continental shelf bottom boundary layer. *J. Geophys. Res.* 106: 9543–9549.

Hilton, R.G., A. Galy, N. Hovius, M.J. Horng, and H. Chen. 2010. The isotopic composition of particulate organic carbon in mountain rivers of Taiwan. *Geochim. Cosmochim. Acta* 74: 3164–3181.

Hilton, R.G., A. Galy, N. Hovius, M.J. Horng, and H. Chen. 2011. Efficient transport of fossil organic carbon to the ocean by steep mountain rivers: an orogenic carbon sequestration mechanism. *Geology* 39: 71–74.

Intergovernmental Panel on Climate Change. 2007. Climate Change 2007: The Physical Science Basis. Contribution of Working Group I to the Fourth Assessment Report of the Intergovernmental Panel on Climate Change [Solomon, S., D. Qin, M. Manning, Z. Chen, M. Marquis, K.B. Averyt, M. Tignor and H.L. Miller (eds.)]. Cambridge, UK: Cambridge University Press.

Jahnke, R.A., C.E. Reimers, and D.B. Craven. 1990. Intensification of recycling of organic matter at the sea floor near ocean margins. *Nature* 348: 50–54.

Jahnke R.A., and D.B. Jahnke. 2000. Rates of C, N, P and Si recycling and denitrification at the US Mid-Atlantic continental slope depocenter. *Deep-Sea Res. I.* 47: 1405–1428.

Johnson, K.S., F.P. Chavez, V.A. Elrod, S.E. Fitzwater, J.T. Pennington, K.R. Buck, and P.M. Walz. 2001. The annual cycle of iron and the biological response in central California coastal waters. *Geophys. Res. Lett.* 28: 1247–1250.

Kieber, R.J., X. Zhou, and K. Mopper. 1990. Formation of carbonyl compounds from UV-induced photodegradation of humic substances in natural waters: fate of riverine carbon in the sea. *Limnol. Oceanogr.* 35: 1503–1515.

Keil, R.G., L.M. Mayer, P.D. Quay, J.E. Richey, J.I. Hedges. 1997. Loss of organic matter from riverine particles in deltas. *Geochim. Cosmochim. Acta* 61: 1507–1511.

Kineke, G.C., K.J. Woolfe, S.A. Kuehl, J.D. Milliman, T.M. Dellapenna, and R.G. Purdon. 2000. Sediment export from the Sepik River, Papua New Guinea: evidence for a divergent sediment plume. *Cont. Shelf Res.* 20: 2239–2266.

Kiriakoulakis, K., S. Blackbird, J. Ingels, A. Vanreusel, and G.A. Wolff. 2011. Organic geochemistry of submarine canyons – The Portuguese Margin. *Deep-Sea Res. Part II Top. Stud. Oceanogr.* 58: 2477–2488.

Liu, J.T., and H.L. Lin. 2004. Sediment dynamics in a submarine canyon: a case of river–sea interaction. *Mar. Geol.* 207: 55–81.

Liu J.P., Z. Xue, K. Ross, H.J. Wang, Z.S. Yang, A.C. Li, and S. Gao. 2009. Fate of sediments delivered to the sea by Asian large rivers: long-distance transport and formation of remote alongshore clinothems. *Sediment. Rec.* 7: 4–9.

Lohnis, F. 1926. Nitrogen availability of green manure. *Soil Sci.* 22: 253–290.

Ludwig, W., and J.L. Probst. 1998. River sediment discharge to the oceans: present-day controls and global budget. *Am. J. Sci.* 298: 265–295.

Ludwig, W., P. AmiotteSuchet, and J.L. Probst. 1996. River discharges of carbon to the world's oceans: Determining local inputs of alkalinity and of dissolved and particulate organic carbon. *Comptes Rendus De L Academie Des Sciences Serie Ii Fascicule a-Sciences De La Terre Et Des Planetes* 323: 1007–1014.

Mackenzie, F.T., and A. Lerman. 2006. Carbon on the Geobiosphere–Earth's Outer Shell. Dordrecht, the Netherlands: Springer.

Mackenzie, F.T., A. Lerman, and A.J. Andersson. 2004. Past and present of sediment and carbon biogeochemical cycling models. *Biogeoscience* 1: 11–32.

Mannino, A. and H.R. Harvey. 2000. Biochemical composition of particles and dissolved organic matter along an estuarine gradient: sources and implications for DOM reactivity. *Limnol. Oceanogr.* 45: 775–788.

McGuire, A.D., L.G. Anderson, T.R. Christensen, S. Dallimore, L. Guo, D.J. Hayes, M. Heimann, T.D. Lorenson, R.W. Macdonald, and N. Roulet. 2009. Sensitivity of the carbon cycle in the Arctic to climate change. *Ecol. Monogr.* 79: 523–555.

McKee, B.A., and M. Baskaran. 1999. Sedimentary processes of the Gulf of Mexico, In: Bianchi, T.S., Pennock, R., and Twilley, R.R. (eds.), Biogeochemistry of Gulf of Mexico Estuaries, pp. 63–81. New York: John Wiley and Sons.

McKee, B.A., R.C. Aller, M.A. Allison, T.S. Bianchi, and G.C. Kineke. 2004. Transport and transformation of dissolved and particulate materials on continental margins influenced by major rivers: benthic boundary layer and seabed processes. *Cont. Shelf Res.* 24: 899–926.

Meade, R.H. 1996. River-sediment inputs to major deltas, In: Milliman, J.D., and Haq, B.U. (eds.), Sea Level Rise and Coastal Subsidence, pp. 63–85. Dordrecht, the Netherlands: Kluwer Academic Publishers.

Meybeck, M. 1981. River transport of organic carbon to the ocean, In: "Flux of organic carbon by rivers to the oceans." Report of a workshop held at the NAS Study Center, Woods Hole, Massachusetts, September 21–25, 1980. CONF-8009140. Washington, DC: United States Department of Energy, pp. 219–269.

Meybeck, M. 1982. Carbon, nitrogen, and phosphorus transport by world rivers. *Am. J. Sci.* 282: 401–450.

Meybeck, M. 1993. Riverine transport of atmospheric carbon: sources, global typology and budget. *Water Air Soil Pollut.* 70: 443–463.

Meyers-Schulte, K.J., and J.I. Hedges. 1986. Molecular evidence for terrestrial component of organic matter dissolved in ocean water. *Nature* 321: 61–63.

Michels, K.H., A. Suckow, M. Breitzke, H.R. Kudrass, and B. Kottke. 2003. Sediment transport in the shelf canyon "Swatch of No Ground" (Bay of Bengal). *Deep-Sea Res. Part II Top. Stud. Oceanogr.* 50: 1003–1022.

Miller, W.L., and M.A. Moran. 1997. Interaction of photochemical and microbial processes in the degradation of refractory dissolved organic matter from a coastal marine environment. *Limnol. Oceanogr.* 42: 1317–1324.

Milligan, T.G., and P.S. Hill, 1998. A laboratory assessment of the relative importance of turbulence, particle composition, and concentration in limiting maximal floc size. *J. Sea Res.* 39: 227–241.

Milliman, J.D., and K.L. Farnsworth. 2011. River Discharge to the Coastal Ocean – A Global Synthesis. Cambridge, UK: Cambridge University Press.

Milliman, J.D., and R.H. Meade. 1983. World-wide delivery of river sediment to the oceans. *J. Geol.* 91: 1–21.

Mitra, S., T.S. Bianchi, L. Guo, and P.H. Santschi. 2000. Terrestrially-derived dissolved organic matter in Chesapeake Bay and the Middle Atlantic Bight. *Geochim. Cosmochim. Acta* 64: 3547–3557.

Mopper, K., X. Zhou, R.J. Kieber, D.J. Kieber, R.J. Sikorski, and R.D. Jones. 1991. Photochemical degradation of dissolved organic carbon and its impact on the ocean carbon cycle. *Nature* 353: 60–62.

Mullenbach, B.L., and C.A. Nittrouer. 2006. Decadal record of sediment export to the deep sea via Eel Canyon. *Cont. Shelf Res.* 26: 2157–2177.

Opsahl, S., and R. Benner. 1997. Distribution and cycling of terrigenous dissolved organic matter in the ocean. *Nature* 386: 480–482.

Opsahl, S., and R. Benner 1998. Photochemical reactivity of dissolved lignin in river and ocean waters. *Limnol. Oceanogr.* 43: 1297–1304.

Puig, P. and A. Palanques. 1998. Temporal variability and composition of settling particle fluxes on the Barcelona continental margin (Northwestern Mediterranean). *J. Mar. Res.* 56: 639–654.

Reimers, C.E., R.A. Jahnke, and D.C. McCorkle. 1992. Carbon fluxes and burial rates over the continental slope and rise off central California with implications for the global carbon cycle. *Global Biogeochem. Cycles* 6: 199–224.

Romankevich, E.A. 1984. Geochemistry of Organic Matter in the Ocean: New York: Springer-Verlag.

Rowe, G.T., and R. Howarth. 1985. Early diagenesis of organic matter in sediments off the coast of Peru. *Deep-Sea Res.* 32: 43–45.

Sampere T.P., T.S. Bianchi, S.G. Wakeham, and M.A. Allison. 2008. Sources of organic matter in surface sediments of the Louisiana Continental margin: effects of major depositional/transport pathways and Hurricane Ivan. *Cont. Shelf Res.* 28: 2472–2487.

Sampere, T.P., T.S. Bianchi, and M.A. Allison. 2011. Historical changes in terrestrially-derived organic carbon inputs to Louisiana Continental margin sediments over the past 150 years. *J. Geophys. Res. (Biogeosciences)* 116: G01016.

Sarmiento, J.L., and E. Sundquist. 1992. Oceanic uptake of anthropogenic CO_2: a new budget. *Nature* 356: 589–593.

Schlünz, B., and R.R. Schneider. 2000. Transport of terrestrial organic carbon to the oceans by rivers: re-estimating flux and burial rates: *Int. J. Earth Sci.* 88: 599–606.

Seitzinger, S., J.A. Harrison, J.K. Bohlke, A.F. Bouwman, R. Lowrance, B. Peterson, C. Tobias, and G.V. Drecht. 2006. Denitrification across landscapes and waterscapes: a synthesis. *Ecol. Appl.* 16: 2064–2090.

Shanmugam, G., and R.J. Moiola. 1985. Submarine fan models: problems and solutions, In: Bouma, A.H., Normark, W.R., Barnes, N.E. (eds.), Submarine Fans and Related Turbidite Systems, pp. 29–34. New York: Springer-Verlag, New York.

Showers, W.J. and D.G. Angle. 1986. Stable isotope characterization of organic accumulation on the Amazon Continental Shelf. *Cont. Shelf Res.* 6: 227–244.

Silverberg, N., B. Sundby, A. Mucci, S. Zhong, T. Arakaki, P. Hall, A. Landen, and A. Tengberg. 2000. Remineralization of organic carbon in eastern Canadian continental margin sediments. *Deep-Sea Res. Part II Top. Stud. Oceanogr.* 47: 699–631.

Smith, S.V., and J.T. Hollibaugh. 1993. Coastal metabolism and the oceanic organic carbon balance. *Rev. Geophys.* 31: 75–89.

Smith, S.V., D.P. Swaney, L. Talaue-McManus, J.D. Bartley, P.T. Sandhei, C.J. McLaughlin, V.C. Dupra, C.J. Crossland, R.W. Buddemeier, B.A. Maxwell, and F. Wulff. 2003. Humans, hydrology, and the distribution of inorganic nutrient loading to the ocean. *BioScience* 53: 235–245.

Stow, D.A.V., D.G. Howell, and C.H. Nelson. 1985. Sedimentary, tectonic, and sea level controls, In: Bouma, A.H., Normark, W.R., and Barnes, N.E. (eds.), Submarine Fans and Related Turbidite Systems, pp. 15–22. New York: Springer-Verlag.

Stubbins, A., R.G.M. Spencer, H. Chen, P.G. Hatcher, K. Mopper, P.J. Hernes, V.L. Mwamba, A.M. Mangangu, J.N. Wabakanghanzi, and J. Six. 2010. Illuminated darkness: molecular signatures of Congo River dissolved organic matter and its photochemical alteration as revealed by ultrahigh precision mass spectrometry. *Limnol. Oceanogr.* 55: 1467–1477.

Syvitski, J.P.M., S.D. Peckham, R. Hilberman, and T. Mulder. 2003. Predicting the terrestrial flux of sediment to the global ocean: a planetary perspective. *Science* 162: 5–24.

Syvitski, J.P.M., C.J. Vörösmarty, A.J. Kettner, and P. Green. 2005. Impact of humans on the flux of terrestrial sediment to the global coastal ocean. *Science* 308: 376–380.

Takahashi, T., S.C. Sutherland, R. Wanninkhof, C. Sweeney, R.A. Feely, D.W. Chipman, B. Hales, G. Friederich, F. Chavez, A. Watson, D.C.E. Bakker, U. Schuster, N. Metzl, H. Yoshikawa-Inoue, M. Ishii, T. Midorikawa, Y. Nojiri, C. Sabine, J. Olafsson, Th. S. Arnarson, B. Tilbrook, T. Johannessen, A. Olsen, Richard Bellerby, A. Körtzinger, T. Steinhoff, M. Hoppema, H.J.W. de Baar, C.S. Wong, B. Delille and N.R. Bates. 2009. Climatological mean and decadal changes in surface ocean pCO_2, and net sea-air CO_2 flux over the global oceans. *Deep-Sea Res. Part II Top. Stud. Oceanogr.* 56: 554–577.

van Loosdrecht, M.C.M., W. Norde, J. Lyklema, and A.J.B. Zehnder. 1990. Hydrophobic and electrostatic parameters in bacterial adhesion. *Aquat. Sci.* 52: 103–114.

Vörösmarty, C.J., and B.J. Peterson. 2000. Macro-scale models of water and nutrient flux to the coastal zone, In: Hobbie, J.E. (ed.), Estuarine Science: A Synthetic Approach to Research and Practice, pp. 43–79. Washington, DC: Island Press.

Walsh, J.J., P.E. Biscaye, and G.T. Csanady. 1988. The 1983–1984 shelf-edge exchange processes (SEEP) – I experiment: hypothesis and highlights. *Cont. Shelf Res.* 8: 435–456.

Walsh, J.P. and C.A. Nittrouer. 2009. Understanding fine-grained river-sediment dispersal on continental margins. *Mar. Geol.* 263: 34–45.

Waterson, E.J., and E.A. Canuel. 2008. Sources of sedimentary organic matter in the Mississippi River and adjacent Gulf of Mexico as revealed by lipid biomarker and d^{13}CTOC analyses. *Org. Geochem.* 39: 422–439.

Weaver, P.P.E., R.B. Wynn, N.H. Kenyon, and J. Evans. 2000. Continental margin sedimentation, with special reference to the northeast Atlantic margin. *Sedimentology* 47: 239–256.

Williams, P.M., and L.I. Gordon. 1970. Carbon-13: carbon-12 ratios in dissolved and particulate organic matter in the sea. *Deep-Sea Res.* 17: 17–27.

Wollast, R. 1998. Evaluation and comparison of the global carbon cycle in the coastal zone and in the open ocean, In: Brink, K.H., and Robinson, A.R. (eds.), The Sea, pp. 213–252. New York: Wiley and Sons.

Xie, H., O.C. Zafiriou, W.-J. Cai, R.G. Zepp, and Y. Wang. 2004. Photooxidation and its effects on the carboxyl content of dissolved organic matter in two coastal rivers in the southeastern United States. *Environ. Sci. Technol.* 38: 4113–4119.

Section II

Water and sediment dynamics from source to sink

2

Water and sediment dynamics through the wetlands and coastal water bodies of large river deltaic plains

M. A. Allison, A. Kolker, and E. Meselhe

1. Introduction

The biogeochemical dynamics of major and trace elements, including nutrients and pollutants, and organic matter, at the coastal-ocean interface of major rivers is controlled in the first order by the physics of water and sediment transport. All of these substances are transported in either (1) solution phase, (2) attached to mineral particles, or (3) as discrete particles (e.g., particulate organic matter), and their vectors from river channel to receiving basin follow the pathways of water flow (surface and groundwater) and sediment transport (suspended and bedload). In the latter half of the Holocene, when transgression of the river-ocean interface slowed, the distribution of sediment-laden river water created the complex network of wetlands, levees, sand ridges, and open water bodies observed today at river mouths that constrain riverine water and sediment transport. The formation of this modern geomorphology began in major river deltas worldwide over a relatively restricted time frame (8,500 to 6,500 yBP; Stanley and Warne 1994). An exception to this is the Ganges-Brahmaputra delta, where shoreline position became stable and deltaic strata formation began at about 11,000 to 10,000 yBP as a result of the enormous sediment flux engendered by a strengthening southwest monsoon over the Himalayan source basin (Goodbred and Kuehl 2000; Kuehl et al. 2005).

Bianchi and Allison (2009) coined the term *large-river delta-front estuary* (LDE) to define the subregion of major river source-to-sink systems where the bulk of riverine sediment accumulates faster than it can be dispersed by marine redistribution processes. The LDE extends inland along the lowland floodplain to the limit of tidal and/or saline intrusion in the adjacent river channel, and seaward, onto the continental shelf. It is also inclusive of funnel-shaped major river mouths lacking a delta-shaped shoreline protuberance, such as the Amazon and Changjiang (Fig. 2.1). The landward limit also approximately coincides with the river channel exiting a relatively narrow alluvial valley incised into uplands, into a low-elevation landscape built of sediments from the river due to reduced water velocities produced by declining water surface slopes as it approaches the land-sea interface; this is often referred to as the *deltaic plain* (Wright 1977). In this context, the terms *deltaic plain* and *lowland floodplain* are synonymous as the constructive, low-elevation subaerial region near the river's mouth that experiences overbank water flow out of the channel, but, as defined using the LDE terminology, the lowland floodplain is the zone inland of the reach of tidal modulation of river velocity and of salinity penetration.

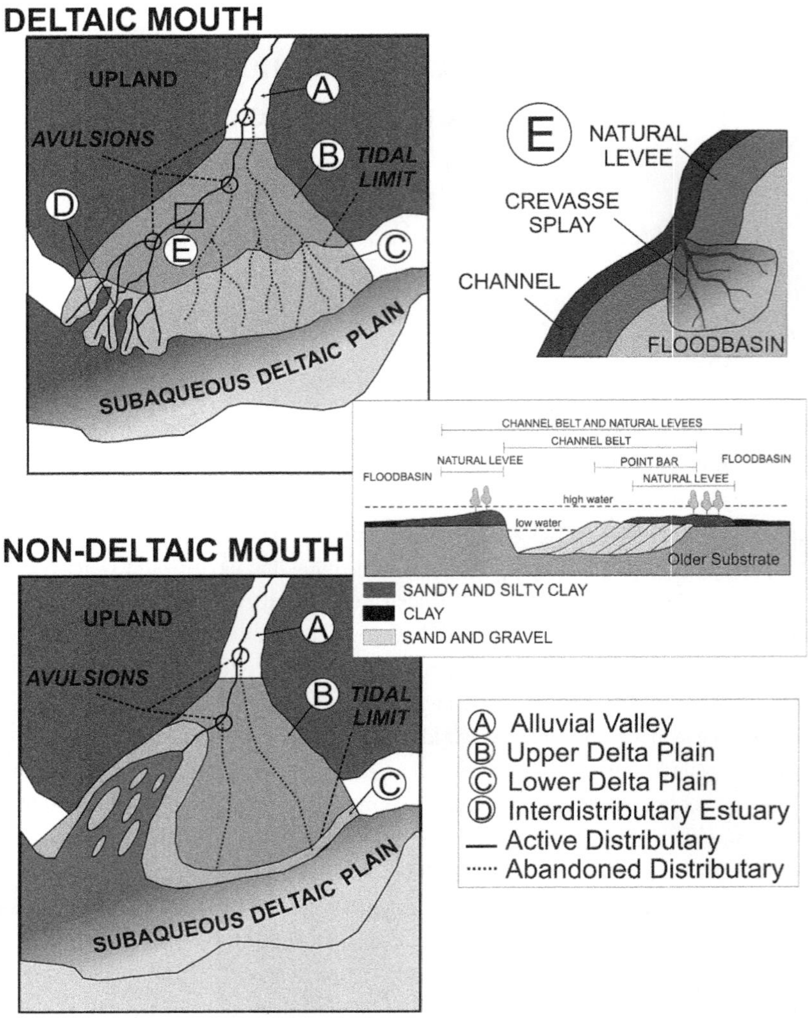

Figure 2.1. Schematic diagram of delta plain components for large rivers with lobate (deltaic) mouths (upper left) and funnel-shaped, estuarine (non-deltaic) mouths (lower left) (adapted from Coleman and Wright 1971). Also shown (right panels) are the components of the deltaic plain from channel to floodbasin (adapted from Berendsen 2005).

The channel's emergence from a confined alluvial valley also approximates the point at which channel bifurcations, known as avulsions, begin to divide the river flow (Slingerland and Smith 2004), providing a second means of distributing water and sediment across the subaerial deltaic plain and lowland floodplain (hereafter referred to together as the deltaic plain) and into inland water bodies (Fig. 2.2). The formation of these channel distributaries appears to be driven by elevation gain in an active region or lobe of the deltaic plain triggering an avulsion and capture of an increasing volume of the river's water discharge down a higher gradient pathway to the ocean (Slingerland

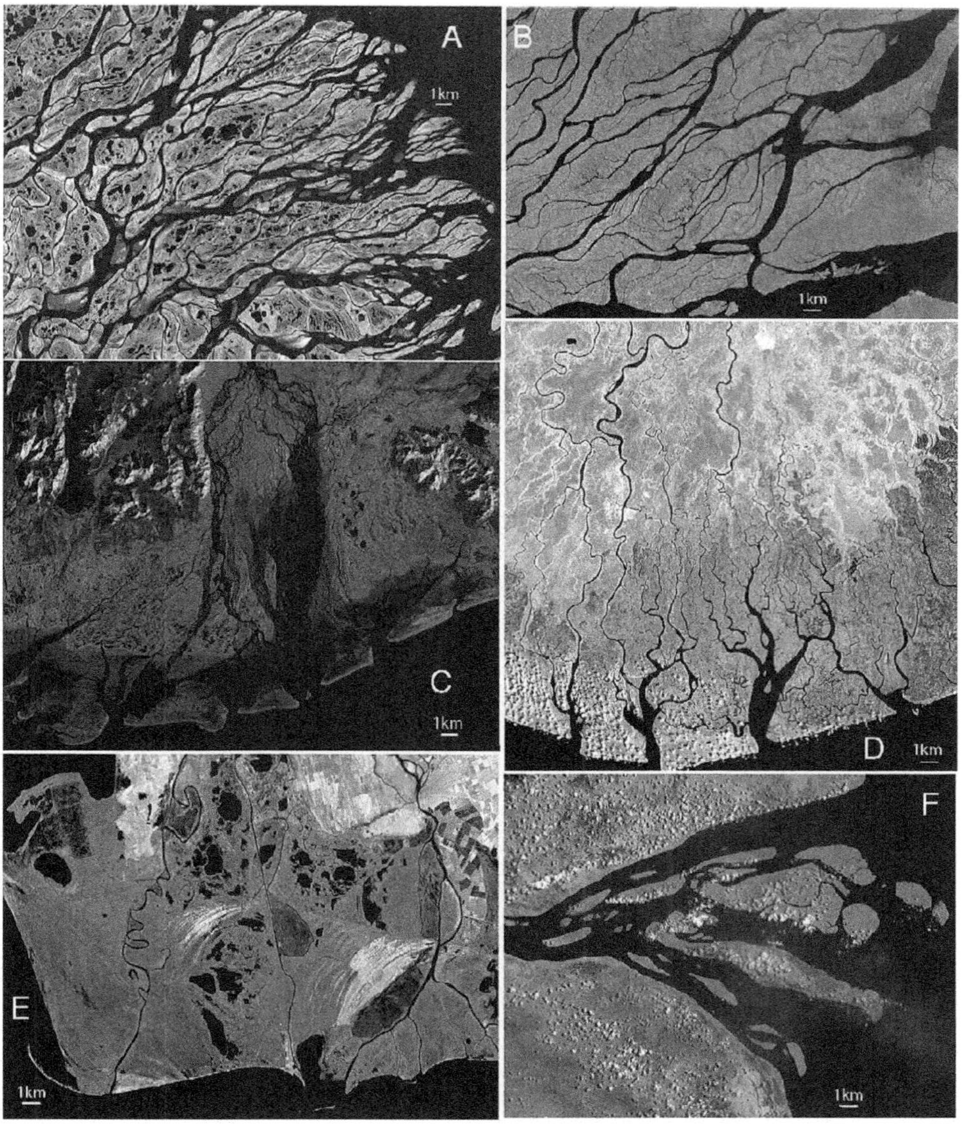

Figure 2.2. Examples of deltaic plains and distributary channel networks on major river deltas including (a) arctic Lena delta, Russia, (b) tropical Orinoco delta, (c) wave- and tide-influenced Copper delta, Alaska, (d) central portion of the tropical wave and tide influenced Niger delta, (e) southern, wave-dominated Danube delta, (f) tide-dominated tropical Fly delta, Papua New Guinea (from Syvitski and Saito 2007).

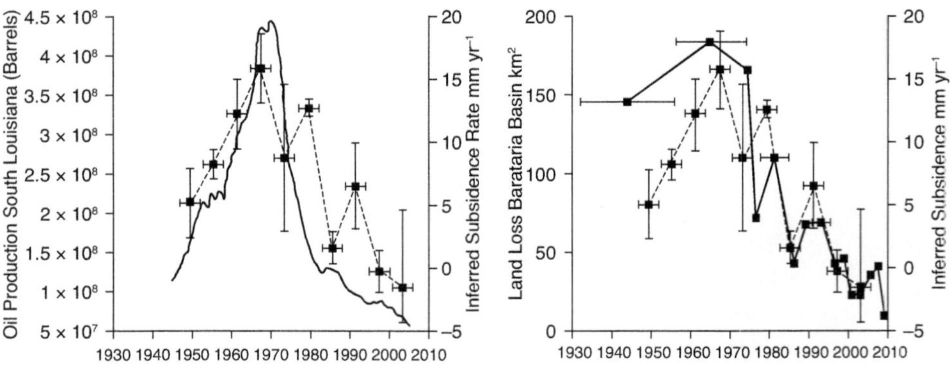

Figure 2.3. Withdrawal of subsurface fluids (e.g., hydrocarbons, groundwater) can increase subsidence rates and land loss in large river deltaic plains. The graph at left shows oil and gas production in the Mississippi deltaic plain from the 1940s plotted against inferred subsidence rate calculated by extracting the local relative sea level component from a long-term tide gauge (e.g., Grand Isle, Louisiana) record in the delta. The graph at right shows the strong direct correlation between subsidence rates and wetland loss rates in one of the major interdistributary basins (e.g., Barataria Basin) that comprise the Mississippi deltaic plain (from Kolker et al. 2011).

and Smith 2004). This process is most evident in sediment-load dominated systems such as the Mississippi, where wave, tide, and current redistribution are less important in controlling where the bulk of riverine sediment accumulates; at least six individual and overlapping, distributary lobe phases have formed in the Mississippi system over the last ~7,500 years (Frazier 1967; Penland et al. 1988).

The subaerial deltaic plain, and intervening open water bodies and their fringing wetlands, are the focus of the present chapter. Given the relatively short time interval in the late Holocene over which overbank and distributary sediment delivery processes have had to construct this zone, its width is highly variable from system to system (Fig. 2.2). An extreme is the Amazon, where tidal processes extend more than 800 km from the river-ocean interface (Vital et al. 1998). In the first order, the extent of the deltaic plain is a function of (1) margin geometry and (2) sediment supply. In most large rivers, the Holocene deltaic plain has aggraded above, and prograded seaward of, older Quaternary deltaic stratigraphic packages (Stanley and Warne 1994). Hence extent of the deltaic plain is controlled by the magnitude of the river's sediment load, which is highly variable from system to system (Syvitski 2011), and the period of time the river has been active in its present location. A second-order control is delta subsidence, which is a limit on subaerial extent (Syvitski et al. 2009). In thick deltaic sequences, particularly in those where rapid aggradation of high water content muddy sediments has taken place in late Holocene distributary lobes, loading-induced compaction can generate subsidence and relative sea level rise at rates of millimeters to centimeters per year, resulting in large-scale wetland loss (Meckel et al. 2007; Tornqvist et al. 2008). Compactional subsidence can be accelerated by the removal of oil, gas, and water from the delta's underlying sediments (Caputo et al. 1970; Morton et al. 2006; Saito et al. 2007; Kolker et al. 2011a; Fig. 2.3). The first-order control of margin geometry, which controls the infilling pattern of Quaternary deltaic sediments above older basement strata, can be highly complex in tectonically active deltas. In the Ganges-Bramaputra, for instance, the deltaic plain is situated atop the tectonically subsiding Bengal Basin (Iman and Shaw 1985; Johnson and Alam 1991). This second subsidence mechanism also reduces subaerial (and subaqueous) elevations, allowing these areas to be continuously creating accommodation space and serving as a modern river sediment depocenter (Goodbred and Kuehl 2000).

In a few large river systems, the magnitude of modern sediment sequestration in various subaerial and subaqueous depositional environments (relative to the input from the drainage basin) is relatively well-constrained. In the Amazon, where human alterations in the drainage basin (e.g., dams) and deltaic plain (e.g., flood control levees) are relatively limited, Nittrouer et al. (1995) calculated that approximately one-third of the sediment passing the last nontidal station on the river at Obidos, Brazil, is sequestered inland of the river-ocean interface. Mertes et al. (1996) and Aalto et al. (2003) suggested that this is primarily a function of overbank flow over the low natural levees constraining the river channel. In the Ganges-Brahmaputra in Bangladesh, which is mainly undammed but has some levee control, overbank flooding is also estimated to be sequestering approximately 32–39% of the late Holocene sediment supply in the subsiding deltaic plain, at rates that increase with proximity to active channels (Allison et al. 1998; Goodbred and Kuehl 1999). Even in the artificially leveed deltaic plain of the Mississippi-Atchafalaya, Allison et al. (2012) reported that 22% of the annual mud (<62.5 μm) suspended sediment load and 80% of the sand load reaching the deltaic plain today is sequestered by a combination of overbank and channel aggradation processes. In the Po River delta plain, Syvitski et al. (2005a) calculated that the combination of delta plain subsidence and artificial levees along distributary channels drive in-channel trapping of 16% of the river's sediment load. This channel bed aggradation is causing a super-elevation of the distributaries above the delta plain today at rates of 4–10 cm y^{-1}. These four examples from diverse systems are sufficient to demonstrate that (1) the deltaic plain and the distributary channels that cross it are a major locale for sequestering sediment and particulate organic matter in large river dispersal systems, and (2) human alterations can impact the locale of sediment sequestration.

Classifications for deltaic plain morphology have been based on the interplay between the magnitude and characteristic grain size delivered by the river and the dominant marine reworking mechanisms (e.g., waves, astronomical tides, coastal currents) that generate erosion at the land-sea interface and along the shorelines of estuarine water bodies in the deltaic plain (Coleman and Wright 1975; Galloway 1975; Wright 1985; Orton and Reading 1993; Postma 1995). More recent efforts to explain individual delta shape and spatial organization of the channel network have attempted to develop quantitative metrics (Edmonds et al. 2010). However, these general classifications of deltaic plain geometry mask the fact that even relatively low-energy, sediment supply–dominated systems such as the Mississippi-Atchafalaya are highly dynamic interfaces on seasonal to interannual timescales.

The delivery of turbid river waters to the deltaic plain during high water events is episodic in all rivers: flood events in large rivers tend to be weeks to several months in duration. Relative to smaller rivers, large systems have basins of enormous size (10^5–10^6 km^2) that span broad climatic zones, which tends to even out the effects of individual precipitation events in the basin (Mulder and Syvitski 1996). The exact timing of fluvial discharge and sediment yield (mass per square kilometers) delivered by the drainage basin to the deltaic plain is a function of many factors, including climate/latitude, basin character and elevation, glaciations, groundwater flow, and human alterations (Beschta 1987; Meybeck and Vörösmarty 2005; Syvitski et al. 2005b; Syvitski and Kettner 2011).

Destructive delta plain physical processes are also highly variable in time and space and from system to system. Erosive tidal currents on wetland and oceanic shorelines, in distributary channels, and in delta plain estuaries are cyclic on diurnal, monthly lunar (spring-neap), and longer timescales. Storm winds generated by seasonal cold air outbreaks (fronts) occur at most latitudes except near the equator, generate large waves at land-water interfaces, intensify and change the vector of coastal currents, and cause meteorological tides (coastal setup and setdown) that submerge the lowest elevation wetlands on the delta plain. Large rivers that debouch into the ocean at latitudes of about 10 to 40° N and S in the Atlantic, Pacific, and Indian Oceans also experience tropical storm landfalls episodically. Because of variability in each storm's size, strength, and direction of approach to a delta plain, their destructive effect is highly individualistic. In general, the strong winds and low barometric pressures produce extreme coastal setup (e.g., storm surge) that can submerge large portions of a delta plain and subject these areas to large-wave orbital and wind-driven current erosive stresses that can remobilize sediments, organic matter, and pore waters from wetland and adjacent estuarine surfaces. On longer timescales, relative sea level rise (RSLR), driven by climate-driven eustatic and subsidence processes, is a destructive process that causes submergence and inland penetration of saline water into freshwater delta plain wetlands (Ericson et al. 2006). A number of recent studies (see, e.g., Blum and Roberts 2009) have modeled this process on individual deltaic plain using elevation data and RSLR rates to predict delta plain land loss in the twenty-first century. Syvitski et al. (2009), in a survey of 33 deltas globally, estimated that flooding of existing delta plain surface area could increase by 50% in the twenty-first century, with the most severe being in deltas that are experiencing rapid subsidence and capture of sediment upstream by dams and reservoirs.

The following sections examine the state of knowledge about how water and dissolved components, sediment, and particulate organic matter are delivered from the channel of large rivers to the deltaic plain and the physical mechanisms that control their dispersal, export, and burial. A major focus is to distinguish overarching versus system-specific controls on water and sediment transport. The mechanisms are examined that disperse water and sediment out of the channel, across wetlands, and into water bodies (e.g., estuaries and lakes) in the deltaic plain, including an examination of the interaction of riverine and marine transport processes and models for simulating these processes.

2. Mechanisms of water and sediment input to the delta plain

The alluvial architecture of large river deltaic plains is composed laterally of contiguous sedimentary facies arranged with increasing distance from the active river distributary channel (Fig. 2.1). In the active distributary channel belt are (1) channel and bar deposits, and (2) raised natural levees lining the channel on both banks. Erosional breaks in the natural levees created in large floods (e.g., crevasses) generate lobate splay deposits delivered by flow channels that deposit riverine sediments into the adjacent interdistributary delta plain. The lower elevation interdistributary basins (referred to here as *floodbasins*) grade toward the land-ocean interface from (1) lakes and swamps on their inland reaches that are indistinguishable from alluvial valley floodplain deposits except by the confining effects of uplands, to (2) estuarine water bodies fringed by increasingly saline marshes and mangrove swamps. These facies are also stacked vertically within the Holocene deltaic plain deposit caused by switching in the loci of river distributary lobe deposition and eventual reoccupation on 10^3 y timescales.

2.1. Avulsion and distributary channel evolution

The three-dimensional alluvial architecture of the deltaic plain is best studied to date in the Rhine-Meuse Delta (see Chapter 25), where Berendsen and Stouthamer (2000; 2001) summarized the data from ca. 200,000 boreholes through the Holocene channel belt and adjacent floodbasins of the delta (Fig. 2.3). Periodically, all or part of an active channel belt is abandoned because of a river switching, or avulsion. *Avulsion* is a process that is controlled by both intrabasinal processes, such as lobe switching and river meandering, and external forcing, such as tectonics, eustasy, and climate change (Beerbower 1964; Allen 1978; Holbrook and Schumm 1999; Stouthamer and Berendsen 2007). The exact conditions for avulsions remain uncertain (Reitz et al. 2010), but Slingerland and Smith (1998) suggested that the occurrence of an avulsion depends on (1) the nature of sediment transport partitioning at the bifurcation, and (2) the ability of channels to change their capacities controlled by stream power and bank cohesion. The latter constrains the ability of a river channel to migrate laterally. Frequency of these avulsions may relate to flow depth and vertical aggradation rate (e.g., sediment supply) in the channel (Jerolmack and Mohrig 2007). Channel avulsion may be favored at locations where vertical aggradation is most rapid, leading to persistent overbank flow (Edmonds et al. 2009). The relationship of avulsions with stream power has led Chatanantavet et al. (2012) to suggest that the association of initial distributary avulsions near the head of the delta plain (Fig. 2.2) is constrained by reduced water surface slopes in the backwater zone of the channel.

The stratigraphic architecture of the deltaic plain is controlled by the number, cross-sectional extent, and downstream connectivity of sandy channel belt "framework" facies (i.e., the channel and natural levees; Figs. 2.1 and 2.4), relative to muddy and organic-rich floodbasin deposits. Mackey and Bridge (1995) modeled the three-dimensional alluvial stratigraphy of channel belts on the deltaic plain as a function of (1) delta plain elevation and downbasin slope, (2) channel and floodbasin deposition rates, (3) compaction, and (4) tectonic subsidence resulting from tilting or faulting. This and earlier studies (Leeder 1978; Bridge and Leeder 1979) assumed that during channel avulsion, rivers relocate randomly or to the lowest point on the delta plain, which is often defined by subsidence patterns. Increases in river discharge on longer timescales are thought to produce zones of closely spaced channel bodies stacked within overbank deposits containing more isolated channel bodies (Demko et al. 2004). Distinct spatial (downbasin) and temporal trends in the alluvial architecture are present in the Rhine-Meuse and Mississippi delta plain that are well-constrained by field studies (Gouw 2008; Gouw and Autin 2008; Bos and Stouthamer 2011). These trends are attributed to variations in available accommodation space (e.g., elevation control), channel belt geometry, overall delta plain geometry, and aggradation rate. In the Anthropocene, distributary channel location and avulsion frequency are often constrained by human alterations. This includes sealing off of minor distributaries to reduce delta plain flood extent and increase navigability of the main distributaries and the use of artificial levees and bank armoring to reduce overbank flow and to prevent avulsion and lateral channel migration.

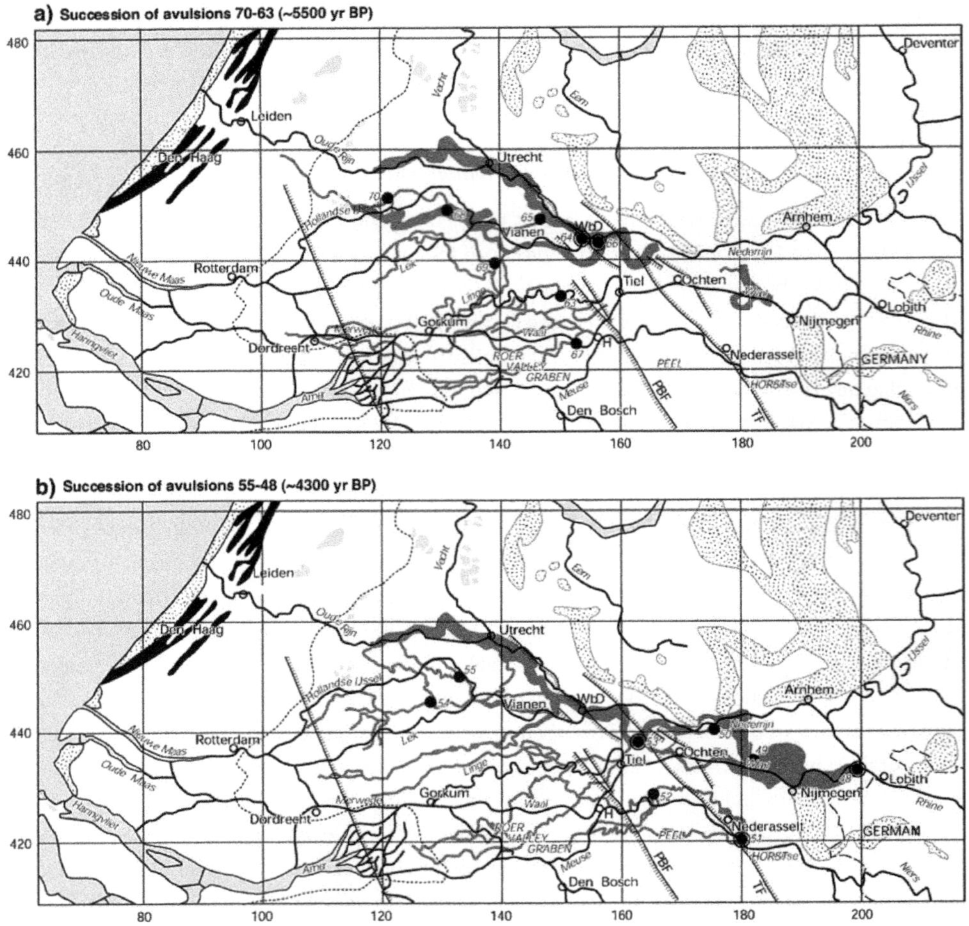

Figure 2.4. Paleogeographic maps of Holocene channel belts emplaced in 5,500 yBP to present in the Rhine-Meuse delta, the Netherlands, along with the location of active channels (from Stouthamer and Berendsen 2007). Points represent the location of channel avulsions at about 5,500 yBP (upper map) and 4,300 yBP (lower).

2.2. Overbank and crevasse processes

Distributary channels deliver surface water and sediment to the deltaic plain through the processes of overbank flow and crevassing, whereas lateral migration of the channel tends to balance sediment accretion due to bar migration with opposite bank erosion (Wolman and Leopold 1957). Overbank flow in delta plains closely resembles the process in the confined alluvial valley further upriver. Because overbank flow tends to draw water from the sediment-poor and finer-grained upper water column, most of the sediment delivered out of channel is mud (Bridge 1993). However, coarse material exiting the river tends to be deposited in the immediate channel margin due to the rapid drop-off in water velocity, forming a natural levee that is thickest and coarsest at the channel margin and thins and fines toward the adjacent floodbasin (Brierley et al. 1997; Gretener and Strömquist 1997; Smith and Pérez-Arlucea 2008). The resulting wedge-shaped ridges (Fig. 2.1) have an elevation, width, and steepness that are controlled by flood height, grain size, stage in evolution, and geometry of the receiving basin (Simm 1995; Cazanacli and Smith 1998; Filgueira-Rivera et al. 2007). Transport of water across the levee and into the adjacent basin is thought to be initially advective and channel perpendicular, but as flow velocities decrease, it increasingly spreads laterally and transports suspended sediment in eddies produced by concentration gradients between riverine and basin water (i.e., turbulent diffusion; James 1985; Pizzuto 1987, Adams et al. 2004; Filgueira-Rivera et al. 2007).

Avulsions through the natural levee that do not result in capture and redirection of channel flow, and that tend to only flow during higher river stages, are referred to as *crevasses*, and the fan- or lobe-shaped deposits produced in the adjacent basin are known as *crevasse splays*. Crevasse splays are sand-rich deposits fining basin-ward that merge with natural levee and floodbasin deposits distally (Coleman 1969). Gautier et al. (2009) documented two large crevasse events that occurred on the Rhône River deltaic plain in the large flood of December 2003. These splays were composed of 81% sand, and 24% of the total splay deposit volumes were derived from incision of the artificial and underlying natural levee and proximal floodplain – the remainder of the sediment was derived from the river. Crevasses can occur throughout the deltaic plain and upriver in the alluvial valley; in the deltaic plain they can debouch into wetlands (swamps and marshes), lakes (Tye and Coleman 1989), or, closer to the mouth, into estuarine floodbasin water bodies (Coleman and Gagliano 1964). Bos and Stouthamer (2011) suggested that the highest proportion of splays in the Rhine-Meuse delta plain is midway between the coast and the delta apex, where the bulk of delta lobe avulsions are concentrated. They also calculate that 7.1% of the total Holocene delta plain sediment volume of the Rhine-Meuse is crevasse splays, inter-fingered with floodbasin, natural levee, and channel belt facies. Crevasse events occur on the delta plains at shorter intervals than lobe switching avulsions: Davis (2000) documented 16 recurring crevasses on the lower 150 km of the Mississippi delta plain from 1770 and 1927, when artificial levees completely constricted the channel.

Crevasse channels can avulse in the same manner as distributary channels and have their own levees and mouth bars at bifurcations (Bridge 1993). This microcosm of the delta plain process distributes water and sediment over the lobate feature (Fig. 2.1). Cahoon et al. (2011) described the evolution of a Mississippi delta splay as four distinct stages: (1) rapid mineral sediment infilling and elevation gain of the receiving area, (2) vegetative colonization when the splay surface becomes emergent, (3) development of a mature wetland community and slowing of mineral sediment input, and (4) decay produced by compactional subsidence. Recent simulations have suggested that the crevassing is favored when the distributary channel mouth bar reaches a critical size and stops prograding seaward, triggering channel bed aggradation that moves upstream and increases overbank flow and bed shear stress on the channel levees (Edmonds et al. 2009).

2.3. Groundwater and rainwater input

Groundwater can play an important role in the hydrology, biogeochemistry, and geology of many deltas, as the flow of groundwater can follow a variety of pathways into and out of the deltaic plain, carrying with it or leaving behind key chemical constituents (e.g., nutrients, dissolved organic carbon).

The magnitude of groundwater flows in deltas is relatively poorly studied to date; however, a recent analysis of hydrological fluxes in the Mississippi River segment crossing the deltaic plain suggests that about 13 km^3 of the 552 km^3 of average annual surface water flux is unaccounted for (flood years 2008–2010), implying a significant loss to groundwater (Allison et al. 2012; Kolker et al. 2013). Groundwater can also be discharged into the coastal ocean, and this submarine groundwater discharge (SGD) can be fresh, brackish, or recirculated seawater (Burnett et al. 2003; Moore 2010). Although most previous studies of SGD have focused on coarse-grained or karstic aquifers, a growing body of studies from the Mississippi, Ganges-Brahmaputra, and Yellow River deltas all suggest that SGD is important in these systems (Moore 1999; Basu et al. 2001; Moore and Krest 2004; Taniguchi et al. 2008; Kolker et al. 2013). Salt waters often intrude into both surface and groundwaters in deltas, a process that naturally occurs during periods of low flow, has been accelerated as a result of fluid withdrawal in anthropogenically impacted system, and may increase with climate change (Mbraheem et al. 1997; Essink et al. 2010).

Given the propensity for saltwater and freshwater mixing in deltas, it is reasonable to think of the subsurface of the deltaic plain as a "subterranean estuary," – a buried mixing zone with intricate particle-water interactions occurring along a spatial heterogeneous gradient (Moore 1999; Burnett et al. 2006; Moore 2010). The factors driving SGD are numerous and can include terrestrially originating hydrological gradients, tidal pumping, thermally driven circulation, and density-driven circulations (Moore 1999; Burnett et al. 2006; Moore 2010). In large river deltas constrained by natural and artificial levees, the hydrological head difference between the river surface and the surrounding wetlands is likely to play a particularly important role in flux to groundwater, as are spatial variations in the permeability of the river channel bottom (Nittrouer et al. 2011).

Although groundwater flow is often orders of magnitude less than surface flow, its geochemical significance can be substantially greater, as ground waters can be enriched in dissolved constituents (Moore 1999; Burnett et al. 2006; Moore 2010). Groundwaters can serve as either sources or sinks for nutrients, contaminants, and many other trace elements and thus have the potential to remediate or accelerate environmental contamination. Ultimately, the fate and transport of dissolved compounds in deltaic plain groundwaters depend on a range of factors that include the solubility of the chemical species at issue, the hydraulic conductivity, the sorption capacity of the sediments, the dissolution and precipitation of Fe-oxides and Fe-sulfides, and the rate of groundwater flow (Burnett et al. 2003; Burnett et al. 2006; Tang and Johannesson 2006; Weinman et al. 2008).

Although it is true that most large river deltas garner most of their water from a large, often distant catchment, local rainfall also can play a role in the surface and groundwater functioning of deltas. Indeed, rainwater can raise water levels on the deltaic plain independent of rise associated with stage in the river channel and can contribute to freshwater lenses in deltas, thereby allowing river water to penetrate deeper into the delta plain that might otherwise be allowable (de Louw et al. 2011). However, this does not always translate into increased sediment transport into deltaic systems, as locally derived water often does not carry high concentrations of sediments relative to that carried by catchment-derived water (Allison et al. 2003). Locally, rainfall can also affect the geomorphology of wetlands when they are exposed, either during periods of low river flow or during low tide. Such activities can lead to local erosion of marsh edges and other unstable features (Torres et al. 2004).

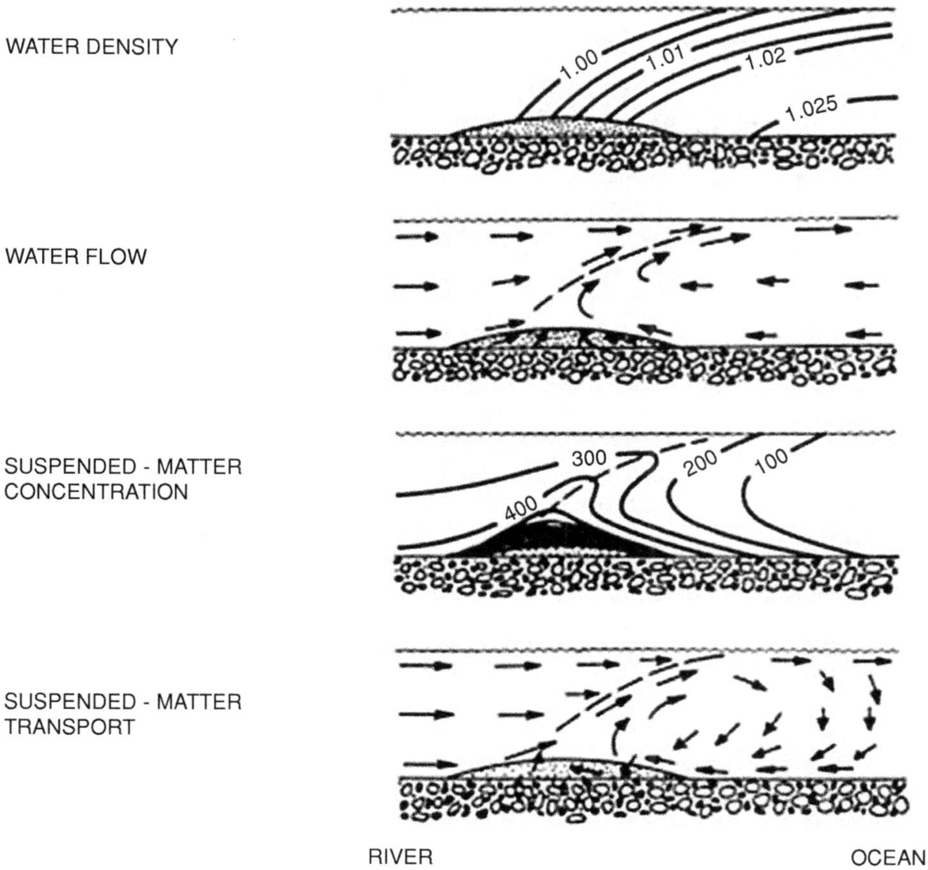

Figure 2.5. Schematic diagram of hydrodynamic and sediment transport conditions in an estuary from the riverine to ocean end-member caused by density-driven circulation. Plots show density, water flow, suspended matter concentration, and suspended matter transport (after NEDECO 1968).

3. Processes and simulation of water and sediment circulation in deltaic plains and estuaries

3.1. Controlling physical processes

The movement of water and sediment transitioning from fluvial channels to the receiving basins, or directly to the ocean, is governed by three main mechanisms: (1) riverine processes, (2) tidal cycle, and (3) wind. The first is the dominant factor governing water and sediment movements within fluvial channels. It diminishes during the transition from the riverine environment across the estuarine environment, to the open marine environment, where tidal and wind-driven processes dominate. Density currents are another mechanism that plays an important role in the transition zone between riverine freshwater flow passing through an estuary and into the ocean (Fig. 2.5). Strong freshwater flow and limited tidal and wind-wave mixing favor development of an upper layer of freshwater (lower density) flowing outward across delta plain water bodies (e.g., estuaries), separated by a pycnocline from a dense, saline bottom layer originating offshore, that moves inland (known as *baroclinic circulation*; Fischer et al. 1979). Greater tidal and wave mixing or reduced freshwater discharge tends to increase mixing of water masses prior to discharge to the shelf.

Generally, this interface between river channel and shelf can be broadly classified as the *estuary*. Estuaries can be categorized based on their size and shape and mixing behavior with riverine and marine end-member waters. Among others, Pritchard (1967) classified estuaries based on their hydrodynamic (e.g., stratified, partially stratified, and well-mixed) or their geomorphological characteristics (e.g., coastal plain, bar built). A broader and more recent definition was provided by the European Environmental Agency (Wolanski 2007), as they coined the term *transitional waters*, described as "bodies of surface water in the vicinity of river mouths which are partly saline in character as a result of their proximity to coastal waters but which are substantially influenced by freshwater flows." It should be noted, however, that not every transitional water is a river mouth, nor does it necessarily exhibit lower salinity. Regardless of the classification, it is important to note that each system is unique in how it interacts with the riverine system, as well as the larger water body (ocean, gulf, sea, etc.) to which it is connected. Each system is also strongly influenced by the variation and seasonality of wind, temperature, and rainfall, as well as by variations in freshwater inflow, sediment load, and sediment grain size distribution.

The physical characteristics of the water flow in such systems is influenced by the small spatial scale turbulence eddies generated by larger scale random motions of water particles and large-scale variability of the mean velocities. These mechanisms enhance the mixing process in estuaries. In the case of confined fluvial channels, the turbulent diffusion (transfer of mass from high-concentration zones to lower concentration zones) is produced by a simple transfer of mass between flow paths, whereas the longitudinal dispersion (the process of scattering of particles due to shear and transverse diffusion) takes place because of the variability in speed among the various streamlines. However, in estuarine water bodies, the advective velocity changes magnitude and direction. These spatial and temporal changes of advection are induced by the tidal cycle, the wind variations, and the Earth's rotation (e.g., Coriolis forcing). Additional factors that can influence the velocity field include the seasonal changes to the tidal cycle, the meteorological patterns, and the riverine inflow.

As mentioned previously, one of the dominant processes that affect water circulation in estuaries is tide. Tide generates a shearing flow against the estuarine floor leading to turbulent mixing. Furthermore, the tidal waves interact with the estuarine bathymetry, generating large-scale currents. The cyclic nature of tides adds to the complexity of the flow and sediment transport patterns in shallow, deltaic plain water bodies in the transition from riverine to estuarine (brackish) environments. The other dominant physical forcing factor controlling water and sediment movement in deltaic plain water bodies is wind. The size and geometry of the water body is critical in determining how dominant wind can be. The larger and more open the water body, the more dominant wind can be as it gets sufficient space (*fetch*) to generate significant currents and to build waves. The direct wind stress on the water surface generates unidirectional circulation currents, especially near-surface, which stimulate mixing and reduce salt stratification. Wind also leads to the formation of waves, whose orbital motion in water bodies further increases mixing. The sum effect of these physical forcing mechanisms imparted by river outflow, tide, and wind on river-derived sediment particles is highly complex temporally and spatially. Suspended particulate (mineral and organic) matter leaving the river channel and transported in shallow water bodies across this estuarine transition are subjected to progressive sorting by size, shape, and mineralogy. Deposition can be ephemeral, with resuspension and further advection on timescales from seconds to interannual, driven by fluctuations in tidal, meteorological, and flood intensity. Further complication is imparted by the exposure to salinity, which can generate (1) particle agglomeration (e.g., flocculation) caused by physicochemical and organic bonding (Dyer 1989 and ref. therein), and (2) sediment depositional focusing caused by density-driven circulation (e.g., turbidity maxima; Postma 1967) (Fig. 2.5). Finally, the dispersal of estuarine waters into and through fringing wetlands can impact sediment dynamics, as is outlined in Section 4. The complexity of these processes has led to simulation approaches to predict system-scale response to these factors.

3.2. Simulation of water and sediment processes

Numerous analytical, physical, and numerical modeling tools have been used to study the water, sediment, and salinity dynamics in estuarine water bodies associated with large rivers (e.g., Horrevoetsa et al. 2004; Chen et al. 2006; Eaton 2007; Gabriel et al. 2007; Phillips and Slattery 2007; Shi 2010; Shi and Lu 2011; Zhou et al. 2011; Bárcena et al. 2012; Chen et al. 2012). The most widely applied analysis and forecast tools today are numerical models. They vary in their degree of complexity and ability to capture the dominant physical processes. The tremendous gain in computational power and resources allowed numerical models to tackle more complex engineering problems, larger spatial domains, and longer duration simulations. For example, modeling turbulence flow was beyond reach (except for simplified applications) until that increase in computing power. Furthermore, numerical models are relatively inexpensive, as they can be used to analyze complex problems regardless of their geographical location. Numerical models are quite capable of modeling complex three-dimensional water flows accurately (e.g., Meselhe et al. 2000; Meselhe and Sotiropoulos 2000; Rego et al. 2010; Kang et al. 2011). Despite the advancement in numerical modeling of turbulence and complex two- and three-dimensional flows, numerical modeling of water constituents and sediment (both cohesive muds and noncohesive sands) is not as fully developed. This can be mainly attributed to the lack of understanding and subsequently the inability to mathematically describe the physical processes that govern the interaction between water and these constituents. For example, the physical processes that govern the interaction between fine sediment (cohesive clay), salinity, and water, to date, are not fully captured in numerical models.

Physical models have also been traditionally used to study estuaries and large riverine systems (Gabriel et al. 2007). The idea behind physical modeling is to create a scaled replica of a natural system (e.g., a river reach, bay). The scaling can be homogeneous, i.e., all dimensions of the natural system are scaled down equally. For example, if a river reach is to be modeled at a 1:100 scale, then the river length, width, and depth are all scaled down by a factor of 100. Care is needed in this scaling process to preserve the main flow features occurring in the natural system. Often, physical models are "distorted" models by scaling different aspects differently: this is typically done to avoid dealing with extremely small flow depth in a laboratory physical model. However, the distortion should not be severe enough to change the natural system's flow characteristics.

There has been a major paradigm shift in the community's use and applications of physical models. Physical models in the past decade or so have focused on two classes of applications: (1) basic research on physical processes (e.g., physical processes of sediment transport and delta growth/decay) that are poorly understood and cannot be accurately modeled numerically (Kim et al. 2009; Connell et al. 2012), and (2) flows in the vicinity of hydraulic structures that include processes that again cannot be captured numerically, e.g., flows around spillways that include aeration mechanisms (Politano et al. 2009). A hybrid approach of coupling numerical and physical models is quite powerful and beneficial (Muste et al. 2001), as each provides complementary information and insights.

3.3. Numerical modeling tools

The focus here is on numerical models due to their lower cost and wider range of applications for examining water and sediment dispersal across deltaic plains. Various types of models can be employed to capture a combination of processes. These include:

- Hydrodynamics (water circulation driven by riverine inflow, tide, wind, and earth motion)
- Density stratification due to salinity
- Sediment (suspended matter) and morphologic (delta growth and dynamics) processes
- Water constituents (e.g., dissolved oxygen, nitrogen, phosphorus)

A brief list of the most common numerical modeling tools employed in deltaic plain simulations follows:

- *Three-Dimensional Circulation Models.* These models (Rego et al. 2010; Kang et al. 2011) calculate the temporal and spatial variability of the flow parameters (e.g., velocities, pressure) in both horizontal directions as well as over the water column (vertical direction). Three-dimensional models vary in degree of complexity. For example, to gain computational efficacy or speed, some assume hydrostatic (linear) distribution of pressure in the vertical direction (Rego et al. 2010). This assumption is acceptable in the absence of strong vertical flow acceleration. Other simplifying assumptions are related to how turbulence fluctuations are being accounted for. The more sophisticated the approach, the higher the computational cost. Overall, three-dimensional models are needed for complex flow patterns, especially if vertical density stratification is present.
- *Two-Dimensional Shallow-Water Circulation Models.* These models are often referred to as *depth-averaged models* (Chen et al. 2012) because they calculate the flow parameters only in the two horizontal directions. As such, the parameters are "depth-averaged," e.g., depth-averaged velocities, or depth-averaged concentrations. Depth-averaged models are more computationally efficient than three-dimensional models, but their applicability is limited to shallow water bodies, or to situations in which vertical density variations are small.
- *Two- and Three-Dimensional Storm Surge Models.* These models are mainly focused on analyzing hurricane storm surge and potential flooding on the deltaic plain (Forbes et al. 2008). One of the most commonly used models is the ADCIRC system (http://adcirc.org). These models analyze tides and wind-driven circulation as well as hurricane storm surge and flooding. Typically to gain

computational speed, these models do not include water constituents (water quality parameters such as salinity or sediment).

- *One-Dimensional Models*: These models are based on equations where the flow parameters are integrated over the entire flow area. As such, they provide variations only along the main-flow direction. For example, along a crevasse splay channel, these models provide cross-sectional average flow parameters. These cross-sectional average parameters can vary temporally and from one location along the channel length to another. One-dimensional models are considerably more computationally efficient than multidimensional models and are capable of simulating hundreds of kilometer transects and allow for performing decadal simulations (Davis 2010).
- *Delta Formation and Land-Building Models*: There are a variety of morphological models that address subaerial delta formation and decay. They range from simple spreadsheet models (Kim et al. 2009) to empirical, spatially explicit models similar to the geomorphological module used in recent environmental planning for restoring the Mississippi Delta (LACPRA 2012), to complex three-dimensional morphological modules such as Delft3D (Edmonds et al. 2010; Edmonds and Slingerland 2009). The latter is based on a complex geomorphological module that includes processes of bottom accretion, channel bifurcation, and lateral bank migration. It is, however, computationally expensive.

The application of numerical models is often driven by the issue of spatial scale: modeled scales range from local (~10^2 km^2), to regional (~10^4 km^2), to continental (~10^6–10^8 km^2), depending on the process(es) of interest. The computational cost (time and computational power required to perform a simulation) increases dramatically as the spatial and temporal coverage expands. Temporally, models can be used to simulate a specific instant in time (steady-state), a single hydrologic (e.g., flood) event, an atmospheric event (e.g., a storm), a season (~weeks), and up to multi-decadal simulations to examine system evolution. To date, and despite the advancement in the computing power in the past two decades, the ability to model large spatial scales for extended temporal duration with a high-resolution model is limited. This is especially true if sediment transport, bottom morphological evolution, and water constituents are included in the numerical simulation.

At the local scale, it is possible to construct high–spatial resolution models to capture the hydrodynamics, sediment, and water quality constituents at the seasonal and annual temporal scales (Chen et al. 2012). At the regional scale, a high-resolution model can be applied, but limitations on the temporal scale will be needed. For example, a high-resolution model at the regional spatial scale can be applied only at the event or at most seasonal temporal scale if the run time is to be kept at a practical level (multi days of simulation time). Figure 2.6 shows an example of an application of the open-source code model Finite Volume Coastal Ocean Model (FVCOM) to the Lower Mississippi River and the surrounding basins. Open source (as opposed to proprietary) indicates that researchers have access to the model formulation and can incorporate adjustment or new knowledge as needed. It is also possible to conduct multi-decadal morphological simulations to mimic land-building and delta dynamic processes on local spatial scales (see Edmonds 2010). However, simplifying assumptions will be made to advance computational speed. These compromises are needed to cope with the current limitations of computing power.

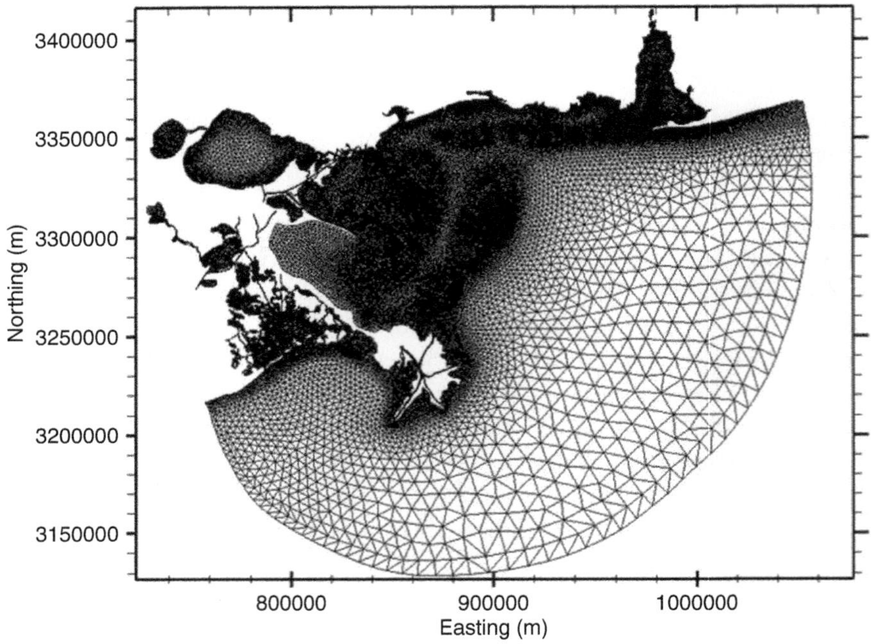

Figure 2.6. Regional FVCOM model grid for the Lower Mississippi River and the receiving basins (from Schindler 2010; Georgiou et al. 2011).

4. The role of wetlands

4.1. Introduction to the ecogeomorphology of deltaic plain wetlands

Wetlands are typically defined as an environment that is intermittently flooded and that has hydric soils that support vegetation (Mitsch and Gosselink 2000). They are among the most widespread of deltaic sub-environments, and they can be found in deltas from the Arctic to the Amazon. Their distribution, morphology, vegetation, and function in a delta are governed by a range of factors that include the flooding regime, soil type, salinity, nutrient availability, and climate (Mitsch and Gosselink 2000; Al-Hilli et al. 2009). These physical and chemical factors can vary in both linear and nonlinear gradients across the delta and can be governed by the morphology of the river channel, shape of the coastal zone, the tidal and wave climates, and the underlying geology.

In wetlands, both in and out of deltas, ecogeomorphological processes play a key role in the development of these systems (Fagherazzi et al. 2004). Ecogeomorphological processes are ones in which the ecology of the area strongly governs the geomorphology and vice versa. There are numerous ecological properties of marshes that have implications for the geomorphology of deltas, including the plant height, stem density, plant morphology, and phenology.

Stem properties are particularly emblematic of the ecogeomorphic relationships that occur in deltaic wetlands, as the density of plant stems in a wetland, their height, and rigidity play a critical role in their ability to trap mineral sediments. This in turn affects the ability of a wetland to accrete in the face of relative sea-level rise in combination with accumulation of in situ organic matter production

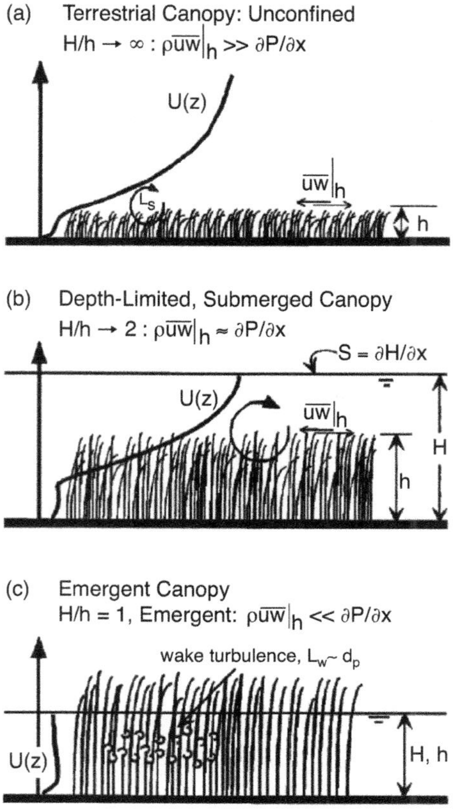

Figure 2.7. Schematic diagram showing the importance of increasing stem height relative to water depth in slowing water currents and trapping sediment. Velocity profiles *U*/(*z*), for three depth ratios (*H/h*), where *H* is water depth, and *h* is canopy height. As the depth ratio decreases (top to bottom) toward the emergent limit, turbulence shifts from predominantly shear dominated to (*Ls*) wake dominated (*Lw*), which signals a decline in turbulence eddy scale (from Nepf and Vivoni 2000).

(Reed 1989; Leonard et al. 1995; Leonard 1997; Nepf and Vivoni 2000; Leonard and Reed 2002; Reed 2002; Nepf 2004; Leonard and Croft 2006). For example, wetland plants can vary from herbaceous plants that are <0.5 m tall to trees that are >30 m tall. This is important because the impact of vegetation on sediment transport is a function of the height of the stems relative to the height of the water (Nepf and Vivoni 2000; Nepf 2004). In cases in which the water depth is much greater than the stem height, flow tends to be driven largely by pressure gradients (Fig. 2.7). As the stem height increases relative to water depth, stresses become increasingly important, and flow vegetation tends to impart turbulence on the flow (Nepf and Vivoni 2000).

The density of stems in a wetland also plays a critical role in effecting spatial patterns of sediment deposition in a wetland, with greater deposition typically found near the front of densely vegetated areas (Coleman 1988; Fagherazzi et al. 2011). The density of stems in deltaic wetlands can range from >100 stems m^{-2} for some grasses to <0.1 stem m^{-2} for bald cypress swamps (Rogers et al. 1996). The arrangement of plant stems can vary from species to species as well. For example, mangroves and large root networks are connected through an above-ground network of branches and stems. Conversely, many herbaceous plants have shoots that are emergent, and the below-ground biomass can consist of roots and rhizomes that are interconnected and form a dense roots mat (i.e., *Spartina patens*) or are not interconnected and are more easily disturbed (i.e., *Sagittaria latifolia*). The strength of wetland plant stems can vary, ranging from grasses that are highly flexible in the environment, to mangroves that are moderate flexible, to cypresses that are effectively rigid (Mitsch and Gosselink 2000).

From this, one can see the difference between the role that marshes (dominated by herbaceous vegetation) and swamps (dominated by woody vegetation) may play in inhibiting hydrodynamic flows. Marshes tend to have flexible stems, low stem heights, and high densities and thus are good at impeding flows of limited water depth and low velocity, but are less effective in deeper water. Conversely, swamps tend to have rigid stems, high stem heights, and low stem densities. As such, swamps are better at impeding flows with greater water depth and higher velocities, but are less effective than marshes in shallow, low-velocity flows. The picture becomes more complex in high-latitude deltaic wetlands (see Chapters 20 and 21), where a thermokarstic topography yields ponds and open bodies of water as ice-rich sediments freeze and thaw (Kokelj and Burn 2005; Burn and Kokelj 2009). These processes can alter sediment erodability and trapping and can promote or inhibit vegetative growth and the retention or flushing of water, depending on the specific conditions of the water body that is formed (Kokelj and Burn 2005; Burn and Kokelj 2009; Schneider et al. 2009).

4.2. Controls on deltaic wetland development

The growth, development, and decay of large river deltaic wetlands are governed by a suite of physical, chemical, geological, and biological controls. Physical processes control the flux of water, sediments, and nutrients into wetlands, as outlined in Part II. Depending on the physical forcing, the movement of water in a deltaic wetland can be unidirectional, bidirectional, or multidirectional. In deltaic wetlands governed by a large river, river-driven flooding is typically a unidirectional process, although river floods can also create smaller secondary features, including eddies and backwater effects. River-driven flooding often produces wetlands that are tear-drop shaped, with higher ground near the natural levees (Coleman and Wright 1975; Roberts 1997). Such higher ground often supports woody vegetation, which has greater stem rigidity and lower stem density than lower areas dominated by grasses, which is often found in the center of such features. Tidal flooding, however, is dominantly a bidirectional process that produces linear-shaped islands (Coleman and Wright 1975). Wave-driven features in deltas typically produce long linear or arcuate features of relatively coarse-grained material that are elevated above sea level (Roberts 1997). These transport processes impact the vegetative community by controlling key habitat components, including the inundation period, the grain size of sediments, the concentration of organic matter in the sediments, and the supply of nutrients to these systems (Coleman and Wright 1975; Coleman 1988; Mitsch and Gosselink 2000). In high-latitude environments, the freeze-thaw cycle of near-surface sediments can retain water and lead to the formations of ponds, both of which have important implications for the development of wetland vegetation (Kokelj and Burn 2005; Burn and Kokelj 2009; Schneider et al. 2009).

Storm-driven flooding can be a multidirectional process. For example, along the coastline adjacent to the Mississippi Delta (see Chapters 3 and 4), water can be transported onshore and into inland water bodies during the approaching phase of a cold front, and offshore during the postfrontal phase, leading to onshore sediment flux into wetlands (Draut et al. 2005) during the prefrontal phase of the storm and mudflat accretion on the postfrontal phase (Roberts et al. 1989; Allison et al. 2000; Draut et al. 2005). Meteorological processes impacting the deltaic shoreline can also affect local sea level (e.g., coastal setup and setdown) and are responsible for centimeters to decimeters scale variability in annual mean sea level that is present in many tide gauges (Kolker et al. 2009; Kolker et al. 2011a; Sturges and Douglas 2011) and can act to accentuate or diminish inundation and sedimentation in coastal wetlands (Reed 1989; Bartholdy et al. 2004; Temmerman et al. 2005; Kolker et al. 2009). Winter storms, tropical cyclones, and other high-energy events can be an important pathway by which sediments are resuspended from bay bottoms and placed on large river coastal wetlands surfaces and transported landward (Reed 1989; Goodbred and Hine 1995; Draut et al. 2005; Ramatchandirane et al. 2010; Pisaric et al. 2011). However, such events can also be erosive, leading to a flux of particles offshore into the coastal zone (Barras 2006). In fact, the frequency of onshore winds and storm setup directly controls the elevation of coastal marshes in many systems. Marshes tend to accrete to, or sometimes even above, the highest astronomical tide levels in onshore-wind–dominated systems, whereas marshes remain below the highest astronomical tide level where tidal forces dominate (Bartholdy et al. 2010). In arctic deltaic wetlands, hydrological processes, and by extension the distribution of vegetation, are strongly linked to thermal processes, including the evolution of ice dams, the freeze-thaw cycle, and the depth of permafrost. Ice rafting also plays a critical role in the creation of ponds and transport of large blocks of sediment (Fig. 2.8) in both high-latitude deltaic and coastal wetlands (Burn and Kokelj 2009; Argow et al. 2011).

These local physical processes are linked to broader scale climate processes that also impact coastal wetlands. For example, changes in climate play a critical role in governing the position of mean sea level: changes in mean sea level can affect the distribution and survival of many wetland plants (Cahoon and Reed 1995; Reed 1995, 2002) and their preservation in the geological record (Kosters 1989). Changes in sea level can also affect the position of the river distributaries that deliver nutrients to marshes, the hydroperiod of wetlands, and the location of wetland depocenters (IPCC 2007). Climate also plays an important role in the temperature and growing period of many wetland plants. Warmer climates have been linked to longer growing periods, which can increase the sediment trapping ability of wetland plants and thus enhance sediment accumulation rates (Kirwan et al. 2009; Kolker et al. 2010). Conversely, warmer year-round temperatures in lower latitudes enhance the period of active bioturbation, decomposition of organic matter, and cycling of chemical constituents (Bertness 1999; Kirwan and Blum 2011). Enhanced rates of mixing and decomposition often lead to marshes that are dominated by inorganic, as opposed to organic, marsh substrates (Nyman et al. 2006). However, increased CO_2 concentrations in some coastal wetlands may contribute to increased below-ground biomass, thereby enhancing wetland accretion (Langley et al. 2009). Future changes in climate may impact the latitudinal boundaries of wetlands, such as mangroves that are limited by winter dieback, with implications for sediment trapping and resistance to sea-level rise and storm erosion (Bakkenes et al. 2002; Walther et al. 2002; Loarie et al. 2008; Comeaux et al. 2012). In arctic

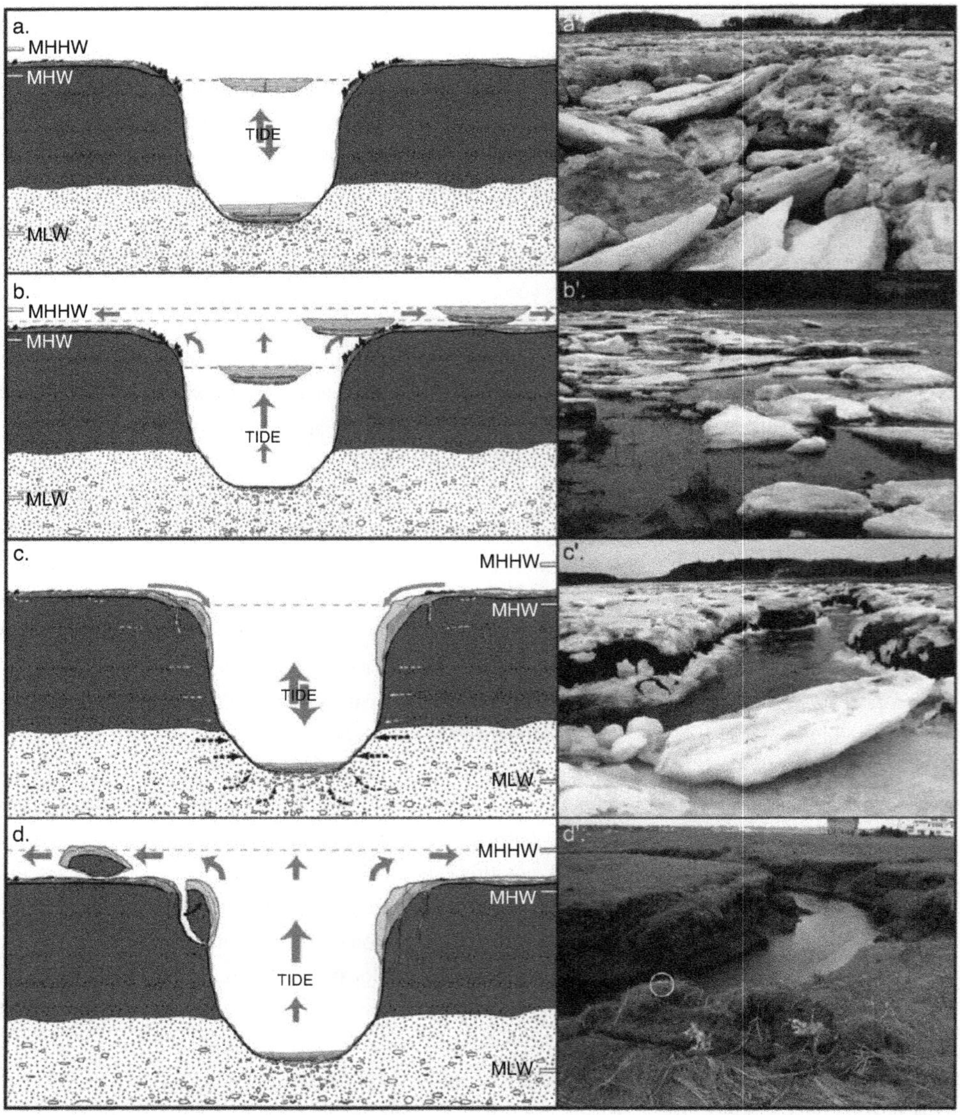

Figure 2.8. Conceptual model of ice raft formation and processes that enlarge the sides and bottom of a wetland tidal creek characteristic of high latitude river-coastal deltaic wetlands (from Argow et al. 2011). (a) floating ice trapped in a tidal channel by insufficient tidal height repeatedly strands at low tide, building up layers of sediment and ice via basal freezing. (b) Spring flood tide allows laden raft to escape the channel and move on to the high-marsh surface, where the raft will be stranded at the turn of the tide. (c) Ice builds on vertical peat banks owing to direct precipitation of ice from tidal waters and from seepage of subsurface flow. Peat becomes weakened under alternating gravitational and buoyant forces. (d) Large cracks develop parallel to the bank and peat blocks may be removed and lifted onto the marsh surface by spring flood waters.

systems, climate also influences the formation of ice, the depth of permafrost, and, through that, the distribution of vegetation (Burn and Kokelj 2009).

The response of wetlands to climate change can be viewed in an ecogeomorphological context (Reed 2002; Fagherazzi et al. 2004). Not only can the rate of relative sea-level rise and the flooding regime impact the health of marsh plants, but these processes can also affect whether and how sediments are bound together by plant roots (DeLaune et al. 1979; Nyman et al. 1990; DeLaune et al. 1994; Day et al. 2008; Feller et al. 2010). Critical geological controls on deltaic wetland development include the rate of subsidence, the supply of sediment to the wetlands, and the geomorphology of both the river and the receiving basin (Reed 2002). Subsidence, along with changes in global sea level, plays a major control on the rate of RSLR, which, as previously mentioned, controls the hydroperiod and the distribution of many wetland plants. In places where sediment supply is not limited, an increase in the rate of RSLR typically leads to an increase in the rate at which sediments are deposited and in which marshes accrete vertically (Orson et al. 1985; Reed 1995, 2002). In places where rates of RSLR are greater than rates at which marshes can accrete, marshes will typically drown (Orson et al. 1985; Reed 1995, 2002). One area where this has been noted is the Mississippi River Delta (see Chapters 17 and 18), where rates of subsidence appear to be linked to rate of fluid production and regional rates of wetland loss (Morton et al. 2002; Meckel et al. 2006; Morton et al. 2006; Meckel 2008; Morton and Bernier 2010; Kolker et al. 2011a). The supply of sediment to deltaic wetlands is governed by patterns of erosion in the uplands and the delivery of sediment to the coastal zone via rivers, which are linked to the regional geology, climate, and patterns of land use (Meade 1996; Meade and Moody 2010). Sediment supply to deltaic wetlands is also linked to marine factors, including longshore currents, onshore transport of sediment, storm setup, and biogenic sediment production (see Chapters 3 and 4). Finally, the geomorphology and hydrodynamics of the river, coupled with internal hydrodynamic processes, play a critical role in determining where avulsions occur (Chatanantavet et al. 2012) and, through that, the areas where deltaic wetlands develop. In some cases, restoration of deltaic wetlands may be achieved by reinitiating the natural hydrodynamic processes that originally built these systems (Paola et al. 2011; Kolker et al. 2012).

The flux and fate of chemical constituents to deltaic wetlands also plays an important role in the life cycle of these systems. Most large rivers carry a large supply of nutrients, such as NO_3^- and PO_4^{3-}, to the coastal zone (Turner et al. 2003; McKee et al. 2004). These nutrients can stimulate plant production, thereby enhancing the ability of marshes to trap sediment and thus stimulating marsh accretion and progradation (Lane et al. 2007; Chapters 17 and 18). However, excess quantities of nutrients may contribute to the eutrophication of coastal wetlands, which can lead to plant diebacks (Darby and Turner 2008). Sulfur, which is present in many deltaic wetlands that receive marine water because SO_4^{2-} is the second most abundant ion in sea water, also plays an important role in the ecogeomorphology of deltaic wetlands. In conditions in which anaerobic respiration dominants, bacteria can reduce SO_4^{2-} to H_2S, which is toxic to marsh plants (Goodman et al. 1959; DeLaune et al. 1994). This is likely to happen in marshes that are either being submerged by RSLR or that have an overly abundant supply of organic matter. This input of SO_4^{2-} also has significant implications for the maintenance of deltaic marshes under conditions of rising sea level, as sulfate drives the decomposition of organic matter via sulfate reduction in salt marshes (Neubauer et al. 2005). In freshwater wetlands, decomposition is more often driven by methanogenesis (Weston et al., 2006).

As sea level rises, salt water intrudes farther upriver, increasing the rate of organic matter decomposition and hastening the inundation of salt marshes in the lower delta plain. These salinity-selective changes, along with the narrowing of typical delta plains, leads to predictable variations in marsh endurance and preferential loss of freshwater marshes and their carbon storage capacity over time (Craft et al. 2009).

4.3. Large river wetlands as particle sources, sinks, and loci of particle transformation

As a result of the factors outlined in the preceding section, deltaic wetlands can function as both sources and sinks for particles to the coastal zone. Wetlands, both in and out of deltas, can function as particle sources when organic parts (e.g., leaves, stems, seeds, and other plant debris) are released into the environment following each year's growing season (Odum and de la Cruz 1967; Gosselink and Kirby 1974; Feijtel et al. 1985; Peterson and Howarth 1987). Many deltaic wetlands across the globe are in a state of degradation, with losses measured in deltas as diverse as the Indus, McKenzie, and the Zambezi, along with many others (Syvitski and Saito 2007; Coleman et al. 2008). In deltaic wetlands that are in an erosional state, the loss of root matter can contribute particles to the coastal zone (Wilson and Allison 2008). Wetland-derived particles can play an important role in the ecology of coastal ecosystems by adding carbon or nutrients to the coastal zone that can contribute to algal blooms and ecosystem metabolism (Bianchi et al. 2011), although wetlands are by no means the only, or even the largest, contributor of carbon to the coastal zone (Howarth 1984; Deegan and Garritt 1997).

Wetlands play a major role as particle traps in the coastal zone. The shoots, stems and leaves of wetland vegetation can reduce water velocities, thereby enhancing sediment deposition (Nepf and Vivoni 2000; Leonard and Reed 2002). Other studies have shown that vegetation, as well other ecosystem engineers (e.g., burrowing crabs, mussels, and oysters), are important for controlling the trapping of sediment and development of topography within marshes (Bertness 1984; Smith and Frey 1985; Newell and Krambeck 1995). One interesting example of the role of vegetation in particle dynamics comes from the lowermost Mississippi deltaic plain (Esposito 2011). A study here found that channels in these wetlands constrained the flow, such that currents were kept relatively high (up to 60 cm s^{-1}) over 5 km from the Mississippi River. Such currents allowed sand to be transported to the front of a developing mouth bar. However, once the flow reached the edge of the developing mouth bar, sediments rapidly fell out of suspension and were deposited on the wetland (Esposito 2011). The particle trapping ability of deltaic wetlands thus serves as a method for removing particle-bound nutrients, contaminants, and metals from the water column and as a means of sequestering these materials in the sediments.

Finally, wetlands play an important role in the transformation of particles in the coastal zone. Wetlands function as a place where particulate organic carbon, nitrogen, and phosphorous are respired and transformed into gaseous and/or dissolved constituents (Mitsch and Gosselink 2000; Wagner et al. 2003; Day et al. 2007). Cai and coworkers have shown that salt marshes are a major source of CO_2 flux to the atmosphere from transformation of organic particles both in the water and within the salt marsh sediments as well (Cai et al. 2003; Cai 2011; Chapter 7). As a locus of primary production, wetlands are also places where dissolved and gaseous carbon, nitrogen, and phosphorus are converted into particles via photosynthesis (Mitsch and Gosselink 2000; Day et al. 2007) and likewise where particulate carbon and nitrogen are converted into gases such as CO_2, and N_2 via respiration, denitrification, and related processes (Baldock et al. 2004). All of these transformations are described in more detail in Chapters 7 and 8.

5. Conclusions

The deltaic plain of major river systems is a complex network of river-dominated channels, estuarine water bodies, wetlands, and shoreline interfaces with the continental shelf, interspersed with infrequently flooded land areas (e.g., levees, beach ridges, relict highlands). The trapping behavior on particulate and dissolved components delivered by the river and derived autochthonously (e.g., wetland plant debris), and the resulting deltaic plain morphology, are highly system specific. As outlined previously, these differences can result from variability in the magnitude, timing, and character of water and sediment delivered by the riverine end member, which is controlled by a wide variety of factors such as catchment precipitation, geology, elevation, and tectonic history. It is also strongly overprinted in many systems by anthropogenic alterations to the catchment and dispersal system as a whole, through damming, agriculture, urbanization, deforestation, and human-induced climate change. Within the deltaic plain, system differences can arise from latitudinal and climatic effects, which control issues such as the type of saline wetlands (mangroves are confined to low latitudes by freeze intolerance, for instance). Astronomical (tides) and meteorological (e.g., wind-driven currents, waves, episodic storms) factors are also highly system-specific in controlling deltaic plain morphology. Antecedent geology also plays a role in controlling subsidence rates, which submerge elevated areas and enlarge inland water bodies, which makes them more susceptible to tidal and meteorological mixing. Here, also, anthropogenic interruption of natural paths of water, sediment, and dissolved component flux from river channel to continental shelf has been severe in many systems, as the fertility of these areas for agriculture and fisheries has made a locus for human civilization for thousands of years. Given these changes, and the low elevation of large river deltaic plains, these are among the most vulnerable natural settings to climate change in the twenty-first century.

References

Aalto, R., L. Maurice-Bourgoin, T. Dunne, D.R. Montgomery, C.A. Nittrouer, and J.L. Guyot. 2003. Episodic sediment accumulation on Amazonian flood plains influenced by El Nino/Southern Oscillation. *Nature* 425: 493–497.

Adams, P.N., R.L. Slingerland, and N.D. Smith. 2004. Variations in natural levee morphology in anastomosed channel flood plain complexes. *Geomorphology* 61: 127–142.

Allen, J.R.L. 1978. Studies in fluviatile sedimentation: an exploratory quantitative model for the architecture of avulsion-controlled alluvial sites. *Sediment. Geol.* 2: 129–147.

Al-Hilli, M.R.A., B.G. Warner, T. Asada, and A. Douabul. 2009. An assessment of vegetation and environmental controls in the 1970s of the Mesopotamian wetlands of southern Iraq. *Wetlands Ecol. Manage.* 17: 207–223.

Allison, M.A., C.R. Demas, B.A. Ebersole, B.A. Kleiss, C.D. Little, E.A. Meselhe, N.J. Powell, T.C. Pratt, and B.M. Vosburg. 2012. A water and sediment budget for the lower

Mississippi-Atchafalaya River in flood years 2008–2010: implications for sediment discharge to the oceans and coastal restoration in Louisiana. *J. Hydrol.* 432: 84–97.

Allison, M.A., and E.A. Meselhe. 2010. The use of large water and sediment diversions in the lower Mississippi River (Louisiana) for coastal restoration. *J. Hydrol.* 387: 346–360.

Allison, M.A., S.R. Khan, S.L. Goodbred, and S.A. Kuehl. 2003. Stratigraphic evolution of the late Holocene Ganges-Brahmaputra lower delta plain. *Sediment. Geol.* 155: 317–342.

Allison, M.A., G.C. Kineke, E.S. Gordon, and M.A. Goni. 2000. Development and reworking of a seasonal flood deposit on the inner continental shelf off the Atchafalaya River. *Cont. Shelf Res.* 20: 2267–2294.

Allison, M.A., Kuehl, S.A., Martin, T.C., and Hassan, A. 1998. Importance of flood-plain sedimentation for river sediment budgets and terrigenous input to the oceans: insights from the Brahmaputra-Jamuna River. *Geology* 26: 175–178.

Argow, B.A., Z.J. Hughes, and D.M. FitzGerald. 2011. Ice raft formation, sediment load, and theoretical potential for ice-rafted sediment influx on northern coastal wetlands. *Cont. Shelf Res.* 31: 1294–1305.

Bakkenes, M., J.R.M. Alkemade, F. Ihle, R. Leemans, and J.B. Latour. 2002. Assessing effects of forecasted climate change on the diversity and distribution of European higher plants for 2050. *Global Change Biol.* 8: 390–407.

Baldock, J.A., C.A. Masiello, Y. Gelinas, Y., and J.L. Hedges. 2004. Cycling and composition of organic matter in terrestrial and marine ecosystems. *Mar. Chem.* 92: 39–64.

Bárcena J.F., A. García, J. García, C. Álvarez, and J.A. Revilla. 2012. Surface analysis of free surface and velocity to changes in river flow and tidal amplitude on a shallow mesotidal estuary: an application in Suances Estuary (Nothern Spain). *J. Hydrol.* 420/1: 301–318.

Barras, J.A. 2006. Land area change in coastal Louisiana after the 2005 hurricanes – a series of three maps. *U.S. Geological Survey Open File Report* 06-1274.

Bartholdy, J., C. Christiansen, and H. Kunzendorf, 2004. Long term variations in backbarrier salt marsh deposition on the Skallingen peninsula – the Danish Wadden Sea. *Mar. Geol.* 203: 1–21.

Bartholdy, A.T., J. Bartholdy, and A. Kroon. 2010. Salt marsh stability and patterns of sedimentation across a backbarrier platform. *Mar. Geol.* 278: 31–42.

Basu, A.R., S. B. Jacobsen, R.J. Poreda, C.B. Dowling, and P.K. Aggarwal. 2001. Large groundwater strontium flux to the oceans from the Bengal basin and the marine strontium isotope record. *Science* 293: 1470–1473.

Beerbower, J.R. 1964. Cyclothems and cyclic depositional mechanisms in alluvial plain sedimentation. In: Merriam, D.F. (ed.), Symposium on Cyclic Sedimentation. *Bull. Kansas Geol. Survey*, 169.2.1.6: pp. 31–42.

Bertness, M.D. 1999. The Ecology of Atlantic Shorelines. Sinauer Associates, Sunderland, MA, p. 417.

Berendsen, H.J.A. 2005. The Rhine-Meuse Delta at a Glance. International Geological Congress Report 2003/2004.

Berendsen, H.J.A. 2000. Late Weichselian and Holocene paleogeography of the Rhine-Meuse delta. *Palaeogeogr. Palaeoclimatol. Palaeoecol.* 161: 311–335.

Berendsen, H. J. A., and E. Stouthamer. 2000. Late Weichselian and Holocene palaeogeography of the Rhine-Meuse delta, the Netherlands. *Palaeogeogr. Palaeoclimatol. Palaeoecol.* 161: 311–335.

Berendsen, H. J. A., and E. Stouthamer. 2001. Palaeogeographic development of the Rhine-Meuse delta, the Netherlands. *Assen, Koninklijke Van Gorcum*, 268 p.

Bertness, M.D. 1984. Ribbed mussels and *Spartina alterniflora* production in a New England salt marsh. *Ecology* 65: 1794–1807.

Beschta, R.L. 1987. Conceptual models of sediment transport in streams. In: Thorne, C.R., Bathurst, J.C., and Hey, R.D. (eds.), Sediment Transport in Gravel-Bed Rivers, pp. 387–419. Chichester: John Wiley & Sons.

Bianchi, T.S., L.A. Wysocki, K.M. Schreiner, T.R. Filley, D.R. Corbett, and A.S. Kolker. 2011. Sources of terrestrial organic carbon in the Mississippi River plume region: Evidence for the importance of coastal marsh inputs. *Aquat. Geochem.* 17: 431–456.

Bianchi, T.S., and M.A. Allison. 2009. Large-river delta-front estuaries as natural "recorders" of global environmental change. *Proc. Natl. Acad. Sci.* 106: 8085–8092.

Blum, M.D., and H.H. Roberts. 2009. Drowning of the Mississippi Delta due to insufficient sediment supply and sea level rise. *Nat. Geosci.* 2(7): 488–491.

Bos, I.J., and E. Stouthamer. 2011. Spatial and temporal distribution of sand-containing basin fills in the Holocene Rhine-Meuse Delta, the Netherlands. *J. Geol.* 119: 641–660.

Bridge, J.S. 2003. Rivers and Floodplains: Forms, Processes, and Sedimentary Record. Malden, Ma: Blackwell.

Bridge, J.S., and M.R. Leeder. 1979. A simulation model of alluvial stratigraphy. *Sedimentology* 26: 617–644.

Brierley, G.J., R.J. Ferguson, and K.J. Woolfe. 1997. What is a fluvial levee? *Sediment. Geol.* 114: 1–9.

Burn, C.R., and S.V. Kokelj. 2009. The environment and permafrost of the Mackenzie delta area. *Permafrost Periglac. Process.* 20: 83–105.

Burnett, W.C., P. Aggarwal, H. Bokuniewicz, J.E. Cable, M.A. Charette, E. Kontar, S. Krupa, K.M. Kulkarni, A. Loveless, W.S. Moore, J.A. Oberdorfer, J. Oliveira, N. Ozyurt, P.P. Povinec, A.M.G. Privitera, R. Rajar, R.T. Ramessur, J. Scholten, T. Stieglitz, M. Taniguchu, and J.V. Turner. 2006. Quantifying submarine groundwater discharge in the coastal zone via multiple methods. *Sci. Total Environ.* 367: 498–543.

Burnett, W.C., H. Bokeniewicz, M. Huettel, W.S. Moore, and M. Taniguchi. 2003. Groundwater and pore water inputs to the coastal zone. *Biogeochemistry* 66: 3–33.

Cahoon, D.R., and D.J. Reed. 1995. Relationships among surface topography, hydroperiod and soil accretion in a deteriorating Louisiana salt-marsh. *J. Coast. Res.* 11: 357–369.

Cahoon, D.R., D.A. White, and J.C. Lynch. 2011. Sediment infilling and wetland formation dynamics in an active crevasse splay of the Mississippi River delta. *Geomorphology* 131: 57–68.

Cai, W.J. 2011. Estuarine and coastal ocean carbon paradox: CO_2 sinks or sites of terrestrial carbon incineration? *Ann. Rev. Mar. Sci.* 3: 123–145.

Cai, W.J., Z. Wang, and Y. Wang. 2003. The role of marsh-dominated heterotrophic continental margins in transport of CO_2 between the atmosphere, the land-sea interface and the ocean. *Geophys. Res. Lett.* 30: 1849.

Caputo, M., L. Pieri, and M. Unghendoli. 1970. Geometric investigation of the subsidence in the Po Delta. *Boll. Geofis. Teor. Appl.* 14: 187–207.

Cazanacli, D., and N.D. Smith. 1998. A study of morphology and texture of natural levees – Cumberland Marshes, Saskatchewan, Canada. *Geomorphology* 25: 43–55.

Chatanantavet, P., M.P. Lamb, and J.A. Nittrouer. 2012. Backwater controls on avulsion location in deltas. *Geophys. Res. Lett.* 39: L01402.

Chen, C., E.A. Meselhe, and M.G. Waldon. 2012. Assessment of mineral concentration impacts from pumped stormwater on an Everglades wetland, Florida, USA – using a spatially-explicit model. *J. Hydrol.* 452/3: 25–39.

Chen S.-L., G.-A. Zhang, S.-L. Yang, and J.Z. Shi. 2006. Temporal variations of fine suspended sediment concentration in the Changjiang River estuary and adjacent coastal waters, China. *J. Hydrol.* 331: 137–145.

Coleman J.M. 1969. Brahmaputra River: channel processes and sedimentation. *Sediment. Geol.* 3: 129–239.

Coleman, J.M. 1988. Dynamic changes and processes in the Mississippi River Delta. *Geol. Soc. of Am. Bull.* 100: 999–1015.

Coleman, J.M., and S.M. Gagliano. 1964. Cyclic sedimentation in the Mississippi River deltaic plain. *Gulf Coast Assoc. Geolog. Soc. Transact.* 14: 67–80.

Coleman, J.M., and L.D. Wright. 1975. Modern river deltas; variability of processes and sand bodies. In: Broussard, M.L. (ed.), Deltas; Models for Exploration, pp. 99–149. Houston, TX: Houston Geological Society.

Coleman, J.M., and L.D. Wright. 1971. *Analysis of Major River Systems and their Deltas, Procedures and Rationale, with Two Examples*. Louisiana State University Coastal Studies Institute Report 95.

Coleman, J.M., O.K. Huh, and D. Braud. 2008. Wetland loss in world deltas. *J. Coastal Res.* 24: 1–14.

Comeaux, R.S., M.A. Allison, and T.S. Bianchi, 2012. Mangrove expansion in the Gulf of Mexico with climate change: implications for wetland health and resistance to rising sea levels. *Estuar. Coast. Shelf Sci.* 96: 81–95.

Connell, S.D., W. Kim, C. Paola, and G.A. Smith. 2012. Fluvial morphology and sediment-flux steering of axial-transverse boundaries in an experimental basin. *J. Sediment. Res.* 82: 310–325.

Craft, C., J. Clough, J. Ehman, S. Joye, D. Park, S. Pennings, H. Guo, and M. Machmuller. 2009. Forecasting the effects of accelerated sea level rise on tidal marsh ecosystem services. *Front. Ecol. Environ.* 7: 73–78.

Darby, F., and R.E. Turner. 2008. Effects of eutrophication on salt marsh root and rhizome biomass accumulation. *Mar. Ecol. Prog. Series* 363: 63–74.

Davis, M. 2010. Numerical Simulation of Unsteady Hydrodynamics in the Lower Mississippi River. Masters of Science Thesis, Civil & Environmental Engineering, Hydraulic Engineering, University of New Orleans, New Orleans, LA.

Davis, D.W. 2000. Historical perspective on crevasses, levees, and the Mississippi River, In: Colten, C.E. (ed.), Transforming New Orleans and its Environs, pp. 84–106. Pittsburgh, PA: University of Pittsburgh Press.

Day, J.W., D.F. Boesch, E.J. Clairain, G.P. Kemp, S.B. Laska, W.J. Mitsch, K. Orth, H. Mashriqui, D.J. Reed, L. Shabman, C.A. Simenstad, B.J. Streever, R.R. Twilley, C.C. Watson, J.T. Wells, and D.F. Whigham. 2007. Restoration of the Mississippi Delta: lessons from Hurricanes Katrina and Rita. *Science* 315: 1679–1684.

Day, J.W., R.R. Christian, D.F. Boesch, A. Yanez-Arancibia, R.R. Twilley, L. Naylor, L. Schaffner, and J.C. Stevenson. 2008. Consequences of climate change on the ecogeomorphology of coastal wetlands. *Ann. Rev. Mar. Sci.* 31: 477–491.

Deegan, L.A. and R.H. Garritt. 1997. Evidence for spatial variability in estuarine food webs. *Mar. Ecol. Prog. Series* 147: 31–47.

DeLaune, R.D., J.A. Nyman, and W.H.J. Patrick. 1994. Peat collapse, ponding and wetland loss in a rapidly submerging coastal marsh. *J. Coast. Res.* 10: 1021–1030.

DeLaune, R.D., R.J. Buresh, and W.H.J. Patrick. 1979. Relationship of soil properties to standing crop biomass of *Spartina alterniflora* in a Louisiana marsh. *Estuar. Coast. Mar. Sci.* 8: 477–487.

de Louw, P.G.B., S. Eeman, B. Siemon, B.R. Voortman, J. Gunnink, E.S. van Baaren, and G.H.P. Oude Essink. 2011. Shallow rainwater lenses in deltaic areas with saline seepage. *Hydrol. Earth Syst. Sci.* 15: 3659–3678.

Demko, T.M., B.S. Currie, and K.A. Nicholl. 2004. Regional paleoclimatic and stratigraphic implications of paleosols and fluvial/overbank architecture in the Morrison Formation (Upper Jurassic), Western Interior, USA. *Sediment. Geol.* 167: 115–135.

Draut, A.E., G.C. Kineke, O.K. Huh, J.M. Grymes, K.A. Westphal, and C.C. Moeller. 2005. Coastal mudflat accretion under energetic conditions, Louisiana chenier-plain coast, USA. *Mar. Geol.* 214: 27–47.

Dyer, K.R. 1989. Sediment processes in estuaries: future research requirements. *J. Geophys. Res.* 94: 14327–14399.

Eaton T. 2007. Analytical estimates of hydraulic parameters for an urbanized estuary – Flushing Bay. *J. Hydrol.* 347: 188–196.

Edmonds, D., and R. Slingerland. 2009. Significant effect of sediment cohesion on delta morphology. *Nat. Geosci.* 3(2): 105–109.

Edmonds, D.A., D.C.J.D. Hoyal, B.A. Sheets, and R.L. Slingerland. 2009. Predicting delta avulsions: implications for coastal wetland restoration. *Geology* 37: 759–762.

Edmonds, D., R. Slingerland, J. Best, D. Parsons, and N. Smith. 2010. Response of river-dominated delta channel networks to permanent changes in river discharge. *Geophys. Res. Lett.* 37: L12404.

Ericson, J.P., C.J. Vörösmarty, S.L. Dingman, L.G. Ward, and M. Meybeck. 2006. Effective sea-level rise and deltas: causes of change and human dimension implications. *Global Planet. Change* 50: 63–82.

Esposito, C.R. 2011. *Differential Sedimentation in a Mississippi River Crevasse Splay*, Masters of Science Thesis, Dept. of Earth of Environmental Sciences, University of New Orleans, New Orleans, LA, p. 88.

Essink, G.H.P.O., E.S. van Baaren, and P.G.B. de Louw. 2010. Effects of climate change on coastal groundwater systems: A modelling study in the Netherlands. *Water Resour. Res.* 46: W00F04.

Fagherazzi, S., M.L. Kirwan, S.M. Mudd, G.R. Guntenspergen, S. Temmerman, A. D'Alpaos, J. van de Koppel, J.M. Rybczyk, E. Reyes, C. Craft, and J. Clough. 2011. Numerical models of salt marsh evolution: ecological, geomorphic an climatic factors. *Rev. Geophys.* 50: 10.1029/2011RG000359.

Fagherazzi, S., M. Marani, and L.K. Blum. 2004. The Ecogeomorphology of Tidal Marshes, Coastal and Estuarine Studies. Washington, DC: American Geophysical Union.

Feller, I.C., C.E. Lovelock, U. Berger, K.L. McKee, S.B. Joye, and M.C. Ball. 2010. Biocomplexity in mangrove ecosystems. *Ann. Rev.Mar. Sci.* 2: 395–417.

Feijtel, T.C., R.D. DeLaune, and W.H.J. Patrick. 1985. Carbon flow in coastal Louisiana. *Mar. Ecol. Prog. Ser.* 24: 255–260.

Filgueira-Rivera M., N.D. Smith, and R.L. Slingerland. 2007. Controls on natural levée development in the Columbia River, British Columbia, Canada. *Sedimentology* 54: 905–919.

Fischer H.B., E. J. List, R.C.Y. Koh, J. Imberger, and N.H. Brooks. 1979. Mixing in Inland and Coastal Waters. San Diego, CA: Academic Press.

Forbes, C., R.A. Luettich, Jr., C. Mattocks, and J.J. Westerink. 2008. A retrospective evaluation of the storm surge produced by Hurricane Gustav (2008): forecast and hindcast results. *Weather Forecast.* 25: 1577–1602.

Frazier, D.E. 1967. Recent deltaic deposits of the Mississippi River – their development and chronology. *Transact. Gulf Coast Assoc. Geolog. Soc.* 17: 287–315.

Gabriel Pavel, L. Josef, and F. Pavel (2007). Hydraulic Research of the Decin Barrage. T.G. Masaryk Water Research Institute, Public Research Institution. *Prague.* ISBN 978-80-85900-73-6.

Galloway, W.E. 1975. Process framework for describing the morphologic and stratigraphic evolution of deltaic depositional systems. In: Broussard, M.L. (ed.), Deltas, Models for Exploration, pp. 87–98. Houston, TX: Houston Geological Society.

Gautier, E., J. Corbonnois, G. Petit, G. Arnaud-Fassetta, D. Brunstein, S. Grivel, G. Houbrechts, and T. Beck. 2009. Multi-disciplinary approach for sediment dynamics study of active floodplains. *Geomorphol. Relief Process Environ.* 1: 65–78.

Georgiou, I.Y. 2011. Deltaic and Estuarine Exchange Processes. Presentation at the Annual Northern Gulf of Mexico (NGOM) Project Science Meeting, U.S. Geological Survey National Wetlands Research Center, May 24–26, Lafayette, LA.

Goodbred, S.L., and A.C. Hine. 1995. Coastal storm deposition: Salt marsh response to a severe extratropical storm, 1993, west-central Florida. *Geology* 23: 679–682.

Goodbred, S.L., and S.A. Kuehl. 2000. The significance of large sediment supply, active tectonism, and eustacy on margin sequence development: Late Quaternary stratigraphy and evolution of the Ganges-Brahmaputra delta. *Sediment. Geol.* 133: 227–248.

Goodbred, S.L., and S.A. Kuehl. 1999. Holocene and modern sediment budgets for the Ganges-Brahmaputra river system: evidence for highstand dispersal to the flood-plain, shelf, and deep-sea depocenters. *Geology* 27: 559–562.

Goodman, P.J., E.M. Braybooks, and J.M. Lambert. 1959. Investigations into 'die-back' in *Spartina townsendii* AGG: The present status of *Spartina townendii* in Britain. *J. Ecol.* 47: 651–677.

Gosselink, J.G., and C.J. Kirby. 1974. Decomposition of salt marsh grass, Spartina alterniflora Loisel. *Limnol. Oceanogr.* 19: 825–832.

Gouw, M.J.P. 2008. Alluvial architecture of the Holocene Rhine-Meuse delta (the Netherlands). *Sedimentology* 55: 1487–1534.

Gouw, M.J.P., and W.J. Autin. 2008. Alluvial architecture of the Holocene Lower Mississippi Valley (USA) and a comparison with the Rhine-Meuse delta (The Netherlands). *Sediment. Geol.* 204: 106–121.

Gretener, B., and L. Strömquist. 1987. Overbank sedimentation-rates of fine-grained sediments – a study of the recent deposition in the lower river Fyrisan. *Geografiska Ann. Series A Phys. Geogr.* 69: 139–146.

Holbrook, J.M., and S.A. Schumm. 1999. Geomorphic and sedimentary response of rivers to tectonic deformation: a brief review and critique of a tool for recognizing subtle epeirogenic deformation in modern and ancient settings: Special issue on Tectonics of Continental Interiors. *Tectonophysics* 305: 287–306.

Horrevoetsa A.C., H.H.G. Savenijea, J.N. Schuurmana, and S. Graasa. 2004. The influence of river discharge on tidal damping in alluvial estuaries. *J. Hydrol.* 294: 213–228.

Howard, J.D., and R.W. Frey. 1985. Physical and biogenic aspects of backbarrier sedimentary sequences, Georgia Coast, USA. *Mar. Geol.* 63: 77–127.

Howarth, R.W. 1984. The ecological significance of sulfur in the energy dynamics of salt marsh and coastal marine sediments. *Biogeochemistry* 1: 5–27.

IPCC, 2007. Climate Change 2007: The Physical Science Basis. Contributions of Working Group I to the Fourth Assessment Report of the Intergovernmental Panel of Climate Change. New York: Cambridge University Press.

Iman, M.B., and H.F. Shaw. 1985. The diagenesis of Neogene clastic sediments from the Bengal Basin, Bangladesh. *J. Sediment. Petrol.* 55: 665–671.

James, C.S. 1985. Sediment transfer to overbank sections. *J. Hydraul. Res.* 23: 435–452

Jerolmack, D. J., and D. Mohrig. 2007. Conditions for branching in depositional rivers, *Geology*, 35(5), 463–466, doi:10.1130/G23308A.1.

Johnson, S.Y., and A.M.N. Alam. 1991. Sedimentation and tectonics of the Sylhet Trough, Bangladesh. *Geol. Soc. Am. Bull.* 103: 1513–1527.

Kang, S., A. Lightbody, C. Hill, and F. Sotiropoulos. 2011. High-resolution numerical simulation of turbulence in natural waterways. *Adv. Water Resour.* 34: 98–113.

Kim, W., D. Mohrig, R. Twilley, C. Paola, and G. Parker. 2009. Is it feasible to build new land in the Mississippi River Delta? *Eos* 90: 373–374.

Kim W., A. Dai, T. Muto, and G. Parker. 2009. Delta progradation driven by an advancing sediment source: coupled theory and experiment describing the evolution of elongated deltas. *Water Resour. Res.* 45: W06428.

Kirwan, M.L., and L.K. Blum. 2011. Enhanced decomposition offsets enhanced productivity and soil carbon accumulation in coastal wetlands responding to climate change. *Biogeosciences* 8: 987–993.

Kirwan, M.L., G.R. Guntenspergen, and J.T. Morris. 2009. Latitudinal trends in *Spartina alterniflora* productivity and the response of coastal marshes to global change. *Global Change Biol.* 15: 1982–1989.

Kokelj, S.V., and C.R. Burn. 2005. Near-surface ground ice in sediments of the Mackenzie Delta, Northwest Territories, Canada. *Permafrost Periglac. Process.* 16: 291–303.

Kolker, A.S., K.H. Johannesson, J.E. Cable, and M.A. Allison. 2013. Pathways and processes associated with the transport of groundwaters in deltaic systems. *Journal of Hydrology* 498: 319–334.

Kolker, A.S., M.D. Miner, and H.D. Weathers. 2012. Depositional dynamics in a river diversion receiving basin: the case of the West Bay Mississippi River Diversion. *Estuar. Coast. Shelf Sci.* 106: 1–12.

Kolker, A.S., M.A. Allison, and S. Hameed. 2011a. An evaluation of subsidence rates and sea-level variability in the Northern Gulf of Mexico. *Geophys. Res. Lett.* 38: L21404.

Kolker, A.S., D.J. Meffert, A. Carbonell, and S.A. Nelson. 2011b. The lessons from a major climate event: Hurricane Katrina and New Orleans. International Panel on Climate Change in Cities First Assessment Report. Cambridge: Cambridge University Press.

Kolker, A.S., M. Kirwan, S.L. Goodbred, and J.K. Cochran. 2010. Global climate changes recorded in coastal wetland sediments: empirical observations lined to theoretical predictions. *Geophys. Res. Lett.* 37: L14706.

Kolker, A.S., S.L. Goodbred, S. Hameed, and J.K. Cochran. 2009. High resolution records of coastal systems responses to short-term sea-level variability. *Estuar. Coast. Shelf Sci.* 84: 493–508.

Kosters, E.C. 1989. Organic-clastic facies relationshops and chronostratigraphy the the Barataria interlobe basin, Mississippi Delta Plain. *J. Sediment. Petrol.* 59: 98–113.

Koutsikopoulos, C., and B. Le Cann. 1996. Physical processes and hydrological structures related to the Bay of Biscay Anchovy. *Sci. Mar.* 60: 9–19.

Krest, J.M., W.S. Moore, and Rama. 1999. ^{226}Ra and ^{228}Ra in the mixing zones of the Mississippi and Atchafalaya Rivers: indicators of groundwater input. *Mar. Chem.* 64: 129–152.

Kuehl, S.A., M.A. Allison, S.L. Goodbred, and H. Kudrass. 2005. Sedimentary processes and products of the modern Ganges-Brahmaputra delta. In: Giosan, L., and Bhattacharya, J. (eds.), River Deltas: Concepts, Models and Examples. Society for Sedimentary Geology (SEPM) Special Publication 83, Houston, Texas, pp. 413–434.

LACPRA. 2012. *Louisiana's Comprehensive Master Plan for a Sustainable Coast.* Baton Rouge, LA: Lousisiana Coastal Protection and Restoration Authority (LACPRA).

Lane, R.R., J.W. Day, B.D. Marx, E. Reyes, E. Hyfield, and J.N. Day. 2007. The effects of riverine discharge on temperature, salinity, suspended sediment and chlorophyll a in a Mississippi delta estuary measured using a flow-through system. *Estuar. Coast. Shelf Sci.* 74: 145–154.

Langley, J.A., K.L. McKee, D.R. Cahoon, J.A. Cherry, and J.P. Megonigal. 2009. Elevated CO_2 stimulates marsh elevation gain, counterbalancing sea-level rise. *Proc. Natl. Acad. Sci.* 106: 6182–6186.

Leeder, M.R. 1978. A quantitative stratigraphic model for alluvium, with special reference to channel deposit density and interconnectedness. *Can. Soc. Petrol. Geologist. Memoir* 5: 587–596.

Leonard, L.A. 1997. Controls of sediment transport and deposition in an incised mainland marsh basin, southeastern North Carolina. *Wetlands* 17: 263–274.

Leonard, L.A., and A.L. Croft. 2006. The effect of standing biomass on flow velocity and turbulence in *Spartina alterniflora* canopies. *Estuar. Coast. Shelf Sci.* 69: 325–336.

Leonard, L.A., and D.J. Reed. 2002. Hydrodynamics and sediment transport through tidal marsh canopies. *J. Coast. Res.* 36: 459–469.

Leonard, L.A., A.C. Hine, and M.E. Luther. 1995. Surficial sediment transport and deposition processes in a *Juncus roemarianus* marsh, west-central Florida. *J. Coast. Res.* 11: 322–336.

Loarie, S.R., B.E. Carter, K. Hayhoe, S. McMahon, R. Moe, C.A. Knight, and D.D. Ackerly. 2008. Climate change and the future of California's endemic flora. *PLoS One* 3: e2502.

Lopes J.F., J.M. Dias, and I. Dekeyse. 2006. Numerical modeling of cohesive sediment transport in the Ria de Aveiro Lagoon, Portugal. *J. Hydrol.* 319: 176–198.

Mackey, S.D., and J.S. Bridge. 1995. Three-dimensional model of alluvial stratigraphy: Theory and applications. *J. Sediment. Res. B Stratigr. Global Stud.* 65: 7–31.

Mbraheem, A.M., M.M. Senosy, and K. Dahab. 1997. Geoelectrical and hydrogeochemical studies for delineating ground-water contamination due to salt water intrusion in the northern part of the Nile Delta, Egypt. *Ground Water* 35: 216–222.

McKee, B.A., R.C. Aller, M.A. Allison, T.S. Bianchi, and G.C. Kineke. 2004. Transport and transformation of dissolved and particulate materials on continental margins influenced by major rivers: benthic boundary layer and seabed processes. *Cont. Shelf Res.* 24: 899–926.

Meade, R.H. 1996. River-sediment inputs to major deltas, In: Milliman, J.D., Haq, B. (eds.), Sea-level Rise and Coastal Subsidence, pp. 63–85. London: Kluwer Academic Publishing.

Meade, R.H., and J.A. Moody. 2010. Causes for the decline of suspended-sediment discharge in the Mississippi River system, 1940–2007. *Hydrol. Process.* 24: 35–49.

Meckel, T.A. 2008. An attempt to reconcile subsidence rates determined from various techniques in southern Louisiana. *Quatern. Sci. Rev.* 27: 1517–1522.

Meckel, T.A., U.S. Ten Brink, and S.J. Williams. 2007. Sediment compaction rates and subsidence in deltaic plains: numerical constraints and stratigraphic influences. *Basin Res.* 19: 19–31.

Meckel, T.A., U.S. Ten Brink, and S.J. Williams. 2006. Current subsidence rates due to compaction of Holocene sediments in southern Louisiana. *Geophys. Res. Lett.* 33: L11403.

Meselhe E.A., and F. Sotiropoulos. 2000. Three dimensional numerical model for open channels with free surface variations. *J. Hydraul. Res.* 38: 115–121.

Meselhe E.A., L.A. Weber, A.J. Odgaard, and T. Johnson. 2000. Numerical modeling for fish diversion studies. *J. Hydraul. Engineer.* 126: 365–374.

Meybeck M., and C.J. Vörösmarty. 2005. Fluvial filtering of land to ocean fluxes: from natural Holocene variations to Anthropocene. *Comptes Rendus* 337: 107–123.

Mertes, L.A.K, T. Dunne, and L.A. Martinelli. 1996. Channel–floodplain geomorphology along the Solimões–Amazon River, Brazil. *Geol. Soc. Am. Bull.* 108: 1089–1107.

Mitsch, W.J., and J.G. Gosselink. 2000. Wetlands. New York: John Wiley and Sons.

Moore, W.S. 1999. The subterranean estuary: A reaction zone of ground water and sea water. *Mar. Chem.* 65: 111–125.

Moore, W.S. 2010. The effect of submarine groundwater discharge on the ocean. *Ann. Rev. Mar. Sci.* 2: 59–88.

Moore, W.S., and J.M. Krest. 2004. Distribution of ^{222}Ra and ^{226}Ra in the plumes of the Mississippi and Atchafalaya River and the Gulf of Mexico. *Mar. Chem.* 86: 105–119.

Morton, R.A., and J.C. Bernier. 2010. Recent subsidence-rate reductions in the Mississippi Delta and their geological implications. *J. Coast. Res.* 26: 555–561.

Morton, R.A., J.C. Bernier, and J.A. Barras. 2006. Evidence of regional subsidence and associated interior wetland loss induced by hydrocarbon production, Gulf coast region. *Environ. Geol.* 50: 261–274.

Morton, R.A., N.A. Buster, and M.D. Krohn. 2002. Subsurface controls on historical subsidence rates and associated wetland loss in southcentral Louisiana. *Transact. Gulf Coast Assoc. Geol. Soc.* 52: 767–778.

Mulder, T., and J.P.M. Syvitski. 1996. Climatic and morphologic relationships of rivers: implications of sea level fluctuations on river loads. *J. Geol.* 104: 509–523.
Muste M., E.A. Meselhe, L. Weber, and A.A. Bradley. 2001. Coupled physical-numerical analysis of flows in natural waterways. *J. Hydraul. Res.* 39: 51–60.
NEDECO. 1965. A Study of the Siltation of the Bangkok Port Channel, Vol. 2, the Field Investigations. The Hague: Netherlands Engineering Consultants.
NEDECO. 1968. Surinam Transportation Study: Report on Hydraulic Investigation. The Hague: Netherlands Engineering Consultants.
Nepf, H.N. 2004. Vegetated flow dynamics, In: Fagherazzi, S., Marani, M., Blum, L.K. (eds.), The Ecogeomorphology of Tidal Marshes, pp. 137–163. Washington, DC: American Geophysical Union.
Nepf, H.M., and E.R. Vivoni. 2000. Flow structure in depth-limited, vegetated flow. *J. Geophys. Res.* 105: 29547–28557.
Neubauer, S.C., K. Givier, S.K. Valentine, and J.P. Megonigal, 2005. Seasonal patterns and plant-mediated controls of subsurface wetland biogeochemistry. *Ecology*, 86(12): 3334–3344.
Newell, S.Y., and C. Krambeck. 1995. Responses of bacterioplankton to tidal inundations of a salt marsh in a flume and adjacent mussel enclosure. *J. Exp. Mar. Biol. Ecol.* 190: 79–95.
Nittrouer, C.A., S.A. Kuehl, R.W. Sternberg, A.G. Figueiredo, and L.E.C. Faria. 1995. An introduction to the geological significance of sediment transport and accumulation on the Amazon continental shelf. *Mar. Geol.* 125: 177–192.
Nittrouer, J.A., D. Mohrig, M.A. Allison, and A-P.B. Peyret. 2011. The lowermost Mississippi River: a mixed bedrock-alluvial channel. *Sedimentology* 58: 1914–1934.
Nyman, J.A., R.J. Walters, R.D. DeLaune, and W.H.J. Patrick. 2006. Marsh vertical accretion via vegetative growth. *Estuar. Coast. Shelf Sci.* 69: 370–380.
Nyman, J.A., R.D. DeLaune, and W.H.J. Patrick. 1990. Wetland soil formation in the rapidly subsiding Mississippi River Deltaic Plain: mineral and organic matter relationships. *Estuar. Coast. Shelf Sci.* 31: 57–69.
Odum, E.P., and A.A. de la Cruz. 1967. Particulate organic detritus in a Georgia salt-marsh estuarine ecosystem, In: Lauff, G.H. (ed.), Estuaries. American Association for the Advancement of Science. Baltimore, MD.
Orson, R., W. Panageotou, and S.P. Leatherman. 1985. Response of tidal salt marshes of the U.S. Atlantic and Gulf coasts to rising sea levels. *J. Coast. Res.* 1: 29–37.
Orton, G.J., and H.G. Reading. 1993. Variability of deltaic processes in terms of sediment supply, with particular emphasis on grain size. *Sedimentology* 40: 475–512.
Paola, C. et al., 2011. Natural Processes in Delta Restoration: Application to the Mississippi Delta. *Ann. Rev. Mar. Sci.* 3: 67–91.
Penland, S., R. Boyd, and J.R. Suter. 1988. Transgressive depositional systems of the Mississippi delta plain: a model for barrier shoreline and shelf sand development. *J. Sediment. Res.* 58: 932–949.
Peterson, B.J., and R.W. Howarth. 1987. Sulfur, carbon and nitrogen isotopes used to trace organic-matter flow in the salt-marsh estuaries of Sapelo Island, Georgia. *Limnol. Oceanogr.* 32(6): 1195–1213.
Phillips, J.D., and M.C. Slattery. 2007. Downstream trends in discharge, slope, and stream power in a lower coastal plain river. *J. Hydrol.* 334: 290–303.
Pisaric, M.F.J., J.R. Thienpont, S.V. Kokelj, H. Nesbitt, T.C. Lantz, S. Solomon, and J.P. Smol. 2011. Impacts of a recent storm surge on an Arctic delta ecosystem examined in the context of the last millenium. *Proc. Natl. Acad. Sci.* 108: 8960–8965.
Pizzuto, J.E. 1987. Sediment diffusion during overbank flows. *Sedimentology* 34: 301–317.

Politano M., P. Carrica, and L. Weber. 2009. A multiphase model for the hydrodynamics and total dissolved gas in tailraces. *Int. J. Multiphase Flow* 35: 1036–1050.

Postma, G. 1995. Sea-level-related architectural trends in coarse grained delta complexes. *Sediment. Geol.* 98: 3–12.

Postma, H. 1967. Sediment transport and sedimentation in the estuarine environment, In: Lauff, G.H. (ed.), Estuaries, pp. 158–179. Washington, DC: American Association for the Advancement of Science.

Pritchard D.W. 1967. Observations of circulation in coastal plain estuaries, In: Lauff, G.H. (ed.), Estuaries, pp. 37–44. Washington, DC: American Association for the Advancement of Science.

Ramatchandirane, C.G., A.S. Kolker, A.D. Ameen, K. Williams, J.P. Donnelly, and L. Giosan. 2010. Does reactivation of Louisiana's Chenier Plain lead to the development of interior wetlands? Assessing the relative roles of storm impacts and riverine deposits, 2010 Fall Meeting, AGU, San Francisco, CA.

Reed, D.J. 2002. Sea-level rise and coastal marsh sustainability: geological and ecological factors in the Mississippi delta plain. *Geomorphology* 48: 233–243.

Reed, D.J. 1995. The response of coastal marshes to sea-level rise: survival or submergence? *Earth Surf. Process. Landforms* 20: 39–48.

Reed, D.J. 1989. Patterns of sediment deposition in subsiding coastal salt marshes: the role of winter storms. *Estuaries* 12: 222–227.

Rego, J., E.A. Meselhe, J. Stronach, and E. Habib. 2010. Numerical modeling of the Mississippi-Atchafalaya Rivers' sediment transport and fate: considerations for diversion scenarios. *J. Coast. Res.* 26: 212–229.

Reitz, M., D. Jerolmack, and J. Swenson 2010. Flooding and flow path selection on alluvial fans and deltas. *Geophys. Res. Lett.* 37(6), L06401, doi:10.1029/2009GL041985.

Roberts, H.H. 1997. Dynamic changes of the Holocene Mississippi River delta plain: the delta cycle. *J. Coast. Res.* 13: 605–627.

Roberts, H.H., O.K. Huh, S.A. Hsu, L.J. Rouse, and D.A. Rickman. 1989. Winter storm impacts on the Chenier Plain coast of southwestern Louisiana. *Transact. Gulf Coast Assoc. Geol. Soc.* 39: 515–522.

Rogers, J.A., S.T. Schwikert, and A. Shapiro-Wenner. 1996. Habitat of wood storks in north and central Florida, USA. *Waterbirds* 19: 1–21.

Saito, Y., N. Chaimanee, T. Jarupongsakul, and J.P.M. Syvitski. 2007. Shrinking megadeltas in Asia: sea-level rise and sediment reduction impacts from case study of the Chao Phraya Delta. *Newslett. IGBP/IHDP Land Ocean Interact. Coast. Zone* 2: 3–9.

Schindler, J. 2010. Estuarine Dynamics as a Function of Barrier Island Transgression and Wetland Loss: Understanding the Transport and Exchange Processes. University of New Orleans Theses and Dissertations. Paper 1260. http://scholarworks.uno.edu/td/1260.

Schneider, J., G. Grosse, and D. Wagner, 2009. Land cover classification of tundra environments inthe Arctic Lena Delta based on Landsat 7 ETM+ data and its application for upscaling of methane emissions. *Remote Sens. Environ.* 113: 380–391.

Shi, J.Z. 2010. Tidal resuspension and transport processes of fine sediment within the river plume in the partially-mixed Changjiang River estuary, China: a personal perspective. *Geomorphology* 121: 133–151.

Shi, J.Z., and L. Li-Feng. 2011. A short note on the dispersion, mixing, stratification and circulation within the plume of the partially-mixed Changjiang River estuary, China. *J. Hydroenviron. Res.* 5: 111–126.

Simm, D.J. 1995. The rates and patterns of overbank deposition on a lowland floodplain. In: Foster, I.D.L., Gurnell, A.M., Webb, B.W. (eds.), Sediment and Water Quality in River Catchments, pp. 247–264. New York: John Wiley and Sons.

Slingerland, R., and N.D. Smith. 1998. Necessary conditions for a meandering river avulsion. *Geology* 26: 435–438.

Slingerland, R., and N.D. Smith. 2004. River avulsions and their deposits. *Ann. Rev. Earth Planet. Sci.* 32: 257–285.

Smith, J.M., and R.W. Frey. 1985. Biodeposition by the ribbed mussel *Guekensia demissa* in a salt marsh, Sapelo Island, Georgia. *J. Sediment. Res.* 55: 817–828.

Smith, N.D., and M. Pérez-Arlucea. 2008. Natural levee deposition during the 2005 flood of the Saskatchewan River. *Geomorphology* 101: 583–594.

Stanley, D.J., and A.G. Warne. 1994. Worldwide initiation of Holocene marine deltas by deceleration of sea-level rise. *Science* 265: 228–231.

Stouthamer, E., and H.J.A. Berendson. 2007. Avulsions: the relative role of allogenic and autogenic processes. *Sediment. Geol.* 198: 309–325.

Sturges, W., and B.C. Douglas. 2011. Wind effects on estimates of sea level rise. *J. Geophys. Res. – Oceans*, 116: C06008.

Syvitski, J.P.M. 2011. Global sediment fluxes to the Earth's coastal ocean. *Appl. Geochem.* 26: S373–S374.

Syvitski, J.P.M., and A.J. Kettner. 2011. Sediment flux and the Anthropocene. *Philosoph. Transact. Royal Soc. A Math. Phys. Engineer. Sci.* 369: 957–975.

Syvitski, J.P.M., and Y. Saito, Y. 2007. Morphodynamics of deltas under the influence of humans. *Global Planet. Change* 57: 261–282.

Syvitski, J.P.M., A.J. Kettner, I. Overeem, E.W.H. Hutton, M.T. Hannon, G.R. Brakenridge, J. Day, C.J. Vörösmarty, Y. Saito, L. Giosan, and R.J. Nicholls. 2009. Sinking deltas due to human activities. *Nat. Geosci.* 2: 681–686.

Syvitski, J.P.M., A.J. Kettner, A. Correggiari, and B.W. Nelson. 2005a. Distributary channels and their impact on sediment dispersal. *Mar. Geol.* 222: 75–94.

Syvitski, J.P.M., C.J. Vörösmarty, A.J. Kettner, and P. Green. 2005b. Impact of humans on the flux of terrestrial sediment to the global coastal ocean. *Science* 308: 376–380.

Tang, J., and K.H. Johannesson, 2006. Controls on the geochemistry of rare earth elements along a groundwater flow path in the Carrizo Sand aquifer, Texas, USA. *Chem. Geol.* 225: 156–171.

Taniguchi, M., T. Ishitobi, J. Chen, S. Onodera, K. Miyaoka, W.C. Burnett, R. Peterson, G. Liu, and Y. Fukushima. 2008. Submarine groundwater discharge from the Yellow River Delta to the Bohai Sea, China. *J. Geophys. Res.* 113: C06025.

Temmerman, S., G. Govers, S. Wartel, and P. Meire. 2005. Spatial and temporal factors controlling short-term sedimentation in a salt and freshwater tidal marsh, Scheldt Estuary, Belgium and SW Netherlands. *Earth Surf. Proc. Land.* 28: 739–755.

Tornqvist, T.E., D.J. Wallace, J.E.A. Storms, et al. 2008. Mississippi Delta subsidence primarily caused by compaction of Holocene strata. *Nat. Geosci.* 1: 173–176.

Torres, R., M.A. Goni, G. Voulgaris, C.R. Lovell, and J.T. Morris. 2004. Effects of low tide rainfall on intertidal zone material processing, In: Fagherazzi, S., Marani, M., and Blum, L.K. (eds), The Ecogeomorphology of Intertidal Marshes. Washington, DC: American Geophysical Union, pp. 93–114.

Turner, R.E., N.N. Rabalais, D. Justic, and Q. Dortch. 2003. Global patterns of dissolved N, P, and Si in large rivers. *Biogeochemistry* 64: 297–317.

Tye, R.S., and J.M. Coleman. 1989. Evolution of Atchafalaya lacustrine deltas, south-central Louisiana. *Sediment. Geol.* 65: 95–112.

van Geen, A., K. Radloff, Z. Aziz, Z. Cheng, M.R. Huq, K.M. Ahmed, B. Weinman, S.L. Goodbred, H.B. Jung, Y. Zheng, M. Berg, P.T.K. Trang, L. Charlet, J. Metral, D. Tisserand, S. Guillot, S. Chakraborty, A.P. Gajurel, and B.N. Upreti. 2008. Comparison of arsenic concentrations in simultaneously-collected groundwater and aquifer particles from Bangladesh, India, Vietnam, and Nepal. *Appl. Geochem.* 23: 3244–3251.

Vital, H., K. Statteger, J. Posewang, J., and F. Theilen. 1998. Lowermost Amazon River: morphology and shallow seismic characteristics. *Mar. Geol.* 152: 277–294.

Wagner, D., S. Kobobe, E.M. Pfieffer, and H.W. Hubberten. 2003. Microbial controls on methane fluxes from a polygonal tundraof the Lena Delta, Siberia. *Permafrost Periglac. Process.* 14: 173–185.

Walther, G., E. Post, P. Convey, A. Menzel, C. Parmesan, T.J.C. Beebee, J. Fromentin, O. Hoegh-Guldberg, and F. Bairlein. 2002. Ecological responses to recent climate change. *Nature* 416: 389–395.

Weston, N.B., R.E. Dixon, and S.B. Joye. 2006. Microbial and geochemical ramifications of salinity intrusion into tidal freshwater sediments. *J. Geophys. Res.* 111: G01009.

Weinman, B., S.L. Goodbred, Y. Zheng, Z. Aziz, M. Steckler, A. van Geen, A.K. Signhvi, and Y.C. Nagar. 2008. Contributions of floodplain stratigraphy and evolution to the spatial patterns of groundwater arsenic in Araihazar, Bangladesh. *Geol. Soc. Am. Bull.* 120: 1567–1580.

Wilson, C.A., and M.A. Allison. 2008. An equilibrium profile model for retreating marsh shorelines in southeast Louisiana. *Estuar. Coast. Shelf Sci.* 80: 483–494.

Wolanski, E. 2007. *Estuarine Ecohydrology*. Amsterdam: Elsevier.

Wolman, M.G., and L.B. Leopold. 1957. River flood plains: Some observations on their formation. *U.S. Geol. Surv. Prof. Pap.* 282-C: 87–109.

Wright, L.D. 1977. Sediment transport and deposition at river mouths: a synthesis. *Geol. Soc. Am. Bull.* 88: 857–868.

Wright, L.D. 1985. River deltas, In: Davis, R.A. (ed.), Coastal Sedimentary Environments, 2nd ed, pp. 1–75. New York: Springer-Verlag.

Zhou N., B. Westrich, S. Jiang, and Y. Wang. 2011. A coupling simulation based on a hydrodynamics and water quality model of the Pearl River Delta, China. *J. Hydrol.* 396: 267–276.

3

Freshwater and sediment dispersal in large river plumes

R. D. Hetland and T. J. Hsu

1. Introduction

Rivers transport freshwater from continents to the ocean. Globally, about 40×10^3 km^3 yr^{-1} of freshwater is transported from the continents to the ocean via rivers, about one-third of the total precipitation over land (Trenberth et al. 2007). According to Syvitski et al. (2005), rivers annually deliver about 12.6 billion metric tons of sediment to the global ocean. There is a geometric increase in the number of smaller rivers, and the size of a river is correlated with river basin area. Many smaller rivers may have significant discharge during storm events, but they are often ungauged, so it is difficult to estimate their net contribution of freshwater and sediment to the ocean. However, it is estimated that the 25 largest rivers (Table 3.1) contribute approximately 50% of the global riverine fresh water and 40% of the global riverine sediment flux (Milliman and Meade 1983; Meade 1996).

There have been studies of river plumes at a wide range of river discharges. Two examples demonstrating discharge extremes are the Teignmouth and the Amazon. Pritchard and Huntley (2006) described a study of the Teignmouth in southern England. At the time of the study, the Teignmouth had a flow rate of 11.7 m^3 s^{-1}. They observed that the plume disperses with each ebb tide pulse, and the extent of the plume's influence is only a few kilometers. Geyer et al. (1996) described the Amazon river plume, with a discharge that ranged from 0.76 to 2.4×10^5 m^3 s^{-1} during the study period. The freshwater discharge is so great that freshwater is forced outside of the river delta, causing processes that usually occur in the estuary to occur over the shelf. The Amazon river plume affects a very large region; water with salinity less than 30 g kg^{-1} can be found more than 1,000 km away from the source (e.g., Molleri et al. 2010).

The magnitude of river discharge is the most dynamic variable when comparing different rivers; the previous two examples of the Teignmouth and Amazon span five orders of magnitude in discharge. Even single rivers typically modulate in discharge by more than an order of magnitude over the course of year. Other aspects associated with the river plume source, for example, estuary mouth width, typically range from a few hundred meters to a few kilometers between different river systems, that is, about one order of magnitude.

There have been a number of recent studies of moderate to large river plume systems, notably, the Columbia (e.g., Horner-Devine et al. 2009; Hickey et al. 2010), Hudson (Chant et al. 2008), Mississippi (Hetland and DiMarco 2008; Bianchi et al. 2010; DiMarco et al. 2010), and the Rhine (de Kok et al. 2001; De Boer et al. 2009) river plumes. Each of these systems has unique characteristics that make direct comparison between them difficult. For example, the Columbia discharges over a

Table 3.1. *The 25 largest rivers with discharge and sediment rates, ordered by water discharge, along with three more rivers that round out the top 10 rivers by sediment discharge*

	River	Water discharge 10^9 m^3 yr^{-1}	Sediment discharge 10^6 t yr^{-1}
Top 25 by water discharge	Amazon, Brazil	6300	1150
	Zaire, Zaire	1250	43
	Orinoco, Venezuela	1200	150
	Ganges-Brahmaputra, Bangladesh	970	1050
	Yangtze (Changjiang), China	900	480
	Yenisey, Russia	630	5
	Mississippi, USA	530	210
	Lena, Russia	510	11
	Mekong, Vietnam	470	160
	Parana/Uruguay, Brazil	470	100
	St. Lawrence, Canada	450	3
	Irrawaddy, Burma	430	260
	Ob, Russia	400	16
	Amur, Russia	325	52
	Mackenzie, Canada	310	100
	Pearl (Xi Jiang), China	300	80
	Salween, Burma	300	100
	Columbia, USA	250	8
	Indus, Pakistan	240	50
	Magdalena, Colombia	240	220
	Zambezi, Mozambique	220	20
	Danube, Romania	210	40
	Yukon, USA	195	60
	Niger, Africa	190	40
	Purari/Fly, New Guinea	150	110
Additional for top 10 sediment discharge	Yellow (Huang He), China	49	1100
	Godavari, India	92	170
	Red (Hunghe)Vietnam	120	160

very shallow bar, such that it interacts with the sea floor long past the estuary mouth. The Hudson discharges into a coastal bight formed by the coast of New Jersey and Long Island; the plume is known to interact with both coasts (Choi and Wilkin 2007; Zhang et al. 2009). The Upper Mississippi has two distributaries, the Lower Mississippi and the Atchafalaya, which enter the ocean through the Mississippi River Delta and Atchafalaya Bay, and carry two-thirds and one-third of the upper Mississippi discharge, respectively. These two locations are more than 200 km apart and discharge into very different coastal environments. Atchafalaya Bay is broad and shallow and is adjacent to a broad, shallow shelf. The Mississippi River Delta has a number of active passes, the largest being Southwest Pass carrying about 30% of the total Delta discharge. The shelf adjacent to the Delta is steep and narrow, and water is discharged very near the shelf edge. All of these river plume systems create a plume with a distinctive pycnocline, separating the surface plume (typically 1–10 m thick) from the deeper ambient shelf water (Murray 1997; Hetland and DiMarco 2008; Zhang et al. 2012). The Rhine is adjacent to a broad, shallow shelf with significant tides, such that the plume is well mixed, and the water column is only periodically stratified with the tide. This is typical for river plumes in the United Kingdom (e.g., the Mersey and Dee river plumes), and such systems are typically referred to as a region of freshwater influence (ROFI; Rippeth et al. 2001; Simpson et al. 1993; 2002).

There have also been a number of studies focusing on small river plume systems, including the Connecticut (Garvine 1974; O'Donnell 1990; O'Donnell et al. 1998), the Merrimack (MacDonald et al. 2007; Hetland and MacDonald 2008), and the Frasier (MacDonald and Geyer 2004). These systems are typically modulated strongly by tides, and studies typically focus on either the strong mixing that occurs near the estuary mouth during ebb tide, or tidal plume frontal properties.

Studies of small-scale plumes typically focus on tidal timescales, whereas studies of larger systems typically focus on weather band to seasonal processes. An exception is the study by Halverson and Pawlowicz (2008), who used observations of salinity mounted on a ferry boat to study differences in plume salinity on timescales ranging from tidal timescales to interannual timescales.

More generally, river plumes are known to affect carbon cycling on the shelf (e.g., Salisbury et al. 2008), and anthropogenic nutrients delivered by large rivers may be the dominant nutrient for shelf ecosystems (e.g., Fennel et al. 2011). Half of the oceanic carbon burial is thought to occur in large river delta systems (Hedges and Keil 1995). Increased residence time within a recirculating bulge feature has been shown to influence local biological activity (Kudela et al. 2010). Turbidity due to the suspended sediment load carried by river plumes can affect biological processes and sea surface temperature (Cahill et al. 2008). Franks and Anderson (1992) showed a relationship between the Kennebec River plume in the Gulf of Maine with the downcoast transport of a toxic dioflagellate. Thus river plumes may affect shelf ecosystem processes in a variety of ways. (See Chapters 1, 2, 4, and 8.)

This chapter describes recent advances in our understanding of river plume dynamics and sediment dynamics within a river plume. This is meant to be a general introduction for chemists and biologists, so we have focused on heuristic explanations in lieu of mathematics. Readers interested in a more rigorous treatment of these issues should refer to the original source literature.

2. The anatomy of a river plume

2.1. Salinity space

The position of a river plume may be affected by tides, wind, bathymetry, and coastline changes. Because of this, river plumes may change shape and position drastically over time. This is problematic when trying to describe the plume in Cartesian coordinates; the plume may be present or absent at a particular position at any given time. Also, there many different processes acting along the path of the plume, and as the plume shifts position, different dynamical regions may be present at any particular point at different times.

One solution to this problem is to describe the plume in salinity coordinates (Hetland 2005). That is, instead of defining a point in the plume by some fixed position in space, a point in the plume is defined by a particular salinity. The primary assumption here is that processes along a particular isohaline, say, the surface contour of 28 g kg^{-1}, will be similar throughout the plume. Although this assumption is not perfect, it is often a reasonable approximation (e.g., Hetland 2005; Hetland and MacDonald 2008). Thus, using salinity coordinates, it is often possible to describe different dynamical regions within the plume by the salinity at which the processes that define that particular region occur.

A cross-section of the plume is shown in Figure 3.1, where the dynamical regions are separated by the dominant vertical mixing processes, and are defined in salinity space, in particular, by a range

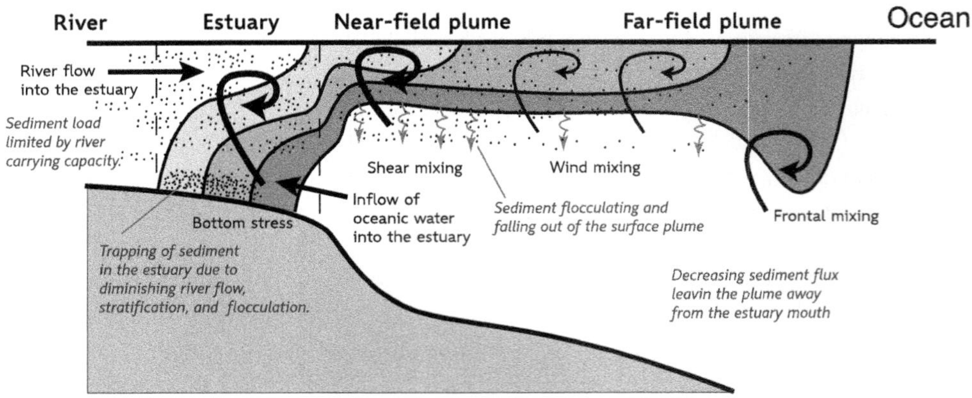

Figure 3.1. An idealized cross-section of a surface-trapped river plume demarks the various dynamical regions and the dominant mixing processes within each region. Note that the diagram is not to scale; the far-field region where wind-mixing is dominant is much larger than the estuarine region or near-field region. (Based on a figure by Hetland 2005.)

of surface salinity values. Figure 3.2 shows the same dynamical regions in plan view, this time highlighting lateral advection processes. These dynamical regions are described in more detail next.

Qualitatively, isohalines may be used to examine the plume in salinity space. As a simple example of looking at the plume in salinity space, the 33 g kg^{-1} isohaline is plotted every month for two years in

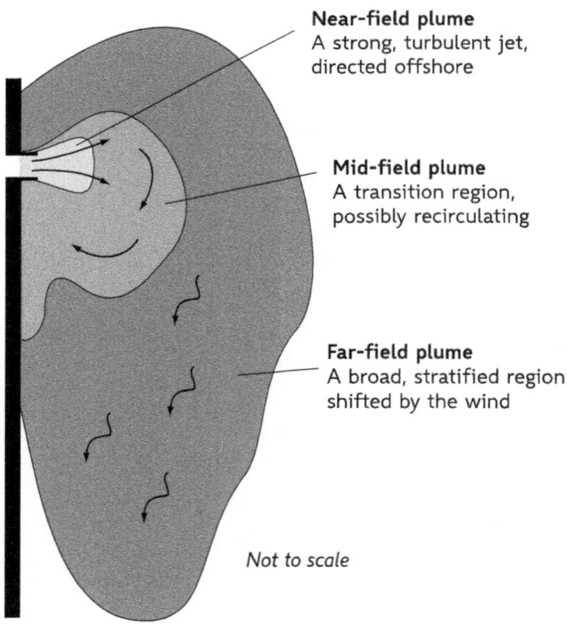

Figure 3.2. The features of a surface-trapped plume are shown in plan view. Here the lateral advection processes are highlighted. Again, the figure is not quite to scale – the near-field is typically much smaller than the far field. See Figure 3.1.

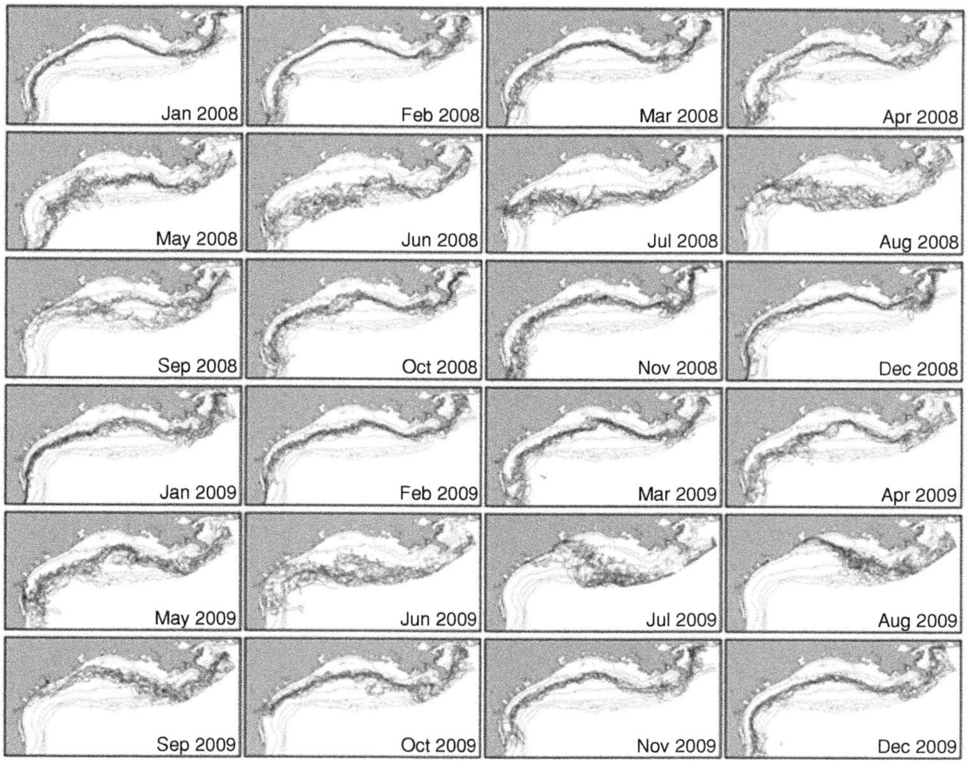

Figure 3.3. Monthly locations of the 33 g kg^{-1} surface salinity contour are shown every three days. Each panel shows a different month in the years 2008 and 2009, which had similar peak discharges, but very different seasonal wind patterns. In particular, summertime winds in 2009 were more upwelling favorable than normal, and this tended to push the plume much further to the east, as compared with 2008.

Figure 3.3. The 33 g kg^{-1} isohaline is used operationally as the boarder of the Mississippi/Atchafalaya river plume, and so this figure shows the shifting boundary of the plume over time. The shifting of the plume is caused by the magnitude of the Mississippi/Atchafalaya discharge and is modulated by seasonal winds. The seasonal winds, in turn, change every year; the plume was pushed much further upcoast in the summer of 2009 by unusually strong summertime upwelling winds as compared with 2008. It is clear from this example that the plume is only occasionally present along the Texas coast and thus a spatial average that defines the plume by some fixed geographic reference point would fail. Using the 33 g kg^{-1} as reference defining the location of the plume for estimating mean properties of chlorophyll within the plume, for example, would be more reasonable. Similar response of a plume to the wind may be seen in Chapter 13.

2.2. The estuary

Although a detailed description of estuarine circulation is beyond the scope of this chapter, it is important because the estuarine outflow forms the source water for a river plume. Mixing within estuaries determines the initial salinity of the river plume. Estuaries may also process other constituents carried by the river water, such as carbon (see Chapters 2 and 16). There are two key factors of estuarine geometry and circulation that influence a river plume: the width of the estuary mouth and the salinity (or density) of the outflowing water.

The density of the estuary is controlled by the amount of mixing within the estuary. The magnitude of the inflowing river water, the mixing within the estuary, and the strength of the estuarine exchange flow at the mouth may be related by Knudsen's relation (Knudsen 1900), which states:

$$Q_f = S_{out} Q_{out} - S_{in} Q_{in} \tag{1}$$

where Q_f is the riverine freshwater transport, Q_{in} and Q_{out} are the transports into and out of the estuary at the mouth, and S_{in} and S_{out} are the characteristic salinities of these transports. Thus the plume source is not a freshwater source with a transport of Q_f, but rather a flow with a transport of Q_{out} and a salinity of S_{out}. Defining the net mixing within an estuary requires a detailed understanding of estuarine dynamics; a recent review by MacCready and Geyer (2010) gives an overview of our understanding of estuarine dynamics.

The other important characteristic for river plume formation is the width of the estuary mouth. This is typically compared with the local, internal deformation radius:

$$R_d = \frac{\sqrt{g\beta S_{out} H}}{f}, \tag{2}$$

defined using the thickness of the estuarine outflow, H, the salinity of the outflowing water, S_{out}, the gravitational acceleration g, the haline contraction coefficient, β, and the Coreollis parameter, f. The internal deformation radius, R_d, characterizes the natural length scale of rotating, stratified flow; a buoyant coastal current, for example, is typically a few deformation radii wide. The ratio of the mouth width, W, to the deformation radius defines the Kelvin number, $K = W/R_d$. A narrow outflow is defined as configurations with a Kelvin number of less than one, a wide outflow greater than one.

Narrow outflows tend to produce strong, directed flow at the mouth, like the Columbia, passes of the Mississippi River Delta, Merrimack, Hudson, and Rhine rivers, to name a few. Because of the strong flow through the estuary mouth, rotational processes are less important, and the outflow resembles a strong jet. Narrow estuary outflows are also characterized by very high mixing near the outflow region, as discussed later in the near-field plume section. Wide outflows tend to come from large bays, such as the Chesapeake, Delaware, and Rio de la Plata, or alluvial estuaries, such as the Elba or Themes. In these estuaries, the flow may be concentrated on the downcoast side of the estuary; that is, the outflow will be concentrated with the coast on the right in the northern hemisphere. In this case, the flow in the estuary may resemble a buoyant coastal current more than a narrow jet. Also, estuarine exchange with the ocean may be more lateral, with inflow on the upcoast side and outflow on the downcoast side, rather than an exchange typical of a narrow estuary with surface buoyant outflow over a deep, salty inflow.

2.3. *The near-field plume*

A near-field plume may occur at the mouth of a narrow estuary. The near-field plume is a region of very strong flow associated with rapid spreading of the plume and very intense mixing (Wright and Coleman 1971; MacDonald and Geyer 2004; MacDonald et al. 2007; Nash et al. 2009; Hetland 2010). The near-field plume is characterized by a single, active layer of freshwater above a layer of more quiescent saline coastal water. The plume is typically thin, ranging from less than a meter to a few meters in thickness, and there can be strong current shear or stratification within the plume. The near-field plume exists only within a few kilometers from the estuary mouth before deceleration due to mixing or turning due to the earth's rotation halts its rapid intrusion into the coastal ocean. The near-field plume is often either modulated by the tides, or, in the case in which there is a reversal in the tidal flow at the estuary mouth, the near-field plume is discharged in tidal pulses. Because of the rapid flow speeds and relatively small size of the near-field plume, residence times of freshwater within this region are typically on the order of only a few hours.

The flow in the near-field is supercritical (see, e.g., Armi and Farmer 1986), meaning that flow has speeds faster than the first-mode baroclinic wave speed (interfacial waves on the pycnocline between the plume and the ambient shelf water). In other words, the flow is so fast that waves cannot propagate upstream. The elevated shear associated with the rapidly moving plume water triggers shear instabilities (Ivey and Imberger 1991; MacDonald and Geyer 2005).

Hetland (2010) described the interaction between lateral spreading and vertical mixing in the near-field plume. Once the brackish estuarine water leaves the confines of the estuary channel, the water will spread laterally on entering the coastal ocean. By conservation of the water volume, the lateral spreading requires the base of the plume to shoal. This shoaling is in turn associated with a drop in the sea level and thus a drop in pressure. By the Bernoulli principle, a drop in pressure corresponds to an acceleration of the flow, which enhances the vertical shear and the associated vertical mixing. Thus there is a competition between the spreading, which tends to accelerate the plume, and mixing, which tends to decelerate the plume. Initially, spreading dominates, and the plume experiences net acceleration as it enters the coastal ocean. Ultimately, though, mixing dominates, and the plume eventually decelerates enough such that it is no longer a jet-like feature, but rather a plume-like feature.

Near-field plumes have much in common with smaller, engineering-scale plumes such as smokestacks and industrial outfalls; a recent classification and review of engineering-scale plumes is given by Jones et al. (2007). The major difference between geophysical scale, near-field river plumes, and engineering-scale plumes is the aspect ratio. Geophysical scale plumes typically have a vertical-to-horizontal aspect ratio of 10^{-3} (e.g., 1 meter thick to 1 kilometer in breadth). Engineering-scale plumes typically have a much larger aspect ratio, say 10^{-1}. Because of this, horizontal mixing is a much more important process in engineering-scale plumes, as large turbulent eddies may develop that are a significant fraction of the plume width, while still retaining an aspect ratio near 1. In a geophysical scale plume, mixing is almost entirely a vertical process, as turbulent eddies cannot have such a small aspect ratio and cannot grow to a significant fraction of the plume width.

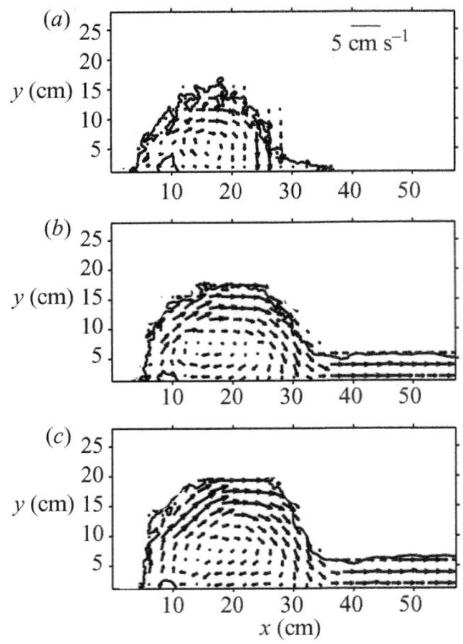

Figure 3.4. An example of a midfield river plume "bulge" and coastal current as observed in a rotating tank experiment by Horner-Devine et al. (2006, their Figure 2, upper left panels) shows a strong recirculating feature near the estuary outflow (the bulge) and a coastal current developing further downcoast. The three frames show the temporal evolution of the bulge, over three rotational periods from the time when the buoyant water inflow was turned on. The bulge continues to grow in time, with a characteristic length scale growing from about three deformation radii to over six, whereas after three rotational periods, the coastal current remains relatively constant in width. This is because the coastal current does not carry all of the water that is introduced through the estuary mouth, resulting in a retention of water in the bulge so that it grows in time.

2.4. The mid-field plume, or the bulge

The mid-field plume is a region of transition between the offshore directed near-field plume and the along-shore directed far-field plume. This region is typically associated with the "bulge," first discussed by Fong and Geyer (2002), who used an idealized numerical model to investigate trapping of water within the bulge. The bulge is also a prominent feature in rotating tank experiments (Avicola and Huq 2003a, 2003b; Horner-Devine et al. 2006; Huq 2009). An example of a bulge from a laboratory experiment is shown in Figure 3.4, where a region of recirculation (the bulge) forms near the estuary outflow before a coastal current forms. Although it was previously thought that the bulge was possibly an unphysical numerical artifact (e.g., Garvine 2001), most researchers now agree that the bulge is a real, if ephemeral, feature of many natural river plumes (e.g., Chant et al. 2008; Horner-Devine et al. 2009; Horner-Devine 2009).

Early numerical studies suggested that the bulge grew over time, and as such, freshwater transport in the coastal current was lower than the freshwater introduced to the coastal ocean by the river (Fong and Geyer 2002). Yankovsky and Chapman (1997) later derived a steady-state balance for the bulge, although momentum balance arguments (Nof and Pichevin 2001) suggested that a steady-state is not possible. However, none of these idealized numerical or analytical studies included the effects of wind stress. It seems that wind may disrupt bulge formation in the near-field plume (Chant et al. 2008; Horner-Devine et al. 2009) and that the bulge forms primarily during calm periods.

2.5. *Far-field plume*

In the absence of wind, the far-field river plume is nearly in geostrophic balance. That is, the along-shore flow generates an onshore force (the Coriolis force) due to the earth's rotation that is balanced by an offshore pressure gradient caused by the high sea-level anomaly at the coast associated with the buoyant water. As such, the buoyant coastal current is directed primarily downcoast, with the coast to the right of the flow direction in the northern hemisphere. The flow is strongest at the front that typically forms at the boundary between the plume and the ambient coastal ocean water. The flow interacts with the bottom if the plume is bottom-advected or is contained entirely within the upper water column if the plume is surface-advected. This classification is refined later and is related to the character of the source water.

Wind can affect the river plume. A downwelling wind will tend to increase the downcoast transport in the river plume, tilt the isopycnals toward vertical, and narrow the surface expression of the plume (Pullen and Allen 2000; Garcia Berdeal et al. 2002). An upwelling wind will decrease along-shore transport and pull the freshwater plume offshore (Fong and Geyer 2001; Lentz 2004). Thus there is an asymmetrical response to the wind, with freshwater moving downcoast during downwelling and offshore during upwelling. Because of this, freshwater is rarely found upcoast of the river source, even when the plume is strongly forced by the wind.

Houghton et al. (2004) described a study where dye was released into the Delaware river plume just before a moderately strong (10 m s^{-1}) upwelling wind event. They followed the dye as it was advected with the plume 28 km offshore and obtained estimates of vertical mixing within the plume. Average mixing rates within the plume were approximately 2×10^{-4} m^2 s^{-2}. They also found evidence of secondary circulation within the plume; the dye patch moved about 50% faster offshore (v_{dye} = 0.30 m s^{-1}) than the plume itself (v_{plume} = 0.21 m s^{-1}). This suggests a convergence at the plume front, where surface plume water subducts frontal water into the plume pycnocline.

Although mixing is generally proportional to the wind speed, upwelling winds are more effective at mixing than downwelling winds (Hetland 2005). This is, in part, because the plume is spread out over a larger area Fong and Geyer (2001) showed with idealized numerical experiments, and Lentz (2004) showed with observations that the wind will tend to mix a plume up to a point, after which mixing is reduced. This comes through calculating the Ekman transport-driven velocity, u_e, of the plume,

$$u_e = \frac{\tau}{\rho_0 f h}, \tag{3}$$

where τ is the wind stress, ρ_0 is a reference density, f is the Coriolis parameter, and h is the plume thickness. This equation may be combined with an equation for the bulk Richardson number,

$$Ri_B = \frac{\Delta \rho g h}{\rho_0 u_e^2}, \tag{4}$$

where $\Delta\rho$ is the density difference between the plume and ambient coastal water, g is the gravitational acceleration, and the other variables are as defined previously. If the bulk Richardson number falls below a critical value, $Ri_c \sim 0.6$, then turbulent mixing will occur; salty ambient shelf water (with a velocity assumed to be near zero) will be entrained into the plume, and thus the plume will increase in density and decrease in velocity. Because mass is conserved, the product $h\Delta\rho$ will be constant in time even as both h and $\Delta\rho$ change, that is,

$$h_{\text{initial}}\Delta\rho_{\text{initial}} = h_{\text{final}}\Delta\rho_{\text{final}} \tag{5}$$

where the initial and final subscripts are the initial and final states of the plume before and after a wind event. Becaue the bulk Richardson number depends linearly on density difference, but on the square of the velocity difference, the bulk Richardson number will increase up to the critical Richardson number, at which point turbulent mixing is suppressed. Combining Equations 3 and 4 with mass conservation, the final thickness of the plume may be calculated (Fong and Geyer 2001; Lentz 2004; Hetland 2005) as

$$h_{\text{final}} = \frac{2\tau}{\rho_0 f}\sqrt{\frac{Ri_c\rho_0}{g h_{\text{initial}}\Delta\rho_{\text{initial}}}}. \tag{6}$$

This result is useful for its quantitative, validated predictions of plume structure under the influence of wind. However, it is also useful conceptually, as it demonstrates how even under upwelling wind conditions, when the plume is most susceptible to mixing, the stratification within the plume is able to limit mixing – for a given wind stress, the plume will mix down to h_{final}, but no further – and is thus able to maintain the cohesion of the plume as a distinct feature. This is also why large river plumes, which tend to be thicker and more stratified, persist over a larger area; they are more able to resist the mixing effects of the wind.

This result assumes that the plume is surface trapped, that it does not interact with the bottom. Larger scale, bottom-advected plumes are also affected by the wind, but the dynamics of this is complicated by general shelf-scale circulation. For example, Whitney and Garvine (2005) discussed wind effects on the Delaware plume in a region where it interacts strongly with the sea floor.

2.6. Very far-field plume

There has not been much research on the ultimate fate of river plumes, the point at which a plume ceases to be a distinctive feature and melds into the broader background shelf or oceanic flow field. Garvine (1999) examined the penetration of a river plume using a very idealized model configuration without wind. He found that the along-shore length scale of the plume was, among other physical parameters, related to the background molecular mixing specified as part of the turbulent closure. In salinity coordinates, this is straightforward to explain. In the absence of other active, turbulent mixing, the mixing is entirely defined by the background mixing. Freshwater must be mixed across any given isopycnal surface; if there is no mixing, the isopycnal will blow up like a balloon. Thus the integrated mixing across the surface must balance the freshwater flux through the surface. Given the same freshwater flux, halving the net mixing across an isohaline surface must double its area to achieve the same net freshwater flux across the isohaline. Thus mixing is one way to destroy a river plume; the amount of mixing required may be calculated by an energy balance, as done by Pritchard and Huntley (2002). This would be an effective way to define the extent of smaller river plumes.

Larger river plumes are less sensitive to vertical mixing processes, as they may already be quite deep. Another mechanism that could potentially be important in defining the spatial extent of large river plumes is evaporation. The relevant scale here would be the area over which a particular river discharge could be evaporated or the area over which integrated precipitation over an area would balance river discharge. For scales much larger than this, the plume is no longer an individual feature, and general circulation combined with the global hydrological cycle must be considered.

There have been a variety of attempts to parameterize freshwater sources from rivers into large-scale climate models. Freshwater fluxes from rivers is thought to be an important factor in determining global thermohaline circulation patterns. Garvine and Whitney (2006) created a model of Delaware Bay to examine ways in which mixing within the estuary might be parameterized in a large-scale model that does not resolve mixing and advective processes within the bay. Hordoir et al. (2008) created a similar sort of box model to parameterize unresolved mixing processes within a river plume. At present, neither methodology is commonly used General Circulation Models. The present approach is to place freshwater from rivers into the uppermost few model cells, where the exact number of cells is set by trial and error. Thus an important goal in river plume research is to understand mixing and advection processes to the point where these processes can be parameterized accurately and efficiently.

3. River plume classification

In the absence of wind, there are five relevant variables that define the estuarine discharge: the width of the estuary mouth, W, the inflowing velocity, u, the thickness of the outflowing water, h, the reduced gravity defined by the density anomaly of the outflowing estuary water, $g' = g\Delta\rho\rho_0^{-1} = g(\rho_0 - \rho_{\text{outflow}})\rho_0^{-1}$, and the Coriolis parameter, f. These dimensional variables may be combined to form a few key nondimensional variables. The width of the estuary mouth is normalized by the deformation radius, R_d, to define the Kelvin number,

$$K = \frac{Wf}{\sqrt{g'h}}. \tag{7}$$

The inverse of the Kelvin number (sometimes written as the square of the inverse) is occasionally used instead; this is called the Burger number. The flow speed is normalized by the internal gravity wave speed to define the Froude number,

$$Fr = \frac{u}{\sqrt{g'h}}. \tag{8}$$

For many problems, the slope of the continental shelf is also an important parameter. This may be calculated through the either the actual slope of continental shelf, α, or through the slope Burger number,

$$S = \alpha\frac{N}{f}, \tag{9}$$

where N is the stratification frequency,

$$N^2 = \frac{g}{\rho_0} \frac{\partial \rho}{\partial z}. \tag{10}$$

The choice of any set of nondimensional numbers is not unique. For example, the Rossby number, $Ro = FrK^{-1} = u(fW)^{-1}$, is a measure of how far the fluid travels in one inertial period, normalized by the estuary mouth width; the Rossby number is often used as a measure of nonlinearity, that is, the potential for instabilities to develop in a flow field. Because the Rossby number is formed by the ratio of the Kelvin and Froude numbers, the Rossby number could replace either in a classification.

As mentioned previously, one way to classify river plumes is by the outflow Kelvin number (Garvine 1995). Kelvin numbers typically range from about 0.1 to 4.0. An example of the low end of this range is the Merrimack River, which has a Kelvin number of about 0.1. The Merrimack has a 300-m estuary mouth defined by jetties, and the ebb tide outflow is nearly fresh during moderate riverine discharges over about 300 m^3 s^{-1} (the freshwater transport at the mouth may be six times this mean value during ebb tides). At the other extreme, the Delaware Bay outflow, a wide estuary formed from a drowned river valley, has a Kelvin number of about 4.0 (Garvine 1995). Outflows with a Kelvin number of less than one typically form narrow, energetic jets directed in a way that is controlled by the coastline and bathymetry of the estuary mouth. In analogy with small-scale, engineering flows, the estuary mouth acts as a nozzle for the near-field plume that forms outside the estuary mouth. In contrast, outflows with a Kelvin number greater than 1 do not form such an energetic jet. The flow here, in the absence of other forcing, will typically hug the downcoast coastline of the estuary (e.g., the southern coast for the Delaware). The flow in this case has more the character of a coastal current rounding a bend than a strong jet entering the coastal ocean.

Yankovsky and Chapman (1997) derived a classification system based on both the Rossby and Burger numbers (again, these two properties may be derived from the Kelvin and Froude numbers). They partition the parameter space into three regions, based in part on the idealized domain that they used, which had a coastal wall with a height of h_0. This might be interpreted in realistic systems as the depth at which the surface and bottom boundary layers separate. They derive two-dimensional scales for the plume, the offshore distance where the pycnocline breaches the surface, and where it intersects the bottom.

The surface front is defined by a scaling for the diameter of the bulge region, based on a "gradient wind" balance. A gradient wind balance is based on a three-way balance between the centripetal acceleration of the flow around the bulge, the pressure gradient force, and the Coriolis force. The first two terms are directed outward from the bulge, and the third is directed inward. Isobe (2005) suggested that this balance may become unstable and might be stabilized by mixing due to tidal currents. The final scaling for the bulge diameter, that is, the offshore position of the surface front, is between 4.2 R_d for a weak outflow ($Fr \ll 1$) and 4.6 R_d for $Fr = 1$.

The bottom front is defined by a geostrophic transport of the plume water with a thermal wind shear across the front. If the bottom front is at a location shallower than its trapping depth, Ekman transport will drive a seaward flow in the bottom boundary, pushing lighter plume water underneath denser shelf water. The outcome of this process is that the pycnocline will be shifted seaward until

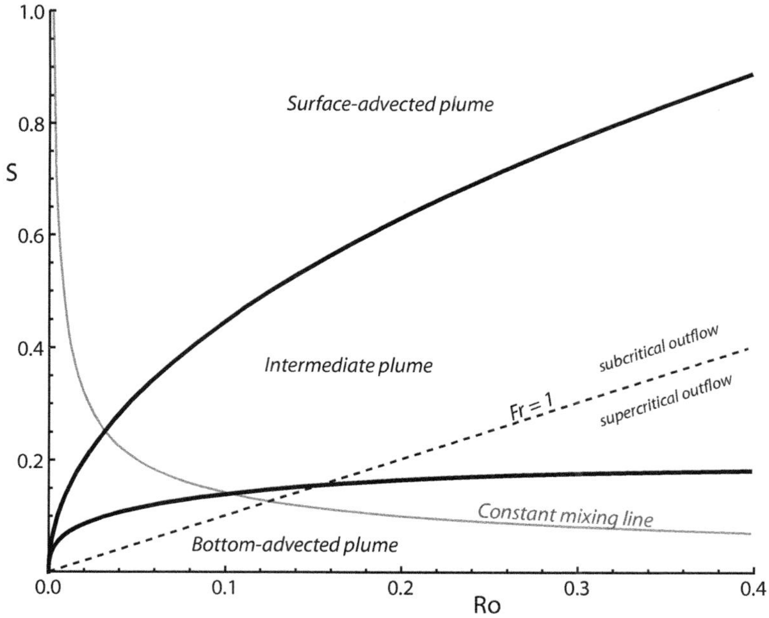

Figure 3.5. This figure shows the plume classification of surface-, intermediate-, and bottom advected plumes, as a function of Burger number, S, and Rossby number, Ro.

the flow at the bottom is 0, and the Ekman transport in the bottom boundary layer is shut down. The depth at which this occurs is

$$h_b = \sqrt{\frac{2Qf}{g'}}, \tag{11}$$

where $Q = Wuh$ is the volume discharge from the estuary, assumed to be equal to the net transport carried within the frontal region.

The surface-advected plume (also commonly referred to as a surface-trapped plume) is one in which the plume never interacts with the bottom, but only touches the coastal wall, where $h_b < h_0$. An intermediate plume is one in which the plume interacts with the bottom, but the location of the bottom front is inshore of the surface front. The bottom-advected plume is one in which the predicted position of the bottom front is seaward of the predicted surface front position. Surface-advected plumes are strongly affected by the dynamics within the bulge region, are typically vertically stratified, and might be more susceptible to wind mixing, and the position of a surface-advected plume is easily shifted by the wind. Bottom-advected plumes are affected by bottom boundary layer processes, even when the bottom boundary layer is a small fraction of the total water column depth, and have significant horizontal stratification. Bottom-advected plumes are characteristic of larger outflows and are thus less sensitive to modification by the wind.

The separation of surface- and bottom-advected plumes based on the trapping depth may be cast in terms of the Burger and Rossby numbers, as shown in Figure 3.5 (assuming a particular shelf slope for the bottom-advected plume). Yankovsky and Chapman (1997) related these properties to the outflow in their idealized numerical experiments. However, in real river plume systems, there can be considerable mixing between the source waters and location on the shelf where the plume is investigated. The light gray line in Figure 3.5 shows valid solutions for a particular riverine freshwater discharge with different amounts of mixing. The plume transitions from surface-advected to bottom-advected as the freshwater is mixed more with ambient shelf waters. Because of this, it is often best to use the classification of surface- and bottom-advected plumes in a diagnostic manner, rather than making predictions based on source water properties.

4. Sediment dispersal

Sediment dispersal via river plumes, often called initial deposition, is the first stage of sediment source to sink (the study of the fate of terrestrial sediment in the coastal ocean; Wright and Nittrouer 1995). Almost all sediment dispersals occur within the continental margin. In fact, for most of the river-shelf systems, initial deposition is quite close to the coast because once river outflows enter the ocean, factors such as diminishing river flow inertia, frictions, and stable density stratification all lead to effective sediment trapping near the river mouth (Wright 1977) (see Chapter 4).

Although studying sediment transport is not equal to transport of organic carbon, through aggregation and enhanced setting, floc aggregates are the main vehicle for the vertical transport of organic carbon, nutrients, and trace metals in the ocean (Santschi et al. 2005). Hence, in this chapter we focus on mechanisms that directly affect the sedimentation process. The initially deposited sediments can be then resuspended by energetic wave and current events and delivered further offshore (Traykovski et al. 2000). However, the resuspension of sediment is directly relevant to the benthic zone processes, and this section only focuses on sedimentation from surface plumes.

4.1. Sediment dynamics

Sediment dispersal is clearly dictated by the hydrodynamics of the river plume in the near field, mid field, and far field discussed in the previous sections. Intuitively, it is straightforward to consider that for fine sediment, sediment particles may simply follow the fluid flow and, meanwhile, settle with their settling velocities because the density of sediment (quartz) is about 2.6 times greater than that of the seawater. Therefore, it is natural to ask the following question: can we model sediment dispersal by simply assuming sediment particles as passive scalar except adding a settling mechanism? If the answer is yes, this is indeed a tempting approach, because it is then sufficient to predict sediment dispersal simply based on our understanding of river plume hydrodynamics. There are several difficulties in this proposition. First, it is not straightforward to estimate sediment settling velocity due to flocculation and hindered settling effects. As the mathematical description of flocculation dynamics is reasonably understood for inorganic sediments, biological effects can significantly change the nature of coagulation and aggregate structures (e.g., Thomsen 2005). Second, when sediment concentration and/or sediment grain size becomes large, increasing collisions and particle inertia do not allow sediments to follow the fluid flow passively, which leads to many interesting fluid mechanic mechanisms better described by a multiphase-phase flow approach.

4.1.1. Buoyancy effect

Consider that an idealized fine sediment particle that has spherical shape settles in still water. The balance between the immersed weight of the particle and drag force in the equilibrium condition gives the generic settling velocity, calculated as

$$W_{s0} = \frac{(s-1)gd^2}{18\nu}, \tag{12}$$

where ν is the kinematic viscosity of the fluid, d is the grain diameter, $s = \rho^s/\rho^f$ is the specific gravity of sediment, with ρ^s representing the sediment density and ρ^f the fluid density. For quartz, the specific gravity is generally specified as $s = 2.65$. Once W_{s0} is calculated, it is often assumed that the velocity of sediment particle can be calculated by the fluid velocity at which particle is located plus the settling velocity W_{s0}. With a more formal analysis of particle-laden flow (Balachandar and Eaton 2010), it can be shown that this approach is appropriate only for dilute flow of fine sediment, for which the particle inertia effect can be neglected. The importance of particle inertia effect is often estimated based on the ratio of particle response time to the timescale of flow turbulence (i.e., Stokes number). It appears that for typical fine sediments that are of interest during the dispersal processes, this ratio is significantly smaller than unity, and sediment inertia effect can be neglected (even for flocs). Therefore, we follow the assumption here that transport of fine sediment can be modeled as passive substance transport with an advection-diffusion equation, except adding a settling term (e.g., Winterwerp 2001; Warner et al. 2008).

It can be also shown theoretically that by considering the difference between the sediment particle velocity and fluid velocity to be only due to settling velocity, the effect of sediment on the hydrodynamics is simply the buoyancy effect, that is, we can model sediment dispersal by adopting the density-driven flow analogy similar to salt or heat transport (e.g., Cantero et al. 2008b). There are two essential features of density-driven flow that the readers need to be aware of. First, the density difference can trigger flow motion as density currents. For example, the river plume itself is a density current that propagate above the seawater. Moreover, when river water is laden with lots of sediments, the bulk density of freshwater-sediment mixture may increase (so does reduced gravity g') and hence modifies the Kelvin number and Froude number shown in Equations 7 and 8. Second, buoyancy effect results in stable density stratification, which can suppress flow turbulence. In this case, sediment-induced stable density stratification can also suppress flow turbulence, which is the main turbulence modulation mechanism for fine sediment (e.g., Winterwerp 2001; Cantero et al. 2008a; Ozdemir et al. 2011).

Following the river plume classification discussed in Section 3, let's consider that a river, characterized by flow velocity u and flow depth h, delivers sediments of settling velocity W_{s0} and bulk sediment mass concentration C_{in} (in grams per liter). Using u and h as the characteristic velocity and length scales, sediment dispersals are essentially controlled by the following three nondimensional parameters. The Reynolds number is written as

$$\mathrm{Re} = \frac{uh}{\nu}. \tag{13}$$

The fall parameter F is a nondimensionalized settling velocity

$$F = \frac{W_{s0}}{u}. \tag{14}$$

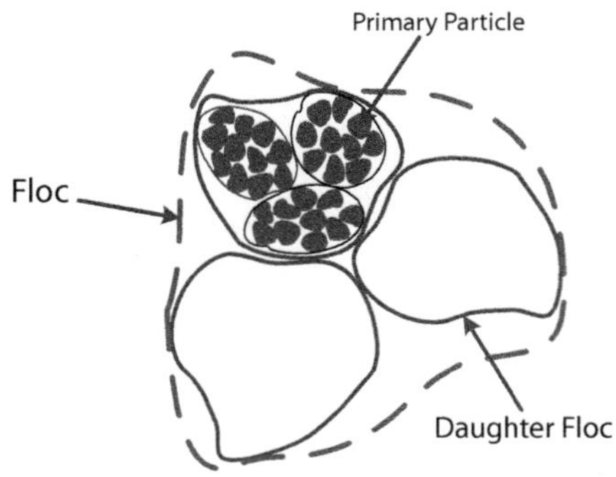

Figure 3.6. A schematic plot of a floc aggregate, which consists of smaller daughter flocs, which are further formed by primary particles.

It is not difficult to image that for the same amount of sediment load, its impact on the hydrodynamics must be dependent on the river discharge. Hence the nondimensionalized sediment load can be expressed as a Richardson number

$$Ri = \frac{(1 - \rho^f/\rho^s)ghC_{in}}{\rho^f u^3}. \tag{15}$$

For typical size and velocity of a river, the Reynolds number is very large (fully turbulent flow). Numerous examples in fluid mechanics have demonstrated that when the Reynolds number is sufficiently large, the flow is fully turbulent, and hence the nondimensional flow quantities become independent of *Re*. Therefore, sediment dispersal is essentially controlled by nondimensionalized settling velocity F and sediment load Ri. In the next several sections, we discuss the effects of settling and sediment loading on sediment dispersal in more detail.

4.1.2. Flocculation and hindered settling

Rivers deliver a significant amount of fine sediment (inorganic) to the ocean. These fine sediments are clay, silt, or very fine sand with grain size d in the range of O(1) μm to several tens of μm. Rivers also deliver significant amount of organic carbon typically in the form of phytoplankton, detritus, and mucilage (Santschi et al. 2005). A main feature of cohesive sediments that makes them distinct from noncohesive sediments is that cohesive sediment particles (primary particle) tend to form complex structures, called *floc aggregate* (e.g., Winterwerp and van Kesteren 2004) (see also Figure 3.6). Floc aggregates are of much larger size and smaller density than that of primary particles. These aggregates further determine the main physical properties of transport, such as settling velocity, permeability of floc, bed strength and rheological stress, and so on (Kranenburg 1994). Because organic materials have their distinct densities and chemical-biological properties, flocs of high organic content also have very different physical properties. In this section, we first review the general mathematical framework of flocculation. Specific closures and empirical coefficients that may be further revised to better include biological effects are also discussed briefly.

Here, we are specifically interested in calculating the settling velocity when fine sediment particles form aggregates. Recall that in Equation 12, we calculate the settling velocity of a sediment particle without flocculation. For floc, the settling velocity is calculated in a similar way but using an equivalent floc size D and floc aggregate density ρ^a,

$$W_{s0} = \frac{(\rho^a/\rho^f - 1)gD^2}{18\nu}. \tag{16}$$

Therefore, to model cohesive sediment transport, it is necessary to first come up with an appropriate parameterization describing aggregate structure so that floc density ρ_a and settling velocity can be readily calculated. In the cohesive sediment literature, pioneering work to understand aggregate structure can be traced back to Krone (1962) and Partheniades (1965). In their work, they noted that the primary particles (clay and silt) form floc, which join together to form floc aggregates (daughter floc). These aggregates combine to form larger aggregates, and so on. Therefore, the concept of orders of aggregation is introduced. This concept, although it may be too idealized, provides a good approximation to model floc structure for cohesive sediment transport. A formal mathematical tool to model such structure is the fractal theory. Here, the basic concept is illustrated by the following example. Consider close packing of spherical particles of uniform diameter d (i.e., the primary particle diameter) in three dimensions. The total number of particles in a given volume of size D can be estimated as

$$N \sim \left[\frac{D}{d}\right]^{n_f}, \tag{17}$$

where $n_f = 3$, and it is called the fractal dimension. Detailed observations of aggregate structure suggest that geometrical packing of floc aggregates is not similar to close packing of uniform spheres described previously. The packing is often quite loose, and the space is not effectively filled in all three directions. Typical value of fractal dimension for floc aggregate in estuary or continental shelf is about $n_f = 2.0 \sim 2.5$ (e.g., Dyer and Manning 1999; Manning and Dyer 1999; Winterwerp and van Kesteren 2004).

Using the concept of fractal and a prescribed fractal dimension, the density of a floc aggregate can be calculated as (see Kranenburg 1994 or Winterwerp and van Kesteren 2004 for more details)

$$\rho^a = \rho^f + \left(\rho^s - \rho^f\right)\left(\frac{D}{d}\right)^{n_f-3}. \tag{18}$$

Hence we can rewrite the calculation for the settling velocity of floc aggregate as

$$W_{s0} = \frac{(\rho^s/\rho^f - 1)gd^{3-n_f}}{18\nu}D^{n_f-1}. \tag{19}$$

Therefore, floc-settling velocity is proportional to floc diameter D to a power of $(n_f - 1)$. Evidently, the relationship between floc-settling velocity and floc diameter depends strongly on the fractal dimension n_f. Moreover, when plotting measured floc diameter and floc-settling velocity in the logarithmic scale, one expects a linear relationship with a slope of $(n_f - 1)$. In laboratory and field experiments, where many realizations of floc size and settling velocity are measured, this relationship becomes a useful method to estimate the fractal dimension of floc. Figure 3.7 shows field data measured in the Tamar estuary (Southern United Kingdom) (Winterwerp et al. 2006). It can be seen that the regression analysis gives a slope of 1.2, suggesting a fractal dimension of the mud investigated in that study to be 2.2. However, we should notice the large scatter in the measured data and moderate

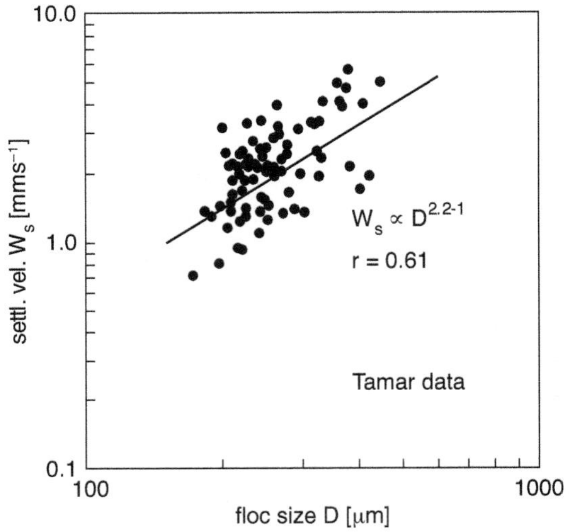

Figure 3.7. Measured floc size plotted with respect to measured settling velocity in the Tamar estuary reported in Winterwerp et al. (2006, their Figure 1, left panel). The slope of the regression line can be used to estimate the fractal dimension.

γ value. Similar data analyses have been applied to other datasets measured in various sites (e.g., Dyer and Manning 1999; Manning and Dyer 1999; Winterwerp and van Kesteren 2004), and the averaged fractal dimension is about $n_f = 2.0$. Without in situ data, many studies simply use $n_f = 2$. Again, recall that although an averaged value of $n_f = 2$ is given, the site-dependent variability can range from $n_f = 1.5 \sim 2.5$.

In reality, floc size ranges from primary particles (O(1) μm) to microflocs (O(10) μm) and to macroflocs (O(100) μm or larger). The assumption that fractal dimension is a constant may be too simplistic. To come up with a more general description of cohesive sediment applicable for a wide range of floc sizes, fractal dimension is suggested to be a variable (Khelifa and Hill 2006; Maggi et al. 2007). Physically, when a floc size is equal to or slightly larger than the size of primary particle, the fractal dimension should be equal or close to 3.0 (i.e., three-dimensional close packing is more likely to occur). However, as the order of aggregation become larger, the packing become less efficient and the fractal dimension decreases. Khelifa and Hill (2006) carried out a comprehensive analysis on published settling velocity data for a wide range of floc sizes measured in the field and laboratory. They proposed that the fractal dimension is dependent on floc size. The fractal dimension is around 2.0 when floc size is as large as 2 mm, and as floc size approaches primary particle diameter, fractal dimension increases to 3.0 via a power law function.

The wide range of fractal dimensions obtained at different sites can be also attributed to biological effects. There is clear evidence suggesting that aggregates with high organic content have a larger porosity, suggesting a lower fractal dimension and a milder increase of settling velocity with floc size (Engel and Schartau 1999; Santschi et al. 2005). The observed increase of aggregate porosity is attributed to enhanced stickiness of aggregates due to the presence of transparent exopolymer particles (TEP). These particles are natural exudates from marine algae and bacteria with high stickiness to

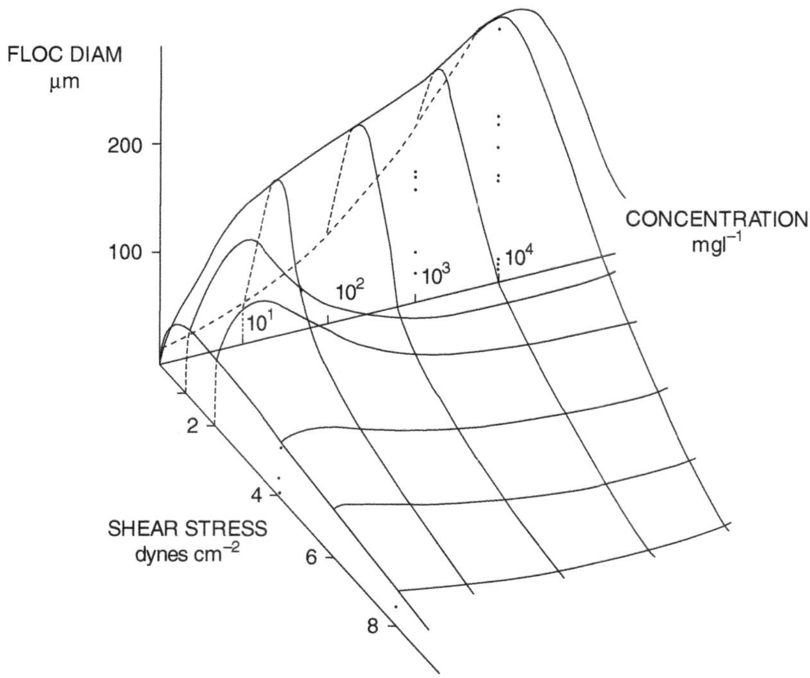

Figure 3.8. A conceptual diagram, adopted from Dyer (1989, Figure 3), illustrating the relationship of floc diameter as a function of sediment concentration and shear stress.

greatly stimulate aggregation (Passow and Alldredge 1995; Santschi et al. 2005). As we discuss next, TEP can increase the efficiency of two particles to adhere after collisions and form large, more porous aggregates (Engel 2000). For example, in the study of Engel and Schartau (1999), all the observed aggregates with TEP presence had fractal dimensions below 2.0 (as low as 1.44). However, Sanford et al. (2005) observed aggregates in Chesapeake Bay estuary with high fecal pellet production. They obtained relatively large settling velocity for a given floc size, suggesting an increased fractal dimension of flocs due to the presence of more compact, high-density fecal pellets (see also Fugate and Friedrichs 2003).

Understanding floc structure is still not sufficient to model cohesive sediment transport. In reality, floc aggregation and breakup processes, called *floc dynamics*, must occur in a turbulent flow (Dyer 1989; Manning and Dyer 1999). Hence floc size, floc density, and floc-settling velocity are all flow-dependent variables that change with the level of turbulence and sediment concentration. In the cohesive sediment literatures, the pioneering study of Dyer (1989) illustrated a conceptual picture (see Figure 3.8) of floc diameter that depends on sediment concentration and turbulent shear stress (one of the several ways to quantify the intensity of turbulence). Both aggregation and breakup of flocs are related to collision of particles. Collisions are more likely to occur when sediment concentration becomes larger and/or flow becomes more turbulent. Hence Figure 3.8 indicates that for a given sediment concentration, floc size initially increases with turbulent shear due to increased collisional frequency. However, floc size then starts to decrease as turbulent shear is further increased. Floc size cannot increase unbounded because turbulent shear can eventually lead to floc breakup. Conversely, under a given turbulent shear stress, floc size initially increases with sediment concentration due to enhanced collisional frequency at higher sediment concentration. Again the increase of floc size cannot be unbounded because for a sufficiently large floc size, further increase in concentration can cause floc breakup. So far, the arguments used to explain the conceptual diagram of Dyer (1989) are based on an equilibrium assumption, that is, the timescale of flocculation is smaller than the residence time of aggregates in the turbulent water column. Winterwerp (1998) carried out a non-equilibrium analysis and demonstrated that his results are still consistent with Dyer (1989). However, Winterwerp (1998) argued that at low shear rate, the timescale of flocculation becomes much larger than the residence time, namely, floc is not in the equilibrium condition, and the ascending branch of floc diameter in Dyer's diagram at low shear rate is the outcome of limited residence time for flocs to grow.

Consistent with the conceptual picture described in Dyer (1989), a phenomenological equation describing the change of averaged floc size due to aggregation (increase of floc size) via particle collision and floc breakup (decrease of floc size) in turbulent flows was proposed by Winterwerp (1998). This mathematical formulation for the variation of floc size can be coupled with standard hydrodynamic and suspended sediment transport models to simulate cohesive sediment transport in tidal boundary layer (e.g., Winterwerp 2002; Son and Hsu 2011). In reality, flocs are polydispersed; namely, at any given time, different sizes of floc coexist. More comprehensive (but complicated) descriptions for polydispersed floc dynamics are based on population balance equations of number density of floc (associated to a given size), with additional source and sink terms representing aggregation and breakup (Sterling et al. 2004, 2005).

The aggregation term in the phenomenological equation of floc dynamics depends on collisional frequency and a coagulation efficiency parameter, which determines the probability of two particles to stay together once they have collided. The coagulation efficiency parameter is often specified empirically, and for inorganic sediment, the coagulation efficiency is less than 0.3. There are also uncertainties in the term describing the breakup process due to the existence of different types of floc breakup mechanism and a lack of good understanding on the floc yield strength. Hence coefficients describing the magnitudes of break-up terms are also specified empirically. Moreover, when the fractal dimension is considered as a variable, floc yield strength may not be assumed as simply an empirical constant (Son and Hsu 2009). However, there is clear evidence from laboratory and field experiments (e.g., Winterwerp 1998; Fettweis et al. 2006) suggesting that the floc size is limited by the smallest turbulent eddy size, called the Kolmogorov length scale, $\nu = (\nu^3/\varepsilon)^{1/4}$, where *nu* is the molecular viscosity of the fluid, and ε is the turbulent kinetic energy dissipation rate (a measure of the intensity of the turbulence). Finally, floc yield strength is known to be highly dependent on the organic content of flocs (e.g., Dyer 1989).

When organic content is significant and aggregates contain greater proportion of TEP, Engel (2000) showed that the stickiness of aggregates is significantly increased, and the coagulation efficiency can be close to 1.0. Large stickiness and high coagulation efficiency during aggregation typically result in large flocs with high porosity and fragility. Hence, comparing with inorganic sediments, when modeling floc dynamics with high organic content and TEP, the fractal dimension should be lower, and empirical coefficients associated with coagulation efficiency and break-up need to be modified accordingly.

Strictly speaking, Equation 16 (or Equation 12) is based on settling of a single particle or a single floc without the interference of other surrounding particles. When sediment (volumetric) concentration is large, the averaged distance among particles/flocs becomes small, and the interference from surrounding particles becomes nonnegligible. When many particles settle concurrently (a cloud of particles), the bulk settling velocity is reduced, and this is called *hindered settling effect.* A hindered settling function is introduced to calculate bulk settling velocity as

$$W_s = \gamma\,(\phi)\,W_{s0}. \tag{20}$$

The hindered settling function is parameterized as a power function of sediment volumetric concentration ϕ (or sometimes sediment mass concentration): $\gamma\,(\phi) = (1 - \phi)^p$, with p depending on particle Reynolds number (see Fredsoe and Deigaard 1992). For fine sediment considered here, p is around 4.5.

Semi-empirical formulations incorporating hindered settling effects, and the bulk features of floc dynamics have been proposed to allow direct calculation of settling velocity of cohesive sediment (e.g., Mehta 1986; Winterwerp et al. 2006). For example, van der Ham and Winterwerp (2001) utilized an algebraic formula suggested by Mehta (1986) (see also van Leussen 1994) to calculate cohesive sediment transport in a tidal channel:

$$W_{s0} = k_1 c^m, \quad c < 1 gL^{-1} \tag{21}$$

$$W_{s0} = W_M\,(1 - k_2 c)^{k_3}, \quad c \geq 1 gL^{-1} \tag{22}$$

$$W_s = W_{s0} \frac{1 + aG}{1 + bG^2}, \tag{23}$$

where k_1, k_2, k_3, m, a, b, and W_M are empirical coefficients associated with different mechanisms. At low-sediment mass concentration $c < 1$ g L^{-1}, aggregation is the dominant process that gives a power-law increase of settling velocity with respect to sediment concentration. Hindered settling starts to become important when sediment mass concentration becomes significantly greater than 1 g L^{-1}. Finally, turbulent shear, which is quantified by a microscale shear parameter $G = \sqrt{\varepsilon/\nu}$, can cause floc breakup and a decrease of settling velocity. These empirical coefficients are calibrated with in situ or laboratory data. However, they are inevitably somewhat site-dependent because of variability in primary particle properties and organic content. In a review provided by (Dyer 1989), the exponent m shown in Equation 21 is shown to range from 0.61 to 2.6.

Because fractal dimension must be smaller than 3.0 for floc, the dependence of floc-settling velocity on floc diameter is a less rapidly increasing function compared with noncohesive sediment. Physically, this reflects the fact that as the floc size grows, the floc density actually decreases, and hence the increase of floc-settling velocity as a function of floc size is mild. As discussed previously, for organic particles with the presence of TEP, the fractal dimension can be lower than 2.0, suggesting an even milder increase of settling velocity on floc diameter. Field observations also indicate that for aggregates with higher organic content, the dependence of floc diameter with respect to sediment concentration is also milder. For example, Winterwerp et al. (2006) obtained the m value in Equation 21 to be only 0.2 (0.44) for the floc observed in Tamar estuary (Scheldt estuary) due to high organic content. Without detailed information on primary particle and floc properties, it is common to simply assume floc-settling velocity to be around 1 mm s^{-1}.

4.2. Riverine input and carrying capacity

Estimating sediment load or river-borne sediment concentration is the first step in modeling sediment dispersal for a given river-shelf system (see Figure 3.1). As we discuss next, sediment discharge is qualitatively proportional to river discharge. Hence, for large rivers, sediment discharge may closely follow the seasonal variation, whereas for small rivers, significant sediment discharge only occurs during more episodic river-flooding events. A universal formula to estimate river-borne sediment load is not possible because the basin area, maximum elevation of the river basin, and level of anthropogenic intervention (damming, farming, and deforestation) are highly variable (Milliman and Syvitski 1992). Hence, for a given river, a power law is often used to relate sediment discharge Q_s with river discharge Q_r (see Geyer et al. 2004 and references therein):

$$Q_s = aQ_r^b, \tag{24}$$

where a and b are site-dependent coefficients to be determined with in situ monitoring data. A similar power law–based formula can be used to relate bulk sediment concentration with river discharge. Other than some of the site-dependent factors mentioned previously, the power law relationship between the sediment discharge and riverine freshwater discharge are in fact consistent with fluid mechanics principles. For example, the total sediment transport rate (bedload and suspended load) is known to follow a power law with respect to flow velocity or bottom stress (e.g., Bagnold 1966; Ribberink 1998). In addition, the power law relationship is also consistent with the concept of carrying capacity. It is well-known that flow turbulence is the main mechanism suspending sediments in a bottom boundary layer. However, sediment-induced density stratification can damp flow turbulence. The sediment-carrying capacity suggests that for a given flow discharge (characterized by flow velocity u and flow depth h) and sediment-settling velocity, there exists a critical concentration, called *saturation concentration*, such that when more sediment is added, the flow cannot sustain such extra sediment load, turbulence is significantly attenuated, and sediments must settle to the bed. Following the three nondimensional parameters discussed in the previous section, for high Reynolds number, it can be shown that the maximum sediment concentration that a river can sustain C_{sat} is calculated as (Winterwerp 2001):

$$C_{sat} = K_s \frac{1}{s-1} \frac{u^3}{ghW_s}, \tag{25}$$

where K_s is an empirical coefficient. Hence, for a given discharge and sediment-settling velocity, there is a maximum sediment load that a river can sustain. The saturation sediment concentration C_{sat} is proportional to flow velocity cubed, consistent with the empirical power law approach.

4.3. The near-, mid-, and far-field dispersal

The main sediment dispersal pathway is generally along the axis of the near-, mid-, and far-field river plume. The first factor in identifying the sediment dispersal pattern is to discern whether the sediment-laden river plume is positively or negatively buoyant (Wright and Nittrouer 1995). Depending on the sediment concentration in a river, the mixture (water and sediment) density of a river plume can become greater than that of the receiving seawater when a large amount of sediment load is delivered by the river. In this case, the river plume becomes a bottom-hopping gravity flow, called *hyperpycnal flow*, with features similar to a turbidity current with reverse buoyancy (e.g., Sequeiros et al. 2009). Depending on the mixing and salinity in the coastal zone, for a negatively buoyant plume to occur, it typically requires a critical sediment concentration of around $35 \sim 45$ g L^{-1} (Mulder and Syvitski 1995). Hence the so-called hyperpycnal flow becomes a rare event that only occurs annually in a few small mountainous rivers (e.g., Milliman and Kao 2005).

In this chapter, we focus on the more commonly observed positively buoyant plume, which is also called *hypopycnal plume*, in which river plume is laden with dilute sediment concentration of no more than O(1) g L^{-1}. In such dilute conditions, the river plume dynamics are approximately independent of sediment dynamics and are well-described by the hydrodynamic theories discussed in Section 2. Conversely, the transport of organic carbon, such as phytoplankton, is more complicated (e.g., Jackson 1990) due to additional source terms related to growth and modification of model coefficients in the aggregation and removal via settling due to increased stickiness. In addition, as light transmission affects the growth, it also depends on the concentration of suspended particulate materials (SPMs). Hence there is a strong coupling between transport of organic particles and light transmission (e.g., Cahill et al. 2008). However, it appears that for both inorganic and organic sediments, the common and key mechanism is the removal rate of SPM from the river plume through settling. Let us consider that a river plume propagates with a vertically averaged velocity u that is of thickness h and assume that fine sediments in the plume are well-mixed with a vertically averaged sediment concentration C. A simple mass balance of sediment in the plume propagation direction (x-direction) gives (Hill et al. 2000; Wu et al. 2006)

$$\frac{\partial C}{\partial t} + u\frac{\partial C}{\partial x} = D_s, \tag{26}$$

where D_s is the deposition rate of sediment off the plume (i.e., removal rate), calculated by

$$D_s = -\frac{CW_e}{h}, \tag{27}$$

where W_e is an "effective" settling velocity. The steady-state solution of this simple formulation provides the sediment concentration distribution along the plume direction x, which is written as:

$$\ln C(x) = \ln C_0 - \frac{W_e x}{hu}, \tag{28}$$

where C_0 is the initial bulk sediment concentration at the river mouth (i.e., $x = 0$). This is a significant simplification over directly solving the three-dimensional advection-diffusion equations of sediment concentration for all the size classes. However, it has the advantage that the sedimentation rate D_s or the effective settling velocity W_e can be estimated based on field-measured sediment concentration in the river plume. A good example using Equations 27 and 28 to inversely estimate W_e using a helicopter-based sampling survey during an Eel River flood event is discussed in detail in Hill et al. (2000). In their formulation, the effect of saltwater entrainment is further incorporated.

It is important to note that the effective settling velocity W_e is different from the settling velocity W_s discussed in equation (12) for primary particles (no flocculation) and Equation 12 for flocs. Because the effective settling shown in Equation 27 is a parameterization of the net vertical flux, it is clear that W_e also includes a contribution from turbulent suspension, which typically gives an upward flux of sediment. In other words, W_e is a conservative estimate (or underestimation) of the actual settling velocity W_s. Hill et al. (2000) reported that their estimate of effective settling velocity of sediment in Eel River plume during flooding is around 0.1 mm s^{-1}. Because the settling velocity estimated based on single-grain settling (no flocculation) is only around 0.04 mm s^{-1}, whereas the actual settling velocity can be larger than 0.1 mm s^{-1} due to turbulent suspension, it is clear that sediments in the Eel River plume settle as flocs, not primary particles. Hence flocculation is probably the most critical mechanism controlling sediment dispersal in a hypopycnal plume. The settling velocity increases as fine sediment become flocculated, which further encourages sedimentation. As a rule of thumb, when fine sediments settle as floc aggregates, the settling velocity is about 0.1 $\sim$ 1 mm s^{-1}.

It should be mentioned that recent field observation of sediment dispersal at the Santa Clara River indicates very rapid sedimentation within 1 km off the river mouth during a flooding event (Warrick et al. 2008). The inversely calculated effective settling velocity is $\mathcal{O}(1)$ cm s^{-1}. A effective settling velocity of $\mathcal{O}(1)$ cm s^{-1} is yet another order of magnitude greater than that can be explained by floc settling ($\mathcal{O}(1)$ mm s^{-1}). To explain such rapid settling, Warrick et al. (2008) conjectured that the observed rapid settling can be due to convective sedimentation because they believe that the sediment concentration in the Santa Clara River plume during flooding can be of several to 10 g L^{-1}. Convective sedimentation is an instability that occurs at the freshwater and saltwater interface due to double-diffusive or settling-driven Rayleigh-Taylor instabilities (e.g., Burns and Meiburg 2012; Yu et al. 2013). Such convective instability triggers the formation of sediment fingers that descend with a speed that exceeds $\mathcal{O}(1)$ cm s^{-1}, an order of magnitude greater than the settling velocity of floc. It has also been observed in the laboratory (e.g., Hoyal et al. 1999; Parsons et al. 2001) and numerical simulation (Snyder and Hsu 2011) that when sediment concentration is of several g L^{-1}, convective sedimentation can occur. Although convective sedimentation is a known process that can be reproduced in the laboratory or numerical simulations, there is no direct evidence from field observation that it occurs, or it is the dominant processes of sediment dispersal in a river plume.

5. Concluding remarks

Although there has been considerable work done to understand some of the nonlinear and complex processes that happen in river plumes, particularly in the near-field plume, and mixing processes associated with upwelling winds, there has not been much work done on the ultimate fate of river plumes. For large river plumes, this can involve the transport and transformation of plume water masses (i.e., increasing salinity through mixing) and other constituents over many hundreds of kilometers. For example, carbon introduced into the coastal ocean by the river plume may be transformed and deposited in shelf sediments or exported off of the shelf. Although there is evidence that most of the carbon burial occurs in large river deltas (Hedges and Keil 1995), it appears that the carbon is heavily modified by processes within plume waters before it is deposited and buried (Chen et al. 2005). For example, the Mississippi/Atchafalaya Rivers may deposit sediments in the river channel or in near-shore environments, before this sediment is later moved to deeper parts of the shelf (Chapter 19).

Also, in large river plumes, evaporation may become an important process that increases the plume salinity and limits the extent of the plume. Most modern surveys of natural river plumes occur at spatial scales over which evaporation is not expected to be important. However, understanding how freshwater in a river plume is modified very far from the source is important to understanding the global water cycle. These processes are also a poorly understood aspect of modern climate models. Freshwater from continents is important in barrier layer formation (e.g., Godfrey and Lindstrom 1989; Lukas and Lindstrom 1991; de Boyer Monteegut et al. 2007) and may alter deep water formation in high latitudes (e.g., Stouffer et al. 2006).

Flocculation is the most critical mechanism determining the removal rate of sediment in the plumes of large rivers. Although the mathematical formulations of flocculation are available, the biological effects have not been incorporated carefully. As there has been significant advancement on how biological effects can change the stickiness and affect the fractal dimension, porosity and settling velocity, and the importance of fecal pellets, these biological-chemical processes need to be better quantified and parameterized in modeling flocculation and settling velocity. Specifically, in the large-scale modeling of plankton population and transport of organic carbon (Jackson 2005), the removal of aggregate via settling is one of the key mechanisms. Hence new findings on the biological effects on settling velocity of aggregate need to be incorporated in the large-scale ecosystem modeling.

References

Armi, L., and D.M. Farmer. 1986. Maximal two-layer exchange through a contraction with barotropic net flow. *J. Fluid Mech.* 164: 27–51.

Avicola, G., and P. Huq. 2003a. The characteristics of the recirculating bulge region in coastal buoyant outflows. *J. Mar. Res.* 61: 435–463.

Avicola, G., and P. Huq. 2003b. The role of outflow geometry in the formation of the recirculating bulge region in coastal buoyant outflows. *J. Mar. Res.* 61: 411–434.

Bagnold, R.A. 1966. Quantifying the contributions of tidal straining and gravitational circulation to residual circulation in periodically stratified tidal estuaries. Geological Survey Professional Paper 422-I.

Balachandar, S., and J.K. Eaton. 2010. Turbulent dispersed multiphase flow. *Annu. Rev. Fluid Mech.* 42: 111–133.

Bianchi, T.S., S.F. DiMarco, J.H. Cowan Jr., R.D. Hetland, P. Chapman, J.W. Day, and M.A. Allison. 2010. The science of hypoxia in the Northern Gulf of Mexico: A review. *Sci. Total Environ.* 408: 1471–1484.

Burns, P., and E. Meiburg. 2012. Sediment-laden fresh water above salt water: linear stability analysis. *J. Fluid Mech.* 691: 279–314.

Cahill, B., O. Schofield, R. Chant, J. Wilkin, E. Hunter, S. Glenn, and P. Bissett. 2008. Dynamics of turbid buoyant plumes and the feedbacks on near-shore biogeochemistry and physics. *Geophys. Res. Lett.* 35, L10605, doi: 10.1029/2008GL033595.

Cantero, M.I., S. Balachandar, A. Cantelli, C. Pirmez, and G. Parker. 2008a. Turbidity current with a roof: direct numerical simulation of self-stratified turbulent channel flow driven by suspended sediment. *J. Geophys. Res.* 114: C03008.

Cantero, M.I., S. Balachandar, and M. Garcia. 2008b. An Eulerian-Eulerian model for gravity currents driven by inertial particles. *Int. J. Multiph. Flow* 34: 484–501.

Chant, R., J. Wilkin, W. Zhang, B. Choi, E. Hunter, R. Castelao, S. Glenn, J. Jurisa, O. Schofield, R. Houghton, J. Kohut, T. Frazer, and M. Moline. 2008. Dispersal of the Hudson River plume in the New York bight: synthesis of observational and numerical studies during LaTTE. *Oceanography* 21: 148–161.

Chen, N., T. Bianchi, and B. McKee. 2005. Early diagenesis of chloropigment biomarkers in the lower Mississippi River and Louisiana shelf: implications for carbon cycling in a river-dominated margin. *Mar. Chem.* 93: 159–177.

Choi, B., and J. Wilkin. 2007. The effect of wind on the dispersal of the Hudson River plume. *J. Phys. Oceanogr.* 37: 1878–1897.

De Boer, G.J., J.D. Pietrzak, and J.C. Winterwerp. 2009. SST observations of upwelling induced by tidal straining in the Rhine ROFI. *Cont. Shelf Res.* 29(1): 263–277.

de Boyer Montegut, C., J. Mignot, A. Lazar, and S. Cravatte. 2007. Control of salinity on the mixed layer depth in the world ocean: 1. General description. *J. Geophys. Res.* 112, C06011, doi: 10.1029/2006JC003953.

de Kok, J., C. de Valk, J. van Kester, E. de Goede, and R. Uittenbogaard. 2001. Salinity and temperature stratification in the Rhine plume. *Estuar. Coast. Shelf Sci.* 53(4): 467–475.

DiMarco, S.F., P. Chapman, N. Walker, and R.D. Hetland. 2010. Does local topography control hypoxia on the eastern Texas-Louisiana shelf? *J. Mar. Sys.* 80: 25–35.

Dyer, K.R. 1989. Sediment processes in estuaries: Future research requirements. *J. Geophys. Res.* 94: 14327–14339.

Dyer, K.R., and A.J. Manning. 1999. Observation of the size, settling velocity and effective density of flocs, and their fractal dimensions. *J. Sea Res.* 41: 87–95.

Engel, A. 2000. The role of transparent exopolymer particles (TEP) in the increase in apparent particle stickiness (α) during the decline of a diatom bloom. *J. Plankton Res.* 22: 485–497.

Engel, A., and M. Schartau. 1999. Influence of transparent exopolymer particles (TEP) on sinking velocity of Nitzschia closterium aggregates. *Mar. Ecol. Prog. Series* 182: 69–76.

Fennel, K., R. Hetland, Y. Feng, and S.F. DiMarco. 2011. A coupled physical-biological model of the Northern Gulf of Mexico shelf: model description, validation and analysis of phytoplankton variability. *Biogeosciences* 8: 1881–1899.

Fettweis, M., F. Francken, V. Pison, and D. Van den Eynde. 2006. Suspended particulate matter dynamics and aggregate sizes in a high turbidity area. *Mar. Geol.* 235: 63–74.

Fong, D.A., and W.R. Geyer. 2001. Response of a river plume during an upwelling favorable wind event. *J. Geophys. Res.* 106: 1067–1084.

Fong, D.A., and W.R. Geyer. 2002. The alongshore transport of freshwater in a surface-trapped river plume. *J. Phys. Oceanogr.* 32: 957–972.

Franks, P.J.S., and D.M. Anderson. 1992. Alongshore transport of a toxic phytoplankton bloom in a buoyancy current: *Afexandnum famaratse* in the Gulf of Maine. *Mar. Biol.* 112: 153–164.

Fredsoe, J., and R. Deigaard. 1992. Mechanics of Coastal Sediment Transport. Singapore: World Scientific.

Fugate, D.C., and C.T. Friedrichs. 2003. Controls on suspended aggregate size in partially mixed estuaries. *Estuar. Coast. Shelf Sci.* 58: 389–404.

Garcia Berdeal, I., B.M. Hickey, and M. Kawase. 2002. Influence of wind stress and ambient flow on a high discharge river plume. *J. Geophys. Res.* 107 (C9), 3130, doi:10.1029/2001JC000932.

Garvine, R.W. 1974. Physical features of the Connecticut River outflow during high discharge. *J. Geophys. Res.* 79: 831–846.

Garvine, R.W. 1995. A dynamical system for classifying buoyant coastal discharges. *Cont. Shelf Res.* 15: 1585–1596.

Garvine, R.W. 1999. Penetration of buoyant coastal discharge onto the continental shelf: a numerical study. *J. Phys. Oceanogr.* 29: 1892–1909.

Garvine, R.W. 2001. The impact of model configuration in studies of buoyant coastal discharge. *J. Mar. Res.* 59: 193–225.

Garvine, R.W., and M.M. Whitney. 2006. An estuarine box model of freshwater delivery to the coastal ocean for use in climate models. *J. Mar. Res.* 64: 173–194.

Geyer, W.R., R.C. Beardsley, S.J. Lentz, J. Candela, R. Limeburner, W.E. Johns, B.M. Castro, and I.D. Soares. 1996. Physical oceanography of the Amazon shelf. *Cont. Shelf Res.* 16: 575–616.

Geyer, W.R., R.P. Signell, D.A. Fong, J. Wang, D.M. Anderson, and B.A. Keafer. 2004. The freshwater transport and dynamics of the western Maine coastal current. *Cont. Shelf Res.* 24: 1339–1357.

Godfrey, JS., and E.J. Lindstrom. 1989. The heat budget of the equatorial western Pacific surface mixed layer. *J. Geophys. Res.* 94: 8007–8017.

Halverson, M.J., and R. Pawlowicz. 2008. Estuarine forcing of a river plume by river flow and tides. *J. Geophys. Res.* 113, C09033, doi:10.1029/2008JC004844.

Hedges, J.I., and R.G. Keil. 1995. Sedimentary organic matter preservation: an assessment and speculative synthesis. *Mar. Chem.* 49: 81–115.

Hetland, R.D. 2005. Relating river plume structure to vertical mixing. *J. Phys. Oceanogr.* 35: 1667–1688.

Hetland, R.D. 2010. The effects of mixing and spreading on density in near-field river plumes. *Dyn. Atmos. Oceans* 49: 37–53.

Hetland, R.D., and S.F. DiMarco. 2008. How does the character of oxygen demand control the structure of hypoxia on the Texas-Louisiana continental shelf? *J. Mar. Sys.* 70: 49–62.

Hetland, R.D., and D.G. MacDonald. 2008. Spreading in the near-field Merrimack River plume. *Ocean Model.* 21: 12–21.

Hickey, B.M., R.M. Kudela, J. Nash, K.W. Bruland, W.T. Peterson, P. MacCready, E.J. Lessard, D.A. Jay, N.S. Banas, A.M. Baptista, E.P. Dever, P.M. Kosro, L.K. Kilcher, A.R. Homer-Devine, E.D. Zaron, R.M. McCabe, J.O. Peterson, P. Orton, J. Pan, and M.C. Lohan. 2010. River influences on shelf ecosystems: Introduction and synthesis. *J. Geophys. Res.* 115: C00B17, doi:10.1029/2009JC005452.

Hill, P.S., T.G. Milligan, and W.R. Geyer. 2000. Controls on effective settling velocity in the Eel River flood plume. *Cont. Shelf Res.* 20: 2095–2111.

Hordoir, R., J. Polcher, J.C. Brun-Cottan, and G. Madec. 2008. Towards a parametrization of river discharges into ocean general circulation models: a closure through energy conservation. *Clim. Dyn.* 31: 891–908.

Horner-Devine, A.R. 2009. The bulge circulation in the Columbia River plume. *Cont. Shelf Res.* 29: 234–251.

Horner-Devine, A.R., D.A. Fong, S.G. Monismith, and T. Maxworthy. 2006. Laboratory experiments simulating a coastal river discharge. *J. Fluid Mech.* 555: 203–232.

Horner-Devine, A.R., D. Jay, P. Orton, and E. Spahn. 2009. A conceptual model of the strongly tidal Columbia River plume. *J. Mar. Sys.* 78: 460–475.

Houghton, R.W., C.E. Tilburg, R.W. Garvine, and A. Fong. 2004. Delaware River plume response to a strong upwelling-favorable wind event. *Geophys. Res. Lett.* 31: L07302.

Hoyal, D.C. J.D., M.I. Bursik, and J.F. Atkinson. 1999. Settling-driven convection: A mechanism of sedimentation from stratified fluids. *J. Geophys. Res.* 104: 7953–7966.

Huq, P. 2009. The role of Kelvin number on bulge formation from estuarine buoyant outflows. *Estuar. Coast.* 32: 709–719.

Isobe, A. 2005. Ballooning of river-plume bulge and its stabilization by tidal currents. *J. Phys. Oceanogr.* 35: 2337–2351.

Ivey, G.N., and J. Imberger. 1991. On the nature of turbulence in a stratified fluid. I: The energetics of mixing. *J. Phys. Ocean.* 21: 650–658.

Jackson, G.A. 1990. A model of the formation of marine algal flocs by physical coagulation processes. *Deep-Sea Res.* 37: 1197.

Jackson, G.A. 2005. Coagulation theory and models of oceanic plankton aggregation, In: I.G. Droppo, G.G. Leppard, S.N. Liss, and T.G. Milligan (eds.), Flocculation in Natural and Engineered Environmental Systems, pp. 271–292. Boca Raton, FL: CRC Press.

Jones, G.R., J.D. Nash, R.L. Doneker, and G.H. Jirka. 2007. Buoyant surface discharges into water bodies. I: flow classification and prediction methodology. *J. Hydraul. Engineer.* 133: 1010–1020.
Khelifa, A., and P.S. Hill. 2006. Models for effective density and settling velocity of flocs. *J. Hydraul. Res.* 44: 390–401.
Knudsen, M. 1900. Ein Hydrographische Lehrsatz. *An. Hydrogr. Maritimen Met.* 28: 316–320.
Kranenburg, C. 1994. The fractal structure of cohesive sediment aggregates. *Estuar. Coast. Shelf Sci.* 39: 451–460.
Krone, R.B. 1962. Flume studies of the transport of sediment in estuarial shoaling processes. Hydraulics Engineering Laboratory, University of California, Berkeley, final report.
Kudela, R.M., A.R. Horner-Devine, N.S. Banas, B.M. Hickey, T.D. Peterson, R.M. McCabe, E.J. Lessard, E. Frame, K.W. Bruland, D.A. Jay, J.O. Peterson, W.T. Peterson, P.M. Kosro, S.L. Palacios, M.C. Lohan, and E.P. Dever. 2010. Multiple trophic levels fueled by recirculation in the Columbia River plume. *Geophys. Res. Lett.* 37: L18607, doi:10.1029/2010GL044342.
Lentz, S. 2004. The Response of Buoyant Coastal Plumes to Upwelling-Favorable Winds. *J. Phys. Oceanogr.* 34: 2458–2467.
Lukas, R., and E. Lindstrom. 1991. The mixed layer of the western equatorial Pacific Ocean. *J. Geophys. Res.* 96: 3343.
MacCready, P., and W. Geyer. 2010. Advances in estuarine physics. *Ann. Rev. Mar. Sci.* 2: 35–58.
MacDonald, D.G., and W.R. Geyer. 2004. Turbulent energy production and en-trainment at a highly stratified estuarine front. *J. Geophys. Res.* 109: C05004.
MacDonald, D.G., and W.R. Geyer. 2005. Hydraulic Control of a Highly Stratified Estuarine Front. *J. Phys. Oceanogr.* 35: 374–387.
MacDonald, D.G., L. Goodman, and R.D. Hetland. 2007. Turbulent dissipation in a near-field river plume: A comparison of control volume and microstructure observations with a numerical model. *J. Geophys. Res.* 112: C07026.
Maggi, F., F. Mietta, and J.C. Winterwer. 2007. Effect of variable fractal dimension on the floc size distribution suspended cohesive sediment. *J. Hydrol.* 343: 43–55.
Manning, A.J., and K.R. Dyer. 1999. A laboratory examination of floc characteristics with regard to turbulent shearing. *Mar. Geol.* 160: 147–170.
Meade, R. 1996. River-Sediment Inputs to Major Deltas. Sea-Level Rise and Coastal Subsidence. London: Kluwer, pp. 63–85.
Mehta, A.J. 1986. Characterization of cohesive sediment properties and transport processes in estuaries. In: Metha (ed.), Lecture Notes on Coastal and Estuarine Studies 14: 290–325.
Milliman, J.D., and S.J. Kao. 2005. Hyperpycnal discharge of fluvial sediment to the ocean: impact of super-typhoon herb (1996) on Taiwanese Rivers. *J. Geol.* 113: 503–506.
Milliman, J.D., and R.H. Meade. 1983. World-wide delivery of river sediment to the ocean. *J. Geol.* 91: 1–21.
Milliman, J.D., and J.P.M. Syvitski. 1992. Geomorphic/tectonic control of sediment discharge to the ocean: the importance of small mountain rivers. *J. Geol.* 100: 525–544.
Molleri, G.S.F., E.M.L.M. Novo, and M. Kampel. 2010. Space-time variability of the Amazon River plume based on satellite ocean color. *Cont. Shelf Res.* 30: 342–352.
Mulder, T., and J.P.M. Syvitski. 1995. Turbidity currents generated at river mouths during exceptional discharges to the world oceans. *J. Geol.* 103: 285–299.
Murray, S.P. 1997. An observational study of the Mississippi-Atchafalaya coastal plume: final report. OCS Study MMS 98–0040.
Nash, J.D., L.F. Kilcher, and J.N. Moum. 2009. Structure and composition of a strongly stratified, tidally pulsed river plume. *J. Geophys. Res.* 114: C00B12, doi:10.1029/2008JC005036.
Nof, D., and T. Pichevin. 2001. The ballooning of outflows. *J. Phys. Oceanogr.* 31: 3045–3058.

O'Donnell, J. 1990. The Formation and Fate of a River Plume: A Numerical Model. *J. Phys. Oceanogr.* 20: 551–569.

O'Donnell, J., G.O. Marmorino, and C.L. Trump. 1998. Convergence and Downwelling at a River Plume Front. *J. Geophys. Res.* 28: 1481–1495.

Ozdemir, C.E., T.-J. Hsu, and S. Balachandar. 2011. A numerical investigation of lutocline dynamics and saturation of fine sediment in the oscillatory boundary layer. *J. Geophys. Res.* 116: C09012, doi:10.1029/2011JC007185.

Parsons, J.D., J.W.M. Bush, and J.P.M. Syvitski. 2001. Hyperpycnal flow formation from riverine outflows with small sediment concentration. *Sedimentology* 48: 465–478.

Partheniades, N. 1965. Erosion and deposition of cohesive soils. *J. Hydraul.* 91: 105–139.

Passow, U., and A.L. Alldredge. 1995. Aggregation of a diatom bloom in a mesocosm: the role of transparent exopolymer particles (TEP). *Deep-Sea Res. Part II Top. Stud. Oceanogr.* 42: 99–109.

Pritchard, M., and D. Huntley. 2002. Instability and mixing in a small estuarine plume front. *Estuar. Coast. Shelf Sci.* 55: 275–285.

Pritchard, M., and D. Huntley. 2006. A simplified energy and mixing budget for a small river plume discharge. *J. Geophys. Res.* 111: C03019, doi:10.1029/2005JC002984.

Pullen, J.D., and J.S. Allen. 2000. Modeling Studies of the Coastal Circulation off Northern California: Shelf Response to a Major Eel River Flood Event. *Cont. Shelf Res.* 20: 2213–2238.

Ribberink, J.S. 1998. Bed-load transport for steady flows and unsteady oscillatory flows. *Coast. Eng.* 34: 59–82.

Rippeth, T., N.R. Fisher, and J.H. Simpson. 2001. The Cycle of Turbulent Dissipation in the Presence of Tidal Straining. *J. Phys. Oceanogr.* 31: 2458–2471.

Salisbury, J.E., D. Vandemark, C.W. Hunt, J. Campbell, W.R. McGillis, and W.H. McDowell. 2008. Seasonal observations of surface waters in two Gulf of Maine estuary-plume systems: Relationships between watershed attributes, optical measurements and surface pCO2. *Estuar. Coast. Shelf Sci.* 77: 245–252.

Sanford, L.P., P.J. Dickhudt, L. Rubiano-Gomez, M. Yates, S.E. Suttles, C.T. Friedrichs, D.C. Fugate, and H. Romine. 2005. Variability of suspended particle concentrations, sizes, and settling velocities, in the Chesapeake Bay turbidity maximum, In: I.G. Droppo, G.G. Leppard, S.N. Liss, and T.G. Milligan (eds.), Flocculation in Natural and Engineered Environmental Systems, pp. 211–236. Boca Raton, FL: CRC Press.

Santschi, P.H., A.B. Burd, J.-F. Gaillard, and A.A. Lazarides. 2005. Transport of materials and chemicals by nanoscale colloids and micro- to macro-scale flocs in marine, freshwater and engineered systems, In: I.G. Droppo, G.G. Leppard, S.N. Liss, and T.G. Milligan (eds.), Flocculation in Natural and Engineered Environmental Systems, pp. 191–209. Boca Raton, FL: CRC Press.

Sequeiros, O.E., A. Cantelli, E. Viparelli, J.D.L. White, M.H. Garca, and G. Parker. 2009. Modeling turbidity currents with nonuniform sediment and reverse buoyancy. *Water Resour. Res.* 45: W06408, doi:10.1029/2008WR007422.

Simpson, J.H., H. Burchard, N.R. Fisher, and T.P. Rippeth. 2002. The semi-diurnal cycle of dissipation in a ROFI: model-measurement comparisons. *Cont. Shelf Res.* 22: 1615–1628.

Simpson, J.H., W.G. Bos, F.F. Schirmer, A.J. Souza, T.P. Rippeth, S.E. Jones, and D. Hydes. 1993. Periodic Stratification in the Rhine ROFI in the North-sea. *Oceanologica Acta* 16: 23–32.

Snyder, P.J., and T.-J. Hsu. 2011. A numerical investigation of convective sedimentation. *J. Geophys. Res.* 116: C09024, doi:10.1029/2010JC006792.

Son, M., and T.-J. Hsu. 2009. The effect of variable yield strength and variable fractal dimension on flocculation of cohesive sediment. *Water Res.* 43: 3582–3592.

Son, M., and T.-J. Hsu. 2011. The effects of flocculation and bed erodibility on modeling cohesive sediment resuspension. *J. Geophys. Res.* 116: C03021, doi:10.1029/2010JC006352.

Sterling Jr., M.C., J.S. Bonner, A.N.S. Ernest, C.A. Page, and R.L. Autenrieth. 2005. Application of fractal flocculation and vertical transport model to aquatic sol-sediment systems. *Water Res.* 39: 1818–1830.

Sterling Jr., M.C., J.S. Bonner, C.A. Page, C.B. Fuller, A.N.S. Ernest, and R.L. Autenrieth. 2004. Modeling crude oil droplet-sediment aggregation in nearshore waters. *Environ. Sci. Technol.* 38: 4627–4634.

Stouffer, R.J., J. Yin, J.M. Gregory, K.W. Dixon, M.J. Spelman, W. Hurlin, A.J. Weaver, M. Eby, G.M. Flato, H. Hasumi, et al. 2006. Investigating the causes of the response of the thermohaline circulation to past and future climate changes. *J. Climate* 19: 1365–1387.

Syvitski, J.P.M., C.J. Vörösmarty, A.J. Kettner, and P. Green. 2005. Impact of humans on the flux of terrestrial sediment to the global coastal ocean. *Science* 308: 376–380.

Thomsen, L. 2005. Organic rich aggregates in the ocean: formation, transport behavior and biochemical composition, In: I.G. Droppo, G.G. Leppard, S.N. Liss, and T.G. Milligan (eds.), Flocculation in Natural and Engineered Environmental System, pp. 237–248. Boca Raton, FL: CRC Press.

Traykovski, P., W.R. Geyer, J.D. Irish, and J.F. Lynch. 2000. The role of wave-induced density-driven fluid mud flows for cross-shelf transport on the Eel River continental shelf. *Cont. Shelf Res.* 20: 2113–2140.

Trenberth, K.E., L. Smith, T. Qian, A. Dai, and J. Fasullo. 2007. Estimates of the global water budget and its annual cycle using observational and model data. *J. Hydrometeorol.* 8: 758–769.

van der Ham, R., and J.C. Winterwerp. 2001. Turbulent exchange of fine sediments in a tidal channel in the Ems/Dollard estuary. Part II. Analysis with a 1DV numerical model. *Cont. Shelf Res.* 21: 1629–1647.

van Leussen, W. 1994. Estuarine Macro Flocs and Their Role in Fine-Grained Sediment Transport. PhD Thesis, Utrecht University, The Netherlands.

Warner, J.C., C.R. Sherwood, R.P. Signell, C.K. Harris, and H.G. Arango. 2008. Development of a three-dimensional, regional, couple wave, current, and sediment-transport model. *Comput. Geosci.* 34: 1284–1306.

Warrick, J.A., J. Xu, M.A. Nobel, and H.J. Lee. 2008. Rapid formation of hyperpycnal sediment gravity currents offshore of a semi-arid California river. *Cont. Shelf Res.* 28: 991–1009.

Whitney, M.M., and R.W. Garvine. 2005. Wind influence on a coastal buoyant outflow. *J. Geophys. Res* 110, doi:10.1029/2003JC002261.

Winterwerp, J.C. 1998. Stratification effects by cohesive and non-cohesive sediment. *J. Hydraul. Res.* 36: 309–326.

Winterwerp, J.C. 2001. Stratification effects by cohesive and non-cohesive sediment. *J. Geophys. Res.* 106: 22559–22574.

Winterwerp, J.C. 2002. On the flocculation and settling velocity of estuarine mud. *Cont. Shelf Res.* 22: 1339–1360.

Winterwerp, J., A.J. Manning, C. Martens, T. de Mulder, and J. Vanlede. 2006. A heuristic formula for turbulence-induced flocculation of cohesive sediment. *Estuar. Coast. Shelf Sci.* 68: 195–207.

Winterwerp, J.C., and W.G.M. van Kesteren. 2004. Introduction to the Physics of Cohesive Sediment in the Marine Environment. New York: Elsevier.

Wright, L.D. 1977. Sediment transport and deposition at river mouth: a synthesis. *Geol. Soc. Am. Bull.* 88: 857–868.

Wright, L.D., and J.M. Coleman. 1971. Effluent expansion and interfacial mixing in the presence of a salt wedge, Mississippi River delta. *J. Geophys. Res.* 76: 8649–8661.

Wright, L.D., and C.A. Nittrouer. 1995. Dispersal of river sediments in coastal seas: six contrasting cases. *Estuaries* 18: 494–508.

Wu, J., L. Ametistova, M. Heron, C.J. Lemckert, and P. Kalangi. 2006. Finite dispersal of a separative nepheloid plume by an internal hydraulic jump in a tropical mountainous river estuary. *J. Geophys. Res.* 111: C11004, doi:10.1029/2005JC003404.

Yankovsky, A.E., and D.C. Chapman. 1997. A simple theory for the fate of buoyant coastal discharges. *J. Phys. Oceanogr.* 27: 1386–1401.

Yu, X., T.-J. Hsu, and S. Balachandar. 2013. Convective instability in sedimentation: Linear stability analysis. *J. Geophys. Res.* 118: 256–272.

Zhang, W., J. Wilkin, and R. Chant. 2009. Modeling the pathways and mean dynamics of river plume dispersal in the New York Bight. *J. Phys. Oceanogr.* 39: 1167–1183.

Zhang, X., M. Marta-Almeida, and R. Hetland. 2012. A high-resolution pre-operational forecast model of circulation on the Texas-Louisiana continental shelf and slope. *J. Operation. Oceanogr.* 5: 19–34.

4

Shelf and slope sedimentation associated with large deltaic systems

J. P. Walsh, D. R. Corbett, A. S. Ogston, C. A. Nittrouer, S. A. Kuehl, M. A. Allison, and S. L. Goodbred, Jr.

1. Introduction

Deltas are key areas for the supply and cycling of sediments and solutes on the Earth's continental margins and sequester about 80% of the carbon buried in marine sediments (Berner 1982). Many large delta systems are home to significant populations, including numerous major cities (e.g., Shanghai [Yangtze], Calcutta, Dhaka [Ganges-Brahmaputra], New Orleans [Mississippi]; Ericson et al. 2006; Syvitski and Saito 2007), and the continental margin surrounding these deltas supports highly productive fisheries and other human resources (e.g., hydrocarbons). Elevated sea levels from eustatic rise, land subsidence, and tropical cyclones are examples of the serious hazards to delta inhabitants and coastal economies in many systems in the present century (Syvitski et al. 2009). The largest deltas on Earth have evolved over geologic timescales (e.g., 10^5–10^6 years), but their present subaerial and subaqueous depositional patterns have formed since the rate of sea-level rise declined at ~7 kya (Stanley and Warne 1994). Their present extent and behavior reflect this development and modifications by humans over the last millennia. Significant anthropogenic modifications were made in particular to the subaerial delta plain (e.g., land-use change, urbanization, channel cutting and damming, artificial levee construction), and these and others influence the modern supply, character, and location of sediment and carbon delivery to the adjacent margin. Delta systems are important locations of material storage and exchange, and deltaic processes extend well beyond the shoreline, carrying materials from terrestrial to marine realms.

Wright (1985) indicates that the term *delta* refers to subaerial and subaqueous river-sediment accumulations near a source stream, including all that is reworked by oceanographic processes. With this in mind, and because a minimal fraction of the sediment load for most rivers is bedload (<5%; Syvitski and Saito 2007), the largest area of sedimentation for the majority of delta systems lies on the adjacent shelf and slope. Margin sedimentation has been well-studied for some of the Earth's largest sediment dispersal systems, and comparison between the Amazon, Ganges-Brahmaputra and Mississippi-Atchafalaya systems in this chapter demonstrates some similarities in the form and function of large deltas. However, contrasting of deltaic systems also shows how they exist in a wide range of geological settings and are influenced by a diversity of ocean conditions. As a result, there are notable differences in behavior in terms of the deposition and burial of sediments and carbon, and many large systems have not been thoroughly studied as a result of their remoteness and size. New insights can be gained from quantitative analysis and modeling of the diversity of systems. This chapter serves as an introduction to the physical forcing and controls on sediment and carbon transport operating in the distal (i.e., open marine) portion of large deltaic systems to provide a context for the following chapters that detail the margin biogeochemical cycling of specific systems. Because of the breadth in the character of deltaic systems, the sedimentary functioning of three well-studied and diverse river dispersal systems is the focus of the present chapter and serves to illuminate the range in delta behavior.

2. Overview of large deltaic systems

Initial efforts to classify deltas to date have focused on sediment supply (riverine), tidal, and wave processes as being the first-order control on their subaerial morphology (Coleman and Wright 1975; Galloway 1975; Wright 1985). Additionally, Orton and Reading (1993) demonstrated how the grain size of the sediment load delivered by the riverine system has an important influence on the depositional sedimentary facies and their configuration. More recent studies of large river sediment dispersal systems have highlighted how margin geomorphology and oceanographic conditions (e.g., waves and currents) are fundamental to determining how and where the sediment load from a river accumulates offshore (e.g., Kuehl et al. 1986; Swift and Thorne 1991; Wright and Nittrouer 1995; Walsh and Nittrouer 2003; Kuehl et al. 2005; Liu et al. 2009; Walsh and Nittrouer 2009; Sheremet et al. 2011). More specifically, factors affecting the amount, timing, and character of sediment input (e.g., Meade et al. 1990; Milliman and Syvitski 1992; Hovius 1998; Syvitski and Milliman 2007) and variability in the shape and dynamics of the receiving basin ultimately affects how and where muddy sediment accumulates (Fig. 4.1). Modeling of river sediment dispersal has made great strides in the last couple decades, enabling a better understanding of the role of ocean processes (e.g., Harris et al. 2005; Overeem et al. 2005; Slingerland et al. 2008), and over that time, sediment-gravity flows have become better recognized for their transport capability in low-gradient shelf systems (Wright and Friedrichs 2006; Hill et al. 2009 and references therein). Today, much effort is being made to develop conceptual, analytical, and numerical simulation models that incorporate the full range of sedimentary processes that shape deltas (e.g., Syvitski 2008; Warner et al. 2008).

Since the mid-1990s, significant research has also focused on the functioning of small mountainous river dispersal systems because of their countless number and considerable combined contribution to the cycling of sediments and solutes in the ocean (Milliman and Syvitski 1992). Yet, the fact remains that large deltaic systems supply a tremendous amount of freshwater and sediment discharge to the oceans. Thirty of the largest sediment-discharging rivers supply 6.9×10^3 Mt y^{-1} or 49% of the estimated global pre-human river sediment load to the sea (Table 4.1; Syvitski et al. 2005), and because these systems individually release such a major volume of water and sediment, they each have recognizable effects on regional biogeochemical processes of continental margins.

Although there are commonalities to many of the largest deltaic systems, they exist in a wide range of geological settings and are influenced by a diversity of ocean conditions (Fig. 4.2). Interestingly, 16 of the 30 listed in Table 4.1 have mouths located in tropical areas, including 8 of the top 10 (excluding the Huanghe or Yangtze); these warm-climate systems have coasts that are home to vast mangrove forests that influence carbon production and cycling (Chapter 15). The sediment discharge of those listed varies over two orders of magnitude (Table 4.1, Fig. 4.2), and the wave and tidal

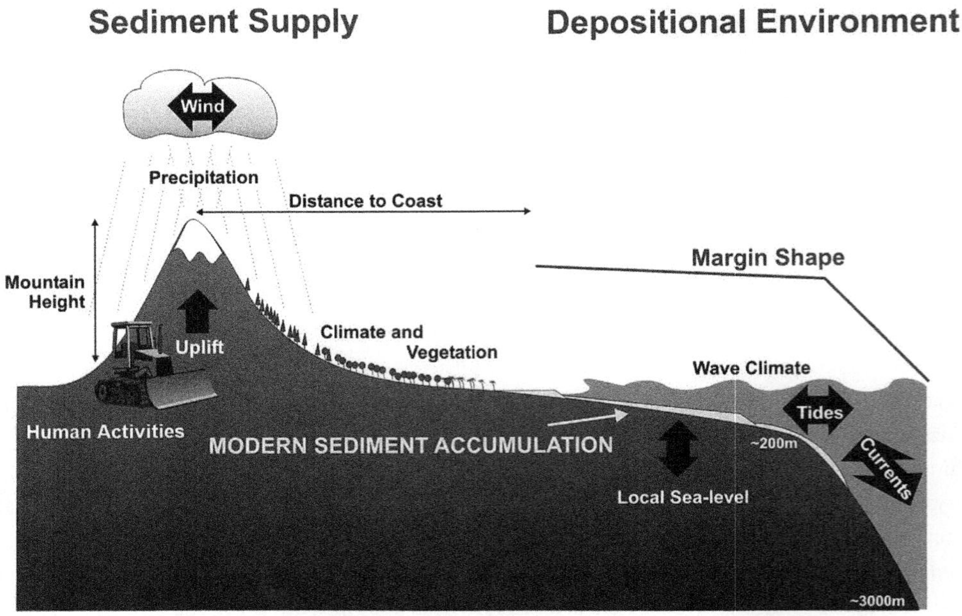

Figure 4.1. Factors affecting continental-margin sedimentation. The many factors can be separated into two broad categories: the terrestrial factors that affect the sediment discharge and the dynamics and properties of the sea that control the accumulation patterns of materials.

regime at the mouths of these systems is markedly different, from 0 to >3-m mean significant wave height and micro- to macrotidal conditions. Walsh and Nittrouer (2009) proposed the use of a hierarchical decision tree as a means to discern the expected sedimentation pattern of fine-grained fluvial dispersal systems. In this scheme, four deltaic types are identified, and large deltas are represented in each of these system types (Table 4.1, Fig. 4.2). These are (listed in frequency order): proximal-accumulation dominated (PAD, 10), subaqueous-delta clinoform (SDC, 8), marine-dispersal dominated (MDD, 9), and canyon captured (CC, 3). PAD systems have much of their load stored immediately adjacent to the distributary mouth(s) (e.g., Mississippi, Po), whereas CC systems transport a large fraction of their sediment supply to the deep-sea via a submarine canyon (e.g., Sepik, Congo). Many of the largest river systems have created substantial shelf sediment deposits (i.e., SDCs), and rivers that discharge into strong wave and/or tide areas spread their load over vast areas, minimizing their geomorphic expression (i.e., MDD). This distribution of class assignments highlights the differing forcing functions, and thus sedimentation behavior expected at large river margins. Regarding biogeochemical cycling, another point to consider is that although continental shelves are the ocean areas most influenced by material emanating from rivers, areas seaward of the shelf break may also experience rapid river-related sedimentation. Modern off-shelf sediment flux depends on the style, volume, and character of sediment input; margin physical processes; and available accommodation space on the shelf, all of which are functions of the shelf geometry and other factors (Milliman and Syvitski 1992; Walsh and Nittrouer 2003). Interestingly, the Amazon, which discharges more sediment than any other river, exports only a minimal fraction of its load beyond the

Table 4.1. *Attributes of large deltaic systems*

Number	River	Sediment discharge (Mt yr^{-1})	Tidal range (m)	Mean significant wave height (m)	Discharge to marine power	Shelf width (km)	System type
1	Amazon	1,194	6	2	0.97	330	SDC
2	Ganges-Brahmaputra	1,100	3.6	1	2.55	231	SDC
3	Huanghe	1,100	0.8	1.5	12	1,105	PAD
4	Yangtze	480	4.5	1.5	0.69	557	SDC
5	Mississippi	398	0.4	0.5	30.9	15	PAD
6	Irrawaddy	256	4.2	1.5	0.42	179	SDC
7	Indus	249	5	3.5	0.21	42	MDD
8	Magdalena	220	1.5	1.5	1.59	11	CC
9	Orinoco	173	1.9	1.3	1.09	143	PAD
10	Godavari	170	3	2	0.43	9	CC
11	Colorado	120	5	0.5	0.15	147	SDC
12	Nile	120	0.4	1.5	1.61	40	PAD
13	Song Hong	109	3.5	1.5	0.25	456	SDC
14	MacKenzie	100	0.2	1.2	2.21	79	PAD
15	Mekong	98	2.6	1.5	0.35	271	SDC
16	Parana	90	4	0.5	0.19	480	PAD
17	Orange	89	1.5	2.5	0.34	155	MDD
18	Purari	80*	3.4*	1.3*	NA	83	MDD
19	Sepik	80*	0.8*	0.9*	NA	3	CC
20	Zhujiang	78	5	1.5	0.09	297	MDD
21	Fly	70	4	1.5	0.13	155	SDC
22	Copper	63*	3.4	3	0.12	81	MDD
23	Danube	67	0	1.5	0.99	31	PAD
24	Krishna	64	3	2	0.17	13	MDD
25	Choshui	63	3.6*	1*	NA	108	MDD
26	Mahanadi	60	3.5	2	0.12	24	MDD
27	Rhone	59	0.5	2	0.45	36	PAD
28	Yukon	54	1.5	2	0.3	699	PAD
29	Tigris–Euphrates	53	3	1	0.18	1008	MDD
30	Amur	52*	0.6*	NA	NA	291	PAD

Data are from Syvitski and Saito (2007) and Walsh and Nittrouer (2009, with *) and references therein.

shelf break (Nittrouer et al. 1995), but some smaller systems (e.g., Sepik River, Papua New Guinea) lose the majority (>90 %) of sediment to deeper waters, perhaps because of gravity flows down a shelf-impinging canyon (Kineke et al. 2000; Walsh and Nittrouer 2003; Kuehl et al. 2004).

Finally, it must be emphasized that most large delta systems have changed substantially over the last few centuries. Numerous papers have demonstrated how human activities have drastically affected river sediment loading (e.g., Syvitski et al. 2005; Syvitski and Saito 2007), and these changes have in turn likely impacted margin sediment accumulation rates (e.g., Sommerfield et al. 1999; Orpin et al. 2006; Allison et al. 2007; Kettner et al. 2007). In this context, it must be realized that the study of delta systems today only provides a modern snapshot of the sedimentary functioning, which includes overprinting by human activities. Longer timescale (geological) investigations, both observational

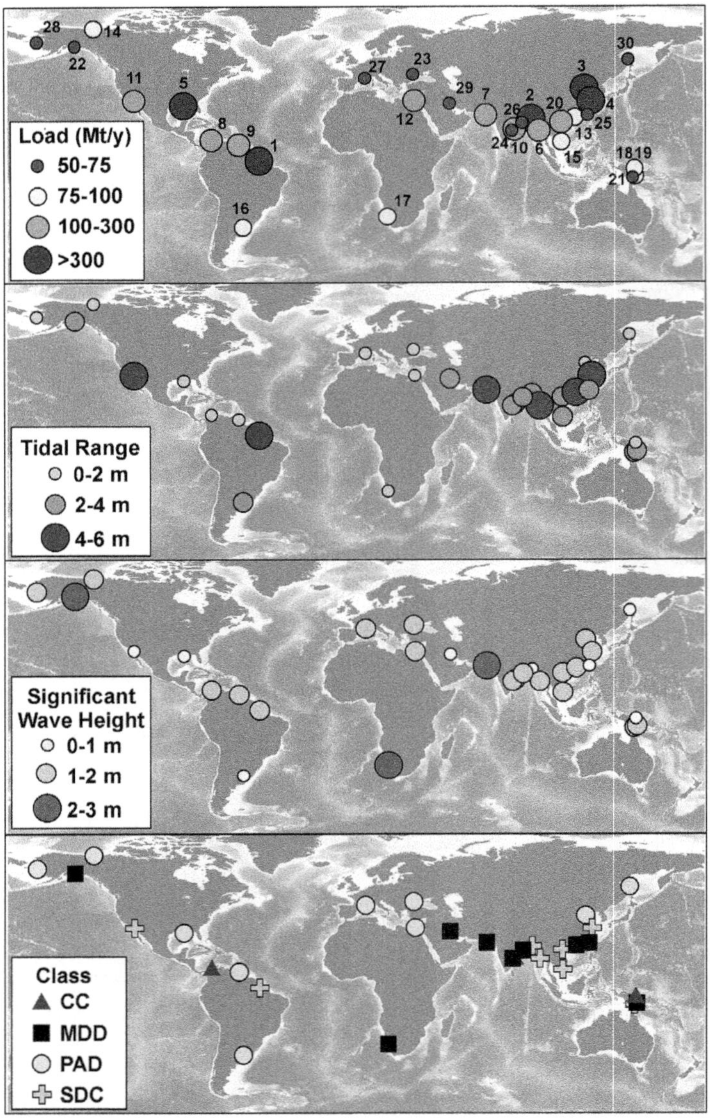

Figure 4.2. Data and classifications for 30 large river-sediment dispersal systems. From top to bottom, panels show river sediment discharge (Mt y^{-1}), tidal range (m), significant wave height (m), and classification (bottom, based on Walsh and Nittrouer 2009). Numbers follow Table 4.1.

and modeling, are essential to providing a deeper perspective on a system's sedimentary evolution (e.g., Fly River/Gulf of Papua, Tcherepanov et al. 2010; Indus River, Giosan et al. 2006; Mississippi River, Roberts 1997; Jerolmack 2009; Yellow River, Liu et al. 2007) and how modern alterations have impacted the natural processes of sediment and carbon burial.

In the following sections, the continental-margin sedimentation of three contrasting large river-sediment dispersal systems (i.e., Amazon, Ganges-Brahmaputra, Mississippi) is reviewed to illustrate the functional differences and highlight the key controlling processes. Recent geochemical research emphasizes the importance of physical processes affecting sedimentary systems to the cycling of organic material (McKee et al. 2004), and these ideas are discussed in detail in the other chapters of this volume.

3. Amazon: a subaqueous delta clinoform system

The Amazon River has the largest catchment in the world, reaching across South America to the Andes Mountains and covering more than 20° of latitude (Fig. 4.3). The adjacent margin has been constructed from millions of years of sedimentation (Damuth and Kumar 1975), and its present shelf is greater than 300 km wide at its mouth, but narrows northwestward to ~100 km near Cabo Cassipore at the Brazil-French Guiana border. Its continental slope is incised by numerous canyons, and the Amazon Cone, the system's deep sea fan, is demarcated by a protrusion in the continental rise, reflecting its re-occupation and construction over many sea-level lowstands (Damuth and Kumar 1975). The Amazon mouth lies at the equator, in the wet tropics, a geographic region defined by constantly high precipitation (>1,500 mm yr^{-1}) and temperatures >20°C (Nittrouer et al. 1996). As a result, the Amazon discharge is received by a coastal zone with relatively high and seasonally stable solar radiation, temperature, and precipitation. These factors along with trade winds from the east and weak Coriolis force are characteristic of tropical areas and have influence on other factors, such as the Rossby radius (which impacts plume behavior), sediment mineralogy, and coastal vegetation (mangroves) (Nittrouer et al. 1995; 1996).

The Amazon freshwater discharge averages 5.7×10^3 km^3 y^{-1} (Oltman 1968), providing almost a fifth of the freshwater to the world's oceans. It ranks first in water and sediment discharge, with a sediment load of 1,194 Mt y^{-1} (Table 4.1; Syvitski and Saito 2007). Interannual water discharge variability is present but minimized (factor of 3–4) because of the vast size of the catchment (Nittrouer et al. 1991). Its runoff also varies seasonally, with the discharge peak in water and sediment about 2.5 and 5 times the minimum, respectively. The sediment discharge peak generally occurs in February prior to the water peak (May–June) due to basin hysteresis (Philander and Pacanowski 1986; Richey et al. 1986; Nittrouer and Demaster 1996).

Sediment dynamics on the Amazon shelf are strongly affected by Atlantic Ocean waves and the regional circulation. The semidiurnal tidal components (M2, S2, N2) are enhanced in this system (Platzman et al. 1981; Cartwright et al. 1991), producing a large (~6 m; Table 4.1) tidal range in the river mouth region and very strong shelf currents (>100 cm s^{-1}; Beardsley et al. 1995; Geyer and Kineke 1995; Geyer et al. 1996; Kineke et al. 1996). The funnel-shaped bathymetry of the river mouth region creates a near-resonant tide and enhances the tidal range (Table 4.1), and a tidal bore, known as the Pororoca (translated as "Roar") is produced. The northward-directed North Brazil Current is induced by impingement of the South Equatorial Current, affecting the outer shelf and slope (Flagg et al. 1986; Philander and Pacanowski 1986). Estimated water transport rates are 10–50 Sv, with a maximum in July–August, and these flows produce currents of >75 cm s^{-1} (Geyer et al. 1996). The strength of the North Brazil Current and shelf circulation vary seasonally and over shorter timescales in response to the regional wind forcing (Geyer et al. 1996). The equatorial position of the Amazon mouth limits the influence from tropical cyclones (Pielke 1990). The moderate wave climate with significant wave heights of 1–2 m is also driven by seasonal trade winds (January–April); waves of >3 m (13-s period) have been observed on the inner shelf (Sternberg et al. 1996).

For about a decade (mid-1980s to mid-1990s), interdisciplinary research efforts, including "A Multidisciplinary Amazon Shelf SEDiment Study" (AmasSeds, 1989–1996), examined the physical oceanography, sedimentation, and related biogeochemical processes in this complex SDC system (see the integrative results in Fig. 4.4; Nittrouer et al. 1996). This research led to a much improved understanding of this system and helped to initiate an era of process-orientated investigation into continental-margin sedimentation near rivers. In the Amazon system, fluvial water discharge peaks in May–June, with low water flow in October–November. Amazon suspended-sediment input is 85–95% silt and clay, derived primarily from the Andes Mountains, and is composed of quartz, kaolinite, illite, montmorillonite, and feldspar (Gibbs 1967; Meade et al. 1985). Sediment from the Amazon has been traced into the Caribbean Sea (Bowles and Fleischer 1985), but much of the sediment load accumulates in a prograding clinoform (i.e., a large sedimentary deposit) on the Amazon shelf (Alexander et al. 1986; Nittrouer et al. 1986; Kuehl et al. 1996).

Allison et al. (1995) documented that northwestward (alongshore) advection of riverine sediment driven by the wave-current regime generates seasonal sedimentation along the mangrove-colonized shoreline. Alongshore supply from the Amazon controls coastal sedimentation and shoreline morphology for ~1,800 km downcoast, to the Orinoco delta (Wells and Coleman 1981; Rine and Ginsburg 1985; Allison and Lee 2004; Anthony et al. 2010).

On the inner shelf (i.e., <40 m water depth) adjacent to the river mouth, deposition (and removal) of decimeter-scale, river-derived sediment layers also occurs seasonally (Kuehl et al. 1995; 1996). In situ measurements indicate how, because of the enormous water discharge of the Amazon River, the estuarine interface is pushed onto the adjacent open shelf, and fluid-mud layers are created by estuarine-like circulation and frontal trapping on the clinoform topset (Kineke and Sternberg 1995). This trapping mechanism has been shown to thicken the seasonal layer (Kuehl et al. 1995), and x-radiographs from time-series coring of these seasonal deposits beneath the benthic estuarine frontal zone have documented significant bed variability on tidal and fortnightly timescales (e.g., Fig. 4.5; Jaeger and Nittrouer 1995). Tidal currents are a key process in varying benthic shear stresses (i.e., removing thick beds during spring tides; Fig. 4.5), and varying the location of the frontal zone where sediment trapping is focused on the inner shelf. However, variations in the inner shelf seasonal layer thickness are correlated with wave climate (Kuehl et al. 1995), and quasi-cyclic ^{210}Pb profiles (Dukat and Kuehl, 1995) suggest that wave forcing is also a significant driver in annual-to-multiannual offshore export from the energetic inner topset region to the outer topset and foreset region. This cross-shelf and cross-clinoform export has been documented directly using seabed tripods that observed seaward transport by gravity-driven fluid-mud flow during a shelf wave event (Cacchione et al. 1995; Sternberg et al. 1996), helping to explain the progradational-aggradational pattern of these subaqueous shelf mud clinoforms. A back-of-the-envelope calculation estimates that an enormous sediment volume is potentially stored in the seasonal layer (3.2–8 × 10^3 Mt, several times the annual

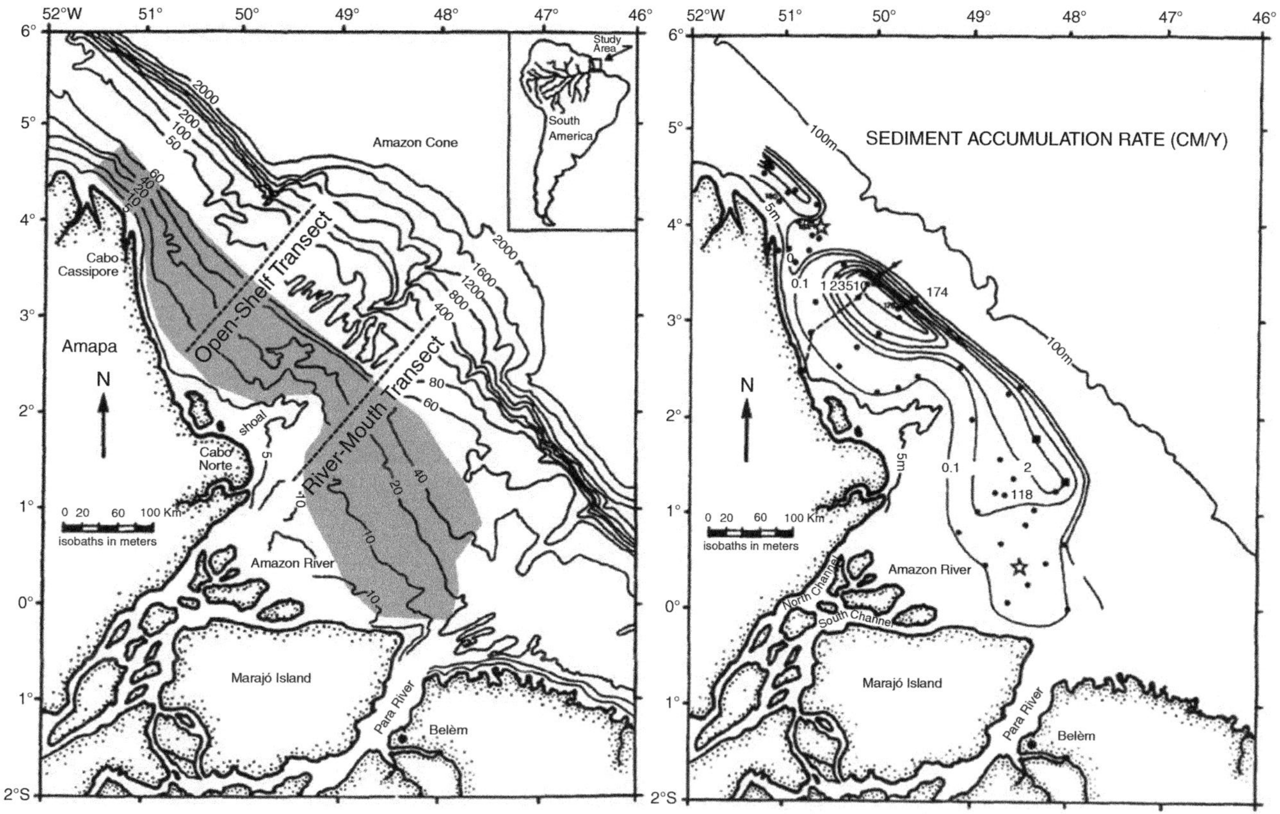

Figure 4.3. Basemap of (left) and sediment accumulation rates on (right, cm yr^{-1}) the Amazon River margin (from Nittrouer et al. 1996 and Kuehl et al. 1986, respectively). Note, much seabed dynamics and sediment accumulation occurs on the subaqueous delta clinoform in the area shaded gray on left.

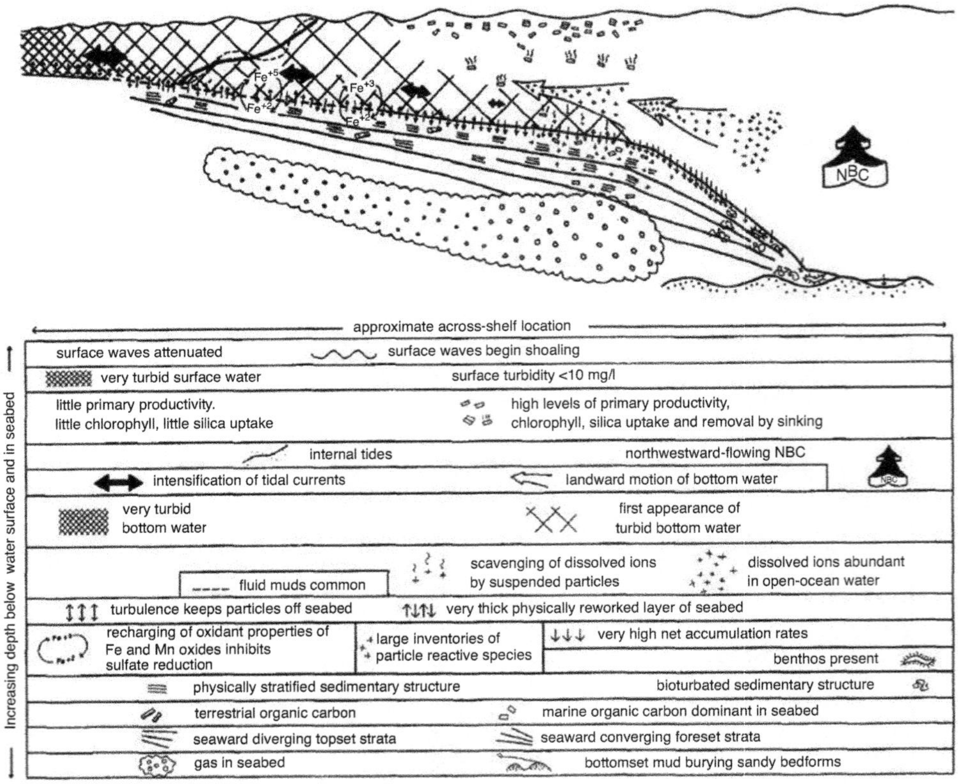

Figure 4.4. The physical and biogeochemical processes occurring on the Amazon subaqueous delta clinoform (Nittrouer et al. 1996). Waves and tidal and subtidal currents are responsible for moving sediment from the river mouth and across the shelf. Accumulation patterns and sedimentary structures reflect the gradient in processes. Although the figure provides two-dimensional perspective, the system certainly has three-dimensional complexity that varies with time.

Amazon River sediment supply to the shelf, Table 4.1), highlighting the connection between shelf sediment dynamics and biogeochemical cycling (Aller, 1998).

Net rates of sediment accumulation on the Amazon subaqueous delta clinoform are low nearshore (<0.1 cm y^{-1}, 5–15 m water depth), increase to a maximum near the topset-forest boundary (~10 cm y^{-1}, ~40 m depth), and decrease seaward to negligible rates on the outer shelf beyond the bottomset region (Kuehl et al. 1986). This accumulation pattern is related to this offshore translation of sediment from the high-energy, high-supply region on the inner-shelf topset to the relatively low-energy and episodic supply-dominated foreset/bottomset areas of the mid-shelf. A similar sedimentation pattern has been noted in other SDC-classed systems (e.g., Ganges-Brahamaputra; Kuehl et al. 1989; Fly/Papua New Guinea, Walsh et al. 2004). Sedimentary structures also vary across the shelf clinoform, with interbedded sands and muds on the topset grading to bioturbated muds (Kuehl et al. 1986), reflecting variation in the benthic community in response to rates of accumulation and physical (erosion-deposition) disturbance (Aller and Aller 1986; Aller and Stupakoff 1996). The

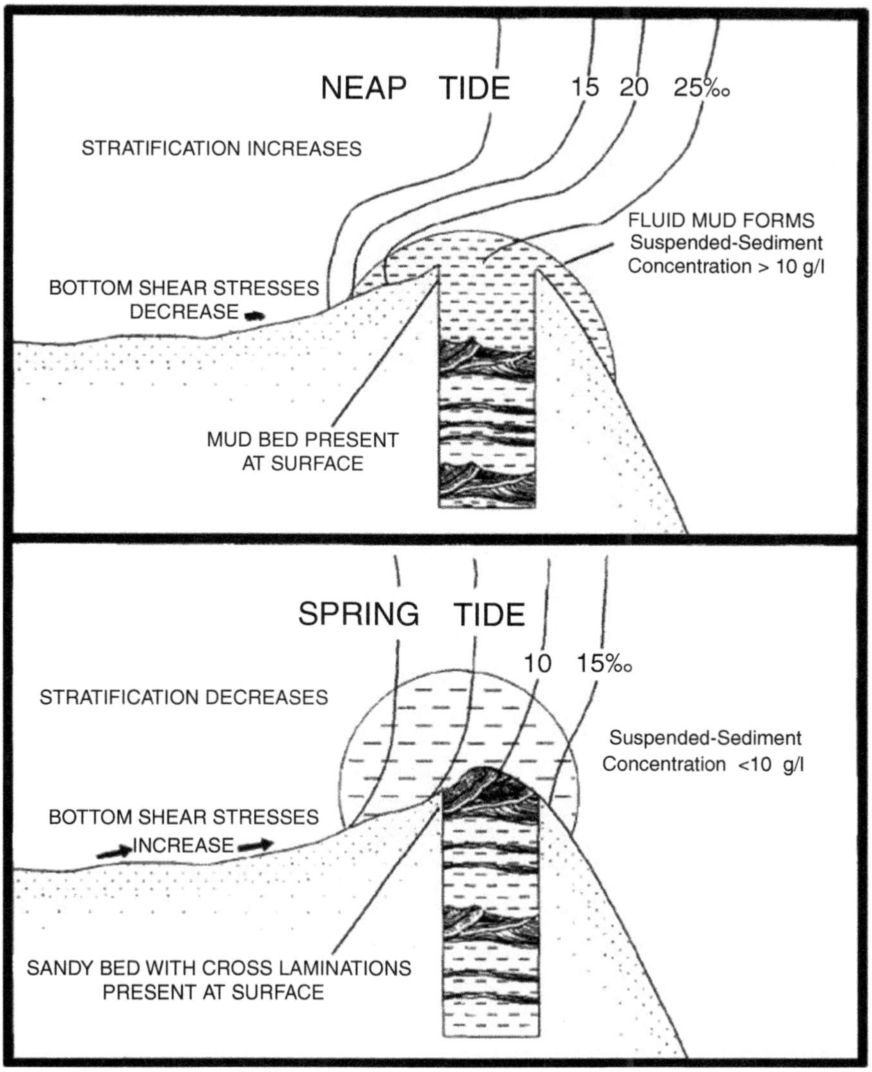

Figure 4.5. Conceptual model of sediment dynamics in response to tidal-regime changes (from Jaeger and Nittrouer 1995; see Fig. 4.3 for location). Associated with weaker currents during neap tides, fluid muds are formed, and sedimentary layers are deposited on the inner-shelf seabed (Kineke et al. 1996). Greater stresses from spring tidal currents disperse the fluid muds and rework the surficial strata.

pattern of modern Amazon sediment accumulation is the result of and control for the system's compound clinoform geometry, with a coastal clinoform (prograding shoreline) building seaward across the topset of the subaqueous delta clinoform, which itself is prograding seaward over relict shelf deposits (Nittrouer et al. 1996). This constructional architecture is achieved because energetic physical processes are dispersing the enormous sediment flux to these lower-energy depocenters and filling available shelf accommodation space through progradation. Today, like earlier high sea-level stands, little Amazon sediment escapes the shelf, and only a thin hemipelagic layer blankets the Amazon Fan (Damuth and Kumar 1975; Nittrouer et al. 1995; Flood and Piper 1997). Consequently, the large 10-km-thick fan that initiated in the Miocene will not likely see appreciable growth until the next sea-level fall (Lopez 2001 and references therein).

4. Ganges-Brahmaputra: a combination of influences

The Ganges and Brahmaputra rivers drain southern and northern areas of the Himalayan Mountains, respectively. Although the watershed is characterized by active tectonics and the river flows through several subsiding basins in the delta plain (Goodbred et al., 2003 and references therein), the mouth is positioned on a trailing-edge margin, albeit with relatively complex morphology. The shelf extends hundreds of kilometers seaward to the south, and the system is an SDC type according to Walsh and Nittrouer (2009), with a large shelf depocenter (Table 4.1). The shelf narrows along the central margin (Fig. 4.6) where the Ganges-Brahmaputra submarine canyon, known as the Swatch of No Ground, incises to within <30 km of the modern shoreline. Sediment passing through this shelf incision has helped build the massive Bengal Fan (Fig. 4.7; Kuehl et al. 2005).

The two large river systems coalesce along with the Meghna River before draining into the Bay of Bengal (Fig. 4.6). The combined watershed is 2×10^6 km^2, and the sediment load is ~1,100 Mt y^{-1} (Syvitski and Saito 2007; Table 4.1), with a distinct difference in the character of the sediment supplied by the two systems, owing to their distinct geological provenance (Huizing 1971; Heroy et al. 2003). Also, because of the high-mountain relief and the high rainfall in the catchment (Sarin et al. 1989), the combined riverine flux of dissolved ions to the ocean is large, 0.13×10^{12} kg y^{-1}, which is about 3% of the global supply. The total area of the delta is approximately 1.4×10^5 km^2, including a large subaerial portion (1.1×10^5 km^2) in southern Bangladesh, and a subaqueous clinoform extends from the intricate shoreline of northernmost Bay of Bengal through the Swatch of No Ground (Kuehl et al. 2005). Discharge from the Ganges-Brahmaputra River is modulated by the southwest monsoon, with the majority (>80%) of water released in the May–November period when discharge is about an order of magnitude higher than in the dry season. The sediment discharge is more focused in July–September when about 95% of load is delivered, with minimal sediment discharged from December–April (Barua et al. 1994). Long-term subsidence in the Bengal Basin is on the order of 1–2 mm y^{-1}, locally 3–5 mm y^{-1}, but overall this accommodation is effectively filled by the immense sediment load (Alam 1996).

Like the discharge, the shelf circulation and wave climate are strongly controlled by seasonal meteorological forcing. A geostrophic counterclockwise circulation is set up on the shelf during the high-discharge wet season, which reverses to a clockwise gyre in the winter period (Tomczak and Godfrey 1994). Wave heights are not well monitored on the Bengal Shelf but are generally highest during onshore winds of the summer monsoon period and considerably attenuated by the broad, shallow inner shelf. Cyclones also regularly influence the shelf and coast with long-period storm waves (e.g., Rogers and Goodbred 2010). The elevated river discharge during the southwest monsoon influences the density structure of the Bay of Bengal, producing a surface water layer of reduced salinity. As a result, upwelling is weakened during this time, and an oxygen-minimum zone

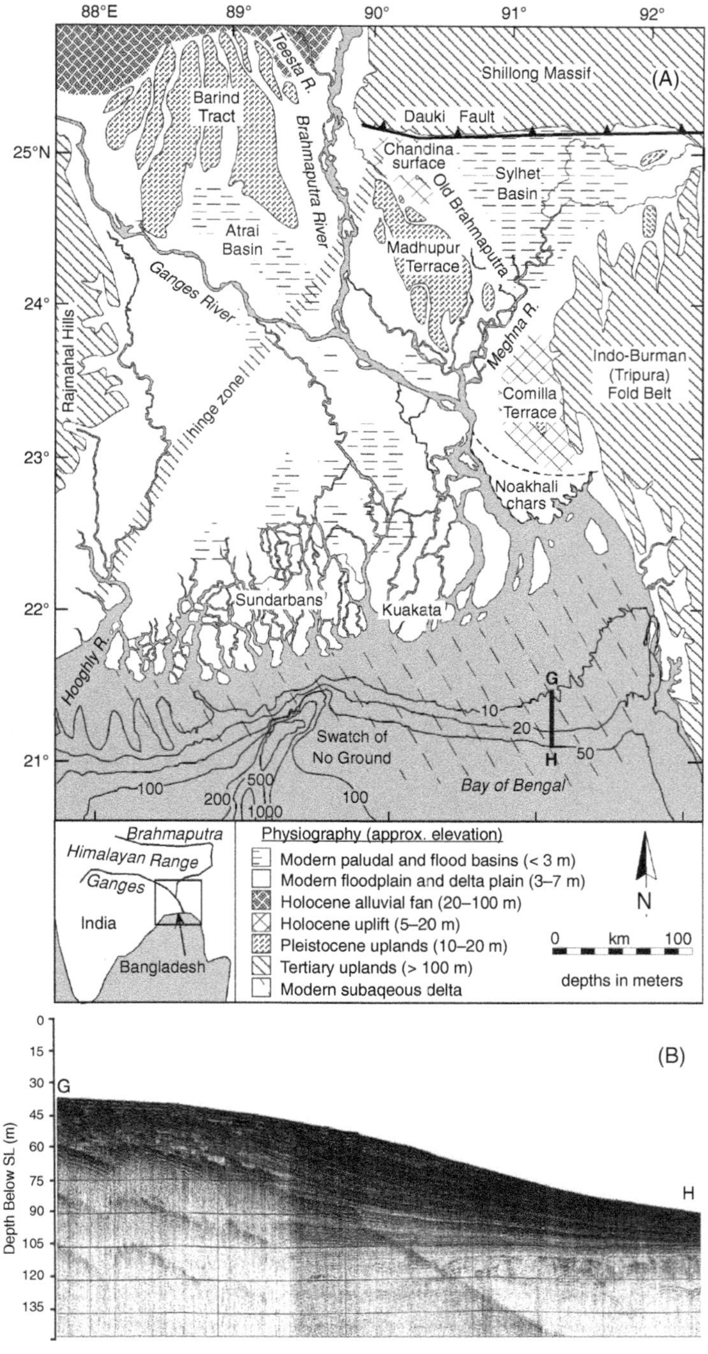

Figure 4.6. (top, A) Basemap for the Ganges-Brahmaputra river sediment dispersal system and (bottom, B) shows a seismic reflection image from line GH (bold solid line in A). Note the area of sediment transport extends from the high peaks of Himalaya to the deep sea (>1,000 m). Goodbred and Kuehl 2000a; Kuehl et al. 2005).

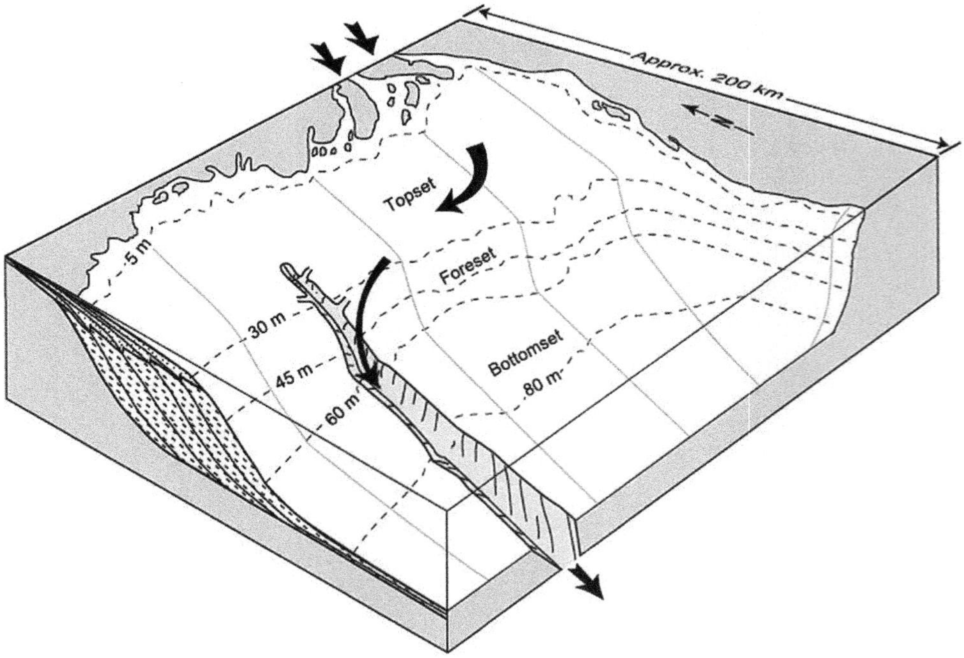

Figure 4.7. Three-dimensional cross-section of the Ganges-Bramaputra shelf sedimentary system. Note, the wide shelf in the east allows sediment to be sequestered in the clinoform, but the Swatch of No Ground intersects the along-shore transport system, capturing sediment into the canyon (Kuehl et al. 2005).

is produced in the lower water column below 60 m on the shelf (Karstensen 1999). Nearly anoxic bottom water conditions have been observed at the head of the Swatch of No Ground (Berner et al. 2003). Strong tidal currents are also very important to the redistribution of sediment in the delta. The tidal range is largely macrotidal, ranging from coastal values of 1.9 m in the west to 5 m in the river mouth region (Barua 1990). The delta region is regularly impacted by tropical cyclones, which produce large waves on the shelf, and storm surges that inundate the extensive mangrove areas of the delta plain (Allison and Kepple 2001). An average of nine tropical cyclones occur per year in the northern Indian Ocean. More than 100 tropical cyclones have made landfall on the Ganges-Brahmaputra delta plain since the late nineteenth century, with most activity in the periods before (April–May) and after (October–November) the monsoon season, although cyclones may occur any time of the year (Singh et al. 2000; Raghavan and Rajesh 2003). These events have had enormous impact on the large population living in the delta: the Bhola Cyclone of 1970 generated a >6 m surge and killed 300,000–500,000 inhabitants (Islam and Peterson 2003).

The network of coastal islands and connecting distributary channels along the actively prograding and relict subaerial delta shoreline (Allison 1998; Allison et al. 2003) is perched atop a thick wedge of fluviodeltaic sediment that has accumulated above transgressive sand deposits over the late Holocene (Goodbred and Kuehl. 2000a, Goodbred et al. 2003; Fig. 4.6). From the channelized innermost shelf to about the 30-m isobath is the low-gradient (<0.1°) topset region of the subaqueous clinoform (Kuehl et al. 1989; Kuehl et al. 1997; Michels et al. 1998). Although this area has had minimal stratigraphic investigation, the strong seabed reflection seen in seismic profiles and numerous grab samples reveal fine to very-fine sand sediment coverage (Kuehl et al. 1997) with ephemeral and localized silt-dominated mud deposition (Kuehl et al. 1989). These heterolithic inner shelf deposits represent the energetic topset region of the advancing subaerial deltaic shoreline. The more steeply dipping (~0.25°) marine foreset beds are mud-dominated and extend from a 20–30 m deep rollover to about 80-m depth in the west and 60 m in the east (Figs. 4.6 and 4.7). The bottomset strata thin seaward to ~100 m depth. The outer shelf comprises relict lowstand and early transgressive shoreline, lagoon, and carbonate shoal sediments (Wiedicke et al. 1999).

The sigmoidal geometry of the inner-to-mid shelf clinoform is similar to those observed in seismic-reflection data near the Amazon and other river deltas (e.g., Fly, Slingerland et al. 2008; Ebro, Diaz et al. 1990; Po, Correggiari et al. 2005; Atchafalaya, Neill and Allison 2005; Indus, Giosan et al. 2006; Mekong, Xue et al. 2010; Yangtze, Liu et al. 2007; Fig. 4.6). Similar to these other systems, the rates of sediment burial on timescales of decades are largest (typically centimeters per year) in the foreset region and slower in topset and bottomset areas (i.e., 5–20 cm yr^{-1}, 1–2 cm yr^{-1}, and 3–5 cm yr^{-1}, respectively, for the Ganges-Brahmaputra system). Such deposition rates on this low gradient shelf have prograded the subaqueous delta front at a rate of 15 m yr^{-1}. Sediment budget calculations show that about one-third of the annual river sediment load is stored in the subaqueous delta (Kuehl et al. 1997; Michels et al. 1998; Suckow et al. 2001).

Widespread acoustically transparent deposits recorded in seismic reflection profiles throughout the outer topset and foreset region have been shown to indicate nonstratified, amorphous sediments that are otherwise lithologically identical to adjacent, well-stratified deposits. The acoustically transparent sediments are interpreted to have formed by the liquefaction of previously stratified deposits (Kuehl et al. 2005) under the influence of tropical storms (Kudrass et al. 1998; Rogers and Goodbred 2010) and major historical earthquakes (Palamenghi et al. 2011). Although acoustically transparent, liquefied sediments are widespread on the rapidly accreting Bengal shelf, the typical runout distances of these flows are only a few hundred meters, and they do not appear to significantly alter shelf morphology.

Like the Amazon and some other systems, seasonally high-energy conditions on the shelf and episodic tropical cyclones create energetic wave-current conditions on the inner shelf (topset) region that leads to frequent resuspension and more distal advection (cross- and alongshore) of river-derived sediments. Although measures of suspended sediment on the inner shelf are sparse, Barua et al. (1994) reported depth-averaged SSC values of 100–1,500 mg L^{-1} during the low-discharge dry season and a 5- to 10-fold increase to 4,000–6,000 mg L^{-1} under high discharge. These latter values are well in the range for inhibited settling (McCave 1984), and given that they are depth-averaged (1 m above seafloor, mid-depth, 1 m below surface), it is possible that near-bed concentrations reached or exceeded that for a fluid mud, as in the Amazon system. Further offshore in northern Bay of Bengal, a peak lithogenic vertical particle flux of 300 mg m^{-2} d^{-1} was measured by sediment traps during the monsoon season (Unger et al. 2003). Widespread physical stratification observed in topset, foreset, and canyon sediments are also thought to indicate regular, rapid event erosion-deposition stratal packets probably from cyclone-induced remobilization (Kuehl et al. 1997; Kudrass et al. 1998;

Michels et al. 1998) Using pre- and post-cyclone geophysical data, Rogers and Goodbred (2010) provided evidence for a large-scale, storm-forced submarine mass movement in the upper canyon region region (<150 m water depth) that may have been an extreme analog to Amazon and the Gulf of Papua observations of fluid-mud flow events that transfer sediments offshore and progading their respective clinoform (Cacchione et al. 1995; Martin et al. 2008).

An especially interesting aspect of the Ganges-Brahmaputra system is that the Ganges-Brahmaputra subaqueous clinoform intersects the Swatch of No Ground shelf incision, causing massive, river-derived sediment accumulation in the canyon head and a conduit for sediment transport to the deep basin (Kuehl et al. 1997; Kottke et al. 2003; Michels et al. 2003). It is believed that episodic, canyon-flushing events occur every few hundred years, resulting in turbidite deposition on the Bengal Fan. This body of research highlights how the Swatch of No Ground is a very active sedimentary system, and this is remarkable considering the high stand in sea level. This aspect of the Ganges-Brahmaputra dramatically distinguishes this system from the other described herein. With about one-third of the load entering, partially accumulating, and transiting through a canyon, this system is notably different than most other major rivers (Fig. 4.2). For example, the Mississippi River also has a sizable canyon that receives sediment, but sediment accumulation rates are modest, suggesting a minimal role in the modern sediment budget (Corbett et al. 2006). However, event forcing (i.e., cyclones and earthquakes) is thought to be critical to sediment redistribution in the Ganges-Brahmaputra system, and in this way it has similarity to the Mississippi (Goñi et al. 2007; Bianchi et al. 2006).

5. Mississippi-Atchafalaya: a proximal accumulation-dominated system

Draining from the Rocky Mountains to the Appalachians, about 41% of the conterminous United States, the Mississippi-Atchafalaya River System (MARS) has the third largest drainage basin in the world (Fig. 4.8). Approximately 60% of the total suspended matter and 66% of the total dissolved load from the conterminous United States is transported by the MARS to the ocean (Presley et al. 1980). The modern PAD dispersal system of the MARS (Walsh and Nittrouer 2009) extends out onto the shelf and is a product of natural and anthropogenic processes, including runoff, land-use impacts, subsidence, sea-level change, and wetland loss (Day et al., 2007). The outflow and sedimentation from the MARS greatly influences coastal processes in the northern Gulf of Mexico (GoM). In this area, the second largest zone of shelf hypoxia in the world develops seasonally (Rabalais et al. 2002).

The Mississippi River has an average freshwater discharge of ~500 km^3 yr^{-1} through its bird-foot-shaped delta (Syvitski and Saito 2007). Water and sediment flow both east (~two-thirds via the main stem of the Mississippi River) onto the eastern Louisiana/Mississippi shelf, and west (~one-third through the Atchafalaya River) onto the eastern Louisiana/Texas continental shelf (Meade and Parker 1985; Kennicutt et al. 1995; Allison et al. 2012). The discharge of the MARS during the high-water months (January–May) is about three times the discharge during the low-water months (July–November; Meade, 1995), and the maximum monthly sediment discharge (March) is five times the minimum monthly discharge (September). Over the last 200 years, the suspended-sediment load of the Mississippi River has decreased by approximately one-half as a result of anthropogenic manipulation, including: (1) deforestation and agriculture; (2) changes in the land-management strategy; and (3) construction of dams, diversions, and levees (Meade and Parker 1985; McKee and

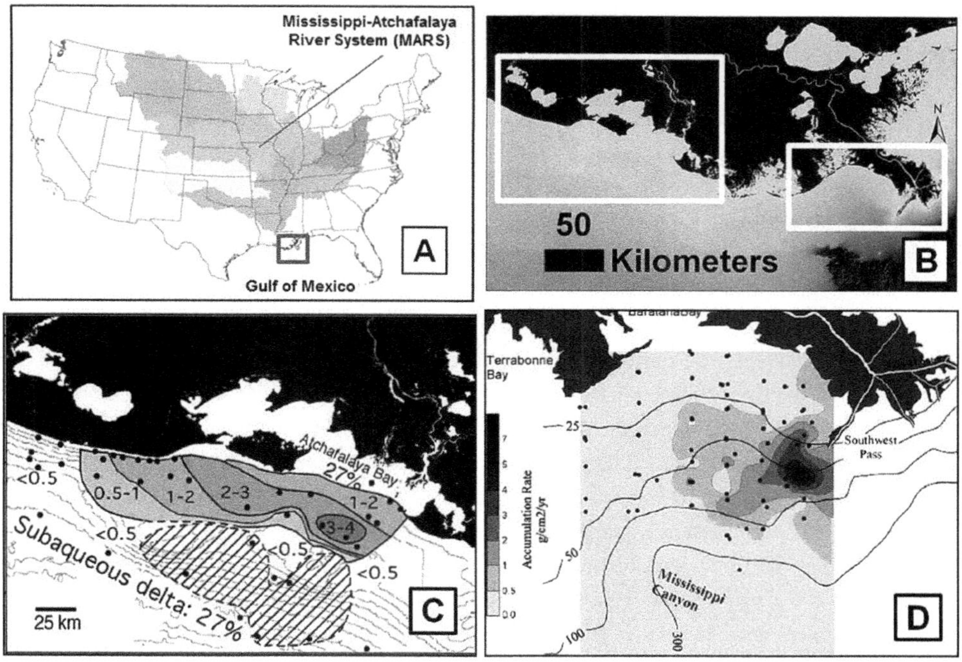

Figure 4.8. (top, A) The drainage basin of the Mississippi-Atchafalaya River System (MARS) (A, figure from the USGS); basemap of Mississippi River margin (B); pattern of sediment accumulation on the shelf near Atchafalaya River mouth (C, Draut et al. 2005) and Southwest Pass (D, Corbett et al. 2006). Note, Southwest Pass is an active distributary of the main stem of the Mississippi River and creates one of the toes of the bird-foot-shaped delta.

Baskaran 1999). Today, the MARS delivers approximately 210 Mt of suspended matter to the northern GoM each year (based on measurements at Tarbert Landing, 495 km upriver from the river mouth; Meade and Moody 2010; Horowitz 2010). This remains a large flux of sediment, but its historical load places it fifth in Table 4.1. Delivery of sediments into the GoM is greatly influenced by temporary and longer term storage and remobilization of fine and coarse sediments in the lower river (Meade et al. 1990; McKee and Baskaran 1999; Galler and Allison 2008, Nittrouer et al. 2008; Allison et al. 2012). Using a quantitative transport model, Demas and Curwick (1988) demonstrated that periods of fine-grained sediment storage occur in the lower river for typically 4–8 months, after discharge drops below 14,000 $m^3\ s^{-1}$ (early summer) and before it exceeds 20,000 $m^3\ s^{-1}$ (mid-winter) as measured at Tarbert Landing. In fact, they observed as much as an 80% decrease and 30% increase in suspended-sediment concentrations during flows between 7,300–7,600 $m^3\ s^{-1}$ and 25,000–28,000 $m^3\ s^{-1}$, respectively. Although this annual storage of fine-grained sediment is released to the MARS margin at high discharge (Galler and Allison 2008), the quantity of coarse and fine sediment (and associated carbon) measured at Tarbert Landing that reaches the MARS margin every year is reduced by at least half by a combination of overbank sedimentation and channel sand storage (Allison et al. 2012).

Circulation patterns on the northern GoM continental self are primarily wind-driven, whereas currents along the shelf break are affected by eddies spun off of the Loop Current (Oey 1995; Cho et al. 1998; Wiseman and Dinnel 1988). Although currents play an important role in transporting sediment, river flow and waves are the dominant influences on initial deposition and sediment remobilization along the Louisiana shelf (Allison et al. 2000; Walker and Hammack 2000; Huh et al. 2001; Jaramillo et al. 2009; Sheremet et al. 2011). Once sediment is delivered to the GoM, the transport of fine sediments by the surface plume is generally limited to <30 km from the river mouth (Corbett et al. 2004, 2007; Geyer et al, 2004). This sediment supply results in rapid accumulation and delta progradation, and early study of this system helped form the foundation of the present concepts of delta sedimentation (e.g., Fisk et al. 1954; Scruton 1960). After initial transport and deposition on the shelf, sediments may be subjected to energetic conditions associated with tropical systems and winter storms, and these events will determine the extent and frequency of sediment remobilization and transport on the shelf. During the winter and early spring, mean wave energy and river discharge are typically the highest measured during the year. Following this period, river discharge decreases, and wave energy typically remains low until mid to late summer. In the fall, the wave energy increases as a result of the augmented frequency of continental weather systems.

Under fair weather conditions, bed shear stresses are insufficient to suspend sediments across most of the shelf, yet during moderate wind events (e.g., frontal passages), enhanced wave orbital motions can induce resuspension (Wright et al. 1997; Allison et al. 2000; Dail et al. 2007). On average, more than five extra-tropical weather systems impact the northern GoM per month between October and April (Baumann 1987) and play an important role in resuspending and advecting sediments on the shelf (Henry 1979; Allison et al. 2000; Booth et al. 2000; Corbett et al. 2004, 2007; Kineke et al. 2006). These systems also affect the sediment transported in the Mississippi River surface plume; typically, the plume is positioned to the west along the shelf during the late spring and summer because of the combined effects of the earth's rotation and the prevailing winds. However, with the passage of continental weather systems during the winter, the river plume is periodically pushed offshore in a southern or even an eastern direction for a number of days. It is noted here that morphology at the Atchafalaya River mouth is dramatically different from that of the Mississippi mainstem (Fig. 4.8C vs. 4.8D), with the former being a shallow bay and shelf and the latter having distributary channels that extend into deeper water. As a result, the plume behavior and ensuing depositional patterns (and subsequent remobilization) are markedly different (Cobb et al. 2008; Xu et al. 2011 and references therein).

Decades of research on this margin have provided significant insights into sedimentation and strata development over a range of time scales. From the Atchafalaya River, sediment accumulation is building a shallow clinoform immediately adjacent to the mouth and prograding this feature along the coast (Fig. 4.8C; Draut et al. 2005; Neill and Allison 2005). This early-stage delta deposit has stratigraphic variability similar to that of larger SDC systems (e.g,. Amazon, Ganges-Brahmaputra). For the mainstem Mississippi River on centennial and decadal timescales, a decrease in accumulation rate and concomitant increase in the influence of bioturbation with distance along the prodelta has been noted (Scruton 1960; Corbett et al. 2006). The highest accumulation rates (>4 cm y^{-1}) are found near Southwest Pass in the area recognized by Corbett et al. (2004) as a zone of rapid deposition of river-borne particles (Fig. 4.8D). Near shore (<20 m depth), radiochemical profiles are suggestive of sediments reworked by waves and currents with little to no accumulation. The remainder of the shelf (distal of Southwest Pass) is dominated by steady-state accumulation beneath a ~10-cm thick mixed layer, with rates typically <0.7 cm y^{-1}. This shelf accumulation west of the mainstem river mouths has been reduced historically, matching the timing and quantity of reduction in sediment supply from the catchment (Allison et al. 2007). From the 20–80-m isobaths, the delta front is characterized by a steeper gradient and bathymetric morphology reflective of gravity-driven mudflow activity (Coleman et al. 1998; Walsh et al. 2006). During Hurricane Camille in 1969, an oil rig was knocked to the seabed from speculated mudflow activity (Bea 1971). In 2005, hurricanes Katrina and Rita appeared to reinitiate mudflow movement, and seabed dynamics may have contributed to significant pipeline damage (Walsh et al. 2006). Beyond the delta front is the prodelta, which, in general, more slowly receives muddy sediment, although sediment-gravity flows produced at shallower depths (Wright et al. 2001; Allison et al. 2005) may be advected to deeper areas during storm conditions (Goñi et al. 2007). The complex MARS sedimentary system highlights how variable margin morphology, even within a single system (e.g., Atchafalaya vs. mainstem Mississippi distributaries), can affect sediment transport and accumulation and thus is critical to understanding shelf and slope geochemical cycling.

Waves and currents associated with tropical systems are known to move sediments along and across the MARS in patterns that mirror the winter frontal pattern, but are much greater in magnitude (Corbett et al. 2004; 2007; Dail et al. 2007). Corbett et al. (2007) suggested that as much as 50% of the sediment delivered annually by the Mississippi River, and initially deposited on the delta, is subsequently remobilized and transported along or across the shelf. Recent research following hurricanes Lili, Ivan, Katrina, and Rita highlight the dramatic role that tropical cyclones can play in cross-shore exchange of sediment (both landward and seaward). Marsh areas can potentially receive much sediment from hurricanes and cause marsh aggradation (e.g., Turner et al. 2006); however, these same events can also result in dramatic wetland reconfigurations and losses (Morton and Barras 2011 and references therein). Based on radionuclide and sedimentological data, shelf event layers associated with tropical cyclones in 2004 and 2005 were clearly identifiable (Walsh et al. 2006; Dail et al. 2007; Goñi et al. 2007; Figure 4.9), and the thickness of storm deposits ranged from 1 to 40 cm (average: 14 ± 9 cm). The majority of the sites near the Mississippi delta showed three (or more) distinct storm layers. However, in water depths shallower than ~20 meters near the Atchafalaya and Mississippi mouths, data suggest resuspension of Katrina deposits leaving only Rita layers (Goñi et al. 2007; Allison et al. 2010). Regardless, the storm deposits and their geochemical signatures indicate massive sediment remobilization, an estimated volume five times greater than the annual MARS sediment supply. Hurricane-related deposition at some locations was equivalent to as much as 70 years of deposition when compared with ^{210}Pb-based rates (Dail et al. 2007). These data dramatically highlighted the importance of stochastic events to across-shelf sediment transport on this margin. Bianchi et al. (2006), Goñi et al. (2007), and Sampere et al. 2008 showed how inorganic and organic sedimentation was also affected in the Mississippi Canyon during episodic events. In sum, morphologically, and in terms of sediment dispersal, the MARS is considerably different than both the Amazon and the Ganges-Brahmaputra, especially considering its intra-system complexity. However, tropical cyclone events and mass movements are critical to sediment redistribution, as was also described for the Ganges-Brahmaputra system. These factors influence carbon cycling in this system, which is discussed in Chapter 19.

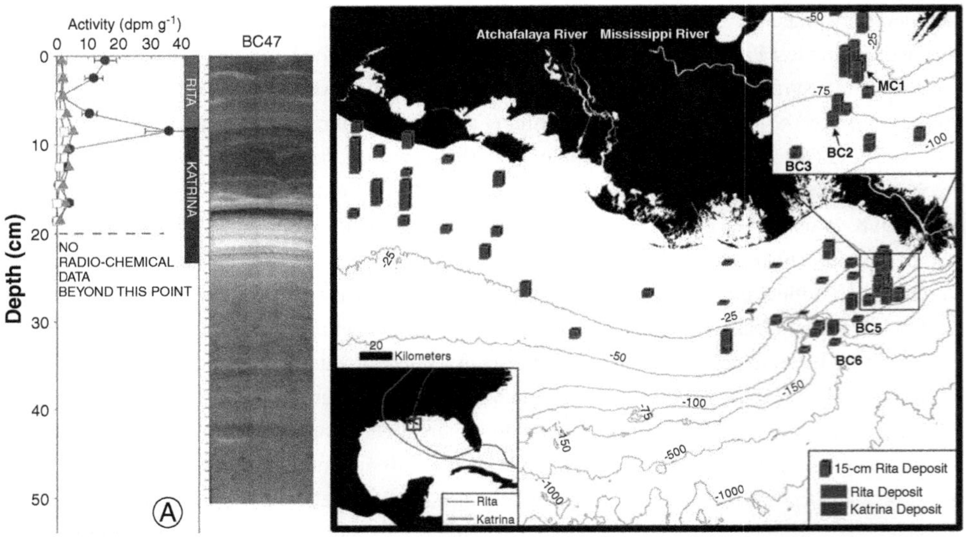

Figure 4.9. Measurements of hurricane deposit thicknesses from Katrina and Rita (Walsh et al. 2006; Goñi et al. 2007). X-radiographs and radiochemical data enabled estimation of the deposit thicknesses (A), and from these data, a map of layer thicknesses was developed (B). Radiochemical and geochemical data indicated that the sediment was primarily resuspended shelf sediment (Goñi et al. 2007).

6. Fluvial and marine controls on large deltaic systems

River sediment dispersal is influenced by an array of processes across the land-ocean interface, ranging from floodplain sedimentation and bank migration in the alluvial valley to transport by turbidity currents down the continental rise. The thick subaerial and subaqueous sedimentary packages associated with large deltaic systems are evidence that the shelf, and even the lower margin, can be dominated by the presence of these enormous point sources of sediment and carbon. Because of the nonlinear sediment rating curve relationship, river sediment loads are strongly governed by floods, and it is common for moderately sized rivers to release the vast majority (potentially >90%) of their annual transport over just a few days (Meade et al. 1990). However, large river systems exhibit less episodicity because the basin integrates and thus mutes the impact of individual weather systems. Instead, these systems can be expected to have more gradual river hydrographs, with flood peak discharges spanning weeks to months. So, unlike small mountainous rivers, large systems are unlikely to have their discharge correlated with energetic oceanic storm processes. Consequently, regardless of the size or strength of fluvial forcing, initial deposition may be ephemeral, and long-term sediment accumulation on the shelf and slope is largely controlled by the wave-current regime and the preexisting margin geometry, as the preceding case studies demonstrate.

Owing to their size, the discharge of large systems represents an integration of weather across large catchments and thus their sediment transport may reflect interannual climate cycles (e.g., El Niño). For example, Richey et al. (1989) argued how flow variation of the Amazon is related to tropical Pacific conditions rather than human activities, and Rogers and Coleman (2003) statistically determined how Pacific and Atlantic ocean-driven signals are recognized to varying degrees in the discharge records of Mississippi River gauges. Milliman and Farnsworth (2011) provided a more in-depth discussion of climate signals and their influence on river discharges. It is logical that recent studies of shelf and deeper sediment deposits on large river margins have been shown to contain responses to sub-annual or longer scale climate variations (e.g., Goodbred and Kuehl 2000b; Kudrass et al. 2001; Saito et al. 2001; Ogston et al. 2008). Thus these deposits potentially provide invaluable sedimentary archives of past climate variability.

Several decades ago it was viewed that resuspension and advection by waves and currents (tidal and subtidal) largely controlled the sediment redistribution from the river mouth to marine seabed (Nittrouer et al. 2009). However, since that time, sediment-gravity flows, especially fluid muds, have become better appreciated as an active cross-shelf redistribution mechanisms on low-gradient shelves fed by large quantities of fine sediments (Wright and Friedrichs 2006; Hill et al. 2009). For large river systems, estuarine processes (by which saline ocean water and freshwater meet and mix) do not always occur within the river distributaries, but at times (or always, in the case of the Amazon; Geyer et al. 1996; Kineke et al. 1996) can be located on the shelf. The estuarine circulation creates a convergence zone at the toe of the salt wedge, allowing the development of a turbidity maximum on the continental shelf and a potential mechanism to form very high near-bed suspended-sediment layers (or fluid muds, if the concentrations are >10 g L^{-1}). Recent studies have implicated sediment-gravity-flow processes as important to the overall fluxes and development of shelf deposits (Traykovski et al. 2007; Hill et al. 2009 and references therein). As described, these dense layers may be formed at sites of lateral convergence, such as at the estuarine turbidity maximum (e.g., Amazon, Kineke et al. 1996). Alternately, waves (Friedrichs et al. 2000) and wave-current interactions in the bottom boundary layer (Wright et al. 2001; Jaramillo et al. 2009; Ma et al. 2010) can resuspend unconsolidated sediment or prevent sediment from leaving suspension, creating very high turbidity in the bottom boundary layer. With suspensions concentrated enough to significantly exceed the density of the ambient seawater (i.e., fluid muds), gravity exerts a force instigating down-slope transport. These flows have been observed on multiple margins, but are likely to be especially important in large river dispersal systems in which there is often a significant source of unconsolidated, fine-grained sediment (Friedrichs et al. 2000). An important aspect of gravity flows is that they do not follow isobath-parallel trajectories, as tends to be the case with advected sediment in coastal flows. Rather, they generally follow the path of greatest (albeit, very low) slope, allowing sediment from large rivers to move directly offshore to help build delta strata. Some fluvial systems occasionally achieve suspended-sediment concentrations sufficiently high to induce gravity-driven flow when the freshwater reaches the sea (Mulder and Syvitski 1995). A direct hyperpycnal plume, as this is called, is not likely to form from most large rivers because of their less flashy nature and ability to store material (Hill et al. 2009), but the Huanghe River is a well-known exception (Wright et al. 1986).

The land-ocean transition (i.e., the sub-aerial delta and inner shelf) is a critical zone in which physical marine processes can potentially overprint the signal of fluvial discharge to the shelf. For example, sediment-transport observations in the Amazon and Fly systems illustrate how neap and spring tide conditions can differ in their impact on fluid-mud formation and transport (e.g., Fig. 4.5). Work in the MARGINS NSF Source to Sink Initiative elucidated how the idealized model of Walsh et al. (2004) based on monsoon-driven physical forcing on the Fly River was too simplistic for this three-dimensionally complex deltaic shelf system (Crockett et al. 2008; Martin et al. 2008;

Ogston et al. 2008). Measurements made at a few sites revealed significant spatial variability in the suspended-sediment concentration (and the conditions necessary to form fluid mud), and work over the span of two years (including an El Niño drought season) highlighted how inter-annual river discharge and oceanographic forcing are important signal modulators. Clearly, macrotidal (and possibly smaller tidal) systems such as the Amazon, Fly, and Ganges-Brahmaputra with extensive shallow areas will experience strong and regular reworking by energetic tidal currents, and these will play a role in modulating the mechanisms (e.g., fluid muds) and patterns of transport and sedimentation. Intensive field investigations on different margins are needed to create a more comprehensive picture of transport and deposition behavior to develop realistic models of these complex bottom-boundary-layer dynamics.

Sequence stratigraphic principles explain how climate cycles and sea-level changes control the forcing mechanisms behind continental margin construction (Vail et al. 1977; Van Wagoneer et al. 1988), but the degree to which single events are recorded in preserved strata as identifiable layers depends on the system and the particular event characteristics. In Table 4.1, the calculated "Discharge to Marine Power Ratio" parameter (see Syvitski and Saito 2007) attempts to provide perspective into the relative dominance of terrestrial and marine processes in various systems. For example, the very high ratio for the MARS suggests dominance of fluvial dynamics, and geological studies and historical classifications support this notion. However, recent work on winter and tropical storms in the last decade has shown that episodically strong waves and currents cause dramatically different sediment dynamics on the MARS shelf than that from flood periods (e.g., Allison et al. 2005; Kineke et al. 2006; Walsh et al., 2006; Dail et al. 2007; Goñi et al. 2007). Thus, even though the MARS is widely considered to be a river-dominated system, its shelf strata may be strongly reshaped episodically by marine processes. Of course, this is important for carbon oxidation as the fate of organic material that reaches the seabed is a function of its transport pathway as well as post-depositional utilization and decomposition (McKee et al. 2004).

7. Geological framework and human influence on large deltaic systems

As described previously and reflected in the discussions of the three case-study systems, the geological setting and its human alteration play a critical role in the functioning of freshwater and sediment dispersal on large river margins. First, they can affect both the amount and timing of sediment influx from the catchment. For example, because discharge scales with drainage basin size, river systems with sizable catchments also generally have sizable sediment discharges, and this occurs despite the fact that yield tends to decrease with basin area (Milliman and Syvitski 1992). Additionally, the fluvial and marine forcing may be largely decoupled because of the size of the basin (e.g., the Andes are far from the Amazon mouth) and the time required for flow routing. Hence shelf sedimentation may be entirely out of phase with meteorological forcing that drives precipitation cycles in the catchment. For example, the Ganges-Brahmaputra and Mississippi River systems have distinct periods of elevated discharge (i.e., the monsoon and late spring rainfall/snowmelt, respectively), and these periods often do not coincide with the most major storms (e.g., the timing of tropical cyclone activity), which have a key control on marine transport. Thus the land-ocean coherency, which is often discussed as important for small mountainous systems (e.g., Kniskern et al. 2011), is less significant for large systems.

There are countless examples of how human activities in the watershed and delta (e.g., channel maintenance and flow diversions) have dramatically affected sediment supply from large rivers (Meade et al. 1990; Syvitski et al. 2005; Day et al. 2007; Syvitski and Saito 2007), but few studies document unequivocal shelf impacts, a point acknowledged by Wheatcroft and Sommerfield (2005) for deforestation impacts on Pacific Northwest shelf sedimentation. However, the right balance of sediment accumulation and mixing is likely present at some sites to resolve detailed records (Allison et al. 2007). In the future, more work is needed to quantitatively measure and/or model (e.g., Kettner et al. 2007) how changes such as shoreline hardening, deforestation and development, and changing land use are impacting sediment exchange and strata development. However, research on the Huanghe (Yellow) River system provides a compelling example of how damming upstream (i.e., the Three Gorges Dam) can affect a large delta and supply to and sedimentation in the shelf system (see Chapters 5 and 6). Measurements demonstrate how hysteresis in the rating curve shifted, and eventually the load dramatically dropped in the delta, instigating the most recent period of widespread delta erosion (Xu and Milliman 2009; Yang et al. 2011). In a similar fashion, the Indus shelf appears to have responded to reduced river sediment loading by eroding along the coast and on the clinoform (Giosan et al. 2006). Detailed time-series mapping studies along with modeling investigation will enable researchers to better understand shelf and slope sediment dynamics.

8. Conclusions

Only 30 large dispersal systems together supply about half of the sediment to the global ocean, highlighting how understanding these huge, dynamic systems with voluminous margin storage is critical for biogeochemical research. Shelf and slope sedimentation has been well-studied for some of the Earth's largest sediment dispersal systems (e.g., the Amazon, Ganges-Brahmaputra, and Mississippi-Atchafalaya), and this body of work demonstrates some notable similarities in the form and function of the shelf-slope region adjacent to large deltas. For example, an estimated one-third of the large river deltas have constructed massive shelf sediment deposits called subaqueous delta clinoforms that have similar morphological character and sedimentation behavior. However, looking at the attributes of 30 large systems, it is clear that a wide range in key variables is represented, and system-specific studies have demonstrated how waves, tides, along-shelf currents, sediment-gravity flows, and many other processes are important to the resulting sediment accumulations to varying degrees in different systems. Thus, owing to their number, remoteness, and size, many large systems have not been thoroughly studied, and more field research is needed on the diversity of delta types. Also, new insight can be gained from quantitative analysis and modeling of these systems, and this aspect is critical to better understanding complex phenomena such as anthropogenic impacts, ecosystem changes, and socioeconomic ramifications from these important marine areas.

Acknowledgments

The authors thank the many scientists, captains, and crews they have been fortunate to work with during their research. Also, interactions with these individuals have enabled success in many research efforts and growth in the knowledge. Research for and writing of this chapter was supported by a

NSF MARGINS Source to Sink to Walsh (OCE-0841092) and an NSF RAPID (OCE-1142496) to Walsh and Corbett.

References

Alam, M. 1996. Subsidence of the Ganges-Brahmaputra delta of Bangladesh and associated drainage, sedimentation and salinity problems, In: J.D. Milliman, J.D., and Haq, B.U. (eds.), Sea-level Rise and Coastal Subsidence; Causes, Consequences, and Strategies. Amsterdam, the Netherlands: Kluwer Academic Publishers.

Alexander Jr, C.R., Nittrouer, C.A. and D.J. DeMaster. 1986. High-resolution seismic stratigraphy and its sedimentological interpretation on the Amazon continental shelf. *Cont. Shelf Res.* 6: 337–357.

Aller, R.C. 1998. Mobile deltaic and continental shelf muds as suboxic, fluidized bed reactors. *Mar. Chem.* 61: 143–155.

Alexander, Jr, C.R., C.A. Nittrouer, and D.J. DeMaster. 1986. High-resolution seismic stratigraphy and its sedimentological interpretation on the Amazon continental shelf. *Cont. Shelf Res.* 6: 337–357.

Aller, J.Y., and R.C. Aller. 1986. General characteristics of benthic faunas on the Amazon inner continental shelf with comparison to the shelf off the Changjiang River, East China Sea. *Cont. Shelf Res.* 6: 291–310.

Aller, J.Y., and I. Stupakoff. 1996. The distribution and seasonal characteristics of benthic communities on the Amazon shelf as indicators of physical processes. *Cont. Shelf Res.* 16: 717–751.

Allison, M.A. 1998. Historical changes in the Ganges-Brahmaputra delta front. *J. Coastal Res.* 14: 1269–1275.

Allison, M.A., and M.T. Lee. 2004. Sediment exchange between Amazon mudbanks and shore-fringing mangroves in French Guiana. *Mar. Geol.* 208: 169–190.

Allison, M.A., C.R. Demas, B.A. Ebersole, B.A. Kleiss, C.D. Little, E.A. Meselhe, N.J. Powell, T.C. Pratt, and B.M. Vosburg. 2012. A water and sediment budget for the lower Mississippi-Atchafalaya River in flood years 2008–2010: implications for sediment discharge to the oceans and coastal restoration in Louisiana. *J. Hydrol.* 432: 84–97.

Allison, M.A., S.R. Khan, S.L. Goodbred, and S.A. Kuehl. 2003. Stratigraphic evolution of the late Holocene Ganges-Brahmaputra lower delta plain. *Sediment. Geol.* 155: 317–342.

Allison, M.A., and E.B. Kepple. 2001. Modern sediment supply to the lower delta plain of the Ganges-Brahmaputra River in Bangladesh. *Geo-Mar. Lett.* 21: 66–74.

Allison, M.A., C.A. Nittrouer, and L.E.C. Faria. 1995. Rates and mechanisms of shoreface progradation and retreat downdrift of the Amazon River mouth. *Mar. Geol.* 125: 373–392.

Allison, M.A., G.C. Kineke, E.S. Gordon, and M.A. Goñi. 2000. Development and reworking of a seasonal flood deposit on the inner continental shelf off the Atchafalaya River. *Cont. Shelf Res.* 20: 2267–2294.

Allison M.A., A. Sheremet, M.A. Goñi, and G.W. Stone. 2005. Storm layer deposition on the Mississippi-Atchafalaya subaqueous delta generated by Hurricane Lili in 2002. *Cont. Shelf Res.* 25: 2213–2232.

Allison, M.A., T.M. Dellapenna, E. S. Gordon, S. Mitra, and S.T. Petsch. 2010. Impact of Hurricane Katrina (2005) on shelf organic carbon burial and deltaic evolution. *Geophys. Res. Lett.* 37: L21605, doi:10.1029/2010GL044547.

Allison, M.A., T.S. Bianchi, B.A. McKee, and T.P. Sampere. 2007. Carbon burial on river-dominated continental shelves: impact of historical changes in sediment loading adjacent to the Mississippi River. *Geophys. Res. Lett.* L01606.

Allison, M.A., A. Sheremet, M.A. Gõni, and G.W. Stone. 2005. Storm layer deposition on the Mississippi-Atchafalaya subaqueous delta generated by Hurricane Lili in 2002. *Cont. Shelf Res.* 25: 2213–2232.

Anthony, E.J., A. Gardel, N. Gratiot, C. Proisy, M.A. Allison, F. Dolique, and F. Fromard. 2010. The Amazon-influenced muddy coast of South America: A review of mud-bank-shoreline interactions. *Earth Sci. Rev.* 103: 99–121.

Barua, D.K. 1990. Suspended sediment movement in the estuary of the Ganges-Brahmaputra-Meghna river system. *Mar. Geol.* 91: 243–253.

Barua, D.K., S.A. Kuehl, R.L. Miller, and W.S. Moore. 1994. Suspended sediment distribution and residual transport in the coastal ocean of the Ganges-Brahmaputra River mouth. *Mar. Geol.* 120: 41–61.

Baumann, R.H. 1987. Physical variables, In: Conner, W.H., Day Jr., J.W. (eds.), The Ecology of Barataria Basin, Louisiana: An Estuarine Profile. U.S. Fish and Wildlife Services, Biological Report 85 (7.13) 8–17.

Bea, R.G. 1971. How sea floor slides affect offshore structures. *Oil Gas J.* 69: 88–92.

Beardsley R.C., C.J. Candela, R. Limeburner, W.R. Geyer, S.J. Lentz, B.M. Castro, D. Cacchione, and N. Carneiro. 1995. The M_2 tide on the Amazon shelf. *J. Geophys. Res.* 100: 2283–2320.

Berner, R.A. 1982. Burial of organic carbon and pyrite sulfur in the modern ocean: Its geochemical and environmental significance. *Amer. J. Sc.* 282: 451–473.

Berner, U., J. Poggenburg, E. Faber, D. Quadfasel, and A. Frische. 2003. Methane in ocean waters of the Bay of Bengal: its sources and exchange with the atmosphere. *Deep-Sea Res. Part II Top. Stud. Oceanogr.* 50: 925–950.

Bianchi, T.S., T.P. Sampere, M.A. Allison, E.A. Canual, D. Corbett, B.A. Mckee, S. Wakeham, and E. Waterson. 2006. Rapid export of organic matter to the Mississippi Canyon. *Eos* 87: 565.

Booth, J.G., R.L. Miller, B.A. McKee, and R.A. Leathers. 2000. Wind-induced bottom sediment resuspension in a microtidal coastal environment. *Cont. Shelf Res.* 20: 785–806.

Bowles, F.E., and P. Fleischer. 1985. Orinocco and Amazon river sediment input to the Eastern Caribbean basin. *Mar. Geol.* 68: 53–72.

Cacchione, D.A., D.E. Drake, R.W. Kayan, R.W Stemberg, G.C. Kineke, and G.B. Tate. 1995. Measurements in the bottom boundary layer on the Amazon subaqueous delta. *Mar. Geol.* 125: 235–257.

Cartwright, D.E., R.D. Ray, and B.V. Sanchez. 1991. Oceanic tide maps and spherical marmonic coefficients from Geosat altimetry. NASA Technical Memorandum 104544. Greenbelt, MD: Goddard Space Flight Center, p. 75.

Cho, K., R.O. Reid, and W.D. Nowlin Jr. 1998. Objectively mapped stream function fields on the Texas-Louisiana shelf based on 32 months of moored current meter data. *J. Geophys. Res.* 103: 35–3524.

Cobb, M., T.R. Keen, and N.D. Walker. 2008. Modeling the circulation of the Atchafalaya Bay system. Part 2: River plume dynamics during cold fronts. *J. Coastal Res.* 244:1048–1062.

Coleman, J.M., and L.D. Wright. 1975. Modern river deltas: variability of processes and sand bodies, In: Broussard, M.L. (ed.), Deltas, Models for Exploration. Houston, TX: Houston Geological Society, pp. 99–149.

Coleman, J.M., H.J. Walker, and W.E. Grabau. 1998. Sediment instability in the Mississippi River delta. *J. Coastal Res.* 14: 872–881.

Corbett, D.C., B. McKee, and D. Duncan. 2004. An evaluation of mobile mud dynamics in the Mississippi River deltaic region. *Mar. Geol.* 209: 91–112.

Corbett, D.R., B. Mckee, and M.A. Allison. 2006. Nature of decadal-scale sediment accumulation on the western shelf of the Mississippi River delta. *Cont. Shelf Res.* 26: 2125–2140.

Corbett, D.R., M.B. Dail, and B.M. McKee. 2007. High-frequency time-series of the dynamic sedimentation processes on the western shelf of the Mississippi river delta. *Cont. Shelf Res.* 27: 1600–1615.

Correggiari, A., A. Cattaneo, and F. Trincardi. 2005. The modern Po Delta system: Lobe switching and asymmetric prodelta growth. *Mar. Geol.* 222: 49–74.

Correggiari, A., F. Trincardi, L. Langone, and M. Roveri. 2001. Styles of failure in late Holocene highstand prodelta wedges on the Adriatic shelf. *J. Sediment. Res.* 71: 218–236.

Crockett, J.S., C.A. Nittrouer, A.S. Ogston, and M.A. Goñi. 2008. Variable styles of sediment accumulation impacting strata formation on a clinoform: Gulf of Papua, Papua New Guinea, In: Bolton, B. (ed.), Anonymous Developments in Earth and Environmental Sciences. Philadelphia: Elsevier, pp. 177–204.

Dail, M.B., D.R. Corbett, and J.P. Walsh. 2007. Assessing the importance of tropical cyclones on continental margin sedimentation in the Mississippi delta region. *Cont. Shelf Res.* 27: 1857–1874.

Damuth J.E., and N. Kumar. 1975. Amazon cone: morphology, sediments, age and growth pattern. *Geol. Soc. Am. Bull.* 86: 863–878.

Day, J.W., D.F. Boesch, E.J. Clairain, G.P. Kemp, S.B. Laska, W.J. Mitsch, K. Orth, H. Mashriqui, D.J. Reed, L. Shabman, C.A. Simenstad, B.J. Streever, R.R. Twilley, C.C. Watson, J.T. Wells, and D.F. Whigham. 2007. Restoration of the Mississippi Delta: Lessons from Hurricanes Katrina and Rita. *Science* 315: 1679–1684.

Demas, C.R., and P.B. Curwick. 1988. Suspended-sediment and associated chemical transport characteristics of the lower Mississippi River, Louisiana. Louisiana Department of Transportation and Development Water Resources Technical Report. No. 45.

Diaz, J.I., C.H. Nelson, J.H. Barber, Jr. and S. Giró. 1990. Late Pleistocene and Holocene sedimentary facies on the Ebro continental shelf. *Mar. Geol.* 95: 333–352.

Draut, A.E., G.C. Kineke, D.W. Velasco, M.A. Allison, and R.J. Prime. 2005. Influence of the Atchafalaya River on recent evolution of the chenier-plain inner continental shelf, northern Gulf of Mexico. *Cont. Shelf Res.* 25: 91–112.

Dukat, D.A., and S.A. Kuehl. 1995. Non-steady-state Pb-210 flux and the use of Ra-228/Ra-226 as a geochronometer on the Amazon Continental-Shelf. *Mar. Geol.* 125: 329–350.

Ericson J.P, C.J. Vorosmarty, S.L. Dingman, L.G. Ward, and M. Meybeck. 2006. Effective sea-level rise and deltas: causes of change and human dimension implications. *Global Planet. Change* 50: 63–82.

Fisk, H.N., E.J. McFarlan, C.R. Kolb, and L.J. Wilbert. 1954. Sedimentary framework of the modern Mississippi delta. *J. Sediment. Petrol.* 24: 76–99.

Flagg, C.N., R.L. Gordon, and S. McDowell. 1986: Hydrographic and current observations on the continental slope and shelf in the western Equatorial Atlantic. *J. Phys. Oceanogr.* 16: 1412–1429.

Flood R.D., and D.JW. Piper. 1997. Amazon Fan sedimentation: the relationship to equatorial climate change, continental denudation, and sea-level fluctuations, In: Flood, R.D., Piper, D.J.W., Klaus, A., and L.C. Peterson (eds.), Proceedings of the Ocean Drilling Program, Scientific Results. Vol. 155, pp. 653–675.

Friedrichs, C.T., L.D. Wright, D.A. Hepworth, and S.C. Kim. 2000. Bottom boundary layer processes associated with fine sediment accumulation in coastal seas and bays. *Cont. Shelf Res.* 20: 807–841.

Galler, J.J., and Allison, M.A. 2008. Estuarine controls on fine-grained sediment storage in the lower Mississippi and Atchafalaya Rivers. *Geol. Soc. Am. Bull.* 120: 386–398.

Galloway, W.E. 1975. Process framework for describing the morphologic and stratigraphic evolution of deltaic depositional systems, In: Broussard, M.L. (ed.), Deltas, Models for Exploration. Houston, TX: Houston Geological Society, pp. 87–98.

Geyer, W.R., R.C. Beardsley, S.J. Lentz, J. Candela, R. Limeburner, W.E. Johns, B.M. Castro, and I.D. Soares. 1996. Physical oceanography of the Amazon shelf, *Continent. Shelf Res.* 5/6: 575–616.

Geyer, W.R., P.S. Hill, and G.C. Kineke. 2004. The transport, transformation and dispersal of sediment by buoyant coastal flows. *Cont. Shelf Res.* 24: 927–949.

Geyer, W.R., and G.C. Kineke. 1995. Observations of currents and water properties in the Amazon frontal zone. *J. Geophys. Res.* 100: 2321–2339.

Gibbs, R.J. 1967. The geochemistry of the Amazon River system; part I, the factors that control the salinity and the composition and concentration of the suspended solids. *Geol. Soc. Am. Bull.* 78: 1203–1232.

Giosan, L., S. Constantinescu, P.D. Clift, A.R. Tabrez, M. Danish, and A. Inam. 2006. Recent morphodynamics of the Indus delta shore and shelf. *Cont. Shelf Res.* 26: 1668–1684.

Goñi, M., Y. Alleau, D.R. Corbett, J.P. Walsh, D. Mallinson, M.A. Allison, E. Gordon, S. Petsch, and T.M. Dellapenna. 2007. The effects of Hurricanes Katrina and Rita on the seabed of the Louisiana shelf. *Sediment. Rec.* 5: 4–9.

Goñi, M.A., E.S. Gordon, N.M. Monacci, R. Clinton, R. Gishewhite, M.A. Allison, and G.C. Kineke. 2006. The effect of Hurricane Lili on the distribution of organic matter in the Inner Louisiana Shelf (Gulf of Mexico, USA). *Cont. Shelf Res.* 26: 2260–2280.

Goodbred, S.L., and S.A. Kuehl. 2000a. The significance of large sediment supply, active tectonism and eustasy on margin sequence development: Late Quaternary stratigraphy and evolution of the Ganges–Brahmaputra delta: *Sediment. Geol.* 133: 227–248.

Goodbred, S.L., and S.A. Kuehl. 2000b. Enormous Ganges–Brahmaputra sediment discharge during strengthened early Holocene monsoon. *Geology* 28: 1083–1086.

Goodbred, S.L., S.A. Kuehl, M.S. Steckler, and M.H. Sarker. 2003. Controls on facies distribution and stratigraphic preservation in the Ganges-Brahmaputra delta sequence. *Sediment. Geol.* 155: 301–316.

Harris, C.K., P. Traykovski, and W.R. Geyer. 2005. Flood dispersal and deposition by near-bed gravitational sediment flows and oceanographic transport: A numerical modeling study of the Eel River shelf, northern California. *J. Geophys. Res.* 110: CP09025.

Henry, W.K. 1979. Some aspects of the fate of cold fronts in the Gulf of Mexico. *Mon. Weather Rev.* 107: 1078–1082.

Heroy, D.C., S.A. Kuehl, and S.L. Goodbred. 2003. Mineralogy of the Ganges and Brahmaputra Rivers: implications for river switching and Late Quaternary climate change. *Sediment. Geol.* 155: 343–359.

Hill, P.S., J.M. Fox, J.S. Crockett, K.J. Curran, C.T. Friedrichs, W.R. Geyer, T.G. Milligan, A.S. Ogston, P. Puig, M.E. Scully, P.A. Traykovski and R.A. Wheatcroft. 2009. Sediment Delivery to the Seabed on Continental Margins, In: Nittrouer, C.A., Austin, J.A., Field, M.E., Kravitz, J.H., Syvitski, J.P.M., and Wiberg, P.L. (eds.), Continental Margin Sedimentation: From Sediment Transport to Sequence Stratigraphy. Oxford, UK: Blackwell Publishing.

Horowitz, A.J. 2010. A quarter century of declining suspended sediment fluxes in the Mississippi River and the effect of the 1993 flood. *Hydrol. Process.* 24: 13–34.

Hovius, N. 1998. Controls on sediment supply by large rivers. SEPM Spec. Publ. 42: 47–69.

Huh, O.K., N.D. Walker, and C. Moeller. 2001. Sedimentation along the eastern Cheneir Plain coast: down drift impact of a delta complex shift. *J. Coastal Res.* 17: 72–81.

Huizing, H.G.J. 1971. A Reconnaissance study of the mineralogy of sand fractions from East Pakistan sediments and soils. *Geoderma.* 6: 109–133.

Islam, T., and R.E. Peterson. 2003. A climatological study on the landfalling tropical cyclones of Bangladesh. American Meteorological Society Annual Meeting Abstracts.

Jaeger, J.M., and C.A. Nittrouer. 1995. Tidal controls on the formation of fine-scale sedimentary strata near the Amazon river mouth. *Mar. Geol.* 125: 259–282.

Jaramillo S., Sheremet, A., Allison, M.A., Holland, K.T., and Reed, A.H. 2009. Wave-mud interactions over the muddy Atchafalaya subaqueous clinoform, Louisiana, United States: wave-supported sediment transport. *J. Geophys. Res. Oceans* 114:C04002.

Jerolmack, D.J. 2009. Conceptual framework for assessing the response of delta channel networks to Holocene sea level rise. *Quat. Sci. Rev*. 28: 1786–1800.

Karstensen, J. 1999. Ueber die Ventilation der Thermokline des Indischen Ozeans: Dissertation, Fachbereich Geowissenschaften, Hamburg, Germany.

Kennicutt M.C., W.W. Schroeder, and J.M. Brooks. 1995. Temporal and spatial variations in sediment characteristics on the Mississippi-Alabama continental-shelf. *Cont. Shelf Res*. 15: 1–18.

Kettner, A.J., B. Gomez, and J.P.M. Syvitski. 2007. Modeling suspended sediment discharge from the Waipaoa River system, New Zealand: the last 3000 years. *Water Resourc. Res*. 43: W07411.

Kineke, G.C., and R.W. Sternberg. 1995. Distribution of fluid muds on the Amazon continental shelf. *Mar. Geol*. 125: 193–233.

Kineke, G.C., R.W. Sternberg, J.H. Trowbridge, and W.R. Geyer. 1996. Fluid-mud processes on the Amazon continental shelf. *Cont. Shelf Res*. 16: 667–696.

Kineke, G.C., K.J. Woolfe, S.A. Kuehl, J. Milliman, T.M. Dellapenna, and R.G Purdon. 2000. Sediment export from the Sepik River, Papua New Guinea: evidence for a divergent sediment plume. *Cont. Shelf Res*. 20: 2239–2266.

Kineke, G.C., E.E. Higgins, K. Hart, and D. Velasco. 2006. Fine-sediment transport associated with cold front passages on the shallow shelf, Gulf of Mexico. *Cont. Shelf Res*. 26: 2073–2091.

Kniskern T., J.A. Warrick, K.L. Farnsworth, R.A. Wheatcroft, and M.A. Goñi. 2011. Coherence of river and ocean conditions along the U.S. west coast during storms. *Cont. Shelf Res*. 31: 789–805.

Kottke, B., T. Schwenk, M. Breitzke, M. Wiedicke, H.R. Kudrass, and V. Spiess. 2003. Acoustic facies and depositional processes in the upper submarine canyon Swatch of No Ground (Bay of Bengal): *Deep-Sea Res. Part II Top. Stud. Oceanogr*. 50: 979–1001.

Kudrass, H.R., K.H. Michels, M. Wiedicke, and A. Suckow. 1998. Cyclones and tides as feeders of a submarine canyon off Bangladesh. *Geology* 26: 715–718.

Kudrass, H.R., F. Hofmann, H. Doose, K. Emeis, and H. Erlenkeuser. 2001. Modulation and amplification of climatic changes in the Northern Hemisphere by the Indian summer monsoon during the past 80 k.y. *Geology* 29: 63–66.

Kuehl, S.A., D.J. DeMaster, and C.A. Nittrouer. 1986. Nature of sediment accumulation on the Amazon continental shelf. *Cont. Shelf Res*. 6: 311–336.

Kuehl, S.A., T.M. Hariu, and W.S. Moore. 1989. Shelf sedimentation off the Ganges-Brahmaputra River system: Evidence for sediment bypassing to the Bengal fan. *Geology* 17: 1132–1135.

Kuehl, S.A, T.D. Pacioni, and J.M. Rine. 1995. Seabed dynamics of the inner Amazon continental shelf: temporal and spatial variability of surficial strata. *Mar. Geol*. 125: 283–302.

Kuehl, S.A., C.A. Nittrouer, M.A. Allison, L.E.C. Faria, D.A. Dukat, J.M. Jaeger, T.D. Pacioni, A.G. Figueiredo, and E.C. Underkoffler. 1996. Sediment deposition, accumulation, and sea bed dynamics in an energetic fine-grained coastal environment. *Cont. Shelf Res*. 16: 787–815.

Kuehl, S.A., B.M. Levy, W.S. Moore, and M.A. Allison. 1997. Subaqueous delta of the Ganges-Brahmaputra river system. *Mar. Geol*. 144: 81–96.

Kuehl, S.A., T.M. Hariu, and W.S. Moore. 1989. Shelf sedimentation off the Ganges-Brahmaputra River system: Evidence for sediment bypassing to the Bengal fan. *Geology* 17: 1132–1135.

Kuehl, S.A., T.M. Hariu, M.W. Sanford, C.A. Nittrouer, and D.J. DeMaster. 1991. Millimeter-scale sedimentary structure of fine-grained sediments: Examples from continental margin environments, In: Bennett, R.H., Bryant, W.R., and Hulbert, M.H. (eds.), Microstructure of Fine-Grained Sediments. New York: Springer-Verlag, pp. 33–45.

Kuehl, S.A., M.A. Allison, S.L. Goodbred, and H. Kudrass. 2005. The Ganges-Brahmaputra Delta, In: Gosian, L., and Bhattacharya, J. (eds.), River Deltas – Concepts, Models, and Examples. SEPM Special Publication No. 83, pp. 413–434.

Kuehl, S.A., G.J. Brunskill, K. Burns, D. Fugate, T. Kiniskern, and L. Meneghini. 2004. Nature of sediment dispersal off the Sepik River, Papua New Guinea: preliminary sediment budget and implications for margin processes. *Cont. Shelf Res.* 24: 2417–2429.

Liu, J., Y. Saito, H. Wang, Z. Yang, and R. Nakashima. 2007. Sedimentary evolution of the Holocene subaqueous clinoform off the Shandong Peninsula in the Yellow Sea. *Mar. Geol.* 236: 165–187.

Liu, J.P., Z. Xue, K. Ross, H.J. Wang, Z.S. Yang, A.C. Li, and S. Gao, 2009. Fate of sediments delivered to the sea by Asian large rivers: Long-distance transport and formation of remote alongshore clinothems. *Sediment. Rec.* 7: 4–9.

Lopez, M. 2001. Architecture and depositional pattern of the Quaternary deep-sea fan of the Amazon. *Mar. Petrol. Geol.* 18: 479–486.

Ma, Y., C.T. Friedrichs, C.K. Harris, and L.D. Wright. 2010. Deposition by seasonal wave- and current-supported sediment gravity flows interacting with spatially varying bathymetry: Waiapu shelf, New Zealand. *Mar. Geol.* 275: 199–211.

Martin, D.P., C.A. Nittrouer, A.S. Ogston, and J.S. Crockett. 2008. Tidal and seasonal dynamics of a muddy inner-shelf environment, Gulf of Papua. *J. Geophys. Res.* 113: F01S07.

McCave, I.N. 1984. Size spectra and aggregation of suspended particles in the deep ocean. *Deep-Sea Res.* 31:329–352.

McKee, B.A., and M. Baskaran. 1999. Sedimentary processes of Gulf of Mexico estuaries, In: Bianchi, T.S., Pennock, J.R., and Twilley, R.R. (eds.), Biogeochemistry of Gulf of Mexico Estuaries. New York: Wiley, pp. 63–85.

McKee, B.A., R.C. Aller, M.A. Allison, T.S. Bianchi, and G.C. Kineke. 2004. Transport and transformation of dissolved and particulate materials on continental margins influenced by major rivers: benthic boundary layer and seabed processes. *Cont. Shelf Res.* 24: 899–926.

Meade, R.H, T.R. Yuzyk, and T.J. Day. 1990. Movement and storage of sediment in rivers of the United States and Canada. In: M.G. Wolman and H.C. Riggs (eds.). The Geology of North America, vol. 1, Surface Water Hydrology. Boulder, CO: Geology Society of America, pp. 255–280.

Meade R. 1995. Contaminants in the Mississippi River. Reston, VA: U.S. Geological Survey Circular 1133.

Meade, R.H. 1996. River-sediment inputs to major deltas, In: Milliman, J., and Haq B. (eds.), Sea-Level Rise and Coastal Subsidence. London: Kluwer Academic Publishing, pp. 63–85.

Meade, R.H., and R. Parker. 1985. Sediment in rivers of the United States, In: National Water Summary 1984: U.S. Geological Survey Water Supply Paper. 2275, pp. 49–60.

Meade, R.H., and J.A. Moody. 2010. Causes for the decline in suspended-sediment discharge in the Mississippi River system, 1940–2007. *Hydrol. Process.* 24: 35–49.

Meade, R.H., T. Dunne, J.E. Richey, U. deM. Santos, and E. Salati. 1985. Storage and remobilization of suspended sediment in the lower Amazon River of Brazil. *Science* 228: 488–490.

Michels, K.H., H.R. Kudrass, C. Hubscher, A. Suckow, and M. Wiedicke. 1998. The submarine delta of the Ganges–Brahmaputra: cyclone-dominated sedimentation patterns. *Mar. Geol.* 149: 133–154.

Michels, K.H., A. Suckow, M. Breitzke, H.-R. Kudrass, and B. Kottke. 2003. Sediment transport in the shelf canyon "Swatch of No Ground" (Bay of Bengal). *Deep-Sea Res. Part II Top. Stud. Oceanogr.* 50: 1003–1022.

Milliman, J.D., and K.L. Farnsworth. 2011. River Discharge to the Coastal Ocean: A Global Synthesis. Cambridge, UK: Cambridge University Press, p. 392.

Milliman, J.D., and J.P.M. Syvitski. 1992. Geomorphic tectonic control of sediment discharge to the ocean – the importance of small mountainous rivers. *J. Geol.* 100: 525–544.

Morton, R.A., and J.A. Barras. 2011. Hurricane impacts on coastal wetlands: a half-century record of storm generated features from southern Louisiana. *J. Coastal Res.* 6A: 27–43.

Mulder, T., and J.P.M. Syvitski. 1995. Turbidity currents generated at river mouths during exceptional discharges to the world oceans. *J. Geol.* 100: 525–544.

Neill, C.F., and M.A. Allison. 2005. Subaqueous deltaic formation on the Atchafalaya Shelf, Louisiana. *Mar. Geol.* 214:411–430.

Nittrouer, C.A., and D.J. Demaster. 1996. The Amazon shelf setting: tropical, energetic, and influenced by a large river. *Cont. Shelf Res.* 16: 553–573.

Nittrouer, C.A., S.A. Kuehl, and D.J. DeMaster. 1986. The deltaic nature of Amazon shelf sedimentation. *Geol. Soc. Am. Bull.* 97: 444–458.

Nittrouer, C.A., D.J. DeMaster, A.G. Figueiredo, and J.M. Rine. 1991. AmasSeds: An interdisciplinary investigation of a complex coastal environment. *Oceanography* 4: 3–7.

Nittrouer, C.A., S.A. Kuehl, R.W. Sternberg, A.G. Figueiredo, and L.E.C. Faria. 1995. An introduction to the geological significance of sediment transport and accumulation on the Amazon continental-shelf. *Mar. Geol.* 125: 177–192.

Nittrouer, C.A., S.A. Kuehl, A.G. Figueiredo, M.A. Allison, C.K. Sommerfield, J.M. Rine, L.E.C. Faria, and O.M. Silveira. 1996. The geological record preserved by Amazon shelf sedimentation. *Cont. Shelf Res.* 16: 817–841.

Nittrouer, C.A., J.A. Austin, M.E. Field, J.H. Kravitz, J.M. Syvitski, and P.L. Wiberg. 2009. Continental Margin Sedimentation: From Sediment Transport to Sequence Stratigraphy. New York: Wiley Blackwell.

Nittrouer, J.A., M.A. Allison, and R. Campanella. 2008. Bedform transport rates for the lowermost Mississippi River. *J. Geophys. Res. Earth Surf. Process.* 113: F03004.

Oey, L. 1995. Eddy- and wind-forced shelf circulation. *J. Geophys.* 100: 8621–8637.

Ogston, A.S., R.W. Sternberg, C.A. Nittrouer, D.P. Martin, M.A. Goñi, and J.S. Crockett. 2008. Sediment delivery from the Fly River tidally dominated delta to the nearshore marine environment and the impact of El Niño. *J. Geophys. Res.* 113: F01S11.

Oltman, R.E. 1968. Reconnaissance Investigations of the Discharge and Water Quality of the Amazon River. U.S. Geological Survey, Circular 552.

Orpin A.R., C. Alexander, L. Carter, S. Kuehl, and J.P. Walsh. 2006. Temporal and spatial complexity in post-glacial sedimentation on the tectonically active, poverty bay continental margin of New Zealand. *Cont. Shelf Res.* 26: 2205–2224.

Orton, G.J., and H.G. Reading. 1993. Variability of deltaic processes in terms of sediment supply, with particular emphasis on grain size. *Sedimentology* 40: 475–512.

Overeem, I., J.P.M. Syvitski, and E.W.H. Hutton. 2005. Three-dimensional numerical modeling of deltas, In: River Deltas; Concepts, Models, and Examples. Special Publication – Society for Sedimentary Geology. 83: 13–30.

Palamenghi, L., T. Schwenk, V. Spiess, and H.R. Kudrass. 2011. Seismostratigraphic analysis with centennial to decadal time resolution of the sediment sink in the Ganges–Brahmaputra subaqueous delta. *Cont. Shelf Res.* 31: 712–730.

Philander, S.G.H. and R.C. Pacanowski. 1986: The mass and heat budget in a model of the tropical Atlantic Ocean. *J. Geophys. Res.* 91: 14212–14220.

Pielke, R.A. 1990. The Hurricane. New York: Routledge, 228 pp.

Platzman, G.W., G.A. Curtis, K.S. Hansen, and R.D. Slater. 1981. Normal modes of the world ocean: II. Description of modes in the period range 8 to 80 hours. *J. Phys. Oceanogr.* 11: 570–603.

Presley, B.J., J.H. Trefry, and R.F. Shokes. 1980. Heavy metal inputs to Mississippi Delta sediments: a historical view. *Water Air Soil Pollut.* 13: 481–494.

Rabalais, N.N., R.E. Turner, and W.J. Wiseman, Jr. 2002. Hypoxia in the Gulf of Mexico, a.k.a. "The Dead Zone." *Ann. Rev. Ecol. Systematics* 33: 235–263.

Raghavan, S., and S. Rajesh. 2003. Trends in tropical cyclone impact: A study in Andhra Pradesh, India. *Bull. Am. Meteorolog. Soc.* 84: 635–644.

Richey, J.E., R.H. Meade, E. Salati, A.H. Devol, C.F. Nordin, and U. dos Santos. 1986. Water discharge and suspended sediment concentrations in the Amazon River: 1982–1984. *Water Resour. Res.* 22: 756–764.

Richey, J.E., C. Nobre, and C. Deser. 1989. Amazon River discharge and climate variability: 1903 to 1985. *Science* 6: 101–103.

Rine, J.M., and R.N. Ginsburg. 1985. Depositional facies of a mud shoreface in Suriname, South America – a mud analogue to sandy shallow-marine deposits. *J. Sed. Petrol.* 55: 633–652.

Roberts, H.H. 1997. Dynamic changes of the Holocene Mississippi River delta plain: The delta cycle. *J. Coastal Res.* 14: 605–627.

Rogers, J.C., and J.S.M. Coleman. 2003. Interactions between the Atlantic Multidecadal Oscillation, El Ninõ/La Ninã, and the PNA in winter Mississippi Valley stream flow. *Geophys. Res. Lett.* 30: 25–1–5.

Rogers, K.G., and S.L. Goodbred, Jr. 2010. Mass failures associated with the passage of a large tropical cyclone over the Swatch of No Ground submarine canyon (Bay of Bengal). *Geology* 38: 1051–1055.

Saito, Y., Z. Yang, and K. Hori. 2001. The Huanghe (Yellow River) and Changjiang (Yangtze River) deltas: a review on their characteristics, evolution and sediment discharge during the Holocene. *Geomorphology* 41: 219–231.

Sampere, T.P., T.S. Bianchi, S.G. Wakeham, and M.A. Allison. 2008. Sources of organic matter in surface sediments of the Louisiana Continental margin: Effects of major depositional/transport pathways and Hurricane Ivan. *Cont. Shelf Res.* 28: 2472–2487.

Sarin, M.M., S. Krishnaswami, K. Dilli, B.L.K. Somayajulu, and W.S. Moore. 1989. Major ion chemistry of the Ganges-Brahmaputra river system: weathering processes and fluxes to the Bay of Bengal. *Geochemica et Cosmochemica Acta.* 53: 997–1009.

Scruton, P.C. 1960. Delta building and the deltaic sequence, In: Shepard, F.P., Phleger, F.B., and van Anderl, T.H. (eds.), Recent sediments northwest Gulf of Mexico. Tulsa, OK: American Association of Petroleum Geologists, pp. 82–102.

Sheremet, A., S. Jaramillo, S.-F. Su, M.A. Allison, and K.T. Holland. 2011. Wave-mud interaction over the muddy Atchafalaya subaqueous clinoform, Louisiana, United States: wave processes. *J. Geophys. Res. Oceans* 116: C06005.

Singh, O.P., T.M. Ali Khan, and Md.S. Rahman. 2000. Changes in the frequency of tropical cyclones over the North Indian Ocean. *Meteorol. Atmosph. Phys.* 75: 11–20.

Slingerland, R., R.W. Selover, A.S. Ogston, T.R. Keen, N.W. Driscoll, and J.D. Milliman. 2008. Building the Holocene clinothem in the Gulf of Papua: An ocean circulation study. *J. Geophys. Res.* 113: F01S14.

Sommerfield, C.K., C.A. Nittrouer, and C.R. Alexander. 1999. ^{7}Be as a tracer of flood sedimentation on the northern California continental margin. *Cont. Shelf Res.* 19: 335–361.

Stanley, D.J., and A.G. Warne 1994. Worldwide initiation of Holocene marine deltas by deceleration of sea-level rise. *Science* 265: 228–231.

Sternberg, R.W., D.A. Cacchione, B. Paulso, G.C. Kineke, and D.E. Drake. 1996. Observations of sediment transport on the Amazon subaqueous delta. *Cont. Shelf Res.* 16: 697–715.

Suckow, A., U. Morgenstern and H.R. Kudrass. 2001. Absolute dating of recent sediments in the cyclone-influenced shelf area off Bangladesh: comparison of gamma spectrometric (^{137}Cs, ^{210}Pb, ^{228}Ra), radiocarbon, and ^{32}Si ages: *Radiocarbon* 43: 917–927.

Swift, D.J.P., and J.A. Thorne. 1991. Sedimentation on continental margins: I. A general model for shelf sedimentation, In: Swift, D.J.P., Tillman, R.W., Oertel, G.F., and Thorne, J.A. (eds.), Shelf Sand and Sandstone Bodies, Geometry, Facies and Sequence Stratigraphy: International Association of Sedimentologists Special Publication 14, pp. 3–31.

Syvitski, J.P.M. 2008. Predictive modeling in sediment transport and stratigraphy. *Comput. Geosci.* 34: 1167–1169.

Syvitski, J.P.M., C.J. Vörösmarty, A.J. Kettner, and P.A. Green. 2005. Impact of humans on the flux of terrestrial sediment to the global coastal ocean. *Science* 308: 376–380.

Syvitski, J.P.M., A.J. Kettner, M.T. Hannon, E.W.H. Hutton, I. Overeem, G.R Brakenridge, J. Day, C. Vörösmarty, Y. Saito, L. Giosan, and R J. Nicholls. 2009. Sinking Deltas. *Nat. Geosci.* 2: 681–689.

Syvitski, J.P.M., and J.D. Milliman. 2007. Geology, geography and humans battle for dominance over the delivery of sediment to the coastal ocean. *J. Geol.* 115: 1–19.

Syvitski J.P.M. and Y. Saito. 2007. Morphodynamics of deltas under the influence of humans. *Global Planet. Change* 57: 261–282.

Tcherepanov, E.N., A.W. Droxler, P.M. Lapoint, K. Mohn, and O.A. Larsen. 2010. Siliciclastic influx and burial of the Cenozoic carbonate system in the Gulf of Papua. *Mar. Pet. Geol.* 27: 533–554.

Tomczak, M., and J.S. Godfrey. 1994. Regional Oceanography: An Introduction: Oxford, UK: Pergamon Press.

Traykovski, P., P.L. Wiberg, and W.R. Geyer. 2007. Observations and modeling of wave-supported sediment gravity flows on the Po prodelta and comparison to prior observations from the Eel shelf. *Cont. Shelf Res.* 27: 375–399.

Turner, R.E., J.J. Baustian, E.M. Swenson, and J.S. Spicer. 2006. Wetland sedimentation from Hurricanes Katrina and Rita. *Science* 314: 449–452.

Unger, D., V. Ittekkot, P. Schaefer, J. Tiemann, and S. Reschke. 2003. Seasonality and interannual variability of particle fluxes to the deep Bay of Bengal: influence of riverine input and oceanographic processes. *Deep-Sea Res. Part II. Top. Stud. Oceanogr.* 50: 897–923.

Vail, P.R., R.M. Mitchum Jr., R.G. Todd, J.M. Widmier, S. Thompson, III, J.B. Sangree, J.N. Bubb, and W.G. Hatlelid. 1977. Seismic stratigraphy and global changes of sea level, In: C.E. Payton (ed.), Seismic Stratigraphy–Applications to Hydrocarbon Exploration. AAPG Memoir, 29. Tulsa, OK: AAPG, pp. 49–211.

Van Wagoner J.C., H.W. Posamentier, R.M. Mitchum, Jr., P.R. Vail, J.F. Sarg JF, T.S. Loutit, and J. Hardenbol. 1988. An overview of the fundamentals of sequence stratigraphy and key definitions, In: Wilgus, C.K., Hastings, B.S., Kendall, C.G., Posamentier, H.W., Ross, C.A., and Van Wagoner, J.C. (eds.), Sea-Level Changes: An Integrated Approach. Soc. Econ. Paleontol. Mineral. Spec. Publ. No. 42, pp. 39–45.

Walker, N.D., and A.B. Hammack. 2000. Impacts of winter storms on circulation and sediment transport: Atchafalaya-Vermilion Bay region, Louisiana, USA. *J. Coastal Res.* 16: 996–1010.

Walsh, J.P., and C.A. Nittrouer. 2003. Contrasting styles of off-shelf sediment accumulation in New Guinea. *Mar. Geol.* 196: 105–125.

Walsh, J.P., and C.A. Nittrouer. 2009. Towards an understanding of fine-grained river-sediment dispersal on continental margins. *Mar. Geol.* 263: 34–45.

Walsh, J.P., C.A. Nittrouer, C.M. Palinkas, A.S. Ogston, R.W. Sternberg and G.J. Brunskill. 2004. Clinoform mechanics in the Gulf of Papua, New Guinea. *Cont. Shelf Res.* 24: 2487–2510.

Walsh, J.P., D.R. Corbett, D. Mallinson, M. Goñi, M. Dail, C. Loewy, K. Marciniak, K. Ryan, C. Smith, A. Stevens, B. Sumners, and T. Tesi. 2006. Mississippi delta mudflow activity and 2005 gulf hurricanes. *Eos* 87: 477–478.

Warner, J.C., C.R. Sherwood, R.P. Signell, C.K. Harris, and H.G. Arango. 2008. Development of a three-dimensional, regional, coupled wave, current, and sediment-transport model. *Comput. Geosci.* 34: 1284–1306.

Wells, J.T., and J.M. Coleman. 1981. Physical processes and fine-grained sediment dynamics, coast of Surinam, South America. *J. Sediment. Petrol.* 51: 1053–1068.

Wheatcroft, R.A., and C.K. Sommerfield. 2005. River sediment flux and shelf accumulation rates on the Pacific Northwest margin. *Cont. Shelf Res.* 25: 311–332.

Wiedicke, M., H.-R. Kudrass, Ch. Hübscher. 1999. Oolitic beach barriers of the last Glacial sea-level lowstand at the outer Bengal shelf. *Mar. Geol.* 157: 7–18.

Wiseman, W.J., and S.P. Dinnel. 1988. Shelf currents near the mouth of the Mississippi River. *J. Phys. Oceanogr.* 18: 1287–1291.

Wright, L.D. 1985. River deltas, In: Davis, R.A. (ed.), Coastal Sedimentary Environments, 2nd ed. New York: Springer-Verlag, pp. 1–75.

Wright, L.D., and C.T. Friedrichs. 2006. Gravity driven sediment transport on continental shelves: a status report. *Cont. Shelf Res.* 26: 2092–2107.

Wright, L.D., and C.A. Nittrouer. 1995. Dispersal of river sediments in coastal seas – 6 contrasting cases. *Estuaries* 18: 494–508.

Wright, L.D., Z.S. Yang, B.D. Bornhold, G.H. Keller, D.B. Prior, and W.J. Wiseman, Jr. 1986. Hyperpycnal plumes and plume fronts over the Huanghe (Yellow River) delta front. *Geo-Mar. Lett.* 6: 97–105.

Wright, L.D., C.R. Sherwood, and R.W. Sternberg. 1997. Field measurements of fairweather bottom boundary layer processes and sediment suspension on the Louisiana inner continental shelf. *Mar. Geol.* 140: 329–345.

Wright, L.D., C.T. Friedrichs, S.C. Kim, and M.E. Scully. 2001. Effects of ambient currents and waves on gravity-driven sediment transport on continental shelves. *Mar. Geol.* 175: 25–45.

Xu, K., and J.D. Milliman. 2009. Seasonal variations of sediment discharge from the Yangtze River before and after impoundment of the Three Gorges Dam. *Geomorphology* 104: 276–283.

Xu, K.H., C.K. Harris, R.D., Hetland, and J.M. Kaihatu. 2011. Dispersal of Mississippi and Atchafalaya sediment on the Texas-Louisiana shelf: Model estimates for the year 1993. *Cont. Shelf Res.* 31: 1558–1575.

Xue, Z., J.P. Liu, D. DeMaster, L.B. Nguyen, and T.K.O Ta. 2010. Late Holocene evolution of the Mekong subaqueous delta, southern Vietnam. *Mar. Geol.* 269: 146–60.

Yang, S.L., J.D. Milliman, P. Li, and K. Xu. 2011. 50,000 dams later: Erosion of the Yangtze River and its delta. *Global Planet. Change* 75: 14–20.

5

Changjiang (Yangtze) and Huanghe (Yellow) Rivers: historical reconstruction of land-use change and sediment load to the sea

H. Wang, Z. Yang, and N. Bi

1. Regional settings of the Changjiang and Huanghe River basins

The total global sediment flux from rivers to the global oceans has been estimated as ~20,000 mt/yr (10^6 metric tons per year), of which ~31.5% is contributed by Asian rivers, particularly those draining the Himalaya–Tibetan Plateau region (Milliman and Meade 1983; Milliman and Syvitski 1992; Syvitski et al. 2005). Among these large rivers, the Changjiang and the Huanghe rivers, annually providing more than 949 km^3/yr of freshwater and 1,560 mt/yr of sediment to the sea (Table 5.1), are of great significance to both regional social community and physical–ecological environments in the East China Shelf Seas such as the Bohai Sea, Yellow Sea, and East China Sea.

1.1. Topographic settings

The rise of the Himalayas–Tibet owing to the collision of India Plate and Eurasian Plate (~55 million year ago) and the corresponding climate shift (e.g., birth of Asian Monsoon) created large rivers in Asia (Tapponnier et al. 2001; Clift et al. 2008). The Changjiang and Huanghe Rivers commonly originate from the Himalaya–Tibet region at an altitude of nearly 4,000–5,000 m above the mean sea level and flow eastward in accord with the terrain and topography (Fig. 5.1). The upper streams of the two rivers are characterized by steep hillslope and narrow gorges, and the rivers enter into the fluvial plain in the middle and lower reaches (Fig. 5.2).

The upper Changjiang is more than 4,300 km long from the source to Yichang and has a total drainage area of about 1.0×10^4 km^2 (Tong and Han 1982). The mainstem of the Jinshajiang is jointed by four major tributaries in this river section, and the rivers are deeply incised into rocky canyons with >1,000 m in elevation difference between the riverbeds and mountain peaks. The channels are about 0.5–1.5 km wide, 5–20 m deep, and vary in slope from $10\sim40\times10^{-5}$, reaching a maximum of 450×10^{-5} (Chen et al. 2001). The middle Yangtze is 950 km in length and has a total drainage area of about 0.68×10^6 km^2. A large tributary (Hanjiang River) joins the mainstem of Changjiang nearby the megacity, Wuhan. Two large lakes connect the Changjiang through narrow outlets in this section: the Dongting Lake and Poyang Lake that are important sinks of the Changjiang sediment in the middle reaches (Yang et al. 2006). In the middle Yangtze, the slope decreases dramatically to $2\sim3\times10^{-5}$, but local slope can be as high as $6\sim8\times10^{-5}$. The channel of middle Changjiang is wider and deeper than the upper course, with the width between 1 and 2 km and the depth between 6 and

Table 5.1. *Information of the Changjiang and Huanghe river systems*

Rivers	Yellow River	Yangtze River
Basin area (10^6 km^2)	0.77	1.94
River length (km)	5,464	6,300
Water discharge (km^3/yr)	49	900
Runoff (mm/yr)	64	464
Sediment load (10^6 t/yr)	1,080	480
Sediment yield (t/km^2/yr)	1,403	247
Annual precipitation (mm)	300	1,190
Average elevation (m)	1,547	748
Maximum elevation (m)	4,200	5,100
Delta area (km^2)	36,272	66,000
Climatic zone	Mountains-arid-temperate	Mountains-temperate
Average population density (person/km^2)	162	224

References: Milliman and Syvitski (1992); Coleman and Huh (2004); World Resources Institute 2005.

15 m. A typical meandering river pattern with many cutoffs prevails in this river reach, where the river exits from the upper rock-confined valley into the Jingjiang fluvial plain. Below the Hukuo, the final 930 km constitutes the lower Changjiang, with a total drainage area of 0.12×10^6 km^2. Several large interior lakes, such as the Chaohu and Taihu, and in association with many tributaries, drain into the lower Yangtze. This segment of the river wanders among plains and hills, on which high-stage water level about 8 m above mean water level is clearly marked. The slope of riverbed decreases to $0.5–1.0\times10^{-5}$, and the channel widens to 2–4 km and deepens to 10–20 m. The river channel can however, be wider than 15 km and as shallow as 6 m in its estuarine region.

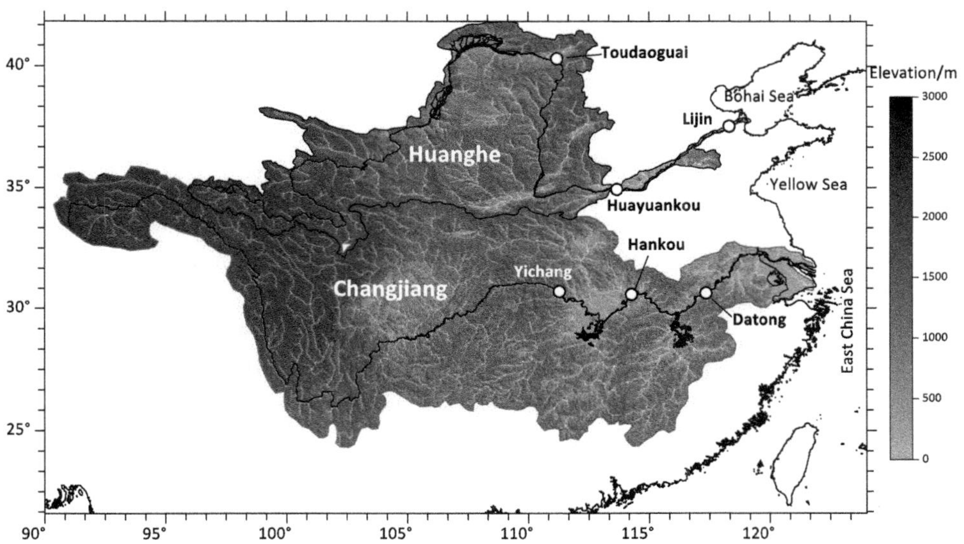

Figure 5.1. Topography of the Changjiang and Huanghe river basin.

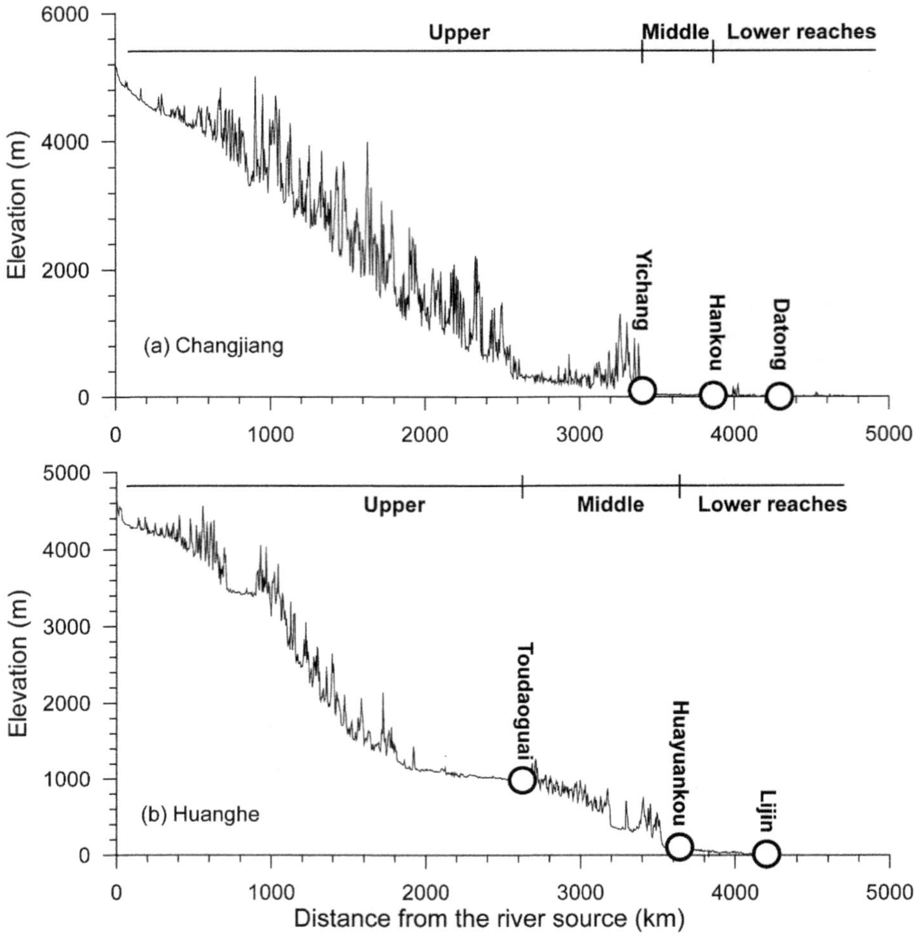

Figure 5.2. Topographic profiles along the mainstream of the Changjiang and Huanghe, illustrating the rivers originating in the high Tibet Plateau and flowing to the fluvial plain in the lower reaches.

The Huanghe originates on the eastern Qinghai–Tibet Plateau, and then flows eastward through northwestern China over a total length of 5,464 km before debouching into the Bohai Sea (Fig. 5.1). It drains a wide basin that covers more than 750,000 km^2 and exhibits a variety of geological and climatic features. The river can be divided into three sections according to its geographical settings (Fig. 5.2): (1) the upper reaches extend over a length of 3,471 km from the river source to Toudaoguai and drain an area of 385,996 km^2; (2) the middle reaches stretch 1,206 km from Toudaoguai to Huayuankou and drain an area of 343,751 km^2, and a considerable number of tributaries join the main stream in the middle reaches; and (3) the lower reaches extend over a length of 786 km from Huayuankou through the flat alluvial plains of Henan and Shandong provinces to the river mouth and drain an area of 22,726 km^2 (Fig. 5.1). The river descends from an altitude of approximately 4,500 m at its source, with variable gradient to 1,000 m at Toudaoguai and then with relatively continuous gradient to 110 m at Huayuankou (Fig. 5.2). Changes from steep to gentler gradients near Lanzhou and Huayuankou correspond to the boundaries between the Qinghai–Tibet Plateau (above Lanzhou), the Loess Plateau (below Lanzhou and including the middle reaches), and the flat alluvial plains of the lower reaches (Fig. 5.2).

1.2. Hydrologic settings

The annual water discharge from the upper reaches recorded at Yichang gauging station is approximately 463 km^3/yr, and the annual sediment load amounts to 470 mt/yr, which constitutes the main source of sediment delivered to the middle and lower reaches. Several major tributaries, including the Jinshajiang, Jialingjiang, Minjiang, and Wujiang, contribute ~86% of the sediment load at Yichang station (Fig. 5.1, Yang et al. 2006). In the middle reaches, the river meanders across the almost flat Jingjiang fluvial plain, forming several large lowland lakes, including Dongting Lake and Poyang Lake. In addition, a major tributary, the Hanjiang, joins the Changjiang at Hankou station (Fig. 5.1). The Hanjiang annually contributed ~73 mt/yr of sediment load to the Changjiang before construction of the Danjiangkou Reservoir in 1968 (Yang et al. 2006). The middle reaches, particularly over the Jingjiang fluvial plain, act as an important flood-diversion area, accounting for at least 100 mt of sediment loss owing to overbank deposition (Yang et al. 2006) during years of catastrophic flooding (e.g., 1954 and 1998). As a result, the annual sediment load at Hankou averages 383 mt/yr, accounting for ~80% of the sediment input from the upper reaches, whereas the mean annual water discharge at Hankou increases to 711 km^3/yr. A large proportion of the lower reaches between Datong station and the river mouth is influenced by the tides of the East China Sea. The mean annual sediment load from the Changjiang to the sea, as recorded at Datong station, is ~413 mt/yr, which is slightly higher than that at Hankou station, whereas the mean annual water discharge at Datong is approximately twice that coming from the upper reaches (records at Yichang station). The sediment load at the three major gauging stations along the main stream indicate that the Changjiang sediment is predominately derived from the upper reaches, where severe soil erosion and frequent landslides have a significant effect, but the water from the upper reaches accounts for only ~50% of that discharging into the East China Sea.

Compared with other large rivers of the world, high sediment load and low water discharge are distinctive characteristics of the Huanghe. The average annual water discharge from the Huanghe to the sea between 1950 and 2005, as recorded at Lijin station, was 32.0 km^3/yr, which represents only 5% of that of the Changjiang (Yang et al. 2006), or 0.5% of that of the Amazon (Milliman and Meade 1983). Most of the river water of the Huanghe originates from the upper reaches, where ice melt and snowfall are the main contributors. The average annual water discharge (1950–2005) at Toudaoguai, near the boundary between the upper and middle reaches, is 22.0 km^3/yr and accounts for 56.4% of the annual discharge at Huayuankou (Table 5.1; Fig. 5.2B). The annual discharge at Lanzhou is somewhat higher than that at Toudaoguai, because water is used for agricultural irrigation along the river between Lanzhou and Toudaoguai (Chen et al. 2005). Downstream from Toudaoguai the water discharge increases as several tributaries of various sizes join the main stream in the middle reaches. The highest annual discharge of the river (39.0 km^3/yr) occurs at Huayuankou. Downstream from Huayuankou the river is confined to a narrow basin characterized by a riverbed raised up to 10 m above the surrounding area by severe siltation. At Lijin, some 100 km upstream from the river mouth (Fig. 5.1), the annual water discharge decreases to 32.0 km^3/yr, which is 82.0% of the discharge at

Huayuankou and is considered to represent the discharge from the Huanghe to the sea. The decrease in discharge between Huayuankou and Lijin is caused mainly by water abstraction for agricultural irrigation and domestic use in the lower reaches (Wang et al. 2006). Because of climate change (e.g., decrease in precipitation related to El Niño–Southern Oscillation [ENSO] events) and increasing water consumption facilitated by dams and reservoirs, discharge from the Huanghe to the sea has decreased continuously over the past half century (Wang et al. 2006) and illustrates a complicated transformation of the river system as a result of both natural and human influences.

The spatial distribution of the sediment load in the Huanghe differs largely from that of the water discharge. Nearly 90% of the sediment load comes from the middle reaches, where the dominant geological feature is the wind-deposited Loess Plateau, which is acknowledged as the region of highest erodibility in the world (Ren and Shi 1986; Huang and Zhang 2004). In contrast, the upper reaches above Toudaoguai provide only 10% of the sediment load. In the upper reaches of the river, sediment load increases from 70 mt/yr at Lanzhou to 100 mt/yr at Toudaoguai, and the average suspended sediment concentration (SSC) increases from 2.3 to 4.4 kg/m^3 along the same stretch of river. In the middle reaches a surprisingly large increase in sediment load to 970 mt/yr at Huayuankou is largely the contribution of the tributaries between Toudaoguai and Tongguan. As the river flows over the flat alluvial plain, considerable quantities of suspended sediment fall from suspension, and the high level of channel deposition in the lower reaches has raised the riverbed 10 m above the surrounding areas to form a unique "suspended river" that creates a high risk of catastrophic flooding in the surrounding area (Chen et al. 2005). Finally, the Huanghe discharges 800 mt/yr (1950–2005) of sediment load to the sea, as recorded at station Lijin.

1.3. Surface lithology of the river basins

Sediment yields of Changjiang and Huanghe differ largely from each other, reflecting differences in their basin sizes and tectonic and geologic features, as well as large difference in regional surficial lithology of the river basins.

The river sediment of Changjiang is mainly derived from soil erosion on hillslopes in the upper reaches, where mixed consolidated sedimentary rocks are dominant (Fig. 5.3, 25% of the basin area), plus input (~12% of the total load before dam construction) from its tributary, the Hanjiang River, in the middle reaches (Yang et al. 2006). Besides the controlling effects of rainfall and human activities, the lithological characteristics are of importance to the sediment yield.

The lithology of the Huanghe River Basin is dominated by loess in the middle reaches that consists of ~41% of the basin area (Fig. 5.3). The average thickness of this semiconsolidated and easily erodible Quaternary formation is estimated as 50~80 m, with maximum of 150~180 m. Therefore, the river has the high sediment yield and sediment load to the ocean, about 90% of which is attributed to summer rainfall on the highly erodible Loess Plateau in its middle reaches.

2. Human impacts on the river basins and land-use change

Many ancient civilizations are developed around large rivers. Like other famous rivers all over the world, the Changjiang and Huanghe have nurtured brilliant civilization in China and thus present

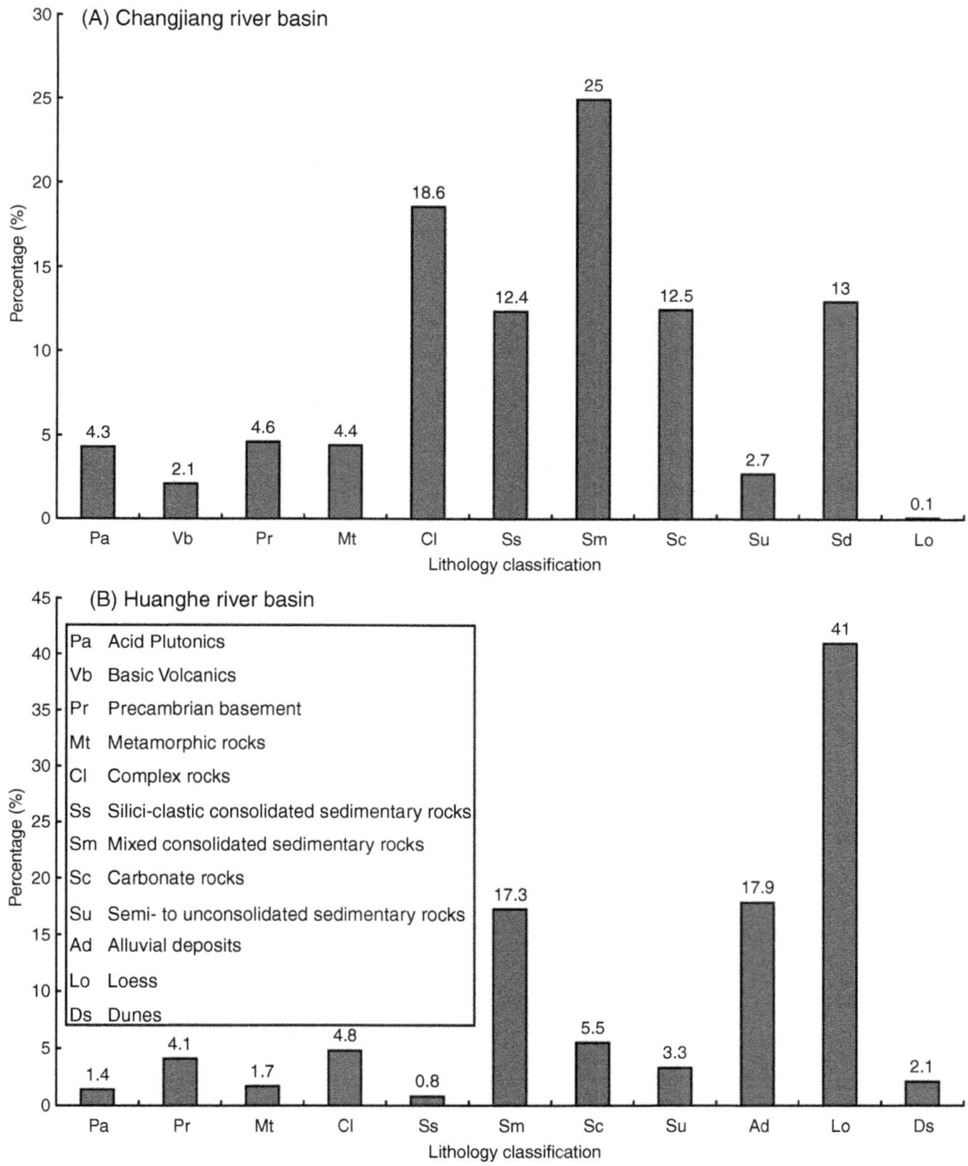

Figure 5.3. Surficial lithology of Changjiang and Huanghe river basins (Data are courtesy of Dr. Hans Durr).

important influence on the Chinese culture. In the Chinese long history, the Changjiang and Huanghe stand for a kind of culture and civilization. The civilizations developed within the two river basins have undoubtedly played a very great role in the development of civilization in China and even across the globe.

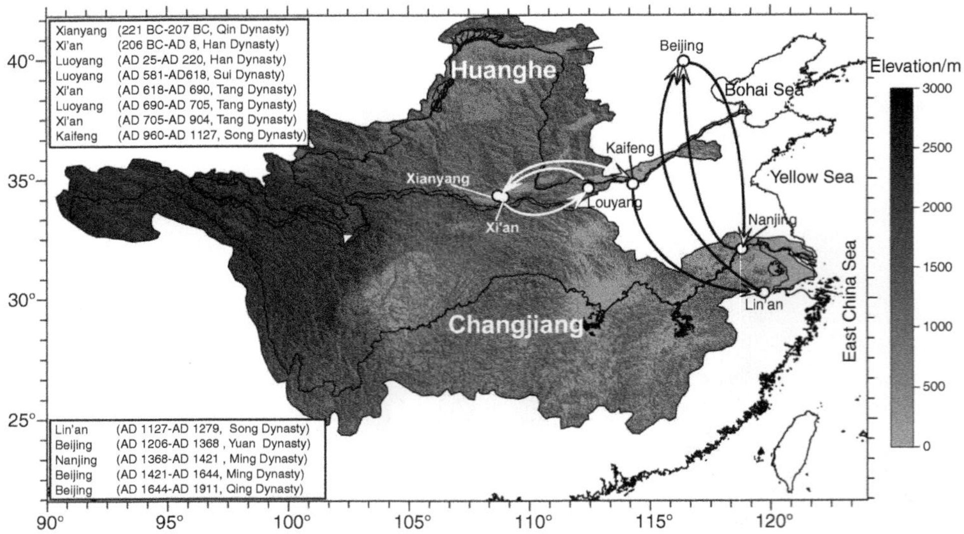

Figure 5.4. Capital migrations of Chinese Dynasties, indicating the human interventions on the Changjiang river basin had been enhanced since A.D. 1127 as the capital moved to the south China. Lin'an is currently the city of Hangzhou.

From the Xia (twenty-first to sixteenth century B.C.) and Shang (sixteenth to eleventh century B.C.) Dynasties to the Tang Dynasty (A.D. 618–907), the Huanghe area, especially the middle and lower part, had always been the center of politics, economy, and culture. Commonly, the Huanghe civilization came into being between the years 4,000 B.C. and 2,000 B.C., which covered more than 2,000 years. Three thousand years ago, about 50% of the Loess Plateau was covered by forests. As the population in the loess region increased, the land cover in the loess region has been destroyed significantly owing to the poor agricultural practices and warfare. As a result, soil erosion and sediment yield has been greatly enhanced. Based on the sedimentary records in the deltaic plain, an abrupt increase in sediment load to 1,000–1,200 mt/yr during 2000–1000 B.P. was identified mostly owing to increased cultivation and deforestation on the Loess Plateau (Milliman et al. 1987; Xue 1993; Ren and Zhu 1994; Liu et al. 2002). This high sediment load persisted during the past 1,000 years and contributed much to the development of the fluvial plain in northern China and the formation of mega-deltas in the Bohai and Yellow seas. These changes are also recorded in pollen in deltaic sediments derived from the Huanghe (Yi et al. 2003a). The drastic increase in sediment yield and consequent siltation in the river bed of the lower reaches characterized by the suspended river bed (see Wang et al. 2007) and overwhelming floods caused fatal disasters to the regional civilization. Since A.D. 1127, the capital of Song Dynasty had to move to South China, such as Lin'an, currently the city of Hangzhou (Song Dynasty) and Nanjing (Ming Dynasty), and North China, such as Beijing (Yuan and Qing Dynasties) (Fig. 5.4).

During the long history of ancient China, the Changjiang river basin had developed the brilliant Liangzhu culture in Zhejiang area, Bashu culture in Sichuan, Jingchu culture in Hubei and Hunan, and Wuyue culture in Jiangsu, Zhejiang, and Shanghai. Data of pollen, algal, fungal spore, and micro-charcoal in the sediment core on the aerial delta indicate that the precise cultural and environmental context of rice cultivation at this earliest known Neolithic site in eastern China, 7,700 cal. yr B.P. (Zong et al. 2007). The culture center was gradually transferred to the south from North China because of the political crises and warfare, bringing a boom in the economy and culture development. Deforestation in the Changjiang river basin is much later compared with deforestation on the Loess Plateau and in the Huanghe river basin (Ren and Zhang 1998; Ren 2000; Yi et al. 2003a). Fossil pollen data from North China and the upper-middle Changjiang basin suggested no obvious forest decline in the Changjiang basin during the middle to late Holocene until very recently (<ka B.P.), whereas a severe decline of arboreal pollen has occurred in the Huanghe catchment since c. 5000 B.P. (Ren 2000). Changes in land use and population pressure possibly explain the difference in the onset of forest decline between the Changjiang and Huanghe river basin. As recorded in the historical literature, a large number of northern people have migrated to the middle and lower Yangtze drainage basin since c. 1.7 cal. ka B.P., and about five million people moved from North China southward, mainly to the Changjiang river basin, including the hilly upper catchment area, during A.D. 1126–1279, when the Song Dynasty was defeated and the move of capital city (Qu and Li 1992; Ge et al. 1993; Fig. 5.4). The sudden increase in population of people from the north not only led to severe deforestation but also increased upland cultivation, including cultivation of millet, and thus resulted in the soil and rock erosion in the Changjiang river basin (Qu and Li 1992). Progradation rates obtained from sediment cores on the Changjiang delta plain suggest that the sediment load from the Changjiang before 2000 yr B.P. was $\sim 240 \times 10^6$ t/yr, only ~50% of the present sediment load (Hori et al. 2001; Saito et al. 2001). Accelerated deposition identified from the subaqueous delta lobe was attributed to the increased deforestation and soil rock erosion associated with human activities during the past c.1.7 ka, and especially since A.D. 1126–1279 (Wang et al. 2010a). Over thousands of years, human activities have modified the landscape of Changjiang river basin (e.g., extensive agricultural reclamations). And the ancient Dujiangyan Irrigation System, constructed in the Minjiang tributary 2,000 years ago but still working well now, essentially changed the input of water and sediment from the Minjiang to the mainstream of Changjiang (Li and Xu 2006). Meanwhile, the constructions of flood diversions in the middle and lower reaches also impacted the geomorphological features of the Changjiang River. By far the levees and embankments have been well developed in the middle and lower reaches to control the floods and prevent from breaching. However, early human activities had been of minor importance to the Changjiang sediment flux to the sea compared with the present situation.

3. Anthropogenic alteration of the rivers since the 1950s

3.1. Variation of the Changjiang sediment flux since the 1950s

Sediment loads recorded at station Yichang in the upper reaches of Changjiang averaged 448 mt/yr during 1950–2008, in contrast to loads of 367 mt/yr at station Hankou in the middle reaches and 398 mt/yr at station Datong in the lower reaches (Fig. 5.5). The Yangtze River has also been strongly fragmented by more than 50,000 dams, most of which are constructed in the tributaries (Yang et al. 2010). The large reservoirs in the Yangtze basin, with a total storage of capacity about 68 km^3, trap most of the sediment derived from the major tributaries in the upper and middle reaches

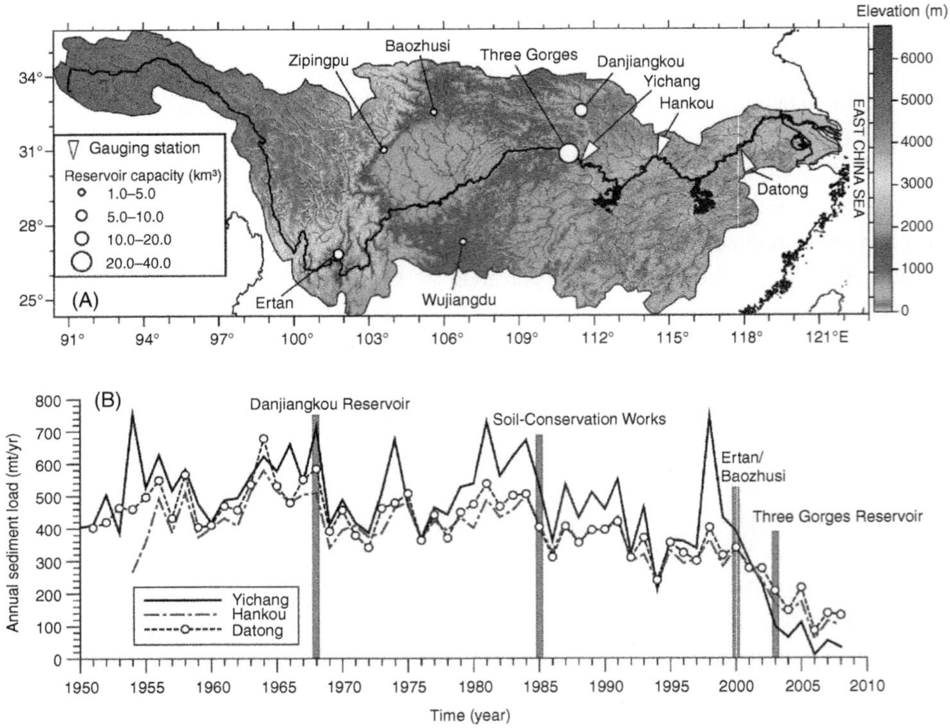

Figure 5.5. The Changjiang river basin: (A) major reservoirs and gauging stations and (B) time-series data (1950s–2008) of annual sediment load recorded at major stations. Since 2002 the sediment loads at Hankou and Datong have become higher than at Yichang, indicating notable sediment supply from the middle and lower channel due to bed scour (from Wang et al. 2011).

(Fig. 5.5A). The Danjiangkou Reservoir (Fig. 5.5A) on the Hanjing River, for instance, caused an abrupt decrease in sediment load from that tributary starting in 1968, affecting sediment records at stations Hankou and Datong in subsequent years (Fig. 5.5B). Soil conservation projects in the mid-1980s caused a distinct decrease in sediment yield from the upper Changjiang River basin, as shown by the sediment records at Yichang, and induced synchronous decreases in sediment load at stations of Hankou and Datong (Fig. 5.5B). For example, conservation measures including terracing, planting of trees, and protecting existent vegetation have been introduced in the Jialing River basin, covering an area of more than 23,000 km^2 during the period 1989–1996 (Zhang and Wen 2004). As the vegetation in the upper reach recovered after soil conservation programs and after the construction of many small dams, sediment yield from the upper reach decreased continuously (Yang et al. 2006). The Three Gorges Reservoir, which began storing water in June 2003, has greatly decreased the sediment delivered to the middle and lower reaches (Fig. 5.5). The sediment loads at Hankou and Datong during 2003–2008 have decreased to 125 and 154 mt/yr, respectively, but remain much higher than at Yichang (61 mt/yr), which means that river bed erosion in the middle and lower reaches has been a major source of sediment delivered to the ocean (Fig. 5.5; Yang et al. 2010).

3.2. Variation of the Huanghe sediment flux since the 1950s

The Huanghe historically has delivered more than 1,000 mt/yr of sediment to the Bohai Sea or the Yellow Sea (during the period 1128–1855), building prominent proximal and distal mud deposits on the continental shelf (Liu et al. 2004; Yang and Liu 2007). The sediment was mostly yielded from the loess region and transported by several major tributaries to the main stream (Fig. 5.6).

The annual sediment load delivered from the upper reach as recorded at station Toudaoguai was ~100 mt/yr (average from 1950 to 2008). Before completion of the Liujiaxia Reservoir in 1968, the average annual sediment load at Toudaoguai was ~170 mt/yr, with slight interannual variability (Fig. 5.6B). After the Longyangxia Reservoir was completed in 1986, the annual sediment load decreased from 110 mt/yr in 1969–1986 to 40 mt/yr in 1987–2008. The sediment loads recorded at stations Huayuankou and Lijin are an order of magnitude higher than at Toudaoguai, reflecting extremely high sediment yields from the Loess Plateau in the middle reaches (Fig. 5.6A). However, these sediment loads were also affected by the large Sanmenxian and Xiaolangdi reservoirs and soil conservation projects in the loess region. Although these reservoirs are smaller than the Longyangxia Reservoir in both storage and dam height, they trapped a considerable amount of sediment derived from the upper and middle reaches (Fig. 5.6). In particular, since operation of the Xiaolangdi Reservoir began in 1999, the sediment loads at Huayuankou and Lijin have decreased to 110 and 150 mt/yr, respectively, only ~10% of the 1950s level. And since 2002, the Water-Sediment Regulation Scheme implemented by the Yellow River Conservancy Committee has continuously caused bed erosion in the lower channel, as indicated by the greater sediment load at Lijin than at Huayuankou (Fig. 5.6B).

In addition to the stepwise decreases in sediment load caused by the operation of large reservoirs, the annual sediment yield from the loess region, as recorded at stations from the seven major tributaries (locations shown in Fig. 5.6A), also decreased at the same rate to the decreasing sediment load at station Tongguan (Fig. 5.7). Since the late 1970s and the early 1980s, the sediment yield from the loess region has been reduced to 760 mt/yr, accounting for ~60% of the average in 1950s and 1960s. The average sediment yield in 2000–2008 even decreased to 360 mt/yr. The decrease in sediment yield from the loess region was mostly contributed to the soil conservation program that started from the late 1970s and has become effective since 1980s. It was estimated that about 20,000 km^2 of terrace were constructed in the middle river basin together with improved land cover (Xu 2003; Syvitski and Kettner 2011). By the end of the 1980s, soil conservation and sediment control works had been introduced in almost 50% of the basin (Mou, 1996), corresponding with the significant decrease in sediment yield from the loess region (Fig. 5.7).

4. Roles of climate change and human activities on river systems

The river basins of Changjiang and Huanghe are climatically dominated by the East Asian Monsoon and both are densely populated (224 and 126 person/km^2, respectively; Wang et al. 2011). Therefore, the river systems are sensitive to the climate change and human activities. The warm phase invariably corresponds with decreased precipitation in the Huanghe River basin and the water discharge from the Changjiang, whereas the cold phase corresponds well with increasing precipitation and water discharge of both river systems (Table 5.2). Before recent human interventions, the changes of river

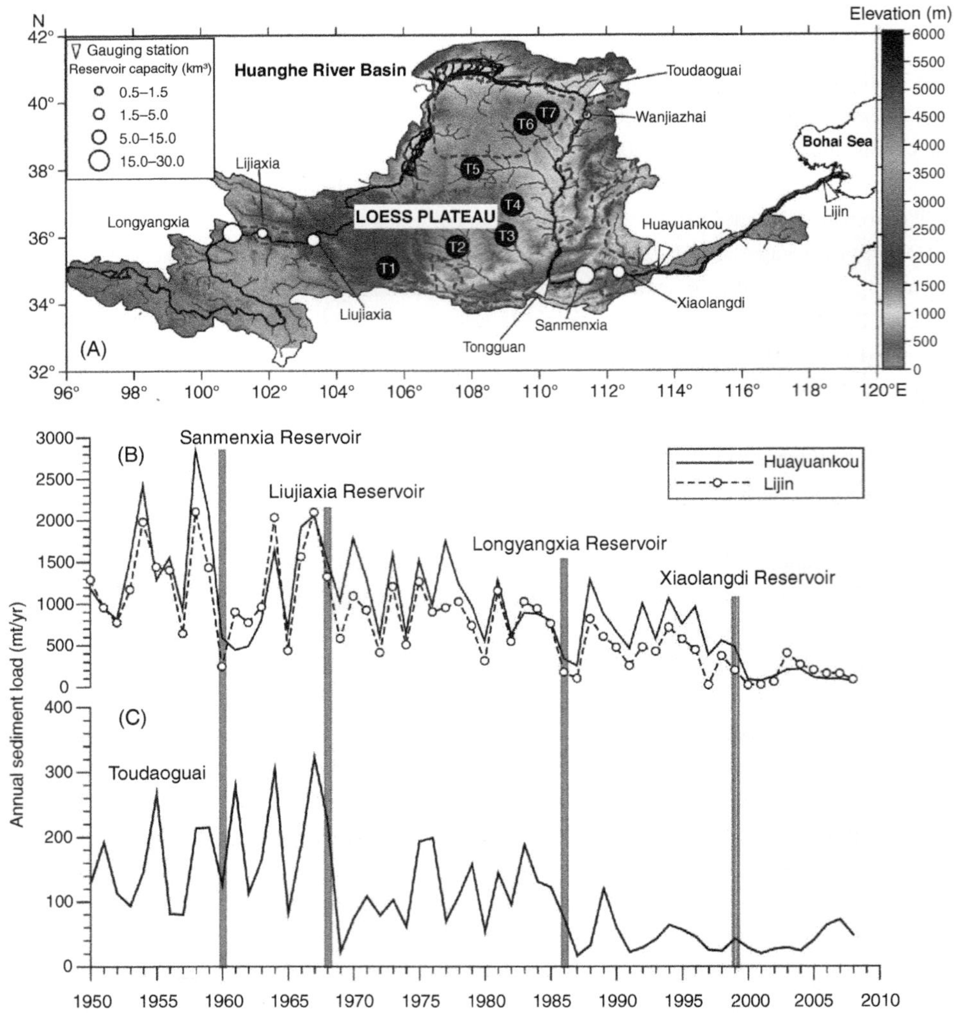

Figure 5.6. The Huanghe river basin: (A) major reservoirs and gauging stations on the mainstream, and time-series data (1950–2008) of annual sediment load recorded at stations of Huayuankou, Lijin (B) and at station Toudaoguai (C) (from Wang et al. 2011). The area surrounded by dashed line in panel (A) indicates the loess region (data from Durr et al. 2005). Major tributaries in panel (A) are: T1 – Weihe River, T2 – Jinghe River, T3 – Beiluohe River, T4 – Yanhe River, T5 – Wudinghe River, T6 – Kuyehe River, and T7 – Huangfuchuan River.

sediment flux to the ocean have been closely related to tectonics and climate change and their effects on river runoff and surface erosion. However, intensive human interventions in the river basin and complex climate change strongly interact to greatly modify river sediment flux to the ocean.

Based on the rating-curve method, Wang et al. (2008) reconstructed the time series data of annual sediment flux from the Changjiang to the sea. The comparisons of reconstructed datasets with climatic

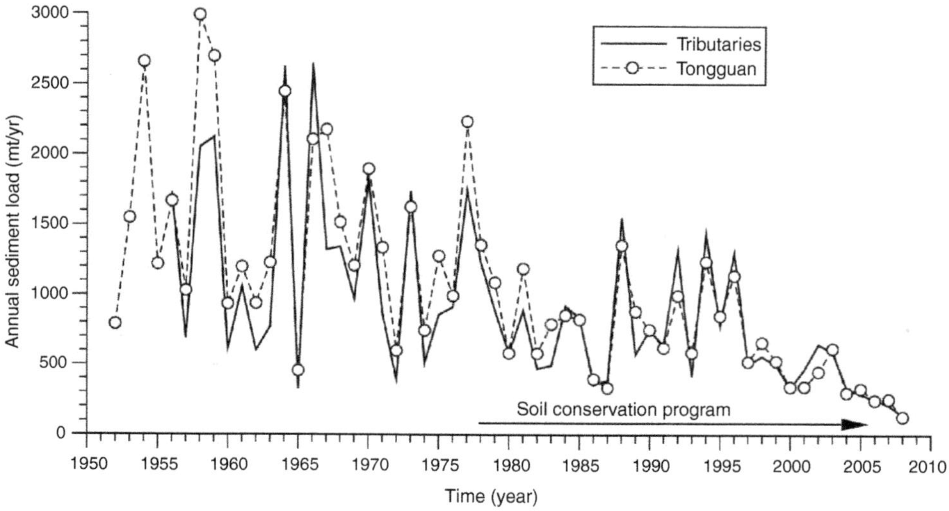

Figure 5.7. Time-series data (1950s–2008) of sediment yield from seven major tributaries (locations shown in Fig. 5.6A) in the loess region and sediment load at station Tongguan (from Wang et al. 2011).

index indicate that the Indian monsoon influences sediment yield in the upper reaches, from where most of the sediment load of the Changjiang is derived, and is therefore of primary importance to the sediment flux from the Changjiang to the sea (Fig. 5.8). It seems that during the period of little human disturbance (1865–1950s), the Changjiang sediment flux to the sea was dominantly influenced by the Indian monsoon, whereas the influence of the East Asian summer monsoon is unclear. Rainfall during the Indian monsoon wet season, which runs from June to September, accounts for about 70%

Table 5.2. *Linkage of the warm/cold phase to the water discharge from the Changjiang River and the precipitation in the Huanghe basin (refer to Wang et al. 2011)*

		Changjiang		Huanghe	
Year	Warm/cold phase	Water discharge (km^3)	Variability (%)	Precipitation (mm)	Variability (%)
1951	Warm	833.9	−6.9	444	0.0
1954	Cold	1,359.0	51.7	508	14.5
1964	Cold	1,038.5	15.9	572	28.8
1965	Warm	876.0	−2.2	356	−19.8
1972	Warm	697.7	−22.1	385	−13.3
1986	Warm	715.2	−20.2	353	−20.6
1987	Warm	837.4	−6.5	418	−5.9
1997	Warm	849.9	−5.1	350	−21.0
1998	Cold	1,244.0	38.8	459	3.5
2006	Warm	688.6	−23.1	407	−8.3

Note: The referred level of annual precipitation in the Huanghe river basin is used as the average of 1950–2008: 444 mm/yr.

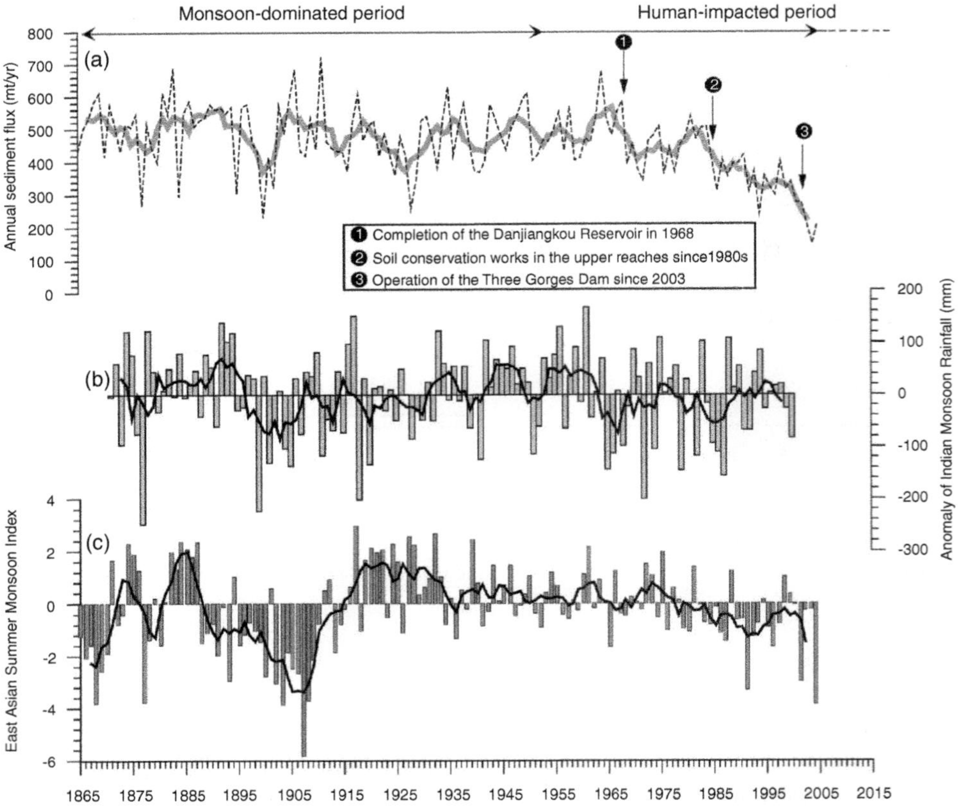

Figure 5.8. (a) Reconstructed annual sediment flux from the Changjiang to the sea (1865–2005); (b) Anomalies of Indian Monsoon Rainfall since 1870 (data available at http://www.iges.org/india/allindia.html); (c) Index of East Asian Summer Monsoon (from Wang et al. 2008). Solid lines in (a), (b), and (c) are 5-year running averages.

of annual rainfall in the upper reaches of the Changjiang. Links between monsoon-related events (e.g., rainfall over East Asia) weakened between 1890 and 1930 but strengthened during the period from 1930 to 1970 (Kripalani and Kulkarni 2001). A strong inverse relationship between Indian monsoon rainfalls and El Niño events has been recognized in several studies (Kumar et al. 1999; Krishnamurthy and Goswami 2000; Sarkar et al. 2004) and implies that before the period of significant human disturbances, annual sediment flux from the Changjiang to the sea was closely associated with climate change (Fig. 5.8). Therefore, the monsoons were the dominant influence on sediment flux to the sea during the period with little human interventions.

Since the 1950s, human activities have become the dominant influence on sediment flux to the sea, leading to stepwise decreases that correspond to major hydrologic events in the recent history of the river basin (Fig. 5.8; Yang et al. 2006). Thus the dominant influence on sediment flux to the sea has shifted from that of monsoons from the 1860s to the1950s, to human activities from the 1950s to the present. Human activities have been of primary importance to the sediment flux from the Changjiang to the sea since the 1950s, particularly since the 1980s. At present, the mechanisms by which the monsoons affect sediment flux to the sea are far from being totally understood. Nevertheless, it seems to be clear that since 1865 monsoons were initially the dominant influence on sediment flux from the Changjiang to the sea and that there was a transition to a period dominated by humans, but the boundary is not clearly identified. Although future global warming is expected to increase the frequency of heavy rainfall and severe drought (Intergovernmental Panel on Climate Change, 2007), the effects of human activities in the Changjiang drainage basin will be more direct and serious for the sediment load of the Changjiang than the effects of global warming.

Wang et al. (2006) compared the ENSO events to the regional precipitation within the Huanghe river basin from 1950 to 2000, the result of which indicated that the ENSO events corresponded to low precipitation in the river basin (Fig. 5.9). Since the mid-1980s, global ENSO events have become stronger and more frequent than ever, and most of the low-level precipitation years in the Huanghe River basin were closely associated with moderate and strong ENSO events (Fig. 5.9). Before the 1970s, the precipitation in the Huanghe River basin had strong interannual variability (>216 mm); however, interannual variation became mild as the strong ENSO events occurred frequently in subsequent years and decreasing trend became evident. The average annual precipitation in the river basin from 1985 to 2000 was only 427.8 mm, about 10% less than that before the 1960s. These correlations between ENSO events and lower regional precipitation indicated that regional precipitation was strongly affected by the global climate system. As a result of the changing pattern of basin-wide precipitation owing to the climate change, both the water discharge and sediment load from the Huanghe to the sea decreased steadily (Fig. 5.9c); meanwhile, human activities, including the soil conservation program in the loess region and operation of large reservoirs on the mainstream, presented important contributions to the decreases of sediment load to the sea (Figs. 5.6 and 5.7).

5. The future of the mega-deltas of Changjiang and Huanghe

Many large deltas in the world are now shrinking because of the increasing coastal erosion and relative sea-level rise induced by the reduction of sediment delivery to the deltas and natural or accelerated subsidence in deltas (e.g., Syvitski 2008; Syvitski et al. 2009). The decrease in sediment flux from both the Changjiang and Huanghe has triggered notable environmental changes in the coastal ocean (e.g., Yang et al. 2010; Wang et al. 2010a). The impacts of decrease in sediment flux are diverse, including morphological changes to deltas and ecological changes due to the involvement of fine-grained sediment in the coastal biogeochemical processes. The mega-deltas of the Changjiang and Huanghe are at high risk because of reductions in the sediment flux from the rivers, which slow down the aggradation rates or increase coastal erosion, thereby contributing to relative sea-level rise and increased risk of coastal flooding (e.g., Syvitski et al. 2009).

The Changjiang delta has prograded more than 200 km into the East China Sea during the last 6,000 years (Saito et al. 2001); however, during the past 60 years, human intervention rather than monsoonal climate has become a primary driving force to the decreasing sediment flux to the ocean. As a consequence, the subaqueous Yangtze delta has recently suffered erosion and coastal wetlands in the intertidal zone have declined (Yang et al. 2006, 2010; Yang and Liu 2007; Fig. 5.10).

In the past 7 kyr, the Huanghe River has discharged more than $1{,}500\times10^9$ tons of sediment to the sea (Liu and Wang 2013, Chapter 6 in this book). Besides the formation of mega-deltas in the southern Bohai Sea and north of the South Yellow Sea, an omega-shaped mud deposit along the coast of

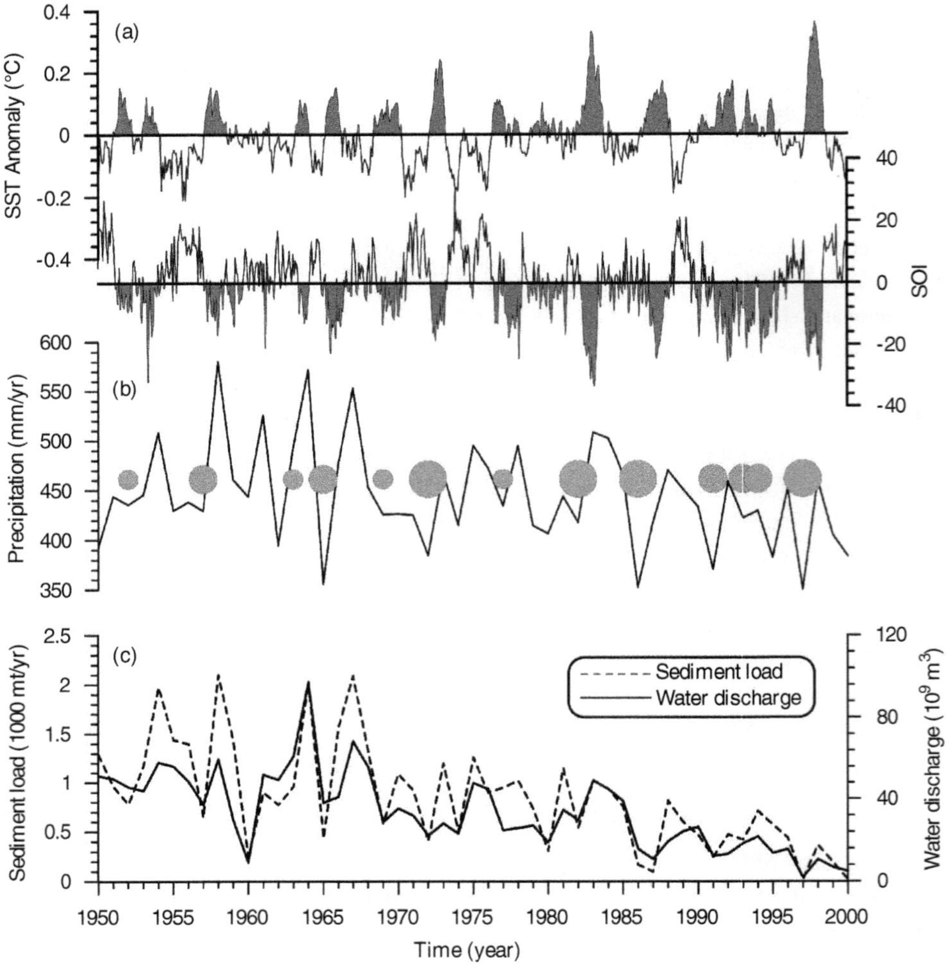

Figure 5.9. Time series of monthly SST anomalies and SOI (South Oscillation Index) (a), annual basin-wide precipitation (b) in the Huanghe drainage basin, sediment load and water discharge to the sea (c) from 1950 to 2000 (from Wang et al. 2006). The bubbles indicate the years of ENSO events, and their sizes correspond to the magnitudes of ENSO events.

Shandong Peninsula has been created as a large portion of the Huanghe sediment has been transported to and accumulated in this region (Liu et al. 2004; Liu and Wang 2013; Chapter 6 in this book). As a result of climate change and human interventions, the sediment flux from the Huanghe to the ocean has been reduced to 150 mt/yr during the last 10 years, approximately 10% of the past levels in 1950s and 1960s (Wang et al. 2007). Channel erosion induced by the Water-Sediment Regulation Scheme has led to coarser sediment outflow to the sea (Wang et al. 2010a). Given the decreasing concentration of suspended sediment (lower than the threshold concentration of hyperpycnal plume) and effect of coarsening sediment particles at the river mouth, the sedimentary regime at the Huanghe River mouth

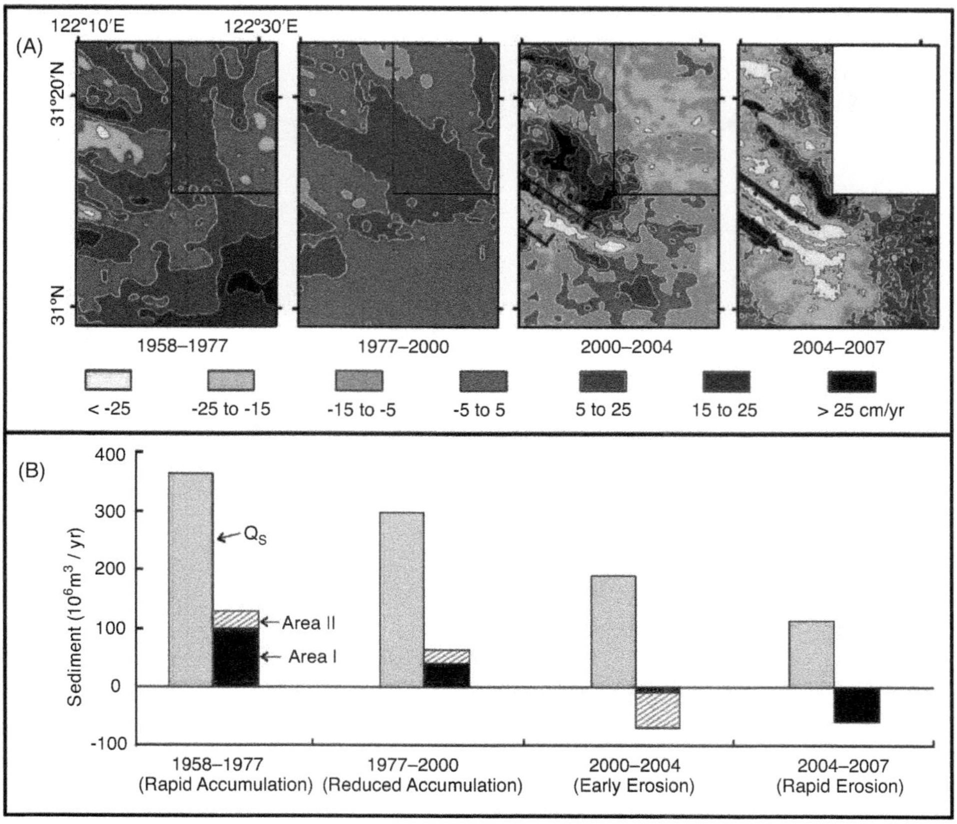

Figure 5.10. (A) Accumulation/erosion in sensitive areas of the Changjiang Delta, 1958–1977, 1977–2000, 2000–2004, 2004–2007; (B) comparison of sediment load at Datong and sediment accumulation/erosion in study areas (modified after Yang et al. 2011).

has been greatly modified as sediment is now dispersed by surface plumes rather than by hyperpycnal plumes as in the past (e.g., Wright et al. 1988, 1990; Wang et al. 2010a, 2010b, 2011). Delta erosion owing to the reduced sediment supply and continued delta subsidence has increased the flood risk along the delta coast (Syvitski et al. 2009). A large portion of the Huanghe delta (~3,400 km^2, see Syvitski et al. 2009) newly formed during the past 100 years is lower than 2 m in elevation compared with the datum of mean sea level (Fig. 5.11) owing to significant compaction, which makes the Huanghe delta much more vulnerable to episodic events such as storm surge and long-term rise of sea level.

6. Conclusion

The Changjiang and Huanghe river systems and their deltas are of great importance to the developments of the ancient Chinese civilization and modern China in the future. On a millennial scale, human interventions in the river basins of the Changjiang and Huanghe have derived high sediment flux to the ocean as a result of soil erosion enhanced by agricultural practices and deforestation; the

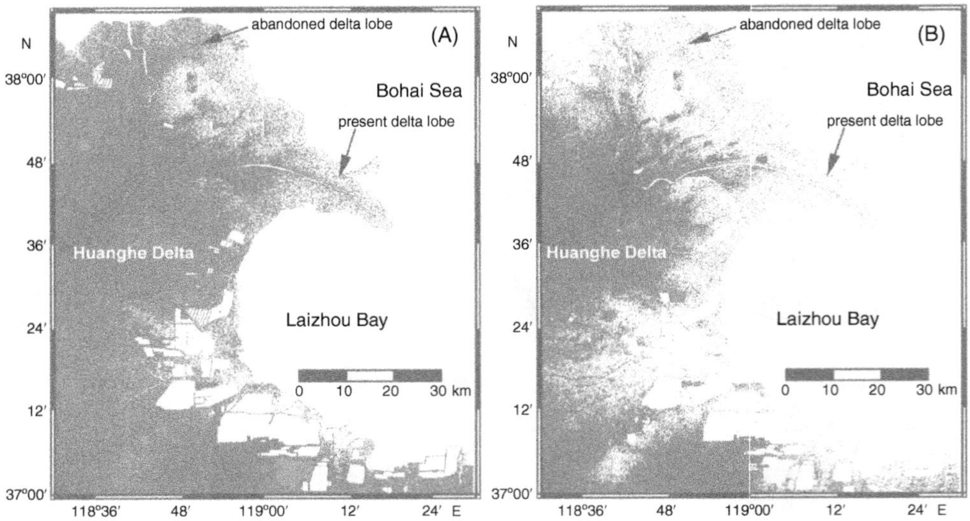

Figure 5.11. Topography of the Huanghe delta (A) and a scenario with 2-m rise of sea level (B). Topographic data are from Shuttle Radar Topography Mission, available at http://dds.cr.usgs.gov/srtm/.

large amount of river sediment (~1,500 mt/yr) has built mega-deltas along the coast of the Bohai Sea, Yellow Sea, and East China Sea (Saito et al. 2001). However, human interventions are decreasing the sediment flux to the regional ocean on the recent centennial scale. New land formation on delta plains has come to a standstill, and delta coastline erosion has commonly replaced progradation. Both the increase of sediment flux from 2000 B.P. and the accelerating decrease from the 1950s are mostly driven by human activities through changing the land use in the river basin that increases or decreases the sediment yield. In particular, hydraulic engineering such as dams and reservoirs since the 1950s with the purposes of flood control and agricultural irrigation also intercepted the sediment derived from the river basins. The prograded mega-deltas fed by high sediment input during the past several thousands of years provided living space and allowed rapid social and economic development for regional community. Now, human activities put these mega-deltas at risk of destruction, and challenges from this regional environmental change continue to grow more severe. The Changjiang and Huanghe delta have been regarded as representatives of deltas in greater peril.

References

Chen, J., F. Wang, M. Meybeck, D. He, X. Xia, and L. Zhang. 2005. Spatial and temporal analysis of water chemistry records (1958–2000) in the Huanghe (Yellow River) basin. *Global Biogeochem. Cycles* 19: GB3016

Chen, Z. Y., J. Li, H. She, and Z. Wang. 2001. Yangtze River of China: historical analysis of discharge variability and sediment flux. *Geomorphology* 41: 77–91.

Clift, P., K. Hodges, D. Heslop, R. Hannigan, H. Long, and G. Calves. 2008. Correlation of Himalayan exhumation rates and Asian monsoon intensity. *Nat. Geosci.* 1: 875–880.

Coleman M., and O.K. Huh. 2004. Major deltas of the world: a perspective from Space. Coastal Studies Institute, Louisiana State University, Baton Rouge, LA. Available at http://www.geol.lsu.edu/WDD/PUBLICATIONS/C&Hnasa04/C&Hfinal04.htm.

Durr, H.H., M. Meybeck, and S.H. Durr. 2005. Lithologic composition of the Earth's continental surfaces derived from a new digital map emphasizing riverine material transfer. *Global Biogeochem. Cycles* 19: GB4S10.

Ge, J., S. Cao, and S. Wu. 1993. Migrations in Historic China. Fuzhou: Fujian People's Publishing House (in Chinese).

Hori, K., Y. Saito, Q.H. Zhao, X.R. Cheng, P.X. Wang, Y. Sato, et al. 2001. Sedimentary facies of the tide-dominated paleo-Changjiang (Yangtze) estuary during the last transgression. *Mar. Geol.* 177: 331–351.

Huang, M.B., and L. Zhang. 2004. Hydrological responses to conservation practices in a catchment of the Loess Plateau, China. *Hydrol. Process.* 18: 1885–1898.

Intergovernmental Panel on Climate Change. 2007. Climate Change 2007: The Physical Science Basis. Contribution of Working Group I to the Fourth Assessment Report of Intergovernmental Panel on Climate Change. Solomon, S., Qin, D., Manning, M., Marquis, M., Averyt, K., Tignor, M. M.B., Miller, H.L., Jr., Chen Z. (eds). New York: Cambridge University Press, pp. 7–8.

Kripalani, R.H., and A. Kulkarni. 2001. Monsoon rainfall variations and teleconnections over South and East Asia. *Int. J. Climatol.* 21: 603–616.

Krishnamurthy, V., and B.N. Goswami. 2000. Indian monsoon–ENSO relationship on interdecadal timescale. *J. Climate* 13: 579–595.

Kumar, K.K., B. Rajagopalan, and A.M. Cane. 1999. On the weakening relationship between the Indian monsoon and ENSO. *Science* 284: 2156–2159.

Li, K., and Z. Xu. 2006. Overview of Dujiangyan Irrigation Scheme of ancient China with current theory. Irrigation and Drainage 55: 291–298.

Liu, J.P., J.D. Milliman, and S. Gao. 2002. The Shandong mud wedge and post-glacial sediment accumulation in the Yellow Sea. *Geo-Mar. Lett.* 21: 212–218.

Liu, J.P., J.D. Milliman, S. Gao, and P. Cheng. 2004. Holocene development of the Yellow River's subaqueous delta, North Yellow Sea. *Mar. Geol.* 209: 45–67.

Liu J.P., and Wang H. 2013. Historical fluxes and fates of the Yellow River derived sediments into the Sea, In: Bianchi T.S., Allison M.A., and Cai, W. (eds), Biogeochemical Dynamics at Large River-Coastal Interfaces: Linkage with Global Climate Change. New York: Cambridge University Press.

Milliman, J.D., and R.H. Meade. 1983. World-wide delivery of river sediment to the oceans. *J. Geol.* 91: 1–21.

Milliman, J.D., Y.S. Qin, M.-E. Ren, and Y. Saito. 1987. Man's influence on the erosion and transport of sediment by Asian rivers: The Yellow River (Huanghe) example. *J. Geol.* 95: 751–762.

Milliman, J.D., and J.P.M. Syvitski. 1992. Geomorphic/tectonic control of sediment discharge to the ocean: the importance of small mountainous rivers. *J. Geol.* 100: 525–544.

Mou, J. 1996. Recent studies of the role of soil conservation in reducing erosion and sediment yield in the loess plateau are of the Yellow River basin, In: Walling, D.E.,Webb, B.W. (eds.), Erosion and Sediment Yield: Global and Regional Perspectives. Proceedings of the Exeter Symposium, July 1996, IAHS Publication, vol. 236. Wallingford, UK: IAHS Press, pp. 541–548.

Qu, G., and J. Li. 1992. Population and Environment of China. Beijing: China Environmental Science Press (in Chinese).

Ren, G. 2000. Decline of the mid- to late Holocene forests in China: Climatic change or human impact? *J. Quatern. Sci.* 15: 273–281.

Ren, G., and L. Zhang. 1998. A preliminary mapped summary of Holocene pollen data for northeast China. *Quatern. Sci. Rev.* 17: 669–688.

Ren, M.E., and T.L. Shi. 1986. Sediment discharge of the Yellow River (China) and its effect on the sedimentation of the Bohai and the Yellow Sea. *Cont. Shelf Res.* 6: 785–810.

Ren, M.E., and X.M. Zhu. 1994. Anthropogenic influences on changes in the sediment load of the Yellow River, China, during the Holocene. *Holocene* 4: 314–320.

Sarkar, S., R.P. Singh, and M. Kafatos. 2004. Further evidences for the weakening relationship of Indian rainfall and ENSO over India. *Geophys. Res. Lett.* 31: L13209.

Saito, Y., Z.S. Yang, and K. Hori. 2001. The Huanghe (Yellow River) and Changjiang (Yangtze River) deltas: A review on their characteristics, evolution and sediment discharge during the Holocene. *Geomorphology* 41: 219–231.

Syvitski, J.P.M. 2008. Deltas at risk. *Sustain. Sci.* 3: 23–32.

Syvitski, J.P.M., and A.J. Kettner. 2011. Sediment flux and the Anthropocene. *Phil. Transact. R. Soc.* 369: 957–975.

Syvitski, J.P.M., A.J. Kettner, I. Overeem, E.W.H. Hutton, M. Hannon, R. Brakenridge, J. Day, C. Vörösmarty, Y. Saito, L. Giosan, and R.J. Nicholls. 2009. Sinking deltas due to human activities. *Nat. Geosci.* 2: 681–686.

Syvitski, J.P.M., C.J. Vörösmarty, A.J. Kettner, and P. Green. 2005. Impact of humans on the flux of terrestrial sediment to the global coastal ocean. *Science* 308: 376–380.

Tapponnier P., Z. Xu, R. Françoise, M. Bertrand, A. Nicolas, W. Gérard, J.S. Yang. 2001. Oblique stepwise rise and growth of the Tibet Plateau. *Science* 294: 1671–1677.

Tong, Z.J., and Q.W. Han. 1982. River Channel Morphology and Evolution Below Danjiangkou Dam, Hanjiang River. Hydrological Bureau of Changjiang River Planning Office, Wuhan, China, pp. 23–36 (Interior report in Chinese).

Wang, H., N. Bi, Y. Saito, Y. Wang, X. Sun, and J. Zhang. 2010a. Recent changes in sediment delivery by the Huanghe (Yellow River) to the sea: causes and environmental implications in its estuary. *J. Hydrol.* 39: 302–313.

Wang, H., N. Bi, Y. Wang, Y. Saito, and Z.S. Yang. 2010b. Tide-modulated Hyperpycnal flows off the Huanghe (Yellow River) mouth, China. *Earth Surf. Process. Landforms* 35: 1315–1329.

Wang, H., Y. Saito, Y. Zhang, N. Bi, X. Sun, and Z. Yang. 2011. Recent changes of sediment flux to the western Pacific Ocean from major rivers in East and Southeast Asia. *Earth-Sci. Rev.* 108: 80–100.

Wang, H., Z.S. Yang, Y. Saito, J.P. Liu, and X. Sun. 2006. Interannual and seasonal variation of the Huanghe (Yellow River) water discharge over the past 50 years: Connections to impacts from ENSO events and dams. *Global Planet. Change* 50: 212–225.

Wang, H., Z.S. Yang, Y. Saito, J.P. Liu, X. Sun, and Y. Wang. 2007. Stepwise decreases of the Huanghe (Yellow River) sediment load (1950–2005): impacts of climate changes and human activities. *Global Planet. Change* 57: 331–354.

Wang, H., Z.S. Yang, Y. Wang, Y. Saito, and J.P. Liu. 2008. Reconstruction of sediment flux from the Chanjiang (Yangtze River) to the sea since the 1860. *J. Hydrol.* 349: 318–332.

Wang Y., H. Wang, N. Bi, and Z. Yang. 2011. Numerical modeling of hyperpycnal flow in an idealized river mouth. *Estuar. Coast. Shelf Sci.* 93: 228–238.

Wang, Z., M. Li, R. Zhang, C. Zhuang, Y. Liu, Y. Saito, J. Xie, and B. Zhao. 2010a. Impacts of human activity on the late-Holocene development of the subaqueous Yangtze delta, China, as shown by magnetic properties and sediment accumulation rates. *Holocene* 21: 1–15.

Wang, Z., H. Xu, Q. Zhan, Y. Saito, Z. He, and J. Xie. 2010b. Lithological and palynological evidence of late Quaternary depositional environments in the subaqueous Yangtze delta, China. *Quatern. Res.* 73: 550–562.

World Population. 2009. Population Reference Bureau. Available at http://www.prb.org/pdf09/09wpds_eng.pdf.

World Resources Institute. 2005. Watersheds of the World 2005. Available at http://earthtrends.wri.org/.

Wright, L.D., W.J. Wiseman, B.D. Bornhold, D.B. Prior, J.N. Suhayda, G.H. Keller, Z.-S. Yang, and Y.B. Fan. 1988. Marine dispersal and deposition of Yellow River silts by gravity-driven underflows. *Nature* 332: 629–632.

Wright, L.D., W.J. Wiseman, Z.S. Yang, B.D. Bornhold, G.H. Keller, D.B. Prior, and J.N. Suhayda. 1990. Processes of marine dispersal and deposition of suspended silts off the modern mouth of the Huanghe (Yellow River). *Cont. Shelf Res.* 10: 1–40.

Xu, J. 2003. Sediment flux into the sea as influenced by the changing human activities and precipitation: example of the Huanghe River, *China. Acta Oceanolog. Sinica* 25: 125–135.

Xue, C. 1993. Historical changes in the Yellow River delta, China. *Mar. Geol.* 113: 321–329.

Yang, S.L., J.D. Milliman, P. Li, and K. Xu. 2011. 50,000 dams later: Erosion of the Yangtze River and its delta. *Global Planet. Change* 75: 14–20.

Yang, Z.S., and J.P. Liu. 2007. A unique Yellow River–derived distal subaqueous delta in the Yellow Sea. *Mar. Geol.* 240: 169–176.

Yang, Z.S., H. Wang, Y. Saito, J.D. Milliman, K. Xu, S. Qiao, and S. Shi. 2006. Dam impacts on the Changjiang (Yangtze) River sediment discharge to the sea: The past 55 years and after the Three Gorges Dam. *Water Resour. Res.* 42: W04407.

Yi, S., Y. Saito, H. Oshima, Y. Zhou, and H. Wei. 2003. Holocene environmental history inferred from pollen assemblages in the Huanghe (Yellow River) delta, China: climatic change and human impact. *Quatern. Sci. Rev.* 22: 609–628.

Zhang, X., and A. Wen. 2004. Current changes of sediment yields in the upper Yangtze River and its two biggest tributaries, China. *Global Planet. Change* 41: 221–227.

Zong, Y., Z. Chen, J.B. Innes, C. Chen, Z. Wang, and H. Wang. 2007. Fire and flood management of coastal swamp enabled first rice paddy cultivation in east China. *Nature* 449: 459–462.

6

Flux and fate of the Yellow (Huanghe) River–derived materials to the sea: impacts of climate change and human activities

P. Liu and H. Wang

1. Introduction

The Yellow River (Huanghe) is the second-longest river in China after the Yangtze and the sixth longest in the world at an estimated length of 5,464 km. It originates in Kunlun Mountains in Qinghai Province in western China, flows eastward through nine provinces, and empties into the Bohai Sea (Fig. 6.1). The Yellow River basin drains 752,000 km^2, which supports more than 100 million people, and historically has been called the "Mother River" and "the cradle of Chinese civilization."

The Yellow River's upper reach starts from its source in the Bayan Har Mountains and ends at Hekou Town of Inner Mongolia just before the river turns sharply to the south (Fig. 6.1). This upper reach portion of the river flows mostly through pastures, swamps, and knolls, resulting in a clear and steady flow of water and minor sediment contribution to the Yellow River overall. Therefore, this segment contributes less than 8% of the river's total sediment load, even though it covers 51.4% of the total basin area. The middle reach starts at Hekou Town and ends at the city of Zhengzhou in Henan Province. In the middle reach, the Yellow River passes through the Loess Plateau (Fig. 6.1), where substantial erosion takes place. The middle section covers 45.7% of the total drainage; however, it contributes 92% of the river's sediment discharge. In the lower reach, from Zhengzhou to the sea, a distance of 786 km, the river is confined within a levee-lined course as it flows northeast across the North China Plain before emptying into the Bohai Sea. This section covers only 3% of the total basin. The low gradient and resulting reduction in river velocities in the lower reach results in channel deposition of large amount of silt, elevating the river bed and creating the famous "river above ground." In the beginning of the lower reach, near Kaifeng city in Henan Province, the Yellow River bed, confined by levees, is perched 10 m above the surrounding floodplain (Wang et al. 2006, 2007; Wikipedia 2012).

The area of the Yellow River's watershed is only one-eighth of that of the Amazon River; however, the middle reach basin is dominated by the highly erodible Loess Plateau farmlands and bare (deforested) land. Based on the land cover data from ESA GlobCover in 2009, 65% of the modern Yellow River basin is cropland and 18% is bare land (Fig. 6.2). Silt input along the middle reaches lends the river a distinctive yellowish brown color, giving the Yellow River its name. However, before A.D. 200, the river was called Dahe, which means "Large River." The river was not muddy at that time (Saito and Yang, 1994). Its sediment discharge was calculated as only 10% of its last hundred-year level (Milliman et al. 1987; Liu et al. 2002). The large amount of clay, silt, and sand discharging

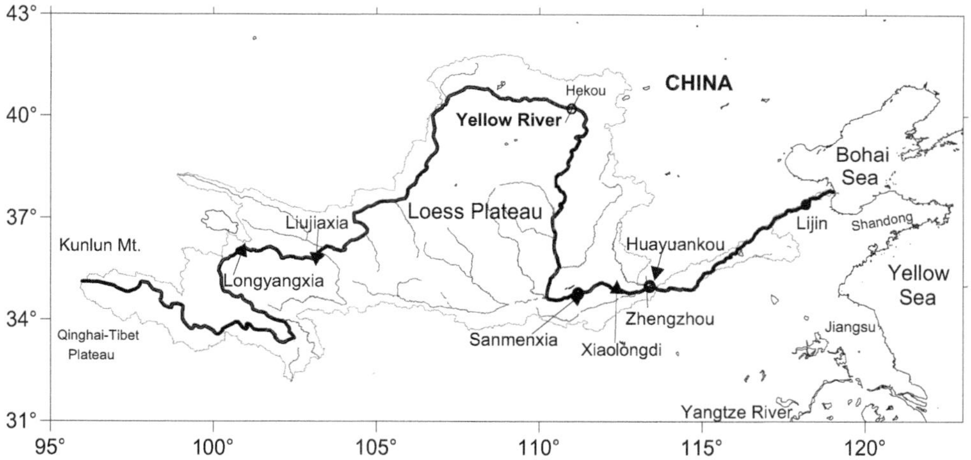

Figure 6.1. The Yellow River watershed in north China. The river originates from the Qinghai-Tibet Plateau in the west, flows through the Loess Plateau in the middle, and empties into the Bohai Sea in the east. The Hydrological Station of Lijin is about 100 km upstream from the river's mouth. Two reservoirs, Longyangxia and Liujiaxia, are located in the upper reaches; the other two, Sanmenxia and Xiaolongdi, are in the middle reaches.

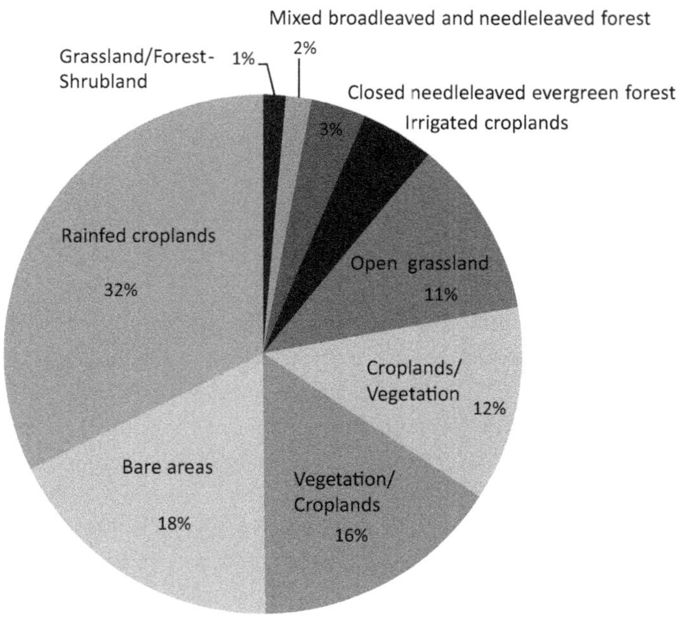

Figure 6.2. The Yellow River basin landcover distribution, which indicates the predominated croplands (65%) and bare areas (18%). Analysis was done in ArcGIS based on the downloadable data from ESA GlobCover 2009.

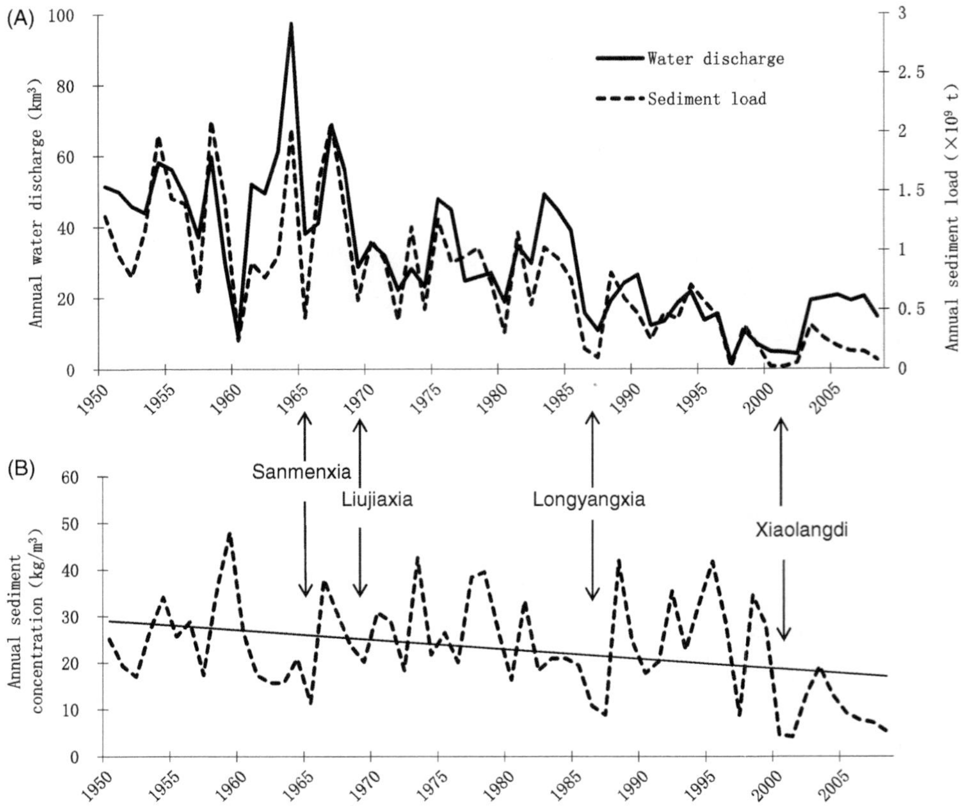

Figure 6.3. (A) Yellow River's annual water discharge and sediment load at Lijin Station from 1950 to 2008; (B) annual sediment concentration at Lijin Station, the line represents a linear trend from 1950 to 2008, showing the overall decreasing after large reservoirs being built.

into the river also makes the modern Yellow River one of the most sediment-laden and turbid rivers in the world. Although it has been decreasing in the last decade, the sediment concentration of the Yellow River averages 20–40 kg/m^3 (Fig. 6.3), compared with 2.8 kg/m^3 for Indus, 0.2 kg/m^3 for the Amazon, and 0.8 kg/m^3 for the Mississippi (Milliman and Farnsworth 2011).

Overall, the historic sediment load of the Yellow River since 1919 has been about 1.6×10^9 tons/yr (Fig. 6.3), of which perhaps at least 25% (0.4×10^9 tons/yr) is deposited in the lower river reach channel, raising the bed about 10 cm/yr (Ministry of Water Resources [MOWR], 2001). The highest recorded annual load of silt discharged into the Yellow River is 3.91×10^9 tons in 1933. The highest suspended sediment concentration was recorded in 1977's flood at 920 kg/m^3. A large portion of this sediment was deposited in the nearly flat and slow-moving lower reaches of the river. As a result of the perching of the channel above the floodplain, during the flood seasons, the river often broke through its confining levees and established a new course. In the past 2000 years, the Yellow River has had at least 26 major course changes. The river mouth on 100-year timescales shifted hundreds of

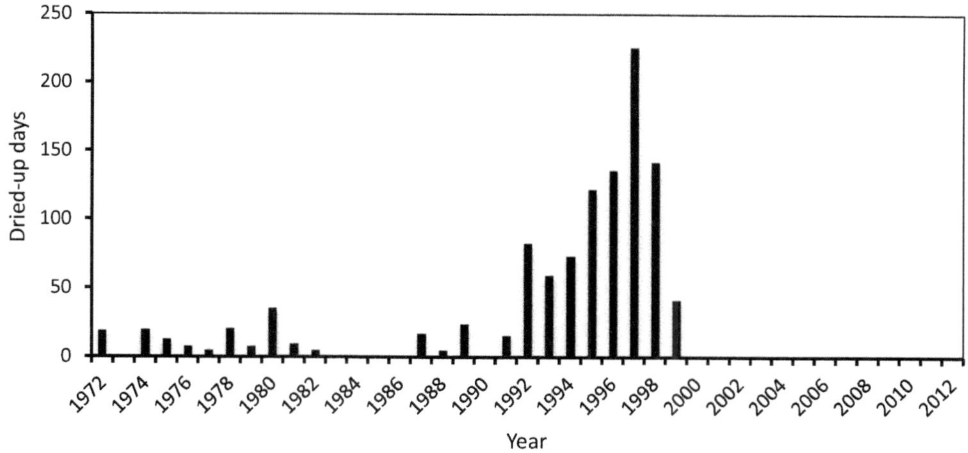

Figure 6.4. The yearly distribution of Yellow River no-flow dried-up days from 1972 to 2012. Starting from late 1999, the YRCC began to manage the river flow, mainly through adjusting the Xiaolangdi Reservior, and since then, the river has never been dried up.

kilometers between Jiangsu and Shandong provinces (Figs. 6.1, and 6.5), causing discharge to vary between the Yellow and Bohai Sea (MOWR, 2001).

Monitoring observations at the Lijin Hydrological Station, about 100 km upstream from the river mouth (Fig. 6.1), show that before the 1980s about 1.1×10^9 tons/yr of sediment was discharged into the Bohai Sea (Qian and Dai 1980). Other than the Amazon and Ganges-Brahmaputra rivers, no other rivers in the world are known to discharge more than 0.5×10^9 tons/yr to the ocean. As a result, the large sediment load of the Yellow River has created a massive flood plain in northern China, two prominent proximal deltas in the Shandong and Jiangsu coasts, and a thick distal subaqueous mud accumulation in the Yellow Sea shelf (Alexander et al. 1991; Liu et al. 2004; Yang and Liu 2007). Therefore, the Yellow River, together with the Yangtze, is thought to account for $\sim$10% of the global sediment flux to the oceans (Milliman and Meade 1983; Milliman and Syvitski 1992).

In the past 30 years, however, due to natural and anthropogenic impacts, the volume of water and sediment discharged by the Yellow River to the Bohai Sea has been steadily decreasing. For example, for several years in the early 1960s, annual water discharge was greater than 90 km^3/yr, and prior to 1970, discharge was consistently greater than 25 km^3/yr (Fig. 6.3A). In contrast, during the last 20 years, annual water discharge has averaged less than 25 km^3/yr (Wang et al. 2007, 2011). Moreover, in 1997, no water flowed into the sea for a total 226 days (Fig. 6.4) (Yang et al. 1998). This decrease in sediment delivery to the coast has caused a series of problems, including delta subsidence, flooding, relative sea-level rise, coastal erosion, and marine ecosystem deterioration (Bianchi and Allison 2009; Syvitski et al. 2009).

In this chapter, we will review the historical changes of Yellow River water and sediment discharges. Included is a discussion of the causes and impact of this variability and the fate of river-derived sediments into the sea, including its distribution, transport processes, and strata thicknesses on adjacent Bohai and Yellow Sea continental shelves.

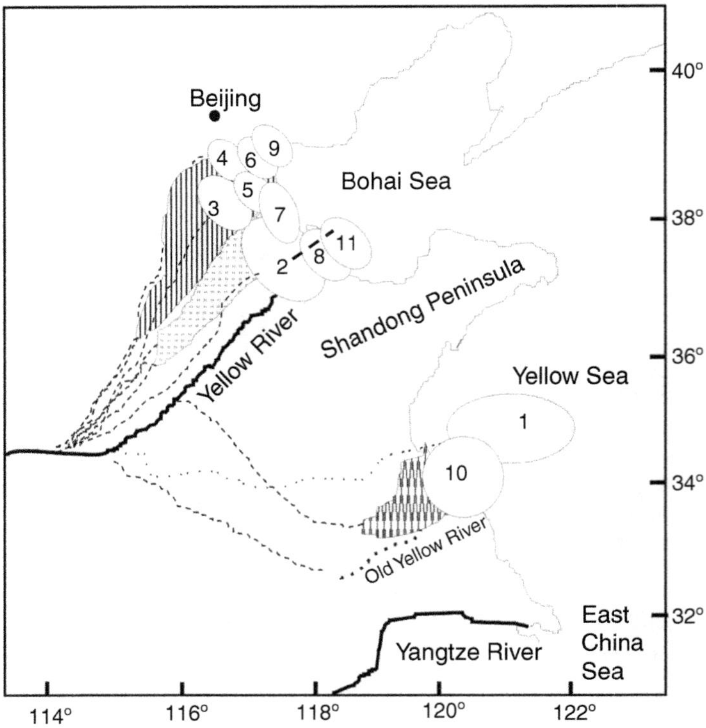

Figure 6.5. Historical courses shifts of the Yellow River lower reaches and the distribution of the super-lobes. (1) 9–7 ka B.P.; (2) 7–5 ka B.P.; (3) 5–4.5 ka B.P.; (4) 4.5–3.4 ka B.P.; (5) 3.4–3ka B.P.; (6) 3 ka B.P.–602 B.C.; (7) 602 B.C.–A.D. 11; (8) A.D. 11–A.D. 1048; (9) A.D. 1048–1128; (10) A.D. 1128–1855; (11) 1855–present. (Data from Milliman et al. 1989, Xue et al. 1993, Saito et al. 1994, and Liu et al. 2002.) The dash line is the across section profile shown in Fig. 6.6.

2. Historical changes of the Yellow River water and sediment discharge

2.1. Sediment fluxes on the millennial scale

Throughout its history, the Yellow River has experienced frequent devastating floods, levee breaching, and course changes, owing to seasonally distributed water discharge, high sediment load, and an elevated river bed (Yang et al. 2000). One of the major results of these changes is the formation of the North China Plain, the largest alluvial plain and lowland area of eastern Asia, covering 410,000 km^2. This fertile soil has undergone intense cultivation since early Chinese history and is considered the cradle of Chinese culture and civilization.

Previous studies shows that in the last 10 kyr (Holocene period), the Yellow River has changed its lower reach course more than 10 times, both regionally (flowing into the Bohai Sea or South Yellow Sea) and locally (e.g., emptying into the northwestern or southwestern Bohai Sea) (Fig. 6.5). In the beginning of the Holocene, at 9–7.5 kyr before present, the Yellow River emptied into the South Yellow Sea (Fig. 6.5, lobe 1), forming a 20-m-thick offshore delta accumulation (Yang 1985; Shi et al. 1986; Milliman et al. 1987, 1989). The estimated annual sediment load to the sea during

Table 6.1. *Holocene sediment and accumulation in the western bohai and yellow seas (see Fig. 6.5 for locations)*

	9–7.5 kyr B.P.	7.0–1.0 kyr B.P.	A.D. 1128–1855	A.D. 1855–Present	9ka–Present
Location	SYS offshore (Lobe 1)	BS nearshore (Lobes 2∼9)	SYS nearshore (Lobe 10)	BS nearshore (Lobe 11)	YS offshore (Distal mud)
Volume (km^3)	200	500	250	108	300
Mass (ton)	240×10^9	600×10^9	300×10^9	140×10^9	360×10^9
Accumulation rate (tons/yr)	0.16×10^9	0.1×10^9	0.43×10^9	0.9×10^9	0.04×10^9
Sediment discharge (tons/yr)	0.23×10^9	0.14×10^9	0.61×10^9	1.3×10^9	

Modified from Liu et al. 2002.

this time was 0.16×10^9 tons/yr (Liu et al. 2002), which agrees quite closely with inferred Yellow River sediment fluxes prior to agricultural activity on the loess plateau of northern China (Milliman et al. 1987; Saito and Yang 1994). Following a northward shift in course around 7–7.5 ka, the Yellow River continued to discharge into the Bohai Sea until A.D. 1128. During this approximately 6-kyr interval, it formed at least eight deltaic superlobes (Fig. 6.5, lobes 2–9) (Xue 1993; Saito et al. 2000), accounting for a total sediment volume/mass of 500 km^3/600×10^9 t (Table 6.1). The annual Yellow River sediment flux during this period would have been about 0.11×10^9 tons/yr (Liu et al. 2002). A section profile crossing lobes 2, 8, and 11 indicates that the initiation of the Yellow River delta in the western Bohai at ∼6–5 ka b.p. formed the first Lijian delta super lobe, about 150 km west of the modern coastline (Fig. 6.6) (Xue. 1993).

From A.D. 1128 to 1855, the Yellow River again discharged southward onto the Jiangsu coast and emptied into the South Yellow Sea (Fig. 6.5, lobe 10). In this 730-yr interval, approximately 250 km^3 of sediment was deposited along the Jiangsu coast, which equates to a mean annual sediment load of 0.43×10^9 tons/yr, reflecting enhanced erosion from increased agricultural activity in the loess hills of northern China. As a result, in the western Bohai Sea, there was a sedimentary facies gap in the delta lobe between 1128–1855 (Fig. 6.6). The last major shift of the Yellow River to the northward

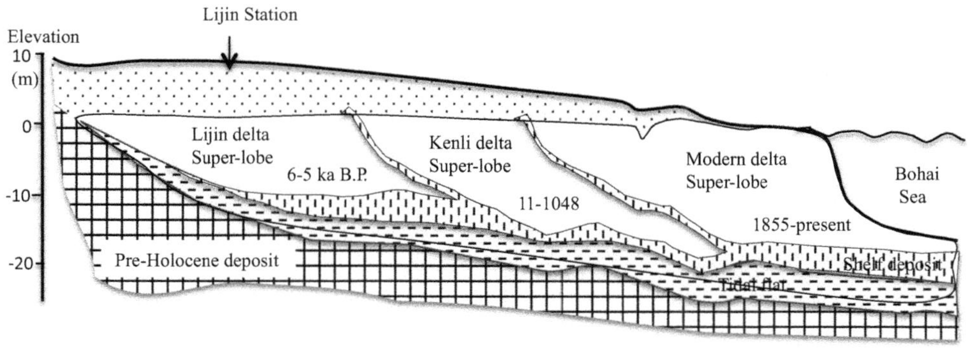

Figure 6.6. A section profile crossing the lobes 2, 8, and 11 shows the time lines of the Yellow River deltaic progradation since the middle Holocene (after Xue et al. 1993).

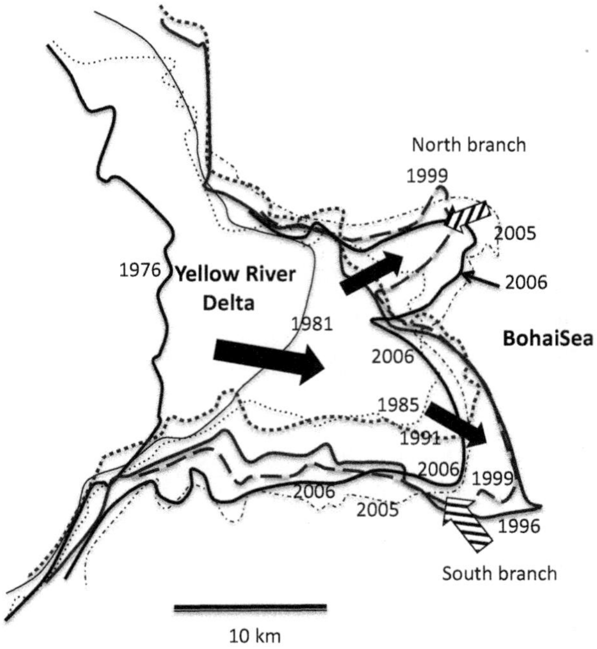

Figure 6.7. Historical coastline changes of the modern Yellow River Delta from 1976 to 2006.

pathway occurred in A.D. 1855 (Fig. 6.5, lobe 11; Fig. 6.6). Assuming an annual riverine sediment load of 0.9×10^9 tons/yr (Milliman and Meade 1983), in the years between 1855 and the present, an estimated 140×10^9 tons (or 108 km^3) (Table 6.1) of sediment passing the Lijin Hydrographic Station (Fig. 6.1) accumulated on the prograding delta and offshore. Overall, the 11 documented shifts in the Yellow River's path in the latter half of the Holocene have collectively deposited nearly 500 km^3 of sediment in the nearshore of the Yellow Sea, 300 km^3 offshore, and about 600 km^3 in the Bohai Sea (Fig. 6.5; Table 6.1).

2.2. Sediment fluxes on the decadal scale

After the last major northward shift of its course from the Jiangsu to the Shandong provinces in 1855, the Yellow River has been discharging into the western Bohai Sea. From 1855 to 1976, the Yellow River created 2058 km^2 of land, with a progradation rate of 24 km^2/yr. From 1976 to 1992, the net land area of accretion was about 364.4 km^2 with a rate of 22.8 km^2/yr (Fig. 6.7) (Li et al. 2000; Li and Chen 2003). However, from 1992 to 2000, the net land progradation was only 37 km^2, with a rate of 4.1 km^2/yr (Chang et al. 2004). Based on historical runoff and sediment load data from 1955 to 1989, and their relationship with deltaic land changes, Xu (2002) concluded that, to keep the delta and coastal zone stable in terms of land loss or gain would require a sediment discharge of about 280×10^6 tons/yr. More recent data (1976–2002), analyzed by Cui et al. (2006), indicated that to prevent the delta and coastal area from retreating, the equilibrium point involves an annual runoff of 13.5 km^3, a sediment load of 350×10^6 t, or a carrying capacity of 19.64 kg/m^3. The Yellow River (Fig. 6.3) at the Lijin Station between 2000 and 2005 had a mean water discharge that was reduced to 12.24 km^3/yr, a sediment load reduced to 160×10^6 tons/yr, and a sediment concentration that was only 10.5 kg/m^3 (Wang et al. 2007).

The Yellow River's oft-cited sediment load of 1.1×10^9 tons/yr given by Qian and Dai (1980) is based on data collected between the 1950s and 1970s, when the river's water and sediment discharge were generally high (Yang et al. 1998) (Fig. 6.3). However, since the 1950s, the Yellow River basin has experienced a series of dramatic changes by both climate shifting and human activities, such as construction of large dams and reservoirs, effective soil conservation, and regional climatic changes. The impacts of the preceding changes have caused stepwise decreases in the annual water and sediment discharges to the sea (Fig. 6.3) (Yang et al. 1998; Walling and Fang 2003; Xu 2003; Wang et al. 2006, 2007, 2010, 2011).

2.3. Causes of the recent decrease of the water and sediment discharge

One of the major causes of the sediment load stepwise decrease is the operation of a series of large reservoirs and dams. From 1960–2000, more than 300 reservoirs had been constructed over the Yellow River basin; 24 of them are considered large reservoirs with individual storage capacities over 1×10^8 m^3. Of these, at least four major reservoirs (Sanmenxia, Liujiaxia, Longyangxia, and Xiaolangdi) have been built over the main channel, two in the upper reach, and two in the middle reach (see Fig. 6.1). These dams have had a dramatic influence on water regulation (floods) and sediment retention (Wang et al. 2006, 2007, 2011). For example, after the first large dam (Sanmenxia) was built in the middle reach in the 1960s, the reservoir had already trapped more than 7.7×10^9 tons of sediment up to 1973 (Fig. 6.3). After the completion of the Liujiaxia Reservoir in 1968, the annual sediment load delivered from the upper reaches decreased from 0.17×10^9 to 0.1×10^9 tons. After the Longyangxia Reservoir was completed in 1986, the annual sediment load from the upper reach decreased again to about 0.04×10^9 tons. Since the completion of the Xiaolangdi Reservoir in 2000, the annual sediment load at Lijin Station has decreased to 0.15×10^9 tons, only ~10% of the 1950s level (Fig. 6.3) (Wang et al. 2011; Wang et al. 2013; Chapter 5).

In addition to the sediment retentions by large dams and reservoirs over the basin, the soil conservation program initiated in the late 1970s and early 1980s has caused a sharp decrease in the drainage basin sediment yield. In the 1950s, the Loess Plateau in the middle reaches discharged an estimated 1.6×10^9 tons of sediment into the river. By the 1970s, this number increased to 2.2×10^9 tons due to the deforestation and agricultural activities (Zheng et al. 1994). However, the sediment load at Lijin station decreased to 0.76×10^9 tons in early 1980s and dropped further to 0.36×10^9 tons in 2000–2008 (Fig. 6.3A) (Wang et al. 2011; Wang et al. 2013; Chapter 5).

At the same time, the decline in water discharge since 1965 (Fig. 6.3A) is most likely caused by a reduction in rainfall and increased use of the river water (Yang et al. 1998). During the past decade, rainfall in the middle and upper reaches of the Yellow River basin has dropped more than 12% compared with the 1950s (National Environment Protection Bureau 1997), but the agricultural irrigation has increased by a factor of five since 1950. In 1999, water diverted from the river served 140 million people and irrigated 74,000 km^2 of land. By 2000, more than 39 km^3 water was used in the Yellow River basin: nearly 80% of that for agriculture, about 12–13% for industry, and 7–8% for drinking water (Yellow River Conservancy Commission 2002). The increased population and rapid economic development in northern China have been a major control on increasing water consumption in the Yellow River basin: total water use increased an estimated 21% between 1990 and 2000 (Helweg 2000).

2.4. Impacts of the recent-year water discharges and sediment loads decreasing

As Figure 6.3 shows, in the past decades, both water and sediment discharge have been regularly decreasing. These decreases have caused some severe impacts on the Yellow River delta environment and coastal stability, such as coastal erosion flooding, saline intrusion, and so forth.

One evident change is the decreasing of the delta progradation rate, or coastline retreat in the delta area with severe erosion. Based on the multiyear satellite images analysis, the historical changes of the modern Yellow River delta coastlines have been reconstructed (Fig. 6.7). Clearly, the Yellow River delta had been progradating much faster between 1976 and 1985, when the annual sediment load was still between 0.5 and 1.0×10^9 tons. After 1985, the annual sediment load basically was below 0.6×10^9 tons, the rate of extending to the sea decreased. In most recent years, the south branch of the modern river delta has gradually retreated since 1996, and the north branch has sharply retreated in the year 2006.

Further analysis of the relationship of the historical Yellow River mean annual sediment load and annual increased deltaic land indicates that, when the annual sediment load is below 300×10^6 tons, there will be a negative gain of the new land (Fig. 6.8). Basically, in the last decade, with the extreme low annual sediment load ($<200\times10^6$ tons), the Yellow River delta has no new land being created; instead, some sections have been sharply eroded, particularly in the south branch.

With the gradual decreasing sediment load by passing the Lijin Station, not only has the delta growth slowed down or stopped, but also the amount of sediment discharged in to the offshore has also gradually decreased (Wang K.R. et al. 2007). Its impacts to the marine environment and sediment depositional and biogeochemical processes need to be studied in the future.

3. Fate of yellow river sediments in the Bohai and Yellow Seas

3.1. Rapid accumulations near the river mouth: proximal deltaic depocenter

Since the postglacial sea level reached its present position at approximately 7 kyr, the modern Yellow River delta has begun to form in the western Bohai Sea (Xue 1993; Saito et al. 2001). The enormous prehistorical Yellow River sediment discharge of $\sim1\times10^9$ tons has created a large subaerial delta plain, with an area of ~8,000 km^2 and up to 20 m thick (Figs. 6.5 and 6.6).

One unique feature of the Yellow River relative to other large rivers, such as the Amazon, Yangtze, and Mississippi, is its extremely high sediment concentration (often greater than 40 g/L, or kg/m^3), which makes it possible to form hyperpycnal flows where the river meets the ocean (Mulder and Syvitski 1995). During the summer flooding season, highly turbid gravity flows have been observed off the Yellow River mouth (Wright et al. 1988, 1990; Wang et al. 2010b), and most of the fluvially

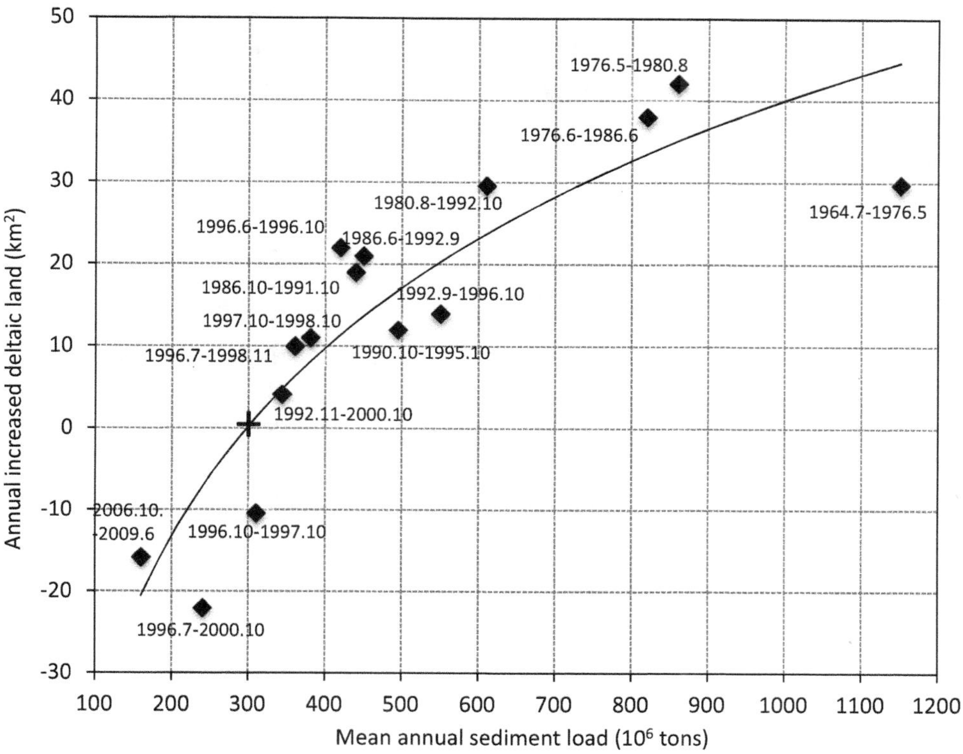

Figure 6.8. The relationship of the historical Yellow River mean annual sediment load and annual increased deltaic land (Data collected from Jiang M.X. et al. 2004; Li X.N. et al. 2001; Li Y.Z. et al. 2012, Wang K.R. et al. 2007). Basically, when the annual sediment load is below 300×10^6 tons, there will be a negative gain of the new land.

derived sediment (70%) appears to remain trapped within 15 km of the modern deltaic shoreline (Qin and Li 1986; Bornhold et al. 1986; Martin et al. 1993; Wright et al. 2001).

In the winter months, the Yellow River delivers only 15% of its annual water discharge and less than 10% of its sediment load (Milliman and Farnsworth 2011; Yang et al. 2011). However, under the prevailing winter monsoon and intensified wave conditions, the previously deposited Yellow River sediment, which mainly accumulates during the summer flood seasons, is partly resuspended and eroded, forming an extensive suspended sediment plume around the Shandong Peninsula (Fig. 6.9). Field data indicate that suspended sediment concentration in the winter of 2006 was 1.7 to 27 times higher than the 2007 summer value, and the suspended sediment flux was 2 to122 times higher in the winter than summer (Yang et al. 2011). This indicates that Yellow River sediment was resuspended and transported offshore beyond the delta, mainly in the winter season. Numerical modeling results also support the concept that a dominant portion of the remobilized sediment is carried out of the Bohai Sea into the Yellow Sea through the Bohai Strait (Jiang et al. 2000, 2004; Li et al. 2010) (Fig. 6.9). Based on shipboard and satellite observations of suspended materials, Bi et al. (2011) concluded that the annual sediment flux through the southern Bohai Strait is about 40×10^6 tons/yr,

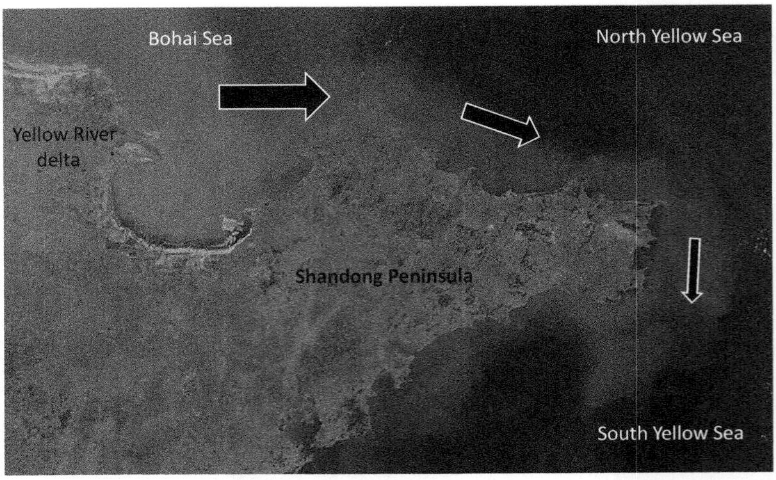

Figure 6.9. NASA Satellite image (February 9, 2004) shows in the winter the suspend matters and plume from the Yellow River in the western Bohai extending all the way into the South Yellow Sea.

80% of which was transported in the winter. This matches very well with the Liu et al. (2004) estimate of 33×10^6 tons/yr from stratigraphic records.

3.2. Longshore transport to the Yellow Sea: distal mud depocenter

In addition to the proximal depocenter on the delta, previous studies have shown that in the past 7,000 years, more than 30% of the total Yellow River sediment has been resuspended and transported out of the Bohai Sea into the Yellow Sea (Liu et al. 2002, 2004; Yang and Liu 2007). The North Yellow Sea has been assumed to be an escape pathway (Milliman et al. 1989; Alexander et al. 1991). Based on limited observations, Qin and Li (1986) indicated that approximately $6–8 \times 10^6$ tons/yr of sediment escapes from the Bohai to the Yellow Sea (Qin and Li 1986; Martin et al. 1993). Based on the seismic surveys and ^{14}C dating of the cores on the Yellow Sea shelf, Liu et al. (2004) concluded that the average sediment accumulation in the Yellow Sea shelf is 33×10^6 tons/yr.

Data from geophysical seismic surveys, coring, and suspended matters observations indicate that the modern Yellow River–derived sediment has transported out of the Baihai Sea, via the North Yellow Sea, and reached the central Yellow Sea, about 700 km from the river mouth (Yang and Liu 2007; Liu et al. 2009). This along-shelf distributed distal clinoform has been deposited since the middle Holocene sea level highstand, mainly by the resuspended Yellow River sediments carried alongshore by the coastal current and interacting with local waves, tides, and upwelling. Around the eastern side of the Shandong Peninsula, there is a 40-m thick distal mud that has accumulated on the middle shelf at a water depth of approximately −40 to −80 m (Figs. 6.10 and 6.11). ^{14}C dating and geochemical features of several deep cores have linked this deposit to the Yellow River and indicate that it has formed since the last sea-level rise and transgression in the middle Holocene (Liu et al. 2007). Over the past 7,000 years, analysis of the size distribution of sediments in this deposit in the Bohai and Yellow Sea indicates that nearly 30% of the Yellow River–derived sediment has been resuspended and transported out of the Bohai Sea into the North and South Yellow Sea. Overall,

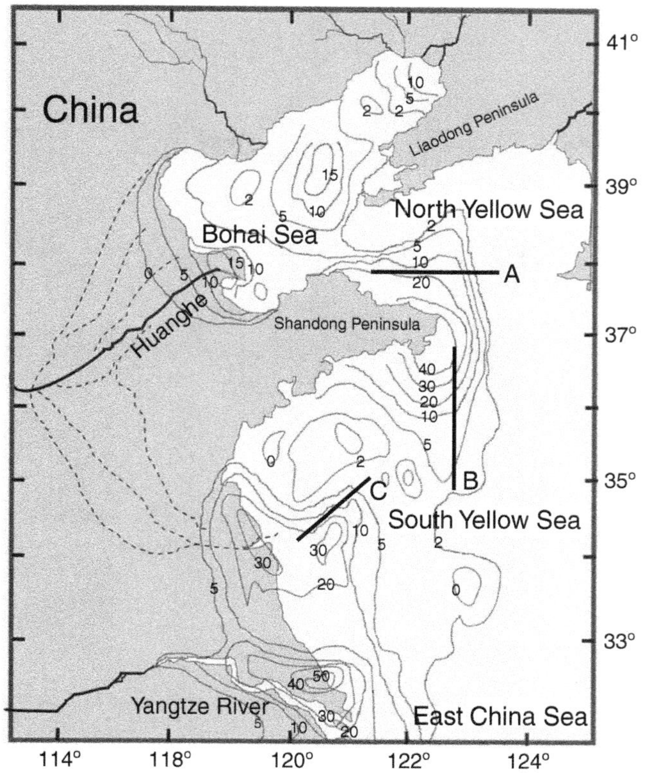

Figure 6.10. Holocene river-derived sediment distribution and thickness (isopach in meters) in the Bohai Sea and the western Yellow Sea; the three solid lines (A, B, C) indicate the locations of three selected seismic profiles in Figure 6.11.

the modern Yellow River–derived sediment could reach the −75 m water depth in the central South Yellow Sea, about 700 km from the river mouth (Figs. 6.10 and 6.12), and a very small fraction of the modern riverine sediment could escape the outer shelf to reach the Okinawa Trough (Yang and Liu 2007; Liu et al. 2009).

4. Conclusions

The Yellow River is one of the most important rivers in the world in terms of water discharge and sediment load to the sea; its basin supports more than 100 million people and vast agricultural enterprises, and its delta region also contains a very rich oil field. Understanding the historical changes of the Yellow River's sediment fluxes and fates of its sediment is critical to China both economically and politically.

In the past 7 kyr, the Yellow River has discharged more than $1{,}500 \times 10^9$ tons of sediment to the sea. In addition to forming a large subaerial and subaqueous delta (>8,000 km^2) proximal to the river mouth(s), at least 30% of the total sediment delivered during this period is estimated to have been transported alongshore out of the Bohai Sea along the Shandong peninsula, ultimately accumulating as a thick distal mud lobe on the South Yellow Sea continental shelf (Figs. 6.10 and 6.12).

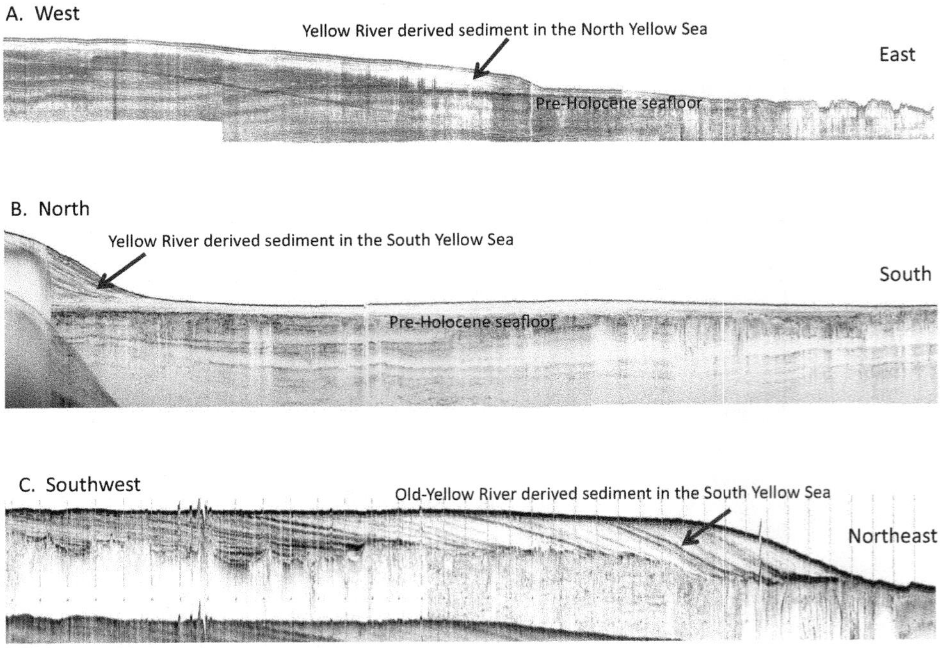

Figure 6.11. Selected seismic profiles indicate the distribution of the Yellow River–derived sediment. Seismic profiles show the Yellow River's longshore-transported distal mud accumulation progrdating from the Bohai Sea to the North Yellow Sea (A), then continue to the South Yellow Sea (B). The Chirp profile show the abandon old Yellow River deltaic deposits in the south Yellow Sea (C), most likely the one formed at A.D. 1128–1855.

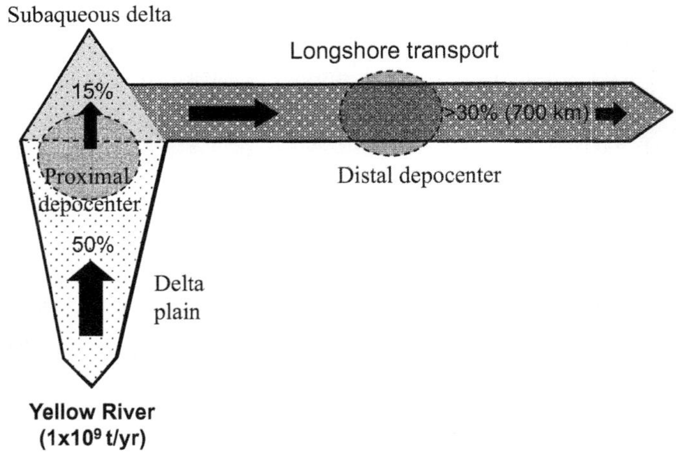

Figure 6.12. A conceptual model of the distribution and fate of the Yellow River–derived sediment to the coast and sea. Two dash circles represent the proximal depocenter near the river mouth and the remote nearshore depocenter hundreds of kilometers away from the river mouth (after Liu et al. 2009).

Over the last 60 years, global and regional climate change, along with human activities, has caused major transformation of the river system in the alluvial valley and at its mouth. The building of large dams and reservoirs, an increased demand for water from growing industry, agriculture, and domestic needs have caused a stepwise decrease in water and sediment loads reaching the sea. As a result, the Yellow River delta is experiencing a paradigm shift from rapid progradation to slow growing or severe erosion (Figs. 6.7 and 6.8). This decrease in sediment delivery to the coast has not yet translated into an observable decrease in alongshore sediment delivery to the distal depositional area, but it will not be surprising to see changes in the rate and nature of materials in the near future. Additional monitoring networks and programs will be necessary in the future to document the impact of human influences on this dynamic fluvio-deltaic system – both onshore and offshore.

References

Alexander, C.R., D.J. DeMaster, and C.A. Nittrouer. 1991. Sediment accumulation in a modern epicontinental-shelf setting: the Yellow Sea. *Mar. Geol.* 98: 51–72.

Bi, N., Z. Yang, H. Wang, D. Fan, X. Sun, and K. Lei. 2011. Seasonal variation of suspended-sediment transport through the southern Bohai Strait. *Estuar. Coast. Shelf Sci.* 93: 239–247.

Bornhold, B.D., Z.S. Yang, G.H. Keller, D.B. Prior, W.J. Wiseman, Q. Wang, L.D. Wright, W.D. Xu, and Z.Y. Zhuang. 1986. Sedimentary framework of the modern Huanghe (Yellow River) Delta. *Geo-Mar. Lett.* 6: 77–83.

Bianchi, T.S., and M.A. Allison. 2009. Large-river delta-front estuaries as natural "recorders" of global environmental change. *Proc. Natl. Acad. Sci.* 106: 8085–8092.

Chang, J., G.H. Liu, and Q.S. Liu. 2004. Analysis on spatio-temporal feature of coastline change in the Yellow River Estuary and its relation with runoff and sand transportation. *Geograph. Res.* 23: 339–346.

Cui, B.L., X.L. Chang, Y.L. Chen, Q. Dong, and W.Q. Li. 2006. The impact of hydrological characteristics of the Yellow River coastline changes in the Yellow River delta. *J. Natl. Res.* 21: 957–964.

Helweg, O.J. 2000. "Water for a growing population: Water supply and groundwater issues in developing countries." *Water Int.* 25: 33–39.

Jiang. M.X., F.D. Yang, R.J. Huo, and J.T. Chen. 2004. On the evolution of the Huanghe River delta and its relation to the riverway and sand entering the sea. *Transact. Oceanol. Limnol.* 3: 6–15.

Jiang, W.S., T. Pohlmann, J. Sundermann, and S.Z. Feng. 2000. A modeling study of SPM transport in the Bohai Sea. *J. Mar. Syst.* 24: 175–200.

Jiang, W.S., T. Pohlmann, J. Sun, and A. Starke. 2004. SPM transport in the Bohai Sea: field experiments and numerical modeling. *J. Mar. Syst.* 44: 175–188.

Li, F.-L., J.-Z. Pang, and M.-X. Jiang. 2000. Shoreline changes of the Yellow River Delta and its environmental geological effect. *Mar. Geol. Quatern. Geol.* 20: 17–21.

Li, F.L., and X.Q. Chen. 2003. Shoreline changes of the Yellow River Delta and its sub-delta Area Forecast. Procedures of International Conference on Estuaries and Coasts. November 9–11, 2003. Hangzhou, China, pp. 246–254.

Li, G., H. Xue, Y. Liu, H. Wang, and H. Liao. 2010. Diagnostic experiments for transport mechanisms of suspended sediment discharged from the Yellow River in the Bohai Sea. *J. Geograph. Sci.* 20: 49–63.

Li, Y.Z., J.B. Yu, G.X. Han, Y.L. Wang, and Z.D. Zhang. 2012. Coastline change detection of the Yellow River Delta by satellite remote sensing. *Mar. Sci.* 36: 99–105.

Liu, J.P., J.D. Milliman, and S. Gao. 2002. The Shandong mud wedge and post-glacial sediment accumulation in the Yellow Sea. *Geo-Mar. Lett.* 21: 212–218.

Liu, J.P., J.D. Milliman, S. Gao, and P. Cheng. 2004. Holocene development of the Yellow River's subaqueous delta, North Yellow Sea. *Mar. Geol.* 209: 45–67.

Liu, J., Y. Saito, H. Wang, Z.G. Yang, and R. Nakashima. 2007. Sedimentary evolution of the Holocene subaqueous clinoform off the Shandong Peninsula in the Yellow Sea. *Mar. Geol.* 236: 165–187.

Liu, J.P., Z. Xue, K. Ross, H. Wang, Z.S. Yang, A.C. Li, and S. Gao. 2009. Fate of sediments delivered to the sea by Asian large rivers: long-distance transport and formation of remote alongshore clinothems. *Sediment. Rec.* 7: 4–9.

Martin, J.M., J. Zhang, M.C. Shi, and Q. Zhou. 1993. Actual flux of the Huanghe (Yellow River) sediment to the western Pacific Ocean. *Neth. J. Sea Res.* 31: 243–254.

Milliman, J.D., and R.H. Meade. 1983. World-wide delivery of river sediment to the oceans. *J. Geol.* 91: 1–21.

Milliman, J.D., Y.S. Qin, M.E. Ren, and Y. Saito. 1987. Man's influence on the erosion and transport of sediment by Asian rivers: the Yellow River (Huanghe) example. *J. Geol.* 95: 751–762.

Milliman, J.D., Y.S. Qin, and Y. Park. 1989. Sediment and sedimentary processes in the Yellow and East China Seas, In: Taira, A., and Masuda, F. (eds.): Sedimentary Facies in the Active Plate Margin. Tokyo, Japan: Terra Scientific Publishing, pp. 233–249.

Milliman, J.D., and J.P.M. Syvitski. 1992. Geomorphic/tectonic control of sediment discharge to the ocean: the importance of small mountainous rivers. *J. Geol.* 100: 525–544.

Milliman, J.D., and K.L. Farnsworth. 2011. River Discharge to the Coastal Ocean: A Global Synthesis. New York: Cambridge University Press.

Ministry of Water Resources. 2001. Integrated water resources management around the Bo Hai Sea–GEF draft concept document. Ministry of Water Resources (MOWR), P. R. China.

Mulder, T., and J.P.M. Syvitski. 1995. Turbidity currents generated at river mouths during exceptional discharges to the world oceans. *J. Geol.* 103: 285–299.

National Environment Protection Bureau (China). 1997. the Yellow River runs dry and its sustainable development (in Chinese). Beijing: China Environment Science Pree, 1997.

Qian, N. and D.Z. Dai. 1980. The problems of river sedimentation and present status of its research in China, In: Proceedings of International Symposium on River Sedimentation. Shanghai, China: Guanghua Press.

Qin, Y.S., and F. Li. 1986. Study of the influence of sediment loads discharged from Huanghe river on sedimentation in the Bohai and Yellow Seas. *Oceanologia et Limnologia Sinica* 27: 125–135.

Saito, Y., and Z.S. Yang. 1995. Historical change of the Huanghe (Yellow River) and its impact on the sediment budget of the East China Sea, In: Tsunogai, S., Iseki, K., Koike, I., and Oba, T. (eds.), Global Fluxes of Carbon and Its Related Substances in the Coastal Sea-Ocean-Atmosphere System. Yokohama, Japan: M&J International, pp. 7–12.

Saito, Y., Z.S. Yang, and K. Hori. 2001. The Huanghe (Yellow River) and Changjiang (Yangtze River) deltas: a review on their characteristics, evolution and sediment discharge during the Holocene. *Geomorphology* 41: 219–231.

Saito, Y., and Z.S. Yang. 1994. The Huanghe River: its water discharge, sediment discharge, and sediment budget. *J. Sed. Soc. Japan* 40: 7–17.

Saito, Y., H. Wei, Y. Zhou, A. Nishimura, Y. Sato, and S. Yokota. 2000. Delta progradation and chenier formation in the Huanghe (Yellow River), delta China. *J. Asian Earth Sci.* 18: 489–497.

Shi, W.B., D.P. Li, X.C. Wang, and Z.X. Zhang. 1986. Shallow seismic surveying in south Huanghai Sea and its geological significance (in Chinese with English abstract). *Mar. Geol. Quatern. Geol.* 6: 87–104.

Syvitski, J.P.M., A.J. Kettner, I. Overeem, E.W.H. Hutton, M. Hannon, R. Brakenridge, J. Day, C. Vörösmarty, Y. Saito, L. Giosan, and R.J. Nicholls. 2009. Sinking deltas due to human activities. *Nat. Geosci.* 2: 681–686.
Wang, H., Z.S. Yang, Y. Saito, J.P. Liu, and X. Sun. 2006. Interannual and seasonal variation of the Huanghe (Yellow River) water discharge over the past 50 years: Connections to impacts from ENSO events and dams. *Global Planet. Change* 50: 212–225.
Wang, H., Z.S. Yang, Y. Saito, J.P. Liu, X. Sun, and Y. Wang. 2007. Stepwise decreases of the Huanghe (Yellow River) sediment load (1950–2005): impacts of climate changes and human activities. *Global Planet. Change* 57: 331–354.
Wang, H., Z.S. Yang, Y. Wang, Y. Saito, and J.P. Liu. 2008. Reconstruction of sediment flux from the Chanjiang (Yangtze River) to the sea since the 1860. *J. Hydrol.* 349: 318–332.
Wang, H., N. Bi, Y. Saito, Y. Wang, X. Sun, and J. Zhang. 2010a. Recent changes in sediment delivery by the Huanghe (Yellow River) to the sea: causes and environmental implications in its estuary. *J. Hydrol.* 39: 302–313.
Wang, H., N. Bi, Y. Wang, Y. Saito, and Z.S. Yang. 2010b. Tide–modulated Hyperpycnal flows off the Huanghe (Yellow River) Mouth, China. *Earth Surf. Process. Landforms* 35: 1315–1329.
Wang, H., Y. Saito, Y. Zhang, N. Bi, X. Sun, and Z. Yang. 2011. Recent changes of sediment flux to the western Pacific Ocean from major rivers in East and Southeast Asia. *Earth-Sci. Rev.* 108: 80–100.
Wang, H., Z. Yang, and N. Bi. 2013. Changjiang and Huanghe Rivers, In: Bianchi T.S., Allison M.A., and Cai, W. (eds.), Historical Reconstruction of Land-Use Change and Sediment Load to the Sea. New York: Cambridge University Press.
Wang K.R., Y.Y. Ru, X.T. Chen, and Z.J. Hou. 2007. Discussion on the dynamic equilibrium problem of the delta coastline of the Yellow River Estuary. *J. Sediment Res.* 6: 66–70.
Wikipedia. 2012. http://en.wikipedia.org/wiki/Yellow_River.
Walling, D.E., and D. Fang. 2003. Recent trends in the suspended sediment loads of the world's rivers. *Global Planet. Change* 39: 111–126.
Wright, L.D., W.J. Wiseman, B.D. Bornhold, D.B. Prior, J.N. Suhayda, G.H. Keller, Z.S. Yang, and Y.B. Fan. 1988. Marine dispersal and deposition of Yellow River silts by gravity-driven underflows. *Nature* 332: 629–632.
Wright, L.D., W.J. Wiseman, Z.S. Yang, B.D. Bornhold, G.H. Keller, D.B. Prior, and J.N. Suhayda. 1990. Processes of marine dispersal and deposition of suspended silts off the modern mouth of the Huanghe (Yellow River). *Cont. Shelf Res.* 10: 1–40.
Wright, L.D., C.T. Friedrichs, S.C. Kim, and M.E. Scully. 2001. Effects of ambient currents and waves on gravity-driven sediment transport on continental shelves. *Mar. Geol.* 175: 25–45.
Xu, J.X. 2002. A study of thresholds of runoff and sediment for the land accretion of the Yellow River delta. *Geograph. Res.* 21: 163–170.
Xu, J.X. 2003. Sediment flux into the sea as influenced by the changing human activities and precipitation: example of the Huanghe River, China. *Acta Oceanolog. Sinica* 25: 125–135.
Xue, C.T. 1993. Historical changes in the Yellow River delta, China. *Mar. Geol.* 113: 321–329.
Yang, Z.G. 1985. Sedimentary and environment in south Huanghai Sea since late Pleistocene (in Chinese with English abstract). *Mar. Geol. Quatern. Geol.* 5: 1–19.
Yang, Z.S., J.D. Milliman, J. Galler, J.P. Liu, and X. Sun. 1998. Yellow River's water and sediment discharge decreasing steadily. *Eos* 79: 589–592.
Yang, D.Y., G. Yu, Y.B. Xie, D.J. Zhan, and Z.J. Li. 2000. Sedimentary records of large Holocene floods from the middle reaches of the Yellow River, China. *Geomorphology* 33: 73–88.
Yang, Z.S., and J.P. Liu. 2007. A unique Yellow River–derived distal subaqueous delta in the Yellow Sea. *Mar. Geol.* 240: 169–176.

Yang, Z., Y. Ji, and N. Bi. 2011. Sediment transport off the Huanghe (Yellow River) delta and in the adjacent Bohai Sea in winter and seasonal comparison. *Estuar. Coast. Shelf Sci.* 93: 173–181.

Yellow River Conservancy Commission. 2002. *Yellow River Basin Planning*. YRCC website. March, 2002 (in Chinese). Available at: http:// www.yrcc.gov.cn/.

Zheng, F.L., K.L. Tang, and H.Y. Bai. 1994. Study on relationship between human's activities and environmental evolution. *Res. Soil Water Conserv.* 1: 36–42 (in Chinese with English abstract).

7

Carbon dioxide dynamics and fluxes in coastal waters influenced by river plumes

W.-J. Cai, C. T. Arthur Chen, and A. Borges

1. Introduction

Globally, estuaries are a source of CO_2 to the atmosphere, releasing a significant fraction of CO_2 that the terrestrial biosphere absorbed from the atmosphere, and thus significantly reduce riverine terrestrial carbon fluxes to the ocean. Although several recent synthesis papers brought the flux lower from ~0.5 to ~0.25 Pg C yr^{-1} (Table 7.1), the flux is still quite significant in consideration of global carbon budget and pathways. Most importantly, the likely uncertainty in estimating air-water CO_2 flux in coastal waters and wetlands is so large (±0.2 Pg C yr^{-1}) that it becomes a major obstacle for a precise assessment of the CO_2 flux at the air earth interface (be it terrestrial or ocean), thus making the effort for further improving the accuracy of estimating air-earth surface CO_2 exchange fluxes less meaningful for climate change research – if such uncertainty in coastal systems is not tackled (Cai 2011).

Previous works have largely focused on inner estuaries within coastal shorelines, and river plumes extending outside the shorelines largely are ignored for a number of reasons. Most importantly, river plumes are not limited geographically, and their areas are not defined clearly. They also have CO_2 values closer to saturation with respect to the atmosphere. These properties are in stark contrast to inner estuaries, which are geographically limited by river banks or shorelines and virtually in all cases have CO_2 values distinctly higher than the atmosphere. Hence river plumes are difficult to quantify and generalize in terms of an annual exchange flux of CO_2 with the atmosphere. River plumes also reside directly on the continental shelves and thus often are included as part of shelves in global syntheses. Freshwater plumes of some large rivers (e.g., Amazon, Congo, Niger) often even extend beyond the continental shelf into the continental slope and ocean basin (Bianchi and Allison 2009). Because of the lack of reliable air-water CO_2 flux estimates and the absence of global estimates of surface areas, river plumes have not been included in previous syntheses of global coastal water CO_2 fluxes (Borges 2005; Borges et al. 2005; Cai et al. 2006; Chen and Borges 2009; Laruelle et al. 2010).

This chapter first briefly reviews the state of knowledge of global (inner) estuarine CO_2 flux and the existing problems. Then, we describe the distribution features of CO_2 parameters in and around large river plumes over continental shelves to discuss factors that control CO_2 distribution in areas dominated by river plumes. We also attempt to estimate air-sea CO_2 flux in river plumes globally and provide a general view on whether inclusion of river plume in the estuarine CO_2 flux assessments would change the currently best known global estuarine CO_2 flux values.

Table 7.1. *Global CO_2 degassing fluxes from inner estuaries to the atmosphere. Here inner estuaries are defined as estuarine areas within shorelines. A positive sign indicates ocean CO_2 loss (degassing) and a negative sign indicates ocean CO_2 uptake (in-gassing)*

	Flux molC/m²/yr	Area 10^6km²	Flux PgC/yr	Reference
Estuaries	35.71	1.40	0.6	Abril and Borges 2004
Estuaries	38.12	0.94	0.43	Borges 2005
Salt marshes	23.45	0.14	0.04	
Mangroves	13.66	0.20	0.04	
Total	33.2	1.28	0.51	
Estuaries	28.62	0.94	0.32	Borges et al. 2005
Salt marshes	21.4	0.14	0.036	
Mangroves	18.66	0.15	0.033	
Total	26.42	1.23	0.39	
Estuaries	32.1	0.943	0.36	Chen and Borges 2009
Salt marshes	30.4	0.384	0.09	
Mangroves	27.1	0.147	0.05	
Total	28.27	1.474	0.50	
Small deltas and estuaries	25.7	0.084	0.026	Laruelle et al. 2010
Tidal systems and embayments	28.5	0.276	0.094	
Lagoons	17.3	0.252	0.052	
Fjords and fjärds	17.5	0.456	0.096	
Total	21.0	1.067	0.268	
Estuaries (both river-dominated and nonriverine coastal lagoons)	20.83	1.05	0.25	Cai. 2011
Estuaries	23.9 ±	1.07	0.26	Chen et al. 2012

Note: Here inner estuaries are defined as estuarine areas within shorelines. A positive sign indicates estuarine CO_2 loss (degassing) and a negative sign indicates estuarine CO_2 uptake (in-gassing).

2. CO_2 degassing flux in inner estuaries

It has long been known that inner estuaries act as sources of CO_2 to the atmosphere (Park et al. 1969) because of their net ecosystem metabolic status being heterotrophic (Odum and Hoskin 1958; Smith and Hollibaugh 1993; Hopkinson and Smith 2005; Borges and Abril 2011). However, it was not until the late 1990s that a regional synthesis of fluxes of CO_2 out of the estuaries was made available by Frankignoulle et al. (1998), who reported that CO_2 emissions from inner estuaries in Europe (average 50 mol $m^{-2}yr^{-1}$) was a significant term in the regional CO_2 budget, equivalent to 5–10% of the total anthropogenic CO_2 emissions from western Europe. The first compilation of global inner estuarine fluxes of CO_2 was that of Abril and Borges (2004) and was based on only a handful of data. The datasets (Borges 2005; Borges et al. 2005) grew gradually, and in 2009 an assessment including 32 estuaries was published (Chen and Borges 2009). This database has been further extended to 60 inner estuaries in 2010 (Laruelle et al. 2010) and 106 inner estuaries in 2012 (Chen et al. 2012).

Although the numbers of inner estuaries included in the CO_2 flux estimation have increased in recent years, all global extrapolations have relied on the surface areas of inner estuaries – roughly estimated four decades ago by Woodwell et al. (1973), with the exception of a recent publication by Laruelle et al. (2010), which based extrapolations on typology (Dürr et al. (2011). The Woodwell et al. (1973) global estimate of surface areas of estuaries was based on the extrapolation of the estuarine area to coastline length for the United States that was extrapolated to the global coastline length. Surprisingly, these two estimates are rather close: 0.94×10^6 km^2 by Woodwell et al. (1973) when intertidal areas (marshes and mangroves) are excluded and 1.10×10^6 km^2 by Dürr et al. (2011). The Dürr et al. (2011) typology is based on a geographical information system approach and also provides surface area estimates by estuarine types. Thus their approach was a major improvement for scaling of CO_2 fluxes, because it was known that carbon was transformed differently according to estuarine types. For example, recent comparisons have been made between stratified fjords and well-mixed macro-tidal estuaries (Borges 2005) and stratified and well-mixed lagoons (Koné et al. 2008). Laruelle et al. (2010) scaled up the inner estuarine CO_2 fluxes using the following four estuarine types: small deltas and estuaries (8% of total surface), tidal systems and embayments (26% of total surface), lagoons (24% of total surface), and fjords and fjärds (43% of total surface).

More recent studies, compiled in Table 7.1, appear to show a convergence of a water-to-air flux of $\sim$0.25 PgC yr^{-1} for the global inner estuaries. However, there are still large uncertainties, as many field expeditions have reported insufficient spatial and temporal coverage and were conducted under highly variable wind mixing and stratification conditions of the water column. The amount of data seems to focus mostly on the upper sections of inner estuaries, which are generally higher in pCO_2. The lower inner estuaries, generally lower in pCO_2, are relatively poorly covered, and the river plumes outside of shorelines are even less studied. Furthermore, coastal estuaries with little freshwater input and surrounded by wetlands generally have lower CO_2 than those of river-dominated systems and have largely been understudied (Borges 2005; Jiang et al. 2008; Cai 2011). There has also been recent controversy on whether the large CO_2 degassing flux in these systems is the result of decomposition of riverine terrestrial organic carbon, or is largely supported by lateral transport from adjacent productive coastal wetlands (Cai 2011). As several recent synthesis papers have discussed CO_2 degassing flux in inner estuarine areas in detail, we refer readers to the literature cited in Table 7.1 and focus next on CO_2 distributions, fluxes, and control mechanisms in river plumes in this chapter.

Frankignoulle et al. (1998) reported air-water CO_2 fluxes in several European river plumes and emphasized that river plumes were characterized by air-water CO_2 fluxes with fluxes being one to two orders of magnitude lower than inner estuarine regions. The river plumes they had studied were, however, limited to rivers of relatively low discharge and had data with little seasonal coverage. On the other end of the spectrum, several studies have provided an estimate of the CO_2 sink of the Amazon River plume and estimated the overall sink in the plume from an extrapolation based on the relationship between the partial pressure of CO_2 (pCO_2) and salinity (Ternon et al. 2000; Körtzinger 2003; Cooley et al. 2007; Lefèvre et al. 2010). Additional data have shown that the relationship between pCO_2 and salinity in the Amazon is remarkably stable on a seasonal basis (Lefèvre et al. 2010). Seasonal CO_2 distributions and air-water fluxes in inner estuaries and river plumes have also been reported for the Changjiang River (Wang et al. 2000; Chen et al. 2008; Chou et al. 2009b; Zhai and Dai 2009) and preliminarily for the Mississippi River (Cai 2003; Dagg et al. 2005; Lohrenz and

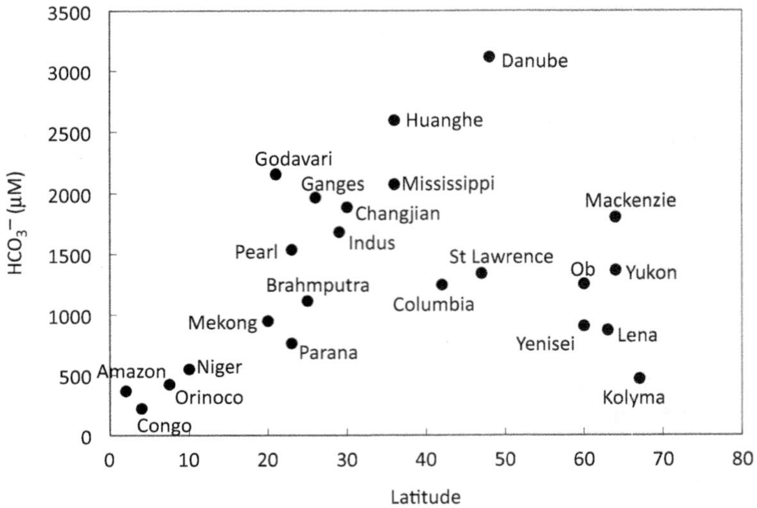

Figure 7.1. Latitudinal distribution of river bicarbonate (HCO_3^-) concentrations. Note that an average latitude value for each drainage basin was assigned to each river. Modified after Cai et al. (2008).

Cai 2006; Lohrenz et al. 2010; Huang 2013). However, the aforementioned information has not been properly incorporated in the global synthesis shown in Table 7.1; an initial attempt is made here to address this problem.

3. Distribution pattern of pCO_2 in river plumes and control mechanisms

3.1. River carbonate contents and the associated buffering capacity changes during river-ocean mixing

Several chapters of this book have discussed the physical and biogeochemical properties of larger river plumes (e.g., Chapter 3), including the Mississippi and Atchafalaya River plumes (e.g., Chapter 18), the Changjiang plume (e.g., Chapter 10), the Pearl River plume (e.g., Chapters 11 and 13), and the Mekong River plume (e.g., Chapter 12). Thus, in this chapter we avoid repeating much of the background information provided in these other chapters. Here, we focus on surface water pCO_2 distribution and air-sea CO_2 flux from a few studied plumes and discuss the relationships between physical and biogeochemical properties of the plumes.

Middle-latitude and subtropical larger rivers (e.g., the Mississippi, Changjiang, and Pearl Rivers) typically have high bicarbonate (HCO_3^-) concentrations and fluxes because of abundant distribution of carbonate minerals in their drainage basins, in contrast to tropical larger rivers with low HCO_3^- concentrations, such as the Amazon and Orinoco Rivers (Cai et al. 2008; Cai et al. 2010). Thus HCO_3^- concentrations in the world's largest rivers do show a distinct latitudinal distribution (Fig. 7.1). Regardless of the particular latitude, river total alkalinity (TAlk) values and dissolved inorganic carbon (DIC) concentrations were generally lowest during peak river discharge and highest

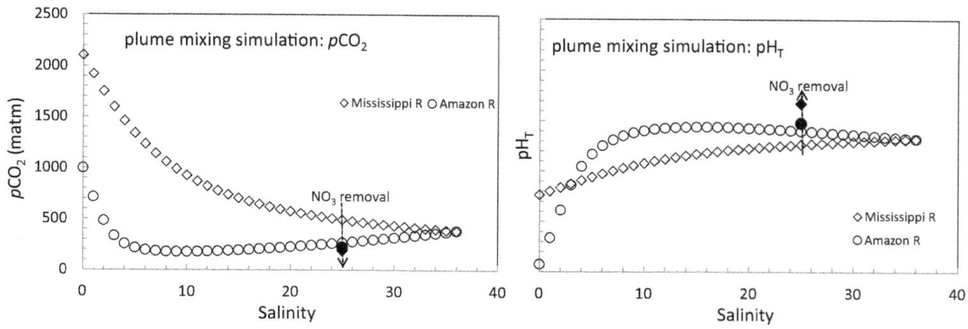

Figure 7.2. Predicted pCO_2 and pH changes during plume mixing for a high carbonate river example (e.g., the Mississippi River) and a low carbonate river example (e.g., the Amazon River) during summer (assuming T = 28°C). Average TAlk values of the Mississippi River and Amazon River were estimated to be 2,300 μmol kg^{-1} and 300 μmol kg^{-1}, respectively. We also assumed a DIC = TAlk + 30 μmol kg^{-1} (Guo et al. 2012) for both rivers, which results in reasonably expected values of pCO_2 during summer. A complete removal of NO_3^- (arbitrarily set to occur at S = 25) leads to pCO_2 as low as 192 μatm and pH as high as 8.34 in the high carbonate river plume ($[NO_3^-]$ = 100 μmol kg^{-1}, which is diluted to 30.6 μmol kg^{-1} at S = 25) and pCO_2 as low as 224 μatm and pH as high as 8.18 in the low carbonate river plume ($[NO_3^-]$ = 15 μmol kg^{-1}, which is diluted to 4.6 μmol kg^{-1} at S = 25). Observed extreme pCO_2 and pH in the Mississippi River plume are similar to the simulations (Guo et al. 2012; Cai unpublished).

during lowest discharge, owing to the dilution effect of weathering products (Cai 2003; Cai et al. 2008; Chen et al. 2008; Bouillon et al. 2012; Guo et al. 2012).

High HCO_3^- content in subtropical to middle-latitude large river waters provides a high acid-base buffering capacity, comparable to that of seawater. For example, the TAlk and DIC concentrations of the Mississippi River are higher than or close to its ocean end-member. Thus pCO_2 and pH changes are expected to be relatively small during their initial stage of mixing with seawater (Fig. 7.2). In sharp contrast, low carbonate river waters in tropical zone have little buffering capacity, and their pCO_2 and pH values are quickly dominated by seawater (i.e., as shown with a rapid decrease in pCO_2 and increase in pH in the Amazon River plume, Fig. 7.2). Therefore, the rapid pCO_2 decrease to a minimum (or pH increase to a maximum) observed in many middle-latitude river plumes (Zhai and Dai 2009; Guo et al. 2012) cannot be attributed simply to thermodynamic changes during river-ocean mixing as was presented in Salisbury et al. (2008). Large rivers in middle-latitudes are also rich in nutrients derived from use of agriculture fertilizers and sewage. Thus coastal waters receiving inputs from large rivers generally have very high biological production rates (i.e., eutrophication). As a result, pCO_2 (or pH) in these river plumes are expected to be very low (or high) (Fig. 7.2); such low pCO_2 and high pH values have been observed in the Mississippi (Guo et al. 2012) and Changjiang River plumes (Chou et al. 2009b; Zhai and Dai 2009; Tseng et al. 2011). Consequently, we suggest that although ocean acidification, which has been linked with CO_2 uptake from the atmospheric (which has increased because of fossil fuel use), has become a major ocean environmental stressor, eutrophication-induced pH increases are a dominant feature in surface water pH in the nutrient-rich river plumes worldwide (Borges and Gypens 2010). However, abundant organic matter exported to

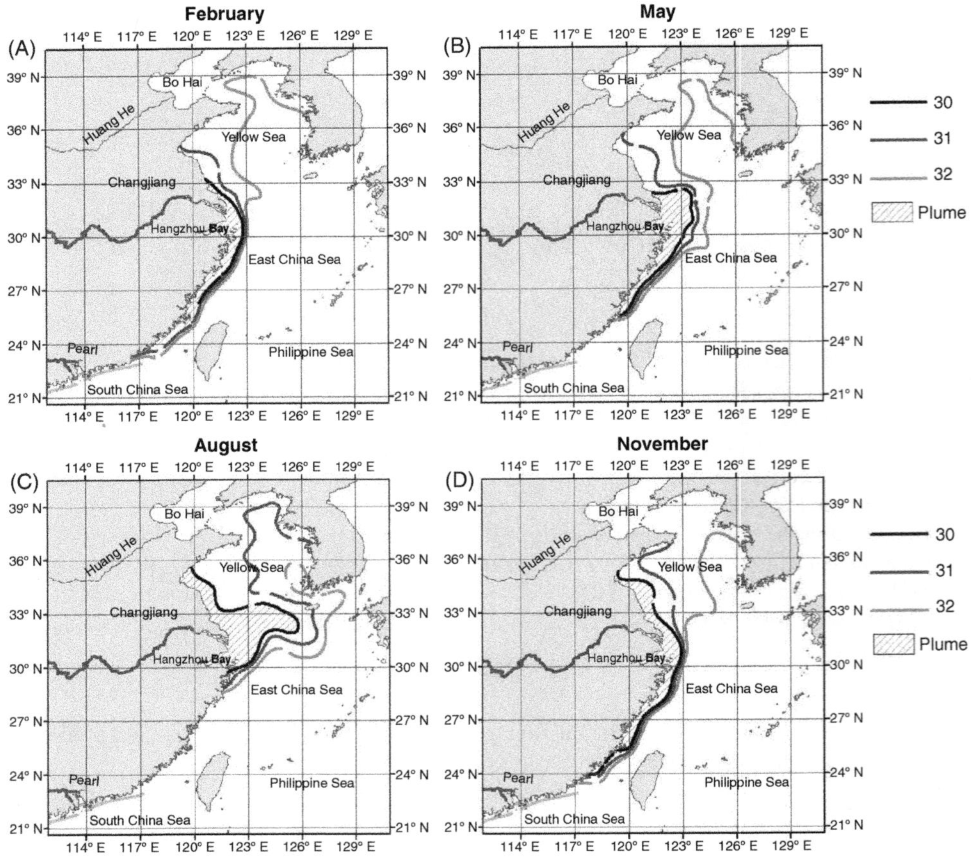

Figure 7.3. The Changjiang River plume in low-discharge February and high-discharge August from Kang et al. (2013). The river plume area was extracted by way of a geographical information system based on data from the Marine Atlas of the North China Sea-Hydrology by W.F. Hon, 2006, China Ocean Press, and the NOAA World Ocean Atlas, 2009 (http://www.nodc.noaa.gov/OC5/SELECT/woaselect/woaselect.html).

the bottom water during water stratified season would lead to extremely low pH and high pCO_2 values in bottom waters as a combined result of enhanced organic matter decomposition, CO_2 uptake from the atmosphere, and decreased buffer capacity in high CO_2 seawaters (Cai et al. 2011).

3.2. CO_2 in the Changjiang River plume

The Changjiang River is the third longest river in the world after the Amazon and Nile Rivers, and it is the fourth largest in terms of discharge after the Amazon, Congo and Orinoco Rivers. The discharge is the highest in July and the lowest in January. The total annual discharge amounts to $950 \times km^3\ yr^{-1}$ with 71% occurring between May and October (Chen et al. 2008). Naturally, the plume is the most extensive in the summer (Fig. 7.3C) and peaks in August, one month after the peak flow.

In August, the plume extends to the northeast because of the southwest monsoon. An S = 32 isoline reaches the Korea Strait and part of the so called Changjiang Diluted Water (CDW), a popular name for the "river plume" in the region, enters the Sea of Japan (Senjyu et al. 2006).

By the fall, the plume occupies a much smaller area owing to a lower discharge (Fig. 7.3D). Because of the northeast monsoon, the CDW extends toward the southwest along the coast of China, with the S = 30 isoline reaching the mid-Taiwan Strait and an S = 32 isoline extending well into the South China Sea as the alongshore current, although several smaller rivers along the coast also contribute to the low salinity coastal current. In February, the plume occupies the smallest area (Fig. 7.3A), before starting to grow again in spring. In May, a bulge of the CDW starts to extend toward to the northeast (Fig. 7.3B).

Time and again, the East China Sea (ECS) has been found to be a sink of CO_2 (Chen and Wang 1999; Tsunogai et al. 1999; Chou et al. 2009a; Zhai and Dai 2009; Tseng et al. 2011). This is in part because the Changjiang River not only exports a significant amount of nutrients to the ECS but also a large quantity of freshwater, which also helps to induce an estuarine type flow. That is, the lighter, fresher water flows out of the shelf on the surface, whereas the subsurface, nutrient-rich waters upwell onto the shelf (Chen and Wang 1999; Chen et al. 2008). The high biological production associated with nutrients brought by the river and this upwelling reduces the pCO_2 in surface waters to below 200 μatm in the area affected by the Changjiang River plume most of the year.

To date, Zhai and Dai (2009) have provided the most comprehensive seasonal sea-to-air CO_2 flux data in the ECS; i.e., 8.8, 4.9, −2.9, and 10.4 mmol m^{-2} d^{-1}, in spring, summer, autumn, and winter, respectively. The area of the plume ($S < 30$) has also been estimated by Kang et al. (2013) as 73×10^3, 88×10^3, 127×10^3, and 96×10^3 km^2, in February, May, August, and November, respectively. Multiplying these fluxes by the corresponding areas results in a total annual absorption of 2 Tg C yr^{-1} for the Changjiang River Plume.

3.3. CO_2 in the Mississippi River plume

The Mississippi-Atchafalaya river system is the largest river in North America, with an annual load of 580 km^3 of freshwater exported to the northern Gulf of Mexico (Milliman and Meade 1983). The Mississippi River splits into two rivers before empting into the Gulf of Mexico, with the Atchafalaya River receiving a combined total of approximately 30% water of the Mississippi and a local river, which has lower HCO_3^- content than the Mississippi. A large part of the freshwater discharge forms a stratified coastal current known as the Louisiana Coastal Current (Wiseman et al. 1997). The size and shape of the plume is strongly affected by wind and current patterns. Persistent southeast winds over the region set up a western flow over the shelf through most of the year except in summer (Ohlmann and Niiler 2005; Hetland and Dimarco 2008). Large-scale circulation patterns often retain the freshwater on the shelf, although the river flow is reduced in summer (Rabalais et al. 1996).

pCO_2 is very high in the lower Mississippi River (Dagg et al. 2005; Cai and Lohrenz 2010). It ranged from about 1,000 μatm in January to 2,200 μatm in July at the Mississippi river mouth and in the Atchafalaya Bay and other coastal bays (based on unpublished data from more than 10 cruises conducted by the Cai laboratory). Once outside the river channels, salinity increases and pCO_2 decreases rapidly, except in the case of extreme flooding conditions near the Atchafalaya Bay

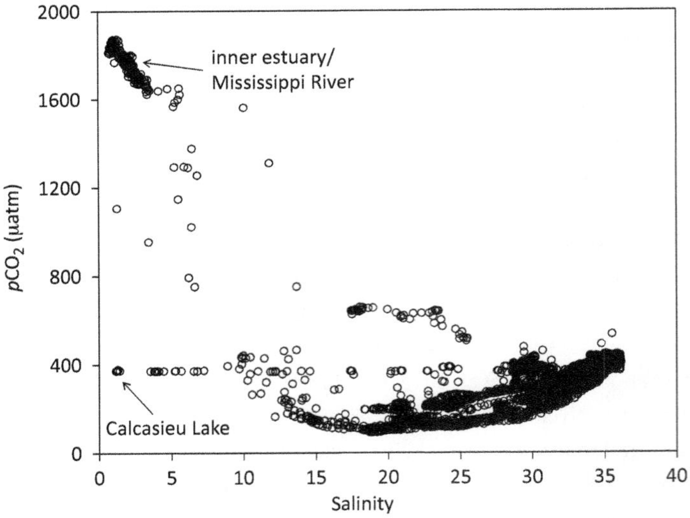

Figure 7.4. pCO_2 vs. salinity plot in the Northern Gulf of Mexico from a June 2006 field survey (unpublished data from Cai laboratory). Very high pCO_2 (over 1,200 µatm) was measured in the main channel of the Mississippi River at the Southwest Pass in eastern Louisiana. Calcasieu Lake, located in south Louisiana, is an isolated coastal bay connected to the shelf water through a coastal water way.

(Bianchi et al., 2013). pCO_2 can be reduced to below 100 µatm with an increase in pH to above 8.5; these values are close to theoretical predictions for a complete biological use of riverine NO_3^- (Fig. 7.2). As expected, intensive biological CO_2 fixation into biomass, owing to very high nutrient concentrations in the river, could reduce pCO_2 to very low values. Air-sea CO_2 flux in the inner shelf during late spring was found to be as high as –10 to –15 mmol m^{-2} d^{-1}. Beyond the mid-field regions of the plume, pCO_2 increases with salinity increase (or distance from shore) and gradually approaches (in spring and winter) or exceeds atmospheric levels (during the summer and fall) in offshore areas (Fig. 7.4). Hence there is not a simple correlation between pCO_2 and salinity in these highly dynamic large river plumes. This is also probably true for other middle-latitude coastal shelves impacted by larger rivers (e.g., see Zhai and Dai 2009 for the Changjiang and Guo et al. 2009 for the Pearl). This feature is in great contrast to that observed in tropical large river plumes (see next section).

Regionally and seasonally averaged based on data from more than 10 cruises in the northern Gulf of Mexico, the inner shelf (<20 m but excluding bays and estuaries) is a sink of atmospheric CO_2 (−3.35 mmol m^{-2} d^{-1}), the middle shelf is usually neutral (–0.08 mmol m^{-2} day^{-1}), and the outer shelf is a weak source of CO_2 (1.46 mmol m^{-2}/d^{-1}). The Louisiana shelf is a CO_2 sink for the atmosphere, with an annual mean air-sea CO_2 flux value of −0.96 mol C m^{-2} yr^{-1}. Total flux in the northern Gulf of Mexico (within a survey area of 39,523 km^2) is −0.46 TgC/yr (Cai laboratory, unpublished results; Huang 2013).

Regarding carbonate system distribution and dynamics, the northern Gulf of Mexico shelf appears as a three end-member mixing system except during the extremely low discharge time (later summer) when the Atchafalaya River has essentially Mississippi River water (Cai 2003; Guo et al. 2012; Huang et al. 2012). In addition, river end-member values can change with time rapidly. Thus TAlk

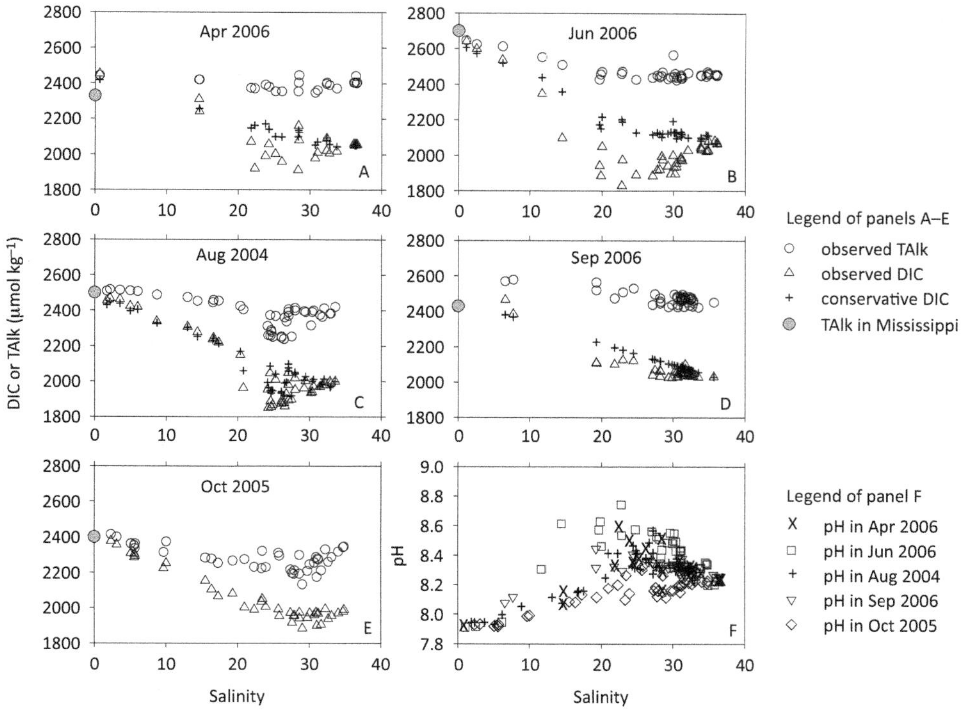

Figure 7.5. DIC and TAlk vs. salinity plots from several cruises in the northern Gulf of Mexico. + indicates predicted DIC values from a conservative mixing model (from salinity and TAlk). The difference between the predicted DIC and observed (and lower) DIC indicates biological uptake. Such mixing values are not given for October 2005, as the properties were completely dominated by mixing after hurricanes Katrina and Rita. Modified after Guo et al. (2012).

does not show simple conservative behavior when examined against salinity in most time of the year (Fig. 7.5). The deviation of TAlk from the linear mixing of the Mississippi River and the ocean would reflect the influences of the third end-member and end-member variations and thus still can be used to separate the biological DIC removal from physical mixing. Thus the difference between the predicted DIC (from salinity and TAlk) and observed (and lower) DIC indicates biological uptake (Fig. 7.5). By dividing the biological uptake by a plume travel time, one can also derive the net community production (NCP) rates. The NCP rates at the maximum DIC removal area (so called mid-salinity zone or mid-field) can be as high as more than 0.3 mol C m^{-2} d^{-1} (or more than 3 gC m^{-2} d^{-1}), highest among all larger river plumes (Cai 2003; Guo et al. 2012). In contrast, DIC removal in the Amazon River plume is much smaller (is nearly invisible in the DIC to salinity plot, see Fig. 7.6 in the next section). We interpret this stark contrast as evidence of a strong anthropogenic nutrient influence on the biological production of coastal ocean waters in areas near population centers. Other factors also matter. For example, the Mississippi River has lower turbidity than the Changjiang River, and thus high biological production and CO_2 removal in the plume has occurred at low salinity zone of 20 to 30 or occasionally even at salinity as low as 12 (Dagg et al. 2008; Huang et al. 2012) in the Mississippi plume, whereas in the Changjiang plume, biological bloom rarely has occurred at salinity less than 28.

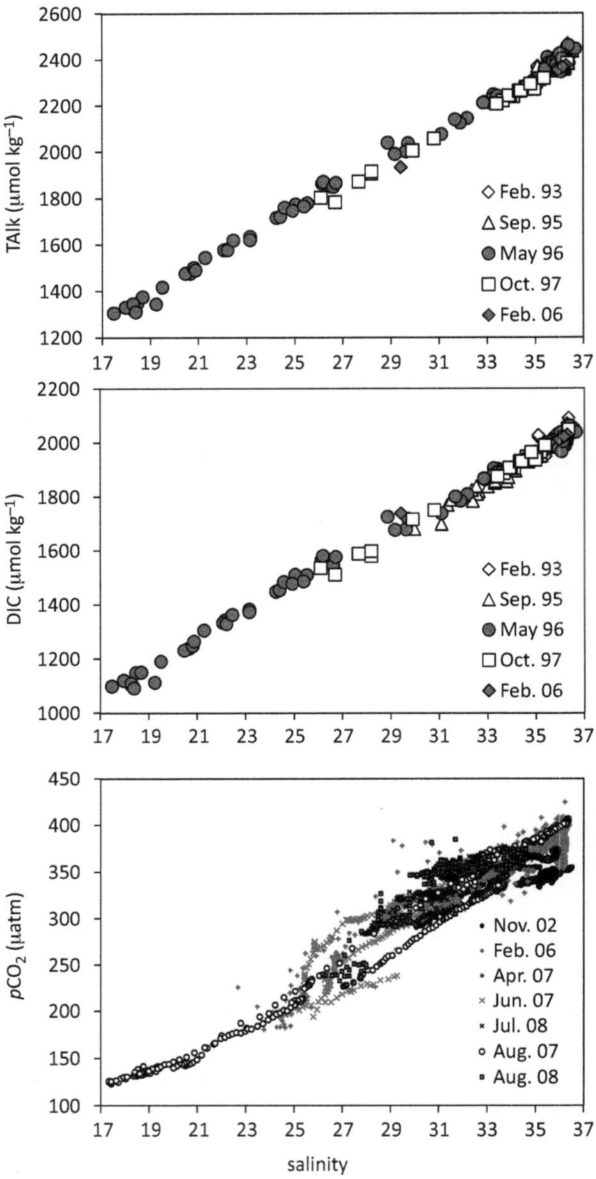

Figure 7.6. TAlk, DIC, and the pCO_2 vs. salinity distributions from the Amazon River plume (reproduced based on data supplied by N. Lefèvre). See Lefèvre et al. (2010) for site and method description and for further discussion.

3.4. CO_2 in the Amazon River plume

The Amazon is the largest river in the world in terms of freshwater discharge (5,520 km^3 yr^{-1}) and watershed size (6.9 10^6 km^2), and accounts for ~30% of fresh water discharge to the Atlantic Ocean (Wisser et al. 2010). Discharge ranges from 2,500 km^3 yr^{-1} (November) to 7,600 km^3 yr^{-1} (late May) (Lentz 1995). The Amazon River mouth is located at the equator, and the plume propagates northwestward toward the Caribbean with the North Brazilian Current, Guyana Current, and Caribbean Current. Part of the Amazon plume propagates eastward (at 3°–10°N) carried by the North Equatorial Counter Current (Müller-Karger et al. 1989). Over the Brazilian shelf, the plume water is on the top 3 to 10 m, and offshore the plume waters reach depths of 40–45 m (Lentz and Limeburner 1995; Hellweger and Gordon 2002). The Amazon River plume can be detected in satellite ocean color data over areas as large as 500,000 km^2 (Muller-Karger et al. 1988; Salisbury et al. 2011).

CO_2 data in the Amazon River plume have been reported by Ternon et al. (2000), Körtzinger (2003), Cooley et al. (2007), and Lefèvre et al. (2010). DIC and TAlk showed quasi-conservative mixing behavior while pCO_2 is below atmospheric equilibrium, in contrast with surrounding waters of the tropical Atlantic Ocean that are above atmospheric equilibrium (Fig. 7.6). The relationship between pCO_2 and sea surface salinity (SSS) is remarkably similar among the different studies, with the lowest pCO_2 value of 125 μatm reported at an SSS value of 18 by Lefèvre et al. (2010). Here, pCO_2 is nearly linearly related to SSS, in great contrast to the complex relationships observed in subtropical and middle-latitude coastal oceans impacted by nutrient-rich large rivers such as in the case of the Mississippi (Fig. 7.4), Changjiang (Zhai and Dai 2009), and the Pearl (Guo et al. 2009). Seasonal variations of pCO_2 were also found to be low in the Amazon River plume (Fig. 7.6), in great contrast with those found in river plumes at subtropical and temperate latitudes, where seasonal oscillations from CO_2 over-saturation (typically in winter or fall) to under-saturation (typically in spring or summer) are observed (Borges and Frankignoulle 2002; Lohrenz and Cai 2006; Chou et al. 2009a; Chou et al. 2009b; Guo et al. 2009; Zhai and Dai 2009; Vandemark et al. 2011; Bozec et al. 2012). Another great contrast between the Amazon plume and some of the other large river plumes is in the DIC (or TAlk) to salinity relationship, which does not show a substantial removal in the Amazon plume (Fig. 7.6), whereas such overall removal is apparent in the Mississippi plume (Fig. 7.5) (Cai 2003; Guo et al. 2012; Huang et al. 2012), in the Pearl River plume (Cai et al. 2004; Dai et al. 2008), and the Changjiang plume (Wang et al. 2000; Chou et al. 2009b).

The low pCO_2 values in the Amazon River plume are due to a combination of thermodynamic effect of mixing of river water with seawater (Fig. 7.2) and biological drawdown within the plume owing to increasing light availability as suspended particles sediment in the delta. Cooley et al. (2007) highlighted the role of diazotroph-supported NCP in sustaining the biological drawdown of CO_2 within the Amazon plume. However, the nearly linear DIC-to-salinity relationship indicates that net biological production rate, at least in a per-volume base, is much lower in the Amazon plume than in the Mississippi and Changjiang plumes. For example, NCP rates estimated from within plume DIC removal were 0.9–1.6 gC m^{-2} d^{-1} for the Amazon (Ternon et al. 2000) and 1–8 gC m^{-2} d^{-1} for the Mississippi plume (Lohrenz et al. 1997; Lohrenz et al. 1999; Guo et al. 2012), whereas the mixed layer depth of the former is nearly 10 times that of the latter. This conclusion is consistent with a comparison of measured biological production rates (see Table 4 of Guo et al. 2012). Finally, we have to point out that all of the mixing models in the Amazon River plume between river and marine end-members ignore processes occurring in the extensive inner estuary (44,200 km^2, Borges 2005) with extensive flooded areas where pCO_2 is extremely dynamic and strongly deviate from conservative mixing (Gwenaël Abril, unpublished manuscript).

Körtzinger (2003) used a pCO_2 versus SSS relationship (SSS >30) and the monthly climatological SSS means of the Word Ocean Atlas 2001 to integrate air-sea CO_2 fluxes. This approach resulted in an overall CO_2 sink of –0.014 Pg C yr^{-1} for the plume area of 2.4×10^6 km^2 (defined as $S < 34.9$) corresponding to an average air-sea CO_2 flux of –0.5 mol m^{-2} yr^{-1}. Using a similar approach but with a more extensive (SSS > 17) and seasonally resolved dataset, and using instead the monthly fields of SSS from the World Ocean Atlas 2005, Lefèvre et al. (2010) estimated a CO_2 flux of –0.35 mol m^{-2} yr^{-1} and a lower Amazon River plume surface area of 1.1×10^6 km^2 (also defined as $S <$ 34.9). Thus Lefèvre et al. (2010) gave an integrated CO_2 flux of –0.005 Pg C yr^{-1} for the Amazon River plume. These fluxes are extremely low compared with the CO_2 emission from the Amazon River itself (0.5 Pg C yr^{-1}, Richey et al. 2002). It should also be noted here that there is no published estimate available for the Amazon River inner estuary.

4. A tentative estimate of global river-plume CO_2 flux

Deriving a global CO_2 flux estimate requires a sufficient number of CO_2 flux measurements adequately accounting for seasonal and spatial variability within a given system and also surface area estimates for scaling the data. In addition, consistent data from multiple systems are preferred for a reliable global extrapolation. The number of projects acquiring CO_2 data in large river plumes has increased steadily over the last decade. Also, high-quality coastal CO_2 data are being banked in the international and public databases such as Carbon Dioxide Information Analysis Center and are being included in large data compilations such as Surface Ocean CO_2 Atlas (SOCAT, Pfeil et al. 2012). Hence, in the near future, it should be possible to estimate the global air-sea CO_2 flux in river plumes with high confidence with the approach outlined previously as the number of studies acquiring CO_2 data in large river plumes increase and become available.

To the best of our knowledge, there is no published global estimate of river plume surface areas. The most straightforward way to achieve this would be to use SSS as a tracer of river plumes. This can be estimated on a global basis using remote sensing products. Direct determination of SSS from AQUARIUS or Soil Moisture and Ocean Salinity satellite missions is probably inadequate because the resolution of the products is ~50 km. Another possibility would be to derive SSS in river plumes from remote sensed colored dissolved organic carbon (CDOM) as applied in several studies in the Amazon (Salisbury et al. 2011), Changjiang (Bai et al. 2013), Mississippi (Lohrenz et al. 2010; Zhu et al. 2011), and Columbia (Palacios et al. 2009). However, algorithms for the retrieval of SSS from remote sensed CDOM are needed on a regional basis (site-specific) in scope, because CDOM values at river end-members are highly variable (Blough and Del Vecchio 2002). Also, CDOM-SSS relationships can seasonally depart from linearity for a variety of CDOM production or removal processes (additional end-members, phytoplankton CDOM release, bacterial utilization or photo-oxidation). Therefore, we take a first-order approximation in this chapter in hope of making some progress regarding this issue.

Table 7.2. *Air-Water CO_2 flux (mol m^{-2} yr^{-1}), surface area (km^2), and freshwater discharge for major river plumes*

	Freshwater discharge (km^3 yr^{-1})	Plume surface (km^2)	Area to discharge ratio	Air-water CO_2 flux (mol m^{-2} yr^{-1})	Reference
Amazon	5520	2,400,000(1)	438	−0.50	Körtzinger (2003)
		1,100,000(1)	199	−0.35	Lefèvre et al. (2010)
Changjiang	950	85,000(2)	90	−1.90	Zhai & Dai (2009)
Mississippi	580	22,000(3)	38	−0.96	Cai unpublished
Loire	18	7,993(4)	–	−1.50	Bozec et al. (2012)

Note: (1)Both Körtzinger (2003) and Lefèvre et al. (2010) defined the Amazon River plume for $S < 34.9$ in the area 3°S–18°N, 30–60°W, but the former used the data from the World Ocean Atlas 2001 and the latter used the World Ocean Atlas 2005. (2)Changjiang River plume area comes from Tseng et al. (2011) based on data in their Fig. 2 and Kang et al. (2013), defined as $S < 30$. (3)The Mississippi River plume is from Kang et al. (2013) defined as $S < 31$ and from W.-J. Cai unpublished. Earlier Green et al. (2006) had provided a very small plume area of 2,200 km^2 immediately outside the Southwest Pass of the Mississippi, which was defined as the area with a suspended particle concentration over 5 mg L^{-1}. If a one-month water residence time is assumed, a plume of 22,000 km^2 or 2,200 km^2 would indicate a mixed layer depth of 2.2 or 22 m, respectively, using the Mississippi River discharge rate, suggesting that the high end plume area is reasonable. However, these two plume area estimates are not necessarily at odds, as the latter is defined as the turbidity plume, whereas the former is defined based on salinity and is thus more consistent with other definition in this chapter. (4)Loire plume area was defined for $S < 34.4$ by Bozec et al. (2012).

Table 7.2 summarizes existing knowledge on air-sea CO_2 fluxes in major river plumes and additional physical data (surface area and freshwater discharge). It appears that all known large river plumes are net sinks for atmospheric CO_2. pCO_2 values below atmospheric equilibrium also have been reported in other major river plumes, such as the Niger River (Körtzinger 2009) and the Congo River (Bakker et al. 1999; Lefèvre 2009); however, integrated air-water CO_2 fluxes have not been computed as data coverage is quite low in these systems. It should also be noted that very small river plumes such as the Scheldt estuary (freshwater discharge < 4 km^3 yr^{-1}) and the Kennebec (freshwater discharge ~8 km^3 yr^{-1}) are a net source of CO_2 to the atmosphere (+1.9 mol m^- yr^{-1} (Borges and Frankignoulle 2002) and +0.9 mol m^- yr^{-1} (Salisbury et al. 2009).

Although there is a general increasing pattern of river plume surface area with discharge (Table 7.2), we are not able to derive a simple relationship from these few data-points. This could be due to a variety of reasons such as inconsistency in the definitions and methods to derive river plume areas (e.g., see notes below Table 7.2) as well as interactions with topography and/or dispersal by wind. Nevertheless, we attempted a crude global extrapolation of CO_2 fluxes for river plumes based on the limited dataset shown in Table 7.2. The intuitive notion is that there should be a general positive relationship between river plume surface area and freshwater discharge.

For our estimate, we removed the Amazon discharge (5,520 km^3 yr^{-1}) from the global freshwater discharge of 37,288 km^3 yr^{-1} (Dai and Trenberth 2002). To this value (31,768 km^3 yr^{-1}) we applied a ratio of river plume surface area (km^2) to freshwater discharge (km^3 yr^{-1}) of 38 (for Mississippi) and 90 (for Changjiang), allowing for computation of a range of surface area for all river plumes, excluding the extremely large Amazon River and the very small Loire River. To this value we added the two Amazon plume area estimates, allowing us to derive a range of surface area of river plumes globally, ranging between 2.31×10^6 km^2 and 5.26×10^6 km^2 (i.e., 9–20% of global continental shelf area; Walsh 1988). If we scale the surface area weighted mean of air-water CO_2 fluxes given in Table 7.2 (-0.49 mol m$^-$ yr^{-1}), this would give a global CO_2 sink ranging between -0.014 and -0.031 PgC yr^{-1}. Alternatively, if we scale the arithmetic mean of air-water CO_2 fluxes given in Table 7.2 (-1.04 mol m$^-$ yr^{-1}), this would provide a global CO_2 sink ranging between -0.029 and -0.066 PgC yr^{-1}. Note that a recent work estimated the annual mean area of global river plumes to be 3.72×10^6 km^2, which is approximately 14% of the total continental shelves area worldwide (Kang et al. 2013). As this work only counted plume areas that were three salinity units lower than the associated ocean salinity, we view the high end of our rough estimate (-0.066 PgC yr^{-1}) to be more reasonable than the lower end estimate.

The tentative global sink of CO_2 in river plumes ranging between -0.014 and -0.066 PgC yr^{-1} is to a large extent driven by the sink in the Amazon River plume (-0.005 to -0.014 PgC yr^{-1}). We notice that this flux only represents 12–26% of the currently best estimated global continental shelf CO_2 sink of ~-0.25 PgC yr^{-1} (Cai et al. 2006; Laruelle et al. 2010) and does not balance the CO_2 source from inner estuaries ($\sim+0.25$ PgC yr^{-1}) (Table 7.1). The global river plume CO_2 sink is much smaller compared with the CO_2 sink in the open ocean ranging between -1.4 PgC yr^{-1} and -2.2 PgC yr^{-1} (Gruber et al. 2009; Takahashi et al. 2009). We acknowledge that this global plume CO_2 sink estimate is very crude. Yet available data converge to the fact that large river plumes are sinks of atmospheric CO_2 (Table 7.2). We suggest that currently the largest unknown term is in the global surface area of large river plumes, which we propose as between 2.31×10^6 km^2 and 5.26×10^6 km^2. However, we do not believe that the true plume can be an order of magnitude higher (23 to 52×10^6 km^2) or as large as 6–15% of the global ocean area (362×10^6 km^2). Only in that scenario, the plume flux can be significant in a global scale, which we view as highly unlikely. However, such plume-induced CO_2 flux can be highly significant in regional air-sea CO_2 budget and represents important biogeochemical processes.

In summary, large river plumes are highly dynamic in terms of CO_2 distributions. They are frequently under-saturated in CO_2 with respect to atmospheric equilibrium owing to biological activity and thermodynamic effect on the marine carbonate system of mixing of river and seawater. The latter effect is more marked in poorly buffered low-latitude rivers owing to lower HCO_3^- concentrations in river water. Tropical river plumes such as the Amazon seem to show more or less seasonally stable pCO$_2$-SSS relationships, whereas in subtropical and temperate (e.g., Mississippi, Changjiang and Pearl, and Loire) systems, pCO$_2$ is more variable seasonally, typically above equilibrium in fall/winter and below equilibrium in spring/summer. Consistently, although DIC-SSS and TAlk-SSS indicate quasi-conservative mixing behavior in the Amazon River, they show a nonconservative behavior in subtropical-temperate systems such as the Mississippi and Changjiang River, where biological CO_2 fixation draws down DIC greatly beyond the prediction from the conservative mixing behavior. Available air-sea CO_2 flux estimates indicate that the Amazon, Changjiang, Mississippi, and Loire river plumes are net sinks of atmospheric CO_2 ranging from -0.4 to -1.9 mol m$^-$ yr^{-1}. Based on a crude estimate of the global extent of river plumes, we have integrated the CO_2 sink. Current best estimate suggests that large river plumes contribute at most to ~25% of the continental shelf CO_2 sink and to ~3% of the global oceanic CO_2 sink.

Acknowledgment

We thank Kang et al. of the Second Institution of Oceanography, Hangzhou, China, for providing an advanced copy of their paper and helping with the Changjiang plume extension information (Figure 7.3). W.-J. Huang and X. Guo helped with plotting Figures 7.4 and 7.5. W.-J. Huang also provided CO_2 flux estimate in the Mississippi River plume. We thank Tom Bianchi for improving the text. A.V.B. is a research associate at the FRS-FNRS. W.-J.C. acknowledges the support of the National Science Foundation (OCE-0752110) and NASA (NNX10AU06G, NNX12AB21G, and NNX12AP84G).

References

Abril, G., and A.V. Borges. 2004. Carbon dioxide and methane emissions from estuaries, In: Tremblay, A., Varfalvy, L., Roehm, C., and Garneau, M. (eds.), Greenhouse Gases Emissions from Natural Environments and Hydroelectric Reservoirs: Fluxes and Processes. Berlin: Springer-Verlag, pp. 187–207.

Bai, Y., D. Pan, W.-J. Cai, X. He, D. Wang, B. Tao, and Q. Zhu. 2013. Remote sensing of salinity from satellite-derived CDOM in the Changjiang River dominated East China Sea. *J. Geophys. Res.: Oceans* 118: 227–243.

Bakker, D. C. E., H. J. W. de Baar, and E. de Jong. 1999. The dependence on temperature and salinity of dissolved inorganic carbon in East Atlantic surface waters. *Mar. Chem.* 65(3–4): 263–280.

Bianchi, T.S., and M.A. Allison. 2009. Large-river delta-front estuaries as natural "recorders" of global environmental change. *Proc. Natl. Acad. Sci.* 106: 8085–8092.

Bianchi, T.S., Garcia-Tigreros, F., Yvon-Lewis, S.A., Shields, M., Mills, H.J., Butman, D., Osburn, C., Raymond, P., Shank, G.C., DiMarco, S.F., Walker, N., Reese, B.K., Mullins-Perry, R., Quigg, A., Aiken, G.R., Grossman, E.L.. 2013. Enhanced transfer of terrestrially derived carbon to the atmosphere in a flooding event. *Geophys. Res. Lett.* 40: 116–122.

Blough, N.V., and R. Del Vecchio. 2002. Chromophoric DOM in the coastal environment, In: Hansell, D.A., and Carlson, C.A. (eds.), Biogeochemistry of Marine Dissolved Organic Matter. Amsterdam: Academic Press, pp. 509–546.

Borges, A.V. 2005. Do we have enough pieces of the jigsaw to integrate CO_2 fluxes in the Coastal Ocean? *Estuaries* 28: 3–27.

Borges, A.V., and G. Abril. 2011. 5.04 – Carbon Dioxide and Methane Dynamics in Estuaries, In: Eric, W., and Donald, M. (eds.), Treatise on Estuarine and Coastal Science. Amsterdam: Academic Press, pp. 119–161.

Borges, A.V., B. Dellile, and M. Frankignoulle. 2005. Budgeting sinks and sources of CO_2 in the coastal oceans: Diversity of ecosystems counts. *Geophys. Res. Lett.* 32: L14601.

Borges, A.V., and M. Frankignoulle. 2002. Distribution and air-water exchange of carbon dioxide in the Scheldt plume off the Belgian coast. *Biogeochemistry* 59: 41–67.

Borges, A.V., and N. Gypens. 2010. Carbonate chemistry in the coastal zone responds more strongly to eutrophication than ocean acidification. *Limnol. Oceanogr.* 55: 346–353.

Bouillon, S., A. Yambélé, R.G.M. Spencer, D.P. Gillikin, P.J. Hernes, J. Six, R. Merckx, and A.V. Borges. 2012. Organic matter sources, fluxes and greenhouse gas exchange in the Oubangui River (Congo River basin). *Biogeosciences* 9: 2045–2062.

Bozec, Y., T. Cariou, E. Macé, P. Morin, D. Thuillier, and M. Vernet. 2012. Seasonal dynamics of air-sea CO2 fluxes in the inner and outer Loire estuary (NW Europe). *Estuar. Coast. Shelf Sci.* 100: 58–71.

Cai, W.-J. 2003. Riverine inorganic carbon flux and rate of biological uptake in the Mississippi River plume. *Geophys. Res. Lett.* 30: 1032.

Cai, W.-J. 2011. Estuarine and coastal ocean carbon paradox: CO_2 sinks or sites of terrestrial carbon incineration? *Ann. Rev. Mar. Sci.* 3: 123–145.
Cai, W.-J., M. Dai, and Y. Wang. 2006. Air-sea exchange of carbon dioxide in ocean margins: a province-based synthesis. *Geophys. Res. Lett.* 33: L12603.
Cai, W.-J., X. Guo, C.-T. A. Chen, M. Dai, L. Zhang, W. Zhai, S.E. Lohrenz, K. Yin, P.J. Harrison, and Y. Wang. 2008. A comparative overview of weathering intensity and HCO_3^- flux in the world's major rivers with emphasis on the Changjiang, Huanghe, Zhujiang (Pearl) and Mississippi Rivers. *Cont. Shelf Res.* 28: 1538–1549.
Cai, W.-J., X. Hu, W.-J. Huang, L. Jiang, Y. Wang, T. H. Peng, and X. Zhang. 2010. Alkalinity distribution in the western North Atlantic Ocean margins. *J. Geophys. Res. Ocean.*, 115(C8), C08014.
Cai, W.-J., Hu, X., Huang, W.-J., Murrell, M.C., Lehrter, J.C., Lohrenz, S.E., Chou, W.-C., Zhai, W., Hollibaugh, J.T., Wang, Y., Zhao, P., Guo, X., Gundersen, K., Dai, M., and Gong, G.-C., 2011. Acidification of subsurface coastal waters enhanced by eutrophication. Nature Geosci 4: 766–770.
Cai, W.-J., and S.E. Lohrenz. 2010. Marginal seas-The Mississippi River plume and adjacent margin in the Gulf of Mexico, In: Liu, K.K., Atkinson, L., Quinones, R., and Talaue-McManus, L. (eds.), A Global Synthesis. Springer.
Cai, W.J., M.H. Dai, Y.C. Wang, W.D. Zhai, T. Huang, S.T. Chen, F. Zhang, Z.Z. Chen, and Z.H. Wang. 2004. The biogeochemistry of inorganic carbon and nutrients in the Pearl River estuary and the adjacent Northern South China Sea. *Cont. Shelf Res.* 24: 1301–1319.
Chen, C.-T.A., and A.V. Borges. 2009. Reconciling opposing views on carbon cycling in the coastal ocean: continental shelves as sinks and near-shore ecosystems as sources of atmospheric CO_2. *Deep Sea Res. Part II Top. Stud. Oceanogr.* 56: 578–590.
Chen, C.-T.A., T.-H. Huang, Y.-H. Fu, Y. Bai, and X. He. 2012. Strong sources of CO_2 in upper estuaries become sinks of CO2 in large river plumes. *Curr. Opin. Environ. Sustain.* 4: 179–185.
Chen, C.-T.A., W. Zhai, and M. Dai. 2008. Riverine input and air-sea CO_2 exchanges near the Changjiang (Yangtze River) Estuary: status quo and implication on possible future changes in metabolic status. *Cont. Shelf Res.* 28: 1476–1482.
Chen, C.T.A., and S.L. Wang. 1999. Carbon, alkalinity and nutrient budgets on the East China Sea continental shelf. *J. Geophys. Res. Oceans* 104: 20675–20686.
Chou, W.-C., G.-C. Gong, D.D. Sheu, C.-C. Hung, and T.-F. Tseng. 2009a. Surface distributions of carbon chemistry parameters in the East China Sea in summer 2007. *J. Geophys. Res.* 114: C07026.
Chou, W.-C., G.-C. Gong, D.D. Sheu, S. Jan, C.-C. Hung, and C.-C. Chen. 2009b. Reconciling the paradox that the heterotrophic waters of the East China Sea shelf act as a significant CO2 sink during the summertime: evidence and implications. *Geophys. Res. Lett.* 36: L15607.
Cooley, S.R., V.J. Coles, A. Subramaniam, and P.L. Yager. 2007. Seasonal variations in the Amazon plume-related atmospheric carbon sink. *Global Biogeochem. Cycles* 21: GB3014.
Dai, A. and K.E. Trenberth 2002. Estimates of freshwater discharge from continents: Latitudinal and seasonal variations. *J. Hydrol.* 3: 660–687.
Dai, M., Zhai, W., Cai, W.-J., Callahan, J., Huang, B., Shang, S., Huang, T., Li, X., Lu, Z., Chen, W., Chen, Z. 2008. Effects of an estuarine plume-associated bloom on the carbonate system in the lower reaches of the Pearl River estuary and the coastal zone of the northern South China Sea. *Cont. Shelf Res.* 28: 1416–1423.
Dagg, M.J., T. Bianchi, B. Mckee, and R. Powell. 2008. Fates of dissolved and particulate materials from the Mississippi river immediately after discharge into the northern Gulf of Mexico, USA, during a period of low wind stress. *Cont. Shelf Res.* 28: 1443–1450.

Dagg, M.J., T.S. Bianchi, G.A. Breed, W.J. Cai, S. Duan, H. Liu, B.A. McKee, R.T. Powell, and C.M. Stewart. 2005. Biogeochemical characteristics of the lower Mississippi River, USA, during June 2003. *Estuaries* 28: 664–674.

Dürr, H.H., G.G. Laruelle, C.M. Van Kempen, C.P. Slomp, M. Meybeck, and H. Middelkoop. 2011. Worldwide typology of nearshore coastal systems: Defining the estuarine filter of riverine inputs to the oceans. *Estuar. Coasts* 34: 441–458.

Frankignoulle, M., G. Abril, A. Borges, I. Bourge, C. Canon, B. DeLille, E. Libert, and J.M. Theate.1998. Carbon dioxide emission from European estuaries. *Science* 282: 434–436.

Green, R., Bianchi, T., Dagg, M., Walker, N., Breed, G., 2006. An organic carbon budget for the Mississippi River turbidity plume and plume contributions to air-sea CO_2 fluxes and bottom water hypoxia. *Estuar. Coast.* 29: 579–597.

Gruber, N., M. Gloor, S.E. Mikaloff Fletcher, Gruber, N., Gloor, M., Mikaloff Fletcher, S.E., Doney, S.C., Dutkiewicz, S., Follows, M.J., Gerber, M., Jacobson, A.R., Joos, F., Lindsay, K., Menemenlis, D., Mouchet, A., Müller, S.A., Sarmiento, J.L., Takahashi, T., 2009. Oceanic sources, sinks, and transport of atmospheric CO_2. *Global Biogeochem. Cycles* 23: GB1005.

Guo, X., W.-J. Cai, W.-J. Huang. Guo, X., Cai, W.-J., Huang, W.-J., Wang, Y., Chen, F., Murrell, M.C., Lohrenz, S.E., Jiang, L.-Q., Dai, M., Hartmann, J., Lin, Q., Culp, R., 2012. Carbon dynamics and community production in the Mississippi River plume. *Limnol. Oceanogr.* 57: 1–17.

Guo, X., M. Dai, W. Zhai, W.-J. Cai, and B. Chen. 2009. CO_2 flux and seasonal variability in a large subtropical estuarine system, the Pearl River Estuary, China. *J. Geophys. Res.* 114: G03013.

Hellweger, F.L., and A.L. Gordon. 2002. Tracing Amazon River water into the Caribbean Sea. *J. Mar. Res.* 60: 537–549.

Hetland, R.D., and S.F. Dimarco. 2008. How does the character of oxygen demand control the structure of hypoxia on the Texas–Louisiana continental shelf? *J. Mar. Syst.* 70: 49–62.

Hopkinson, C.S., and E.M. Smith. 2005. Estuarine respiration: an overview of benthic, pelagic, and whole system respiration, In: del Giorgio, P., and Williams, P. (eds.), Respiration in Aquatic Ecosystems. New York: Oxford University Press, pp. 122–147.

Huang, W.-J. 2013. Inorganic carbon distribution and dynamics in the Mississippi River plume on the northern Gulf of Mexico. PhD dissertation, Univ. of Georgia.

Huang, W.J., W.J. Cai, R.T. Powell, S.E. Lohrenz, Y. Wang, L.Q. Jiang, and C.S. Hopkinson. 2012. The stoichiometry of inorganic carbon and nutrient removal in the Mississippi River plume and adjacent continental shelf. *Biogeosciences* 9: 2781–2792.

Jiang, L.-Q., W.-J. Cai, and Y. Wang. 2008. A comparative study of carbon dioxide degassing in river- and marine-dominated estuaries. *Limnol. Oceanogr.* 53: 2603–2615.

Kang, Y., D. Pan, Y. Bai, X. He, X. Chen, C.-T. Chen, and D. Wang (2013), Areas of the global major river plumes, *Acta Oceanol. Sin.* 32(1): 79–88.

Koné, Y.J.M., G. Abril, K.N. Kouadio, B. Delille, and A.V. Borges. 2008. Seasonal Variability of Carbon Dioxide in the Rivers and Lagoons of Ivory Coast (West Africa). *Estuar. Coasts* 32: 246–260.

Körtzinger, A. 2003. A significant CO_2 sink in the tropical Atlantic Ocean associated with the Amazon River plume. *Geophys. Res. Lett.* 30: 2287.

Körtzinger, A. 2009. The outer Amazon Plume: An atmospheric CO_2 sink, In: K.-K. Liu, L. Atkinson, R. Quinones and L. Talaue-McManus (eds.), Carbon and Nutrient Fluxes in Continental Margins, Heidelberg: Springer, pp. 450–453.

Laruelle, G.G., H.H. Dürr, C.P. Slomp, and A.V. Borges. 2010. Evaluation of sinks and sources of CO_2 in the global coastal ocean using a spatially-explicit typology of estuaries and continental shelves. *Geophys. Res. Lett.* 37: L15607.

Lefèvre, N. 2009. Low CO_2 concentrations in the Gulf of Guinea during the upwelling season in 2006. *Mar. Chem.* 113(1–2): 93–101.

Lefèvre, N., D. Diverrès, and F. Gallois. 2010. Origin of CO_2 undersaturation in the western tropical Atlantic. *Tellus B* 62: 595–607.

Lentz, S.J. 1995. Seasonal variations in the horizontal structure of the Amazon Plume inferred from historical hydrographic data. *J. Geophys. Res.* 100: 2391–2400.

Lentz, S.J., and R. Limeburner. 1995. The Amazon River Plume during AMASSEDS: Spatial characteristics and salinity variability. *J. Geophys. Res.* 100: 2355–2375.

Lohrenz, S.E., and W.-J. Cai. 2006. Satellite ocean color assessment of air-sea fluxes of CO_2 in a river-dominated coastal margin. *Geophys. Res. Lett.* 33.

Lohrenz, S.E., W.-J. Cai, F. Chen, X. Chen, and M. Tuel. 2010. Seasonal variability in air-sea fluxes of CO2 in a river-influenced coastal margin. *J. Geophys. Res.* 115: C10034.

Lohrenz, S.E., G.L. Fahnenstiel, D.G. Redalje, G.A. Lang, X.G. Chen, and M.J. Dagg. 1997. Variations in primary production of northern Gulf of Mexico continental shelf waters linked to nutrient inputs from the Mississippi River. *Mar. Ecol. Prog. Series* 155: 45–54.

Lohrenz, S.E., G.L. Fahnenstiel, D.G. Redalje, G.A. Lang, M.J. Dagg, T.E. Whitledge, and Q. Dortch. 1999. Nutrients, irradiance, and mixing as factors regulating primary production in coastal waters impacted by the Mississippi River plume. *Cont. Shelf Res.* 19: 1113–1141.

Milliman, J.D., and R.H. Meade. 1983. World-wide delivery of sediment to the oceans. *J. Geol.* 91: 1–21.

Müller-Karger, F.E., C.R. Mcclain, T.R. Fisher, W.E. Esaias, and R. Varela. 1989. Pigment distribution in the Caribbean sea: observations from space. *Prog. Oceanogr.* 23: 23–64.

Muller-Karger, F.E., C.R. Mcclain, and P.L. Richardson. 1988. The dispersal of the Amazon's water. *Nature* 333: 56–59.

Odum, H.T., and C.M. Hoskin. 1958. Comparative studies of the metabolism of Texas Bays. Publications of the Institute of Marine Science, University of Texas 5: 16–46.

Ohlmann, J.C., and P.P. Niiler. 2005. Circulation over the continental shelf in the northern Gulf of Mexico. *Prog. Oceanogr.* 64: 45–81.

Palacios, S.L., T.D. Peterson, and R.M. Kudela. 2009. Development of synthetic salinity from remote sensing for the Columbia River plume. *J. Geophys. Res.* 114: C00B05.

Park, P.K., L.I. Gordon, S.W. Hager, and M.C. Cissel. 1969. Carbon dioxide partial pressure in the Columbia River. *Science* 166: 867–868.

Pfeil, B., A., Olsen, D.C.E., Bakker, Pfeil, B., Olsen, A., Bakker, D.C.E., Hankin, S., Koyuk, H., Kozyr, A., Malczyk, J., Manke, A., Metzl, N., Sabine, C.L., Akl, J., Alin, S.R., Bellerby, R.G.J., Borges, A., Boutin, J., Brown, P.J., Cai, W.J., Chavez, F.P., Chen, A., Cosca, C., Fassbender, A.J., Feely, R.A., González-Dávila, M., Goyet, C., Hardman-Mountford, N., Heinze, C., Hood, M., Hoppema, M., Hunt, C.W., Hydes, D., Ishii, M., Johannessen, T., Jones, S.D., Key, R.M., Körtzinger, A., Landschützer, P., Lauvset, S.K., Lefèvre, N., Lenton, A., Lourantou, A., Merlivat, L., Midorikawa, T., Mintrop, L., Miyazaki, C., Murata, A., Nakadate, A., Nakano, Y., Nakaoka, S., Nojiri, Y., Omar, A.M., Padin, X.A., Park, G.H., Paterson, K., Perez, F.F., Pierrot, D., Poisson, A., Ríos, A.F., Santana-Casiano, J.M., Salisbury, J., Sarma, V.V.S.S., Schlitzer, R., Schneider, B., Schuster, U., Sieger, R., Skjelvan, I., Steinhoff, T., Suzuki, T., Takahashi, T., Tedesco, K., Telszewski, M., Thomas, H., Tilbrook, B., Tjiputra, J., Vandemark, D., Veness, T., Wanninkhof, R., Watson, A.J., Weiss, R., Wong, C.S., Yoshikawa-Inoue, H., 2012. A uniform, quality controlled Surface Ocean CO2 Atlas (SOCAT). *Earth Syst. Sci. Data Discuss.* 5: 735–780.

Rabalais, N., R. Turner, D. Justić, Q. Dortch, W. Wiseman, and B. Sen Gupta. 1996. Nutrient changes in the Mississippi River and system responses on the adjacent continental shelf. *Estuar. Coasts* 19: 386–407.

Richey, J.E., J.M. Melack, A.K. Aufdenkampe, V.M. Ballester, and L.L. Hess. 2002. Outgassing from Amazonian rivers and wetlands as a large tropical source of atmospheric CO2. *Nature* 416: 617–620.

Salisbury, J., M. Green, C. Hunt, and J. Campbell. 2008. Coastal acidification by rivers: a threat to shellfish? *Eos* 89: 513–514.

Salisbury, J., D. Vandemark, C. Hunt, J. Campbell, B. Jonsson, A. Mahadevan, W. McGillis, and H. Xue. 2009. Episodic riverine influence on surface DIC in the coastal Gulf of Maine. *Estuar. Coast. Shelf Sci.* 82: 108–118.

Salisbury, J., D. Vandemark, J. Campbell, C. Hunt, D. Wisser, N. Reul, and B. Chapron. 2011. Spatial and temporal coherence between Amazon River discharge, salinity, and light absorption by colored organic carbon in western tropical Atlantic surface waters. *J. Geophys. Res.* 116: C00H02.

Senjyu, T., H. Enomoto, T. Matsuno, and S. Matsui. 2006. Interannual salinity variations in the Tsushima Strait and its relation to the Changjiang discharge. *J. Oceanogr.* 62: 681–692.

Smith, S.V., and J.T. Hollibaugh. 1993. Coastal metabolism and the oceanic organic-carbon balance. *Rev. Geophys.* 31: 75–89.

Takahashi, T., S.C. Sutherland, R. Wanninkhof, Takahashi, T., Sutherland, S.C., Wanninkhof, R., Sweeney, C., Feely, R.A., Chipman, D.W., Hales, B., Friederich, G., Chavez, F., Sabine, C., Watson, A., Bakker, D.C.E., Schuster, U., Metzl, N., Yoshikawa-Inoue, H., Ishii, M., Midorikawa, T., Nojiri, Y., Körtzinger, A., Steinhoff, T., Hoppema, M., Olafsson, J., Arnarson, T.S., Tilbrook, B., Johannessen, T., Olsen, A., Bellerby, R., Wong, C.S., Delille, B., Bates, N.R., de Baar, H.J.W., 2009. Climatological mean and decadal change in surface ocean pCO_2, and net sea-air CO_2 flux over the global oceans. *Deep Sea Res. Part II Top. Stud. Oceanogr.* 56: 554–577.

Ternon, J.F., C. Oudot, A. Dessier, and D. Diverres. 2000. A seasonal tropical sink for atmospheric CO2 in the Atlantic ocean: the role of the Amazon River discharge. *Mar. Chem.* 68: 183–201.

Tseng, C.-M., K.K. Liu, G.C. Gong, P.Y. Shen, and W.J. Cai. 2011. CO2 uptake in the East China Sea relying on Changjiang runoff is prone to change. *Geophys. Res. Lett.* 38: L24609.

Tsunogai, S., S. Watanabe, and T. Sato. 1999. Is there a "continental shelf pump" for the absorption of atmospheric CO_2? *Tellus* 51B: 701–712.

Vandemark, D., J.E. Salisbury, C.W. Hunt, S.M. Shellito, J.D. Irish, W.R. McGillis, C.L. Sabine, and S.M. Maenner. 2011. Temporal and spatial dynamics of CO_2 air-sea flux in the Gulf of Maine. *J. Geophys. Res.* 116: C01012.

Walsh, J.J. 1988. On the Nature of Continental Shelves. San Diego: Academic Press.

Wang, S.L., C.T.A. Chen, G.H. Hong, and C.S. Chung. 2000. Carbon dioxide and related parameters in the East China Sea. *Cont. Shelf Res.* 20: 525–544.

Wiseman, W.J., N.N. Rabalais, R.E. Turner, S.P. Dinnel, and A. Macnaughton. 1997. Seasonal and interannual variability within the Louisiana coastal current: stratification and hypoxia. *J. Mar. Syst.* 12: 237–248.

Wisser, D., B.M. Fekete, C.J. Vörösmarty, and A.H. Schumann. 2010. Reconstructing 20th century global hydrography: a contribution to the Global Terrestrial Network-Hydrology (GTN-H). *Hydrol. Earth Syst. Sci.* 14: 1–24.

Woodwell, G.M., P.H. Rich, and C.A.S. Hall. 1973. Carbon in estuaries, In: Woodwell, G.M., and Pecan, E.V. (eds.), Carbon and the Biosphere. Springfield, VA: Atomic Commission.

Zhai, W., and M. Dai. 2009. On the seasonal variation of air – sea CO_2 fluxes in the outer Changjiang (Yangtze River) Estuary, East China Sea. *Mar. Chem.* 117: 2–10.

Zhu, W., Q. Yu, Y.Q. Tian, R.F. Chen, and G.B. Gardner. 2011. Estimation of chromophoric dissolved organic matter in the Mississippi and Atchafalaya river plume regions using above-surface hyperspectral remote sensing. *J. Geophys. Res.* 116: C02011.

8

Impacts of watershed processes on exported riverine organic carbon

N. Blair and E. L. Leithold

1. Introduction

The organic geochemical signature of terrestrial organic matter is well recognized in continental margin sediments, and its variation in the sedimentary record has long been used as an indicator of climate change through time. Rivers are the dominant source of terrestrial organic carbon (OC) to the global ocean (Hedges 1992), but the complexity and variability of riverine OC limits our ability to understand its behavior in the marine environment (Chapter 19). Understanding the processes that influence fluvial OC composition and flux would enrich interpretations of the contemporary and paleoenvironmental sedimentary records.

Predicting the future response of the coupled watershed-ocean system to anthropogenic-induced change and understanding the feedback between its biogeochemical cycling of carbon and climate will obviously require an understanding of the riverine OC signature as an integrated signal of watershed processes as well. The anthropogenic impact on river sediment and OC has been significant. Humans have accelerated soil erosion rates over nature by nearly an order of magnitude (Syvitski et al. 2005). Whereas many individual river-ocean dispersal systems have felt the impact of elevated sediment fluxes from land use, the net global riverine flux to the ocean has been greatly attenuated by dams (Syvitski et al. 2005). Today as much terrestrial OC is deposited behind reservoirs as is delivered to the ocean (Syvitski et al. 2005). Surface stocks of soil organic matter have been greatly depleted via agriculture (Lal 2003); however, it is uncertain how much of the depletion represents a net transfer to the atmosphere as opposed to translocation downstream by erosion (Kuhn et al. 2009). The engineered river has seen dramatic impacts on floodplain and delta sediment dynamics (Chapters 2, 16, 23, and 25). This in turn must influence the carbon biogeochemistry of and export from these systems. Ascertaining the role of the coupled watershed-ocean system in the global carbon cycle thus becomes an imperative if we are to gain a predictive capability for future environmental conditions.

Riverine OC is partitioned between dissolved and particulate phases (Hedges et al. 1997; Burdige 2005; Chapters 15 and 20–22). The contribution of riverine dissolved OC (DOC) to the sedimentary organic geochemical record is virtually unconsidered in the literature. Some DOC flocculates as a result of estuarine mixing, although its fate thereafter is unknown. Large estuaries act as effective DOC filters (Chapter 15). In general, riverine DOC is assumed to be degraded in the marine water column either microbially or photolytically (Hedges et al. 1997; Chapter 22).

Accordingly, the focus of this chapter is to summarize the roles that watershed properties play in influencing the nature of fluvial particulate OC (POC) exported to the ocean and its fate in the seabed. Controls of flux, mode of delivery, and POC composition can be traced to interplay of watershed lithology, tectonics, climate, hydrology, geomorphology, and ecology. These same collective factors conspire in both direct and indirect ways with marine processes to fractionate and disperse the fluvial POC, as well as to promote either degradation or preservation in the seabed.

The chapter is organized around descriptions of the dominant sources of river POC: terrestrial vegetation, soils, the deeper regolith, and in-channel production. Examples of interactions between watershed and POC characteristics are provided in each subsection. A summary of conceptual models that attempt to synthesize the dominant interactions to predict exported POC compositions is provided as a conclusion.

2. POC Sources

2.1. Vegetation

The synergy between terrestrial ecosystems and environmental factors has considerable influence over the nature of the OC produced and exported from watersheds. The impacts of climatic drivers on ecosystem composition, plant response and associated biochemical signatures, are the foundation of the interpretation of much of the organic geochemical record (Smith and Freeman 2006; Makou et al. 2007; Schouten et al. 2007; Eglinton and Eglinton 2008). Other external factors, such as soil chemistry (Wright and Fridley 2010) and altitude can influence community composition and organic geochemical signatures (Boom et al. 2002; Polissar et al. 2009). Plant succession driven by environmental perturbations such as climate and sea level change (Johnson et al. 2007; Pauli et al. 2007; Galy et al. 2008b; Nichols et al. 2010), large weather events, volcanic eruptions, fires (Turner et al. 1997), sediment trapping on deltas (Chapter 2), and human land use (Jansen et al. 2007) adds a dynamic time-dependent component to the organic carbon signature.

One of the most often used tracers of terrestrial organic carbon in the marine environment is the $^{13}C/^{12}C$ ratio or $\delta^{13}C$ value (Farquhar et al. 1989), which is defined as:

$$\delta^{13}C = 10^3 \times (R_{sample} - R_{std})/R_{std}, \quad (1)$$

where R = the $^{13}C/^{12}C$ ratio of the sample or the standard (historically PeeDee Belemnite or the more recently calibrated Vienna PDB). The $\delta^{13}C$ values of terrestrial vegetation rest in part on genomic and phenomic controls (Ainsworth et al. 2008) in addition to external environmental factors. To a first approximation, the $\delta^{13}C$ of a plant is dependent on the $^{13}C/^{12}C$ ratio of the CO_2 it sees, the intrinsic isotope effects associated with carboxylating enzymes, and the leakiness or carboxylating inefficiency of the plant (Farquhar et al. 1989). C3 plants are less efficient in their C-fixation than C4 counterparts, and thus they exhibit a greater C-isotope discrimination. The differences in $\delta^{13}C$ values of C3 ($\sim -23‰$ to $-35‰$) and C4 plants ($\sim -9‰$ to $-13‰$) (O'Leary 1981; Farquhar et al. 1989; Collister et al. 1994) is one of the most robust carbon isotopic signatures known. This isotopic indicator of plant physiology has been used to track the inputs of C4 tropical grasses and crops (especially corn or maize) against the otherwise more typical C3 background as a result of climatic

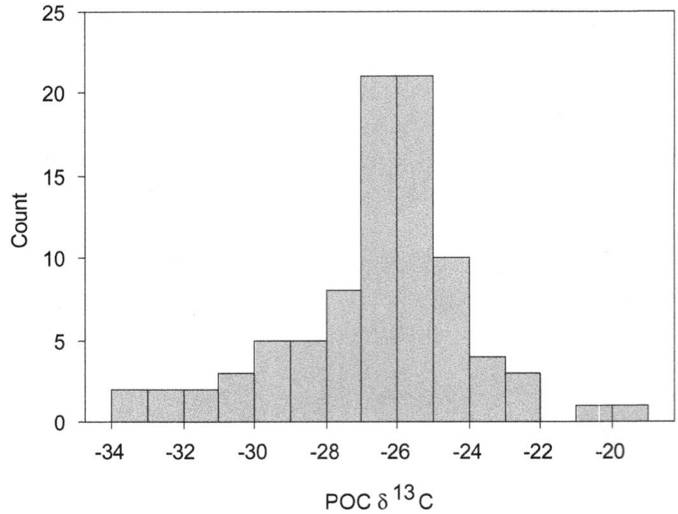

Figure 8.1. Frequencies of riverine exported POC $\delta^{13}C$ values. Rivers include the Amazon, Delaware, Haringvliet, Hudson, Ishikari, Mackenzie, Meuse, Mohawk, Parker, Rhine, Santa Clara, Susquehanna, Tokachi, York, Waipaoa, Waiapu, and Zengjiang (Alam et al. 2007; Blair et al. 2010; Gao et al. 2007; Goñi et al. 2005; Komada et al. 2004; Leithold et al. 2006; Mayorga et al. 2005; Megens et al. 2001; Nagao et al. 2005; Raymond and Bauer 2001; Raymond et al. 2004).

and anthropogenic forcings (Bernardes et al. 2004; Galy et al. 2008b; Blair et al. 2010; Carr et al. 2010; Chapters 14 and 16). The $\delta^{13}C$ distribution of riverine POC reflects the global dominance of the C3 metabolism (Fig. 8.1).

Resolving environmental forcings on plant $^{13}C/^{12}C$ content is inherently difficult because of the complex interplay of processes (Korner et al. 1991). Even so, studies of global C3-vegetation $\delta^{13}C$ patterns indicate that water availability is an important control (Arens et al. 2000; Diefendorf et al. 2010). Water stress leads to decreased CO_2 leakiness as plants attempt to conserve internal moisture levels via manipulating stomatal conductance. The end result is that the magnitude of the discrimination against ^{13}C-substituted CO_2 relative to the ^{12}C species diminishes with decreasing water availability; that is, the difference between the $\delta^{13}C$ values of the plant and atmospheric CO_2 ($\Delta\,\delta$) decreases. Other factors of relevance to watershed studies include latitudinal and altitudinal effects that appear to act through temperature and atmospheric pressure controls (Korner et al. 1991). There appear to be modest positive correlations between temperature and/or pressure and $\Delta\,\delta$. The trapping of ^{13}C depleted CO_2 from soil respiration within forest canopies generates a $^{13}C/^{12}C$ gradient with height (Da Silveira et al. 1989). Changes in atmospheric pCO_2 levels also influence vegetation $\delta^{13}C$ values (Jahren et al. 2008).

Vascular plants produce distinctive biomarkers that have been exploited in river margin studies. Long (C25-C35), straight-chained (normal or n–) alkanes with a chain-length distribution favoring the odd carbon numbers are among the extractable lipids found in protective leaf waxes (Table 8.1). Long-chained even C-numbered fatty acids and alcohols (n-C22-C32) are also common vascular

Table 8.1. *Indicators of terrestrial organic carbon*

Terrestrial OC biomarkers	Indicator/parameter	Citations
Bulk OC		
C/N elemental ratio	20–500	(Hedges et al. 1997)
$\delta^{13}C$	C3 plants: −29±6‰ C4 plants: −11±2‰	(Collister et al. 1994; Farquhar et al. 1989; O' Leary 1981)
Extractable lipids		
n-alkanes	Odd numbered C25–C35	(Eglinton and Hamilton 1967)
n-alcohols	Even numbered C22–C32	ibid.
n-fatty acids	Even numbered C22–C32	ibid.
Pentacyclic triterpenoids	Taraxerol, α-amyrin, lupeol	(Medeiros and Simoneit 2008)
Tetracyclic triterpenoids (phytosterols)	Stigmasterol, β-sitosterol	ibid.
Macromolecular components		
Lignin	Vanillyl (V), syringyl (S), cinnamyl (C) phenols	(Hedges and Mann 1979)
Cutins and Suberans	Hydroxy mono- and dioic substituted fatty acids	(Goñi and Hedges 1990)
Sporopollenin	Pollen and lower plant spores	(Derenne and Largeau 2001)
Tannins	Polyhydroxy aromatics (e.g. gallic acid)	(Nierop and Filley 2008)
Diagenetically altered (soil) OC		
Glycerol dialkyl glycerol tetraethers	Branched and Isoprenoid Tetraether Index (BIT)	(Hopmans et al. 2004)
3,5-Dihydroxyybenzoic acid	3,5-Bd/V phenol ratio	(Gordon and Goñi 2003)

plant indicators. Other extractable lipids include the phytosterols, stigmasterol and β-sitosterol, and various less-well studied pentacyclic triterpenoids (Table 8.1).

Vascular plants produce abundant non-extractable macromolecular organic species that must be hydrolyzed and/or partially oxidized prior to detection and analysis (Table 8.1). Of these, lignin is the most abundant vascular plant-specific biomarker. Analysis is typically accomplished via either an oxidation using alkaline CuO (Hedges and Mann 1979) or a transmethylation reaction using tetramethyl ammonium hydroxide (Clifford et al. 1995), whereby both methods release the phenolic building blocks of the lignin. Distinctions between woody/nonwoody sources (forests vs. grassland; Leopold et al. 1982; Castaneda et al. 2009) and gymnosperms/angiosperms can be resolved by studying specific phenol ratios. The same analytical methods release monomers from cutins, suberins, and tannins (Table 8.1). The abundance of lignin and cellulose in vascular plants, both N-free materials, elevates elemental C/N ratios to characteristic values greater than 20 (Hedges et al. 1997).

Compound-specific isotopic measurements provide an additional dimension to the interpretation of the organic geochemical record in soils and sediments. $\delta^{13}C$ measurements of terrestrial lipids and lignin have been used to resolve C3 and C4 terrestrial plant sources from each other and from marine organic carbon (Collister et al. 1994; Goñi and Eglinton 1996). The radiocarbon contents of vascular plant fatty acids and lignin provide measures of residence time within the watershed (Galy and Eglinton 2011; Chapter 14). Stable H-isotopic compositions (δD) of plant alkanes have been used to detect changes in precipitation patterns and paleoaltimetry (Smith and Freeman 2006; Smith et al. 2007; Polissar et al. 2009).

2.2. Soils

Viewed from the perspective of the marine organic geochemist, soil organic matter is often treated as a single and almost discrete source of diagenetically altered material recognized by its N-enrichment relative to the original plant OC (Hedges and Oades 1997; Sollins et al. 2009), oxidized lignin (Kuzyk et al. 2008; Chapter 22), microbial biomarkers (Cooke et al. 2008a; Cooke et al. 2008b), and ^{14}C-age (Blair et al. 2010; Drenzek et al. 2009; Chapters 14 and 16). In reality, the soil OC is an inherently complex suite of mixtures of young and old material with ranges of diagenetic states that are reflective of soil particle history within the watershed. Black carbon from combustion, kerogen from sedimentary rocks, and petrogenic (metamorphic) graphite can all enter the soil pool (Blair et al. 2003; Gonzalez-Perez et al. 2004; Bouchez et al. 2010) as intrinsically recalcitrant fractions. Anthropogenic materials, such as pesticides, pharmaceuticals, and fossil fuel byproducts, are incorporated into soils (Chefetz et al. 2008; Gao et al. 2009; Larsbo et al. 2009; Komárek et al. 2010).

Above-ground plant litter and below-ground biomass are the primary sources of the soil OC pool. The diagenesis of this material to form what is generically referred to as soil organic matter has been reviewed previously (Hedges and Oades 1997). As plant-derived material is heterotrophically cycled within the soil, a transition from large OC fragments with adhered soil particles to organically glued clay aggregates occurs (Hedges and Oades 1997; Sollins et al. 2007). The aggregation process may play a role in the stabilization of soil OC by limiting exposure to oxidants, microbially released exoenzymes and/or water (Jastrow and Miller 1997; Lehmann et al. 2007; Bachmann et al. 2008; Schmidt et al. 2011). Disaggregration processes, such as via tillage, re-expose the sequestered OC, thereby accelerating its oxidation (Golchin et al. 1995; Angers and Chenu 1997). Sorption via inorganic minerals and especially iron oxides also stabilizes the OC (Haider 1999; Eusterhues et al. 2003; Aufdenkampe and Hedges 2005; Basile-Doelsch et al. 2007; Kaiser and Guggenberger 2007).

Some macromolecular polymers from plants and microbial sources, such as lignin, cutin, and chitin, have been argued to be recalcitrant and thus to be important contributors to the soil OC pool (Poirier et al. 2003; Lorenz et al. 2007; Rillig et al. 2007). Recent studies indicate these compounds can turn over rapidly in soils and thus are conditionally reactive (Schmidt et al. 2011). Their persistence appears to rely on the same sequestration mechanisms outlined previously, albeit they may be more prone to preservation under anoxic conditions.

There are numerous environmental parameters that can influence the quantity and nature of the soil OC. Aside from the OC inputs, precipitation patterns and temperature exert control over soil OC composition primarily by influencing diagenetic processes. The extent of degradation tends to increase with temperature (Schimel et al. 1994). Oscillating iron redox cycles can increase OC mobility (Thompson et al. 2006) and, by inference, reactivity. Mineralogical controls on soil column permeability and particle surface area can influence OC content and composition (Bachmann et al. 2008).

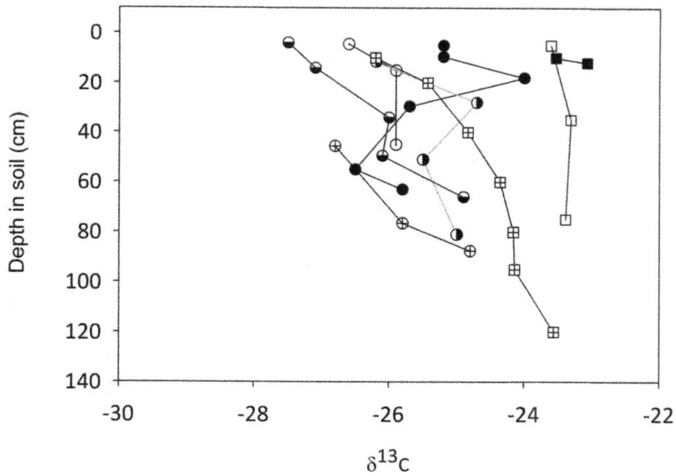

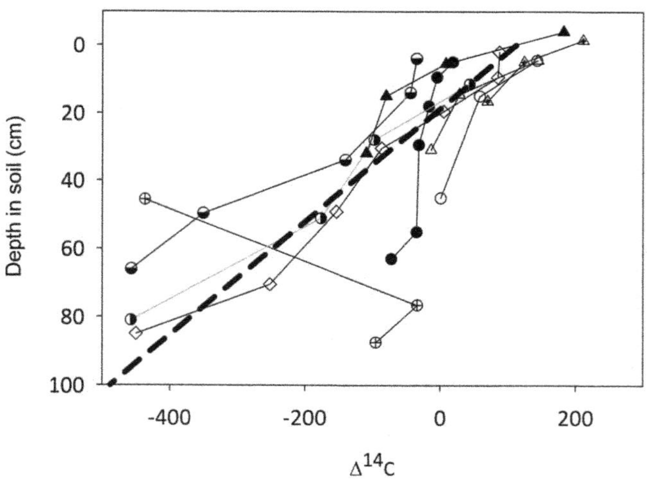

Figure 8.2. $\delta^{13}C$ (top) and $\Delta^{14}C$ (bottom) values of soil POC. Soils are from the North Island of New Zealand (Blair et al. 2010).

The stable and radio-carbon isotopic compositions of soil OC reflect source inputs, diagenetic processing, vertical mixing, and in the case of the radiocarbon, ^{14}C-decay (Ehleringer et al. 2000; Trumbore 2009). Surface horizons, and especially the upper 10 cm or less, capture the isotopic compositions of the vegetation inputs. An enrichment of ^{13}C (more positive $\delta^{13}C$ values) with depth is often seen (Fig. 8.2), and this has been attributed primarily to the release of ^{13}C-depleted CO_2 via respiration (Ehleringer et al. 2000; Bostrom et al. 2007). Mixing of surface-supplied terrestrial OC with marine sedimentary rock-derived parent material, which tends to be relatively ^{13}C-enriched, could generate the same isotopic gradients, however. ^{14}C decay decreases radiocarbon content with depth or age of soil (Fig. 8.2). The ^{14}C-depth gradient is attenuated via vertical mixing associated with the penetration of the rhizosphere into lower soil horizons, burrowing animals, and the downward percolation of dissolved OC. Selective respiration of young (^{14}C-rich) and reactive fractions, much as seen in marine sediments (Blair and Aller 2012), should also be an important process. Deviations from these general isotopic trends indicate that non–steady-state OC inputs overwhelm the diagenetic signal.

Hypothetically, the complexity of soil OC composition should increase with distance from the headwater region as soil is lost from upland areas and transported to downstream storage (Aufdenkampe et al. 2007; Hedges et al. 2000). Downstream sediment reservoirs, such as floodplains, integrate and degrade OC from upland sources as they develop their own OC signatures derived from colonizing vegetation and soil formation (Bechtold and Naiman 2009; Zehetner et al. 2009; Blair et al. 2010; Bouchez et al. 2010). This, and the many factors discussed previously, will produce a broad range of soil OC sources within a watershed.

2.3. The deeper regolith and geomorphic processes

Sedimentary rocks, and especially mudstones, can contribute significant quantities of ancient OC to the riverine pool (Kao and Liu 1996; Leithold and Blair 2001; Masiello and Druffel 2001; Blair et al. 2003; Gomez et al. 2003; Komada et al. 2004; Komada et al. 2005; Leithold et al. 2006; Galy et al. 2008a; Hilton et al. 2008a; Blair et al. 2010; Chapter 14). The relative importance of sedimentary rock C tends to scale with watershed sediment yield (Meybeck 1993; Ludwig et al. 1996; Komada et al. 2004; Leithold et al. 2006). In a study of small rivers along the Pacific Rim, ~8–80% of the riverine POC was attributed to rock carbon with the highest contributions associated with the muddiest rivers (Leithold et al. 2006). Rock OC itself is variable in composition, being derived from a range of sources and having traveled diverse diagenetic and catagenetic paths (Vandenbroucke and Largeau 2007). This aspect of riverine OC is understudied and has been barely exploited (Blair et al. 2010). Metamorphic rocks may contribute petrogenic graphite (Dickens et al. 2004). Igneous rock should make minimal contributions to the organic carbon pool.

The balance of contemporary, aged soil and rock OC delivered to rivers is influenced by the sizes of those pools in the watershed and sediment generation and transport processes. Surface erosion, such as sheet wash and rilling, will preferentially delivery younger material. This process occurs over all but the lowest hydrographic regimes and is thus a chronic source. As it is often dominated by plant debris, it can be easily missed in bulk sediment budgets because it is a low-volume, though high–OC content source (Page et al. 2004; Blair et al. 2010). Gullies and channel banks are also chronic sediment sources and each deliver aged soil C and in some cases, ancient C to the channel (Gomez et al. 2003; Blair et al. 2010). Mass wasting processes that excavate deeper into hillslopes will access older soils and rock sources and generate high sediment yields. Commonly there is interplay between lithology, tectonics, climate, and anthropogenic disturbances that contributes to deeper incision of the regolith (Hilton et al. 2008b; Kao and Milliman 2008). Steep landscapes underlain by soft, and often tectonically crushed, sedimentary rock are prone to mass wasting, especially during intense precipitation events (Hicks et al. 2000; Reid and Page 2003; Goldsmith et al. 2008; Hilton et al. 2008b). Human land use, and especially deforestation and road building, decreases the precipitation threshold for failures. Earthquakes can also destabilize hillslopes and trigger landslides and debris flows (Dadson et al. 2004; Hovius et al. 2011). Carbon sources from soils and rocks are blended as these materials are stored down-slope, such as in slide debris tails (Hilton et al. 2008a). Storage in and remobilization from down-slope sedimentary reservoirs appear to buffer changes in OC composition as a function of hydrograph and precipitation event intensity (Hilton et al. 2008a; Blair et al. 2010).

2.4. In-channel and lowland sources and processes

The view of the river channel acting as a pipeline of terrestrial OM to the ocean has evolved to one of the fluvial system as a reactor (Bouillon et al. 2009). In reality, both the pipeline and reactor models are apt descriptions of rivers, albeit under different conditions. Small, mountainous rivers during high flow events behave more as pipelines because in-channel residence times are as short as hours to days, the time it takes a pulse of floodwater to move from headwaters to mouth (Goldsmith et al. 2008). Under more tranquil low-flow conditions, riverine primary production becomes more important within the system, especially from a stream ecosystem perspective (Power 1991; Hilton et al. 2006; Bouillon et al. 2009). On a mass flux basis, however, in-channel production is considered to be a minor contribution to the OM exported by the high-sediment yield systems (Blair et al. 2003; Leithold et al. 2006; Hilton et al. 2008a).

Rivers transition to reactor behavior as they become larger, and the low gradient portion of the channel begins to dominate the hydrograph. Recycling of OC leads to biogeochemical "spiraling" of materials as they are transported downstream (Newbold et al. 1982; Richey et al. 1990). Even when primary production occurs in-channel, many rivers may exhibit net heterotrophic behavior because external inputs of OC from terrestrial sources contribute to remineralization (Duarte and Prairie 2005). Light availability, as influenced by factors such as water column turbidity, turbulence, riparian shading, and season, exerts control over the aquatic production (Vannote et al. 1980; Young and Huryn 1999; Salmaso and Zignin 2010). Sunlight exposure also leads to photochemical degradation of DOC (Amon and Benner 1996b). Nutrient levels stimulate primary production when light is not limiting (Hilton et al. 2006). Observations of elevated phytoplankton biomass in the Mississippi River may be the result of lower suspended loads resulting from soil conservation measures and increased agricultural nutrient inputs (Chapter 16).

Tracking in-channel sources of OC into the marine environment is difficult and not commonly attempted because of the similarity of biomarkers from freshwater and marine aquatic sources (Waterson and Canuel 2008; Schmidt et al. 2010). Whereas marine OC has a relatively well-constrained carbon stable isotope composition (~ -19 to -24‰), the same is not true for stream and river sources because of highly variable dissolved inorganic carbon (DIC) isotopic compositions and other factors such as water velocity (Fry and Sherr 1984; Peterson and Fry 1987; Finlay et al. 2002). The use of isotopic measurements to distinguish freshwater and marine aquatic sources will depend on the characteristics of the particular system.

Anthropogenic organic compounds, such as biocides, pharmaceuticals, and petroleum products, are delivered to waterways via atmospheric deposition, surface runoff, groundwater inputs, and direct, point source injection (Ackerman et al. 2008; Loos et al. 2009; Maria and Maria 2009; Blocksom et al. 2010). These materials are often trace pollutants (Kronimus and Schwarzbauer 2007; Loos et al.

2009); however, petroleum releases and waste treatment plant inputs can perturb bulk parameters such as $\Delta^{14}C$ (Spiker and Rubin 1975; Griffith et al. 2009).

3. Climate impacts on riverine OC

Numerous effects of anthropogenic climate change on riverine POC are anticipated in the next century and in some cases may already be manifest. Warming temperatures as well as increased frequency and magnitude of heavy precipitation events (IPCC 2007; Smith et al. 2009; Trenberth 2011) may potentially cause shifts in the amount and character of riverine POC in diverse settings.

Amplified climatic warming of the Arctic, where permafrost contains half the global soil OC (SOC) pool (Tarnocai et al. 2009), is anticipated to have significant impacts on fluvial OC. The OC carried by Arctic rivers today is dominantly DOC (Rachold et al. 2004; McGuire et al. 2009), which radiocarbon analyses indicate is predominately young, recently fixed carbon (Guo et al. 2007; Raymond et al. 2007; Chapters 20–22). POC in these rivers, in contrast, has mean radiocarbon ages of several thousands of years and is derived from permafrost and peats (Goñi et al. 2005; Guo and Macdonald 2006; Guo et al. 2007; Drenzek et al. 2009; Vonk et al. 2010; Chapters 20–22). Warming of the Arctic is expected to increasingly lead to release of this old material into fluvial systems via mechanisms including thermokarst and river bank erosion (Bowden et al. 2008; Gustafsson et al. 2011). POC export from Siberian Rivers is expected to increase along with a projected 30–122% increase in sediment flux by 2100 (Gordeev 2006; Gordeev and Kravchishina 2009). Vonk et al. (2010) presented evidence that pre-aged soil OC released from deep mineral soils and discharged from the Kalix River undergoes little degradation after discharge and transport across Bothian Bay on the Barents Sea. In contrast, a younger pool of remobilized SOC from upper soil horizons and shallow peatlands in the Kalix catchment degrades rapidly in the coastal water column and would be expected to add greenhouse gases to the atmosphere.

Mass wasting in some mountainous and upland river catchments is anticipated to increase as a result of temperature and precipitation changes. In high mountains, permafrost melting and glacial retreat may increase the frequency of rockfalls, landslides, and debris flows (Huggel et al. 2012; Stoffel and Huggel 2012), potentially increasing contributions of rock carbon to fluvial sediment loads. At lower elevations, shallow, rain-triggered landslides are expected to become more common as intensity-duration thresholds are exceeded with greater frequency (Jakob and Lambert 2009; Jomelli et al. 2009; Crozier 2010). Research in New Zealand and Taiwan (Hilton et al. 2012; Hilton et al. 2011) points to the increasingly important role that such storm-driven landslides will likely play in delivering OC from biomass and soils to mountainous rivers.

Along with changing land use patterns, climate change may also have significant impacts on POC sources in lowland environments. Increased temperatures can be anticipated to accelerate the turnover of OC in floodplain soils, leading to more extensive oxidation of rock and soil C delivered from upland sources (Bouchez et al. 2011). The impact of changing precipitation patterns on fine-grained sediment sources to the upper Mississippi River has been noted (Belmont et al. 2011). In that area, increased river discharge has led to a shift in the dominant source of sediment from agricultural erosion to erosion of stream banks and bluffs. An increased supply of aged soil carbon to the channel network would be expected as a result of this fluvial incision.

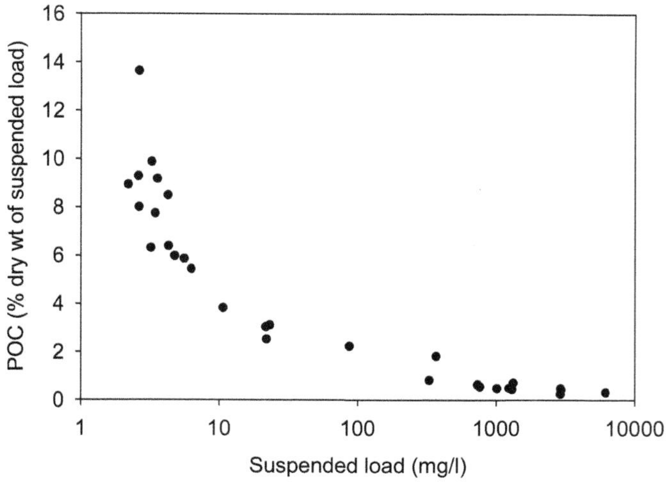

Figure 8.3. POC concentration as a function of total suspended load for Ishikari, Ganga, Brahamaputra, and Zengjiang Rivers (Alam et al. 2007; Galy et al. 2007; Gao et al. 2007).

4. Predicting the exported riverine organic signature

The complexity of the sources of OC to rivers makes characterizing the composition and reactivity of fluvial export a challenging endeavor. The task cannot be accomplished with measurements alone because of the large number of rivers worldwide and the need to sample many as a function of hydrograph. Models that fill in between measured datasets are necessary to obtain a comprehensive perspective (Blair et al. 2004; Wheatcroft et al. 2010). The relationship between POC concentration and total suspended solids (TSS) is an example of how a simple empirical model could be brought to bear on the problem (Fig. 8.3). As noted by Meybeck (1993) and Ludwig et al. (1996), POC concentrations approach an asymptotic value of approximately 0.5% on a dry weight basis at high sediment loads. As hypothesized by Meybeck (1993) and subsequently demonstrated using carbon isotopic measurements (Kao and Liu 1996; Leithold and Blair 2001; Masiello and Druffel 2001; Blair et al. 2003; Gomez et al. 2003; Komada et al. 2004; Komada et al. 2005; Galy et al. 2008a; Hilton et al. 2008a; Blair et al. 2010), the relationship is the result of mixing between modern biogenic OC sources (primarily plant debris) and lithogenic sediment (Fig. 8.3). The asymptotic concentration of ~0.5% C reflects the background of sedimentary rock-derived organics. Thus by simply knowing the suspended load in a river, one could predict in a general sense not only POC concentration, but also the approximate balance of reactive (modern) and unreactive (ancient) material. This approach does not consider the contribution of aged soil C (Drenzek et al. 2009).

More process-driven models have been proposed that are based on relatively simple and intuitive concepts. As described by river continuum conceptual model (Vannote et al. 1980), coarse POC (CPOM) derived from predominantly riparian sources is degraded to fine-grained (FPOM) and ultra-fine material (UPOM) as the POC moves from headwaters to higher order streams. More generally and without reference to stream order, the size-reactivity continuum model similarly argues that large

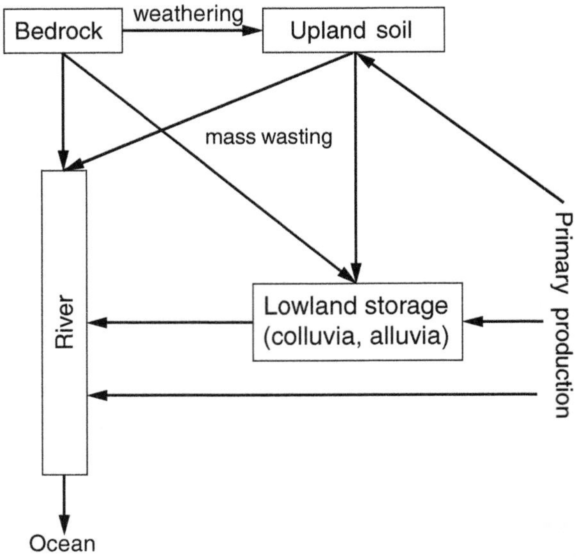

Figure 8.4. Watersheds may be viewed as a collection of coupled bioreactors that mix ingredients and degrade and/or modify them. The POC exported to the marine environment is an integration of the products from the bioreactors, which include upland and lowland soils as well as the river channel. The extent of overprinting or replacement of POC signatures from an upstream location will depend on multiple factors including climate and residence time within the downstream bioreactor.

OC particles (i.e., plant debris) are degraded to smaller, less reactive fractions (Amon and Benner 1996a; Hedges and Oades 1997). Taken together, one would logically predict that POC particle size and reactivity should decrease with distance down river, and the extent of that decrease should scale with the size of the watershed. Differential transport of the riverine OC fractions must also be considered. As an example, the regional chromatography model is built on the supposition that different fractions of OC, and especially those dissolved or fine-grained, will migrate through the watershed at different rates and thus possess varying ages, degrees of degradation, and reactivity (Hedges et al. 1986; Hedges et al. 2000).

None of the models completely explained organic geochemical patterns when put to a test in the Amazon basin (Hedges et al. 2000). Where the models most fail is in the description of the relative behaviors of coarse and fine-grained POC and DOC. This failure results in part because of an unstated but implicit assumption that the various fractions of OC within a sample are products of a single diagenetic continuum. This is not the case because mass wasting processes, such as those described earlier, expose and mix organics that range in age from the contemporary to millions of years. Multiple diagenetic lineages exist within samples. Particle genealogy and history matter.

The generation and diagenetic processing of multi-cycle POC, material derived from multiple C-cycles, is the conceptual driver behind the bedrock-to-burial (or residence time) model (Blair et al. 2004; Fig. 8.4). In it, POC evolves as particulates move from upland to the ocean by the mixing and turnover of sources in bio-reactive storage reservoirs such as soils. As an example, deposition

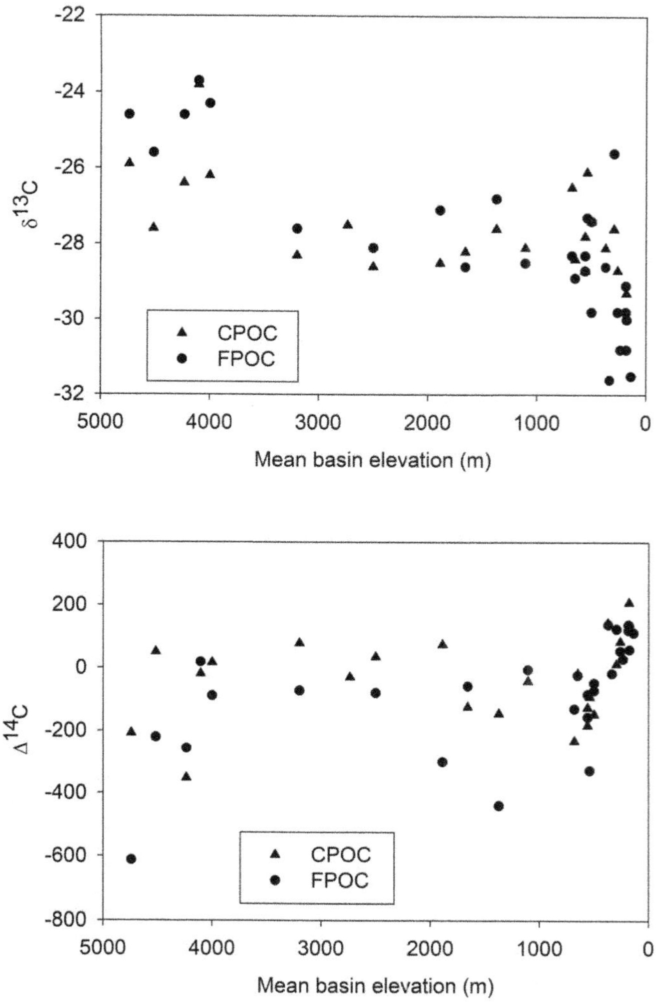

Figure 8.5. $\delta^{13}C$ (top) and $\Delta^{14}C$ (bottom) values of coarse (63–2,000 μm) and fine (0.1–63 μm) POC from Amazon River and tributaries as functions of mean basin elevation in meters. The changes in isotopic composition reflect particle history as upland POC sources (vegetation, soils, and eroded sedimentary rock) are slowly overprinted and replaced by OC generated at lower elevations. Data are from Mayorga et al. 2005.

in a floodplain of sedimentary POC derived from upland sources will lead to the gradual loss of the upland signature via heterotrophic reactions and replacement with that derived from the floodplain biome. The evolution of the Mississippi delta splay via sediment infilling, colonization by vegetation, development of a mature plant community and slowing of sediment input, and finally decay and subsidence illustrates the time-dependent, dynamic nature of a riverine bioreactor (Chapter 2).

The evolution of riverine carbon isotope values is evident in the Amazon River watershed where the upland signatures are replaced by those from the floodplain (Fig. 8.5). Bypassing or traveling rapidly

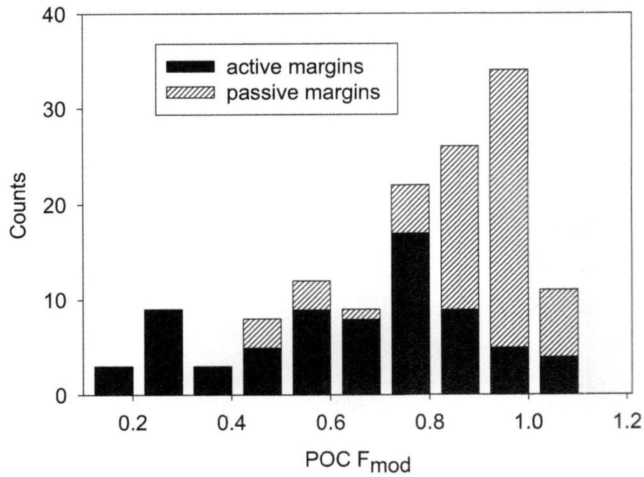

Figure 8.6. Frequencies of river POC ^{14}C-contents (F_{mod}). ^{14}C-contents span almost the entire range of possible values reflecting mixing of ancient (sedimentary rock-derived, $F_{mod} = 0$) and more modern (contemporary and aged soil) OC. Small mountainous rivers aligned along active margins tend to export the most ^{14}C-depleted material because of rapid mechanical weathering of sedimentary rock. The ancient OC derived from the rocks is slowly lost during storage in larger, passive margin systems (Blair and Aller 2012).

through the floodplain reactor preserves the upstream POC signal. Residence time in the bio-active reservoirs thus becomes an important parameter. In a conceptually parallel treatment, river DOC reactivity has been described to be a function of sources, hydrologic flow paths, and the residence time in reactive reservoirs (Battin et al. 2008).

Riverine POC ^{14}C provides a test of the residence time model. Rivers export POC with radiocarbon contents (expressed as fraction modern C, F_{mod}) that span almost all permissible values (Fig. 8.6). Hypothetically, one might expect that exported POC would age or become more ^{14}C-depleted as a function of watershed size because of greater storage opportunities. A compilation of riverine POC radiocarbon measurements does not reveal the hypothetical relationship when only bulk measurements are considered (Blair and Aller 2012; Fig. 8.7). As described earlier, riverine POC is commonly a mixture of materials derived from contemporary, aged soil and rock C sources, and these mixtures can vary as a function of hydrograph and suspended load (Fig. 8.8). Consequently bulk POC radiocarbon signals can be highly variable for an individual river and not directly related to watershed area. The mixture compositions can be characterized in terms of rock C concentrations and ^{14}C-content of the non-rock (soil + vegetation) component, and these two parameters can be extracted from data using plots of ^{14}C-content versus %C (Fig. 8.9). The rock C content of a sediment sample equals the %C value at $F_{mod} = 0$. The F_{mod} of the non-rock C is the asymptotic value at large %C contents. Based on our compilation, the global mean rock C concentration in riverine sediments is $0.4 \pm 0.05\%$, a value similar to that estimated by Meybeck (1993) of ~0.5%. The global exported F_{mod} of the non-rock component is 1.00 ± 0.01.

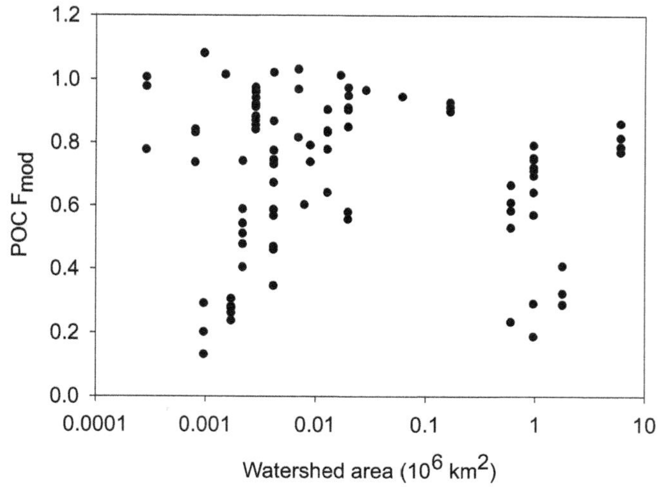

Figure 8.7. Riverine POC ^{14}C-content (F_{mod}) versus the watershed area for global rivers (Kao and Liu 1996; Masiello and Druffel 2001; Megens et al. 2001; Raymond and Bauer 2001; Komada et al. 2004; Raymond et al. 2004; Goñi et al. 2005; Nagao et al. 2005; Leithold et al. 2006; Alam et al. 2007; Galy et al. 2007; Gao et al. 2007; Blair et al. 2010). Vertical groups of data are from the same river and represent the variability resulting from a changing hydrograph and sediment load.

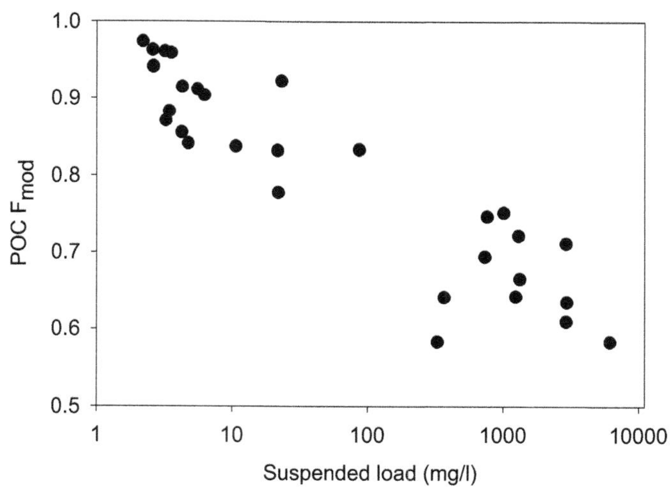

Figure 8.8. The relation between riverine POC ^{14}C-content (F_{mod}) and suspended load for the Ishikari, Ganga, Brahamaputra, and Zengjiang Rivers (Alam et al. 2007; Galy et al. 2007; Gao et al. 2007). ^{14}C-free fossil C dilutes the radiocarbon content of the POC at high suspended sediment loads.

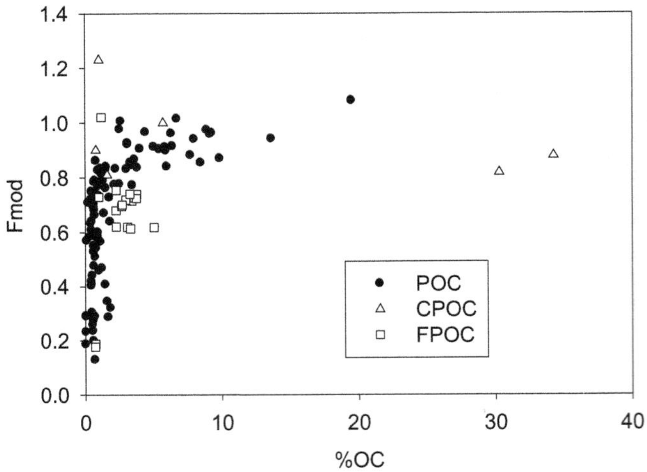

Figure 8.9. Riverine POC ^{14}C-content (F_{mod}) as a function of %OC in suspended load (Ishikari, Ganga, Brhamaputra, Lower Megha, Zengjiang, Santa Clara, Mackenzie, Waipaoa, Waiapu, Langyang, Eel, Noyo, Navarro, Suislaw, Tokachi, Rhine, Strickland, Fly, Amazon, Piracica). Data include bulk POC and coarse (plant debris, CPOC) and fine (lithogenic-dominated, FPOC) fractions (Krusche et al. 2002; Alin et al. 2008). Additional citations are in captions of Figures 8.1 and 8.7.

It is somewhat problematic to estimate the two parameters for individual rivers because in many cases radiocarbon C and %C measurements sufficient to fully define a mixing curve do not exist. A mixing model can be utilized, however, by manipulating the following mass balance equations,

$$F_r c_r = F_k c_k + F_t c_t \tag{2}$$

$$c_r = c_k + c_t \tag{3}$$

where F_r, is the fraction modern C of the riverine POC and F_k, F_t are the F_{mod} values of the constituent rock (kerogen) and non-rock (terrestrial) OC end-members. POC concentrations (%; 100 × mass OC/mass dry sediment) for the river suspended sediment and the material derived from the two sources are given by c_r, c_k, and c_t, respectively.

Equations (2) and (3) can be combined and rearranged to

$$F_r c_r = F_t c_r + c_k(F_k - F_t). \tag{4}$$

Plotting $F_r c_r$ as a function of c_r will reveal a linear relationship with slope F_t if the end-member isotopic compositions are reasonably invariant. In principle it is possible to produce linear relationships from random associations of F_r and c_r using Equation 4. Artifacts are avoided by using datasets that exhibit a relationship between F and c consistent with two-end-member mixing and by pursuing additional modes of validation such as by the radiocarbon analysis of particle separates (Blair et al. 2010).

The effect of watershed size can be seen when the rock C concentration and non-rock F_{mod} are resolved. Rock C concentrations are lowest in the largest watersheds, and this is presumably due to the slow oxidation of the POC derived from sedimentary rocks on prolonged storage in lowland soils (Fig. 8.10). However, the youngest non-rock components are associated with the smallest

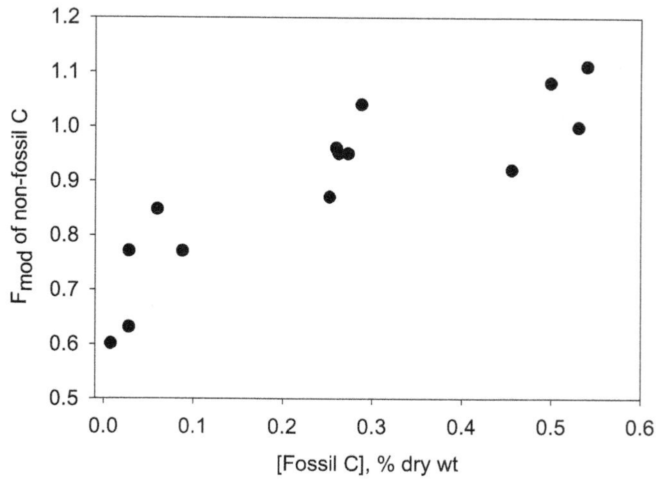

Figure 8.10. The fossil (rock-derived) OC content of riverine POC versus the ^{14}C-content of the non-fossil component. Small rivers, such as the Ishikari of Japan and Langyang of Taiwan, transport the highest concentrations of fossil C and the youngest non-fossil fraction. The Amazon and Ganges-Brahmaputra systems export the lowest concentrations of fossil C and the oldest non-fossil component.

watersheds as a result of short residence times (Fig. 8.10). Some of the smallest watersheds (Ishikari and Langyang) appear to export a nearly biomodal mixture of rock and contemporary C, whereas the largest systems (Amazon, Ganges, Brahmaputra) delivery more complex mixtures (Chapters 14 and 16).

Watershed size is not the only variable that can influence storage and POC radiocarbon contents. Climate, tectonics, and geologic history can be factors. The Waipaoa River of New Zealand's North Island is an example in which tephric soils accumulated on bedrock and alluvial terraces that formed during a dryer, cooler period of the Last Glacial Maximum (Berryman et al. 2000). Destabilization of these soils has been accelerated by anthropogenic deforestation, which has increased river sediment loads five-fold (Foster and Carter 1997). Thus in spite of the small size of the watershed (2,200 km^2), the Waipaoa transports a non-rock component with a mean age of ~1,000 years (Blair et al. 2010) that must result from relict storage. Non–steady-state processes will thus complicate interpretation of riverine POC signatures.

5. Summary

The particulate organic carbon exported to the ocean results from a complex integration of sources and processes. Watersheds may be viewed as coupled arrays of biogeochemical reactors that leak out product via the river. Within these reactors, organic fractions with vastly different ages and reactivities are cycled. Contemporary POC derived from extant ecosystems turns over rapidly within soils. The composition of the contemporary fraction is controlled dominantly by recent climate and anthropogenic drivers such as land use. Aged soil C (~10^2–10^4 yrs) is derived from the diagenetic processing of ecosystem inputs, and its age is reflective of its modest reactivity. Retardation of POC oxidation is accomplished in part by aggregation of soil particles. Disaggregation in the marine environment is expected to reactivate this pool of C. Fossil (petrogenic) C originates from the sedimentary rocks that commonly contribute to the bedrock underlying watersheds. Rapid tectonic uplift of mudstones coupled with erosion that is accelerated by seismic events (earthquakes), storms, and/or by human activities can cause this ancient ($>10^6$ yrs) and highly unreactive source to be an important component of riverine POC. The relative contribution of these three general sources varies among watersheds. One hypothetical end-member environment is that of the active margin small mountainous river from which a mixture of fossil rock-derived C and a non-rock fraction that is skewed toward the modern is exported. Large passive margin rivers are the counter example in which the fossil component is small and the non-rock fraction is dominated by contemporary and aged soil pools. These differences will greatly influence the behavior of the river POC when delivered to the ocean (Blair and Aller 2012). Both climate and land use change are expected to accentuate the delivery of older terrestrial POC fractions to the global ocean, although rock C fluxes from large watersheds may decrease because warmer temperatures could accelerate the oxidation of the ancient material (Bouchez et al. 2010).

Acknowledgments

Numerous scientific colleagues, including students and support personnel, assisted over the years in the river research that helped provide the perspectives for this review, and especially those involved in the Amasseds, Strataform, and MARGINS S2S field programs. Particular thanks to C. Nittrouer who headed several of these efforts. Essential assistance was provided by our colleagues in New Zealand, including Noel Trustrum, Mike Page and Hannah Brackley of GNS, Mike Marden of Landcare and Alan Palmer of Massey University. The research was funded by NSF Grants EAR-0222584 and OCE-0646159.

References

Ackerman, L.K., A.R. Schwindt, S.L. Massey Simonich, et al. 2008. Atmospherically Deposited PBDEs, Pesticides, PCBs, and PAHs in Western U.S. National Park fish: concentrations and consumption guidelines. *Environ. Sci. Technol.* 42: 2334–2341.

Ainsworth, E.A., C. Beier, C. Calfapietra, et al. 2008. Next generation of elevated CO_2 experiments with crops: a critical investment for feeding the future world. *Plant Cell Env.* 31: 1317–1324.

Alam, M.J., S. Nagao, T. Aramaki, Y. Shibata, and M. Yoneda. 2007. Transport of particulate organic matter in the Ishikari River, Japan during spring and summer. *Nucl. Instrum. Methods Phys. Res. Sect. B* 259: 513–517.

Alin, S.R., R. Aalto, M.A. Goñi, J.E. Richey, and W.E. Dietrich. 2008. Biogeochemical characterization of carbon sources in the Strickland and Fly rivers, Papua New Guinea. *J. Geophys. Res.* 113: F01S05.

Amon, R.M.W., and R. Benner. 1996a. Bacterial utilization of different size classes of dissolved organic matter. *Limnol. Oceanogr.* 41: 41–51.

Amon, R.M.W., and R. Benner. 1996b. Photochemical and microbial consumption of dissolved organic carbon and dissolved oxygen in the Amazon River system. *Geochim. Cosmochim. Acta* 60: 1783–1792.

Angers, D.A., and C. Chenu. 1997. Dynamics of soil aggregation and C sequestration, In: R. Lal, Kimble, J.M., Follett, R.F., Stewart, B.A. (eds.), Soil Processes and the Carbon Cycle, pp. 199–206. Boca Raton, FL: CRC Press.

Arens, N.C., A.H. Jahren, and R. Amundson. 2000. Can C3 plants faithfully record the carbon isotopic composition of atmospheric carbon dioxide? *Paleobiology* 26: 137–164.

Aufdenkampe, A.K., and J.I. Hedges. 2005. Is the characteristic composition of mineral-associated organic matter the consequence of sorption? *Abstr. Pap. Am. Chem. S.* 230: U1796–U1796.

Aufdenkampe, A.K., E. Mayorga, J.I. Hedges, et al. 2007. Organic matter in the Peruvian headwaters of the Amazon: compositional evolution from the Andes to the lowland Amazon mainstem. *Org. Geochem.* 38: 337–364.

Bachmann, J., G. Guggenberger, T. Baumgartl, et al. 2008. Physical carbon-sequestration mechanisms under special consideration of soil wettability. *J. Plant Nutr. Soil Sci. Zeitschrift Fur Pflanzenernahrung Und Bodenkunde* 171: 14–26.

Basile-Doelsch, I., R. Amundson, W.E.E. Stone, et al. 2007. Mineral control of carbon pools in a volcanic soil horizon. *Geoderma* 137: 477–489.

Battin, T.J., L.A. Kaplan, S. Findlay, et al. 2008. Biophysical controls on organic carbon fluxes in fluvial networks. *Nat. Geosci.* 1: 95–100.

Bechtold, J.S., and R.J. Naiman. 2009. A quantitative model of soil organic matter accumulation during floodplain primary succession. *Ecosystems* 12: 1352–1368.

Belmont, P., K.B. Gran, S.P. Schottler, et al. 2011. Large shift in source of fine sediment in the upper Mississippi River. *Environ. Sci. Technol.* 45: 8804–8810.

Bernardes, M.C., L.A. Martinelli, A.V. Krusche, et al. 2004. Riverine organic matter composition as a function of land use changes, Southwest Amazon. *Ecol. Appl.* 14: S263–S279.

Berryman, K., M. Marden, D. Eden, C. Mazengarb, Y. Ota, and I. Moriya. 2000. Tectonic and paleoclimatic significance of Quaternary river terraces of the Waipaoa River, east coast, North Island, New Zealand. *New Zeal. J. Geol. Geophys.* 43: 229–245.

Blair, N.E., and R.C. Aller. 2012. The fate of terrestrial organic carbon in the marine environment. *Annu. Rev. Mar. Sci.* 4: 401–423.

Blair, N.E., E.L. Leithold, and R.C. Aller. 2004. From bedrock to burial: the evolution of particulate organic carbon across coupled watershed-continental margin systems. *Mar. Chem.* 92: 141–156.

Blair, N.E., E.L. Leithold, H. Brackley, N. Trustrum, M. Page, and L. Childress. 2010. Terrestrial sources and export of particulate organic carbon in the Waipaoa sedimentary system: problems, progress and processes. *Mar. Geol.* 270: 108–118.

Blair, N.E., E.L. Leithold, S.T. Ford, K.A. Peeler, J.C. Holmes, and D.W. Perkey. 2003. The persistence of memory: the fate of ancient sedimentary organic carbon in a modern sedimentary system. *Geochim. Cosmochim. Acta* 67: 63–73.

Blocksom, K.A., D.M. Walters, T.M. Jicha, J.M. Lazorchak, T.R. Angradi, and D.W. Bolgrien. 2010. Persistent organic pollutants in fish tissue in the mid-continental great rivers of the United States. *Sci. Tot. Environ.* 408: 1180–1189.

Boom, A., R. Marchant, H. Hooghiemstra, and J.S.S. Damste. 2002. CO2- and temperature-controlled altitudinal shifts of C-4- and C-3-dominated grasslands allow reconstruction of palaeoatmospheric pCO_2. *Palaeogeogr. Palaeoclimatol. Palaeoecol.* 177: 151–168.

Bostrom, B., D. Comstedt, and A. Ekblad. 2007. Isotope fractionation and ^{13}C enrichment in soil profiles during the decomposition of soil organic matter. *Oecologia* 153: 89–98.

Bouchez, J., O. Beyssac, V. Galy, J. Gaillardet, C. France-Lanord, L. Maurice, P. Moreira-Turq. 2010. Oxidation of petrogenic organic carbon inthe Amazon floodplain as a source of atmospheric CO_2. *Geology* 38: 255–258.

Bouillon, S., G. Abril, A.V. Borges, et al. 2009. Distribution, origin and cycling of carbon in the Tana River (Kenya): a dry season basin-scale survey from headwaters to the delta. *Biogeosciences* 6: 2475–2493.

Bowden, W.B., M.N. Gooseff, A. Balser, A. Green, B.J. Peterson, and J. Bradford. 2008. Sediment and nutrient delivery from thermokarst features in the foothills of the North Slope, Alaska: Potential impacts on headwater stream ecosystems. *J. Geophys. Res. Biogeosci.* 113, G02026, doi:10.1029/2007JG000470.

Burdige, D.J. 2005. Burial of terrestrial organic matter in marine sediments: a re-assessment. *Global Biogeochem. Cycles* 19, Gb4011 doi:10.1029/2004gb002368.

Carr, A.S., A. Boom, and B.M. Chase. 2010. The potential of plant biomarker evidence derived from rock hyrax middens as an indicator of palaeoenvironmental change. *Palaeogeogr. Palaeoclimatol. Palaeoecol.* 285: 321–330.

Castaneda, I.S., J.P. Werne, T.C. Johnson, and T.R. Filley. 2009. Late Quaternary vegetation history of southeast Africa: the molecular isotopic record from Lake Malawi. *Palaeogeogr. Palaeoclimatol. Palaeoecol.* 275: 100–112.

Chefetz, B., T. Mualem, and J. Ben-Ari. 2008. Sorption and mobility of pharmaceutical compounds in soil irrigated with reclaimed wastewater. *Chemosphere* 73: 1335–1343.

Clifford, D.J., D.M. Carson, D.E. Mckinney, J.M. Bortiatynski, and P.G. Hatcher. 1995. A new rapid techniques for the characterization of lignin in vascular plants – thermchemolysis with tetramethylammonium hydroxide (TMAH). *Org. Geochem.* 23: 169–175.

Collister, J.W., G. Rieley, B. Stern, G. Eglinton, and B. Fry. 1994. Compound-specific $\delta^{13}C$ analyses of leaf lipids from plants with differing carbon-dioxide metabolisms. *Org. Geochem.* 21: 619–627.

Cooke, M.P., H.M. Talbot, and P. Farrimond. 2008a. Bacterial populations recorded in bacteriohopanepolyol distributions in soils from Northern England. *Org. Geochem.* 39: 1347–1358.

Cooke, M.P., H.M. Talbot, and T. Wagner. 2008b. Tracking soil organic carbon transport to continental margin sediments using soil-specific hopanoid biomarkers: a case study from the Congo fan (ODP site 1075). *Org. Geochem.* 39: 965–971.

Crozier, M.J. 2010. Deciphering the effect of climate change on landslide activity: a review. *Geomorphology* 124: 260–267.

Da Silveira, L., L. Sternberg, S.S. Mulkey, and S.J. Wright. 1989. Ecological interpretation of leaf carbon isotope ratios: influence of respired carbon dioxide. *Ecology* 70: 1317–1324.

Dadson, S.J., N. Hovius, H. Chen, et al. 2004. Earthquake-triggered increase in sediment delivery from an active mountain belt. *Geology* 32: 733–736.

Derenne, S., and C. Largeau. 2001. A review of some important families of refractory macromolecules: composition, origin, and fate in soils and sediments. *Soil Sci.* 166: 833–847.

Dickens, A.F., Y. Gelinas, C.A. Masiello, S. Wakeham, and J.I. Hedges. 2004. Reburial of fossil organic carbon in marine sediments. *Nature* 427: 336–339.

Diefendorf, A.F., K.E. Mueller, S.L. Wing, P.L. Koch, and K.H. Freeman. 2010. Global patterns in leaf ^{13}C discrimination and implications for studies of past and future climate. *Proc. Nat. Acad. Sci.* 107: 5738–5743.

Drenzek, N.J., K.A. Hughen, D.B. Montluçon, et al. 2009. A new look at old carbon in active margin sediments. *Geology* 37: 239–242.

Duarte, C.M., and Y.T. Prairie. 2005. Prevalence of heterotrophy and atmospheric CO2 emissions from aquatic ecosystems. *Ecosystems* 8: 862–870.

Eglinton, G., and R.J. Hamilton. 1967. Leaf epicuticular waxes. *Science* 156: 1322.

Eglinton, T.I., and G. Eglinton. 2008. Molecular proxies for paleoclimatology. *Earth Planet. Sci. Lett.* 275: 1–16.

Ehleringer, J.R., N. Buchmann, and L.B. Flanagan. 2000. Carbon isotope ratios in belowground carbon cycle processes. *Ecol. Appl.* 10: 412–422.

Eusterhues, K., C. Rumpel, M. Kleber, and I. Kogel-Knabner. 2003. Stabilisation of soil organic matter by interactions with minerals as revealed by mineral dissolution and oxidative degradation. *Org. Geochem.* 34: 1591–1600.

Farquhar, G.D., J.R. Ehleringer, and K.T. Hubick. 1989. Carbon isotope discrimination and photosynthesis. *Annu. Rev. Plant Physiol. Plant Mol. Biol.* 40: 503–537.

Finlay, J.C., S. Khandwala, and M.E. Power. 2002. Spatial scales of carbon flow in a river food web. *Ecology* 83: 1845–1859.

Foster, G., and L. Carter. 1997. Mud sedimentation on the continental shelf at an accretionary margin – Poverty Bay, New Zealand. *New Zeal. J. Geol. Geophys.* 40: 157–173.

Fry, B., and E.B. Sherr. 1984. $\delta^{13}C$ measurements as indicators of carbon flow in marine and fresh-water ecosystems. *Contrib. Mar. Sci.* 27: 13–47.

Galy, V., and T. Eglinton. 2011. Protracted storage of biospheric carbon in the Ganges-Brahmaputra basin. *Nat. Geosci.* 4: 843–847.

Galy, V., C. France-Lanord, O. Beyssac, P. Faure, H. Kudrass, and F. Palhol. 2007. Efficient organic carbon burial in the Bengal fan sustained by the Himalayan erosional system. *Nature* 450: 407-U406.

Galy, V., C. France-Lanord, and B. Lartiges. 2008a. Loading and fate of particulate organic carbon from the Himalaya to the Ganga-Brahmaputra delta. *Geochim. Cosmochim. Acta* 72: 1767–1787.

Galy, V., L. François, C. France-Lanord, et al. 2008b. C4 plants decline in the Himalayan basin since the Last Glacial Maximum. *Quat. Sci. Rev.* 27: 1396–1409.

Gao, Q.Z., Z. Tao, G.R. Yao, J. Ding, Z.F. Liu, and K.X. Liu. 2007. Elemental and isotopic signatures of particulate organic carbon in the Zengjiang River, southern China. *Hydrol. Processes* 21: 1318–1327.

Gao, Y., Y. Zeng, Q. Shen, W. Ling, and J. Han. 2009. Fractionation of polycyclic aromatic hydrocarbon residues in soils. *J. Hazard. Mat.* 172: 897–903.

Golchin, A., P. Clarke, J.M. Oades, and J.O. Skjemstad. 1995. The effects of cultivation on the composition of organic-matter and structural stability of soils Aust. *J. Soil Res.* 33: 975–993.

Goldsmith, S.T., A.E. Carey, W.B. Lyons, S.-J. Kao, T.-Y. Lee, and J. Chen. 2008. Extreme storm events, landscape denudation, and carbon sequestration: Typhoon Mindulle, Choshui River, Taiwan. *Geology* 36: 483–486.

Gomez, B., N.A. Trustrum, D.M. Hicks, K.M. Rogers, M.J. Page, and K.R. Tate. 2003. Production, storage, and output of particulate organic carbon: Waipaoa River basin, New Zealand. *Water Resour. Res.* 39, 1161 doi:10.1029/2002wr001619.

Goñi, M.A., and T.I. Eglinton. 1996. Stable carbon isotopic analyses of lignin-derived CuO oxidation products by isotope ratio monitoring gas chromatography mass spectrometry (irm-GC-MS). *Org. Geochem.* 24: 601–615.

Goñi, M.A., and J.I. Hedges. 1990. Potential applications of cutin-derived CuO reaction-products for discriminating vascular plant sources in natural environments. *Geochim. Cosmochim. Acta* 54: 3073–3081.

Goñi, M.A., M.B. Yunker, R.W. Macdonald, and T.I. Eglinton. 2005. The supply and preservation of ancient and modem components of organic carbon in the Canadian Beaufort Shelf of the Arctic Ocean. *Mar. Chem.* 93: 53–73.

Gonzalez-Perez, J.A., F.J. Gonzalez-Vila, G. Almendros, and H. Knicker. 2004. The effect of fire on soil organic matter – a review. *Environ. Int.* 30: 855–870.

Gordeev, V.V. 2006. Fluvial sediment flux to the Arctic Ocean. *Geomorphology* 80: 94–104.

Gordeev, V.V., and M.D. Kravchishina. 2009. River flux of dissolved organic carbon (DOC) and particulate organic carbon (POC) to the Arctic Ocean: what are the consequences of the global changes? In: J.C.J. Nihoul and A.G. Kostianoy (eds.), Influence of Climate Change on the Changing Arctic and Sub-Arctic Conditions. NATO Science for Peace and Security Series C – Environmental Security, pp. 145–160.

Gordon, E.S., and M.A. Goñi. 2003. Sources and distribution of terrigenous organic matter delivered by the Atchafalaya River to sediments in the northern Gulf of Mexico. *Geochim. Cosmochim. Acta* 67: 2359–2375.

Griffith, D.R., R.T. Barnes, and P.A. Raymond. 2009. Inputs of fossil carbon from wastewater treatment plants to US rivers and oceans. *Environ. Sci. Technol.* 43: 5647–5651.

Guo, L.D., and R.W. Macdonald. 2006. Source and transport of terrigenous organic matter in the upper Yukon River: evidence from isotope ($\delta^{13}C$, $\Delta^{14}C$, and $\delta^{15}N$) composition of dissolved, colloidal, and particulate phases. *Global Biogeochem. Cycles* 20, Gb2011 doi:10.1029/2005gb002593.

Guo, L.D., C.L. Ping, and R.W. Macdonald. 2007. Mobilization pathways of organic carbon from permafrost to arctic rivers in a changing climate. *Geophys. Res. Lett.* 34, L13603 doi:10.1029/2007gl030689.

Gustafsson, O., B.E. Van Dongen, J.E. Vonk, O.V. Dudarev, and I.P. Semiletov. 2011. Widespread release of old carbon across the Siberian Arctic echoed by its large rivers. *Biogeosciences* 8: 1737–1743.

Haider, K. 1999. From dead organic residues to humus. *J. Plant Nutr. Soil Sci.* 162: 363–371.

Hedges, J.I. 1992. Global biogeochemical cycles – progress and problems. *Mar. Chem.* 39: 67–93.

Hedges, J.I., W.A. Clark, P.D. Quay, J.E. Richey, A.H. Devol, and U.D. Santos. 1986. Compositions and fluxes of particulate organic material in the Amazon River. *Limnol. Oceanogr.* 31: 717–738.

Hedges, J.I., R.G. Keil, and R. Benner. 1997. What happens to terrestrial organic matter in the ocean? *Org. Geochem.* 27: 195–212.

Hedges, J.I., and D.C. Mann. 1979. Characterization of plant-tissues by their lignin oxidation-products. *Geochim. Cosmochim. Acta* 43: 1803–1807.

Hedges, J.I., E. Mayorga, E. Tsamkis, et al. 2000. Organic matter in Bolivian tributaries of the Amazon River: A comparison to the lower mainstream. *Limnol. Oceanogr.* 45: 1449–1466.

Hedges, J.I., and J.M. Oades. 1997. Comparative organic geochemistries of soils and marine sediments. *Org. Geochem.* 27: 319–361.

Hicks, D.M., B. Gomez, and N.A. Trustrum. 2000. Erosion thresholds and suspended sediment yields, Waipaoa River Basin, New Zealand. *Water Resour. Res.* 36: 1129–1142.

Hilton, J., M. O'Hare, M.J. Bowes, and J.I. Jones. 2006. How green is my river? A new paradigm of eutrophication in rivers. *Sci. Tot. Environ.* 365: 66–83.

Hilton, R.G., A. Galy, and N. Hovius. 2008a. Riverine particulate organic carbon from an active mountain belt: importance of landslides. *Global Biogeochem. Cycles* 22, Gb1017 doi:1010.1029/2006gb002905.

Hilton, R.G., A. Galy, N. Hovius, M.-C. Chen, M.-J. Horng, and H. Chen. 2008b. Tropical-cyclone-driven erosion of the terrestrial biosphere from mountains. *Nat. Geosci.* 1: 759–762.

Hilton, R.G., A. Galy, N. Hovius, S.J. Kao, M.J. Horng, and H. Chen. 2012. Climatic and geomorphic controls on the erosion of terrestrial biomass from subtropical mountain forest. *Global Biogeochem. Cycles* 26, Gb3014 doi:10.1029/2012gb004314.

Hilton, R.G., P. Meunier, N. Hovius, P.J. Bellingham, and A. Galy. 2011. Landslide impact on organic carbon cycling in a temperate montane forest. *Earth Surf. Processes Landforms* 36: 1670–1679.

Hopmans, E.C., J.W.H. Weijers, E. Schefuss, L. Herfort, J.S.S. Damste, and S. Schouten. 2004. A novel proxy for terrestrial organic matter in sediments based on branched and isoprenoid tetraether lipids. *Earth Planet. Sci. Lett.* 224: 107–116.

Hovius, N., A. Galy, R.G. Hilton, et al. 2011. Erosion-driven drawdown of atmospheric carbon dioxide: The organic pathway. *Appl. Geochem.* 26: S285–S287.

Huggel, C., J.J. Clague, and O. Korup. 2012. Is climate change responsible for changing landslide activity in high mountains? *Earth Surf. Processes Landforms* 37: 77–91.

Intergovernmental Panel on Climate Change (IPCC). 2007. Climate Change 2007, In: S. Solomon, Qin, S.D., Manning, M., Chen, Z., Marquis, M., Averyt, K.B., Tignor, M., and Miller, H.L. (eds.), The Physical Science Basis: contribution of Working Group I to the Fourth Assessment Report of the Intergovernmental Panel on Climate Change, pp. 145–160. New York: Cambridge University Press.

Jahren, A.H., N.C. Arens, and S.A. Harbeson. 2008. Prediction of atmospheric $\delta^{13}CO_2$ using fossil plant tissues. *Rev. Geophys.* 46, Rg1002 doi:10.1029/2006rg000219.

Jakob, M., and S. Lambert. 2009. Climate change effects on landslides along the southwest coast of British Columbia. *Geomorphology* 107: 275–284.

Jansen, B., K.G.J. Nierop, F.H. Tonneijck, F.W.M. Van Der Wielen, and J.M. Verstraten. 2007. Can isoprenoids in leaves and roots of plants serve as biomarkers for past vegetation changes? A case study from the Ecuadorian Andes. *Plant Soil* 291: 181–198.

Jastrow, J.D. and R.M. Miller. 1997. Soil aggregate stabilization and carbon sequestration: feedbacks through organomineral associations, In: R. Lal, Kimble, J.M., Follett, R.F., Stewart, B.A. (eds.), Soil Processes and the Carbon Cycle, pp. 207–223. Boca Raton, FL: CRC Press.

Johnson, B.J., K.A. Moore, C. Lehmann, C. Bohlen, and T.A. Brown. 2007. Middle to late Holocene fluctuations of C-3 and C-4 vegetation in a Northern New England Salt Marsh, Sprague Marsh, Phippsburg Maine. *Org. Geochem.* 38: 394–403.

Jomelli, V., D. Brunstein, M. Deque, M. Vrac, and D. Grancher. 2009. Impacts of future climatic change (2070–2099) on the potential occurrence of debris flows: a case study in the Massif des Ecrins (French Alps). *Clim. Change* 97: 171–191.

Kaiser, K., and G. Guggenberger. 2007. Sorptive stabilization of organic matter by microporous goethite: sorption into small pores vs. surface complexation. *Eur. J. Soil Sci.* 58: 45–59.

Kao, S.J., and K.K. Liu. 1996. Particulate organic carbon export from a subtropical mountainous river (Lanyang Hsi) in Taiwan. *Limnol. Oceanogr.* 41: 1749–1757.

Kao, S.J., and J.D. Milliman. 2008. Water and sediment discharge from small mountainous rivers, Taiwan: the roles of lithology, episodic events, and human activities. *J. Geol.* 116: 431–448.

Komada, T., E.R.M. Druffel, and J. Hwang. 2005. Sedimentary rocks as sources of ancient organic carbon to the ocean: an investigation through $\Delta\ ^{14}C$ and $\delta^{13}C$ signatures of organic compound classes. *Global Biogeochem. Cycles* 19, Gb2017 doi:10.1029/2004gb002347.

Komada, T., E.R.M. Druffel, and S.E. Trumbore. 2004. Oceanic export of relict carbon by small mountainous rivers. *Geophys. Res. Lett.* 31: L07504.

Komárek, M., E. Cadková, V. Chrastný, F. Bordas, and J.-C. Bollinger. 2010. Contamination of vineyard soils with fungicides: a review of environmental and toxicological aspects. *Environ. Int.* 36: 138–151.

Korner, C., G.D. Farquhar, and S.C. Wong. 1991. Carbon isotope discrimination by plants follows latitudinal and altitudinal trends. *Oecologia* 88: 30–40.

Kronimus, A., and J. Schwarzbauer. 2007. Non-target screening of extractable and non-extractable organic xenobiotics in riverine sediments of Ems and Mulde Rivers, Germany. *Environ. Pollut.* 147: 176–186.

Krusche, A.V., L.A. Martinelli, M.R. Williams, et al. 2002. Composition of particulate and dissolved organic matter in a disturbed watershed of southeast Brazil (Piracicaba River basin). *Water Res.* 36: 2743–2752.

Kuhn, N.J., T. Hoffmann, W. Schwanghart, and M. Dotterweich. 2009. Agricultural soil erosion and global carbon cycle: controversy over? *Earth Surf. Processes Landforms* 34: 1033–1038.

Kuzyk, Z.Z.A., M.A. Goñi, G.A. Stern, and R.W. Macdonald. 2008. Sources, pathways and sinks of particulate organic matter in Hudson Bay: evidence from lignin distributions. *Mar. Chem.* 112: 215–229.

Lal, R. 2003. Soil erosion and the global carbon budget. *Environ. Int.* 29: 437–450.

Larsbo, M., J. Stenström, A. Etana, E. Börjesson, and N.J. Jarvis. 2009. Herbicide sorption, degradation, and leaching in three Swedish soils under long-term conventional and reduced tillage. *Soil Tillage Res.* 105: 200–208.

Lehmann, J., J. Kinyangi, and D. Solomon. 2007. Organic matter stabilization in soil microaggregates: implications from spatial heterogeneity of organic carbon contents and carbon forms. *Biogeochemistry* 85: 45–57.

Leithold, E.L., and N.E. Blair. 2001. Watershed control on the carbon loading of marine sedimentary particles. *Geochim. Cosmochim. Acta* 65: 2231–2240.

Leithold, E.L., N.E. Blair, and D.W. Perkey. 2006. Geomorphologic controls on the age of particulate organic carbon from small mountainous and upland rivers. *Global Biogeochem. Cycles* 20: Gb3022, doi:10.1029/2005gb002677.

Leopold, E.B., R. Nickmann, J.I. Hedges, and J.R. Ertel. 1982. Pollen and lignin records of late Quaternary vegetation, Lake Washington. *Science* 218: 1305–1307.

Loos, R., B.M. Gawlik, G. Locoro, E. Rimaviciute, S. Contini, and G. Bidoglio. 2009. EU-wide survey of polar organic persistent pollutants in European river waters. *Environ. Pollut.* 157: 561–568.

Lorenz, K., R. Lal, C.M. Preston, and K.G.J. Nierop. 2007. Strengthening the soil organic carbon pool by increasing contributions from recalcitrant aliphatic bio(macro)molecules. *Geoderma* 142: 1–10.

Ludwig, W., J.L. Probst, and S. Kempe. 1996. Predicting the oceanic input of organic carbon by continental erosion. *Global Biogeochem. Cycles* 10: 23–41.

Makou, M.C., K.A. Hughen, L. Xu, S.P. Sylva, and T.I. Eglinton. 2007. Isotopic records of tropical vegetation and climate change from terrestrial vascular plant biomarkers preserved in Cariaco Basin sediments. *Org. Geochem.* 38: 1680–1691.

Maria, G., and C. Maria. 2009. Bioaccumulation dynamics of a PCB low-level discharge in a riverine pathway downstream the release point. *Chem. Biochem. Eng. Q.* 23: 121–134.

Masiello, C.A., and E.R.M. Druffel. 2001. Carbon isotope geochemistry of the Santa Clara River. *Global Biogeochem. Cycles* 15: 407–416.

Mayorga, E., A.K. Aufdenkampe, C.A. Masiello, et al. 2005. Young organic matter as a source of carbon dioxide outgassing from Amazonian rivers. *Nature* 436: 538–541.

McGuire, A.D., L.G. Anderson, T.R. Christensen, et al. 2009. Sensitivity of the carbon cycle in the Arctic to climate change. *Ecol. Monogr.* 79: 523–555.

Medeiros, P.M., and B.R.T. Simoneit. 2008. Multi-biomarker characterization of sedimentary organic carbon in small rivers draining the Northwestern United States. *Org. Geochem.* 39: 52–74.

Megens, L., J. Van Der Plicht, and J.W. De Leeuw. 2001. Temporal variations in C-13 and C-14 concentrations in particulate organic matter from the southern North Sea. *Geochim. Cosmochim. Acta* 65: 2899–2911.

Meybeck, M. 1993. C, N, P and S in rivers: From sources to global inputs, In: F.T.M.R. Wollast, and L. Chou (eds.), Interactions of C, N, P and S Biogeochemical Cycles and Global Change, pp. 163–193. New York: Springer-Verlag.

Nagao, S., T. Usui, M. Yamamoto, M. Minagawa, T. Iwatsuki, and A. Noda. 2005. Combined use of $\Delta^{14}C$ and $\delta^{13}C$ values to trace transportation and deposition processes of terrestrial particulate organic matter in coastal marine environments. *Chem. Geol.* 218: 63–72.

Newbold, J.D., P.J. Mulholland, J.W. Elwood, and R.V. O'Neill. 1982. Organic-carbon spiralling in stream ecosystems. *Oikos* 38: 266–272.

Nichols, J., R.K. Booth, S.T. Jackson, E.G. Pendall, and Y. Huang. 2010. Differential hydrogen isotopic ratios of Sphagnum and vascular plant biomarkers in ombrotrophic peatlands as a quantitative proxy for precipitation–evaporation balance. *Geochim. Cosmochim. Acta* 74: 1407–1416.

Nierop, K.G.J., and T.R. Filley. 2008. Simultaneous analysis of tannin and lignin signatures in soils by thermally assisted hydrolysis and methylation using ^{13}C-labeled TMAH. *J. Anal. Appl. Pyrolysis* 83: 227–231.

O' Leary, M.H. 1981. Carbon isotope fractionation in plants. *Phytochemistry* 20: 553–567.

Page, M., N. Trustrum, H. Brackley, and T. Baisden. 2004. Erosion-related soil carbon fluxes in a pastoral steepland catchment, New Zealand. *Agri. Ecosyst. Env.* 103: 561–579.

Pauli, H., M. Gottfried, K. Reier, C. Klettner, and G. Grabherr. 2007. Signals of range expansions and contractions of vascular plants in the high Alps: observations (1994–2004) at the GLORIA*master site Schrankogel, Tyrol, Austria. *Global Change Biol.* 13: 147–156.

Peterson, B.J., and B. Fry. 1987. Stable isotopes in ecosystem studies. *Annu. Rev. Ecol. Syst.* 18: 293–320.

Poirier, N., S. Derenne, J. Balesdent, A. Mariotti, D. Massiot, and C. Largeau. 2003. Isolation and analysis of the non-hydrolysable fraction of a forest soil and an arable soil (Lacadee, southwest France). *Eur. J. Soil Sci.* 54: 243–255.

Polissar, P.J., K.H. Freeman, D.B. Rowley, F.A. Mcinerney, and B.S. Currie. 2009. Paleoaltimetry of the Tibetan Plateau from D/H ratios of lipid biomarkers. *Earth Planet. Sci. Lett.* 287: 64–76.

Power, M.E. 1991. Shifts in the effects of tuft-weaving midges on filamentous algae. *Am. Midland Nat.* 125: 275–285.

Rachold, V., Eicken, H., Gordeev, V.V., Grigoriev, M.N., Hubberten, H.-W., Lisitzin, A.P. Schevchenko, V.P., and Schirrmeister, L. 2004. Modern terrigenous organic carbon input to the Arctic Ocean, In: R. Stein, Macdonald, R.W. (eds.), *The Organic Carbon Cycle in the Arctic Ocean*, pp. 33–55. New York: Springer.

Raymond, P.A., and J.E. Bauer. 2001. Use of ^{14}C and ^{13}C natural abundances for evaluating riverine, estuarine, and coastal DOC and POC sources and cycling: a review and synthesis. *Org. Geochem.* 32: 469–485.

Raymond, P.A., J.E. Bauer, N.F. Caraco, J.J. Cole, B. Longworth, and S.T. Petsch. 2004. Controls on the variability of organic matter and dissolved inorganic carbon ages in northeast US rivers. *Mar. Chem.* 92: 353–366.

Raymond, P.A., J.W. McClelland, R.M. Holmes, et al. 2007. Flux and age of dissolved organic carbon exported to the Arctic Ocean: a carbon isotopic study of the five largest arctic rivers. *Global Biogeochem. Cycles* 21, Gb4011 doi:10.1029/2007gb002934.

Reid, L.M., and M.J. Page. 2003. Magnitude and frequency of landsliding in a large New Zealand catchment. *Geomorphology* 49: 71–88.

Richey, J.E., J.I. Hedges, A.H. Devol, et al. 1990. Biogeochemistry of carbon in the Amazon River. *Limnol. Oceanogr.* 35: 352–371.

Rillig, M.C., B.A. Caldwell, H.A.B. Wosten, and P. Sollins. 2007. Role of proteins in soil carbon and nitrogen storage: controls on persistence. *Biogeochemistry* 85: 25–44.

Salmaso, N., and A. Zignin. 2010. At the extreme of physical gradients: phytoplankton in highly flushed, large rivers. *Hydrobiologia* 639: 21–36.

Schimel, D.S., B.H. Braswell, E.A. Holland, et al. 1994. Climatic, edaphic, and biotic controls over storage and turnover of carbonm in soils. *Global Biogeochem. Cycles* 8: 279–293.

Schmidt, F., K.-U. Hinrichs, and M. Elvert. 2010. Sources, transport, and partitioning of organic matter at a highly dynamic continental margin. *Mar. Chem.* 118: 37–55.

Schmidt, M.W.I., M.S. Torn, S. Abiven, et al. 2011. Persistence of soil organic matter as an ecosystem property. *Nature* 478: 49–56.

Schouten, S., J.W.H. Weijers, F. Peterse, M.T.J. Van Der Meer, and J.S.S. Damste. 2007. Development and application of novel organic proxies. *Geochim. Cosmochim. Acta* 71: A905–A905.

Smith, F.A., and K.H. Freeman. 2006. Influence of physiology and climate on delta D of leaf wax n-alkanes from C-3 and C-4 grasses. *Geochim. Cosmochim. Acta* 70: 1172–1187.

Smith, F.A., S.L. Wing, and K.H. Freeman. 2007. Magnitude of the carbon isotope excursion at the Paleocene-Eocene thermal maximum: the role of plant community change. *Earth Planet. Sci. Lett.* 262: 50–65.

Smith, J.B., S.H. Schneider, M. Oppenheimer, et al. 2009. Assessing dangerous climate change through an update of the Intergovernmental Panel on Climate Change (IPCC) "reasons for concern". *Proc. Natl. Acad. Sci.* 106: 4133–4137.

Sollins, P., M.G. Kramer, C. Swanston, et al. 2009. Sequential density fractionation across soils of contrasting mineralogy: evidence for both microbial- and mineral-controlled soil organic matter stabilization. *Biogeochemistry* 96: 209–231.

Sollins, P., C. Swanston, and M. Kramer. 2007. Stabilization and destabilization of soil organic matter – a new focus. *Biogeochemistry* 85: 1–7.

Spiker, E.C., and M. Rubin. 1975. Petroleum pollutants in surface and groundwater as indicated by the carbon-14 activity of dissolved organic carbon. *Science* 187: 61–64.

Stoffel, M., and C. Huggel. 2012. Effects of climate change on mass movements in mountain environments. *Prog. Phys. Geog.* 36: 421–439.

Syvitski, J.P.M., C.J. Vorosmarty, A.J. Kettner, and P. Green. 2005. Impact of humans on the flux of terrestrial sediment to the global coastal ocean. *Science* 308: 376–380.

Tarnocai, C., J.G. Canadell, E.a.G. Schuur, P. Kuhry, G. Mazhitova, and S. Zimov. 2009. Soil organic carbon pools in the northern circumpolar permafrost region. *Global Biogeochem. Cycles* 23, Gb2023 doi:10.1029/2008gb003327.

Thompson, A., O.A. Chadwick, S. Boman, and J. Chorover. 2006. Colloid mobilization during soil iron redox oscillations. *Environ. Sci. Technol.* 40: 5743–5749.

Trenberth, K.E. 2011. Changes in precipitation with climate change. *Climate Res.* 47: 123–138.

Trumbore, S. 2009. Radiocarbon and soil carbon dynamics. *Annu. Rev. Earth Planet. Sci.* 37: 47–66.

Turner, M.G., V.H. Dale, and E.H. Everham, III. 1997. Fires, hurricanes, and volcanoes: comparing large disturbances. *BioScience* 47: 758–768.

Vandenbroucke, M., and C. Largeau. 2007. Kerogen origin, evolution and structure. *Org. Geochem.* 38: 719–833.

Vannote, R.L., G.W. Minshall, K.W. Cummins, J.R. Sedell, and C.E. Cushing. 1980. River continuum concept. *Can. J. Fish. Aquat. Sci.* 37: 130–137.

Vonk, J.E., B.E. Van Dongen, and O. Gustafsson. 2010. Selective preservation of old organic carbon fluvially released from sub-Arctic soils. *Geophys. Res. Lett.* 37, L11605 doi:10.1029/2010gl042909.

Waterson, E.J., and E.A. Canuel. 2008. Sources of sedimentary organic matter in the Mississippi River and adjacent Gulf of Mexico as revealed by lipid biomarker and $\delta^{13}C$ TOC analyses. *Org. Geochem.* 39: 422–439.

Wheatcroft, R.A., M.A. Goñi, J.A. Hatten, G.B. Pasternack, and J.A. Warrick. 2010. The role of effective discharge in the ocean delivery of particulate organic carbon by small, mountainous river systems. *Limnol. Oceanogr.* 55: 161–171.

Wright, J.P., and J.D. Fridley. 2010. Biogeographic synthesis of secondary succession rates in eastern North America. *J. Biogeogr.* 37: 1584–1596.

Young, R.G., and A.D. Huryn. 1999. Effects of land use on stream metabolism and organic matter turnover. *Ecol. Appl.* 9: 1359–1376.

Zehetner, F., G.J. Lair, and M.H. Gerzabek. 2009. Rapid carbon accretion and organic matter pool stabilization in riverine floodplain soils. *Global Biogeochem. Cycles* 23, Gb4004 doi:4010.1029/2009gb003481.

9

Black carbon in coastal and large river systems

S. Mitra, A. R. Zimmerman, G. Hunsinger, and W. R. Woerner

1. Introduction

Black carbon (BC) is defined as the highly condensed carbonaceous products of organic matter combustion (Novakov 1984; Goldberg 1985). It can be formed either via the condensation of combustion gases (i.e., soot), or as the residues of incomplete combustion of any type of biomass. BC in the environment has recently received increasing attention with the growing awareness of its potential importance to the short- and particularly the long-term C cycle and accompanying climate feedbacks. Although BC in the terrestrial environment has been extensively investigated, BC transport to and storage in aquatic environments has received far less attention. In this chapter, the evolution of BC is discussed, from its creation in watersheds to its fate in the oceans, with specific focus on its crossing of the terrestrial-aquatic interface.

The two major possible sources of BC to rivers or the ocean are from fossil fuel and biomass burning. Subsequent to its generation, BC may be transported to the ocean either by atmospheric deposition or after incorporation into soils and sediments, followed by its erosion and transport via fluvial systems (Fig. 9.1). Although these are important, examination and quantification of fluvial BC in large rivers is likely an important route for BC transfer to the oceans deserving further research attention for several reasons. First, despite the importance of BC and its fluvial discharge, there are a limited number of studies addressing BC discharge from the world's major rivers and limited measurements of BC in aquatic environments as a whole. Second, when measured, BC has been shown to comprise a substantial portion of sedimentary organic matter across different river systems and their coastal margins globally (Masiello and Druffel 2001; Mitra et al. 2002; Elmquist et al. 2008; Lohmann et al. 2009). Third, the quantity of BC identified in riverine particulate organic carbon (POC) is approximately twice the amount of BC calculated to be buried in ocean sediments (Druffel 2004), suggesting a missing sink, unidentified transformation processes, or uncertainty in the estimates associated with fluvial and coastal BC sources or oceanic sinks. Finally, as marine sedimentary rock is by far, the largest reservoir of organic matter on Earth (Hedges 1992), it might follow that it is also the largest reservoir for BC on the Earth. However, unlike other organic matter, BC is only created in the terrestrial sphere. Thus, quantifying the efficiency of the terrestrial-to-marine transfer of BC is particularly important toward understanding how biomass combustion is related to climate change and the C cycle. This is because, as a relatively refractory pool of carbon (e.g., Masiello 2004), the formation of BC from biomass combustion, followed by its deposition and burial,

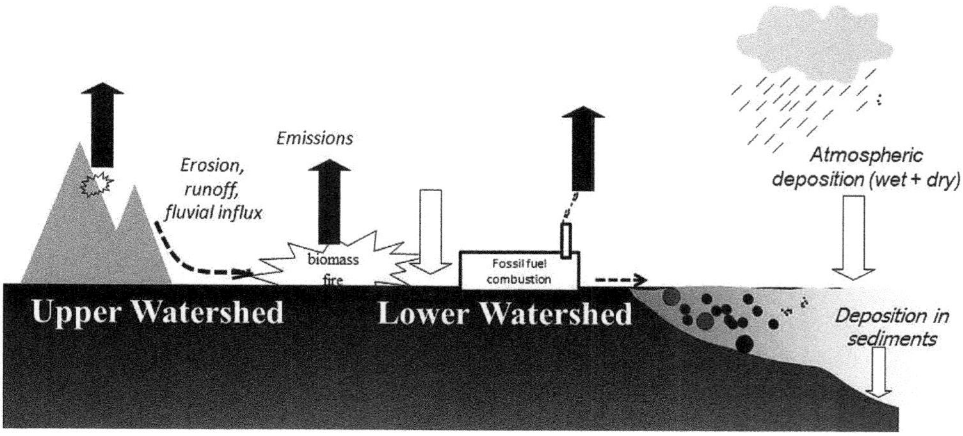

Figure 9.1. Conceptual model of reservoirs and fluxes of particulate black carbon in the terrestrial and aquatic environments. Large black arrows depict BC emissions processes, white arrows depict deposition processes. Small dashed arrow depicts erosion, runoff, and fluvial influx which may occur throughout a watershed.

represents a transfer of carbon from the short- to the long-term carbon cycling pools (i.e., long term sequestration – Fig. 9.2).

Different aspects of the biogeochemical cycling of BC have been reviewed previously (e.g., Schmidt and Noack 2000; Forbes et al. 2006; Preston and Schmidt 2006). The objective of this chapter is to provide a review of studies of BC sources, fluxes, and reservoirs, with emphasis on its fluvial transport and discharge into the ocean. It will be shown that, to the extent that can be determined, in many ways, the fluvial discharge of BC and its burial in the ocean is similar to that of terrestrial organic matter, which is discussed in other chapters of this book (Chapters 4 and 8); whereas in other respects, its cycling is quite distinct. Indeed, the result of the comparisons made in this chapter suggest that

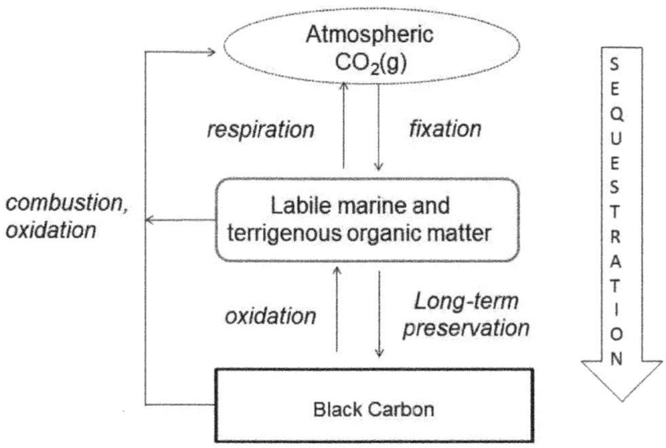

Figure 9.2. The relationship between biomass-derived BC and the global carbon cycle.

unlike bulk terrigenous organic matter, the riverine delivery of refractory BC, irrespective of isolation method used, may in fact be in balance with that found in marine sediments.

It is now widely recognized that atmospheric BC is an important driver of climate, only second or third in importance behind CO_2 in its atmosphere warming effect (Forster et al. 2007). And unlike CO_2 and methane, the reduction of BC emission to the atmosphere may be an easier goal to implement in efforts toward climate change mitigation (Jacobson 2001). Under future climate change scenarios, the threat of wildfires will increase (IPCC 2007) and, as society's dependence on fossil fuels continues to grow, concomitant increases in fossil fuel and aerosol BC emissions are expected. In addition, individuals and companies may soon intentionally produce BC (biochar) from waste biomass in large quantities in efforts to sequester C and improve soil fertility (e.g., Lehmann 2007). The fate of this increasing BC supply is not known, so knowledge of riverine fluxes and aquatic BC reservoir processes are needed now more than ever.

2. Sources of black carbon

The main sources of BC to the environment are fossil fuel (coal, oil, gasoline, and diesel) and biomass combustion. Except in the case of natural fires, these fluxes are driven largely by society's present and historical energy needs. Natural gas combustion is not a significant contributor to BC formation (Goldberg 1985); thus we do not consider it further in this chapter. Complete combustion of these organic materials results in the production of CO_2, an important greenhouse gas. However, the efficiency of combustion is dependent on several variables, including heating time, temperature, and type of fuel, but primarily the availability of oxygen. A sub-stoichiometric reaction of organic matter with oxygen, or what is known as pyrolysis, results in the formation of particulate, carbon rich, and O-H-S-N–poor compounds (Stanmore et al. 2001). For example, in the following equation,

$$C_mH_n + yO_2 = 2yCO + (n/2)H_2 + (m\text{-}2y)C \quad (1)$$

BC will be formed when m exceeds the value of 2y (Goldberg 1985). It should be noted that kinetics of the reaction will often determine the ratio of m:2y.

Several terms have been used to identify the residuals (or subsets of the residuals) of incomplete combustion of organic matter, including char, biochar, charcoal, carbonaceous spherules, soot, fusain, elemental carbon, carbon black, oxidation resistant elemental carbon, and graphitic carbon. It may be best to conceptualize these types of BC as existing along a continuum of combustion, with the abundance of each of these products dependent on combustion conditions (Fig. 9.3) (Masiello 2004; Hammes et al. 2007). On the left side of this figure are the residues remaining from incomplete combustion of biomass, including slightly charred biomass and charcoal; these constitute the larger size class of pyrogenic BC. In contrast, soot and graphitized C, is a mix of amorphous microscopic particles that contain BC and other carbon and/or sulfur-containing products of incomplete combustion. During combustion, residual char and charcoal may form simultaneously with soot, depending on fuel composition and efficiency of combustion. For example, BC from diesel combustion has a higher proportion of soot compared to other incomplete combustion products. In contrast, smoke plumes from wildfires may appear brown, blue, or gray due to the reduced combustion efficiency of wildfires,

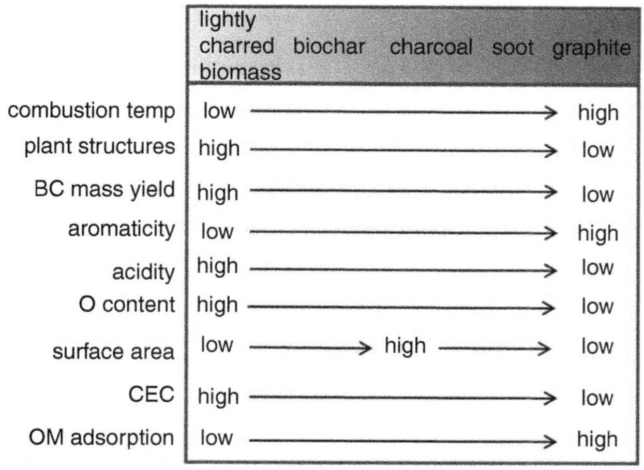

Figure 9.3. The combustion continuum. Modified from Masiello (2004) and Hedges et al. (2000).

which produce relatively greater amounts of non-condensed or partially-condensed organic carbon (sometimes called brown carbon) and other pyrolysis aerosols in comparison with soot (Bachmann 2009).

Laboratory charring experiments of individual components of wood (e.g., cellulose, lignin, pectin), as well as bulk wood and grass, offer some clues to understanding the controls on and variations in BC composition. In general, initial heating of organic matter results in dewatering and dehydration to phenols and aromatics. These are followed by additional chemical transformations that occur at higher temperatures and longer duration of heating leading to formation of more condensed refractory organic matter. For example, Czimczik et al. (2002) showed that at temperatures below 500 °C, BC formed via smoldering hardwood or softwood yielded small clusters of condensed C, lacking a high proportion of graphitic OM. Thus levels of condensed soot-type BC may be low in ecosystems dominated by low-temperature smoldering fires, especially when quantified by isolation methods that rely on detection of condensed aromatic structures (Czimczik and Schmidt 2002).

At formation temperatures beyond 450 °C, progressive depolymerization, loss of functional groups, aromatization, dehydrogenation reactions, and removal of substituents occurs, resulting in larger sheets of fused aromatic rings (Baldock and Smernik 2002; Knicker 2007; Arranz et al. 2009; Keiluweit et al. 2010; Zimmerman and Gao 2013). These condensed structures would likely be very resistant to biotic or abioitic degradation, as would the graphitic turbostatic crystallites that have been observed to form at temperatures >600 °C and in soot (Kercher and Nagle 2003; Keiluweit et al. 2010). In contrast to charcoal, which is irregularly shaped and larger, the basic units of soot are spherical or nearly spherical particles with diameters often in the range of 20–30 nanometers, corresponding to about one million carbon atoms (Haynes and Wagner 1981). Soot may be formed from the nucleation of gaseous carbonaceous species, such as acetylene and polycyclic aromatic hydrocarbons (PAHs) around uncharged radicals, which occur during intermolecular collisions. Subsequently, condensation reactions of gas phase species lead to the appearance of the first recognizable soot particles during the cooling process (Haynes and Wagner 1981). For some solid fuels, soot is also formed from the pyrolysis of tarry materials ejected during devolatilization. These gases then subsequently dehydrogenate and partially oxidize into soot spherules (Stanmore et al. 2001).

2.1. Atmospheric BC

The atmosphere is an important source of BC influx to rivers and the ocean. Moreover, the presence of BC aerosols in the atmosphere has been shown to affect hydrological cycles. For example, the radiative forcing of most aerosols is negative (i.e., leading to cooling of the Earth's atmosphere with increased concentrations) (Penner 2001), but BC aerosol's radiative forcing is positive as BC absorbs sunlight and warms the atmosphere and may enhance evaporation (Hansen et al. 2000; Jacobson 2001). These radiative effects have been shown to alter regional atmospheric stability and vertical circulation, while affecting large-scale circulation and hydrologic cycles, causing significant regional climate effects such as enhanced regional precipitation (Menon et al. 2002).

The aeolian transport of BC aerosols and their deposition on snow and ice are intimately linked via a positive feedback loop in the climate-carbon cycle. The long-range transport of BC aerosols to snow and ice in remote areas reduces albedo (Warren and Wiscombe 1979; Warren and Wiscombe 1981; Warren and Wiscombe 1985), resulting in increased ice melt in polar and boreal areas (Robock 1984; Clarke and Noone 1985; Barry 1996). For example, a concentration of 15 mg kg^{-1} of BC in snow reduced its albedo by 1% (Warren and Wiscombe 1980). In some cases, the BC remaining at the surface of glacier snow increased melting by 50% over areas of ice that were not impacted by BC deposition (Conway et al. 1996). Such BC-driven snow melting has been shown to contribute to historical rapid Tibetan glacier retreat, suggesting that BC deposition may significantly affect some of the world's largest freshwater supplies as well as sea level globally (Hansen and Nazarenko 2004; Xu et al. 2009). Furthermore, some recent studies have alluded to anthropogenic aerosols, rich in fossil carbon, contributing to much of the ancient organic matter in glacial runoff (Stubbins et al. 2012a). As a result, increased BC emission may ultimately enhance melting of ice from polar and boreal areas, leading to increased freshwater transport of BC into the oceans.

There are several factors that affect the concentration of BC aerosols in the atmosphere and their subsequent deposition. Geographically, modern day emissions of BC are highest in the tropics and East Asia (Ramanathan and Carmichael 2008), suggesting that watersheds and rivers in this area may be enriched in BC. Also, aerosol BC is found above both land and ocean, with concentrations being highest near urbanized areas and lowest above remote areas (Clarke et al. 1984). Present day estimates of annual global BC emission range from 6 to 24 Teragrams (1 Tg = 10^{12} g) BC from fossil fuel emission (Penner and Eddleman 1993; Novakov et al. 2003) and 50 to 260 Tg BC from biomass burning (Kuhlbusch and Crutzen 1995; Fernandes et al. 2007), suggesting both processes contribute to aerosol BC loading. However, another recent estimate of annual fossil fuel and biomass BC emission together (8 Tg y^{-1}; Bond et al. 2004), suggests that there is still some uncertainty in the scientific community regarding the amount of BC emitted into the atmosphere.

The residence time of BC aerosols in the atmosphere is determined by initial size distribution, concentration of ambient particles, frequency and duration of precipitation, and wind conditions (Ogren and Charlson 1983; Ogren et al. 1984). The largest BC particles are not likely to become or stay airborne and as such will deposit proximally to the source of combustion (Clark 1988). The majority of aerosol BC is found in the smaller 0.01–1.0 μm size fraction (Clarke et al. 2006) and will have atmosphere residence times of days to weeks (Clark 1988) (Fig. 9.3). Despite its importance locally and regionally, dry deposition does not contribute very much to the total atmospheric deposition of aerosol organic matter (Bidleman 1988). Additional studies at the regional level have shown that wet deposition accounted for 52–99% of aerosol BC deposition and thus is considered to be the most important deposition mechanism by which atmospheric BC enters the ocean (Suman et al. 1997). The most comprehensive estimate for atmospheric BC deposition to the ocean to date, suggests that dry and wet deposition of BC to the ocean are 2 and 10 TgC y^{-1}, respectively (Jurado et al. 2008). Next we address how this estimate compares with other BC sources into the ocean, namely that derived through soil erosion and fluvial influx.

2.2. Soil BC

Given that humans have been burning biomass since ~9 kya (Carcaillet et al. 2002), the widespread occurrence of natural fires (when unsuppressed), and the emission products associated with both processes, the occurrence of BC in nearly all soils is not surprising. Although there is still uncertainty regarding the most appropriate method to measure BC (see later section), calculation of pyrogenic OM contribution to soil organic carbon (SOC) using solid state ^{13}C nuclear magnetic resonance spectroscopy, acid potassium dichromate digestion, and benzenepolycarboxylic acid concentrations are becoming more common. Using these methods of determination, concentration of BC in most soils generally ranges 0.2–5 weight % of soil and represents between 2% and 15% of the soil TOC, with values of about 10% SOC most common (e.g., Bird et al. 1999; Hammes et al. 2007; Cusack et al. 2012). Soils thought to have anthropic inputs of pyrogenic OM such as Amazonan *terra preta* and higher latitude grassland OM-rich soils such as Chernozems and Mollisols have been found to have BC contents as great as 45% SOC (Chasar et al. 2000; Czimczik et al. 2002; Forster et al. 2007; Schmidt et al. 2002).

Much as for soil OM as a whole, there is considerable debate regarding the primary controls on soil BC concentrations. As with soil OM, the possible controls can be reduced to either input or preservation-related mechanisms. Although some studies have found soil BC stocks to be related to fire frequency or intensity, other studies have failed to find such relationships. For example, the BC content of an Andean volcanic agricultural soil that had yearly burns of wheat and barley residues was not different from one that had experienced no burning (Rivas et al. 2012). In contrast, a nearby forest with no recent burning had significantly less BC content (0.5% SOC) than the topsoil of a site affected by wildfire 3 years previously (7% SOC, Rivas et al. 2012). In studies of Siberian Scots pine forest soils, increases in BC of up to 40% were found in some soils and no lasting effect on others (Czimczik et al. 2003; Czimczik et al. 2005).

Despite all these uncertainties, wildfires have been estimated to produce approximately 0.04–0.2 Pg soil BC yr^{-1} (Preston and Schmidt 2006). Given these global estimates, as well as those for specific areas, workers have concluded that charcoal soil stocks should be several times greater than actually observed (Baldock et al. 2004; Ohlson and Tryterud 2000). Part of the difficulty in linking BC production to BC soil stocks is that BC may not be as fully inert as was previously assumed.

The observation of strong correlations between SOC and BC levels (Hammes et al. 2007; Cusack et al. 2012) suggests that similar processes may control their distribution and perhaps export, including degradation, translocation within soil, dissolution, and erosion. Both BC degradation and dissolution (soil as a dissolved BC source) are discussed further later.

Although horizontal and vertical movement of particulate soil BC by erosion and downward translocation, respectively, are infrequently quantified, there is a growing body of indirect evidence that these processes represent significant losses of soil BC at short timescales. First is the very presence of BC in aquatic environments, despite its exclusive production on land. Second, studies examining soil profiles in which the history of fire incidence is known have calculated wildly varying loss rates, leading many to conclude that process other than BC degradation, such as erosion, may play a dominant role (e.g., Nguyen et al. 2008; Zimmerman and Gao 2013). Finally, a few studies have attempted to follow all BC losses from a BC-amended soil in the field. Working in Colombia over two years, only a small portion of the BC losses could be accounted for, leading the researchers to conclude that 20–53% of the applied BC must have been lost by surface erosion (Major et al. 2010). Another study found that 7–55% of pyrogenic C added to the soil via a surficial grass fire was lost to erosion, and another 23–46% was moved vertically to greater soil depths (Rumpel et al. 2009). In addition, observations of low BC concentrations in upper layers of soils subject to frequent burning, but increased BC with depth, suggests mobility of BC within the soil profile (Dai et al. 2005; Dick et al. 2005; Knicker et al. 2012).

Perhaps it is not surprising that BC should make its way from soils into rivers, given that natural fires often occur on unstable slopes and may denude the landscape of erosion-deterring vegetation (Fig. 9.1). The passage of fire across soil can also lead to an increase in soil hydrophobicity and a concomitant decrease in water infiltration capacity, the net result of which is increased surface runoff and erosion, which are magnified in proportion to the intensity of the fire (Conedera et al. 2009). In addition, BC particles may be preferentially exported compared with other soil OM owing to their predominantly small size and low density (Skjemstad et al. 1999; Hammes et al. 2007). Exported BC was found to be most closely associated with the 20–50 μm soil fraction (Rumpel et al. 2009). Certainly, BC exported from soil versus storage or loss via other mechanisms will vary with such factors as climate, topography, soil type, and hydrological regime.

2.3. Dissolved BC

Despite the fact that fluvial BC has been typically regarded as particles found in suspended sediments or bedload, the recently proposed presence of BC in the dissolved phase is certainly reasonable. The definition of particulate versus dissolved classes of organic matter is operational, with POC often defined as organic material retained on filters with nominal size cutoffs (e.g., 0.45, 0.7, or 1 μm). Thus it is conceivable that microparticulate BC escapes the analytical window of what is defined as particulate material. Furthermore, material that passes through such filters may remain suspended in the water column for substantial periods of time, leading to the observed age offset between POC and BC (Masiello and Druffel 1998; Abiven et al. 2011). Also, desorption from soot BC aerosols at room temperature has yielded condensed aromatic molecules in the water soluble fraction, whereas oxidation of particulate BC causes it to become more water soluble (Chugtai et al. 1991;

Decesari et al. 2002; Wozniak et al. 2008). Thus leaching of carbon from the particulate to the dissolved phase, aided by chemical oxidation or biodegradation, may be significant.

Researchers have claimed to identify substantial amounts of pyrogenic substances in the dissolved phase of organic matter in rainwater, rivers, and the ocean (Kim et al. 2004; Mannino and Harvey 2004; Lee 2005; Hockaday et al. 2006; Dittmar 2008). Some of the recent studies that have detected the presence of BC in the sub-particulate phase are summarized in Table 9.1. What is clear from the comparison of these studies is that dissolved BC may comprise trace amounts up to 7% of the total DOC in an aquatic system. The studies in Table 9.1 suggest that a median value of ~5% is appropriate for the fraction of total DOC that is BC. Since DOC in the oceans is estimated at 7×10^5 Tg C (Hansell and Carlson 1998), the ocean's reservoir of dissolved BC can be approximated at 4×10^4 Tg. Using a value of 35,000 km^3 y^{-1} for the water discharged from all rivers to the ocean (Chapter 3), with a median concentration of all dissolved BC concentrations listed in Table 9.1 (10 μg L^{-1}), an estimate for the residence time of dissolved BC is calculated to be ~100,000 years. As the median age of bulk DOC has been measured to be far younger (~6,000 years; Williams and Druffel 1987), our simple calculations suggest that dissolved BC is certainly old enough to bias the age of bulk DOC. That said, several studies have found dissolved BC to be rapidly photooxidized (Stubbins et al. 2010; Stubbins et al. 2012b). It is likely that, as with particulate BC, dissolved BC is composed of materials with a continuum of compositions and labilities. Clearly, additional investigations into the chemical composition, age, and environmental fate of dissolved BC in the ocean are warranted.

3. Isolation, quantification, and source discrimination methods for black carbon

There are as many methods to isolate BC as there are BC practitioners. Furthermore, each method of BC isolation is often uniquely suited to the interests of the researcher. For example, the atmospheric community often quantifies BC by measurement of light transmittance or reflectance on filters as it is concerned about BC's radiative properties (Smith et al. 1975; Rosen and Novakov 1977; Novakov 1984; Horvath 1993; Liousse et al. 1993; Martins et al. 1998a; Martins et al. 1998b; Bond 2001; Currie et al. 2002; Moosmuller et al. 2009). But, isolation, quantification, and source discrimination methods for aquatic and sediment-associated BC fall into categories of visual/optical, oxidation-based (chemical and thermal), spectroscopic approaches, molecular proxy-based, and some combination of these. Specifically, quantification of BC after its isolation involves either direct optical methods (e.g., counting) or quantification of C by high temperature oxidation in the presence of a metal catalyst to CO_2. These BC isolation methods are briefly discussed later. The reader is also referred to several additional methodological review studies for further information (Schmidt and Noack 2000; Currie et al. 2002; Nguyen et al. 2004; Hammes et al. 2007).

The methods used to isolate BC in aquatic systems can result in large differences in the amount of BC quantified. Furthermore, each method for BC isolation has its inherent limitations based on the fact that each BC isolation method may be optimized for a specific portion of the BC continuum (Fig. 9.3). Also, there is no measure of accuracy for BC analysis. Methodological limitations can be attributed to (1) a lack of a common definition for BC, (2) failure to detect a broad range of pyrogenic carbonaceous residue, and (3) false-positive identification of non-BC carbonaceous substances such as soil geopolymers, kerogen, or other petrogenic carbon.

Table 9.1. *Compilation of dissolved or colloidal black carbon data measured in aquatic systems*

Geographical region	Isolation method	Comments on isolation method	Amount of BC in dissolved or colloidal phase (μg BC L^{-1})	Dissolved or colloidal BC/DOC (%)	Reference
Delaware Bay, Chesapeake Bay, USA	CTO-375 on HMW DOM	Some charring observed but not quantified	45–1,038	4–7	(Mannino and Harvey 2004)
Stream in New Jersey, USA	FT ICR-MS of aromatic compounds in DOC	Non-quantitative	NA	NA	(Kim et al. 2004)
Rio Negro, Amazon River, Brazil	FT ICR-MS on HMW DOM	Non-quantitative	NA	NA	(Kim et al. 2004)
River-to-ocean transect, north eastern Gulf of Mexico	Benzenepolycarboxylic acid quantification on cartridge isolated DOM via HPLC		9,600–44,000	0.9–2.6	(Dittmar 2008)
Southern Ocean	Benzenepolycarboxylic acid quantification on cartridge isolated DOM via HPLC	Functionality confirmed using FT ICR-MS	7.3–9.6	~2	(Dittmar and Paeng 2009)
Open Ocean (Sargasso Sea, North Central Pacific)	Radiocarbon age of benzenepolycarboxylic acid on cartridge-isolated DOM	Ultrafiltered material	0.96–6	0.5–3.5	(Ziolkowski and Druffel 2010)
North Atlantic Deep Water	Benzenepolycarboxylic acid quantification on cartridge isolated DOM	Functionality confirmed using FT ICR-MS	0.66–13	0.2–2.1	(Stubbins et al. 2012a)
Lake Tahoe, USA	SP2-FIA[1] on whole water	Restricted to ~60–400 nm particle size range	0.02–0.45	0.01–0.02	(Bisiaux et al. 2011)
Amazon River	Benzenepolycarboxylic acid quantification on cartridge isolated DOM	Aromaticity confirmed using FT ICR-MS	0–240	≤7	(Dittmar et al. 2012)

Notes:
[1] flow injection, ultrasonic nebulization, laser induced incandescence photometer.
NA – not available.

3.1. Visual methods for BC quantification

Paleoecological studies have benefited tremendously from quantification of the historical occurrence of fires made by enumeration and quantification on a per unit area basis of macroscopic or microscopic charcoal in the sedimentary record. Several recent studies that have comprehensively quantified historical charcoal deposition globally have suggested that both geographical and climatological differences contribute to BC loading (Carcaillet et al. 2002; Power et al. 2008; Bowman et al. 2009; Marlon et al. 2009a; Marlon et al. 2009b). Although light microscopy can identify larger particles that are clearly charcoal (i.e., jet black, opaque, angular) some particles can be challenging to identify due to their ambiguous color and/or shape (Patterson et al. 1987). Distinguishing between carbonized plant tissue and much smaller carbonaceous residue of combustion of fossil fuels and other sources is controversial, though important to climate studies. For example, using scanning electron microscopy, Kralovec et al. (2002) proposed that larger BC particles (>20 μm) isolated downcore in Lake Erie sediments were either (1) porous and spheroidal (oil-derived), (2) elongate with length-to-width ratios greater than three (wood-derived), or (3) irregular with a length to width ratio between 1 and 2 (coal-derived). An attempt to independently validate the BC isolation methods in that study using historical energy consumption data was only partially successful (Kralovec et al. 2002), suggesting that this approach may be limited in its application.

3.2. Chemical and thermal BC quantification methods

Quantification of BC in natural soils and sediments requires the ability to distinguish between inorganic carbon, non-pyrogenic OC and pyrogenic carbon. Chemical and thermal oxidation methods entail the removal of the former two, followed by quantification of residual carbon abundance. Prevalent chemical and thermal methods and their variables currently in use are noted in Table 1 of Nguyen et al. (2004). Subtle differences such as isolation temperature (Elmquist et al. 2006; Elmquist et al. 2004) or chemical oxidant reaction time (Masiello et al. 2002) can lead to substantial differences in BC yield for the same natural sample or reference material (Masiello et al. 2002; Elmquist et al. 2004; Elmquist et al. 2006). Thus it is critical that any BC isolation/quantification method be applied carefully and precisely. To that end, it is advisable to incorporate well-characterized available standard reference materials into any method of BC isolation such as those used in the BC Ring Trial (Hammes et al. 2007).

The thermal and chemical oxidation methods used most often for analysis of sedimentary BC are (1) acid-dichromate oxidation (AD) and (2) chemo-thermal oxidation at 375 °C (CTO-375). The AD technique involves thorough demineralization of a sample via hydrochloric acid (HCl) and hydrofluoric acid (HF), followed by a lengthy potassium dichromate (KCr_2O_7) and sulfuric acid oxidation of the demineralized residue (Wolbach and Anders 1989). The technique is based on the kinetics of the reduction reaction of an acidic solution of dichromate in conjunction with that solution's simultaneous oxidation of reduced organic matter. The dichromate in acid oxidizes labile and refractory carbon ranging from fresh biomass to kerogen and even labile BC, depending on the duration of oxidation time (Wolbach and Anders 1989). It is also assumed that each fraction of OM present in the sample responds differently to the oxidant, which is available in excess. The oxidation of carbon in the residue during the etching procedure is then modeled as a first-order exponential loss process, with a "fast" rate constant corresponding to labile non-BC, intermediate rate constants corresponding to kerogens, and a "slow" rate constant corresponding to BC (e.g., elemental carbon) (Wolbach and Anders 1989) with 400 hours being an adequate representation of the onset of the slower rate constant (Masiello 1999). It is commonly and incorrectly assumed in this approach, the BC can be approximated by carbon that is left in the residue after the oxidation reaction. However, that is not the case, as proper application of this technique relies on knowledge of the chemical kinetics of sample oxidation (Masiello et al. 2002). One criticism of wet-chemical methods is that they may underestimate BC as a result of losses during decantation or other transfers of the samples (Elmquist et al. 2004; Elmquist et al. 2006). However, other studies have not observed a difference in BC content of Standard Reference Material 1649a (urban dust) isolated by both the CTO-375 and the AD method (Mitra et al. 2009), suggesting that the BC losses during the washing and transfer steps can be minimized with careful handling. Although the AD method has been used to isolate fluvial BC in numerous studies (see Table 9.2), the toxicity of the reagents and the lengthy duration of the oxidation process are somewhat disadvantageous.

The CTO-375 technique for isolating BC has probably been the technique most widely applied to soils and sediments, owing to its ease of use. This method originally consisted of thermal oxidation of OC in the presence of excess air and was conducted at 375 °C followed by release of carbonates via in situ acidification in 1 M HCl (Gustafsson et al. 1997). The residual sedimentary carbon is then quantified by CHN elemental analysis (Gustafsson et al. 1997). Later testing of this technique indicated that this method only quantified soot BC (Nguyen et al. 2004). Several modifications have been made to this technique over the years. For example, the temperature as well as the point at which acidification should be done have both been modified since the initial study (Elmquist et al. 2004; Elmquist et al. 2006). One problem with thermal techniques is that they can yield false positives and overestimation of pyrogenic OM via charring. For example, nitrogen-rich macromolecules seem to be charred during thermal oxidation (Accardi-Dey 2003; Flores-Cervantes 2008), leading to quantities of soot BC in coastal planktonic organic matter isolates (Gelinas et al. 2001; Accardi-Dey 2003), bovine serum albumin (Flores-Cervantes 2008), and melanoidin (Hammes et al. 2007). It also appears that pollen is refractory to chemical and thermal treatment associated with the CTO-375 (Gustafsson et al. 1997). The reader is advised to carefully review the evolution of this procedure before proceeding and, in particular, evaluate its limitations alongside its simplicity. Several studies quantifying soot BC using the CTO technique are also listed in Table 9.2.

Recent advances in spectroscopic methods for identification of BC, which are based on molecular structure such as aromaticity, have yielded intriguing results about chemically and thermally isolated pools of BC. Spectroscopic methods such as ^{13}C nuclear magnetic resonance (NMR) and near-edge X-ray absorption fine structure (NEXAFS) have been used to demonstrate the presence of aromatic, graphitic, and amorphous moieties in chemically and thermally isolated BC (Lehmann et al. 2005; Dickens et al. 2006; Haberstroh et al. 2006). This suggests that material that is chemically or thermally isolated as BC may in fact be extremely heterogeneous in functional groups and composition.

3.3. Isolation of BC in the dissolved phase

Researchers have used a variety of techniques to isolate and quantify BC in the dissolved or colloidal phase in aquatic systems. For example, Mannino and Harvey (2004) quantified BC in

Table 9.2. *Values of particulate black carbon and organic carbon concurrently measured in river systems globally*

River	BC (mg/g)	TOC[1] (mg/g)	BC/TOC (%)	Type of BC[2]	Sample type	Reference
Santa Clara[3]	3.0	35.6	8.5	AD	Surface water TSS	(Masiello and Druffel 2001)
	7.1	16.0	44.3	AD		
	1.0	11.5	8.8	AD		
Ganges-Brahmaputra	2.9	5.7	50.9	AD	Surface water TSS	Mitra et al., unpublished
Mississippi	3.8	28.0	16.0	Soot	Surface water TSS	(Mitra et al. 2002)
	4.2	25.0	16.8	Soot	Bottom water TSS	
	0.5	32.0	1.5	Soot	Surface water TSS	
	0.8	25.0	3.0	Soot	Bottom water TSS	
AVERAGE BC IN TSS	2.9 ± 2.2	22.4 ± 10.3	18.7 ± 18.7			
Changjiang River	0.6	2.6	24.6	Soot	Surface sed (0–10cm)	(Kang et al. 2009)
	1.4	5.4	25.6	Soot		
	0.6	2.7	21.9	Soot		
	1.2	5.4	21.9	Soot		
	0.2	2.5	7.2	Soot		
	0.7	6.2	11.9	Soot		
	1.9	6.9	27.1	AD	Surface sed (0–2 cm)	(Li et al. 2012b)
	2.2	4.1	52.9	AD		
	2.2	4.7	46.2	AD		
	1.7	5.9	28.8	AD		
	2.2	6.3	34.1	AD		
	1.5	5.2	29.2	AD		
	2.3	6.7	34.0	AD		
Congo Fan	0.8	33.0	2.4	Soot	Surface sed (0–1 cm)	(Lohmann et al. 2009)
	0.9	20.2	4.5	Soot		
	0.9	13.3	6.8	Soot		
Amazon fan	1.2	3.7	32.4	Soot	Surface sed (0–1 cm)	(Lohmann et al. 2009)
	1.1	7.1	15.5	Soot		
	1.1	5.7	19.3	Soot		
	1.1	5.6	19.6	Soot		
Ob	0.5	9.2	5.1	Soot	Surface sed (0–2 cm)	(Elmquist et al. 2008)
Yenisey	1.2	19.4	6.2	Soot		
Lena	0.2	5.7	3.0	Soot		
Indigirka	1.4	18.1	7.7	Soot		

(continued)

Table 9.2 (*continued*)

River	BC (mg/g)	TOC[1] (mg/g)	BC/TOC (%)	Type of BC[2]	Sample Type	Reference
Kolyma	1.5	19.2	7.8	Soot		
Yukon	1.2	12.4	9.7	Soot		
Mackenzie	1.0	20.2	5.0	Soot		
AVERAGE BC IN SEABED SEDIMENTS[4]	1.2 ± 0.6	9.5 ± 7.5	18.9 ± 13.7			
OVERALL AVERAGE	1.6 ± 1.3	12.5 ± 9.7	18.9 ± 14.7			

Notes:

[1] TOC values correspond to the same time period as that of BC collection.

[2] Type of BC "AD" refers to isolation by acid-dichromate, "soot" refers to that isolated using CTO-375 or a similar chemothermal oxidation method. See original references for additional method details.

[3] Although not a large river, BC concentrations in the Santa Clara were included due to the overall paucity of information on BC in suspended sediments for large rivers globally.

[4] Although many stations were sampled within each of these studies, values of BC in seabed were only included if they were obtained directly at the mouth of that river system.

high-molecular-weight dissolved organic matter from the Delaware Bay using the CTO-375 method. However, quantities of BC in algal exudates, as well as the modern radiocarbon ages of the BC isolate (Mannino personal communication), suggest that their results were influenced by some charring of carbon during their sample processing. Indeed, the level of charring in the blanks used in that study created BC that was similar to the concentrations of BC estimated to exist in the open ocean (Mannino and Harvey 2004). As noted previously, selective charring of nitrogen-rich organic matter has been observed to occur via the CTO-375 technique (Gelinas et al. 2001; Accardi-Dey 2003). Thus, the thermal oxidation method may not be appropriate for isolating BC in coastal or oceanic DOM containing nitrogenous organic matter.

Evidence of BC in the dissolved phase is determined by the presence of condensed polyaromatic moieties in water, which has been revolutionized by Fourier Transform Ion Cyclotron Mass Spectrometry (FTICR-MS) (Kim et al. 2004; Hockaday et al. 2006; Koch and Dittmar 2006; Hockaday et al. 2007). This technique, although not quantitative, offers extremely high-mass resolving capacity of DOM (~200,000 mass units). To quantify dissolved BC, several researchers have relied on the fact that dissolved and unsubstituted PAHs can be oxidized in the lab with nitric acid into various benzenepolycarboxylic acids (BPCA). These BPCAs can then be quantified as a proxy for dissolved BC using liquid or gas chromatography (Glaser et al. 1998; Brodowski et al. 2005). However, it should be understood that the similarity of functionalized water-soluble condensed aromatics in natural waters to the degradative products of BC does not necessarily confirm their common origin. For example, functionalized polyaromatics such as those used as proxies for dissolved BC may originate from oil-weathering (Aeppli et al. 2012) or in situ thermogenic processes known to occur in hydrothermal seeps in the ocean (Dittmar and Paeng 2009). Furthermore, both improper laboratory acidification during the procedure and the presence of darker soil pigments have been shown to yield non-pyrogenic BPCA. Thus the use of BPCAs as a proxy for dissolved BC should also be done with caution.

One novel method of quantifying colloidal BC suspended in the water column involves using flow injection analysis (FIA) coupled to a single particle (SP) intercavity laser-induced incandescence photometer (Bisiaux et al. 2011). The FIA-SP2 method determines BC from the wavelength-resolved incandescent light emitted by individual BC particles heated to their boiling point (3,700–4,300 K) via laser. As a result, the method is free of positive interferences from humic acids, charcoal, mineral graphite, or PAHs and is insensitive to particle morphology. However, the method is restricted to a particle size range of ~60–400 nm, and thus larger colloidal BC would be omitted from being quantified.

3.4. Source discrimination of BC

Most BC quantification methods do not provide information about the BC parent material of mode of combustion. Nonetheless, this source information is useful in understanding BC's role in the global carbon cycle or effect on climate. Stable and radiogenic isotopes of BC as well as molecular proxies of combustion have been used to determine its sources. If minimal isotopic fractionation of the natural isotopes of carbon occurred during (1) combustion of fuel, (2) post combustion preservation, and (3) while carrying out a specific BC isolation technique, stable carbon isotopic signatures ($\delta^{13}C$) of residual BC might be used to distinguish BC from different sources. However, evidence suggests there is carbon isotopic fractionation of both fossil fuels and biomass, both during combustion as well as diagenesis. For example, the isotopic signature of fossil fuels is a function of their physical state (i.e., solid, liquid, or gas), and fractionation during combustion leads to isotopic depletion during the production of combustion gases, but is generally positive during the process of particle formation (Widory 2006). Similarly, isotopic shifts during the heating process of biomass resulted in fractionation of C-13 by as much as 2–4‰ (Gilhooly et al. 1996; Turekian et al. 1996). Also, up to 8‰ depletion in $\delta^{13}C$ was observed in C4-derived chars from natural burning without a similar isotopic change in chars from wood or C3 grasses, owing to the presence of occluded phytoliths in C4-derived chars (Krull et al. 2003). Residual BC may also contain some nitrogen, oxygen, hydrogen, and sulfur, in which case their isotopic signature may be used to fingerprint the source of BC. Because these elements comprise such a minor percentage of its composition, they are often not considered practical in BC source discrimination (Bird and Ascough 2012).

Radiocarbon (C-14) abundance can be used to elucidate the extent to which BC was produced from the combustion of fossil sources or modern biomass (Reddy et al. 2002; Gustafsson et al. 2009). Although living biomass incorporates C-14 from the atmosphere, it is deficient in ancient organic material. This method was applied in a most elegant manner to the infamous "brown cloud" or annual haze of combustion-derived aerosols over the Indian Ocean (Gustafsson et al. 2009). In that study, the C-14 value of aerosol BC was quantitatively used to conclude that 50% of the BC was derived from biomass combustion. Unfortunately, ancient petrogenic and paleosol carbon, which have been found in stream and river waters (Raymond and Bauer 2001), are also likely to be depleted in radiocarbon. Thus improper accounting for such recycled carbon may lead to overestimates of carbon sequestration in marine sediments. Although physical gravimetric separation conducted prior to oxidation may be effective in separating out some petrogenic material (Dickens et al. 2004b), more robust confirmation of BC source is generally necessary, as each source of carbon (fossil fuel or biomass) will have different net effects on the global carbon cycle.

Some of the uncertainties associated with chemical and thermal BC isolation methods of aquatic sediments have been overcome by quantifying trace organic chemical molecules cogenerated during combustion, such as selected polycyclic aromatic hydrocarbons (PAHs) (Gustafsson and Gschwend 1997; Louchouarn et al. 2007; Mitra et al. 2009) and levoglucosan, a byproduct of cellulose combustion (Elias et al. 2001; Kuo et al. 2011). Isomer ratios of high-molecular-weight and particle-reactive PAHs are often used to discriminate between sources of combustion (Yunker et al. 2002), but have also been used to corroborate the abundance and provenance of sedimentary BC (Mitra et al. 2002; Mitra et al. 2009). Another similar technique along these lines employs pyrene fluorescence quenching as a proxy for BC concentrations in natural sediments (Flores-Cervantes et al. 2009b). This method relies on the concept that BC has a high sorption affinity for PAHs such as pyrene and has yielded promising results for its quantification in coastal aquatic ecosystems (Flores-Cervantes et al. 2009b). In summary, it is recommended that any attempt at BC quantification should be coupled to multiple methods of its source discrimination.

4. Black carbon in large river systems

Despite the importance of the fluvial delivery of BC in the global carbon cycle as shown previously, only a few studies have characterized or quantified BC in the suspended river and coastal bed sediments associated with major rivers. The results of those studies are summarized in Table 9.2. Quantification of BC discharge from measurements of BC in riverine suspended sediments (e.g., Masiello and Druffel 2001; Mitra et al. 2002) are associated with uncertainties in that BC in suspended sediments can vary dramatically within and among river watersheds and likely vary with river size and discharge (Masiello and Druffel 2001). The amount of BC transported as bedload is also unknown. Additionally, erosion of soil from fire-affected areas can introduce non-BC soil-derived geopolymers, which may produce false-positive measurements of BC suspended in the water column. Other researchers have attempted to estimate BC discharge by quantifying BC in the seabed adjacent to major river systems (Elmquist et al. 2008; Lohmann et al. 2009) but this approach also has inherent limitations. For example, BC may be preferentially transported away from river deltas. And bed sediments adjacent to the mouth of large rivers have been termed "fluidized bed reactors," which are extremely efficient hot spots for remineralization of carbon (Aller 1998). As a result, quantities of BC in these areas may be disproportionately low and unrepresentative of global fluvial quantities entering the oceans.

4.1. Fluvial BC

Existing fluvial sedimentary BC concentration data were collected and parsed into estimates for BC in riverine suspended sediments as well as bed sediments (Table 9.2). In comparing these numbers, a few salient trends are apparent. First, irrespective of the isolation method used or geographic location, BC concentrations per mass of sediment are generally greater in suspended sediments than in bed

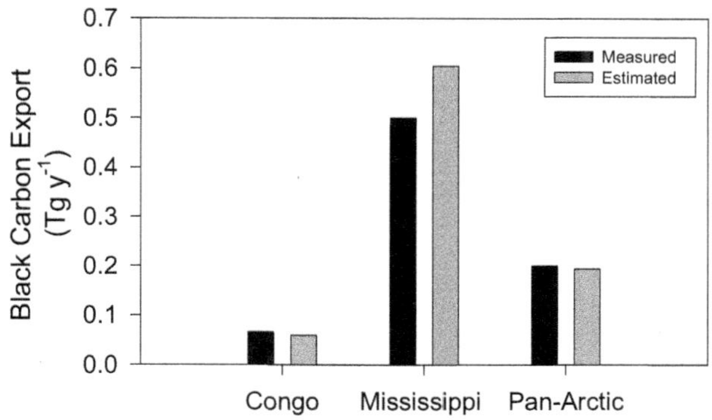

Figure 9.4. Measured and estimated BC exported from a selected subset of major river systems from Table 9.3. Measured values for the Congo, Mississippi, and Pan-Arctic Rivers are from Lohmann et al. 2009, Mitra et al. 2002, and Elmquist et al. 2008, respectively. Pan-Arctic rivers BC export is calculated as the sum of the well-studied Yenisey, Ob, Lena, and MacKenzie rivers.

sediments. Second, the concentration of POC, when quantified concurrently with BC, is significantly higher in suspended sediments than bed sediments. As a result, BC-to-TOC concentration ratios appear similar in all the fluvial systems in Table 9.2. This suggests that non-BC sedimentary organic carbon is transported, sorted, or degraded by similar processes as BC across coastal systems. For these reasons, we suggest that the average value of 1.6 mg BC per gram dry weight (gdw) of sediment (Table 9.2), integrated across suspended and bed sediments, be used as a convenient representative concentration of fluvial BC entering the oceans. To calculate global particulate BC exported from large rivers for which BC has not yet been quantified, the value of 1.6 $mggdw^{-1}$ was multiplied by suspended sediment discharged from each large river listed in Table 9.3. A similar approach has been used to calculate global fluvial discharge of sedimentary POC (Burdige 2005). Estimated values of BC were then compared with known values of BC in a subset of the large river systems listed in Table 9.3, as determined by others. This comparison shows that estimated values of fluvial BC export are similar to experimentally-determined values of soot BC for the Congo, the Mississippi, and selected Pan-Arctic rivers (Fig. 9.4). Part of the reason that this estimation method agrees well with existing values of soot BC may be that averaging BC/TSS across different methods and across a larger number of rivers may eliminate differences owing to methodological artifacts. Alternatively, the comparison may be biased by the fact that most of the values and the 1.6 mg gdw^{-1} in Table 9.2 are derived from refractory BC, isolated using the AD and CTO-375 procedures.

The estimates for each river's BC discharge noted in Table 9.3 are graphically depicted in Figure 9.5, where the area of each arrow is proportional to the estimated annual BC discharge. It is apparent from Figure 9.5 that not only is the Amazon the dominant large river for BC discharge, but the BC discharge of the four major South American rivers (Amazon, Magdalena, Orinoco, and Paraná) is nearly equal to the total BC discharge of every other major river in the study combined (2.64 vs. 2.72 Tg y^{-1}, respectively). Perhaps this is due to the compound effects of residual widespread biomass burning (Dittmar et al. 2012) exacerbated by the lengthy watershed residence time of the basin, coupled with the sheer enormity of that river system's discharge. Nonetheless, the fluvial BC

Table 9.3. *Sediment, particulate organic carbon, black carbon, and polycyclic aromatic hydrocarbon discharge from large rivers*

River	Qs ($T\ yr^{-1}$)	Qav (m^{-3}/s)	Yield ($T\ km^{-2}\ yr^{-1}$)	TSS ($kg\ m^{-3}$)	TSS ($MT\ yr^{-1}$)	POC ($mg\ L^{-1}$)	POC reference	BC discharged[1] ($Tg\ y^{-1}$)	tPAH ($ng\ g^{-1}$)	b[a]p ($ng\ g^{-1}$)	b[a]p discharged ($kg\ y^{-1}$)	PAH reference
Amazon	1.2×10^9	2.0×10^5	193	0.19	1190.4	2.83	(Richey et al. 1990)	1.91	701	NA		(Budzinski et al. 1997; Lohmann et al. 2009)
Mississippi	4.0×10^8	1.7×10^4	124	0.11	61.3	16.90	(Bianchi et al. 2007)	0.60	5300	372	22,817	(Maruya et al. 1997; Overton et al. 2004; Santschi et al. 2001)
Ganges-Brahmaputra	3.1×10^8	3.1×10^4	701	0.32	312.8	2.50	(Aucour et al. 2006)	0.47	2938	148	46,300	(Ahmad et al. 1996; Domìnguez et al. 2010)
Indus	2.5×10^8	3.2×10^3	261	1.05	105.0	8.82	(Ludwig et al. 1996)	0.38	NA	NA		
Magdalena	2.2×10^8	7.2×10^3	868	0.63	143.3	NA		0.33	NA	NA		
Orinoco	1.7×10^8	3.4×10^4	180	0.06	65.7	1.50	(Paolini 1995)	0.26	NA	NA		
Godavari	1.7×10^8	3.3×10^3	611	2.03	213.2	7.20	(Balakrishna and Probst 2005)	0.26	NA	NA		
Nile	1.2×10^8	3.5×10^3	43	1.4	153.8	3.85	(Ludwig et al. 1996)	0.18	1229	94	14,459	(Badawy and Emababy 2010)
Irrawaddy	1.1×10^8	1.4×10^4	614	0.26	111.2	3.59	(Bird et al. 2008)	0.17	133	0.6	67	(Chernova et al. 2001)
Yangtze	1.0×10^8	2.8×10^4	251	0.13	119.3	1.62	(Wang et al. 2012)	0.16	502	14.6	1,741	(Bouloubassi et al. 2001; Li et al. 2012a)

MacKenzie	1.0×10^{8}	1.0×10^{4}	58	0.39	128.0	6.36	(Macdonald et al. 1998)	0.15	1200	66	8,450	(Yunker et al. 1993; Yunker et al. 2011)
Mekong	9.8×10^{7}	1.7×10^{4}	123	0.08	44.2	2.01	(Ellis et al. 2012)	0.15	1379	30	1,326	(Keenan et al. 2010)
Paraná	9.0×10^{7}	1.5×10^{4}	31	0.18	82.3	1.69	(Ludwig et al. 1996)	0.14	2120	179	14,739	(Cataldo et al. 2001; Colombo et al. 2006)
Congo[2]	3.9×10^{7}	3.7×10^{4}	8.8	0.03	35.0	1.70	(Coynel et al. 2005)	0.06	227	NA		(Lohmann et al. 2009)
Huanghe	3.3×10^{7}	1.5×10^{3}	1122	2.52	117.7	15.26	(Wang et al. 2012)	0.05	5734	180	21,189	(Jiao et al. 2012; Ma et al. 2008; Wu et al. 2001; Xu et al. 2007)
Amur	2.4×10^{7}	1.1×10^{4}	28[3]	0.07	25.3	0.24	(Meybeck and Ragu 1997)	0.04	1100			(Nemirovskaya 2007)
Lena	2.0×10^{7}	1.7×10^{4}	8	0.02	10.8	0.99	(Lobbes et al. 2000)	0.03	97			(Elmquist et al. 2008)
Yenisey	4.9×10^{6}	1.8×10^{4}	5[3]	0.008	5.0	0.98	(Unger et al. 2005)	0.01	636	4.3	22	(Fernandes and Sicre 1999); (Dahle et al. 2003)
Ob	3.5×10^{6}	1.3×10^{4}	6.4[3]	0.009	3.8	0.66	(Unger et al. 2005)	0.01	376	6.9	26	(Fernandes and Sicre 1999); (Dahle et al. 2003)

Notes:

[1] A value of 5.4 Tg y^{-1} is estimated as the sum of BC discharged.

[2] Although the Congo's POC load is low, its expansive watershed is in an area subject to high biomass burning (Cachier and Ducret 1991; Delmas et al. 1991), and 2–7% of its POC is made up of soot BC.

[3] Milliman and Meade 1983.

Unless otherwise noted, physical variables for rivers are from Syvitski and Saito (2007). Qs: suspended load, Qav: average discharge Yield, TSS: total suspended sediments, POC: particulate organic carbon, BC: black carbon, b[a]p: benzo[a]pyrene. NA – not available.

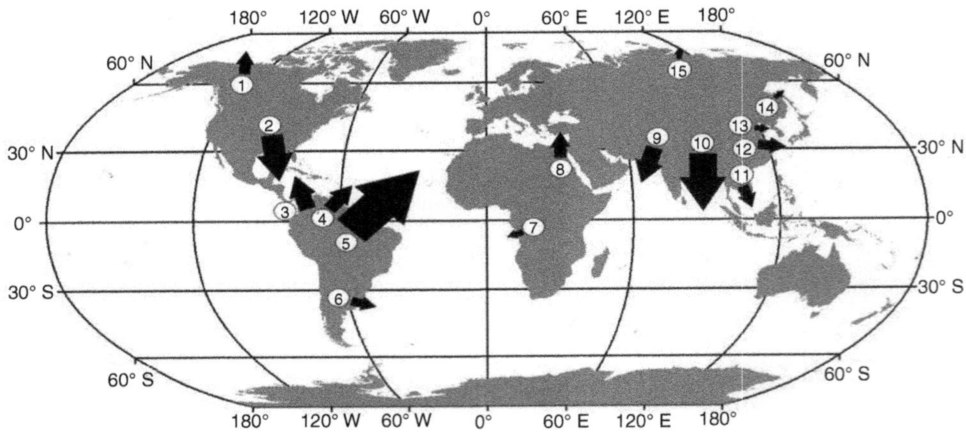

Figure 9.5. World map with arrows depicting fluvial BC discharge from major rivers noted in Table 9.3. Names of rivers are: (1) Mackenzie, (2) Mississippi, (3) Magdalena, (4) Orinoco, (5) Amazon, (6) Parana, (7) Congo, (8) Nile, (9) Indus, (10) Ganges – Brahmaputra, Irrawaddy, and Godavari, (11) Mekong, (12) Yangtze, (13) Huanghe, (14) Amur, and (15) Lena, Ob, and Yenisey.

discharged from South America seems to supersede that of any other continent. However, as we note in another section later, the quantities of BC measured in the seabed adjacent to the Amazon River are surprisingly low.

Although quantifying fluvial BC discharge is not a straightforward process, the BC discharge of some large river systems as determined with our simple BC discharge model agree well with the measured values (Fig. 9.3). As noted in other chapters of this book, the large rivers listed in Table 9.3 are viewed as some of the most influential ones globally in terms of either water or sediment discharge (e.g., see Chapter 3). Thus we assume their collective contribution to BC discharge to the ocean to be the most important, globally.

Although collectively, small mountainous high-yield river systems (SMR) provide 40–70% of the global sedimentary delivery to the ocean (Milliman and Syvitski 1992; Leithold et al. 2006), they are not included in the global BC discharge estimates in Table 9.3. This may be justified given that their watersheds are smaller. Nonetheless, we do discuss their potential role in fluvial BC discharge later.

4.2. Global pyrogenic PAH discharge and comparison with BC

As noted earlier, PAHs co-evolve with BC during combustion (Lima et al. 2005). During biomass and fuel combustion, the liberation of hydrogen associated with fuel combustion en route to its formation of BC produces aromatic molecules (Chakraborty and Long 1967). Several researchers have isolated PAHs concurrently with BC in an attempt to confirm the pyrogenic origin of BC quantified (Mitra et al. 2002; Louchouarn et al. 2007; Kuo et al. 2008; Flores-Cervantes et al. 2009b). Here we attempt to validate our approach for calculating global BC discharge for the river systems listed in Table 9.3 by making a comparison between the discharge of benzo[*a*]pyrene (a high-molecular-weight particle-reactive PAH) and the estimated values of large river BC discharge depicted in Fig. 9.5. The

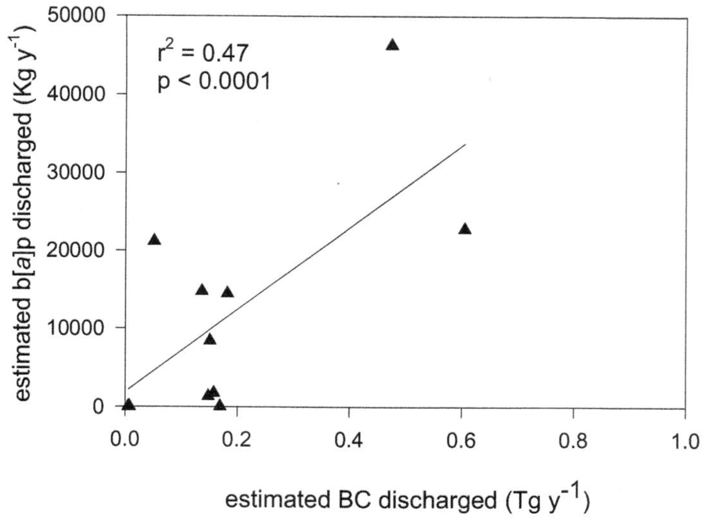

Figure 9.6. Relationship between estimated values of benzo[*a*]pyrene and black carbon discharged in rivers listed in Table 9.3. Estimated values are the product of benzo[*a*]pyrene and black carbon concentrations in sediment multiplied by total suspended sediment discharged for each river system.

significant correlation between the two values ($r^2 = 0.47$; $p < 0.005$ – Fig. 9.6) suggests that the fluvial particulate BC estimated using our approach is pyrogenic and not substantially influenced by non-pyrogenic refractory material (e.g., petrogenic C or soil-derived polymers). In contrast, river discharge of total PAH and BC was not significantly correlated ($r^2 < 0.01$), which is a reasonable result, because not all PAHs are pyrogenic. For example, low-molecular-weight and alkylated PAHs would not be expected to be coeval with combustion-derived BC (Lima et al. 2005).

The global sources and sinks of BC to the ocean can be summarized as follows. First, the most recent estimate of atmospheric BC deposition to the ocean was combined with our estimated fluvial discharge summary (Table 9.3). It was assumed that the most recent value for atmospheric deposition of BC into the world's oceans (12 Tg y^{-1}) is also the most reliable (Jurado et al. 2008), because it is similar in magnitude to the most updated and comprehensive estimate of the amount of fossil fuel and biomass BC emitted annually throughout the world (8 Tg y^{-1}; Bond et al. 2004). This value for annual BC emission globally is much lower than the often cited 50–270 Tg y^{-1} for atmospheric BC emission strictly attributed to biomass burning (Kuhlbusch et al. 1996). Use of the larger range for BC emission calculated by Kuhlbusch et al. (1996) has yielded higher and perhaps erroneous estimates for the magnitude of BC introduced into the ocean compared with the lower amounts found in ocean sediments (Druffel 2004; Masiello 2004).

The value for the global estimate for large river delivery of BC into the ocean calculated above is 5.4 Tg y^{-1} (sum of BC discharged column in Table 9.3). If this fluvial estimate is increased by 40% to incorporate SMR contributions, the total influx of particulate BC to the oceans from rivers may be as high as 8 Tg y^{-1}. This value is comparable to the 12 Tg y^{-1} supplied by atmospheric deposition (Jurado et al. 2008) and implies a total flux of ~20 Tg y^{-1} BC to the ocean. It should be noted, however, that these estimates pertain mainly to more refractory BC (e.g., soot) and may not include other more labile portions of the BC continuum such as charcoal.

5. Fate of black carbon

Despite efforts to understand the short- and long-term cycling of BC derived from both natural and anthropogenic sources, the role of the coastal zone in sequestering fluvial BC is still uncertain. As noted earlier, long-term burial of biomass-derived BC in sediments is a net long-term sink for atmospheric CO_2. In contrast, deposition of fossil fuel BC amounts to reburial of ancient carbon and does not promote net carbon sequestration. It is important to note that understanding of the role of BC in the carbon cycle is confounded by the timescales of the records in which it is quantified in various reservoirs. For example, variations in aeolian BC production and deposition have been observed at seasonal-to-annual time scales, whereas the marine sedimentary record can only be reconstructed at decadal or longer resolution. Consequently, in many areas of the open ocean where sedimentation is low (mm 1,000 y^{-1}) variations in BC will be dominated by BC derived from natural forest fire sources as opposed to modern anthropogenic sources. In contrast, coastal areas with higher sedimentation rates are more likely to record BC inputs from both anthropic fossil fuel and biomass combustion. Next, we discuss sedimentary BC after it leaves a river and consider its fate both on continental margins as well as the deep ocean.

5.1. Storage BC in the ocean

Globally, continental margins (shelf, slope and rise), estimated to be responsible for about 80–90% of the ocean's OC storage (Berner 1982; Hedges and Keil 1995), are the largest reservoirs of marine OC. This is likely to be true for the majority of sedimentary BC as well, which is thought to be deposited in continental shelf regions where water depths are less than 200 m (Suman et al. 1997). The amount and ultimate fate of sedimentary BC on continental margins will vary as a function of the concentration and flux of delivered BC, climate, tectonics, biological and physical mixing, and geomorphology of the margin. Additionally, throughout the Quaternary, there is evidence that stochastic events such as large storms may be drivers of the geochemical record in coastal environments (Goni et al. 2007; Rogers and Goodbred 2010). Most of these drivers have not been comprehensively addressed in existing studies of bulk sedimentary carbon, let alone in BC studies.

We address this data gap by compiling existing estimates of BC deposition in coastal and open ocean sediments (Table 9.4) to determine whether any salient trends can be attributed to any of the state variables noted previously. Furthermore, this compendium may help determine whether there is imbalance in BC delivery and deposition in the ocean. The data in Table 9.4 are listed geographically from high-to-low latitude. What is particularly striking is that the highest concentrations and largest fluxes of BC to coastal sediments do not seem to be related to either latitude or to proximity to large river mouths. And, once again, there does not appear to be a trend in BC concentrations or fluxes related to its quantification method. Rather, the highest fluxes of sedimentary BC listed in Table 9.4 (Swedish Continental Shelf, Chesapeake Bay, coastal East China Sea, and the Palos Verdes Shelf) are

Table 9.4. *Deposition of BC in coastal and open ocean sediments*

Geographical region	Isolation method	BC concentration in sediments ($mg\ g^{-1}$)	BC Flux ($\mu g\ cm^{-2}\ y^{-1}$)	Reference[1]
Coastal Areas				
Saanich Inlet,BC, Canada	Chemical treatment, IR spectrometry	0.55	74	(Griffin and Goldberg 1975)
Pan Arctic Shelf	CTO – 375	0.5–1.5	0.13–5.6	(Elmquist et al. 2008)
Swedish Continental Shelf	CTO – 375	2–2.8	270–8100	(Sanchez-Garcia et al. 2012)
Gulf of Maine, USA	CTO-375, pyrene fluorescence quenching	0.11–0.17; 0.1–15[2]	100–200; 80–170	(Flores-Cervantes et al. 2009a; Gustafsson and Gschwend 1998)
Puget Sound, Washington, USA	GBC[3]	0.1–0.3	30–1,040	(Kuo et al. 2011)
Washington Coast, USA	GBC[3]	0.14–0.71	1.2–3.1	(Dickens et al. 2004a)
Chesapeake Bay, USA	AD	1.1–7.5	760–5,400	(Mitra et al. 2009)
Coastal East China Sea	AD	1.5–2.2	210–1300	(Li et al. 2012b)
Santa Barbara Basin	Chemical treatment, IR spectrometry; AD	0.29; 2.2	26; 35	(Griffin and Goldberg 1975, Masiello and Druffel 2003)
Gulf of Panama	Chemical treatment, IR spectrometry	NQ	109–354	(Suman 1986)
Continental rise adjacent to Nicaragua	Chemical treatment, IR spectrometry	NQ	23–135	(Suman 1986)
Eastern South Atlantic Shelf	CTO-375	0.8–1.6	0.5–7.8	(Lohmann et al. 2009)
Western South Atlantic Shelf	CTO -375	0.4–1.7	0.6–2.6	(Lohmann et al. 2009)
Palos Verdes Shelf	CTO -375	1.2–1.4	700–1000	(Gustafsson and Gschwend 1998)
Open Ocean				
South China Sea	AD	0.01–1.9	NQ	(Jia et al. 2003)
Pelagic Pacificand Atlantic	Chemical treatment, IR spectrometry	0.01–1	0.002–0.2	(Smith et al. 1973)
Pelagic Pacific	Chemical treatment, IR spectrometry	0.44	0.057	(Griffin and Goldberg 1975)
Pelagic Pacific	Chemical treatment, IR spectrometry	0.001–3	< 0.0014–3.6	(Herring 1985)
Equatorial Pacific	GBC[3]	0.092	0.05	(Dickens et al. 2004a)

Notes:

[1] The following studies of BC in the deep ocean sediments were reviewed but were not included in the table because neither concentration nor sediment mass flux were provided: (Bird and Cali 1998; Masiello and Druffel 1998; Thevenon et al. 2004).

[2] Concentrations in $\mu g\ L^{-1}$.

[3] The GBC (graphitic black carbon) technique is a thermal oxidation technique with additional organic acid pretreatment specifically designed to minimize charring of occluded organic matter (Gelinas et al. 2001).

NQ – unable to be quantified.

adjacent to fluvial systems affected by urbanization within their watershed, suggesting that fossil fuel combustion primarily drives the bulk of sedimentary BC influx into coastal margins today. Without higher temporal or spatially resolved data, little more can be said about the additional controls on the BC in the coastal sedimentary record.

Based on the calculations discussed in Section 4.2, the global flux of BC to the ocean is estimated at ~20 Tg y^{-1}, with the reasonable assumption that 40% of this is contributed by rivers and 60% by the atmosphere. Although there exists the methodological issue of whether this estimate is mainly for soot or includes all BC, we have shown that estimates don't vary greatly across detection methods that employ chemical and thermal oxidation (Tables 9.2 and 9.4). Assuming a figure of about 90% of total seaward flux of particulate BC to be deposited on continental margins (Suman et al. 1997), BC storage in these regions can be estimated to be 18 Tg y^{-1}. This leads to a delivery estimate of ~2 Tg of BC y^{-1} deposited annually in the deep ocean at water depths greater than 200 m. Assuming the deep ocean has an area of 3.34×10^8 km^2 (Suman et al. 1997), we calculate an expected average BC flux of 0.60 μg cm^{-2} y^{-1}. This falls within the range of measured open ocean flux values from 0.001 to 3.6 μg cm^{-2} y^{-1} (Table 9.4). Using a similar approach, 18 Tg y^{-1} delivered to continental shelves globally can be scaled to the total area of continental shelves (2.75×10^7 km^2) to yield an expected average flux of 65 μg cm^{-2} y^{-1}, which is on the lower end, but within the range of values for BC burial flux in coastal regions globally that are listed in Table 9.4. In summary, our compilation of data for the annual budget of refractory BC suggests that its atmosphere and riverine delivery and deposition to surface sediments in the coastal and deep ocean are in balance. However, our estimates are not able to provide an evaluation of the role of labile pools of BC in the global carbon cycle. In light of recent studies quantifying BC degradation in the environment, data on the production, transport, and fate of these labile pools such as char and charcoal are sorely needed.

5.2. Degradation of BC

There is a growing awareness that, although BC is a relatively refractory OM form, it is not completely inert in the environment. There are a number of lines of evidence that suggest that some pools of BC degrade via abiotic and biotic processes. Physical processes that may enhance BC degradation include UV exposure, leaching/solubilization, volatilization and consumption by later fires, erosion, resuspension and transport in the seabed, and biological processes that include bioturbation and ingestion by macrofauna or remineralization by microbes.

The first type of evidence showing BC oxidation comes from studies measuring CO_2 efflux during in vitro laboratory incubations of laboratory-produced pyrogenic OM. These experiments have recorded pyrogenic C losses ranging about 0.5–5% over a few months to a few years, with some larger losses for materials produced at lower temperatures (<400 °C, Baldock and Smernik 2002; Hamer et al. 2004; Nguyen and Lehmann 2009; Nguyen et al. 2010; Zimmerman 2010; Cross and Sohi 2011; Bruun et al. 2012). About half of these losses can be attributed to abiotic mechanisms (Zimmerman 2010), and greater losses have been observed in the presence of labile sources of OM (so-called positive priming; Cross and Sohi 2011; Zimmerman et al. 2011). These observed losses translate to half-lives of 10 to several hundred years. However, exponential decreases in rates of C loss over time suggest that the power model of OM decomposition (Middleburg 1989) might be more appropriately applied, leading to the calculation of BC half-lives in the thousand to million year range (Zimmerman 2010).

Second, measurements of BC distributions in soil profiles with known pyrogenic OM input histories lead to calculations of apparent BC field loss rates ranging from zero loss (completely stable) to BC half-lives of 3 years. Discrepancies among these studies are likely due to uncertainties that include (1) the likelihood that BC particles are mobile in the soil column and lost via erosion, (2) differing BC degradation rates in different climates and soil types, (3) poorly known BC input rates, and (4) differing BC quantification methods. Only a few studies have examined BC in marine sediment columns, and none of these show progressive BC losses with time of burial (Masiello and Druffel 1998; Middleburg et al. 1999; Masiello and Druffel 2003). However, observations of 64% greater BC contents in abyssal plain sediments exposed to oxygen over 10–20 thousand years versus anoxic sediments suggest BC half-lives of $\sim$10 ky (Middleburg et al. 1999). In addition, smaller BC-nonBC OM ^{14}C-age offsets in oxygenated versus anoxic Santa Monica Basin sediments (Masiello and Druffel 2003) also indicate that sedimentary remineralization of BC does occur, at least in the presence of oxygen.

Next, mass balance considerations suggest the occurrence of large-scale BC remineralization at some location between its production and final marine burial. For example, given estimated BC production rates, over the past 7,000 years, soil OC would become 25–125% BC in just 20,000 years (Masiello and Druffel, 2003). Calculations presented in this chapter also suggest losses between soils and coastal deposition. For example, globally averaged BC concentration in TSS of 2.9 mg g^{-1} versus 1.2 mg g^{-1} in deltaic sediment beds suggest about half the BC is lost during transport (although this could also result from preferential transport processes). This is despite the fact that many of these estimates are based on BC measurement by chemical and thermal oxidation methods that isolate relatively refractory BC, perhaps only soot. However, at this time, estimates of BC production, river export, and sediment sink fluxes within the same geographical area are too few and too widely ranging to use for predicting global BC mineralization rates.

One area where BC may be readily degraded is on broad continental shelves. These environments, which are often physically dynamic due to periodic resuspension, contain sediments frequently subjected to redox oscillations, which have been shown to enhance BC degradation (Nguyen and Lehmann 2009). These areas have been termed "fluidized bed reactors" (Aller 1998) and have been shown to be hot spots of terrestrial OM remineralization. In addition, continuous addition of labile planktonic debris may co-metabolize the degradation of OM, which has also been shown to enhance BC degradation (i.e., positive priming; Hamer et al. 2004; Zimmerman et al. 2011). This may explain the seemingly contradictory observations of large amounts of BC export by the Amazon and other South American rivers (calculated in this chapter) and relatively low concentrations of BC in Amazon shelf sediments (Lohmann et al. 2009).

Finally, losses to BC may occur by solubilization or leaching either in the soil or as part of the suspended or deposited sediment. Although laboratory leaching experiments have recorded only small losses of BC ranging from 0.04% to 1% (Hammes et al. 2007; Abiven et al. 2011; Katsoyiannis et al. 2011; Zimmerman and Gao 2013), aged charcoals seem to exhibit leach C rates similar to those of fresh ones, and detection of mobilized BC in rivers draining Brazil's Atlantic forest even 40 years after widespread burning (Dittmar et al. 2012) suggests C losses that occur by this method may be continuous. And there is evidence from both laboratory microbial incubations showing up to 70% C loss over 1 year (Zimmerman and Gao 2013) and photo-degradation experiments showing a 20-fold decrease in coastal and North Atlantic deepwater dissolved BC (Stubbins et al. 2012b; Stubbins et al. 2010) that losses via leaching followed by mineralization may be a primary sink for BC.

6. Conclusions

Many recently developed analytical techniques show great promise toward identification and quantification of pyrogenic carbon. Yet, even with the advent of these techniques, further study of BC in the environment is warranted to fully understand it from a source-to-sink perspective. This is because several important questions about the global cycling of BC still remain. For example, how long does BC remain in soil, and what processes move it to river systems? Does BC spend very little or a substantial amount of time in transit through river systems and how is BC transformed there? To what extent does BC comprise the bedload versus suspended load of a river system? As there are few studies of the transport and fate of BC within river systems, these questions cannot yet be answered. Because few studies concurrently measure charred OM charcoal, and refractory BC, questions also remain about the fate and transport of labile pools of BC shown in the left side of Figure 9.3. Also, the presence of dissolved BC in rainwater as well as in fluvial and ocean systems begs the question, "What are the mechanisms by which dissolved BC cycles in the ocean?" Lastly, although it seems clear that continental margins are repositories for the bulk of sedimentary BC, a comprehensive understanding of BC fate and degradation in these areas is lacking. With ever-increasing changes in land use and anthropic impacts to climate, now more than ever, it is necessary to address these data gaps in order to understand the role of BC in the global carbon cycle and its possible use for climate mitigation.

References

Abiven, S., Hengartner, P., Schneider, M.P.W., Singh, N., and Schmidt, M.W.I. 2011. Pyrogenic carbon soluble fraction is larger and more aromatic in aged charcoal than in fresh charcoal. *Soil Biol. Biochem.* 43: 1615–1617.

Accardi-Dey, A. 2003. Black Carbon in Marine Sediments: Quantification and Implications for the Sorption of Polycyclic Aromatic Hydrocarbons. PhD Dissertation, Massachusetts Institute of Technology, Cambridge, MA.

Aeppli, C., Carmichael, C.A., Nelson, R.K., Lemkau, K.L., Graham, W.M., Redmond, M.C., Valentine, D.L., and Reddy, C.M. 2012. Oil weathering after the Deepwater Horizon disaster led to the formation of oxygenated residues. *Environ. Sci. Technol.* 46: 8799–8807.

Ahmad, S., Ajmal, M., and Nomani, A.A. 1996. Organochlorines and polycyclic aromatic hydrocarbons in the sediments of Ganges River (India). *Bull. Environ. Contam. Toxicol.* 57: 794.

Aller, R. 1998. Mobile deltaic and continental shelf muds as suboxic, fluidized bed reactors. *Mar. Chem.* 61: 143–155.

Arranz, J.M.D., Gonzalez-Vila, F.J., Lopez-Capel, E., Manning, D.A.C., Knicker, H. Gonzalez-Perez, J.A. 2009. Structural properties of non-combustion-derived refractory organic matter which interfere with BC quantification. *J. Anal. Appl. Pyrolysis* 85: 399–407.

Aucour, A.M., France-Lanord, C., Pedoja, K., Pierson-Wickmann, A.C., and Sheppard, S.M.F. 2006. Fluxes and sources of particulate organic carbon in the Ganga-Brahmaputra river system. *Global Biogeochem. Cycles* 20: GB2006, doi:10.1029/2004GB002324.

Bachmann, J. 2009. Black Carbon: A Science/Policy Primer. Arlington, VA: Pew Center on Global Climate Change. Available at http://www.c2es.org/docUploads/black-carbon-12-16-09.pdf.
Badawy, M.I., Emababy, M.A. 2010. Distribution of polycyclic aromatic hydrocarbons in drinking water in Egypt. *Desalination* 251: 34–40.
Balakrishna, K., and Probst, J.L. 2005. Organic carbon transport and C/N ratio variations in a large tropical river: Godavari as a case study, India. *Biogeochemistry* 73: 457–473.
Baldock, J.A., Masiello, C.A., Gelinas, Y., and Hedges, J.I. 2004. Cycling and composition of organic matter in terrestrial and marine ecosystems. *Mar. Chem.* 92: 39–64.
Baldock, J.A., and Smernik, R.J. 2002. Chemical composition and bioavailability of thermally, altered *Pinus resinosa* (Red Pine) wood. *Org. Geochem.* 33: 1093–1109.
Barry, R.G. 1996. The parameterization of surface albedo for sea ice and its snow cover. *Prog. Phys. Geograph.* 20: 63–79.
Berner, R. 1982. Burial of organic carbon and pyrite sulfur in the modern ocean and its geochemical and environmental significance. *Am. J. Sci.* 282: 451–473.
Bianchi, T.S., Wysocki, L.A., Stewart, M., Filley, T.R., and McKee, B.A. 2007. Temporal variability in terrestrially-derived sources of particulate organic carbon in the lower Mississippi River and its upper tributaries. *Geochim. Cosmochim. Acta* 71: 4425–4437.
Bidleman, T.F. 1988. Atmospheric processes – wet and dry deposition of organic compunds are controlled by their vapor particle partitioning. *Environ. Sci. Technol.* 22: 361–367.
Bird, M.I., and Ascough, P.L. 2012. Isotopes in pyrogenic carbon: a review. *Org. Geochem.* 42: 1529–1539.
Bird, M.I., and Cali, J.A. 1998. A million-year record of fire in sub-Saharan Africa. *Nature* 394: 767–769.
Bird, M.I., Moyo, C., Veenendaal, E.M., Lloyd, J., and Frost, P. 1999. Stability of elemental carbon in a savanna soil. *Global Biogeochem. Cycles* 13: 923–932.
Bird, M.I., et al. 2008. A preliminary estimate of organic carbon transported by the Ayeyarwady (Irrawaddy) and Thanlwin (Salween) Rivers of Myanmar. *Quat. Int.* 186: 113–122.
Bisiaux, M.M., et al. 2011. Stormwater and fire as sources of black carbon nanoparticles to Lake Tahoe. *Environ. Sci. Technol.* 45: 2065–2071.
Bond, T.C. 2001. Spectral dependence of visible light absorption by carbonaceous particles emitted from coal combustion. *Geophys. Res. Lett.* 28: 4075–4078.
Bond, T.C., Streets, D.G., Yarber, K.F., Nelson, S.M., Woo, J.-H., and Klimont, Z. 2004. A technology-based global inventory of black and organic carbon emissions from combustion, *J. Geophys. Res. Atmos.* 109, doi:10.1029/2003JD003697.
Bouloubassi, I., Fillaux, J., and Saliot, A. 2001. Hydrocarbons in surface sediments from the Changjiang (Yangtze River) Estuary, East China Sea. *Mar. Pollution Bull.* 42: 1335–1346.
Bowman, D.M.J.S., et al. 2009. Fire in the Earth system. *Science* 324: 481–484.
Brodowski, S., Rodionov, A., Haumaier, L., Glaser, B., and Amelung, W. 2005. Revised black carbon assessment using benzene polycarboxylic acids. *Org. Geochem.* 36: 1299–1310.
Bruun, E.W., Ambus, P., Egsgaard, H., and Hauggaard-Nielsen, H. 2012. Effects of slow and fast pyrolysis biochar on soil C and N turnover dynamics. *Soil Biol. Biochem.* 46: 73–79.
Budzinski, H., et al. 1997. Identification of polycyclic aromatic hydrocarbons in sediments from the Amazon Fan: occurrence and diagenetic evolution, *Proc. Ocean Drilling Progr. Sci. Results. NSF* 555–564.
Burdige, D.J. 2005. Burial of terrestrial organic matter in marine sediments: a re-assessment. *Global Biogeochem. Cycles* 19, doi:10.1029/2004GB002368.
Cachier, H., and Ducret, J. 1991. Influence of biomass burning on equatorial African rains. *Nature* 352: 228–230.
Carcaillet, C., et al. 2002. Holocene biomass burning and global dynamics of the carbon cycle. *Chemosphere* 49: 845–863.

Cataldo, D., Colombo, J., Boltovskoy, D., Bilos, C., and Landoni, P. 2001. Environmental toxicity assessment in the Paran· river delta (Argentina): simultaneous evaluation of selected pollutants and mortality rates of *Corbicula fluminea* (Bivalvia) early juveniles. *Environ. Pollution* 112: 379–389.

Chakraborty, B.B., and Long, R. 1967. Gas chromatographic analysis of polycyclic aromatic hydrocarbons in soot samples. *Environ. Sci. Technol.* 1: 828–834.

Chasar, L.S., Chanton, J.P., Glaser, P.H., Siegel, D.I., and Rivers, J.S. 2000. Radiocarbon and stable carbon isotopic evidence for transport and transformation of dissolved organic carbon, dissolved inorganic carbon, and CH_4 in a northern Minnesota peatland. *Global Biogeochem. Cycles* 14: 1095–1108.

Chernova, T., et al. 2001. The composition and the source of hydrocarbons in sediments taken from the tectonically active Andaman Backarc Basin, Indian Ocean. *Mar. Chem.* 75(1): 1–15.

Chugtai, A.R., Jassim, J.A., Peterson, J.H., Stedman, D.H., and Smith, D.M. 1991. Spectroscopic and solubility characteristics of oxidized soots. *Aerosol Sci. Technol.* 15: 112–126.

Clark, J.S. 1988. Particle motion and the theory of charcoal analysis: source area, transport, deposition, and sampling. *Quatern. Res.* 30: 67–80.

Clarke, A.D., and Noone, K.J. 1985. Soot in the Arctic snowpack – a cause for perturbations in radiative-transfer. *Atmos. Environ.* 19: 2045–2053.

Clarke, A.D., Owens, S.R., and Zhou, J.C. 2006. An ultrafine sea-salt flux from breaking waves: implications for cloud condensation nuclei in the remote marine atmosphere. *J. Geophys. Res. Atmos.* 111, doi:10.1029/2005JD006565.

Clarke, A.D., Weiss, R.E., and Charlson, R.J. 1984. Elemental carbon aerosols in the urban, rural, and remote-marine troposphere and in the stratosphere – inferences from light-absorption data and consequences regarding radiative transfer. *Sci. Total Environ.* 36: 97–102.

Colombo, J.C., et al. 2006. Sources, vertical fluxes, and equivalent toxicity of aromatic hydrocarbons in coastal sediments of the Rio de la Plata Estuary, Argentina. *Environ. Sci. Technol.* 40: 734–740.

Conedera, M., et al. 2009. Reconstructing past fire regimes: methods, applications, and relevance to fire management and conservation. *Quatern. Sci. Rev.* 28: 555–576.

Conway, H., Gades, A., and Raymond, C.F. 1996. Albedo of dirty snow during conditions of melt. *Water Resour. Res.* 32: 1713–1718.

Coynel, A., Seyler, P., Etcheber, H., Meybeck, M., and Orange, D. 2005. Spatial and seasonal dynamics of total suspended sediment and organic carbon species in the Congo River. *Global Biogeochem. Cycles* 19, doi:10.1029/2004GB002335.

Cross, A., and Sohi, S.P. 2011. The priming potential of biochar products in relation to labile carbon contents and soil organic matter status. *Soil Biol. Biochem.* 43: 2127–2134.

Currie, L.A., et al. 2002. A critical evaluation of interlaboratory data on total, elemental, and isotopic carbon in the carbonaceous particle reference material, NIST SRM 1649a. *J. Res. Natl. Inst. Standards Technol.* 107: 279–298.

Cusack, D.F., Chadwick, O.A., Hockaday, W.C., and Vitousek, P.M. 2012. Mineralogical controls on soil black carbon preservation. *Global Biogeochem. Cycles* 26: 10.1029/2011GB004109.

Czimczik, C.I., Preston, C.M., Schmidt, M.W.I., and Schulze, E.D. 2003. How surface fire in Siberian Scots pine forests affects soil organic carbon in the forest floor: stocks, molecular structure, and conversion to black carbon (charcoal). *Global Biogeochem. Cycles* 17: 10.1029/2002GB001956.

Czimczik, C.I., Preston, C.M., Schmidt, M.W.I., Werner, R.A., and Schulze, E.-D. 2002. Effects of charring on mass, organic carbon, and stable carbon isotopic composition of wood. *Org. Geochem.* 33: 1207–1223.

Czimczik, C.I., Schmidt, M.W.I., and Schulze, E.D. 2005. Effects of increasing fire frequency on black carbon and organic matter in Podzols of Siberian Scots pine forests. *Eur. J. Soil Sci.* 56: 417–428.

Dahle, S., Savinov, V.M., Matishov, G.G., Evenset, A., and Naes, K. 2003. Polycyclic aromatic hydrocarbons (PAHs) in bottom sediments of the Kara Sea shelf, Gulf of Ob and Yenisei Bay. *Sci. Total Environ.* 306: 57.

Dai, X., Boutton, T.W., Glaser, B., Ansley, R.J., and Zech, W. 2005. Black carbon in a temperate mixed-grass savanna. Soil Biol. Biochem. 37: 1879–1881.

Decesari, S., et al. 2002. Water soluble organic compounds formed by oxidation of soot. *Atmos. Environ.* 36: 1827–1832.

Delmas, R.A., Loudjani, P., Podaire, A., and Menaut, J. 1991. Biomass burning in Africa: an assessment of annually burned biomass, In: Levine, J. (ed.), Global Biomass Burning. Cambridge, MA: MIT Press.

Dick, D.P., et al. 2005. Characteristics of soil organic matter of different Brazilian Ferralsols under native vegetation as a function of soil depth. *Geoderma* 124: 319–333.

Dickens, A., Gelinas, Y., Masiello, C.A., Wakeham, S.G., and Hedges, J.I. 2004a. Reburial of fossil organic carbon in marine sediment. *Nature* 427: 336–339.

Dickens, A.F., et al. 2006. Solid-state C-13 NMR analysis of size and density fractions of marine sediments: insight into organic carbon sources and preservation mechanisms. *Geochim. Cosmochim. Acta* 70: 666–686.

Dickens, A.F., Gelinas, Y., and Hedges, J.I. 2004b. Physical separation of combustion and rock sources of graphitic black carbon in sediments. *Mar. Chem.* 92: 215–223.

Dittmar, T. 2008. The molecular level determination of black carbon in marine dissolved organic matter. *Org. Geochem.* 39: 396–407.

Dittmar, T., et al. 2012. Continuous flux of dissolved black carbon from a vanished tropical forest biome. *Nat. Geosci.* 5: 618–622.

Dittmar, T., and Paeng, J. 2009. A heat-induced molecular signature in marine dissolved organic matter. *Nat. Geosci.* 2: 175–179.

Domìnguez, C., et al. 2010. Quantification and source identification of polycyclic aromatic hydrocarbons in core sediments from Sundarban Mangrove Wetland, India. *Arch. Environ. Contam. Toxicol.* 59: 49–61.

Druffel, E.R.M. 2004. Comments on the importance of black carbon in the global carbon cycle. *Mar. Chem.* 92: 197–200.

Elias, V.O., Simoneit, B.R.T., Cordeiro, R.C., and Turcq, B. 2001. Evaluating levoglucosan as an indicator of biomass burning in Carajas, Amazonia: a comparison to the charcoal record. *Geochim. Cosmochim. Acta* 65: 267–272.

Ellis, E.E., Keil, R.G., Ingalls, A.E., Richey, J.E., and Alin, S.R. 2012. Seasonal variability in the sources of particulate organic matter of the Mekong River as discerned by elemental and lignin analyses. *J. Geophys. Res.* 117: G01038.

Elmquist, M., Cornelissen, G., Kukulska, Z., and Gustafsson, O. 2006. Distinct oxidative stabilities of char versus soot black carbon: implications for quantification and environmental recalcitrance. *Global Biogeochem. Cycles* 20: 10.1029/2005GB002629.

Elmquist, M., Gustafsson, O., and Andersson, P. 2004. Quantification of sedimentary black carbon using the chemothermal oxidation method: an evaluation of ex situ pretreatments and standard addition approaches. *Limnol. Oceanogr. Methods* 2: 417–427.

Elmquist, M., Semiletov, I., Guo, L., and Gustafsson, O. 2008. Pan-Arctic patterns in black carbon sources and fluvial discharges deduced from radiocarbon and PAH source apportionment markers in estuarine surface sediments. *Global Biogeochem. Cycles* 22: 10, doi:1029/2007GB002994.

Fernandes, M., and Sicre, M.A. 1999. Polycyclic aromatic hydrocarbons in the Arctic: Ob and Yenisei estuaries and Kara Sea shelf. *Estuar. Coastal Shelf Sci.* 48: 725–737.

Fernandes, S.D., Trautmann, N.M., Streets, D.G., Roden, C.A., and Bond, T.C. 2007. Global biofuel use, 1850–2000. *Global Biogeochem. Cycles* 21, doi:10.1029/2006GB002836.

Flores-Cervantes, D.X. 2008. Black Carbon in the Gulf of Maine: New Insights into Inputs and Cycling of Combustion-Derived Organic Carbon. PhD Dissertation, Massachusetts Institute of Technology, Boston, Massachusetts, USA.

Flores-Cervantes, D.X., Plata, D.L., MacFarlane, J.K., Reddy, C.M., and Gschwend, P.M. 2009a. Black carbon in marine particulate organic carbon: inputs and cycling of highly recalcitrant organic carbon in the Gulf of Maine. *Mar. Chem.* 113: 172–181.

Flores-Cervantes, D.X., Reddy, C.M., and Gschwend, P.M. 2009b. Inferring black carbon concentrations in particulate organic matter by observing pyrene fluorescence losses. *Environ. Sci. Technol.* 43: 4864–4870.

Forbes, M.S., Raison, R.J., and Skjemstad, J.O. 2006. Formation, transformation, and transport of black carbon (charcoal) in terrestrial and aquatic ecosystems. *Sci. Total Environ.* 370: 190–206.

Forster, P., et al. 2007. Changes in atmospheric constituents and in radiative forcing, In: Solomon, S., et al. (ed.), Climate Change 2007 – The Physical Science Basis. Cambridge: Cambridge University Press.

Gelinas, Y., Prentice, K.M., Baldock, J.A., and Hedges, J.I. 2001. An improved thermal oxidation method for the quantification of soot/graphitic carbon in sediments and soils. *Environ. Sci. Technol.* 35: 3519–3525.

Gilhooly, W.P., et al. 1996. Stable carbon isotopic analysis of charcoal from single plant sources, In: Levine, J. (ed.), Biomass Burning and Global Change. Cambridge, MA: MIT Press.

Glaser, B., Haumaier, L., Guggenberger, G., and Zech, W. 1998. Black carbon in soils: the use of benzenecarboxylic acid as specific markers. *Org. Geochem.* 29: 811–819.

Goldberg, E. 1985. Black Carbon in the Environment: Properties and Distribution. New York: John Wiley & Sons.

Goni, M., Alleau, Y., Crobett, R., and Walsh, J. 2007. The effects of Hurricanes Katrina and Rita on the seabed of the Louisiana shelf. *J. Sediment. Rec.* 5: 1.

Griffin, J., and Goldberg, E.D. 1975. The fluxes of elemental carbon in coastal marine sediments. *Limnol. Oceanogr.* 20: 456–463.

Gustafsson, O., and Gschwend, P.M. 1997. Soot as a strong partitioning medium for polycyclic aromatic hydrocarbons in aquatic systems. Molecular markers in environmental geochemistry, ACS Symposium Series Vol. 671. Washington, DC: American Chemical Society.

Gustafsson, O., and Gschwend, P.M. 1998. The flux of black carbon to surface sediments on the New England continental shelf. *Geochim. Cosmochim. Acta* 62: 465–472.

Gustafsson, O., Haghseta, F., Chan, C., Macfarlane, J., and Gschwend, P. 1997. Quantification of the dilute sedimentary soot phase: implications for PAH speciation and bioavailability. *Environ. Sci. Technol.* 31: 203–209.

Gustafsson, O., et al. 2009. Brown clouds over South Asia: biomass or fossil fuel combustion? *Science* 323: 495–498.

Haberstroh, P.R., et al. 2006. Chemical composition of the graphitic black carbon fraction in riverine and marine sediments at sub-micron scales using carbon X-ray spectromicroscopy. *Geochim. Cosmochim. Acta* 70: 1483–1494.

Hamer, U., Marschner, B., Brodowski, S., and Amelung, W. 2004. Interactive priming of black carbon and glucose mineralisation. *Org. Geochem.* 35: 823–830.

Hammes, K., et al. 2007. Comparison of quantification methods to measure fire-derived (black/elemental) carbon in soils and sediments using reference materials from soil, water, sediment and the atmosphere. *Global Biogeochem. Cycles* 21, doi:10.1029/2006GB002914.

Hansell, D.A., and Carlson, C.A. 1998. Deep ocean gradients in dissolved organic carbon concentrations. *Nature*. 395: 263–266.

Hansen, J., and Nazarenko, L. 2004. Soot climate forcing via snow and ice albedos. *Proc Natl Acad. Sci*. 101: 423–428.

Hansen, J., Sato, M., Ruedy, R., Lacis, A., and Oinas, V. 2000. Global warming in the twenty-first century: an alternative scenario. *Proc. Natl Acad. Sci*. 97: 9875–9880.

Haynes, B.S., and Wagner, H.G. 1981. Soot formation. *Progr. Energy Combustion Sci*. 7: 229–273.

Hedges, J.I. 1992. Global biogeochemical cycles: progress and problems. *Mar. Chem*. 39: 67–93.

Hedges, J.I., and Keil, R.G. 1995. Sedimentary organic matter preservation: an assessment and speculative synthesis. *Mar. Chem*. 49: 81–115.

Herring, J.R. 1985. Charcoal fluxes into sediments of the North Pacific Ocean: The Cenozoic record of burning, In: Sundquist, E.T., Broecker, W.S. (eds.), The Carbon Cycle and Atmospheric CO_2: Natural Variations Archean to Present. Washington, DC: American Geophysical Union.

Hockaday, W.C., Grannas, A.M., Kim, S., and Hatcher, P.G. 2006. Direct molecular evidence for the degradation and mobility of black carbon in soils from ultrahigh-resolution mass spectral analysis of dissolved organic matter from a fire-impacted forest soil. *Org. Geochem*. 37: 501–510.

Hockaday, W.C., Grannas, A.M., Kim, S., and Hatcher, P.G. 2007. The transformation and mobility of charcoal in a fire-impacted watershed. *Geochim. Cosmochim. Acta* 71: 3432–3445.

Horvath, H. 1993. Atmospheric light absorption – a review. *Atmos. Environ. Part A Gen. Top*. 27: 293–317.

IPCC. 2007. Fourth Assessment Report (AR4). Cambridge University Press.

Jacobson, M.Z. 2001. Strong radiative heating due to the mixing state of black carbon in atmospheric aerosols. *Nature* 409: 695–698.

Jia, G., Peng, P.A., Zhao, Q., and Jian, Z. 2003. Changes in terrestrial ecosystem since 30Ma in East Asia: stable isotope evidence from black carbon in the South China Sea. *Geology* 31: 1093–1096.

Jiao, W., et al. 2012. PAHs in surface sediments from coastal and estuarine areas of the northern Bohai and Yellow Seas, China. *Environ. Geochem. Health* 1–12.

Jurado, E., Dachs, J., Duarte, C.M., and Simo, R. 2008. Atmospheric deposition of organic and black carbon to the global oceans. *Atmos. Environ*. 42: 7931–7939.

Kang, Y., et al. 2009. Black carbon and polycyclic aromatic hydrocarbons (PAHs) in surface sediments of Chinaís marginal seas. *Chinese J. Oceanol. Limnol*. 27: 297–308.

Katsoyiannis, A., Sweetman, A.J., and Jones, K.C. 2011. PAH Molecular diagnostic ratios applied to atmospheric sources: a critical evaluation using two decades of source inventory and air concentration data from the UK. *Environ. Sci. Technol*. 45: 8897–8906.

Keenan, H.E., Bangkedphol, S., Sakultantimetha, A., and Songsasen, A. 2010. The ecological complexity of the Thai-Laos Mekong River: II. Metals and polyaromatic hydrocarbons (PAHs) monitoring, modelling and environmental fate. *J. Environ. Sci. Health Part A* 45: 1674–1680.

Keiluweit, M., Nico, P.S., Johnson, M.G., and Kleber, M. 2010. Dynamic molecular structure of plant biomass-derived black carbon (biochar). *Environ. Sci. Technol*. 44: 1247–1253.

Kercher, A.K., and Nagle, D.C. 2003. Microstructural evolution during charcoal carbonization by X-ray diffraction analysis. *Carbon* 41: 15–27.

Kim, S., Kaplan, L.A., Benner, R., and Hatcher, P.G. 2004. Hydrogen-deficient molecules in natural riverine water samples-evidence for the existence of black carbon in DOM. *Mar. Chem*. 92: 225–234.

Knicker, H. 2007. How does fire affect the nature and stability of soil organic nitrogen and carbon? A review. *Biogeochemistry* 85: 91–118.

Knicker, H., Nikolova, R., Dick, D.P., and Dalmolin, R.S.D. 2012. Alteration of quality and stability of organic matter in grassland soils of Southern Brazil highlands after ceasing biannual burning. *Geoderma* 181: 11–21.

Koch, B.P., and Dittmar, T. 2006. From mass to structure: an aromaticity index for high-resolution mass data of natural organic matter. *Rapid Comm. Mass Spectrom.* 20: 926–932.

Kralovec, A.C., Christensen, E.R., and Van Camp, R.P., 2002. Fossil fuel and wood combustion as recorded by carbon particles in Lake Erie sediments 1850–1998. *Environ. Sci. Technol.* 36: 1405–1413.

Krull, E.S., et al. 2003. ^{13}C-depleted charcoal from C4 grasses and the rold of occluded carbon in phytoliths. *Org. Geochem.* 34: 1337–1352.

Kuhlbusch, T.A., and Crutzen, P.J. 1995. Toward a global estimate of black carbon in residues of vegetation fires representing a sink of atmospheric CO_2 and a source of O_2. *Global Biogeochem. Cycles* 9: 491–501.

Kuhlbusch, T.A.J., et al. 1996. Black carbon formation by savanna fires: measurements and implications for the global carbon cycle. *J. Geophys. Res.* 101: 23651–23665.

Kuo, L.J., Herbert, and B.E., Louchouarn, P. 2008. Can levoglucosan be used to characterize and quantify char/charcoal black carbon in environmental media? *Org. Geochem.* 39: 1466–1478.

Kuo, L.J., et al. 2011. Combustion-derived substances in deep basins of Puget Sound: historical inputs from fossil fuel and biomass combustion. *Environ. Pollution* 159: 983–990.

Lee, E.J. 2005. Studies of dissolved black carbon and organic matter utilizing electrospray ionization mass spectrometry. MS Thesis. Columbus, OH: The Ohio State University.

Lehmann, J. 2007. A handful of carbon. *Nature* 447: 143–144.

Lehmann, J., et al. 2005. Near-edge X-ray absorption fine structure (NEXAFS) spectroscopy for mapping nano-scale distribution of organic carbon forms in soil: application to black carbon particles. *Global Biogeochem. Cycles* 19, doi:10.1029/2004GB002435.

Leithold, E., Blair, N., and Perkey, D. 2006. Geomorphologic controls on the age of particulate organic carbon from small mountainous and upland rivers. *Global Biogeochem. Cycles* 20, doi:10.1029/2005GB002677.

Li, B., et al. 2012a. Spatial distribution and source apportionment of PAHs in surficial sediments of the Yangtze Estuary, China. *Mar. Pollution Bull* 64: 636–643.

Li, X., et al. 2012b. Composition, abundance and age of total organic carbon in surface sediments from the inner shelf of the East China Sea. *Mar. Chem* 145–147: 37–52.

Lima, A.L.C., Farrington, J.W., and Reddy, C.M. 2005. Combustion-derived polycyclic aromatic hydrocarbons in the environment – a review. *Environ. Forens.* 6: 109–131.

Liousse, C., Cachier, H., and Jennings, S.G. 1993. Optical and thermal measurements of black carbon aerosol content in different environments – variation of the specific attenuation cross-section (sigma). *Atmos. Environ. Part A Gen. Top.* 27: 1203–1211.

Lobbes, J.M., Fitznar, H.P., and Kattner, G. 2000. Biogeochemical characteristics of dissolved and particulate organic matter in Russian rivers entering the Arctic Ocean. *Geochim. Cosmochim. Acta* 64: 2973–2983.

Lohmann, R., et al. 2009. Fluxes of soot black carbon to South Atlantic sediments. *Global Biogeochem. Cycles* 23, doi:10.1029/2008GB003253.

Louchouarn, P., et al. 2007. Elemental and molecular evidence of soot- and char-derived black carbon inputs to New York City's atmosphere during the 20th century. *Environ. Sci. Technol.* 41: 82–87.

Ludwig, W., Amiotte Suchet, P., and Probst, J.L. 1996. River discharges of carbon to the world's oceans: determining local inputs of alkalinity and of dissolved and particulate organic carbon. *Sciences de la terre et des planètes (Comptes rendus de l'Académie des sciences)* 323: 1007–1014.

Ma, Y. et al. 2008. Distribution, sources, and potential risk of polycyclic aromatic hydrocarbons (PAHs) in drinking water resources from Henan Province in middle of China. *Environ. Monitor. Assess.* 146: 127–138.

Macdonald, R., et al. 1998. A sediment and organic carbon budget for the Canadian Beaufort Shelf. *Mar. Geol.* 144: 255–273.

Major, J., Lehmann, J., Rondon, M., and Goodale, C. 2010. Fate of soil-applied black carbon: downward migration, leaching and soil respiration. *Global Change Biol.* 16: 1366–1379.

Mannino, A. 2000. Discussion at USGS Reston on Radiocarbon Ages of BC in Delaware River HMW DOM.

Mannino, A., and Harvey, H.R. 2004. Black carbon in estuarine and coastal ocean dissolved organic matter. *Limnol. Oceanogr.* 491: 735–740.

Marlon, J.R., et al. 2009a. Climate and human influences on global biomass burning over the past two millennia (vol 1, pg 697, 2008). *Nat. Geosci.* 2: 307–307.

Marlon, J.R., et al. 2009b. Wildfire responses to abrupt climate change in North America. *Proc. Natl. Acad. Sci.* 106: 2519–2524.

Martins, J.V., et al. 1998a. Effects of black carbon content, particle size, and mixing on light absorption by aerosols from biomass burning in Brazil. *J. Geophys. Res. Atmos.* 103: 32041–32050.

Martins, J.V., Hobbs, P.V., Weiss, R.E., and Artaxo, P. 1998b. Sphericity and morphology of smoke particles from biomass burning in Brazil. *J. Geophys. Res. Atmos.* 103: 32051–32057.

Maruya, K., Loganathan, B., Kannan, K., McCumber-Kahn, S., and Lee, R. 1997. Organic and organometallic compounds in estuarine sediments from the Gulf of Mexico (1993–1994). *Estuar. Coasts* 20: 700–709.

Masiello, C.A. 1999. Radiocarbon measurements of black carbon in sediments and a small river. PhD Dissertation. University of California, Irvine.

Masiello, C.A. 2004. New directions in black carbon organic geochemistry. *Mar. Chem.* 92: 201–213.

Masiello, C.A., and Druffel, E.R.M. 1998. Black carbon in deep sea sediments. *Science* 280: 1911–1913.

Masiello, C.A., and Druffel, E.R.M. 2001. Carbon isotope geochemistry of the Santa Clara River. *Global Biogeochem. Cycles* 15: 407–416.

Masiello, C.A., and Druffel, E.R.M. 2003. Organic and black carbon C-13 and C-14 through the Santa Monica Basin sediment oxic-anoxic transition. *Geophys. Res. Lett.* 30, doi:10.1029/2002GL015050.

Masiello, C.A., Druffel, E.R.M., and Currie, L.A. 2002. Radiocarbon measurements of black carbon in aerosols and ocean sediments. *Geochim. Cosmochim. Acta* 66: 1025–1036.

Menon, S., Hansen, J.E., Nazarenki, L., and Luo, Y. 2002. Climate effects of black carbon aerosols in China and India. *Science* 297: 2250–2253.

Meybeck, M., and Ragu, A. 1997. Presenting the GEMS-GLORI, a compendium of world river discharge to the oceans, Freshwater Contamination (Proceedings of Rabat Symposium S4, Aprii-May 1997). IAHS Publication No. 243, PANGAEA.

Middleburg, J.J. 1989. A simple rate model for organic matter decomposition in marine sediments. *Geochim. Cosmochim. Acta* 53: 1577–1581.

Middleburg, J.J., Nieuwenhuize, J., and van Breugel, P. 1999. Black carbon in marine sediments. *Mar. Chem.* 65: 245–252.

Milliman, J.D., and Syvitski, J.P.M. 1992. Geomorphic-tectonic control of sediment discharge to the ocean: the importance of small mountainous rivers. *J. Geol.* 100: 525–544.

Mitra, S., Bianchi, T.S., McKee, B.A., and Sutula, M. 2002. Black carbon from the Mississippi River: Quantities, sources, and potential implications for the global carbon cycle. *Environ. Sci. Technol.* 36: 2296–2302.

Mitra, S., Zimmerman, A.R., Hunsinger, G.B., Willard, D., and Dunn, J.C. 2009. A Holocene record of climate-driven shifts in coastal carbon sequestration. *Geophys. Res. Lett.* 36: 10.1029/2008GL036875. L05704

Moosmuller, H., Chakrabarty, R.K., and Arnott, W.P. 2009. Aerosol light absorption and its measurement: a review. *J. Quant. Spectrosc. Radiat. Transfer* 110: 844–878.

Nemirovskaya, I. 2007. Hydrocarbons in the water and bottom sediments of a region with continuous petroleum contamination. *Geochem. Int.* 45: 638–651.

Nguyen, B.T., and Lehmann, J. 2009. Black carbon decomposition under varying water regimes. *Org. Geochem.* 40: 846–853.

Nguyen, B.T., Lehmann, J., Hockaday, W.C., Joseph, S., and Masiello, C.A. 2010. Temperature sensitivity of black carbon decomposition and oxidation. *Environ. Sci. Technol.* 44: 3324–3331.

Nguyen, B.T., et al. 2008. Long-term black carbon dynamics in cultivated soil. *Biogeochemistry* 89: 295–308.

Nguyen, T.H., Brown, R.A., and Ball, W.P. 2004. An evaluation of thermal resistance as a measure of black carbon content in diesel soot, wood char, and sediment. *Org. Geochem.* 35: 217–234.

Novakov, T. 1984. The role of soot and primary oxidants in atmospheric chemistry. *Sci. Total Environ.* 36: 1–10.

Novakov, T., et al. 2003. Large historical changes of fossil-fuel black carbon aerosols. *Geophys. Res. Lett.* 30: 57–61.

Ogren, J.A., and Charlson, R.J. 1983. Elemental carbon in the atmosphere: cycle and lifetime. *Tellus B* 35B: 241–254.

Ogren, J.A., Groblicki, P.J., and Charlson, R.J. 1984. Measurement of the removal rate of elemental carbon from the atmosphere. *Sci. Total Environ.* 36: 329–338.

Ohlson, M., and Tryterud, E. 2000. Interpretation of the charcoal record in forest soils: forest fires and their production and deposition of macroscopic charcoal. *Holocene* 10: 519–525.

Overton, E., Ashton, B., and Miles, M. 2004. Historical polycyclic aromatic and petrogenic hydrocarbon loading in Northern Central Gulf of Mexico shelf sediments. *Mar. Poll. Bull.* 49: 557–563.

Paolini, J. 1995. Particulate organic carbon and nitrogen in the Orinoco River (Venezuela). *Biogeochemistry* 29: 59–70.

Patterson, W.A., Edwards, K.J., and Maguire, D.J. 1987. Microscopic charcoal as a fossil indicator of fire. *Quatern. Sci. Rev.* 6: 3–23.

Penner, J.E., and Eddleman, H. 1993. Towards the development of a global inventory for black carbon emissions. *Atmos. Environ.* 27A: 1277–1295.

Penner, J.E. 2001. Climate Change 2001: The Scientific Basis. Cambridge, UK: Cambridge University Press.

Power, M.J., et al. 2008. Changes in fire regimes since the Last Glacial Maximum: an assessment based on a global synthesis and analysis of charcoal data. *Clim. Dynam.* 30: 887–907.

Preston, C.M., and Schmidt, M.W.I. 2006. Black (pyrogenic) carbon: a synthesis of current knowledge and uncertainties with special consideration of boreal regions. *Biogeosciences* 3: 397–420.

Ramanathan, V., and Carmichael, G. 2008. Global and regional climate changes due to black carbon. *Nat. Geosci.* 1: 221–227.

Raymond, P.A., Bauer, J.E., 2001. Riverine export of aged terrestrial organic matter to the North Atlantic Ocean. *Nature* 409: 497–500.

Reddy, C.M., et al. 2002. Radiocarbon as a tool to apportion the sources of polycyclic aromatic hydrocarbons and black carbon in environmental samples. *Environ. Sci. Technol.* 36: 1774–1782.

Richey, J.E., et al. 1990. Biogeochemistry of carbon in the Amazon River. *Limnol. Oceanogr.* 35: 352–371.

Rivas, Y., Matus, F., Rumpel, C., Knicker, H., and Garrido, E. 2012. Black carbon contribution in volcanic soils affected by wildfire or stubble burning. *Org. Geochem.* 47: 41–50.

Robock, A. 1984. Snow and ice feedbacks prolong effects of nuclear winter. Nature 310: 667–670.

Rogers, K.G., and Goodbred, S.L. 2010. Mass failures associated with the passage of a large tropical cyclone over the Swatch of No Ground submarine canyon (Bay of Bengal). *Geology* 38: 1051-U123.

Rosen, H., and Novakov, T. 1977. Raman scattering and the characterisation of atmospheric aerosol particles. *Nature* 266: 708–710.

Rumpel, C., Ba, A., Darboux, F., Chaplot, V., and Planchon, O. 2009. Erosion budget and process selectivity of black carbon at meter scale. *Geoderma* 154: 131–137.

Sanchez-Garcia, L., Cato, I., and Gustafsson, O. 2012. The sequestration sink of soot black carbon in the Northern European Shelf sediments. *Global Biogeochem. Cycles* 26: 10.1029/2010GB003956.

Santschi, P., Presley, B., Wade, T., Garcia-Romero, B., and Baskaran, M. 2001. Historical contamination of PAHs, PCBs, DDTs, and heavy metals in Mississippi River Delta, Galveston Bay and Tampa Bay sediment cores. *Mar. Environ. Res.* 52: 51–79.

Schmidt, M.W.I., and Noack, A.G. 2000. Black carbon in soils and sediments: analysis, distribution, implications, and current challenges. *Global Biogeochem. Cycles* 14: 777–793.

Schmidt, M.W.I., Skjemstad, J.O., and Jager, C. 2002. Carbon isotope geochemistry and nanomorphology of soil black carbon: black chernozemic soils in central Europe originate from ancient biomass burning. *Global Biogeochem. Cycles* 16, doi:10.1129/2002GB001939.

Skjemstad, J.O., Taylor, J.A., and Smernik, R.J. 1999. Estimation of charcoal (Char) in soils. *Comm. Soil Sci. Plant Anal.* 30: 2283–2298.

Smith, D.M., Griffin, J.J., and Goldberg, E.D. 1973. Elemental carbon in marine sediments: a baseline for burning. Nature 241: 268–270.

Smith, D.M., Griffin, J.J., and Goldberg, E.D. 1975. Spectrometric method for quantitative-determination of elemental carbon. *Anal. Chem.* 47: 233–238.

Stanmore, B.R., Brilhac, J.F., and Gilot, P. 2001. The oxidation of soot: a review of experiments, mechanisms and models. *Carbon* 39: 2247–2268.

Stubbins, A., et al. 2012a. Anthropogenic aerosols as a source of ancient dissolved organic matter in glaciers. *Nat. Geosci.* 5: 198–201.

Stubbins, A., Niggemann, J., and Dittmar, T. 2012b. Photo-lability of deep ocean dissolved black carbon. *Biogeosciences* 9: 1661–1670.

Stubbins, A., et al. 2010. Illuminated darkness: molecular signatures of Congo River dissolved organic matter and its photochemical alteration as revealed by ultrahigh precision mass spectrometry. *Limnol. Oceanogr.* 55: 1467–1477.

Suman, D.O. 1986. Charcoal production from agricultural burning in Central Panama and its deposition in the sediments of the Gulf of Panama. *Environ. Conserv.* 13: 51–60.

Suman, D.O., Kuhlbusch, T.A.J., and Lim, B. 1997. Marine sediments: a reservoir for black carbon and their use as spatial and temporal records of combustion, In: Clark, J., Cachier, H., Goldammer, J.G., Stocks, B. (eds.), Sediment records of biomass burning and global change. NATO ASI Series. Berlin: Springer-Verlag.

Thevenon, F., Bard, E., Williamson, D., and Beaufort, L. 2004. A biomass burning record from the West Equatorial Pacific over the last 360 ky: methodological, climatic and anthropic implications. *Palaeogeogr. Palaeoclimatol. Palaeoecol.* 213: 83–99.

Turekian, V.C., et al. 1996. Bulk and compound-specific isotopic characterization of the products of biomass burning: laboratory studies, In: Levine, J. (ed.), Biomass Burning and Global Change. Cambridge, MA: MIT Press.

Unger, D., Gaye-Haake, B., Neumann, K., Gebhardt, A.C., and Ittekkot, V. 2005. Biogeochemistry of suspended and sedimentary material in the Ob and Yenisei rivers and Kara Sea: amino acids and amino sugars. *Cont. Shelf Res.* 25: 437–460.

Wang, X.C., Ma, H.Q., Li, R.H., Song, Z.S., and Wu, J.P. 2012. Seasonal fluxes and source variation of organic carbon transported by two major Chinese Rivers: the Yellow River and Changjiang (Yangtze) River. *Global Biogeochem. Cycles* 26, doi:10.1029/2011GB004130.

Warren, S.G., and Wiscombe, W.J. 1979. New model for the spectral albedo of snow. *Bull. Am. Meteorol. Soc.* 60: 846–846.

Warren, S.G., and Wiscombe, W.J. 1980. A model for the spectral albedo of snow. II. Snow containing atmospheric aerosols. *J. Atmos. Sci.* 37: 2734–2745.

Warren, S.G., and Wiscombe, W.J. 1981. Comment on radiative properties of snow for clear sky solar-radiation. *Cold Regions Sci. Technol.* 5: 177–180.

Warren, S.G., and Wiscombe, W.J. 1985. Dirty snow after nuclear war. *Nature* 313: 467–470.

Widory, D. 2006. Combustibles, fuels and their combustion products: a view through carbon isotopes. *Combust. Theory Model.* 10: 831–841.

Williams, P.M., and Druffel, E.R.M. 1987. Radiocarbon in dissolved organic matter in the Central North Pacific Ocean. *Nature* 330: 246–248.

Wolbach, W.S., and Anders, E. 1989. Elemental carbon in sediments: Determination and isotopic analysis in presence of kerogen. *Geochim. Cosmochim. Acta* 53: 1637–1647.

Wozniak, A.S., Bauer, J.E., Sleighter, R.L., Dickhut, R.M., and Hatcher, P.G. 2008. Technical Note: Molecular characterization of aerosol-derived water soluble organic carbon using ultrahigh resolution electrospray ionization Fourier transform ion cyclotron resonance mass spectrometry. *Atmos. Chem. Phys.* 8: 5099–5111.

Wu, Y., Zhang, J., Mi, T., and Li, B. 2001. Occurrence of *n*-alkanes and polycyclic aromatic hydrocarbons in the core sediments of the Yellow Sea. *Mar. Chem.* 76: 1–15.

Xu, B.Q., et al. 2009. Black soot and the survival of Tibetan glaciers. *Proc. Natl Acad. Sci.* 106: 22114–22118.

Xu, J., et al. 2007. Polycyclic aromatic hydrocarbons in the surface sediments from Yellow River, China. *Chemosphere* 67: 1408–1414.

Yunker, M.B., Macdonald, R.W., Cretney, W.J., Fowler, B.R., and McLaughlin, F.A. 1993. Alkane, terpene and polycyclic aromatic hydrocarbon geochemistry of the Mackenzie River and Mackenzie Shelf: riverine contributions to Beaufort Sea coastal sediment. *Geochim. Cosmochim. Acta* 57: 3041–3061.

Yunker, M.B., Macdonald, R.W., Snowdon, L.R., and Fowler, B.R. 2011. Alkane and PAH biomarkers as tracers of terrigenous organic carbon in Arctic Ocean sediments. *Org. Geochem.* 42: 1109–1146.

Yunker, M.B., et al. 2002. PAHs in the Fraser River basin: a critical appraisal of PAH ratios as indicators of PAH source and composition. *Org. Geochem.* 33: 489–515.

Zimmerman, A. 2010. Abiotic and microbial oxidation of laboratory-produced black carbon (biochar). *Environ. Sci. Technol.* 44: 1295–1301.

Zimmerman, A.R., and Gao, B. 2013. The Stability of Biochar in the Environment, In: Ladygina, N., Rineau, F. (eds.), Biochar and Soil Biota. Boca Raton, FL: CRC Press.

Zimmerman, A.R., Gao, B., and Ahn, M.-Y. 2011. Positive and negative carbon mineralization priming effects among a variety of biochar-amended soils. *Soil Biol. Biochem.* 43: 1169–1179.

Ziolkowski, L., and Druffel, E. 2010. Aged black carbon identified in marine dissolved organic carbon. *Geophys. Res. Lett* 37: L16601.

Section III

Eastern hemisphere systems

10

Carbon biogeochemistry in the continuum of the Changjiang (Yangtze) River watersheds across the East China Sea

J. Zhang, Y. Wu, G. L. Zhang, and Z. Y. Zhu

1. Introduction

The biogeochemistry (e.g., carbon cycle) in the coastal environment is regulated by the processes that bridge the terrestrial ecosystem with the coastal ocean, with additional linkages with the atmosphere. Moreover, the fate of carbon is intrinsically linked to the dynamics and transformation of other biologically essential and bio-mediated elements, such as plant nutrients and micronutrients, which affect elemental cycling through changes in chemical speciation and participation between intra- and subcellular processes. Biogeochemical processes in various scales are important in controlling the flux between different subsystems of Earth by the mediation of primary production as it transitions from eutrophic coastal to oligotrophic open oceanic waters and vice versa.

Our knowledge to date shows that given the rather small portion of surface area and water volume of coastal relative to global oceans, carbon fixation and its impact on the sequestration of atmospheric CO_2 in the coastal ocean is of critical importance in global change science. Coastal environments receive significant amounts of land-source inputs (e.g., river), whereby the biogeochemistry of carbon, as well as plant nutrients and micronutrients, are affected by changing inputs from terrestrial influx (i.e., chemical elements associated with fresh water and sediment loads) and open ocean waters (e.g., Kuroshio in the Western Pacific Ocean at mid-latitudes), which are under the influence of seasonal monsoon and climate change effects in East Asia.

The Changjiang (Yangtze River) (note: "jiang" means "river" in Chinese) takes its water sources from the Tibetan Plateau at elevation of ca. 6,500 m with a river course of 6,400 km and watersheds of 1.8×10^6 km^2. The upstream region (i.e., first 2,000 km from the river source) of the Changjiang runs through mountainous areas with an elevation to length ratio, or slope of river bed (H/L), of 1–2‰. The downstream region (i.e., 2,000 km upstream from the river mouth) of river course is essentially composed of the fluvial plains with H/L of ca. 0.1‰. Historically, the annual water discharge of Changjiang amounts to ca. 1.0×10^{12} m^3 with a sediment load of ca. 0.5×10^9 tons yr^{-1}, of which the bed load is ca. 10%. Over the last two decades, the annual water discharge has varied within ca. 10% of the long-term (e.g., 50–100 years) average, but the seaward sediment load has dropped by 30–40%, especially after the construction of Three Gorges Dam in 2003 (Yang et al. 2006, 2007).

Semidiurnal tides dominate in the Changjiang Estuary, with a tidal range of up to 4–5 m, which affects 600–650 km of the river course. The dispersal of riverine effluents and circulation of the East China Sea (ECS) is affected by the monsoon climate, with northeasterly winds in winter and southwesterly winds in summer. In summer, the dispersal of freshwater plumes from Changjiang is mainly eastward and northeastward and can reach as far as shelf break and Tsusima Strait. In winter, riverine influx is constrained to the west of ECS shelf, and water flows southwards along the coast of East China under the influence of the winter monsoon. The influence of open-ocean on the circulation of ECS is mainly through the Taiwan Strait, meso-scale processes at shelf break, and incursion of Kuroshio Subsurface Water across the shelf (Su 1998).

Monitoring data for water and sediment loads in the watersheds of Changjiang can be traced in the literature back to the first part of the twentieth century, with measurements of chemical parameters starting to be incorporated at major hydrographic and gauge stations along the river course after 1950s. Systematic observations in the coastal ECS started in late 1950s, when China carried out the first national survey of oceanography from the north (i.e., Bohai) to the south (i.e., South China Sea). Moreover, research data bloomed after the 1980s after China launched the economic innovation on the mainland. Those data in national reports and peer-reviewed literature formed the basis of this study in terms of a historical review of research progress and examination of the system evolution.

In this chapter, we review the knowledge of carbon biogeochemistry based on the data from the Changjiang Watersheds across the ECS to the Kuroshio; data examined here include dissolved and particulate chemical species and their linkages with biomass at low trophic levels using plant pigments as chemical biomarkers. Then, we use data from the literature and our own studies to integrate information along a continuum from the watersheds of Changjiang to the entire ECS. Although this study is focused on carbon, other chemical elements are included that have an influence on carbon biogeochemistry. Given that biogeochemical cycles of carbon in the coastal ocean are strongly modulated by the dynamics of the ecosystem as well as physical driving forces, we attempt to link the biogeochemistry with hydrographic processes in one perspective, along with ecosystem (e.g., food-web structure) dynamics on another side.

2. Characters of the land-source influx and open boundary processes

2.1. Dissolved inorganic carbon and CH_4

Among all the dissolved inorganic carbon (DIC) species in the Changjiang, HCO_3^- is the dominant species, which accounts for 55–60% of the total ion equivalent concentrations, together with Ca^{2+} (Li and Zhang 2003). Long-term monthly variation of the HCO_3^- and CO_2 concentrations in the Changjiang at Datong Hydrographic Station (ca. 625 km upstream from the river mouth) in the period 1960s to 2000 are shown in Figure 10.1. HCO_3^- concentrations illustrated obvious seasonal variations, with highest values in winter and lowest values in spring (Fig. 10.1). This trend of variation reflects a dilution of the weathering-derived HCO_3^- in the drainage basin (Cai et al. 2008). Although the monthly water discharge of Changjiang reaches the maximum in summer, the effect of dilution can be compensated by dissolution of carbonates as a result of an intensification of soil erosion and elevated pCO_2 during flood season (Chen et al. 2002). Hence the low concentration was not observed during but before July each year. There was no clear trend with regard to interannual variations in the riverine HCO_3^- in the period of 1960s to 2000, with concentrations between 1.2 and 2.4 mM and an annual average of 1.7 mM (Wang et al. 2007). The annual DIC flux to the East China Sea was estimated 19.5×10^6 tons yr^{-1} (Liu et al. 2002).

Figure 10.2 shows the long-term annual mean calculated pCO_2 in the Changjiang at Datong Hydrographic Station for the period of 1960s to 2000, which indicated a clear trend of decline

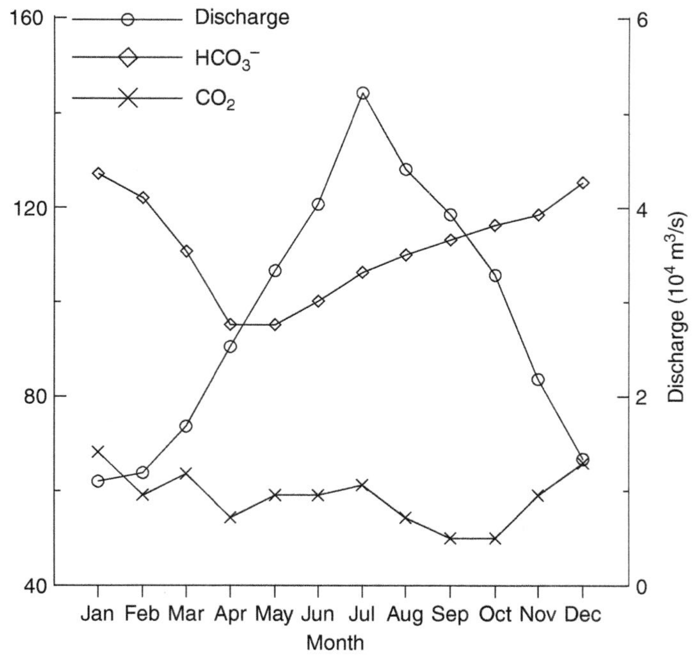

Figure 10.1. Long-term mean monthly discharge, CO_2, and HCO_3^- concentrations in the Changjiang at Datong Hydrographic Station for the period of 1960s–2000. Data are from Liu et al. (2002).

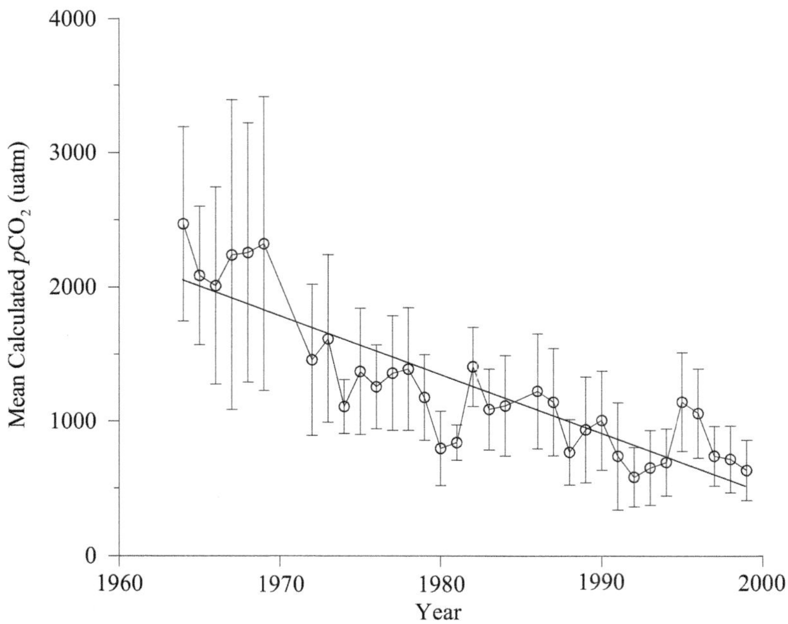

Figure 10.2. Long-term annual mean calculated pCO_2 at Datong Hydrographic Station of the Changjiang for the period of 1960s–2000. Data are from Wang et al. (2007).

Table 10.1. *Compilation of surface DIC, TAlk, pH, and* pCO_2 *in the Changjiang Estuary and the East China Sea*

Region	Study period	DIC (μmol kg^{-1})	TAlk (μmol kg^{-1})	pH	pCO_2 (μatm)	References
Upper Estuary	May 2004		1.1–1.6 mM	7.8–8.1	1000–3050	Li et al. 2006
	Aug. 2003		1630–1950	7.81–7.98	400–1510	Zhai et al. 2007
	Oct. 2005	1695–1750	1700–1740		620–2160	
	Dec. 2005	1939–1995	1920–1970	7.92–8.04	410–1275	
	Apr. 2006	1575–1840	1575–1810	7.76–8.03	770–3120	
Lower Estuary	Aug. 2003		2120–2230	8.1–8.4	181–712	Zhai et al. 2007; Chen et al. 2008
	Aug. 2004	1808–2997	1805–3155	7.69–8.52		Gao et al. 2008
	Oct. 2005		1800–1865		248–775	Zhai et al. 2007
	Dec. 2005	1930–2080	2250–2270		356–450	
	Apr. 2006				560–800	
East China Sea	Feb.–Mar 1993	1945–2093*	2276–2337*			Tsunogai et al. 1997
	Oct. 1993	1904–2141*	2273–2372*			
	May 1996	1970–2250*	2195–2280		257.8–370.1	Peng et al. 1999
	Apr–May 1995		2300–2480*	8.04–8.22	270–380	Wang et al. 2000
	Aug. 2003				236–445	Shim et al. 2007
	May 2004				240–431	
	Oct. 2004				326–517	
	Nov. 2005				340–510	
	Jul. 2007	1773–2023	2120–2269	7.929–8.338	160–615	Chou et al. 2009a
	Jan. 2008	1941–2081	2240–2286	7.905–8.091	306–390	Chou et al. 2011

* Data normalized to salinity of 35.

since the 1960s. This decrease was attributed to changes on the trophic relationships and damming activities in combination in the Changjiang Drainage Basin (Wang et al. 2007). However, pCO_2 in the Changjiang (i.e., >1000 μatm) was still much higher than that in the atmosphere (i.e., 360 ppm), and the degassing flux from the river was estimated to be 14.2 mol m^{-2} yr^{-1} in the 1990s (Wang et al. 2007).

Table 10.1 summarizes the previous published data on carbonate parameters in the Changjiang Estuary and ECS. Surface pCO_2 was 350–3,500 μatm in the upper part of the Changjiang Estuary (Li et al. 2006; Zhai et al. 2007), which showed a reduction down to the river mouth. However, there was a CO_2 peak near the Huangpujiang outlet due to the confluence of fresh waters with extremely high pCO_2 (Zhai et al. 2007). Furthermore, pCO_2 rapidly dropped from 600–1,200 μatm to 120–620 μatm in the lower part of Changjiang Estuary (Fig. 10.3), which could still be higher than the atmospheric pCO_2 level (Zhai et al. 2007; Zhai and Dai 2009). Surface pCO_2 in the Changjiang Estuary showed significant seasonal variations, with the lowest values usually in summer, medium values in winter and spring, and the highest values in fall (Zhai et al. 2007; Zhai and Dai 2009). For the ECS, surface pCO_2 values were generally lower than the atmospheric pCO_2 level in most seasons except the fall, in which surface waters may be slightly oversaturated with CO_2 and act as a weak source of atmospheric CO_2 (Shim et al. 2007; Zhai and Dai 2009; Chou et al. 2011). The highest influx (i.e., –13.7±5.3 mmol C m^{-2} d^{-1}) of atmospheric CO_2 taken by the ECS was observed in winter (Chou et al. 2011), followed

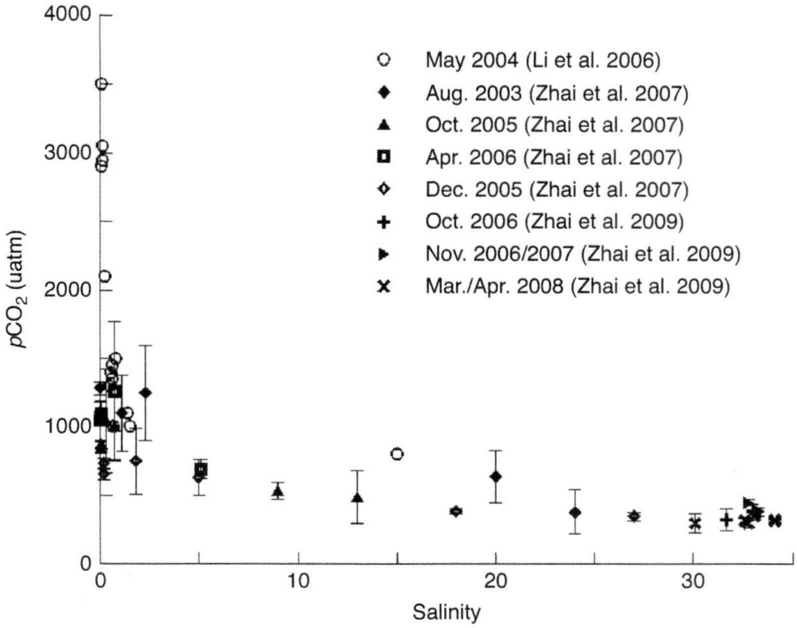

Figure 10.3. Relationship between surface pCO_2 and salinity in the Changjiang Estuary.

by spring (–5.0 to –8.8 mmol C m^{-2} d^{-1}) (Wang et al. 2000; Shim et al. 2007; Zhai and Dai 2009) and summer (–6.3±3.1 mmol C m^{-2} d^{-1}) (Chou et al. 2009a).

When examining variations of the pCO_2 values obtained in 1990s–2000s (Fig. 10.4), some appreciable characters appear. Tsunogai et al. (1997) reported a mean ΔpCO_2 of −63±6 μatm (n=139) in February 1993 and –44±20 μatm (n=176) in October 1993 for the shelf zone to the west of 125.5 °E along the PN Section (note: PN section starts off the Changjiang Estuary and ends southeasterly at Ryukyu Islands across the Changjiang plume front and shelf waters until the Okinawa Trough). Zhai and Dai (2009) reported a mean ΔpCO_2 of −37±17 μatm in January 2006, 4±32 μatm in November 2006, and 64±22 μatm in November 2007 for the inner shelf of the ECS (i.e., 29.5–32.5 °N and 122.5–124.5 °E). Chou et al. (2011) documented an average ΔpCO_2 of −30±18 μatm (n=197) in January 2008 for the inner shelf of the ECS (i.e., 30–31.5 °N and 123–125.5 °E). These results suggest that ΔpCO_2 values in the shelf area off the Changjiang Estuary in both fall and winter increased slightly from the 1990s to the 2000s; this was attributed to the enhanced accumulation of respired CO_2 in bottom waters in summer owing to the coastal eutrophication and hypoxia in combination.

Overall, DIC in the Changjiang Estuary behaved conservatively in the mixing zone between fresh and marine end-members. DIC in surface waters increased gradually from <1,700 μmol kg^{-1} at the riverine end-member to 1,870 μmol kg^{-1} in the lower part of Changjiang Estuary in spring and autumn, whereas DIC varied between 1930 and 2080 μmol kg^{-1} in winter (Zhai et al. 2007). DIC in surface waters of the ECS ranged between 1,773 and 2,023 μmol kg^{-1} with an average of 1903±56 μmol kg^{-1} in July 2007 (Chou et al. 2009a), and ranged between 1,941 and 2,081 μmol kg^{-1} with an average of 2,001±36 μmol kg^{-1} in January 2008 (Chou et al. 2011). The salinity normalized DIC (i.e., DIC×35/S) for the shelf region of ECS in January 2008 was 2,041±27 μmol kg^{-1} for February

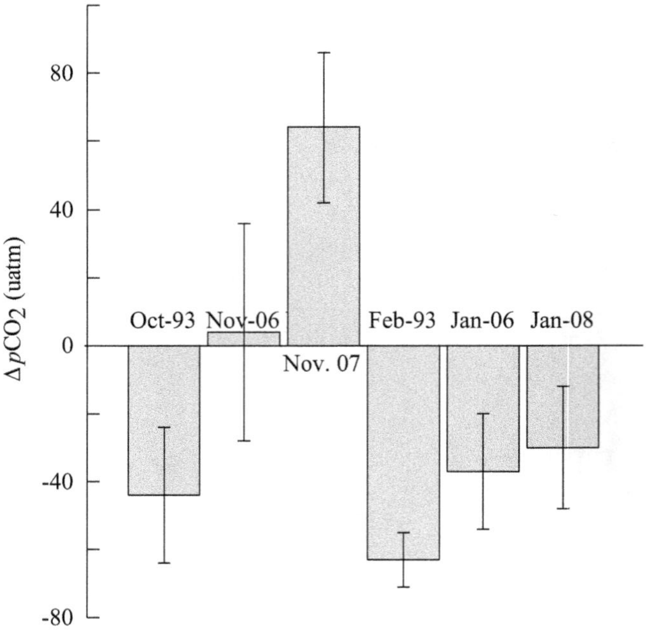

Figure 10.4. Comparison of ΔpCO_2 data for the East China Sea Shelf waters in literature. Data are from Tsunogai et al. 1997, Zhai and Dai 2009, and Chou et al. 2011.

1993 (cf. Tsunogai et al. 1997). The spatial distribution trends of DIC for summer and winter are quite different (Fig. 10.5). In winter, high DIC values occurred mainly in the inner shelf of water depth <50 m along the coast, whereas lower DIC occurred in the region further offshore at water depth >100 m, influenced by the Kuroshio (Chou et al. 2011). In summer, low DIC was observed in

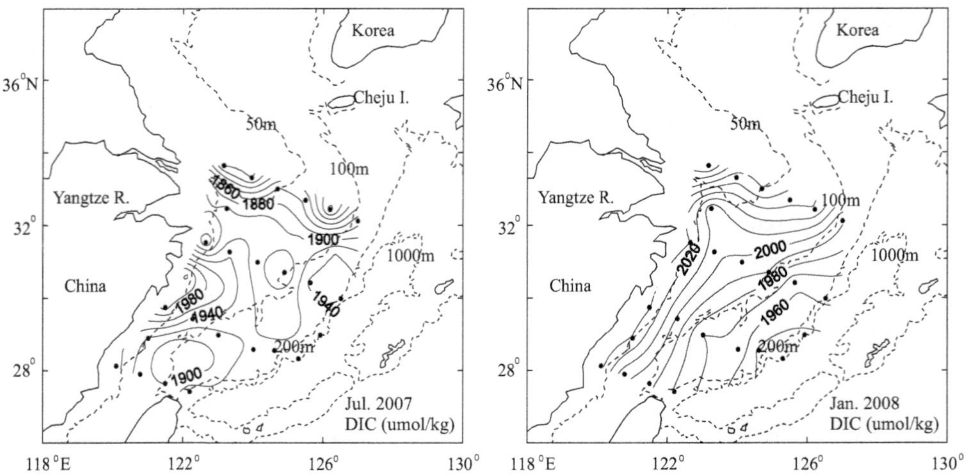

Figure 10.5. Horizontal distribution of DIC in the ECS in July 2007 and January 2008. Data are from Chou et al. 2009a-b and 2011.

Table 10.2. *Compilation of surface CH_4 in the Changjiang Estuary and the East China Sea*

Region	Study period	Stations	CH_4 (nM)	CH_4 saturation (%)	CH_4 flux (μmol m^{-2} d^{-1})	References
Changjiang Estuary	May 2002	10	13.17±4.86	525±196	21.1±9.6[a]; 41.1±18.7[b]	Zhang et al. 2008a
	June 2006	11	12.41±10.37	544±456	22.2±22.9[a]; 43.2±44.4[b]	
	Aug. 2005	35	19.59±16.22	876±686	55.4±52.6[a]; 108.9±103.4[b]	
	Oct. 2006	3	26.79±25.16	1284±1200	67.0±67.9[a]; 130.6±132.4[b]	
	Nov. 2002	18	7.39±4.00	300±144	12.7±3.6[a]; 25.0±18.9[b]	
	Aug.–Sep. 2009		2.9–67.3 (7.3±6.2)	145–2934 (365±303)	22.1±38.4[a]; 34.3±59.6[b]	Zhao et al. 2011b
East China Sea	Feb. 1993	12	2.70–3.27 (2.90±0.22)			Tsurushima et al. 1996
	Oct. 1993	14	2.14–4.02 (2.92±0.73)			
	June 1994	10	2.37–3.64 (2.75±0.38)			
	Aug. 1994	10	1.77–4.76 (2.49±0.72)			
	Sep.–Oct. 1994	14	2.48–7.58 (3.49)	127–254		Zang 1998
	Apr. 2001	29	3.24±0.59	141±23.6	1.63±1.67[a]; 2.77±2.71[b]	Zhang et al. 2004
	May 2002	18	5.05±2.49	216±101	5.4±4.7[a]; 10.5±9.2[b]	Zhang et al. 2008a
	Sep. 2003	21	2.75–49.0 (9.5±11.0)	148–2490 (487±555)	0.02–250.4 (20.9±54.8)[a] 0.02–440.6 (36.6±95.7)[b]	Zhang et al. 2008b
	June 2006	10	6.22±3.52	289±180	9.3±8.8[a]; 18.1±17.1[b]	
	Aug. 2005	5	6.85±3.77	348±187	13.3±11.8[a]; 26.2±23.1[b]	
	Oct. 2006	24	7.09±3.62	350±179	13.9±10.0[a]; 27.0±19.5[b]	
	Nov. 2002	12	3.53±0.68	161±32	3.7±1.9[a]; 7.3±3.8[b]	

[a] Flux was estimated by LM86 equation; [b] flux was estimated by W92 equation.

the northern part of the ECS owing to the influence of waters from Yellow Sea, whereas high DIC was observed in the coastal upwelling area further to the south (Chou et al. 2009a).

Table 10.2 summarizes the previously published methane data in the Changjiang Estuary and ECS. Figure 10.6 showed the monthly variations of dissolved CH_4 and water discharge at station Xuliujing (ca. 125 km upstream from the river mouth) in the lower Changjiang from June 2007 to December

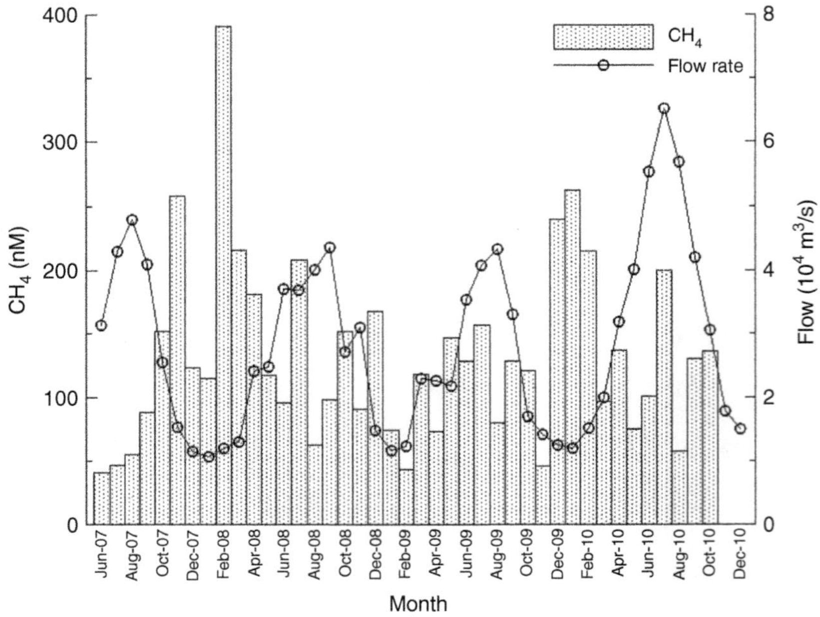

Figure 10.6. Monthly CH_4 concentrations at Xuliujing and water discharges at Datong Hydrographic Station of Changjiang from June 2007 to December 2010. Data were from Zhang et al. 2008a and Zhao et al. 2011a.

2010. Monthly dissolved CH_4 in the lower reaches of Changjiang ranged from 41.0 to 390.6 nM with an average of 131.7±70.9 nM and showed obvious seasonal as well as interannual variations. Annual average riverine input flux of CH_4 to the Changjiang Estuary was estimated 121.2×10^6 mol yr^{-1}. However, as shown in Figure 10.7, riverine CH_4 influx can be rapidly modified via microbial oxidation and air-sea exchange in the upper part of the estuary at salinity of 0 to10‰ (Zhang et al. 2008a).

Methane in the ECS shows a concentration gradient from Changjiang Estuary to the Kuroshio Surface Water with considerable spatial and temporal variations (Zhang et al. 2004, 2008b). Elevated CH_4 concentrations occur in the water column near the river mouth but decrease rapidly offshore ward (Fig. 10.8). In the open ECS, where the influence by Kuroshio can be significant, low CH_4 concentrations of 2–3 nM were observed for the surface waters (water depth: >100 m). Over the ECS shelf, CH_4 concentrations were 3–8 nM and tended to increase from surface to the near-bottom waters, which suggested CH_4 release from the bottom sediments (Fig. 10.8). The net CH_4 loss through exchange between ECS shelf and Kuroshio can be estimated from water volume transport ratios and the methane concentrations associated with the inflow and outflow. Here the water budget used by Wong et al. (2000) is applied to our preliminary estimation, namely about 0.5 Sv of Kuroshio Surface Water and 0.5 Sv of the upwelling Kuroshio Subsurface Water is imported into the ECS shelf, whereas 1.1 Sv of the ECS shelf water is exported to the Kuroshio and the Japan/East Sea. Assuming the concentrations of 2.28 and 2.14 nM to be representative for the inflow water from the Kuroshio

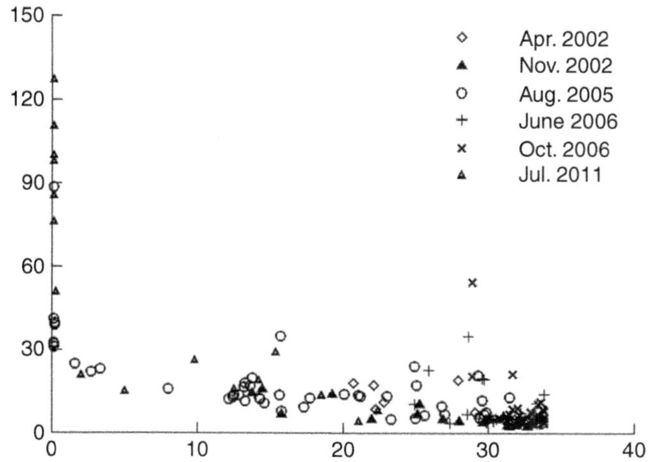

Figure 10.7. Relationship between surface CH_4 and salinity in the Changjiang Estuary and its adjacent coastal areas. Data for July 2011 are from our unpublished observations and other data are from Zhang et al. 2008a.

Surface and Subsurface Waters, respectively, and 3.68 nM for the outflow of ECS shelf water in April 2001, the net CH_4 exported from the ECS shelf was ca. 58.0×10^6 mol yr^{-1} in April 2001 (Zhang et al. 2004). Assuming the concentrations of 2.82 and 2.52 nM to be representative for the inflow water from the Kuroshio Surface and Subsurface Waters, respectively, and 5.81 nM for the outflow of ECS shelf water, the net CH_4 exported from the ECS shelf is ca. 117.3×10^6 mol yr^{-1} in September 2003. Hence the amount of CH_4 exported from the ECS to the NW Pacific Ocean showed strong seasonality, with a mean of 88×10^6 mol yr^{-1} averaged between spring and fall.

2.2. Dissolved organic carbon and particulate organic carbon

Several studies have addressed the flux of dissolved organic compound (DOC) and particulate organic compound (POC) of the Changjiang to the ECS (cf. Gan et al. 1983; Dagg et al. 2004; Lin et al. 2007; Wu et al. 2007). Table 10.3 lists the annual flux of DOC and POC of the Changjiang based on different observatories; the significantly higher values of DOC flux observed in earlier studies, for example, before 1980s, may in part, be due to the problems of sample collection and chemical analysis (e.g., method). Moreover, sampling strategy (e.g., sampling frequency and sampling site choice) may also lead to the inconsistency in flux estimations. The data collected since 1997 are quite consistent except for the flood period of 1998. The total flux of organic carbon (OC) of the Changjiang amounts to $2\text{–}4 \times 10^6$ tons yr^{-1} in 2003–2010, about 1% of global riverine OC flux, and equals to the value of Mississippi River (Dagg et al. 2004). The noticeable character of the Changjiang is the increase in the ratio of DOC to POC fluxes (i.e., 0.45 in 1997 and 2003 and 0.65 in 2010). Such kind of change should be considered in evaluating the riverine impact on the biogeochemistry of adjacent coastal ocean.

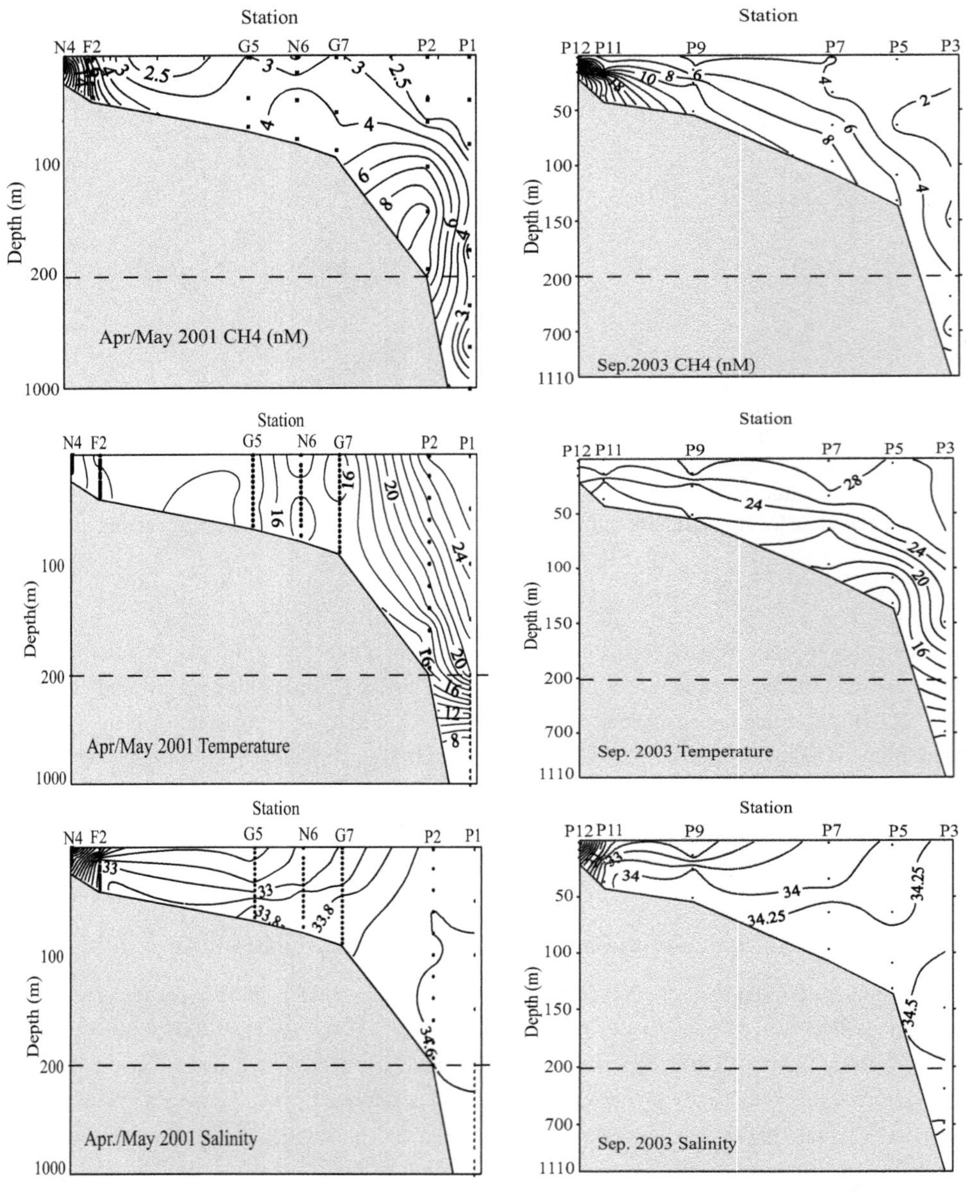

Figure 10.8. Vertical distributions of CH_4 (nmol/L), salinity, temperature (°C) along the PN Section. Data are from Zhang et al. 2004 and 2008b.

The turbidity maximum zone (TMZ) is one of the distinct hydrodynamic characteristics of the Changjiang Estuary. It exists in the area of surface water salinity at 5–15‰ with extreme high total suspended matter (TSM) of 1–10 g L^{-1} in near-bottom waters (Shen and Pan 2001). The distribution of DOC against salinity indicated that no conservative behavior can be found in the Changjiang

Table 10.3. *Comparison of annual flux ($\times 10^6$ tons yr^{-1}) data of DOC and POC from the Changjiang in this study with previous works in literature*

Year/Period	DOC flux	POC flux	References
1958–84	8.5	2.98	Wang 1989
1983	6		Gan et al. 1983
1986	1.8	4.5	Cauwet and Mackenzie 1993
1997	0.9	2.2	Wu et al. 2007
1998–6–9	5.5	9	Duan 2000
2003	1.32	2.69	Lin 2007
2004	1.2	1.63	Lin 2007
2005	1.47	2.21	Lin 2007
2006	1.03	1.04	Lin 2007
2007	1.25	1.86	This study
2008	1.07	1.23	This study
2009	1.11	1.56	This study
2010	1.54	2.55	This study

Estuary (Fig. 10.9), and an increase in DOC concentration takes place in the region of the salinity 10–15‰, where significantly higher POC is also observed. Positive correlation of POC and TSM suggested the dominance of an estuarine pool of OC. Table 10.4 presents the values observed of organic matter in the Changjiang Estuary (Lin et al. 2009); concentrations tend to decrease from river to the coast, but extremely high values of TSM and POC can be found in the turbidity maximum zone. The modification of DOC and POC could be tracked with the ratio of DOC/POC along the

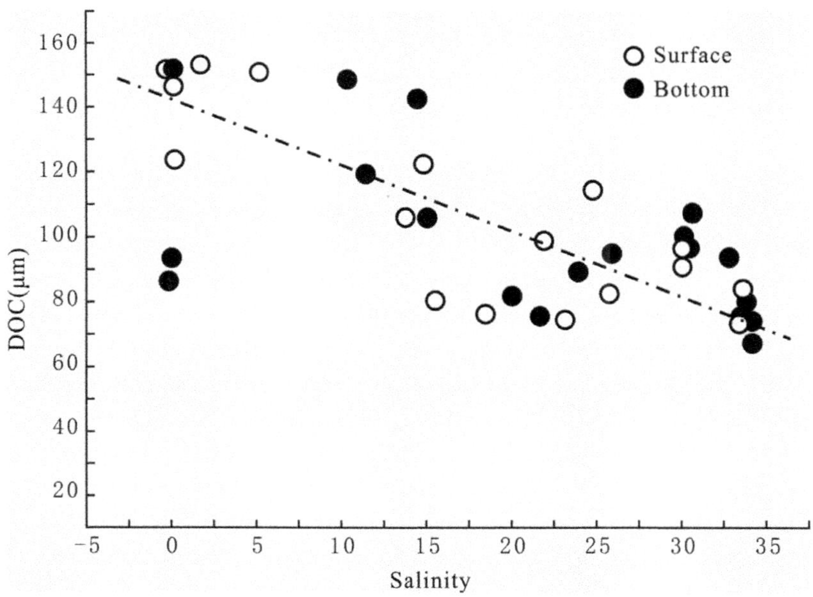

Figure 10.9. Distribution of DOC against salinity in the Changjiang Estuary. Data are from Lin 2007.

Table 10.4. *Composition of Organic Carbon (OC) in the main stream and Estuary of Changjiang and adjacent coastal waters*

		TSM (mg L^{-1})		DOC (μM)		POC (μM)		
Region	Layer	mean	s.d.	mean	s.d.	mean	s.d.	DOC/POC
River	Surface	543	275	147	16	168	490	0.88
	Near-bottom	489	294	86	6	174	388	0.49
TMZ	Surface	277	272	120	33	156	296	0.77
	Near-bottom	1715	203	120	29	1014	1090	0.12
Coast	Surface	16.8	28.6	87.7	13	19.4	110	4.52
	Near-bottom	164	253	88.2	13	131	189	0.67

Note: Data are compiled from Lin et al. 2009.

estuarine transects, for example, the lowest DOC/POC ratio was observed in TMZ, which indicated the dominance of POC, although the increase of DOC values was also observed in this area.

In the ECS, different water masses have variable DOC concentrations: highest values (ca. 110 μM) were found off the Changjiang Estuary and lowest values (ca. 60 μM) were found for the subsurface waters of Kuroshio, with DOC of ~75 μM in the surface water of Kuroshio and China coastal water (Hung et al. 2003). Figure 10.10 shows that in three cruises off the Changjiang Estuary in 2006, DOC values were ca. 100 μM, with a 10-μM difference from surface to the near-bottom waters (Lin 2007).

Vertical profiles of DOC and COC (i.e., colloidal organic carbon) were studied by Hung et al. (2000), and DOC was usually high in surface waters and then decreased slightly with water depth. The so-called truly dissolved fraction (i.e., <1 kDa) of OC ranged from 79.6% in surface and 89.2% in near-bottom waters in coastal environment, and from 60.3% at 100 m to 70.5% at 1,000 m depth in Kuroshio; the observed COC fractions (i.e., 30–40%) in the Kuroshio Waters (Hung et al. 2000) were in agreement with those in the NW Pacific Ocean (i.e., 30–37%) found by Ogawa and Ogura (1992).

Several studies were undertaken along the PN Section with regard to DOC and POC dynamics (Table 10.5). DOC and POC abundance generally decreased seaward from the Changjiang Estuary to Kuroshio Surface Water (Hung et al. 2000; Wu et al. 2003), and POC showed more significant spatial and temporal variabilities than DOC, especially in the region close to the Changjiang Estuary. Hung and Gong 2007 reported that the POC was 50–60 μg L^{-1} at the shelf break and ca. 90 μg L^{-1} at mid-shelf and then increased to 150 μg L^{-1} in the area affected by the Changjiang effluent plumes. High POC values measured in the water column coincided with rather elevated concentration of Chl-a, including the subsurface Chl-a maximum (Hung and Gong 2007). The high concentrations of POC in the near-bottom waters in the inner and mid-shelves resulted from resuspension of bottom sediments, particularly during rough sea conditions (Fig. 10.10). In this area, the layer of mobile mud sediments above the seabed can be of critical importance on the decomposition and transport of organic matter, as found from the Mississippi/Atchafalaya margins (Bianchi et al., Chapter 19 of this volume). Besides mixing induced by winds, strong tidal dynamics could also induce resuspension in coastal waters and hence change the POC profile (Hung et al. 2000; Zhu et al. 2006).

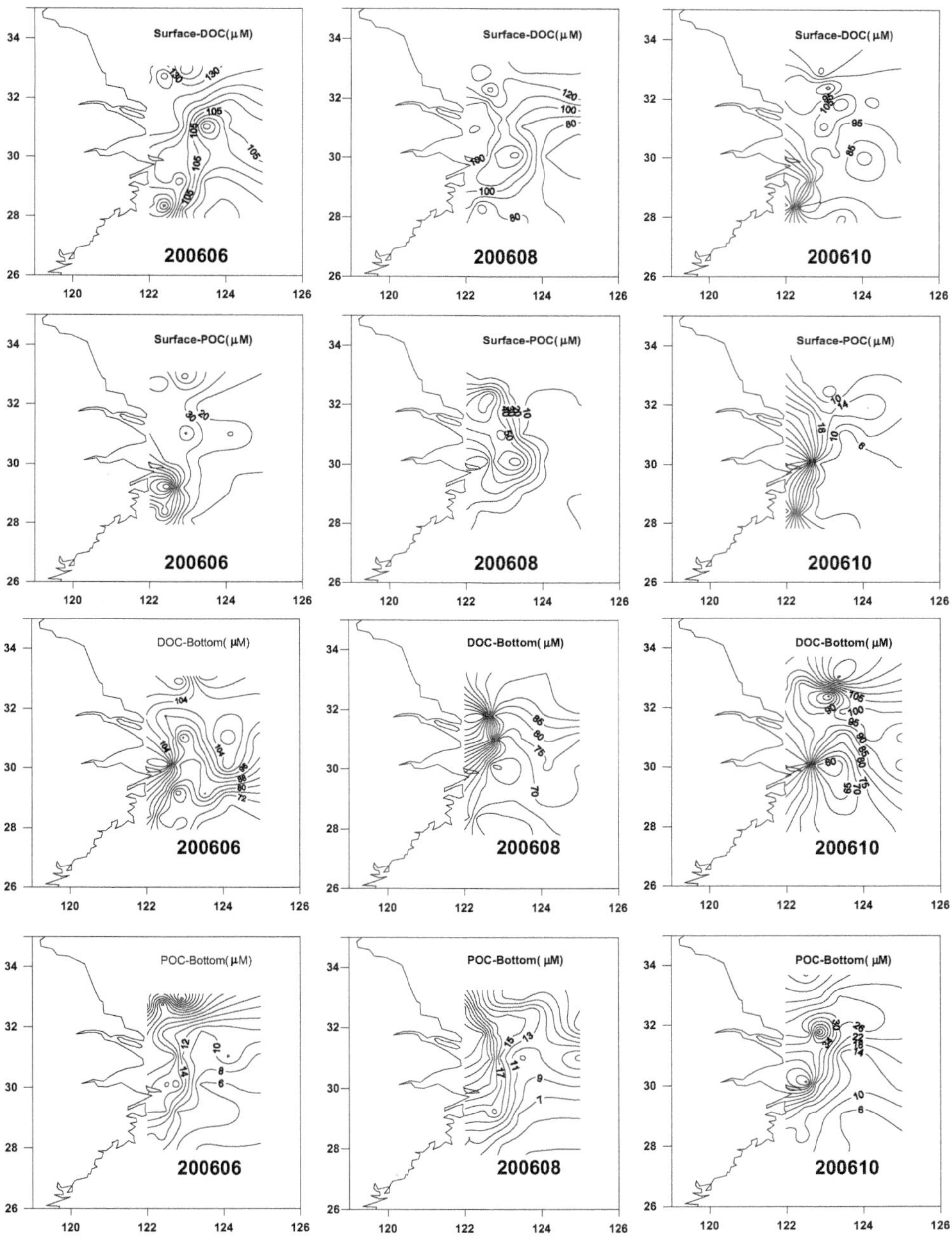

Figure 10.10. Spatial distribution of DOC and POC in three cruises of June, August, and October 2006. Data are from Lin 2007.

Table 10.5. *Distribution of DOC and POC in the East China Sea, illustrated along the PN section (Lin 2007)*

Region	DOC (μM)	POC (μM)	Sampling period	Depth	References
PN-A		15–38.3	01–1993	Near-bottom	Hoshika et al. 2003
PN-A		14.2–50	07/08–1994	Near-bottom	Hoshika et al. 2003
PN-A	65–72		04–1996		Ogawa et al. 2003
PN-A	110–120		06/07–1998		Hung et al. 2003
PN-A		25.0–71.1	01–2000		Zhu et al. 2006
PN-A	71.6–101	1.58–71.7	09–2002		Lin 2007
PN-B		2.5–8.33	02–1993	Near-bottom	Hoshika et al. 2003
PN-B		38.3–190	01–1993	Near-bottom	Hoshika et al. 2003
PN-B		8.33–33.3	04–1994		Huang et al. 1997
PN-B		5.83–23.3	07/08–1994	Near-bottom	Hoshika et al. 2003
PN-B	60–70		04–1996		Ogawa et al. 2003
PN-B	72–85		06/07–1998	Surface	Hung et al. 2003
PN-B		1.42–17.5	01–2000		Zhu et al. 2006
PN-B	63.0–88.0	1.25–10.8	09–2002		Lin 2007
PN-C		2.5–18.3	01–1993		Hoshika et al. 2003
PN-C		8.33	04–1994		Huang et al. 1997
PN-C		2.5–12.5	07/08–1994		Hoshika et al. 2003
PN-C		8.33	01–1994		Huang et al. 1997
PN-C	60–75		04–1996		Ogawa et al. 2003
PN-C	75–85		06/07–1998		Hung et al. 2003
PN-C		0.25–0.81	01–2000		Zhu et al. 2006
PN-C	41.5–77.6	0.15–4.67	09–2002		Lin 2007

PN-A: the region mainly impacted by Changjiang diluted water; PN-B: the broad shelf region; PN-C: the region impacted by Kuroshio Surface and Subsurface Waters.

Time series data illustrated that near to the Changjiang Estuary, tidal dynamics had remarkable influence on POC distributions, with elevated values of POC and chlorophyll *a* occurring twice a day, following the pattern of TSM in the water column (Zhu et al. 2006). Stations across the ECS shelf showed strong stratification with a thermocline at the depth of 40 m, and POC concentration was found to be 10–20 $\mu g\ L^{-1}$ in surface waters and 5–10 $\mu g\ L^{-1}$ beneath the thermocline (Zhu et al. 2006).

2.3. Pigments

In the water column, sources of pigments are mainly related to in situ primary production; in the estuary under strong terrestrial influence, potential direct inputs such as detritus of higher plants should also be noted, however. In coastal environment (e.g., estuary) where the land-source influx is of concern, direct contribution from phytoplankton pigments to organic carbon is usually ignored (Lee et al. 2000), as pigments themselves are low in concentration (i.e., on the order of $\mu g\ L^{-1}$) compared with bulk riverine organic carbon pool (i.e., both DOC and POC are usually on the order of mg L^{-1}). There is another aspect of phytoplankton-represented organic carbon (e.g. freshwater pigments), which is derived from in situ primary production in the drainage basin of the Changjiang.

Three Gorges Reservoir (TGR) has resulted in an increase in algal blooms in tributaries and the main stream of Changjiang, following an increase in residence time of water and accumulation of riverine nutrients. Hydrographic data from the Daninghe (note: "he" means "river" in Chinese) in TGR in 2004–2008 showed that Chl-a concentrations are $\geq 30\ \mu g\ L^{-1}$ (Wu et al. 2010). Survey of 39 lakes in the Changjiang Drainage Basin showed that Chl-a in the water column ranged from 0.74 to 200 $\mu g\ L^{-1}$ (cf. Zeng et al. 2011). In the Taihu (note: "hu" means "lake" in Chinese) from the lower reaches of Changjiang Watersheds, Chl-a showed a vertically well-mixed profile, and phytoplankton blooms (e.g., Chl-a concentration $>30\ \mu g\ L^{-1}$) occurred at an interval of ~50 days (Wu and Chen 2004). Again in the Taihu, pigment decreased quickly within the upper 50 cm of bottom sediments from ~0.15 $\mu g\ g^{-1}$ OC to <0.05 $\mu g\ g^{-1}$ OC, corresponding to a sedimentary record of ca. 5,000 years (Qu et al. 2000), which is comparable to the data from Guchenghu with a sediment core of 20 m further upstream and in the middle reaches (Ma et al. 1996). In the Zhangduhu from the middle reaches of Changjiang, chlorophyll and its derivatives in the sediment cores ranged from 0.5 to1.0 $\mu g\ g^{-1}$ OC over last 100 years (Wang et al. 2006). Further down the river, pigment concentrations are relatively low at Xuliujing all year around, and Chl-a concentration remains at a level of 0.01–3.5 $\mu g\ L^{-1}$, with an average of 0.88 $\mu g\ L^{-1}$.

In the Changjiang Estuary, Chl-a is influenced by strong terrestrial signature, and the surface water chlorophylls are low (e.g., $<1\ \mu g\ L^{-1}$), with the spring the highest season for Chl-a (Liu et al. 2008). Chlorophylls maxima are usually found in the estuarine front zone off Changjiang, where salinity increases and TSM quickly decreases (Ning et al. 2004; Song et al. 2009; Zhu et al. 2009). For example, in spring, surface Chl-a concentrations are found to be well correlated with Zeu/Zmix ratio ($r^2 = 0.83$), with Zeu standing for the depth of euphotic layer and Zmix for the depth of surface mixed layer (Zhu et al. 2009). In the shelf region of ECS, chlorophylls concentration and light (i.e., radiation) and nutrients are closely coupled. It should be noted that that phytoplankton structure was highly dynamic off the Changjiang Estuary and was presumably influenced by the location of the frontal zone.

Application of CHEMTAX revealed that nano-phytoplankton played an important role in estuarine and coastal zone, whereas in the open shelf of ECS, pico-phytoplankton can contribute more than 60% of the phytoplankton biomass (Furuya et al. 2003; Chen et al. 2006), which is in agreement with size-fractionized studies (Huang et al. 2006; Sun and Tian 2011). Chlorophytes in the coastal waters and open East China Sea contained a relatively low level of lutein (Furuya et al. 2003), which is usually a key diagnostic carotenoids (Mackey et al. 1996).

In the bottom sediments, Chl-a generally decreased with depth, and the most elevated sedimentary pigments can be found in region close to the Changjiang Estuary (Li et al. 2004; Bao et al. 2010). The chemical stability of organic compounds has a great influence on the concentration of preserved sedimentary records of pigments. For example, lutein shows a notable concentration in bottom sediments (e.g., 8 nmol g OC^{-1}), which is comparable to a more abundant fucoxanthin in the water column (Bao et al. 2010).

Based on phytoplankton OC to chlorophyll *a* ratio (i.e., OC_{phyto}:Chl-*a*), one can estimate the phytoplankton-derived organic carbon. However, as OC_{phyto}:Chl-*a* ratio is highly related to the structure of phytoplankton community, the calculation can be biased when the community structure changes, for example, in the continuum of Changjiang Watersheds and ECS. Using such a simple approach, an estimate can be made with empirical ratios of 25–250 for OC_{phyto}:Chl-*a* (Steele and Baird 1962; Banse 1977). Given the Chl-a concentration of 0.88 μg L^{-1}, phytoplankton-derived organic carbon would then be 22–220 μg L^{-1}, or 1.8–18 μM in the Changjiang at the river mouth. Considering that the mean POC concentration at Xuliujing is 75.8 μM (Lin et al. 2007), contribution from phytoplankton to the POC pool would be from 2% to 24%, or up to one-fifth of the Changjiang POC input into the ECS is from riverine in situ primary production.

There are some works trying to distinguish between higher plants and green algae based on lutein versus Chl-b ratios (i.e., Lut:Chl-*b*) (e.g., Bianchi and Findlay 1990), owing to the notable input of OC from higher plants. In the open ECS, however, terrestrial influence is much less; with overwhelming marine phytoplankton versus negligible contribution from higher plants, variation of the ratio of OC_{phyto}:Chl-*a* is most likely driven by the change in phytoplankton community structure, instead of contribution from higher plants from land sources (West et al. 2011).

3. Major pathways of carbon cycle and influence factors

3.1. DIC and CH_4

Aqueous photosynthesis and respiration rates are low in the upper part of Changjiang Estuary, and the DIC production is also limited (Zhai et al. 2007). High biological production fueled by allochthonous nutrients in the lower estuary and the shelf of ECS consumes a large amount of DIC and draws down the concentration of dissolved CO_2 through the biological pump. In spring, DIC in the ECS was found to be mainly influenced by temperature (Li and Wang 2011). Chou et al. (2009b) demonstrated that summer stratification is the key factor maintaining the CO_2 sink status in the ECS shelf waters. In summer, biological production overwhelmed community respiration in surface layers above the pycnocline. High biological production consumes a large amount of DIC in drawing down the surface fCO_2 below the saturation level. Community respiration exceeded primary production in the lower part of water column, and DIC is regenerated and stored in the subsurface layer through respiration. The CO_2 fixed by photosynthesis is partly transported from the ECS shelf to open oceans via the "continental shelf pump," as suggested by Tsunogai et al. (1999). However, Chou et al. (2009b) suggested that photosynthetically fixed CO_2 in summer might not be completely exported from the ECS shelf. Instead, CO_2 may be stored in the deep water, which can be readily released into the atmosphere through strong vertical mixing events that occur during the storm period in summer and/or during the prevailing period of northeast monsoon and seasonal cooling in early autumn. Temperature was the dominant factor controlling the spatial variation of carbon parameters in the shelf of ECS in winter, when the water column is well-stratified (Chou et al. 2011).

CH_4 is produced in an environment in which organic matter accumulates and oxygen is absent in organisms (e.g., guts of copepods) and respiration is over the production at the ecosystem level (e.g., hypoxic conditions). High CH_4 was found from the subsurface waters of the ECS (Zhang et al. 2004, 2008a), suggesting in situ CH_4 production from microenvironments such as digestive tracts and freshly released fecal pellets (Reeburgh 2007). Estuarine and coastal areas off the Changjiang are characterized by a high load of organic matter and mineral particles, which provide conditions for the production of CH_4 by methanogenesis in the water column and bottom sediments as well. A mean

Table 10.6. *Estimated contributions (%) of POC derived from Phytoplankton in the East China Sea (Lin 2007)*

Time	Depth	POC (μM)	Chl a (mg m^{-3})	Phytoplankton (%)
06–2006	Surface	31.7	2.38	8.15
	Middle depth	22	2.18	10.7
	Near-bottom	13.1	0.48	3.98
08–2006	Surface	24.3	3.43	15.3
	Middle depth	14.2	2.65	20.2
	Near-bottom	12.6	1.9	16.4
10–2006	Surface	13.4	3.06	24.8
	Middle depth	11.9	2.1	19.2
	Near-bottom	19.6	1.5	8.31
Average				14.1

CH_4 emission rate of 1.97 μmol m^{-2} d^{-1} was found from the bottom sediments of the Changjiang Estuary in spring (Zhang et al. 2008b), but it could be much higher because of hypoxic characters of near-bottom waters and higher temperature in summer. Distributions of CH_4 in the water column of the ECS are influenced by the Changjiang freshwater discharge and intrusion of Kuroshio Waters in combination (Zhang et al. 2004, 2008b). As discussed previously, exchange between the shelf water and the Kuroshio causes a net loss of CH_4 from the East China Sea, which may account for more than 80% of the total riverine input via the Changjiang.

3.2. DOC and POC

How far terrestrial materials along with Changjiang effluent plumes could be transported offshore and across the broad shelf is still under discussion. Different tracer techniques have been used to elucidate the fact, and it was reported that the riverine materials could be transported out of the estuary as far as ca. 250 km, with significant spatial and temporal variability (Wu et al. 2003; Ren et al. 2010).

The combined human activities and extreme weather conditions (i.e., draught vs. flood periods) in the watersheds have induced distinctive temporal fluctuations of water discharge and sediment load of the Changjiang. The average carbon fixation rate was about three times higher in flooding than nonflooding periods (Gong et al. 2011), which presumably affects the higher tropic levels and eventually fish catches as well. However, it should also be remembered that the flood-enhanced productivity and stratification could favor the development of seasonal hypoxia in the ECS.

In the shelf area of the ECS, high correlation between POC and *Chl-a* was observed in previous studies (Chang et al. 2003; Lin 2007), which suggested that POC could be mainly derived from in situ production. Based on the correlation of POC and Chl-a, the converted ratio (f) was set for 13 between biological production and Chl-a concentration (Chang et al. 2003); other factors could change the *f*-ratio in dynamic environment, such as phytoplankton structure, nutrients, and turbidity as well (Geider 1993). The percentage of contribution of phytoplankton to POC was estimated by this ratio as listed in Table 10.6 and elucidated that higher contributions were observed from surface and

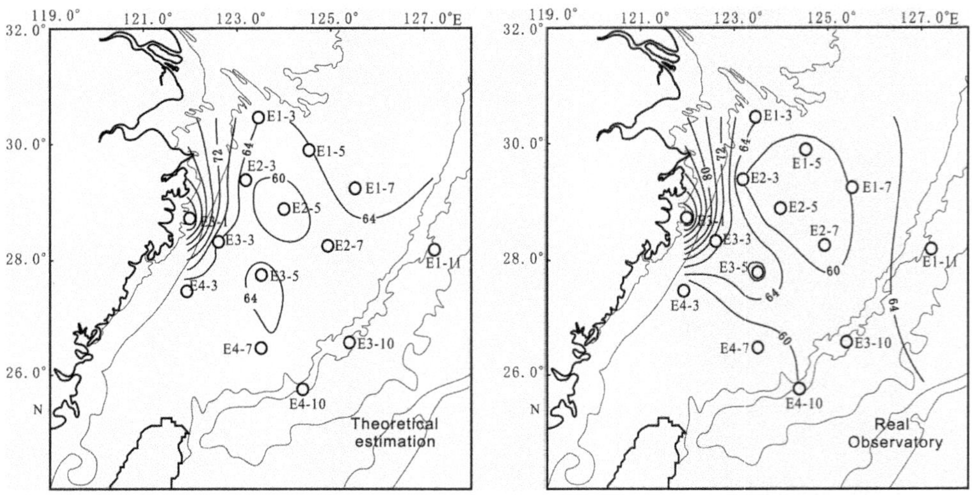

Figure 10.11. Comparison between model calculation by Ra-isotopes (left panel) and observations (right panel) for DOC distributions in the East China Sea. Data are from Tian 2008.

subsurface waters (Lin 2007). It is reported that DOC released from phytoplankton is an important source for heterotrophic bacteria organic matter, but it contributes ≤2% to new primary production in the open ocean (Hansell and Carlson 1998). Although the data from various cruises indicated an increase of DOC along with higher Chl-a in surface waters of ECS shelf, it is still difficult, if not impossible, to make the estimation of contribution of in situ primary production to the DOC pool (Lin 2007).

Recent studies have revealed the positive relationships between DOC and salinity in coastal environment (Hung et al. 2000; Ogawa et al. 2003; Lin 2007), because the concentration of DOC was highly related with the density in the ocean. Zhang et al. (2007) used ^{226}Ra plus hydrographic properties (i.e., salinity and temperature) to trace the mixing character of four different water masses in the ECS shelf. Tian (2008) simulated DOC distributions based on the mixing model by Zhang et al. (2007) and compared them with the observed DOC data in February 2007 and found that model output and observational data matched each other with a difference of ±7 μM on average; the maximum deviation of model output from observations was ca. 10% (Fig. 10.11).

In the coastal environment, re-suspension of bottom sediments is one of the important sources of POC in near-bottom waters (Hoshika et al. 2003; Zhu et al. 2006). In early studies, it was understood that the distribution of POC in the ECS in autumn was mainly derived from re-suspension (Liu et al. 1998). Resuspension is not only limited to the region close to the Changjiang Estuary; in the winter season, due to intensification of the Yellow Sea Coastal Current under northerly monsoon, the near-bottom water POC was also higher than in surface waters in mid-shelf region of ECS (Zhao et al. 2003). Hoshika et al. (2003) calculated the residual current of 0.052 m s^{-1} east to northeast at 10 m above the bottom at water depth of 115 m, which was considered important to carry fine suspended particles across over the ECS shelf. Resuspension and elevated POC levels were observed, and the

Table 10.7. *Comparison of organic Carbon Degradation Rates (K) between East China Sea and other world regions (Lin 2007)*

Region	Water type	Initial DOC level (μM)	K (d^{-1})	References
East China Sea Coast	Coastal water	85–120	0.038–0.048	Hung et al. 2003
Kuroshio	Open ocean waters	75–85	0.009	
East China Sea Shelf	Shelf waters	80	0.017–0.047	
North Atlantic Shelf	Shelf waters		0–0.4	Kirchman et al. 1991
Georges Bank Shoal	Shelf waters	92	0.0024	Hopkinson et al. 1997
North Pacific Ocean	Open ocean waters		0.005	Ogura 1972
Satilla River	Riverine waters	333	0.0012	Moran et al. 1999
Taiwan Strait	Shelf waters	50–74	0.047	Lin 2007
East China Sea	Shelf waters	50–98	3.6	Lin 2007
Changjiang	Riverine waters	119	<0.005	This study
Kuroshio	Open ocean waters	55	0.004	This study

cross-shelf export of POC through nephiloide mechanism was one of the important across-shelf pathways (Zhu et al. 2006).

Tropical cyclones in the NW Pacific Ocean take place ca. 25–30 times each year with total number of stormy days for three to four weeks in adjacent marginal seas (Zhu et al. 2006). Peng and Hu (1997) found that total suspended matter from the Changjiang could be dumped into the deep ocean by winter wind-driven vertical circulations with strong resuspension events, and the elevated POC value observed in 2003 cruise was six times as high as the typical values in the non-storm period (Zhu et al. 2006). One of the recent studies compared the POC and primary production before and after typhoon Marakot, which passed over Taiwan on August 8, 2009, and the export ratio (i.e., POC flux relative to primary production) increased significantly following the typhoon event, which was accompanied by changes of phytoplankton community structure (cf. Gong et al. 2011).

The removal of DOC from the water column could be regulated by photodegradation and microbial degradation (Benner and Opsahl 2001). Compared with the photo-bleaching processes in the fresh water system, the rate of photodegradation in the open ocean was much slower (Mopper et al. 1991). Utilization of DOC in the ECS was related to the chemical and biological properties of water masses, such as original concentrations of organic matter, nutrients level and biological species, and so forth. The highest utilization rate of DOC was found in coastal waters, and then in the shelf waters that mix with Kuroshio waters in various proportions (Hung et al. 2003). The incubation experiments designed to understand the biological utilization of DOM with variable water masses showed the similar results compared with the previous studies (Table 10.7). In the experiments, riverine samples had limited bioavailability, whereas Taiwan Strait Warm Water had similar behavior of DOC degradation as compared with the shelf waters (Lin et al. 2007).

Vertical flux of POC from euphotic waters is highly related to the physical setting of suspended particles, in situ primary production, aggregation of particles, consumption by predators, and so forth (Olli et al. 2002; Iseki et al. 2003). The POC export flux determined directly by sediment trap data at 65, 90, 120, and 140 m depth in the ECS ranged from 54 to 20 mg C m^{-2} d^{-1}; the ^{234}Th-derived

POC export flux varied from 67 to 26 mg C m^{-2} d^{-1}, using the ratio of POC/^{234}Th for sinking particles (Hung and Gong 2007). Hung and Gong (2007) pointed out that estimation of POC export flux based on the ratio of POC/^{234}Th for the intermediate (i.e., 10–50 μm) and small (i.e., 1.0–10 μm) size fractions of particles and ^{234}Th inventory in the water column are generally in agreement with the POC export fluxes measured by drifting sediment traps. With regard to the POC export flux in Kuroshio Region, data are still limited, and hence the spatial and temporal variations of POC export flux in the ECS as whole are not well understood.

3.3. Carbon fate inferred from pigments

Based on OC_{phyto}:Chl-*a* ratio, we reach an estimate of phytoplankton-derived carbon to be one-fifth in the POC export flux from the Changjiang. In the ECS, previous works reported a more pertinent OC_{phyto}:Chl-*a* ratio (Chang et al. 2003); in the region across the eutrophic coastal environment to the oligotrophic Kuroshio waters, the OC_{phyto}:Chl-*a* ratios are estimated to be 18 g g^{-1} for coastal environment, 67 g g^{-1} at mid-shelf, and 94 g g^{-1} for the Kuroshio. Considering the reported surface chlorophyll *a* (Furuya et al. 2003; Gong et al. 2003) and POC concentrations in the ECS (Zhu et al. 2006), phytoplankton-represented organic carbon contributes 20%, 45%, and 60% in the surface waters of coastal zone, middle shelf of ECS, and Kuroshio Region, respectively.

The organic carbon produced by phytoplankton via photosynthesis is the main pathway for inorganic carbon to be fixed as organic compounds, including pigment molecules. In the ECS and the Changjiang Estuary, primary production is highly variable in temporal and special dimensions. For instance, in the Changjiang Estuary and adjacent coastal waters, primary production can be 939 mg C m^{-2} d^{-1} in summer, whereas in winter, the primary production can drop to ca. 300–400 mg C m^{-2} d^{-1} (Gong et al. 2003). In the open waters of the ECS, where Kuroshio plays an important role in regulating photosynthesis, primary production shows low but consistent concentrations, with a mean value of 395 mg C m^{-2} d^{-1} (Gong et al. 2003). For the entire shelf of the ECS, the primary production is estimated ca. 145 g C m^{-2} yr^{-1} (Gong et al. 2003).

Because of the critical role of primary production in carbon cycles in the marine ecosystem, factors that affect the composition and structure of phytoplankton community (e.g., pigment cells) modify the pathway and efficiency of carbon transfer from low to higher trophic levels of the food web. In the lower reaches and upper estuary of the Changjiang, the water column is highly turbid with elevated nutrients concentrations (e.g., inorganic nitrogen), where the limiting factor for phytoplankton growth is the radiation (i.e., light) (Zhu et al. 2009). In the offshore region further down the estuary, where the turbidity decreases and nutrients become depleted, nutrients turn to be the limiting factor in photosynthesis. Moreover, dinoflagellates are the main phytoplankton groups off the Changjiang Estuary in spring, which is taken over by the diatoms in summer. For example, in May 2011, there was a large area of dinoflagellates blooms (i.e., *Noctiluca scintillans*) off the Changjiang Estuary and adjacent coasts along the Zhejiang Province, whereas in August 2011, the same area was found to have diatoms blooms, mainly *Thalassiosira pseudonana* and *Skeletonema costatum*.

It is known that planktonic blooms off the Changjiang Estuary have increased notably in last two decades (Zhou et al. 2003). Furthermore, flux and molar ratio among plant nutrients in the river can be influenced by the anthropogenic activities (e.g., the Three Gorges Dam), which can lead to a shift of phytoplankton community in the adjacent coastal areas in the ECS (Zhang et al. 1999).

In the ECS, nutrients and temperature are the two limiting factors for phytoplankton growth. In the coastal zone, diatoms are usually the main primary production contributors, whereas in the Kuroshio Region, prochlorophytes and cyanobacteria become very common (Pan et al. 2005). Patchy characters and high concentrations of fucoxanthin in Kuroshio Surface Water indicate that diatom-rich water from the shelf can be transported off the ECS shelf (Furuya et al. 2003). The subsurface chlorophyll *a* maximum in the continental shelf of ECS, which is repeatedly observed, is probably due to the influences of Taiwan Strait Warm Water and Kuroshio waters, which are the main sources of plant nutrients for the open shelf of the ECS. Being invisible to remote sensing from space, the role of the subsurface Chl-a maximum needs more attention in terms of both carbon cycle and primary production studies.

4. Biogeochemical budgets and feedbacks to open ocean and atmosphere

The ECS is regarded as an important carbon sink in time scale of 10–100 years, that is the period of sediment deposition recorded by radio nuclides such as ^{210}Pb and ^{137}Cs and so forth, because of the considerable amount of carbon accumulations in bottom sediments and the outflow of dissolved carbon to the NW Pacific Ocean (Deng et al. 2006). It was reported that the DOC export was more than twice the combined riverine fluxes (i.e., 155 Gmol C yr^{-1}) from Changjiang and Huanghe (Yellow River), whereas the POC export flux from the ECS was considerably smaller than the combined riverine inputs (i.e., 750–875 Gmol C yr^{-1}) (Cauwet and Mackenzie 1993). However, much of the riverine POC can be trapped in the area close to the river mouth and/or in the adjacent coastal zone (Deng et al. 2006). This is in contrast to the Bay of Bengal, where the burial efficiency of the terrestrial organic carbon from the Ganges-Brahmaputra River System could be as high as close to 100% (Galy et al., Chapter 14 of this volume). A summary of carbon budget for the ECS shelf was elaborated by Deng et al. (2006), and it was shown that overall, 80% of the carbon inventory of ECS can be exported out of the shelf area in the form of dissolved carbon (i.e., DOC and/or DIC). In the shelf region of ECS, carbon burial in the form of organic matter and carbonates amounts to $12–17 \times 10^6$ tons C yr^{-1} (Deng et al. 2006).

Because the major export species of carbon is in the form of dissolved organic carbon, a box model was elaborated based on a series of observations to evaluate export of DOC out off the ECS shelf at seasonal resolutions (Lin 2007). Figure 10.12 presents the end-member contributions and fluxes of DOC using up-to-date data. It is revealed that much higher DOC export out of the ECS shelf happens in winter (i.e., 87.4 kmol s^{-1}) than in summer (i.e., 9.2 kmol s^{-1}). The role of Kuroshio waters and Taiwan Strait Warm Water in regulating DOC chemistry of ECS was highlighted in comparison with the contribution of riverine influx, notably from the Changjiang. The exchange of ECS with the Yellow Sea was also evaluated in the model, and there is significant difference between summer and winter, as shown in Figure 10.12. Contributions from atmospheric pathway played an important role in DOC delivery in summer, which was considerably decreased in winter, owing to abundant precipitations in the summer monsoon period. The estimation of DOC annual flux by Hung et al.

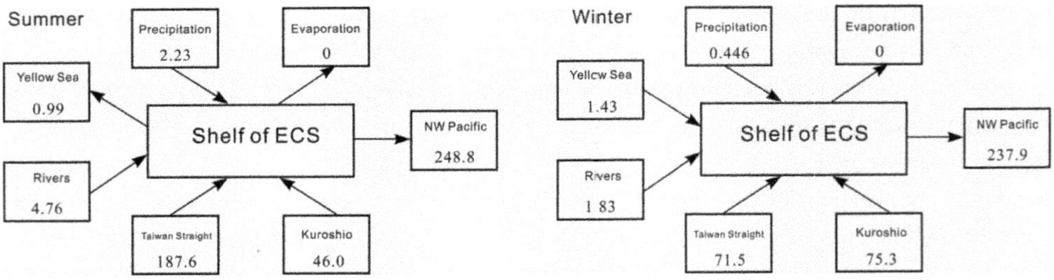

Figure 10.12. A simplified budget calculation of DOC (kmol s^{-1}) for the East China Sea Shelf, with case studies for summer (left panel) and winter (right panel) scenarios. Data are from Lin 2007.

(2003) was 1.9±1.5 Tg C yr^{-1} (1 Tg=10^{12} g), which is similar to the summer results shown in Figure 10.12, but much lower than the values in winter, presumably owing to underestimation of the impact from Kuroshio. Nevertheless, the budget estimations in Figure 10.12 need to be refined with more detailed information to constrain the variability of end-members. For example, uncertainty of fluxes from atmospheric deposition can be large owing to limited observational data in temporal and spatial dimensions, and contributions from sediment-water interface exchange and submarine groundwater discharges are still poorly quantified owing to very limited data sets.

The ECS receives about 2×10^7 tons DIC each year mainly as bicarbonate ion from the discharge of Changjiang (Liu et al. 2002). Although both decomposition of organic carbon from biological pump within the coastal waters and riverine source contribute to the DIC in the ECS (Tsunogai et al. 1997; Peng et al. 1999), gas exchange is the dominant source of DIC in the ECS. The ECS was found to act as a year-round CO_2 sink except in the fall (Tsunogai et al. 1999; Shim et al. 2007). Tsunogai et al. (1999) estimated the annual air-sea CO_2 influx from the ECS to be 8.0 mmol C m^{-2} d^{-1}; Chou et al. (2011) recently recalculated the flux and determined the annual mean air-sea CO_2 of the ECS to be 2.7 mmol C m^{-2} d^{-1}, which is 66% less. Figure 10.13 summarizes the DIC fluxes across different boundaries. The ECS receives 20 Tg C yr^{-1} supplied from rivers as DIC, and 9.1 Tg C yr^{-1} from atmosphere as CO_2. There is a net loss of DIC (i.e., 53.9 Tg C yr^{-1}) from the shelf to the ocean, which is about twice the total flux from the atmosphere and rivers. A high biological production is therefore required to support such an offshore transport of dissolved inorganic carbon.

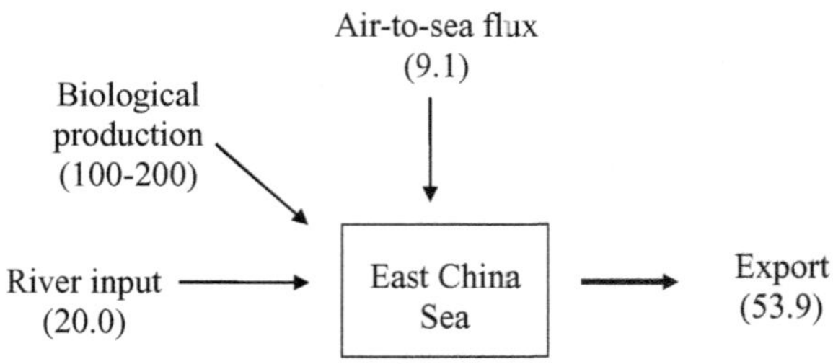

Figure 10.13. A simplified budget calculation for the dissolved inorganic carbon (Tg C/yr) in the East China Sea.

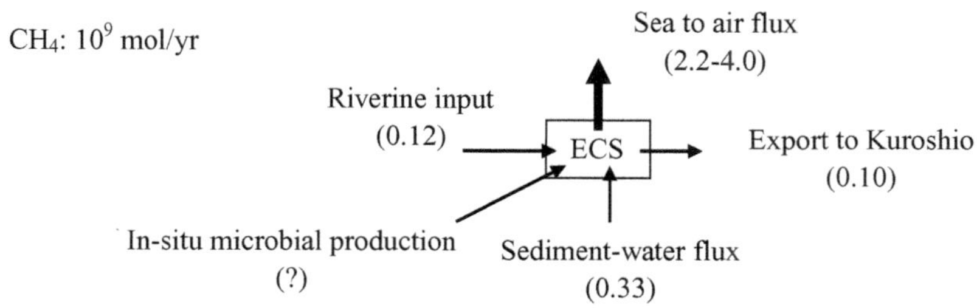

Figure 10.14. Preliminary budgets for dissolved CH_4 in the East China Sea.

The main pathways for CH_4 budgets in the ECS include sea-to-air release and export to the NW Pacific Ocean (e.g., Kuroshio). Total CH_4 input from Changjiang is estimated to be 0.12×10^9 mol yr^{-1}. The CH_4 input from sediments to the ECS is estimated to be 0.33×10^9 mol yr^{-1}, which is more significant to the water column CH_4 inventory than the riverine input (Fig. 10.14). The water exchanges between the shelf and the Kuroshio result in a net export of methane (i.e., 0.10×10^9 mol yr^{-1}) from the ECS shelf, implying a significant and net source of methane to the ocean interior (Fig. 10.14). The sea-to-air fluxes of CH_4 showed obvious seasonal variations with a mean of 7.7 ± 6.2 and 14.0 ± 11.6 μmol m^{-2} d^{-1} by LM86 and W92 equations, respectively. CH_4 emission from ECS was estimated to be 2.2–4.0×10^9 mol yr^{-1}. From the CH_4 emission data, the contribution of the ECS (i.e., about 0.2% of the total ocean surface area) to the global sea-to-air CH_4 flux can be estimated, which is about 0.3% relative to the global oceanic CH_4 emission of 11–18 Tg yr^{-1} (Bange et al. 1994), and suggests that the ECS shelf is an important source of CH_4 flux and hence hot spot in global scale. An in situ microbial CH_4 production of $(1.85–3.65) \times 10^9$ mol yr^{-1} is needed to balance the sources and sinks of CH_4 in the ECS, suggesting that microbial production could play an important role in methane biogeochemistry.

5. Evidence of historical records

Pigments are involved in photosynthesis, which is a critical process for the inorganic carbon to be transformed and fixed as organic matter, and hence they affect the carbon biogeochemistry and food-web dynamics of the ecosystem. Therefore, pigments are useful biomarkers for historical reconstruction in terms of the carbon cycle, particularly in the coastal environment (Bianchi and Canuel 2011). Although there has been work reporting preserved pigments over a time scale up to 20×10^3 years in marine sediments (Watts et al. 1977), care must be taken when applying pigment compounds in historical reconstructions, as pigments are liable to various factors (e.g., microbial degradation and consumption in benthic food-web) in both water column and seabed before they are eventually preserved in bottom sediments; even after sedimentation, processes of the early diagenesis can still modify the composition of pigments (Leavitt 1993). As labile compounds, pigments are also sensitive to illumination and temperature, which can accelerate the biodegradation. In the Changjiang Estuary and adjacent ECS shelf, positive factors for pigments preservation include elevated water column primary production, high sedimentation rate, and relatively shallow water depth (i.e., $\leq$50 m). In summer, seasonal oxygen depletion in the near-bottom waters are also thought to benefit pigment preservation.

Given that early diagenetic processes of organic materials in bottom sediments function with time, the content of organic molecules decreases exponentially with time span after deposition. Hence a simple model can usually be applied to describe the level of organic matter in early diagenesis (cf. Berner 1980):

$$Y = Y_0 \times e^{-k \times t} \tag{10.1}$$

where Y_0 represents the concentration of a given type of organic molecules in the surface of bottom sediments, Y is the concentration of the same kind of organic matter in the sediment cores after time t of deposition, and k is the rate of organic molecules degradation. In Equation 10.1, time span t after deposition can be linked to the position H (i.e., length) of sediment cores by $t = H/\omega$, if the deposition rate (ω) is known by radio nuclides measurements (e.g., ^{7}Be, ^{210}Pb and ^{137}Cs).

Degradation of pigments can be divided into three stages: first stage is in the water column; before pigments are deposited on to the sea bottom, they are under a quick degradation process in the water column, usually with a half-life period on the order of days. The second stage is in the surface sediments, with a half-life period on the orders of years, and the final stage (i.e., stage 3) is in the deep part of sediment cores, with a half-life period of even over hundreds of years (Leavitt 1993). The data interpretation would be biased if the historical reconstruction is based on sediment record of pigments from different stages (e.g., stage 2 vs. stage 3).

The Changjiang Estuary is characterized by seasonal disturbance of bottom sediments (e.g., different monsoon regimes involved) and high sedimentation rate (ca. 1–5 cm yr^{-1}). Therefore, a few centimeters of sediment core length may cover a record of several years and up to a decade. In this time scale, sedimentary pigments are still mainly at degradation stage 2, which is supported by observed sedimentary chlorophyll *a* half-life of 0.5 to 1 year (unpublished data), in comparison with the half-life period of 0.05–5 years in the Louisiana shelf of the Gulf of Mexico (Chen et al. 2005). The observed sedimentary Chl-a profiles are usually showing a pattern that increases from the bottom to the top of sediment core (Fig. 10.15). A similar trend can also be observed for pheophytin a, although stability of which is higher relative to Chl-a (Fig. 10.15).

A series of heavy floods occurred in the middle and lower part of Changjiang Watersheds in the summer of 1998 with freshwater discharge up to 82.3×10^3 m^3 s^{-1}. In the sedimentary cores from the offshore of Changjiang Estuary, a corresponding maximum of pigments was also identified (Fig. 10.16). Influenced by floods in 1998, detritus of higher plants can also be potential sources for the preserved sedimentary pigment pool. With regard to the diagnostic pigment compounds, a similar distribution pattern can be found for lutein and zeaxanthin, with similar chemical stabilities (Fig. 10.16). Similar distribution patterns of lutein and zeaxanthin in the sediment cores indicate that these two pigments are from the same origin (Bao et al. 2010). Again, contents of zeaxanthin and lutein can be higher relative to fucoxanthin and peridinin, but it is still hard to identify the most abundant phytoplankton groups considering the different stabilities of these pigment compounds (Reuss 2005). Moreover, stable nitrogen isotopes (i.e., $\delta^{15}N$) show a decreasing trend from the bottom to the top of the sediment cores (Fig. 10.17), which is in agreement with the increasing application of chemically fixed N-fertilizers in agriculture of the Changjiang Watersheds, corresponding to the

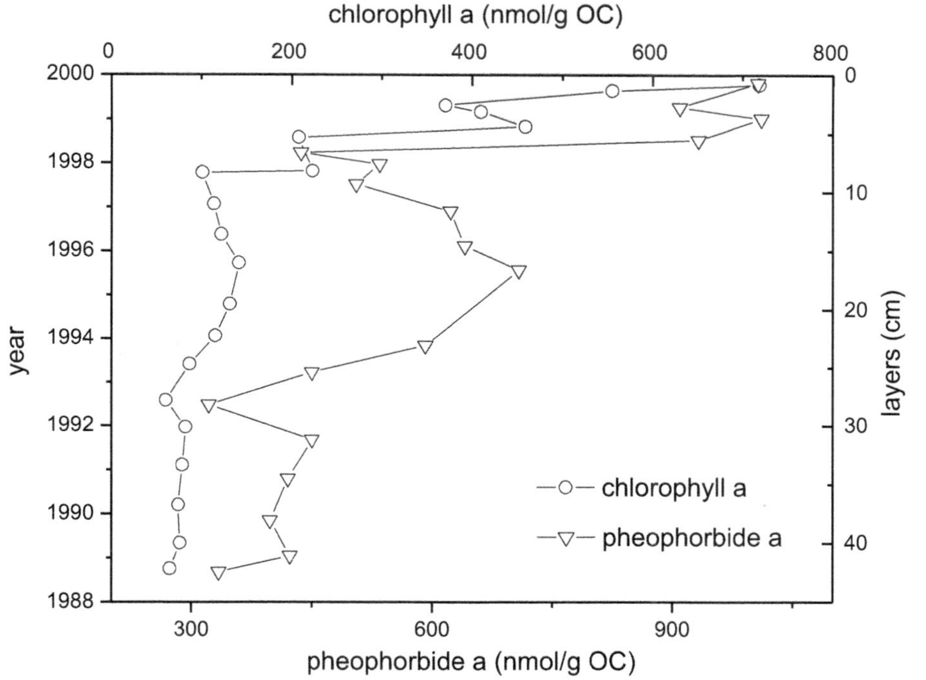

Figure 10.15. Typical chlorophyll *a* and pheophorbide a profile in a sediment core off the Changjiang Estuary detected by fluorometer. Data are from Li et al. 2004.

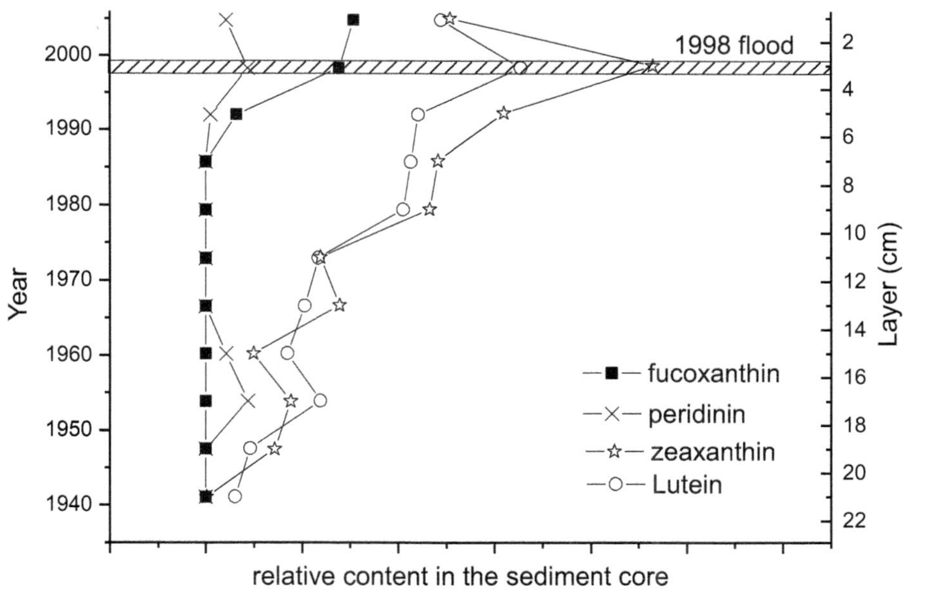

Figure 10.16. Preserved diagnostic pigments in the sediment core off the Changjiang Estuary. Data are from Li et al. 2011.

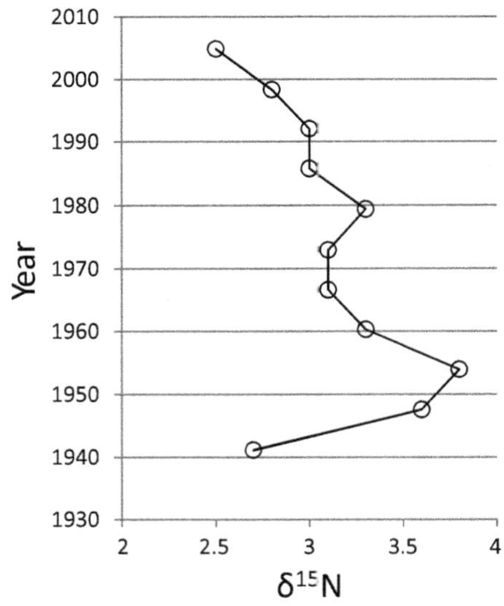

Figure 10.17. $\delta^{15}N$ variation trend in the sediment cores off the Changjiang Estuary. Data are from Li et al. 2011.

fact that dissolved inorganic nitrogen has continuously increased since 1950s in the Changjiang (Zhang et al. 1999; Yan and Zhang 2003), followed by an increase in the eutrophication and higher frequency of algal blooms in the Changjiang Estuary and ECS (Fig. 10.18). Li et al. (2011) showed that increases in the frequency of hypoxia based on the study of foraminifera off the Changjiang was correlated with pigments concentrations and historical changes in nutrients in sediment records. A similar reconstruction using pigments was successful off the Mississippi River and adjacent coastal environment (Chen et al. 2001).

6. Biogeochemical provinces of the Changjiang estuary – ECS in the perspective of carbon biogeochemistry

Although the seasonal and interannual variabilities are found in the ECS, the strong gradients of carbon properties occur from the eutrophic coastal environment to the oligotrophic Kurishio waters. Indeed, the observed CH_4/CO_2 ratio (i.e., nmol L^{-1}:ppm) along the PN Section shows almost linear reduction from inner shelf of ECS to the interior of Kuroshio in surface, either in autumn or in winter but with different gradients (Tsurushima et al. 1996; Tsunogai et al. 1997). For instance, the surface water concentration ratio of CH_4 (in nmol L^{-1}) to CO_2 (in ppm) range from 0.010–0.015 in the coast to 0.006–0.008 in the Kuroshio Region, and such a cross-shelf gradient in CH_4/CO_2 between coast and open ocean can be twice as high in autumn than in winter, showing a coupled feedback to atmosphere (Fig. 10.19). Again, the available data along the PN Section confirm that ECS is essentially a strong source of CH_4 to atmosphere, whereas it behaves as a sink of atmospheric CO_2 across the broad shelf, at least for the winter season (Fig. 10.20).

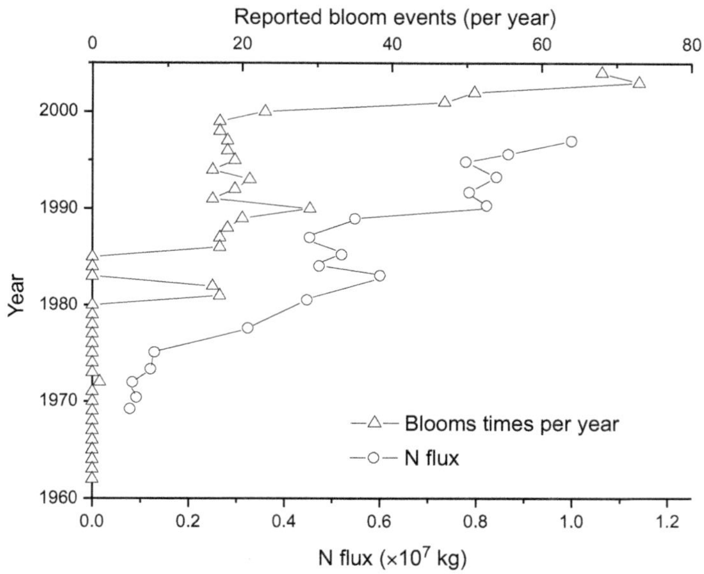

Figure 10.18. Annual nitrogen flux measured at the Datong Hydrographic Station of the Changjiang and reported bloom events in coastal environment off the Changjiang Estuary. Data are from Yan and Zhang 2003 and Wang 2006.

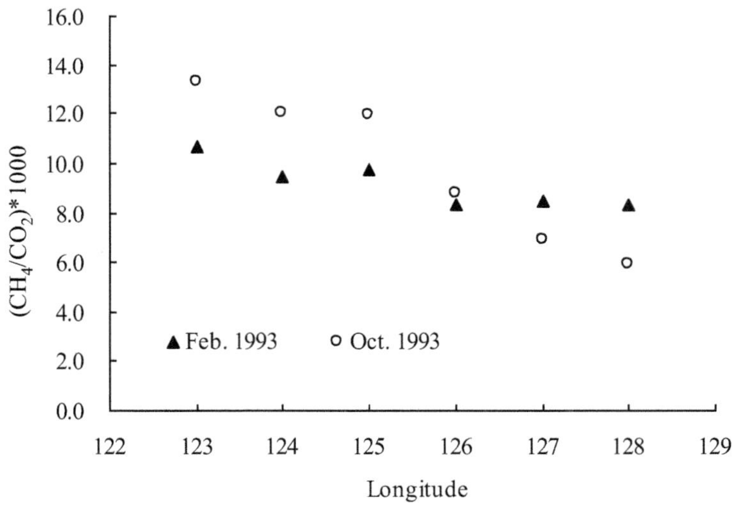

Figure 10.19. CH_4/CO_2 ratio along the PN Section during the cruises in February 1993 and October 1993. Surface CO_2 data were from Tsunogai et al. (1997) in ppm and surface dissolved CH_4 data were from Tsurushima et al. (1996) in nmol/L.

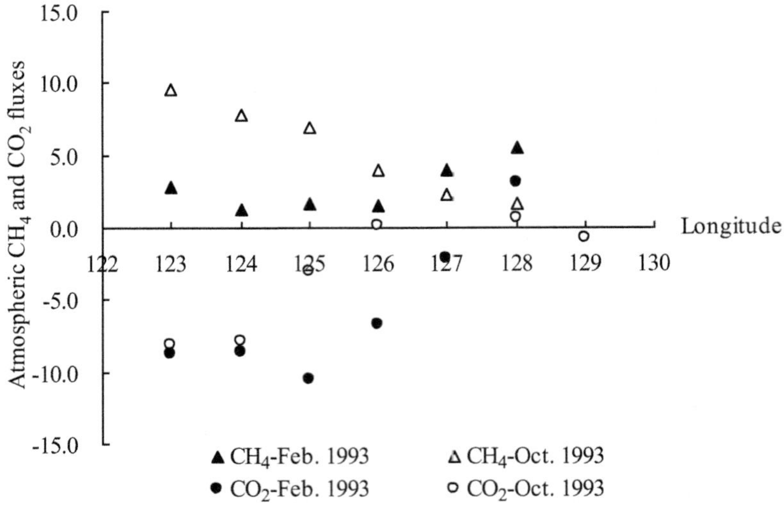

Figure 10.20. CH_4 fluxes (μmol m^{-2} d^{-1}) and CO_2 fluxes (mmol m^{-2} d^{-1}) along the PN Section during the cruises in February 1993 and October 1993. Surface CO_2 data were from Tsunogai et al. (1997), and a constant CO_2 gas exchange velocity of 3.5 m/d was used to calculate the atmospheric CO_2 fluxes. Surface dissolved CH_4 data were from Tsurushima et al. (1996), and a constant CH_4 gas exchange velocity of 5.1 m/d was used to calculate the atmospheric CH_4 fluxes.

With regard to the organic matter in the water column, distribution of organic pool (i.e., POC+DOC) shows different provinces between the coastal environment and the open Kuroshio waters. Briefly, the DOC/POC ratio in surface waters indicates an important contribution of particulate form to the organic carbon pool in the coastal province with DOC/POC <4 to 5, whereas in the province of open Kuroshio waters, DOC/POC can be as high as 20–25 (Fig. 10.21). The low DOC/POC ratio in coastal waters is presumably from the terrestrial sources (e.g., river) with high TSM (e.g., up to 1–10 g L^{-1} in estuarine TMZ) but low POC concentration (i.e., >0.5%). In the open shelf and Kuroshio Surface Water, TSM is as low as 1–5 mg L^{-1}, with POC as high as 10–20%. Similarly, the Chl-a to POC ratio in the water column can be as low as 0.004 in the area affected by the high turbidity plumes off the Changjiang; it then increases up to 0.01 or even higher in the offshore waters, corresponding to the high biomass of phytoplankton (Fig. 10.22).

In summary, owing to the complex structure of water circulation and monsoon climate, the carbon biogeochemistry of the East China Sea shows the characters of biogeochemical sub-provinces. In the coastal environment and off the estuary, distribution of carbon species (i.e., inorganic and organic forms) illustrates the influences from the land-sources influx. On one side, the terrestrial source of nutrients stimulates the coastal eutrophication, which may reduce the partial pressure of CO_2 in the water column but with strong feedback to atmosphere with CH_4 emission. Nevertheless, the heterotrophic nature of high turbidity and eutrophic waters in the coastal environment off the Changjiang may induce consumption of dissolved oxygen and emission of CO_2 as well, particularly in summer, followed by coastal hypoxia in near-bottom waters. The open shelf waters of the ECS as well as Kuroshio Surface Water have oligotrophic character, where the primary production and

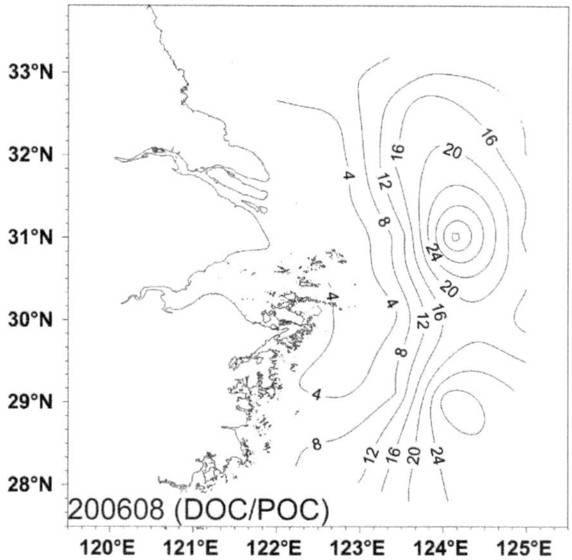

Figure 10.21. Ratio of DOC to POC in surface waters off the Changjiang Estuary and in the East China Sea Shelf in August 2006.

carbon fixation respond to the external supply of plant and micronutrients and the system functions autotrophy. Again in the open ECS and Kuroshio Region, vertical mixing (e.g., convection in winter) of nutrient-poor surface with nutrient-enriched deep waters have an important role in regulating the planktonic community and hence the CO_2 uptake. The ECS behaves as a barrier and/or channel,

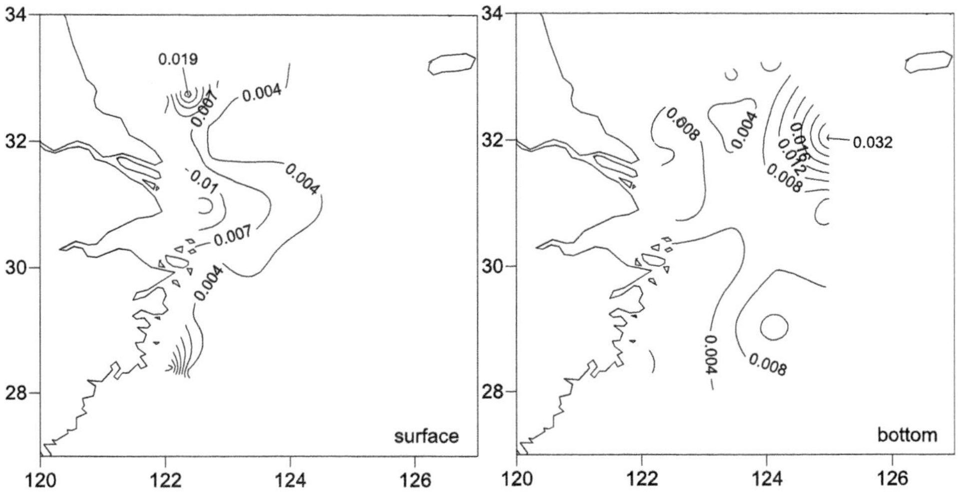

Figure 10.22. Ratio of *Chl-a* to POC in region off the Changjiang Estuary and in the East China Sea Shelf in August 2006.

depending on the species of interest, through which terrestrial and oceanic materials are mixed and modified via biogeochemical processes and eventually deposited into the seabed, fed back to atmosphere, and exported to the ocean interior.

7. Concluding remarks and future work

In the coastal ocean, carbon biogeochemistry is interactively linked to the influxes from land and open ocean. The ecosystems function in response to the changes of climate in different temporal and spatial dimensions, with strong feedbacks to the atmosphere through emission and uptake of greenhouse gases (e.g., CO_2 and CH_4), and to the global ocean through the outflow of materials (e.g., continental pump), and to bottom sediments through deposition and burial of organic and inorganic particles.

In the East China Sea, the major external sources of carbon include riverine influx of dissolved and particulate forms, mainly from the Changjiang, and the Kuroshio, including Taiwan Strait Warm Water. In this study, we summarize the knowledge of carbon biogeochemistry from the Changjiang (Yangtze River) Watersheds across the ECS until Kuroshio, and the data include DIC, CO_2, CH_4, DOC, and POC, as well as photosynthetic pigments. Briefly, the historical data show that DIC concentration in the Changjiang has dropped over the last three to four decades, whereas the flux ratio of DOC to POC tends to increase, particularly after the construction of Three Gorges Dam. In the ECS, the observational data indicate that the ecosystem functions essentially as a CO_2 sink but a CH_4 source to atmosphere and exports a considerable amount of inorganic and organic carbon to the NW Pacific Ocean. Following the loss of plant nutrients from the watersheds (e.g., Changjiang), the adjacent coastal environment has suffered from eutrophication, with a change in photosynthetic plankton community structure revealed by the pigment compositions of sediment cores, which has an important consequence on the carbon biogeochemistry of the ECS.

With regard to the dynamics of carbon in the Changjiang Estuary and ECS, the response to the variability of land-derived materials should be continuously explored, taking into consideration ever-increased human disturbances in the watershed and adjacent coastal environment, such as change in the flux of land-derived carbon and modification of the ratio between carbon and other plant nutrients. At this point, our knowledge of bioavailability of terrestrial organic carbon is still in its infancy, which is, however, closely linked with the function of the ecosystem (e.g., micro-loop) in the marine recipient. The role of submarine groundwater discharges in the coastal and shelf region needs to be evaluated because a preliminary study indicated that the contribution of ground water to the CDOM (i.e., colored dissolved organic matter) composition of marine environment was underestimated (Gao et al. 2011). The interaction of inorganic species with CDOM and/or terrestrial humus is worthy of study as well. Isotopic compositions of POC and DOC (i.e., $\delta^{13}C$ and $\Delta^{14}C$) combined with biomarkers should be applied for the discrimination of sources and pathways of carbon in the dynamic ECS, which will be very helpful to understand the differences and similarities in sources and biogeochemical behaviors of DOC and POC in comparison with other world oceans. In correspondence to the changing impact (e.g., shift of nutrients ratios and eutrophication), there is already report of increasing dinoflagellates blooms in recent years relative to two to three decades ago (Tang et al. 2006; Zhou et al. 2008). Confirmation of phytoplankton community structure shift and its consequence on carbon cycles needs further studies. Finally, to understand better the transport and fluxes of organic matter from terrestrial and riverine systems to the continental margin and to link this to open ocean and atmosphere, it is prerequisite to assess the temporal variability (i.e., intra- and interannual scales) in isotopic compositions, sources, and residence times of DOC and POC (Bauer et al. 2002).

Although the open ECS and Kuroshio Surface Water are still in the oligotrophic condition, the coastal environment adjacent to the China mainland has suffered from an increase in eutrophication affected by land-source input of plant nutrients, particularly nitrogen. Thus carbon biogeochemistry in the coastal waters of ECS is regulated by the change in ecosystem function that alters from autotrophy to heterotrophy with strong impact on the atmosphere (e.g., CO_2 vs CH_4), as well as change in the ratio of labile to refractory organic matter in the water column and the deposition flux to the sea bottom. It appears that the meso-scale processes (e.g., front and eddy) play a critical role in the interactions (e.g., exchange flux) between eutrophic coastal to oligotrophic open oceanic waters, which has, however, not been studied in depth before.

Acknowledgments

This work was financially supported by the Ministry of Science and Technology of China (No: 2011CB409801/2) and Natural Science Foundation of China (No: 41021064). Prof. Tom Bianchi is acknowledged for the syntax correction and suggestions on the draft manuscript of this work.

References

Banse, K. 1977. Determining the carbon-to-chlorophyll ratio of natural phytoplankton. Mar. Biol. 41: 199–212.

Bange, H.W., U.H. Bartell, S. Rapsomanikis and M.O. Andrae. 1994. Methane in the Baltic and North Seas and a reassessment of the marine emissions of methane. Global Biogeochem. Cycles 8: 465–480.

Bao, H.Y., Y. Wu, Z.Y. Zhu and J. Zhang. 2010. Distribution of pigments in E5 core sediments in the Changjiang Estuary. Mar. Environ. Sci. 29: 314–316 (in Chinese).

Bauer, J.E., E.R.M. Druffel, D.M. Wolgast and S. Griffin. 2002. Temporal and regional variability in sources and cycling of DOC and POC in the northwest Atlantic continental shelf and slope. Deep-Sea Res. Part II Top. Stud. Oceanogr. 49: 4387–4419.

Berner, R.A. 1980. Early Diagenesis: A Theoretical Approach. Princeton University Press, Princeton.

Benner, R. and S. Opsahl. 2001. Molecular indicators of the sources and transformations of dissolved organic matter in the Mississippi river plume. Organic Geochem. 32: 597–611.

Bianchi, T.S. and E.A. Canuel. 2011. Chemical Biomarkers in Aquatic Ecosystems, Princeton, NJ: Princeton University Press.

Bianchi, T.S. and S. Findlay. 1990. Plant pigments as tracers of emergent and submergent macrophytes from the Hudson River. Can. J. Fish. Aquat. Sci. 47: 492–494.

Cai, W.J., X. Guo, C.A. Chen, M. Dai, L. Zhang, S.E. Lohrenz, W. Zhai and Y. Wang. 2008. A comparative overview of weathering intensity and HCO_3^- flux in the world's largest rivers with emphasis on the Changjiang, Huanghe, Pearl and Mississippi Rivers. Cont. Shelf Res. 28: 1538–1549.

Cauwet, G. and F.T. Mackenzie. 1993. Carbon inputs and distribution in estuaries of turbid rivers: the Yangtze and Yellow rivers (China). Mar. Chem. 43: 235–246.

Chang, J., F.K. Shiah, G.C. Gong and K.P. Chiang. 2003. Cross-shelf variation in carbon-to-chlorophyll *a* ratios in the East China Sea, summer 1998. Deep-Sea Res. Part II Top. Stud. Oceanogr. 50: 1237–1247.

Chen, C.-T.A., W. Zhai and M. Dai. 2008. Riverine input and air-sea CO_2 exchanges near the Changjiang (Yangtze River) Estuary: Status and implication on possible future changes in metabolic status. Cont. Shelf Res. 28: 1476–1482.

Chen, J.S., F.Y. Wang, X.H. Xia and L. Zhang. 2002. Major element chemistry of the Changjiang (Yangtze River). Chem. Geol. 187: 231–255.

Chen, J.X., B.Q. Huang, Y. Liu, Z.R. Cao and H.S. Hong. 2006. Phytoplankton community structure in the transects across East China Sea and Northern South China Sea, determined by analysis of HPLC photosynthetic pigment signatures. Advance Earth Sci. 21: 738–746 (in Chinese).

Chen, N., T.S. Bianchi and B.A. McKee. 2005. Early diagenesis of chloropigment biomarkers in the lower Mississippi River and Louisiana shelf: implications for carbon cycling in a river-dominated margin. Mar. Chem. 93: 159–177.

Chen, N.H., T.S. Bianchi, B.A. McKee and J.M. Bland. 2001. Historical trends of hypoxia on the Louisiana shelf: application of pigments as biomarkers. Org. Geochem. 32: 543–561.

Chou, W.C., G.C. Gong, D.D. Sheu, C.C. Hung and T.F. Tseng. 2009a. The surface distributions of carbon chemistry parameters in the East China Sea in summer 2007. J. Geophys. Res. 114: C07026.

Chou, W.C., G.C. Gong, D.D. Sheu, S. Jen, C.C. Hung and C.C. Chen. 2009b. Reconciling the paradox that the heterotrophic waters of the East China Sea shelf act as a significant CO_2 sink during the summertime: evidence and implications. Geophys. Res. Lett. 36: L15607.

Chou, W.C., G.C. Gong, C.M. Tseng, D.D. Sheu, C.C. Hung, L.P. Chang and L.W. Wang. 2011. The carbonate system in the East China Sea in winter. Mar. Chem. 123: 44–55.

Dagg, M., R. Benner and S. Lohrenz. 2004. Transformation of dissolved and particulate materials on continental shelves influenced by large rivers: plume processes. Cont. Shelf Res. 24: 833–858.

Deng, B., J. Zhang and Y. Wu. 2006. Recent sediment accumulation and carbon burial in the East China Sea. Global Biogeochem. Cycles 20: GB3014.

Duan, S.W. 2000. The distribution and transportation of nutrients and carbon in the Yangtze River (in Chinese). Ph.D. Dissertation, Institute of Geography, Chinese Acad. Sci., Beijing.

Furuya, K., M. Hayashi, Y. Yabushita and A. Ishikawa. 2003. Phytoplankton dynamics in the East China Sea in spring and summer as revealed by HPLC-derived pigment signatures. Deep-Sea Res. Part II Top. Stud. Oceanogr. 50: 367–387.

Gan, W.B., H.M. Chen and Y.F. Han. 1983. Carbon transport by the Yangtze (at Nanjing) and Huanghe (at Jinan) Rivers, People's Republic of China. pp. 459–470. In: E. T. Degens, S. Kempe, H. Soliman (Eds.), Transport of Carbon and Minerals in Major World Rivers, Part 2, SCOPE/UNEP, Vol. 55. University of Hamburg, Germany.

Gao, L., D.D. Fan, C.X. Sun, D.J. Li and J.G. Cai. 2011. Optical characterization of CDOM in a marsh-influenced environment in the Changjiang (Yangtze River) Estuary. Environ. Earth Sci. 64: 643–658.

Gao, X.-L., J.-M. Song, X.-G. Li, N. Li and H.-M.Yuan. 2008. Dissolved inorganic carbon in surface waters around the Changjiang Estuary and Hangzhou Bay in summer. Mar. Sci. 32: 61–67 (in Chinese).

Geider, R.J. 1993. Quantitative phytoplakton physiology: implications for primary production and phytoplankton growth. ICES Mar. Sci. Symposia 197: 52–62.

Gong, G.C., Y.H. Wen, B.W. Wang and G.J. Liu. 2003. Seasonal variation of chlorophyll *a* concentration, primary production and environmental conditions in the subtropical East China Sea. Deep-Sea Res. Part II Top. Stud. Oceanogr. 50: 1219–1236.

Gong, G.C., K.K. Liu, K.P. Chiang, T.M. Hsiung, J. Chang, C.C. Chen, C.C. Hung, W.C. Chou, C.C. Chung, H.Y. Chen, F.K. Shiah, A.Y. Tsai, C.H. Hsieh, J.C. Shiao, C.M. Tseng, S.C. Hsu, H.J. Lee, M.A. Lee, I.I. Lin and F.J. Tsai. 2011. Yangtze River floods enhance coastal ocean phytoplankton biomass and potential fish production. Geophys. Res. Lett. 38: L13603.

Hansell, D.A. and C.A. Carlson. 1998. Deep-ocean gradients in the concentration of dissolved organic carbon. Nature 395: 263–267.

Hopkinson, C.S., B. Fry and A.L. Nolin. 1997. Stoichiometry of dissolved organic matter dynamics on the continental shelf of the northeastern U.S.A. Cont. Shelf Res. 17: 473–489.

Hoshika, A., T. Tanimoto, Y. Mishima, K. Iseki and K. Okamura. 2003. Variation of turbidity and particle transport in the bottom layer of the East China Sea. Deep-Sea Res. Part II Top. Stud. Oceanogr. 50: 443–455.

Huang, B.Q., Y. Liu, J.X. Chen, D.Z. Wang, H.S. Hong, R.H. Lü, L.F. Huang, Y.A. Lin and H. Wei. 2006. Temporal and spatial distribution of size-fractionized phytoplankton biomass in East China Sea and Huanghai Sea. Acta Oceanol. Sin. 28: 156–164 (in Chinese).

Huang, Z.Q., T.B. Fu and Y.H. Zhang. 1997. Distribution characteristics of POC in sea water in East China Sea. J. Oceanogr. Taiwan Strait 16: 145–152 (in Chinese).

Hung, C.C. and G.C. Gong. 2007. Export flux of POC in the main stream of the Kuroshio. Geophys. Res. Lett. 34: L18606.

Hung, J.J., C.H. Chen, G.C. Gong, D.D. Sheu and F.K. Shiah. 2003. Distributions, stoichiometric patterns and cross-shelf exports of dissolved organic matter in the East China Sea. Deep-Sea Res. Part II Top. Stud. Oceanogr. 50: 1127–1145.

Hung, J.J., P.L. Lin and K.K. Liu. 2000. Dissolved and particulate organic carbon in the southern East China Sea. Cont. Shelf Res. 20: 545–569.

Iseki, K., K. Okamura and Y. Kiyomoto. 2003. Seasonality and composition of downward particulate fluxes at the continental shelf and Okinawa Trough in the East China Sea. Deep-Sea Res. Part II Top. Stud. Oceanogr. 50: 457–473.

Kirchman, D.L., Y. Suzuki, C. Garside and H.W. Ducklow. 1991. High turnover rates of dissolved organic carbon during a spring phytoplankton bloom. Nature 352: 612–614.

Leavitt, P.R. 1993. A review of factors that regulate carotenoid and chlorophyll deposition and fossil pigment abundance. J. Paleolimnol. 9: 109–127.

Lee, C., S.G. Wakeham and J.I. Hedges. 2000. Composition and flux of particulate amino acids and chloropigments in equatorial Pacific seawater and sediments. Deep-Sea Res. Part I 47: 1535–1568.

Li, N. and J.-T. Wang. 2011. Dissolved inorganic and organic carbon in the north of East China Sea (ECS) coastal waters in spring. Mar. Sci. 35: 5–10 (in Chinese).

Li, J.Y. and J. Zhang. 2003. Variations of solid content and water chemistry at Nantong station and weathering processes of the Changjiang watershed. Resour. Environ. Yangtze Basin 12: 363–369 (in Chinese).

Li, X.N., S.M. Liu, R.H. Lü, J. Zhang and L. Zou. 2004. An analysis of chlorophyll in the sediments of East China Sea and Yellow Sea. J. Ocean Univ. China 34: 603–610 (in Chinese).

Li, X.X., T.S. Bianchi, Z.S. Yang, L.E. Osterman, M.A. Allison, S.F. DiMarco and G.P. Yang. 2011. Historical trends of hypoxia in Changjiang River estuary: Applications of chemical biomarkers and microfossils. J. Mar. Syst. 86: 57–68.

Li, Y., L.J. Zhang, Z. Su, X.L. Wang and J.M. Pan. 2006. Rapid variation of pCO_2 and its influencing factors at the initial mixing stage of fresh water and saltwater in the Changjiang River Estuary. J. Ocean Univ. China 36: 295–298 (in Chinese).

Lin, J. 2007. Distributions of dissolved organic carbon and particulate organic carbon in the Changjiang Estuary and its adjacent area. MSc Dissertation, East China Normal University, Shanghai.

Lin, J., Y. Wu, J. Zhang, S.L. Yang and Z.Y. Zhu. 2007. Seasonal variation of organic carbon fluxes in the Yangtze River and influence of Three-Gorges engineering. China Environ. Sci. 27: 246–249 (in Chinese).

Lin, J., Y. Wu, J. Zhang and Z.Y. Zhu. 2009. Impact of turbidity maximum zone on organic carbon distribution in Changjiang Estuary. Channel Sci. 30: 151–159 (in Chinese).

Liu, W.C., R. Wang and C.L. Li. 1998. C/N ratios of particulate organic matter in the East China Sea. Oceanol. Limnol. Sin. 29: 467–470 (in Chinese).

Liu, X.C., H.T. Shen and Q.H. Huang. 2002. Concentration variation and flux estimation of dissolved inorganic nutrient from the Changjiang River into its estuary. Oceanol. Limnol. Sin. 33: 332–340 (in Chinese).

Liu, Z.L., T. Zhang, Y.L. Liu, J.M. Pan, Z.Y. Chen and X.Y. Liu. 2008. The distributions of chlorophyll a concentration in the water and the surface sediments in the Changjiang River estuary during spring of 2004. Mar. Sci. 26: 1–7 (in Chinese).

Ma, Y., S.M. Wang and H.X. Pan. 1996. Significance of diatom and pigment in the study on palaeoenvironmental evolution. J. Lake Sci. 8: 16–26 (in Chinese).

Mackey, M.D., D.J. Mackey, H.W. Higgins and S.W. Wright. 1996. CHEMTAX – a program for estimating class abundances from chemical markers: Application to HPLC measurements of phytoplankton. Mar. Ecol. Progr. Ser. 144: 265–283.

Mopper, K., X. Zhou, R.J. Kieber, D.J. Kieber, R.J. Sikorski and R.D. Jones. 1991. Photochemical degradation of dissolved organic carbon and its impact on the oceanic carbon cycle. Nature 353: 60–62.

Moran, M.A., W.M. Sheldon and J.E. Sheldon. 1999. Biodegradation of riverine dissolved organic carbon in five estuaries of the southeastern United States. Estuaries 22: 55–64.

Ning, X.R., J.X. Shi, L.M. Cai and C.G. Liu. 2004. Biological productivity front in the Changjiang Estuary and the Hangzhou Bay and its ecological effects. Acta Oceanol. Sin. 26: 96–106 (in Chinese).

Ogawa, H. and N. Ogura. 1992. Comparison of two methods for measuring dissolved organic carbon in sea water. Nature 356: 696–698.

Ogawa, H., T. Usui and I. Koike. 2003. Distribution of dissolved organic carbon in the East China Sea. Deep-Sea Res. Part II Top. Stud. Oceanogr. 50: 353–366.

Ogura, N. 1972. Rate and extent of decomposition of dissolved organic matter in surface seawater. Mar. Biol. 13: 89–93.

Olli, K., W.R. Riser, P. Wassmann, T. Ratkova, E. Arashkevich and A. Pasternak. 2002. Seasonal variation in vertical flux of biogenic matter in the marginal ice zone and the central Barents Sea. J. Mar. Syst. 38: 189–204.

Pan, L.A., L.H. Zhang, J. Zhang, J.M. Gasol and M. Chao. 2005. On-board flow cytometric observation of picoplankton community structure in the East China Sea during the fall of different years. FEMS Microbiol. Ecol. 52: 243–253.

Peng, M.C. and D.X. Hu. 1997. Outflow of suspended materials from Changjiang River Mouth, Chinese J. Oceanol. Limnol. 15: 258–263.

Peng, T.H., J.J. Hung, R. Wanninkhof and F.J. Millero. 1999. Carbon budget in the East China Sea in spring. Tellus B 51: 531–540.

Qu, W.C., B. Xue, M.D. Dickman, S.M. Wang, C.X. Fan, R.J. Wu, P.Z. Zhang, J.F. Chen and Y.H. Wu. 2000. A 14000-year record of paleoenvironmental change in the western basin of China's third largest lake, Lake Taihu. Hydrobiologia 432: 113–120.

Reeburgh, W.S. 2007. Oceanic methane biogeochemistry. Chem. Rev. 107: 486–513.

Ren, J.L., J. Zhang, D.D. Li, Y. Cheng and S.M. Liu. 2010. Behavior of dissolved inorganic arsenic in the Yellow Sea and East China Sea. Deep-Sea Res. Part II Top. Stud. Oceanogr. 57: 1035–1046.

Reuss, N. 2005. Sediment pigments as biomarkers of environmental change. Ph.D. Dissertation, National Environmental Research Institute, Roskilde, Denmark.

Shen, H.T. and D.A. Pan. 2001. Turbidity Maximum Zone of Changjiang Estuary. Beijing: China Ocean Press (in Chinese).

Shim, J., D. Kim, Y.C. Kang, J.H. Lee, S.T. Jang and C.H. Kim. 2007. Seasonal variations in pCO_2 and its controlling factors in surface seawater of the northern East China Sea. Cont. Shelf Res. 27: 2623–2636.

Song, S.Q., J. Sun and Z.M. Yu. 2009. Vertical pattern of chlorophyll a in the Yangtze River estuary and its adjacent waters. Chin. J. Plant Ecol. 33: 369–379 (in Chinese).

Steele, J.H. and I.E. Baird. 1962. Further relations between primary production, chlorophyll, and particulate carbon. Limnol. Oceanogr. 7: 42–47.

Su, J.L. 1998. Circulation dynamics of the China Seas north of 18 °N. pp. 483–505. In: A.R. Robinson and K.H. Brink (eds.), The Sea, Vol. 11. New York: John Wiley & Sons.

Sun, J. and W. Tian. 2011. Phytoplankton in Yangtze River estuary and its adjacent waters in spring in 2009: Species composition and size-fractionated chlorophyll a. Chin. J. Applied Ecol. 22: 235–242 (in Chinese).

Tang, D., B. Di, G. Wei, I.H. Ni, I. Oh and S. Wang. 2006. Spatial, seasonal and species variations of harmful algal blooms in the South Yellow Sea and East China Sea. Hydrobiologia 568: 245–253.

Tian, L.X. 2008. Distributions of organic carbon in the East China Sea, Yellow Sea, northeastern Hainan rivers and coasts. MSc Dissertation, East China Normal University, Shanghai.

Tsunogai, S., S. Watanabe, J. Nakamura, T. Ono and T. Sato. 1997. A preliminary study of carbon system in the East China Sea. J. Oceanogr. 53: 9–17.

Tsunogai, S., S. Watanabe and T. Sato. 1999. Is there a "continental shelf pump" for the absorption of atmospheric CO_2? Tellus B 51: 701–712.

Tsurushima, N., S. Watanabe and S. Tsunogai. 1996. Methane in the East China Sea water. J. Oceanogr. 52: 221–233.

Wang, B.D. 2006. Cultural eutrophication in the Changjiang (Yangtze River) plume: History and perspective. Estuar. Coast. Shelf Sci. 69: 471–477.

Wang, F.S., Y. Wang, J. Zhang, H. Xu and X. Wei. 2007. Human impact on the historical change of CO_2 degassing flux in river Changjiang. Geochem. Trans. 8: 7.

Wang, M.Y. 1989. Research on carbon, nitrogen, phosphorus and sulfur transportation from Changjiang. p. 122–131. In: Background Values Study of Chemical Elements in Water Environment. Beijing: Press of Beijing.

Wang, S.-L., C.-T.A. Chen, G.-H. Hong and C.-S. Chung, 2000. Carbon dioxide and related parameters in the East China Sea. Cont. Shelf Res. 20: 525–544.

Wang, Y.Z., H.O. Qiu, A.H. Peng, Z.Y. Yang, M. Yang and Y.S. Gu. 2006. Determination of pigments and their environmental interpretations in the sediments of Zhangdu Lake in Wuhan. Chin. J. Anal. Lab. 25: 32–35 (in Chinese).

Watts, C.D., Maxwell, J.R and H. Kjosen. 1977. The potential of carotenoids as environmental indicators. In: R. Campos and J. Goñi (eds.), Advances in Organic Geochemistry, pp. 391–413. Madrid: Enadimsa.

West, A.J., C.W. Lin, T.C. Lin, R.G. Hilton, S.H. Liu, C.T. Chang, K.C. Lin, A. Galy, R.B. Sparkes and N. Hovius. 2011. Mobilization and transport of coarse woody debris to the oceans triggered by an extreme tropical storm. Limnol. Oceanogr. 56: 77–85.

Wong, G.-T.-F., S.-Y. Chao, Y.-H. Li and F.-K. Shiah. 2000. The Kuroshio edge exchange processes (KEEP) study an introduction to hypotheses and highlights. Cont. Shelf Res. 20: 335–347.
Wu, G.Y., X.A. Liu and D. Wan. 2010. The temporal and spatial distribution characteristic of algae blooms in the Daninghe River. Environ. Monitor. China 26: 69–74 (in Chinese).
Wu, S.C. and W.M. Chen. 2004. The periodic change of phytoplankton biomass in Taihu Lake. China Environ. Sci. 24: 151–154 (in Chinese).
Wu, Y., J. Zhang, S.M. Liu, Z.F. Zhang, H.T. Chen and H. Xiong. 2007. Sources and distribution of carbon within the Yangtze River system. Estuar. Coast. Shelf Sci. 71: 13–25.
Wu, Y., J. Zhang, D.J. Li, H. Wei and R.X. Lu. 2003. Isotope variability of particulate organic matter at the PN section in the East China Sea. Biogeochemistry 65: 31–49.
Yan, W.J. and S. Zhang. 2003. How do nitrogen inputs to the Changjiang basin impact the Changjiang River nitrate: A temporal analysis for 1968–1997. Global Biogeochem. Cycles 17: 1091.
Yang, S.L., M. Li, S.B. Dai, Z. Liu, J. Zhang and P.X. Ding. 2006. Drastic decrease in sediment supply from the Yangtze River and its challenge to coastal wetland management. Geophys. Res. Lett. 33: L06408.
Yang, S.L., J. Zhang and X.J. Xu. 2007. Influence of the Three Gorges Dam on downstream delivery of sediment and its environmental implications, Yangtze River. Geophys. Res. Lett. 34: L10401.
Zang, J.Y. 1998. Dissolved methane in the sea water of the East China Sea, Acta Oceanol. Sin. 20: 52–59 (in Chinese).
Zeng, T.H., G.X. Liu and Z.Y. Hu. 2011. Estimation of phytoplankton primary production of lakes in the middle and lower reaches of the Yangtze River. Resour. Environ. Yangtze Basin 20: 717–722 (in Chinese).
Zhai, W. and M. Dai. 2009. On the seasonal variation of air-sea CO_2 fluxes in the outer Changjiang (Yangtze River) Estuary, East China Sea. Mar. Chem. 117: 2–10.
Zhai, W., M. Dai and X. Guo. 2007. Carbonate system and CO_2 degassing fluxes in the inner estuary of Changjiang (Yangtze) River, China. Mar. Chem. 107: 342–356.
Zhang, G.L., J. Zhang, S.M. Liu, J.L. Ren, J. Xu and F. Zhang. 2008a. Methane in the Changjiang (Yangtze River) Estuary and its adjacent marine area: Riverine input, sediment release and atmospheric fluxes. Biogeochemistry 91: 71–84.
Zhang, G.L., J. Zhang, J.L. Ren, J.B. Li and S.M. Liu. 2008b. Distributions and sea to air fluxes of methane and nitrous oxide in the North East China Sea in Summer. Mar. Chem. 110: 42–55.
Zhang, G.L., J. Zhang, Y.B. Kang and S.M. Liu. 2004. Distributions and fluxes of dissolved methane in the East China Sea and the Yellow Sea in spring. J. Geophy. Res. 109: C07011.
Zhang, J., Z.F. Zhang, S.M. Liu, Y. Wu, H. Xiong and H.T. Chen. 1999. Human impacts on the large world rivers: Would the Changjiang (Yangtze River) be an illustration? Global Biogeochem. Cycles 13: 1099–1105.
Zhang, L., Z. Liu, J. Zhang, G.H. Hong, Y. Park and H.F. Zhang. 2007. Reevaluation of mixing among multiple water masses in the shelf: An example from the East China Sea. Cont. Shelf Res. 27: 1969–1979.
Zhao, J.S., H.W. Ji and Z.G. Guo. 2003. The vertical distribution of particulate organic carbon in the typical areas of the East China Sea in winter. Mar. Sci. 27: 59–63 (in Chinese).
Zhao, J., G.L. Zhang, Y. Wu and J. Yang. 2011a. Distributions and fluxes of methane in the Changjiang. Environ. Sci. 32: 18–25 (in Chinese).
Zhao, Y.C., G.L. Zhang and S.M. Liu. 2011b. Distributions and fluxes of methane in the Changjiang Estuary at the end of August 2009. Mar. Sci. 35: 81–89 (in Chinese).
Zhou, M.J., T. Yan and J.Z. Zou. 2003. Preliminary analysis of the characteristics of red tide areas in Changjiang River estuary and its adjacent area. Chin. J. Applied Ecol. 14: 1031–1038 (in Chinese).

Zhou, M.J., Z.L. Shen and R.C. Ren. 2008. Responses of a coastal phytoplankton community to increased nutrient input from the Changjiang (Yangtze) River. Cont. Shelf Res. 28: 1483–1489.

Zhu, Z.Y., J. Zhang, Y. Wu and J. Lin. 2006. Bulk particulate organic carbon in the East China Sea: Tidal influence and bottom transport. Prog. Oceanogr. 69: 37–60.

Zhu, Z.Y., W.M. Ng, S.M. Liu, J. Zhang, J.C. Chen and Y. Wu. 2009. Estuarine phytoplankton dynamics and shift of limiting factors: A study in the Changjiang (Yangtze River) Estuary and adjacent area. Estuar. Coast. Shelf Sci. 84: 393–401.

11

Dynamics of phytoplankton blooms and nutrient limitation in the Pearl River (Zhujiang) estuarine coastal waters

K. Yin, J. Xu, Z. Lai, and P. J. Harrison

1. Introduction

The Pearl River Delta region (Fig. 11.1) is located on the south coast of China and is one of the most developed regions in China in recent years. The region, including Hong Kong, has 26.5 million people and many primary and secondary industries. The Pearl River receives a high load of anthropogenic nutrients from increased activities in agriculture (Neller and Lam 1994), fish dike farming (Ruddle and Zhong 1988), and sewage effluent (Hills et al. 1998; Dai et al. 2006; Lu et al. 2009) due to the population increase and economic development in southern China and the Pearl River delta region (Lin 1997; Xu et al. 2012).

The Pearl River is the second largest river in China next to the Changjiang (Yangtze) River, and it is the 13th largest river in the world by discharge volume. The Pearl River stretches for 2,214 km and drains an area of 452,000 km^2. The Pearl River consists largely of three rivers: the Xi Jiang (West River), Bei Jiang (North River), and the Dong Jiang (East River). The three rivers branch into many small rivers in a network and merge into eight river "gates" before entering the South China Sea. The yearly average Pearl River discharge is 10,520 m^3 s^{-1}, with 20% occurring during the dry season in October–March and 80% during the wet season in April–September, with the monthly maximum in July reaching as high as 22,200 m^3 s^{-1} (Yin et al. 2000). Four of the river gates, which enter into the Pearl River Estuary (PRE), discharge about 53% of the total freshwater coming from the Pearl River (Lee et al. 2006; Harrison et al. 2008; Chapter 13).

Concentrations and loadings of nutrients (nitrogen and phosphorus) have increased to high levels during the last few decades. The annual mean total nitrogen concentration is 164 μmol L^{-1}, with dissolved inorganic nitrogen (DIN), particulate nitrogen (PN) and dissolved organic nitrogen (DON) being 114, 19, and 31 μmol L^{-1}, respectively, and 3.2 and 0.7 μmol L^{-1} for total phosphorus (TP) and orthophosphate (DIP), respectively. The annual fluxes from the Pearl River to the coastal waters of the South China Sea are 4.1×10^{10} mol y^{-1} for TN, with DIN, DON, and PN being 2.9×10^{10}, 0.69×10^{10}, and 0.51×10^{10} mol y^{-1}, respectively, and 7.7 and 1.8×10^{8} mol y^{-1} for TP and DIP (Lu et al. 2009). The Pearl River delta receives high rainfall, ranging between 1,600 and 2,300 mm. Zhang et al. (1999) estimated that 1 mm of rainfall carries about 2×10^{4} mol NH_4-N, 1×10^{4} mol NO_3-N, and about 4.8–6.9 $\times10^{7}$ mol DIN.

Anthropogenic loading of nutrients in rivers has increased in many coastal waters over the last few decades. The increased nutrients significantly promote primary productivity in receiving coastal

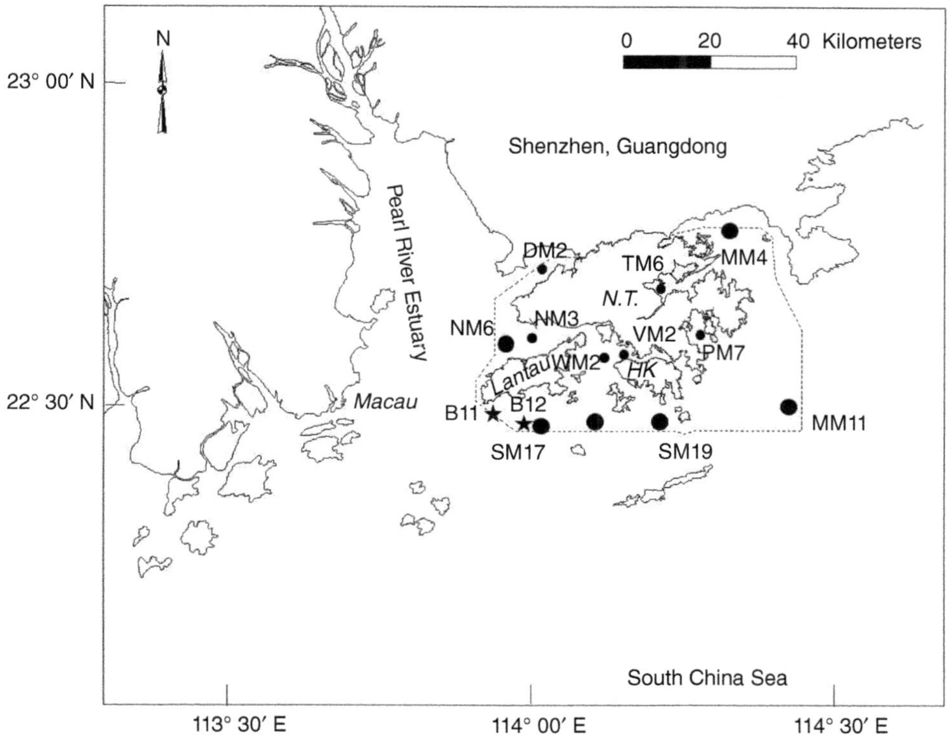

Figure 11.1. Map showing the main Pearl River estuary. On its east side are the Hong Kong territorial waters surrounded by the dashed line. DM2, NM6, NM3, SM17, SM19, MM11, and MM4 are sampling stations regularly monitored by the Hong Kong Environmental Protection Department. B11 and B12 are sampling stations for the cruise conducted in 1998.

waters and cause algal blooms and hypoxic waters in the bottom (Cloern 2001). The number and the extent of dead zones (hypoxic bottom waters) have increased in coastal waters around the world (Diaz and Rosenberg 2008). The typical example of a dead zone is in the northern Gulf of Mexico, which has been receiving an increasing load of anthropogenic nutrients from the Mississippi River (Turner et al. 2005; Chapter 16). Large-scale harmful algal blooms (HABs) and hypoxic waters have also become a seasonal phenomenon in the Changjiang River estuary and East China Sea (Zhou et al. 2008). In contrast, in the Mekong River, anthropogenic impacts are less obvious. Nutrients are much lower than the Pearl River with no strong seasonal cycle, and productivity and chl *a* are low (1–4 $\mu g\ L^{-1}$) (Chapter 12). However, not every estuarine coastal system receiving anthropogenic nutrients exhibits eutrophication symptoms (Cloern 2001). In the Pearl River estuary and adjacent coastal waters, events such as algal blooms and bottom hypoxic waters usually occur episodically and on a small scale, suggesting that the degree of eutrophication is not as severe as expected compared with other large estuarine coastal waters such as the Changjiang River–East China Sea system and the Mississippi River–Northern Gulf of Mexico system (Chapter 16).

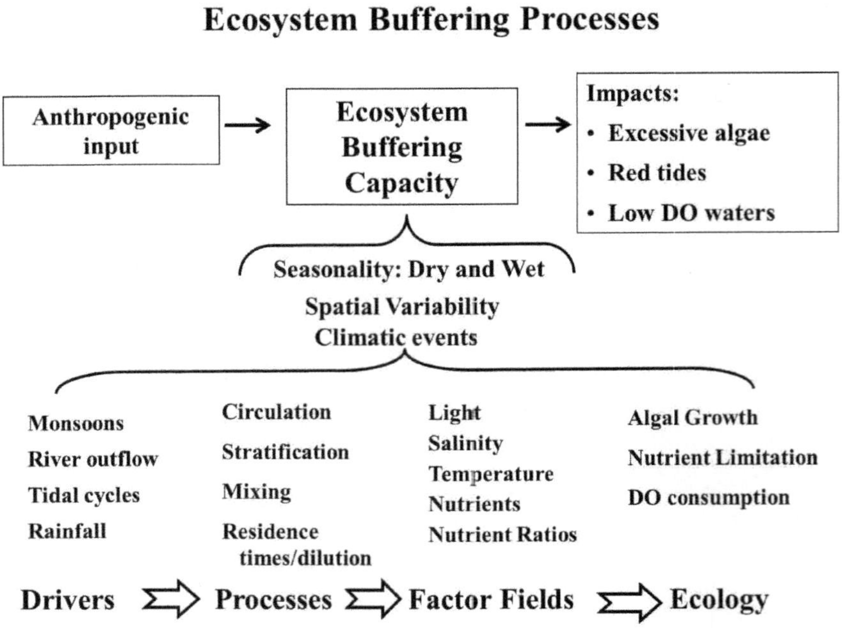

Figure 11.2. Schematic showing the ecosystem-buffering processes.

An estuary has an environmental assimilative capacity to buffer against enrichment of nutrients in the estuary, as illustrated in Fig. 11.2. We refer to it as the "buffering capacity." Just as a chemical buffer system has an extensive capacity to reduce changes before it finally collapses, an estuarine ecosystem has an ecosystem buffering capacity to resist eutrophication symptoms/impacts such as algal blooms and hypoxia. The capacity for the level of phytoplankton biomass and consumption of dissolved oxygen is defined by the most limiting nutrient; however, the buffering processes are regulated by physical drivers and processes that determine the distribution fields of growth factors and biological activities (Fig. 11.2). In the Pearl River estuarine coastal system, the buffering capacity varies seasonally and is driven by winds, freshwater outflow, and tides as well as climate variability, through physical processes of circulation, mixing/stratification, and dilution. The physical processes determine the distribution (field) of light conditions and nutrients and other chemical constituents, and in turn, they drive phytoplankton growth and bloom dynamics and microbial activities such as bacterial respiration and subsequently dissolved oxygen in the bottom water.

2. General geographic and oceanographic settings of the Pearl River estuary

The Pearl River estuary (PRE) is a subtropical region in the northwestern part of the South China Sea and forms part of the western waters of Hong Kong. Thus the estuarine coastal system is directly influenced by three water regimes: freshwater outflow from the Pearl River, coastal waters from the China Coastal Current from the northeast, and oceanic waters from the South China Sea. The Kuroshio current invades through Luzon Strait and moves westward into the South China Sea. In

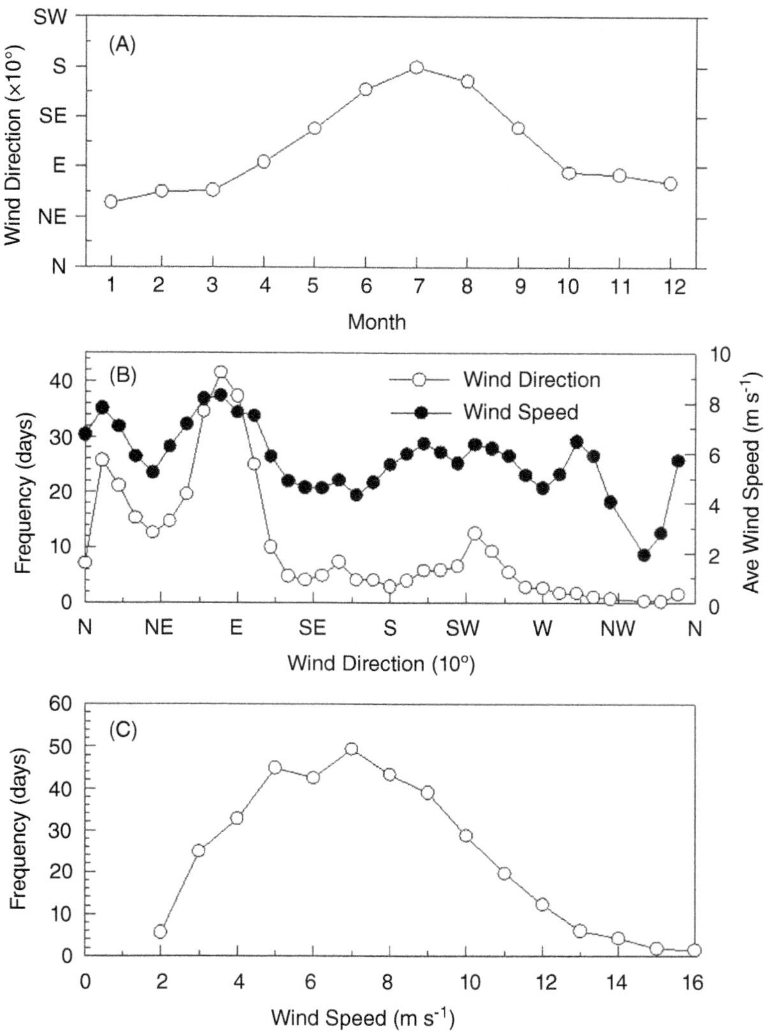

Figure 11.3. (A) Monthly mean wind direction during 1990–2000; (B) Mean frequency (days) and wind speed and wind directions, (C) mean frequency (days) of the wind speed.

winter when the invading Kuroshio current meets the continental shelf break, it turns to the south, forming a southward cold water tongue (Yin et al. 1999). Between the Kuroshio and China Coastal Current, there is the South China Sea warm current (Guan 1994), which is supported by geological records (Liu et al. 2010).

Seasonal monsoons and the Pearl River discharge as well as rainfall are important in determining coastal currents, circulation, stratification, mixing, and residence time in the Pearl River estuary and adjacent coastal waters (Lee et al. 2006; Kuang et al. 2011; Chapter 13). Northeasterly monsoon winds prevail in winter and southwesterly winds blow in summer, as shown in Fig. 11.3A. The frequency

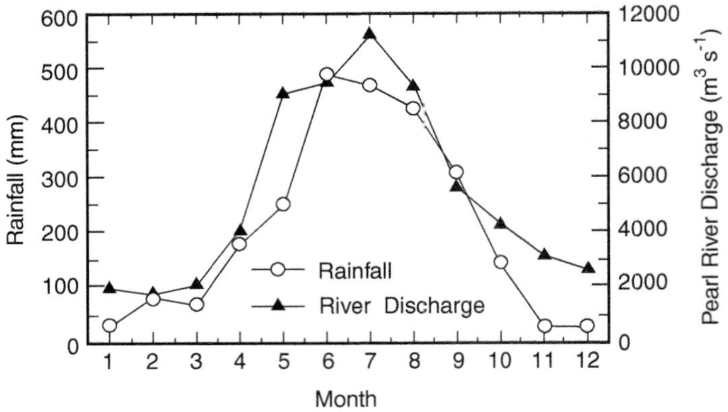

Figure 11.4. Rainfall during 10 years (1990–2000) in Hong Kong and freshwater outflow of the Pearl River.

distribution of the wind direction is not balanced over the year, with easterly winds prevailing most of the year (Fig. 11.3B). Westerly and northwesterly winds on average are rare events. Winds are stronger in the eastern and northern (10°) direction than in other directions, ranging mostly between 4 and 10 m s^{-1} (Fig. 11.3C).

There is a rainy season during spring and summer in the region. Mean annual rainfall during 1990–1998 was 2,480 mm, with 2,110 mm (85%) falling during April–September and maximal rainfall occurs in June, July, and August (Fig. 11.4). Rainfall not only adds freshwater to the water column and strengthens the stratification, but it also adds nutrients directly to the photic zone.

The PRE and coastal waters are very dynamic and complex due to advection, mixing, and the interaction of freshwater from the PRE, typhoons in summer and early fall, runoff, and oceanic water from summer upwelling (Su, 2004; Harrison et al. 2008). These processes have a large influence on the spatial and temporal variability of the water column structure and water quality (Chapter 13).

3. Processes in estuarine and coastal waters

One of the most important factors buffering against eutrophication appears to be the exchange of water with the open ocean (Jickells 1998; National Research Council 2000; Cloern 2001). When the estuarine water is frequently exchanged with the open ocean or flushes rapidly, the estuarine ecosystem appears to be more resistant to nutrient over-enrichment, whereas estuaries with low exchange rates or flushing rates appear to be particularly vulnerable to eutrophication. When water exchange and mixing are slower than biological activities (e.g. growth, productivity, and respiration), the dynamics of the most limiting nutrient become an important mechanism controlling the amount of algal biomass produced. Climate change often results in a change in ocean circulation, exchange between the open ocean and coastal waters, and stratification/mixing, which leads to a change in residence time (Harley et al. 2006).

The Pearl River estuarine coastal waters progress through seasonal changes in water masses, circulation, and stratification owing to seasonal freshwater outflow and monsoonal winds. In winter, the China coastal current brings coastal waters from the northeast, which becomes dominant in the PRE coastal waters (Williamson 1970; Watts 1973; Liu et al. 2010), as observed in satellite images as well (Yin et al. 1999; Gan et al. 2009). Because of the NE monsoon winds, there is downwelling of coastal waters along the south China coast (Liu et al. 2010). In summer, the coastal water is subjected to upwelling owing to the southwest monsoon winds and strong estuarine circulation due to Pearl River outflow (Yin 2002, 2003). Therefore, residence times of coastal waters are shorter in summer than in winter because of the combined effects of upwelling and estuarine circulation. It appears that seasonal monsoons serve as a mixing and flushing mechanism in two ways: (1) reducing potential seasonal eutrophication effects during summer by diluting the N-rich surface riverine water with N-low deep water during summer upwelling; and (2) reducing long-term (years) accumulation of organic matter in the sediments owing to nutrient enrichment in the region by transport and dispersion offshore (Yin 2002; Liu et al. 2009).

The Pearl River outflow results in a typical estuarine circulation with an estuarine plume flowing outward at the surface and a salt wedge (oceanic waters) invading at the bottom. This estuarine-like circulation is confined within the Pearl River estuary in winter and extends to the open adjacent coastal areas in the northern South China Sea in summer (Dong et al. 2004). The southwest monsoon in summer results in upwelling and strengthens this estuarine-like circulation in coastal waters beyond the estuary (Harrison et al. 2008; Gan et al. 2009; Chapter 13). At the surface, there are the riverine, estuarine, and coastal plumes. The riverine plume is a thin freshwater layer resulting directly from the river outflow, which is regulated by tidal cycles, being dammed/restrained during higher high water and released during lower low water. The estuarine plume is a mixture of freshwater and seawater when it is in the estuary (geographically bounded by the land margin between Hong Kong and Macau). As the estuarine plume extends into the coastal waters of the northwestern South China Sea, it covers a relatively large area, including the eastern Hong Kong coastal waters (Watts 1971; Wang et al. 2002; Gan et al. 2009, 2010). This water body is referred to as the coastal plume. The riverine, estuarine, and coastal plumes are separated by riverine, estuarine, or coastal fronts (Dong et al. 2004). The size and extent of the plumes vary with freshwater discharge and winds. In winter, when river outflow is small and winds are strong, the distinctions between these plumes are not obvious: the estuarine plume is small and is confined within the estuary, and thus a coastal plume may be absent. In summer, the riverine plume dominates on the eastern side of the Pearl River estuary; the estuarine plume tends to move out along the east side of the estuary due to SW monsoon winds and the Coriolis effect and turns to the east, and it becomes the coastal plume (Wong et al. 2003, 2004; Lee et al. 2006). Driven by the SW monsoon, the coastal plume continues to move eastward, covering a large area of the northern South China Sea and moving across Hong Kong southern waters to the east of Hong Kong (Gan et al. 2009, 2010; Shu et al. 2011; Chapter 13).

Physical conditions of the water column clearly show marked spatial and temporal differences based on regular monitoring stations in Hong Kong (Fig. 11.1; Environmental Protection Department 2010). The coastal plume spreads out in a west-to-east direction from the Pearl River estuary to the eastern waters of Hong Kong and produces a pronounced gradient in salinity and nutrients along a transect from the estuarine-coastal plume to oceanic waters, especially in summer. Temporal and

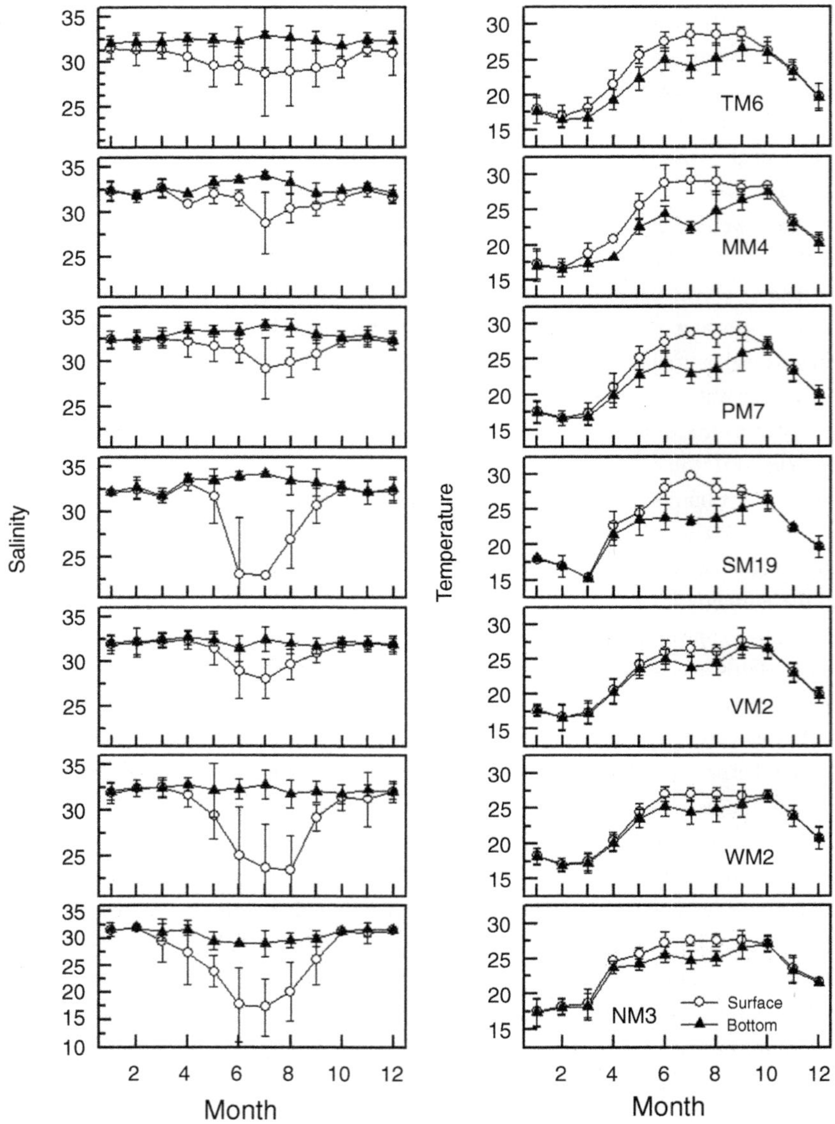

Figure 11.5. Average monthly change in salinity and temperature at 7 stations during 1990–2000 (data from EPD water quality monitoring program).

spatial gradients are evident in salinity. Surface salinity decreases to the lowest values during July in Hong Kong waters, and it is lower in the west than in the east (Fig. 11.5). Surface temperature increases to a maximum in June and remains relatively constant through July and August as well as September. Bottom salinity increases to maximum in July at all stations except for NM3. An increase in salinity in the bottom water is due to upwelling in these semi-enclosed bays (TM6, MM4, and

PM7) and coastal waters (SM19, VM2, and WM2), whereas a decrease in salinity at the surface in July (Fig. 11.5) is due to high rainfall. Haline stratification (a difference in salinity of >1.0 between the surface and bottom) starts in March in the Pearl River estuary (NM3), one month later (in April) in eastern waters (TM6, MM4, PM6) and southern waters (SM19), and even later in June near Victoria Harbour (VM2). Thermal stratification (a difference in temperature of >1°C) starts two months later than haline stratification in May at NM3 and in June at WM2, but earlier in March in the eastern waters (TM6 and MM4), in April at PM7 and SM19 and in June at VM2 (Fig. 11.5). The strength of the haline or thermal stratification indicates the stability of the water column, which influences the formation of phytoplankton blooms.

4. Spatial and seasonal variability of phytoplankton biomass

The Pearl River discharge of freshwater determines nutrient concentrations and potential phytoplankton biomass in the estuarine-influenced waters because the northern South China Sea is oligotrophic. In an estuary, there is usually a regional maximum of phytoplankton biomass (Mann and Lazier 2006), which is important in determining not only the spatial distribution of the exhaustion of nutrients, but also the spatial fate of phytoplankton biomass, which sinks and consumes dissolved oxygen in the bottom water (Lohrenz et al. 1997). There is often a progression of the phytoplankton maximum downstream or upstream of the estuarine/coastal plumes associated with seasonal changes in river flow. Upstream in the Pearl River estuary, the dilution rate is too fast to allow phytoplankton blooms, and in addition, phytoplankton growth rates are often limited by light in these turbid waters. As the river outflow spreads horizontally downstream along the widened estuary or into open coastal waters, the dilution rate decreases, the flow velocity decreases, and suspended sediments settle out, which improves light penetration. As a result, phytoplankton blooms can occur in a region downstream and form a regional phytoplankton maximum. The region of the phytoplankton maximum is often where a nutrient becomes depleted (other nutrients are still in excess) and limits the level of algal biomass at the phytoplankton maximum. Spatial and temporal variability of the regional maximum is largely controlled by various physical forces such as river outflow, tidal cycles, and wind mixing in estuaries (Cloern 1996).

4.1. River outflow controlled variability

River flow regulates dilution rate, light penetration, and nutrient concentrations. In the PRE, when the freshet starts during the rainy season, the riverine and estuarine plumes dominate (Yin 2002; Lee et al. 2006), the dilution rate is high in the estuary, and the euphotic zone (1% light depth) is shallow (Yin et al. 2004). As a result, phytoplankton growth is light-limited, and they grow slowly relative to the dilution and vertical mixing and cannot accumulate in the euphotic zone in the estuary, even though nutrients are high. Therefore, there is no phytoplankton bloom in the estuary. As the estuarine plume moves out into the open coastal waters, it becomes the coastal plume. The horizontal dilution decreases, and the water column becomes more stable. The light penetration deepens as salinity increases (Yin et al. 2004). As a result, the phytoplankton usually bloom at intermediate salinities, and the most limiting nutrient determines the phytoplankton biomass. Phytoplankton productivity is dependent on light penetration across the coastal plume. In June 1998, the regional maximum of

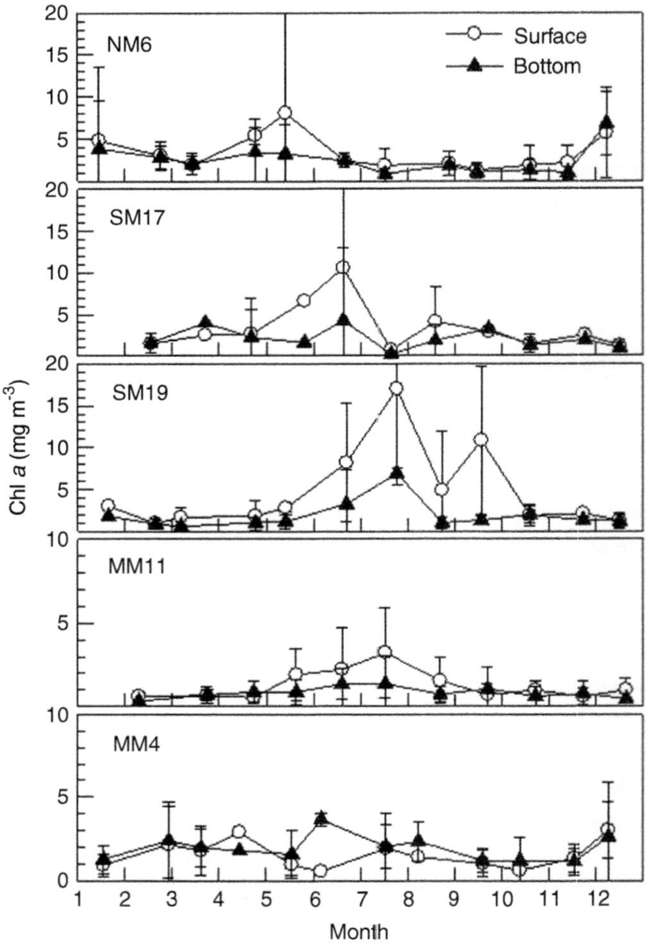

Figure 11.6. Monthly average chl *a* in Hong Kong waters at 5 stations during 1990–2000.

chlorophyll was in waters south of Lantau Island, but progressed to waters south of Hong Kong Island by July (Yin et al. 2000). Therefore, the largest blooms tend to occur in waters south of Hong Kong Island (Yin et al. 2001; Zhou et al. 2012). Analysis of a 12-year time series of chl *a* indicates that there is a temporal and spatial progression of the chl *a* maximum around Hong Kong (Yin 2002, 2003). Chl *a* concentrations in the PRE (at NM6) are low during March and increase during April and May, remain <3 μg L^{-1} during June–November, and increase during December and January, with the highest monthly value being <10 μg L^{-1} (Fig. 11.5) because residence time increases in the dry season. In the waters south of Hong Kong (at SM17), chl *a* concentrations increase during May and June (ca. 10 μg L^{-1}), one month later than in the estuary at NM6, and remain <3 μg L^{-1} during other months. However, the increase in chl *a* occurs later (June and July) at SM19, and the maximum chl *a* is >15 μg L^{-1} (Fig. 11.6). Chl *a* concentrations at MM11 in the eastern waters reach a maximum around July, whereas the chl *a* concentrations in the inner eastern waters (MM4) that are not influenced by the PR remain low <3 $\mu g\ L^{-1}$ and do not show the same fluctuations observed at other stations (Fig. 11.6). In many instances, the chl *a* concentrations at the bottom increase at the same time as chl *a* at the surface, probably because of vertical mixing and sinking of blooms.

The appearance of the regional maximum in phytoplankton biomass and productivity in the vicinity of the PRE agrees with observations in many other estuaries. In the Mississippi River, the mean monthly discharge is minimal in September, starts to increase rapidly in January-February, and reaches a maximum in April because of snow melting (Ho and Barrett 1977). In the Mississippi River estuary and northern Gulf of Mexico, there is a "down plume" progression of phytoplankton biomass and productivity, with the displacement of the plume further away from the river mouth as the river discharge increases. The chl *a* maximum in fall (e.g., October 1983) occurred in the Mississippi estuary when river flow was very low (Fox et al. 1987) and moved to the river mouth in the spring (February 19–March 10, 1991) (Hitchcock and Whitledge 1992). The maximum primary productivity was also near the river mouth in April 1988 but was pushed along with the chl *a* maximum further downstream in summer (July 17–August 10, 1990) as the maximum of the Mississippi River discharge persisted from early April to late June (Lohrenz et al. 1992; Chapter 16). Typically, maximum phytoplankton biomass and productivity occur at intermediate salinities and coincide with the non-conservative decrease in nutrients along the salinity gradient (Lohrenz et al. 1999). River-controlled dynamics of phytoplankton blooms have been well studied in the Chesapeake Bay estuary. The mean annual cycle of freshwater flow in the Susquehanna River is characterized by a spring maximum during March–April and a summer minimum during August–September (Schubel and Pritchard 1986). As a result, the chl *a* and primary productivity maxima were found to retreat upstream between March and June (Fisher et al. 1988; Malone et al. 1988). Evidence that river flow regulates the progression of phytoplankton biomass and productivity can also be found in many other estuaries, including San Francisco Bay (Cloern et al. 1983, 1985; Peterson et al. 1985), the Hudson River estuary (Malone 1977; Malone and Chervin 1979; Bowman and Iverson 1978), the St. Lawrence estuary (de Lafontaine et al. 1991), the Adriatic Sea receiving Po River discharge (Revelente and Gilmartin 1976), and the Mekong River and estuary (Chapter 12). Recent studies also showed very high biomass of chl *a* and high POC in the plume of the Changjiang River in the East China Sea (Gong et al. 1996; Chapter 10).

4.2. Tidal cycle regulated variability

There are several effects of tidal cycles in estuaries. Tidal ranges in the Pearl River estuary are small, about 1–2 m (Mao et al. 2004; Lee et al. 2006). The salt wedge moves up and down with the tidal excursion. Phytoplankton that sink to depth downstream can be brought back into the estuary in the salt wedge and entrained up into the surface layer and act as seed populations during estuarine circulation and upwelling in summer. Hence, a spatial cycle occurs in the estuary (Yin et al. 1995a, 1995b). Higher chl *a* biomass was observed at the bottom than at the surface in the middle of the estuary during the flood tide (Yin et al. 2004). Higher phytoplankton biomass below the euphotic zone at the bottom of the estuary has been observed in other estuaries, for example, in the Chesapeake

Bay (Malone et al. 1988). This tidally regulated estuarine circulation of phytoplankton biomass has implications not only for phytoplankton production, but also for nutrient regeneration and DO consumption, which contributes to the formation of low oxygen bottom water in an estuary (Cloern 1996).

4.3. Wind event driven variability

Winds are important in influencing the water column structure, the circulation, and primary productivity in the estuary (Mann and Lazier 2006; Yin et al. 1997; Huisman et al. 1999). The coastal plume in open waters moves horizontally depending on the wind direction. Under southwest monsoon winds in summer, the coastal plume moves eastward across the waters of Hong Kong, which significantly influences these waters. However, when westerly winds blow, they push the coastal plume westward, holding the estuarine plume in the estuary, and at the same time, cause vertical mixing in the water column. As a result, the estuarine plume may stay in the estuary until the water level pressure builds up to force an outflow or the wind relaxes. Water column mixing produces a deeper mixed layer depth. When northeast to easterly winds were strong during July 22–26, 1999, a high phytoplankton biomass was observed in the deeper mixed layer west of Lantau Island owing to the retention of the estuarine plume inside the estuary, which allowed phytoplankton to bloom. Such a wind-induced episodic event also occurred during July 2000. After a strong east wind that blew at 7.2–10.5 m s^{-1}, the water column was homogeneously mixed one day later in waters south of Hong Kong and eastward (Yin et al. 2004).

Two typhoons passed over Hong Kong waters and provided an opportunity to document the changes in nutrients and chl *a* in late summer 2003 (Zhou et al. 2012). Before the typhoon, there was a pronounced algal bloom (chl *a* >10 μg L^{-1}) with the near depletion of phosphate and silicate in Lamma Channel. After the typhoon, chl *a* decreased by five-fold because of strong wind-induced vertical mixing, dilution, and rapid flushing. As a result of the heavy torrential rainfall, the increase in the Pearl River discharge quickly reset the typical summer strong haline stratification. A new algal (diatom) bloom was produced a few days after the typhoon when vertical mixing in the water column was reduced. As the river discharge and rainfall brought high nitrogen to the area, the new algal bloom became P and Si limited. Hence, the mixing and dilution effects of typhoons are relatively short-lived in coastal ecosystems that are dominated by high river discharge. Therefore, a typhoon represents a quick mixing and flushing event, and it can be added to the list of other factors such as spring tidal mixing, the Pearl River discharge, summer upwelling, and shallow water depth that make Hong Kong waters relatively resistant to eutrophication impacts (Zhou et al. 2012).

Phytoplankton biomass and productivity were high in the Pearl River estuary and its adjacent coastal waters, comparable with other large eutrophic estuarine waters. In the northern Gulf of Mexico, which receives discharge from the eutrophic Mississippi River, high phytoplankton biomass and productivity were observed to be >30 mg chl *a* m^{-2} and 10 g C m^{-2} d^{-1} in April, May, and July–August (Lohrenz et al. 1999). At the continental shelf beyond the Amazon River, the maximum phytoplankton biomass and productivity were 25 μg chl *a* L^{-1} and 8 g C m^{-2} d^{-1} respectively. In the Chesapeake Bay, the maximal chl *a* biomass was 126 mg m^{-2} in the euphotic zone, but was extremely high in the whole water column (>2,000 mg chl *a* m^{-2}) (Malone et al. 1988). The maximum in primary productivity

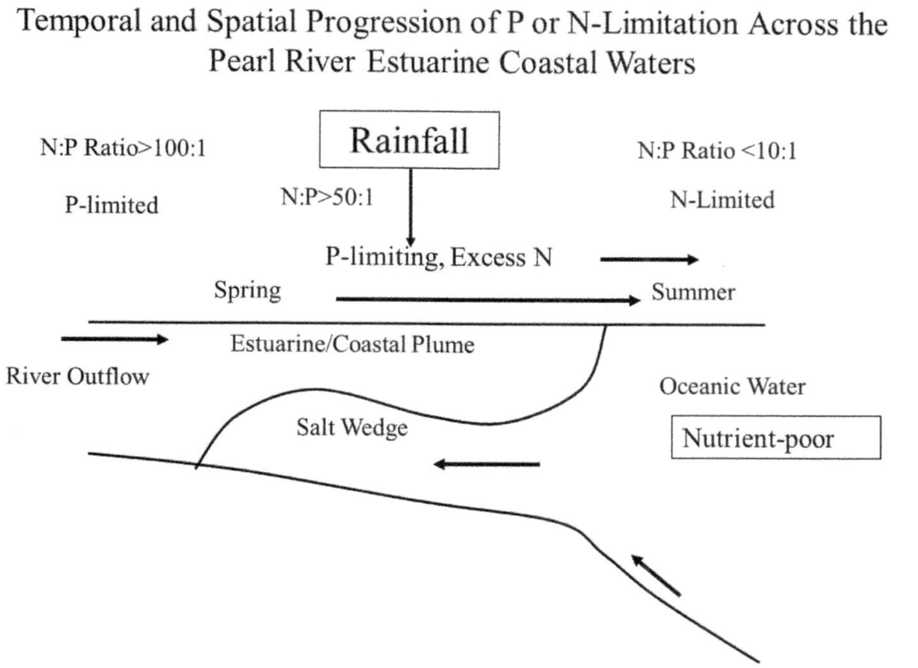

Figure 11.7. Conceptual model of the temporal and spatial shift in nutrient limitation across the Pearl River estuarine coastal plume.

did not exceed 4.2 g C m^{-2} d^{-1} (Malone et al. 1988). Chl *a* in San Francisco Bay reached as high as 60 μg L^{-1}, but primary productivity did not exceed 3.5 g C m^{-2} d^{-1}. Maximum phytoplankton biomass and productivity were relatively low in other estuaries, including the St. Lawrence estuary/Gulf of St. Lawrence (Sinclair 1978; Tremblay et al. 1997), Zaire River estuary (Cadee 1978), and Danube River estuary (Humborg et al. 1997). The Huang He (Yellow River) in northern China showed very low primary productivity of 0.51 mg C m^{-2} d^{-1} (Turner et al. 1990). In the Changjiang River estuary, chl *a* biomass reached 30–110 mg m^{-2} in the coastal plume (Gong et al. 1996), but primary productivity was relatively low (1.5 g C m^{-2} d^{-1}) (Ning et al. 1988).

5. Dynamics of nutrients and nutrient limitation variability

The Pearl River contains very high N concentrations, but low P concentrations, and therefore, ambient N:P ratios are very high (>100:1). Conversely, N:P ratios are <16:1 in the oligotrophic South China Sea. Therefore, salinity and N:P ratios >16:1 are good indicators of waters that are influenced by the Pearl River estuarine waters (Xu et al. 2008). High N:P ratios result in P limitation in the Pearl River–influenced waters, whereas low N:P ratios (Yin et al. 2000, 2004, 2008) indicate N limitation in South China Sea waters that are not influenced by the PR (Yin et al. 2000, 2001; Wu et al. 2003; Xu et al. 2008). There is a spatial shift from P limitation to N limitation along a transect from the estuarine-influenced waters to the SCS waters (Yin et al. 2001), as shown in Fig. 11.7.

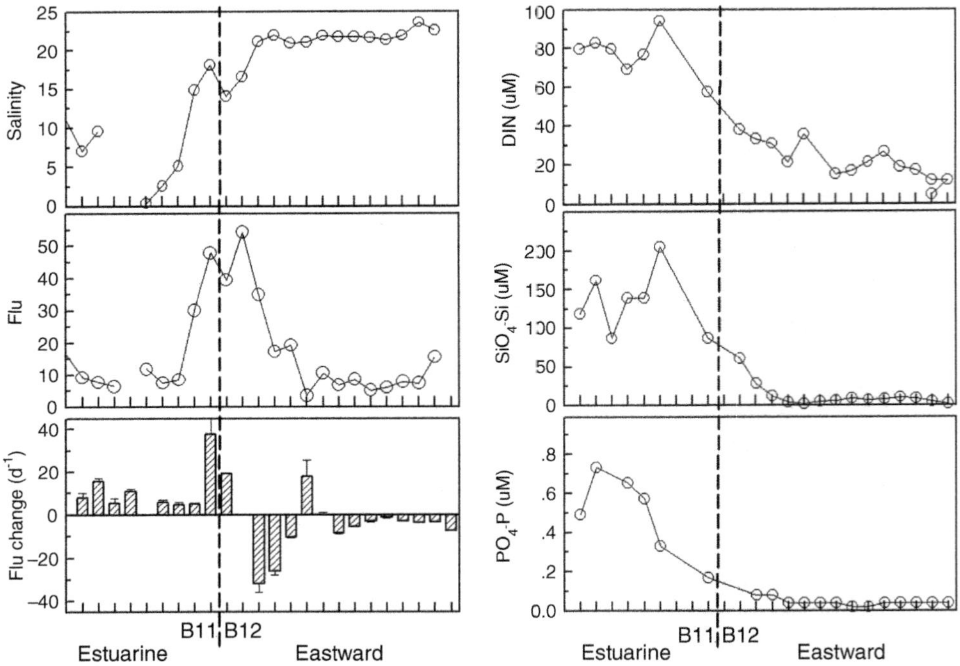

Figure 11.8. Spatial changes in salinity, in-situ fluorescence (Flu), a change in fluorescence after a 24 h incubation, NO_3, SiO_4, and PO_4 during a west to east transect across the Pearl River estuarine coastal front to eastern waters in the southern waters of Hong Kong during a cruise in July 1998.

The shift usually occurs at the inner edge of the coastal plume front. For example, in 1999, concentrations of NO_3 were very high (ca. 90 μmol L^{-1}) upstream of the PRE, and PO_4 was low in surface waters (<0.5 μmol L^{-1}), resulting in high N:P ratios >200:1 (Fig. 11.8). Much of the riverine NO_3 was not utilized in the estuary until depletion at the edge of the coastal plume on the east side of Hong Kong (Yin and Harrison 2008). Beyond the coastal plume to the east of Hong Kong, the N:P ratios were <5:1, indicating potential N limitation. N and P limitation was demonstrated in nutrient addition experiments (Yin et al. 2001, 2004, 2008; Xu et al. 2009). Using ratios of nutrients to infer the potential limiting nutrient is valid for Hong Kong waters as a comparison of several methods to determine potential and actual limitation agreed well with each other (Xu et al. 2009). A spatial shift in nutrient limitation in estuarine regions is common in other estuaries. For example, in the estuarine waters of the eastern Gulf of Finland, P limitation was found in the inner estuary, whereas the open eastern Gulf was mainly N limited (Pitkanen and Tamminen 1995). There was evidence of a shift from phosphorus to nitrogen limitation (Lohrenz et al. 1999), or nitrogen and silicon co-limitation in the Mississippi River–influenced region of the northern Gulf of Mexico (Dortch and Whitledge 1992; Chapter 16 in this volume) and in the Chesapeake Bay (Fisher et al. 1999). This spatial shift in nutrient limitation has important implications for nutrient pollution control and coastal management of Hong Kong waters (Yin and Harrison 2007).

6. Potential effects of climate change on eutrophication

Climate change may lead to a change in ecosystem structure or exacerbate the eutrophication impacts. A recent example is the climate change–induced shift in biological communities in the San Francisco Bay (Cloern et al. 2007). The abrupt change in the biological communities was first detected as increasing phytoplankton biomass and the occurrences of new seasonal blooms that began in 1999, overriding the influence of changes in the input of nutrients. These community changes in San Francisco Bay across three trophic levels followed a state change in the California current system in the form of a sudden increased in upwelling intensity, amplified primary production, and strengthened southerly water flows. These diagnostic features of the East Pacific "cold phase" led to strong recruitment and immigration of juvenile flatfish and crustaceans into estuaries where they feed and develop. Therefore, interdecadal oceanic regime shifts can propagate into estuaries and coastal waters, altering their community structure and efficiency of transforming land-derived nutrients into algal biomass. This example demonstrates the importance of climate change in the ecosystem-buffering processes against eutrophication, as climate change affects many drivers and processes that are variable and interact with each other (Fig. 11.2).

In the Pearl River estuarine coastal waters, episodic events of blooms and hypoxia often occur during the summer, but they do not form severe seasonal eutrophication symptoms over a large scale (coastal scale). One of the reasons is that bloom formation and bottom hypoxia can be interrupted by strong wind events, as demonstrated by Zhou et al. (2012). It has been suggested that the frequency of typhoons striking the China coast per year is on the rise (Lin et al. 2003; Cao et al. 2007). An average of about 30 tropical cyclones form in the western tropical Pacifice very year, and this area has the highest number of typhoons among all the ocean basins (Chan et al. 2004). Strong wind events can pose frequent interruptions to the formation of bottom hypoxia.

There are strong signs of climate change in Hong Kong. Air temperature has been increasing steadily at a rate of 0.17°C per decade during 1947–1988 and at an accelerated rate of 0.34°C per decade during 1989–2006 (Yeung 2006). The water temperature has also increased in Deep Bay, on the east side of the PRE (Xu et al. 2010), and other areas (Ho et al. 2008) during the last several decades. The annual average wind speed was reported to decrease by 1.69 km per decade during 1986–2004 (Yeung 2006). The El Niño Southern Oscillation (ENSO) has been suggested to lead to changes in temperature and rainfall patterns and to influence the monsoons (Leung and Wu 2004). The El Niño years are generally warmer than average and have high rainfall, whereas La Niña effects are less pronounced (Chang 1999). Rainfall in southern China underwent a regime shift from wet to dry in the mid 1970s when the Pacific Decadal Oscillation switched from the cool to the warm phase (Leung and Wu 2005; Zhou et al. 2007). Over the last two decades, rainfall has significantly increased in Hong Kong (Xu et al. 2010). These observations suggest that climate change may exert strong forcing on the Pearl River estuarine coastal waters. Generally, climatic forcing exerts an influence on marine ecosystems by altering physical factors and processes that affect phytoplankton production dynamics (Miller et al. 2004). A good example is Deep Bay, where there has been an increase in chl *a* over the last 21 years (Xu et al. 2010). Because nutrients are very high, the increase in chl *a* was considered to be due to the climatic effect of the significant increase in rainfall (11 mm y^{-1}) over the last 45 years, which has increased stratification and reduced light limitation (Xu et al. 2010).

Currently we have a limited understanding of how large-scale climatic variability affects marine ecosystems, particularly for coastal ecosystems, where there are multiple natural stresses in addition to anthropogenic influences (IMBER 2005). ENSO is the dominant mode of climatic forcing affecting the western equatorial Pacific over interannual time scales and subsequently can result in variations in rainfall, winds, thermocline depth, and oceanic circulation (Fiedler 2002). Variations in these physical parameters can consequently exert effects on biological activities, including a change in primary productivity to changes in the population dynamics of fish, seabirds, and marine mammals (Fiedler 2002). Hong Kong is linked to the SCS, which connects to the western Pacific Ocean through Luzon Strait between Taiwan and the Philippines. The dominant climatic forcing factor in the western Pacific over interannual time scales is ENSO. The intensity of the El Niño in 1997–1998 was one of the strongest of the century. A series of red tides occurred in Hong Kong territorial waters between mid-March and mid-April 1998, killing the majority of the caged fish. The events coincided with the dramatic change in the oceanographic conditions in the northern portion of the South China Sea between 1997 and 1998 from March to mid-April (Yin et al. 1999). It was believed that a large pool of warmer (but saltier) water moved northward underneath the Hong Kong coastal water, behaving like a high-density wedge and the NE monsoon-induced downwelling moved against the wedge, which led to a trapping effect of the coastal waters of the south China coast. Given the local eutrophic conditions of the China coast, an outbreak of HABs occurred over a coast-wide scale (~400 km) in winter 1997 and spring 1998 (Yin et al. 1999). The decrease in El Niño intensity in the SCS in 1998 lagged behind that of the western Pacific Ocean (Wang et al. 2002). As a result, the number of red tides in 1998 was somewhat higher than average (Agriculture, Fisheries and Conservation Department 2004). This observation suggests that El Niño does affect Hong Kong waters. Based on a long time series from 1986 to 2006, the 12-month averaged chl *a* was significantly correlated with MEI (multiple ENSO index), and the relationship was negative at DM2, VM2, and PM1 (Shaw 2008), as shown in Fig. 11.9. The correlation was higher at DM2 and VM2 than PM1. The three stations in three different water zones have different nutrient conditions and oceanographic processes (Yin et al. 2010). DM2 in Deep Bay is contaminated by the Shenzhen River and has very high nutrients, and VM2 in Victoria Harbour is near the main domestic sewage outfall, and nutrients are rarely limiting phytoplankton growth (Yin et al. 2010; Xu et al. 2010, 2012). Nutrients at PM1 in Port Shelter are low and often become limiting. Residence times are very short in Victoria Harbour (2 days) and Deep Bay (9 days), whereas residence times are longer in Port Shelter (25 days) (Arega and Lee 2000). In Port Shelter with long residence times, the already low nutrients become a limiting factor for phytoplankton growth. In Deep Bay with turbid waters, light is the limiting factor (Xu et al. 2010). In Victoria Harbour, residence time controls the level of algal biomass, and therefore, processes that lead to longer residence times would favor an increase in phytoplankton biomass. The strong El Niño in 1998 appeared to result from a change in the SCS circulation and hence led to warmer waters and longer residence times and consequently an outbreak of red tides in Hong Kong (Yin et al. 1999). Longer residence times mean that phytoplankton biomass would remain longer and accumulate in eutrophic waters such as Deep Bay and Victoria Harbour. This explains why the correlation between chl *a* and MEI is stronger in Deep Bay and Victoria Harbour than in Port Shelter, which already has a very long residence time.

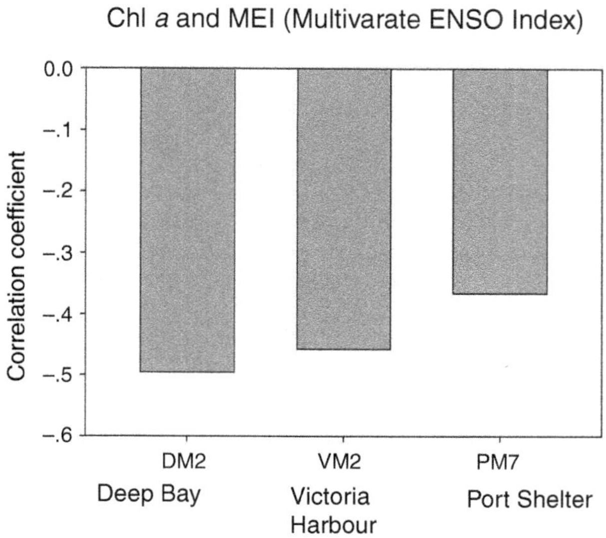

Figure 11.9. Correlation between chl *a* and the multivariate ENSO index (MEI) at 3 different stations representing 3 zones for data between 1986 and 2005.

7. Conclusions

The Pearl River is one of the largest rivers in the world, collects a large load of anthropogenic nutrients from the densely populated watershed, and discharges them into the northern South China Sea. Dissolved inorganic nitrogen in the Pearl River is as high as the Changjiang and Mississippi Rivers, but the coastal waters in the northern South China Sea have not formed coast-wide seasonal hypoxia as in the East China Sea and northern Gulf of Mexico. The ecosystem-buffering capacity of the Pearl River estuary–South China Sea system is large, and hence it reduces eutrophication symptoms such as high biomass HABs and bottom hypoxia. The major physical driving forces are monsoonal winds and typhoon events. Phosphorus is the most limiting nutrient and is so low that it limits the level of phytoplankton biomass and subsequently limits the level of dissolved oxygen consumption in deeper waters.

Acknowledgments

This research was supported by China MOST Project 2012CB956004, NSFC 41176129, and ARC DP110103155. Jie Xu was supported by Chinese Academy of Sciences "Hundred Talents program" grant (Y35L041001) and Paul Harrison was supported by the University Grants Council of Hong Kong AoE project (AoE/P-04/04–4-II). Zhigang Lai was supported by NSFC Project 41206005. We thank the Environmental Protection Dept of Hong Kong for use of its water quality data.

References

Agriculture, Fisheries and Conservation Department. 2004. Red Tides/Harmful Algal Blooms in Hong Kong, 2002 and 2003. Hong Kong Government, p. 27.

Arega, F., and J.H.W. Lee. 2000. Long-term circulation and eutrophication model for Tolo Harbour, Hong Kong. Water Quality Ecosys. Modeling 1: 169–192.
Bowman, M.J., and R.L. Iverson. 1978. Estuarine and plume fronts. pp. 87–104. *In* M.J. Bowman and W.E. Esaias (eds), Oceanic fronts in coastal processes. Springer-Verlag, New York.
Bricker, S.B., C.G. Clement, D.E. Pirhalla, S.P. Orlando, and D.G.G. Farrow. 1999. National estuarine eutrophication assessment: Effects of nutrient enrichment in the nation's estuaries. Special Projects Office and the National Centers for Coastal Ocean Science, National Ocean Service, National Oceanic and Atmospheric Administration, Silver Spring, MD.
Cadee, G.C. 1978. Primary productivity and chlorophyll in the Zaire River estuary and plume. Neth. J. Sea. Res., 12: 368–381.
Cao, X.C., Q.Z. Yuan, J.L. Yang, and H.Q. Yi. 2007. Features of the tropical cyclones landing on China in 2005. Quarterly J. Appl. Meteorol. 18: 412–416 (in Chinese, with English Abstract).
Chan, J.C.L., K.S. Liu, S.E. Ching, and E.S.T. Lai. 2004. Asymmetric distribution of convection associated with tropical cyclones making landfall along the south China coast. Mon. Weather. Rev. 132: 2410–2420.
Chang, W.L. 1999. Some impacts of El Niño and La Niña events on the weather of Hong Kong. *In* Workshop on the impact of the El Niño Southern Oscillation (ENSO) / La Niña on meteorology and hydrology in the typhoon committee area, Macau.
Cloern, J.E., A.E. Alpine, B.E. Cole, R.L.J. Wong, J.F. Arthur, and M.D. Ball. 1983. River discharge controls phytoplankton dynamics in the northern San Francisco Bay estuary. Estuar. Coast. Shelf Sci. 16: 415–429.
Cloern, J.E., B.E. Cole, R.L.J. Wong, and A.E. Alpine. 1985. Temporal dynamics of estuarine phytoplankton: a case study of San Francisco Bay. Hydrobiologia 129: 153–176.
Cloern, J.E. 1996. Phytoplankton bloom dynamics in coastal ecosystems: A review with some general lessons from sustained investigation of San Francisco Bay, California. Rev. Geophys. 34: 127–168.
Cloern, J.E. 2001. Our evolving conceptual model of the coastal eutrophication problem. Mar. Ecol. Prog. Ser. 210: 223–253.
Cloern, J.E., A.D. Jassby, J.K. Thompson, and K.A. Hieb. 2007. A cold phase of the East Pacific triggers new phytoplankton blooms in San Francisco Bay. Proc. Natl. Acad. Sci. 104: 18561–18565.
Dai, M., X. Guo, W. Zhai, L. Yuan, B. Wang, L. Wang, P. Cai, T. Tang, and W.J. Cai. 2006. Oxygen depletion in the upper reach of the Pearl River estuary during a winter drought. Mar. Chem. 102: 159–169.
de Lafontaine, Y., S. Demers, and J. Runge. 1991. Pelagic food web interactions and productivity in the Gulf of St. Lawrence: a perspective. Can. Spec. Publ. Fish. Aquat. Sci. 113: 99–123
Diaz, R.J., and R. Rosenberg. 2008. Spreading dead zones and consequences for marine ecosystems. Science 321: 926–929.
Dong, L., J. Su, L. A. Wong, Z. Cao and J. Chen. 2004. Seasonal variation and dynamics of the Pearl River Plume. Cont. Shelf Res. 24:1761–1777.
Dortch, Q., and T.E. Whitledge. 1992. Does nitrogen or silicon limit phytoplankton production in the Mississippi River plume and nearby regions? Cont. Shelf Res. 12: 1293–1309.
Environmental Protection Department. 2011. Marine Water Quality in Hong Kong Government. 2010.The Government of Hong Kong Special Administration Region, China, Hong Kong Government Printer.
Fiedler, P.C. 2002. Environmental change in the eastern tropical pacific ocean: Review of ENSO and decadal variability. Mar. Ecol. Prog. Ser. 244: 265–283.

Fisher, T.R., L.W. Harding Jr, D.W. Stanley, and L.G. Ward. 1988. Phytoplankton, nutrients and turbidity in the Chesapeake, Delaware and Hudson estuaries. Estuar. Coast. Shelf. Sci. 27: 61–93.

Fisher, T.R., A.B. Gustafson, K. Sellner, R. Lacouture, L.W. Haas, R.L. Wetzel, R. Magnien, D. Everitt, B. Michaels, and R. Karrh. 1999. Spatial and temporal variation of resource limitation in Chesapeake Bay. Mar. Biol. 133: 763–778.

Fox, L.E., F. Lipschultz, L. Kerkhof, and S. Wofsy. 1987. A chemical survey of the Mississippi Estuary. Estuaries 10: 1–12.

Gan, J.P., L. Li, D.X. Wang, and XG. Guo. 2009. Interaction of a river plume with a coastal upwelling in the northern South China Sea. Cont. Shelf Res. 29: 728–740.

Gan, J., Z. Lu, M. Dai, A. Y. Y. Cheung, H. Liu, and P. Harrison. 2010. Biological response to intensified upwelling and to a river plume in the northeastern South China Sea: A modeling study. J. Geophys. Res. 115: C09001. doi:10.1029/2009 JC005569.2010.

Gong, G.C., Y.L.L. Chen, and K.K. Liu. 1996. Chemical hydrography and chlorophyll *a* distribution in the East China Sea in summer: implications in nutrient dynamics. Cont. Shelf Res. 16: 1561–1590.

Guan, B.X. 1994. Patterns and structures of the currents in Bohai, Huanghai and East China Seas. pp. 17–26. *In* D. Zhou, Y.B. Liang and C.K. Zeng (eds.), Oceanology of China Seas. Kluwer Academic, London.

Harley, C.D.G., A.R. Hughes, K.M. Hultgren, B.G. Miner, C.J.B. Sorte, C.S. Thornber, L.F. Rodriguez, L. Tomanek, and S.L. Williams. 2006. The impacts of climate change in coastal marine systems. Ecol. Lett. 9: 228–241.

Harrison, P.J., K. Yin, J.H.W. Lee, J. Gan, and H. Liu. 2008. Physical-biological coupling in the Pearl River Estuary. Cont. Shelf Res. 28: 1405–1415.

Hills, P., L. Zhang, and J. Liu. 1998. Transboundary pollution between Guangdong Province and Hong Kong: Threats to water quality of the Pearl River Estuary and their implications for environmental policy planning. J. Environ. Plan. Manag. 41: 375–396.

Hitchcock, G., and T. Whitledge. 1992. Nutrient/pigment variability in the Mississippi River plume and adjacent waters. Proceedings of workshop Louisiana Universities Marine Consortium October 1991: Nutrient enhanced coastal ocean productivity.

HKO. 2003. Global warming – the Hong Kong connection, August. Available at: http://www.hko.gov.hk/wxinfo/news/2003/pre0801e.htm.

Ho, C.L., and B.B. Barrett. 1977. Distribution of nutrients in Louisiana's coastal waters influenced by Mississippi River. Estuar. Coast. Shelf Sci. 5: 173–195.

Ho, A.Y.T., J. Xu, K. Yin, X. Yuan, L. He, Y. Jiang, J.H.W. Lee, D.M. Anderson, and P. J. Harrison. 2008. Seasonal and spatial dynamics of nutrients and phytoplankton biomass in Victoria Harbour and its vicinity before and after sewage abatement. Mar. Poll. Bull. 57: 313–324.

Ho, AYT., J. Xu, K. Yin, Y. Jiang, X. Yuan, L. He, D.M. Anderson, J.H.W. Lee, and P.J. Harrison. 2010. Phytoplankton biomass and production in subtropical Hong Kong waters: Influence of the Pearl River outflow. Estuar. Coast. 33: 170–181.

Hoyos, C.D., P.A. Agudelo, P.J. Webster, and J.A. Curry. 2006. Deconvolution of the factors contributing to the increase in global hurricane intensity. Science 312: 94–97.

Huisman, J., P. van Oostveen, and F.J. Weissing. 1999. Critical depth and critical turbulence: Two different mechanisms for the development of phytoplankton blooms. Limnol. Oceanogr. 44: 1781–1787.

Humborg, C., V. Ittekot, A. Cociosu, and B. Bdungen. 1997. Effect of Danube River dam on Black Sea biogeochemistry and ecosystem structure. Nature 386: 385–388.

IMBER. 2005. Science plan and implementation strategy, IGBP Report No. 52, Stockholm.

Jickells, T.D. 1998. Nutrient biogeochemistry of the coastal zone. Science 281: 217–222.

Kuang, C.P., H.W.J. Lee, P.J. Harrison, and K. Yin. 2011. Effect of wind speed and direction on summer tidal circulation and vertical mixing in Hong Kong waters. J. Coastal Res. 27: 74–86.

Lee, J.H.W., P.J. Harrison, C.P. Kuang, and K.D. Yin. 2006. Eutrophication dynamics in Hong Kong coastal waters: Physical and biological interactions. pp. 187–206. *In* E. Wolanski (ed.), The Environment in Asia Pacific Harbors. Springer, The Netherlands.

Leung, Y.K., and M.C. Wu. 2004. The effect of ENSO and East Asian monsoon on the annual rainfall in Hong Kong, China. *In* Sixth joint meeting of seasonal prediction on East Asian summer monsoon. Guilin, China.

Leung, Y.K., M.C. Wu, and K.K. Yeung. 2007. Recent decline in typhoon activity in the South China Sea. International conference on climate change, Hong Kong, China, pp. 29–31.

Lin G.C.S. 1997. Red capitalism in South China: growth and development of the Pearl River Delta. UBC Press, Vancouver, Canada.

Lin, I., W.T. Liu, C.C. Wu, G.T.F. Wong, C. Hu, Z.Q. Chen, W.D. Liang, Y. Yang, and K.K. Liu, 2003. New evidence for enhanced ocean primary production triggered by tropical cyclone. Geophys. Res. Lett. 30: 1718.

Liu, C., H. Xia, and D. Wang. 2010. The observation and analysis of eastern Guangdong coastal downwelling in the winter of 2006. Acta. Oceanol. Sin. 32: 1–9.

Liu, J.P., Z. Xue, K. Ross, H.J. Wang, Z.S. Yang, A.C. Li, and S. Gao. 2009. Fate of sediments delivered to the sea by Asian large rivers: long-distance transport and formation of remote alongshore clinothems. Sediment. Rec. 7: 4–9.

Liu, J., R. Xiang, M. Chen, Z. Chen, W. Yan, and F. Liu. 2011. Influence of the Kuroshio current intrusion on depositional environment in the northern South China Sea: Evidence from surface sediment records. Mar. Geol. 285: 59–68.

Lohrenz, S.E., D.G. Redalje, G.L. Fahnenstiel, and G.A. Lang. 1992. Nutrient/pigment variability in the Mississippi River plume and adjacent waters. *In* NOAA Coastal Ocean Program Office, Texas A&M University, and Sea Grant College Program (eds.), Nutrient Enhanced Coastal Ocean Productivity: Proceedings of workshop, Louisiana Universities Marine Consortium, October 1991.

Lohrenz, S.E., G.L. Fahnenstiel, D.G. Redalje, G.A. Lang, X. Chen, and M.J. Dagg. 1997. Variations in primary production of northern Gulf of Mexico continental shelf waters linked to nutrient inputs from the Mississippi River. Mar. Ecol. Prog. Ser. 155: 45–54.

Lohrenz. S.E., G.L. Fahnenstiel., D.G. Redalje, G.A. Lang, M.J. Dagg, T.E. Whitledge, and Q. Dortch. 1999. Nutrients, irradiance, and mixing as factors regulating primary production in coastal waters impacted by the Mississippi River plume. Cont. Shelf Res. 19: 1113–1141

Lu, F.H., H.G. Ni, F. Liu, and E.Y. Zeng. 2009. Occurrence of nutrients in riverine runoff of the Pearl River Delta, South China. J. Hydrol. 376: 107–115.

Malone, T.C. 1977. Environmental regulation of phytoplankton productivity in the lower Hudson estuary. Estuar. Coast Mar. Sci. 5: 157–171

Malone, T.C., and M.B. Chervin. 1979. The production and fate of phytoplankton size fractions in the plume of the Hudson River, N.Y. Bight. Limnol. Oceanogr. 24: 683–696

Malone, T.C., L.H. Crocker, S.E. Pike, and B.W. Wendler. 1988. Influences of river flow on the dynamics of phytoplankton production in a partially stratified estuary. Mar. Ecol. Prog. Ser. 48: 235–249.

Mann, K.H., and J.R.N. Lazier. 2006. Dynamics of Marine Ecosystems: Biological-Physical Interactions in the Oceans, p. 496, 3rd ed. Blackwell Scientific, London.

Mantua, N. J., and Hare, S. R. 2002. The Pacific decadal oscillation. J. Oceanogr. 58: 35–44.

Mao, Q., P. Shi, K. Yin, J. Gan and Y, Qi. 2004. Tides and tidal currents in the Pearl River Estuary. Cont. Shelf Res. 24: 1797–1808.

Miller, A.J., M.A. Alexander, G.J. Boer, F. Chai, K. Denman, D.J. Erickson, R. Frouin, A.J. Gabric, E.A. Laws, M.R. Lewis, Z.Y. Liu, R. Murtugudde, S. Nakamoto, D.J. Neilson, J.R. Norris, J.C. Ohlmann, R.I. Perry, N. Schneider, K.M. Shell, and A. Timmermann. 2003. Potential feedbacks between Pacific Ocean ecosystems and interdecadal climate variations. B. Am. Meteorol. Soc. 84: 617–633

Neller, R.J. and K.C. Lam 1994. The environment. *In* Y.M. Yeung and D.K.Y. Chu (eds.), Guangdong: Survey of a Province Undergoing Rapid Change. The Chinese University of Hong Kong Press, Hong Kong.

Ning, X., D. Vaulot, Z. Liu, and Z. Liu. 1988. Standing stock and production of phytoplankton in the estuary of the Changjiang (Yangtze River) and the adjacent East China Sea. Mar. Ecol. Prog. Ser. 49: 141–150.

National Research Council. 2000. Clean Coastal Waters: Understanding and Reducing the Effects of Nutrient Pollution, p. 405. National Academy Press, Washington, DC.

Peterson, D.H., R.E. Smith, S.W. Hager, D.D. Harmon, R.E. Herndon, and L.E. Schemel. 1985. Interannual variability in dissolved inorganic nutrients in Northern San Francisco Bay estuary. Hydrobiol. 129: 37–58.

Pitkanen, H., and T. Tamminen. 1995. Nitrogen and phosphorus as production limiting factors in the estuarine waters of the eastern Gulf of Finland. Mar. Ecol. Prog. Ser. 129: 283–294.

Revelante, N. and M. Gilmartin. 1976. The effects of Po River discharge on phytoplankton dynamics in the northern Adriatic Sea. Mar. Biol. 34: 259–271.

Ruddle, K., and G. Zhong. 1988. Integrated Agriculture-Aquaculture in South China: The Dike-Pond System of the Zhujiang Delta, p. 173. Cambridge University Press, Cambridge.

Schlüter, L. 1998. The influence of nutrient addition on growth rates of phytoplankton groups, and microzooplankton grazing rates in a mesocosm experiment. J. Exp. Mar. Biol. Ecol. 228: 53–71.

Shaw, E. 2008. Response of phytoplankton to climatic variability in Hong Kong coastal waters. Proceedings of the 13th international conference on harmful algae, November 3–7 2008, Hong Kong.

Sinclair, M. 1978. Summer phytoplankton variability in the lower St. Lawrence estuary. J. Fish. Res. Board. Can. 35: 1171–1185.

Shu, Y., D. Wang, J. Zhu, S. Peng. 2011. The 4-D structure of upwelling and Pearl River plume in the northern SouthChina Sea during summer 2008 revealed by a data assimilation model. Ocean Model. 36: 228–241.

Su, J. 2004. Overview of the South China Sea circulation and its influence on the coastal physical oceanography outside the Pearl River Estuary. Cont. Shelf Res. 24: 1745–1760.

Tremblay, J.É., L. Legendre and J.C. Therriault. 1997. Size-differential effects of vertical stability on the biomass and production of phytoplankton in a large estuarine system. Estuar. Coast. Shelf Sci. 45: 415–431.

Turner, R.E., N.N. Rabalais and Z.Z. Nan. 1990. Phytoplankton biomass, production and growth limitations on the Huanghe (Yellow) River continental shelf. Cont. Shelf Res. 10: 545–571.

Turner, R.E., N.N. Rabalais, E.M. Swenson, M. Kasprzak, and T. Romaire. 2005. Summer hypoxia in the northern Gulf of Mexico and its prediction from 1978 to 1995. Mar. Environ. Res. 59: 65–77.

Wang, D.X, Q. Xie, Y. Du, W. Wang, and J. Chen. 2002. The 1997–1998 warm event in the South China Sea. Chin. Sci. Bull. 47: 1221–1227.

Watts, J.C.D. 1971. A general review of the oceanography of the northern sector of the South China Sea. Hong Kong Fish. Bull. 2: 41–50.

Watts, J.C.D. 1973. Further observation on the hydrology of the Hong Kong territorial waters. Hong Kong Fish. Bull. 3: 9–35.

Webster, P.J., G.J. Holland, J.A. Curry, and H.R. Chang. 2005. Changes in tropical number, duration and intensity in a warming environment. Science 309: 1844–1846.

Williamson, G.R. 1970. The hydrography and weather of the Hong Kong fishing grounds. Hong Kong Fish. Bull. 1: 43–49.

Wong, L. A., J. Chen, H. Xue, L. Dong, J. Su and G. Heinke. 2003. A model study of the circulation in the Pearl River Estuary (PRE) and its adjacent coastal waters: 1. Simulations and comparison with observations. J. Geophys. Res. 108: C3156.

Wong, L. A., J. C. Chen and L. X. Dong. 2004. A model of the plume front of the Pearl River Estuary, China and adjacent coastal waters in the winter dry season. Cont. Shelf Res. 24: 1779–1795.

Wu, J., S.W. Chung, L.S. Wen, K.K. Liu, Y.L.L. Chen, H.Y. Chen, and D.M. Karl. 2003. Dissolved inorganic phosphorus, dissolved iron, and *Trichodesmium* in the oligotrophic South China Sea. Global Biogeochem. Cy. 17: 1008–1016.

Xu, J., A.Y.T. Ho, K. Yin, X. Yuan, D.M. Anderson, J.H.W. Lee, and P.J. Harrison. 2008. Temporal and spatial variations in nutrient stoichiometry and regulation of phytoplankton biomass in Hong Kong waters: Influence of the Pearl River outflow and sewage inputs. Mar. Poll. Bull. 57: 335–348.

Xu, J., K. Yin, A.Y.T. Ho, J.H.W. Lee, D.M. Anderson and P.J. Harrison. 2009. Regulation of nutrient limitation in Hong Kong waters inferred from comparison of nutrient ratios, nutrient enrichment bioassays, and ^{33}P turnover times. Mar. Prog. Ecol. Ser. 388: 81–97.

Xu, J, K. Yin, J.H.W. Lee, H. Liu, A.Y.T. Ho, X. Yuan, and P.J. Harrison. 2010. Long-term and seasonal changes in nutrients, phytoplankton biomass and dissolved oxygen in Deep Bay, Hong Kong. Estuar. Coast. 33: 399–416.

Xu, J., K. Yin, J.H.W. Lee, D.M. Anderson, X. Yuan, A.Y.T. Ho and P.J. Harrison. 2012. Resistance of Hong Kong waters to nutrient enrichment: assessment of the role of physical processes in reducing eutrophication. J. Oceanogr. 68: 545–560.

Yeung, KH. 2006. Issues related to global warming – myths, realities and warnings. Fifth conference on catastrophe insurance in Asia, June 20–21 2006, Hong Kong, Keynote Address.

Yin, K., P.J. Harrison, S. Pond, and R.J. Beamish. 1995a. Entrainment of nitrate in the Fraser River plume and its biological implications. I. Effects of salt wedge. Estuar. Coast. Shelf Sci. 40: 505–528.

Yin. K., P.J. Harrison, S. Pond, and R.J. Beamish. 1995b. Entrainment of nitrate in the Fraser River plume and its biological implications. II. Effects of spring vs. neap tides and river discharge. Estuar. Coast. Shelf Sci. 40: 529–544.

Yin, K., R.H. Goldblatt, P.J. Harrison, P.J. Clifford, M.A. St. John, and R.J. Beamish. 1997. Importance of wind and river discharge in influencing nutrient dynamics and phytoplankton production in summer in the central Strait of Georgia. Mar. Ecol. Prog. Ser. 161:173–183.

Yin, K., P.J. Harrison, J. Chen, W. Huang, and P.Y. Qian. 1999. Red tides during spring 1998 in Hong Kong: is El Niño responsible? Mar. Ecol. Prog. Ser. 187: 289–294.

Yin, K., P.Y. Qian, J.C. Chen, D.P.H. Hsieh, and P.J. Harrison. 2000. Dynamics of nutrients and phytoplankton biomass in the Pearl River estuary and adjacent waters of Hong Kong during summer: Preliminary evidence for phosphorus and silicon limitation. Mar. Ecol. Prog. Ser. 194: 295–305.

Yin, K., P.Y. Qian, M.C.S. Wu, J.C. Chen, L.M. Huang, X.Y. Song, and W.J. Jian. 2001. Shift from P to N limitation of phytoplankton biomass across the Pearl River estuarine plume during summer. Mar. Ecol. Prog. Ser. 221: 17–28.

Yin, K. 2002. Monsoonal influence on seasonal variations in nutrients and phytoplankton biomass in coastal waters of Hong Kong in the vicinity of the Pearl River Estuary. Mar. Ecol. Prog. Ser. 245: 111–122.

Yin, K. 2003. Influence of monsoons and oceanographic processes on red tides in Hong Kong waters. Mar. Ecol. Prog. Ser. 262: 27–41.

Yin, K., P.Y. Qian, L. Huang, J. Zhang, M.C.S. Wu, W. Jian, X. Song, and J. Chen. 2004. Effect of wind events on phytoplankton blooms in the Pearl River Estuary during summer. Cont. Shelf Res. 24: 1909–192.

Yin, K., and P.J. Harrison. 2007. Influence of the Pearl River estuary and vertical mixing in Victoria Harbor on water quality in relation to eutrophication impacts in Hong Kong waters. Mar. Poll. Bull. 54: 646–656.

Yin, K., and P.J. Harrison. 2008. Nitrogen enrichment in subtropical Pearl River estuarine coastal waters. Cont. Shelf Res. 28: 1435–1442.

Yin, K., J. Xu, and P.J. Harrison. 2010. A comparison of eutrophication processes in three Chinese subtropical semi-enclosed embayments. pp. 367–397. *In* M. Kennish and H.W. Paerl (eds.), Coastal Lagoons:Critical Habitats of Environmental Change. CRC Press, Taylor and Francis Publisher, New York.

Zhang, J., Z.G. Yu, J.T. Wang, J.L. Ren, H.T. Chen, H. Xiong, L.X. Dong, and W.Y. Xu. 1999. The subtropical Zhujiang (Pearl River) Estuary: Nutrient, trace species and their relationship to photosynthesis. Estuar. Coast. Shelf Sci. 49: 385–400.

Zheng, G.M., and D.L. Tang. 2007. Offshore and nearshore chlorophyll increases induced by typhoon winds and subsequent terrestrial rainwater runoff. Mar. Ecol. Prog. Ser. 333: 61–74.

Zhou, W., X. Wang, T.J. Zhou, and J.C.L. Chan. 2007. Interdecadal variability of the relationship between the East Asian winter monsoon and ENSO. Meteorol. Atmos. Phys. 98: 283–293.

Zhou, M.J., Z.L. Shen, and R.C. Yu. 2008. Response of a coastal phytoplankton community to increased nutrient input from the Changjiang (Yangtze) River. Cont. Shelf Sci. 28: 1483–1489.

Zhou, W., K. Yin, P.J. Harrison, and J.H.W. Lee. 2012. Changes in water quality caused by two late summer typhoons in Hong Kong waters dominated by high river discharge. Estuar. Coast. Shelf Sci. 111: 35–47.

12

The Mekong River and its influence on the nutrient chemistry and matter cycling in the Vietnamese coastal zone

M. Voss, D. Bombar, J. W. Dippner, D. Nhu Hai, N. Ngoc Lam, and N. Loick-Wilde

1. Introduction

Coastal upwelling systems provide the productivity on which the largest fisheries in the world are based, like off Peru or Namibia. Most of these economically important systems are located along eastern boundary current systems and are rarely found at western boundaries of the oceans. An exception is the coastline off Vietnam, where monsoon driven upwelling occurs between March and September (Dippner et al. 2007).

Moreover, fluvial inputs carry nutrients to the coastal systems and thus support primary and secondary production (Milliman and Maede 1983). Studies in the Amazon plume, which extends thousands of kilometers into the Atlantic Ocean, have revealed interesting links between nutrient supply and plankton populations. Nutrients fuel phytoplankton growth in these plumes as well as a symbiosis between diatoms of the genus *Hemiaulus* and an endosymbiont of the genus *Richelia*, which fix significant amounts of nitrogen (Carpenter et al. 1999; Foster et al. 2007). However, the combination of a coastal upwelling system, which is also impacted by a large river, is hardly found in tropical oceans elsewhere, except in the South China Sea (SCS) off Vietnam. Moreover, the region experiences pronounced seasonal differences in precipitation and wind strength so that the nutrient input varies in time and space. To study the productivity and biogeochemistry in this coastal system potentially provides novel insights into the role of various N-sources. Moreover, the monsoon activity and El Nino phenomena are impacted by climate change and climate variability and may influence the upwelling regime.

Vietnam has a 3,400 km long coastline stretching between 8.23° and 21.39° north along the eastern coastline of the Indochina peninsula. The coastal fishery is an essential economic component and makes Vietnam the world's fourth largest seafood exporter after China, Norway, and Thailand (FAO 2009). Not only the fishing activity but also the aqua-culture of shrimp and catfish have changed the landscape and impacted the biogeochemical fluxes along the coast, where large algae blooms became a common phenomenon (Liu et al. 2009).

Although several marine institutions and research groups exist, most internationally published papers about the SCS dealt with numerical modeling (Shaw and Chao 1994; Metzger and Hurlburt 1996; Wu et al. 1998; Hu et al. 2000; Morimoto et al. 2000; Qu 2000; Isobe and Namba 2001; Yang et al. 2002) and the analysis and interpretation of satellite images (Liu and Xie 1999; Hwang and Chen 2000; Xie et al. 2003; Kuo et al. 2004; Zhao and Tang 2007). Coastal monitoring programs and a coastal stations network were established, but dedicated science projects concerning in situ observations in the Vietnamese upwelling area were rarely performed. Measurements and data existed (e.g., from the three-lateral monitoring activity Joint Oceanographic Marine Scientific Research Expedition [JOMSRE] between China, Philippines, and Vietnam), but unfortunately only cooperative research reports were published (Pham et al. 2002). Under these settings, an interdisciplinary bilateral Vietnamese-German (VG) research cooperation was run from 2003–2008, focusing on the "Land-Ocean-Atmospheric Interactions in the Coastal Zone of Southern Vietnam." It was well known that a central upwelling area was located directly in front of the coast of Khanh Hoa Province; however, its impact on the productivity remained unclear. Moreover, blooms of diazotrophs and other sometimes harmful algae were reported from the waters off southern Vietnam (Nguyen and Doan-Nhu 1996) but hardly related to the hydrochemistry of the area. We therefore carried out seven cruises in different seasons to generate an improved understanding of the physics of the local upwelling combined with specific studies of nitrogen cycling in different monsoon seasons and under the influence of the River Mekong. Moreover, the trophic transfer of nitrogen to higher trophic levels was investigated. Here we give an overview on the project achievements considering the existing literature and our own observation during the period 2003–2008.

2. Climatology of the Southeast Asian monsoon region

Climatic variations in the atmosphere and in the upper ocean of the SCS are primarily controlled by the East-Asian monsoon, which follows closely the climatic variations in the equatorial central Pacific. The south-west (SW) summer monsoon lasts from June to September. The stronger northeast (NE) monsoon appears during winter from November to March and is known to vary with the Southern Oscillation Index (Liu et al. 2002). Both are separated by a relatively calm period in April and May, called intermonsoon. The annual cycle of wind stress fields (Hellerman and Rosenstein 1983) shows that NE monsoon prevails over the whole region with an averaged magnitude of 9 m s^{-1} in winter. During summer, weaker SW monsoon of about 6 m s^{-1} dominates over most parts of the SCS, which is characterized by a spatial asymmetry. The inhomogeneous wind forcing over this region results in complex circulation patterns.

Investigations of large-scale structures by Zhang (2000) show a teleconnection between the western tropical Pacific and the East Asian region, which has significant impact on the East Asian summer climate. The intensity of the East-Asian monsoon, especially the SW monsoon, is driven by the interannual variability of the El Niño Southern Oscillation (ENSO), the variability of the Inter-Tropical Convergence Zone (ITCZ) (Dippner et al. 2013), and the variability of Indian Ocean Dipole (IOD) (Liu et al. 2012). They influence the intensity of precipitation and therefore the runoff of River Mekong as well as the intensity of upwelling along the Vietnamese coast.

3. Hydrography of the Vietnamese coastal waters

SCS is the largest marginal sea in the tropics, with a maximum depth of more than 5,000 m. It covers an area from the equator to 23°N and from 99°E to 121°E. The deep central basin is bordered by two broad shelf regions with water depth less than 100 m (Fig. 12.1). The northern shelf, including the area south

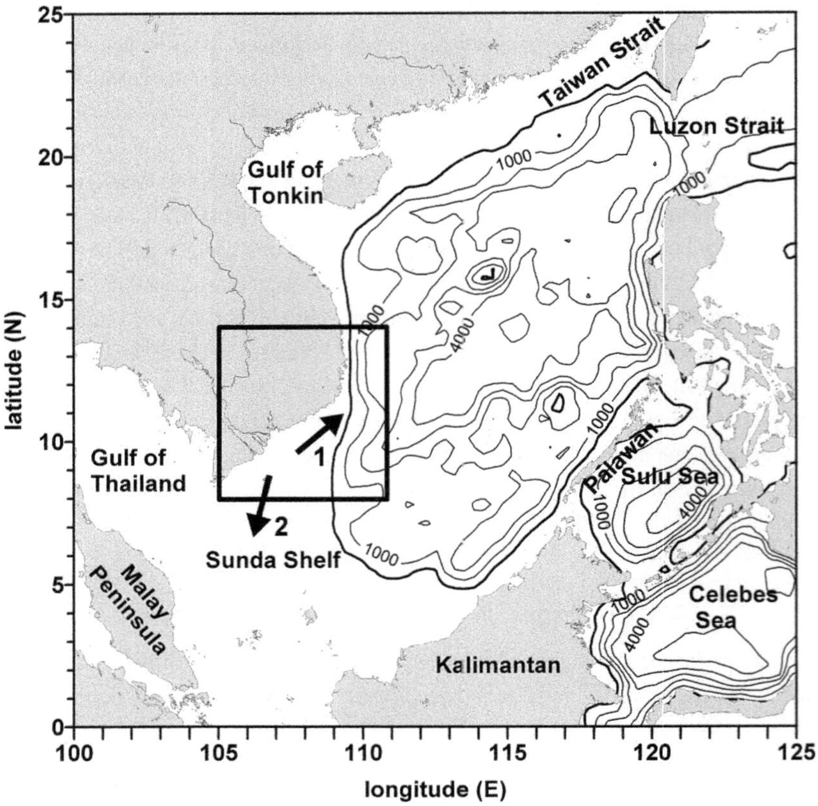

Figure 12.1. Map and topography of the South China Sea. The contour interval is 1,000 m. In addition, the shelf edge (200 m contour line) is marked by a bold line. The rectangle marks the area of investigation. Arrow 1 indicates river plume flow during the SW monsoon; arrow 2 indicates the direction along the shelf edge typical for all other seasons.

of China and the Gulf of Tonkin, extends from Taiwan southwestward to 15°N. The southern shelf consists of the Gulf of Thailand and the Sunda Shelf between the Malay Peninsula and Kalimantan. In the south, the Sunda Shelf is connected to the Java Sea and the Indian Ocean through the Strait of Malacca. In the east, the Philippines and Palawan Islands separate SCS from the Pacific Ocean. The main exchange with the East China Sea and the Pacific Ocean is through the Strait of Taiwan and the Strait of Luzon, where a strong inflow of the Kuroshio Current occurs during NE monsoon.

The seasonal circulation pattern can be separated into a northern and a southern circulation in the SCS. The circulation in the southern SCS is driven by the monsoon only, forming a cyclonic gyre during winter and an anticyclonic gyre during summer. In the northern part of the SCS, the circulation is forced by the monsoon and the inflow through the Strait of Taiwan and the Strait of Luzon (Hu et al. 2000). During winter, both cyclonic gyres merge to a basin-wide cyclonic circulation pattern, which is part of the SCS trough flow, an important heat and freshwater conveyor in the transition between Pacific and Indian Ocean (Qu et al. 2005; Qu et al. 2006). Also in summer, a cyclonic gyre exists that interacts with the anticyclonic gyre in the south to form a dipole circulation cell, which is therefore associated with an eastward offshore current off the coast of Vietnam (Wu et al. 1998). After an ENSO event, upwelling is suppressed and no offshore transport occurs. This clearly has consequences for the production in the upwelling area as well as on the extension of the Mekong River plume.

4. The Mekong River and river plume

The Mekong is one of the largest rivers in the world, among the 10 largest in terms of length and runoff. A sediment transport rate of app. 130×10^6 t y^{-1} was suggested (Milliman and Maede 1983) and is similar to geological estimates, which yield a mean amount of 145×10^6 t y^{-1} over the last 3,000 years (Ta et al. 2002). The Mekong flows 4,300 km from north to the southeast, starting in the Dangla mountains in China; flows through east Tibet, Myanmar, Thailand, and Laos; and forms a vast delta of approximately 70,000 km^2 in Cambodia and South Vietnam before it drains into the SCS (Fig. 12.2). During the past 4,000 years, the delta has propagated more than 200 km around the Cambodian border (Ta et al. 2002) and grew south-eastward approximately 70–80 km from land into the sea (Schimanski and Stattegger 2005). The river forms a delta with two major channels called Bassac River and Mekong River, which are dominated by mangrove forests along their river mouth and coast. The estuary is tide-dominated, and changes in the water level can be experienced from the actual coastline to 390 km (Dippner et al. 2007). Annual loads of nitrogen and phosphorus measured close to the river mouth are approximately 2,700 and 900 t y^{-1} (Yoshimura and Takeuchi 2007) and data from the web (www.gemstat.org/) report an annual mean of approximately 35 μmol L^{-1} NO_3^- and 3 μmol L^{-1} PO_4^{3-}, which is much lower than, for example,the concentrations found in the Yellow river (Huang He), which drains into the Yellow Sea (Chapter 11) Runoff as observed at Stung Treng shows a strong monsoon-driven seasonality (Fig. 12.2). The maximum runoff entering the SCS occurs during SW monsoon, with 40,000 m^3 s^{-1} $\pm$11,000 m^3 s^{-1}, and is highly correlated to South China Sea Climate Index (SCSCI) and to Southern Oscillation Index (SOI) and IOD. During NE winter monsoon, the runoff is in the order of 5,000 m^3/s, with less than 2,000 m^3/s during spring inter-monsoon. During the low flow period, the plume of the Mekong River propagates to the south onto the Sunda Shelf due to Coriolis force (Fig. 12.1).

The runoff as well as the plume propagation into the SCS is also strongly controlled by the interannual variability of the East Asian monsoon system. During SW monsoon, the plume propagates parallel to the coast in NE direction forced by the wind (Fig. 12.1). However, depending on the wind speed, two slightly differing pathways are possible. During a strong SW monsoon, the Mekong plume propagates directly into the central SCS after passing Ca Na Bay offshore. During weak SW monsoons, the plume moves northward after passing Ca Na Bay and propagates into the upwelling area in front of the coast of Khanh Hoa Province. Both situations have been documented by in situ observations (Voss et al. 2006) and by satellite images (Dippner and Loick-Wilde 2011).

5. Biochemistry and productivity in the Mekong plume and coastal waters

5.1. Phytoplankton species composition

Despite the enormous productivity, phytoplankton species composition in the Mekong estuary was hardly studied until the 1980s. An intensive investigation from 1978–1980 documented 342 phytoplankton species with an average density of 248 cells L^{-1}. The study was based on 15 cruises in

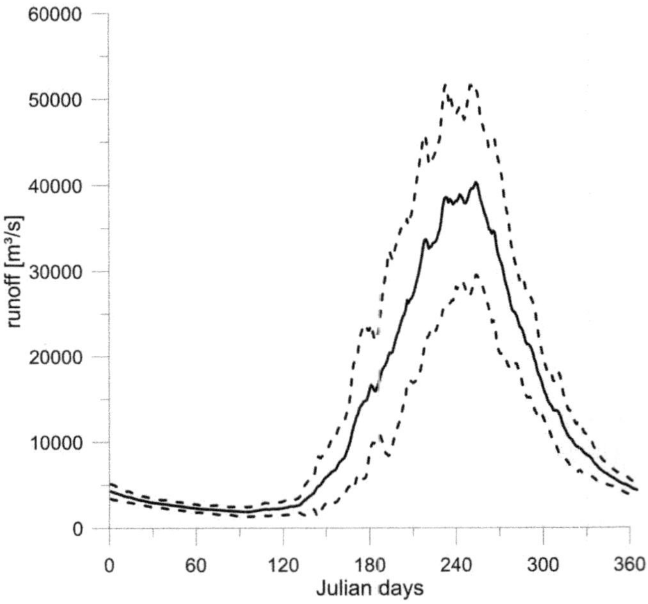

Figure 12.2. Long-term seasonal mean of Mekong runoff ± standard deviation measured daily at Stung Treng.

different seasons along the coast of Vietnam, from Provinces of Ninh Thuan to Ca Mau (Nguyen-Van et al. 1995). The results were extended by another survey during May 1999, which reported 357 phytoplankton species with diatoms as dominating group (159 species) followed by dinoflagellates (161 species). Diatoms were dominated in the surface waters of the Mekong plume close to shore by the medium-size species *Thalassionema nitzschioides* and small-size diatoms *Cyclotella meneghiniana* (Shamsudin et al. 2001). Species of the cyanobacteria *Trichodesmium* were reported further offshore. At the chlorophyll maximum, the diatoms *Thalassionema frauenfeldii* dominated the entire central Vietnam and Mekong estuary area (Boonyapiwat 2001).

A concerted sampling program in the Vung Tau area, just north of the Mekong estuary, documented 241 species (Table 12.1) during 1998–2010, whereas the Mekong delta hosted 334 species; among them were 96 dinoflagellate species, 226 species of diatoms (Table 12.1), and 13 other species, including freshwater chlorophytes, cyanobacteria, and dictyochophyceans. An analysis based on *nifH* gene diversity during a cruise in the intermonsoon season revealed relatively low diversity with a mixture of estuarine and oceanic fingerprints, but recorded the presence of *Trichodesmium* as well as α-proteobacteria, symbiotic species, and the unicellular cyanobacteria UCYN A and B (Moisander et al. 2008).

In general, the species composition was significantly influenced by the amount of freshwater input to the coastal system. The freshwater green algae were only found during the rainy SW monsoon season and disappeared during the dry season. Moreover, higher numbers of cyanobacteria species and lower number of dinoflagellates were present during the SW monsoon (Table 12.1). One example for the changes in the dinoflagellate group is the genus *Ceratium* that consists of many coastal

Table 12.1. *Number of phytoplankton species recorded in Mekong estuary during 1998–2010*

	Vung Tau 1998–2010[1]	September 2008[2]	September 2009[2]	April 2010[2]	Total
Total	241	165	180	233	334
Dinoflagelates	62	37	48	72	96
Cyanobacteria	2	5	3	3	5
Chlorophytes	2	3	4	0	5
Diatoms	173	119	123	157	226
Centric diatoms	*138*	*91*	*98*	*102*	*160*
Pennate diatoms	*37*	*29*	*27*	*56*	*68*
Dictyochophycean	2	1	2	1	2

Source: [1]Nguyen-Chi et al. (2011); [2]Unpublished data of Dept. of Marine Plankton (Institute of Oceanography).

(e.g., *Ceratium furca*, *C. fusus*) and oceanic species (e.g., *Ceratium ranipes*). During the rainy season, only 13 marine species were found, whereas in the intermonsoon, 23 species were recorded. The sensitivity toward salinity is also visible in, for example, the occurrence of the oceanic species *Podolampas elegans*, which occurs offshore during the SW monsoon and close to the coast during the intermonsoon. Other abundant species at Vung Tau station are *Ceratium furca*, *Thalassionema frauenfeldii*, *Ditylum brightwellii*, *Ditylum sol*, and species in genera *Pseudo-nitzschia*, *Chaetoceros*, and *Cyclotella*. Rare species found once in 10 years were *Alexandrium tamutum*, *Asteromphalus flabellatus*, *Chaetoceros muelleri*, and *Coscinodiscus granii*.

There is a trend of decreasing similarity in years 2002–2010 compared with the previous year 2001 (from 80% to 65% similarity), indicating a possible shift of species composition/diversity (Nguyen-Chi et al. 2011). Moreover, the number of species found during the spring inter-monsoon was 95 and thus significantly higher than during SW monsoon, with 60 species (Doan-Nhu and Nguyen-Ngoc 2008; Nguyen-Chi et al. 2011).

In the Mekong estuary, a similar pattern of salinity-dependent species distribution was reported (Nguyen-Van et al. 1995; Grosse et al. 2010; Nguyen-Chi et al. 2011). During the SW monsoon, highest cell density were found further away from the river mouth with up to 880×10^3 cells L^{-1}, whereas during lower runoff abundances of phytoplankton were lower, with only 200×10^3 cells L^{-1} close to river mouth (Doan-Nhu and Nguyen-Ngoc 2008; Nguyen-Chi et al. 2011). Diatoms were the dominating group in both seasons, *Chaetoceros* spp. cell numbers varied between 33×10^3 cells L^{-1} during intermonsoon to 705×10^3 cells L^{-1} in the SW monsoon season. At all times, the diatom species *Thalassionema frauenfeldii*, *Pseudo-nitzschia spp.*, and *Chaetoceros diversus* were found. In the rainy season, dominant species were the dinoflagellate *Ceratium furca*, or diatoms such as *Rhizosolenia imbricata*, *Hemiaulus sinensis*, whereas during the dry season, *Chaetoceros lorenzianus*, *Chaetoceros peruvianus*, *Guinardia cylindrus*, *Proboscia alata*, and *Pseudosolenia calcar-avis* were abundant (Grosse et al. 2010; Nguyen-Chi et al. 2011). Freshwater species were again only abundant close to the river mouth (*Pediastrum* sp., *Scenedesmus* sp., and *Phacus* sp.) (Nguyen-Chi et al. 2011).

There are 21 species of potentially harmful planktonic microalgal in the Mekong estuaries, whereas 34 species are documented in the northern part of Mekong estuaries, including the region of Ninh Thuan to Binh Thuan (Table 12.2). In the Mekong estuary, not many species of potentially harmful

Table 12.2. *A comparison in diversity of potentially harmful planktonic microalgae in Ninh Thuan-Binh Thuan waters and Mekong estuaries*

	Taxa	Ninh Thuan- Binh Thuan[1]	Vung Tau – Mekong estuaries[2]
	Diatoms		
1	*Pseudo-nitzschia brasiliana*	+	+
2	*P. calliantha*	+	+
3	*P. cuspidata*	+	+
4	*P. delicatissima*	+	+
5	*P.* cf. *granii*	+	+
6	*P. inflatula*	+	–
7	*P. micropora*	+	–
8	*P. multistriata*	+	–
9	*P. pungens*	+	+
10	*P.* cf. *sinica*	+	+
11	*P. subraudulenta*	+	–
	Cyanobacteria		
1	*Trichodesmium erythraeum*	+	+
2	*T. thiebautii*	+	–
	Dinophyceae		
1	*Alexandrium affine*	+	+
2	*A. compressum*	+	–
3	*A. fraterculus*	+	–
4	*A. insuetum*	+	–
5	*A. leei*	+	+
6	*A. minutum*	+	+
7	*A. pseudogonyaulax*	+	+
8	*A. tamarense*	+	+
9	*A. tamiyavanichi*	+	+
10	*A. tamutum*		+
11	*Dinophysis caudata*	+	+
12	*D. hastata*	+	+
13	*D. miles*	+	+
14	*D. mitra*	+	+
15	*D. rotundata*	+	–
17	*Lingulodinium polyedricum*	+	–
18	*Noctiluca scintillans*	+	+
19	*Prorocentrum lima*	+	–
20	*P. cordatum*	+	–
21	*P. rhathymum*	+	–
	Haptophytes		
1	*Phaeocystis globosa*	+	+
	Number of species	**34**	**21**

Source: [1]Chen et al. 2004, Nguyen-Ngoc et al. 2004, Nguyen-Ngoc and Ho-Van 2011; [2]Nguyen-Chi et al. 2011, Nguyen-Ngoc et al. 2004.
+, Presence; –, absence.

algae have been recorded, and those were recorded in the southern region in low-salinity and high-turbidity waters (Nguyen-Ngoc et al. 2004).

The haptophyte *Phaeocystis globosa* was first observed in 2002 in the coastal waters and since then regularly during the southwest monsoon. The blooms have the potential to devastate flora and fauna along the coast of south-central Vietnam (Doan-Nhu et al. 2003; Doan-Nhu and Nguyen-Ngoc 2008). The origin of the *Phaeocystis globosa* bloom was unclear until a study based on satellite imagery and the wind regime formulated the hypothesis that this species first appears in the Mekong River estuary and from there propagates to the north, a phenomenon that is hydrographically plausible (Dippner et al. 2011). Moreover, Doan-Nhu et al. (2010) suggested a scenario in which the species is closely associated with the strong development of diatoms, or dinoflagellate blooms of *Noctiluca scintillans*.

Dinophysis species, which are well known for their potential toxicity, occur at high densities in waters above 20°C and salinity from 5 to 35. Dinophysis species often live in high-salinity waters; *Dinophysis caudata and D. miles* were found abundantly in the saline part of the Mekong estuary. *D. caudata* is found throughout the year in the Vietnamese coastal waters from brackish to marine (salinity 15–35) conditions, reaching highest densities in the Mekong estuary in July–October (Nguyen-Ngoc et al. 2004). *Alexandrium* were found only rarely in net samples under variable salinities (4–34) and temperatures (20–30°C), but most commonly in water temperatures around 30° C (Nguyen-Ngoc et al. 2004). Other dinoflagellate species such as *Ceratium furca, C. trichoceros*, and *Peridinium quinquecorne* appear in high density to change the color of the water, but adverse effects have not yet been documented. Likewise, blooms of *Noctiluca scintillans* and *Alexandrium pseudogonyaulax* were noted near shrimp ponds in Nha Trang Bay without harmful effects.

In the past ten years, *Trichodesmium erythaeum* has formed regular blooms during the southwest monsoon season, as evidenced by surveys. It appears in Vietnamese coastal regions, from Da Nang located in the middle of the country, to Kien Giang, situated in the very south. The first observation is from March 1999 (Nguyen 1999), and interestingly, the highest number of filaments of 5,000 L^{-1} was recorded in the upwelling region of the south central coast north of the Mekong River system, which fits well to other in situ observations, as described later.

5.2. Primary production and nitrogen fixation

Phytoplankton blooms are common features in estuaries, and for temperate systems such as the Chesapeake Bay or San Francisco Bay, much attention has been paid to the underlying mechanisms of elevated phytoplankton production, especially in the context of eutrophication research (Harding 1994; Cloern 1996; Kemp et al. 2005). In contrast, comparatively few studies have been conducted in subtropical and tropical estuaries (Murrell et al. 2007; Ho et al. 2010) and none on the impact of Mekong runoff on the biogeochemistry in the SCS prior to investigations of the Vietnamese-German project (Grosse et al. 2010; Bombar et al. 2011).

The SCS is an extremely oligotrophic marginal sea where phytoplankton production appears to be primarily limited by nitrogen availability (Wu et al. 2003; Chen et al. 2004; Chen 2005). The Mekong discharges into this system with nutrient concentrations of 30–40 μmol L^{-1} of nitrate plus nitrite ($NO_3^- + NO_2^-$), about 1 μmol L^{-1} of PO_4^{3-}, but more than 100 μmol L^{-1} of $Si(OH)_4$ (Grosse et al. 2010; www.gemstat.org). Thus the river inputs represent an important nutrient source, and the

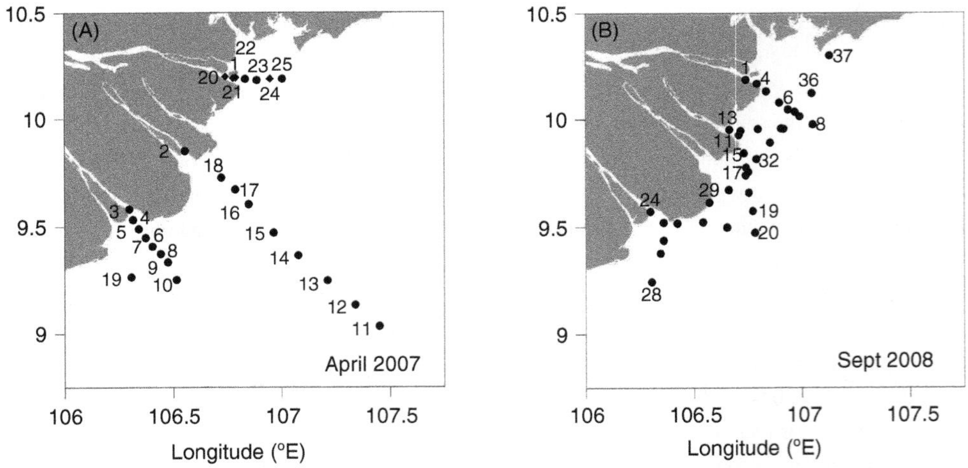

Figure 12.3. Map of the Mekong River Estuary with all stations sampled in (A) April 2007 and (B) September 2008. From: Grosse, J., Bombar, D., Doan, H.N., Nguyen, L.N., Voss, M. 2010. The Mekong River plume fuels nitrogen fixation and determines phytoplankton species distribution in the South China Sea during low- and high-discharge season. Limnology and Oceanography 55: 1668–1680.

biological responses have been clearly observed by ocean color satellite sensors as high chlorophyll patches off the river mouth (Liu et al. 2002).

Our in situ observations in close proximity to the Mekong outflow were conducted in April 2007, coinciding with inter-monsoon and the river's lowest annual outflow, and in September 2008, SW-monsoon, during the river's peak discharge (Fig. 12.3). Concentrations of nitrate, phosphate, and silicate were only up to 30 μmol L^{-1}, 1.0 μmol L^{-1}, and 130 μmol L^{-1} in April 2007, respectively (Grosse et al. 2010). As can be expected, salinity and nutrient distributions varied between these monsoon seasons (Fig 12.4). Interestingly, however, concentrations of $NO_3^- + NO_2^-$ and PO_4^{3-} in mesohaline (14–32) waters in April 2007 did not differ much from concentrations at salinities as low as 0 in September 2008, whereas for $Si(OH)_4$ concentrations, there was indeed a clear difference (Fig. 12.4). This would suggest that $NO_3^- + NO_2^-$ and PO_4^{3-} concentrations in Mekong freshwater were much higher in April 2007 compared with September 2008, but our measurements showed that they were about equal (Grosse et al. 2010). Possibly this observation rather points to the effect of local (close to the river mouth) inputs of $NO_3^- + NO_2^-$ and PO_4^{3-} from aquaculture facilities and sewage in April 2007, whereas $Si(OH)_4$ is not contained in such anthropogenic sources and thus showed a more conservative mixing behavior.

Primary production rates of up to 0.64 μmol C L^{-1} h^{-1} were measured in river-influenced waters in April 2007, and slightly higher rates of up to 0.68 μmol C L^{-1} h^{-1} were measured in September 2008 (Grosse et al. 2010). The region covered by our station grids extended into open sea waters only in April 2007, and there, primary production rates did not exceed 0.05 μmol C L^{-1} h^{-1}. These low rates are in good agreement with earlier measurements in the open SCS in July 2003–2004, which included sampling of surface waters down to subsurface chlorophyll maxima (Table 12.3). Thus the

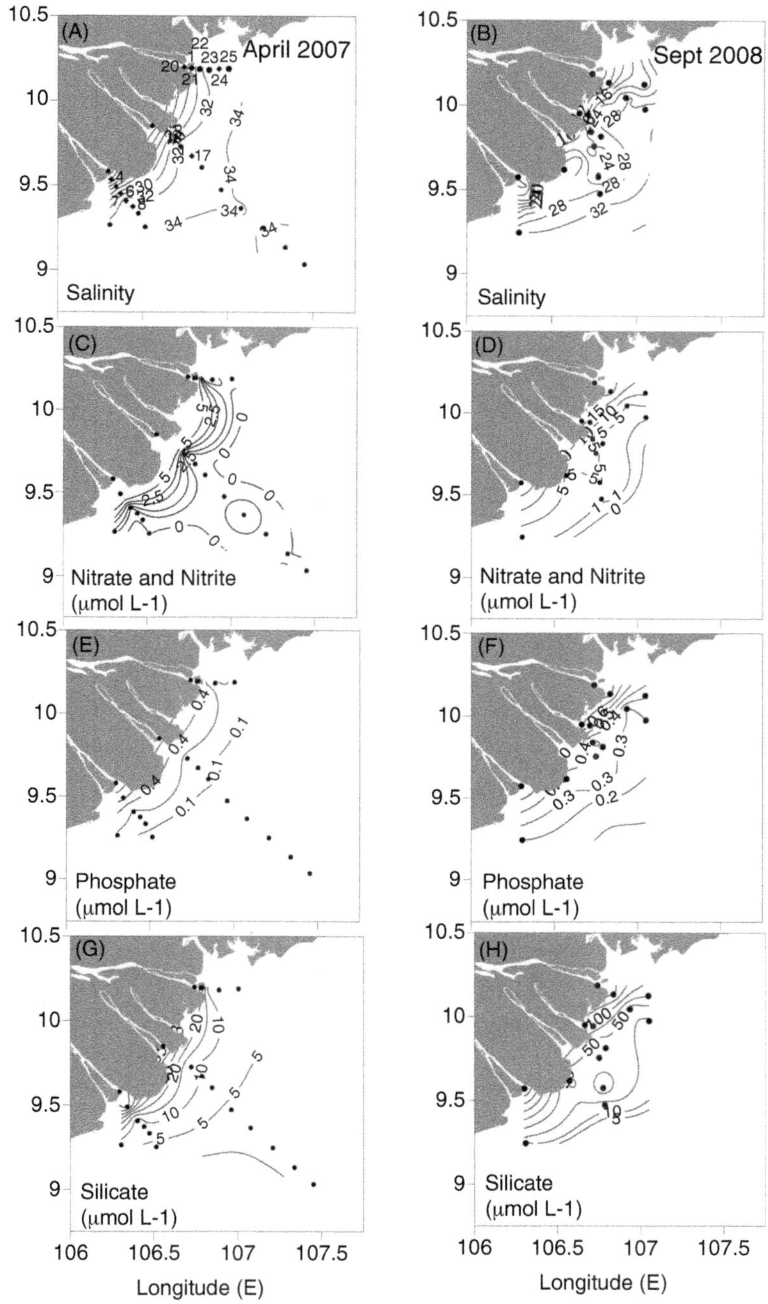

Figure 12.4. Salinity and nutrient distribution of the Mekong estuary. Left and right column show results from April 2007 and September 2008, respectively. 0.1 μmol L^{-1} denotes the detection limit for all nutrients in this study. The dashed lines denote the 32 salinity boundary. From: Grosse, J., Bombar, D., Doan, H.N., Nguyen, L.N., Voss, M. 2010. The Mekong River plume fuels nitrogen fixation and determines phytoplankton species distribution in the South China Sea during low- and high-discharge season. Limnology and Oceanography 55: 1668–1680.

Table 12.3. *Maximal volumetric rates of pelagic primary productivity measured at various locations in the SCS*

Location	Date	Maximal rate of primary production (μmol C L^{-1} h^{-1})	References
Vietnamese upwelling area, 40-m depth	July 2003	0.14	Bombar et al. 2010
Vietnamese upwelling area, 10-m depth	July 2004	0.4	Bombar et al. 2010
SCS offshore waters, 30-m depth	July 2003	0.07	Bombar et al. 2010
SCS offshore waters, 50-m depth	July 2004	0.1	Bombar et al. 2010
Mekong River plume, transitional surface waters (salinity of 32–33.5)	April 2007	0.64	Grosse et al. 2010
Mekong River Plume, mesohaline surface waters	September 2008	0.68	Grosse et al. 2010

rates measured in the Mekong plume are 7–14 times higher than typical SCS offshore rates, and they exceed maximal rates of around 0.4 μmol C L^{-1} h^{-1} encountered in the highly productive Vietnamese upwelling area around 12°N (Table 12.3).

The Mekong plume extends far offshore into the central SCS during SW monsoon, and therefore, it appears likely that during the spatially limited observation in September 2008, we only sampled a developing phytoplankton bloom. Within the sampling area, nutrient concentrations were high throughout, and chlorophyll *a* concentrations did not exceed 1.3 μg L^{-1}, which is much lower than in April 2007, when concentrations of up to 3.9 μg L^{-1} were measured. We assume that in September 2008 the peak in algal biomass and primary production must have occurred further offshore within the plume, and consequently,the measured rates do not represent the potential maxima in discharge-induced primary production.

Nitrogen (N_2) fixation rates in the Mekong plume waters fall among the highest ever measured in the marine pelagial, with up to 22.77 nmol N L^{-1} h^{-1} in April 2007 and up to 5.05 nmol N L^{-1} h^{-1} in September 2008 (Fig. 12.5) (Grosse et al. 2010). The rates would satisfy up to 47% of the N demand of primary productivity, assuming Redfield stoichiometry. Elevated N_2 fixation and high abundances of diazotroph microorganisms have been found not only in the Mekong plume, but also in several river plumes around the world (Carpenter et al. 1999; Foster et al. 2007; Foster et al. 2009), suggesting that there might be similar favorable factors sustaining the diazotroph planktonic microorganisms. During both cruises to the Mekong area, ratios of $NO_3^- + NO_2^-$ to PO_4^{3-} (N:P) in surface waters sharply declined with increasing salinity (Fig. 12.5). Sampling stations having N:P ratios lower than the typical oceanic Redfield ratio (16:1) were the ones characterized by the highest rates of N_2 fixation. The low N:P ratios indicate that N_2 fixation could be triggered by a PO_4^{3-} surplus

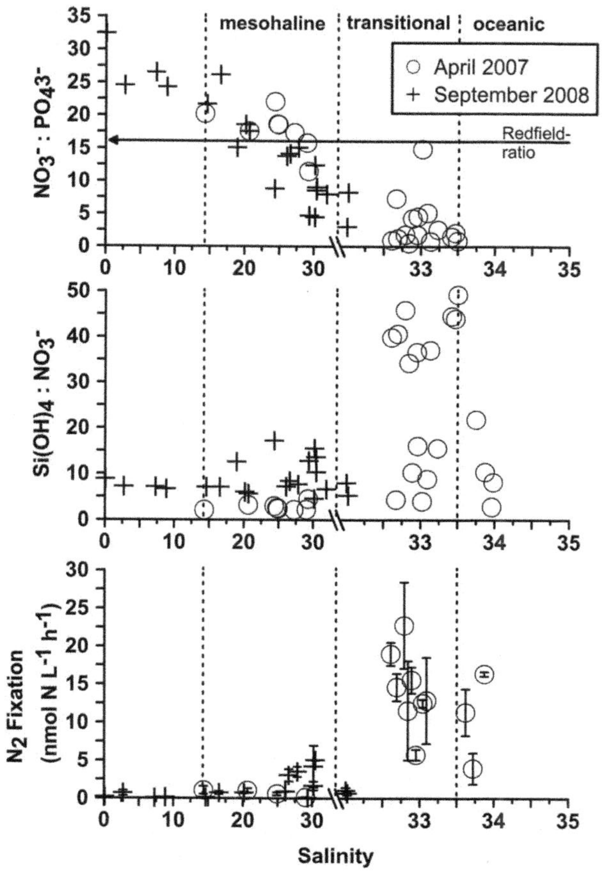

Figure 12.5. Salinity zonation mesohaline, transition, and oceanic with the nutrient ratios N:P, Si:N, and rates of N_2 fixation in Mekong estuary for April 2007 and September 2008. (D. Bombar, 2010 PhD Thesis, University of Rostock, Germany).

relative to phytoplankton nutrient requirements, a mechanism that was also suspected for the Amazon plume system (Subramaniam et al. 2008).

Coincident with this PO_4^{3-} surplus, concentrations of $Si(OH)_4$ were high in the Mekong plume (Fig. 12.5), suggesting that nutrient conditions particularly favored the growth of symbioses between different diatom species and the heterocystous cyanobacterium *Richelia intracellularis* (diatom-diazotroph associations, DDAs), which were first shown to appear as massive blooms in the Amazon plume system (Carpenter et al. 1999; Foster et al. 2007). Based on microscopy, diatom species including *Rhizosolenia* spp., *Chaetoceros* spp., *Guinardia* spp., *Hemiaulius* spp., and *Bacteriastrum* spp., which potentially hosted diazotroph symbionts, were indeed highly abundant in April 2007 as well as September 2008 in the Mekong plume. The presence and activity of the *Richelia intracellularis* symbiont was further confirmed in April 2007 by analyzing the distribution and expression of the nitrogenase (*nifH*) gene by quantitative polymerase chain reaction (PCR) (Bombar et al. 2011).

However, this molecular approach also revealed that *Trichodesmium* spp. was an at least equally important diazotroph in the Mekong River plume. *NifH* gene copies and transcripts of unicellular cyanobacterial diazotrophs (groups B and C) and a γ-proteobacterial phylotype were exclusively found at oceanic stations, confirming earlier studies that showed that the nutrient gradients within tropical river plumes are responsible for unequal distribution patterns of different marine diazotrophs (Foster et al. 2007). The diversity of the diazotroph community was assessed by a nested PCR approach targeting *nifH* and subsequent DNA sequencing of representative *nifH* clones from clone libraries. The recovered sequences in large part confirmed the presence of well-known oceanic diazotrophs, which were found previously in the area by the same method (Moisander et al. 2008). However, several *nifH* sequences represented new lineages of heterotrophic proteobacteria and diazotrophs belonging to *nifH* cluster III (including anaerobic sulfate reducers). These previously uncharacterized prokaryotes were possibly responsible for the unexpectedly high N_2 fixation (1.13 ± 1.04 nmol N L^{-1} h^{-1}) (Fig. 12.5) that was found in parts of the plume with lowest salinity and highest nutrient concentrations, but this awaits further confirmation. Such organisms could reside within anoxic microzones of suspended particles. Overall, it was shown that the Mekong River plume could be a "hot spot" of N_2 fixation, similar to the Amazon plume in the Atlantic. Given the possibility that tropical rivers might generally enhance oceanic N_2 fixation in a similar way by supplying otherwise limiting nutrients to ocean, it appears necessary to further study whether these sites substantially contribute to global oceanic N_2 fixation.

5.3. Food web and zooplankton nutrition

Nitrogen and carbon cycling in the microbial and herbivore food web in the upwelling area off Vietnam were studied by means of stable isotopes of carbon (^{13}C) and nitrogen (^{15}N) to identify the difference between spring (April/May) inter-monsoon (SpIM) and southwest Monsoon (SWM) seasons (Loick et al. 2007a; Loick et al. 2007b). In the previous paragraph, we discussed the role of N_2 fixation in the coastal area. This process has significant effects on the distribution of nitrogen stable isotopes in the planktonic food, since N_2 fixation introduces lower $\delta^{15}N$ values (Montoya et al. 2002).

The approach to study the food web included the separation of trophic levels by size fractionation following the classification approach by Sieburth et al. (1978) and the stable isotope approach by Peterson and Fry (1987). The microbial food web, including pico-, nano-, and microplankton, was collected by filtration of whole water samples from Niskin bottles onto filters with a pore size of 0.7 μm (Sameoto et al. 2000). The micro- and mesoplankton was separated with different plankton net hauls including a 55-μm and 200-μm mesh size (Sameoto et al. 2000). The net hauls were divided into six size fractions: 55–166 μm, 166–330 μm, 330–1,000 μm, 1,000–1,500 μm, 1,500–2,500 μm, and >2,500 μm.

The stable carbon isotope data ($\delta^{13}C$) showed that there was a large difference of up to 3‰ in the $\delta^{13}C$-signature between the filter samples and the net samples (Fig. 12.6) during both seasons. The trophic fractionation of carbon is very small, and a deviation by 1‰ already indicates a difference by one trophic level (Peterson and Fry 1987). The 1–3‰ gap between filter plankton (−24‰ to −23‰) and net-plankton data (−21‰ to −22‰) indicated the lack of one to two trophic levels in the size

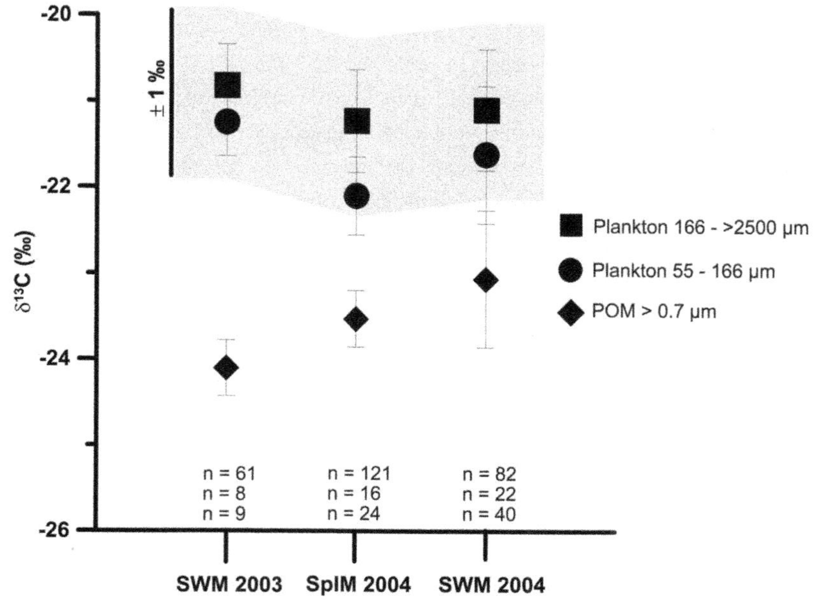

Figure 12.6. Average $\delta^{13}C$ values of particulate organic matter (POM) from filter samples, net-plankton size fraction 55–166 μm, and net-plankton size fractions 166–>2,500 μm. Number of samples are given above the x-axis for each respective season and plankton size fraction for size fraction 166–>2,500 μm (top), size fraction 55–166 μm (middle), and >0.7 μm (POM, bottom). Dotted area indicates the ±1‰ interval for potential food source of plankton in size fraction 166–2,500 μm according to Peterson and Fry (1987).

fraction <55 μm and therefore in the nano- and microplankton. The only study that systematically analyzed $\delta^{13}C$ values in plankton size classes, however, covering smaller size range of pico-, nano-, micro-, and mesoplankton was published from Mediterranean samples (Rau et al. 1990). They filtered 140 l of seawater from the Mediterranean Sea into <3 μm, 3–8 μm, 8–20 μm, 20–150 μm, and >150 μm size fractions during four seasons (Rau et al. 1990). Their results showed that particles larger than 20 μm were always greater than −22‰, whereas in the size fractions between <3 μm and 20 μm, values of −24‰ and −23‰ were observed. The size fraction between 8 and 20 μm was dominated by diatoms, naked oligotrichous ciliates, dinoflagellates, and tintinids (Rau et al. 1990). In contrast, these organisms were absent in size fraction 3–8 μm, which suggests an effective separation of groups belonging to four different trophic levels (Rau et al. 1990). The minicell recapture predator-prey and model results identified four trophic levels in plankton between 1 and 12 μm (Wikner and Hagström 1988):(1) pico- and small nanoflagellates of 1–5 μm (e.g., *Ochromonas* sp., and *Bodo* sp.), (2) nanoflagellates of 5–8 μm (e.g., *Chryptomonas* sp. and *Monas* sp.), (3) nanoflagellates of 8–10 μm (e.g., *Chromulina* sp.), and (4) ciliates and flagellates of 10–12 μm (e.g., *Cyclidium* sp.). Although we lack a qualitative analysis of nano- and microzooplankton off Vietnam, a similar food web structure is plausible from the stable carbon isotope values from filter and net-plankton. We concluded that our POM filters were dominated by small pico- and nanoplankton up to a size of 20 μm, whereas the net plankton may have been dominated by micro- and mesoplankton.

Carbon cycling in marine planktonic food webs is generally separated in two different pathways, the microbial and the herbivore pathway (Legendre and Le Fèvre 1991; Pesant et al. 2000). The difference in $\delta^{13}C$ of 1–3‰ between POM (equivalent to the microbial) and net plankton (equivalent to the herbivore) indicate that pico- and nanoplankton may have been too small to be directly grazed by zooplankton with 166 to >2,500 μm because copepods feed at a predator-to-prey size ratio of 18 (Hansen et al. 1994). Predators in the size fractions 166–1,000 μm would therefore feed on particles between 9 μm and 45 μm, a range dominated by flagellates, copepods, and cladocerans (Calbet and Landry 1999; Hansen et al. 1994). This is in close accordance with our data so that the 166 to >2,500 μm size fraction feeds on plankton of 55–166 μm, which is equivalent to a predator-to-prey size ratio between 6 and 45. We assume that carbon cycling was decoupled between the microbial and herbivore food web. Consequently, mesozooplankton may have a negligible direct grazing impact on the microbial food web, which is in consensus with observed coexistences of pico-and nanoplankton populations and larger-sized phytoplankton as observed in the oceans (Kiørboe 1993) and in the central Pacific ocean (Calbet and Landry 1999).

In the SCS, the role of diazotrophy for the food web manifested in a difference of 2‰ $\delta^{15}N$ in samples from within and outside the upwelling area. Increasing $\delta^{15}N$ values in bulk and amino acids were found with the size of the zooplankton (Fig. 12.7) (Loick et al. 2007b). Enrichments in $\delta^{15}N$ with increasing plankton size have been mainly attributed to trophic fractionation effects (Fry and Quinones 1994; Rolff 2000), whereas the lower $\delta^{15}N$ values outside the upwelling center have been attributed to the input of nitrogen from N_2 fixation (Montoya et al. 2002; Loick et al. 2007a). The specific $\delta^{15}N$ isotope signature of N_2 of 0‰ can be found almost unaltered in the total nitrogen of the N_2-fixing organisms (Montoya et al. 2002). If the diazotroph nitrogen is assimilated by zooplankton, it lowers the total nitrogen $\delta^{15}N$ value of zooplankton significantly below 5‰ (Owens 1987). This unique signal identifies diazotroph nitrogen in zooplankton in areas with high N_2 fixation such as the tropical North Atlantic (Montoya et al. 2002) or the southern SCS (Loick et al. 2007a). The observed increases in $\delta^{15}N$ between the consecutive size fractions of average 0.2‰ ($\pm$ 0.6, $n = 263$) were much smaller than the reported 2–3.5‰ increase between consumer and diet (Peterson and Fry 1987). One reason may be the insufficient resolution of trophic levels and the overlap within one size fraction (Fry and Quinones 1994). Nutritional stress can cause increases in $\delta^{15}N$ of up to 0.4‰ in *Daphnia magna*, which may be significantly correlated with the C:N ratio of the food (Adams and Sterner 2000). During decreasing N content in food from a C:N ratio of 6 to 25, the corresponding $\delta^{15}N$ values of the daphnids feeding on the algae increased from 1‰ to 4‰ (Adams and Sterner 2000). In the SCS we found no such correlation. However, the general finding of increase in $\delta^{15}N$ with size along trophic levels from the Atlantic ocean (Fry and Quinones 1994; Montoya et al. 2002) was confirmed by Loick (2007a). Amino acids such as glutamic acid, alanine, aspartic acid, isoleucine, leucine, proline, and valine are up to 7‰ enriched in the consumer compared with the diet as is and reflects the trophic increase in amino acid specific $\delta^{15}N$ much clearer than in the bulk $\delta^{15}N$. This was first shown by McClelland (2002) and confirmed for the SCS (Loick et al. 2007b).

The transfer of new nitrogen from N_2 fixation into larger zooplankton size fractions and higher trophic levels was a distinct feature during SWM season. We found a lowering in $\delta^{15}N$ in all zooplankton size fractions at offshore sites (Fig. 12.7). Its occurrence outside the upwelling area

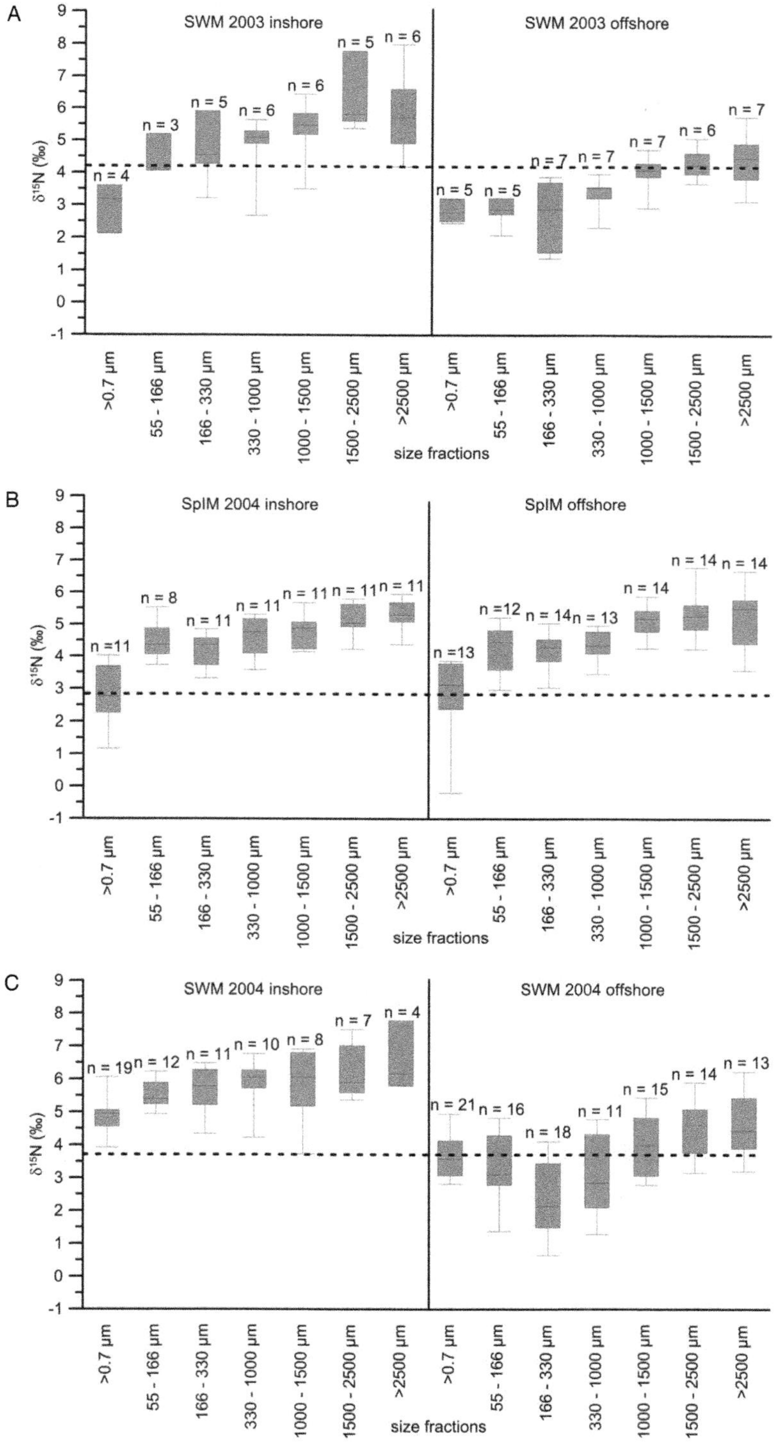

Figure 12.7. $\delta^{15}N$ patters of particulate organic matter (POM >0.7) and six net-plankton size fractions (55–>2,500) from the upper 100 m during (A) southwest monsoon 2003, (B) spring inter-monsoon 2004, and (C) southwest monsoon 2004. The full lines separate inshore (<42 km off the coast) and offshore (>42 km off the coast) stations. Dashed lines give the integrated concentration weighed $\delta^{15}N$ value of sub-thermocline nitrate for each season, Note: Inshore means within the upwelling area as identified by Dippner et al. (2007).

during SWM season probably was related to enhanced N_2 fixation rates and fit to the northward deflected Mekong River plume where favorable conditions for diazotrophs prevailed. The current view is that nitrogen from nitrogen fixation is passed on to diatoms, dinoflagellates, and their grazers via ammonium from decomposed cyanobacteria and dissolved organic nitrogen (DON) exudates, although diazotrophs also may be directly grazed by some mesozooplankton (Loick-Wilde et al. 2012). It is not clear in what form nitrogen from N_2 fixation was transferred into the herbivore food chain, but the significant differences in $\delta^{13}C$ between POM and net plankton rule out a direct consumption of cyanobacteria.

Based on the assumption that size fraction 55–166 μm was dominated by phytoplankton, the median $\delta^{15}N$ values in this size fraction from stations outside the upwelling area (2.9‰ and 3.1‰ for SWM 2003 and 2004, respectively) show that a mixture of the two ultimate N sources (dissolved N_2 and NO_3^-) may be a mixture of nitrate in the source upwelling water ($\delta^{15}N=3.5 \pm 0.3$‰ and 3.7 ± 0.6‰ for SWM 2003 and 2004, respectively) and regenerated N from N_2 fixers (−1‰, Montoya et al. 2002). According to the rule of proportion, the $\delta^{15}N$ in phytoplankton is presumably composed of 9% and 13% N from N_2 fixation during SWM 2003 and 2004, respectively. This is 10% higher than the estimated contribution of diazotroph nitrogen for primary production (Voss et al. 2006) and may reflect the accumulation of diazotroph nitrogen in zooplankton tissue owing to slower turnover times.

Correlations between the $\delta^{15}N$ values of essential versus nonessential amino acids in zooplankton from the SCS (Fig. 12.8) and the tropical north Atlantic at sites of enhanced N_2 fixation point to spatial differences in the animals' nitrogen metabolism. Animals from areas with high N_2 fixation may be different from ones where nitrate is the principal new nitrogen source (Loick et al. 2007a). Starvation led to an increase of amino acid $\delta^{15}N$ in, for example, leucine from 12‰ to 16‰ within 72 hours (N. Loick-Wilde, unpublished data). The underlying mechanism may be that enzymatic isotope fractionation leads to a preferential catabolism of the lighter, faster reacting ^{14}N isotope, leaving an enhanced ^{15}N pool in the residual amino acids (Macko et al. 1987; Gannes et al. 1998). An increase of the $\delta^{15}N$ in leucine resulting from starvation rather than trophic fractionation therefore may have caused the observed positive correlation of $\delta^{15}N$ of some amino acids at stations where food may have been limited because of nitrate limitation of phytoplankton (Fig. 12.8), which has been described for tropical rivers in general (Chapter 15). This would mean that zooplankton outside the upwelling area catabolized body proteins to meet their energy demand. However, this is a tentative interpretation, and more data are needed to better understand underlying mechanisms.

Recent experimental data showed that zooplankton quickly responded to a changing food environment by selective incorporation of nitrogen into amino acids, of which essential isoleucine received most of the offered diazotroph nitrogen (nitrogen from N_2 fixation) (Loick-Wilde et al. 2012). At present, the combination of molecule-specific nitrogen isotope analysis using natural nitrogen stable isotope abundances and stable nitrogen isotope tracers offers one of the best approaches to directly follow different nitrogen sources into functional molecules such as essential and nonessential amino acids in zooplankton tissue. Considering the key position of amino acids in zooplankton metabolism, this approach can elucidate outstanding new insights into the animals' physiological and biochemical responses to a changing environment.

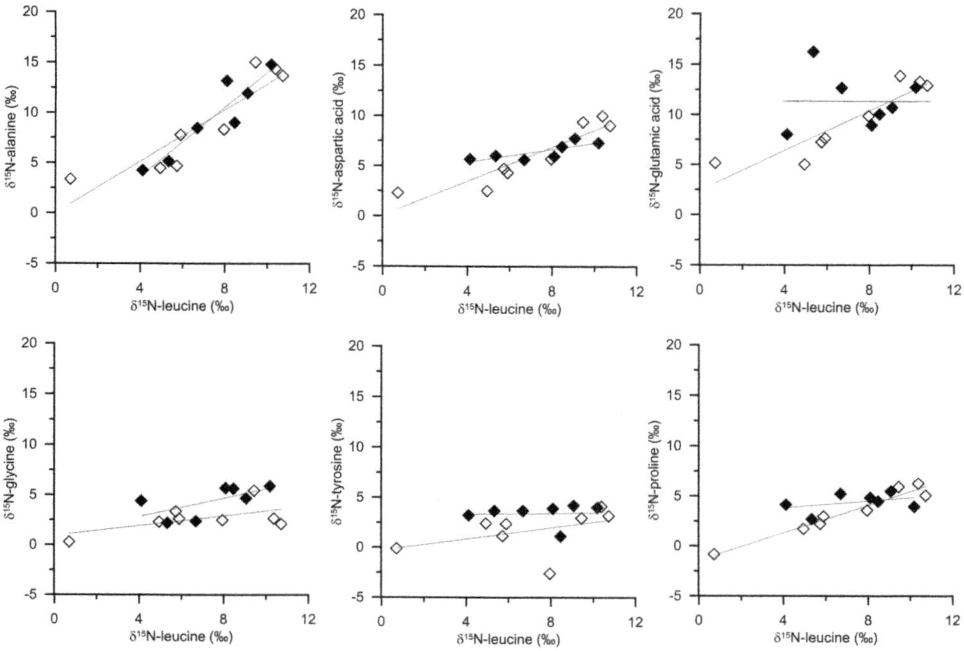

Figure 12.8. Correlation of $\delta^{15}N$ of different amino acids from increasing plankton size fractions (55 μm to >2,500 μm) from stations inside (filled symbols) and outside the upwelling area from SWM 2003 and 2004. Data points are composed from six different plankton size fractions from SWM 2003 and 2004. Black lines give the linear correlations of the different ratios. From Loick, N., Gehre, M., Voss, M. 2007b. Stable nitrogen isotopes in essential versus non-essential amino acids of different plankton size fractions. Isotopes in Environmental and Health Studies 43: 281–293.

6. How does this compare with other tropical estuaries?

Estuaries vary considerably in their physical and hydrographic properties, such as tidal stirring, depth, and magnitude and timing of freshwater runoff. The coastal waters of the SCS off Vietnam are characterized by strong tides of up to 4.5 m (Hoa et al. 2007), which cause a deep intrusion of ocean waters upstream into the Mekong River during inter-monsoon, the low discharge season. In April 2007, waters with a salinity of 0 were therefore found approximately 30 km upstream. The lowest salinity at sampling stations in the river mouth was 14.3 (Sta. 3, Fig. 12.3), and from there, salinities gradually increased to maximal values of around 34 at stations farthest offshore (Fig. 12.3). During SW monsoon, the discharge is much higher so that during our investigation in September 2008, freshwater was found in the river arms at some stations near the outflow. These pronounced seasonal differences in discharge volume basically represent large flushing events and as such should prevent to a large degree the build-up of nutrient enrichment and eutrophication within the estuary, unlike estuaries that are poorly flushed or have only short and/or episodic flushing events (Lin et al. 2006).

Similar to the Mekong, the discharge volume of the Pearl River, which drains into the northern SCS, is primarily determined by seasonal monsoon precipitation (Ho et al. 2010). However, the coastal ocean near the Pearl River outflow already receives high nutrient inputs from sewage effluent throughout the year (e.g., from Hong Kong) and from intense agriculture and aquaculture in the Pearl River delta (Yin and Harrison 2008). This has resulted in high concentrations of dissolved nutrients, with up to 100 μmol L^{-1} of nitrate in summer months when discharge is highest. Furthermore, the proportions of the dissolved nutrients appear shifted, because phosphorus concentrations on average remain <2 μmol L^{-1} and silicate concentrations remain <100 μmol L^{-1} (Yin and Harrison 2008). The consequence is nitrogen over-enrichment in estuarine waters and widespread phosphorus limitation of phytoplankton biomass and nitrogen utilization in the river estuary and the SCS (Yin and Harrison 2008). In contrast, the available data on nutrient concentrations (our scarce data and on gemstat.org) in Mekong freshwater do not show large seasonal differences in nutrient concentrations, and there does not seem to be a clear increasing trend in nutrient concentrations over several years (www.gemstat.org). This might indicate that the Mekong system is still relatively pristine as compared with the Pearl River, and/or that Mekong wetlands and remaining mangrove forests still act as an effective buffer, absorbing the increasing nitrogen loading. This is also suggested by the observed nutrient ratios in this system: N:P ratios at salinities of 0 were never higher than around 33, and Si:N ratios were around 10 (Grosse et al. 2010). We did not measure concentrations of ammonium, but concentration values given on www.gemstat.org rarely exceed 10 μmol L^{-1}, which would bring Si:N ratios down to around 5.

7. Future of the Mekong estuary and its catchment

In 1957, Cambodia, Laos, Thailand, and Vietnam started coordinating the large-scale exploitation of the natural resources of the Mekong River system by forming the Mekong Committee, which represents the United Nation's first direct involvement in international river basin planning. Since 1995, this intergovernmental body has been called the Mekong River Commission (MRC), and its main mission statement is listed as "the joint management of shared water resources and sustainable development of the Mekong River" (www.mrcmekong.org). Clearly, the biggest ongoing change in the Mekong estuary and its catchment is due to the exploitation of the hydroelectric power potential, its use as an irrigation source, and increasing ship traffic and increasing sewage loading from growing cities and aquaculture facilities. The hydroelectric potential of more than 20 GW and an irrigation potential of more than 5 million hectares have immense importance for the economic development of these countries. The population of the Lower Mekong Basin is predicted to grow to more than 100 million by 2025. Besides fishery, which is the primary source of protein for most people in the area, one of the largest business sectors is aquaculture. This sector, according to the MRC website, has grown in "phenomenal" proportions in recent years, with 1.9 million tons production in 2008, which is five times more than in 2000, and of which about 1.6 million tons originate from the Mekong Delta in Vietnam.

The combined effects of these environmental changes should ultimately result in eutrophication of the coastal zone, but there is large uncertainty about the eutrophication status of the Mekong and about how this will change the planktonic ecosystem. As mentioned previously, the available data on nutrient concentrations in Mekong freshwater do not show large seasonal differences in nutrient concentrations, and there does not seem to be a clear increasing trend in nutrient concentrations over the years. Possibly extreme damming would so much alter the timing and magnitude of discharge that signs of eutrophication would manifest themselves at an increased pace (Chapter 2).

According to the data presented, nutrient concentrations and ratios are decisive for the selection of certain functional phytoplankton groups such as diazotrophs, and thus it appears likely that increased nutrient loading and especially intense nitrogen over-enrichment would have strong effects on the structuring of the phytoplankton communities in the estuary, as has been discussed for other estuarine systems (Conley et al. 1993; Howarth and Marino 2006).

8. Conclusion

The biochemistry of the Mekong River and the adjacent coastal seas of Vietnam show distinct features unique to the region, such as the very strong and only local upwelling cell at 12°N, the important role of nitrogen fixation of various groups of phytoplankton, and their significant role for the food web. Presumably the productivity and wealth of marine food depends strongly on the Mekong River. The relationships between factors and controlling features are still far from being understood. The fast economic growth and increasing human impact in the Mekong drainage basin suggest that there is not much time left to understand the system before it turns into a different state. Decreases in sediment load and increases in nutrient input will alter the coastal system. The ecological role of the mangrove forests as filters is rapidly degraded. The impact on the people and their life will presumably be substantial.

Acknowledgment

The German-Vietnamese cooperation has been financed by the German Research Foundation, the Federal Ministry for Economic Cooperation and Development, and the Vietnamese Ministry of Science and Technology, which is greatly acknowledged. Drs. Nguyen Ngoc Lam and Doan Nhu Hai thank the National Foundation of Science and Technology Development for funding the Project No. 106.13.35.09. The observations were carried out on board of *MV Nghien Cuu Bien* and of *FS Sonne*. We are indebted to the captain and crew of both ships for their hospitality and support. Most of the cruises and other work would not have happened without the continuous support of our project coordinator, Dr. Rolf Peinert. The topography of the South China Sea and adjacent areas was provided by Delft Hydraulics, which is greatly acknowledged. Special thanks to Jeff Richey for the runoff data of Mekong River.

References

Adams, T. S., and R. W. Sterner. 2000. The effect of dietary nitrogen content on trophic level ^{15}N enrichment. Limnology and Oceanography 45: 601–607.

Bombar, D., J. W. Dippner, H. N. Doan, L. N. Ngoc, I. Liskow, N. Loick-Wilde, M. Voss, et al. 2010. Sources of new nitrogen in the Vietnamese upwelling region of the South China Sea. Journal of Geophysical Research C: Oceans 115: 1–14.

Bombar, D., P. H. Moisander, J. W. Dippner, R. A. Foster, M. Voss, B. Karfeld, J. P. Zehr. 2011. Distribution of diazotrophic microorganisms and *nifH* gene expression in the Mekong River plume during intermonsoon. Marine Ecology Progress Series. 424: 39–52.

Boonyapiwat, S. 2001. Species Composition, Abundance and Distribution of Phytoplankton in the Thermocline Layer in the South China Sea, Area IV: Vietnamese Waters, pp. 292–309. Proceedings of the SEAFDEC Seminar on Fishery Resources in the South China Sea, Area IV: Vietnamese Waters.

Calbet, A., and M. R. Landry. 1999. Mesozooplankton influences on the microbial food web: direct and indirect trophic interactions in the oligotrohic open-ocean. Limnology and Oceanography 44: 1370–1380.

Carpenter, E. J., J. P. Montoya, J. Burns, M. R. Mulholland, A. Subramaniam, and D. G. Capone. 1999. Extensive bloom of a N_2-fixing diatom/cyanobacterial association in the tropical Atlantic Ocean. Marine Ecology Progress Series 185: 273–283.

Chen, Y. F. L. 2005. Spatial and seasonal variations of nitrate-based new production and primary production in the South China Sea. Deep-Sea Research Part I-Oceanographic Research Papers 52: 319–340.

Chen, Y. L. L., H. Y. Chen, D. M.Karl, and M. Takahashi. 2004. Nitrogen modulates phytoplankton growth in spring in the South China Sea. Continental Shelf Research 24: 527–541.

Cloern, J. E. 1996. Phytoplankton bloom dynamics in coastal ecosystems: a review with some general lessons from sustained investigation of San Francisco Bay, California. Reviews of Geophysics 34: 127–168.

Conley, D. J., C. L. Schelske, and E. F. Stoermer. 1993. Modification of the biogeochemical cycle of silica with eutrophication. Marine Ecology Progress Series 101: 179–192.

Dippner, J. W., D. Bombar, N. Loick-Wilde, M. Voss, and A. Subramaniam. 2013. Comment on "Current separation and upwelling over the southeast shelf of Vietnam in the South China Sea" by Chen et al. J. Geophys. Res. 118: 1618–1623.

Dippner, J. W., and N. Loick-Wilde. 2011. A redefinition of water masses in the Vietnamese upwelling area. Journal of Marine Systems 84: 42–47.

Dippner, J. W., L. Nguyen-Ngoc, H. Doan-Nhu, and A. Subramaniam. 2011. A model for the prediction of harmful algae blooms in the Vietnamese upwelling area. Harmful Algae 10: 606–611.

Dippner, J. W., K. V. Nguyen, H. Hein, T. Ohde, and N. Loick. 2007. Monsoon Induced Upwelling off the Vietnamese Coast. Ocean Dynamics 57: 46–62.

Doan-Nhu, H., N. N. Lam, and J. W. Dippner. 2010. Development of Phaeocystis globosa blooms in the upwelling waters of the South Central coast of Viet Nam. Journal of Marine Systems 83: 253–261.

Doan-Nhu, H., and L. Nguyen-Ngoc. 2008. Phytoplankton in Binh Thuan coastal waters in 1998–2001, pp. 221–236. Proceedings of National Conference "Bien Dong-2007," Sept. 12–14, 2007. Viet Nam Academy of Science and Technology.

Doan-Nhu, H., L. Nguyen-Ngoc, C. Nguyen, T. Ho-Van, and M. A. Nguyen-Thi. 2003. Plankton assemblages during the late bloom of Haptophyte Algae in Binh Thuan province, Southern central Vietnam, in July 2002. Collection of Marine Research Works 13: 105–118.

Food and Agriculture Organization. 2009. Food and Agriculture Organization of the United Nations, Statistical Yearbook: world food and agriculture, Vol. 4, 2009. Available at:http://www.fao.org/economic/the-statistics-division-ess/publications-studies/statistical-yearbook/fao-statistical-yearbook-2009/c-international-trade/en/.

Foster, R. A., A. Subramaniam, C. Mahaffey, E. J. Carpenter, D. G. Capone, and J. P. Zehr. 2007. Influence of the Amazon River plume on distributions of free-living and symbiotic cyanobacteria in the western tropical north Atlantic Ocean. Limnology and Oceanography 52: 517–532.

Foster, R. A., A. Subramaniam, and J. P. Zehr. 2009. Distribution and activity of diazotrophs in the Eastern Equatorial Atlantic. Environmental Microbiology 11: 741–750.

Fry, B., and R. Quinones. 1994. Biomass spectra and stable isotope indicators of trophic level in zooplankton of the northwest Atlantic. Marine Ecology Progress Series 112: 201–204.
Gannes, L. Z., C. Martinez Del Rio, and P. Koch. 1998. Natural abundance variations in stable isotopes and their potential uses in animal physiological ecology. Comparative Biochemistry and Physiology Part A 119: 725–737.
Grosse, J., D. Bombar, H. N. Doan, L. N. Nguyen, and M. Voss. 2010. The Mekong River plume fuels nitrogen fixation and determines phytoplankton species distribution in the South China Sea during low- and high-discharge season. Limnology and Oceanography 55: 1668–1680.
Hansen, B., P. K. Bjørnsen, and P. J. Hansen. 1994. The size ratio between planktonic predators and prey. Limnology and Oceanography 39: 395–403.
Harding, L. W. 1994. Long-term trends in the distribution of phytoplankon in Chesapeake Bay-Roles of light, nutrients and stream flow. Marine Ecology Progress Series 104: 267–291.
Hellerman, S., and M. Rosenstein. 1983. Normal monthly wind stress over the world ocean with error estimates. Journal of Physical Oceanography 13: 1093–1104.
Ho, A. Y. T., J. Xu, K. Yin, Y. Jiang, X. Yuan, L. He, D. M. Anderson, H. W. Lee Joseph, and P. J. Harrison. 2010. Phytoplankton Biomass and Production in Subtropical Hong Kong Waters: Influence of the Pearl River Outflow. Estuaries and Coasts 33: 170–181.
Hoa, L. T. V., N. H. Nhan, E. Wolanski, T. T. Cong, and H. Shigeko. 2007. The combined impact on the flooding in Vietnam's Mekong River delta of local man-made structures, sea level rise, and dams upstream in the river catchment. Estuarine Coastal and Shelf Science 71: 110–116.
Howarth, R. W., and R. Marino. 2006. Nitrogen as the limiting nutrient for eutrophication in coastal marine ecosystems: evolving views over three decades. Limnology and Oceanography 51: 364–376.
Hu, J., H. Kawamura, H. Hong, and Y. Qi. 2000. A review on the currents in the South China Sea: seasonal circulation, South China Sea warm current and Kuroshio intrusion. Journal of Oceanography 56: 607–624.
Hwang, C., and S. A. Chen. 2000. Circulations and eddies over the South China Sea derived from TOPEX/Poseidon altimetry. Journal of Geophysical Research 105: 23943–23965.
Isobe, A., and T. Namba. 2001. The circulation in the upper and intermediate layers of the South China Sea. Journal of Oceanography 57: 93–104.
Kemp, W. M., W. R. Boynton, J. E. Adolf, D. F. Boesch, W. C. Boicourt, G. Brush, J. C. Cornwell, T. R. Fisher, P. M. Glibert, J. D. Hagy, L. W. Harding, E. D. Houde, D. G. Kimmel, W. D. Miller, R. I. E. Newell, M. R. Roman, E. M. Smith, and J. C. Stevenson. 2005. Eutrophication of Chesapeake Bay: historical trends and ecological interactions. Marine Ecology Progress Series 303: 1–29.
Kiørboe, T. 1993. Turbulence, phytoplankton cell size, and the structure of pelagic food webs. Advances in Marine Biology 29: 1–72.
Kuo, N. J., Q. Zheng, and C. R. Ho. 2004. Response of Vietnam coastal upwelling to the 1997–1998 ENSO event observed by multisensor data. Remote Sensing of Environment 89: 106–115.
Legendre, L., and J. Le Fèvre (eds.). 1991. From Individual Plankton Cells to Pelagic Ecosystems and to Global Biogeochemical Cycles. Springer Verlag.
Lin, H.-J., X.-X. Dai, K.-T. Shao, H.-M. Su, W.-T. Lo, H.-L. Hsieh, L.-S. Fang, and J.-J. Hung. 2006. Trophic structure and functioning in a eutrophic and poorly flushed lagoon in southwestern Taiwan. Marine Environmental Research 62: 61–82.
Liu, C. L., D. L. Tang, S. F. Wang, and L. Nguyen-Ngoc. 2009. Variations in the dominant algal bloom-forming species in the western South China Sea from 1993 to 2007. African Journal of Marine Science 31: 373–380.
Liu, K. K., S. Y. Chao, P. T. Shaw, G. C. Gong, C. C. Chen, and T. Y. Tang. 2002. Monsoon-forced chlorophyll distribution and primary production in the South China Sea: observations and a

numerical study. Deep-Sea Research Part I-Oceanographic Research Papers 49: 1387–1412.

Liu, W. T., and X. Xie. 1999. Space-based observations of the seasonal changes of South Asian monsoons and oceanic response. Journal of Geophysical Research 109: C07012.

Liu, X., X. J. Wang, and Y. D. Cheng. 2012. Abnormal upwelling and chlorophyll-a concentration off South Vietnam in summer 2007. Journal of Geophysical Research 117C7: C07021, doi: 10.1029/2012JC008052.

Loick-Wilde, N., J. Dutz, A. Miltner, M. Gehre, J. P. Montoya, and M. Voss. 2012. Incorporation of nitrogen from N2 fixation into amino acids of zooplankton. Limnology and Oceanography 57: 199–210.

Loick, N., J. W. Dippner, H. N. Doan, I. Liskow, and M. Voss. 2007a. Pelagic nitrogen dynamics in the Vietnamese upwelling area according to stable nitrogen and carbon isotope data. Deep-Sea Research Part I-Oceanographic Research Papers 54: 596–607.

Loick, N., M. Gehre, and M. Voss. 2007b. Stable nitrogen isotopes in essential versus non-essential amino acids of different plankton size fractions. Isotopes in Environmental and Health Studies 43: 281–293.

Macko, S. A., M. L. Fogel, P. E. Hare, and T. C. Hoering. 1987. Isotopic fractionation of nitrogen and carbon in the synthesis of amino acids by microorganisms. Chemical Geology (Isotope Geoscience Section) 65: 79–92.

McClelland, J. W., and J. P. Montoya. 2002. Trophic relationships and the nitrogen isotopic composition of amino acids in plankton. Ecology 83: 2173–2180.

Metzger, E. J., and H. E. Hurlburt. 1996. Coupled dynamics of the South China Sea, the Sulu Sea, and the Pacific Ocean. Journal of Geophysical Research 101: 12331–12352.

Milliman, J. D., and R. H. Maede. 1983. World-wide delivery of river sediment to the oceans. Journal of Geology 91: 1–21.

Moisander, P. H., R. A. Beinart, M. Voss, and J. P. Zehr. 2008. Diversity and abundance of diazotrophic microorganisms in the South China Sea during intermonsoon. ISME Journal 2: 954–967.

Montoya, J. P., E. J. Carpenter, and D. G. Capone. 2002. Nitrogen-fixation and nitrogen isotope abundances in zooplankton of the oligotrophic North Atlantic. Limnology and Oceanography 47: 1617–1628.

Morimoto, A., K. Yoshimoto, and T. Yanagi. 2000. Characteristics of sea surface circulation and eddy field in the South China Sea revealed by satellite altimetric data. Journal of Oceanography 56: 331–344.

Murrell, M. C., J. D. Hagy, III, E. M. Lores, and R. M. Greene. 2007. Phytoplankton production and nutrient distributions in a subtropical estuary: importance of freshwater flow. Estuaries and Coasts 30: 390–402.

Nguyen-Chi, T., L. Nguyen-Ngoc, and M. A. Nguyen-Thi. 2011. Phytoplankton at Vung Tau monitoring station in 2001–2010. Vietnam Academy of Science and Technology. Viet Nam Journal of Science and Technology 49(6B): 57–64. In Vietnamese with English abstract.

Nguyen, N.L., and H. Doan-Nhu. 1996. Harmful marine phytoplankton in Vietnam waters, In: Yasumoto, T., Oshima, and Y. Fukuyo (eds.), Harmful and Toxic algae blooms. Paris: UNESCO – IOC, 586 pp.

Nguyen-Ngoc, L., H. Doan-Nhu, P. Andersen, V. T. Ho, J. Skov, V. T. Chu, and T. B. L. Do. 2004. Occurrence of potentially toxic algae in Vietnamese coastal waters. In: Larsen, J., and Nguyen-Ngoc, L. (eds.), Potentially Toxic Microalgae of Vietnamese Waters. Opera Botanica 140: 159–180.

Nguyen-Ngoc, L., and T. Ho-Van. 2011. Processes of formation of red tide and developmental ecology of toxic algae in specific coastal waters and their impact to marine resources. In: Le,

D. T., Tran, N., Dang, V. B., and Nguyen, M. H. (eds.), Collection of the Main Results of Marine Science and Technology Program for the Development of Economic and Social Sustainable Development. Hanoi: Ministry of Science and Technology, pp. 3–123. In Vietnamese.

Nguyen-Van, K., C. Nguyen, and L. N. Ngoc. 1995. Reviews of studies on phytoplankton in the sea waters of Vietnam during 70 years, 1924–1994. Collection of Marine Research Works 6: 81–90.

Nguyen, N. L. 1999. Trichedesmium erythraeum bloom. IOC Newsletter: Harmful Algae News 19: 13.

Owens, N. J. P. 1987. Natural variations in 15N in the marine environment. Advances in Marine Biology 24: 390–451.

Pesant, S., L. Legendre, M. Gosselin, P. Bjornsen, L. Fortier, J. Michaud, and T. Nielsen 2000. Pathways of carbon cycling in marine surface waters: The fate of small-sized phytoplankton in the Northeast Water Polynya. Journal of Plankton Research 22: 779–801.

Peterson, B. J., and B. Fry. 1987. Stable isotopes in ecosystem studies. Annual Review of Ecological Systems 18: 293–320.

Pham, V. T., T. A. Nguyen, and T. D. Hoang. 2002. Some remarks on the distribution of nutrients along the transect Nha Trang-Luzon (Vietnamese-Philippines cooperative investigation, JOMSRE II, May 2000). Collection of Marine Research Works 12: 91–102.

Qu, T. 2000. Upper-layer circulation in the South China Sea. Journal of Physical Oceanography 30: 1450–1460.

Qu, T., Y. Du, G. Meyers, A. Ishida, and D. Wang. 2005. Connecting the tropical Pacific with Indian Ocean through South China Sea. Geophysical Research Letters 32: L24609.

Qu, T., Y. Du, and H. Sasaki. 2006. South China Sea throughflow: a heat and freshwater conveyor. Geophysical Research Letters 33: L23617.

Rau, G. H., J.-L. Teyssie, F. Rassoulzadegan, and S. W. Fowler. 1990. 13C/12C and 15N/14N variations among size-fractionated marine particles: implications for their origin and trophic relationships. Marine Ecology Progress Series 59: 33–38.

Rolff, C. 2000. Seasonal variation in δ13C and δ15N of size-fractionated plankton at a coastal station in the northern Baltic proper. Marine Ecology Progress Series 203: 47–65.

Sameoto, D., P. Wiebe, J. Runge, L. Postel, J. Dunn, C. Miller, and S. Coombs. 2000. Collecting Zooplankton. In: Harris, R. P., Wiebe, P. H., Lenz, J., Skjoldal, H. R., and Huntley, M. (eds.), ICES, Zooplankton Methodology Manual. London: Academic Press, pp. 55–81.

Schimanski, A., and K. Stattegger. 2005. Deglacial and Holocene evolution of the Vietnam shelf: stratigraphy, sediments and sea-level change. Marine Geology 214: 365–387.

Shamsudin, L., K. Mohamad, S. Noraslizans, and M. Kasina. 2001. Nanoplankton distribution and abundance in the Vietnamese waters of the South China Sea, pp. 198–232. Proceedings of the SEAFDEC Seminar on Fisheries Resources in the South China Sea, Area IV: Vietnamese Waters.

Shaw, P. T., and S. Y. Chao. 1994. Surface circulation in the South China Sea. Deep Sea Research Part I: Oceanographic Research Papers 41: 1663–1683.

Sieburth, J. M., V. Smetacek, and J. Lenz. 1978. Pelagic ecosystem structure: heterotrophic compartments of the plankton and their relationships to plankton size fractions. Limnology and Oceanography 23: 1256–1263.

Subramaniam, A., P. L. Yager, E. J. Carpenter, C. Mahaffey, K. Björkman, S. Cooley, A. B. Kustka, J. P. Montoya, S. A. Sanudo-Wilhelmy, R. Shipe, and D. G. Capone 2008. Amazon River enhances diazotrophy and carbon sequestration in the tropical North Atlantic Ocean. Proceedings of the National Academy of Sciences of the United States of America 105: 10460–10465.

Ta, T. K. O., V. L. Nguyen, M. Tateishi, I. Kobayashi, S. Tanabe, and Y. Saito. 2002. Holocene delta evolution and sediment discharge of the Mekong River, southern Vietnam. Quaternary Science Reviews 21: 1807–1819.

Voss, M., D. Bombar, N. Loick, and J. W. Dippner. 2006. Riverine influence on nitrogen fixation in the upwelling region off Vietnam, South China Sea. Geophysical Research Letters 33: L07604.

Wikner, J., and Å. Hagström. 1988. Evidence for a tightly coupled nanoplanktonic predator-prey link regulating the bacterivores in the marine environment. Marine Ecology Progress Series 50: 137–145.

Wu, C. R., P. T. Shaw, and S. Y. Chao. 1998. Seasonal and interannual variations in the velocity field of the South China Sea. Journal of Oceanography 54: 361–372.

Wu, J. F., S.-W. Chung, L.-S. Wen, K.-K. Liu, Y.-L. Chen, H.-Y. Chen, D. M. Karl. 2003. Dissolved inorganic phosphorus, dissolved iron, and Trichodesmium in the oligotrophic South China Sea. Global Biogeochemical Cycles 17: 8–1–8–10.

Xie, S. P., Q. Xie, D. Wang, and W.T. Liu. 2003. Summer upwelling in the South China Sea and its role in regional climate variations. Journal of Geophysical Research 108: 3261, doi:10.1029/2003JC001867, C8.

Yang, H., Q. Liu, Z. Liu, D. Wang, and X. Liu. 2002. A general circulation model study of the dynamics of the upper ocean circulation of the South China Sea. Journal of Geophysical Research 107: 3085.

Yin, K., and P. J. Harrison. 2008. Nitrogen over enrichment in subtropical Pearl River estuarine coastal waters: possible causes and consequences. Continental Shelf Research 28: 1435–1442.

Yoshimura, C., and K. Takeuchi. 2007. Estimation of nutrient runoff process in the Mekong river basin using a distributed hydrological model. Journal of Japan Society of Hydrology and Water Ressources 20: 493–504.

Zhang, R. 2000. Features of the East Asian monsoon during El Niño Episode. CLIVAR Exchanges 5: 21–22.

Zhao, H., and D. L. Tang. 2007. Effect of 1998 El Niño on the distribution of phytoplankton in the South China Sea. Journal of Geophysical Research 112: C02017.

13

Physical dynamics and biogeochemistry of the Pearl River plume

M. Dai, J. Gan, A. Han, H. S. Kung, and Z. Yin

1. Introduction

River plumes, typical of large freshwater discharges, may extend into the adjacent continental shelf hundreds of kilometers away from the estuarine mouth and become critical areas of land-ocean interaction both physically and biogeochemically (Hickey et al. 1998; Nash and Moum 2005; Dagg et al. 2008; Dai et al. 2008a; Chen and Borges 2009; Gan et al. 2010; Cao et al. 2011; Bianchi et al. 2012; Han et al. 2012).

From a physical dynamic point of view, buoyancy input from freshwater discharge forms gravitational circulation, changes the course of flow direction, and modulates the mixing intensity in the estuary. After exiting into the ambient shelf, river plume often yields a right-tilted (in the northern hemisphere) quasi-stationary bulge of buoyant freshwater and associated circular currents over the shelf at the entrance to the estuary (Chao and Boicourt 1986; Zu and Gan 2008). With the existence of ambient coastal currents, the fate and characteristics of the plume, as well as the coastal currents themselves, are largely controlled by the interaction between the plume and coastal currents (Fong and Geyer 2002; Gan et al. 2009a). The plume insulates surface coastal water from the water below and amplifies the efficiency of wind forcing near the surface (Lentz 2001)and may even generate internal waves (Nash and Moum 2005). At the same time, the lateral density gradient or pressure gradient formed between the buoyant plume and ambient seawater geostrophically alters the intensity of the wind-driven currents (Chao 1988; Gan et al. 2009b). Therefore, the interaction between the plume and coastal circulations affect not only the advection but also the turbulence mixing on the shelf, thereby affecting significantly the biogeochemistry therein. At the same time, river plumes are often loaded with carbon, nutrients, and sediments (Dagg et al. 2004; Mckee et al. 2004). Both the high nutrient discharge within the river plume and the low turbidity of its lower reach are favorable for phytoplankton growth and very often result in enhanced biological activity (Gaston et al. 2006 and references therein). River plumes are thus frequently sites of phytoplankton blooms and intensive carbon uptake in coastal seas.

It is therefore clear that river plumes, which are initiated by large river discharge, transported, and modulated by the estuarine and adjacent shelf circulation, make dynamical and biogeochemical links of the land-ocean interactions. However, the complexity of the interaction between the plume dynamics and estuarine/shelf circulation along with the associated biogeochemical alteration therein makes it challenge to elucidate their processes and mechanism. To make quantitative assessment of these processes adds to the challenge. This chapter provides an overview of the basic characteristics of the Pearl River, Pearl River Estuary (PRE), and their adjacent northern South China Sea (NSCS). We focus on the physical and biogeochemical characteristics of the river plume off the PRE over the subtropical shelf sea in the NSCS. We emphasize the coupled physical-biogeochemical processes in this extremely dynamic and complex system impacted by river plumes. We also offer approaches to distinguish the biogeochemical rates from the complex water mass transport and/or mixing. Although this chapter is site-specific based on the regional studies, we are attempting to demonstrate that such integration between physical dynamics and biogeochemistry is a key to understanding these systems of similar nature, and the approach we have exemplified should have applicability to many coastal ocean settings in the world.

2. Basics of the Pearl River, estuary, and the shelf

2.1. The Pearl River

2.1.1. Basics

The Pearl River, or Zhujiang in Chinese, is an extensive river system in southern China, ranking as the third longest river in China after the Yangtze River and the Yellow River. It is mainly composed of three tributaries, the West River (Xijiang), the North River (Beijiang), and the East River (Dongjiang), all of which share a common delta, the Pearl River Delta (PRD) (Fig. 13.1A). Both the North and East Rivers originate from Jiangxi Province with a length of 573 and 562 km, respectively (Table 13.1). Originating from Yunnan province, the West River is the largest tributary of the Pearl River system, with a length of 2,214 km (Table 13.1).

The 450,000 km^2 Pearl River basin drains the majority of the south central (Guangdong and Guangxi provinces), as well as parts of the southwest (Yunnan, Guizhou, Hunan, and Jiangxi provinces) of China, and the northeast of Vietnam. The catchment areas of the West River, the North River, and the East River are 351,500, 44,700, and 25,300 km^2, respectively (Table 13.1). The entire drainage basin of the Pearl River is located south of 27°N. With a subtropical climate, the area has a long summer (wet season) and a short winter (dry season). The average annual rainfall is 1,470 mm (Dai et al. 2008a).

The West River basin is characterized by a "karst" landscape and thus is high in carbonate mineral content of ~80% (Cai et al. 2008 and references therein). The total ion content is 176.5 mg L^{-1} in the West River (Table 13.1). The long-term average concentration of HCO_3^- is modest, being 118.3 mg L^{-1} (1939 $\mu mol\ L^{-1}$) in the West River (Table 13.1), whereas the specific HCO_3^- flux (1279×10^3 mol km^{-2} yr^{-1}) is highest among all of the world large rivers because of the high weathering rate at its drainage basin (Cai et al. 2008). The West River is also characterized by high inorganic nitrogen concentration (DIN, NO_3+NO_2+NH_4, ~126 $\mu mol\ L^{-1}$) and moderate silicate concentration ($Si(OH)_4$ ~120 $\mu mol\ L^{-1}$) (Table 13.1).

The North River is in the intermediate range in terms of carbonate content, with [HCO_3^-] of 87.1 mg L^{-1} (1,428 $\mu mol\ L^{-1}$) and the total ion content of 131.9 mg L^{-1} (Table 13.1). Nutrient concentrations in the North River are slightly higher than those of the West River, with DIN of 151.8 $\mu mol\ L^{-1}$, PO_4 (DIP) of 0.41 $\mu mol\ L^{-1}$, and $Si(OH)_4$ of 133.2 $\mu mol\ L^{-1}$, respectively (Table 13.1).

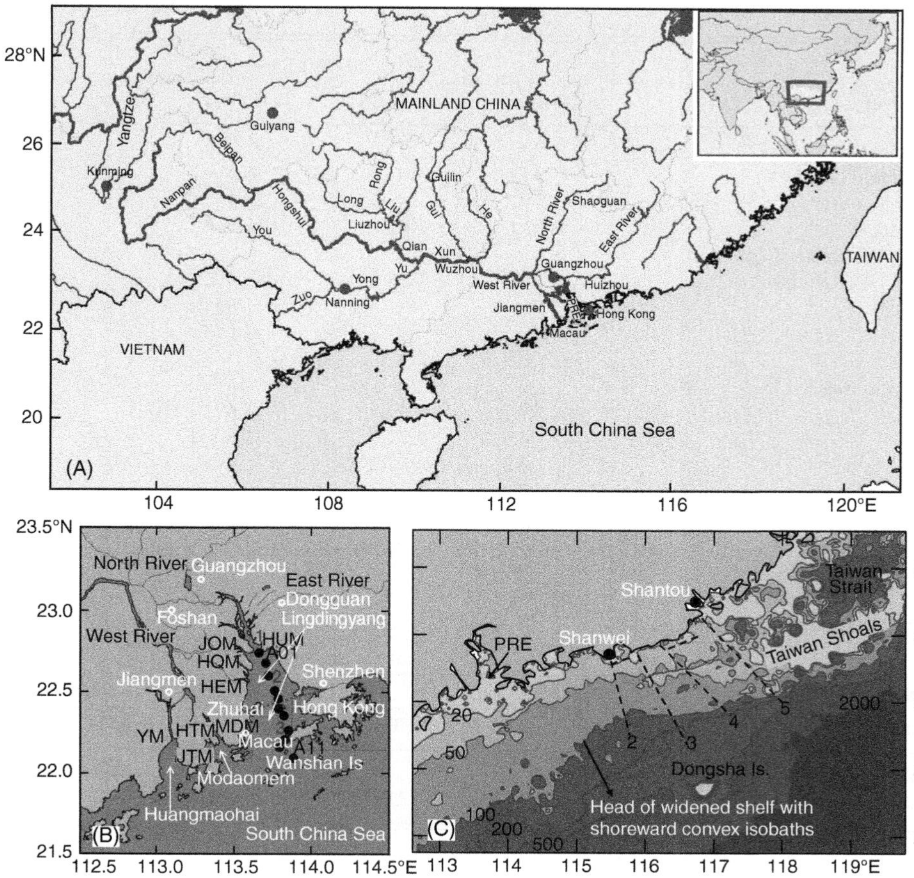

Figure 13.1. Maps of the Pearl River drainage basins (A), the Pearl River estuarine system (B, revised from Figure 1 in Guo et al. (2009), and the northern South China Sea (NSCS) shelf (C, revised from Figure 1 in Gan et al. (2010). Zones 1, 2, and 3 in Panel B denote upper, mid, and lower Pearl River estuary. Stations located near Humen to outer Lingdingyang Island are also shown in Panel B (A01 to A11). Also shown in Panel C is the topography (in meters, solid lines) in the NSCS. The selected cross-shelf sections (dashed lines) are marked by white numbers. The location of shoreward convex isobaths exists at the head of the widened shelf about half degree southwest of Shanwei.

In the eastern basin of the Pearl River, granites are abundant. The East River thus has characteristic higher silicate concentration (173.1 μmol L^{-1}) but lower bicarbonate ion concentration (31.8 mg L^{-1} or 521 μmol L^{-1}) and total ion content (52.7 mg L^{-1}) (Table 13.1). Dissolved organic carbon, dissolved inorganic nitrogen, and phosphate are similar in all of the three tributaries of the Pearl River system.

On an annual basis, the West River discharges 66.8×10^6 ton yr^{-1} of suspended sediment into the SCS, which accounts for ~90% of the total sediment flux of the Pearl River. The summation of the

Table 13.1. *Basic characteristics and hydrochemistry of the Pearl River system and its estuaries*

	Major tributaries				
Pearl River	West River	North River	East River	Others	Total
Length[a] (km)	2214	573	562	–	
Basin area[b] (10^3 km^2)	351.5	44.7	25.3	–	421.5
Discharge[c] (10^9 m^3 yr^{-1})	219.7	42.1	23.4	–	285.2
Landscape[d]	Karst	Karst	Granite		
Chemical parameters[e] (μmol L^{-1})					
DIN (NO_3+NO_2+NH_4)	125.9	151.8	145.4		
PO_4 (DIP)	NA	0.41	1.59		
$Si(OH)_4$	119.6	133.2	173.1		
DOC	108.9	82.6	86.7		
Major ions[f] (mg L^{-1})					
Ca^{2+}	29.7	23.1	5.6		
Mg^{2+}	5.0	2.7	1.6		
Na^++K^+	8.6	7.6	6.8		
Cl^-	3.3	1.9	2.8		
$SO_4{}^{2-}$	10.4	9.4	3.9		
$HCO_3{}^-$	118.3	87.1	31.8		
Total dissolved solids	176.5	131.9	52.7		
Total dissolved solids flux[f] (10^6 t yr^{-1})	35.2	5.3	1.3	–	41.8
Suspended sediment flux[g] (10^6 t yr^{-1})	66.8	5.4	2.5	–	74.7

	Sub-estuaries							
Sub-estuaries	Lingdingyang				Modaomen		Huangmaohai[#]	
Outlets[*]	HUM	JOM	HQM	HEM	MDM	JTM	HTM	YM
Surface area[h] (km^2)	1180				350		440	
Discharge[a] (10^9 m^3 yr^{-1})	57.8	54.1	20.0	35.0	88.4	18.9	19.4	18.8

[a] Chinese Bays and Estuaries Records Compilation Committee (1998).
[b] Cai et al. (2004).
[c] Dai et al. (2009).
[d] Zhao (1990).
[e] X. Guo unpublished data (Jan. 2010); Sampled at 111.33°E, 23.47°N in the West River, 112.96°E, 23.56°N in the North River, and 114.29°E, 23.16°N in the East River.
[f] Zhang et al. (2007).
[g] Zhang et al. (2011).
[h] Wong and Cheung (2000).
[*] HUM: Humen; JOM: Jiaomen; HQM: Hongqimen; HEM: Hengmen; MDM: Modaomen; JTM: Jitimen; HTM: Hutiaomen; YM: Yamen.
[#] August 2005, unpublished data from X. Guo.

Pearl River has a total suspended sediment flux of 74.7×10^6 ton yr^{-1}, including the contribution from the North (5.4×10^6 ton yr^{-1}) and East (2.5×10^6 ton yr^{-1}) Rivers (Table 13.1). Such a sediment flux from the Pearl River system only accounts for ~0.4–0.5% of the global total flux to the ocean (15–20 Gt yr^{-1}, Zhang et al. 2011).

Regarding the total dissolved solids load, the Pearl River has a total flux of 41.8×10^6 ton yr^{-1}, 84% of which is delivered by the West River (35.2×10^6 ton yr^{-1}), and only accounts for ~1% of the global total flux ($3,843 \times 10^6$ ton yr^{-1}, Zhang et al. 2007 and references therein). Similarly, the North (5.3×10^6 ton yr^{-1}) and East (1.3×10^6 ton yr^{-1}) Rivers make minor contributions to the total flux of the Pearl River (Table 13.1, Zhang et al. 2007).

2.1.2. Precipitation and river discharge

According to Dai et al. (2009), the West River has a long-term average discharge of 219.7×10^9 m^3 yr^{-1}, and ranks as the 23rd world largest river in terms of water discharge. Among the three major tributaries that merge into the PRD, the West River contributes ~76% of the total discharge. The summation of the Pearl River system has a total discharge of 285.2×10^9 m^3 yr^{-1}, including the contribution of the East and North Rivers, which makes the Pearl River system the 17th largest river in the world (Table 13.1).

There is no significant interannual variation in water discharge during this long-term period between 1948 and 2004. Exception occurred, however, to the periods of relatively high flows around the 1950s and the later 1990s and low flows in the later 1980s (Fig. 13.2A). To explore the causes behind the discharge trends, Figure 13.2A shows the trends during the same period in precipitation and air temperature (both from Dai et al. 2007). Water discharge is significantly correlated with precipitation, suggesting that precipitation change is a major cause for the discharge trends and large interannual to decadal variations. Widespread decreases in precipitation in the mid-1960s, 1990, and 2005 coincide with the decrease in runoff in these periods, whereas the increase in precipitation in the years 1959, 1969, 1975, 1980, 1995, 1997, and 2000 are consistent with runoff increases. Although the droughts in mid-1960s, 1990, and 2005 are reflected by the increase in air temperature, the high flows period corresponds to the decreased air temperature.

Monthly average discharge of the West River distributes asymmetrically and is characterized distinguishably by peak values in July. Large upward trend is from April to September, during which 80% of the annual water discharge takes place, and downward trend is from October to March, indicating the flood/wet season in summer and dry season in winter (Fig. 13.2B).

In addition to the significant monthly changes in the water discharge, there often occurs synoptic at a weekly time scale, typically forced by heavy precipitation at monsoonal season (Fig. 13.2C), which very much initiates the river plumes demonstrated in later sections (see Section 3).

2.2. Pearl River estuary

2.2.1. Basics

The Pearl River system empties into the SCS through three sub-estuaries, Lingdingyang, Modaomen, and Huangmaohai, via eight major outlets, namely Humen (HUM), Jiaomen (JOM), Hongqimen (HQM), Hengmen (HEM), Modaomen (MDM), Jitimen (JTM), Hutiaomen (HTM), and Yamen (YM) (Fig. 13.1B).

Lingdingyang, traditionally referred to as the PRE, is a funnel-shaped sub-estuary with a surface area of 1,180 km^2 (Table 13.1). Generally, the Lingdingyang sub-estuary is divided into two parts by two islands around the latitude of 22°25′ N. The northern part is designated Inner Lingdingyang; the southern part, outer Lingdingyang (Han 1998).

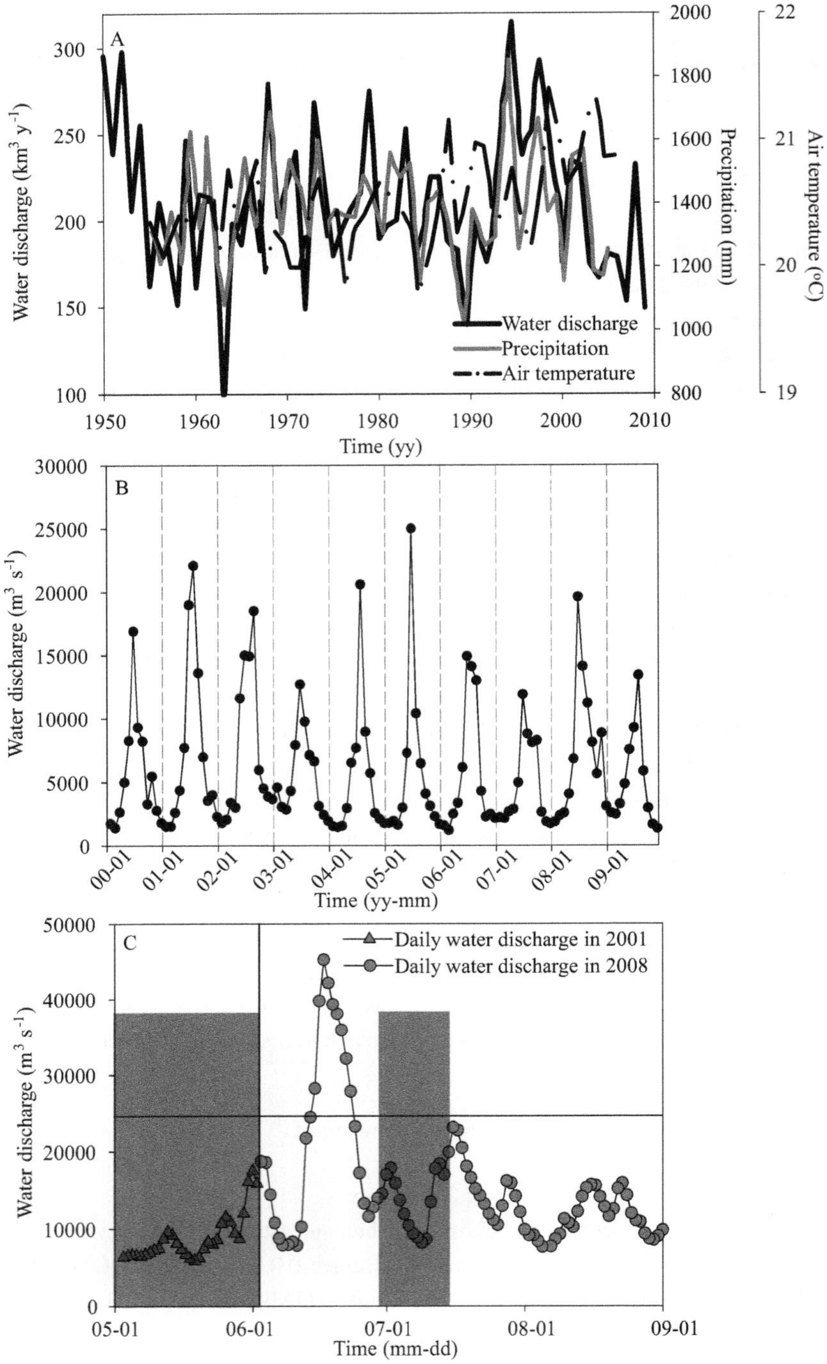

Figure 13.2. River discharge of the Pearl River system at different time scales. (A) Long-term variability of the freshwater discharge of the West River, the largest tributary of the Pearl River, which accounts for ~76% of the total freshwater discharge. Data in 1950–1984 recorded at Wuzhou gauge station are from Dai et al. (2009). Data in 1985–1999 recorded at Gaoyao gauge station are from Dai et al. (2007). Data between 2000–2009 (Wuzhou) are from xxfb.hydroinfo.gov.cn/. Data of precipitation and air temperature are collected from Dai et al. (2007). (B) Monthly average discharge of the West River in 2000–2009. (C) Daily water discharge of the West River in May 2001 and June–August 2008. The gray bar indicates values during the cruise periods in May 13–June 3, 2001, and in June 29–July 15, 2008; both cases are illustrated in this chapter showing the discharge initiated river plumes. The solid line shows the generally observed maximum value in the wet season.

The topography of the Lingdingyang sub-estuary has mixed features of channels, shoals, and tidal flats (Figure 13.1B). The depth of the sub-estuary varies from 0 to 30 m. Humen is one of the outlets at the northern end of the PRE. Two deep channels with varying depths are located in the eastern half of the sub-estuary, providing important pathways for seawater intrusion and freshwater outflow. The east channel has a water depth of about 10 m, and the west channel is shallower. These geographic and topographic features exert dynamic influences on tidal cycles, water circulation, and the water column structure. Consequently, they affect water quality and estuarine ecosystems.

The four eastern outlets (HUM, JOM, HQM, and HEM) collect about 50–55% of the Pearl River freshwater from the East and North Rivers, as well as some branches of the West River, and discharge their waters into the Lingdingyang and eventually into the continental shelf of the NSCS.

Lingdingyang is a heavily perturbed area and surrounded by several metropolis, such as Guangzhou, Shenzhen, and Hong Kong. These metropolis have populations of several to >10 million and annual sewage discharge of ~700–1,000 million tons (Bu and Ye 2007).

The Modaomen sub-estuary receives most of the freshwater of the West River through the MDM and JTM outlets, which account for about 28% (Cheung et al. 2000) of the total discharge into the sea south of Macau. The Modaomen sub-estuary is an arc-like siltation zone with its apex at the MDM and JTM outlets, with a surface area of 350 km^2 (Table 13.1). The MDM outlet is very shallow, with a water depth of 1–2 m (Guo et al. 2009).

Huangmaohai, with a surface area of 440 km^2, is also a funnel-shaped sub-estuary similar to Lingdingyang (Table 13.1). The Huangmaohai sub-estuary collects the discharge from two branches of the West River and a local river (the Tanjiang) through the HTM and YM outlets.

In contrast to the Lingdingyang, the Huangmaohai and Modaomen sub-estuaries are surrounded by relatively less populated cities such as Jiangmen and Zhuhai. Populations of these two cities are 4 and 1.5 million and the sewage discharges are 150 and 120 million tons, respectively (Bu and Ye 2007).

2.2.2. Circulation in the estuary

2.2.2.1. Gravitational circulation

As a semi-enclosed coastal water body that connects with the Pearl River discharge at its upper reach and with the adjacent NSCS shelf sea at its lower reach, PRE has an estuarine circulation that is largely forced by tides and the river influx. Tides are mainly semi-diurnal (M2) and diurnal (K1) around PRE region and have ~1.0 m magnitude inside the PRE. They are amplified and modulated as they propagate back and forth in the estuary with spatially variable water depths. It strengthens vertical shear of the currents, reduces the vertical stability of the water column, and introduces stronger vertical mixing. Tides form a counterclockwise tidal residual circulation (Mao et al. 2004) and affect the estuarine circulation in both tidal and subtidal frequencies. Although the freshwater from Pearl River discharge pushes seawater beyond the river mouth, it forms the thermohaline forcing between the buoyant river water and dense seawater, leading to a gravitational circulation in the PRE. The circulation pattern is thus determined by the relative strength between freshwater volume R and tidal volume V. In the wet season, R is strong or the ratio R/V is large. The circulation in PRE exhibits generally as a salt wedge estuary (Fig. 13.3A), in which the freshwater flushes seaward at the upper layer and seawater directs landward at the lower layer. It generates a density front that wedges

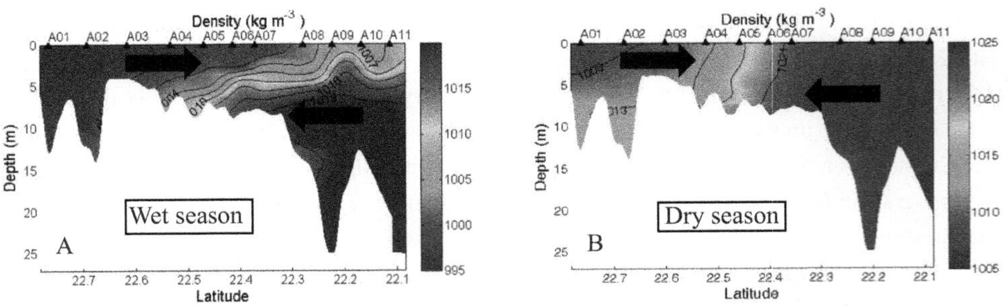

Figure 13.3. Water density along the north-south central axis of the Pearl River estuary (Lingdingyang subestuary) as a function of water depth observed during the wet season (A: August 2010) and during dry season (B: December 2010). The estuarine circulation is marked by solid arrows.

landward from the surface near the estuary mouth toward the bottom near mid-estuary. An upward transport of mass and salt (entrainment) enhances the estuarine circulation, and the vertical velocity shear near the front creates instability. The river discharge decreases dramatically in dry seasons as the ratio *R/V* becomes smaller (Fig. 13.3B). With the additional strong northeasterly wind-stirring mixing, the circulation in the PRE is generally characterized as a slightly stratified estuary, in which turbulence is strong and water column is vigorously mixed in dry seasons.

2.2.2.2. Subtidal circulation

Besides the periodical motion of tidal flow within the time scale of tidal period, the PRE is greatly controlled by the subtidal currents that vary over the period beyond the tidal period. In fact, the subtidal currents, which are controlled by both tidal and subtidal forcing, play a dominant role in the net material transport in estuary. Unlike classical estuaries that have relatively small spatial scale with gravitational circulation, the subtidal circulation in the PRE may be controlled by the intrusion of local wind-driven shelf current, besides the tidal and freshwater discharge.

The southwesterly and northeasterly monsoonal winds prevail in NSCS during the wet and the dry seasons, respectively (Fig. 13.4). The magnitude of seasonally averaged wind stress is about 0.1 Pa in the dry season (Fig. 13.4B) and about 0.025 Pa in the wet season (Fig. 13.4A). They direct surface current westward and southeastward inside the PRE and form respective upwelling and downwelling circulations (see later) on the adjacent shelf. The wind-driven surface currents can be identified by the orientation of the river plume inferred from chlorophyll *a* concentrations of satellite remote sensing (Figs. 13.4C and 13.4D) and by surface currents.

The plume mainly tilts southeastward inside the PRE and directs eastward by coastal current over the shelf in the wet season. It attaches along the west bank inside the PRE and flows westward after exiting the estuary in the dry season. The seasonal variation is mainly governed by the subtidal forcing induced by seasonal monsoon and the volume of the river discharge (Zu and Gan 2008). In the dry season, relatively strong wind and weak river discharge form westward moving of the freshwater. In the wet season, opposite conditions reduce the wind effect inside the estuary. Zu and Gan (2012) showed

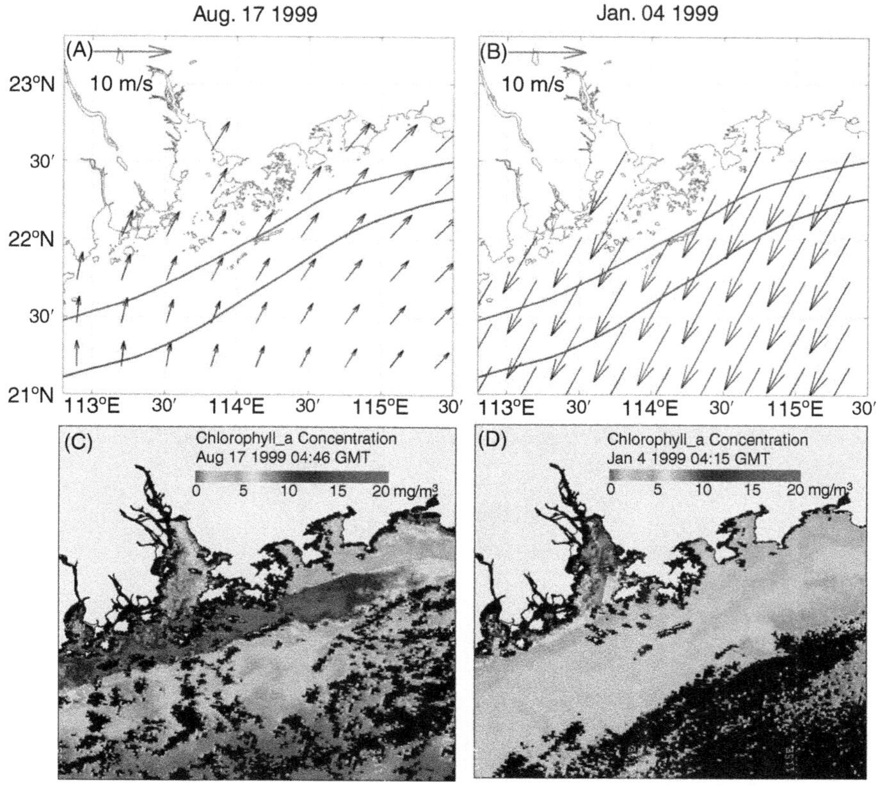

Figure 13.4. Vectors of wind speed (A and B) and surface chlorophyll *a* (Chl *a*) concentrations (C and D) in the Pearl River estuary observed in different seasons, in the wet season on August 7, 1999 (A, C) and in the dry season on January 4, 1999 (B, D). Chl *a* data are from SeaWiFS data, and the wind speed data are from NCEP reanalysis product.

that the current directs eastward inside the estuary when southwesterly wind reaches $\sim>0.05$ Pa in the wet season.

2.2.2.3. Intra-tidal circulation

The strengths of tides, river discharge, and associated mixing jointly control the advancing/retreating of seawater/river water in the PRE at intra-tidal time scale. The results obtained from the numerical results of Zu and Gan (2012) were used to demonstrates the dynamic response to forcing during different tidal phases (Fig. 13.5). The model was forced with observed wind, river discharge, and tides in the wet season of July 2000, and the results on July 27 were presented.

At Phase 1 at the peak of ebbing, the strong southward currents occupied over the entire estuary; it lowered the water level in the PRE and the adjacent shelf, as tides retreated toward the South China Sea. Relatively high elevation was formed by the buoyant river discharge at the head of the estuary, which tended to force water southward. The opposite condition occurred at Phase 4 at the peak of

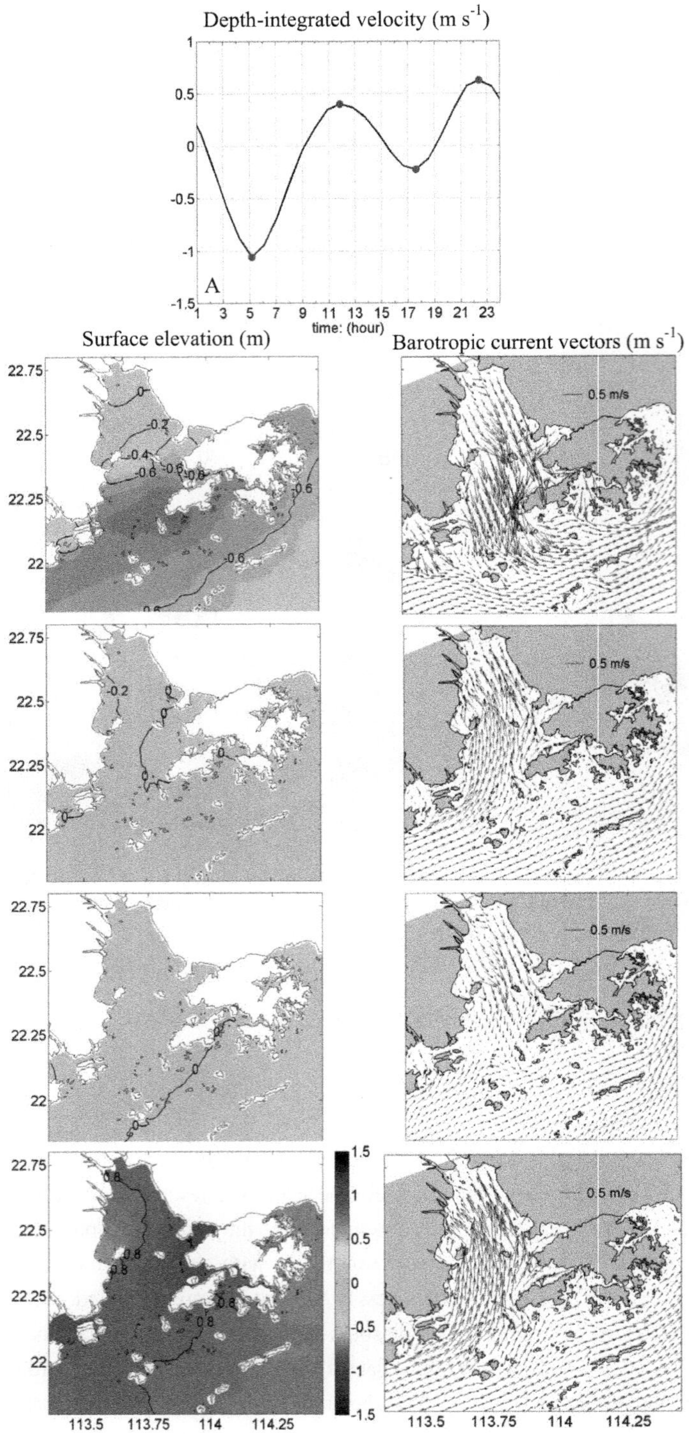

Figure 13.5. Variations of surface elevation (m, left column) and barotropic current vectors (m s^{-1}, right column) at four different tidal phases on July 27, 2000. The top panel (A) shows northward (>0)-southward (<0) depth-integrated velocity (m s^{-1}) at a station around Lingding Island in the middle of the Pearl River estuary.

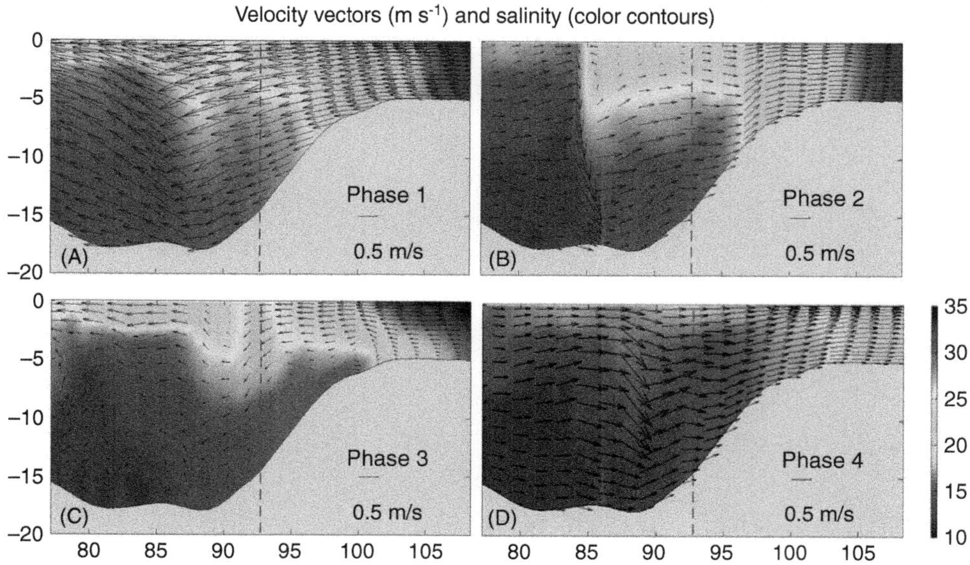

Figure 13.6. Velocity vectors (m s^{-1}) and salinity (color contours) along the axial section A (Fig. 13.1B) during the four tidal phases. The vertical velocity was artificially enlarged such that the cross-estuary current can be identified. The red dashed line indicates the location of entrance of the estuary. The x-axis is the grid number (<800 m per grid) from the southern boundary of the model domain. The red dashed line is the location of estuary entrance.

flooding, whereas the conditions at Phases 2 and 3 showed the similar features with the respective weak flooding and ebbing events between Phases 1 and 4. During all phases, the tidal currents were amplified in the estuary, and the northeastward monsoon-driven currents prevailed over the shelf. The tidal effect was relatively weak over the shelf, and the shelf currents east of Hong Kong tended to be enhanced during ebbing current.

Perhaps the intra-tidal circulation in the estuary can be more clearly seen from the velocity and salinity distribution along the axial section A (Fig. 13.1B). During Phases 1 and 4, Figures 13.6A and13.6D show the respective seaward and landward flows, dominated by tidal currents, in the weakly stratified estuary and strongly stratified adjacent shelf. Water column appeared more stratified at the ebbing phases as fresh river discharge flushing out of the estuary in the upper layer. The interesting response of circulation to tides occurred during Phases 2 and 3 (Figs. 13.6B and 13.6C), in which the *R/V* ratio was comparable and the velocity in the water column was vertically sheared in the shelf and in the estuary as well at Phase 3. The intruded seawater, built up by the flooding current before Phase 3 and entrained from the lower layer, was pushed seaward in the upper layer, which created a highly unstable two-layer water column in the estuary. At Phase 2, the landward advancing seawater met the freshwater from the prior ebbing current and generated a salinity front ~8 km south of the estuarine entrance. The convergence in the front pushed the surface water downward while the flooding current advanced seawater landward in the lower layer.

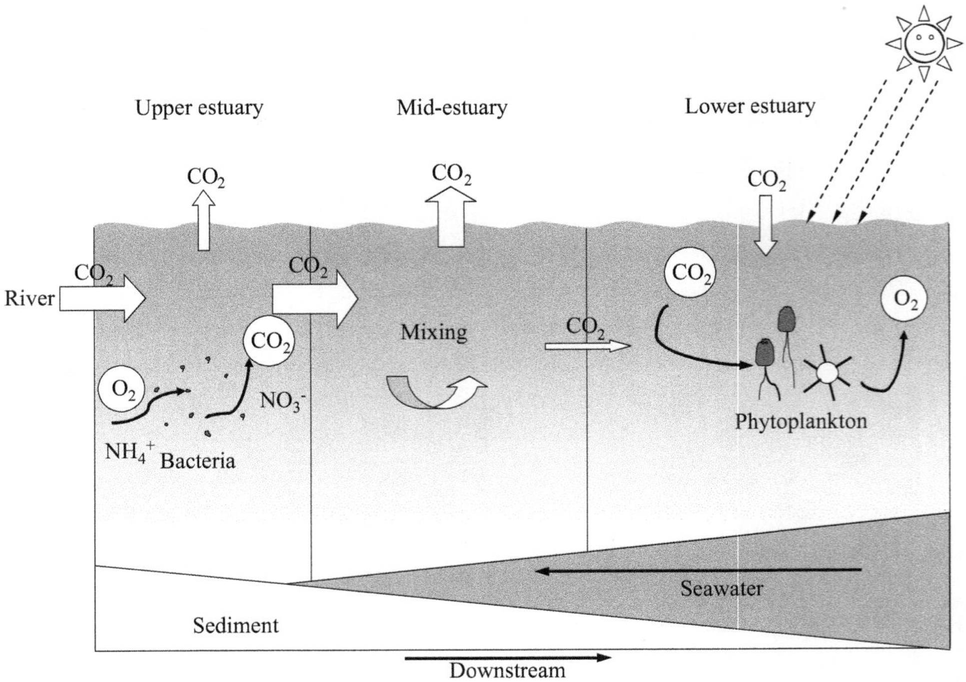

Figure 13.7. Conceptual paradigm describing the controls on biogeochemistry in the Pearl River Estuary (from Guo et al. 2009). The upper estuary was dominated by oxic respiration of organic matter and nitrification; the mid-estuary was controlled by mixing between freshwater and seawater; the lower estuary was dominated by net community productivity.

In addition, Zu and Gan (2012) found that the salt water intrusion has a distinct spring-neap variation, as the distribution of the salinity gradient changes from a sharp front, separating the freshwater inshore with seawater offshore, during spring tide into a highly stratified water column during neap tide. The landward intrusion of salt water caused by tides cannot monotonically increase/decrease with the flooding/ebbing currents, but changes with the competing effects of tidal mixing and river discharge. The classical two-layer circulation is only one of the circulation modes in the PRE. Subtidal currents, such as wind-driven currents, tend to intensify/weaken the intensities of cross estuary-shelf circulation during the different tidal phases.

2.2.3. Hydrology and biogeochemistry of the PRE

The hydrology of the PRE as illustrated by the spatial distribution of salinity within the estuary is largely reflective of the river discharge pattern and the estuarine circulation. As previously described (Guo et al. 2009), we divided the PRE into three zones for ease of discussion, namely upper estuary upstream of Human outlet, mid-estuary in the Inner Lingdingyang and Huangmaohai (here as the case of Inner Lingdingyang), and lower estuary in the Outer Lingdingyang and its adjacent northern shelf waters of NSCS (Fig. 13.1B and Fig. 13.7).

As shown in Figure 13.8, the salinity near Humen Outlet (distance = 0 in Fig. 13.9) was 0–~4.2 in the wet season as in the cases of August 2005 (summer) and April 2007 (spring), with the freshwater

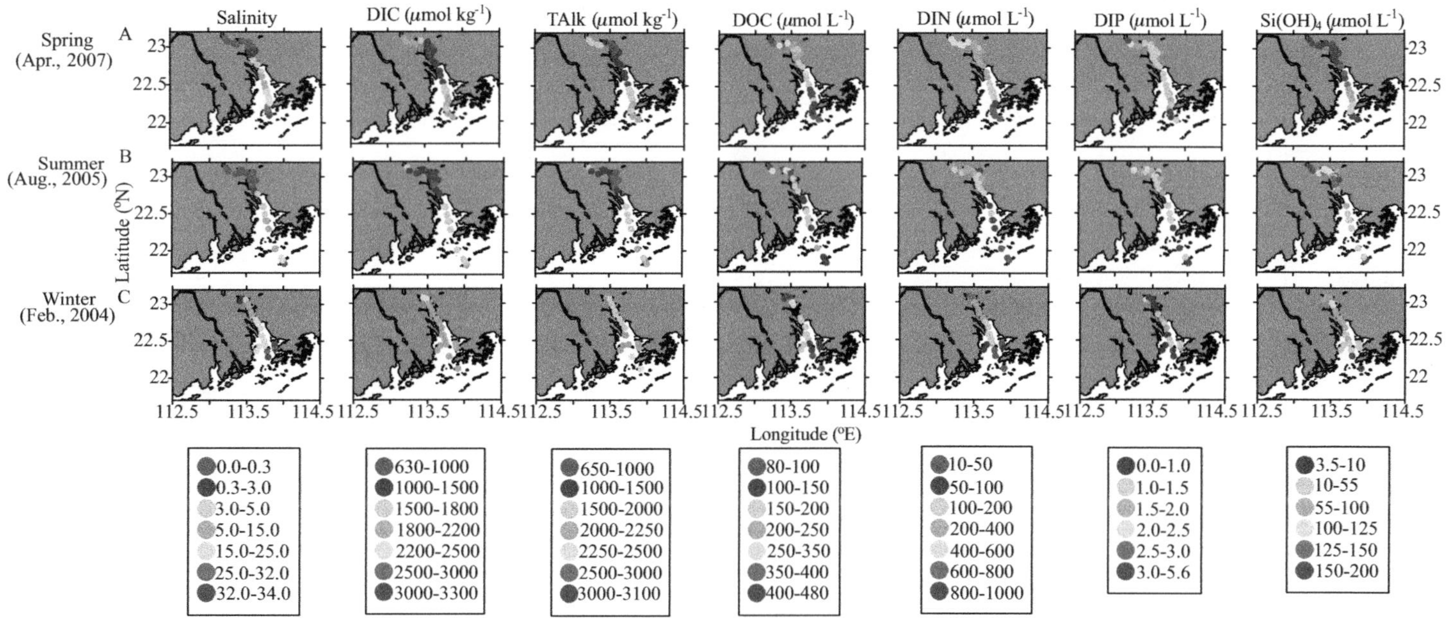

Figure 13.8. Surface distributions (≤ 5 m) of salinity, dissolved inorganic carbon (DIC) (μmol kg^{-1}), TAlk (μmol kg^{-1}), dissolved inorganic nitrogen (DIN, NO_3+NO_2+NH_4) (μmol L^{-1}), dissolved inorganic phosphorus (DIP) (μmol L^{-1}), and silicate ($Si(OH)_4$) (μmol L^{-1}) in the Pearl River Estuary in spring (April 2007) (A), wet season (summer, August 2005) (B), and dry season (winter, February 2004) (C). Data of salinity, DIC, and TAlk in the dry season [February 2004] are from Dai et al., 2006 and Guo et al. 2008. Data of salinity, DIC, and TAlk in the wet season [August 2005] are from Guo et al. 2009. Data of salinity in spring [April 2007] are from Guo et al. 2009.

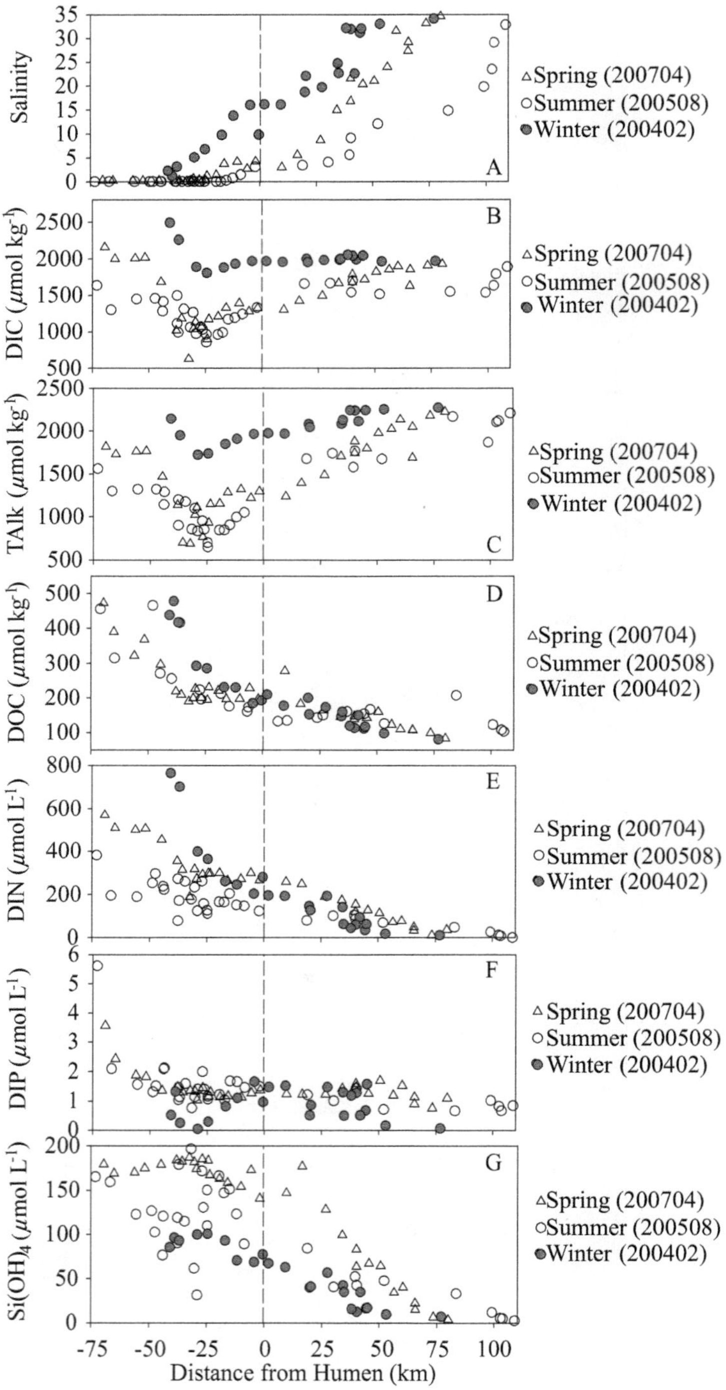

Figure 13.9. Salinity (A), DIC (μmol kg^{-1}) (B), TAlk (μmol kg^{-1}) (C), DOC (μmol L^{-1}) (D), DIN (μmol L^{-1}) (E), DIP (μmol L^{-1}) (F), and $Si(OH)_4$ (μmol L^{-1}) (G) vs. distance from Humen along the sampling transects in Lingdingyang sub-estuary during spring (April 2007), summer (August 2005), and winter (February 2004). The broken vertical lines represent the location of Humen. Positive numbers denote downstream and negative values are upstream of Humen. Data of salinity, DIC, and TAlk in the dry season [February 2004] are from Dai et al., 2006 and Guo et al. 2008. Data of salinity, DIC, and TAlk in the wet season [August 2005] are from Guo et al. 2009. Data of salinity in spring [April 2007] are from Guo et al. 2009.

end-member located at ~20 km upstream of Humen, whereas the salinity was 10–15 in the dry season in February 2004 (winter) when the zero salinity was located at ~40 km upstream of Humen. In the mid-estuary of the Lingdingyang sub-estuary, the spatial distribution of salinity was highly variable between seasons (Figs. 13.8 and 13.9), with high values in winter (~15.0–~34.0) as compared with spring (~4.0–~20.0) and summer (~3.0–~15.0), reflective of the complexity of river discharge and estuarine circulations discussed earlier. In the lower estuary, average salinity was 30.0–34.0 in the dry season in February 2004, whereas it was 15.0–33.0 in the wet season in August 2005. The salinity in the lower estuary in spring was between that in summer and winter, with average values of 22.0–34.0.

The distributions of carbon and nutrients in the PRE and their controls have been examined in a number of studies (e.g., Cai et al. 2004; Zhai et al. 2005; Dai et al., 2006; Dai et al. 2008a; Dai et al. 2008b; Guo et al. 2008; He et al., 2010; Cao et al. 2011; Han et al. 2012; Yin et al. 2012), which are briefly summarized here.

Generally, among the three zones of the PRE, the upper estuary is biogeochemically characterized by extremely high nutrients, notably NH_4, high DOC, and high pCO_2, but depleted O_2 (Guo et al. 2009), although a significant seasonal variation occurs (Figs. 13.8 and 13.9). The mid-estuary is dominated by mixing between freshwater and seawater. Consequently, nutrients and DOC behave conservatively or apparently conservatively and display less seasonal variations. In the lower estuary, nutrients and DOC decrease and are controlled by net community production owing to low turbidity.

In the upper estuary, DIC concentration is very high in winter, with the value of ~2,500 μmol kg^{-1} at the freshwater end-member, and decreases rapidly to ~1,800 μmol kg^{-1} at ~25 km upstream of Humen. In contrast, in spring and summer, DIC is ~1,250–2,250 μmol kg^{-1} at ~40–70 km upstream of Humen and with the minimum values of ~550–1,100 μmol kg^{-1} at ~25 km upstream of Humen. The distribution patterns of TAlk are similar to those of DIC, with values of ~2,146 μmol kg^{-1} in winter and 1,560–1,816 μmol kg^{-1} in summer and spring at the freshwater end-member. At ~25 km upstream of Humen, the minimum TAlk are 688, 647, and 1,722 μmol kg^{-1} in spring, summer, and winter, respectively. DOC is also enriched in the upper estuary freshwater end-member as high as ~480 μmol L^{-1} in all seasons, which might be influenced profoundly by the wastewater input from upstream cities, and decreases rapidly to ~200 μmol L^{-1} in the vicinity of Humen. The upper estuary also has a very high DIN concentration in the freshwater end-member (760 μmol L^{-1} at 40 km upstream of Humen) in winter, but lower concentrations of 570 and 380 μmol L^{-1} in spring and summer, respectively. A highly variable DIN is observed in summer at 30 km upstream of Humen, but not in winter, which is apparently influenced by large branch inputs of the East River. A remarkable feature of the Pearl River estuary is that NH_4^+ is the dominant species of inorganic nitrogen. There exists a year-round pattern of dramatic decrease in NH_4^+, increase in NO_3^-, and insignificant change in NO_2^- in the upper estuary, which is dominated by the nitrification. This process has been elaborated by Dai et al. (2008b) and is not discussed here. DIP is overall at a level of 1.0 μmol L^{-1} over the wide sampling distance in all seasons, except 3.5 and 5.6 μmol L^{-1} in freshwater end-member in spring and summer, respectively. $Si(OH)_4$ concentration is always enriched in the upper estuary in all seasons. In the freshwater end-member, $Si(OH)_4$ concentrations are ~179, 165, and 85 μmol L^{-1} in spring, summer, and winter, respectively. Note that there is a $Si(OH)_4$ peak at ~30 km upstream of Humen, and $Si(OH)_4$ displays great variable in the upper estuary zone, especially in spring and summer. This might be related to the increasing branch inputs from the East River, whose landscape is granite.

In the mid-estuary of the Inner Lingdingyang, DIC concentration increases from Humen, and is higher in winter, with the value of ~1,900 μmol kg^{-1}, but lower in spring and summer, with the value of ~1,300–1,500 μmol kg^{-1}. Similarly, TAlk also increases from Humen and is higher in winter (~1,900 μmol kg^{-1}) than that in spring and summer (~1,300–1,600 μmol kg^{-1}). DOC decreases from Humen to the Inner Lingdingyang and displays an apparently conservative behavior, with concentration in winter slightly higher than that in spring and summer. DIN in the mid-estuary is also controlled by mixing and decreases from ~180–300 μmol L^{-1} around Humen to ~100–220 μmol L^{-1} in the Inner Lingdingyang, with higher concentrations in spring and winter. As mentioned previously, DIP remains at the level around 1.0 μmol L^{-1} in all seasons. $Si(OH)_4$ decreases rapidly from ~140–170 μmol L^{-1} at Humen to ~60–100 μmol L^{-1} in the Inner Lingdingyang in spring and summer, and from ~70 μmol L^{-1} to ~30–40 μmol L^{-1} in winter.

In the lower estuary, DIC concentration increases downstream, reaching 1,933 μmol kg^{-1} in spring and 1890 μmol kg^{-1} in summer. Also, TAlk keeps increasing in the lower estuary, with the values of ~2,210–2,280 μmol kg^{-1} in all seasons. DOC continues decreasing to 80–100 μmol L^{-1} at the downstream 100 km away from Humen. DIN decreases rapidly to below ~50–100 μmol L^{-1} in all seasons. DIP is still at the level of 1.0 μmol L^{-1} in all seasons. $Si(OH)_4$ decreases from ~60–100 μmol L^{-1} in the Inner Lingdingyang to 2.5–30 μmol L^{-1} in the Outer Lingdingyang in spring and summer, and from ~30–40 μmol L^{-1} to 6.9 μmol L^{-1} in winter.

2.2.4. Conceptual summary and about the mixing behavior: Conservative and nonconservative

PRE is such a complex estuarine system that the application of the classic two-end member mixing model should be done with caution. When taking into account the mixing scheme in the upper estuary (upstream of Human), the East River has distinct end-member values (e.g., of DIC/DOC) because of the drainage characteristics and the different extent of the local material sources (see Table 13.1 and description in Section 2.1.1). As a result, the mixing curve in the upstream of Humen should adopt a three end-member mixing scheme. An example of this has been demonstrated by Guo et al. (2008), which considers the highly variable end-member concentrations apparently influenced by different tributaries with different drainage basin chemistry and anthropogenic influences.

When considering the mixing scheme downstream of Humen, in particular in the mid-estuary, a two-end member mixing model may be applicable depending on the target chemical elements, which may or may not be different in other outlets, all of which discharge into the Lingdingyang. In this mixing dominated zone, most of the chemical parameters appear to be conservative at salinity >5 in winter. In summer, biological uptake of nutrients (DIN and $Si(OH)_4$) and DIC occurs in the outer estuary and inner shelf areas where salinity is 12–25 as indicated by the nonconservative mixing line and higher DOC concentration (Fig. 13.10).

2.3. Northern South China Sea Shelf

2.3.1. Basics

The shelf over NSCS stretches from the northwest to the southeast of mainland China and from the coast to roughly the 200 m isobath with an area of about 1.2×10^6 km^2 (Fig. 13.1C). The shelf

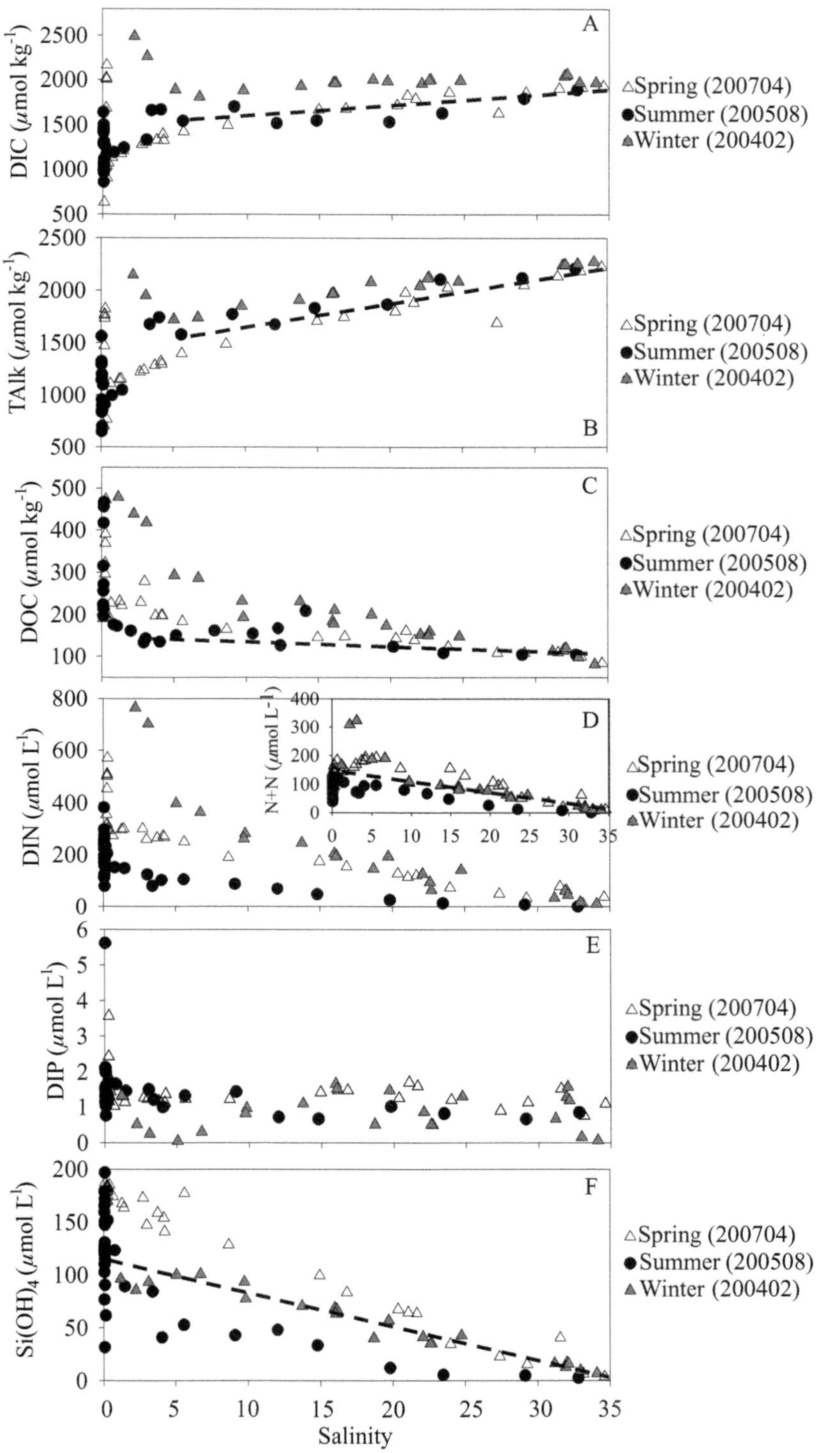

Figure 13.10. Salinity distributions of DIC (μmol kg^{-1}) (A), TAlk (μmol kg^{-1}) (B), DOC (μmol L^{-1}) (C), DIN (N+N: NO_3+NO_2) (μmol L^{-1}) (D), DIP (μmol L^{-1}) (E), and $Si(OH)_4$ (μmol L^{-1}) (F) in the Pearl River Estuary. Dashed lines indicate the conservation mixing line in summer. Data of salinity, DIC, and TAlk in the dry season [February 2004] are from Dai et al., 2006 and Guo et al. 2008. Data of salinity, DIC, and TAlk in the wet season [August 2005] are from Guo et al. 2009. Data of salinity in spring [April 2007] are from Guo et al. 2009.

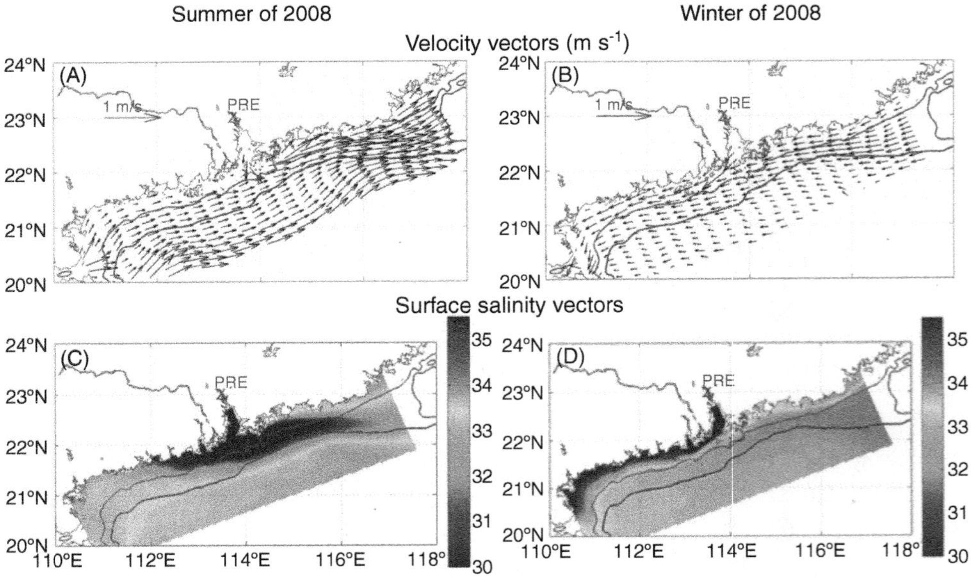

Figure 13.11. Seasonally averaged surface velocity vectors (A, B) and surface salinity in the wet (June, July, and August, C) and dry (December, January, and February, D) seasons of 2008. The red and blue contour lines are the 30 m and 50 m isobaths, respective.

topography in the NSCS is characterized by the complex coastline variation in the nearshore region and by the existence of a prominent eastward widened shelf formed by an abrupt offshore extension of isobaths east of the PRE and bounded by the 50 m isobath at its southern edge (Fig. 13.1C). A shallow bank, the Taiwan Shoals, is located between the 50-m isobath in the south and the 30-m isobath in the north at the eastern end of the widened shelf. Shelf circulation over the NSCS is mainly dominated by monsoonal wind-driven shelf circulation over the unique variable shelf topography (Gan et al. 2009a) and under the influence of buoyancy from river plume (Gan et al. 2009b).

2.3.2. Shelf circulation in the northern South China Sea

2.3.2.1. Upwelling circulation in summer (wet season)

In summer, coastal upwelling driven by strong prevailing southwesterly monsoon winds occurs over the near-shore NSCS (Gan et al. 2009a) and interplays with the buoyant river plume (Gan et al. 2009b) in the wet season. These two physical processes largely shape the nutrient dynamics and influence phytoplankton growth and the associated biological production (Gan et al. 2010; Cao et al. 2011; Han et al. 2012).

The coastal upwelling circulation in the NSCS is characterized by a strong upwelling jet regulated by the variable isobaths and coastline over the shelf with water depth less than 50 m (Fig. 13.11A). A distinct intensified upwelling occurred as a result of the unique widened shelf topography east of PRE at 115.5°E, as shown by the amplified alongshore current and associated cross-isobath transport at depths from a numerical simulation (Gan et al. 2009a). It is noteworthy that the intensified upwelling over the widened shelf is a common phenomenon occurring to many shelf seas around the world, which may invoke the geostrophically amplified shoreward advection of dense deep waters over the widened shelf and strong and efficient upslope dense water advection in the bottom boundary layer by the converging isobaths at the head of the widened shelf.

The intensified cross-isobath transport at depths also existed near the entrance of PRE and at the lee of Hong Kong Island arising from local topographic effect (Gan and Allen 2002). These hotspots of intensified upwelling circulation shaped the river plume over the shelf east of PRE and enhanced the estuary (or bay)-shelf exchange rate.

2.3.2.2. Downwelling circulation in winter (dry season)

Forced by northeasterly monsoon during winter, or the dry season, the surface currents over the shelf direct southwestward, roughly following the shelf topography as a result of geostrophy (Fig. 13.11B). With the bottom frictional effect and the effect arising from the flow-topography (effect) interaction (Gan et al. 2009a, 2013), the current near the bottom deviated from the isobaths and directed seaward. Relatively strong seaward cross-isobath transport existed at the head of the widened shelf at 115.5°E and in the waters off PRE. Similar to the conditions in summer, the alone-shore component of downwelling circulation controlled the fate of the plume in both near and far fields after river water exited the estuary. The cross-shore component, particularly in the places where its magnitude was amplified, tended to move waters at depths over the inner shelf seaward and suppressed seaward expansion of the river plume in the surface.

3. Physical dynamics and biogeochemistry of the plume

3.1. Plume over the Shelf

The Pearl River freshwater exits the PRE and forms a river plume over the broad continental shelf in the northern part of the SCS. The nature of the plume over the shelf is governed by its intrinsic dynamics as well as by the wind-driven circulation over the shelf.

The plume first formed a bulge after leaving the estuary. In the absence of the coastal current, the bulge expanded seaward and reached a quasi-stationary when the total freshwater from the river discharge and in the bulge balanced each other. It generated a positive surface elevation anomaly and isohaline. The plume attached the coastline when the river discharge was small or overshot into the open shelf water when the discharge was large. It touched with the bottom in the near field or stayed in the upper part of the water column in the far field, respectively. Plume often propagated like a first mode baroclinic wave that resulted from the density difference between the river and seawater, as described by inviscid theory (Chao and Boicourt 1986; Garvin 1987; Rennie et al. 1999). In PRE, the plume is, however, subject to the control of earth rotation, tides, and wind-driven coastal circulation (Zu et al. 2008; Gan et al. 2009b).

After entering the shelf, the discharge moved westward under the Coriolis force, forming a freshwater bulge at the western side of the estuary entrance, when the coastal current was weak or absent. The plume re-hugged the coastline as it moved westward and tended to form an anti-cyclonic circulation (Zu et al. 2008). The structure of the plume can be greatly modified by the tidal forcing. The seaward movement of the plume was deterred under the influence of tide, leading to more freshwater piling up at the head of the PRE, and forming a larger surface tilt (Fig. 13.5). The associated seaward pressure gradient increases and contributes to the formation of a stronger southward moving jet in the upper part of the estuary.

Over the shelf off PRE, the strongest forcing that controls the plume is the wind-driven coastal circulation. The plume swung westward and eastward during the wet and dry seasons, respectively (Figs. 13.11C and 13.11D). In the dry season, the co-effect of rotation and wind-driven current confined the plume to the western side of the estuary. Inside this strong and slender plume, the advection term was comparable to the Coriolis term, as the current is highly nonlinear, with a large value of the ratio of relative vorticity to planetary vorticity (ζ/f) (Zu and Gan 2008). The river plume reached Hainan Island under the northeasterly driven shelf current (Fig. 13.11B). In the wet season, strong river discharge generated a strong plume and extended over a large area in the NSCS. It yielded a bulge of plume water near the entrance of the estuary and extended westward when upwelling wind relaxed or reversed direction. A fraction of the plume emanated from the outer part of the bulge, detached from the coast and the bottom, advected eastward with its central axis approximately directed 22.1°N over the shelf, and gradually turned toward the offshore side of the upwelling jet (Fig. 13.11C). It formed a widening and deepening buoyant plume over the shelf.

Unlike the plume in the absence of the coastal current, the freshwater in the outer part of the bulge flows downstream at the speed of the current (Gan et al. 2009b), rather than the first baroclinic wave as in Chao and Boicourt (1986) or Rennie et al. (1999). In this plume-current system, the fraction of the discharged freshwater volume accumulated in the bulge reached a steady state, and the volume of newly discharged freshwater was transported downstream by the upwelling current. There was no further plume water accumulation in the bulge afterward, and all newly discharged freshwater advected downstream. The coastal current was close enough to the bulge at the entrance of the PRE that it limited the growth of the bulge (Gan et al. 2009b). With the existence of wind-driven currents in the ambient coastal water, the fate and characteristics of the plume, as well as the currents, were controlled by the interaction between the plume and wind-driven circulation.

3.2. Plume effect on the shelf circulation

The coastal current is profoundly influenced by the stratification in the water column (Allen et al. 1995; Lentz 2001). With the increase of vertical stratification by the plume, the mixed layer thins, and the role of wind stress in the surface Ekman layer is enhanced. As a result, the intensity of surface alongshore currents and cross-shelf circulation is amplified, whereas the coastal wind-driven jet is located farther from shore.

The modulation of upwelling circulation by the buoyant plume over the NSCS shelf during the wet season can be seen from numerical results (Fig. 13.11C) obtained on day 30, with the model being forced with an upwelling favorable wind (Gan et al. 2009b). The less dense plume water and the seawater formed density fronts at the lateral edges of the plume, particularly on its northern flank, where the upwelled dense water over the inner shelf met the lighter plume water offshore. Enhancement of stratification by the plume thinned the surface frictional layer and enhanced the cross-shelf circulation in the upper water column such that the surface Ekman current and compensating

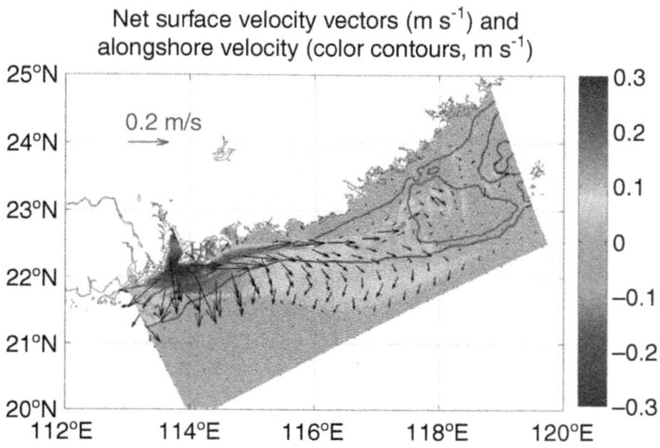

Figure 13.12. The simulated result of net surface velocity vectors (m s^{-1}) and alongshore velocity (color contours, m s^{-1}) induced by river plume 30 days after the onset of upwelling. The red and blue contour lines are the 30 m and 50 m isobaths, respectively.

flow beneath the plume were amplified, whereas the shoaling of the deeper dense water in the upwelling region changed minimally (Gan et al. 2009b). The pressure gradient generated between the buoyant plume and ambient seawater accelerated the wind-driven current along the inshore edge of the plume but retarded it along the offshore edge (Fig. 13.12).

During the dry season, the buoyant water attached along the coastline west of PRE and was expected to enhance the westward coastal current. Zu and Gan (2008) found that the co-effect of the river buoyancy and the surface Ekman transport generated a larger cross-shelf pressure gradient and resulted in a much stronger coastal current over the shelf off PRE in the dry season. Consequently, the magnitudes of the saltier water inflow on the eastern side and the freshwater outflow on the western side of the estuary were strengthened.

3.3. Biogeochemistry of the river plumes

The Pearl River plume has a profound impact on the biogeochemistry of the NSCS, primarily owing to the abundant nutrients that the plume carries to the shelf system. As being examined, the scale of the plume is clearly determined by the runoff in the upper stream, whereas the spatial pattern of the plume on the shelf is modulated by the shelf circulation manifested particularly by the upwelling.

The southwestern (SW) monsoon typically begins in April–May, and is followed by a rainy season, when potential river plumes may be formed under flood upstream. Here we present two cases studies of the Pearl River plume and its impact on the nutrient and carbon biogeochemistry. The first case is under relatively low river discharge observed in the PRE and the NSCS in May 2001 (see details in Dai et al. 2008a). In May 2001 (Fig. 13.2C), SW winds on May 1, 6–8, 13–14, and 19–22 induced significant precipitation on May 1–4, 8–9, 16–18, and 21–22. River discharge recorded showed a steady increase from 8,000 m^3 s^{-1} to as high as ~20,000 m^3 s^{-1}. Another case is in summer of 2008 (Fig. 13.2C); as detailed in Han et al. (2012), continuous heavy rain caused water discharge peaked at ~43,000 m^3 s^{-1} on June 16 and was down to ~22,000 m^3 s^{-1} on July 15. Such river discharges were

MODIS color index

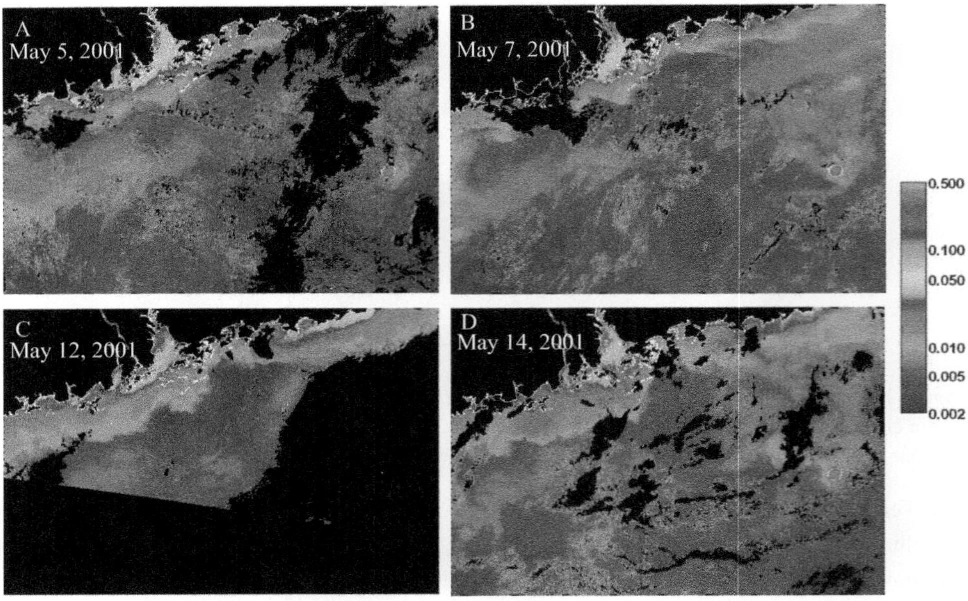

Figure 13.13. MODIS color index (CI) in the Pearl River estuary in May 2001 showing the daily average on May 5 (A), 7 (B), 12 (C), and 14 (D), 2001. This color index represents the MODIS ocean color derived empirically. The approach has applicability even under severe sun glint. The color index is significantly correlated with Chl *a*. See details in Hu (2011).

much higher than the annual mean water discharge of about 6,700 $m^3 s^{-1}$, or the monthly long-term average value of ~14,000 $m^3 s^{-1}$ from June to August.

3.3.1. Case of May 2001

Figure 13.13 shows the average MODIS ocean color index (CI) in PRE and in the adjacent SCS on May 5, 7, 12, and 14, 2001, which clearly suggests that the Pearl River plume stretched from the PRE and flowed southwestward. Notably, the plume expansion can be observable from the images of May 12 and 14 as compared with that on May 7. This is consistent with the field observation on May 8–9 (Fig. 13.2, see Fig. 2 in Dai et al. 2008a), suggesting again the plume development following the high precipitation upstream. The plume clearly brought a significant amount of nutrients into the region, as demonstrated by an increase in inorganic nitrogen concentration. At the river end during the survey, high $Si(OH)_4$ (~150 μmol L^{-1}) and NO_3 (75–120 μmol L^{-1}) concentration were observed at levels very similar to those observed during summertime measurement (Figs. 13.8 and 13.9). The consequence of such delivery of nutrients was the phytoplankton bloom observed associated with the river plume. For example, a several-fold increase (from ~0.1–0.2 mg m^3 to a maximum level of 1.8 mg m^3, an order of magnitude higher than pre-bloom conditions) in biomass (Chl *a*) was observed. In addition to increased Chl *a*, significant drawdown of pCO_2 from ~350 to ~200 μatm,

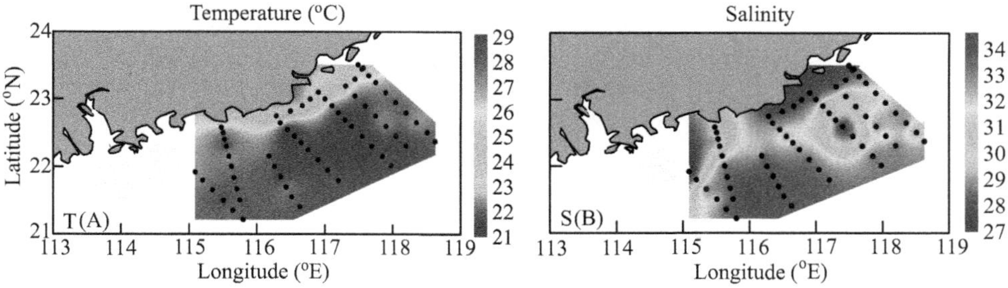

Figure 13.14. Surface distribution (≤ 5 m) of temperature (°C) (A) and salinity (B) in the northern South China Sea shelf in summer 2008 (from Fig. 2 in Han et al. 2012).

biological uptake of DIC (decreased from ~1,660 to ~1,500–1,560 μmol kg^{-1}), and associated enhancement of DO (saturation from ~95% to ~120–130%) and pH (~8.2–~8.6) were also observed.

Net DIC drawdown associated with the plume-induced bloom was assessed to be of 100–150 μmol kg^{-1} and TAlk increase of 0–50 μmol kg^{-1} (from ~2,030–2,080 to ~2,030 μmol kg^{-1}). For an average surface water depth of 5 m, a very high apparent biological CO_2 consumption rate (net community production, NCP) of 70–110 mmol m^{-2} d^{-1} was estimated. This value is 2–6 times higher than the estimated air-sea exchange rate (~18 mmol m^{-2} d^{-1}). POC concentrations in the surface waters reached 30–40 μmol L^{-1}, which was also an order of magnitude higher than the value in the pre-bloom period of ~5 μmol L^{-1}.

3.3.2. Case of August 2008

Following a continuous heavy rain in the upstream Pearl River as described previously, a strong plume that extended more than 300 km from the PRE mouth in summer 2008 was observed according to the temperature and salinity distributions (Fig. 13.14). This plume area was characterized by high nutrient concentrations (~0.1–14.2 μmol L^{-1} for DIN, ~0.02–0.10 μmol L^{-1} for DIP, and ~0.2–18.9 μmol L^{-1} for $Si(OH)_4$) and by low DIC (< 1740 μmol kg^{-1}) and TAlk (<2010 μmol kg^{-1}).

In contrast, the near shore area (upwelling) had high nutrients (0.8–6.4 μmol L^{-1} for DIN, ~0.20–0.37 μmol L^{-1} for DIP, and 5.7–19.2 μmol L^{-1} for $Si(OH)_4$) and high DIC and TAlk (higher than ~1940 μmol kg^{-1} and ~2210 μmol kg^{-1}) apparently sourced from subsurface nutrient-replete waters through wind-driven coastal upwelling, higher than those in the outer shelf surface seawater. The consequence of biomass contributed by nutrient-enriched plume and upwelling was also significant, which was expatiated in Cao et al. (2011) and Han et al. (2012).

3.3.3. Comparison between 2001 and 2008

To put the preceding two cases into comparison, we see the similarity in between in terms of the precipitation-initiated enhanced river discharge and the subsequent river plumes formed. Table 13.2 demonstrates that the discharge of the Pearl River determined the river plume intensity (plume expansion). For example, the much higher river discharge in 2008 (Fig. 13.2C) induced a much larger plume extension away from the PRE mouth relative to that in 2001. However, the concentrations of nutrients, DIC, and TAlk in both cases were similar around the Pearl River plume bulge under

Table 13.2. *Comparison of the river plume between 2001 and 2008*

	Case in May 2001	Case in June–Jul. 2008
Discharge ($m^3\ s^{-1}$)	8,000 – ~20,000	peaked at ~43,000 and was down to ~22,000
Plume direction/ extension	southwest	northeast
Plume area/extent	~109 km away from the PRE mouth	~400 km away from the PRE mouth to the southern Taiwan Strait
Nutrient	NO_3: 75–120 μmol L^{-1} at the river end	NO_3+NO_2: from ~100 μmol L^{-1} at near null salinity to ~8.0–1.5 μmol L^{-1} at the river mouth
(Carbon) DIC/TAlk (μmol kg^{-1})	DIC: ~1,660 – ~1,500–1,560; TAlk: ~2,030 – ~2,030–2,080	DIC: <1,740 TAlk: <2,010
NCP (mmol C $m^{-2}\ d^{-1}$)	~70–110	~36±19

Data are from Dai et al. (2008) for the case of 2001, and from Cao et al. (2011) and Han et al. (2012) for the case of 2008.

different discharge conditions. Nevertheless, the conservation of these chemical parameters might be variable in upper-mid-estuary during flood and/or after flood period in wet seasons, which has been illustrated by Han et al. (2012). In addition, the directions of the plume extension were different in the two cases.

Biological responses to both plume cases were also significant. However, the NCP value in the case of 2008 was much lower than in May 2001 on the NSCS shelf. This is primarily related to the plume extension. The location of the plume-induced bloom observed in the May 2001 case was limited to nearshore at the mouth of the PRE, whereas significant DIC removal extended to the far reaches of the plume in the 2008 case (Cao et al. 2011). In addition, the river plume in summer of 2008 was additionally impacted by the coastal upwelling over the NSCS shelf (see later).

4. Coupling the physical dynamics and biogeochemistry

Deconvolution of physical dynamics and biogeochemistry in complex systems such as river plumes is not an easy task, in particular when quantitative assessment is to be made. Numerical modeling is certainly a sophisticated approach that heavily relies on the rightness of the physical dynamics that the model can resolve. Alternatively, mass-balance–based end-member mixing model is a relatively straightforward way to use as far as the end-member values can be defined and quantified (Cao et al. 2011; Han et al. 2012). Here we first demonstrate how to establish the end-member mixing between different water masses and its subsequent application to quantify the biologically mediated processes.

4.1. Mixing of different water masses

The mixing model used to differentiate biogeochemical rates on top of conservative physical mixing between different water masses involves essentially water masses and their end-member concentrations of targeted chemicals, estimation of concentrations under conservative mixing without biogeochemical alteration, and finally comparing the difference between the predicted concentrations

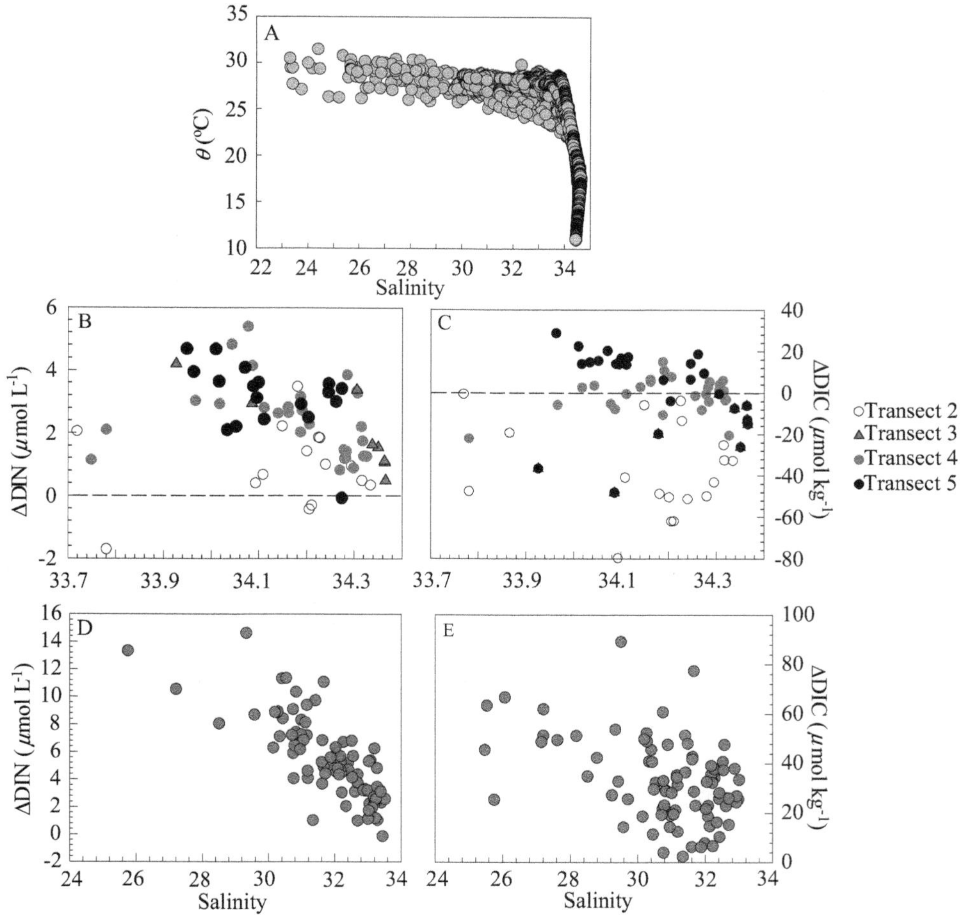

Figure 13.15. (A) The potential temperature (θ) (°C) vs. salinity scheme in the northern South China Sea (NSCS) collected from the conductivity-temperature-depth recorder per meter dataset in the whole water column (redrawn from Shu et al. 2011 and Han et al. 2012). θ–S plots exhibit three water masses: river plume, SCS surface water, and SCS subsurface water. Biological mediated dissolved inorganic nitrogen (NO_3+NO_2) (ΔDIN) (μmol L^{-1}) (B) and biological mediated dissolved inorganic carbon (ΔDIC) (μmol kg^{-1}) (C) vs. salinity in coastal upwelling and in river plume (D, E) on the NSCS shelf in summer 2008. Data in Panels B, D were collected from Han et al. 2012; Panels C, E were redrawn from Cao et al. 2011).

and actually measured concentration, which estimates the biogeochemical alteration of the targeted chemicals.

First, we have to identify the water masses that can be easily derived based on the θ-S (potential temperature-salinity) diagram. Taking the summer 2008 case as an example (Fig. 13.15A), we identified three primary water masses, namely, plume water, surface SCS water, and subsurface SCS water. Second, based on the mass balance equations for potential temperature, salinity, and the water fractions originating from the three end-members, we may resolve the fractional contribution of each water mass and predict nutrient or carbon concentrations if no biogeochemical alteration occurred to the system. The mass balance equations are as follows:

$$\theta_{RI} F_{RI} + \theta_{SW} F_{SW} + \theta_{SUB} F_{SUB} = \theta_{in\ situ} \tag{13.1}$$

$$S_{RI} F_{RI} + S_{SW} F_{SW} + S_{SUB} F_{SUB} = S_{in\ situ} \tag{13.2}$$

$$F_{RI} + F_{SW} + F_{SUB} = 1 \tag{13.3}$$

where $\theta_{in\ situ}$ and $S_{in\ situ}$ represent the potential temperature and salinity in the samples; the subscripts *RI*, *SW*, and *SUB* denote the three different sources: Pearl River plume, the SCS surface water, and the subsurface water; and F_{RI}, F_{SW}, and F_{SUB} represent the fractions in the in situ water samples contributed by the three end-members, which were calculated from the potential temperature and salinity.

The conservative nutrient concentrations of DIN (DIN°), DIP (DIP°), and $Si(OH)_4$ ($Si(OH)_4{}^{\circ}$) and DIC (DIC°) by mixing of the end-members can then be derived as:

$$DIN^{\circ} = DIN_{RI} F_{RI} + DIN_{SW} F_{SW} + DIN_{SUB} F_{SUB} \tag{13.4}$$

$$DIP^{\circ} = DIP_{RI} F_{RI} + DIP_{SW} F_{SW} + DIP_{SUB} F_{SUB} \tag{13.5}$$

$$Si(OH)_4^{\circ} = Si(OH)_{4RI} F_{RI} + Si(OH)_{4SW} F_{SW} + Si(OH)_{4SUB} F_{SUB} \tag{13.6}$$

$$DIC^{\circ} = DIC_{RI} F_{RI} + DIC_{SW} F_{SW} + DIC_{SUB} F_{SUB} \tag{13.7}$$

where DIN_{RI}, DIN_{SW}, DIN_{SUB}, DIP_{RI}, DIP_{SW}, DIP_{SUB}, $Si(OH)_{4RI}$, $Si(OH)_{4SW}$, $Si(OH)_{4SUB}$, DIC_{RI}, DIC_{SW}, and DIC_{SUB} are the concentrations of the three end-members for DIN, DIP, $Si(OH)_4$, and DIC.

Third, we may calculate the difference between the prediction based on conservative mixing and the field-measured values denoted as "Δ", which reflected the amount of nutrients produced (negative) or removed (positive) associated with biological processes:

$$\Delta DIN = DIN^{\circ} - DIN_{\text{in situ}} \tag{13.8}$$

$$\Delta DIP = DIP^{\circ} - DIP_{\text{in situ}} \tag{13.9}$$

$$\Delta Si(OH)_4 = Si(OH)_4^{\circ} - Si(OH)_{4\text{in situ}} \tag{13.10}$$

$$\Delta DIC = DIC^{\circ} - DIC_{\text{in situ}} \tag{13.11}$$

where $DIN_{\text{in situ}}$, $DIP_{\text{in situ}}$, $Si(OH)_{4\text{in situ}}$, and $DIC_{\text{in situ}}$ represent the nutrient concentrations measured during the cruise.

Based on the preceding model estimation, we derived that most of the ΔDIN and ΔDIP in coastal upwelling are positive, suggesting DIN and DIP consumption during the upslope advection of the subsurface water from a depth of ~100 m until it outcropped nearshore (Fig. 13.15B, ΔDIP is not shown here). However, a few negative ΔDIN and ΔDIP along the coast likely indicated DIN and

DIP additions sourced from organic matter degradation. Such regenerated DIN and DIP might have been consumed very quickly by phytoplankton given the oligotrophic nature of the ambient water, the process of which was however difficult to elucidate. We may further estimate the net community production based on the ΔDIN in coastal upwelling to be 54±24 mmol C m^{-2} d^{-1}.

In the plume regions, overall ΔDIN and ΔDIP were positive, indicating net biological uptake (Fig. 13.15D). Based on the modeled consumption of individual nutrients, the nutrient uptake ratio was 16.7 in coastal upwelling area and agreed well with the classic Redfield ratio, and was 61.3±8.7 in river plume, which an alternative non-DIP source likely contributed (see Fig. 10 in Han et al. 2012).

ΔDIC in coastal upwelling displayed the combination of DIC addition and uptake along the upwelling current (Fig. 13.15C), indicating two processes of DIC regeneration and biological consumption. In addition, for stations involved in Shantou upwelling region during the entire sampling period, the average ΔDIC value was 8±9 μmol kg^{-1}, and the consequent NCP for upwelled waters on the NSCS shelf was estimated to be 23±26 mmol C m^{-2} d^{-1}. Positive ΔDIC in river plume were observed, indicating the exclusively biological removal of DIC (Fig. 13.15E). The ΔDIC displayed an overall trend decreasing with the increasing salinity, and the consequent NCP for the river plume was 36±19 mmol C m^{-2} d^{-1}. Δ$Si(OH)_4$ displayed similar behaviors to that of ΔDIC (see Figs. 8 and 9 in Han et al. 2012). We also found that both the DIC-derived NCP and nutrient-derived production were very well agreed in between and also were comparable to the primary production arrived at using numerical simulations (Gan et al. 2010), which suggests that our three end-member mixing model was able to estimate the biological alteration on top of the physical mixing between different water masses.

Taking together the physical dynamics and the simple mass balance, we were able to come up with the following mechanistic understanding of carbon and nutrients in such a complex system under the influence of both river plumes and coastal upwelling (Fig. 13.16). The upslope advection of subsurface waters intensified the cross-shelf advection off Shanwei and then subsequently transported northeastward by the upwelling coastal current and outcropped at the lee of the coastal cape off Shantou. Pearl River plume enhanced upwelling wind-driven current near surface, whereas it was advected eastward. Among these physical processes, organic matter appeared to be remineralized, and DIC and nutrients were regenerated along with the upslope advection of subsurface waters toward Shanwei. In the inner shelf along the upwelling coastal current from Shanwei to Shantou, there appeared a northward enhancement trend of DIC and nutrient consumption rates, although regeneration and consumption of DIC and nutrients (notably for $Si(OH)_4$) coexisted. DIN and DIP consumption followed the Redfield stoichiometry. In the plume areas, net consumption of nutrients and DIC were obvious, with an apparent non-Redfield DIN:DIP uptake ratio (Han et al. 2012).

4.2. Coupled physical-biogeochemical model

With a coupled three-dimensional physical model and a nitrogen-based dissolved inorganic nitrogen, phytoplankton, zooplankton, and detritus (NPZD) ecosystem model and field measurements, Gan et al. (2010) conducted a process-oriented study of the biological response to upwelling and

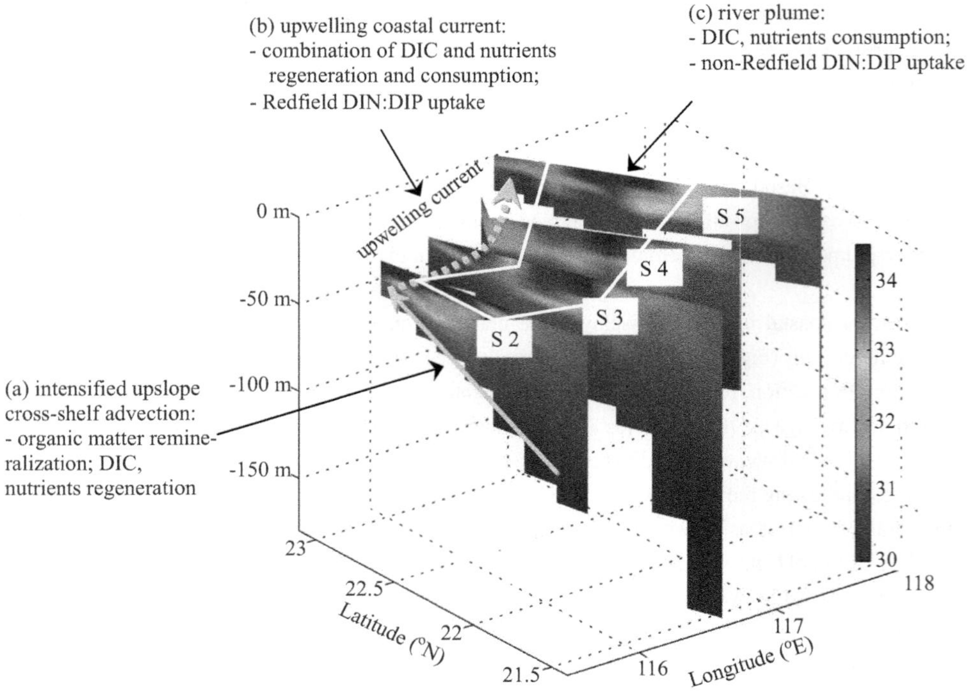

Figure 13.16. Conceptual scheme illustrating nutrients (DIN, DIP, and $Si(OH)_4$) and DIC dynamics under the co-influence of both the river plume and coastal upwelling over the northern South China Sea shelf from Han et al. (2012). Symbols S2 to S5 represent the transect numbers 2–5 as shown in Figure 13.1C.

Pearl River plume in the NSCS during the wet season. They identified the two high chlorophyll centers that are typically observed over the NSCS shelf and stimulated by nutrient enrichment from intensified upwelling over the widened shelf and from the river plume. The nutrient enrichment has strong along-shore variability involving the variable cross-isobath nutrient transport between the middle and the inner widened shelf during the upwelling and an eastward expansion of the nutrient-rich plume. Only a relatively small portion of upwelled nutrient-rich deep water from the outer shelf reaches the inner shelf, where algal blooms occur. Nutrient enrichment in the plume stretches over a broad extent of the shelf and produces significant biomass on the NSCS shelf. The spatial dislocation and temporal variation of NO_3, phytoplankton (P), and zooplankton (Z) biomasses are found in the plume waters, in which Z and N have the largest and the smallest eastward and seaward extensions with P stays between them, respectively. This is jointly controlled by the eastward advection of the plume and by the different growth rates and growth time lag in P and Z.

Frequently observed subsurface chlorophyll maximum (SCM) contributes substantial biomass to the waters over the continental shelf of the NSCS. Based on the coupled physical-biological numerical model and validated by field measurements, Lu et al. (2010) showed the influences of physical forcing processes, the upwelling circulation and plume, on the SCM. They found that the depth and intensity of the SCM are spatially variable regulated by the variable upwelling circulation and associated plume distribution over the NSCS shelf. Cross-shore component of upwelling circulation shoals and weakens the SCM toward the coast as a result of the upwelling of high-nutrient, low-chlorophyll deep water. The enhanced upwelling-favorable wind weakens the intensity of the SCM due to dilution by the enhanced mixing. In the vast region of NSCS that is covered by the plume, the SCM weakens because of the substantial reduction of photosynthetic active radiation (PAR) in the water column beneath the plume.

5. Summary and perspectives

We have demonstrated in this chapter that river discharge, wind forcing, tides, gravitational circulation, and shelf current intrusion jointly govern the circulation and water mass properties in the PRE, which largely regulates the physical and biogeochemical characteristics of freshwater discharged onto the continental shelf or of the plume in the NSCS. The motion of the plume and thus the cross-gradient material transport over the shelf are determined by plume dynamics of density front and stratification effect on the frictional transport, besides by the predominant wind-driven circulation. Therefore, the biogeochemical characteristics and evolutions in the PRE plume over the shelf are the synthesized response of biogeochemical dynamics in the plume and in the ambient shelf water to the physical forcing dynamics of the time-dependent, three-dimensional shelf circulation in the NSCS.

The dynamics of the Pearl River plume have predominant roles in stimulating primary production in the NSCS through delivery of a large quantity of nutrients into the shelf system otherwise oligotrophic in nature. We also demonstrated the approaches to differentiate the biogeochemical rate on top of complex physical dynamics of plume systems, which are often superimposed with many other coastal processes. One of the typical schemes is the coastal upwelling, which is often observable in the western boundary under southwest monsoon. We have shown that the plume is a very efficient reactor for biogeochemical processes as seen by high nutrient and inorganic carbon consumption rates, which apparently supported the enhanced biological productivity.

Site-specifically, coupled physical-biogeochemical processes have been much less studied for the river plume in other seasons in this particular system, for example, in the winter time. What is basically known is that the Pearl River plume is much less prominent in the dry season, when the northeastern monsoon prevails, which modulates the plume to the western part of the estuary.

Although this chapter has emphasized the interactive physical and biogeochemical processes, we must point out that the PRD has been one of the most rapid developing regions in the world during the past 30 years. As a consequence, the PRE has been experiencing intense anthropogenic disturbance (see also Chapter 11). For example, the PRE currently receives an annual wastewater discharge of recently $\sim 5{,}000 \times 10^6$ ton yr^{-1} from upstream cities such as Guangzhou, Foshan, and Dongguan (see the Environmental Status Bulletins of Guangdong Province, China; www.gdepb.gov.cn/). Agricultural activities have given rise to increasingly high levels of pollution from fertilizers and pesticides. Associated with these waste/fertilizer discharges, the nutrients and inorganic carbon system might be greatly influenced in the lower reaches of the estuary, even to the NSCS shelf through the river plumes and estuary-shelf circulation, the effect of which on the shelf ecosystem system has however never been carefully examined. Added in more complexity is global change, which may well be reflected in regional changes in temperature, CO_2, and circulation, the effect of this global scale multiple drivers are yet to be assessed for the PRE and NSCS regions.

Acknowledgment

This research was funded by the National Basic Research Program of China (973 Program) through grant 2009CB421204 and 2009CB421208, and by the National Natural Science Foundation of China (NSFC) through grants 41121091 and 41130857. Authors are grateful to Tingting Zu, Xiaoyan Wu, Linlin Liang, and Wei Huang for helping with the data processing. Harish Gupta is thanked for his assistance in preparing Figure 13.1A. Chuanmin Hu and Shaoling Shang are appreciated for generating the RS images. We thank Xianghui Guo and Biyan He for their suggestions. JPG is supported by project Hong Kong GRF grants of 601008 and 612412.

References

Allen, J. S., P. A. Newberger, and J. Federiuk. 1995. Upwelling circulation on the oregon continental shelf. part 1: response to idealized forcing. J. Geophys. Res. 25:1843–1866.

Bu, X., and J. Ye. 2007. 广东省统计年鉴 2007. p. 641, China Stat. Press, Beijing. [Gongdong Statistical Yearbook-2007].

Cai, W. J., M. H. Dai, Y. C. Wang, W. D. Zhai, T. Huang, S. T. Chen, F. Zhang, and Z. Z. Chen. 2004. The biogeochemistry of inorganic carbon and nutrients in the Pearl River estuary and the adjacent Northern South China Sea. Cont. Shelf Res. 24:1301–1319.

Cai, W. J., X. H. Guo, C. T. A. Chen, M. H. Dai, L. J. Zhang, W. D. Zhai, S. E. Lohrenz, K. D. Yin, P. J. Harrison, and Y. C. Wang. 2008. A comparative overview of weathering intensity and HCO_3^- flux in the world's major rivers with emphasis on the Changjiang, Huanghe, Zhujiang (Pearl) and Mississippi Rivers. Cont. Shelf Res. 28:1538–1549.

Cao, Z. M., M. H. Dai, N. Zheng, D. L. Wang, Q. Li, W. D. Zhai, F. F. Meng, and J. P. Gan. 2011. Dynamics of the carbonate system in a large continental shelf system under the influence of both a river plume and coastal upwelling. J. Geophys. Res. 116: G02010.

Chao, S. Y., and W. C. Boicourt. 1986. Onset of estuarine plumes. J. Phys. Oceanogr. 16:2137–2149.

Chao, S. Y. 1988. River-forced estuarine plumes. J. Phys. Oceanogr. 18:72–88.

Chen, C. T. A., and A. V. Borges. 2009. Reconciling opposing views on carbon cycling in the coastal ocean: continental shelves as sinks and near-shore ecosystems as sources of atmospheric CO_2. Deep-Sea Res. II 56:578–590.

Cheung, K. C., X. P. Nie, C. Y. Lan, and M. H. Wong. 2000. Aimen, Modaomen and Pearl River estuaries of the Pearl River Delta, pp. 45–52. In Dupra, S. V., J. I. M. C.Smith, and C. J.Crossland (eds.), V. LOICZ Reports and Studies. LOICZ, Texel, the Netherlands.

Chinese Bays and Estuaries Records Compilation Committee. 1998. 中国海湾志. 重要河口. Vol.14. Ocean Press. [Chinese Bays and Estuaries Records, Important Estuaries]

Dagg, M., R. Benner, S. Lohrenz, and D. Lawrence. 2004. Transformation of dissolved and particulate materials on continental shelves influenced by large rivers: plume processes. Cont. Shelf Res. 24:833–858.

Dagg, M., M. H. Dai, P. J. Harrison, and C. Rabouille. 2008. Introduction. Cont. Shelf Res. 28: V–VI.

Dai, A. G., T. T. Qian, K. E. Trenberth, and J. D. Milliman. 2009. Changes in continental freshwater discharge from 1948 to 2004. J. Climate. 22:2773–2792.

Dai, M. H., X. H. Guo, W. D. Zhai, L. Y. Yuan, B. W. Wang, L. F. Wang, P. H. Cai, T. T. Tang, and W. J. Cai. 2006. Oxygen depletion in the upper reach of the Pearl River estuary during a winter drought. Mar. Chem. 102:159–169.

Dai, M. H., W. D. Zhai, W. J. Cai, J. Callahan, B. Q. Huang, S. L. Shang, T. Huang, X. L. Li, Z. M. Lu, W. F. Chen, and Z. Z. Chen. 2008a. Effects of an estuarine plume-associated bloom on the carbonate system in the lower reaches of the Pearl River estuary and the coastal zone of the northern South China Sea. Cont. Shelf Res. 28:1416–1423.

Dai, M. H., L. F. Wang, X. H. Guo, W. D. Zhai, Q. Li, B. Y. He, and S. J. Kao. 2008b. Nitrification and inorganic nitrogen distribution in a large perturbed river/estuarine system: the Pearl River Estuary, China. Biogeosciences 5:1227–1244.

Dai, S. B., S. L. Yang, and A. M. Cai. 2007. 51 年来珠江流域输沙量的变化. Acta Geogr. Sin. 62(5):545–554. [Variation of sediment discharge of the Pearl River basin from 1955 to 2005]

Fong, D. A., and W. R. Geyer. 2002. The alongshore transport of freshwater in a surface-trapped river plume. J. Geophys. Res. 32:957–972.

Gan, J. P. and J. S. Allen. 2002. A modeling study of shelf circulation off northern California in the region of the Coastal Ocean Dynamics Experiment: Response to relaxation of upwelling winds. J. Geophys. Res. 107: 3123.

Gan, J. P., A. Y. Y. Cheung, X. G. Guo, and L. Li. 2009a. Intensified upwelling over a widened shelf in the northeastern South China Sea. J. Geophys. Res. 114: C09019.

Gan, J. P., L. Li, D. X. Wang, and X. G. Guo. 2009b. Interaction of a river plume with coastal upwelling in the northeastern South China Sea. Cont. Shelf Res. 29:728–740.

Gan, J. P., Z. M. Lu, M. H. Dai, A. Y. Y. Cheung, H. B. Liu, and P. J. Harrison. 2010. Biological response to intensified upwelling and to a river plume in the northeastern South China Sea: A modeling study. J. Geophys. Res. 115: C09001.

Gan, J. P., H. S. Ho, and L. L. Liang. 2013. Dynamics of intensified downwelling circulation over a widened shelf in the northeastern South China Sea. J. Phys. Oceanogr. 43: 80–94, doi: 10.1175/JPO-D-12-02.1.

Garvin, R. W. 1987. Estuary plumes and fronts in shelf waters: a layer model. J. Phys. Oceanogr. 17:1877–1896.

Gaston, T. F., T. A. Schlacher, and R. M. Connolly. 2006. Flood discharges of a small river into open coastal waters: plume traits and material fate. Estuar. Coast. Shelf S. 69:4–6.

Guo, X. H., W. J. Cai, W. D. Zhai, M. H. Dai, Y. C. Wang, and B. S. Chen. 2008. Seasonal variations in the inorganic carbon system in the Pearl River (Zhujiang) estuary. Cont. Shelf Res. 28:1424–1434.

Guo, X. H., M. H. Dai, W. D. Zhai, W. J. Cai, and B. S. Chen. 2009. CO_2 flux and seasonal variability in a large subtropical estuarine system, the Pearl River Estuary, China. J. Geophys. Res. 114: G03013.

Han, A. Q., M. H. Dai, S. J. Kao, J. P. Gan, Q. Li, L. F. Wang, W. D. Zhai, and L. Wang. 2012. Nutrient dynamics and biological consumption in a large continental shelf system under the influence of both a river plume and coastal upwelling. Limno. Oceanogr. 57:486–502.

Han, W. Y. 1998. 南海海洋化学. Science Press, Beijing, China. [Marine Chemistry in the South China Sea]

Harrison, P. J., K. D. Yin, J. H. W. Lee, J. P. Gan, and H. B. Liu. 2008. Physical-biological coupling in the Pearl River Estuary. Cont. Shelf Res. 28:1405–1415.

He, B. Y., M. H. Dai, W. D. Zhai, L. F. Wang, K. J. Wang, J. H. Chen, J. R. Lin, A. Q. Han, and Y. P. Xu. 2010. Distribution, degradation and dynamics of dissolved organic carbon and its major compound classes in the Pearl River estuary, China. Mar. Chem. 119:52–64

Hickey, B., L. Pietrafesa, D. Jay, and W. Boicourt. 1998. The Columbia river plume study: subtidal variability in the velocity and salinity fields. J. Geophys. Res. 103:10339–10368.

Hu, C. M. 2011. An empirical approach to derive MODIS ocean color patterns under severe sun glint. Geophys. Res. Lett. 38: L01603.

Lentz, S. J. 2001. The influence of stratification on the wind-driven cross-shelf circulation over the north Carolina shelf. J. Phys. Oceanogr. 23:2749–2760.

Lu, Z. M., J. P. Gan, M. H. Dai, and A. Y. Y. Cheung. 2010. The influence of coastal upwelling and a river plume on the subsurface chlorophyll maximum over the shelf of the northeastern South China Sea. J. Mar. Syst. 82:35–46

Mckee, B. A., R. C. Aller, M. A. Allison, T. S. Bianchi, and G. C. Kineke. 2004. Transport and transformation of dissolved and particulate materials on continental margins influenced by major rivers: benthic boundary layer and seabed processes. Cont. Shelf Res. 24:899–926.

Mao, Q., P. Shi, K. D. Yin, J. P. Gan, and Y. Qi. 2004. Tides and tidal currents in the Pearl River Estuary. Cont. Shelf Res. 24:1797–1808.

Nash, J. D., and J. N. Moum. 2005. River plumes as a source of large-amplitude internal waves in the coastal ocean. Nature, 437:400–403.

Rennie, S., J. Largier, and S. J. Lentz. 1999. Observations of a pulsed buoyancy current downstream of Chesapeake Bay. J. Geophys. Res. 104:18227–18240.

Shu, Y. Q., J. Zhu, D. X. Wang, and X. Xiao. 2011. Assimilating remote sensing and in situ observations into a coastal model of northern South China Sea using ensemble Kalman filter. Cont. Shelf Res. 31:S24–S36.

Wong, M. H., and K. C. Cheung. 2000. Pearl River Estuary and Mirs Bay, South China. The Estuarine Systems of the South China Sea Region: carbon, nitrogen, phosphorus flux workshop report: 7.

Zhai, W. D., M. H. Dai, W. J. Cai, Y. C. Wang, and Z. Wang. 2005. High partial pressure of CO_2 and its maintaining mechanism in a subtropical estuary: the Pearl River estuary, China. Mar. Chem. 93:21–32.

Zhang, S. R., X. X. Lu, D. L. Higgitt, C. T. A. Chen, H. G. Sun, and J. T. Han. 2007. Water chemistry of the Zhujiang (Pearl River): natural processes and anthropogenic influences. J. Geophys. Res. 112: F01011.

Zhang, W., S. Mu, Y. Zhang, and K. Chen. 2011. Temporal variation of suspended sediment load in the Pearl River due to human activities. Int. J. Sediment. Res. 26:487–497.

Zhao, H. T. 1990. 珠江志. Ocean Press, Beijing, China. [Evolution of the Pearl River Estuary]

Zu, T. T., and J. P. Gan. 2012. A numerical study of coupled estuary-shelf circulation around the Pearl River Estuary during summer: responses to variable winds, tides and river discharge. J.Geophys. Res. (submitted).

Zu, T. T., and J. P. Gan. 2008. Process-oriented study of the circulation and river plume in the Pearl River Estuary: response to the wind and tidal forcing. Advances in Geosciences 12: 225–242, b672-v12-ch16, Ocean Science Series.

Zu, T. T., J. P. Gan, and S. Y. Erofeeva. 2008. Numerical study of the tide and tidal dynamics in the South China Sea. Deep-Sea Res. I 55:137–154.

14

The evolution of carbon signatures carried by the Ganges-Brahmaputra river system: a source-to-sink perspective

V. Galy, C. Hein, C. France-Lanord, and T. I. Eglinton

1. Introduction to the Ganges-Brahmaputra river system

The Himalayan mountain range is drained southward by a dozen Trans–Himalayan rivers originating in southern Tibet, which collect most of the Himalayan runoff and supply the Ganges-Brahmaputra (G-B) rivers in the East-West Indo-Gangetic floodplain (Fig. 14.1). The G-B basin is composed of two sub-basins: the Ganges basin to the West and the Brahmaputra basin to the East. In both basins, the transition between the Himalayan range – characterized by steep slopes and high incision – and the mostly flat, aggradating floodplain is very abrupt as the slope of the rivers decrease by several orders of magnitude within a few kilometers. The Himalayan parts of these two basins display comparable geomorphology, but their floodplains exhibit sharp contrasts. The Ganges floodplain is 200–300 km wide and very flat, with the Ganga meandering far south of the Himalayan front, whereas the Brahmaputra floodplain is only 50–100 km wide, owing to the presence of the Shillong plateau and Burmese range to the south (Fig. 14.1). Consequently, the Brahmaputra is steeper and more channelized and remains a braided river until its confluence with the Ganges. The Ganges and Brahmaputra merge in Bangladesh to form the Lower Meghna River that eventually discharges into the northern Bay of Bengal.

Geologically, the Himalayan range is divided in four superimposed units (e.g., Le Fort 1989). From north to South: the Tethyan Sedimentary Series (TSS) consists of variably metamorphosed clastic sediments and carbonates; the High Himalaya Crystalline formations (HHC) are composed of orthogneisses, highly metamorphosed paragneisses, and some marbles; the Lesser Himalaya (LH) is composed of metamorphosed metasediments, including minor black shales and limestones; and finally, the Siwaliks forms the frontal range and is composed of sandstones and conglomerate that are uplifted Mio-Pliocene sediments originally deposited in the paleo-floodplain (Fig. 14.1). To the south of the range, the floodplain is entirely composed of Quaternary fluvial sediment. Owing to intense fluvial migration and floodplain aggradation, most surface deposits are Holocene.

Climate in the G-B drainage-basin complex is largely dominated by the Indian monsoon, a coupled ocean-atmosphere-land climate system characterized by annual turnover of both atmospheric circulation and associated precipitation in the tropics and subtropics. During the summer, stronger heating of the tropical landmasses (principally the Tibetan Plateau) than of the equatorial ocean creates a low-pressure system above the continent and a corresponding high-pressure system above the ocean. This differential heating is eventually responsible for a reversal of the classic Hadley atmospheric

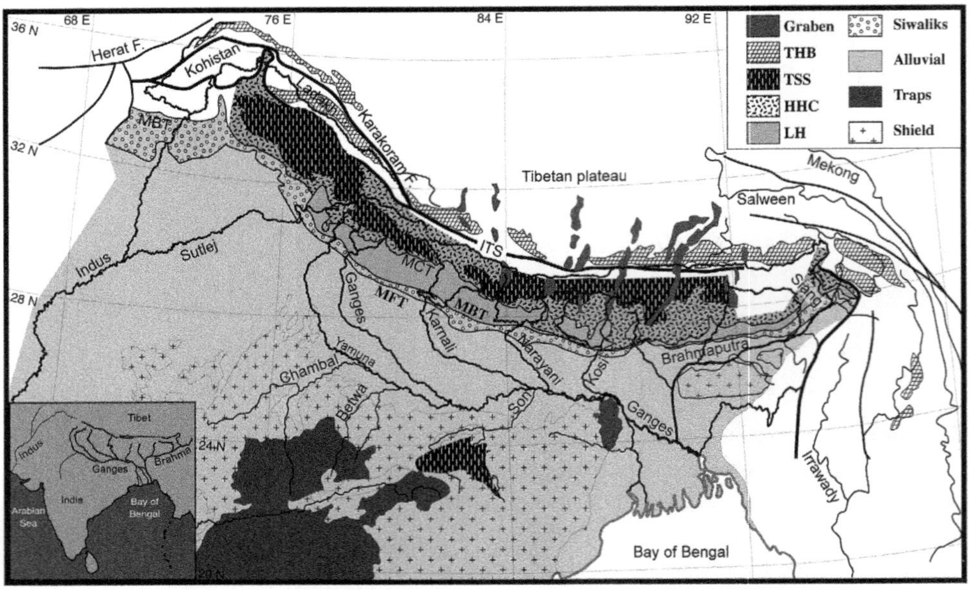

Figure 14.1. Map of the Ganges-Brahmaputra system showing the main physiographic and lithologic units, as well as the main rivers. THB: Trans-Himalaya Batholith; TSS: Thetisian Sedimentary Series; HHC: High Himalaya Crystalline, LH: Lesser Himalaya. Modified from Galy et al. (2010).

circulation, with moist air traveling from the equatorial ocean to the tropical continental masses. Thus, during the boreal summer, monsoon winds bring moisture from the Indian Ocean over to South Asia, generating very intense precipitation over Western India, the Bay of Bengal, and the southern flank of the Himalayas. In the winter, the cooling of the Tibetan plateau and the expansion of the Siberian high-pressure zone causes the reversal of the monsoon winds, leading to a dry, cool season in the G-B system.

As a result of this strong climate seasonality, river flow in the G-B system also displays large and systematic seasonal variations. Similar to the Changjiang and Huanghe rivers (Chapters 5 and 6), the hydrographs of the Ganges, Brahmaputra, and their large trans-Himalayan tributaries are controlled by monsoon rainfall and characterized by two main stages: high discharge during the summer monsoon period (July–September) and low discharge during the remainder of the hydrologic cycle, with abrupt transitions between these two stages (Fig. 14.2). The hydrographs of Ganges and Brahmaputra are typically offset by several weeks as a result of later arrival of the summer monsoon in the Ganges basin. Time series surveys indicate that >95% of the sediment flux of the Ganges and Brahmaputra Rivers is transported during the summer monsoon period (Islam and Jaman 2006). The combined sediment discharge from these rivers peaks at ca. 1.7×10^5 m^3/s from a boreal winter base level closer to 0.2×10^5 m^3/s (Kudrass et al. 1998). This sediment is composed of a bedload fraction (up to half of the total load; Galy and France-Lanord 2001) that is dominated by fine- to medium-grained sand (Palamenghi et al. 2011); suspended sediments are primarily composed of silt and clay with a fine sand fraction ranging from 15 to 40% (Thorne et al. 1993; Lupker et al. 2011).

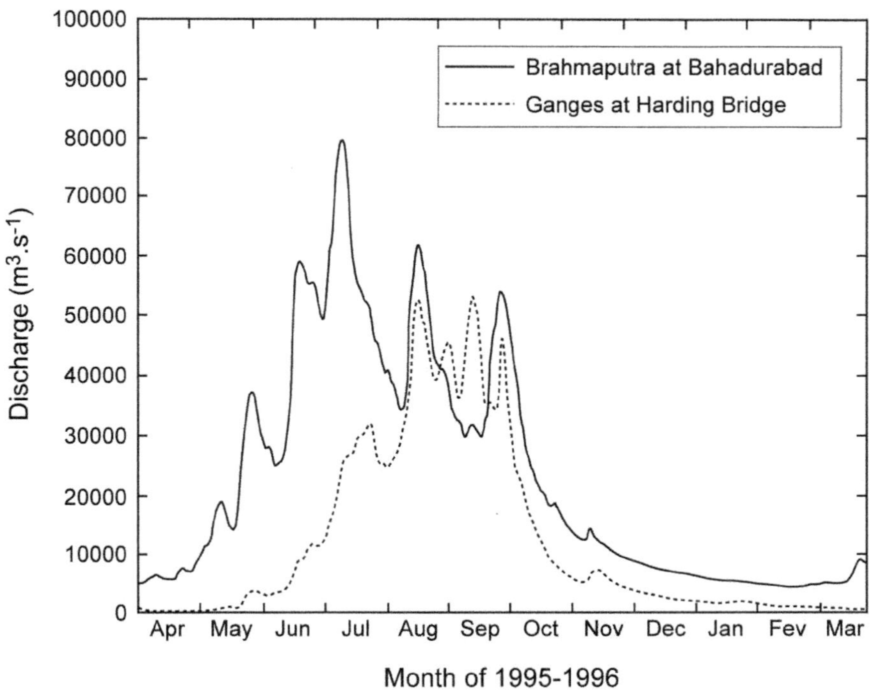

Figure 14.2. Hydrograph of the Ganges and Brahmaputra Rivers in Bangladesh for the hydrological year 1995–1996. Data from RSP (1996).

The G-B basin is one of the most densely populated regions of the world. As a consequence, present-day vegetation in the floodplain as well as in the lower part of the Himalayan range is largely controlled by human activities (i.e., deforestation and farming). In the Himalayan range, forested areas still predominate, although this region has also been severely affected by deforestation during the past few decades. The vegetation distribution is however largely controlled by elevation, resulting in the dominance of alpine scrub grassland above 3,000–3,500 m (Dobremez 1976). In the floodplain, the vegetation is composed of a mixture of crops and tropical grasslands. The main crops are rice, wheat, millet, maize, and sugar cane. The contribution of C4 crops, such as millet, maize, and sugar cane, to the total production is variable but significant, reaching ca. 80% in the state of Uttar Pradesh (Aucour et al. 2006).

Human activities also disturb water and sediment flow. Irrigation, especially in the western part of the Gangetic plain, uses a large proportion of the river flow during the low water stage. The mainstream of the Ganges is blocked by a large dam (the Faraka Dam) at the India-Bangladesh border, and most of the water is diverted toward Calcutta during the dry season, resulting in the Ganges being almost completely dry in Bangladesh during that time of year. However, in stark contrast to human perturbations seen on other major river systems (e.g., Chapter 6) floodgates on the Faraka Dam remain open during the summer monsoon, resulting in only a modest perturbation to the Ganges River flow.

2. Reworking and deposition of fluvial sediment and organic carbon in the Bay of Bengal

Sediment from the G-B river system is eventually deposited in the Bay of Bengal, the site of the world's largest delta (1.4×10^5 km^2; Kuehl et al. 2005), the world's largest submarine fan ($\sim 3 \times 10^6$ km^2; Curray 1994), and the world's largest reservoir of terrigenous sediments ($\sim 2.9 \times 10^{16}$ tons; Curray 1994). Roughly one-third of the sediment discharged from the G-B rivers is deposited in each the subaerial delta, the subaqueous delta, and the canyon-fan system (Fig. 14.3; Kuehl et al. 1997; Michels et al. 1998; Goodbred and Kuehl 1999).

The subaerial delta composes most of the land area of Bangladesh. It is a broad, flat alluvial lowland composed of swamps, marshes, former river channels, levees, flood plains, and river bars (Umitsu 1993; Kuehl et al. 1997). It is dominated by relatively coarse sediments (sand and silt; Umitsu 1993), a unique feature for a delta that is related to the coarse sediment load of the G-B Rivers (Kuehl et al. 2005; Chapter 4). Formation of the subaerial delta was controlled by eustatic sea-level rise with strong influences from variations in the tectonic and sediment-supply regimes across the delta (Goodbred 2003). The "upper delta plain" is marked by distributary development and channel avulsion (Kuehl et al. 2005) and is thus dominated by sandy alluvial deposits (Goodbred 2003). However, along the tectonically active northeast region, fine-grained floodplain deposits are most common (Goodbred 2003). The southern or "lower" delta plain is dominated by a series of relict distributary channels of the Ganges River that subdivide this mangrove-dominated region (the Sundarbans) into a series of shore-normal elongate islands and peninsulas (Kuehl et al. 2005). Onshore transport of shoreface sediment reworked by tides and storms accounts for a significant fraction of the sediments deposited in the most distal part of the lower delta plain (Paul et al. 2011). This region has been heavily influenced by Holocene sea-level change and modern tidal fluxes and thus contains a mix of fine- and coarse-grained facies (Allison et al. 2003; Goodbred 2003). Ecogeomorphologic feedbacks associated with the tide-dominated wetlands in the lower delta plain likely play a major, though understudied, role in the cycling of sediment, carbon, and nutrients in the lower subaerial delta and proximal Bay of Bengal (Chapters 2 and 15).

Nearshore reaches of the subaqueous delta are dominated by a series of subaqueous shoals that extend to 8 m depth seaward of peninsulas and islands that separate subaerial distributary channels (Kuehl et al. 1997; Kuehl et al. 2005). These extend out onto the topset beds of the delta, which are dominated by fining-upward sequences of very fine sand and silt (Kuehl et al. 1997; Kuehl et al. 2005). Topsets have an average gradient of <0.1 degrees and extend to ca. 30 m water depth (Kuehl et al. 1997). The delta front is composed of foresets that are silt dominated, dip at 0.20–0.27 degrees, and extend to ca. 55–75 m depth, varying along shore (Kuehl et al. 1997; Michels et al. 1998). Sedimentation rates range up to 6–8 cm/yr (Kuehl et al. 1989; Michels et al. 1998), and progradation occurs at a rate of ca. 12–17 m/yr (Michels et al. 1998). However, sedimentation is not uniform across the delta (Chapter 4): the eastern foresets, proximal to the mouth of the Lower Meghna, have generally been found to be more sensitive to discharge (monsoon floods), whereas sediment transport and deposition along the western side is driven largely by monsoon wind–driven currents (Palamenghi et al. 2011; Chapter 4). Finally, the distal reaches of the delta are composed of bottomset beds, lowstand delta deposits, and outer shelf deposits (Hübscher et al. 1998). These regions extend at a slope of ca. 0.022 degrees to the shelf break at ca. 150 m (Kuehl et al. 1997; Kuehl

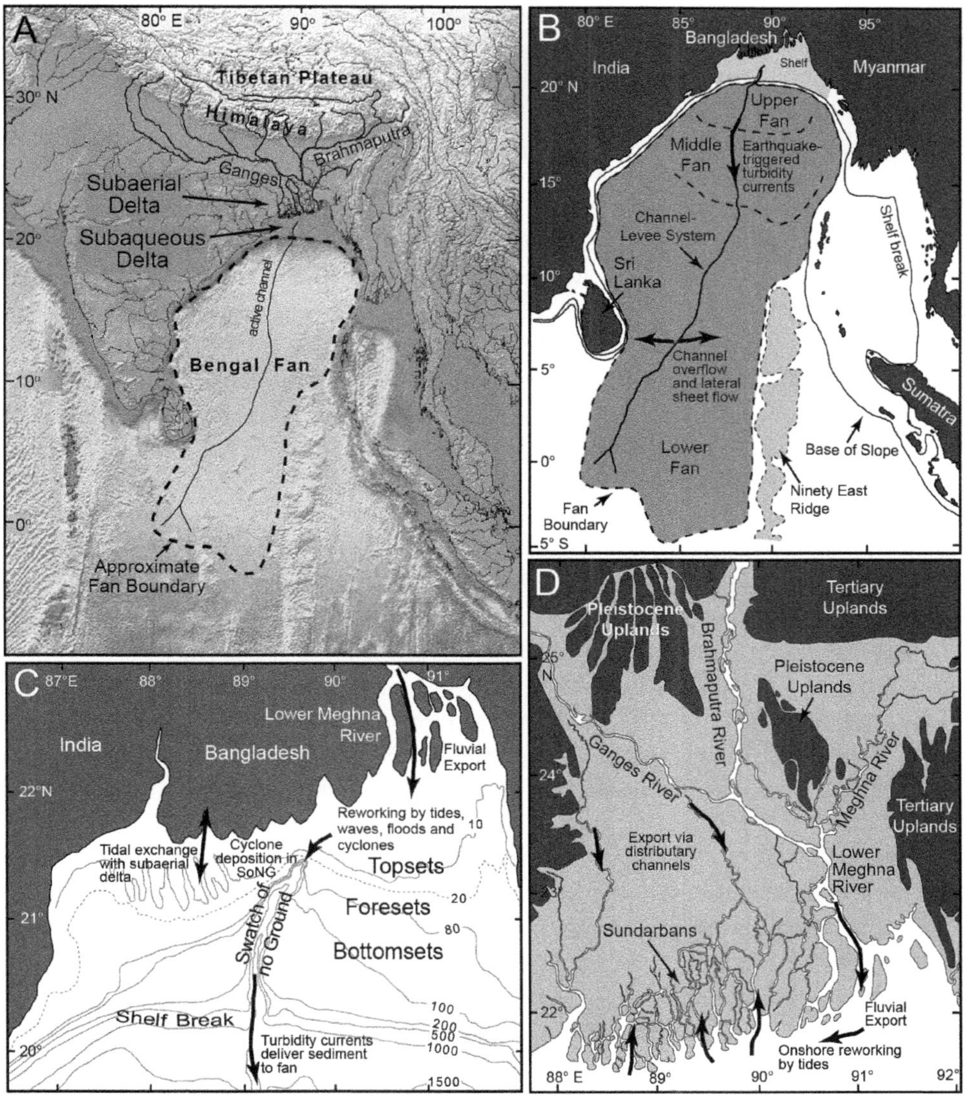

Figure 14.3. Three major depocenters for sediment and organic carbon from the Ganges and Brahmaputra Rivers. Major pathways and mechanisms of sediment reworking and deposition are indicated by arrows in B–D. (A) Overview map showing locations of the Bengal Fan, the subaqueous delta, and the subaerial delta. Over the Holocene, each has received approximately one-third of the sediment delivered by the river system. Figure modified from Galy et al. (2007) and Galy et al. (2008c). (B) Morphology of the Bengal Fan, extending from the shelf break to near 5° S latitude. Sediment is delivered to the fan dominantly via turbidity currents that originate in the upper Swatch of No Ground (SoNG) and travels along a single-channel channel-levee system. Modern sediment is transferred to channel-distal regions of the fan via channel overflow and lateral sheet flow. Figure modified from Galy et al. (2008a). (C) Morphology of the subaqueous delta, located on the shelf, adjacent to the mouth of the Lower Meghna River. Sediment is reworked from the mouth and along the shelf via tides, waves, flood events and tropical cyclones. Nearly 10% of modern fluvial sediment is eventually reworked into the SoNG, where it is temporarily stored prior to export to the Bengal fan via the channel-levee system. Other sediment is deposited along the prograding front of the delta, and the rest is reworked onshore to the subaerial delta via tidal exchange. Figure modified from Kudrass et al. (1998) and Galy (2007). (D) Physiographic map of the subaerial delta showing extent of Pleistocene and Tertiary upland deposits (dark gray) and Holocene deltaic and floodplain deposits (light gray). Sediment and organic carbon reach the delta directly via distributary channels from the Ganges River and via shoreface reworking by tides and storms. Figure modified from Kuehl et al. (2005).

et al. 2005). The proximal side of this region is dominated by foreset slumps. Sedimentation rates are generally <2 cm/yr (Suckow et al. 2001; Kuehl et al. 2005).

Proximal to the India/Bangladesh border, the subaqueous delta is incised by the Swatch of No Ground (SoNG), a 160-km long deep submarine canyon whose head lies in 20 m water depth and within 30 km of the present coastline. It directly intersects the prograding forest beds of the submarine delta (Kuehl et al. 2005). The SoNG is 20–30 km wide, with wall slopes ranging from 5–10 degrees (Kuehl et al. 2005) and maintains an average gradient of 8.2 m/km to its outlet in ~1,400 m depth (Curray et al. 2003). The head of the SoNG is the presumed location of the confluent G-B river mouth at the sea-level lowstand corresponding to the Last Glacial Maximum (21 ka). Following the Younger Dryas (ca. 11.5–12.8 ka), rapid sea-level rise resulted in the flooding of the shelf and a gradual reduction in direct sediment supply to the middle Bengal Fan; little terrestrial sediment has *directly* entered the canyon since ~6.5 ka (Curray et al. 2003). Nonetheless, the highest rates of modern sediment accumulation are found near the head of the SoNG: estimates range from >15 to nearly 100 cm/yr (Kuehl et al. 1997; Kudrass et al. 1998; Michels et al. 1998; Pierson-Wickmann et al. 2001; Suckow et al. 2001; Michels et al. 2003).

Offshore of the submarine delta lies the Bengal Fan (Fig. 14.3). This is the largest submarine fan in the world: it is 1,000 km wide, ca. 16.5 km thick, and extends offshore for approximately 3,000 km (Curray et al. 2003). Unlike most other submarine fans at major river mouths (Amazon, Mississippi, Indus; Chapters 4, 17, and 19), the Bengal Fan has been active throughout the Holocene and until present (Hübscher et al. 1997; Weber et al. 1997; Chapter 4). Much of this fan was built through a series of migrating, avulsing, and intersecting channel-levee systems (Curray et al. 2003; Schwenk et al. 2005). However, at present, there is only one active fan-valley system connected to the submarine canyon (Curray et al. 2003) that delivers sediment to the distal reaches of the fan.

Whereas much of the sediment that reaches the distal lower delta plain likely does so through reworking in the nearshore (Paul et al. 2011), direct deposition in the delta via overbank flooding (Chapter 2) has been estimated to be responsible for 32–39% of the late Holocene sediment supply (Allison et al. 2003; Goodbred and Kuehl 1999). Sediment and organic carbon (OC) that reach the mouth of the Lower Mengha River are discharged as a river plume (Chapter 3) that extends across the entire shallow shelf of the Bay of Bengal during the strong monsoon season. It is then reworked in both along- and cross-shore directions by freshwater fluxes, longshore currents, storm waves, and tidal fluxes (Kuehl et al. 1989; Segall and Kuehl 1992; Barua et al. 1994; Kuehl et al. 1997; Michels et al. 1998; Michels et al. 2003). Sediment and OC are largely transported to the southwest entrained in the large freshwater plume that is modified by tidal fluxes (Kuehl et al. 1989; Barua et al. 1994; Michels et al. 1998; Chapter 3). Transport in the central portion of the shelf, west of the river outlet, is dominated by tidal cyclicity and storm resuspension (Kuehl et al. 1989; Segall and Kuehl 1992; Barua et al. 1994; Kudrass et al. 1998; Michels et al. 2003; Chapter 4). Approximately 9% of the annual river sediment load is eventually reworked into the head of the SoNG (Michels et al. 2003). Sediment cores from this region (Kudrass et al. 1998; Michels et al. 1998; Michels et al. 2003) contain fine-grained muddy sedimentary sequences with interspersed 2- to 15-cm thick, fining-upward, interbedded sand and silt layers with distinct basal contacts that are interpreted as tempestites (Kudrass et al. 1998). These units are deposited by settling from suspension clouds that are mobilized and transported by tropical cyclone–induced downwelling that, on crossing the deeper water of the SoNG, lose their ability to transport coarser particles (Kudrass et al. 1998; Michels et al. 2003).

Over centennial to millennial timescales, the SoNG serves as the primary conduit for the bypassing of large volumes of sediment to the lower deltaic plain and Bengal Fan (Coleman 1969; Segall and Kuehl 1992; Curray et al. 2003) via a nonbifurcating, 2,500-km-long, 13-km-wide channel-levee system (Fig. 14.2; Weber et al. 1997). This process occurs dominantly via mass wasting (likely earthquake-triggered) and subsequent turbidity currents (Weber et al. 1997; Kudrass et al. 1998; Curray et al. 2003; Kottke et al. 2003; Michels et al. 2003). In this manner, large volumes of sediment are conveyed to the middle and lower Bengal Fan via lateral sheet flow and channel overflow (Curray et al. 2003). Elsewhere in the delta, acousticallytransparent seismic units within the foreset beds have been identified as evidence for the likely role of mass wasting associated with earthquakes in the subaqueous delta that could also deliver sediment to more proximal sections of the fan (Kuehl et al. 2005; Steckler et al. 2008; Palamenghi et al. 2011).

3. Characterization of organic carbon signatures of the Ganges and Brahmaputra rivers and the Bengal Fan

To geochemically characterize the G-B fluvial system, Himalayan rivers were sampled repeatedly over the past decade, at or near the outflow of the range, whereas mainstream rivers were sampled at several locations in the floodplain (Fig. 14.1). Most sampling was performed during the peak monsoon period (July–August), although sampling during late monsoon conditions (September) in 2008 was also conducted. Suspended sediments were sampled at different depths in the river, from surface to bottom, to fully represent the sediment heterogeneity typical of large rivers stemming from hydrodynamic-sorting processes. Bedload sediments were also dredged at the bottom of the rivers. This procedure was often coupled with current velocity measurements using an Acoustic Doppler Current Profiler, allowing the integration of the sediment variability over the whole river channel.

In the Ganges basin, all of the major Trans–Himalayan tributaries of the Ganges at their outlet of the Himalaya in Nepal were sampled: the Yamuna, Ganges, Karnali, Narayani, and Kosi (Fig. 14.1). The mainstream of the Ganges was sampled in India at several stations along its course in the floodplain. Finally, sampling has been repeated over the course of five monsoon seasons at Harding Bridge (Bangladesh) just before the confluence of the Ganges with the Brahmaputra.

A similar approach was applied for the Brahmaputra basin. Major Trans–Himalayan tributaries (Siang, Subansiri, Manas, and Tista) were sampled at the outlet of the Himalayan range or in the floodplain before their confluence with the Brahmaputra and the mainstream along its course in the floodplain (Singh and France-Lanord 2002). Again, sampling has been repeated over five monsoon seasons, here at Sirajganj (Bangladesh) before its confluence with the Ganges.

The mixing of the Ganges and Brahmaputra – named the Padma River – was sampled in Bangladesh at Mawa (ca. 65 km downstream of the confluence). Finally, the confluence of the Padma and Meghna Rivers – named the Lower Meghna River – was sampled at four different locations from Bohla to Chor Mankia (at the coastline).

The Bengal Fan sedimentary units are documented by subsurface sediments cored in the shelf, in the mid-fan active channel-levee system and in the distal part of the fan during *R/V Sonne* cruises

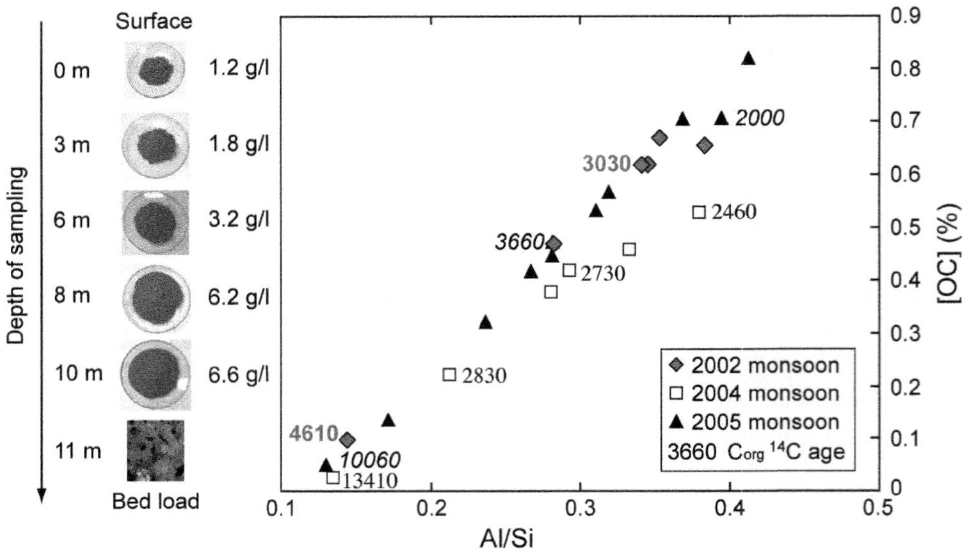

Figure 14.4. Example of vertical depth profile of suspended and bed sediments from the Ganges-Brahmaputra river system. Total suspended sediment concentration with increasing water depth is shown on the left. Clay-rich surface load has high Al/Si and OC content, whereas quartz-rich bottom load has low Al/Si and OC content. For each depth profile, the bulk ^{14}C age increases with increasing depth, reflecting increasing proportion of ^{14}C-free petrogenic OC. Data are from Galy et al. (2007) and Galy et al. (2008b).

SO93 and SO188 in 1994 and 2006, respectively (Ittekkot et al. 2003; Palamenghi et al. 2011; Kudrass et al. unpublished).

4. Carbon sources, transformations, and storage in the Himalayan range

During transport, river detritus experiences strong hydrodynamic forcing that results in mineral and grain-size sorting. These sorting processes are in turn responsible for very large sediment heterogeneity in the river channel. Quartz and other coarse, rounded, and dense particles tend to increase with water depth and in the bedload, whereas fine and tabular phyllosilicates are more evenly distributed throughout the river channel (Galy et al. 2008b; Lupker et al. 2011; Lupker et al. 2012). The OC content of the river sediments is also highly variable but remains low and never exceeds 1% during the monsoon period when >95% of sediment discharge occurs. In the river channel, total particulate OC concentrations are depth-dependent, with surface suspended particles having the highest values and bed sediments the lowest (Fig. 14.4). Accounting for the heterogeneity of the detrital load is thus essential for deriving accurate flux estimates.

Although this degree of heterogeneity complicates attempts to develop a comprehensive assessment of OC cycling in the system, strong geochemical relationships exist that permit extrapolation of trends. Specifically, total OC (TOC) is positively and tightly correlated with the Al to Si ratio (Al/Si) (Fig. 14.4), implying that the latter can be used as a proxy for mineral and grain-size sorting. OC concentrations linearly increase with the relative proportion of clay and fine-grained minerals (as indicated by increasing proportions of Al relative to Si), suggesting that OC and mineral particles have similar responses to sorting processes in the river.

OC loading – defined as Al/Si-normalized OC concentrations – are remarkably uniform at the outflow of the Himalayan range. Specifically, the eight large trans-Himalayan rivers display second-order variability of OC loading. For instance, in the Ganges basin, the Narayani and Kosi Rivers have lower OC loading than the Karnali and Ganges Rivers. These small differences likely reflect different efficiency of OC export that might be related, for instance, to variable suspended-sediment yield. However, temporal variability of OC loading – either inter- or intra-annual – cannot be ruled out because of the lack of actual time-series survey data.

Stable carbon isotopes ($\delta^{13}C$) have long been used as a biogeochemical tool to derive information on the source of OC due to isotope contrasts resulting from photosynthetic C fixation mechanisms employed by different plant types (Deines 1980). In river sediments, the $\delta^{13}C$ value of bulk OC is a proxy for its provenance, because river algae, C3, and C4 plants have distinct isotope compositions.

In large rivers at the outflow of the Himalayan range, bulk OC $\delta^{13}C$ values range between −19.9 and −26.3‰. Most of this variability is due to isotopic heterogeneity in OC-poor bed sediments and reflects contributions from petrogenic OC with variable $\delta^{13}C$ in the Himalayan bedrocks. In suspended sediments, the $\delta^{13}C$ values display a narrower range, resulting from dilution of the petrogenic C component by OC derived from the terrestrial biosphere (Galy et al. 2008a; Galy and Eglinton 2011). Using the average Al/Si ratio of suspended sediments at the outflow of the Himalayan range (0.22; Lupker et al. 2012), the relationship between $\delta^{13}C$ values and Al/Si ratios can be used to constrain the average isotopic composition of exported sediments. The variability described by the eight large trans-Himalayan rivers is significant (−23.7‰ to −26.3‰; Fig. 14.5). However, there is no systematic difference between Ganges and Brahmaputra tributaries: the flux-weighted average $\delta^{13}C$ value for the Ganges tributaries is −25.1‰, whereas that for those of the Brahmaputra is −24.8‰. Overall, the flux-weighted composition of trans-Himalayan rivers at the outflow of the range is −24.9‰. This isotope signature suggests only minor inputs of algal matter (which is likely to be in the −30‰ to −35‰ range; Mook and Tan 1991), consistent with high sediment loads and very dynamic transport regime that inhibits primary production within the Himalayan rivers. In addition, the mean $\delta^{13}C$ value reveals that OC is largely dominated by C3 plant inputs, consistent with vegetation patterns in the Himalayan region, where forest and high altitude ecosystems dominate (Dobremez 1978; Blasco et al. 1996). These interpretations of bulk OC isotopic composition are further supported by $\delta^{13}C$ values of organic molecules specifically produced by vascular plants, such as long-chain n-alkanes and n-alkanoic acids derived from epicuticular leaf waxes (Galy et al. 2011).

However, OC transfer from the G-B basin to the Bay of Bengal is not instantaneous. Some fraction of OC synthesized in the basin is "aged" in various terrestrial reservoirs – such as soils – prior to export by the river system. The radiocarbon (^{14}C) content of OC in G-B river sediments has been shown to provide an estimate of the time elapsed between OC synthesis and its export to the Bay of Bengal, i.e., the integrated average age of the OC in the basin. Riverine particulate OC is, however, known to be highly heterogeneous (Hedges et al. 1986; Blair et al. 2004; Aufdenkampe et al. 2007; Galy et al. 2008b; Chapters 8, 16, and 22), thus complicating the interpretation of radiocarbon signatures.

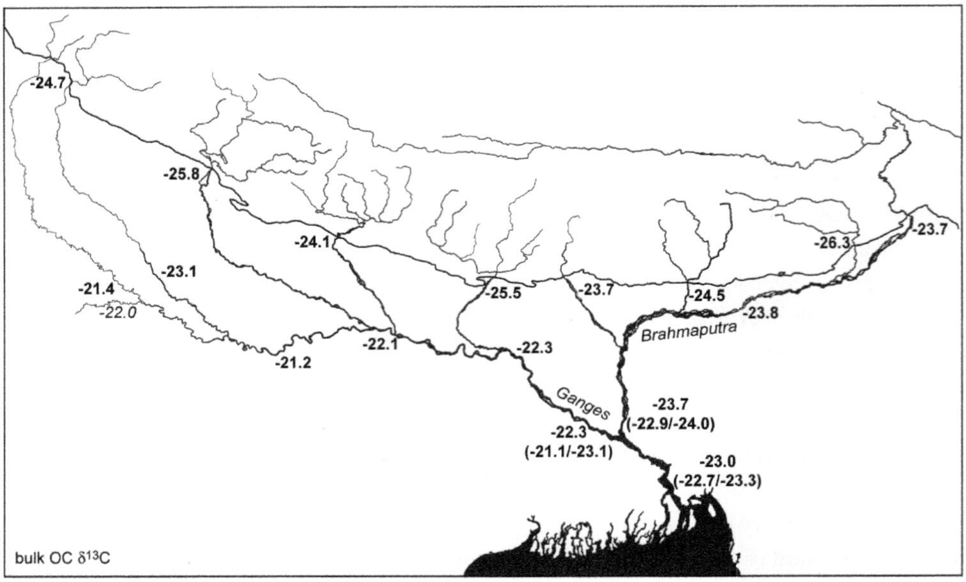

Figure 14.5. Simplified map of the Ganges-Brahmaputra system showing the evolution of bulk organic carbon $\delta^{13}C$ values from the Himalayan range to the delta. Bold numbers represent depth-integrated $\delta^{13}C$ values; italic numbers represent $\delta^{13}C$ values of surface suspended sediments.

The presence of ancient (^{14}C-free) carbon derived from rock erosion (petrogenic carbon) dilutes the apparent ^{14}C content of the terrestrial biospheric carbon component (Blair et al. 2004; Drenzek et al. 2007; Galy et al. 2008a; Drenzek et al. 2009; Bouchez et al. 2010; Hilton et al. 2011; Chapter 8), whereas the ^{14}C content of carbon stemming from autotrophic and heterotrophic processes within the river can vary markedly depending on the sources of dissolved inorganic carbon and degree of exchange with the atmosphere (Raymond et al. 2004; Mayorga et al. 2005). However, by taking advantage of the natural particle segregation that occurs with depth in large rivers, bulk OC ^{14}C measurements have been used to calculate the concentration of petrogenic carbon in these sediments (Galy et al. 2008a).

In trans-Himalayan tributaries, petrogenic carbon contributions were determined to be 0.03–0.17% (Galy et al. 2008a; Galy and Eglinton 2011; Rosenheim and Galy 2012). Correcting bulk ^{14}C measurements for this contribution, it was determined that bulk biospheric carbon exported from the trans-Himalayan tributaries ranged from 450 to 17,000 years old, with an average residence time of the overall pool of ca. 2,000 years (Galy and Eglinton 2011). These residence times contrast sharply with estimates of the residence time of individual vascular plant–derived OC compounds in the same river sediments, which vary from 50 to 1,300 years (Galy and Eglinton 2011). These systematically younger ages for vascular plant biomarkers reveal the existence of one or more components that are characterized by longer residences within the drainage basin, whose mixing with labile vegetation-derived OC explains the overall residence time of the bulk biospheric carbon. Using bulk and molecular ^{14}C data, it was estimated that the slow cycling, refractory component needed to explain the average age of the biospheric carbon is at least 15,000 years old (Galy and Eglinton, 2011). This refractory biospheric C is likely sourced from old soils that have accumulated in semi-arid parts of the catchments, north of the Himalayan range, in Tibet (Galy and Eglinton 2011).

Suspended sediments exiting the Himalayan range thus represent a complex mixture of OC, with a portion of this comprised of fossil carbon from sedimentary rock erosion, another comprised of refractory soil OC sourced in the upper reaches of the catchments, and the remainder comprised of C3 vegetation and C3-dominated soil OC from lower elevations within the Himalayan range.

5. Carbon sources, transformations, and storage in the Ganges floodplain

In the Gangetic floodplain, Trans-Himalayan tributaries as well as the Ganges mainstream show relatively similar OC loadings. Averaged over seven samplings (accounting for five monsoon seasons), the OC loading of the Ganges in the delta (0.36% OC at Al/Si of 0.23) is statistically identical to the flux-weighted average OC loading of its four main trans-Himalayan tributaries (0.32% OC at Al/Si of 0.22). However, second-order spatial and temporal variations of OC loading exist. The most striking difference is observed for the Ganges in the delta. Samples collected from three late monsoon field programs (September 2008) are characterized by significantly higher OC loading than samples collected during the previous peak monsoon intervals (July and August 2002, 2004, 2005, 2007). This might suggest an increase of OC loading during the monsoon, but could also reflect interannual variability. Clearly, time-series sampling over several monsoon seasons would be required to adequately document temporal variability of OC loading in these large river systems (see Chapters 16 and 22).

Within the delta, mean $\delta^{13}C$ values of sediments exported by the Ganges (i.e., Al/Si = 0.23; Lupker et al. 2011) vary between −21.2‰ and −23.1‰, with an average $\delta^{13}C$ value over seven sampling campaigns of −22.3‰ (Fig. 14.5). These values are up to 3.9‰ (2.8‰ on average) higher than those of Himalayan river-suspended sediments. The $\delta^{13}C$ value of long-chain n-alkanes and n-alkanoic acids (specific biomarkers of vascular plants) show very similar enrichment between the outflow of the Himalaya and the delta. This isotopic offset is thus interpreted to result from significant contributions from ^{13}C-enriched C4-plant OC to the Ganges suspended sediments. As the OC loading in suspended sediments is similar at the outflow of the range and in the floodplain, the $\delta^{13}C$ shift is considered to reflect removal, and subsequent replacement of C3-derived OC with C4 plant OC during transit through the Ganges floodplain (Keil et al. 1997; Aufdenkampe et al. 2007; Galy et al. 2008b). Interestingly, the average residence time of bulk biospheric OC is identical within uncertainties at the outflow of the Himalayas and in the delta (ca. 1,950 and 2,000 years, respectively). In contrast, vascular plant biomarkers appear to be younger in the delta than at the outflow of the range (ca. 50 and 250 years, respectively). Taken together, these observations suggest selective exchange and preservation processes acting on both labile biospheric OC and old refractory OC during floodplain transit. Although more work is clearly needed to fully constrain the behavior of different pools of OC in the Ganges floodplain, our results demonstrate the usefulness of coupling bulk and compound-specific isotopic measurements to characterize the fate of OC in large river systems.

6. Carbon sources, transformations, and storage in the Brahmaputra floodplain

Averaged over eight sampling programs (spanning five monsoon seasons), the OC loading of the Brahmaputra in the delta (0.33% OC at Al/Si of 0.23) is slightly higher than the average OC loading of its four main trans-Himalayan tributaries (0.26% OC at Al/Si of 0.22). However, contrary to the Ganges basin, the sediment load of trans-Himalayan tributaries of the Brahmaputra has only been relatively poorly characterized. Therefore, the average OC loading of all eight sampled tributaries at the outflow of the range (0.31% OC) provides a more robust point of comparison. Similar to the Ganges basin, OC loading appears to remain essentially constant during sediment transfer through the Brahmaputra floodplain. Second-order temporal variations of OC loading of the Brahmaputra in the delta mimic those observed for the Ganges. In particular, the three late monsoon samplings (September 2008) are once again characterized by significantly higher OC loading than the previous peak monsoon samplings (July and August 2002, 2004, 2005, 2007). The concomitant OC loading increase in the Ganges and Brahmaputra basin suggests a regional (e.g., monsoon variability), rather than local, control of OC export.

The average $\delta^{13}C$ composition of OC in Brahmaputra suspended sediments varies between −22.9‰ and −24.0‰ (average −23.7‰). On average, these compositions are ca. 1.5‰ more depleted than OC carried by the Ganges in the delta (Fig. 14.4) and ca. 1.2‰ less depleted than that of the Himalayan rivers at the outflow of the range. A similar isotopic shift is observed in $\delta^{13}C$ value of vascular plants biomarkers, suggesting that removal and subsequent replacement of C3 plant–derived OC with C4 plant OC also occurs during transit through the Brahmaputra floodplain. In contrast to the Ganges, the dominance of C3 vegetation in the Brahmaputra floodplain diminishes the isotopic contrast between mountain and floodplain vegetation, which might explain why Brahmaputra sediments in the delta have more ^{13}C-depleted compositions than Ganges sediments. However, less efficient OC replacement in the Brahmaputra than in the Ganges basin would also be compatible with geomorphologic differences between these two catchments: the Brahmaputra floodplain is much narrower than the Ganges floodplain and, within the floodplain, the braided Brahmaputra River is steeper than the meandering Ganges River. Older ^{14}C ages of vascular plant biomarkers in the Brahmaputra basin (ca. 800 years on average) further suggest that OC replacement might be less efficient than in the Ganges floodplain.

Another striking difference between the river basins is the residence time of bulk biospheric OC: the Brahmaputra (ca. 4,000 years on average) is double that of the Ganges. This difference reflects greater proportions of old refractory OC in the Brahmaputra basin, in line with the large contribution of sediments from Tibet (through the Siang River) in this basin (Galy and Eglinton 2011). This observation clearly implies that OC carried by large rivers is a mixture of OC sourced in different parts of the catchment, with labile components preferentially sourced and entrained in the lower reaches of the river system.

7. Carbon export from the G-B basin to the Bengal Fan

Bulk and molecular isotopic compositions of OC delivered by the Lower Meghna to the Bay of Bengal lie in between the compositions of Ganges and Brahmaputra sediments, suggesting that, to the first order, OC behaves conservatively in the lower reaches of the G-B river system. For instance, the average $\delta^{13}C$ composition of OC in Lower Meghna suspended sediments is −23.0‰, which is the exact average of Ganges and Brahmaputra average compositions (−22.3‰ and −23.7‰, respectively). This composition illustrates the significance of the C4 plant contribution in this river system, which is not common for very large river catchments. On average, the residence time of bulk biospheric OC discharged to the Bay of Bengal is ca. 3,000 years, illustrating the significance of C storage in soils prior to export by the G-B rivers. Such a long residence time is somewhat surprising for a tropical/subtropical river system, especially when considering the very high physical erosion rates characteristic of the G-B system. This illustrates the sensitivity of the biospheric OC reservoir to external forcing such as climate change and human activities.

Averaged over five sampling periods (five different monsoon seasons), the OC loading of the Lower Meghna is similar to that of Ganges and Brahmaputra in the delta (0.33% OC at Al/Si of 0.23). The average petrogenic C concentration is 0.025%, i.e., <10% of the total OC. The corresponding biospheric OC concentration is thus 0.30%. Although this biospheric OC concentration is quite low, multiplying this value by the enormous suspended sediment flux for this system (1.15×10^{9} t/yr; River Survey Project 1996) yields a corresponding biospheric OC flux of 2.9×10^{11} mol/yr. This flux does not take into account the contribution of bed sediments. On average, these bed sediments have low OC content, with an average value of 0.05%. When corrected for the petrogenic C concentration, the biospheric OC content of bed sediments drops to ca. 0.025%, illustrating the disproportionate role of suspended-sediment load in terms of biospheric OC delivery to the ocean. The flux of bed sediment cannot be directly measured, but geochemical mass balance and geomorphologic arguments constrain the sum of floodplain sequestration and bedload transport to 10–50% of the suspended sediment flux (Galy and France-Lanord 2001). Accounting for these sediment fluxes, it has been estimated that the G-B system delivers 3.0×10^{11} mol/yr of biospheric OC to the Bay of Bengal (Galy et al., 2008b; Galy et al., 2011b).

8. Fate of OC in the Bengal Fan

Bengal Fan sediments have highly variable characteristics (grain size and mineralogy) owing to hydrodynamic sorting during the transport along the shelf and within the turbiditic system. Nevertheless, as in rivers, although OC contents are also variable, they remain tightly correlated with Al/Si ratios.

The relationship between TOC and Al/Si defined by Bengal Fan sediments is statistically identical to that of the G-B river sediments (Fig. 14.6), suggesting identical OC loading in G-B river sediments and in modern Bengal Fan sediments (Galy 2007). Several biomarker and isotope studies have shown that the OC in both modern and ancient Bengal Fan sediments is overwhelmingly dominated by terrestrial inputs with negligible marine OC (Cochran et al. 1989; Poynter and Eglinton 1990; Meyers and Dickens 1992; France-Lanord and Derry 1994; Freeman and Colarusso 2001; Galy 2007; Galy et al. 2008c). In addition, a series of seven core top samples from the shelf and channel-levee system have bulk OC ^{14}C contents and $\delta^{13}C$ values highly compatible with the signature of Lower Meghna sediments (Fig. 14.7). On average, sediments recently deposited in the Bengal Fan are characterized by petrogenic C concentration and biospheric OC residence time identical with the average composition

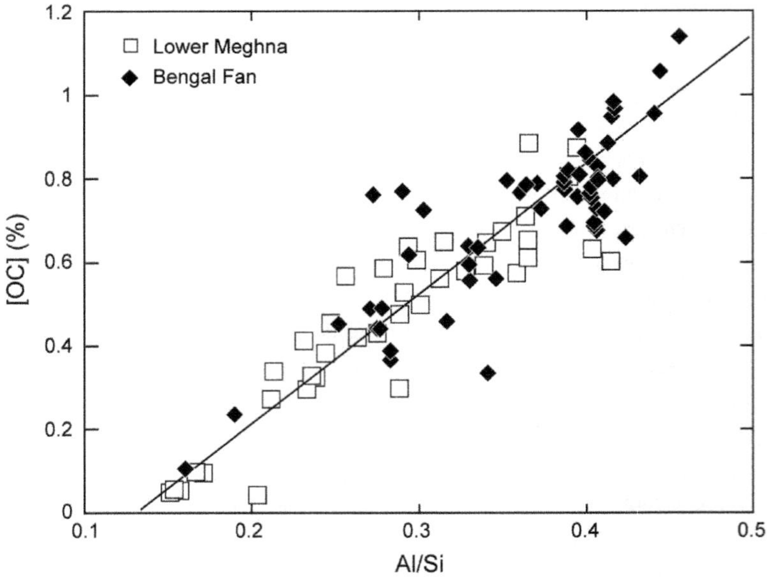

Figure 14.6. OC content of Bengal Fan sediments as a function of Al/Si with best linear fit. Fan sediments define a trend comparable to that defined by sediments from the Lower Meghna River (open squares), suggesting they have similar organic carbon loading. Data are from Galy et al. (2007), Galy et al. (2008b), and this study.

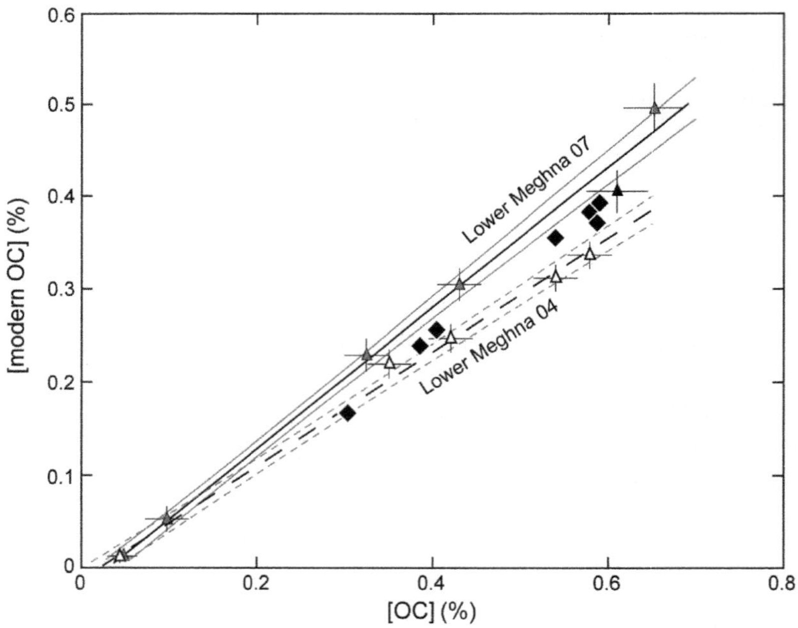

Figure 14.7. Bulk OC radiocarbon content of Lower Meghna sediments (triangles) and Bengal fan core top sediments (diamonds). This type of diagram shows the product of the radiocarbon composition by the OC content (or absolute modern organic carbon content) as a function of OC content. River and Bengal fan sediments are characterized by similar petrogenic carbon concentration and biospheric OC residence time. River sediment ^{14}C data are from Galy et al. (2008a) and Galy and Eglinton (2011).

of sediments delivered by the Lower Meghna River. This key observation implies that, in the Bengal Fan, OC is not significantly affected by marine OC addition or by selective degradation of labile biospheric OC.

Together, these observations imply that terrestrial OC preservation in Bengal Fan sediments is exceptionally high, with the proportion of OC delivered to the sedimentary system that is actually buried in sediments (the burial efficiency) approaching 100%. Although high terrestrial organic carbon export efficiencies have been reported for small mountainous rivers (Goni et al. 2006; Chapter 8), this situation stands in sharp contrast with other large deltaic systems such as those of the Amazon and Mississippi rivers, where terrestrial OC burial efficiencies typically do not exceed 30%, as most of terrestrial OC is replaced by marine OC prior to burial (Hedges et al. 1997; Burdige 2005; Burdige 2007; Chapter 19).

The very high OC burial efficiency in the Bengal Fan likely derives from the transport dynamic, with rapid sediment accumulation on the delta shelf. In shallowwater, where OC oxidation is theoretically most efficient, OC is protected from oxidation by quick burial under a thick layer of fresh sediments. Typical O_2 penetration depths in such sediments are only a few centimeters (Cai and Sayles 1996): the same order of magnitude as the Bengal shelf annual accumulation. Therefore, OC exposure time to O_2 is on the order of only a few years, insufficient for effective OC oxidation.

Some of the OC deposited on the shallow shelf is transferred, along with associated sediments, to the deep-water zone by turbiditic currents and deposited in the channel-levee system. Identical OC loadings in shelf, active channel-levee, and mid-deep fan sediments (Galy et al. 2007) indicate that OC oxidation during sediment transfer from shelf to deep ocean is negligible. In addition, river discharge and precipitation are responsible for a low salinity anomaly and maintain stratified waters in the Bay of Bengal (Broecker et al. 1980; Benner et al. 2003). With high productivity in surface water sustained by high nutrient flux delivered by the G-B, the stratification of the Bay of Bengal waters favors an intense respiration of marine OC in the surface waters. This process consumes O_2 and generates an extended O_2 minimum zone and generally low O_2 concentrations in the Bay of Bengal (Broecker et al. 1980; Benner et al. 2003). In Bengal Fan sediments, terrestrial OC oxidation is thus limited by both short exposure to, and low availability of, O_2. Whereas a recent study of sediments accumulated in the Ganges floodplain over the last glacial-interglacial cycle showed significant degradation of OC in this setting (Agrawal et al. in review), this low degree of preservation in floodplain sediments contrasts with high burial efficiency of essentially the same OC and sediments in the Bengal Fan. This further suggests that depositional conditions exert a first-order control on OC preservation, whereas OC chemical composition or OC-mineral associations do not seem to play a significant role.

9. Conclusions

The evolution of stable-isotopic composition ($\delta^{13}C$) of OC from the outflow of the Himalayan range to the delta of Bangladesh shows that a significant fraction of OC derived from the Himalaya is oxidized and replaced by OC derived from the floodplain. Bulk and compound-specific ^{14}C data further suggest a selective replacement of biospheric OC during floodplain transit combined with selective preservation of old refractory OC.

The OC loading and radiocarbon content of recent Bengal Fan sediments is comparable to that of G-B river sediments. Biomarker abundance and $\delta^{13}C$ values show that OC is dominated by terrestrial inputs. The terrestrial OC burial efficiency is close to 100%, which strongly contrasts with other large deltaic systems on Earth, where ~70% of terrestrial OC is typically oxidized and replaced by marine OC prior to burial (Chapter 19). This extreme burial efficiency is sustained by high erosion rates in the Himalaya that generate high sedimentation rates and low oxygen availability in the Bay of Bengal.

As a result of ongoing global warming, annual temperatures are predicted to increase significantly (2–4°) over the Himalayas (Intergovernmental Panel on Climate Change 2007). This might destabilize ancient OC stored in soils through enhanced microbial decomposition of OC (Kirschbaum 1995). Destabilization of ancient OC stored in permafrost in response to global warming has been proposed as a major positive feedback (Zimov et al. 2006). This process could also be at play in low latitude catchments, such as the Ganges-Brahmaputra basin. However, increasing glacial melt and intensification of the Indian summer monsoon are other possible consequences of global warming. This would result in higher runoff and physical erosion and hence more efficient transfer and burial of biospheric OC in the Bengal fan. This enhanced sequestration of biospheric OC would represent a negative feedback, thereby counteracting the microbial destabilization of soil OC. The respective magnitudes of these two mechanisms affecting the C cycle in opposite directions are currently unknown, as projections of future temperature and precipitations in the Himalayas are still very uncertain. This illustrates the need for better climate simulations – especially precipitations – and the formulation of long-term time-series monitoring of river biogeochemistry.

References

Agrawal, S., V. Galy, P. Sanyal, and T. Eglinton. In review. C4 plants expansion in the Ganga plain during the last glacial cycle: insights from isotopic composition of vascular plant biomarkers. Organic Geochemistry.

Allison, M. A., S. R. Khan, S. L. Goodbred, and S. A. Kuehl. 2003. Stratigraphic evolution of the late Holocene Ganges–Brahmaputra lower delta plain. Sedimentary Geology 155: 317–342.

Aucour, A.–M., C. France–Lanord, K. Pedoja, A.–C. Pierson–Wickmann, and S. M. F. Sheppard. 2006. Fluxes and sources of particulate organic carbon in the Ganga–Brahmaputra river system. Global Biogeochemical Cycles 20: 1–12.

Aufdenkampe, A. K., E. Mayorga, J. I. Hedges, et al. 2007. Organic matter in the Peruvian headwaters of the Amazon: compositional evolution from the Andes to the lowland Amazon mainstem. Organic Geochemistry 38: 337–364.

Barua, D. K., S. A. Kuehl, and R. L. Miller. 1994. Suspended sediment distribution and residual transport in the coastal ocean off the Ganges–Brahmaputra river mouth. Marine Geology 120: 41–61.

Benner, U., J. Poggenburg, E. Faber, D. Quadfasel, and A. Frische. 2003. Methane in ocean waters of the Bay of Bengal: its sources and exchange with the atmosphere. Deep Sea Research Part II: Topical Studies in Oceanography 50: 925–950.

Blair, N. E., E. L. Leithold, and R. C. Aller. 2004. From bedrock to burial: the evolution of particulate organic carbon across coupled watershed–continental margin systems. Marine Chemistry 92: 141–156.

Blasco, F., M. F. Bellan, and M. Aizpuru. 1996. A vegetation map of tropical continental Asia at scale 1.5 million. Journal of Vegetation Science 7: 623–634.
Bouchez, J., O. Beyssac, V. Galy, et al. 2010. Oxidation of petrogenic organic carbon in the Amazon floodplain as a source of atmospheric CO2. Geology 38: 255.
Broecker, W. S., J. R. Toggweiler, and T. Takahashi. 1980. The Bay of Bengal – A major nutrient source for the deep Indian Ocean. Earth and Planetary Science Letters 49: 506–512.
Burdige, D. J. 2005. Burial of terrestrial organic matter in marine sediments: A re–assessment. Global Biogeochemical Cycles 19: 1–7.
Burdige, D. J. 2007. Preservation of organic matter in marine sediments: Controls, mechanisms and an imbalance in sediment organic carbon budgets? Chemical Reviews 107: 467–485.
Cai, W.–J., and F. L. Sayles. 1996. Oxygen penetration depths and fluxes in marine sediments. Marine Chemistry 52: 123–131.
Cochran, J. R., D. A. V. Stow, C. Auroux, et al. 1989. Leg 116 Distal Bengal Fan. Proceedings of the Ocean Drilling Program, Initial Reports. College Station, TX: Ocean Drilling Program.
Coleman, J. M. 1969. Brahmaputra river: channel processes and sedimentation. Sedimentary Geology 3: 129–239.
Curray, J. R. 1994. Sediment volume and mass beneath the Bay of Bengal. Earth and Planetary Science Letters 125: 371–383.
Curray, J. R., F. J. Emmel, and D. G. Moore. 2003. The Bengal Fan: morphology, geometry, stratigraphy, history and processes. Marine and Petroleum Geology 19: 1191–1223.
Deines, P. 1980. The isotopic composition of reduced organic carbon, pp. 329–406. *In* P. Fritz and J. C. Fontes (eds.), Handbook of Environmental Isotope Geochemistry. Elsevier.
Dobremez, J. F. 1976. Le Népal: écologie et biogéographie. Paris: Editions du CNRS.
Dobremez, J. F. E. A. 1978. Carte écologique du Népal 1/250 000. University of Grenoble.
Drenzek, N. J., D. B. Montlucon, M. B. Yunker, R. W. Macdonald, and T. I. Eglinton. 2007. Constraints on the origin of sedimentary organic carbon in the Beaufort Sea from coupled molecular 13C and 14C measurements. Marine Chemistry 103: 146–162.
Drenzek, N., K. Hughen, D. B. Montlucon, et al. 2009. A new look at old carbon in active margin sediments. Geology 37: 239.
France–Lanord, C., and L. A. Derry. 1994. $\delta^{13}C$ of organic carbon in the Bengal Fan: source evolution and transport of C3 and C4 plant carbon to marine sediments. Geochimica et Cosmochimica Acta 58: 4809–4814.
Freeman, K. H., and L. A. Colarusso. 2001. Molecular and isotopic records of C4 grassland expansion in the late Miocene. Geochimica et Cosmochimica Acta 65: 1439–1454.
Galy, V. 2007. Source, transport et enfouissement du carbone organique lors de l'erosion continentale: l'exemple du systeme himalayen. Institut National Polytechnique de Lorraine.
Galy, A., and C. France–Lanord. 2001. Higher Erosion rates in the Himalaya: geochemical constraints on riverine fluxes. Geology 29: 23–26.
Galy, V., O. Beyssac, C. France–Lanord, and T. I. Eglinton. 2008a. Recycling of graphite during Himalayan erosion: a geological stabilization of carbon in the crust. Science 322: 943–945.
Galy, V., C. France–Lanord, and B. Lartiges. 2008b. Loading and fate of particulate organic carbon from the Himalaya to the Ganga–Brahmaputra delta. Geochimica et Cosmochimica Acta 72: 1767–1787.
Galy, V., L. François, C. France–Lanord, et al. 2008c. C4 plants decline in the Himalayan basin since the Last Glacial Maximum. Quaternary Science Reviews 27: 1396–1409.
Galy, V., and T. I. Eglinton. 2011. Protracted storage of biospheric carbon in the Ganges–Brahmaputra basin. Nature Geoscience 4: 843–847.

Galy, V., T. Eglinton, C. France-Lanord, and S. Sylva. 2011. The provenance of vegetation and environmental signatures encoded in vascular plant biomarkers carried by the Ganges–Brahmaputra rivers. Earth and Planetary Science 304: 1–12.

Galy, V., C. France-Lanord, O. Beyssac, P. Faure, H. Kudrass, and F. Palhol. 2007. Efficient organic carbon burial in the Bengal fan sustained by the Himalayan erosional system. Nature 450: 407–410.

Goni, M. A., N. Monacci, R. Gisewhite, A. Ogston, J. Crockett, and C. Nittrouer. 2006. Distribution and sources of particulate organic matter in the water column and sediments of the Fly River Delta, Gulf of Papua (Papua New Guinea). Estuarine, Coastal and Shelf Science 69: 225–245.

Goodbred, S. L. 2003. Response of the Ganges dispersal system to climate change: a source–to–sink view since the last interstade. Sedimentary Geology 162: 83–104.

Goodbred, S. L., and S. A. Kuehl. 1999. Holocene and modern sediment budgets for the Ganges–Brahmaputra river system: evidence for highstand dispersal to flood–plain, shelf, and deep–sea depocenters. Geology 27: 559–562.

Hedges, J. I., W. A. Clark, P. D. Quay, J. E. Richey, A. H. Devol, and U. D. M. Santos. 1986. Compositions and fluxes of particulate organic material in the Amazon River. Limnology & Oceanography 31: 717–738.

Hedges, J. I., R. G. Keil, and R. Benner. 1997. What happens to terrestrial organic matter in the ocean? Organic Geochemistry 27: 195–212.

Hilton, R. G., A. Galy, N. Hovius, and M. J. Horng. 2011. Efficient transport of fossil organic carbon to the ocean by steep mountain rivers: An orogenic carbon sequestration mechanism. Geology 39: 71–74.

Hübscher, C., M. Breitzke, K. Michels, H. R. Kudrass, V. Spiess, and M. Wiedicke. 1998. Late Quaternary seismic stratigraphy of the eastern Bengal Shelf. Marine Geophysical Research 20: 57–71.

Hübscher, C., V. Spiess, M. Breitzke, and M. E. Weber. 1997. The youngest channel–levee system of the Bengal Fan: results from digitalsediment echosounder data. Marine Geology 141: 125–145.

Intergovernmental Panel on Climate Change. 2007. Climate change 2007: The physical science basis. *In* S. Solomon et al. (eds.), Contribution of working group I to the fourth assessment report of the Intergovernmental Panel on Climate Change. Cambridge University Press.

Islam, G. M. T., and S. T. Jaman. 2006. Modeling sediment loads in the lower Ganges, Bangladesh. Water Management 159: 87–94.

Ittekkot, V., H. R. Kudrass, D. Quadfasel, and D. Unger. 2003. Bay of Bengal. Deep Sea Research Part II: Topical Studies in Oceanography 50: 853–1053.

Keil, R. G., L. M. Mayer, P. D. Quay, J. E. Richey, and J. I. Hedges. 1997. Loss of organic matter from riverine particules in deltas. Geochimica et Cosmochimica Acta 61: 1507–1511.

Kirschbaum, M. 1995. The temperature dependence of soil organic matter decomposition, and the effect of global warming on soil organic C storage. Soil Biology and Biochemistry 27: 753–760.

Kottke, B., T. Schwenk, M. Breitzke, M. Wiedicke, H. R. Kudrass, and V. Spiess. 2003. Acoustic facies and depositional processes in the upper submarine canyon Swatch of No Ground (Bay of Bengal). Deep Sea Research Part II: Topical Studies in Oceanography 50: 979–1001.

Kudrass, H. R., K. H. Michels, M. Wiedicke, and A. Suckow. 1998. Cyclones and tides as feeders of a submarine canyon off Bangladesh. Geology 26: 715–718.

Kuehl, S. A., M. A. Allison, S. L. Goodbred, and H. Kudrass. 2005. The Ganges–Brahmaputra delta. Special Publication–SEPM 83: 413.

Kuehl, S. A., T. M. Hariu, and W. S. Moore. 1989. Shelf sedimentation off the Ganges–Brahmaputra river system: Evidence for sediment bypassing to the Bengal fan. Geology 17: 1132–1135.

Kuehl, S. A., B. M. Levy, W. S. Moore, and M. A. Allison. 1997. Subaqueous delta of the Ganges–Brahmaputra river system. Marine Geology 144: 81–96.

Le Fort, P. 1989. The Himalayan orogenic segment. *In* A. M. C. Sengör (ed.), Tectonic evolution of the Tethyan regions. Proceedings of the NATO ASI meeting, Istanbul, October 1985.

Lupker, M., C. France-Lanord, J. Lavé, et al. 2011. A Rouse–based method to integrate the chemical composition of river sediments: application to the Ganga basin. Journal of Geophysical Research – Earth Surface 116: 1–24.

Lupker, M., C. France-Lanord, V. Galy, et al. 2012. Predominant floodplain over mountain weathering of Himalayan sediments (Ganga basin). Geochimica et Cosmochimica Acta 84: 410–432.

Mayorga, E., A. K. Aufdenkampe, C. A. Masiello, et al. 2005. Young organic matter as a source of carbon dioxide outgassing from Amazonian rivers. Nature 436: 538–541.

Meyers, P. A., and G. R. Dickens. 1992. Accumulation of organic Matter in sediments of the Indian Ocean: A synthesis of results from scientific deep sea drilling, pp. 295–309. *In* R. A. Duncan, D. K. Rea, R. B. Kidd, U. von Rad and J. K. Weissel (eds.), Synthesis of Results from Scientific Drilling in the Indian Ocean. Geophysical Monograph. American Geophysical Union.

Michels, K. H., H. R. Kudrass, C. Hübscher, A. Suckow, and M. Wiedicke. 1998. The submarine delta of the Ganges–Brahmaputra: cyclone–dominated sedimentation patterns. Marine Geology 149: 133–154.

Michels, K. H., A. Suckow, M. Breitzke, H. R. Kudrass, and B. Kottke. 2003. Sediment transport in the shelf canyon "Swatch of No Ground" (Bay of Bengal). Deep Sea Research Part II: Topical Studies in Oceanography 50: 1003–1022.

Mook, W. G., and F. C. Tan. 1991. Stable carbon isotopes in rivers and estuaries, pp. 246–264. *In* E. T. Degens, S. Kempe and J. E. Richey (eds.), Biogeochemistry of Major World Rivers.

Palamenghi, L., T. Schwenk, V. Spiess, and H. R. Kudrass. 2011. Seismostratigraphic analysis with centennial to decadal time resolution of the sediment sink in the Ganges–Brahmaputra subaqueous delta. Continental Shelf Research 31: 712–730.

Paul, M., L. Reisberg, N. Vigier, and C. France–Lanord. 2011. Behavior of osmium at the freshwater–saltwater interface based on Ganga derived sediments from the estuarine zone. Geochemistry, Geophysics, Geosystems 12: Q12023.

Pierson–Wickmann, A.–C., L. Reisberg, C. France–Lanord, and H. Kudrass. 2001. Os–Sr–Nd results from sediments in the Bay of Bengal: Implications for sediment transport and the marine Os record. Paleoceanography 16: 435–444.

Poynter, J., and G. Eglinton. 1990. Molecular composition of three sediments from Hole 717c: the Bengal Fan, pp. 155–161. *In* J. R. Cochran, D. A. V. Stow and et al. (eds.), Proc. ODP, Sci. Res. Ocean Drilling Program.

Raymond, P. A., J. E. Bauer, N. F. Caraco, J. J. Cole, B. Longworth, and S. T. Petsch. 2004. Controls on the variability of organic matter and dissolved inorganic carbon ages in northeast US rivers. Marine Chemistry 92: 353–366.

Rosenheim, B. E., and V. Galy. 2012. Direct measurement of riverine particulate organic carbon age structure. Geophys. Res. Lett. 39: L19703.

River Survey Project. 1996. Spatial representation and analysis of hydraulic and morphological data. Water Resources Planning Organization (WARPO).

Schwenk, T., V. Spiess, M. Breitzke, and C. Hubscher. 2005. The architecture and evolution of the Middle Bengal Fan in vicinity of the active channel–levee system imaged by high–resolution seismic data. Marine and Petroleum Geology 22: 637–656.

Segall, M. P., and S. A. Kuehl. 1992. Sedimentary processes on the Bengal continental shelf as revealed by clay–size mineralogy. Continental Shelf Research 12: 517–541.

Singh, S., and C. France–Lanord. 2002. Tracing the distribution of erosion in the Brahmaputra watershed from isotopic compositions of stream sediments. Earth and Planetary Science Letters 252: 645–662.

Steckler, M. S., S. H. Akhter, and L. Seeber. 2008. Collision of the Ganges–Brahmaputra Delta with the Burma Arc: implications for earthquake hazard. Earth and Planetary Science Letters 273: 367–378.

Suckow, A., U. Morgenstern, and H.–R. Kudrass. 2001. Abslolute dating of recent sediments in the cyclone–influenced shelf area off Bangladesh: comparison of Gamma spectrometric (^{137}Cs, ^{210}Pb, ^{228}Ra), radiocarbon, and ^{32}Si ages. Radiocarbon 43: 917–927.

Thorne, C. R., A. P. G. Russell, and M. K. Alam. 1993. Planform pattern and channel evolution of the Brahmaputra river, Bangladesh. *In* J. L. Best and C. S. Bristow (eds.), Braided rivers. Special Publication. Geological Society.

Umitsu, M. 1993. Late Quaternary sedimentary environments and landforms in the Ganges Delta. Sedimentary Geology 83: 177–186.

Weber, M. E., M. H. Wiedicke, H. R. Kudrass, C. Hübscher, and H. Erlenkeuser. 1997. Active growth of the bengal Fan during sea–level rise and highstand. Geology 25: 315–318.

Zimov, S. A., E. a. G. Schuur, and F. S. Chapin. 2006. Permafrost and the global carbon budget. Science 312: 1612–1613.

15

Carbon and nutrient fluxes across tropical river-coastal boundaries

D. M. Alongi, S. Bouillon, C. Duarte, A. Ramanathan, and A. I. Robertson

1. Introduction

The tropical coastal zone is a crucial nexus in the global ocean, being part of the engine for natural and anthropogenic climate change and for oceanic circulation of heat, salt, and water vapor. Although only one-quarter of the world's continental margins lie within the tropics (Jahnke 2010), most of the world's water and sediment is discharged from tropical rivers (Milliman and Farnsworth 2011). The tropical coastal ocean is thus the major conduit for exchange of dissolved and particulate elements and gases between land, sea, and atmosphere. Highly variable salinities, strong tidal fronts, complex coastal boundary layers, high salinity plugs, and tidal trapping and time lags in estuarine circulation, low mean tidal amplitudes, high rates of precipitation and evaporation, permanently stratified thermoclines and haloclines, and migrating fluid mudbanks are just some of the environmental characteristics peculiar to or dominant in tropical coastal waters that have important consequences for biogeochemical cycling (Nittrouer et al. 1995).

Tropical coastal wetland habitats include freshwater marshes and bogs, peatlands, salt marshes, tidal algal mats, and freshwater forests, but mangroves and seagrasses are most often the dominant form of wetland along tropical coastlines (Mitra et al. 2003). The biogeochemical links between tropical mangroves and freshwater wetlands are underappreciated and have not been quantified (Ewel 2010). Most studies of material exchange have focused on the connections between mangrove estuaries and adjacent nearshore waters, rather than on connectivity between mangroves, seagrass meadows, and tidal flats (Davis et al. 2009; Bouillon and Connolly 2009; Kennedy et al. 2010).

Mangroves and seagrasses are highly productive and are a vital ecological and economic resource (Hogarth 2007). They provide many essential ecosystem services, such as food and fuel; habitats for a variety of fauna including fish, mammals, and invertebrates (Nagelkerken et al. 2008); some protection from waves and currents; and a net sink for coastal carbon and other elements. Mangroves, and to a lesser extent, seagrasses, are among the most carbon-rich biomes (Alongi 2009; Duarte et al. 2010) Proportionally large below-ground pools of roots help to stabilize these habitats from tides, waves, winds, and storms, and fine root material is a major contributor to total sediment carbon (e.g., Castaneda-Moya et al. 2011). Both vegetation types facilitate the accumulation of fine particles, fostering rapid rates of sediment accretion and subsequent carbon burial. Mangroves account for only ~1% of carbon sequestration by the world's forests owing to their small global area (<140,000 km^2), but when considered as coastal habitat, they account for 14% of carbon sequestration by the

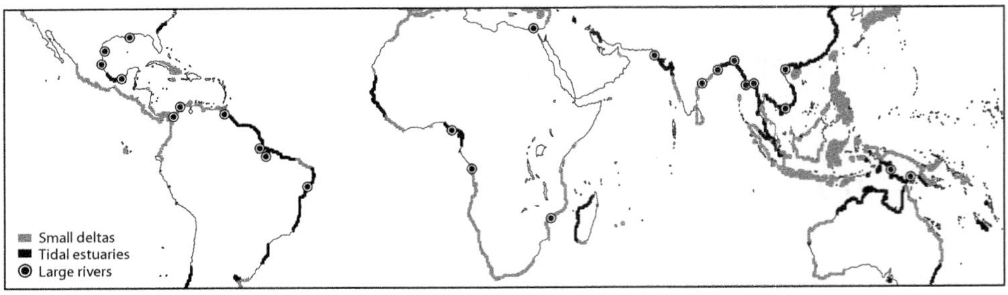

Figure 15.1. Global map of the distribution of large rivers, small deltas, and coastlines dominated by tidal estuaries, throughout the tropical coastal ocean. Modified from Dürr et al. (2011).

global coastal ocean (Alongi 2012). For seagrasses, recent estimates similarly suggest that seagrass meadows are significant coastal carbon sinks, with a global carbon pool possibly as high as 20 Pg C, with high losses owing to erosion and remineralization of nearly 300 Tg C y^{-1} (Fourqurean et al. 2012). Both ecosystems thus play a considerable role in biogeochemical connectivity along the tropical coastal continuum.

In this chapter, we examine the role mangroves and seagrasses play in coupled land-ocean-atmosphere processes and, on a larger scale, to what extent coastal zones in the tropics play – and will continue to play – a role in ocean biogeochemistry. Global change is already altering processes at the river-coast-atmosphere interface (Barange et al. 2010), so here we offer some insights into how increases in sea-level, temperature, and river and land use and changes in atmospheric and ocean chemistry may diminish, accelerate, or leave untouched rates and pathways of carbon and nutrient cycling in tropical coastal seas.

2. Types and diversity of tropical coasts

The world's coasts are composed of many types and sizes of habitats, and a number of typologies exist for coastal ecosystems (Longhurst 1998; Alongi 2004; Dürr et al. 2011). Tropical systems have more generally been typed based on seasonal patterns of precipitation, that is, the "wet" and "dry" tropics. There are basic similarities among classifications, but the scheme of Dürr et al. (2011) is used here because it is most relevant in that it defines coastal areas based on the extent of river inputs to the ocean. Dürr et al. (2011) distinguished seven coastal types: small deltas, tidal systems, lagoons, fjords, large rivers, and karstic and arheic coasts; the latter two types approximate the classic "dry" tropics designation and are not considered further. Small and large river/tidal systems are classified as active filters and lagoons,and karstic and arheic coasts are classified as limited filters and are clearly differentiated on the basis of freshwater discharge (climate) and degree of openness to the ocean (geomorphology). Consequently, this chapter focuses on small and large river deltas and tidally-dominated bays, estuaries, and lagoons (Fig. 15.1). Coastal lagoons that do not receive significant inflow from rivers are not considered here; lagoonal carbon and nutrient dynamics have been reviewed by Marbá et al. (2006), Bouillon and Connolly (2009), Davis et al. (2009), Kennish and Paerl (2010), and Adame and Lovelock (2011).

3. Cross-boundary exchanges

Many estimates have been made over the past two decades regarding the contribution of coastal ecosystems to the global C, N, and P cycles (Wollast 1993; Mackenzie et al. 2002; Smith et al. 2005; Jahnke 2010; also see Chapter 7), and all budgets point to coastal carbon and nutrient fluxes disproportionate to the small area of the world's continental margins. Tropical river-coastal margins deliver most of the carbon and nutrient supplied from land to sea (Davis et al. 2009; Bouillon and Connolly 2009; see also Chapters 12, 14, and 22), with the percentage of global ocean fluxes of dissolved organic and inorganic C, N, and P in the tropics ranging from 65% to 80% and of particulate C, N, and P ranging from 65% to 70% (Seitzinger et al. 2005).

Few data for tropical rivers have either been available or used to extrapolate to all tropical systems in most global estimates (Jahnke 2010). More detailed analysis of carbon fluxes from tropical rivers was made recently by Huang et al. (2012), in which data from 175 rivers were used. Their analysis indicates global water discharge from tropical rivers of 38,540 $km^3\ y^{-1}$ (Fig. 15.2A), within the range of earlier estimates (Milliman and Farnsworth 2011), but their estimate of global sediment discharge of 12,610 mt y^{-1} (Fig. 15.2B) is somewhat lower than earlier approximations (Syvitski et al. 2005). Tropical rivers provide 0.21 Pg C y^{-1} or 55 to 64% of the total global riverine DIC flux (Fig. 15.2C), with the highest yield from Asia, where the percentage of carbonate rock is highest. PIC flux is difficult to estimate owing to lack of data; a crude estimate of 0.051 Pg C y^{-1} (Fig. 15.2D) suggests that about 30% of the global PIC flux is discharged from the tropics. Equatorial rivers have lower DIC but higher DOC concentrations than those at other latitudes; the latter can be attributed to carbon-rich African soils with abundant kaolinite and the prevalence of peat and black-water rivers in tropical Asia and Americas (Kapos et al. 2008). Total DOC discharge (0.136 Pg C y^{-1}) equates to 63–66% of total global flux (Fig. 15.2E). Most tropical POC discharge occurs in the Americas and Asia, with total POC discharge from the tropics estimated at 0.131 Pg C y^{-1} (Fig. 15.2F) or 64–70% of the global flux. Although there is much information on nitrogen and phosphorus concentrations in tropical rivers and estuaries, there are very little data on N and P discharge rates (Davis et al. 2009) or the contribution of wetlands to rates of nutrient export.

3.1. Small to medium river deltas

Wetlands, especially mangroves, tidal flats, and seagrass meadows, front most tropical river deltas, but only a handful of studies have evaluated the contribution of these habitats to river-estuary-ocean biogeochemistry. The lack of information is unfortunate, as small, wetland-fringed deltas are found along the Pacific coast of Central America to northern South America, along the Atlantic coast and of most of the Indian Ocean coast of southern Africa, much of the Indian coast, the Queensland coast of Australia, and among the thousands of islands in Southeast Asia, including the northern coast of Papua New Guinea (Fig. 15.1). The most extensive information of the biogeochemical role of wetlands in small deltaic systems comes from Java, the Sepik River on the north coast of Papua New Guinea, Hinchinbrook Channel on the Queensland coast, and both coasts of India; there are also some flux data from small deltas in the LOICZ case studies, but few identify the contribution of wetlands to C, N, or P export (Ong et al. 2001; Wattayakorn et al. 2001).

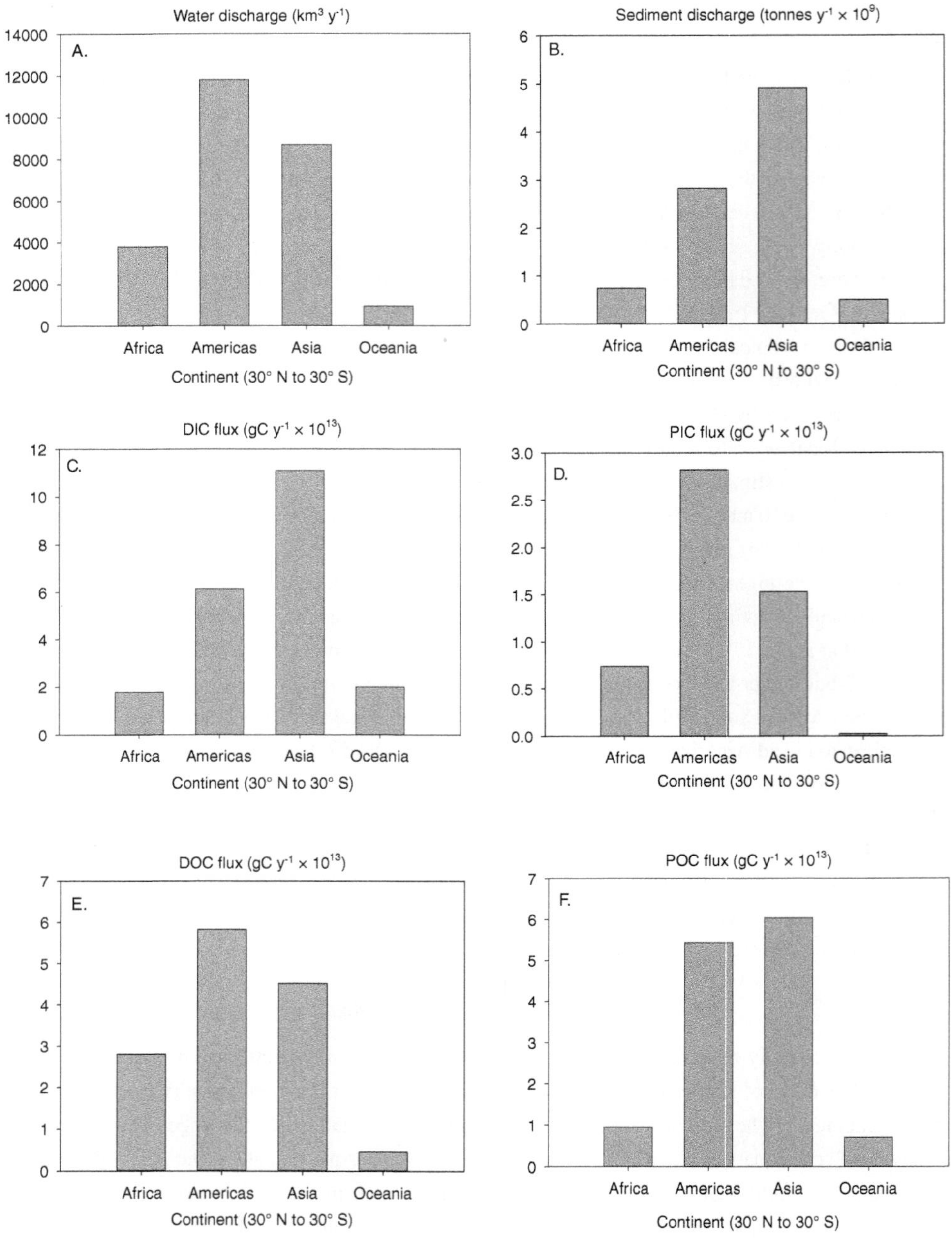

Figure 15.2. Global discharge from tropical rivers of (A) water, (B) sediment, (C) DIC, (D) PIC, (E) DOC, and (F) POC from the four continents. Data from Huang et al. (2012).

Although the contribution of wetlands has not been identified in most studies of small deltas, most do suggest a strong seasonal pattern in relation to the onset and decline of the monsoon season. For instance, biogeochemical budgets for the Tapi River delta in Bandon Bay, Thailand, show that nutrient export is greater in the wet season, dominated mostly by NH_4^+. Most material appears to be metabolized within the bay, indicating that any mangrove organic matter exported from the delta is retained close to the coast (Wattayakorn et al. 2001). Efficiency of recycling of C, N, and P is high within the bay, and the presence of benthic algal mats on mudflats and mangroves at the river mouth can account for most nitrogen fixation. Similarly, fluxes of carbon and nutrients from the Tsengwen delta, Taiwan, are highest in the wet season and in flood periods, although total fluxes are declining due to damming of the river and historically low rainfall (Hung and Huang 2005). Like most tropical estuaries, nutrient and POC behavior in the Tsengwen is nonconservative, but net ecosystem metabolism is positive, with net uptake of CO_2 estimated as 17.8 g C m^{-2} y^{-1}, despite high rates of microbial activity and rapid particle scavenging and deposition in sediments within the estuary.

A strong monsoonal signal is also a dominant feature of the Brantas River delta on Java in Indonesia (Aldrian et al. 2008), where DIC concentrations are high throughout the year, mainly as a result of carbonate weathering. A carbon budget for the estuary indicates that upstream portions are a carbon source, but that middle sections are a carbon sink. Outwelling of carbon and nutrients from the Brantas River to the adjacent Madura Strait is strong year-round with agricultural runoff being a major source of nutrients, but a major part of fluvial input appears to be retained in the inner estuary (Jennerjahn et al. 2004). Like the other small deltaic systems, most labile matter is metabolized within the estuary, and most outwelled material appears to be relatively old and refractory; terrestrial material (including wetlands) may be a significant OM source mainly in the wet season. Although seagrasses were not identified in this study, it is possible that seagrass meadows offshore may utilize river-derived nitrogen and phosphorus, as was found to occur off the south coast of Sulawesi (Stapel et al. 1996; Alongi et al. 2008).

Within the Red River delta, biogeochemical studies have been able to pinpoint the contribution of mangroves to carbon and nutrient dynamics (Wösten et al. 2003; Tue et al. 2012). A LOICZ-CARARET budget model shows that mangroves are sinks for nitrogen and phosphorus. However, marine phytoplankton is the dominant source of organic matter at the delta mouth, whereas phytoplankton and mangrove detritus co-dominates in large and small creeks up-estuary. Overall, the mangrove contribution to the POM pool is up to 76% during low tides, but only 27% at high tides; thus mangrove export of detritus to the creeks occurs mostly during ebb tides.

Arguably, the best-studied small delta is the Hinchinbrook Channel in north Queensland, Australia. The Hinchinbrook Channel is a tropical delta with a barrier (Hinchinbrook Island), with a long channel dominated by mangroves, seagrass beds, and tidal flats and fed mostly by the Herbert River. A mass balance model of organic carbon (Fig. 15.3) shows that mangroves, seagrasses, benthic microalgae on tidal flats, and phytoplankton contribute 55.5%, 6.6%, 1.1%, and 10%, respectively, to total C_{ORG} inputs in the channel, with the Herbert River contributing the remaining 26.8% (Alongi et al. 1998). Much of the organic carbon in adjacent coastal sediment is of mangrove origin, supporting the budget calculation that mangroves are the largest contributor to the coastal zone; much of this material is trapped within the immediate coastal zone (<20 m depth) by a coastal boundary layer, which only breaks down during significant flood events in the summer wet season.

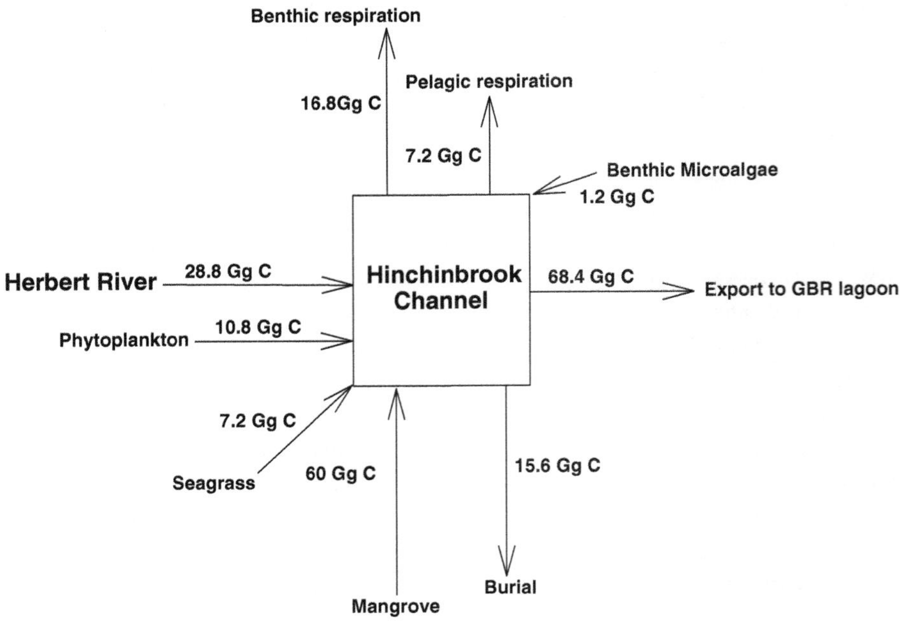

Figure 15.3. Carbon budget of Hinchinbrook Channel. Data from Alongi et al. (1998).

It appears that small-to medium-sized river deltas in the tropics can export significant quantities of carbon derived from mangroves, seagrasses, algae, and river phytoplankton to the adjacent coastal zone, depending on a number of drivers such as climate, tidal amplitude, ratio of wetland to total catchment area, and wetland productivity. However, quantitative estimates of nitrogen and phosphorus export and/or import are poorly constrained, although we can estimate some export of these and other elements assuming known C:N:P ratios of wetland organic matter. It is also readily apparent from these studies that the bulk of wetland material exported to the coastal zone is refractory, the net result of high rates of mineralization of labile fractions (with subsequent efflux of CO_2 to the atmosphere) within these deltas.

3.2. Large river deltas

The exemplar of large tropical rivers is the Amazon, and it is not surprising that most knowledge about tropical coastal biogeochemistry comes from the north Brazilian coast and adjacent shelf. Freshwater wetlands abound within the Amazon Basin, but an analysis of stable isotope patterns points toward upland trees as the major source of carbon to the lower Amazon (Devol and Hedges 2001). The contribution of Amazon-derived materials to offshore carbon and nutrient cycles is well known, but the proportion of this material that is derived from wetlands has not been determined, owing to complex water circulation and sediment transport and to the rich mixture of detritus from a wide variety of sources (DeMaster and Aller 2001).

From the Amazon River mouth to the Orinoco delta to the northwest, the 1,600-km-long Guianas coast receives about 20% of the 1.2×10^9 tons y^{-1} of suspended particulate matter discharged from the river. Most of this material is diverted along a NW path, where complex and irregular episodes of deposition, erosion, and resuspension result in migrating, shore-attached mud banks. Along the north Brazilian coast, both northwest and southeast of the Amazon, extensive mangrove forests develop most fully. Mangroves, mostly *Avicennia germinans*, opportunistically colonize the most stable portions of mud bank along the Guianas coast. The distribution of mangroves changes rapidly, resulting in pioneer forests, young forests, young mature forests, senescent forests, and "recolonized cemetery" forests (Marchand et al. 2003). The organic matter in the soils of these forests is a blend of allochthonous refractory material derived from the Amazon and autochthonous reactive material originating from algal mats and mangrove trees (Marchand et al. 2003).

The contribution of mangroves to coastal biogeochemistry has been assessed southeast of the Amazon (Dittmar et al. 2001, 2006; Jennerjahn and Ittekot 2002), where mangroves fringe the coast between the Caeté, Gurupi, Maracaçumé, and Turiaçu Rivers. Mangroves comprise only about 6% of the catchment area along this portion of the coast, but their contribution to the estuarine DOC and POC pool (2.7×10^{10} mol C y^{-1}) exceeds by several times the total export from the hinderland (Dittmar et al. 2001). Transport of terrestrially-derived DOC appears to be inhibited by a geochemical barrier within the estuaries. Most of the export occurs during the rainy season. A further study (Dittmar et al. 2006) shows that mangroves are a significant source of DOC to the open ocean off northern Brazil. One-third of mangrove DOC is destroyed by UV light, with the remainder being refractory. Extrapolating to the world ocean, Dittmar et al. (2006) calculated that mangroves can account for $>$10% of the terrestrially-derived, refractory DOC to the ocean. Based on data in the literature and partly from their own data acquired along the north Brazilian coast, Jennerjahn and Ittekot (2002) similarly calculated that mangroves account for $\approx$11% of the total terrestrial carbon input in the ocean and 15% of total carbon buried in marine sediments. The contribution of mangrove-derived nitrogen and phosphorus from the Brazilian mangrove-fringed coast is unknown (Koch et al. 2010), although we do know that 28% of P and 20 to 50% of N supplied to the Amazon shelf is derived from rivers (DeMaster and Aller 2001).

Other large tropical rivers with significant coastal mangroves and seagrass beds, such as the Mekong and Irrawaddy deltas (also see Chapter 12), are known to export large quantities of terrestrially-derived C, N, and P to the coastal ocean. The contribution of coastal wetlands has not been quantified, although it may be considerable, if not spatially limited, to the pro-delta front. For instance, off the Irrawaddy delta, isotopic signatures show a rapid decline in terrestrially-derived organic matter from delta mouth to the open sea (Ramaswamy et al. 2008).

Like other large rivers, the Godavari River of east India exports refractory dissolved and particulate materials to the sea. Although the upper catchment delivers significant amounts of DIC further downstream (Sarin et al. 2002), the source of organic matter in the lower catchment is mainly soil in the wet season and phytoplankton in the dry season; heavy erosion owing to deforestation and agricultural practices lead to high OM loading to the extent that the Godavari exports 756×10^9 g C y^{-1} of POC to the Bay of Bengal, representing about 78% of POC derived from the main channel; the remainder is metabolized within the river, deposited on the floodplains, and trapped behind dams (Balakrishna and Probst 2005). A more detailed analysis of the Godavari system

(Borges et al. 2003; Bouillon et al. 2003) indicates that carbonate dissolution dominates DIC dynamics in the oligohaline and mesohaline sections of the river. The partial pressure of CO_2 is low upriver but is high and variable in mangrove creeks within the lower delta. These mangrove creeks are active sites of metabolism. A major pathway for mangrove carbon in this system may be as eventual release as CO_2 to the atmosphere, rather than burial in sediments or export to the Bay of Bengal (also see Chapter 14). Ganguly et al. (2011) recently observed that land-sea breezes, atmospheric stability, monsoonal variability, and net heterotrophy, in addition to metabolic processes, are the major drivers behind CO_2 flux from mangrove waters to the atmosphere.

In the Sundarbans, which is bordered by one of the largest continuous tracts of mangrove in the world, biogenic controls on CO_2 exchange between wetland and atmosphere predominate. Mangrove waters of the Sundarbans are heterotrophic, with high rates of mineralization both in sediments and the water column, leading to supersaturation of CO_2 and net release to the atmosphere (Biswas et al. 2004). There are seasonal variations linked to the monsoons, but overall, the mangroves supply nearly 14 kg C ha^{-1} y^{-1} to the atmosphere. Similar net release rates of CO_2 from mangrove waters have been calculated for other large mangrove-dominated rivers, including creeks and secondary waterways within the Mekong delta (Koné and Borges 2008; also see Chapter 12). Mangrove waters may also be a major source of nitrogenous gases. In the Hooghly River, Biswas et al. (2005) measured both emission and gaseous dry deposition of ammonia at rates of 1.8×10^3 and 2.3×10^3 kgN km^{-2} y^{-1}, respectively, although there is net gaseous ammonia deposition in the mangroves at a rate of 490 kg N km^{-2} y^{-1}. Imported flux from land to mangrove forest averages 4.3×10^6 kg N y^{-1}, out of which 33% is deposited in coastal waters (Biswas et al. 2005). In the Hooghly estuary, an offshoot of the Ganges within the western portion of the Sundarbans, annual fluxes of DIC, DIN, and DIP have been calculated as 2.8×10^6 t, 65.8×10^3 t, and 12.8×10^3 t, respectively, and although not quantified, derived mostly from the extensive mangrove forests lining the lower estuary (Mukhopadhyay et al. 2006). The net export of these carbon and nitrogen species contribute significantly to new production in the northern Indian Ocean (Singh and Ramesh 2011; also see Chapter 14).

Being net heterotrophic, the mangrove waterways of the Hooghly River are also a source of CH_4, with emission rates ranging from 1.97 to 134.6 μmol m^{-2} d^{-1}, and stimulated by decomposition of post-monsoon, phytoplankton blooms and sewage from the millions of people living within the catchment. Methane release is higher from mangrove soils (18.36 mmol m^{-2} d^{-1}). Extrapolating from both water column and soil estimates, Biswas et al. (2007) estimated that the Sundarbans releases CH_4 at an area average rate of 47.3×10^5 mol km^{-2} y^{-1}.

Like many tropical river deltas, the vegetation within the Fly River delta in Papua New Guinea is dominated by mangroves, with community composition governed by the salinity regime. Upstream forests are dominated by the mangrove palm *Nypa fruticans*, which give way in the mid delta to *Bruguiera parviflora* and *B. gymnorrhiza*, replaced in the lower delta by *Rhizophora stylosa* and *Avicennia marina*; at the forefront of these forested islands are accreting mud banks that are major sites of sediment, carbon, and nutrient storage (Alongi 1991). A mass balance for organic carbon in the Fly Delta (Fig. 15.4) shows that the Fly River contributes 1.7 Tg C y^{-1} to the Fly Delta and mangroves contribute 0.6 Tg C y^{-1}; subtracting respiratory losses (1.0 TgC y^{-1}) and sedimentation on accreting mud banks (0.6 Tg C y^{-1}) leaves 0.7 Tg C y^{-1}available for export to the adjacent Gulf of Papua. Export of mangrove POC and DOC (0.3 Tg C y^{-1}) equates to >43% of total organic carbon (C_{ORG}) export. Organic carbon being mineralized in offshore sediments 2 to 5 km from the delta

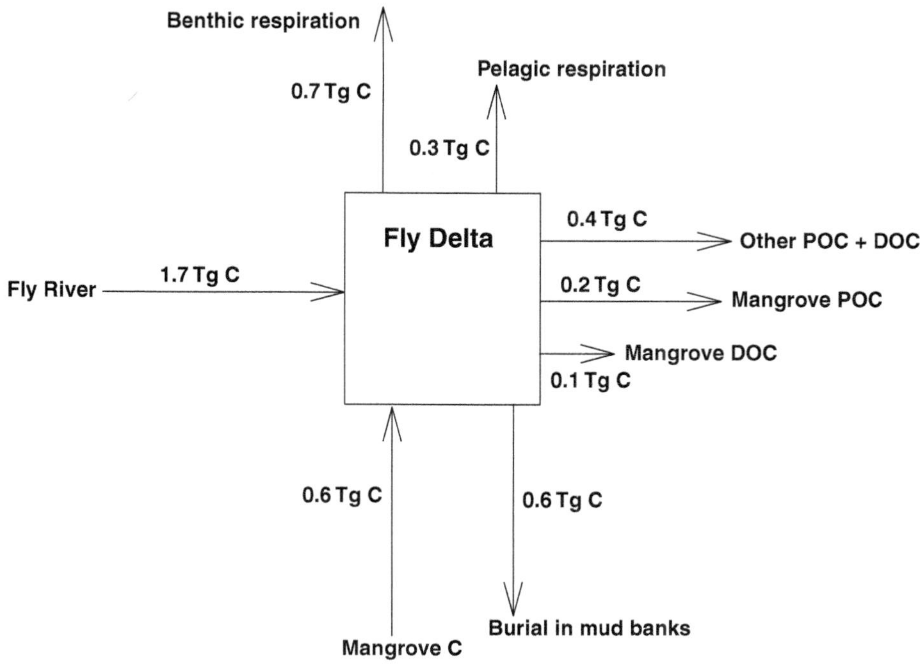

Figure 15.4. Carbon budget for the Fly Delta, Papua New Guinea. Data from Robertson and Alongi (1995) and updated with sediment accumulation data in Walsh and Ridd (2009).

mouth is ≈40% terrestrial and ≈60% marine, with a clear decrease in terrestrial influence further offshore (Aller and Blair 2008). Robertson and Alongi (1995) noted large quantities of macrodetritus floating offshore and sinking to the adjacent deep Coral Sea, but overall, these amounts are minor compared with marine-derived C_{ORG} and terrestrial fine particulate organic matter. Limited transport from the Fly Delta is supported by isotopic signatures (Bird et al. 1995) in Gulf sediments. However, several other large deltas discharge into the Gulf of Papua to the extent that 40% of the Gulf seabed contains at least 75% terrestrially-derived carbon; terrestrial material escapes to the deep ocean. The convergence of several large rivers into the Gulf of Papua translates to significant total export of terrestriallyderived C_{ORG}.

For the vast majority of tropical river-ocean margins there is, however, very little data or understanding of the mechanisms of DIC export. This is unfortunate as two recent global models (Bouillon et al. 2007a; Alongi 2009) of mangrove carbon indicate that up to 50% of net mangrove ecosystem production is unaccounted for, most probably lost as DIC from the forest floor via lateral or groundwater transport to adjacent coastal waters. A similar mechanism has been proposed for temperate coastal ecosystems (Cai 2011). This C loss would explain the supersaturated state of coastal waters and rapid rates of degassing of CO_2 to the atmosphere, especially in tropical deltaic environments where carbon preservation is low (Blair and Aller 2012).

It is clear from the examples provided that, like smaller river deltas, wetlands in large deltas also make a significant contribution to biogeochemical fluxes across the land-sea boundary. The relative contribution will, of course, vary depending on factors such as tidal regime, the proportion of wetland to total delta area, wetland productivity, the magnitude of freshwater flow and its seasonality, coastal water circulation and geomorphology, as well as physicochemical conditions in delta waters. Like small deltas, large deltaic systems export mostly refractory materials, retaining and metabolizing labile matter across the stream-river-estuary-nearshore continuum.

3.3. Tidally-dominated creeks, estuaries, and lagoons

Tides are an important physical feature of nearly all tropical river deltas, but some coastal ecosystems are devoid of a prograding delta, being transgressive and often characterized by meso- or macro-tides and a low-gradient intertidal zone available for wetland colonization. There are, of course, rivers that bring large volumes of sediment to wave-dominated coasts, and it is in these areas where complex landforms develop within which extensive wetlands can exist. Thus tidally-dominated, non-deltaic systems actually span a continuum of geomorphologic types.

Tidally-dominated systems have been extensively studied in the tropics, so a large database is available of the exchange of carbon and nutrients with adjacent coastal waters. A recent review by Adame and Lovelock (2011) for mangrove-dominated creeks, estuaries, and lagoons found that such systems export carbon and nutrients to the coastal zone. Most carbon is exported as litter (Fig. 15.5A), with lesser amounts of small C particles (Fig. 15.5B) and dissolved organic carbon (Fig. 15.5C) also exported, but with greater variation among systems. Both nitrogen (Fig. 15.6A) and phosphorus (Fig. 15.6B) fluxes are highly variable among systems, although removal of one outlier showing net import (Vietnam) indicates that most systems export N and P, in either dissolved or particulate form. Flux rates relate best to precipitation and tidal range; more flushing downstream by tides or rainfall or both results in more material exported. Export of carbon and nutrients from tropical catchments is greater on a per area basis than the global average. Tropical riverine export averages 49.5 and 19.2 g C m^{-2} y^{-1} for POC and DOC (Fig. 15.5), 5.98 g N m^{-2} y^{-1} for total N (Fig. 15.6A), and 2.46 g P m^{-2} y^{-1} for total P (Fig. 15.6B). These values are from 5 to 40 times greater than the global averages of 2.4 and 3.7 g C m^{-2} y^{-1} for POC and DOC (Alvarez-Cobelas et al. 2012), 0.93 g N m^{-2} y^{-1} for total N (Alvarez-Cobelas et al. 2008), and 2.46 g P m^{-2} y^{-1} for total P (Alvarez-Cobelas et al. 2009), demonstrating the greater intensity of tropical river discharge. Furthermore, the tropical C:N:P molar ratio of 72:5:1 (excluding litter C for comparison) is greatly different from the C:N:P ratio of 251:33:1 based on the global average, indicating less nitrogen in relation to carbon but C:P and N:P stoichiometry less than the Redfield ratio in tropical waters.

Export from tidally-dominated systems is delivered to adjacent bays and the inner continental shelf, and like deltaic systems, most of this material is either processed or deposited close inshore. The exported material is metabolized mostly by bacteria, but in many cases, there is significant flux of material between estuaries and inshore and vegetation further offshore. For instance, in Gazi Bay, Kenya, carbon derived from seagrass beds is imported during flood tides into mangrove creeks. Some mangrove-derived material gets transported to the adjacent seagrass beds where most carbon is rapidly metabolized, with little permanent burial (Hemminga et al. 1994). Rapid, but spatially restricted, transfer of materials between the mangroves and seagrasses is facilitated by trapping of estuarine water within the bay; thus the connection is mostly through river plumes and tidal effects. Dissolved carbon dominates, however, and there is a strong but declining mangrove signature from within the

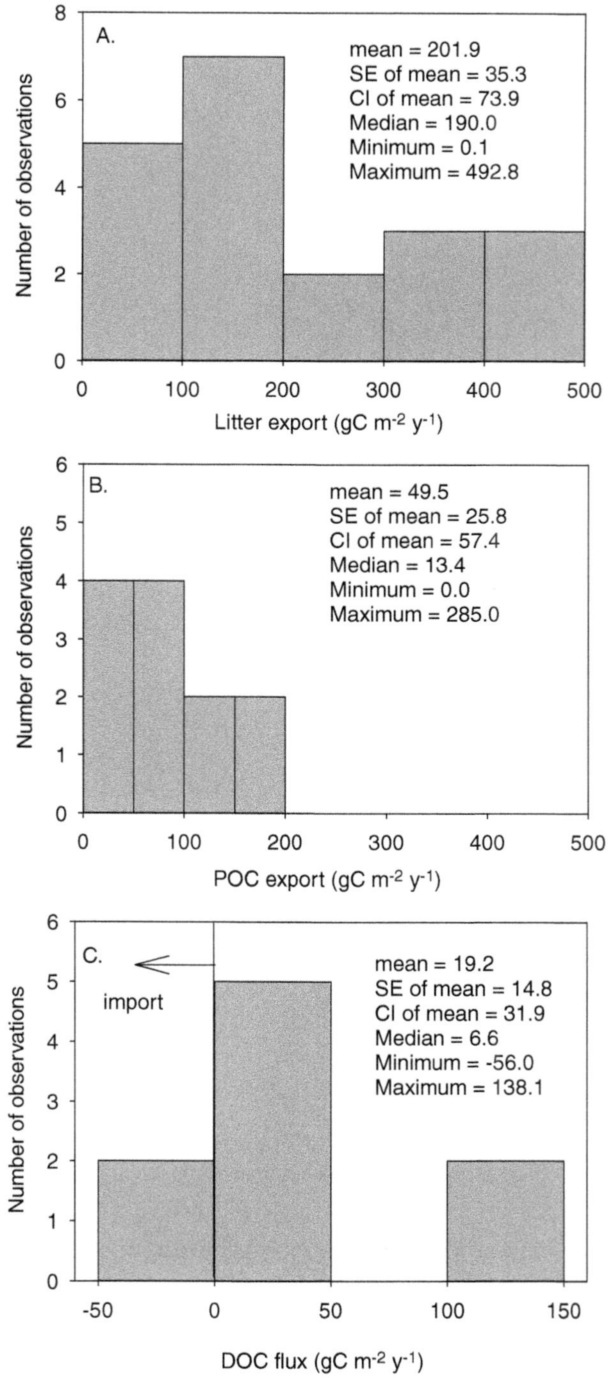

Figure 15.5. Histogram and descriptive statistics of rates of (A) litter export, (B) POC export, and (C) flux of DOC from mangrove-dominated creeks, estuaries, and lagoons. Data from Adame and Lovelock (2011).

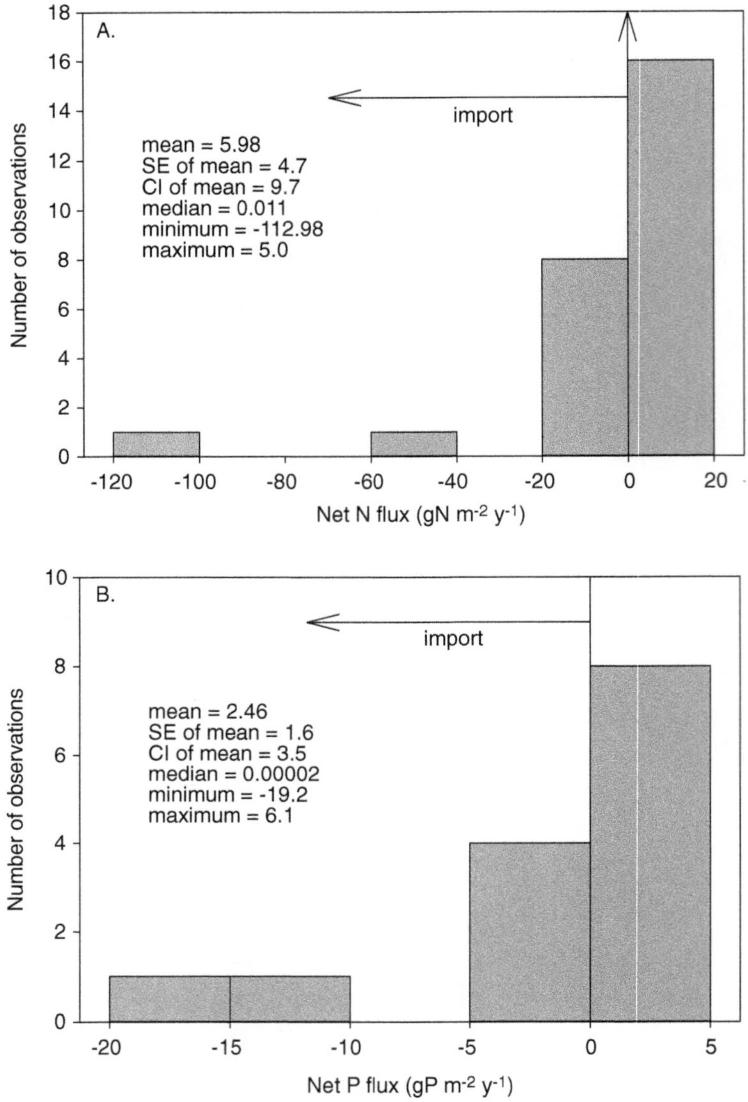

Figure 15.6. Histogram and descriptive statistics of rates of (A) net N flux, and (B) net P flux from mangrove-dominated creeks, estuaries, and lagoons. Data from Adame and Lovelock (2011).

creeks to reefs further offshore (Bouillon et al. 2007b). The pattern of DIC is very different, with mangrove-lined creeks being a source of CO_2 to the atmosphere owing to supersaturated conditions fostered by high metabolic rates. Rapid outgassing occurs when these CO_2-rich waters reach open water conditions close to the mangrove forest boundary, but within a short distance the effect of dense seagrass vegetation becomes apparent, the latter apparently acting as a sink of CO_2, utilizing exported DIC for growth and production. In other African coastal areas, the connection between mangroves and seagrasses is similar, with net export of mangrove-derived matter to closely adjacent seagrass beds (Mohammed et al. 2001). There is little or no evidence that mangrove material is exported to seagrass beds a few kilometers offshore.

Flux patterns of nitrogen and phosphorus to seagrass beds adjacent to tidal estuaries show that upland runoff can be important to seagrass growth and development. For example, in the southern Everglades, discharge during the wet season accounts for nearly all N and P export to Florida Bay. Storms also contribute to the nutrient economy of the bay, as hurricanes and storms account for much of the nutrient exchange between mangroves and seagrasses (Sutula et al. 2003; Davis et al. 2004). Like mangroves, tropical seagrass beds retain and accumulate fine particles, as input of N and P from sediments can provide most of the nutrients required for their nutrition (Davis et al. 2009). Seagrass-derived material is exported to adjacent habitats, including mangroves and coral reefs, providing a vital link along the land-river-estuary-ocean continuum.

Some tidally-dominated estuaries are turbid, with at most only small patches of seagrass even offshore. For instance, the mangrove-lined estuaries along the west coast of peninsular Malaysia are nearly devoid of seagrass owing to high turbidity maintained by strong tidal currents and strong winds (Ong et al. 2000; Alongi et al. 2004). In the Merbok estuary, simple LOICZ single-box models show that the estuary can be either a sink or source of N and P depending on nutrient species (export: DON; sink for DIN) and on the extent of rainfall, with nonconservative behavior of both elements year-round. Carbon fluxes were high, as this estuary is net heterotrophic because of a high organic load. In the Matang estuary further north, a carbon budget (Alongi et al. 2004) shows that most carbon is lost via tree respiration (73%), with only 8% exported to the Straits of Malacca as DOC and POC. Nitrogen cycling in the estuary is dominated by nitrification-denitrification pathways and high rates of DIN uptake by the trees. Sediment N is efficiently (67–81%) mineralized, suggesting that any export of N is refractory in nature.

On the opposite side of the peninsula in southern Thailand, a mass balance for carbon in Sawi Bay, a mangrove-fringed embayment, indicates rapid rates of DOC and POC exchange between the creeks, Sawi Bay, and the Gulf of Thailand (Alongi et al. 2000). The major source of carbon is mangrove-derived (86% of total input). Tree and pelagic respiration are responsible for major losses (34% and 46%, respectively), but there is some net export to the Gulf. Modeling studies of N and P indicate that most nutrients are trapped (most N is denitrified) within the estuarine section of the bay, although there is some export of nitrogen and phosphorus, mostly as DON and DIP (Wattayakorn et al. 2000).

Regardless of tidal regime, most tidally-dominated creeks and estuaries, even semi-restricted lagoons (Young et al. 2005) and small estuaries receiving inputs from tropical blackwater rivers (Yoshioka et al. 2002; Alkhatib et al. 2007), show some export of C, N, and P and considerable CO_2 emissions to the atmosphere (Boonphakdee and Fujiwara 2008; Ralison et al. 2008; Miyajima et al. 2009; Souza et al. 2009). The greater the tidal range, the greater the amount of C, N or P exported. In macrotidal estuaries, such as Darwin Harbour, the 7–9 m tidal range results in a very large fraction of dissolved and particulate estuarine materials being exported to the open sea (Burford et al. 2008). However, much of the carbon in mangroves lining the estuary appears to be retained within the forests or respired in creek waters. Although new nitrogen from the open ocean is a significant flux pathway, phytoplankton communities showed nitrogen limitation.

4. Anthropogenic disturbances and global change

Most of the world's tropical rivers and associated wetlands are not in a pristine state, predisposing these estuarine ecosystems to be less resilient and resistant to further global change. As suggested by Hamilton (2010), the most important changes at the tropical river-coastal boundary in the future are likely to be (1) a shift in temperature leading to a shift in plant and animal species composition; (2) an increase in rates of microbial growth resulting in localized depletion of oxygen and stratification of water bodies; and (3) a shift in hydrological regimes leading to shifts in ecosystem structure and function due to mainly to shifts in the salinity regime. These impacts would be in concert with other anthropogenic drivers such as deforestation, other land-use change, and changes in retention and withdrawal of freshwater.

For centuries, many tropical catchments have suffered water diversion or damming and wastes from humans and agriculture, especially in historically high-density regions, such as tropical South and Southeast Asia (Mimura 2006). More than 23% of the world's rivers are currently influenced by dam construction, resulting in increased water storage and hence increased water residence time (Vörösmarty et al. 1997). During the past 50 years, more than 100 billion tons of sediment and 1–3 billion tons of carbon have been sequestered in reservoirs; soil erosion has resulted in increases in sediment transport by global rivers, yet has reduced the flux of sediment reaching the coastal zone (Syvitski et al. 2005). In the Paía River of Colombia, water diversion has resulted in (1) coastal erosion along part of the coastal delta; (2) transgressive barrier islands with exposed peat soils in the surf zone; (3) closing of abandoned inlets and formation of ebb tidal delta; (4) breaching events on barrier islands; and (5) channel accretion. Within the past 20 years, tidal flats have advanced seaward and freshwater vegetation has advanced downstream, with a decline in mangrove area and an increase in episodes of defoliation (Restrepo and Kettner 2012).

Potential impacts of reservoirs relevant in the context of sediment and nutrient delivery to coastal ecosystems include the retention of sediment (Vörösmarty et al. 1997) and nutrients (Humborg et al. 2000; Harrison et al. 2009), changes in quantity and timing of downstream river flow and flooding patterns (Maingi and Marsh 2002), and downstream changes in river sediment balance (Tamooh et al. 2012). The Tana River delta in Kenya is a good example of a tropical delta characterized by strong sediment retention and alteration of water chemistry due to damming (Maingi and Marsh 2002). Here, the Masinga Dam results in the retention of $\approx$50% of the DIN pool, $\approx$72% of the DIP pool, and $\approx$40% of the dissolved silicate pool (Bouillon et al. 2009). The POC pool in reservoir waters is dominated by ^{13}C-depleted phytoplankton that are rapidly metabolized downstream (Tamooh et al. 2012). Furthermore, relatively old $^{14}C_{POC}$ and ^{14}C-depleted DIC signatures in the Tana delta indicate carbonate dissolution by organic or sulfuric acids, as chemical weathering is intense within the waterways, with most labile organic material being rapidly metabolized within the estuary.

Current environmental problems that are unlikely to abate in the near future include: habitat destruction, sewage, industrial and agricultural pollution, coastal erosion, dams, and siltation owing to deforestation and catchment erosion. Megadeltas formed by the Ganges-Brahmaputra-Meghna, Irrawaddy, Indus, and Mekong rivers have been greatly affected by environmental change, especially by increases in the construction of dams, pollution, and encroachment of rapidly growing cities

(Woodroffe et al. 2006). These alterations have resulted in declining runoff of water and sediment to the coastal zone.

Although the concomitant export of particulate organic matter to coastal seas has been declining, and despite high retention levels of nutrients in reservoirs, there has been a 50% increase in DIN and DIP delivery to South and Southeast Asian coastal waters (Seitzinger et al. 2010). Increased delivery of dissolved nutrients is one cause of observed increases in the incidence of red tides and eutrophication of tropical Asian coastal waters (Mimura 2008). However, not all perturbed systems have suffered loss of control of their biogeochemical functions. For instance, the mangrove-fringed Segara Anakan Lagoon on Java, Indonesia, receives significant inputs of nutrients from an agriculturally dominated catchment (Jennerjahn et al. 2009), yet nutrient pollution and eutrophication in the lagoon is low to moderate because rapid flushing and short residence time of lagoon water results in rapid export of land-derived nutrients to the sea. Similarly, Yasin et al. (2010) estimated that riverine export of nutrients has increased markedly in African rivers during the period 1970–2000, most markedly for DIP, where the increase has reached 80% above pre-1970 levels. This increase in nutrient inputs is related to a shift from natural to anthropogenic nutrient sources. Nevertheless, Yasin et al. (2010) estimated that riverine export of sediments and particulate C, N, and P are expected to decrease substantially in the coming decades.

Problems associated with climate change, such as sea-level rise, have been with us for centuries, but an acceleration of these changes on perturbed tropical river-coastal boundaries will drive changes (and the pace of change) to a new level in the future. Because of current trends in sea-level rise, rivers are already experiencing an increase in flooding events as well as greater seawater encroachment upstream. Regional expression of sea-level rise will be geographically variable, because deltas have highly variable rates of subsidence (Woodroffe et al. 2006).

Two well-documented cases of the response of tropical deltas (with a declining sediment load) to sea-level rise are the Sundarbans (Rahman et al. 2011) and the Irrawaddy Delta (Furuichi et al. 2009). From 1973 to 2010, the direction and extent of erosion and accretion in the Sundarbans varied throughout different time periods; overall, accretion rate has declined because of declining sediment transport, but the erosion rate has remained high because of rising sea-level, with the net result being a net erosion of coastal land (mostly inhabited by mangroves). The Irrawaddy Delta has suffered a similar fate, although variations in river discharge are partly the result of the influence of El Niño-Southern Oscillation. Nevertheless, discharge has decreased in the last 100 years, whereas sealevel has risen, resulting in a net loss of wetlands within the lower delta (Furuichi et al. 2009).

In some areas, sea-level rise accompanied by warmer temperatures is resulting in a shift in vegetation types, such as the expansion of mangroves into heretofore salt marsh habitat (Comeaux et al. 2012), and to the decline in the ability of wetlands to keep pace with the rise in sea-level (Kirwan and Blum 2011). Some climate-related changes will be neutral or even positive, but most foreseen changes will be negative in terms of ecosystem functioning, especially considering the ongoing decline of mangroves, seagrasses, and coral reefs (Mimura 2008). Areas considered to be most vulnerable are Africa, South and Southeast Asia, the Caribbean, the small tropical islands of the Indian and Pacific Oceans, and the Atlantic coast of Central and South America (Nicholls et al. 1999).

There has been considerable interest recently in utilizing the ability of tropical seagrass meadows (Fourqurean et al. 2012) and mangroves (Donato et al. 2011) to accumulate carbon to offset anthropogenic CO_2 emissions and habitat losses. Commonly referred to as "blue carbon," replanting and restoration of these intertidal ecosystems would, in theory, recoup some carbon lost to the atmosphere (McLeod et al. 2011). Considerable discussion in the literature (Hopkinson et al. 2012) now concerns the magnitude and capacity of seagrass meadows (Duarte et al. 2010) and mangroves (Alongi 2012) to sequester carbon.

Although climate-driven change in carbon uptake is potentially largest in tropical land areas (Raddatz et al. 2007), the decline in wetlands coupled with the decline in freshwater and sediment runoff and increases in coastal erosion and pollution suggest that tropical river-coastal boundaries will shift more strongly to a state of net heterotrophy, making these boundaries a greater source of CO_2 to the atmosphere.

5. Conclusions

Most of the world's water and sediment is discharged from tropical rivers, making the tropical coastal ocean the major conduit for exchange of dissolved and particulate elements and gases between land, sea, and atmosphere. Tropical riverine fluxes of dissolved C, N, and P range from 65% to 80% and from 65% to 70% for particulate C, N, and P of total riverine flux to the global ocean. Carbonate weathering, a strong monsoonal signal, rapid rates of accumulation and metabolism, and high recycling efficiencies are dominant features of both small and large tropical river-coastal boundaries. Mangroves, tidal flats, and seagrasses front most deltas, but mangroves are the main contributor of material fluxes to the coastal zone; most material is refractory and trapped within the immediate coastal zone by complex circulation mechanisms. A major carbon pathway is the release of CO_2 from supersaturated coastal waters. Most tropical rivers are under an intensifying anthropogenic influence, with increases in the construction of dams, pollution, and encroachment of rapidly growing cities, predisposing them to be less resilient and resistant to future change. These alterations have resulted in declining delivery of water and sediment to the coastal zone. Tropical rivers are already experiencing an increase in flooding events as well as greater seawater encroachment upstream. Some climate-related changes will be neutral or positive, but most changes will be negative, especially considering the ongoing decline of mangroves, seagrasses, and coral reefs. The decline in wetlands coupled with declines in freshwater and sediment delivery and increases in coastal erosion and pollution suggest that tropical river-coastal areas will shift more strongly to a state of net heterotrophy. Tropical coastal waters and sediments will therefore become increasingly important sources of greenhouse gases to the atmosphere.

References

Adame, M.F., and C.E. Lovelock. 2011. Carbon and nutrient exchange of mangrove forests with the coastal ocean. Hydrobiologia 663:23–50.

Aldrian, E., C-T.A. Chen, S. Adi, Prihartanto, et al. 2008. Spatial and seasonal dynamics of riverine carbon fluxes of the Brantas catchment in East Java. J. Geophys. Res. 113: G03029.

Alkhatib, M., T.C. Jennerjahn, and J. Samiaji. 2007. Biogeochemistry of the Dumai River estuary, Sumatra, Indonesia, a tropical blackwater river. Limnol. Oceanogr. 52:2410–2417.

Aller, R.C., and N.E. Blair. 2008. Early diagenetic remineralization of sedimentary organic C in the Gulf of Papua deltaic complex (Papua New Guinea): Net loss of terrestrial C and diagenetic fractionation of C isotopes. Geochim. Cosmochim. Act. 68:1815–1825.

Alongi, D.M. 1995. The role of intertidal mudbanks in the diagensis and export of dissolved and particulate materials from the Fly Delta, Papua New Guinea. J. Exp. Mar. Biol. Ecol. 149:81–107.

Alongi, D.M. 2004. Ecosystem types and processes, pp. 317–352. *In* A.R. Robinson, J. McCarthy, and B.J. Rothschild (eds.), The Sea, vol. 13. Harvard.

Alongi, D.M. 2009. The Energetics of Mangrove Forests. Springer.

Alongi, D.M. 2012. Carbon sequestration in mangrove forests. Carbon Manage. 3:313–322.

Alongi, D.M., L.A. Trott, M.C. Undu, and F. Tirendi. 2008. Benthic microbial metabolism in seagrass meadows along a carbonate gradient in Sulawesi, Indonesia. Aquat. Microb. Ecol. 51: 141–152.

Alongi, D.M., T. Ayukai, G.J. Brunskill, B.F. Clough, et al. 1998. Sources, sinks, and export of organic carbon through a tropical, semi-enclosed delta (Hinchinbrook Channel, Australia). Mangr. Salt Marsh. 2:237–242.

Alongi, D.M., G. Wattayakorn, T. Ayukai, B. Clough, et al. 2000. An organic carbon budget for mangrove-fringed Sawi Bay, southern Thailand. Phuket Mar. Biol. Cent. Sp. Publ. 22: 79–85.

Alongi, D.M., A. Sasekumar, V.C. Chong, J. Pfitzner, et al. 2004. Sediment accumulation and organic material flux in a managed mangrove ecosystem: Estimates of land-ocean-atmosphere exchange in peninsular Malaysia. Mar. Geol. 208:383–402.

Alvarez-Cobelas, M., D.G. Angeler, and S. Sánchez-Carrillo. 2008. Export of nitrogen from catchments: a worldwide analysis. Environ. Pollut. 156:261–269.

Alvarez-Cobelas, M., S. Sánchez-Carrillo, D.G. Angeler, and R. Sánchez-Andrés. 2009. Phosphorus export from catchments: a global view. J. N. Am. Benthol. Soc. 28:805–820.

Alvarez-Cobelas, M., D.G. Angler, S. Sánchez-Carrillo, and G. Almendros. 2012. A worldwide view of organic carbon export from catchments. Biogeochemistry 107:275–293.

Balakrishna, K., and J.L. Probst. 2005. Organic carbon transport and C/N ratio variations in a large tropical river: Godavari as a case study, India. Biogeochemistry 73:457–473.

Barange, M., J.G. Field, R.P. Harris, E.E. Hofmann, R.I. Perry, and F.E. Werner (eds.). 2010. Marine Ecosystems and Global Change. Oxford.

Bird, M.I., G.J. Brunskill, and A.R. Chivas. 1995. Carbon-isotope composition of sediments from the Gulf of Papua. Geo-Mar. Lett. 15:153–159.

Biswas, H., S.K. Mukhopadhyay, T.K. De, S. Sen, and T.K. Jana. 2004. Biogenic controls on the air-water carbon dioxide exchange in the Sundarban mangrove environment, northeast coast of Bay of Bengal, India. Limnol. Oceanogr. 49:95–101.

Biswas, H., A. Chatterjee, S.K. Mukhopadhya, T.K. De, et al. 2005. Estimation of ammonia exchange at the land-ocean boundary condition of Sundarban mangrove, northeast coast of Bay of Bengal, India. Atmos. Environ. 39:4489–4499.

Biswas, H., S.K. Mukhopadhyay, S. Sen, and T.K. Jana. 2007. Spatial and temporal patterns of methane dynamics in the tropical mangrove dominated estuary, NE coast of Bay of Bengal, India. J. Mar. Syst. 68:55–64.

Blair, N.E., and R.C. Aller. 2012. The fate of terrestrial organic carbon in the marine environment. Ann. Rev. Mar. Sci. 4:401–423.

Boonphakdee, T., and T. Fujiwara. 2008. Temporal variability of nutrient budgets in a tropical river estuary: the Bangpakong River Estuary, Thailand. Environ. Asia 1:7–21.

Borges, A.V., S. Djenidi, G. Lacroix, J. The'ate, B. Delille, and M. Frankignoulle. 2003. Atmospheric CO_2 flux from mangrove surrounding waters. Geophys. Res. Lett. 30: 1558.

Bouillon, S., and R.M. Connolly. 2009. Carbon exchange among tropical coastal ecosystems, pp. 45–70. *In* I. Nagelkerken (ed.), Ecological connectivity among tropical coastal ecosystems. Springer.

Bouillon, S., M. Frankignoulle, F. Dehairs, B. Velimirov, et al. 2003. Inorganic and organic biogeochemistry in the Gautami-Godavari estuary (Andhra Pradesh, India) during the pre-monsoon: The local impact of extensive mangrove forests. Glob. Biogeochem. Cy. 17, doi:10.1029/2002GB002026.

Bouillon, S., A.V. Borges, E. Castañeda-Moya, K. Diele, et al. 2007a. Mangrove production and carbon sinks: a revision of global budget estimates. Glob. Biogeochem. Cy. 22: GB2013.

Bouillon, S., F. Dehairs, B. Velimirov, G. Abril, et al. 2007b. Dynamics of organic and inorganic carbon across contiguous mangrove and seagrass systems (Gazi Bay, Kenya). J. Geophys. Res. 112: G02018.

Bouillon, S., G. Abril, A.V. Borges, F. Dehairs, et al. 2009. Distribution, origin and cycling of carbon in the Tana River (Kenya): a dry season basin-scale survey from headwaters to the delta. Biogeosciences 6:2475–2493.

Burford, M.A., D.M. Alongi, A.D. McKinnon, and L.A. Trott. 2008. Primary production and nutrients in a tropical macrotidal estuary, Darwin Harbour, Australia. Estuar. Coast.Shelf Sci. 79:440–448.

Cai, W-J., 2011. Estuarine and coastal ocean carbon paradox: CO_2 sinks or sites of terrestrial carbon incineration? Ann. Rev. Mar. Sci. 3:123–146.

Castaneda-Moya, E., R.R. Twilley, V.H. Rivera-Monroy, B.D. Marx, C. Coronado-Molina, and S.M.L. Trott. 2011. Patterns of root dynamics in mangrove forests along environmental gradients in the Florida coastal Everglades. Ecosystems 14:1178–1195.

Comeaux, R.S., M.A. Allison, and T.S. Bianchi. 2012. Mangrove expansion in the Gulf of Mexico with climate change: Implications for wetland health and resistance to rising sea levels. Estuar. Coast. Shelf Sci. 96:81–95.

Davis III, S.E., J.E. Cable, D.L. Childers, C. Coronado-Molina, et al. 2004. Importance of episodic storm events in controlling ecosystem structure and function in a Gulf Coast estuary. J. Coastal Res. 20:1198–1208.

Davis III, S.E., D. Lirman, and J.R. Wozniak. 2009. Nitrogen and phosphorus exchange among tropical coastal ecosystems, pp. 9–43. *In* I. Nagelkerken (ed.), Ecological connectivity among tropical coastal ecosystems. Springer.

DeMaster, D.J., and R.C. Aller. 2001. Biogeochemical processes on the Amazon Shelf: Changes in dissolved and particulate fluxes during river/ocean mixing, pp. 328–357. *In* M.E. McClain, R.L.Victoria, and J.E. Richey (eds.), The Biogeochemistry of the Amazon Basin. Oxford.

Devol, A.H., and J.I. Hedges. 2001. Organic matter and nutrients in the mainstream Amazon River, pp. 275–306. *In* M.E. McClain, R.L.Victoria, and J.E. Richey (eds.), The Biogeochemistry of the Amazon Basin. Oxford.

Dittmar, T., R.J. Lara, and G. Kattner. 2001. River or mangrove? Tracing major organic matter sources in tropical Brazilian coastal waters. Mar. Chem. 73:253–271.

Dittmar, T., N. Hertkorn, G. Kattner, and R.J. Lara. 2006. Mangroves, a major source of dissolved organic carbon to the ocean. Glob. Biogeochem. Cy. 20: GB1012, doi: 10.1029/2005GB002570.

Donato, D.C., J.B. Kauffman, D. Murdiyarso, S. Kurnianto, M. Stidham, and M. Kanninen. 2011. Mangroves among the most carbon-rich forests in the tropics. Nat. Geosci. 4: 293–297.

Duarte, C.M., N. Marbá, E. Gacia, J.W. Fourqurean, et al. 2010. Seagrass community metabolism: Assessing the carbon sink capacity of seagrass meadows. Glob. Biogeochem. Cy. 24: GB4032.

Dürr, H.H., G.G. Laruelle, C.M. van Kempen, C.P. Slomp, et al. 2011. Worldwide typology of nearshore coastal systems: defining the estuarine filter of river inputs to the oceans. Estuaries Coasts 34:441–458.

Ewel, K.C. 2010. Appreciating tropical coastal wetlands from a landscape perspective. Front. Ecol. Environ. 8:20–26.

Fourqurean, J.W., C.M. Duarte, H. Kennedy, N. Marbá, et al. 2012. Seagrass ecosystems as a significant global carbon stock. Nat. Geosci. 5:505–509.

Furuichi, T., Z. Win, and R.J. Wasson. 2009. Discharge and suspended sediment transport in the Ayeyarwady River, Myanmar: centennial and decadal changes. Hydrol. Process. 23:1631–1641.

Ganguly, D., M. Dey, C. Chowdhury, A.A. Pattnaik, et al. 2011. Coupled micrometeorological and biological processes on atmospheric CO_2 concentrations at the land-ocean boundary, NE coast of India. Atmos. Environ. 45:3903–3910.

Hamilton, S.K. 2010. Biogeochemical implications of climate change for tropical rivers and floodplains. Hydrobiologia 657:19–35.

Harrison, J.A., R.J. Maranger, R.B. Alexander, A.E. Giblin, P. Jacinthe, E. Mayorga, S.P. Seitzinger, D.J. Sobota, and W.W. Wollheim. 2009. The regional and global significance of nitrogen removal in lakes and reservoirs. Biogeochemistry 93:143–157.

Hemminga, M.A., F.J. Slim, J. Kazungu, G.M. Ganssen, et al. 1994. Carbon outwelling from a mangrove forests with adjacent seagrass beds and coral reefs (Gazi Bay, Kenya). Mar. Ecol. Prog. Ser. 106:291–301.

Hogarth, P. 2007.The Biology of Mangroves and Seagrasses. Oxford: Oxford University Press.

Hopkinson, C.S., W-J. Cai, and X. Hu. 2012. Carbon sequestration in wetland dominated coastal systems- a global sink of rapidly diminishing magnitude. Curr. Opinion Environ. Sustain. 4: 186–194.

Huang, T-H.,Y-H. Fu, P-Y. Pan, and C-T.A. Chen. 2012. Fluvial carbon fluxes in tropical rivers. Curr. Opin. Environ. Sustain. 4:1–8.

Humborg, C., D.J. Conley, L. Rahm, F. Wulff, A. Cociasu, and V. Ittekot. 2000. Silicon retention in river basins: far-reaching effects on biogeochemistry and aquatic food webs in coastal marine environments. Ambio 29:45–50.

Hung, J-J., and M-H. Huang. 2005. Seasonal variations of organic carbon and nutrient transport through a tropical estuary (Tsengwen) in southwestern Taiwan. Environ. Geochem. Health 27:75–95.

Jahnke, R.A. 2010. Global synthesis, pp. 597–615. *In* K-K. Liu, L. Atkinson, R. Quiñones, and L. Talaue-McManus (eds.), Carbon and Nutrient Fluxes in Continental Margins: A Global Synthesis. Springer.

Jennerjahn, T., and V. Ittekkot. 2002. Relevance of mangroves for the production and deposition of organic matter along tropical continental margins. Naturwissenschaften 89: 23–30.

Jennerjahn, T., B. Nasir, and I. Pohlenga. 2009. Spatio-temporal variation of dissolved inorganic nutrients related to hydrodynamics and land use in the mangrove-fringed Segara Anakan Lagoon, Java, Indonesia. Reg. Environ. Change 9:259–274.

Jennerjahn, T., V. Ittekot, S. Klöpper, S. Adi, et al. 2004. Biogeochemistry of a tropical river affected by human activities in its catchment: Brantas River estuary and coastal waters of Madura Strait, Java, Indonesia. Estuar. Coast.Shelf Sci. 60:503–514.

Kapos, V., C. Ravilious, A. Campbell, B. Dickson, et al. 2008. Carbon and biodiversity: a demonstration atlas. UNEP-WCMC.

Kennedy, H.A., J. Beggins, C.M. Duarte, J.W. Fourqurean, et al. 2010. Seagrass sediments as a global carbon sink: Isotopic constraints. Global Biogeochem. Cy. 24: GB4026.

Kennish, M. J., and H.W. Paerl. 2010. Coastal lagoons: critical habitats of environmental change. CRC.

Kirwan, M.L., and L.K. Blum. 2011. Enhanced decomposition offsets enhanced productivity and soil carbon accumulation in coastal wetlands responding to climate change. Biogeosciences 8:978–993.

Koch, B.P., T. Dittmar, and R.J. Lara. 2010. The biogeochemistry of the Caeté mangrove-shelf system, pp. 45–67. *In* U. Saint-Paul and H. Schneider (eds.), Mangrove Dynamics and Management in North Brazil. Springer.

Koné, Y.J-M., and A.V. Borges. 2008. Dissolved inorganic carbon dynamics in the waters surrounding forested mangroves of the Ca Mau Province (Vietnam). Estuar. Coast. Shelf Sci. 77:409–421.

Longhurst, A. 1998. Ecological Geography of the Sea. Academic.

Mackenzie, F.T., L.M. Ver, and A. Lerman. 2002. Coastal zone biogeochemical dynamics under global warming, pp. 27–40. *In* W. Ernst (ed.), Frontiers in Geochemistry: Organic, Solution, and Ore Deposit Geochemistry, Vol. 2. Bellwether.

Maingi, J.K., and S.E. Marsh. 2002. Quantifying hydrologic impacts following dam construction along the Tana River, Kenya. J. Arid. Environ. 50:53–79.

Marbá, N., M. Holmer, E. Gacia, and C. Barrón. 2006. Seagrass beds and coastal biogeochemistry, pp. 135–157. *In* A.W.D. Larkum, R.J. Orth and C.M. Duarte (eds.), Seagrasses: Biology, Ecology and Conservation. Springer.

Marchand, C., E. Lallier-Vergés, and F. Baltzer. 2003. The composition of sedimentary organic matter in relation to the dynamic features of a mangrove-fringed coast in French Guiana. Estuar. Coast. Shelf Sci. 56:119–130.

McLeod, E., G. Chmura, S. Bouillon, R. Saim, et al. 2011. A blueprint for blue carbon: toward an improved understanding of the role of vegetated coastal habitats in sequestering CO_2. Front. Ecol. Environ. 9:552–560.

Milliman, J.D., and K.L. Farnsworth. 2011. River Discharge to the Coastal Ocean: A Global Synthesis. Cambridge.

Mimura, N. 2006. State of the environment in the Asia and Pacific coastal zones and effects of global change, pp. 17–38. *In* N. Harvey (ed.), Global Change and Integrated Coastal Management: The Asia-Pacific Region. Springer.

Mimura, N. (ed.). 2008. Asia-Pacific Coasts and Their Management: States of Environment. Springer.

Mitra, S., R. Wassmann, and P.L.G. Vlek. 2003. Global inventory of wetlands and their role in the carbon cycle. ZEF- Discussion Papers on Development Policy No. 64, Center for Development Research, Bonn.

Miyajima, T., Y. Tsuboi, Y. Tanaka, and I. Koike. 2009. Export of inorganic carbon from two Southeast Asian mangrove forest to adjacent estuaries as estimated by the stable isotope composition of dissolved inorganic carbon. J. Geophy. Res. 114: G01024.

Mohammed, S.M., R.W. Johnstone, B. Widen, and E. Jordelius. 2001. The role of mangroves in the nutrient cycling and productivity of adjacent seagrass communities, Chwaka Bay, Zanzibar, pp. 205–226. *In* D. Richmond and J. Francis (eds.), 20th annual conference on advances in marine science in Tanzaa. Inst. Mar. Sci., Univ. Dar es Salaam ana, Zanzibar.

Mukhopadhyay, S.K., H. Biswas, T.K. De, and T.K. Jana. 2006. Fluxes of nutrients from the tropical River Hooghly at the land-ocean boundary of Sundarbans, NE coast of Bay of Bengal, India. J. Mar. Syst. 62:9–21.

Nagelkerken, I., S.J.M. Blaber, S. Bouillon, P. Green, et al. 2008. The habitat function of mangroves for terrestrial and marine fauna: a review. Aquat. Bot. 89:155–185.

Nicholls, R.J., F.M.J. Hoozemans, and M. Marchand. 1999. Increased flood risk and wetland losses due to global sea-level rise: regional and global analyses. Global Environ. Change 9:S69–S87.

Nittrouer, C.A., G.J. Brunskill, and A.G. Figueiredo. 1995. Importance of tropical coastal environments. Geo-Mar. Lett. 15:121–126.

Ong, J.E., W.K. Gong, A. bin Rahim, C.H. Chang, et al. 2001. Carbon and nutrient fluxes and socio-economic studies in the Merbok mangrove estuary, pp. 1–57. *In* L. Talaue-McManus, H.H.Kremer, and J.I. Marshall Crossland (eds.), Biogeochemical and human dimensions of coastal functioning and change in Southeast Asia. LOICZ Reports and Studies No. 17. Texel, the Netherlands.

Raddatz, T.J., C.H. Reick, W. Knorr, J. Kattge, et al. 2007. Will the tropical land biosphere dominate the climate-carbon cycle feedback during the twenty-first century? Clim. Dyn. 29:565–574.

Rahman, A.F., D. Dragoni, and B. El-Masri. 2011. Response of the Sundarbans coastline to sea-level rise and decreased sediment flow: A remote sensing assessment. Remote Sens. Environ. 115:3121–3128.

Ramaswamy, V., B. Gaye, P.V. Shirodkar, P.S. Rao, et al. 2008. Distribution and sources of organic carbon, nitrogen and their isotopic signatures in sediments from the Ayeyarwady (Irrawaddy) continental shelf, northern Andaman Sea. Mar. Chem. 111:137–150.

Restrepo, J.D., and A. Kettner. 2012. Human induced discharge diversion in a tropical delta and its environmental implications: the Patía River, Colombia. J. Hydrol. 424–425:124–142.

Robertson, A.I., and D.M. Alongi. 1995. Role of riverine mangrove forests in organic carbon export to the tropical coastal ocean: A preliminary mass balance for the Fly Delta (Papua New Guinea). Geo-Mar. Lett. 15:134–139.

Sarin, M.M., A.K. Sudheer, and K. Balakrishna. 2002. Significance of riverine carbon transport: A case study of a large tropical river, Godavari (India). Sci. China 45:97–108.

Seitzinger, S.P., J. A. Harrison, E. Dumont, A.H.W. Beusen, et al. 2005. Sources and delivery of carbon, nitrogen and phosphorus to the coastal zone: An overview of Global Nutrient Export from Watersheds (NEWS) models and their application. Glob. Biogeochem. Cy. 19: GB4S01.

Seitzinger, S.P., E. Mayorga, A.F. Bouwman, C. Kroeze, et al. 2010. Global river nutrient export: A scenario analysis of past and future trends. Global Biogeochem. Cy. 24: GB0A08.

Singh, A., and R. Ramesh. 2011. Contribution of riverine dissolved inorganic nitrogen flux to new production in the coastal Northern Indian Ocean: An assessment. Int. J. Oceanogr. 2011, doi:10.1155/2011/983561.

Smith, S.V., R.W. Buddemeier, F. Wulff, D.P. Swaney, et al. 2005. C,N, and P fluxes in the coastal zone, pp. 95–143. *In* C.J. Crossland, H.H. Kremer, H.J. Lindeboom, J.I.M. Crossland, and M.D.A. Le Tissier (eds.), Coastal Fluxes in the Anthropocene. Springer.

Souza, M.F.L., V. Gomes, S.S. Freitas, R.C.B. Andrade, et al. 2009. Net ecosystem metabolism and nonconservative fluxes of organic matter in a tropical mangrove estuary, Piauí River (NE of Brazil). Estuar. Coast. 32:111–122.

Stapel, J., T.L. Aarts, B.H.M. van Duynhoven, J.D. de Groot, et al. 1996. Nutrient uptake by leaves and roots of the seagrass *Thalassia hemprichii* in the Spermonde Archipelago, Indonesia. Mar. Ecol. Prog. Ser. 134:195–206.

Sutula, M.A., B.C. Perez, E. Reyes, D.L. Childers, et al. 2003. Factors affecting spatial and temporal variability in material exchange between the Southern Everglades wetlands and Florida Bay (USA). Estuar. Coast. Shelf Sci. 57:757–781.

Syvitski, J.P.M., C.J. Vörösmarty, A.J. Kettner, and P. Green. 2005. Impact of humans on the flux of terrestrial sediment to the global coastal ocean. Science 308:376–380.

Tamooh, F., K. Van den Meersche, F. Meysman, T.R. Marwick, et al. 2012. Distribution and origin of suspended sediments and organic carbon pools in the Tana River Basin, Kenya. Biogeosci. Discuss. 9:2523–2566.

Tue, N.T., T.D. Quy, H. Hamaoka, M.T. Nhuan, et al. 2012. Sources and exchange of particulate organic matter in an estuarine mangrove ecosystems of Xuan Thuy National Park, Vietnam. Estuar. Coast. 35:1060–1068.

Vörösmarty, C.J., M. Meybeck, B. Fekete, K. Sharma, P. Green, and J.P.M. Syvitski. 2003. Anthropogenic sediment retention: major global impact from registered river impoundments. Global Planet. Change 39:169–190.

Walsh, J.P., and P.V. Ridd. 2009. Processes, sediments, and stratigraphy of the Fly River Delta, pp. 153–176. In B. Bolton (ed.), The Fly River Papua New Guinea: Environmental studies in an impacted tropical river system. Developments in Earth and Environmental Sciences Series No 9. Elsevier.

Wattayakorn, G., T. Ayukai, and P. Sojisuporn. Material transport and biogeochemical processes in Sawi Bay, southern Thailand. Phuket Mar. Biol. Cent. Sp. Publ. 22:63–77.

Wattayakorn, P. Prapong, and D. Noichareon. 2001. Biogeochemical budgets and processes in Bandon Bay, Suratthani, Thailand. J. Sea Res. 46:133–142.

Wollast, R. 1993. Interactions of carbon and nitrogen cycles in the coastal zone, pp. 195–210. *In* R. Wollast, F.T. Mackenzie, and L. Chou (eds.), Interactions of C, N, P, and S Biogeochemical Cycles and Global Change. Springer.

Woodroffe, C.D., R.J. Nicholls, Y. Saito, Z. Chen, et al. 2006. Landscape variability and the response of Asian megadeltas to environmental change, pp. 277–314. *In* N. Harvey (ed.), Global Change and Integrated Coastal Management: The Asia-Pacific Region. Springer.

Wösten, J.H., P. de Willigen, N.H. Tri, T.V. Lien, et al. 2003. Nutrient dynamics in mangrove areas of the Red River Estuary in Vietnam. Estuar. Coast. Shelf Sci. 57:65–72.

Yasin, J.A., C. Kroeze, and E. Mayorga. 2010. Nutrient export by rivers to the coastal water of Africa: past and future trends. Global Biogeochem. Cy. 24: GB0A07.

Yoshioka, T., S. Ueda, T. Miyajima, E. Wada, et al. 2002. Biogeochemical properties of a tropical swamp forest ecosystem in southern Thailand. Limnol. 3:51–59.

Young, M., M.E. Gonneea, J. Herrera-Silveira, and A. Paytan. 2005. Export of dissolved and particulate carbon and nitrogen from a mangrove-dominated lagoon, Yucatan Peninsula, Mexico. Int. J. Ecol. Environ. Sci. 31:189–202.

Section IV

Western hemisphere systems

16

Sediment, organic carbon, nutrients, and trace elements: sources, transport, and biogeochemical cycles in the lowermost Mississippi River

S. Duan, M. A. Allison, T. S. Bianchi, B. A. McKee, A. M. Shiller, L. Guo, and B. E. Rosenheim

1. Introduction

The Mississippi River, the largest river system in North America, drains an area of 3.22×10^6 km^2, nearly 40% of the landmass of the contiguous United States. It is 6,275 km long from its source in the Missouri and discharges at an annual average rate of 17,000 m^3/s, ranking fourth longest and tenth largest water discharge among the world's rivers (Kammerer 1990; http://pubs.usgs.gov/of/1987/ofr87-242/). The Mississippi River is divided into the Upper Mississippi (UMR) and the Lower Mississippi (LMR) at its confluence with the Ohio River, where the UMR begins in the headwaters of Lake Itasca, Minnesota, and merges with the Missouri River upstream of the Ohio confluence. Below the Ohio confluence, the LMR stretches 1,600 km to the Gulf of Mexico. The waterway's flow through the UMR is moderated by 43 major dams and thousands of smaller impoundments that serve multiple purposes, including power generation, water storage for various uses, recreation, and commercial navigation. The LMR is a meandering river in a broad floodplain from Cairo, Illinois, to the Gulf that is constrained throughout by earthen levees and dikes to control flooding, limit lateral channel migration, and secure a navigation channel for barges. Several major tributaries enter the LMR prior to reaching the Gulf, including the White River, Arkansas River, Big Black River, Yazoo River, and Red River. The lowermost Mississippi River, which is often mentioned in this chapter, refers to the section from the Old River Control Structures (ORCS, Fig. 16.1), to Head of Passes, and including the Atchafalaya and Wax Lake outflows. The ORCS is a river management system in central Louisiana that regulates the flow of water leaving the Mississippi into the Atchafalaya River, thereby preventing the Mississippi river from changing course. The LMR (Mississippi pathway) section below ORCS includes long-term monitoring stations for water and sediment discharge and water quality at St. Francisville, Louisiana, and Tarbert Landing, Mississippi. The LMR (including the Atchafalaya River pathway) is the dominant source of freshwater, sediment, terrestrial organic carbon, and nutrients to the northern Gulf of Mexico (NGOM), nourishing 12,000 km^2 of coastal wetlands, 40% of the salt marsh in the contiguous United States, and providing one-fifth of the total U.S. fisheries harvest (www.gulfbase.org/facts.php).

Several environmental issues have arisen in the last few decades, including decreases in suspended sediment and dissolved silica and excess input of nutrients, trace elements, and organic contaminants (Turner and Rabalais 1991; Barber et al. 1995; Shiller 1997; Meade and Moody 2010) (see Chapter 18). These changes pose a significant risk to the health of the northern Gulf of Mexico in that they

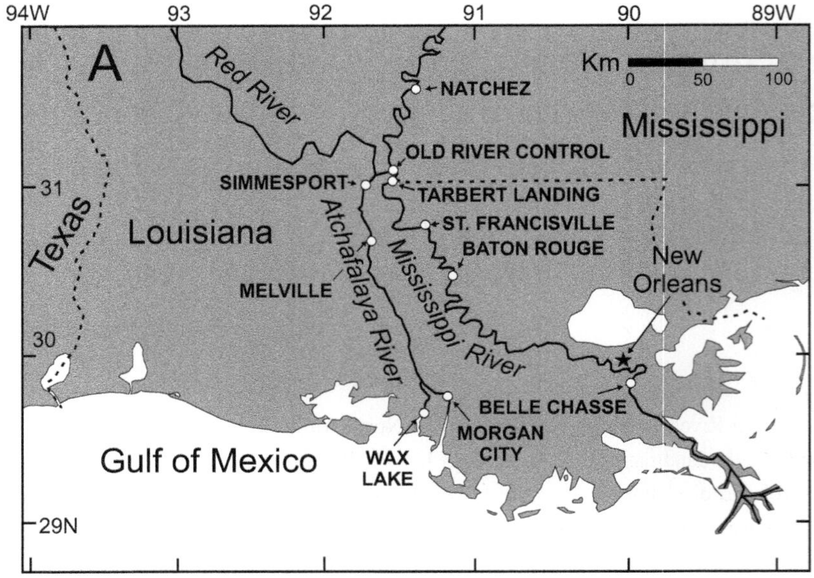

Figure 16.1. Map of the lowermost Mississippi and Atchafalaya Rivers, showing the location of existing water, sediment, and water quality monitoring stations. Figure modified from Allison et al. 2012.

may be linked to deltaic wetland loss, harmful algal blooms in coastal water bodies, and seasonal hypoxia on the adjacent continental shelf. This chapter aims to summarize the present level of understanding about loads and seasonal changes in transporting fresh water, suspended sediment, organic matter, trace metals, and nutrients (nitrogen, phosphorus, and dissolved silica) from the LMR to the Gulf of Mexico. Changes in river flow and suspended sediment are summarized in Section 2, followed by organic matter in Section 3, and trace metals in Section 4. Section 5 reviews the long-term changes and seasonal pattern in nutrient export fluxes, and future changes in organic carbon and nutrient fluxes with land use and climate changes are discussed in the final section.

2. Water and sediment dynamics in the lower Mississippi River

Water movement through the lower Mississippi-Atchafalaya reach is coordinated by the U.S. Army Corps of Engineers through the use of (1) earthen guide levees along the Mississippi and Atchafalaya channels, (2) channel improvements and bank stabilization, and (3) floodways and spillways to divert floodwaters out of the Mississippi channel above New Orleans, Louisiana (see Chapter 2). Levee heights decrease downstream to reflect decreasing stage ranges between high and low discharge, from historic stage extremes at Tarbert Landing of 0.91 and 18.1 m above National Geodetic Vertical Datum (NGVD) to −0.2 and 2.8 m at Venice, Louisiana (river kilometer 18). Downriver of Baton Rouge, Louisiana (river kilometer 372), levees are generally within 1 km of the channel, creating only narrow belts of vegetated land inside the levees (known as batture) that flood during high discharges. Upriver of Baton Rouge, flood protection levees do not closely follow the highly meandering river channel, which allows for significant overbank water storage and sediment sequestration during floods, including exchange of carbon and metals between floodplain sediment and suspended load (Allison et al. 2012). In the Atchafalaya River, the flood control levees that form the Atchafalaya Basin are set back from the channel to create a 30- to 50-km-wide flood basin. However, the upper Atchafalaya basin (upriver of about 68 km below ORCS) has a second set of lower earthen levees that constrain the channel to a width of less than 5 km. Lateral channel migration is constrained throughout the lowermost Mississippi channel through the use of concrete mats lining the channel shallows as well as bankline stone (rip-rap) revetments on the cut bank side of the channel. In the Atchafalaya, bankline stabilization is limited to the same upper basin reach as the secondary levees. In large floods, when discharge in the Mississippi channel below ORCS reaches 35,400 m^3/s, additional water up to 7,079 m^3/s can be passed through the Bonnet Carré Spillway above New Orleans (river kilometer 204 to 208) into Lake Pontchartrain (Fig. 16.1). This has occurred on 10 occasions since 1930, including 2008 and 2011. Discharges above 42,475 m^3/s cannot be passed through Bonnet Carré and are accommodated through the Morganza Spillway and West Atchafalaya Floodway upriver of Baton Rouge, and pass westward into the Atchafalaya Basin. The flood control system, which was fully in place by 1963, means that the LMR channel below New Orleans today never experiences a maximum water discharge above 35,400 m^3/s.

The Mississippi hydrograph has a large seasonal and interannual variability, with high discharge of $>$28,300 m^3/s below ORCS occurring between January and May, and typically with several individual peaks of 1- to 2-week duration (Fig. 16.2). The Atchafalaya hydrograph closely mirrors the Mississippi's given that 60–93% of the water discharge originates in the Mississippi rather than the Red River (Mossa 1996). Recent data seem to suggest that late summer is the time when the two rivers are most similar in chemical properties (Guo, X.H., et al. 2012). Mean Mississippi discharges during the high-water months are about three times the discharge during low water months (Meade 1995). Below ORCS, the timing of the sediment discharge precedes the water discharge maxima as a result of hysteresis, with a more pronounced offset in large flood years (Mossa 1996; Allison et al. 2012). Tides penetrate approximately 200 km (from Head of Passes, or HOP) into the Mississippi River during low discharge (September–November) and as little as 50 km during high discharge (January–May). Currents (0.8 to 1.6 m/s) in the tidal reach are dominated by the river outflow throughout the year; tidal currents are generally too weak to reverse the flow, except briefly during hurricane storm surge influx. A salt wedge is present throughout the discharge cycle and is confined at higher discharges to the Southwest (SW) Pass channel (Fig. 16.1) rather than the shallower Mississippi River passes to the east (e.g., South Pass, Pass a Loutre). At discharges below about 8,500 m^3/s, the saltwater wedge emerges into the main channel above the HOP (Soileau et al. 1989). The thalweg (or the lowest points) of the Mississippi River channel exceeds the depth of the surface of the Gulf of Mexico, allowing for upstream migration of the salt front for ca. 200 km above HOP. The migration of the salt front is constrained and controlled primarily by the volume of water discharge, although flow duration, wind velocity and direction, tides, and water temperature all are contributing factors (Soileau et al. 1989; Galler and Allison 2008). No salt wedge penetration has been observed upriver of the main Atchafalaya entrance into Atchafalaya Bay (Galler and Allison 2008).

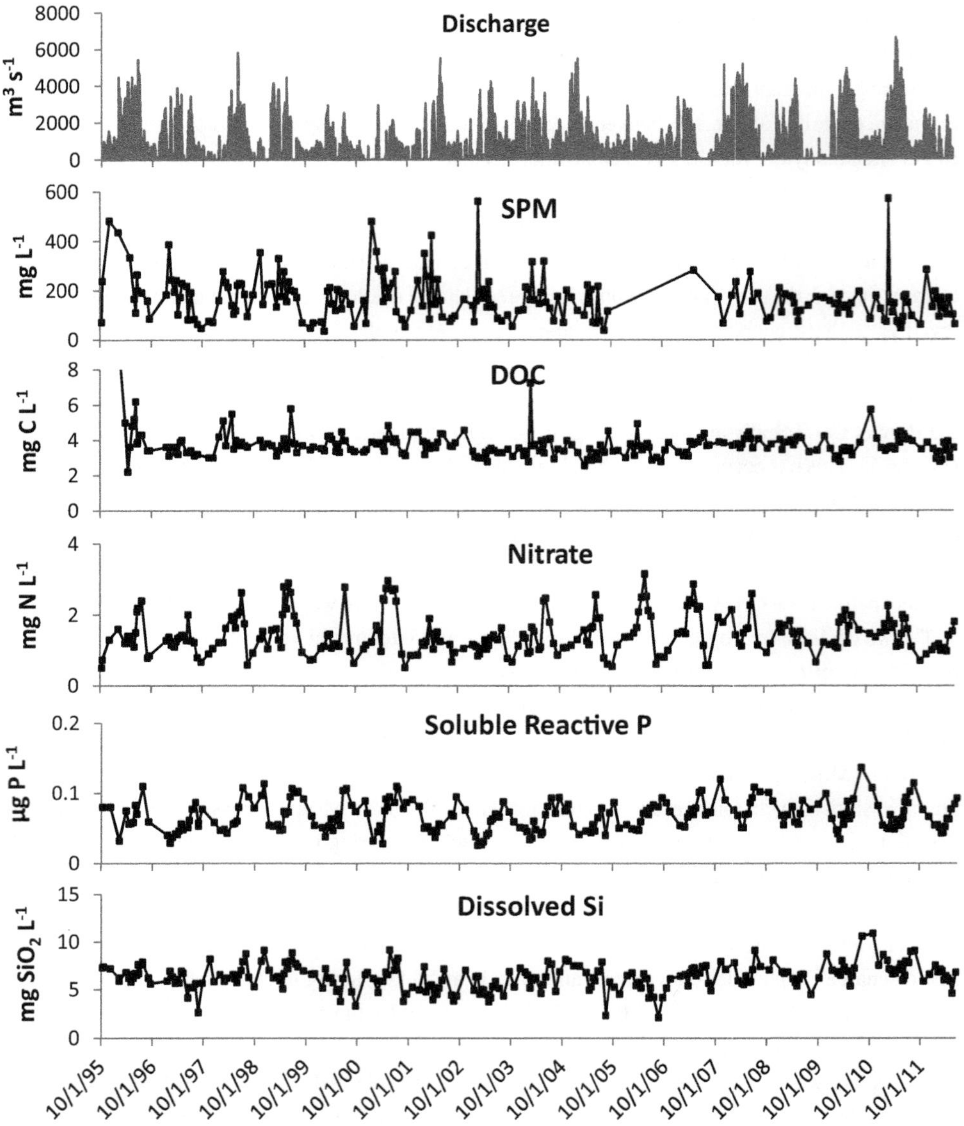

Figure 16.2. Temporal changes in flow, concentrations of SPM, DOC, nitrate, SRP, and dissolved silica in the lower Mississippi River at Tarbert Landing, Mississippi. Concentration and discharge data are from http://water.usgs.gov/nasqan/ and http://www.mvn.usace.army.mil/, respectively.

During low discharge, significant trapping of fine-scale suspended particles ($<62.5\ \mu m$) occurs on the channel bed in both the freshwater tidal reach of the Mississippi associated with the reduced water slope approaching the Gulf (Demas and Curwick 1988; Dagg et al. 2005) and associated with salt wedge trapping driven by estuarine circulation (Galler and Allison 2008; Allison and Meselhe 2010). Periods of net storage and remobilization of sediments in the lower Mississippi can be predicted based on daily water discharge rates measured at Tarbert Landing (Demas and Curwick 1988). Significant deposition and net storage of river sediments occurs below Tarbert Landing during periods of river discharge less than 14,000 m^3/s, and significant resuspension and net remobilization of sediments occurs during periods of river discharge greater than 20,000 m^3/s (Demas and Curwick 1988). As a result, periods of sediment storage are typically 4–8 months in the lower Mississippi River. Although these processes are broadly tied to water discharge, the concentration of fine suspended particles in the river can increase by a factor of 4–5 for a period of days during the rising limb of individual spikes in discharge, attributed to hysteresis in supply from the drainage basin and to the reworking of fines stored on the bed at lower discharges (Allison et al. 2012). Coarse (sand) throughput in suspension of the lowermost river reach is more closely tied to water discharge and the increase in water surface slopes and hence stream power during large floods (Nittrouer et al. 2011; Ramirez and Allison in review). Large woody particles are transported along with the sand in these events (Bianchi et al. 2007). Approximately 10% of the total annual sand transport in the reach below New Orleans is calculated to take place through bedload transport in migrating dunes (Nittrouer et al. 2008).

Total Mississippi River suspended loads reaching ORCS have been falling since at least the mid-twentieth century because of a combination of damming, river control, and soil conservation measures, and this reduction is ongoing (Horowitz 2010; Meade and Moody 2010). In a sediment budget for the Mississippi and Atchafalaya Rivers downriver of ORCS in 2008–2010, Allison et al. (2012) demonstrated that, in addition to the historical declines in sediment load, both basins are efficient at storing sediment input from the upper basin (44% of the Mississippi+Red total suspended particulate matter input and 80% of the sand). This is attributed to a combination of overbank storage onto the lowland floodplain and delta plain and bed aggradation of sand. In the Mississippi River downriver of New Orleans, bed aggradation is exacerbated by reduced stream power associated with the multiple natural and artificial water exits. These exits also remove water upriver of the deepwater exits downriver of HOP, so that Allison et al. (2012) calculated that only 19% of the total suspended load (and 1.4% of the sand) present in the channel immediately below ORCS is debouched below HOP.

3. Sources, transport, and transformation of organic carbon

The annual fluxes of dissolved and particulate organic carbon (DOC and POC) introduced to the Gulf of Mexico by the Mississippi River have been estimated by several studies (Leeneer 1982; Trefry et al. 1994; Bianchi et al. 2004, 2007; Duan and Bianchi 2006; Duan et al. 2007b; Guo et al. 2008; Shen et al. 2012) since the first attempted by Malcolm and Durum (1976). Total organic carbon (TOC) load varies from 3.23×10^9 kg to 5.76×10^9 kg per year, with a DOC/TOC ratio changing from 0.34 to 0.80 (Table 16.1). The values from the studies differ for several reasons. Analytically, for instance, there are fundamental differences in terms of filter pore size (definition of particulate versus dissolved), ranging from 0.2 μm to 0.7 μm. For POC, the difference may be also attributed to the depth in the water column samples that samples were collected, e.g., surface vs. subsurface and grab vs. integrated samples. Large differences in particulate matter concentration (and thus POC) between surface and bottom waters have been reported in the LMR by Sutula et al. (2004). Most importantly, there are significant differences in the timing and location where samples were collected

Table 16.1. *Comparisons of annual fluxes of DOC and POC exported by the LMR system*

References	TOC (10^9kg)	DOC (10^9kg)	POC (10^9kg)	DOC/ TOC	Year/Location	Sampling Frequency
Malcolm and Durum 1976	3.72	1.72	2.00	46%	1969/ Belle Chasse	Monthly
Leeneer 1982	4.33	3.48	0.85	80%	Belle Chase	Not known
Trefry et al. 1994	5.74	1.96	3.78	34%	1974–1991/HOP	11 times
Bianchi et al. 2004, 2007	4.03	3.10	0.93	77%	1998–1999/New Orleans	6 times
Duan and Bianchi 2006, 2007	3.23	2.03	1.20	63%	2002–2003/Baton Rouge	Monthly
Guo et al. 2008		1.95–2.10			2006–2007/ Baton Rouge	Monthly
Shen et al. 2012		2.70			1996–2010/St. Francisville and Melville	Biweekly

to calculate fluxes. A longitudinal change in POC concentration (a 43% decline) has been observed in the LMR from Baton Rouge to Head of Passes (Dagg et al. 2005), and there may be an even greater variability in concentration on an interannual basis. To capture this variability, which is the product of basin-scale hydrology, more frequent sampling (weekly to biweekly) is carried out during high flows by U.S. Geological Survey's National Water-Quality Assessment Program, using the Load Estimate (LOADEST) model (Runkel et al. 2004) to calculate loads and fluxes.

Recent studies (Duan et al. 2007b, Guo 2010; Shen et al. 2012; Fig. 16.2) have shown that there is a relatively small temporal variability of DOC concentrations in the LMR, with an average concentration of 301±28 μM (<10% variability) during the periods of 2006–2009. Likely, this minor seasonal variability was controlled by spatial variability of a larger integrative signal from the watershed as well as in situ DOM processing (Duan et al. 2007b). Therefore, DOC fluxes can be quantitatively derived using either (1) the slope value of the correlation between daily DOC flux and daily freshwater discharge, or (2) the USGS LOADEST model (Guo, L.D. et al. 2012a) (see USGS at http://waterdata.usgs.gov/la/nwis). In contrast to DOC, seasonal variability of POC is positively correlated with total suspended particulate matter (SPM; Duan and Bianchi 2006). So, the POC flux of the LMR can be approximated from the SPM load (Trefry et al. 1994). Presently, no long-term record of POC fluxes in the lower Mississippi River is available and is not estimated in this chapter. However, considering that SPM load of the LMR has decreased by half in the last century (Meade and Moody 2010), as a result of construction of high and low dams in the primary tributaries, a long-term decrease in POC load of the LMR system is likely.

The Mississippi River system integrates river discharge draining extensive C_4 dominated grasslands (Missouri and Red River drainages, western portion of basin) and extensive C_3 dominated ecosystems (Ohio drainage, eastern portion of basin) (e.g., Onstad et al. 2000). Onstad et al. (2000) examined elemental, stable carbon isotope ($\delta^{13}C$) and lignin-phenol compositions of suspended particulate organic matter (POM) of the Mississippi and primary tributaries. Results show $\delta^{13}C$ values from –18.5% to –26.4‰, in a spatial pattern that corresponds to the relative abundances of C_3 and C_4

plants in the drainage basins. The variations in POM sources are supported by the composition of lignin phenols, which show that cinnamyl/vanillyl (C/V) ratios (the indicator for nonwoody sources) are higher in the Missouri River that drains a larger area of grassland than the Ohio River. For the main stem of the LMR, the composition of lignin phenols (e.g., C/V ratios) and $\delta^{13}C$ value of POM indicated of a mixture of woody and nonwoody carbon sources from the upper stream tributaries (Bianchi et al. 2002; Wissel and Fry 2005; Bianchi et al. 2007). However, the $\delta^{13}C$ value of POM in the LMR varied substantially in measurements made by prior studies (e.g., Onstad et al. 2000; Kendall et al. 2001), and studies of the seasonal variability of riverine organic matter composition are needed to precisely estimate the terrigenous fraction of POM exported to offshore sediments (Goñi et al. 1997).

Algal organic matter is generally considered a minor component in a large, mud-dominated river such as LMR, owing to turbulence, waters depths that can reach 60 m in the thalweg, and limited light penetration (Vannote et al. 1980). However, a two-year monthly monitoring study showed that algal biomass in the LMR is higher than expected (Duan and Bianchi 2006). The examination of its relationship with physical and chemical controls suggests that this unexpected phytoplankton biomass is not from in situ phytoplankton production. This hypothesis was supported by a Lagrangian experiment along the lower river below Baton Rouge to HOP that showed no evidence on phytoplankton growth along the LMR (Dagg et al. 2005). Conversely, phytoplankton biomass is high in the UMR (Baker and Baker 1979) and the Missouri River (Knowlton and Jones 2000), indicating a potential upstream phytoplankton source. This unexpectedly high phytoplankton biomass in the UMR is likely a result of decreasing SPM and increasing nutrient concentrations that has been documented over the past few decades as resulting from the constructions of high and low dams and chemical fertilizer applications in the UMR mainstem and associated tributaries. The high concentrations of phytoplankton biomass in the LMR indicate that certain large river systems may represent potential sources of labile organic matter that has been largely ignored in many global ocean budgets. Furthermore, the dense diatom-dominated phytoplankton population in the Mississippi River (Duan and Bianchi 2006) may also be linked with a decrease in dissolved silicate load in the LMR in the last few decades (Turner and Rabalais 1991).

Recent work (Rosenheim et al. 2013) comparing the radiocarbon age distribution of POC during different river discharges in the LMR illustrates a shift to older ages during high discharge (Fig. 16.3). The age increase of POC is observed across all parts of the age spectrum (Figs. 16.3a and 16.3b). It is likely that the source of the aged POC is floodplain sediment rather than bedload sediment. Samples taken at 0.9D (where D is the total depth of the river at the sampling location) show higher content of sand particles sourced from the bed during high discharge, but a similar POC age distribution as the surface during high discharge. Possible sources of the aged POC may include anthropogenic (e.g., fossil fuel combustion) and natural rock sources (e.g., shales) (Mitra et al. 2002), and ramped pyrolysis isotope data illustrate that the older material is generally isotopically lighter (Fig. 16.3c). Previous work (Mitra et al. 2002) showed that black carbon (residual carbon remaining after vapor-phase acidification followed by thermal oxidation at 375 °C for 24 h) in the LMR was an order of magnitude higher during high flow (16% of POC) than base flow (1.9% of POC), and approximately 27% of the black carbon discharged from the LMR originated from fossil fuel combustion. Furthermore, the younger POC fraction becomes smaller relative to the older POC fraction at high discharge

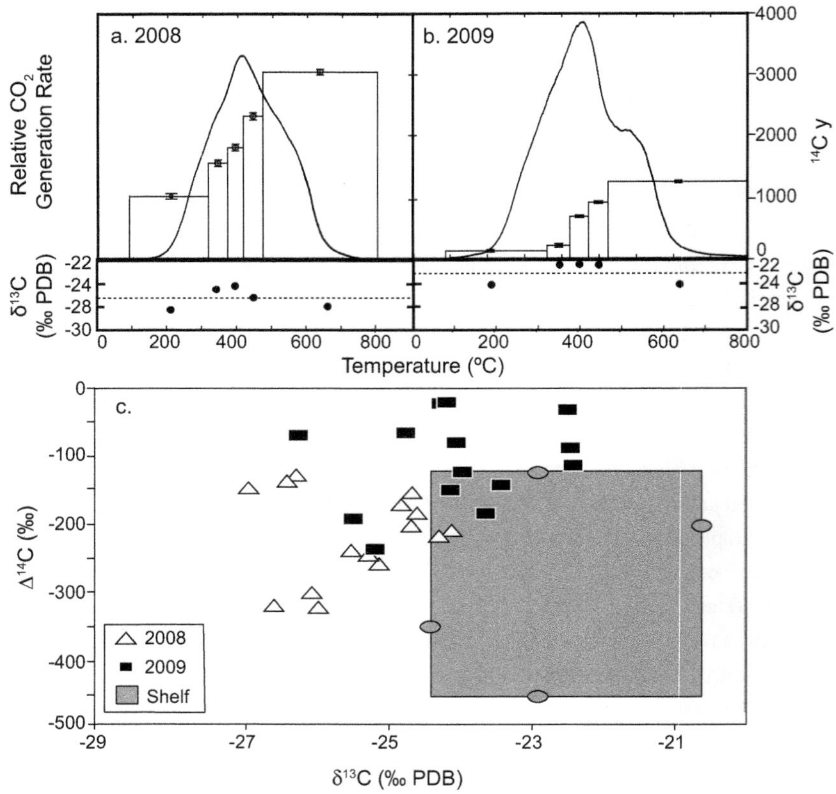

Figure 16.3. Ramped pyrolysis thermograph for LMR POC in (a) spring 2008 (52,400 m^3/s discharge, and (b) spring 2009 (28,100 m^3/s). Black curves correspond to the generation of CO_2 during pyrolysis of the POC (Rosenheim et al. 2008; Rosenheim et al. 2013 for more details). Uneven bars show the temperature interval integrated and the radiocarbon age of the individual dated samples (right axis). Stable isotope data are shown in the bottom panel of (a) and (b), integrated over the same temperature intervals. During high discharge, the full age spectrum is older, and material is generally isotopically lighter. Plotting $\Delta^{14}C$ vs. $\delta^{13}C$ (c), it is evident that the older material during higher discharge sourced from C_3 vegetation and carbon derivatives, whereas the younger material during the more normal spring discharge of 2009 had a more C_4 signal. All values are slightly younger and isotopically lighter than sedimentary organic matter found in shelf sediments adjacent to the river (Goñi et al. 1996; Gordon and Goñi 2004). Figure modified from Rosenheim et al., 2013.

and near-mean discharge as the POC enters the Gulf of Mexico (Rosenheim et al. 2013). This shift has been attributed to more favorable remineralization of younger, more chemically labile POC, and it suggests differential deposition of older POC in deltaic and deepwater deposits (Rosenheim et al. 2013). Overall, during both high and low water discharge conditions, there is neither a predominance of very young (modern and post-modern) nor very old (>30,000 ^{14}C y) POC (Fig. 16.3a, b, Fig. 16.4). This is an indication of the ability of a large river system to attenuate contributions of POC from two primary sources – old, refractory rock-bound carbon, and young, fresh primary production and soil carbon – through remineralization and storage in a large developed floodplain. Such large

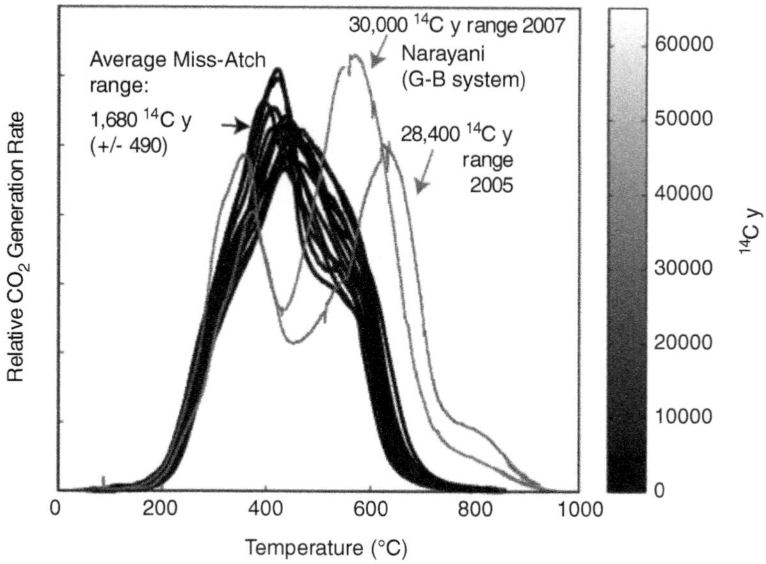

Figure 16.4. Comparison of compiled Mississippi and Atchafalaya ramped pyrolysis data (Fig. 16.3, Rosenheim and Galy, 2012) to ramped pyrolysis data from the Narayani River tributary to the Ganges River. Generation of CO_2 proceeds along slightly different trajectories depending on discharge and other factors (Rosenheim et al. 2013) in the Mississippi/Atchafalaya system. However, the differences within the Mississippi/Atchafalaya system are small compared with the data from the smaller, rapidly incising Narayani River. Age difference (^{14}C y) is shown by grayscale color bar that represents the measureable range in ^{14}C ages; the ranges for two Narayani samples taken during the 2005 and 2007 monsoons are more than one order of magnitude higher than the average Mississippi/Atchafalaya age range. This comparison shows that the large basin and long length of the Mississippi River system yields mixing and reworking of the bulk POC compared with smaller, steeper basins that consist of more easily definable end-member POC sources. Modified from Rosenheim and Galy, 2012.

integrative river systems are predicted to differ from small, steep basins in this manner (Blair and Aller 2012), and these data from the Mississippi River system support this prediction quantitatively. Indeed, compared to a small, mountainous tributary of the Ganges River (India), the Narayani River, the compilation of 2008 and 2009 Mississippi-Atchafalaya ramped pyrolysis radiocarbon data show differences in the pyrolysis characteristics of POC (related to stability of the POC) and the overall age range (Fig. 16.4, from Rosenheim and Galy 2012) (see Chapter 14).

The composition of dissolved organic matter (DOM) and high molecular weight (HMW) DOM (the fraction with molecular size >1,000 Dalton) of the LMR reflects both origin and in situ transformation of organic matter. Relative to that of large natural rivers (Hedges et al. 2000; Guo et al. 2003; Guo and Macdonald 2006), HMW DOM from the LMR is relatively lower in C/N ratio (12.5 to 23) but enriched in $\delta^{13}C$ (−27.1 to −25.8‰) and $\delta^{15}N$ values (+4.7 to +11.5‰) (Bianchi et al. 2004; Duan et al. 2007b; Guo et al. 2009), highlighting the effect of human impact on sources (e.g., less humic substances from forest) (Duan et al. 2007b). Effect of land use on HMW DOM sources is also reflected in the functionality of carbon as determined by solid C-13 nuclear magnetic resonance

(NMR) spectra and lignin biomarkers. For instance, the HMW DOM from the LMR is significantly higher in carboxyl C but lower in aromatic C (or % aromaticity) and lignin-phenol yield than that of the Pearl River (Duan et al. 2007a, 2007b) – a small blackwater river ~200 km east of the LMR that is much less disturbed. Significant differences in LMR organic matter composition are present when comparing HMW DOM with POM (Onstad et al. 2000; Kendall et al. 2001; Wissel and Fry 2005; Duan and Bianchi 2006; Mayer et al. 1998). Possible reasons include differences in source as well as the fractionation between POM and DOM: C-12 enriched humic substances are leached into water, whereas N is absorbed on particles (Hedges et al. 2000). Moreover, DOM composition of the LMR displays a pronounced temporal variability, with more carbohydrates and lignin occurring during high-discharge periods (Bianchi et al. 2004; Duan et al. 2007a, 2007b; Duan and Bianchi 2007). Such a seasonal pattern suggests that there is an export of watershed DOM from anthropogenic and soil sources during high flows. The labile organic matter (e.g., carbohydrates) during the high flows provides a food and energy source for heterophoric bacteria, whose production is also highest during the high-flow period (Duan et al. 2007b).

Rivers act not only as a transporter of organic matter, but as an active transformer as well. Thus in situ processing may be important in controlling temporal and spatial changes in riverine DOM abundance and composition. Results of Lagrangian experiments on longitudinal changes in organic matter in the Mississippi River have been reported in Leenheer et al. (1995), Dagg et al. (2005), and Duan and Bianchi (2007). These results constrain OM transformation in the LMR. Surveys of spatial variability of total DOC from headwaters of the Mississippi River by USGS (Leenheer et al. 1995) display reductions in DOC by 6.5–8.1% in the LMR above Baton Rouge, Louisiana, and a slight decrease in DOC (by 3%) is present along the last 390 km of the LMR (Dagg et al. 2005). These decreases are more likely attributable to in situ DOC mineralization, because of minor tributary input from the LMR watershed and isolation from floodplain by levees (Allison et al. 2012). DOC in situ mineralization also displays seasonal variability, with more DOC loss taking place during summer (April–August) than in other seasons (Duan et al. 2007b). The large loss of DOC during the summer months indicates that mechanisms for the loss are related to water temperature or/and high solar insolation. It is still not clear whether photochemical oxidation or bacterial decay is more important for the losses, although in large turbid rivers, such as the Mississippi River, bacterial decay is more likely than photochemical oxidation (Duan et al. 2007b). In contrast to the small spatial changes in DOC (by 3%), POM shows more drastic decreases (POC by 42% and PN by 57%) within the lowermost MR during falling discharge (Dagg et al. 2005). This drastic decrease was likely attributable to particle sedimentation with decreasing flow. Examinations of amino acids and pigment biomarkers (Dagg et al. 2005; Duan and Bianchi 2007) gave more detailed information on POM transport and transformation during this sedimentation event, including a switch of phytoplankton taxonomy, bacterial organic matter degradation, and short-term sedimentation and resuspension events (see Chapter 19).

The LMR system also receives organic matter inputs from Atchafalaya floodplains and an independent tributary (e.g., Red River). It has been recognized that the organic matter concentration and composition of the Atchafalaya River are different from that of the LMR main stem, with more chromophoric component in the Atchafalaya River than the LMR (e.g., Stolpe et al. 2010). This difference was found to influence organic matter distribution in the northern GOM (e.g., Chen and Gardner 2004). A recent study by Shen et al. (2012) examines the floodplain influence on dissolved organic matter composition and export from the Mississippi-Atchafalaya River system to the Gulf of Mexico. They found that concentrations of DOC, lignin phenols, amino acids, neutral sugars, and chromophoric DOM (CDOM) absorption coefficients and yields of labile organic matter are higher in the Atchafalaya than in the LMR. Significantly lower syringyl to vanillyl phenol ratios (S:V) in the Atchafalaya River indicate substantial contributions from watershed gymnosperms. The organic matter inputs from Atchafalaya River are so large that about 35% of the DOC and 44% of the lignin phenols, amino acids, and neutral sugars exported from the Mississippi-Atchafalaya River system is carried by the Atchafalaya River (30% for water flow) (Shen et al. 2012).

4. Trace metal geochemistry and complexion with organic matter

Historical data on dissolved trace elements in the Mississippi River are limited due to problems of sample contamination in older datasets (e.g., Shiller and Boyle 1987; Taylor and Shiller 1995). Trefry et al. (1986) provided some of the first reliable dissolved trace element measurements in the LMR, reporting dissolved concentrations of Cd, Cu, Cr, Fe, Mn, Ni, and Pb in four samples collected at various flow stages. Their results, though limited, are in general accord with subsequent reports that also used trace element clean sampling and analysis techniques. Interestingly, their dissolved Pb data are about 10-fold higher than results reported a decade later (Shiller 1997), which could reflect declining contamination of the environment with Pb from leaded gasoline. Interannual variability of dissolved trace element concentrations is more poorly understood. Hydrological variability may be important on the annual timescale, though interestingly, a major flood event had little impact on dissolved concentrations (Shiller 1997).

Shiller and Boyle (1987) presented dissolved concentrations of eight trace elements in the LMR (Cd, Cu, Cr, Fe, Mo, Ni, V, and Zn), emphasizing consistent trends with discharge as well as significantly lower levels than were being reported by national monitoring programs. This was later elaborated by Shiller (1997)in a 27-month study of a dozen dissolved trace elements in the LMR. That study found that hydrological factors were generally minimal in affecting seasonality of dissolved trace element concentrations, except in the cases of Ba and U. These two elements mix fairly conservatively and have significantly contrasting concentrations in the different major tributaries of the Mississippi River. The study of Grzymko et al. (2007) confirmed this to be the case for both dissolved U and the dissolved $^{234}U/^{238}U$ ratio in the LMR. Likewise, Xu and Marcantonio (2007) examined dissolved strontium and $^{87}Sr/^{86}Sr$ ratios in the LMR and also suggested variable tributary contributions as the likely cause of temporal changes Sr concentrations and isotopic composition.

Shiller's (1997) study also suggested that for many elements, redox processes were likely of prime importance in determining seasonal dissolved concentration variability in the LMR. This includes upstream processes such as seasonal flushing of element-depleted reducing sources as well as in-stream processes affecting the dynamic balance between local oxidation and reduction. Some support for this has come from observations of temperature-dependent microbial Mn oxidation in the LMR (Shiller and Stephens 2005), as well as from seasonal variability of dissolved rare earth elements in the Mississippi River (Shiller 2002).

Although composition and fluxes of dissolved organic matter and trace elements through the Mississippi River have separately received some attention, the relationship between the DOM and trace elements as well as the role of DOM in governing the abundance, speciation, fate, and transport of trace elements remain poorly understood. Dissolved and colloidal organic matter can effectively bind trace elements and will alter the distribution and partitioning of trace metals and thus their bioavailability and toxicity in aquatic systems (Santschi et al. 1999; Guo et al. 2001). Using flow field flow fractionation coupled online with ICP-MS, Stolpe et al. (2010) examined the continuous size spectra of colloidal metals in the size range of 0.5–40 nm, including Fe, Mn, Cu, Zn, Pb, and U, in the lower Mississippi and Atchafalaya Rivers. They found that DOM composition and abundance in the river waters significantly regulated the partitioning of trace metals between dissolved and colloidal phases and their colloidal size spectra. In general, colloidal size spectra of Cu and U in the Mississippi and Atchafalaya Rivers resembled those detected for CDOM or UV_{254} absorbance, with a major peak in the 0.5–4 nm size range. However, the colloidal size spectra of Mn and Pb also had an additional peak in the 5–25 nm size range, similar to the size spectra of colloidal Fe (Stolpe et al. 2010).

In rivers with higher DOC concentrations and more terrestrial fulvic acids, the smaller organic colloids in the size range of <4 nm are abundant and mostly composed of fulvic acids and tend to bind most metals with high affinity to DOM. The middle-sized colloids (i.e., 5–40 nm size range), are composed of Fe-rich colloids that bind mostly with P, Mn, Zn, and Pb. In addition, the larger sized colloids are found to contain protein-rich colloidal organic matter and be enriched in Fe, P, Mn, and Pb (Stolpe et al. 2010).

Only limited work to date has examined the dissolved-particulate partitioning of trace elements in the Mississippi River, with Trefry et al. (1986) finding that the elements they studied were largely in the particulate phase. Shiller (1997) largely confirmed this, with the exception of a few very soluble elements such as Mo and U.

5. Nutrient export from the Mississippi River to the Gulf of Mexico

Studies on long-term changes in nutrient loading of the LMR system can be traced back to the early 1990s (e.g., Turner and Rabalais 1991), showing that the mean annual concentration and flux of nitrogen from the LMR has doubled or even tripled since the 1950s (e.g., Lohrenz et al. 1997; Goolsby et al. 2000) (see Chapter 18). The increase in annual nitrate flux to the GOM can be largely explained by three factors: increased fertilizer use (Turner and Rabalais 1991), annual variability in precipitation and increased streamflow (Zhang and Schilling 2006; Raymond et al. 2008), and the year-to-year variability in the amount of nitrogen available in the soil-ground water system for leaching to streams (Donner and Scavia 2007). Later estimates showed that nearly all of this increase occurred between 1970 and 1983 (e.g., Goolsby and Battaglin 2001), and there has been a minimal decrease (~6%) since then (Goolsby and Battaglin 2001; Turner et al. 2007). This minimal decrease has been primarily attributed to a reduction (by 47%) in total Kjehldahl nitrogen (including ammonium and organic nitrogen; http://water.usgs.gov/nasqan/), which may be related to discharge of domestic, municipal, or industrial wastewater. Within the total nitrogen flux of 1.57×10^6 tons/y (for years 1980–1996), approximately 61% was nitrate, 37% organic N, and 2% ammonium as N. This TN flux composition is very close to that observed in another N-enriched large river – the Changjiang (Duan et al. 2008), showing the importance of both nitrate and organic N. Mayer et al. (1998) alsohighlighted the importance of organic matter in transporting nitrogen from the LMR to the

GOM. Compared with the large change in nitrogen, the increases in the concentration and loading of total phosphorus have been relatively small (Rabalais et al. 1996; Turner et al. 2007). The difference in P and N variability with time has caused the long-term change in N/P ratio of the LMR water (Turner et al. 2007). Within nutrient species, there are large uncertainties in concentration of DOP that need to be figured out in future. Current values of DOP concentration vary considerably between studies from 0.19–0.24 μM (Cai and Guo 2009), 0.3–0.5 μM (USGS data; Duan et al. 2010), 0.5–0.6 μM (Sutula et al. 2004), to 2.1–2.4 μM (Rinker and Powell 2006) – up to an order of magnitude difference.

In contrast to nitrogen, dissolved silica (DSi) loading of the Mississippi River has decreased by half since the 1950s (Turner and Rabalais 1991; Rabalais et al. 1996) but showed only minor changes since the late 1980s (Turner et al. 2007). Two hypotheses were proposed by Turner and Rabalais (1991) for this DSi loss: (1) P stimulated freshwater diatom uptake in the streams and lakes, and (2) DSi attachment to, or deposition with, sediments behind dams or in reservoirs. The contrasting long-term changes in N, P, and Si loadings have resulted in shifts of the stoichiometry of dissolved nutrients in the Mississippi River (Turner and Rabalais 1991; Justic et al. 1995; Rabalais et al. 1996; Turner et al. 2007). As a result, the ratios of N:Si, N:P, and Si:P of the LMR (1.1, 15, and 14) closely approximate the Redfield ratios (Justic et al. 1995). This shift has changed the stoichiometric nutrient balance of northern GOM and may affect the coastal phytoplankton productivity and the ecosystem as a whole (Justic et al. 1995). In addition to DSi, fluxes of biogenic silica from the Mississippi River have been estimated for the water year 2008–2010 (Yang et al. 2011b). The averaged flux of biogenic silica could be up to 39% of the dissolved silicate flux, which is relatively higher than the global value of 30% (Conley 1997).

Nutrient concentrations and fluxes from the LMR system to the GOM also have pronounced seasonal variations (Fig. 16.2; Turner and Rabalais 1991; Lohrenz et al. 1997; Duan and Bianchi 2006; Turner et al. 2007; Duan et al. 2010, 2011). It was generally assumed that seasonal patterns of nutrient fluxes of the Mississippi River follow that of the river flow, with a peak typically occurring in March and lowest value in October. However, the peaks of nitrate and soluble reactive phosphorus (SRP) fluxes occur in early summer (June), as a result of highest nitrate and SRP concentrations in river water during this season (Lohrenz et al. 1997; Duan and Bianchi 2006; Duan et al. 2011). Only the fluxes of dissolved silica and dissolved organic nitrogen (not shown) follow the seasonal pattern of river flow, owing to minor seasonal changes in their concentrations. Nutrient elemental ratios also vary seasonally, owing to differences in seasonal pattern in N, P, and Si. In general, Si and P are in shortest supply during the spring high flow and N is more likely to be limiting during the rest of the year (Duan and Bianchi 2006; Turner et al. 2007).

The impact of limiting nutrient input to the northern GOM is a topic presently being debated by the scientific community. The resulting conclusion will directly affect the strategies of watershed nutrient management for the Mississippi River basin. In general, N is considered as the limiting nutrient for annual phytoplankton production (Turner et al. 2007), but nutrient limitation in the northern Gulf varies seasonally and spatially (e.g., Lohrenz et al. 1997; Dagg et al. 2007). Nutrient enrichment bioassays in the Mississippi River plume (Sylvan et al. 2006) show that phosphorus limits phytoplankton growth on the Louisiana shelf during the period of hypoxia (May–July), whereas N is a limiting nutrient in September. This is consistent with seasonal change in N/P ratio of the

LMR (highest in April), if there is a two-month delay from the lower river to the shelf (Dinnel and Wiseman 1986). This consistence on nutrient stoichiometric balance highlights the coupling between river inputs and coastal water nutrient dynamics.

Many factors may contribute to the seasonal patterns in riverine nutrient concentrations and fluxes of the LMR system to the GOM. Learning the factors affecting seasonal variations in nutrient concentration will contribute to the controls for eutrophication in the northern GOM. Duan et al. (2010) suggested that the pronounced peaks in nitrate in early summer are largely due to fertilizer application in the Mississippi River watershed, because this nitrate peak did not exist before the 1970s, when chemical fertilizers were not widely used (Turner and Rabalais 1991). Mass balances in seasonal scale (Duan et al. 2010) further show that the nitrate flux peaks in June are primarily attributed to the inputs from the UMR, the tributary with highest nitrate concentration. For the seasonality of SRP, the mechanism is found to be "tributary-by-tributary" dilution (Duan et al. 2011), where the SRP-enriched UMR and Missouri River water, with a flow peak at early summer, is gradually diluted by other SRP-depleted tributaries (the Ohio and other rivers in the LMR) as the river flows downstream toward the Gulf. In contrast to nitrate and SRP, a seasonal summer–fall maximum in DSi concentration occurred earlier in the twentieth century, but it is no longer evident (Turner and Rabalais 1991). This change agrees with the long-term loss of DSi in the Mississippi River, as a result of dam construction and enhanced growth of diatoms in lakes and reservoirs (Turner and Rabalais 1991).

Three types of studies have examined in situ recycling and nutrient dynamics in the LMR system. The first was Lagrangian experiment (e.g., Dagg et al. 2005), where a parcel of water in the LMR was sampled every two hours during its four-day transit. The result shows that there was no measurable loss of NO_3, SRP, or DSi. Similar results are observed in the other Lagrangian experiments (e.g., Sutula et al. 2004) in a section below New Orleans, and by Battaglin et al. (2001) in a 10-day study from Thebes, Illinois, to Belle Chasse, Louisiana. The minor changes in nutrient concentration in the Lagrangian experiments are consistent with results with the second approach – nutrient mass balances (e.g., Duan et al. 2010), showing that only 1.9% of nitrate and 3.5–4.9% of SiO_2 and SRP were lost or gained on an annual basis. Finally, measurements of nutrient fluxes from bed sediment when a fraction of the total SPM is deposited seasonally in the LMR (Sutula et al. 2004) also displayed no significant net effect of diagenesis on the changes in major P phase and SRP release. Possibly, dissolved inorganic nutrient pools in the LMR are so large that the fluxes originating from nutrient transformation are minor or undistinguishable. Compared with minor changes in dissolved inorganic nutrients, mass balances showed retention of particulate nitrogen and phosphorus in the LMR (Duan et al. 2010). Losses of particulate N and P occur during the falling hydrograph limb (March–July), suggesting a hydrological control on sedimentation and suspension of river particles. Mass balances in the LMR (Duan et al. 2010) also indicate loss of dissolved organic nitrogen and phosphorus in summer (similar to DOC), which can likely be ascribed to temperature-enhanced bacterial transformation, and to a lesser extent, to enhanced photochemical oxidization.

Excess nutrient inputs from the LMR are considered as one of the main reasons for eutrophication and related ecosystem problems in the northern GOM. To reduce the size of the hypoxic zone and improve water quality in the Mississippi River basin, the EPA Science Advisory Board Panel recommends a dual nutrient strategy targeting at least a 45% reduction in riverine total nitrogen flux and at least a 45% reduction in riverine total phosphorus flux (http://yosemite.epa.gov/sab/). For nitrogen, it seems that the greatest emphasis would be placed on reducing flux in the upper Mississippi watershed in spring (Duan et al. 2010), the time period most correlated with the size of the hypoxic zone (Rabalais et al. 1996). For the LMR system, however, the effect of nutrient cutoff is likely limited because the LMR basin is not the main nutrient source, and nutrient retentions within large rivers are minor because of decreasing retention capability with stream size (Alexander et al. 2000).

Here, we briefly review the studies on using or modifying existing manmade or natural resources of the LMR (e.g., wetlands, Lake Pontchartrain, Old River Control Structure) to remediate the seasonal hypoxia in the northern Gulf of Mexico. As previously discussed, the LMR is separated from its floodplain by levees and channelization, and nutrient retentions within the LMR floodplain wetland are limited. Thus diversions of the Mississippi River water to floodplain wetlands seem to be a possible solution for nutrient removal. However, large-scale managed input of nutrient-enriched Mississippi waters into wetlands and open waters has been controversial since its implementation in coastal Louisiana (Lane et al. 2003; Turner et al. 1997; Day et al. 2007). Presently there is no clear consensus on whether restoring wetlands with sediment from the river will also enhance the capacity of nitrate removal. The effects of opening of the Bonnet Carré Spillway (BCS) on water quality of Lake Pontchartrain have been reported by Turner et al. (2002) and McCorquodale et al. (2009). Occasional openings of the BCS for flood protection could triple the annual average loading to the lake within 1–2 months (Turner et al. 2002), and a net transfer from inorganic to algal biomass occurred during the 1997 diversion. High nutrient loadings from the BCS can cause algal blooms in the northwest quadrant of Lake Pontchartrain (McCorquodale et al. 2009). At present, the contribution of BCS-diverted water to Lake Pontchartrain on nutrient removals from the LMR has not been measured.

A number of recent studies have suggested that the Atchafalaya River discharge may play a strong role in the development of the northern Gulf hypoxia (Dagg et al. 2007; Hetland and DiMarco 2008; Bianchi et al. 2010; DiMarco et al. 2010). The Atchafalaya River discharges into a shallow and broad, gently sloping shelf (Fig. 16.1). The freshwater from the Atchafalaya tends to have a longer residence time on the shelf because it must traverse the entire shelf before crossing the shelf break and entering the deep water Gulf. Coastal topography near Atchafalaya River mouth also contributes to semi-isolation of the river plume, leading to horizontal and vertical structure of hydrograph, stability, and hypoxia in this region (Hetland and DiMarco 2008; DiMarco et al. 2010). Therefore, almost all freshwater and nutrients delivered by the Atchafalaya River discharge remains on the shelf. As a result, the Atchafalaya River, which carries 30% of total system flow, may have a disproportionately larger effect (estimated at 50%; Day et al. 1995) on Gulf hypoxia. In contrast, the Mississippi River "injects" its nutrient-enriched freshwater into the open Gulf near the shelf break, because the deepwater pass discharge points of the LMR (HOP) are approximately 80 km seaward of most of the Louisiana coastal zone. Additionally, more than half the water available at Tarbert Landing is discharged upriver of these passes through natural and manmade outlets and loss to groundwater (Allison et al. 2012). As a result, only 43–53% of the Mississippi River discharge is carried westward to the present hypoxia zone (Dinnel and Wiseman 1986; Etter et al. 2004). A modeling study also suggests that sedimentation from the Mississippi River plume directly supplies a maximum of 23%

of the organic C required for development and maintenance of hypoxia (Green et al. 2006). It thus appears that future decrease in flow percentage from the Atchafalaya River could mitigate the Gulf coastal hypoxia.

6. The future of the biogeochemistry of the lower Mississippi River: the role of global warming, land-use change, and extreme climates

The LMR is facing many uncertainties regarding the future of fluxes of water, sediment, organic matter, trace elements, and nutrients into the GOM. Such uncertainties are due to intertwined effects of global warming, land-use change, extreme climates, and other human impacts. Miller and Russell (1992) suggested that increased warming will increase runoff by 75% for the world's major rivers. Future global warming is expected to cause significant shifts in rainfall patterns and storm intensity (Watson et al. 1996), which in turn will have a direct impact on the export of material from terrestrial environments. Additionally, land-use change (e.g., urbanization) may drive changes in local climate (e.g., temperature increase) and flow pattern (e.g., Raymond et al. 2008). These climate and land-use changes in the watershed can be expected to alter biogeochemical cycles in the LMR and the northern GOM.

The intensification of agriculture in the central United States is commonly cited as the primary cause of the increase in nitrogen (N) flux of the Mississippi River since the 1950s. For instance, land-use changes were found to be more important than changes in climate and plant CO_2 fertilization to increases in Mississippi River water and carbon export from the watershed over the past 50 years (Yang et al. 2011a). The amount of chemical fertilizer used during the last few decades is positively correlated to nitrogen flux of the LMR (Turner and Rabalais 1991), and results from Alexander et al. (2008) show that agricultural sources in the watersheds contribute more than 70% of the delivered N and P. In addition, agricultural land use is found to decrease structural complexity of fluvial DOM and increase the amount of microbially derived DOM (Wilson and Xenopoulos 2009). Beside direct influences, land use change and accompanying agricultural activities in the MR basin during the last 60 years also have been found to increase MR flow (Zhang and Schilling 2006), via less evapotranspiration and more groundwater recharge, and is thus likely to increase export of organic matter and nutrients from watershed into the river. In contrast to the large area of cropland, urban land use accounts for only a small percentage (0.6%) of land cover of the Mississippi River watershed. But, urban watersheds are generally hot spots for reactive P and ammonium nitrogen (Antweiler et al. 1996; Alexander et al. 2008). In the future, there are still many factors that may affect change in the fluxes of organic matter and nutrients of the LMR. Factors that may increase fluxes include population increase and associated increased need for fertilizer applications and urbanization, whereas better management of fertilizer application in agriculture and watershed/stream restorations might be expected to decrease the fluxes.

With continued high N inputs and without major changes in basin land-use, climate variability can be considered as the controlling factor for the variability in nutrient and organic matter leaching from land and transport through the river system (Justic et al. 2005; Donner and Scavia 2007). For example, the mean annual N flux of the Mississippi River has changed little since the early 1980s, but large year-to-year variations in N flux occur because of variations in precipitation (Goolsby and

Battaglin 2001). Precipitation amounts across the Corn Belt (a region of the Midwestern United States where corn has been the predominant crop and nitrogen load comes from) in the previous November–December and in March–April–May are together a strong predictor ($r^2 = 0.68$) of the spring nitrate flux by the river in any particular year (Donner and Scavia 2007). During wet years, the N flux can increase by 50% or more due to flushing of nitrate that has accumulated in the soils and unsaturated zones in the basin. Watershed nitrogen removal also varied with precipitation and stream flow. Results from a HYDRA aquatic transport model (Donner et al. 2004) suggested an almost two-fold variability in the percentage of N removed in the Mississippi River system (mainly watershed and small tributaries) and a three-fold range in the associated N_2O emissions, with the lowest percentage removed (10–33%) occurring in the wettest years. So, climate variability and change in precipitation should be considered in the management of nutrient export by the Mississippi River. Climate variability and associated changes in freshwater and nutrient input have been used to model the coastal eutrophication and frequency of hypoxia in the northern Gulf of Mexico (Justic et al. 2003, 2005).

It is expected that, as global warming continues, both watershed runoff and storm intensity will increase, and extreme climate events will become frequent for most areas in the world. According to modeling studies of the Mississippi River (Justic et al. 2003; Donner et al. 2004), both the fluxes and the interannual variability of the fluxes of water, sediment, and nutrients (and likely dissolved organic matter) will increase as well. In addition to precipitation pattern and intensity, warming may directly influence carbon, nitrogen, and phosphorus cycles in land and aquatic environments. Lab and field experiments with temperature manipulation have shown that increasing temperature can cause increases in DOC flux, CO_2 efflux, nitrification, and denitrification rate and release rate of phosphate from sediment into water column (e.g., Duan and Kaushal 2013). It is likely the Mississippi River system will continue to see greater fluxes of nutrients and organic matter with increasing global temperature.

7. Conclusions

In the last century, existing records indicate that water flux from the Mississippi catchment has been relatively stable, although climate model projections suggest this may change in the twenty-first century. However, there has been an increased capture of Mississippi water by the Atchafalaya distributary, culminating in the construction of the Old River Control structures in 1963 and the fixing of water discharge of the Atchafalaya at 30% of the total Mississippi+Red water flux. This means that there has been a shift in the exit point of a significant proportion of the water down the western (Atchafalaya) pathway, which has significant implications for the processing of organic carbon, nutrients, and trace elements in the lowermost (delta plain) basin prior to release to the shelf. Sediment flux to the margin (and, associated POC flux) has changed more dramatically in the last century owing to anthropogenic changes in the catchment (e.g., damming, river control works, soil conservation measures) and in the delta plain (e.g., artificial levees, channel aggradation). The net effect has been to reduce the sediment flux to the shelf by more than half and to selectively reduce the coarse grain flux (e.g., sand and large particle POC) by an even greater extent. Recent research suggests that localized storage in the delta plain by channel aggradation and overbank flooding is more significant than previously thought for this highly engineered lowermost river section.

Prior estimates of DOC and POC fluxes vary considerably. POC composition indicated of a mixture of woody C_3 plants from Ohio, nonwoody C_4 sources from the Missouri grassland, and gymnosperms inputs from Atchafalaya wetlands. Algal inputs to the LMR are higher than expected, but the phytoplankton biomass seems from sources of upstream tributaries. POC shifts to older ages during high flows, and the younger POC fraction becomes less relative to the older fraction as the POC enters the GOM. DOC composition reflects effects of both watershed land-use and organic matter fractionation between water and particles. Up to 10% of DOC is lost during transport through the LMR, and the transformation is strongest in summer, whereas the longitudinal changes in POC are much more drastic.

Only limited data exist on dissolved trace element concentrations in the Mississippi River owing to past problems of contaminated samples. Reliable data from the past 25 years show consistent trends and reveal that many of the more reactive elements show seasonality reflective of the effects of redox processes occurring both upstream within the watershed as well as in stream locally. Some other more soluble elements show temporal variability indicative of changes in contributions from tributaries with different trace element concentrations. Limited analyses of colloidal material in the river suggest two main classes of material, carbon-rich and iron-rich, which affect the partitioning and transport of many trace elements. Clearly, there is still much to learn about trace element concentrations, speciation, and fluxes in the Mississippi River.

Nitrogen concentration and flux has substantially increased in the twentieth century, in contrast to minor changes in phosphorus and considerable reduction of dissolve silica, owing to land-use change (e.g., crop cultivation) and dam constructions.Future changes in watershed precipitation and global warming may affect nutrient export. The changes in nutrient sources and spatial distributions (e.g., more from UMR) have caused shifts in seasonal pattern of nutrient fluxes and nutrient limitation in the northern GOM. Transformations of dissolved inorganic nutrient in the LMR are undistinguishable from its large pools.

References

Alexander, R.B., R.A. Smith, G.E. Schwarz, E.W. Boyer, J.V. Nolan, and J.W. Brakebill. 2008. Differences in phosphorus and nitrogen delivery to the Gulf of Mexico from the Mississippi River Basin. *Environment Sciences & Technology* 42: 822–830.

Alexander, R.B., R.A. Smith, and G.E. Schwarz. 2000. Effect of stream channel size on the delivery of nitrogen to the Gulf of Mexico. *Nature* 403: 758–761.

Allison, M. A., C. R. Demas, B.A. Ebersole, B.A. Kleissd, C. D. Littlec, E.A. Meselhee, N.J. Powellf, T.C. Prattc, and B.M. Vosburgg. 2012. A water and sediment budget for the lower Mississippi-Atchafalaya River in flood years 2008–2010: Implications for sediment discharge to the oceans and coastal restoration in Louisiana. *Journal of Hydrology* 432: 84–97.

Allison, M.A., and E.A. Meselhe. 2010. The use of large water and sediment diversions in the lower Mississippi River (Louisiana) for coastal restoration. *Journal of Hydrology* 387: 346–360.

Antweiler, R.C., D.A. Goolsby, and H.E. Taylor. 1996. Nutrients in the Mississippi River, 1987–92, *In* R. H. Meade (ed.), Contaminants in the Mississippi River, *U.S. Geological Survey Circular* 1133, Reston, Virginia, pp. 73–85.

Baker, A.L., and K.K. Baker. 1979. Effects of temperature and current discharge on the concentration and photosynthetic activity of the phytoplankton in the upper Mississippi River. *Freshwater Biology* 9: 191–198.

Barber, L.B. II, J.A. Leenheer, W.E. Pereira, T.I. Noyes, G.K. Brown, C.F. Tabor, and J.H. Writer. 1995. Organic Contamination of the Mississippi River from Municipal and Industrial Wastewater. *In* R.H. Meade (ed.), Contaminants in the Mississippi River, *U.S. Geological Survey Circular* 1133, Reston, Virginia.

Battaglin, W.A., Kendall, C., Chang, C.C.Y., Silva, S.R., and Campbell, D.H. 2001. Chemical and isotopic evidence of nitrogen transformation in the Mississippi River, 1997–98. *Hydrological Processes* 15: 1285–1300.

Bianchi, T.S., J.J. Galler, and M.A. Allison. 2007. Hydrodynamic sorting and transport of terrestrially derived organic carbon in sediments of the Mississippi and Atchafalaya Rivers. *Estuarine Coastal and Shelf Science* 73: 211–222.

Bianchi, T. S., S.F. DiMarco, J.H. Cowan Jr, R.D. Hetland, P. Chapman a, J.W. Day, and M.A. Allison. 2010. The science of hypoxia in the Northern Gulf of Mexico: A review. *Science of the Total Environment* 408: 1471–1484.

Bianchi, T.S., S. Mitra, and B.A. McKee. 2002. Sources of terrestrially-derived organic carbon in lower Mississippi River and Louisiana shelf sediments: implications for differential sedimentation and transport at the coastal margin. *Marine Chemistry* 77: 211–223.

Bianchi, T.S., T. Filley, K. Dria, and P.G. Hatcher. 2004. Temporal variability in sources of dissolved organic carbon in the lower Mississippi River. *Geochimica et Cosmochimica ACTA* 68: 959–967.

Blair, N.E., and R.C. Aller. 2012. The fate of terrestrial organic carbon in the marine environment. *Annual Reviews in Marine Science* 4: 401–423.

Cai, Y., and L. Guo. 2009. Abundance and variation of colloidal organic phosphorus in riverine, estuarine, and coastal waters in the northern Gulf of Mexico. *Limnology and Oceanography* 54: 1393–1402.

Chen, R.F., and G.B. Gardner. 2004. High-resolution measurements of chromophoric dissolved organic matter in the Mississippi and Atchafalaya River plume regions. *Marine Chemistry* 89: 103–125.

Conley, D.J. 1997. Riverine contribution of biogenic silica to the oceanic silica budget. *Limnology and Oceanography* 42: 774–777.

Dagg, M.J., T.S. Bianchi, G.A. Breed, W.J. Cai, S. Duan, H. Liu, B.A. McKee, R.T. Powell, and C.M. Stewart. 2005. Biogeochemical characteristics of the lower Mississippi River, USA, during June 2003. *Estuaries* 28: 664–674.

Dagg, M., J. Ammerman, R. Amon, W. Gardner, R. Green and S. Lohrenz. 2007. Review of water column processes influencing hypoxia in the northern Gulf of Mexico. *Estuaries and Coasts* 30: 735–752.

Day, J.W., R.R. Twilley, R.F. Shaw, C.J. Madden, B.A. McKee, M.J. Dagg, D.L. Childers, R.C. Rainie, and L.J. Rouse. 1995. The influence of Atchafalaya River discharge on Fourleague Bay, Louisiana, USA. *In*: K. Dyer (ed.), Changes in Fluxes in Estuaries: Implications from Science to Management. Olsen and Olsen. pp. 151–160.

Day, J.W., Jr., et al. 2007. Restoration of the Mississippi delta: Lessons learned from Hurricanes Katrina and Rita. *Science* 315: 1679–1684.29.

Demas, C., and P. Curwick. 1988. Suspended Sediment and Associated Chemical Transport Characteristics of the Lower Mississippi River, Louisiana. Louisiana Dept. of Transportation: Baton Rouge, LA.

DiMarco, S.F., P. Chapman, N. Walker, and R.D. Hetland. 2010. Does local topography control hypoxia on the eastern Texas-Louisiana shelf? *Journal of Marine Systems* 80: 25–35.

Dinnel, S., and W.J. Wiseman Jr. 1986. Freshwater on the Louisiana shelf. *Continental Shelf Research* 6:765–784.

Donner, S.D., C.J. Kucharik, and J.A. Foley. 2004. Impact of changing land use practices on nitrate export by the Mississippi River. *Global Biogeochemical Cycles* 18: GB1028.

Donner, S.D., and D. Scavia. 2007. How climate controls the flux of nitrogen by theMississippi River and the development of hypoxia in the Gulf of Mexico. *Limnology and Oceanography* 52: 856–861.

Duan S.W., R.W.M. Amon, T.S. Bianchi and P.H., Santschi. 2011. Temperature Control on Soluble Reactive Phosphorus in the Lower Mississippi River? *Estuaries and Coasts* 34: 78–89.

Duan S.W., T.S. Bianchi, and P.H., Santschi, R.W.M. Amon. 2010. Effects of tributary inputs on nutrient export from the Mississippi and Atchafalaya Rivers to the Gulf of Mexico. *Marine and Freshwater Research* 61: 1029–1038.

Duan, S., and T.S. Bianchi, 2007. Particulate and Dissolved Amino Acids in the Lower Mississippi and Pearl Rivers (USA). *Marine Chemistry* 107: 214–229.

Duan, S.W., and S.S. Kaushal. 2013. Warming increases carbon-nutrient fluxes from sediments in streams across land use. *Biogeosciences* 10: 1193–1207.

Duan, S.W., and T.S. Bianchi. 2006. Seasonal changes in the abundance and composition of plant pigments in particulate organic carbon in the lower Mississippi and Pearl Rivers (USA). *Estuaries and Coasts* 29: 427–442.

Duan, S.W., T. Liang, L.J. Wang, X.M. Zhang, and X.B. Chen. 2008. Seasonal Changes in Nitrogen and Phosphorus Transport in the Lower Changjiang River before the Construction of the Three Gorges Dam. *Coastal and Shelf Science* 79: 239–250.

Duan, S.W., T. S. Bianchi, A.M. Shiller, K. Dria, P.G. Hatcher and K. R. Carman. 2007b. Variability in the bulk composition and abundance of dissolved organic matter in the lower Mississippi and Pearl rivers (USA). *Journal Geophysical Research – Biosciences* 112: G02024.

Duan, S.W., T.S. Bianchi, and T.P. Sampere. 2007a. Temporal variability in the composition and abundance of terrestrially-derived dissolved organic matter in the lower Mississippi and Pearl Rivers. *Marine Chemistry* 103: 172–184.

Etter, P.C., M.K. Howard, J.D. Cochrane. 2004. Heat and freshwater budgets of the Texas-Louisiana shelf. *Journal of Geophysical Research – Oceans* 109: C02024.

Galler, J.J., and M.A. Allison. 2008. Estuarine controls on fine-grained sediment storage in the Lower Mississippi and Atchafalaya Rivers. *Geological Society of America Bulletin* 120: 386–398.

Goñi, M.A., K.C. Ruttenberg, and T. I. Eglinton. 1997. Sources and contribution of terrigenous organic carbon to surface sediments in the Gulf of Mexico. *Nature* 389: 275–278.

Goolsby, D.A., and W.A. Battaglin. 2001. Long-term changes in concentrations and flux of nitrogen in the Mississippi River Basin, USA. *Hydrological Processes* 15: 1209–1226.

Goolsby, D.A., W.A. Battaglin, B.T. Aulenbach, and R.P. Hooper. 2000. Nitrogen flux and sources in the Mississippi River Basin. *Science of the Total Environment* 248: 75–86.

Green, R.E., T.S. Bianchi, M.J. Dagg, N.D. Walker and G.A. Breed. 2006. An organic carbon budget for the Mississippi River turbidity plume and plume contributions to air-sea CO2 fluxes and bottom-water hypoxia. *Estuaries* 29: 579–597.

Grzymko, T.J., F. Marcantonio, B.A. McKee, and C.M. Stewart. 2007. Temporal variability of uranium concentrations and 234U/238U activity ratios in the Mississippi River and its tributaries. *Chemical Geology* 243: 344–356.

Guo, L. 2010. Isotopic composition and export fluxes of organic carbon species from the Lower Mississippi River. Invited presentation at the 2010 AGU Fall Meeting, San Francisco, CA, December 13–17, 2010, abstract # OS22A-01.

Guo, L. B. J. Hunt, P. H. Santschi, and S.M. Ray. 2001. Effect of dissolved organic matter on the metal uptake by American oysters. *Environmental Science & Technology* 35: 885–893.

Guo, L., D.M. White, C. Xu, and P.H. Santschi. 2009. Chemical and isotopic composition of HMW-DOM from the Mississippi River plume. *Marine Chemistry* 114: 63–71.

Guo, L., Y. Cai, C. Belzile, and R. Macdonald. 2012a. Sources and export fluxes of inorganic and organic carbon and nutrient species from the seasonally ice-covered Yukon River. *Biogeochemistry* 107: 187–206.

Guo, L.D., J.K. Lehner, D.M. White, and D.S. Garland. 2003. Heterogeneity of natural organic matter from the Chena River, Alaska. *Water Research* 37: 1015–1022.

Guo, L.D., and R.W. Macdonald. 2006. Source and transport of terrigenous organic matter in the upper Yukon River: Evidence from isotope (delta C-13, Delta C-14, and delta N-15) composition of dissolved, colloidal, and particulate phases. *Global Biogeochemical Cycles* 20: GB2011.

Guo, L.D., Y. Cai, and X. Wang. 2008. Export fluxes of dissolved organic and inorganic carbon and nitrogen from the Mississippi River. EOS, Trans AGU, 89, 2008 Ocean Science Meeting, Orlando, FL March 2–7, 2008, abstract # OS108–1366.

Guo, X.H., W.J. Cai, W.J. Huang, Y.C. Wang, F.Z. Chen, M.C. Murrell, S. Lohrenz, L.Q. Jiang, M.H. Dai, J. Hartmann and R. Culp. 2012b. Carbon dynamics and community production in the Mississippi River plume. *Limnology and Oceanography* 57: 1–17.

Hedges, J. I., M. Emilio, T. Elizabeth, M.E. McClain, A. Aufdenkampe, P. Quay, J.E. Richey, R. Benner, S. Opsahl, B. Black, T. Pimentel, J. Quintanilla, L. Maurice. 2000. Organic matter in Bolivian tributaries of the Amazon River: A comparison to the lower mainstream. *Limnology and Oceanography* 45: 1449–1466.

Hetland, R.D., and S.F. DiMarco. 2008. How does the character of oxygen demand control the structure of hypoxia on the Texas–Louisiana continental shelf? *Journal of Marine Systems* 70: 49–62.

Horowitz, A.J. 2010. A quarter century of declining suspended sediment fluxes in the Mississippi River and the effect of the 1993 flood. *Hydrological Processes* 24: 13–34.

Justic, D., N.N. Rabalais, and R.E. Turner. 1995. Stoichiometric nutrient balance and origin of coastal eutrophication. *Marine Pollution Bulletin* 30: 41–46.

Justic, D., N.N. Rabalais, and R.E. Turner. 2005. Coupling between climate variability and coastal eutrophication: Evidence and outlook for the northern Gulf of Mexico. *Journal of Sea Research* 54: 25–35.

Justic, D., Turner, R.E., and Rabalais, N.N. 2003. Climatic influences on riverine nitrate flux: implications for coastal marine eutrophication and hypoxia. *Estuaries* 26: 1–11.

Kammerer, J.C. 1990. Largest Rivers in the United States. US Geological Survey Fact Sheet OFR 87–242 rev. http://pubs.usgs.gov/of/1987/ofr87-242/.

Kendall, C., S.R. Silva, and V.J. Kelly. 2001. Carbon and nitrogen isotopic compositions of particulate organic matter in four large river systems across the United States. *Hydrological Processes* 15: 1301–1346.

Knowlton, M.F., and J.R. Jones. 2000. Seston, light, nutrients and chlorophyll in the lower Missouri River, 1984–1998. *Journal of Freshwater Ecology* 15: 283–297.

Lane, R.R., H.S. Mashriqui, G.P. Kemp, J.W. Day, J.N. Day, and A. Hamilton. 2003. Potential nitrate removal from a river diversion into a Mississippi delta forested wetland. *Ecological Engineering* 20: 237.

Leenheer, J. 1982. United States Geological Survey Data Information Service. *In*: E. T. Degens (ed.), *Transport* of Carbon and Minerals in Major World Rivers, Pt. I, Mitt. Geol.-Paläont. Inst. Univ. Hamburg, SCOPE/UNEP Sonderbd, 52. pp. 355–356.

Leenheer, J. A., T. I. Noyes, and P. A. Brown. 1995. Data on natural organic substances in dissolved, colloidal, suspended-silt, and -clay and bed-sediment phases in the Mississippi River and some of its tributaries, 1991–1992. *US Geological Survey Water Resources Investigations Report* 94–4191, Denver, CO.

Lohrenz, S.E., G.L. Fahnenstiel, D.G. Redalje, G.A. Lang, X. Chen, and M.J. Dagg. 1997. Variations in primary production o f northern Gulf of Mexico continental shelf waters linked to nutrient inputs from the Mississippi River, *Marine Ecological Progress Series* 155: 435–454.

Malcolm, R., and W. Durum. 1976. Organic carbon and nitrogen concentrations and annual organic carbon load of six selected rivers of the United States. Water-Supply Paper 1817-F. U.S. Geological Survey.

Mayer, L.M., R.G. Keil, S.A. Macko, S.B. Joye, K.C. Ruttenberg, and R.C. Aller. 1998. Importance of suspended particulates in riverine delivery of bioavailable nitrogen to coastal zones. *Global Biogeochemical Cycles* 12: 573–579.

McCorquodale, J.A., R.J. Roblin, I.Y. Georgiou, and K. Haralampides. 2009. Salinity, Nutrient, and Sediment Dynamics in the Pontchartrain Estuary. *Journal of Coastal Research: Special Issue* 54: 71–87.

Meade, R.H. 1995. Setting – geology, hydrology, sediments, and engineering of the Mississippi River. *In* R.H. Mead (ed), Contaminants in the Mississippi River, 1987–92: *U.S. Geological Survey Circular* 1133, pp. 13–30.

Meade, R.H., and J.A. Moody. 2010. Causes for the decline of the suspended-sediment discharge in the Mississippi River System, 1940–2007. *Hydrological Processes* 24: 35–49.

Miller, J.R., and G.L. Russell. 1992. The impact of global warming on river runoff. *Journal of Geophysical Research-Atmospheres* 97: 2757–2764.

Mitra, S., T.S. Bianchi, B.A. McKee, and M. Sutula. 2002. Black carbon from the Mississippi River: Quantities, sources, and potential implications for the global carbon cycle. *Environmental Science & Technology* 36: 2296–2302.

Mossa, J. 1996. Sediment dynamics in the lowermost Mississippi River. *Engineering Geology* 45: 457–479.

Nittrouer, J.A., D. Mohrig, M.A. Allison, and A.-P. B. Peyret. 2011. The lowermost Mississippi River: a mixed bedrock-alluvial channel. *Sedimentology* 58: 1914–1934.

Nittrouer, J.A., M.A. Allison, and R. Campanella. 2008. Bedform transport rates for the lowermost Mississippi River. *Journal of Geophysical Research-Earth Surface* 113: F03004.

Onstad, G.D., D.E. Canfield, P.D. Quay, and J.I. Hedges. 2000. Sources of particulate organic matter in rivers from the continental USA: Lignin phenol and stable carbon isotope compositions. *Geochemica et Cosmochimica Acta* 64: 3539–3546.

Rabalais, N.N., R.E. Turner, D. Justic, Q. Dortch, W.J. Wiseman, Jr., and B.K. SenGupta. 1996. Nutrient changes in the Mississippi River and system responses on the adjacent continental shelf. *Estuaries* 19: 386–407.

Ramirez, M.T. and Allison, M.A., in review. Suspension of bed-material sand over lateral bars in the lower Mississippi River, southeastern Louisiana. *Journal of Geophysical Research – Earth Surface Processes.*

Raymond, P.A., N. Oh, R.E. Turner and W. Broussard. 2008. Anthropogenically enhanced fluxes of water and carbon from the Mississippi River. *Nature* 451: 449–452.

Rinker, K.R. and R.T. Powell. 2006. Dissolved organic phosphorus in the Mississippi River plume during spring and fall 2002. *Marine Chemistry* 102 SI: 170–179.

Rosenheim, B.E. and V. Galy. 2012. Direct measurement of riverine particulate organic carbon age structure. *Geophysical Research Letters* 39: L19703.

Rosenheim, B.E., K.M. Roe, B.J. Roberts, A.S. Kolker, M.A. Allison, and K.H. Johannesson. 2013. River discharge influences on particulate organic carbon age structure in the Mississippi/Atchafalaya River System. *Global Biogeochemical Cycles* 27: 154–166.

Runkel, R.L., Crawford, C.G., and Cohn, T.A. 2004. Load Estimator (LOADEST): A FORTRAN Program for Estimating Constituent Loads in Streams and Rivers: U.S. Geological Survey Techniques and Methods Book 4, Chapter A5.

Santschi, P.H., L. Guo , J.C. Means, and M. Ravichandran. 1999. Natural organic matter binding of trace metal and trace organic contaminants in estuaries. *In*: T.S. Bianchi, J.R. Pennock, and R. Twilley (eds.), Biogeochemistry of Gulf of Mexico Estuaries. John Wiley & Sons, pp. 347–380.

Shen, Y., C.G. Fichot, and R. Benner. 2012. Floodplain influence on dissolved organic matter composition and export from the Mississippi–Atchafalaya River system to the Gulf of Mexico. *Limnology & Oceanography* 57: 1149–1160.

Shiller, A.M. 2002. Seasonality of dissolved rare earth elements in the Lower Mississippi River. *Geochemistry, Geophysics, Geosystems* 3: 1068.

Shiller, A.M. 1997. Dissolved trace elements in the Mississippi river: Seasonal, interannual, and decadal variability. *Geochimica et Cosmochimica Acta* 61: 4321–4330.

Shiller, A.M., and T.H. Stephens. 2005. Microbial manganese oxidation in the lower Mississippi River: methods and evidence. *Geomicrobiology Journal* 22: 117–125.

Shiller, AM., and E.A. Boyle. 1987. Variability of dissolved trace metals in the Mississippi River. *Geochimica et Cosmochimica Acta* 51: 3273–3277.

Soileau, C.W., B.J. Garrett, and B.J. Thibodeaux. 1989. Drought Induced Saltwater Intrusion on the Mississippi River, Proceedings of Coastal Zone '89 Conference, ASCE.

Stolpe, B., L. Guo, A. Shiller, and M. Hassellov. 2010. Size and composition of colloidal organic matter and trace elements in the Mississippi River, Pearl River and the northern Gulf of Mexico as characterized by flow field-flow fractionation. *Marine Chemistry* 118: 119–129.

Sutula, M., T.S. Bianchi, and B.A. McKee. 2004. Effect of seasonal sediment storage in the lower Mississippi River on the flux of reactive particulate phosphorus to the Gulf of Mexico. *Limnology and Oceanography* 49: 2223–2235.

Sylvan, J.B., Q. Dortch, D.M. Nelson, A.F.M. Brown, W. Morrison, and J.W. Ammerman. 2006. Phosphorus limits phytoplankton growth on the Louisiana shelf during the period of hypoxia formation. *Environmental Science & Technology* 40: 7548–7553.

Taylor, H.E., and A.M. Shiller. 1995. The Mississippi River Methods Comparison Study: implications for water quality monitoring of dissolved trace elements. *Environmental Science and Technology* 29: 1313–1317.

Trefry, J.H., S. Metz, T.A. Nelsen, R.P. Trocine, and B.J. Eadie. 1994. Transport of particulate organic carbon by Mississippi River and its fate in the Gulf of Mexico. *Estuaries* 17: 839–849.

Trefry, J.H., T.A. Nelson, R.P. Trocine, S. Metz, and T.W. Vetter. 1986. Trace metal fluxes through the Mississippi River Delta system. Rapp. *P.-v. Réun. Cons. int. Explor. Mer.* 186: 277–288.

Turner, R.E., and M.E. Boyer. 1997. Mississippi River diversions, coastal wetland restoration/creation and an economy of scale. *Ecological Engineering* 8: 117–128.

Turner, R.E., and N.N. Rabalais. 1991. Changes in Mississippi River water quality this century. *Bioscience* 41:140–147.

Turner, R.E., N.N. Rabalais, R.B. Alexander, G. McIsaac, and R.W. Howarth. 2007. Characterization of nutrient, organic carbon, and sediment loads and concentrations from the Mississippi River into the Northern Gulf of Mexico. *Estuaries and Coasts* 30: 773–790.

Turner, R.E., Q. Dortch, D. Justic, and E.M. Swenson. 2002. Nitrogen loading into an urban estuary: Lake Pontchartrain (Louisiana, U.S.A.). *Hydrobiologia* 487: 137–152.

Vannote R.L., G.W. Minshall, K.W. Cummins, J.R. Sedell, C.E. Cushing. 1980. The river continuum concept. *Canadian Journal of Fishery and Aquatic Science* 37: 130–137.

Watson, R.T., M.C. Zinyowera, R.H. Moss, and D.J. Dokken, eds. 1996. Climate Change 1995-Impacts, Adaptations and Mitigation of Climate Change: Scientific-Technical Analyses. Contribution of Working Group II to the Second Assessment Report of the Intergovernmental Panel on Climate Change. Cambridge: Cambridge University Press.

Wilson, H.F., and M.A. Xenopoulos. 2009. Effects of agricultural land use on the composition of fluvial dissolved organic matter. *Nature Geoscience* 2: 37–41.

Wissel, B., and B. Fry. 2005. Sources of particulate organic matter in the Mississippi River, USA. *In*: A.D. Buijse, F. Klijn, R.S.E.W. Leuven, H. Middelkoop, F. Schiemer, J.H. Thorp & H.P. Wolfert (eds.), Rehabilitating large regulated rivers. *Arch. Hydrobiol. Suppl.* 155 (Large Rivers 15). E. Schweizerbart'sche Verlagsbuchhandlung, Stuttgart.

Xu, Y., and F. Marcantonio. 2007. Strontium isotope variations in the lower Mississippi River and its estuarine mixing zone. *Marine Chemistry* 105: 118–128.

Yang, Q., H. Tian, M. Liu, X. Li, G. Chen. 2011a. Modeling Nutrients Export in responses to Land Use Changes in Mississippi River Basin during 1901–2008. American Geophysical Union, Fall Meeting, abstract #B13G-0653.

Yang, W., H. He, and L. Guo. 2011b. Interannual variations and export fluxes of biogenic silica from the Mississippi River. Presentation at the 2011 NGI annual meeting, May 17–19, 2011, Mobile, AL.

Zhang, Y.K., and K.E. Schilling. 2006. Increasing streamflow and baseflow in Mississippi River since the 1940 s: Effect of land use change. *Journal of Hydrology* 324: 412–422.

17

Climate change effects on the ecology of the Mississippi River Delta

J. M. Visser, W. P. Broussard III, G. P. Shaffer, and J. W. Day, Jr.

1. Climate change

The consensus in the scientific community is that human activity is affecting global climate (Intergovernmental Panel on Climate Change [IPCC] 2007), and climate change will significantly alter coastal and wetland ecosystems of the Mississippi delta (Titus et al. 1991; Wilkinson and Buddemeier 1994; Tarasona et al. 2001; Poff et al. 2002; Twilley et al. 2001; Ning et al. 2003; Day et al. 2005). Global climate change will interact with, and magnify, other human stresses on the Mississippi Delta and the goods and services it provides (see Day et al. 2008). Climate change impacts include temperature increase, accelerated sea-level rise, changes in rainfall and freshwater discharge, and changes in the frequency and intensity of tropical storms (Day and Templet 1989; Scavia et al. 2002; Day et al. 2005; Yáñez-Arancibia and Day 2005; Hoyos et al. 2006; Day et al. 2008).

Both precipitation and evaporation will likely change in the Mississippi Delta and Basin (Day et al. 2005), changing the rate of freshwater input. Because freshwater input affects so many critical parameters and processes in coastal ecosystems, this will be an important way that climate change alters these ecosystems. The rate at which water moves through a system affects many estuarine ecosystem processes. This rate of movement will also be affected by sea-level rise, which may impede drainage.

Global warming is leading to accelerated eustatic sea-level rise. The IPCC (2007) predicted that sea level will rise by about 40 cm by 2100 (20–80 cm in the twenty-first century), but recent reports suggest the rise will be a meter or more (Rahmstorf 2007; Vermeer and Rahmstorf 2009). Increasing sea level will strongly impact wetlands of the Delta (Blum and Roberts 2009) through increased inundation of the wetlands, although wetland accretion can offset sea-level rise up to a point, depending on sediment supply and rate of change (Morris et al 2002; Kirwan et al. 2010).

Finally a number of reports suggest that global warming will lead to increases in the total and the number of category 4 and 5 storms (Emanuel 2005; Webster et al. 2005; Hoyos et al. 2006). Storms can have both negative (erosion losses) and positive (sediment input) affects on the wetlands in the Mississippi River Delta (MRD) (Conner et al. 1989; Guntenspergen et al. 1995; Turner et al. 2006a, Howes et al. 2010).

2. Ecology of the Mississippi River Delta

The Delta is generally divided into areas of growth (active delta lobes) and areas of deterioration (abandoned lobes) that represent different stages of the transgressive phase of the delta cycle (Penland

Table 17.1. *Summary of the different habitats in the Mississippi deltaic plain*

Habitat type	Area[a] (hectares)	Soil bulk density[b] (g cm^{-3})	Soil organic matter[b] (%)	Inundation events[c] (number per year)	Duration of inundation events[c] (hours)
Deltaic Marsh	95,862	0.65–1.04	3.7–7.1		
Forested Wetlands	344,000	0.30–0.41	21.7–47.9		
Fresh Marsh	210,538	0.06–0.28	19.3–81.7	32	115
Oligohaline Marsh	139,819	0.11–0.43	17.2–66.6	32	29
Mesohaline Marsh	169,293	0.16–0.31	29.4–48.6	75–125	28–50
Saline Marsh	248,188	0.29–0.53	23.9–25.8	150–263	16–27

[a] Marsh based on Sasser et al. (2008), forested wetlands from Chambers et al. (2005).
[b] From Wang et al. (2012).
[c] From Gosselink (1984).

et al. 1988). Areas with active river inputs are highly influenced by the physical forcing of annual floods. Soils of the recently colonized mudflats in these areas are highly mineral with little organic matter (Table 17.1). Salinity incursions can occur during the low discharge months (September–November) and are generally associated with tropic disturbances and the passage of cold fronts that occur during those months (Holm and Sasser 2001).

The abandoned lobes are characterized by a series of vegetation zones first mapped by Chabreck (1972) that run roughly parallel to the coast and have different salinity, hydrology, and soil conditions (Table 17.1). These vegetation zones range from saline marshes along the coast to tidal freshwater forested wetlands in the upper parts of the basins. Near the coast, saline marshes that are largely dominated by *Spartina alterniflora*, but also have areas dominated by *Avicennia germinans* or *Juncus roemerianus*, have mineral soils and are flooded almost daily by a microtidal regime. In the mesohaline and oligohaline marshes, *Spartina patens* dominates the largest areas, with organic soils that have fewer and slightly longer flooding events. In the fresh marsh, *Panicum hemitomon* dominates the most area, but large areas are also dominated by *Typha domingensis* or *Sagittaria lancifolia*. In the fresh marsh, inundation is dominated by wind-driven events, and therefore there are fewer events, but they last for five days on average. In addition, a large percentage (>70%) of the *Panicum hemitomon* marshes in the MRD occur on peat mats that float up and down with the water level, with a soil that is always saturated but never inundated (Sasser et al. 1995).

Over the past half century, there has been an enormous loss of coastal wetlands in the delta, with a total loss of about 2,551 km^2 between 1956 and 2006 (Barras et al. 2008). Erosion along large water bodies and along the Gulf of Mexico accounts for approximately 30% of the loss (Penland et al. 2002). Most (57%) of the wetland loss is due to elevation deficits (Penland et al. 2002). Wetlands in the area experience relative sea-level rise (RSLR), a combination of subsidence plus eustatic sea-level rise (ESLR). Subsidence in the area is driven by many processes that occur at different spatial and temporal scales (Yuill et al. 2009). Wetlands must grow vertically at a rate equal to the rate of RSLR if they are to survive in the long term. If there is an elevation deficit (surface elevation gain < RSLR), the site will become progressively more waterlogged, and vegetation will become stressed and eventually die (Mendelssohn and Morris 2000). The elevation deficit is increasing as sediment

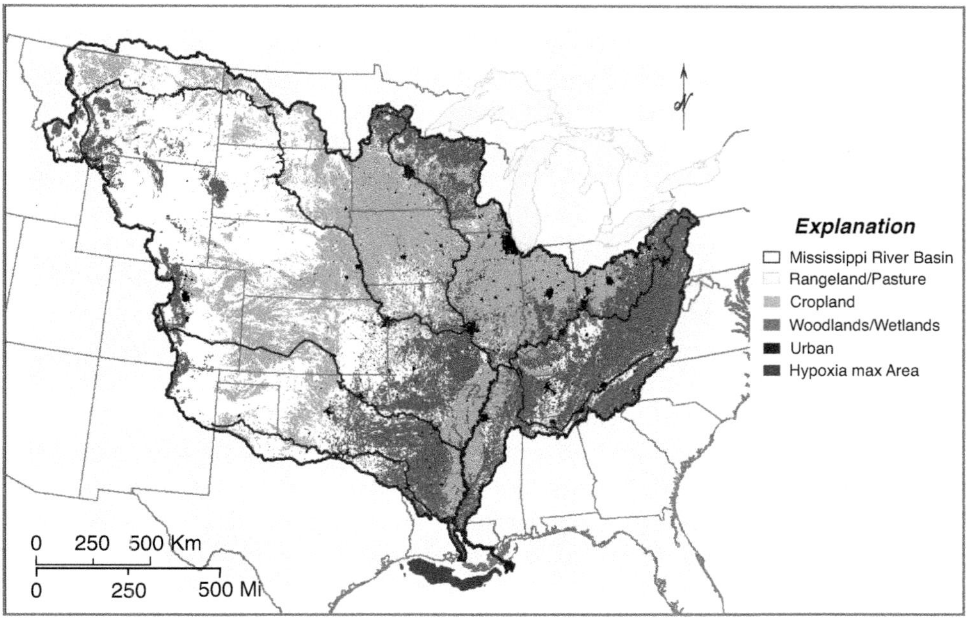

Figure 17.1. Mississippi River Basin depicting 2001 land cover classifications and the 2001 extent of bottom water hypoxia in the Northern Gulf of Mexico.

delivery to the delta has decreased owing to a reduced sediment load in the river (Kesel 1988; Meade 1995) and flood control levees on the Mississippi River that have eliminated riverine input from most of the wetlands (Boesch et al. 1994; Day et al. 2000, 2007).

3. Antropogenic impacts on the Mississippi River Basin

The Mississippi River is the largest river in North America. The average discharge of the Mississippi River and its main distributary, the Atchafalaya River, into the northern Gulf of Mexico averages approximately 22,600 m^3/s. This represents nearly 40% of total terrestrial discharge into coastal oceans from the continental United States.

The Mississippi River Basin (MRB) is the fourth largest watershed in the world. It covers roughly 3 million km^2 (Figure 17.1). Agricultural croplands account for 35% of land area within the MRB, with grasslands, savannas, and shrubland covering 29%; evergreen and deciduous forests covering 22%; and urban/industrial lands covering 14% of the total area (Revenga et al. 1998). This large watershed contributes to a dynamic fluvial system that is subject to many different biomes, anthropogenic influences, and regional climates.

Nitrogen enrichment of the biosphere is an anthropogenic impact of global concern (Vitousek et al. 1997). The global production rate of biologically available nitrogen (N) in terrestrial ecosystems has doubled since 1960, primarily resulting from human activity that consumed nearly eight times more nitrogenous fertilizer in 2003 than in 1960 (Millennium Ecosystem Assessment [MEA] 2005). This human-derived N fixation now produces more biologically available N than all other natural sources combined (Galloway et al. 1995). Excessive nutrient loading has become one of the most dominant drivers of ecosystem change in the last 40 years for both terrestrial and aquatic ecosystems and is projected to increase in the twenty-first century (MEA 2005).

Very little water quality data exist in the MRB from the first half of the twentieth century. Some alkalinity and nitrate-N records are available, allowing for centennial analyses of water quality trends (Raymond et al. 2008; Broussard and Turner 2009; Stets et al. in press). Though phosphorus and silica concentrations are known causes of changes in aquatic ecosystems (Redfield 1958; Goolsby et al. 1999), nitrogen appears to be the limiting nutrient for eutrophication in the Northern Gulf of Mexico (Mississippi River/Gulf of Mexico Watershed Nutrient Task Force 2001) and a nutrient of great concern. The N flux from the MRB increased nearly 300% from 1955–1970 to 1980–1999, almost entirely as a result of increases in nitrate concentration accompanying the intensification of agricultural practices (Goolsby and Battaglin 2001; Turner and Rabalais 2003; Broussard and Turner 2009). The total N flux is approximately 61% nitrate, 2% ammonia, and the remaining 37% is dissolved and particulate organic nitrogen. All of the increase in total N that has occurred since 1972 can be attributed to the increase in nitrate. Stream flows in the MRB generally increased during the 1955–1996 period and this could account for part of the increase in nitrate flux (Goolsby 2000).

Sprague et al. (2011) found that flow-adjusted trends in the concentration of nitrogen and phosphorus from 1993 to 2003 were not significant at a majority of nationwide sampling sites (Sprague et al. 2009). These findings indicate that human activities such as changes in nutrient inputs or the implementation of management practices did not contribute to net changes in nutrient concentrations in most sampled streams. Human activities did, however, contribute to more increases than decreases in the remaining streams (Sprague et al. 2009; Dubrovsky et al. 2010). Furthermore, an observed increase in concentrations at low streamflows in the MRB is a strong indication that increasing nitrate concentrations in groundwater are having a substantial effect on river concentrations in the basin. As a result, conservation practices designed to reduce infiltration to groundwater may help with managing nitrate in these rivers (Sprague et al. 2011).

Several studies (Goolsby et al. 1999; Goolsby et al. 2000; McIsaac et al. 2001; McIsaac et al. 2002; Broussard and Turner 2009) have pointed to an intensification of agricultural activity and other anthropogenic influences in the MRB as drivers of N flux to the Northern Gulf of Mexico. In many estuaries and coastal marine environments, biologically available N limits primary production. Consequently, anthropogenic additions of N to these systems contribute to a process of cultural eutrophication and periodic hypoxia and anoxia in lower portions of the water column (Rabalais et al. 1999; Diaz, 2001; Goolsby and Battaglin 2001; McIsaac et al. 2002). The size of the bottom water hypoxic zone in the Northern Gulf of Mexico has grown to the size of New Jersey, primarily because of nutrient loading from the Mississippi River, and particularly because of the increased nitrate loading (Rabalais et al. 2007; Turner et al. 2008; Environmental Protection Agency [EPA] 2008). To date, little consistent progress has been made in controlling nitrate concentration and flux in the MRB, whereas concentration and flux are increasing in some parts of the basin. For example, N flux from the MRB increased only 9% from 1980 to 2008, but the sub-basins of the upper MRB above Clinton, Iowa, and the Missouri River Basin above Hermann, Missouri, increased 67% and

57%, respectively (Sprague et al. 2011). Reaching the Hypoxia Task Force goal of reducing nitrate flux by 30% to reduce the size of hypoxia in the Northern Gulf of Mexico will require substantial interstate cooperation and federally backed initiatives (Mississippi River/Gulf of Mexico Watershed Nutrient Task Force [MRGOMWNTF] 2008; Broussard et al. 2012).

Modern production agriculture in much of the central and eastern United States would not be possible without the extensive drainage network that has been built up since the late 1800s (U.S. Department of Agriculture [USDA] 1987). Because of its ubiquitous nature, extended installation history, and incomplete maps or surveys of subsurface drains, the current extent of drained cropland in the United States is poorly known. Some of the largest riverine N fluxes in the continental United States have been observed in agricultural regions with extensive artificial subsurface drainage, commonly called tile drainage (McIsaac and Hu 2004). McIsaac and Hu (2004) reported that during 1977–1997, net N input (NNI) to the tile-drained region had increased to 27 kg N ha^{-1} $year^{-1}$, and riverine N flux was approximately 100% of this value. In the non–tile-drained region, NNI was approximately 23 kg N ha^{-1} $year^{-1}$ and riverine N flux was between 25% and 37% of this value (5–9 kg N ha^{-1} $year^{-1}$).

Although tile drainage increased the transport efficiency of NNI to streams, it is surprising that riverine N flux was equal to or greater than NNI in tile-drained regions. This is probably partially due to reduced in-field denitrification. Tile drainage reduces water saturation in fields and therefore reduces the anoxic conditions that promote denitrification. Meisinger and Randall (1991) suggested that denitrification rates should be approximately 30–40% lower in fields with tile drainage as compared with those without. Additionally, tile drainage reduces the interaction between drainage water and riparian zones, which can serve as significant sinks of N (Osborne and Kovacic 1993). The USDA (1987) estimated that in 1985 there were approximately 11 million ha of tile-drained croplands in the MRB. This represents 4% of the total land area and 16% of the area used for annual crop production in the watershed. The larger riverine N flux from tile-drained regions may be a result of low denitrification rates, ongoing depletion of soil organic N, and/or greater biological N fixation inputs. Further research is needed to more accurately quantify these fluxes in tile-drained settings. Mitsch et al. (2001) reported that changes in agricultural practices in conjunction with creation or restoration of about 2 million ha of wetlands strategically placed in the agricultural landscape could significantly reduce nitrogen loading to streams in the middle Mississippi basin and the Gulf of Mexico.

Before 1850, the Mississippi River system exported an estimated 338 million metric tons per year of sediment to coastal Louisiana and the Northern Gulf of Mexico. Peak flux occurred between 1870s and 1890s at 348 million metric tons per year. During the last two decades (1987–2006), this transport has averaged 145–157 million metric tons per year (Meade and Moody 2010; Tweel and Turner 2012). The conversion of prairie and forest to an agricultural landscape following European settlement, and subsequent nineteenth and twentieth century land-use changes in the MRB, is mirrored in several large-scale changes in the river system, including increases in offshore nutrient concentrations (Turner and Rabalais 1994), suspended sediment and nutrient deposition along the Upper Mississippi River (Engstrom et al. 2009), and overbank sedimentation in the Upper Mississippi Valley (Knox 1987), and the rate of enlargement of the Mississippi River Bird-Foot Delta (Tweel and Turner 2012). The reservoir volume in the Missouri River Basin increased in two large increments between 1935–1937 and 1952–1958. The decline in sediment concentration was coincidental with the increase in reservoir volume.

Reductions in sediment flux after the 1960s, however, do not correspond to a reduction in discharge (Meade and Moody 2010). Changes in the relationship between water discharge and suspended-sediment concentration suggests that the Mississippi River has been transformed from a transport-limited to a supply limited system. Dams alone cannot explain this observation. Other anthropogenic activities such as meander cutoffs, river-training structures, bank revetments, and soil erosion controls have trapped sediment, eliminated sediment sources, or protected sediment that was once available for transport. Addressing the numerous smaller engineering structures and soil-retention works throughout the Mississippi system would be necessary to increase sediment flux to the Louisiana coastal zone (Meade and Moody 2010).

Substantial hydrologic modification has occurred throughout the main stem of the Mississippi River through the construction of wing-dams, levees, and lock-and-dam structures designed to maintain a navigable channel. Major tributaries of the Mississippi River including the Ohio, Illinois, Missouri, and Arkansas Rivers also have been extensively dammed for navigation, flood control projects, irrigation, and municipal water supplies.

Virtually the entire length of the Lower Mississippi River channel is surrounded by levees. The floodplain inside the levees is <10% of the original natural floodplain for this part of the river (Benke and Cushing 2005). The river channel itself has been extensively modified for flood control, including armored shorelines, artificial channel cutoffs, and revetments to prevent the channel from meandering.

The Atchafalaya River is the main distributary of the Mississippi River. Since 1963, the U.S. Army Corps of Engineers has regulated the amount of water flowing from the Mississippi into the Atchafalaya River through the Old River Control Structure. This structure is managed to ensure that not more than 30% of the total flow of the Mississippi and Red River enters the Atchafalaya (Benke and Cushing 2005).

Further upstream, Carlisle et al. (2010) reported that streamflow magnitudes were altered in most (86%) of the assessed streams. Minimum flows were the most frequently altered, being inflated or diminished in 74% of streams. Maximum flows were altered in 54% of streams and diminished in most cases. The type and severity of streamflow alterations were associated with climate. In arid climates, minimum and maximum flows were severely diminished, being less than half of expected magnitudes in most (~70%) monitored streams. Maximum flow magnitudes in wet climates were also commonly diminished, being less than three-fourths of expected magnitudes in most (>60%) monitored streams. In contrast, minimum flows in wet climates were commonly inflated, being >25% higher than expected magnitudes in about half of monitored sites (Carlisle et al. 2010).

Pronounced differences in streamflow alteration between arid and wet climates are partly due to distinctive management of watershed hydrology. The tendency for diminished flow magnitudes in arid climates is indicative of consumptive water uses causing net streamflow loss. The primary use of water in arid climates is for irrigated agriculture (Pimentel et al. 1997), but interbasin transfers and groundwater withdrawal for other uses also reduce streamflows (Jackson et al. 2001). Management of watershed hydrology in wet climates, in contrast, is often focused on flood control. This is most often achieved through small impoundments or large reservoirs that remove flood peaks and release the water later. During normally low flow periods, this management technique can result in inflated minimum flows and diminished maximum flows (Magilligan and Nislow 2005).

4. Anthropogenic impacts on the Louisiana coastal zone

Since European settlement in the MRD, humans have gradually changed the hydrology to protect themselves from spring floods of the Mississippi River. As early as 1718, the first levees were constructed to protect the New Orleans population from a river that overflowed nearly six months of the year (Davis 1990). A state law in 1770 required everyone given a tract of land to settle to build a levee to protect the land within three years (Davis 1990). However, in the late 1890s, benefits of completing the levees along the entire length of the river was questioned (Cortwell 1897). In 1928, the protectionist and navigation interest succeeded in the demand for complete levees along the river with the directive from the U.S. Congress to the Army Corps of Engineers to develop a comprehensive flood protection program for the Mississippi River. Although providing flood protection, the levee system separated the river from most of the deltaic plain, and this is a major cause of wetland loss in the delta (Day et al. 2007).

As river flooding diminished and population expanded, interest developed for reclamation of wetlands (Harrison and Kolmorgen 1947). In the mid 1800s, reclamation efforts were built in the vicinity of New Orleans with the expectation of selling crops produced on the reclaimed land to the city (Harrison and Kolmorgen 1947). However, many reclamation efforts failed as levees broke during hurricanes and land subsided as soils were dewatered and organic matter decomposed. By 1985, as much as 30% of the wetland area of the MRD was intentionally impounded (Day et al. 1990), with 21% of these reclaimed areas converted to open water, 58% remaining as developed lands, and 21% as undeveloped land. Penland et al. (2002) estimated that the land loss owing to failed land reclamation is only 2% of the overall land loss in the MRD, but that estimate does not take into consideration wetlands that were converted to agricultural and urban use.

During the early to mid-1900s, a series of navigation channels were dug that connected the upland areas with the coast (generally north-south) in straight lines, and another channel provided an inland route for shipping from east to west (Good et al. 1995). As ships became larger, these channels as well as the major rivers were deepened and channels widened (Good et al. 1995). As a result of wakes from the ship traffic, erosion of the banks along these channels is widespread (Good et al. 1995). The Mississippi River Gulf Outlet (MRGO) is perhaps the most infamous of these channels. Not only did construction of the channel lead to widespread wetland loss as a result of salinity intrusion and hydrologic alteration, the channel facilitated massive flooding of the New Orleans area during hurricanes Betsy in 1965 and Katrina in 2005 (Shaffer et al. 2009). Other canals have had similar impacts (Day et al. 2000). In the mid 1900s, oil exploration in the delta resulted in a patchwork of canals that interfere with natural marsh hydrology, especially in combination with the navigation channels. Marsh loss in the delta directly owing to canal dredging and spoil placement has been estimated between 13% and 22%, whereas indirect loss owing to these canals range from 10% to 25% of the loss (Deegan et al. 1984; Penland et al. 2002). Oil and gas exploration also increased local subsidence rates, which may be the primary reason for the indirect loss (Morton et al. 2006).

5. Climate change effects on the Mississippi River Basin

It has been suggested that human-influenced global warming may be partly responsible for increases in heavy precipitation because atmospheric water-holding capacity increases roughly exponentially with temperature. It appears that human-induced increases in greenhouse gases have contributed to an observed intensification of heavy precipitation events over approximately two-thirds of those parts of the Northern Hemisphere land areas where such data exist (Min et al. 2011). This is projected to lead to increased precipitation over much of the upper basin of the Mississippi (Pachauri and Reisinger 2007), and some models indicate that mean discharge of the Mississippi may increase by 40% by 2100 (Day et al. 2005).

Although increasing precipitation may drive increases in discharge, other factors such as land use and land cover influence the proportion of precipitation that eventually drains to surface streams. Several studies have documented increases in discharge for many gauging stations in the MRB (Lins and Slack 1999; Lins and Slack 2005; Raymond et al. 2008). However, discharge increases are only partly explained by concomitant precipitation increases. A larger percentage of the increase in discharge, particularly in agricultural watersheds, can be explained by changes in discharge at average precipitation (Raymond et al. 2008). An increase in the discharge at average precipitation demonstrates that anthropogenic modifications of the watershed are changing the relationship between precipitation and discharge, resulting in an increase in discharge during a normal precipitation year. Some studies argue that increases in greenhouse gases, especially CO_2, should increase discharge, and the discharge:precipitation ratio, by decreasing evapotranspiration (Betts et al. 2007; Gedney et al. 2006). Others show that although discharge in forested watersheds is highly correlated with precipitation, there is a strong correlation between the percentage of a watershed in agriculture and increases in riverine discharge (Raymond et al. 2008). The relationship between agricultural land cover and the change in discharge at average precipitation demonstrates that agricultural practices are increasing river discharge and altering the water budget of the Mississippi River (Zhang and Schilling 2006; Schilling 2005).

In addition to changes in runoff and groundwater recharge, which ultimately affect riverine discharge, it is important to consider the effects of land cover on regional climate. It is important to include land cover change in forcing scenarios for future climate change studies and accurate climate model predictions (Feddema et al. 2005). Although the effects of land cover tend to offset with respect to global average temperatures, they can significantly alter regional climate outcomes associated with global warming.

6. Climate change effects on biochemical cycling in the Mississippi Delta

Several aspects of climate change will impact biogeochemical processes in the Delta. These aspects include temperature, freshwater inflow, sea-level rise, tropical-storm frequency, and intensity.

An increase in temperature will lead to increased rates of metabolic processes, including rates of biogeochemical processing as well as to longer growing seasons for both emergent vegetation and plankton communities. An example of the impact of temperature on biogeochemical processes is the rate of nitrogen uptake. When diverted river water was measured as it flowed over marshes, water temperature increased 5–7 °C as water flowed as a sheet over several kilometers of marsh (Day et al. 2009). Turbidity owing to suspended sediments decreased from 20–30 NTU to zero and NO_3 decreased from 60–80 μM to undetectable. The increase in temperature enhanced NO_3 uptake. At temperatures greater than 10 °C, there was high N_2 production, reflecting high denitrification rates (250–300 mmol m^{-2} h^{-1}), whereas at temperatures lower than 10 °C, rates were very low.

Changes in rainfall and freshwater input affect both salinity and the replacement time of water in a system. Salinity is inversely related to freshwater input. Inorganic nitrogen is rapidly reduced as water flows through the marshes (Lane et al. 1999; Day et al. 2009). Estuarine sediments are generally net sources of PO_4 with flux rates highly correlated with temperature (Nixon et al. 1980), but cases of estuaries acting as net sinks for PO_4 have also been reported (Callender and Hammond 1982; Froelich 1988). These contradictory findings may be because of PO_4 sorption by clay and detrital organic particles at high concentrations, whereas at lower concentrations PO_4 is released into the water, maintaining moderate ambient concentrations (Jitts 1959; Patrick and Khalid 1974). Cyclic aerobic and anaerobic conditions in the top several centimeters of the soil column affect the sorption and release of PO_4, with PO_4 being released during anaerobic conditions (Patrick and DeLaune 1977), possibly exacerbating hypoxic events. These sorption-desorption processes provide a buffering mechanism for phosphorus in estuaries (Sharp et al. 1982; Madden et al. 1988). Lane et al. (2004) reported that there were significant reductions in nutrient concentrations, and DSi:DIN ratios increased and N:P ratios decreased as diverted river water passed through a Louisiana estuary. Studies in Louisiana and elsewhere have shown high rates of denitrification when river water with high nitrate concentrations flows into estuaries (Jenkins and Kemp 1984; Smith et al. 1985; Boynton et al. 1995; Nixon et al. 1996; Bachand and Horne 2000; Brock 2001). It has been suggested that an increase in the DSi:DIN ratio would likely reduce the potential for harmful algal blooms (Officer and Ryther 1980; Dortch et al. 2001). Nutrient reduction and changes in nutrient ratios are related to loading rate. Thus the interaction of temperature and freshwater input affects P, N, and Si dynamics. Changing freshwater input also affects the turbidity maximum, which could impact the location and rates of phosphorus dynamics because of variations in the amount of both phosphorus and suspended clay particles input by the Mississippi to the coastal zone.

Changes in salinity can affect a number of biogeochemical processes. The interaction between changes in freshwater input and accelerated sea-level rise will impact the salinity gradient in estuaries. Methane release from flooded soils is impacted by salinity. Because the presence of sulfate suppresses methane formation, increased salinity will lead to lower methane formation. DeLaune et al. (1983) reported on methane release from wetlands along a salinity gradient in the MRD. Methane emission was inversely related to salinity and sulfate concentration. Methane production was 4.3, 73, and 160 g C m^{-2} yr^{-1} in salt, brackish, and fresh marshes, respectively.

Rising sea level can affect biogeochemical processes in two basic ways: changing the salinity gradient and loss of wetlands. The impacts of changing salinity were addressed previously. The impact of wetland loss on biogeochemistry is related to the importance of wetlands in so many coastal biogeochemical processes. Coastal wetlands have high levels of productivity and sequester large amounts of organic carbon. Loss of wetlands results in the release of large amounts of stored carbon (DeLaune and White 2012). Many of the biogeochemical processes occurring in wetlands (uptake and release of nutrients, denitrification, methane production) also occur in the sediments

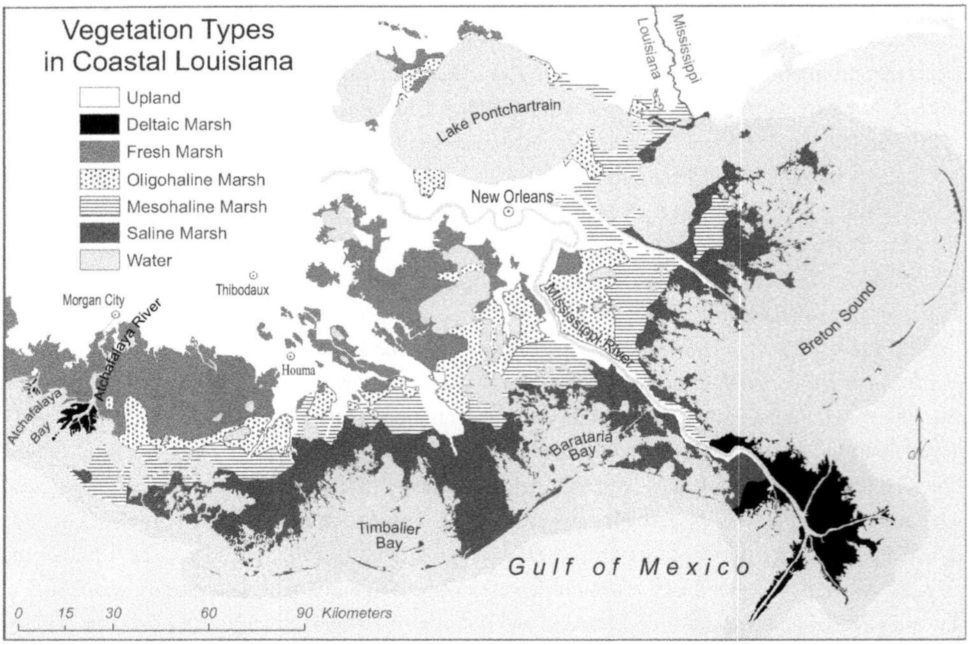

Figure 17.2. Distribution of marsh types in the Mississippi River Deltaic Plain (Modified from Sasser et al. 2008).

of shallow water bodies. If wetlands are lost, then water exchange in estuarine bays could increase significantly, thus leading to less time for nutrient processing and other biogeochemical processes.

Changes in the frequency and intensity of hurricanes impacts biogeochemical processes indirectly. Strong mixing of sediments occurs with storm passage, but this is transitory. There is high sediment deposition in wetlands during hurricanes (Baumann et al. 1984; Conner et al. 1989). These sediments can provide nutrients and iron as well as bulk up sediments to make wetland soils stronger (Day et al. 2011). Prolonged waterlogging owing to freshwater input can lead to weaker soils and make wetlands more susceptible to hurricanes (Howes et al. 2010). Morton and Barras (2011) reported that hurricanes commonly disturb fresher, more organic soils, and wetland recovery is related to hurricane frequency. Thus more frequent stronger hurricanes may retard or prevent wetland recovery.

7. Effects of climate change on delta vegetation

Species composition in the deltaic plain follows the estuarine gradient from saline marshes at the coast to fresh tidal marshes and swamps toward the upland boundary (Sasser et al. 2008; Fig. 17.2). However, many environmental factors that affect plant communities such as inundation frequency and duration, soil organic matter, soil fertility, and acidity are highly correlated with the estuarine gradient (Palmisano 1970; Grace and Pugesek 1997; Table 17.1). Over the last decade, significant changes in species composition have occurred in coastal marshes of the MRD.

The first map of the vegetation of this area was published by O'Neill (1949). This map focused on vegetation as potential habitat for muskrats and described the vegetation zones from the coast inland as follows:

- Salt to brackish marshes near the coast that are poor muskrat habitat and contain *Juncus roemerianus*, *Spartina patens*, and *Spartina alterniflora*, with occasional clumps of *Cladium jamaiscence*.
- Brackish *Schoenoplectus americanus* marshes that are burned every other year to control *Spartina patens*. This is the best muskrat habitat. At higher elevations, *Spartina cynosuroides* dominates.
- Floating *Schoenoplectus americanus* marshes that are almost as productive as the attached (brackish) marshes when they are healthy, but frequently are completely denuded and take a long time to recover.
- Fresh floating marshes dominated by *Panicum hemitomon*.
- Fresh attached marshes dominated by *Panicum hemitomon* with *Typha* spp., *Schoenoplectus californicus*, *Cladium jamaiscence*, *Sagittaria platyphylla*, and *Althernanthera philoxeroides*.

O'Neil described some strips along the Missisissippi River natural levees near New Orleans as intermediate between fresh and brackish as visually dominated by *Cladium jamaiscence*, *Phragmites australis*, *Typha* spp., and *Schoenoplectus californicus*, with an understory of *Schoenoplectus americanus*, *Spartina patens*, *Sagittaria lancifolia*, and *Spartina cynosuroides*. The marshes at the mouth of the Mississippi River were described by O'Neil as a mixture of *Typha* spp., *Phragmites australis*, *Schoenoplectus pungens*, *Panicum repens*, *Zizaniopsis milliacea*, *Spartina alterniflora*, and *Sagittaria latifolia*. O' Neil also mentioned that *Althernanthera philoxeroides* and *Eichornia crassipes* were rapidly invading throughout the coastal marshes and that that the recently introduced nutria (*Myocastor coypu*) had chosen freshwater marshes as their preferred habitat.

Chabreck (1972) documented the first comprehensive vegetation survey of the Louisiana coastal zone in 1968. These data were used to describe nine vegetation types within the deltaic plain (Visser et al. 1998). This analysis splits O'Neill's salt to brackish marshes into small strips of *Avicennia germinans* along coastal bay rims, saltmarshes dominated by *Spartina alterniflora* with patches of *Juncus roemerianus*, and brackish marshes that are a mixture of *Spartina alterniflora*, *Spartina patens*, and *Distichlis spicata*. O'Neil's brackish *Schoenoplectus americanus* marshes have converted to low species diversity *Spartina patens* marshes. O'Neil's floating *Schoenoplectus americanus* marshes are divided into a highly diverse *Spartina patens* dominated marsh with *Schoenoplectus americanus* as a subdominant, interspersed with a highly diverse mixture of *Sagittaria lancifolia*, *Cyperus* spp., and *Bacopa monnieri*. Fresh marshes remain dominated by *Panicum hemitomon*, but Visser et al. (1998) separates areas that are dominated by *Sagittaria lancifolia* with *Panicum hemitomon* as a codominant and areas along the natural levee flanks that are dominated by *Zizaniopsis milliacea*.

Since 1968, the fresh marshes have seen a dramatic change, with thin floating-mat marshes formed by *Eleocharis baldwinii* and *Hydrocotyle* spp. replacing vast expanses of *Panicum hemitomon* marshes (Visser et al. 1999). Most likely this change is due to grazing by nutria. A transplant study (Sasser et al. 2005) showed that *Panicum hemitomon* can grow when transplanted into the thin-mat marshes when it is protected from grazing. The shift to more saline community types during the 2000 drought (Visser et al. 2002) has reversed itself, but demonstrates that shifts up the basin can occur rapidly when salinity increases inland, as is expected with sea-level rise.

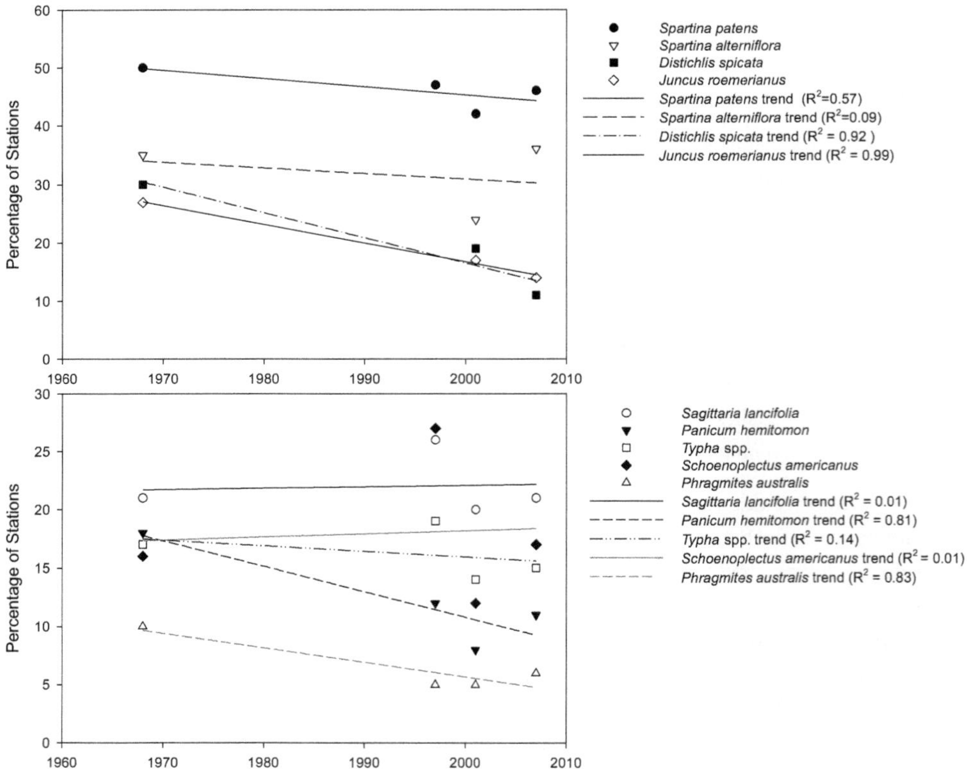

Figure 17.3. Historic changes in the most common plant species in the Mississippi River Deltaic Plain. Top graph shows trends in the brackish to saline species. Bottom graph shows changes in fresh to intermediate species.

Analysis of the occurrence on the coast-wide vegetation surveys for some of the most common species over the last four decades shows that *Spartina alterniflora*, *Sagittaria lancifolia*, *Typha* spp., and *Schoenoplectus americanus* have been relatively stable (Fig. 17.3). In contrast, *Distichlis spicata*, *Juncus roemerianus*, and *Panicum hemiton* have seen significant declines. Although *Spartina patens* and *Phragmites australis* have slightly decreased over time *Spartina patens*, *Distichlis spicata* and *Juncus roemerianus* are generally more sensitive to flooding and may reflect the increasing inundation with sea-level rise. The slight decrease over time in *Phragmites australis* is as of yet unexplained, as this is a very flood-tolerant and somewhat salt-tolerant species.

Except where delta building is occurring, the seaward migration of salt marsh vegetation is already constrained by flood duration, as is its productivity. Therefore, relative sea-level rise can only result in northerly migration of salt marsh, if it can successfully replace less salt tolerant habitat types. If so, no decrease in primary production should occur. However, fresh marshes and swamps subject to saltwater intrusion often transition to open water rather than to more salt-tolerant vegetation (Shaffer et al. 2009).

The MRD contains about 344,000 ha of baldcypress – water tupelo (*Taxodium distichum – Nyssa Aquatica*) swamp, most of which is in a state of deterioration (Chambers et al. 2005; Shaffer et al.

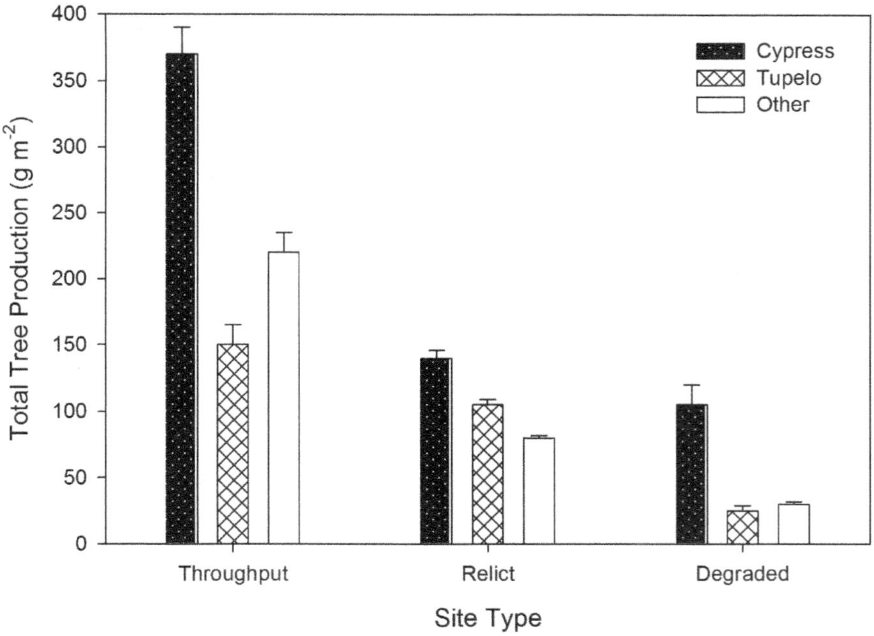

Figure 17.4. Primary production of swamp forest trees shown as a function of hydrology. Total primary production differed significantly between habitat types ($F2, 126 = 14.13$, $p < 0.00001$) decreasing from Throughput to Relic to Degraded habitats.

2009b). Within the forested wetlands, inundation plays an important role, with diverse bottomland hardwoods at higher elevations and low-diversity baldcypress–water tupelo swamps at lower elevations (Wharton et al 1982). As inundation increases, adult trees can survive for long periods, but regeneration can be low and is highly dependent on light availability (Denslow and Battaglia 1999). Modeling by Conner and Brody (1989) forecasted the slow decrease in basal area of *T. distichum* and *N. aquatica* in the swamp forest and replacement of bottomland hardwood species with swamp forest species as a result of increasing inundation. In some areas, there is a clear trajectory of swamp to marsh to open water (Shaffer et al. 2009; Fig. 17.4) with greater than 30% of the monitored swamp suffering mortality over the past 12 years. The primary cause of mortality in interior swamp is impoundment with stagnant nutrient-poor water, whereas swamp more proximal to the lake is deteriorating from saltwater intrusion (Shaffer et al. 2009). Continued sea-level rise is expected to exacerbate both of these unless restoration measures are initiated to increase freshwater throughput; without these restoration measures, much of the swamp in coastal Louisiana is expected to experience complete mortality with the next 20 years.

Fortunately, baldcypress–water tupelo swamps are amongst the most flood-tolerant wetland habitat types on earth, but only under reliable, nutrient-rich, sheet-flow conditions. The two major restoration measures that could sustain swamps are river diversions and assimilation wetlands. Assimilation wetlands are wetlands used for tertiary treatment of municipal wastewater. One assimilation wetland has sustained healthy swamp with elevation gains that exceeded sea level rise for more than 50 years

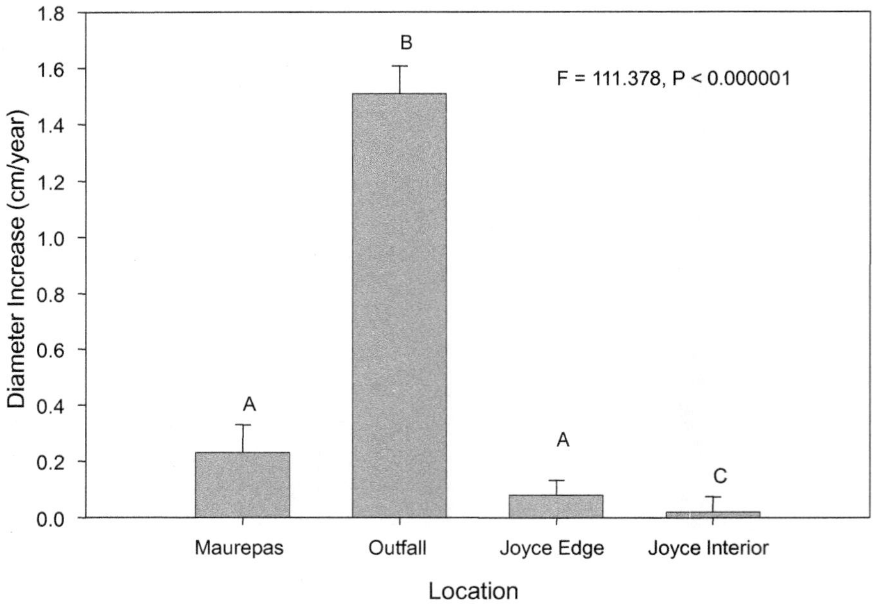

Figure 17.5. Diameter increase of mature baldcypress growing in a reference swamp (Maurepas), the immediate outfall area of an assimilation wetland, swamp at edge of assimilation wetland (Joyce edge), and interior of swamp adjacent to the assimilation wetland (Joyce interior).

(Day et al. 2004). Moreover, mature baldcypress at another coastal Louisiana assimilation wetland has growth rates five-fold greater than that of the adjacent swamp, despite an increase in average water level of greater than 20 cm (Fig. 17.5, Shaffer et al. in prep.).

The predicted increase in both frequency and intensity of hurricanes with climate change should amplify and expand restoration efforts directed at baldcypress–water tupeloforests of the MRD, as this habitat type is extremely resistant and resilient to hurricane damage (Williams et al. 1999; Chambers et al. 2007; Shaffer et al. 2009). Baldcypress–water tupelo swamps fared far better than other forest types in Hurricanes Camille (Touliatos and Roth 1971), Andrew (Doyle et al. 1995), and Hugo (Gresham et al. 1991; Putz and Sharitz 1991). In addition, fresh, oligohaline, and brackish marshes suffered vastly greater loss in Hurricanes Katrina and Rita than did baldcypress–water tupelo swamps (Barras 2006). Hurricane Katrina caused wind throw of an estimated 320 million bottomland hardwood trees, whereas contiguous swamps remained largely intact (Chambers et al. 2007).

Hurricanes also may select for recruitment of invasive species, such as *Triadica sebiferum* (Chambers et al. 2007; Chapman et al. 2008), which have high growth rates, yet are short-lived and themselves highly susceptible to wind throw from subsequent hurricane events. In addition to enhancing the dominance of invasive species, hurricanes could facilitate landward migration of vegetation relatively tolerant to sea-level rise by negative impacts to competitive dominants that would otherwise reduce invasive species through effects such as shading.

Souther-Effler (2004) produced several findings that may help predict future interactions of biotic and abiotic factors affecting forests impacted by rising sea levels. First, from a controlled study utilizing 2–4-year-old water tupelo saplings exposed to flooding, low-level (3 ppt) salinity, and insect herbivory, it was ascertained that defoliation reduced sapling productivity except when salt stress was an overriding factor. Salinity alone, in excess of 3 ppt over a prolonged period, was most detrimental and, when coupled with permanent flooding, resulted in high rates of sapling mortality. Second, the presence of nutrient enhancement, as one would find in a river diversion or assimilation wetland scenario, ameliorated the effects of baldcypress leafroller defoliation on sapling growth. Water tupelo growth, even with forest tent caterpillar defoliation, also was higher at nutrient-rich sites as long as the trees were not severely degraded by salinity. Thus it appears that insects will continue to act in concert with other stress factors to enhance the degradation of many forested wetlands unless depth and duration of flooding is reduced, or restoration measures are implemented to provide an influx of nutrient-rich fresh water.

The combination of salinity and flooding stress has greater effects than either alone, and the negative impacts increase with increasing salinity (Conner 1994; Allen et al. 1996). There is substantial intraspecific variability in salt tolerance within baldcypress species, suggesting that more salt-tolerant strains can be developed (Allen 1994; Allen et al. 1994; Krauss et al. 1998; Pezeshki et al. 1995; Krauss et al. 2000).

A series of models developed to evaluate coastal restoration projects (Louisiana Coastal Protection and Restoration Agency [CPRA] 2012) forecasts the effect of three different climate change scenarios (Table 17.2). Results from the vegetation model do not show a strong change in the salinity gradient in the next 50 years (Fig. 17.6). However, the model forecasts a decline in swamp-forest, which is most rapid under the less optimistic scenario and is mostly replaced by floating fresh marsh vegetation. Under all scenarios there is a slight expansion of saline marsh, but this comes at the expense of brackish and intermediate salinity vegetation types. This modeling effort included no trend in local rainfall and evaporation (Table 17.2), and therefore the freshwater inflow at the upland boundary is maintained, although with more variation in the less optimistic scenario.

7.1. Synergistic interaction of human impacts and climate change on the Mississippi River Delta

The area of bottom-water hypoxia in the northern Gulf of Mexico during the late spring and early summer is one of the largest in the world (Rabalais et al. 2002). Its size is influenced by the flux of nitrate from the MRB (Turner et al. 2006), which is strongly influenced by changes in streamflow, precipitation, and land-use practices that influence runoff (Donner and Scavia 2007; Goolsby and Battaglin 2001; McIsaac et al. 2001; Broussard and Turner 2009). This climate-driven variability in nitrate flux has been shown to be one of the primary factors influencing interannual variability in the size of the hypoxic zone (Donner and Scavia 2007; MRGOMWNTF 2008). As such, it is important to account for random variations in regional climate patterns and general trends in global climate patterns when assessing the impact of conservation practices on nutrient flux. Understanding climate variability and future changes will be paramount to this objective.

The complex interaction of nutrient concentrations, runoff coefficients from differing land cover classes, and changing precipitation patterns makes future projections challenging. Changing surface runoff patterns driven by changes in rainfall and land use could alter river, lake, and coastal ecosystems.

Table 17.2. *Future scenarios used for the 2012 master plan models (CPRA 2012)*

Factors	Plausible range over 50 years	Moderate value	Less optimistic value
Sea-level rise	0.12 m to 0.65 m of sea-level rise over 50 years	0.27 m of sea-level rise over 50 years	0.45 m of sea-level rise over 50 years
Subsidence (varies spatially)	0 to 35 mm/yr	0 to 19 mm/yr	0 to 25 mm/yr
Storm intensity	0% to +30%	+ 10% of current storm intensities	+ 20% of current storm intensities
Storm frequency	−20% to +10%	Current storm frequency (one Category 3 or greater storm every 19 years)	+ 2.5% of current storm frequency (one Category 3 or greater storm every 18 years)
River discharge / sediment load	−7% to + 14% (annual mean discharge, adjusted for seasonality)	534,000 cubic feet per second (annual mean)	509,000 cubic feet per second (−5% annual mean)
River nutrient concentration (nitrogen and phosphorus)	−45% to +20%	−12% of current concentrations	Current concentrations
Rainfall (varies spatially)	Historical monthly range	Variable percentage of historical monthly mean	Variable percentage of historical monthly mean
Evapotranspiration (varies spatially)	+/− 1 standard deviation of historical monthly range	Mean monthly values of the historical record	+0.4 Standard Deviation from historical mean monthly values
Marsh collapse threshold	Swamp salinity: 4–7 ppt Fresh marsh salinity: 6–8 ppt Intermediate marsh inundation: 31–38 cm Brackish marsh inundation: 20–26 cm Saline marsh inundation: 16–23 cm	Swamp: 6 ppt Fresh: 7 ppt Intermediate: 34 cm Brackish: 23 cm Saline: 21 cm	Swamp: 5 ppt Fresh: 7 ppt Intermediate: 33 cm Brackish: 21 cm Saline: 18 cm

If, for example, rainfall and runoff from urban and agricultural areas increase, the flux of fertilizer and other contaminants into lakes and coastal waters will likely increase, exacerbating the risk of eutrophication (Mullholland et al. 1997).

Changes in flooding patterns also could significantly affect river ecosystems. The pulsing erosion and deposition of floodplains is essential to the health of river ecosystems. If climate change increases the likelihood of intense rainfall events, humans may reduce flood risks by channelization of rivers and building dams, levees, and reservoirs. Maintaining floodplain habitats, however, may be a functional

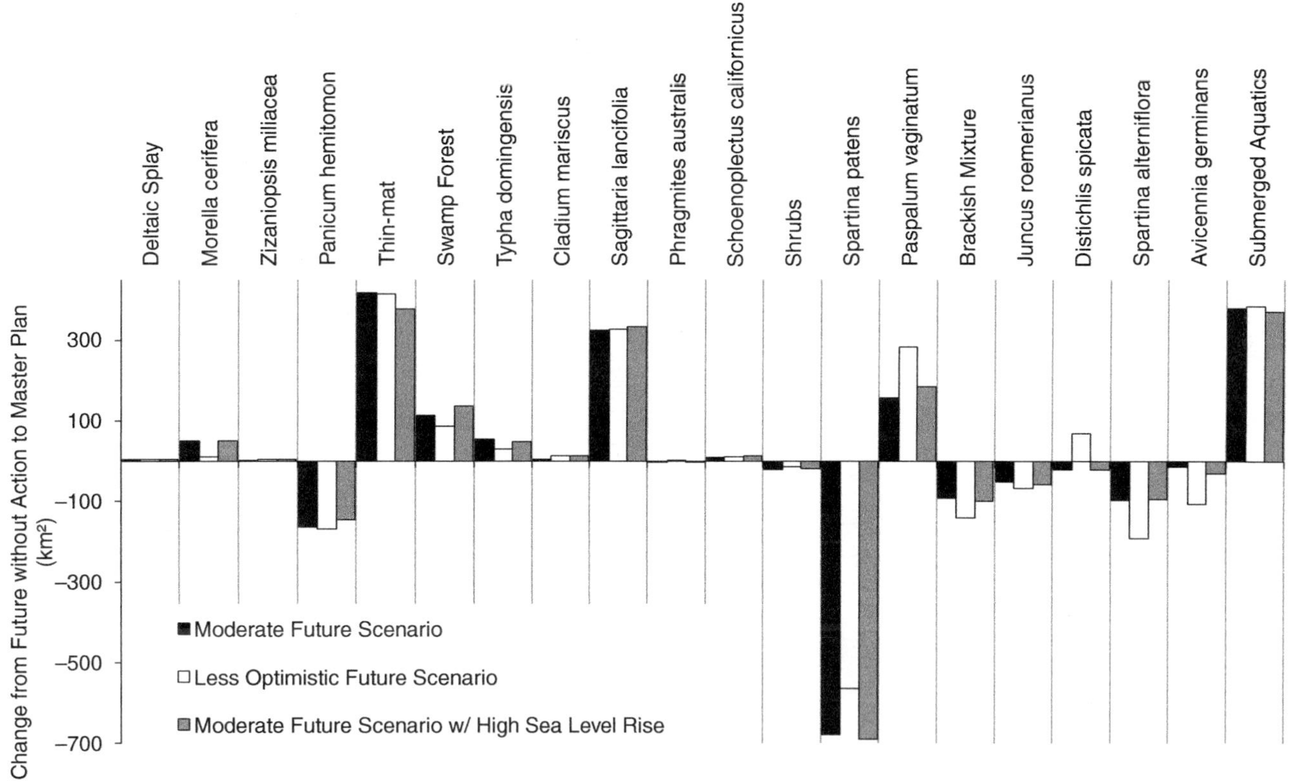

Figure 17.6. Projected change in emergent vegetation in the Mississippi River delta under three different future climate scenarios. Vegetation types are ordered from fresh to saline, except for submerged aquatic vegetation.

alternative that could receive floodwaters, slow the overbank flows, and maintain ecologically viable habitat (Twilley et al. 2001).

Understanding potential change and future scenarios is challenging. Two diverse opinions can be observed in the Hadley and Canadian models with regard to precipitation, evaporation, subsequent runoff, and riverine discharge. Discharge in natural watersheds is driven by the balance between precipitation and evaporation or evapotranspiration (Wolack and McCabe 1999). If evaporation and evapotranspiration increases in a changing climate due to increased temperature and productivity from CO_2 enrichment, freshwater runoff from regional rivers should decrease. If increases in precipitation exceed those losses, then runoff should increase. To this end, the Canadian model projects decreased stream runoff, whereas the Hadley model projects increased regional runoff in the next century. The Canadian model predicts drier soils and more drought-like conditions in the MRB, whereas the Hadley model predicts that much of the region will experience increased rainfall. The timing of peak runoff may change if snow cover decreases and snowmelt occurs earlier in the spring. Human factors also are important in runoff projections, such as the massive flood-control systems that have been engineered along the Mississippi River, which can significantly alter the relationship between rainfall upstream and runoff far downstream in the Gulf Coast. Several other case-specific studies suggest that precipitation and runoff from the greater MRB is likely to increase in the next century (Miller and Russel 1992; Giorgi et al. 1994; Justic et al. 1996). A 20% increase in discharge from the Mississippi River to the Northern Gulf of Mexico could increase the risk of the hypoxic "dead zone" and should be considered in restoration efforts to reintroduce riverine freshwater back to the natural delta system.

Although specific projections are impossible to make, changes in precipitation, evapotranspiration, and runoff will be extremely important for the MRD and the many ecosystems along the Gulf Coast that rely on freshwater from the Mississippi. The discrepancies in model results suggest that we must consider the effects of either an increase or a decrease in regional rainfall in the assessment of ecosystem impacts (Twilley et al. 2001).

It is generally accepted that the introduction of mineral sediment will build wetlands in the Mississippi Delta. However, questions remain about the unintended consequences or collateral effects of nutrients in the Mississippi River on the wetlands, effects that may contravene the benefits of sediment, at least in fragile floating marsh systems (Swarzenski et al. 2008; Howes et al. 2010). In estuaries where relative sea-level rise is rapid, as in the subsiding Mississippi Delta, the concentration of suspended sediment in floodwater and mineral sedimentation are critical to sustaining healthy wetlands (Day et al. 2000). There is strong evidence that freshwater and sediment diversions can be effective and beneficial for restoring wetlands in the Delta, especially if designed to introduce high levels of sediment. Additional research is needed before we fully understand the consequences of nutrients, particularly of nitrate, on the preservation of organic matter in sediment, but the need for action to restore the wetlands is urgent.

The cost of restoring the wetlands of the MRD recently has been estimated at $50 billion (CPRA 2012). The American Carbon Registry and Tierra Resources estimate that a substantial portion of this cost (up to $15 billion) may be provided through carbon sequestration credits (tierraresources.com, methodology WR-MF 2012). By using reliable sources of nutrient-rich freshwater, parts of degraded coastal wetlands can be converted to sustainable forested wetlands. These sources include river diversions, assimilation wetlands, and non-point sources such as storm water runoff. Rates of carbon sequestration may be as much as five-fold higher than in terrestrial systems because (1) net carbon loss reverses as wetlands convert from carbon sources in degraded wetlands to carbon sinks in restored wetlands, (2) burial of aboveground material and belowground production must match sea level rise to achieve sustainable elevations, (3) rates of decomposition are much lower in hydric soils than aerobic soils, and (4) species such as *Taxodium distichum* achieve maximal rates of above- and belowground production under fresh, nutrient-rich, throughput conditions. In addition, canopy species of forested wetlands live for centuries and are resistant to fire and hurricanes.

References

Allen, J. A. 1994. Intraspecific variation in the response of baldcypress (*Taxodiumdistichum*) seedlings to salinity. Dissertation. Louisiana State University.

Allen, J. A., J. L. Chambers, and D. McKinney. 1994. Intraspecific variation in the response of *Taxodiumdistichum* seedlings to salinity. Forest Ecol. Manag. 70: 203–214.

Allen, J. A., S. R. Pezeshki, and J. L. Chambers. 1996. Interaction of flooding and salinity stress on baldcypress (*Taxodiumdistichum*) Tree Physiol. 16: 307–313.

Bachand, P. A. M., and A. J. Horne. 2000. Denitrification in constructed free–water surface wetlands: I. Very high nitrate removal rates in a macrocosm study. Ecol. Eng. 14: 9–15.

Barras, J. A. 2006. Land area change in coastal Louisiana after the 2005 hurricanes – a series of three maps: U.S. Geological Survey Open–File Report 06-1274.

Barras, J. A., J. C. Bernier, R. A. Morton. 2008. Land area change in coastal Louisiana – a multidecadal perspective (from 1956 to 2006). Scientific Investigations Map 3019. USGS.

Baumann, R. H., J. W. Day, and C. A. Miller, 1984. Mississippi deltaic wetland survival: sedimentation vs. coastal submergence. Science 224: 1093–1095.

Benke, A. C., and C. E. Cushing. 2005. Rivers of North America. Elsevier Academic Press.

Betts, R. A., O. Boucher, M. Collins, P. M. Cox, P. D. Falloon, N. Gedney, D. L. Hemming, C. Huntingford, C. D. Jones, D. M. Sexton, and M. J. Webb. 2007. Projected increase in continental runoff due to plant responses to increasing carbon dioxide. Nature 448: 1037–1041.

Blum, M. D., and Roberts, H. H., 2009. Drowning of the Mississippi Delta due to insufficient sediment supply and global sea-level rise. Nat. Geosci. 2: 488–491.

Boesch, D. F., M. N. Josselyn, and A. J. Mehta, J. T. Morris, W. K. Nuttle, C. A. Simmenstad, and D. J. P. Swift. 1994. Scientific assessment of coastal wetland loss, restoration and management in Louisiana. J. Coast. Res. Special Issue 20.

Boynton, W. R., J. H. Garber, R. Summers, and W. M. Kemp. 1995. Inputs, transformations, and transport of nitrogen and phosphorus in Chesapeake Bay and selected tributaries. Estuaries 18: 285–314.

Brock, D. A. 2001. Nitrogen budget for low and high freshwater inflows, Nueces Estuary, Texas. Estuaries 24: 509–521.

Broussard, W., and R. E. Turner. 2009. A century of changing land-use and water-quality relationships in the continental US. Front. Ecol. Environ. 7: 302–307.

Broussard, W. P., III, R. E. Turner RE, and J. V. Westra. 2012. Do federal farm policies influence surface water quality? Agr. Ecosyst. Environ. 158: 103–109.

Callender, E., and D. Hammond. 1982. Nutrient exchange across the sediment-water interface in the Potomac River Estuary. Estuar. Coast. Shelf Sci. 15: 395–413.

Carlisle, D. M., D. M. Wolock, and M. R. Meador. 2010. Alteration of streamflow magnitudes and potential ecological consequences: a multiregional assessment. Front. Ecol. Environ. 9: 264–270.

Chabreck, R. H. 1972. Vegetation, water and soil characteristics of the Louisiana coastal region. LA AES Bull. 664.
Chambers, J. L., W. H. Conner, J. W. Day, S. P. Faulkner, E. S. Gardiner, M. S. Hughes, R. F. Keim, S. L. King, K. W. McLeod, C. A. Miller, J. A. Nyman, and G. P. Shaffer. 2005. Conservation, Protection and Utilization of Louisiana's Coastal Wetland Forests. Final Report to the Governor of Louisiana from the Coastal Wetland Forest Conservation and Use Science Working Group. http://www.coastalforestswg.lsu.edu/.
Chambers, J. Q., J. L. Fisher, H. Zeng, E. L. Chapman, D. B. Baker, and G. C. Hurtt. 2007. Hurricane Katrina's carbon footprint on U.S. Gulf coast forests. Science 318: 1107.
Chapman, E. L., J. Q. Chambers, K. F. Ribbeck, D. B. Baker, M. A. Tobler, H. C. Zeng, and D. A. White. 2008. Hurricane Katrina impacts on forest trees of Louisiana's Pearl River basin. Forest Ecol. Manag. 256: 883–889.
Conner, W.H. 1994. The effect of salinity and waterlogging on growth and survival of baldcypress and Chinese tallow seedlings. J. Coastal Res. 10: 1045–1049.
Conner, W. H., and M. Brody. 1989. Rising water levels and the future of southeastern Louisiana swamp forests. Estuaries 12: 318–323.
Conner, W. H., J. W. Day, R. H. Baumann, and J. W. Randall. 1989. Influence of Hurricanes on Coastal Ecosystems along the Northern Gulf Coast. Wetl. Ecol. Manag. 1: 45–56.
Cortwell, D. L. 1897. The delta of the Mississippi River. Natl. Geogr. Mag. 8: 351–354.
CPRA (Louisiana Coastal Protection and Restoration Agency). 2012. Louisiana's Comprehensive Master Plan for a sustainable coast. http://www.coastalmasterplan.louisiana.gov/2012–master–plan/final–master–plan/.
Davis, D. W. 1990. Living on the edge: Louisiana's marsh, estuary, and barrier island population. Trans. Gulf Coast Assoc. Geol. Soc. 40: 147–159.
Day, J. W. Jr., and P. H. Templet. 1989. Consequences of sea–level rise: Implications from the Mississippi Delta. Coast. Manag. 17: 241–257.
Day, R. H., R. K. Holz, and J. W. Day. 1990. An inventory of wetland impoundments in the coastal zone of Louisiana, USA: Historical trends. Environ. Manage. 14: 229–240.
Day, J. W., Jr., G. P. Shaffer, L. D. Britsch, D. J. Reed, S. R. Hawes, and D. Cahoon. 2000. Pattern and process of land loss in the Mississippi Delta: a spatial and temporal analysis of wetland habitat change. Estuaries 23: 425–438.
Day, J. W. J.–Y. Ko, J. Rybczyk, D. Sabins, R. Bean, G. Berthelot, C. Brantley, L. Cardoch, W. Conner, J.N. Day, A. J. Englande, S. Feagley, E. Hyfield, R. Lane, J. Lindsey, J. Mistich, E. Reyes, and R. Twilley. 2004. The use of wetlands in the Mississippi Delta for wastewater assimilation: A review. J. Ocean Coastal Manag. 47: 671–691.
Day, J. W., J. Barras, E. Clairains, J. Johnston, D. Justic, G. P. Kemp, J. Y. Ko, R. Lane, W. J. Mitsch, G. Steyer, P. Templet, and A. Yáñez–Arancibia. 2005. Implications of global climatic change and energy cost and availability for the restoration of the Mississippi Delta. Ecol. Eng. 24: 253–265.
Day, J. W., Jr., D. F. Boesch, E. J. Clarain, G. P. Kemp, S. B. Laska, W. J. Mitsch, K. Orth, H. Mashriqui, D. J. Reed, L. Shabman, C. A. Simenstad, B. J. Streever, R. R. Twilley, C. C. Watson, J. T. Wells, and D. F. Whigham. 2007. Restoration of the Mississippi Delta: lessons learned from hurricanes Katrina and Rita. Science 315: 1679–1684.
Day, J., R. Christian, D. Boesch, A. Yáñez–Arancibia, J. Morris, R. Twilley, L. Naylor, L. Schaffner, and C. Stevenson. 2008. Consequences of climate change on the ecogeomorphology of coastal wetlands. Estuaries Coasts 37: 477–491.
Day, J. W., J. E. Cable, J. H. Cowan, R. DeLaune, B. Fry, H. Mashriqui, D. Justic, P. Kemp, R. R. Lane, J. Rick, S. Rick, L. P. Rozas, G. Snedden, E. Swenson, R. R. Twilley, and B. Wissel.

2009. The impacts of pulsed reintroduction of river water on a Mississippi Delta coastal basin. J. Coast. Res. 54: 225–243.

Day, J. W, G. P. Kemp, D. J. Reed, D. R. Cahoon, R. M. Boumans, J. M. Suhayda, and R. Gambrell. 2011. Vegetation death and rapid loss of surface elevation in two contrasting Mississippi delta salt marshes: The role of sedimentation, autocompaction and sea-level rise. Ecolog. Eng. 37: 229–240.

Deegan, L. A., H. M. Kennedy, and C. Neill. 1984. Natural factors and human modifications contributing to marsh loss in Louisiana's Mississippi River Deltaic Plain. Environ. Manage. 8: 519–528.

Delaune, R. D., R. H. Baumann, and J. G. Gosselink. 1983. Relationships among vertical accretion, apparent sea–level rise and land loss in a Louisiana Gulf Coast marsh. J. Sediment. Petrol. 53: 147–157.

DeLaune, R., and J. White. 2012. Will coastal wetlands continue to sequester carbon in response to an increase in global sea-level rise? A case study of the rapidly subsiding Mississippi river deltaic plain. Climate Change. 110: 297–314.

Denslow, J. S., and L. L. Battaglia. 1999. Stand composition and structure across a changing hydrologic gradient: Jean Lafitte National Park, Louisiana, USA. Wetlands 22: 738–752.

Diaz, R. J. 2001. Overview of hypoxia around the world. J. Environ. Qual. 30: 275–281.

Donner, S. D., and D. Scavia. 2007. How climate controls the flux of nitrogen by the Mississippi River and the development of hypoxia in the Gulf of Mexico. Limnol. Oceanogr. 52: 856–861.

Dortch, Q., N. N. Rabalais, R. E. Turner, and N. A. Qureshi. 2001. Impacts of changing Si/N ratios and phytoplankton species composition, pp. 37–48. In N. N. Rabalais, and R. E. Turner, R.E. (eds.), Coastal Hypoxia: Consequences for Living Resources and Ecosystems. American Geophysical Union.

Doyle, T. W., B. D. Keeland, L. E. Gorham, and. D. J. Johnson. 1995. Structural impact of Hurricane Andrew on forested wetlands of the Atchafalaya Basin in south Louisiana. J. Coast. Res. 21: 354–364.

Dubrovsky, N. M., K. R. Burow, G. M. Clark, J. M. Gronberg, P. A. Hamilton, K. J. Hitt, D. K. Mueller, M. D. Munn, B. T. Nolan, L. J. Puckett, M. G. Rupert, T. M. Short, N. E. Spahr, L. A. Sprague, and W. G. Wilber. 2010. The quality of our nation's waters – nutrients in the nation's streams and groundwater, 1992–2004. USGS Circular 1350.

Emanuel. K. 2005. Increasing destructiveness of tropical cyclones over the last 30 years. Nature 436: 686–688.

Engstrom, D. R., J. E. Almendinger, and J. A. Wolin. 2009. Historical changes in sediment and phosphorus loading to the upper Mississippi River: mass–balance reconstructions from the sediments of Lake Pepin. J. Paleolimnol. 41: 563–588.

Environmental Protection Agency (EPA). 2008. Hypoxia in the Northern Gulf of Mexico: an update by the EPA Science Advisory Board. EPA Science Advisory Board Report 08–003. http://www.epa.gov/sab.

Feddema, J. J., K. W. Oleson, G. B. Bonan, L. O. Mearns, L. E. Buja, G. A. Meehl, and W. M. Washington. 2005. The importance of land-cover change in simulating future climates. Science 310: 1674–1678.

Froelich, P. 1988. Kinetic control of dissolved phosphate in natural rivers and estuaries: A primer on the phosphate buffer mechanism. Limnol. Oceanogr. 4: 649–668.

Galloway, J. N., W. H. Schlesinger, H. Levy II, A. Michaels, and J. L. Schnoor. 1995. Nitrogen fixation: anthropogenic enhancement–environmental response. Global Biogeochem. Cycles 9: 235–252.

Gedney, N., P. M. Cox, R. A. Betts, O. Boucher, C. Huntingford, and P. A. Stott. 2006. Detection of a direct carbon dioxide effect in continental river runoff records. Nature 439: 835–838.

Giorgi, F., C. S. Brodeur, and G. T. Bates. 1994. Regional climate change scenarios over the United States produced with a nested regional climate model. J. Climate 7: 375–400.

Good, B., J. Buchtel, D. Meffert, J. Radford, K. Rhinehart, and R. Wilson. 1995. Louisiana's major coastal navigation channels. Louisiana Department of Natural Resources http://www.ocpr.louisiana.gov/crm/D%;20R%;20S%;20Reports/General/La.%;20Navigation%;20Channels/Channels_1.pdf.

Goolsby, D. A. 2000. Nitrogen in the Mississippi River Basin – estimating sources and predicting flux to the Gulf of Mexico. USGS Fact Sheet 135–00. U.S. Department of the Interior.

Goolsby, D. A., and W. A. Battaglin. 2001. Long–term changes in concentrations and flux of nitrogen in the Mississippi River Basin, USA. Hydrol. Process. 15: 1209–1226.

Goolsby, D. A., W. A. Battaglin, B. T. Lawrence, R. S. Artz, B. T. Aulenback, R. P. Hooper, D. R. Keeney, and G. J. Stensland. 1999. Flux and sources of nutrients in the Mississippi–Atchafalaya River basin: topic 3 report of the integrated assessment on hypoxia in the Gulf of Mexico. Coastal Ocean Program Decision Analysis Series Number 17. NOAA Coastal Ocean Program.

Goolsby, D. A., W. A. Battaglin, B. T. Aulenback, and R. P. Hooper. 2000. Nitrogen flux and sources in the Mississippi River Basin. Sci. Tot. Environ. 248: 75–86.

Gosselink, J. G. 1984. The ecology of delta marshes of coastal Louisiana: a community profile. U.S. Fish and Wildlife Service FWS/OBS–84/09.

Grace, J. B., and B. H. Pugesek. 1997. A structural equation model of plant species richness and its application to a coastal wetland. Am. Nat. 149: 436–460.

Gresham, C. A., T. M. Williams, and D. J. Lipscomb. 1991. Hurricane Hugo wind damage to Southeastern U.S. coastal forest tree species. Biotropica 23: 420–426.

Gunternspergen, G. R., D. R. Cahoon, J. Grace, G. D. Steyer, S. Fournet, M. A. Townson, and A. L. Foote. 1995. Disturbance and recovery of the Louisiana coastal marsh landscape from the impacts of Hurricane Andrew. J. Coastal Res. 21: 324–339.

Harrison, R. W., and W. M. Kollmorgen. 1947. Past and prospective drainage reclamations in the coastal marshlands of the Mississippi River Delta. J. Land Public Utility Econ. 23: 297–320.

Holm, G. O., and C. E. Sasser. 2001. Differential salinity response between two Mississippi river subdeltas: Implications for changes in plant composition. Estuaries 24: 78–89.

Howes, N. C., D. M. FitzGerald, Z. J., Hughes, I. Y. Georgiou, A. Kulp, M. D. Miner, J. M. Smith, and J. A. Barras. 2010. Hurricane-induced failure of low salinity wetlands. Proc. Natl. Acad. Sci. 107: 14014–14019.

Hoyos, C., P. Agudelo, P. Webster, and J. Curry. 2006. Deconvolution of the factors contributing to the increase in global hurricane intensity. Sci. 312: 94–97.

Intergovernmental Panel on Climate Change. 2007. Climate Change 2007: The Scientific Basis, Contribution of Working Group 1 to the Third Assessment Report. Cambridge University Press, Cambridge, UK.

Jackson, R. B., S. R. Carpenter, C. N. Dahm, D. M. McKnight, R. J. Naiman, S. L. Postel, and S. W. Running. 2001. Water in a changing world. Ecol. Appl. 11: 1027–1045.

Jenkins, M. C., and Kemp, W. M., 1984. The coupling of nitrification and denitrification in two estuarine sediments. Limnol. Oceanogr. 29: 609–619.

Jitts, H. R. 1959. The adsorption of phosphate by estuarine bottom deposits. Aust. J. Mar. Fresh. Res.10: 7–21.

Justic, D., N. N. Rabalais, and R. E. Turner. 1996. Effects of climate change on hypoxia in coastal waters: a doubled CO_2 scenario for the northern Gulf of Mexico. Limnol. Oceanogr. 41: 992–1003.

Kesel, R. H. 1988. The decline of the suspended load of the lower Mississippi River and its influence on adjacent wetlands. Environ. Geol. Water Sci. 11: 271–281.

Kirwan, M. L., G. R. Guntenspergen, A. D' Alpaos, J. T. Morris, S. M. Mudd, and S. Temmerman. 2010. Limits on the adaptability of coastal marshes to rising sea level. Geophys. Res. Lett. 37: L23401.

Knox, J. C. 1987. Historical valley floor sedimentation in the Upper Mississippi Valley. Ann. Assoc. Am. Geogr. 77: 224–244.

Krauss, K. W., J. L. Chambers, and J. A. Allen. 1998. Salinity effects and differential germination of several half–sib families of baldcypress from different seed sources. New Forests 15: 53–68.

Krauss, K.W., J. L. Chambers, J. A. Allen, D. M. Soileau, Jr., and A. S. DeBosier. 2000. Growth and nutrition of baldcypress families planted under varying salinity regimes in Louisiana, USA. J. Coastal Res. 16: 153–163.

Lane, R. R., J. W. Day, and B. Thibodeaux. 1999. Water quality analysis of a freshwater diversion at Caernarvon, Louisiana. Estuaries 22: 327–336.

Lane, R. R., J. W. Day, D. Justic, E. Reyes, B. Marx, J. N. Day, and E. Hyfield. 2004. Changes in stoichiometric Si, N and P ratios of Mississippi River water diverted through coastal wetlands to the Gulf of Mexico. Estuar. Coast. Shelf Sci. 60: 1–10.

Lins, H. F., and J. R. Slack. 1999. Streamflow trends in the United States. Geophys. Res. Lett. 26: 227–230.

Lins, H. F., and J. R. Slack. 2005. Seasonal and regional characteristics of US streamflow trends in the United States from 1940 to 1999. Phys. Geogr. 26: 489–501.

Magilligan, F. J., and K. H. Nislow. 2005. Changes in hydrologic regime by dams. Geomorphology 71: 61–78.

McIsaac, G. F., M. B. David, G. Z. Gertner, and D. A. Goolsby. 2001. Nitrate flux in the Mississippi River. Nature 414: 166–167.

McIsaac. G. F., M. B. David, and G. Z. Gertner. 2002. Relating net nitrogen input in the Mississippi River Basin to nitrate flux in the lower Mississippi River: a comparison of approaches. J. Environ. Qual. 31: 1610–1622.

McIsaac, G.F., and X. Hu. 2004. Net N input and riverine N export from Illinois agricultural watersheds with and without extensive tile drainage. Biogeochemistry 70: 251–271.

Madden, C. J., J. W. Day, and J. M. Randall. 1988. Freshwater and marine coupling in estuaries of the Mississippi River deltaic plain. Limnol. Oceanogr. 33: 982–1004.

Millennium Ecosystem Assessment. 2005. Ecosystems and Human Well-Being: Synthesis. Island Press.

Meade, R. H. 1995. Setting:geology, hydrology, sediments, and engineering of the Mississippi River. In R. H. Meade (ed.), Contaminants in the Mississippi River. USGS Circular 1133. http://pubs.water.usgs.gov/cir1133.

Meade, R. H., and J. A. Moody. 2010. Causes for the decline of suspended–sediment discharge in the Mississippi River system, 1940–2007. Hydrol. Process. 24: 35–49.

Mendelssohn, I. A., and J. T. Morris. 2000. Eco–physiological constraints on the primary productivity of *Spartina alterniflora*. pp. 59–80. In M. P. Weinstein and D. A. Kreeger (eds.), Concepts and Controversies of Tidal Marsh Ecology. Kluwer Academic.

Meisinger, J. J., and G. W. Randall. 1991. Estimating crop nitrogen budgets for soil crop systems, pp. 85–124. In R. F. Follett, D. R. Keeney, and R. M. Cruse (eds), Managing Nitrogen for Groundwater Quality and Farm Profitability. Soil Science Society of America.

Miller, J. R., and G. L. Russel. 1992. The impact of global warming on river runoff. J. Geophys. Res. 97: 2757–2764.

Min, S., X. Zhang, F. Zwiers, and G. Hegerl. 2011. Human contribution to more-intense precipitation extremes. Nature 470: 378–381.

Morris, J. T., P. V. Sundareshwar, C. T. Nietch, B. Kjerve, and D. R. Cahoon. 2002 Responses of coastal wetlands to rising sea level. Ecol. 83: 2869–2877.

Morton, R. A., and J. A. Barras. 2011. Hurricane impacts on coastal wetlands: A half-century record of storm-generated features from southern Louisiana. J. Coastal Res. 27: 27–43.

Morton, R. A., J. C. Bernier, and J. A. Barras. 2006. Evidence of regional subsidence and associated interior wetland loss induced by hydrocarbon production, Gulf Coast region, USA. Environ. Geol. 50: 261–274.

Mississippi River/Gulf of Mexico Watershed Nutrient Task Force (MRGOMWNTF). 2008. Gulf Hypoxia Action Plan 2008 for Reducing, Mitigating, and Controlling Hypoxia in the Northern Gulf of Mexico and Improving Water Quality in the Mississippi River Basin.

Mulholland, P. J., G. R. Best, C. G. Coutant, and G. M. Hornberger. 1997. Effects of climate change on freshwater ecosystems of the Southeastern United States and the Gulf Coast of Mexico. Hydrol. Process. 11: 949–970.

Ning, Z. H., R. E. Turner, T. Doyle, and K. Abdollahi (eds.). 2003. Integrated assessment of the climate change impacts on the Gulf Coast region. U.S. EPA and USGS.

Nixon, S. W., J. R. Kelly, B. N. Furnas, C. A. Oviatt, and S. S. Hale. 1980. Phosphorus regeneration and the metabolism of coastal marine bottom communities, pp. 219–242. In K. R. Tenore and B. C. Coull (eds.), Marine Benthic Dynamics. University of South Carolina Press.

Nixon, S. W., J. W. Ammerman, L. P. Atkinson, V. M. Berounsky, G. Billen, W. C. Boicourt, W. R. Boynton, T. M. Church, D. M. Ditoro, R. Elmgren, J. H. Garber, A. E. Giblin, R. A. Jahnke, N. J. P. Owens, M. E. Q. Pilson, S. P. Seitzinger. 1996. The fate of nitrogen and phosphorus at the landesea margin of the North Atlantic Ocean. Biogeochemistry 35: 141–180.

Officer, C. B., and J. H. Ryther. 1980. The possible importance of silicon in marine eutrophication. Mar. Ecol. Prog. Ser. 3: 83–91.

O'Neil, T. 1949. The muskrat in the Louisiana coastal marshes. Louisiana Department of Wildlife and Fisheries.

Osborne, L. L., and D. A. Kovacic. 1993. Riparian vegetated buffer strips in water quality restoration and stream management. Freshwater Biol. 29: 243–258.

Pachauri, R. K, and A. Reisinger (eds.). Climate Change 2007: Synthesis Report. Contribution of Working Groups I, II and III to the Fourth Assessment Report of the Intergovernmental Panel on Climate Change. IPCC.

Palmisano, A. W. 1970. Plant community-soil relationships in Louisiana coastal marshes. Dissertation. Louisiana State University, Baton Rouge.

Patrick, W. H. Jr., and R. A. Khalid. 1974. Phosphate release and sorption by soils and sediments: effect of aerobic and anaerobic conditions. Science 186: 53–55.

Patrick, W. H., and R. D. DeLaune. 1977. Chemical and biological redox systems affecting nutrient availability in the coastal wetlands. Geosci. Man. 18: 131–137.

Penland, S., R. Boyd, J. R. Suter. 1988. Transgressive depositional systems of the Mississippi Delta plain; a model for barrier shoreline and shelf sand development. J. Sediment. Res. 58: 932–949.

Penland, S., A. D. Beall, and L. D. Britsch III. 2002. Geologic classification of coastal land loss between 1932 and 1990 in the Mississippi River Delta Plain, southeastern Louisiana. Gulf Coast Assoc. Geo. Soc. Trans. 52: 799–807.

Pezeshki, S. R., R. D. DeLaune, and H. S. Choi. 1995. Gas exchange and growth of baldcypress seedlings from selected U.S. Gulf Coast populations: responses to elevated salinities. Can. J. Forest Res. 25: 1409–1415.

Pimentel, D., J. Houser, E. Preiss, O. White, H. Fang, m L. Mesnick, T. Barsky, S. Tariche, J. Schreck, and S. Alpert. 1997. Water resources: agriculture, the environment, and society. BioScience 47: 97–106.

Poff, N. L., M. M. Brinson, and J. W. Day. 2002. Aquatic ecosystems & global climate change: potential impacts on inland freshwater and coastal wetlands ecosystems in the United States. Pew Center on Global Climate Change.

Putz, F. E., and R. R. Sharitz. 1991. Hurricane damage to old-growth forest in Conaree Swamp National Monument, South Carolina, USA. Can. J. Forest Res. 21: 1765–1770.

Rabalais, N. N., R. E. Turner, D. Justić, Q. Dortch, and W. J. Wiseman. 1999. Characterization of hypoxia: topic 1 report for the integrated assessment of hypoxia in the Gulf of Mexico. Coastal Ocean Program Decision Analysis Series No. 15. NOAA Coastal Ocean Program.

Rabalais, N. N, R. E. Turner, and W. J. Wiseman Jr. 2002. Gulf of Mexico hypoxia, a.k.a. "The dead zone". Annu. Rev. Ecol. Syst. 33: 235–263.

Rabalais, N. N., R. E. Turner, B. K. SenGupta, D. F. Boesch, P. Chapman, and M. C.Murrell. 2007. Hypoxia in the Northern Gulf of Mexico: does the science support the plan to reduce, mitigate, and control hypoxia? Estuar. Coasts 30: 753–772.

Rahmstorf, S. 2007. A semi–empirical approach to predicting sea-level rise. Sci. 315: 368–370.

Raymond, P. A., N.-W. Oh, R. E. Turner, W. Broussard. 2008. Anthropogenically enhanced fluxes of water and carbon from the Mississippi River. Nature 451: 449–452.

Redfield, A. C. 1958. The biological control of chemical factors in the environment.Am. Sci. 46: 205–221.

Revenga, C., S. Murray, J. Abramovitz, and A. Hammond. 1998. Watersheds of the World: Ecological Value and Vulnerability. World Resources Institute.

Sasser, C. E., J. G. Gosselink, E. M. Swenson, C. M. Swarzenski, and N. C. Leibowitz. 1996. Vegetation, substrate and hydrology in floating marshes in the Mississippi river delta plain wetlands, USA. Vegetatio 122: 129–142.

Sasser, C. E., G. O. Holm, J. M. Visser, and E. M. Swenson. 2005. Thin-mat floating marsh enhancement demonstration project TE-36. Louisiana Department of Natural Resources. http://lacoast.gov/reports/co/TE-36%;20FINAL%;20REPORT.pdf.

Sasser, C .E., J. M. Visser, E. Mouton, J. Linscombe, and S. B. Hartley. 2008. Vegetation types in coastal Louisiana in 2007: U.S. Geological Survey Open-File Report 2008–1224, 1 sheet, scale 1: 550,000.

Scavia, D., J. C. Field, D. F. Boesch, R. W. Buddemeier, V. Burkett, D. R. Cayan, M. Fogarty, M. A. Harwell, R. W. Howarth, C. Mason, D .J. Reed, R. C. Royer, A. H. Sallenger, and J. G. Titus. 2002. Climate change impacts on U.S. coastal and marine ecosystems. Estuaries 25: 149–164.

Schilling, K. E. 2005. Relation of baseflow to row crop intensity in Iowa. Agric. Ecosyst. Environ. 105: 433–438.

Shaffer, G. P., W. B. Wood, S. S. Hoeppner, T. E. Perkins, J. A. Zoller, and D. Kandalepas. 2009. Degradation of baldcypress–water tupelo swamp to marsh and open water in southeastern Louisiana, USA: an irreversible trajectory? J. Coastal Res. 54: 152–165.

Sharp, J., C. Culberson, and T. Church. 1982. The chemistry of the Delaware estuary. General considerations. Limnol. Oceanogr. 27: 1015–1028.

Smith, C. J., R. D. DeLaune, W. H. Patrick. 1985. Fate of riverine nitrate entering an estuary: I. Denitrification and nitrogen burial. Estuaries 8: 15–21.

Souther-Effler, R. F. 2004. Interactions of herbivory and multiple abiotic stress gents on two wetland tree species in southeast Louisiana. Dissertation, Louisiana State University.

Sprague, L. A., D. K. Mueller, G. E. Schwarz, and D. L. Lorenz. 2009. Nutrient trends in streams and rivers of the United States, 1993–2003: USGS Scientific Investigations Report 2008–5202.

Sprague, L. A., R. M., Hirsch, and B. T. Aulenbach. 2011. Nitrate in the Mississippi River and its tributaries, 1980 to 2008: are we making progress? Environ. Sci. Technol. 45: 7209–7216.

Swarzenski, C. M., T. W. Doyle, B. Fry, and T. G. Hargis. 2008. Biogeochemical response of organic-rich freshwater marshes in the Louisiana delta plain to chronic river water influx. Biogeochem. 90: 49–63.

Tarasona, J., W. E. Arntz, and E. C. Maruenda (eds.) 2001. El Niño en América Latina: Impactos Biológicos y Sociales. El Niño in Latin America: Biologic and Social Impacts. Consejo Nacional de Ciencia y Tecnología.

Titus, J., R. Park, S. Leatherman, J. Weggel, M. Green, P. Mausel, S. Brown, C. Gaunt, M. Trehan, and G. Yohe. 1991. Greenhouse effect and sea–level rise: potential loss of land and the cost of holding back the sea. Coast. Manage. 19: 171–204.

Touliatos, P., and E. Roth. 1971. Hurricanes and trees – Ten lessons from Camille. J. Forest. 69: 285–289.

Turner, R. E., and N. N. Rabalais. 1994. Coastal eutrophication near the Mississippi river delta. Nature 368: 619–621.

Turner, R. E., and N. N. Rabalais. 2003. Linking landscape and water quality in the Mississippi River Basin for 200 years. BioScience 53: 563–572.

Turner, R. E., J. J. Baustian, E. M. Swenson, and J. S. Spicer. 2006a. Wetland sedimentation from hurricanes Katrina and Rita. Sci. 314: 449–452.

Turner, R. E., N. N. Rabalais, and D. Justic. 2006b. Predicting summer hypoxia in the northern Gulf of Mexico: Riverine N, P, and Si loading. Mar. Pollut. Bull. 52: 139–148.

Turner, R. E., N. N. Rabalais, and D. Justic. 2008. Gulf of Mexico hypoxia: alternate states and a legacy. Environ. Sci. Tech. 42: 2323–2327.

Tweel, A. W., and R. E. Turner. 2012. Watershed land use and river engineering drive wetland formation and loss in the Mississippi River birdfoot delta. Limnol. Oceanogr. 57: 18–28

Twilley, R. R., E. J. Barron, H. L. Gholz, M. A. Harwell, R. L. Miller, D. J. Reed, J. B. Rose, E. H. Siemann, R. G. Wetzel, and R. J. Zimmerman. 2001. Confronting climate change in the Gulf Coast region: prospects for sustaining our ecological heritage. Union of Concerned Scientists and Ecological Society of America.

U.S. Department of Agriculture (USDA). 1987. Farm drainage in the United States: history, status and prospects. USDA Misc. Pub. No. 1455.

Vermeer, M., and S. Rahmstorf. 2009. Global sea level linked to global temperature. Proc. Nat. Acad. Sci. 106: 21527–21532.

Visser, J. M., C. E. Sasser, R. H. Chabreck, and R. G. Linscombe. 1998. Marsh vegetation types of the Mississippi River Deltaic Plain, USA. Estuaries 21: 818–828.

Visser, J. M., C. E. Sasser, R. H. Chabreck, and R. G. Linscombe. 1999. Long-term vegetation change in Louisiana tidal marshes, 1968–92. Wetlands 19: 168–175.

Visser, J. M., C. E. Sasser, R. H. Chabreck, and R. G. Linscombe. 2002. The impact of a severe drought on the vegetation of a subtropical estuary. Estuaries 25: 1185–1196.

Vitousek, P. M., H. A. Mooney, J. Lunchenco, and J. M. Melillo. 1997. Human domination of the Earth's ecosystems. Sci. 277: 494–499.

Wang, H., Steyer G. D., Piazza S. C., Holm G. O., Stagg C. L., Rybczyk J. M., Fischenich C. J., Couvillion B. R., Boustany R. G., Fischer M. R., and Sharp L. A. 2012. Horizontal and Vertical Variability in Soil Bulk Density and Organic Matter Across Coastal Louisiana Wetlands Detected by the Coast-Wide Reference Monitoring System (CRMS) – Wetlands. Poster presented at the State of the Coast Conference, June 25–27, New Orleans, LA.

Webster, J., G. J. Holland, J. A. Curry, and H.-R. Chang. 2005. Changes in tropical cyclone number, duration, and intensity in a warming environment. Science 309: 1844–1846.

Wharton, C. H., W. M. Kitchens, E. C. Pendleton, and T. W. Sipe. 1982. Ecology of bottomland hardwood swamps of the southeast: a community profile. FWS/OBS–81/37 U.S. Dept. of the Interior.

Williams, K., Z. S. Pinzon, R. P. Stumpf, and E. A. Raabe. 1999. Sea-level rise and coastal forests on the Gulf of Mexico. USGS–99–441.

Wilkinson, C. R., and R. W. Buddemeier. 1994. Global climate change and coral reefs: implications for people and reefs. Report of the UNEP–IOC–ASPEI–IUCN Global task Team on the Implications of Climate Change on Coral Reefs. IUCN.

Wolack, D. M., and G. J. McCabe. 1999. Simulated effects of climate change on mean annual runoff in the conterminous United States. In Proceedings, Specialty Conference on Potential Consequences of Climate Variability and Change to Water Resources of the United States. American Water Resources Association.

Yáñez-Arancibia, A., and J. W. Day. 2005. Ecosistemas vulnerables, riesgo ecológico y el record 2005 de huracanes en el Golfo de México y Mar Caribe. URL: http://www.ine.gob.mx/download/huracanes2005.pdf.

Yuill, B., D. Lavoie, and D. J. Reed. 2009. Understanding subsidence processes in coastal Louisiana. J. Coast. Res. 54: 23–36.

Zhang, Y. K., and K. E. Schilling. 2006. Increasing streamflow and baseflow in Mississippi River since the 1940s: effect of land use change. J. Hydrol. 324: 412–422.

18

Nutrient and carbon dynamics in a large river-dominated coastal ecosystem: the Mississippi-Atchafalaya River system

S. E. Lohrenz, W.-J. Cai, S. Chakraborty, K. Gundersen, and M. C. Murrell

1. Introduction

Large rivers have major impacts on biogeochemical processes in adjacent coastal margins. These systems introduce large quantities of freshwater and terrestrially derived organic and inorganic constituents into coastal environments. Large river-influenced coastal systems are characteristically high in biological productivity and exhibit intense biogeochemical processing of materials (Dagg et al. 2004; Mckee et al. 2004). Large rivers are also sensitive to both natural and anthropogenic impacts on watersheds and thus are subject to changing climatic conditions (Raymond and Cole 2003; Justic et al. 2005; Bianchi and Allison 2009).

The extent of the impact of large rivers on coastal ecosystems has been debated regarding whether river influence is primarily regionally localized or can extend well beyond the coastal margins to the open ocean (Longhurst 1993, 1995; Müller-Karger et al. 1995; Hu et al. 2004; Raymond et al. 2008). However, a variety of evidence suggests that the impact of large river inputs to coastal margins can extend over large regions of the ocean well beyond the immediate area of discharge (Lee et al. 1993; Ortner et al. 1995; Lie et al. 2003; Del Vecchio and Blough 2004; Cooley and Yager 2006). Moreover, there is increasing evidence that human impacts on watersheds contribute substantially to changes in nutrient and carbon export, with associated consequences for coastal water quality and ecosystem function (Rabouille et al. 2001; Rabalais et al. 2009). Large river outflow regions typically coincide with productive fisheries (Darnell 1990; Grimes 2001; Cowan et al. 2008) and thus are believed to contribute to environmental conditions favorable for fish production and habitat. However, our understanding of the basis for linkages between high primary productive and production at higher trophic levels remains limited, as does the ability to model and predict the behavior of these systems (Holt et al. 2009). In addition, large rivers are sites of massive fluxes of carbon, nutrients, and sediments to ocean margins (Bianchi et al. 2007a; Cai and Lohrenz 2010). Although the fate of these materials is not fully understood, much of the terrestrially derived organic material is believed to be remineralized in coastal margins (Berner 1989; Hedges and Kiel 1995). Paradoxically, river-influenced regions, and coastal margins in general, can also be strong sinks for atmospheric carbon dioxide (Chen 2004; Chen and Borges 2009; Cai 2011) (also see Chapter 7).

Although the outflow regions of many large rivers have undergone extensive study, questions remain regarding the processing and fate of materials entering the coastal margins and the overall extent of river influences on coastal and ocean processes. Here, we examine nutrient and carbon dynamics in the Mississippi-Atchafalaya River–influenced coastal ecosystem of the northern Gulf of Mexico. Specifically, we focus on nutrient and carbon dynamics and associated linkages to primary and secondary productivity in the river-influenced region of the northern Gulf of Mexico. We subsequently review the relationships between productivity, oxygen dynamics, and recurring hypoxia and consider implications of projected changes in climate for these relationships. Finally, we examine recent work describing the potential enhancement of ocean acidification in regions of river-influenced hypoxia.

2. Overview of Mississippi River and northern Gulf of Mexico coastal ecosystem

The Mississippi-Atchafalaya river system (MARS) is among the 10 largest rivers in the world with respect to water and sediment discharge and the largest in the North American continent (Milliman and Meade 1983; Meade 1995; Dagg et al. 2004; Meade and Moody 2010). The Mississippi-Atchafalaya River basin drains about 41% of the conterminous United States (Goolsby et al. 2001), and its watershed extends into Canada. The combination of dams, other engineering structures, and erosion abatement along the length of the Mississippi River and its largest tributary, the Missouri River, has resulted in a long-term decline in the sediment load delivered by the river (Bianchi and Allison 2009; Meade and Moody 2010). In contrast, agricultural practices have contributed to increased riverine nitrate-nitrogen concentrations (Goolsby et al. 2000; Broussard and Turner 2009; Raymond et al. 2012). In addition, increases in bicarbonate and water fluxes over the past 50 years in the Mississippi River have been attributed to anthropogenic drivers (Raymond and Cole 2003; Raymond et al. 2008). The MARS also represents a large and concentrated source of terrestrial organic carbon to the North American continental margin (Bianchi et al. 2004; Wang et al. 2004; Bianchi et al. 2007b; Cai and Lohrenz 2010).

The outflow of the Mississippi River occurs largely through a series of passes in the Belize ("birdfoot") delta (Fig. 18.1). Flow through the Atchafalaya River is managed so as to deliver 30% of the combined flows of the Red River and Mississippi River (Ford and Nyman 2011; see also Chapter 16). Discharge of the MARS varies seasonally, with highest discharge generally occurring in the late spring (Fig. 18.2). The freshwater discharge from the major passes in the Belize delta can be large enough and have high enough velocity that a distinct freshwater layer can extend well out onto the shelf and beyond. In contrast, the Atchafalaya plume undergoes substantial mixing with adjacent shelf water and does not generally retain a distinct jet feature. The Atchafalaya and Mississippi Rivers also differ in the extent to which exchanges occur between the adjacent watershed and the river along its lower reaches. The Mississippi River is highly engineered and constrained by levees. In contrast, the lower reaches of the Atchafalaya are relatively unmanaged and undergo substantial interactions with the river. These very different dynamics influence the composition of organic matter (Shen et al. 2012) and nutrients (see next section).

The Mississippi-Atchafalaya basin will likely continue to undergo substantial anthropogenic and natural changes. The contributed upland areas of the Gulf of Mexico have experienced large-scale land cover and land use change since the mid-1800s (Chen et al. 2006) and will likely undergo further rapid development in the coming years (Alig et al. 2003). Terrestrial land cover and land use are major factors influencing the delivery of sediment (Vörösmarty et al. 2003), organic carbon (Meybeck et al. 2003), dissolved inorganic carbon (DIC) (Cai et al. 2008). and other constituents from land to rivers and eventually to the ocean (Ciais et al. 2008). Moreover, changes in climate will contribute further

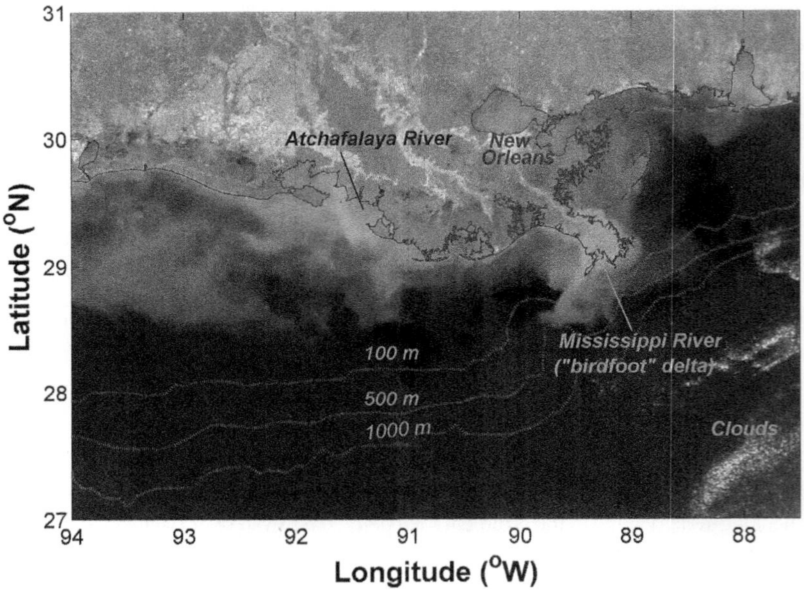

Figure 18.1. Grayscale image of the Mississippi and Atchafalaya outflow regions from the Moderate-Resolution Digital Imaging Spectrometer (MODIS) on the Aqua satellite from April 14, 2009 (image data acquired from NASA, http://oceancolor.gsfc.nasa.gov).

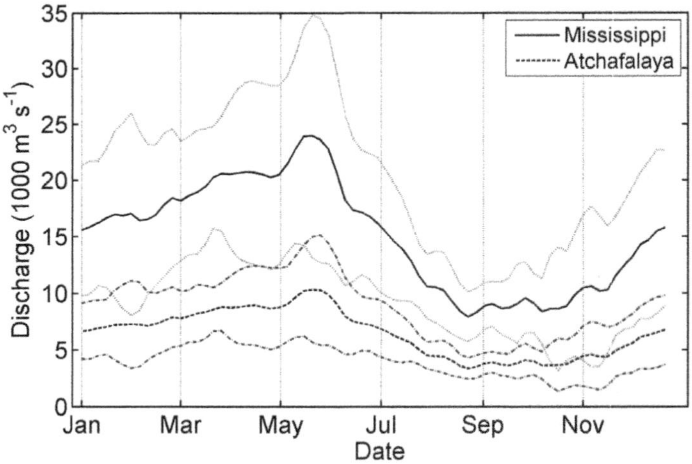

Figure 18.2. Mean discharge from the Mississippi and Atchafalaya rivers for the period 2003–2011 (U.S. Army Corps of Engineers). Discharge data were from Tarbert Landing, MS (Mississippi River, Gage ID 01100), and Simmesport, LA (Atchafalaya River, Gage ID 3045) sites. See Chapter 16 for map showing monitoring sites. Dotted (Mississippi) and dash-dot (Atchafalaya) lines indicate one standard deviation above or below the means.

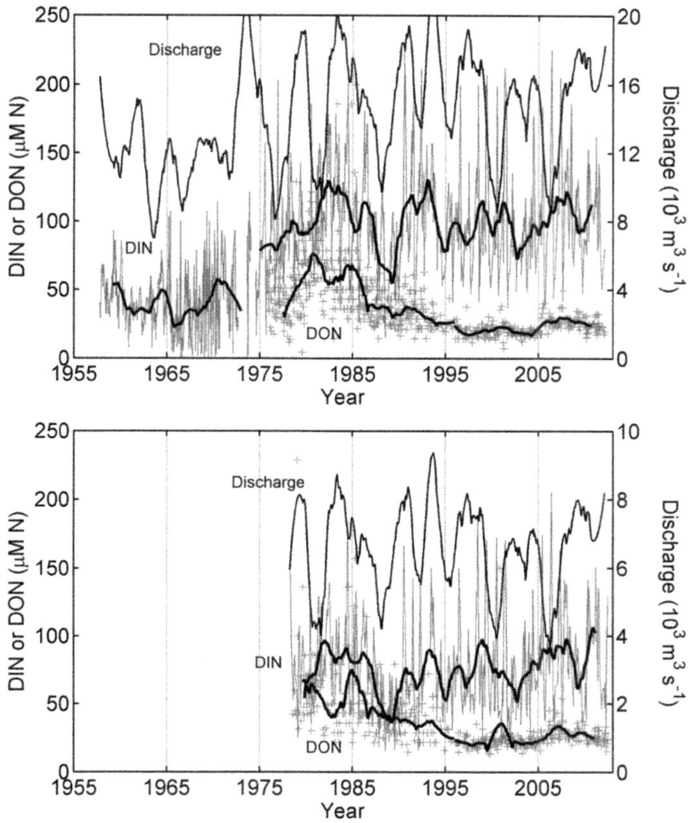

Figure 18.3. Time series of DIN (light gray line) and DON (crosses) for the lower Mississippi River at St. Francisville, LA (upper panel), and Atchafalaya River at Melville, LA (lower panel), are shown in relation to river discharge (upper black lines in each panel). See Chapter 16 for map showing monitoring sites. Note difference in discharge scale between the upper and lower panels. The river discharge was smoothed using a Savitsky-Golay second-order polynomial filter with a 35-month window. Thick black lines represent smoothed trend lines of the nutrient concentration data using the same method. Nutrient data were provided courtesy of United States Geological Survey from the St. Francisville (water quality station number 07373420) and Melville (water quality station number 07381495) NASQAN sites. See map in Chapter 16 for locations of monitoring sites.

to changes in watershed processes that impact biogeochemical processes and ecosystem dynamics on coastal margins (Justić et al. 2005; see also Chapters 16 and 17). In the following section, we focus on changes in nutrients and carbon delivery from the river to the continental margin and subsequently examine the ecosystem responses to these large inputs of terrestrially derived materials.

3. Retrospective analysis of changes in nutrients and carbon in the Mississippi and Atchafalaya Rivers

Nitrogen, specifically nitrate+nitrate (hereafter referred to as DIN) concentrations in the Mississippi River have increased over the past century (Fig. 18.3, upper panel), with the most dramatic increases seen between 1970 and 1983 and large year-to-year fluctuations since the 1980s (Goolsby et al. 1999; Goolsby et al. 2000; Lohrenz et al. 2008; Cai and Lohrenz 2010). The historical record for DIN concentrations extends only back to the late 1970s for the Atchafalaya River (Fig. 18.3, lower panel). DIN concentrations ranged significantly higher in the Mississippi River as compared with the Atchafalaya River over the period 1980–2011 (Mann-Whitney *U* test, $p<0.001$), possibly a reflection of the more channelized nature of the lower Mississippi River. Dissolved organic nitrogen (DON) concentrations for both the Mississippi and Atchafalaya rivers were comparable and were characterized by a decrease from the early 1980s to the present (Fig. 18.3).

Total phosphorus (Total P) as well as orthophosphate concentrations were significantly (Mann-Whitney *U* test, $p<0.001$) higher over the period from 1980–2011 for the Atchafalaya River (Fig. 18.4, lower panel) as compared with the Mississippi River (Fig. 18.4, upper panel). The higher phosphorus concentrations are likely a consequence of differences in the hydrology of the two river systems. The less channelized nature of the Atchafalaya River and the contributing Red River would result in greater exchange between river and adjacent wetlands and associated sediments, thereby contributing to higher levels of particle reactive phosphorus. In addition to the higher range in phosphorus concentrations in the Atchafalaya River, there has been an increasing trend in both TP (Kendall's tau=0.17; $p<0.001$) and orthoP (Kendall's tau=0.13; $p=0.0011$) since 1990. Reasons for this are not clear, but may be attributed to the fact that the Atchafalaya River delta system is accreting, and consequently, the interaction between river and the surrounding watershed is increasing.

The long-term changes in DIN concentrations in the Mississippi River (Fig. 18.3) have resulted in an increase in flux of nitrogen into the Gulf of Mexico (Fig. 18.5 and Goolsby et al. 1999). The mean daily N flux for the period from 2000–2011 determined using the Weighted Regressions on Time, Discharge, and Season (WRTDS) method (Hirsch et al. 2010) was 1.73 10^6 kg N d^{-1} (0.63 Tg y^{-1}) from the Mississippi River and 0.60 10^6 kg N d^{-1} (0.22 Tg y^{-1}) or a total MARS DIN flux of 0.85 Tg y^{-1}. This is slightly lower than the estimate by Goolsby et al. (1999) of 0.95 Tg y^{-1} for the period from 1980–1996. A substantial fraction of this increase included dissolved organic nitrogen during the late 1970s and early 1980s, but this has since been reduced (Fig. 18.3). Estimated annual flux of DON from the MARS in this study was 0.19 Tg y^{-1}, down from the value of 0.38 reported by Goolsby et al. (1999). The earlier increases were attributed to human-related activities, largely agricultural fertilizer use and other sources contributing to net anthropogenic nitrogen inputs (NANI) (Goolsby and Battaglin 2001; Mcisaac et al. 2002; Turner and Rabalais 2003; Donner et al. 2004; Turner and Rabalais 2004; Alexander et al. 2008; Donner and Kucharik 2008; Raymond et al. 2012). Raymond et al. (2012) provided evidence that in addition to fertilizer inputs, increased use of certain agricultural practices such as row cropping and tile drainage have increased water throughput, thereby decreasing nitrogen residence time and processing in the watershed. In addition, over the last 50 years, there has been a long-term trend of increasing discharge from the Mississippi River (Raymond et al. 2012), which is also illustrated in Figure 18.5.

Nutrient ratios have been shown to vary seasonally in this system, with high N:P ratios associated with higher discharge (Lohrenz et al. 1999; Turner et al. 2007; Lohrenz et al. 2008; Cai and Lohrenz 2010). Ratios of DIN to orthophosphate in the Mississippi River (Fig. 18.6, upper panel) tend to

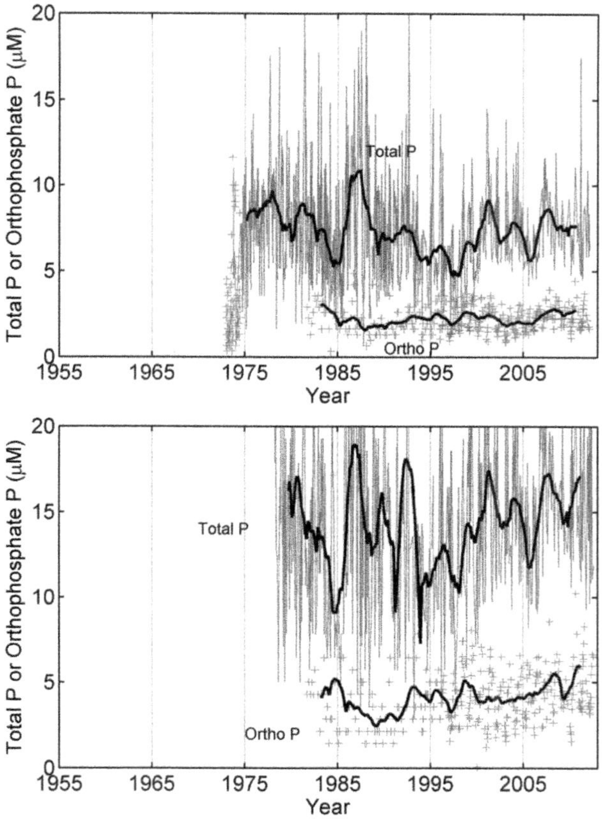

Figure 18.4. Time-series of Total P (gray lines) and orthophosphate (crosses) for the lower Mississippi River at St. Francisville LA (upper panel) and Atchafalaya River at Melville, LA (lower panel). Thick black lines represent smoothed trend lines of the nutrient concentration data using a Savitsky-Golay second-order polynomial filter with a 35-month window. Nutrient data were provided courtesy of United States Geological Survey. There was a statistically significant increasing trend for both Total P ($r^2 = 0.0.0173$; N = 301; p = 0.022) and orthophosphate ($r^2 = 0.0.0349$; N = 276; p = 0.0018) in the Atchafalaya River since 1990.

exceed the Redfield ratio of 16:1, which is representative of the compositional ratios typically found in marine organisms (Redfield 1958). The ratio of DIN to Total P, however, fluctuated around the Redfield N:P value of 16:1. In contrast, ratios of DIN to orthophosphate in the Atchafalaya River (Fig. 18.6, lower panel) were comparable to the Redfield N:P ratio, whereas the ratio of DIN to Total P tended to be lower than the Redfield N:P. The differences in nutrient ratios between these river systems probably reflect a higher degree of biological processing of nutrients within the less channelized Atchafalaya River as compared with the Mississippi River.

Studies examining long-term changes in ratios document a decrease in dissolved silica concentrations relative to dissolved nitrogen and phosphorus, with values approaching Redfield proportions

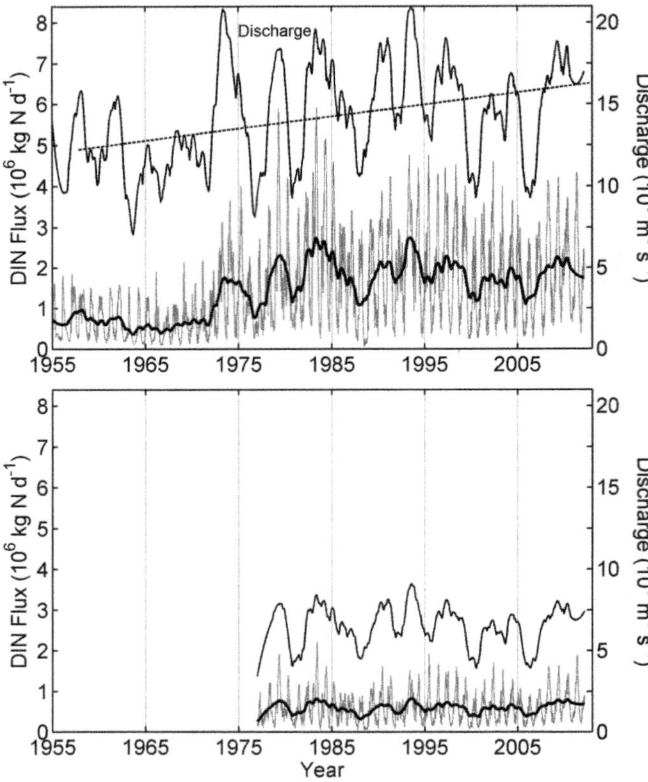

Figure 18.5. Upper panel: Mississippi River discharge at Tarbert Landing (U.S. Army Corps of Engineers) in comparison with dissolved inorganic nitrogen flux (gray line) based on measurements at St. Francisville. DIN flux was estimated by using the Weighted Regressions on Time, Discharge, and Season (WRTDS) method (Hirsch et al. 2010). The upper black line is river discharge smoothed as in Fig. 18.3. The heavy black line is the smoothed DIN flux as in Fig. 18.3. The dashed line is a linear regression fit to the discharge data ($r^2 = 0.135$; $p = 0.001$). The slope of the regression was 74.8 m^3 s^{-1} yr^{-1}. The statistical significance of this trend was also indicated by Kendall's tau of 0.11 ($p < 0.001$). Lower panel: Atchafalaya River discharge at Simmesport, LA, in comparison with dissolved inorganic nitrogen flux based on measurements at Melville.

(Justic et al. 1995; Rabalais et al. 1996; Goolsby et al. 1999). It has been speculated that such changes may have consequences for phytoplankton community composition, as reduced silicate levels may favor non–diatom-dominated communities and alter the dynamics of coastal food webs (Rabalais et al. 1996; Turner et al. 1998). However, Goolsby et al. (1999) reported that there had not been a significant decrease in annual flux of dissolved silica for the period 1955–1996. The lack of a significant trend despite long-term decreases in dissolved silica concentrations was argued to be a consequence of increased streamflow resulting in a dilution of silicate concentrations. Estimates of mean dissolved Si fluxes from the MARS (Table 18.1) determined in this study for the period 2000–2011 (1.91 Tg y^{-1}) were slightly lower than that reported by Goolsby et al. (1999) for the period 1980–1996 (2.32 Tg y^{-1}),

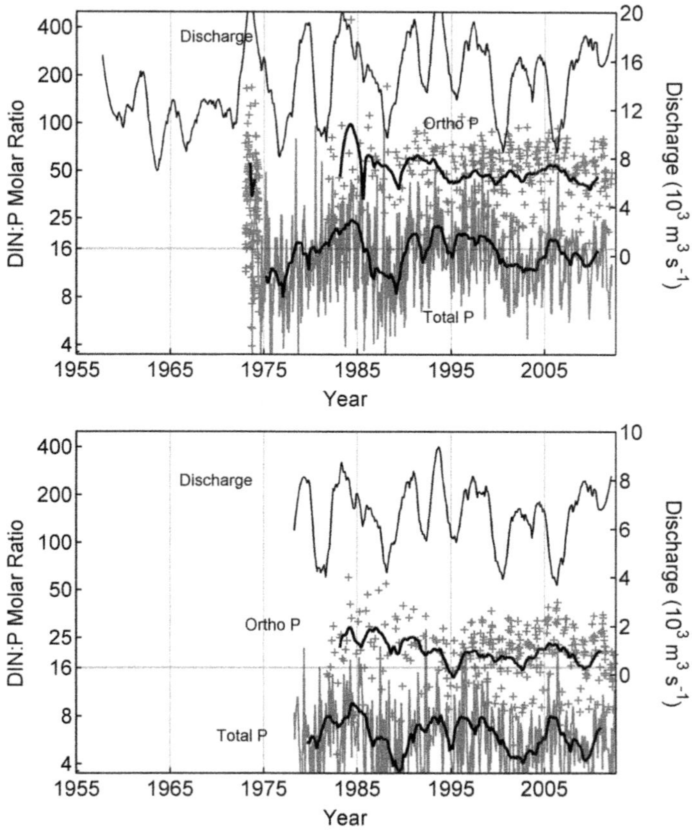

Figure 18.6. Molar ratios of dissolved inorganic nitrogen to either orthophosphate (crosses) or total phosphorus (gray lines) for the Mississippi River at St. Francisville (upper panel) and the Atchafalaya River at Melville (lower panel). The upper black lines are river discharge (measured at Tarbert Landing for Mississippi River and Simmesport for Atchafalaya) smoothed as in Fig. 18.3. The horizontal dotted lines correspond to the Redfield molar N:P ratio of 16:1.

but these differences were not significant and may be due to the use of slightly different methods for estimating fluxes.

The Mississippi and Atchafalaya rivers are a large source of both inorganic and organic carbon to the northern Gulf of Mexico coastal region (Table 18.1). Estimates of organic carbon inputs from the river were reviewed by Cai and Lohrenz (2010). POC flux was tightly coupled to total suspended matter loads, which have been documented to have decreased over time as a result of the damming and other engineering and erosion control activities in the Mississippi basin (Walling 2006; Meade and Moody 2010).

Thus, presumably, there has also been a decrease in POC flux over the past century. Annual particulate organic carbon (POC) flux for the Mississippi and Atchafalaya rivers was estimated at 3.36 Tg C y^{-1} by Cai and Lohrenz (2010) using a long-term average sediment loading of 210 metric tons per year (Meade and Parker 1985) and an average %POC of 1.6 (Trefry et al. 1994). This estimate

Table 18.1. *Average concentrations and annual fluxes of organic and inorganic C, N, and P delivered to the Gulf of Mexico by the Mississippi-Atchafalaya River system (MARS) (in 10^{12} g y^{-1} or Tg y^{-1})*

Chemical	Concentration	Annual flux Tg y^{-1}	Reference
TSM	~360 mg/L[a]	210	Meade and Parker (1985)
	~297 mg/L[a]	172	Meade and Moody (2010)
POC	1.6% of TSM	2.75–3.36	Trefry et al. (1994)
		1.2[b]	Duan and Bianchi 2006
DOC	0.28 mM	2.0	Trefry et al. (1994)
	0.33 mM	2.3	Benner and Opsahl (2001)
	0.49 mM	3.4	Bianchi et al. (2004)
	0.34 mM	2.7	Shen et al. (2012)
PIC	0.15% of TSM	0.26–0.32	Trefry et al. (1994)
DIC	2.408 mM	17	Cai (2003); Guo et al. (2012)
TAlk	2.378 mM	1.38×10^{12} mol y^{-1}	Cai (2003); Raymond and Cole (2003); Guo et al. (2012)
Total N		1.57	Goolsby et al. (1999)
	1.24	This study	
$NO_3 + NO_2$		0.95	Goolsby et al. (1999); Howarth et al. (1996)
	0.85	This study	
Ammonium		0.03	Goolsby et al. (1999)
	0.018	This study	
DON		0.38	Goolsby et al. (1999)
	0.19	This study	
PON		0.20	Goolsby et al. (1999)
	0.45[c]	Trefry et al. (1994)	
	0.18	This study	
Total P		0.136	Goolsby et al. (1999)
	0.20	This study	
PO_4		0.042	Goolsby et al. (1999)
	0.054	This study	
Particulate P		0.095	Goolsby et al. (1999)
Si-dissolved		2.32	Goolsby et al. (1999)
	1.91	This study	

[a] TSM, total suspended matter. Average concentrations estimated by dividing the average sediment flux by the average water flux; total water discharge used for flux calculation is 580 km^3 y^{-1} (from Meade and Moody 2010), which is slightly higher than the value of 530 km^3 y^{-1} used by Cai and Lohrenz (2010).

[b] Duan and Bianchi (2006) reported their flux (0.8 Tg C/yr) for the Mississippi River alone, and thus is scaled up. A 50% increase is applied here for the entire Mississippi-Atchafalaya River System according to a 2 to 1 ratio of water discharge

[c] Based on a 8.5 weight ratio of C:N from Trefry et al. (1994); the Goolsby et al. (1999) values were derived from USGS river monitoring sites, whereas the Trefry et al. (1994) values came from a monitoring station immediately outside the river mouth.

of POC flux is more than twice that of 1.1 Tg C y^{-1}, estimated by scaling up to the entire MARS the value of 0.8 TgC y^{-1} from the Duan and Bianchi (2006) field study in the lower Mississippi River during the period in 2001–2003. Meade and Moody (2010) recently concluded that sediment delivery from the MARS has decreased from 400 metric tons per year before 1900 to about 172 metric tons per year during the last two decades, which would result in an adjusted average POC flux of 2.75 Tg C/yr for the more recent time. However, another recent study by Allison et al. (2012) suggested that sediment discharge from the MARS was as high as 200–260 metric tons per year. Thus we adopt, somewhat arbitrarily, an average POC flux of 3.1 (range from 2.75 to 3.36) Tg C y^{-1} for the sake of discussion. Suspended loads also fluctuate seasonally, with higher loads occurring during periods of higher discharge (Cai and Lohrenz 2010).

In contrast to the seasonal variability in POC loading, dissolved organic carbon showed less seasonal variability, with an average concentration of about 0.33 mM (Benner and Opsahl 2001). This value was lower than the value of 0.49 mM reported for the lower Mississippi River by Bianchi et al. (2004) but higher than that of 0.28 mM by Trefry et al. (1994). Thus an annual flux of 2.3 Tg C y^{-1} was given based on average DOC concentration and an average discharge of 580 km^3 (Cai and Lohrenz 2010). A more recent study based on five cruises during 2009–2010 suggested an average DOC concentration of 0.30 mM in the Mississippi River and 0.44 mM in Atchafalaya River and an average annual DOC export flux of 2.7 Tg C y^{-1} from the MARS (Shen et al. 2012). Again, here we somewhat arbitrarily adopt an average DOC flux of 2.5 (from 2.3–2.7) Tg C y^{-1}. Thus the total annual organic carbon flux from the Mississippi was estimated here as 5.6 Tg C y^{-1} (see Chapter 16).

The quantity and composition of particulate and dissolved organic matter has also been altered as a result of damming, increases in nutrient loading, and other human activities (Bianchi et al. 2004; Sampere et al. 2011). In addition, substantial differences in composition and export of DOC between the Mississippi and Atchafalaya rivers have been attributed to differences in floodplain processes (Shen et al. 2012). There are arguments that terrestrial material exported from coastal marshes may be a significant source of carbon-supporting food webs in the Mississippi plume region (Dagg et al. 2008; Bianchi et al. 2011).

Finally, even with the range of uncertainties, we note that the ratio of inorganic nutrient to total organic carbon (TOC = POC+DOC) delivered by the MARS to the Gulf of Mexico is much higher than the global average as a result of agricultural use of fertilizer in the drainage basin. By multiplying the DIN loading by the Redfield C:N ratio of 6.6:1 and equating it to autotrophic production, Cai (2011) surmised that globally this autotrophic loading is only ~1:15 when compared with the riverine TOC or heterotrophic loading. Thus, globally riverine TOC loading is far in excess of the autotrophic production derived from riverine DIN inputs. In contrast, the ratio of (DIN × 6.6) to TOC is much closer to 1 in the MARS, making it more likely that the system is autotrophic or metabolically balanced.

A long-term increasing trend in river export of alkalinity and inorganic carbon from the Mississippi River has been reported and attributed to increasing rainfall and human-related changes in land cover and land-use activities that can affect weathering rates (Cai 2003; Raymond and Cole 2003; Cai et al. 2008; Raymond et al. 2008; Cai and Lohrenz 2010). As noted previously for nitrogen export, the increased discharge occurs disproportionately in agricultural watersheds, where practices have led to reduced water residence times and increased weathering and throughput of inorganic carbon. This is a vivid example of the profound impact of human activities on both terrestrial and coastal ocean

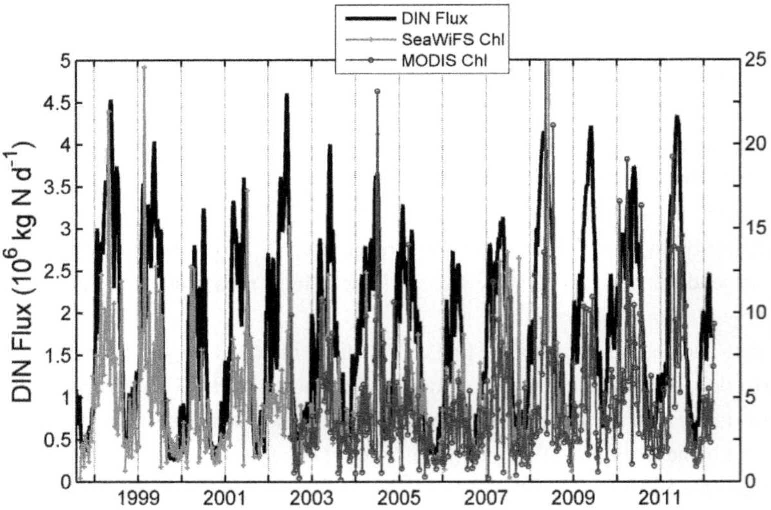

Figure 18.7. Time series of DIN flux and SeaWiFS and MODIS Aqua chlorophyll a products from 1997 to 2012 for the region delimited by 29.0N, 89.5W and 28.5N, 90.0W. The SeaWiFS and MODIS Aqua chlorophyll values are the averages over the specified region of the 9 km 8-day composite product obtained using the NASA GIOVANNI satellite archive (http://oceancolor.gsfc.nasa.gov).

biogeochemical processes, in our so called "Anthropocene" era (Crutzen and Stoermer 2000; Steffen et al. 2005).

4. Nutrient enhancement of primary and secondary productivity in the outflow region

Early descriptions of elevated chlorophyll concentrations in the river outflow region were provided by the work of Riley (1937), who attributed this to high phosphorus inputs from the river. Since then there have been numerous studies characterizing phytoplankton primary production in the Mississippi River outflow plume (Thomas and Simmons 1960; Lohrenz et al. 1990; Redalje et al. 1994; Prasad et al. 1995; Lohrenz et al. 1997; Lohrenz et al. 1999), as well in the northern Gulf of Mexico region (Sklar and Turner 1981; Biggs and Sanchez 1997; Chen et al. 2000; Lehrter et al. 2009). These studies consistently reported elevated levels of primary production in shelf waters proximal to the Mississippi River and implicate the high inputs of nutrients as a major driver. Seasonal relationships between N loading and net community production have also been reported for the Louisiana coastal region (Justic et al. 1993). The pattern of enhanced productivity and distinct phytoplankton community assemblages has been found to extend well out into the oligotrophic waters of the Gulf of Mexico (Wawrik and Paul 2004). Additional evidence for linkages between river inputs and biological response comes from a comparison of riverine dissolved inorganic nitrogen (DIN) flux and satellite estimates of chlorophyll (Lohrenz et al. 2008, and Fig. 18.7).

A classical pattern of elevated biomass and productivity is characteristically observed at intermediate salinities along the river-ocean mixing gradient (Lohrenz et al. 1999, references therein). This has been attributed to conditions described at the "optimal growth zone" (Demaster et al. 1996).

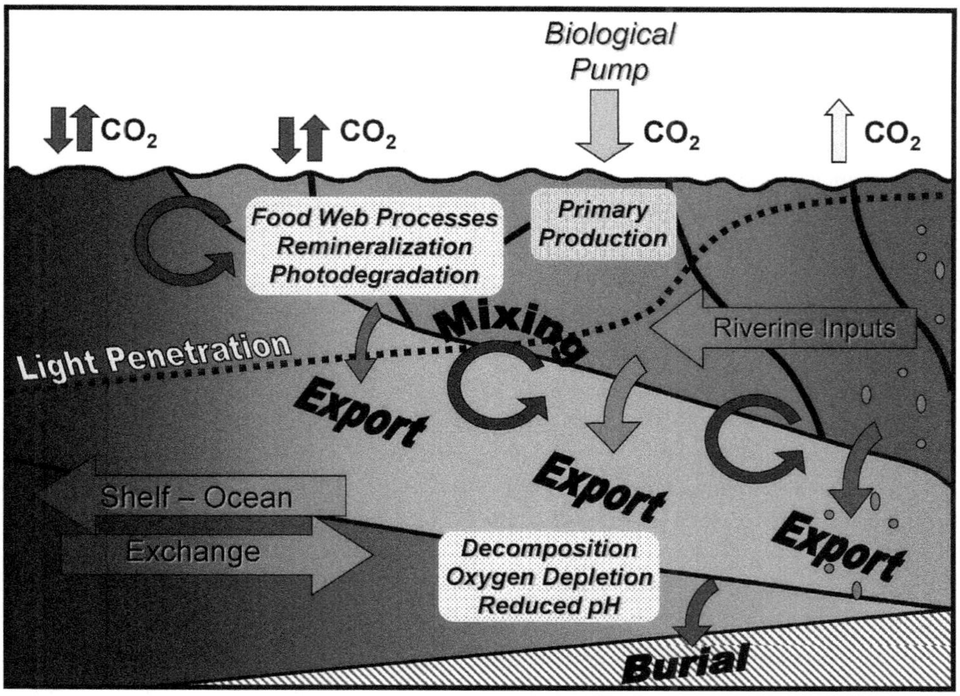

Figure 18.8. Diagram illustrating biogeochemical processes along river-ocean mixing gradient.

A conceptual illustration of this is provided in Fig. 18.8. High levels of suspended particulates and colored dissolved organic matter (CDOM) contribute to low light conditions at and near the river end-member. As particles settle and CDOM is diluted or degraded, light penetration increases and leading to increased productivity. Eventually, as nutrients are consumed or diluted, production becomes limited by availability of nutrients.

The nature of which nutrients limit productivity across terrestrial, freshwater, and marine systems has been actively considered (Elser et al. 2007). The conventional view for marine ecosystems has been that nitrogen limits growth and accumulation of biomass (Smith 1984; Howarth 1988; Rabalais 2002). There is considerable evidence that, in addition to nitrogen, limitation by phosphorus and silicon may occur in certain regions of river-influenced continental shelf waters in the vicinity of the Mississippi River outflow (Dortch and Whitledge 1992; Smith and Hitchcock 1994; Nelson and Dortch 1996; Lohrenz et al. 1999; Sylvan et al. 2006). The mode of nutrient limitation has spatial and seasonal patterns, with P-limitation more prevalent in polyhaline regions, particularly in spring during high river discharge, and N-limitation more likely to occur at higher salinity regions and particularly in summer and fall.

Our observations of systematically lower N:P ratios in the Atchafalaya River as compared with the Mississippi River (Fig. 18.6) provide a basis for speculation that limitation by phosphorus may be less common in the Atchafalaya outflow region. Patterns of nutrient limitation have been shown to be spatially localized and temporally variable in coastal marine ecosystems (Malone et al. 1996;

Conley et al. 2009), and consequently, efforts to manage coastal eutrophication must necessarily adopt a dual strategy for controlling both nitrogen and phosphorus inputs. Rabalais (2002) noted, however, that regarding issues of eutrophication and hypoxia, the primary concern is accumulation of biomass as opposed to limitation of growth and argued that nitrogen is the primary limiting factor for biomass accumulation in many coastal ecosystems.

The quantitative importance of nutrient sources other than river inputs is poorly constrained. There have been considerations of nutrient contributions via upwelling from offshore (Sahl et al. 1993; Chen et al. 2000), although such contributions are likely to be more important in the vicinity of the shelf break. Both upwelling (Han et al. 2012) and groundwater sources (Liu et al. 2012) are potentially important as nutrient sources in the Pearl River and northern South China Sea shelf ecosystem. Other sources include direct atmospheric deposition to Gulf of Mexico waters, which has been reported to be relatively small (Goolsby et al. 1999; Turner and Rabalais 1999), although deposition in the watershed has been estimated to a major fraction (16%) of overall riverine contributions to nitrogen flux (Alexander et al. 2008). Nitrogen fixation has been observed to be important in the outer reaches of the Amazon River plume (Cooley and Yager 2006; Subramaniam et al. 2008) and the Mekong River plume (Grosse et al. 2010), but is less well characterized for the Mississippi and Atchafalaya River system.

A variety of studies have linked elevated productivity to nutrient inputs (Lohrenz et al. 1997; Rabalais et al. 2002a; Lohrenz et al. 2008; Lehrter et al. 2009) and subsequently to hypoxic conditions in bottom waters characteristically observed in the summer and fall (Rabalais et al. 2000, 2002b; Boesch et al. 2009). However, there has been debate regarding the extent to which hypoxia is attributable to nutrient inputs as opposed to other factors such as enhanced stratification (Bianchi et al. 2008; Hetland and Dimarco 2008; Bianchi et al. 2010; Feng et al. 2012). Various modeling approaches have also been used to describe the dynamics of primary production in the plume outflow region and associated linkages to hypoxia (Bierman et al. 1994; Chen et al. 1997; Justic et al. 2002; Scavia et al. 2003; Scavia et al. 2004; Green et al. 2008; Fennel et al. 2011).

In addition to the high primary production in plume waters, there are numerous observations of enhanced bacterial and secondary production at intermediate salinities (Chin-Leo and Benner 1992; Biddanda et al. 1994; Gardner et al. 1994; Amon and Benner 1998). Bacterially mediated transformations of nitrogen have also been shown to be important in this system, including high rates of nitrification (Pakulski et al. 1995; Pakulski et al. 2000) and ammonium recycling (Cotner and Gardner 1993; Wawrik et al. 2004). A conceptual model showing relationships between nutrient inputs, phytoplankton production, and coupling to heterotrophic processes was provided by Dagg and Breed (2003). Such linkages were also reviewed in Dagg et al. (2004) and (2007).

In conjunction with the relatively high bacterial activity, terrestrial carbon appears to be rapidly degraded along the river-ocean mixing gradient (Benner and Opsahl 2001; Bianchi et al. 2004). This has been partially attributed to a high proportion of labile organic matter that is more readily mineralized, but has also been attributed to microbial degradation of terrestrial organic carbon linked to photo-oxidation processes (Benner and Opsahl 2001). More recently, arguments have been made that co-metabolism or "priming" of microbial communities contribute to more efficient mineralization of terrestrial organic matter in river-influenced systems (Bianchi 2011). Organic carbon signatures show a rapid shift to marine signatures due to a combination of removal of the terrestrial organic carbon and in situ production of marine organic carbon (Wang et al. 2004). Autochthonous inputs related to freshwater phytoplankton production in the lower river may also be an important factor contributing to a shift in composition of organic carbon (Bianchi et al. 2004).

5. Oxygen and respiration patterns in subsurface waters

Processes influencing oxygen dynamics in the Mississippi-Atchafalaya River system outflow region have been the subject of considerable study and debate. Rabalais and coworkers have maintained a time series of hypoxia off the Louisiana coast (Rabalais et al. 2001; Turner et al. 2012), and their findings have contributed to a major policy effort to mitigate hypoxia through nutrient reduction in the watershed (Rabalais et al. 2007).

As noted previously, there are clear linkages between nutrient inputs and elevated primary production, and this higher productivity is believed to contribute to increased export of organic carbon into subsurface waters. The subsequent decomposition of this material contributes to an associated consumption of oxygen that leads to hypoxia (Fig. 18.8). Such a pattern is generally accepted as the basis for a global increase in the frequency and severity of hypoxia in coastal waters (Diaz 2001; Diaz and Rosenberg 2008; Rabalais et al. 2010).

A variety of modeling approaches have been used to predict hypoxia in the northern Gulf of Mexico (Bierman et al. 2003; Justic et al. 2003; Scavia et al. 2003; Scavia et al. 2004; Turner et al. 2006; Justic et al. 2007; Greene et al. 2009). However, in some cases, these models rely largely on empirical relationships between hypoxia, oxygen consumption, and freshwater and nutrient loading. Such models may fail to account for variability in meteorological forcing (Feng et al. 2012; Turner et al. 2012) and to explicitly address mechanisms for organic carbon inputs. Although limited measurements of vertical export of organic carbon in the river outflow region have been made (Redalje et al. 1994; Qureshi 1995), the sources and mechanisms of delivery of organic matter fueling respiration in bottom waters remain poorly understood. An inverse model to describe carbon flow through the food webs (Breed et al. 2004) was coupled with a physical model to develop an organic carbon budget of the plume (Green et al. 2006). Modeled estimates of autochthonous organic carbon export from the plume accounted for only a 23% of the oxygen demand required for hypoxic conditions. Non-riverine organic matter inputs as well as active sediment processes may also play a role in mediating the temporal and spatial patterns in hypoxia (Bianchi et al. 2008; Bianchi et al. 2010). Such terms must be considered in the context of physical variability in this complex system (Bianchi et al. 2010; Feng et al. 2012).

Fundamental differences exist between the Mississippi River plume and Atchafalaya plume regions in the physics of the plume and the relative importance of benthic versus water column respiration in contributing to hypoxia, with benthic processes playing a more dominant role in the Atchafalaya system (Hetland and Dimarco 2008; Bianchi et al. 2010). However, Murrell and Lehrter (2010) conducted extensive analyses of both sediment and water column respiration rates in the northern Gulf of Mexico, as well as reviewing findings from previous studies, and they concluded that water column respiration, even when only that below the pycnocline was considered, was consistently the major sink for dissolved oxygen. This suggests that water column organic matter is a major substrate for respiration. This view was further reinforced by a comparison of rates of water column community respiration and phytoplankton productivity that was consistent with net heterotrophy over much of the shelf (Murrell et al. 2013), a condition attributed to input of organic carbon from inshore coastal and riverine sources. Recent estimates by Guo et al. (2012) of net community production based on DIC removal in the Mississippi River plume also supports this argument that organic carbon production in and export from the river plume are the primary driving force for heterotrophy and oxygen consumption in the subsurface and surrounding environments. Clearly, further study is needed to discern sources of organic inputs and mechanisms driving oxygen consumption.

Coupled physical-biological models are now progressing to a point where, with reasonable further development, they can be used to examine the complex interactions among, physics, biogeochemistry, and ecology and consequences for hypoxia in the northern Gulf of Mexico ecosystem (Fennel et al. 2011). However, such work must be complemented by continued efforts to observe and understand physical and biogeochemical processes influencing transformation and fluxes of carbon and nutrients.

6. Carbon system properties and air-sea flux of carbon dioxide

The high biological activity characteristically associated with large river ecosystems can contribute to a strong biological pump and associated draw down of inorganic carbon (cf., Fig. 18.8). This pattern has been observed for the Mississippi River (Cai 2003; Lohrenz and Cai 2006; Cai and Lohrenz 2010; Lohrenz et al. 2010) and is also evident in other major river systems, including the Amazon (Ternon et al. 2000; Kortzinger 2003; Cooley and Yager 2006; Cooley et al. 2007; Subramaniam et al. 2008), Changjiang or Yangtze (Tsunogai et al. 1999; Chen et al. 2008; Zhai and Dai 2009), and the Pearl River and South China Sea (Cai et al. 2004; Dai et al. 2008; Guo et al. 2009; Han et al. 2012). The pattern of river-enhanced drawdown has been shown to be seasonally variable and subject to varying contributions of biological and physicochemical forcing. Salisbury et al. (2008) argued that low-alkalinity river outflows could lower the pH of coastal waters and potentially produce conditions unfavorable for shellfish. However, for the Mississippi River system, higher alkalinity in river waters coupled with high biological activity create conditions leading to higher pH and supersaturated carbonate saturation state values (Guo et al. 2012). Using a three-end-member mixing model to estimate net community production (NCP) along the river-ocean mixing gradient, Guo et al. (2012) found highest NCP rates at intermediate salinities consistent with the conceptual model shown in Figure 18.8 and the carbon budget model of Green et al. (2006). In addition, net community production rates were comparable to independent ^{14}C-derived estimates of primary production (Lohrenz et al. 1997; Lohrenz et al. 1999) and earlier estimates of DIC removal by Cai (2003).

Ocean acidification owing to the increase of CO_2 in the atmosphere and its subsequent uptake by the ocean has become an acute human-induced ocean stressor in recent years and has rightfully received wide attention because of its potentially very negative impacts on marine life and possible alteration of biogeochemical processes (Doney 2010). When this stressor acts together with coastal eutrophication, as in the case of a nutrient-replete large river plume, contrary to the prediction (Salisbury et al. 2008), the high rates of autotrophic carbon fixation can result in higher pH values and thus suppress effects of ocean acidification, as shown by Guo et al. (2012) and Cai et al. (Chapter 7) for the Mississippi plume and (Chou et al. submitted) in the Changjiang. However, conditions in bottom waters are a different story. A recent analysis by Cai et al. (2011) examined the consequences of export of organic matter into bottom waters and its decomposition to CO_2 combined with the potential for introduction of offshore waters with elevated CO_2. They concluded that the multiple stressors (ocean acidification and respiration derived acidification) have compounding effects that result in an additional lowering of pH. They projected that conditions would approach undersaturation of $CaCO_3$ over the next century, thus posing a threat to marine life.

7. Conclusions and implications of climate- and human-related drivers of long-term change

It is clear that human-related activities have a profound impact on biogeochemical processes in the Mississippi and Atchafalaya rivers. The consequences of human activity in the watershed is coupled with continued anticipated increases in atmospheric carbon dioxide and climate-related effects on precipitation, storm frequency and intensity, sea level, and water temperatures. Borges and Gypens (2010) have argued that changes in nutrient delivery will have a stronger impact on carbonate chemistry in coastal waters, with increasing coastal eutrophication potentially countering the effects of ocean acidification. Increasing levels of atmospheric carbon dioxide will likely lead to carbon management efforts in terrestrial ecosystems. We suggest here that the two most important coastal ocean stressors, ocean acidification and eutrophication, have shifted large river-dominated coastal ocean waters to more extreme pH conditions, with higher pH in surface waters and much lower pH in hypoxic bottom waters. Furthermore, conditions will further deteriorate unless both eutrophication and anthropogenic CO_2 release are curtailed.

As noted here, a better understanding is needed of factors influencing the delivery of terrestrial inputs of carbon and nutrients in coastal ecosystems, as well as the consequences of these inputs for ecosystem processes ranging from primary production, community structure, and food web processes to biogeochemical processing and impacts on carbon and oxygen dynamics. These factors ultimately play key roles in the integrity of coastal ocean habitats and their ability to support marine life. From the concepts presented here, it is apparent that there is a critical need for coordinated management of both terrestrial and coastal ecosystems, as they are intimately linked, and actions affecting processes in watersheds will have consequences for carbon cycling and ecosystem processes in coastal ecosystems. Such efforts will require extensive, sustained observations of land and ocean properties coupled with improved models of these environments and their interactions.

Acknowledgments

We acknowledge the support of the National Science Foundation (OCE-0752254 to Lohrenz and OCE-0752110 to Cai) and the National Aeronautics and Space Administration (NNX10AU06G, NNX12AB21G, and NNX12AP84G to Lohrenz and Cai).

References

Alexander, R. B., R. A. Smith, G. E. Schwarz, E. W. Boyer, J. V. Nolan, and J. W. Brakebill. 2008. Differences in phosphorus and nitrogen delivery to the gulf of Mexico from the Mississippi river basin. Environmental Science & Technology 42: 822–830.

Alig, R. J., A. J. Plantinga, S. Ahn, and J. D. Kline. 2003. Land use changes involving forestry in the United States: 1952 to 1997, with projections to 2050, Gen. Tech. Rep. PNW-GTR-587. U.S. Department of Agriculture, Forest Service, Pacific Northwest Research Station.

Allison, M. A., C. R. Demas, B. A. Ebersole, B. A. Kleiss, C. D. Little, E. A. Meselhe, N. J. Powell, T. C. Pratt, and B. M. Vosburg. 2012. A water and sediment budget for the lower Mississippi–Atchafalaya River in flood years 2008–2010: Implications for sediment discharge to the oceans and coastal restoration in Louisiana. Journal of Hydrology 432–433: 84–97.

Amon, R. M. W., and R. Benner. 1998. Seasonal patterns of bacterial abundance and production in the Mississippi River plume and their importance for the fate of enhanced primary production. Microbial Ecology 35: 289–300.

Benner, R., and S. Opsahl. 2001. Molecular indicators of the sources and transformations of dissolved organic matter in the Mississippi river plume. Organic Geochemistry 32: 597–611.

Berner, R. A. 1989. Biogeochemical cycles of carbon and sulfur and their effect on atmospheric oxygen over phanerozoic time. Palaeogeography, Palaeoclimatology, Palaeoecology 75: 97–122.

Bianchi, T., L. Wysocki, K. Schreiner, T. Filley, D. Corbett, and A. Kolker. 2011. Sources of terrestrial organic carbon in the Mississippi plume region: Evidence for the importance of coastal marsh inputs. Aquat. Geochem. 17: 431–456.

Bianchi, T. S. 2011. The role of terrestrially derived organic carbon in the coastal ocean: A changing paradigm and the priming effect. Proceedings of the National Academy of Sciences of the United States of America 108: 19473–19481.

Bianchi, T. S., and M. A. Allison. 2009. Large-river delta-front estuaries as natural "recorders" of global environmental change. Proceedings of the National Academy of Sciences of the United States of America 106: 8085–8092.

Bianchi, T. S., S. F. DiMarco, M. A. Allison, P. Chapman, J. H. Cowan, Jr., R. D. Hetland, J. W. Morse, and G. Rowe. 2008. Controls and consequences of hypoxia on the Louisiana Shelf (USA): beyond the nutrient-centric view. EOS 89: 236–237.

Bianchi, T. S., S. F. DiMarco, J. H. Cowan, Jr., R. D. Hetland, P. Chapman, J. W. Day, and M. A. Allison. 2010. The science of hypoxia in the Northern Gulf of Mexico: A review. Science of the Total Environment 408: 1471–1484.

Bianchi, T. S., T. Filley, K. Dria, and P. G. Hatcher. 2004. Temporal variability in sources of dissolved organic carbon in the lower Mississippi River. Geochimica Et Cosmochimica Acta 68: 959–967.

Bianchi, T. S., J. J. Galler, and M. A. Allison. 2007a. Hydrodynamic sorting and transport of terrestrially derived organic carbon in sediments of the Mississippi and Atchafalaya Rivers. Estuarine, Coastal and Shelf Science 73: 211–222.

Bianchi, T. S., L. A. Wysocki, M. Stewart, T. R. Filley, and B. A. Mckee. 2007b. Temporal variability in terrestrially-derived sources of particulate organic carbon in the lower Mississippi River and its upper tributaries. Geochimica Et Cosmochimica Acta 71: 4425–4437.

Biddanda, B., S. Opsahl, and R. Benner. 1994. Plankton respiration and carbon flux through bacterioplankton on the Louisiana shelf. Limnology and Oceanography 39: 1259–1275.

Bierman, J., V. J., S. C. Hinz, D. Zhu, J. Wiseman, W. J., N. N. Rabalais, and R. E. Turner. 2003. Forecasting hypoxia in the Gulf of Mexico: Responses to nutrient loadings from the Mississippi River Basin, pp. 91–96. *In* N. J. Valette-Silver and D. Scavia (eds.), Ecological Forecasting: New Tools for Coastal and Ecosystem Management. NOAA Technical Memorandum NOS NCCOS 1, NOAA Coastal Ocean Program.

Bierman, V. J., Jr., S. C. Hinz, D.-W. Zhu, W. J. Wiseman, Jr., N. N. Rabalais, and R. E. Turner. 1994. A preliminary mass balance model of primary productivity and dissolved oxygen in the Mississippi River plume/Inner Gulf Shelf region. Estuaries 17: 886–899.

Biggs, D. C., and L. L. Sanchez. 1997. Nutrient enhanced primary productivity of the Texas-Louisiana continental shelf. Journal of Marine Systems 11: 237–247.

Boesch, D. F., W. R. Boynton, L. B. Crowder, R. J. Diaz, R. W. Howarth, L. D. Mee, S. W. Nixon, N. N. Rabalais, R. Rosenberg, J. G. Sander, D. Scavia, and R. E. Turner. 2009. Nutrient Enrichment Drives Gulf of Mexico Hypoxia. EOS 90: 117–128.

Borges, A. V., and N. Gypens. 2010. Carbonate chemistry in the coastal zone responds more strongly to eutrophication than to ocean acidification. Limnology and Oceanography 55: 346–353.

Breed, G. A., G. A. Jackson, and T. L. Richardson. 2004. Sedimentation, carbon export and food web structure in the Mississippi River plume described by inverse analysis. Marine Ecology Progress Series 278: 35–51.

Broussard, W., and R. E. Turner. 2009. A century of changing land-use and water-quality relationships in the continental US. Frontiers in Ecology and the Environment 7: 302–307.

Cai, W.-J. 2003. Riverine inorganic carbon flux and rate of biological uptake in the Mississippi River plume. Geophysical Research Letters 30: 1032.

Cai, W.-J. 2011. Estuarine and coastal ocean carbon paradox: CO2 sinks or sites of terrestrial carbon incineration? Annual Review of Marine Science 3: 123–145.

Cai, W.-J., X. Guo, C.-T. A. Chen, M. Dai, L. Zhang, W. Zhai, S. E. Lohrenz, and Y. Wang Y 2008. A comparative overview of weathering intensity and HCO3- flux of the world's major rivers with emphasis on the Changjiang, Huanghe, Pearl and Mississippi rivers. Continental Shelf Research 28: 1538–1549.

Cai, W.-J., X. Hu, W.-J. Huang, M. C. Murrell, J. C. Lehrter, S. E. Lohrenz, W.-C. Chou, W. Zhai, J. T. Hollibaugh, Y. Wang, P. Zhao, X. Guo, K. Gundersen, M. Dai, and G.-C. Gong. 2011. Acidification of subsurface coastal waters enhanced by eutrophication. Nature Geoscience 4: 766–770.

Cai, W.-J., and S. E. Lohrenz. 2010. The Mississippi River Plume and Adjacent Margin in the Gulf of Mexico, pp. 406–422. *In* K.-K. Liu, L. Atkinson, R. Quiñones and L. Talaue-McManus (eds.), Carbon and Nutrient Fluxes in the Continental Margins: A Global Synthesis. Springer.

Cai, W. J., M. H. Dai, Y. C. Wang, W. D. Zhai, T. Huang, S. T. Chen, F. Zhang, Z. Z. Chen, and Z. H. Wang. 2004. The biogeochemistry of inorganic carbon and nutrients in the Pearl River estuary and the adjacent Northern South China Sea. Continental Shelf Research 24: 1301–1319.

Chen, C.-T. A. 2004. Exchanges of carbon in the coastal seas, pp. 341–351. *In* C. B. Field and M. R. Raupach (eds.), The Global Carbon Cycle: Integrating Humans, Climate, and the Natural World. Scientific Committee on Problems of the Environment (SCOPE). Island Press.

Chen, C.-T. A., W. Zhai, and M. Dai. 2008. Riverine input and air-sea CO_2 exchanges near the Changjiang (Yangtze River) Estuary: Status quo and implication on possible future changes in metabolic status. Continental Shelf Research 28: 1476–1482.

Chen, C., D. A. Wiesenburg, and L. Xie. 1997. Influence of river discharge on biological production in the inner shelf: A coupled biological and physical model of the Louisiana-Texas Shelf. Journal of Marine Research 55: 293–320.

Chen, C. T. A., and A. V. Borges. 2009. Reconciling opposing views on carbon cycling in the coastal ocean: Continental shelves as sinks and near-shore ecosystems as sources of atmospheric CO_2. Deep-Sea Research Part Ii-Topical Studies in Oceanography 56: 578–590.

Chen, H., H. Tian, M. Liu, J. Melillo, S. Pan, and C. Zhang. 2006. Effect of land-cover change on terrestrial carbon dynamics in the southern United States. J. Environ. Qual. 35: 1533–1547.

Chen, X., S. E. Lohrenz, and D. A. Wiesenburg. 2000. Distribution and controlling mechanisms of primary production on the Louisiana-Texas continental shelf. Journal of Marine Systems 25: 179–207.

Chin-Leo, G., and R. Benner. 1992. Enhanced bacterioplankton production and respiration at intermediate salinities in the Mississippi River plume. Marine Ecology Progress Series 87: 87–103.

Chou, W.-C., G.-C. Gong, and W.-J. Cai. Submitted. Seasonality of CO2 in coastal oceans altered by increasing anthropogenic nutrient delivery from large rivers: evidence from the Changjiang-East China Sea system. Biogeosciences Discussions.

Ciais, P., A. V. Borges, G. Abril, M. Meybeck, G. Folberth, D. Hauglustaine, I. A. Janssens. 2008. The impact of lateral carbon fluxes on the European carbon balance. Biogeosciences 5: 1259–1271.

Conley, D. J., H. W. Paerl, R. W. Howarth, D. F. Boesch, S. P. Seitzinger, K. E. Havens, C. Lancelot, and G. E. Likens. 2009. Controlling eutrophication: Nitrogen and phosphorus. Science 323: 1014–1015.

Cooley, S. R., V. J. Coles, A. Subramaniam, and P. L. Yager. 2007. Seasonal variations in the Amazon plume-related atmospheric carbon sink. Global Biogeochem. Cycles 21: GB3014.

Cooley, S. R., and P. L. Yager. 2006. Physical and biological contributions to the western tropical North Atlantic Ocean carbon sink formed by the Amazon River plume. Journal of Geophysical Research-Oceans 111: C08018.

Cotner, J. B. J., and W. S. Gardner. 1993. Heterotrophic bacterial mediation of ammonium and dissolved free amino acid fluxes in the Mississippi River plume. Marine Ecology Progress Series 93: 75–87.

Cowan, J. H., Jr., C. B. Grimes, and R. F. Shaw. 2008. Life history, history, hysteresis, and habitat changes in Louisiana's coastal ecosystem. Bulletin of Marine Science 83: 197–215.

Crutzen, P. J., and E. F. Stoermer. 2000. The Anthropocene. Global Change 41: 17–18.

Dagg, M., R. Sato, H. Liu, T. S. Bianchi, R. Green, and R. Powell. 2008. Microbial food web contributions to bottom water hypoxia in the northern Gulf of Mexico. Continental Shelf Research 28: 1127–1137.

Dagg, M. J., J. W. Ammerman, R. M. W. Amon, W. S. Gardner, R. E. Green, and S. E. Lohrenz. 2007. A review of water column processes influencing hypoxia in the northern Gulf of Mexico. Estuaries and Coasts 30: 735–752.

Dagg, M. J., R. Benner, S. Lohrenz, and D. Lawrence. 2004. Transformation of dissolved and particulate materials on continental shelves influenced by large rivers: plume processes. Continental Shelf Research 24: 833–858.

Dagg, M. J., and G. A. Breed. 2003. Biological effects of Mississippi River nitrogen on the northern gulf of Mexico – a review and synthesis. J. Mar. Syst. 43: 133–152.

Dai, M., W. Zhai, W.-J. Cai, J. Callahan, B. Huang, S. Shang, T. Huang, X. Li, Z. Lu, W. Chen, and Z. Chen. 2008. Effects of an estuarine plume-associated bloom on the carbonate system in the lower reaches of the Pearl River estuary and the coastal zone of the northern South China Sea. Continental Shelf Research 28: 1416–1423.

Darnell, R. M. 1990. Mapping of the biological resources of the continental shelf. American Zoologist 30: 15–21.

Del Vecchio, R., and N. V. Blough. 2004. Spatial and seasonal distribution of chromophoric dissolved organic matter and dissolved organic carbon in the Middle Atlantic Bight. Marine Chemistry 89: 169–187.

Demaster, D. J., W. O. Smith, Jr., D. M. Nelson, and J. Y. Aller. 1996. Biogeochemical processes in Amazon shelf waters: chemical distributions and uptake rates of silicon, carbon and nitrogen. Continental Shelf Research 16: 617–643.

Diaz, R. J. 2001. Overview of hypoxia around the world. Journal of Environment Quality 30: 275–281.

Diaz, R. J., and R. Rosenberg. 2008. Spreading dead zones and consequences for marine ecosystems. Science 321: 926–929.

Doney, S. C. 2010. The growing human footprint on coastal and open-ocean biogeochemistry. Science 328: 1512–1516.

Donner, S. D., and C. J. Kucharik. 2008. Corn-based ethanol production compromises goal of reducing nitrogen export by the Mississippi River. Proceedings of the National Academy of Sciences of the United States of America 105: 4513–4518.

Donner, S. D., C. J. Kucharik, and J. A. Foley. 2004. Impact of changing land use practices on nitrate export by the Mississippi River. Global Biogeochemical Cycles 18: GB1028, doi:10.1029/2003GB002093.

Dortch, Q., and T. E. Whitledge. 1992. Does nitrogen or silicon limit phytoplankton production in the Mississippi River plume and nearby regions? Continental Shelf Research 12: 1293–1309.

Duan, S. W., and T. S. Bianchi. 2006. Seasonal changes in the abundance and composition of plant pigments in particulate organic carbon in the lower Mississippi and Pearl Rivers. Estuaries and Coasts 29: 427–442.

Elser, J. J., M. E. S. Bracken, E. E. Cleland, D. S. Gruner, W. S. Harpole, H. Hillebrand, J. T. Ngai, E. W. Seabloom, J. B. Shurin, J. E. Smith. 2007. Global analysis of nitrogen and phosphorus limitation of primary producers in freshwater, marine and terrestrial ecosystems. Ecology Letters 10: 1135–1142.

Feng, Y., S. F. Dimarco, and G. A. Jackson. 2012. Relative role of wind forcing and riverine nutrient input on the extent of hypoxia in the northern Gulf of Mexico. Geophysical Research Letters 39: L09601.

Fennel, K., R. Hetland, Y. Feng, and S. Dimarco. 2011. A coupled physical-biological model of the Northern Gulf of Mexico shelf: model description, validation and analysis of phytoplankton variability. Biogeosciences Discussions 8: 121–156.

Ford, M., and J. Nyman. 2011. Preface: an overview of the Atchafalaya River. Hydrobiologia 658: 1–5.

Gardner, W. S., et al. 1994. Mineralization of organic material and bacterial dynamics in Mississippi River plume water. Estuaries 17: 816–828.

Goolsby, D. A., and W. A. Battaglin. 2001. Long-term changes in concentrations and flux of nitrogen in the Mississippi River Basin, USA. Hydrological Processes 15: 1209–1226.

Goolsby, D. A., W. A. Battaglin, B. T. Aulenbach, and R. P. Hooper. 2000. Nitrogen flux and sources in the Mississippi River Basin. Science of the Total Environment 248: 75–86.

Goolsby, D. A., W. A. Battaglin, B. T. Aulenbach, and R. P. Hooper. 2001. Nitrogen Input to the Gulf of Mexico. Journal of Environmental Quality 30: 329–336.

Goolsby, D. A., W. A. Battaglin, G. B. Lawrence, R. S. Artz, B. T. Aulenbach, R. P. Hooper, D. R. Keeney, and G. Stensland J. 1999. Flux and Sources of Nutrients in the Mississippi–Atchafalaya River Basin: Topic 3 Report for the Integrated Assessment on Hypoxia in the Gulf of Mexico. NOAA Decision Analysis Series. U.S. Department of Commerce, NOAA Coastal Ocean Program.

Green, R. E., T. S. Bianchi, M. J. Dagg, N. D. Walker, and G. A. Breed. 2006. An organic carbon budget for the Mississippi River turbidity plume and plume contributions to air-sea CO2 fluxes and bottom water hypoxia. Estuaries and Coasts 29: 579–597.

Green, R. E., G. A. Breed, M. J. Dagg, and S. E. Lohrenz. 2008. Modeling the response of primary production and sedimentation to variable nitrate loading in the Mississippi River plume. Continental Shelf Research 28: 1451–1465.

Greene, R. M., J. C. Lehrter, and J. D. Hagy. 2009. Multiple regression models for hindcasting and forecasting midsummer hypoxia in the Gulf of Mexico. Ecological Applications 19: 1161–1175.

Grimes, C. B. 2001. Fishery production and the Mississippi River discharge. Fisheries 26: 17–26.

Grosse, J., D. Bombar, H. N. Doan, L. N. Nguyen, and M. Voss. 2010. The Mekong River plume fuels nitrogen fixation and determines phytoplankton species distribution in the South China Sea during low and high discharge season. Limnology and Oceanography 55: 1668–1680.

Guo, X, W.-J. Cai, W.-J. Huang, Y. Wang, F. Chen, M. C. Murrell, S. E. Lohrenz, L.-Q. Jiang, M. Dai, J. Hartmann, Q. Lin, and R. Culp. 2012. Carbon dynamics and community production in the Mississippi River plume. Limnology and Oceanography 57: 1–17.

Guo, X., M. Dai, W. Zhai, W.-J. Cai, and B. Chen. 2009. CO2 flux and seasonal variability in a large subtropical estuarine system, the Pearl River Estuary, China. Journal of Geophysical Research 114: G03013.

Han, A. Q., M. H. Dai, S. J. Kao, J. P. Gan, Q. Li, L.F. Wang, W. D. Zhai, and L. Wang. 2012. Nutrient dynamics and biological consumption in a large continental shelf system under the influence of both a river plume and coastal upwelling. Limnology and Oceanography 57: 486–502.

Hedges, J. I., and R. G. Kiel. 1995. Sedimentary organic matter preservation: An assessment and speculative synthesis. Marine Chemistry 49: 81–115.

Hetland, R. D., and S. F. Dimarco. 2008. How does the character of oxygen demand control the structure of hypoxia on the Texas-Louisiana continental shelf? Journal of Marine Systems 70: 49–62.

Hirsch, R. M., D. L. Moyer, and S. A. Archfield. 2010. Weighted Regressions on Time, Discharge, and Season (WRTDS), with an application to Chesapeake Bay River inputs. Journal of the American Water Resources Association 46: 857–880.

Holt, J., J. Harle, R. Proctor, S. Michel, M. Ashworth, C. Batstone, I. Allen, R. Holmes, T. Smyth, K. Haines, D. Bretherton, and G. Smith. 2009. Modelling the global coastal ocean. Philosophical Transactions of the Royal Society A: Mathematical, Physical and Engineering Sciences 367: 939–951.

Howarth, R. W. 1988. Nutrient limitation of net primary production in marine ecosystems. Annual Review of Ecology and Systematics 19: 89–110.

Howarth, R. W., G. Billen, D. Swaney, A. Townsend, N. Jaworski, K. Lajtha, J. A. Downing, R. Elmgren, N. Caraco, T. Jordan, F. Berendse, J. Freney, V. Kudeyarov, P. Murdoch, and Z. L. Zhu. 1996. Regional nitrogen budgets and riverine N&P fluxes for the drainages to the North Atlantic Ocean: Natural and human influences. Biogeochemistry 35: 75–139.

Hu, C. M., E. T. Montgomery, R. W. Schmitt, and F. E. Muller-Karger. 2004. The dispersal of the Amazon and Orinoco River water in the tropical Atlantic and Caribbean Sea: Observation from space and S-PALACE floats. Deep-Sea Research Part II-Topical Studies in Oceanography 51: 1151–1171.

Justic, D., V. J. Bierman, D. Scavia, and R. D. Hetland. 2007. Forecasting Gulf's hypoxia: The next 50 years? Estuaries and Coasts 30: 791–801.

Justic, D., N. N. Rabalais, and R. E. Turner. 1995. Stoichiometric nutrient balance and origin of coastal eutrophication. Marine Pollution Bulletin 30: 41–46.

Justic, D., N. N. Rabalais, and R. E. Turner. 2002. Modeling the impacts of decadal changes in riverine nutrient fluxes on coastal eutrophication near the Mississippi River Delta. Ecological Modelling 152: 33–46.

Justic, D., N. N. Rabalais, and R. E. Turner. 2003. Simulated responses of the Gulf of Mexico hypoxia to variations in climate and anthropogenic nutrient loading. Journal of Marine Systems 42: 115–126.

Justic, D., N. N. Rabalais, and R. E. Turner. 2005. Coupling between climate variability and coastal eutrophication: Evidence and outlook for the northern Gulf of Mexico. Journal of Sea Research 54: 25–35.

Justić, D., N. N. Rabalais, and R. E. Turner. 2005. Coupling between climate variability and coastal eutrophication: Evidence and outlook for the northern Gulf of Mexico. Journal of Sea Research 54: 25–35.

Justic, D., N. N. Rabalais, R. E. Turner, and W. J. Wiseman, Jr. 1993. Seasonal coupling between riverborne nutrients, net productivity and hypoxia. Marine Pollution Bulletin 26: 184–189.

Kortzinger, A. 2003. A significant CO_2 sink in the tropical Atlantic Ocean associated with the Amazon River plume. Geophysical Research Letters 30: 2287.

Lee, T. N., G. Podesta, E. Williams, J. Splain, and W. Johnson. 1993. Low salinity water in the Straits of Florida, pp. 65–70. *In* M. J. Dowgiallo (ed.), Coastal Oceanographic Effects of Summer 1993 Mississippi River Flooding. Special NOAA Report. NOAA Coastal Ocean Office/National Weather Service.

Lehrter, J. C., M. C. Murrell, and J. C. Kurtz. 2009. Interactions between freshwater input, light, and phytoplankton dynamics on the Louisiana continental shelf. Continental Shelf Research 29: 1861–1872.

Lie, H.-J., C.-H. Cho, J.-H. Lee, and S. Lee. 2003. Structure and eastward extension of the Changjiang River plume in the East China Sea. Journal of Geophysical Research 108: 3077.

Liu, Q., M. Dai, W. Chen, C. A. Huh, G. Wang, Q. Li, and M. A. Charette. 2012. How significant is submarine groundwater discharge and its associated dissolved inorganic carbon in a river-dominated shelf system? Biogeosciences 9: 1777–1795.

Lohrenz, S. E., and W.-J. Cai. 2006. Satellite ocean color assessment of air-sea fluxes of CO_2 in a river-dominated coastal margin. Geophysical Research Letters 33: L01601.

Lohrenz, S. E., W.-J. Cai, F. Chen, X. Chen, and M. Tuel. 2010. Seasonal variability in air-sea fluxes of CO_2 in a river-influenced coastal margin. Journal of Geophysical Research 115: C10034.

Lohrenz, S. E., M. J. Dagg, and T. E. Whitledge. 1990. Enhanced primary production at the plume oceanic interface of the Mississippi River. Continental Shelf Research 10: 639–664.

Lohrenz, S. E., G. L. Fahnenstiel, D. G. Redalje, G. A. Lang, X. G. Chen, and M. J. Dagg. 1997. Variations in primary production of northern Gulf of Mexico continental shelf waters linked to nutrient inputs from the Mississippi River. Marine Ecology-Progress Series 155: 45–54.

Lohrenz, S. E., G. L. Fahnenstiel, D. G. Redalje, G. A. Lang, M. J. Dagg, T. E. Whitledge, Q. Dortch. 1999. Nutrients, irradiance, and mixing as factors regulating primary production in coastal waters impacted by the Mississippi River plume. Continental Shelf Research 19: 1113–1141.

Lohrenz, S. E., D. G. Redalje, W.-J. Cai, J. Acker, and M. Dagg. 2008. A retrospective analysis of nutrients and phytoplankton productivity in the Mississippi River plume. Continental Shelf Research 28: 1466–1475.

Longhurst, A. 1993. Seasonal cooling and blooming in tropical oceans. Deep-Sea Research 40: 2145–2165.

Longhurst, A. 1995. Interpreting CZCS images of the Amazon Plume: Reply to comments by F. E. Muller-Karger, P.L. Richardson and D. McGillicuddy. Deep-Sea Research 42: 2139.

Malone, T. C., D. J. Conley, T. R. Fisher, P. M. Glibert, L. W. Harding, and K. G. Sellner. 1996. Scales of nutrient-limited phytoplankton productivity in Chesapeake Bay. Estuaries 19: 371–385.

Mcisaac, G. F., M. B. David, G. Z. Gertner, and D. A. Goolsby. 2002. Relating net nitrogen input in the Mississippi River basin to nitrate flux in the lower Mississippi River: A comparison of approaches. Journal of Environmental Quality 31: 1610–1622.

Mckee, B. A., R. C. Aller, M. A. Allison, T. S. Bianchi, and G. C. Kineke. 2004. Transport and transformation of dissolved and particulate materials on continental margins influenced by major rivers: benthic boundary layer and seabed processes. Continental Shelf Research 24: 899–926.

Meade, R., and R. Parker. 1985. National Water Summary. U.S. Geological Survey Water Supply Paper 2275: 49–60.

Meade, R. H. 1995. Setting: Geology, hydrology, sediments, and engineering of the Mississippi River. *In* R. H. Meade (ed.), Contaminants in the Mississippi River, 1987–92, Circular 1133. United States Geological Survey.

Meade, R. H., and J. A. Moody. 2010. Causes for the decline of suspended-sediment discharge in the Mississippi River system, 1940–2007. Hydrological Processes 24: 35–49.
Meybeck, M., L. Laroche, H. H. Durr, and J. P. M. Syvitski. 2003. Global variability of daily total suspended solids and their fluxes in rivers. Global and Planetary Change 39: 65–93.
Milliman, J. D., and R. H. Meade. 1983. World-wide delivery of river sediment to the oceans. The Journal of Geology 91: 1–21.
Müller-Karger, F. E., P. L. Richardson, and D. Mcgillicuddy. 1995. On the offshore dispersal of the Amazon's Plume in the North Atlantic: Comment on the paper by A. Longhurst, "Seasonal cooling and blooming in the tropical oceans". Deep-Sea Research 42: 2127–2131,2133–2137.
Murrell, M., and J. Lehrter. 2010. Sediment and lower water column oxygen consumption in the seasonally hypoxic region of the Louisiana continental shelf. Estuaries and Coasts: 1–13.
Murrell, M. C., R. S. Stanley, J. C. Lehrter, and J. D. Hagy III. 2013. Plankton community respiration, net ecosystem metabolism, and oxygen dynamics on the Louisiana continental shelf: Implications for hypoxia. Continental Shelf Research 52: 27–38.
Nelson, D. M., and Q. Dortch. 1996. Silicic acid depletion and silicon limitation in the plume of the Mississippi River: evidence from kinetic studies in spring and summer. Marine Ecology Progress Series 136: 163–178.
Ortner, P. B., T. N. Lee, P. J. Milne, R. G. Zika, M. E. Clarke, G. P. Podesta, P. K. Swart, P. A. Tester, L. P. Atkinson, and W. R. Johnson. 1995. Mississippi River flood waters that reached the Gulf Stream. Journal of Geophysical Research 100: 13595–13601.
Pakulski, J. D., R. Benner, R. Amon, B. Eadie, and T. Whitledge. 1995. Community metabolism and nutrient cycling in the Mississippi River plume: evidence for intense nitrification at intermediate salinities. Marine Ecology Progress Series 117: 207–218.
Pakulski, J. D., R. Benner, T. Whitledge, R. Amon, B. Eadie, L. Cifuentes, J. Ammerman, and D. Stockwell. 2000. Microbial metabolism and nutrient cycling in the Mississippi and Atchafalaya River plumes. Estuarine Coastal and Shelf Science 50: 173–184.
Prasad, K. S., S. E. Lohrenz, D. G. Redalje, and G. L. Fahnenstiel. 1995. Primary production in the Gulf of Mexico coastal waters using "remotely-sensed" trophic category approach. Continental Shelf Research 15: 1355–1368.
Qureshi, N. A. 1995. The role of fecal pellets in tile flux of carhon to the sea floor on a river-influenced continental shelf subject to hypoxia. Ph.D. Louisiana State University.
Rabalais, N. N. 2002. Nitrogen in aquatic ecosystems. Ambio 31: 102–112.
Rabalais, N. N., R. J. Diaz, L. A. Levin, R. E. Turner, D. Gilbert, and J. Zhang. 2010. Dynamics and distribution of natural and human-caused hypoxia. Biogeosciences 7: 585–619.
Rabalais, N. N., R. E. Turner, R. J. Diaz, and D. Justic. 2009. Global change and eutrophication of coastal waters. ICES Journal of Marine Science 66: 1528–1537, doi:10.1093/icesjms/fsp047.
Rabalais, N. N., R. E. Turner, Q. Dortch, D. Justic, V. J. Bierman, and W. J. Wiseman. 2002a. Nutrient-enhanced productivity in the northern Gulf of Mexico: past, present and future. Hydrobiologia 475: 39–63.
Rabalais, N. N., R. E. Turner, B. K. Sen Gupta, D. F. Boesch, P. Chapman, and M. C. Murrell. 2007. Hypoxia in the northern Gulf of Mexico: Does the science support the plan to reduce, mitigate, and control hypoxia? Estuaries and Coasts 30: 753–772.
Rabalais, N. N., R. E. Turner, and W. J. Wiseman. 2001. Hypoxia in the Gulf of Mexico. Journal of Environmental Quality 30: 320–329.
Rabalais, N. N., R. E. Turner, and W. J. Wiseman. 2002b. Gulf of Mexico hypoxia, aka "The dead zone". Annual Review of Ecology and Systematics 33: 235–263.
Rabalais, N. N., R. E. Turner, and W. J. Wiseman, Jr. 2000. Hypoxia in the Gulf of Mexico. Journal of Environmental Quality 30: 320–329.

Rabalais, N. N., W. J. Wiseman, R. E. Turner, B. K. Sengupta, and Q. Dortch. 1996. Nutrient changes in the Mississippi River and system responses on the adjacent continental shelf. Estuaries 19: 386–407.

Rabouille, C., F. T. Mackenzie, and L. M. Ver. 2001. Influence of the human perturbation on carbon, nitrogen, and oxygen biogeochemical cycles in the global coastal ocean. Geochimica Et Cosmochimica Acta 65: 3615–3641.

Raymond, P. A., and J. J. Cole. 2003. Increase in the export of alkalinity from North America's largest river. Science 301: 88–91.

Raymond, P. A., M. B. David, and J. E. Saiers. 2012. The impact of fertilization and hydrology on nitrate fluxes from Mississippi watersheds. Current Opinion in Environmental Sustainability 4: 212–218.

Raymond, P. A., N. H. Oh, R. E. Turner, and W. Broussard. 2008. Anthropogenically enhanced fluxes of water and carbon from the Mississippi River. Nature 451: 449–452.

Redalje, D. G., S. E. Lohrenz, and G. L. Fahnenstiel. 1994. The relationship between primary production and the vertical export of particulate organic matter in a river impacted coastal ecosystem. Estuaries 17: 829–838.

Redfield, A. C. 1958. The biological control of chemical factors in the environment. American Scientist 46: 205–221.

Riley, G. A. 1937. The significance of the Mississippi River drainage for biological conditions in the northern Gulf of Mexico. Journal of Marine Research 1: 60–74.

Sahl, L. E., W. J. Merrell, and D. C. Biggs. 1993. The influence of advection on the spatial variability of nutrient concentrations on the Texas-Louisiana continental shelf. Continental Shelf Research 13: 233–251.

Salisbury, J., M. Green, C. Hunt, and J. Campbell. 2008. Coastal acidification by rivers: a threat to shellfish? EOS 89: 513–514.

Sampere, T. P., T. S. Bianchi, and M. A. Allison. 2011. Historical changes in terrestrially derived organic carbon inputs to Louisiana continental margin sediments over the past 150 years. Journal of Geophysical Research 116: G01016.

Scavia, D., D. Justic, and V. J. Bierman. 2004. Reducing hypoxia in the Gulf of Mexico: Advice from three models. Estuaries 27: 419–425.

Scavia, D., N. N. Rabalais, R. E. Turner, D. Justic, and W. J. Wiseman. 2003. Predicting the response of Gulf of Mexico hypoxia to variations in Mississippi River nitrogen load. Limnology and Oceanography 48: 951–956.

Shen, Y., C. G. Fichot, and R. Benner. 2012. Floodplain influence on dissolved organic matter composition and export from the Mississippi-Atchafalaya River system to the Gulf of Mexico. Limnology and Oceanography 57: 1149–1160.

Sklar, F. H., and R. E. Turner. 1981. Characteristics of phytoplankton production off Barataria Bay in an area influenced by the Mississippi River. Contributions in Marine Science 24: 93–106.

Smith, S. M., and G. L. Hitchcock. 1994. Nutrient enrichments and phytoplankton growth in the surface waters of the Louisiana Bight. Estuaries 17: 740–753.

Smith, S. V. 1984. Phosphorus versus nitrogen limitation in the marine environment. Limnology and Oceanography 29: 1149–1160.

Steffen, W., A. Sanderson, P. Tyson, J. Jäger, P. Matson, B. Moore, F. Oldfield, K. Richardson, H. J. Schellnhuber, B. L. Turner, and R. J. Wasson RJ. 2005. The Anthropocene Era: How Humans are Changing the Earth System Global Change and the Earth System, pp. 81–141. Global Change – The IGBP Series (closed). Springer Berlin Heidelberg.

Subramaniam, A., P. L. Yager, E. J. Carpenter, C. Mahaffey, K. Björkman, S. Cooley, A. B. Kustka, J. P. Montoya, S. A. Sañudo-Wilhelmy, R. Shipe, and D. G. Capone. 2008. Amazon River enhances diazotrophy and carbon sequestration in the tropical North Atlantic Ocean.

Proceedings of the National Academy of Sciences of the United States of America 105: 10460–10465.

Sylvan, J. B., Q. Dortch, D. M. Nelson, A. F. M. Brown, W. Morrison, and J. W. Ammerman. 2006. Phosphorus limits phytoplankton growth on the Louisiana shelf during the period of hypoxia formation. Environmental Science & Technology 40: 7548–7553.

Ternon, J., C. Oudot, A. Dessier, and D. Diverres. 2000. A seasonal tropical sink for atmospheric CO_2 in the Atlantic ocean: the role of the Amazon River discharge. Marine Chemistry 68: 183–201.

Thomas, W. H., and E. G. Simmons. 1960. Phytoplankton production in the Mississippi Delta, pp. 103–116. *In* F. Shepard (ed.), Recent Sediments, Northwest Gulf of Mexico. American Association of Petrologists.

Trefry, J. H., S. Metz, T. A. Nelsen, R. P. Trocine, and B. J. Eadie. 1994. Transport of particulate organic carbon by the Mississippi River and its fate in the Gulf of Mexico. Estuaries 17: 839–849.

Tsunogai, S., S. Watanabe, and T. Sato. 1999. Is there a "continental shelf pump" for the absorption of atmospheric CO2? Tellus B 51: 701–712.

Turner, R. E., N. Qureshi, N. N. Rabalais, Q. Dortch, D. Justic, R. F. Shaw, and J. Cope. 1998. Fluctuating silicate:nitrate ratios and coastal plankton food webs. Proceedings of the National Academy of Sciences of the United States of America 95: 13048–13051.

Turner, R. E., and N. N. Rabalais. 1999. Suspended particulate and dissolved nutrient loadings to Gulf of Mexico estuaries, pp. 89–107. *In* T. S. Bianchi, J. R. Pennock and R. R. Twilley (eds.), Biogeochemistry of Gulf of Mexico estuaries. John Wiley & Sons.

Turner, R. E., and N. N. Rabalais. 2003. Linking landscape and water quality in the Mississippi river basin for 200 years. Bioscience 53: 563–572.

Turner, R. E., and N. N. Rabalais. 2004. Suspended sediment, C, N, P, and Si yields from the Mississippi River Basin. Hydrobiologia 511: 79–89.

Turner, R. E., N. N. Rabalais, R. B. Alexander, G. Mcisaac, and R. W. Howarth. 2007. Characterization of nutrient, organic carbon, and sediment loads and concentrations from the Mississippi River into the Northern gulf of Mexico. Estuaries and Coasts 30: 773–790.

Turner, R. E., N. N. Rabalais, and D. Justic. 2006. Predicting summer hypoxia in the northern Gulf of Mexico: Riverine N, P, and Si loading. Marine Pollution Bulleti 52: 139–148.

Turner, R. E., N. N. Rabalais, and D. Justić. 2012. Predicting summer hypoxia in the northern Gulf of Mexico: Redux. Marine Pollution Bulletin 64: 319–324.

Vörösmarty, C. J., M. Meybeck, B. Fekete, K. Sharma, P. Green, and J. P. M. Syvitski. 2003. Anthropogenic sediment retention: major global impact from registered river impoundments. Global and Planetary Change 39: 169–190.

Walling, D. E. 2006. Human impact on land–ocean sediment transfer by the world's rivers. Geomorphology 79: 192–216.

Wang, X. C., R. F. Chen, and G. B. Gardner. 2004. Sources and transport of dissolved and particulate organic carbon in the Mississippi River estuary and adjacent coastal waters of the northern Gulf of Mexico. Marine Chemistry 89: 241–256.

Wawrik, B., and J. H. Paul. 2004. Phytoplankton community structure and productivity along the axis of the Mississippi River plume in oligotrophic Gulf of Mexico waters. Aquatic Microbial Ecology 35: 185–196.

Wawrik, B., J. H. Paul, D. A. Bronk, D. John, and M. Gray. 2004. High rates of ammonium recycling drive phytoplankton productivity in the offshore Mississippi River plume. Aquatic Microbial Ecology 35: 175–184.

Zhai, W., and M. Dai. 2009. On the seasonal variation of air – sea CO2 fluxes in the outer Changjiang (Yangtze River) Estuary, East China Sea. Marine Chemistry 117: 2–10.

19

Sedimentary carbon dynamics of the Atchafalaya and Mississippi River Delta system and associated margin

T. S. Bianchi, M. Goñi, M. A. Allison, N. Chen, and B. A. McKee

1. Introduction

Approximately 87% of Earth's land surface is connected to the ocean by rivers (Ludwig and Probst 1998). Rivers are the active interface between terrestrial and oceanic environments, providing connectivity between the two largest global sinks for atmospheric CO_2 (Cai 2011).

The annual flux of carbon from rivers is approximately 1 Pg C, which is of the same order as the net ocean-atmosphere exchange (Mackenzie and Lerman 2006 and references therein). The riverine carbon flux is comprised of an organic fraction (0.5 Pg C y^{-1}) containing roughly equal fractions of dissolved and particulate matter, and an inorganic fraction (Richey 2004).

In modern marine environments, riverine delivery of sedimentary organic matter (SOM) to continental margins is an important means by which terrigenous OM is introduced to the global ocean, influencing global biogeochemical cycles and the ocean's ability to sequester atmospheric CO_2 (see Chapter 1). Yet, there remains considerable uncertainty in our ability to adequately quantify carbon exchange from land to the coastal ocean, particularly in large river delta-front estuaries (LDE) (Bianchi and Allison 2009, and references therein), and in our understanding of the processes influencing the fate of terrigenous carbon in coastal and deep sea sediments (Berner 1982; Sarmiento and Sundquist 1992; Hedges and Keil 1995; Schlünz and Schneider 2000; Burdige 2007). The Mississippi-Atchafalaya River system is one of the best studied LDEs to date both in terms of basic carbon dynamic processes and the impact of human modifications. The material flux from LDEs has a profound impact on the biogeochemistry of the world's ocean and, through this river-ocean interaction, plays an important role in global biogeochemical cycles of elements such as carbon and nitrogen.

2. Sediment and carbon transport dynamics

2.1. Sediment input and deposition in the Atchafalaya River (AR) and Mississippi River (MR) LDE

The MR has the largest drainage basin area in North America (3.3×10^6 km^2; Milliman and Syvitski 1992) and the seventh largest water discharge and suspended load among the world's rivers (Milliman and Meade 1983). Approximately 60% of the total suspended matter and 66% of the total dissolved materials transported from the conterminous United States to the ocean are carried by the MR

(Presley et al. 1980). The MR delivers approximately 2.1×10^{14} g of suspended matter to the northern Gulf of Mexico (GOM) shelf each year (Milliman and Meade 1983; Meade and Parker 1985). In the final, low-gradient section in southern Louisiana, the river's bedload is made up of 90% fine sand, whereas the disaggregated suspended load consists of about 65% clay and 35% silt at low discharge, with a contribution of up to 23% sand at higher discharges (Coleman 1988; Allison and Meselhe 2010; Nittrouer et al. 2011). The MR water and sediment discharge hydrograph exhibits a large seasonal and interannual variability, with high discharge (>30,000 m^3 s^{-1}) typically occurring between January and May, with a series of individual peaks that last 1–2 weeks. Mean water discharges during the high-water months can be expected to be about three times the discharges during the low water months (Meade 1995). Periods of net storage and remobilization of sediments in the lower Mississippi can be predicted based on daily water discharge rates measured at Tarbert Landing, Mississippi (United States), with significant deposition and net storage of river sediments occurring during periods of river discharge less than 14,000 m^3 s^{-1} and significant resuspension and net remobilization of sediments occurring during periods of river discharge greater than 20,000 m^3 s^{-1} (Demas and Curwick 1988). As a result, periods of sediment storage are typically 4–8 months in the lower Mississippi River. In the final, tidal reach of the river below Baton Rouge, Louisiana, the timing of the sediment and water discharge maxima to the GOM are offset, with discharge of fines increasing by a factor of 4–5 for a few days during the rising limb of individual spikes in water discharge that last 1–2 weeks (Mossa 1996; Allison et al. 2012). This phenomenon is attributed to hysteresis in the sediment supply from the upper basin, and to resuspension of fines stored on the channel bed in low discharge in the ca. 200-km-long tidal and estuarine reach in Louisiana (Dagg et al. 2005; Galler and Allison 2008).

The suspended sediment load input to the lowermost MR has decreased by approximately one-half since the 1800s, the majority of that occurring since 1950 owing to construction of dams and river control works and soil conservation measures (Horowitz 2010; Meade and Moody 2010). That load is divided between two distributaries, the AR and the MR, that are the product of deltaic lobe switching in the late Holocene on 10^3 year timescales (Frazier 1967; Penland et al. 1988). Water discharge into the AR distributary is fixed at 30% of the total MR and Red River discharge by a series of control structures at Old River Control (ORC). Allison et al. (2012) examined sediment budgets downriver of ORC for three flood years (2008–2010) and determined that both distributaries were efficient at sequestering sediment by overbank and channel aggradation processes and that the MR distributary experiences additional water and sediment loss through natural and artificial exits upriver of the deepwater passes to the GOM. They calculated that only 60% of the fines (<62.5 μm; 28% AR, 32% MR) and 15% of the sand (9% AR, 6% MR) supplied to the lowermost river system in suspension by the combined Mississippi-Red exits the deepwater passes of the MR or enters the GOM through Atchafalaya Bay, with the rest being retained in bays and nearshore areas.

The dual discharge points of the MR-AR produce shelf depocenters separated by about 200 km and of distinct depositional style. Southwest Pass (SWP) of the MR carries about 64% of the MR water flux through the deepwater passes (Allison et al. 2012). Sand delivered to SWP and the other deepwater MR passes is initially deposited within four channel widths seaward of the distributary mouth to produce a prograding distributary-mouth-bar (Wright and Coleman 1974; Coleman 1988). Most of the fine-grained sediments are delivered farther afield to a fan-shaped prodelta and deposited at shelf water depths of >30 m (Coleman 1981; Adams et al. 1987). Corbett et al. (2006) utilized ^{210}Pb geochronology to examine decadal-scale shelf sediment accumulation in the prodelta west of SWP. They identified three zones: (1) a zone proximal to SWP in 30–100-m water depths where non–steady-state profiles indicated rapid accumulation (>2.5 cm y^{-1}), (2) a nearshore (<20 m water depth) zone with limited to no modern accumulation, and (3) a more distal zone seaward of the 20-m isobath dominated by steady-state accumulation (<0.7 cm y^{-1}) beneath a ~10-cm-thick biologically mixed layer. Allison et al. (2007) demonstrated that the historical decline in sediment supply from the basin is reflected in declining ^{210}Pb-derived burial rates in the MR shelf depocenter. Water and sediment is discharged by the AR enter shallow (2–4 m deep) Atchafalaya Bay, where much of the sand is deposited in bay head deltas that have been prograding into the Bay since the large flood of 1973 (Shlemon 1975; Van Heerden and Roberts 1988; Roberts 1998). Much of the fine-grained sediment is either stored in the Bay, or on the shelf in water depths of <8 m, associated with a subaqueous clinoform (Neill and Allison 2005). ^{210}Pb geochronology in Neill and Allison (2005) indicates that this clinoform is prograding and aggrading rapidly: accumulation rates exceed 3 cm y^{-1} on the foreset region. Further westward, the clinoform merges with a progradational shore-face and mudflat deposit along the Chenier Coast of Louisiana (Draut et al. 2005). The active (Plaquemine-Balize) lobe of the delta has prograded in the last ~1,300 years (Tornqvist et al. 1996) close to the shelf break.

2.2. Mechanisms and timing of sediment remobilization on the MR and AR shelf

The Mississippi-Atchafalaya delta is a model for the geologic evolution of deltas controlled primarily by the constructional processes of sediment input (Coleman and Wright 1975; Galloway 1975; Wright 1985; Orton and Reading 1993; Postma 1995) because of its limited diurnal astronomical tide (30–40 cm amplitude), 10 cm s^{-1} currents in <10 m water depth, and a relatively low energy wave climate (Wells and Roberts 1980; Kemp 1986; Walker and Hammack 2000). However, a number of recent studies have indicated that the shelf depocenters episodically experience rapid deposition and resuspension with major implications for carbon dynamics and burial; these are driven by (1) riverine loading, (2) winter frontal storms, and (3) tropical storms (see Chapters 2 and 4).

Corbett et al. (2007) utilized the particle reactive radiotracer ^{7}Be (half-life of 53 days) as an indicator of MR-derived sediment on the shelf adjacent to SWP. They found a significant correlation of monthly ^{7}Be inventories in the seabed with MR discharge at a station proximal to SWP (10 km away, 35 m water depth) in the zone of high burial rates. At more distal stations, this was not observed, and seasonal ^{7}Be inventories were correlated with wave orbital velocities, which they suggest was a supply from seafloor resuspension owing to winter (December–April) frontal storms. Allison et al. (2000) observed a similar ^{7}Be relationship on the Atchafalaya shelf depocenter: seabed inventories on the subaqueous *mud* clinoform remained low in the December–March high-wave energy (resuspension) period despite rising AR discharges and increased 3–v5-fold in April–May, coincident with high discharge and declining frontal storm energies. These studies suggest that plume-supplied river sediment is confined to proximal areas on the MR and AR depocenters and that supply to more distal areas is more closely linked to seasonal forcing.

The Louisiana coastal zone has a complex coastal circulation regime where the wind-driven and westward-flowing Louisiana Coastal Current (Cochrane and Kelly 1986; Mezic et al. 2010) interacts with the multiple freshwater plumes emanating from the MR and AR mouths and the brackish plumes advected by tides through the passes in the adjacent shallow inter-distributary bays (Wiseman et al. 1997; Wang and Justic 1999). In winter and early spring (November–April), the coastal zone of the northern Gulf is swept periodically (about 3–10-day intervals) by atmospheric cold fronts accompanied in the prefrontal phase by energetic onshore swells (8–10-second period, 1–2-m wave height) and winds (30–80 km h^{-1}). On the innermost shelf (less than approximately 10 m water depth; Sheremet et al. 2011) and in adjacent shallow estuaries, wave-induced bottom turbulence and direct wind stress breaks down water column circulation and resuspends significant quantities of sediment (Walker and Hammack 2000; Allison et al. 2000; Jaramillo et al. 2009). As wave activity decreases at the end of the storm, and the direction of propagation of short-wave fields rotates to align with postfrontal seaward winds, hindered settling of resuspended sediment leads to the formation and seaward advection of episodic wave-supported sediment gravity flows (term coined by Parsons et al. 2007) on the AR clinoform with a duration of about 12 hours and thickness less than 30 cm (Jaramillo et al. 2009; Safak et al. 2010; Sheremet et al. 2011). These frontal events also create a regular, 1–2-m magnitude coastal setup and set-down cycle that exchanges water between the estuaries and shelf and floods the marshes (and beaches) that fringe the estuaries (see Chapter 3). These events are significantly larger than those induced by the astronomical tides but less than those caused by tropical storms (Walker and Hammack 2000).

In the relatively low-energy summer period (June–October), the coastal circulation pattern is complicated by salinity-induced stratification stretching along the inner shelf from the MR mouth to the Texas border (Wiseman et al. 1997; Hetland and DiMarco 2008; DiMarco et al. 2010). Periodically, tropical cyclones break down this stratification and create massive seafloor resuspension and cross-shore exchange of sediment between bay, shelf, and wetland environments. Studies of recent hurricanes suggest that these larger episodic events can resuspend sediment from shelf environments to ~40 m water depth (Allison et al. 2010) with storm surge (e.g., coastal setup) to 5 m or more locally. Studies of a number of recent northern GOM hurricanes (e.g., Lili, Ivan, Katrina-Rita, and Ike) (Sheremet et al. 2005; Allison et al. 2005, 2010; Goff et al. 2010) indicate that, in addition to the flood surge that submerges wetlands, the subsequent ebb surge can advect inner shelf sediments and particulate organic matter sourced from wetland-bay-shelf settings and redeposit them across the entire shelf and even into deeper waters settings such as Mississippi Canyon (e.g., Goñi et al. 2006, 2007; Mead and Goñi 2006; Sampere et al. 2008).

2.3. Linkages between sediments and carbon dynamics across the shelf/slope

River systems can be divided into three strongly interactive domains: (1) the basin, extending from upland streams to the tidally influenced lower river; (2) the river-ocean interface, extending from the tidally influenced freshwater portion of the river to the river mouth; and (3) the margin, which extends from the inner shelf, adjacent to the river mouth, to the shelf-slope break/canyon (see Chapter 4). The importance of LDEs to global organic carbon (OC) burial (Hedges and Keil 1995) is evidenced by the tremendous magnitude of material fluxes to these regions.

Approximately 80% of the total OC preserved in marine sediments occurs in "terrigenous-deltaic" (e.g., the Margin of the LDE) regions near river mouths (Romankevich 1984; Berner 1989). Yet, despite the importance of these environments, there remains a fundamental lack of understanding about OC dynamics operating within these regions. This lack of understanding results from the high spatial and temporal variability in the sources of OC: (1) primary production by phytoplankton and (2) discharge of terrestrially derived OC (OC_{terr}) by rivers. The details of this source variability in OC and its impact on OC diagenesis for the Mississippi-Atchafalaya margin are discussed in Sections 3.1 and 3.2 that follow. A second factor contributing to our lack of understanding about OC dynamics in coastal regions is the complex architecture of margin environments, particularly the deeper, canyon regions associated with large rivers, and the potential role they play in OC dynamics as depo-centers for OC burial and an export pathway for delivery of OC to the deep sea.

Canyons are deep incisions of the continental shelf and slope and remain one of the understudied regions of the continental margin with respect to OC dynamics. One reason for this is their complex wall and floor terrain, which results in high spatial variability and makes detailed studies difficult. Despite a paucity of detailed studies, canyons are known to be: (1) characterized by high faunal biomass, (2) important conduits for delivery and burial of OC in the oceans, and (3) "fast-track corridors" for sediment and carbon transported from land to the deep sea (Weaver et al. 2004; Liu and Lin 2004; Tesi et al. 2008, 2010). The proximity and orientation of river mouth-canyon systems relative to alongshore sediment transport pathways leads to differences in the capture of riverine sediment and OC between LDEs.

Understanding the connectivity between coastal systems and the deep sea has received considerable attention in recent years. Many LDEs are surmised to export a relatively small volume of river-derived particulate matter seaward of the shelf break either (1) because of: (1) their location on wide, passive continental margins where deltaic sedimentation is confined to the inner shelf or (2) because they are located where shelf-dominated coastal currents occur (McKee et al. 2004). This has led to the prevailing dogma that sediments reach the continental slope through hemipelagic processes (Shanmugam et al. 1985; Stow et al. 1985). However, increasingly it is being recognized that some LDEs (e.g., Sepik, Ganges-Brahmaputra, Eel, Rhone, Po) are characterized by direct export of large volumes of sediment and OC to the lower continental margin owing to: (1) a narrow (active continental margin) shelf (Kineke et al. 2000; Mullenbach and Nittrouer 2006); (2) progradation of the deltaic clinoform in the late Holocene near to the shelf edge (Coleman et al. 1998); (3) density-driven, cold shelf water (winter) advection (Puig and Palanques 1998); or (4) landward incision of the associated submarine canyon so that it intercepts along-shelf transport pathways (Johnson et al. 2001; Michels et al. 2003). The margin adjacent to the Mississippi-pathway discharge point possesses two of these characteristics (2 and 4), making it a candidate for large-scale OC export. The recent recognition of active export of river and shelf sediment to the deep sea on LDEs still lacks a rigorous examination of its implications for global carbon cycling, particularly the role of submarine canyons, where enhanced axial downslope flows can rapidly disperse OC out to abyssal depths.

The main pathway of freshwater exiting the river (Southwest Pass) and the net shelf dispersal pathway (driven by a coastal current) are toward the west, where the Mississippi Canyon incises the mid-to-outer shelf (Fig. 19.1). ^{210}Pb sediment accumulation rates on the shelf show a tongue of higher accumulation extending outward from the depocenter on the delta front toward the canyon head; this setting may allow for the rapid transport of shelf-derived primary production to the canyon floor (Bianchi et al. 2006). In fact, recent work has shown that labile OM, likely produced by in situ

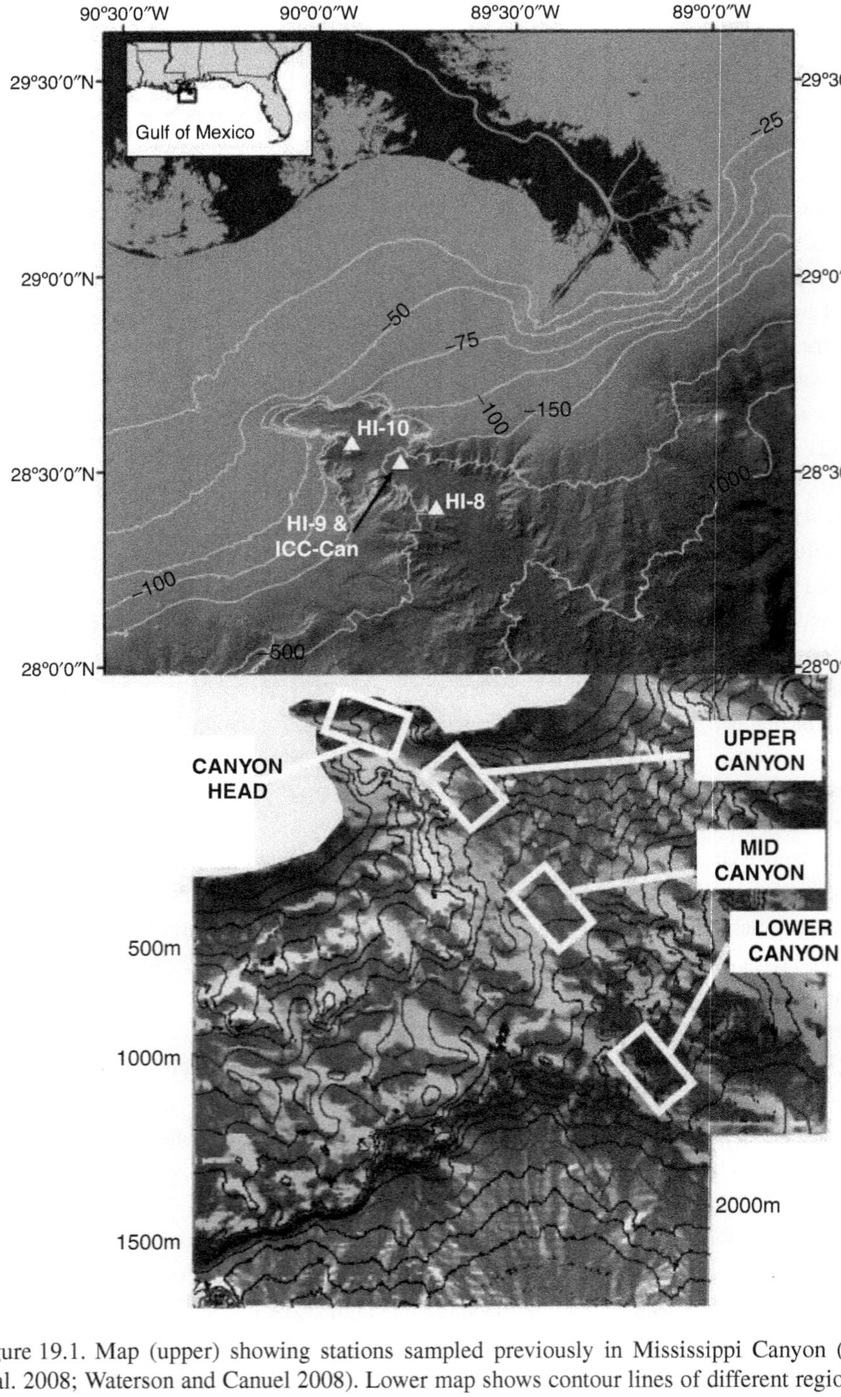

Figure 19.1. Map (upper) showing stations sampled previously in Mississippi Canyon (Sampere et al. 2008; Waterson and Canuel 2008). Lower map shows contour lines of different regions of the Mississippi Canyon.

diatom production in the MRP, is rapidly transported to the Mississippi Canyon (Bianchi et al. 2006; Sampere et al. 2008; Waterson and Canuel 2008). This work suggests that productive LDEs, such as the Mississippi, may be important conduits for transporting fixed carbon from highly productive MRP waters on the shelf to deeper benthic communities.

Given that ^{210}Pb sediment accumulation rates on the shelf apron of the upper canyon are relatively low (<0.5 g cm^{-2} y^{-1}), the question arises about sources and pathways of riverine and shelf sedimentary OC (SOC) into the canyon. Although direct plume deposition seems unlikely given what is known about rapid settling of river-derived particulate matter (PM) from the plume, other active processes can rapidly redistribute fresh riverine PM initially deposited on the shallow delta and adjacent inner shelf. These include wind-driven (down-welling) transport of benthic boundary layers (BBL) (Baker and Hickey 1986; DiMarco and Reid 1998) into the canyon head or resuspension of PM by winter frontal and tropical storms (Allison et al. 2000; Walsh et al. 2006; Sampere et al. 2008; Allison et al. 2010). Episodic mass movements are also well-known along the SW Pass delta front (Coleman et al. 1998) and have been suggested to be a major mechanism for spin-off turbidity currents and nepheloid layers on the continental margin (Mulder and Alexander 2001). This, and recent seismic evidence of active downlope sediment advection at the canyon head, suggest that the canyon head and the western wall of the upper canyon may be the preferential pathway for SOC export into the Mississippi Canyon (Allison et al. unpublished).

2.4. Sedimentary processes in upper Mississippi Canyon

Borings and seismic examination show that upper Mississippi Canyon has been a depocenter for enormous quantities of sediment on Holocene timescales: up to 550 m since 12.5 kyBP (Coleman and Roberts 1988; Goodwin and Prior 1989). Cores collected in the canyon axis during our July 2003 ICC study show evidence, including radiotracer profiles, mud turbidite layers, and terrigenous sand layers, that modern sediment accumulation remains high (higher than the adjacent shelf margin; Bianchi et al. 2006). High-resolution (CHIRP) sub-bottom transects collected along the east wall of the canyon head during a post-Katrina shelf study in October 2005 (Allison et al. 2010) and a remapping of Exclusive Economic Zone (EEZ) survey multi-beam bathymetric data obtained from NOAA show examples of all the major acoustic facies characteristic of active downslope sediment transport. Similar features were observed in Swatch of No Ground Canyon (SoNG), where up to 30% of the Ganges-Brahmaputra sediment load is intercepted (Kuehl et al. 1997; Kottke et al. 2003). These include multiple canyon floor thalwegs, active growth faults on the canyon and gully walls, areas of deformed surficial strata characteristic of rotational sliding, and surficial and shallow-buried, spatially extensive, and acoustically transparent event layers. The latter are interpreted as debris flows and turbidites in SoNG Canyon. In the head of Mississippi Canyon, these layers are decimeters to several meters thick.

Evidence of active sediment delivery in the upper section of Mississippi Canyon contrasts with seafloor mapping and seismic stratigraphic analysis, which indicates the latest phase of lobe formation on Mississippi Fan dates from the late Pleistocene (Stelting et al. 1986; Twichell et al. 1991; Nelson et al. 1992). Prior to recent efforts, this was accepted for the Bengal Fan as well. However, it has recently been shown in piston cores (Weber et al. 1997; Schwenk et al. 2003) that gravity-driven

Table 19.1. *Major hurricanes (category 3 at landfall) to strike Central Louisiana or Mississippi coastlines from 1851 to 2003 (from Sampere et al. 2011)*

Year	Month	State/Category[a]	Central Pressure (mb)	(mph)	Name
1852	Aug	MS, 3; LA, 2	961	115	"Great Mobile"
1855	Sept	LA, 3; MS, 3	950	127	"Middle Gulf Shore"
1856	Aug	LA, 4	934	150	"Last Island"
1860	Aug	LA, 3; MS, 3	950	127	n/a
1879	Sept	LA, 3	950	127	n/a
1893	Oct	LA, 4; MS, 2	948	132	"Chenier Caminanda"
1909	Sept	LA, 3; MS, 2	952	121	"Grand Isle"
1915	Sept	LA, 4	931	n.d	"New Orleans"
1916	July	MS, 3	948	n.d	n/a
1926	Aug	LA, 3	955	n.d	n/a
1934	June	LA, 3	962	n.d	n/a
1947	Sept	MS, 3; LA, 3	940	n.d	n/a
1964	Oct	LA, 3	950	n.d	Hilda
1965	Sept	LA, 3	948	n.d	Betsy
1969	Aug	LA, 5; MS, 5	909	n.d	Camille
1974	Sept	LA, 3	952	n.d	Carmen
1985	Sept	AL, 3; MS, 3	959	115	Elena
1992	Aug	LA, 3	922	167	Andrew

Data from NOAA's best-track database (HURDAT). MS: Mississippi; LA: Louisiana; n.d.: no reliable data available.

[a] Saffir-Simpson Scale.

transport is active on the upper Bengal Fan and that up to 7 m of overspill turbidite deposits on channel margin levees have been deposited since 6,000 yBP.

2.5. Effects of hurricanes on sediment and OC transport

Mobilization of large amounts of sediments and deposition of storm layers have been observed in shelf sediments adjacent to the Mississippi-Atchafalaya for Hurricanes Ivan, Katrina, and Rita (Goñi et al. 2006; Mead and Goñi, 2006; Dail et al. 2007; Goñi et al. 2007). Overall, the total accumulation of TOC associated with such events was of the same magnitude as the combined annual inputs from the MR and AR (Goñi et al. 2007). The mechanisms for hurricane-induced mobilization and dispersal include wave-induced resuspension of large amount of sediment from the shelf during storm passage and seaward transport by ebb surge flows, perhaps involving cross-shelf gravity or mud flows (Allison et al. 2005; Walsh et al. 2006). However, only the most severe storms appear capable of redistributing material to distal portions of the shelf and slope (Dail et al. 2007). Recent work showed that Hurricane Rita completely removed sediments deposited following Katrina at near shelf sites on the Louisiana Shelf (Goñi et al. 2007). Sampere et al. (2011) identified 18 major hurricanes that made landfall as category 3 storms in eastern Louisiana and/or Mississippi since 1850 to 2003 that could have potentially produced hurricane layers preserved in the sedimentary record (Table 19.1).

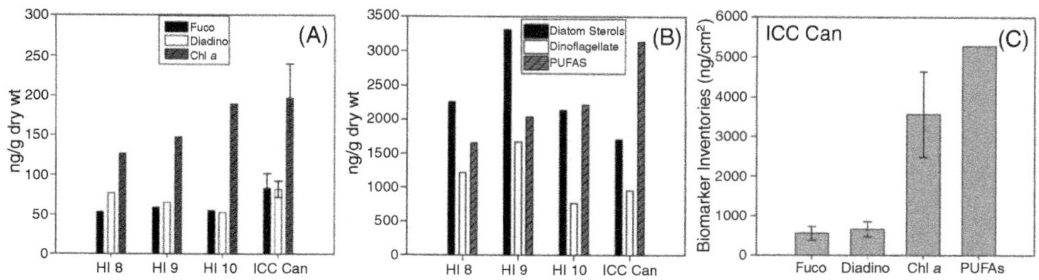

Figure 19.2. (A) Concentrations (ng g^{-1} dry wt. sediment) of algal pigment biomarkers fucoxanthin (Fuco), diadinoxanthin (Diadino), and chlorophyll-*a* (Chl *a*) in surface sediments from 4 stations in the Mississippi Canyon in July 2003 (ICC-Can) and October 2004 (HI-8, HI-9, and HI-10). Diadinoxanthin and fucoxanthin are biomarkers for diatoms. (B) Concentrations (ng g^{-1} dry wt. sediment) of diatom sterols, dinosterol (from dinoflagellates), and polyunsaturated fatty acids (PUFAs) in surface sediments from 4 stations in the Mississippi Canyon. (C) Inventories of algal pigments and PUFAs (ng cm^{-2}) at Station ICC-Can collected in July 2003; downcore samples at the 3 stations collected in October 2004 were not available for biomarker analyses (adapted from Bianchi et al. 2006).

Based on these sediment profiles, we speculate that nearshore areas do preserve hurricane layers on up to decadal timescales, but that these signals are frequently not apparent due to remobilization and bioturbation-induced degradation. The lack of distinct laminations in deeper core X-radiographs confirm that hurricane layers are often lost on century-time scales (Sampere et al. 2011).

Tropical storms contribute to the offshore transport of sediment and associated organic matter to the deeper regions of the northern GOM. These hurricane-induced inputs include plankton-derived detritus from the shelf that contribute labile organic matter to sites that would otherwise be considered oligotrophic (e.g., Lohrenz et al. 1999). For example, biomarkers of labile algal organic matter (e.g., algal pigments and polyunsaturated fatty acids [PUFA]) were observed in surface sediments at all four stations in the Mississippi Canyon (Fig. 19.2a and 19.2b; Bianchi et al. 2006; Waterson and Canuel 2008). Like ^{7}Be, these biomarkers reflected recent deposition of labile OC (Canuel and Martens 1996). These "fresh" biomarkers were likely derived from sediments redistributed from the productive Louisiana shelf region – where diatom productivity and fluxes to sediments are high. Moreover, it is likely that macrobenthic polychaete communities, such as those observed in the canyon in 2003 (Wei et al. 2010), are supported by shelf-derived sediments rich in diatom phytodetritus. Thus, despite the predominant notion that canyon sediments are typically unstable and lack adequate food resources to support significant macrobenthic communities, as described earlier, new evidence indicates inputs of labile carbon resources from productive shelf environments.

Hurricanes can modify these linkages between and transport of fixed labile carbon from highly productive plume waters in LDEs on the shelf to fuel deeper benthic communities. As mentioned earlier, Sampere et al. (2008) showed that the ^{7}Be inventory at HI-9 was higher than at ICC-Can, its nearest site (Fig. 19.1). Because ^{7}Be inventories integrate depositional input over time (250–300 days), passage of Hurricane Ivan between sampling at ICC-Can (July 2003) and HI-9 (October 2004) may explain this difference. This is consistent with high rates of sediment accumulation (i.e., 6–19 times the typical annual accumulation) at the 50-m isobath and deeper observed following the passage of Hurricanes Katrina and Rita (Walsh et al. 2006). The macrobenthos observed at Station ICC-Can in July 2003 was absent at Station HI-9, or the other canyon stations in October 2004 (Corbett et al. 2006), suggesting that these deeper macrobenthic communities had not yet recovered following the large depositional events (Corbett et al. 2006). Further work is needed to better understand the temporal and spatial dynamics of inputs of labile shelf-derived carbon and deep macrobenthic communities before and after hurricane events.

Gravity cores collected at ICC-Can also show evidence of mud turbidites at decimeters below the tube-dwelling community, indicating episodic hyperpycnal down-canyon transport at lower frequency intervals than storms and likely scouring of epibenthic communities (Wei et al. 2010). Although there is a 25-m difference in water depth between Station ICC-Can and HI-9, the sites are immediately adjacent to one another on the canyon floor (Fig. 19.1). Inputs of ^{7}Be and Chl-*a* to these sites indicate that these stations received similar inputs from river/shelf sources, which would not be the case if delivery of material was via bottom-hugging hyperpycnal transport. Furthermore, although algal pigments and PUFAs decreased in concentration in samples collected before and after Hurricane Ivan, sterols (a more stable class of lipid biomarkers) increased in concentration (Fig. 19.2). A potential mechanism for these differences involves redistribution of sediments from the delta to the Canyon during Hurricane Ivan. Pigment and PUFA inventories at Station ICC-Can confirmed that labile phytodetritus was advected to the canyon, most likely during passage of winter fronts and sporadically by hurricanes (Corbett et al. 2004, 2006).

3. Sources and decay of organic matter: linkages with nutrients and transport processes

3.1. Controls on phytoplankton sources and carbon diagenesis across a depositional gradient

The nutrient flux/cycle, primary production, and OC deposition on the Mississippi/Atchafalaya deltaic margin are dynamic, varying on spatial and temporal scales linked to the variable freshwater input by the MR and AR (see Chapter 18). Previous studies have revealed a statistically significant relationship between annual Mississippi River nitrogen load and primary production in the plume region near the Mississippi delta (Lohrenz et al. 1997; Lehrter et al. 2009) and with the spatial extent of the hypoxic area in the summer (Turner et al. 2005; Greene et al. 2009). However, this relationship does not apply to the entire Louisiana shelf (Quigg et al. 2011). Recent studies have shown that phytoplankton productivity is limited by availability of phosphorus in the spring and early summer across the Louisiana shelf extending west of the AR. (Ammerman 1992; Dortch and Whitledge 1992; Ammerman et al. 1995; Lohrenz et al. 1999; Rabalais et al. 2002, 2007; Sylvan et al. 2006; Turner et al. 2007; Quigg et al. 2011). Spring silicate (Si) limitation has also been documented by numerous studies (Dortch and Whitledge 1992, Redalje et al. 1994, Lohrenz et al. 1999), in addition to light availability as limiting factors for phytoplankton production (Quigg et al. 2011). The maximum of primary production occurs in the intermediate salinity region, where dilution and settling of riverine particulate material improve the light penetration in the water column and facilitate the initial consumption of the riverine N and P in the plume waters (Rowe and Chapman 2002; Dagg et al. 2004; Lehrter et al. 2009; Fennel et al. 2011). Primary production across the Louisiana shelf is highly dynamic, with rates ranging from as low as 0.5 g C m^{-2} d^{-1} in winter months up to 10 g C m^{-2} d^{-1} during the summer (Lohrenz et al. 1990; Redalje et al. 1994; Lohrenz et al. 1999).

DIC removal rates also suggests net community production (NCP) at over 3 g C m^{-2} d^{-1} in mid salinity plume water during the summer (Cai 2003; Guo et al. 2012).

Changes in nutrient loading (nitrogen, P, Si) by the MR over the past 50 years have altered N/P/Si ratios on the adjacent margin, such that P and Si limitation of phytoplankton may now be important (Justic et al. 1995; Rabalais et al. 1996, 2002; Sylvan et al. 2006, 2007). Increased N inputs from the MR to the Louisiana coast are believed to be the main reason for P and Si limitation (Dagg et al. 2007). Sylvan et al. (2006) provided evidence that phytoplankton growth on the Louisiana shelf near the MR plume area was strongly limited by P in May and July, whereas N was limiting in September of 2001. Quigg et al. (2011) extended the study area further west to include the shelf region, adjacent to the AR and Terrebonne Bay, and found evidence for early summer P limitation in this extensive part of the margin. This change in limiting factors can also have an impact on the deposition of organic matter derived from algal sources through changes in (1) the magnitude of primary production; (2) the composition of phytoplankton, which affect the efficiency of vertical settling of phytoplankton; and (3) the nutritional value of phytoplankton biomass, which can impact zooplankton dynamics (e.g., grazing and formation of fecal pellets).

According to the modeling efforts of Dagg and Breed (2003), nitrogen availability in the MR plume controls the structure of the phytoplankton community and interactions between different size classes of both phytoplankton and zooplankton. Larger cells generally account for a greater fraction of productivity at low and intermediate salinities within the plume. Redalje et al. (1994) showed that the >8 μm components of the phytoplankton community were responsible for 34–50% of the integrated phytoplankton production in the mid-field region. Wysocki et al. (2006) studied phytoplankton community composition using photosynthetic pigments as biomarkers. They found that diatoms comprised the largest fraction of the total chlorophyll *a* pool for all shelf regions in both April and October and were more dominant during high river discharge. Other studies showed that the abundance and competitive dominance of diatoms species throughout the year on the Louisiana shelf is largely a function of the continuous supply of nutrients from the MR (Dortch and Whitledge 1992; Eadie et al. 1994; Redalje et al. 1994; Bode and Dortch 1996; Lohrenz et al. 1997; Chen et al. 2000; Rabalais et al. 2002). Even within MR plume waters, small shifts in phytoplankton community structure can be observed, with other phytoplankton taxa (cryptophytes, dinoflagellates, and cyanobacteria) becoming more abundant during low discharge period (October).

In the MR plume (MRP) region immediately adjacent to the river mouth and inner shelf, a fraction of the phytoplankton might be derived from riverine diatoms (Wysocki et al. 2006; Sampere et al. 2011), because in situ marine primary production is light-limited in this turbid region. As described in earlier work, a high abundance of freshwater diatoms in the lower river, which appear to be in a more senescent state (based on pigment characteristics), may be derived from more locations upstream. For example, in the Missouri River tributary, large phytoplankton blooms are found in dam reservoirs, owing to greater light availability in waters where sediments have settled out of the water column (Duan and Bianchi 2006). These senescent freshwater diatoms in the lower MR may enter the inner shelf region and likely settle very quickly as phytodetritus to surface sediments in shallow waters (e.g., <20 m isobath), as shown by Wysocki et al. (2006). These freshwater inputs phytodetritus are clearly not likely to be as important to secondary production in the MRP as the higher in situ marine primary production of more saline waters of the mid-field regions of the plume. Nevertheless, these

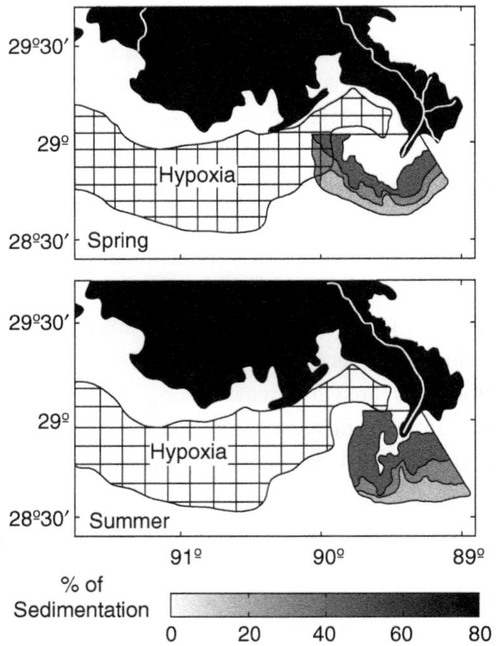

Figure 19.3. Maps showing the location of the 1996 hypoxic region (hatched; Rabalais et al. 1999) relative to plume regions of variable sedimentation in the spring and summer. The Atchafalaya River and Bay are located in the upper left-hand corner of each plot. The highest plume sedimentation occurs in salinity subregion 2, accounting for 63% of total plume sedimentation in the spring and summer. On average, lower sedimentation in salinity sub-regions 1, 3, and 4 in both seasons accounted for 1%, 28%, and 8% of total sedimentation, respectively (adapted from Green et al. 2006).

river-derived diatoms can still serve as an additional resource to the very shallow reaches of inner MRP region (where marine productivity is low), which is something that has been largely ignored.

In an attempt to construct the first carbon model of the MRP, seasonal variability in organic carbon (OC) budgets were estimated using a physical-biological model for the MR turbidity plume (Green et al. 2006). The plume volume was calculated from mixed layer depth and area in each of four salinity subregions based on an extensive set of cruise data and satellite-derived suspended sediment distributions. These physical measurements were coupled with an existing food web model to determine seasonally dependent budgets for labile (reactive on time scales of days to weeks) OC in each salinity subregion. This work also showed that there is a broader region of phytodetrital deposition in spring versus summer owing to differences in the greater extent of nutrient distribution MRP during higher flow in the spring (Fig. 19.3). This model was also used to calculate O_2 demand for the development of regional hypoxia, and the budgets indicated that sedimentation of autochthonous OC from the immediate plume contributed only 23% of the O_2 demand necessary for establishment of hypoxia in the region, which typically extends to the Texas border. It should be noted that this model did not account for the aforementioned inputs of freshwater phytodetritus, because this model was largely focused on linkages between riverine nutrient inputs and in situ production in the MRP.

Zooplankton grazing is typically the most important immediate fate of marine phytoplankton (Banse 1992). Three groups dominate zooplankton grazing of phytoplankton in the northern GOM: protozoa, gelatinous zooplankton, and copepods. Protozoan populations respond quickly when phytoplankton food is abundant. Liu and Dagg (2003) showed that microzooplankton grazing rates were low at near-field stations around the largest discharge point of the MR, peaked at locations with intermediate salinities, and then declined at their far-field station to west of the plume with high salinity. Once again the regions of highest phytodetrial deposition (e.g., Fig. 19.3) in the MRP will also be linked with zooplankton grazing activities. In the near-field region, there is close coupling of heterotrophic and autotrophic populations, particularly in the <20-μm size fraction (Bode and Dortch 1996; Liu and Dagg 2003), and protozoa consume large numbers of phytoplankton. For example, Fahnenstiel et al. (1995) reported that microzooplankton grazing rates on cells <20 μm averaged 82% of algal growth rates during summer. In contrast, no microzooplankton grazing was observed on cells >20 μm (Fahnenstiel et al. 1995), indicating that chain-forming diatoms are not subject to the same degree of protozoan grazing mortality as smaller cells in the near- and mid-field regions. Once again, this work does not consider the importance of senescent phytodetrital material coming into the turbid near-field shallow reaches of the MRP, because most of the high in situ productivity occurs in the mid-plume regions, where grazing is also going to be most prolific.

Mesozooplankton grazing enhances the decline in plume phytoplankton by adding an additional grazing mortality, especially to the large phytoplankton cells. The copepod community also consumes significant portions of the phytoplankton stock and productivity in the northern GOM. Depending on the time of year, between 14% and 62% of the daily algal production can be consumed by copepods in the mid- or far-field regions of the MRP (Dagg 1995). Liu and Dagg (2003) showed that copepod grazing was small compared with microzooplankton grazing, but that copepods enhanced the rate of phytoplankton decline in the far field, especially the large phytoplankton. Gelatinous zooplankton, primarily larvaceans, salps, and doliolids, often form large swarms in coastal regions of the northern GOM. High concentrations of the larvacean, *Oikopleura dioica*, are commonly found in the vicinity of the MRP (Dagg 1995; Dagg et al. 1996), and this organism is an important component of the grazer community. For example, during May 1992, *O. dioica* populations filtered a mean of 20% of the upper 5 m daily at stations within the MRP (Dagg et al. 1996), which likely contributes significantly to the flux of OC to sediments.

3.2. Effects between OC sources and transport rates on diagenesis

In the shelf region immediately west of the MR mouths, burial rates of organic carbon are between 25 and 300 g C m^2 yr^{-1} (Eadie et al. 1994; McKee and Twilley 1996). Several observations suggest that much of the particulate OC (POC) from the Mississippi River and/or biologically fixed on the Louisiana shelf is carried along-shelf, decomposed, buried, or transported to deeper regions in the GOM (Bianchi et al. 2006; Sampere et al. 2008). Vertical fluxes of OC in the MRP as high as

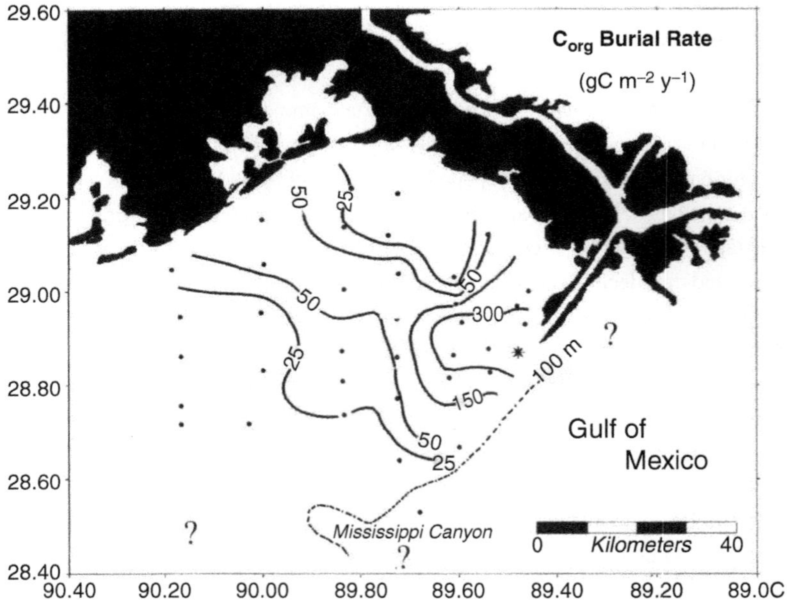

Figure 19.4. Bulk organic carbon burial rates in the MRP region.

1.80 g C m^{-2} d^{1} have been observed during spring (Redalje et al. 1994), but are lower during other seasons (0.29–0.95 g C m^{-2} d^{-1}) and away from the immediate plume (0.18–0.40 g C m^{-2} d^{-1}). In fact, the highest C burial rates in sediments (ca. 300 g C m^{-2} y^{-1}) occur near the mouth of Southwest Pass (Fig. 19.4) and generally decrease with distance from the river mouth (McKee and Twilley 1996). Much of the in situ plume productivity supporting carbon flux is comprised of diatoms (Lohrenz et al. 1999), as reflected in pigment concentrations of surface sediments below the plume (Wysocki et al. 2006). In the immediate vicinity of riverine discharge, up to 70% of carbon (and presumably also nitrogen) burial is from terrestrial sources. This component decreases to <5% at far-field sites (McKee and Twilley 1996; Gordon and Goñi, 2004), where marine carbon dominates burial. Burial rates of organic carbon have increased since 1930, but the increase is attributable primarily to marine carbon, not terrestrial carbon (Eadie et al. 1994). Higher rates of flux to the bottom are due to greater water column productivity supported by increased NO_3^- loading. Eadie et al. (1994) demonstrated there has been a linear relationship between increased NO_3^- loading and increased organic carbon burial since 1950.

Although marine organic matter might represent a large portion of incoming flux to the surface sediment, the potential for long-term burial is low. Aller and Blair (2004) proposed that modern marine organic matter makes up a small but relatively reactive component of sedimentary organic matter that decomposes in deltas. Marine organic matter instead appears to serve primarily as short-lived "kindling" that enhances the metabolism of terrigenous organic matter (Aller and Blair 2004). Mayer et al. (2007) also pointed out that marine production on the Louisiana continental shelf may dominate the incoming organic flux, but not achieve dominance in terms of the standing stocks measured in the sediments. Work by Mead and Goñi (2006) showed that marine-derived lipids in surface sediments off the Atchafalaya Shelf were easily hydrolyzed by chemical treatments, whereas terrestrial lipids were much resistant to hydrolysis as a result of their association with mineral matrices. Thus mineral protection appears to play an important role on the preservation of terrestrial and, likely, marine organic matter in this margin.

Usually non-LDE continental slope sediments have only modest early diagenesis taking place in the first few tens of centimeters below the sediment-water interface and have oxygen penetration depths of several centimeters that tend to increase with increasing water depth (e.g., Cai and Sayles 1996). This reflects increased oxygen exposure time and extent of organic matter degradation (e.g., Hartnett et al. 1998; Hedges et al. 1999). Notable exceptions are sediments found beneath highly productive overlying waters that have a high rain rate of POC, resulting in rapid and extensive diagenesis near the sediment-water interface where formation of high concentrations of authigenic minerals is commonly observed (e.g. Rowe and Howarth 1985; Hedges and Keil 1995; Arthur et al. 1998). However, there is now clear evidence from measurements of diagenetic processes that lateral downslope transport of OM can lead to significantly greater benthic metabolism and associated diagenesis than can be explained by POC input from overlying waters (e.g., Jahnke et al. 1990; DeMaster et al. 1994; Jahnke and Jahnke 2000). Studies such as those by Silverberg et al. (2000) and Arnosti and Holmer (2003) have also demonstrated that the intensity of diagenetic activity is often not well predicted by sediment TOC content and characteristics in continental margin sediments. These earlier studies provide sound examples of the importance of measuring diagenetic processes in association with the study of OM transport and deposition in slope environments.

The rapid transport of labile SOC down the Mississippi Canyon can potentially have a profound influence on the geochemistry of the sediments. This is because these sediments may behave in a manner similar to shallow water shelf sediments near the MR delta, where bacterially driven cycling of reactive manganese and iron and formation of sedimentary iron sulfides takes place. Based on studies of shelf sediments west of the MR delta (Morse and Rowe 1999; Rowe et al. 2002), a non–steady-state diagenetic model was developed to investigate the significance of various diagenetic processes (Morse and Eldridge 2007). Preliminary evidence suggests that the diagenetic activity in the canyon sediments may be considerably higher than in adjacent slope sediments at equivalent water depths. Lin and Morse (1991) found that, at the same depth below the sediment-water interface, the extent of sulfate reduction was about three times greater in canyon sediment than nearby slope sediment (both ~500 m water depth), even though the slope sediment had about 15% greater terrestrially derived OC. Also, using microelectrodes, Morse and Rowe (1999) found evidence for enhanced diagenetic activity in Mississippi Canyon sediments compared with other northern GOM sediments at equivalent water depths.

A particularly intriguing aspect of the diagenesis of sedimentary OM in Mississippi Canyon is that it is likely, in major part, derived from the mobile muds on the Louisiana shelf and consequently may behave in a similar manner. The pioneering work of Aller and his associates (e.g., Aller et al. 1996; Aller 1998; Aller and Blair 2006) has shown the importance of mobile muds as an "incinerator" of OC, and the proposed diagenetic studies will contribute to determining whether the process continues down the canyon. Complementing this body of work, Arzayus and Canuel (2004) showed that degradation of more refractory organic matter was facilitated in mobile, physically mixed estuarine sediments.

3.3. Transport and sources of algal organic carbon across the shelf/slope

Wysocki et al. (2006) showed that the dominant source of phytoplankton to the surface sediments on the inner Louisiana shelf was diatoms. For example, sediment chlorophyll *a* (Chl-*a*) was elevated within the depositional path of the MR during April 2000 (high flow) and near the river mouth during October 2000 (low flow), along with fucoxanthin and diatoxanthin, the dominant carotenoids in diatoms (Wysocki et al. 2006). In another study, within the lower MR, Duan and Bianchi (2006) observed Chl-*a* concentrations within the range found on the shelf (0.8–23.6 nM) and also high fucoxanthin, indicating higher riverine diatom biomass (up to 86% of total biomass) than previously reported. Although these studies were not conducted simultaneously, they suggest that elevated pigment and carbon concentrations on the inner shelf may be derived from river diatoms – as well as marine shelf production. Recent work by Sampere et al. (2008) showed that dominant algal groups present in the lower MR sediments included diatoms, cryptomonads, green algae, and cyanobacteria, with diatoms dominating the shelf/slope and canyon sites (Sampere et al. 2008). Chl-*a* degradation products, including steryl chlorin esters (SCE), chlorophyll-*a* epimer (chl-a′), and chlorophyll-*a* allomer (chl-a allo), were not present in lower MR sediments, but were abundant on the shelf, reflective of diatom inputs from the river and new shelf production (Wysocki et al. 2006). SCE made up a large portion of the total chloropigments at most shelf sites, comprising 72% and 67% of the Chl-*a* decay products at the Mid and Bar 3 Stations, respectively; at the 95 m and 110 m Stations, SCE made-up 45% and 49% of Chl-*a* decay products, respectively. Together, these results suggest the importance of algal OM from the lower river to inner shelf, as well as from the estuaries to the inner shelf (50 isobath). Moreover, the increase in pheopigments and SCEs along the westward, canyon, and estuarine-shelf transects showed that labile phytodetritus is exported from the bays to the shelf and that production near the mouth of the river is transported, likely in mobile muds, westward along the shelf and south to the Mississippi Canyon (Bianchi et al. 2006).

3.4. Transport and sources of terrestrial OC sources across the shelf-slope

Terrestrial inputs of OC to continental margin waters of the northern GOM are high compared with other coastal margins of the United States because of significant discharge from one of the world's largest river systems – the combined inputs of the AR and MR (Hedges and Parker 1976; Malcolm and Durum 1976; Eadie et al. 1994; Trefry et al. 1994; Bianchi et al. 1997). Bulk OC and carbon isotope measurements, made on suspended river particulates and seabed samples within the MR dispersal system (Eadie et al. 1994; Trefry et al. 1994), have shown that river particulates are ^{13}C-depleted (relative to marine OC) and that 60–80% of the OC buried on the shelf adjacent to the river is marine in origin (as determined by the enriched ^{13}C values of the sediments). Subsequently, it was suggested that much of the terrestrial OC (OC_{terr}) delivered to the shelf is from a mixture of C_3 and C_4 plants and materials from eroded soils in the northwestern grasslands of the MR drainage basin (Goñi et al. 1997, 1998). These investigators also concluded that lignin-depleted, aged organic matter with isotopic signatures from C_4 plants (^{13}C-enriched values similar to MOC) was transported greater distances offshore because of its association with small-sized (e.g., clays) particles (Goñi et al. 1997, 1998; Onstad et al. 2000; Gordon et al. 2001). In contrast, lignin-rich, ^{13}C-depleted

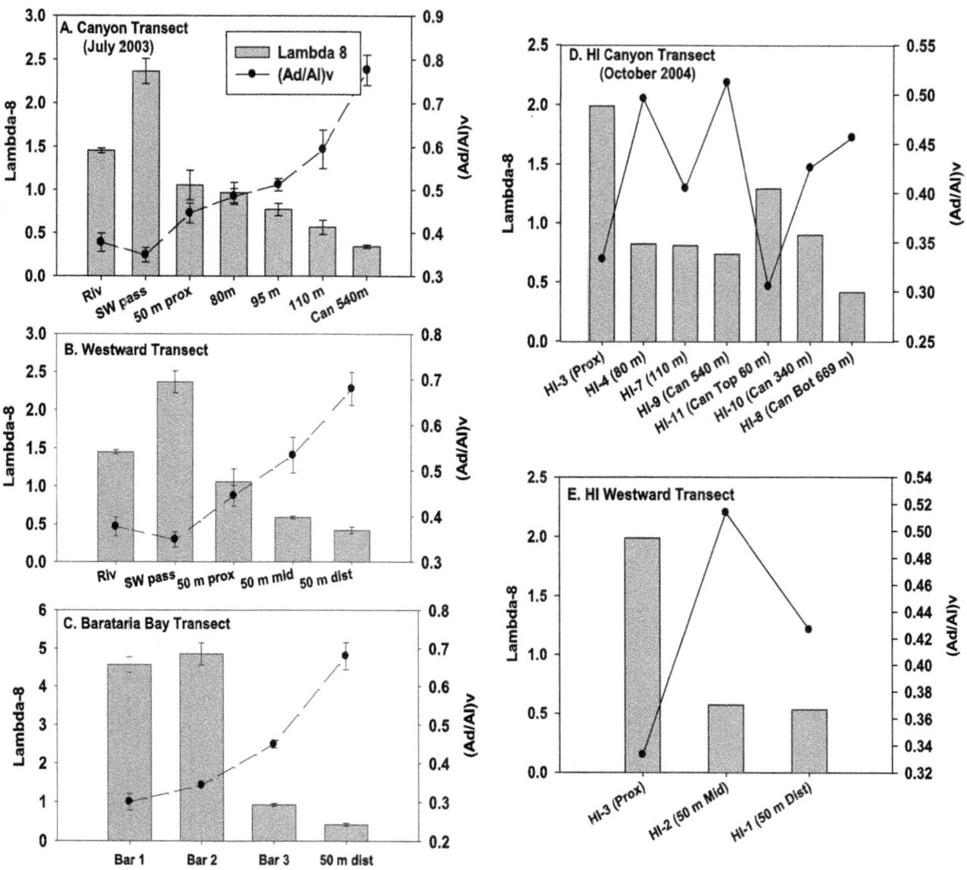

Figure 19.5. Lambda 8 (mg lignin/100 mg OC), (Ad/Al)v (vanillic acid/vanillin ratio) in surface sediments (0–1 cm) for 3 transects sampled during July 2003 (A, B, and C) and 2 transects sampled during October 2004 (D and E) (adapted from Sampere et al. 2008).

woody angiosperm material of coarser size preferentially settles within the lower MR and in the proximal portion of the dispersal system on the shelf (Bianchi et al. 2002; Gordon and Goñi 2003). These findings partially explain the spatial trends in lignin-phenol biomarkers (Fig. 19.5) and suggest that a greater fraction of total OC may be represented by OC_{terr} further offshore, presumably because of selective removal of the more labile, algal fraction of carbon during transport (e.g., Sampere et al. 2008). This hypothesis is supported by recent data showing that Chl-*a* decreased from 0.9 to 0.1 μmol g OC^{-1}, whereas total pheopigments increased from 1.0 to 2.1 μmol g OC^{-1} in nearshore to offshore sites, respectively (Chen et al. 2003). However, if transport is rapid, the more labile organic components can also "survive" offshore transport as far out as the Mississippi Canyon, as discussed earlier (Bianchi et al. 2006; Waterson and Canuel 2008; Sampere et al. 2008).

Lignin-derived CuO oxidation products, which are uniquely derived from terrestrial land plants (e.g., Hedges and Ertel 1982), have been used extensively to understand the sources and distribution of OCterr along the Mississippi/Atchafalaya margins (e.g., Hedges and Ertel 1976; Goñi et al. 1997; 1998; Gordon and Goñi 2003, 2004; Bianchi et al. 2006, 2007a; Sampere et al. 2008). These studies show that the contributions of lignin phenols to overall sedimentary OC are highest in the inner shelf regions and decrease with distance from land. The composition of the lignin phenols, including tracers of oxidation state (i.e., acid:aldehyde ratios) and source (i.e., cinnamyl:vanillyl ratios), display marked changes that indicate that degraded organic matter from vascular plant sources becomes relatively more important in distal sites across the margin. Examination of the content and composition of lignin products among samples of different size, surface area, and density (e.g., Gordon and Goñi 2003; Sampere et al. 2008) shows that there are significant contrasts in the lignin content and compositional signatures among particles with different hydrodynamic properties. The sorting of these particles during offshore transport contributes to the spatial patterns of lignin distribution that is also affected by degradation. Hence, for example, coarse vascular plant detritus, enriched in lignin phenols, contributes to the high lignin contents observed in the lower MR and inner-shelf regions (e.g., Bianchi et al. 2002; Gordon and Goñi 2003). Low-density, organic-rich particles with a marsh lignin signature appear to be degraded significantly during transport (e.g., Sampere et al. 2008), whereas lithogenic-rich fine particles containing lignin derived from aged soil sources are distributed further offshore with much less alteration (e.g., Goñi et al. 1998; Gordon and Goñi 2004; Sampere et al. 2008).

Overall, these observations are consistent with reworking of terrestrial OM in the shallow waters and mobile mud deposits west of SW pass and in the inner Atchafalaya margin. However, a significant fraction of this terrestrial OM survives decay in the inner shelf and is exported offshore along the Mississippi Canyon and across the shelf to the open slope (e.g., Sampere et al. 2008; Gordon and Goñi 2004). Lipid compositions from surface sediments showed that approximately 17–34% of the fatty acid and sterols at even the most distal slope and canyon sites, further supporting that these deeper regions could be an important sink for terrigenous carbon (Waterson and Canuel 2008).

The transport processes responsible for the dispersal of particles across both the Mississippi and Atchafalaya margins result in marked contrasts in the texture and inorganic carbon content of sediments along their shelf and slope (including the Mississippi Canyon). For example, Gordon and Goñi (2004) showed that offshore sediments in both the open slope and along the Mississippi Canyon were characterized by markedly lower ^{210}Pb-based accumulation rates and contained sediments with significantly higher carbonate contents and greater specific mineral surface areas than those found in the shelf (Fig. 19.6). Such a pattern suggests that hydrodynamic sorting during transport favors the offshore dispersal of clay-rich particles, and the decreased supply of terrigenous materials in the most offshore locations results in a higher proportion of biogenic carbonate relative to the lithogenic component.

Because fluvial particles of different grain sizes can have marked differences in the composition of its associated OM (e.g., Hedges et al. 1986; data from the Mississippi/Atchafalaya), differential transport of lithogenic materials can significant affect the distribution of terrestrial OM on this margin. In their study of surface sediments from the Mississippi and Atchafalaya margins, Gordon and Goñi (2004) argued that these physical transport processes are partially responsible for the distinctly higher fossil OC contents and lower yields of terrestrial biomarkers (e.g., lignin-phenols) observed among stations across the shelf-slope transects sampled (Fig. 19.6). In addition to the preferential offshore

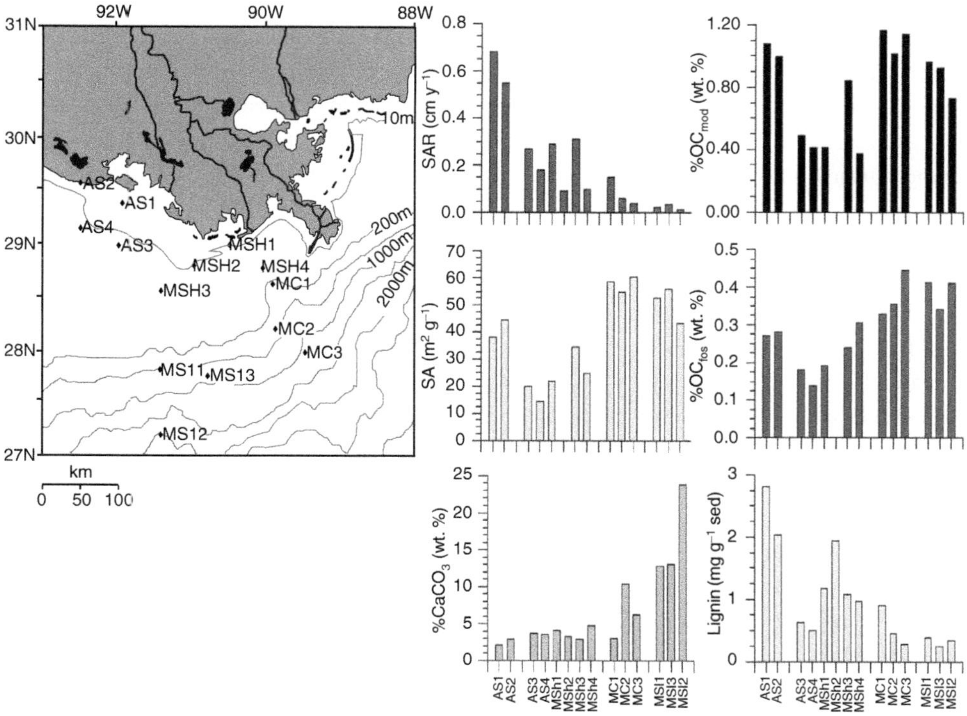

Figure 19.6. Pb-210 based accumulation rates and compositional characteristics of surface sediments across the Atchafalaya/Mississippi margin collected in March–August 1998 (after Gordon and Goñi 2004).

transport and deposition of highly altered, aged terrestrial OM associated with fine-grained (i.e., high surface area) lithogenic particles, there is evidence for elevated contributions of marine OM to sediments at the offshore slope and canyon stations. Higher $CaCO_3$ and modern OC contents reflect the input of marine OM at these locations. The effects of physical sorting of particles of different sizes and compositions during cross-margin transport also need to be taken into account when evaluating the fate of both terrestrial and marine OM in these systems.

Terrestrial-derived OC in surface sediments in the MR plume region is highly coupled to seasonal changes in river discharge. Based on the isotopic and lignin analyses, the character of the terrestrially derived material was very similar to that found in the POC of the lower MR (Bianchi 2007) and did not change much seasonally. This suggests that there is minimal predepositional decay of lignin as it is transported from the lower MR to surface sediments in the MR plume region. Based on lignin Ad/Al ratios, the terrestrial material in this region is also comparatively less degraded than terrigenous OM on the rest of the shelf, consistent with previous sedimentological work that suggested that material from this region is exported to and subsequently reworked in mobile muds across the shelf. This is consistent with the recent work that shows similar transport pathways along and across the shelf in mobile muds during storm events, to as far as the Mississippi Canyon (Bianchi et al. 2006; Sampere et al. 2008).

Wetlands are being lost from Louisiana at a rapid rate, estimated to be about 70 $km^2\ y^{-1}$ (Britsch and Dunbar 1993), as a result of subsidence, decreased sediment availability, and construction of canals (Bass and Turner 1997; Webb et al. 1995). This rate spiked in 2005, when 562 km^2 of land was lost in the period surrounding Hurricanes Katrina and Rita (Barras 2006). In fact, recent work has shown that sediment and organic matter are exported from marshes to open water bay areas in Louisiana with erosional marsh shoreline retreat and down-cutting (Wilson and Allison 2008). At average wetland loss rates for the period 1932–1999 in these areas, this translates to a yield of particulate organic matter to Barataria Bay of 4.2×10^9 kg during this time period, whereas the yield of inorganic material was 24.7×10^9 kg. For Breton Sound, 5.3×10^9 kg of organic material was yielded, whereas 25.9×10^9 kg of inorganic material was yielded (Wilson and Allison 2008). Other carbon isotope and iodine data from the MR margin has shown evidence of episodic organic carbon deposition in shelf sediments of older wetland plant residues (which are known to contain high iodine concentrations) (Santschi et al. 2006).

A molecular and stable isotope-based mixing model defining riverine, marsh, and marine organic carbon suggested that the highest OC inputs to the shelf in spring were from marine sources (55–61% marine OC), whereas riverine OC had the highest (63%) in fall. The spring increase is likely due to lower inputs of marine OC at this time compared with the fall season (Bianchi et al. 2011). This model also indicated that marsh inputs, ranging from 19% to 34% and 3% to 15% of the OC in spring and fall, respectively, were significantly more important sources of OC on the inner Louisiana shelf than previously suggested. In previous research conducted on the Atchafalaya shelf, west of the site studied by Bianchi et al. (2011), Gordon and Goñi (2003) used a similar model to quantify estimates of terrestrial and marine contributions to SOC. In that study, large variations in the character of OC were observed, with one site dominated by coarse, woody, C_3 material, and other samples more representative of well-degraded OC likely derived from river inputs of soil OC. Overall, they found that ca. 65–77% of the OM signal on the Atchafalaya shelf was terrestrial soil-derived, with marine OC increasing from 21% to 34% offshore. Their estimate of the marine contribution is somewhat less than the estimate from that of Bianchi et al. (2011), which ranged from 38% up to 60%. This difference was consistent with previous work in these areas in which chlorophyll *a* concentrations were found to decrease westward along the shelf from the MR (Rabalais et al. 1996; Chen et al. 2000).

Finally, a key difference in the two modeling approached was that Gordon and Goñi (2003) did not consider marsh end-members in their treatment of the data and focused on the distinction between soil OC and vascular plant debris. In contrast to their work, Bianchi et al. (2011) did include the salt marshes as an end-member source, but did not differentiate between soil-derived and vascular plant-derived OC, instead considering that the river had integrated the soil and plant materials from the various drainage basin sources. Because in large complex river systems such as the Mississippi-Atchafalaya, the distinctions between different sources of OM from regions of the watershed with distinct vegetation, geologic, and hydrologic conditions are integrated into the overall OC signal, additional modeling approaches are needed to better constrain source inputs in these regions.

4. Concluding remarks

The Mississippi/Atchafalaya margin encompasses a large region of the northern Gulf of Mexico that displays high connectivity between land and ocean. The system is characterized by highly dynamic physical and geologic processes, which drive intense biogeochemical cycles, including production, degradation, and preservation of terrestrial and marine OM. For example, the presence of mobile muds has a significant impact on the decay and transport of organic matter in this region. Because of its location, the Mississippi/Atchafalaya margin has been and continues to be impacted by both direct human activities (e.g., enhanced nutrient loading in the watershed, changes in freshwater and sediment discharges, pollution associated with fossil fuel exploration) and forcings related to climate change (e.g., increases in sea level, hurricane magnitude, and frequency). The challenges ahead in this period of accelerated change include developing the understanding required to predict how these future changes will affect the biogeochemistry and ecosystem services of the region and the role that the system plays in the connectivity between land and the deep sea.

References

Adams, C. E., J. P. Swift, and J. M. Coleman. 1987. Bottom currents and fluviomarine sedimentation on the Mississippi prodelta shelf: February–May 1984. J. Geophys. Res. 92: 14595–14609.

Aller, R. C. 1998. Mobile deltaic and continental shelf muds as suboxic, fluidized bed reactors. Mar. Chem. 61: 143–155.

Aller, R. C., and N. E. Blair. 2006. Carbon remineralization in the Amazon - Guianas mobile mudbelt: a sedimentary incinerator. Cont. Shelf Res. 26: 2241–2259.

Aller, R. C., and N. E. Blair. 2004. Early diagenetic remineralization of sedimentary organic C in the Gulf of Papua deltaic complex (Papua New Guinea): net loss of terrestrial C and diagenetic fractionation of C isotopes. Geochim. Cosmochim. Acta. 68: 1815–1825.

Aller, R. C., N. E. Blair, Q. Xia, and P. D. Rude. 1996. Remineralization rates, recycling and storage of carbon in Amazon shelf sediments. Cont. Shelf Res. 16: 753–786.

Allison, M. A., T. S. Bianchi, B. A. McKee, and T. P. Sampere. 2007. Carbon burial on river-dominated continental shelves: impact of historical changes in sediment loading adjacent to the Mississippi River. Geophys. Res. Lett. L01606.

Allison, M. A., T. M. Dellapenna, E. S. Gordon, S. Mitra, and S. T. Petsch. 2010. Impact of Hurricane Katrina (2005) on shelf organic carbon burial and deltaic evolution. Geophys. Res. Lett. 37: L21605.

Allison, M. A., C. R. Demas, B. A. Ebersole, B. A. Kleiss, C. D. Little, E. A. Meselhe, N. J. Powell, T. C. Pratt, and B. M. Vosburg. 2012. A water and sediment budget for the lower Mississippi-Atchafalaya River in flood years 2008–2010: implications for sediment discharge to the oceans and coastal restoration in Louisiana. J. Hydrol. 432: 84–97.

Allison, M. A., G. C. Kineke, E. S. Gordon, and M. A. Goñi. 2000. Development and reworking of an annual flood deposit on the inner continental shelf off the Atchafalaya River. Cont. Shelf Res. 20: 2267–2294.

Allison, M. A., M. T. Lee, A. S. Ogston, and R. C. Aller. 2000. Origin of Amazon mudbanks along northeastern coast of South America. Mar. Geol. 163: 241–256.

Allison, M. A., and E. A. Meselhe. 2010. The use of large water and sediment diversions in the lower Mississippi River (Louisiana) for coastal restoration. J. Hydrol. 387: 346–360.

Allison, M. A., A. Sheremet, M. A. Gõni, and G. W. Stone. 2005. Storm layer deposition on the Mississippi-Atchafalaya subaqueous delta generated by Hurricane Lili in 2002. Cont. Shelf Res. 25: 2213–2232.

Ammerman, J. W., W. B. Glover, R. H. S. Ruvalcaba, and M. J. D. MacRae. 1995. Continuous underway measurement of microbial enzyme activities in surface waters of the Mississippi River plume and the Louisiana shelf, pp. 1–8. *In* D. K. Atwood, W. F. Graham and C. B.

Grimes (eds.), Nutrient-enhanced coastal ocean productivity, proceedings of 1994 synthesis workshop. Louisiana Sea Grant College Program, Baton Rouge.

Ammerman, J. W. 1992. Seasonal variation in phosphate turnover in the Mississippi River Plume and the Inner Gulf Shelf: Rapid summer turnover, pp. 69–75. *In* T. J. Bright (eds.), Proceedings of the 1991 NECOP synthesis workshop. Texas A&M Sea Grant, Galveston.

Arnosti, C., and M. Holmer. 2003. Carbon cycling in a continental margin sediment: contrasts between organic matter characteristics and remineralization rates and pathways. Estuar. Coast. Shelf Sci. 58: 197–208.

Arthur, M. A., B. B. Sageman, W. E. Dean, R. L. Slingerland, L. R. Kump, and T. S. White. 1998. Transgression, advection of oxygen-depleted water, and eutrophication of the mid-Cretaceous Western Interior Seaway of North America, paper presented at the Annual Meeting of the Geological Society of America, Toronto, Ontario, October 26–29.

Arzayus, K. M., and E. A. Canuel. 2004. Organic matter degradation in sediments of the York River estuary: Effects of biological vs. physical mixing. Geochim. Cosmochim. Acta. 69: 455–463.

Baker, E. T., and B. M. Hickey. 1986. Contemporary sedimentation processes in and around an active West Coast submarine canyon. Mar. Geol. 71: 15–34.

Banse, K. 1992. Grazing, temporal changes of phytoplankton concentrations, and the microbial loop in the open sea, pp. 409–440. *In* P. G. Falkowski and A. D. Woodhead (eds.), Primary Productivity and Biogeochemical Cycles in the Sea. Plenum, New York.

Barras, J. A. 2006. Land area change in coastal Louisiana after the 2005 hurricanes – a series of three maps. Open Files Report 06-1274, U.S. Geological Survey.

Bass, A. S., and R. E. Turner. 1997. Relationships between salt marsh loss and dredged canals in three Louisiana estuaries. J. Coast. Res. 13: 895–903.

Berner, R. A. 1982. Burial of organic carbon and pyrite sulfur in the modern ocean: its geochemical and environmental significance. Am. J. Sci. 282: 451–473.

Berner, R. A. 1989. Biogeochemical cycles of carbon and sulfur and their effect on atmospheric oxygen over Phanerozoic time. Global Planet. Change. 75: 97–122.

Bianchi, T. S. 2007. *Biogeochem. Estuar.* Oxford University Press, 720 pp.

Bianchi, T. S. 2011. The role of terrestrially derived organic carbon in the coastal ocean: A changing paradigm and the priming effect. Proc. Nat. Acad. Sci. 108: 19473–19481.

Bianchi, T. S., and M. A. Allison. 2009. Large-river delta-front estuaries as natural "recorders" of global environmental change. Proc. Nat. Acad. Sci. 106: 8085–8092.

Bianchi, T. S., C. D. Lambert, P. H. Santschi, and L. Guo. 1997. Sources and transport of land-derived particulate and dissolved organic matter in the Gulf of Mexcio (Texas shelf/slope): The use of lignin-phenols and loliolides as biomarkers. Org. Geochem. 27: 65–78.

Bianchi, T. S., S. Mitra, B. A. McKee. 2002. Sources of terrestrially-derived organic carbon in lower Mississippi River and Louisiana shelf sediments: implications for differential sedimentation and transport at the coastal margin. Mar. Chem. 77: 211–223.

Bianchi, T. S., T. Sampere, M. Allison, E. A. Canuel, B. A. McKee, S. Wakeham, and B. Waterson. 2006. Rapid export of organic matter to the Mississippi Canyon. EOS. 87: 565, 572–573.

Bianchi, T. S., L. A. Wysocki, K. M. Schneider, T. R. Filley, D. R. Corbett, and A. Kolker. 2011. Sources of terrestrial organic carbon in the Louisiana shelf (USA): Evidence for the importance of coastal marsh inputs. Aquat. Geochem. 17: 431–456.

Bode, A., Q. Dortch. 1996. Uptake and regeneration of inorganic nitrogen in coastal waters influenced by the Mississippi River: spatial and seasonal variations. J. Plankton Res. 18: 2251–2268.

Britsch, L. D., and J. B. Dunbar. 1993. Land loss rates – Louisiana coastal-plain. J. Coast. Res. 9: 324–338.

Burdige, D. J. 2007. Preservation of organic matter in marine sediments: Controls, mechanisms, and an imbalance in sediment organic carbon budgets? Chem. Rev. 107: 467–485.

Cai, W. J., and F. L. Sayles. 1996. Oxygen penetration depths and fluxes in marine sediments. Mar. Chem. 52: 123–131.

Cai, W.-J. 2011. Estuarine and coastal ocean carbon paradox: CO_2 sinks or sites of terrestrial carbon incineration? Ann. Rev. Mar. Sci. 3: 123–145.

Canuel, E. A., and C. S. Martens. 1996. Reactivity of recently deposited organic matter: Degradation of lipid compounds near the sediment-water interface. Geochim. Cosmochim. Acta. 60: 1793–1806.

Chen, N., T. S. Bianchi, J. M. Bland, and B. A. McKee. 2003. Novel carotenoid chlorin esters in Louisiana shelf sediments: formation and transformation of CCEs. Geochim. Cosmochim. Acta. 67: 2027–2042.

Chen, X., S. E. Lohrenz, D. A. Weisenburg. 2000. Distribution and controlling mechanisms of primary production on the Louisiana-Texas continental shelf. J. Mar. Sys. 25: 179–207.

Cochrane, J. D., and F. J. Kelly. 1986. Low-frequency circulation on the Tx–La continental shelf. J. Geophys. Res. 91: 10,645–10,659.

Coleman, J. M. 1981. Deltas: Processes of Deposition and Models for Exploration. Burgess Publishing Company, Minneapolis, MN.

Coleman, J. M. 1988. Dynamic changes and processes in the Mississippi River delta. Geol. Soc. Am. Bull. 100: 999–1015.

Coleman, J. M. and H. H. Roberts. 1988. Sedimentary development of the Louisiana continental shelf related to sea level cycles: Part I – Sedimentary sequences. Geo. Mar. Lett. 8: 63–108.

Coleman, J. M., H. H. Roberts, and G. W. Stone. 1998. Mississippi River delta: An overview. J. Coast. Res. 14: 698–716.

Coleman, J. M., and L. D. Wright. 1975. Modern river deltas; variability of processes and sand bodies, pp. 99–149. *In* M. L. Broussard (eds.), Deltas; Models for Exploration. Houston Geological Society.

Corbett, D. R., M. Dail, and B. A. McKee. 2007. High-frequency time-series of the dynamic sedimentation processes on the western shelf of the Mississippi River Delta. Cont. Shelf Res. 27: 1600–1615.

Corbett, D. R., B. A. McKee, and M. A. Allison. 2006. Nature of decadal-scale sediment accumulation in the Mississippi River deltaic region. Cont. Shelf Res. 26: 2125–2140.

Corbett, D. R., B. A. McKee, and D. Duncan. 2004. An evaluation of mobile mud dynamics in the Mississippi River deltaic region, Mar. Geol. 209: 91–112.

Dagg, M. J. 1995. Copepod grazing and the fate of phytoplankton in the northern Gulf of Mexico. Cont. Shelf Res. 15: 1303–1317.

Dagg, M. J., J. W. Ammerman, R. M. W. Amon, W. S. Gardner, and R. E. Green, and S. E. Lohrenz. 2007. A review of water column processes influencing hypoxia in the northern Gulf of Mexico. Estuar. Coast. 30: 735–752.

Dagg, M.J., R. Benner, S. Lohrenz, and D. Lawrence. 2004. Transformation of dissolved and particulate materials on continental shelves influenced by large rivers: plume processes. Cont. Shelf Res. 24: 833–858.

Dagg, M. J., T. S. Bianchi, G. A. Breed, W. J. Cai, S. Duan, H. Liu, B. A. McKee, R. T. Powell, and C. M. Stewart. 2005. Biogeochemical characteristics of the lower Mississippi River, USA, during June 2003. Estuaries. 28: 664–674.

Dagg, M. J., and G. A. Breed. 2003. Biological effects of Mississippi River nitrogen on the northern gulf of Mexico – a review and synthesis. J. Mar. Sys. 43: 133–152.

Dagg, M. J., E. P. Green, B. A. McKee, P. B. Ortner. 1996. Biological removal of fine-grained lithogenic particles from a large river plume. J. Mar. Res. 54: 149–160.

Dail, M. B., D. R. Corbett, and J. P. Walsh. 2007. Assessing the importance of tropical cyclones on continental margin sedimentation in the Mississippi delta region. Cont. Shelf Res. 27: 1857–1874.

Demas, C., and P. Curwick. 1988. Suspended Sediment and Associated Chemical Transport Characteristics of the Lower Mississippi River, Louisiana. Louisiana Dept. of Transportation: Baton Rouge, LA.

DeMaster, D. J., R. H. Pope, L. A. Levin, and N. E. Blair. 1994. Biological mixing intensity and rates of organic carbon accumulation in North Carolina slope sediments. Deep-Sea Res. II. 41: 735–753.

DiMarco, S. F., P. Chapman, N. D. Walker, and R. D. Hetland. 2010. Does local topography control hypoxia on the eastern Texas-Louisiana shelf? J. Mar. Sci. 80: 25–35.

DiMarco, S. F., and R. O. Reid. 1998. Characterization of the principal tidal current constituents on the Texas–Louisiana shelf. J. Geophys. Res. 103: 3093–3110.

Dortch, Q., and T. E. Whitledge. 1992. Does nitrogen or silicon limit phytoplankton production in the Mississippi River plume and nearby regions? Cont. Shelf Res. 12: 1293–1309.

Draut, A. E., G. C. Kineke, D. W. Velasco, M. A. Allison, and R. J. Prime. 2005. Influence of the Atchafalaya River on recent evolution of the chenier plain inner continental shelf, northern Gulf of Mexico. Cont. Shelf Res. 25: 91–112.

Duan, S. W., and T. S. Bianchi. 2006. Seasonal changes in the abundance and composition of plant pigments in particulate organic carbon in the lower Mississippi and Pearl rivers (USA). Estuar. Coast. 29: 427–442.

Eadie, B. J., B. A. McKee, M. B. Lansing, J. A. Robbins, S. Metz, and J. H. Trefrey. 1994. Records of nutrient enhanced coastal ocean productivity in sediments from the Louisiana continental shelf. Estuar. 17: 754–765.

Fahnenstiel, G. L., M. J. McCormick, G. A. Lang, D. G. Redalje, S. E. Lohrenz, M. Markowitz, B. Wagoner, and H. J. Carrick. 1995. Taxon-specific growth and loss rates for dominant phytoplankton populations from the northern Gulf of Mexico. Mar. Ecol. Prog. Ser. 117: 229–239.

Fennel, K., R. Hetland, Y. Feng, and S. DiMarco. 2011. A coupled physical-biological model of the Northern Gulf of Mexico shelf: Model description, validation and analysis of phytoplankton variability. Biogeosci. Discuss. 8: 121–156.

Frazier, D. E. 1967. Recent deltaic deposits of the Mississippi River–their development and chronology. Trans. Gulf Coast. Assoc. Geol. Socs. 17: 287–315.

Galler, J. J., and M. A. Allison. 2008. Estuarine controls on fine-grained sediment storage in the lower Mississippi and Atchafalaya Rivers. Geol. Soc. Amer. Bull. 120: 386–398.

Galloway, W. E. 1975. Process framework for describing the morphologic and stratigraphic evolution of deltaic depositional systems. pp. 87–98. *In* M. L. Broussard (eds.), Deltas, Models for Exploration. Houston Geological Society, Houston.

Goff, J. A., M. A. Allison, and S. P. S. Gulick. 2010.Offshore transport of sediment during cyclonic storms: Hurricane Ike 2008. Texas Gulf Coast, USA. Geology 38: 351–354.

Goñi, M. A., Y. Alleau, R. Corbett, J. P. Walsh, D. Mallinson, M. Allison, E. Gordon, S. Petsch, and T. M. Dellapenna. 2007. The effects of hurricanes Katrina and Rita on the seabed of the Louisiana Shelf. Sediment. Rec. 5: 4–9.

Goñi, M. A., N. Monacci, R. Gisewhite, A. Ogston, J. Crockett, and C. Nittrouer. 2006. Distribution and sources of particulate organic matter in the water column and sediments of the Fly River delta, Gulf of Papua (Papua New Guinea). Estuar. Coast. Shelf Sci. 69: 225–245.

Goñi, M. A., K. C. Ruttenberg, and T. I. Eglinton. 1997. Sources and contribution of terrigenous organic carbon to surface sediments in the Gulf of Mexico: Nature. 389: 275–278.

Goñi, M. A., K. C. Ruttenberg, and T. I. Eglinton. 1998. A reassessment of the sources and importance of land-derived organic matter in surface sediments from the Gulf of Mexico. Geochim. Cosmochim. Acta. 62: 3055–3075.

Goodwin, R. H., and D. B. Prior. 1989. Geometry and depositional sequences of the Mississippi Canyon, Gulf of Mexico. J. Sediment. Petrol. 59: 318–329.

Gordon, E. S., and M. A. Goñi. 2003. Sources and distribution of terrigenous organic matter delivered by the Atchafalaya River to Sediments in the northern Gulf of Mexico. Geochim. Cosmochim. Acta. 67: 2359–2375.

Gordon, E. S., and M. A. Goñi. 2004. Controls on the distribution and accumulation of terrigenous organic matter in sediments from the Mississippi and Atchafalaya river margin. Mar. Chem. 92: 331–352.

Gordon, E. S., M. A. Goñi, Q. N. Roberts, G. C. Kineke, and M. A. Allison. 2001. Organic matter distribution and accumulation on the inner Louisiana shelf west of the Atchafalaya River. Cont. Shelf Res. 21: 1691–1721.

Green, R. E., T. S. Bianchi, M. J. Dagg, N. D. Walker, and G. A. Breed. 2006. An organic carbon budget for the Mississippi River turbidity plume and plume contributions to air–sea CO2 fluxes and bottom-water hypoxia. Estuaries 29: 579–597.

Greene, R. M., J. C. Lehrter, and J. D. Hagy III. 2009. Multiple regression models for hindcasting and forecasting midsummer hypoxia in the Gulf of Mexico. Ecol. Appl. 19: 1161–1175.

Guo, X., W.J. Cai, W.J. Huang, Y. Wang, F. Chen, M.C. Murrell, S.E. Lohrenz, L. Jiang, M. Dai, J. Hartmann, Q. Lin, and R. Culp. 2012. Carbon dynamics and community production in the Mississippi River plume. Limnol. Oceanogr. 57: 1–17.

Hartnett, H. E., R. G. Keil, J. I. Hedges, and A. H. Devol. 1998. Influence of oxygen exposure time on organic carbon preservation in continental margin sediments. Nature 391: 572–574.

Hedges, J. I. and Ertel, J. R. 1982. Characterization of lignin by gas capillary chromatography of cupric oxide oxidation-products. Anal. Chem. 54(2): 174–178.

Hedges, J. I., W. A. Clark, P. D. Quay, J. E. Richey, A. H. Devol, and U. de M. Santos. 1986. Compositions and fluxes of particulate organic material in the Amazon River. Limnol. Oceanogr. 31: 717–738.

Hedges, J. I., F. S. Hu, A. H. Devol, H. E. Hartnett, E. Tsamakis, and R. G. Keil. 1999. Sedimentary organic matter preservation: A test for selective degradation under oxic conditions. Am. J. Sci. 299: 529–555.

Hedges, J. I., and R. G. Keil. 1995. Sedimentary organic matter preservation: an assessment and speculative synthesis. Mar. Chem. 49: 81–115.

Hedges, J. I., and P. L. Parker. 1976. Land-derived organic matter in surface sediments from the Gulf of Mexico. Geochim. Cosmochim. Acta. 40: 1019–1029.

Hetland, R. D., and S. F. DiMarco. 2008. The effects of bottom oxygen demand in controlling the structure of hypoxia on the Texas–Louisiana continental shelf. J. Mar. Syst. 70: 49–62.

Horowitz, A. J. 2010. A quarter century of declining suspended sediment fluxes in the Mississippi River and the effect of the 1993 flood. Hydrol. Proc. 24: 13–34.

Jahnke, R. A., and D. B. Jahnke. 2000. Rates of C, N, P and Si recycling and denitrification at the US Mid-Atlantic continental slope depocenter. Deep-sea Res. I 47:1405–1428.

Jahnke, R. A., C. E. Reimers, and D. B. Craven. 1990. Intensification of recycling of organic matter at the sea floor near ocean margins. Nature 348: 50–54.

Jaramillo, S., A. Sheremet, M. A. Allison, K. T. Holland, and A. H. Reed. 2009. Wave-mud interactions over the muddy Atchafalaya subaqueous clinoform, Louisiana, United States: wave-supported sediment transport. J. Geophys. Res.- Oceans. 114: C04002.

Johnson, C. L., L. E. Webb, S. A. Graham, M. S. Hendrix, and G. Badarch. 2001. Sedimentary and structural records of late Mesozoic high-strain extension and strain partitioning, East Gobi basin, southern Mongolia, pp. 413–434. *In* Hendrix, M.S., Davis, G.A. (eds.), Paleozoic and Mesozoic tectonic evolution of central and eastern Asia. Geological Society of America Memoir 194.

Justic, D., N. N. Rabalais, and R. E. Turner. 1995. Stoichiometric nutrient balance and origin of coastal eutrophication. Mar. Poll. Bull. 30: 41–46.

Kemp, G. P. 1986. Mud deposition at the shoreface: Wave and sediment dynamics on the Chenier plain of Louisiana, Ph.D. thesis. Louisiana State Univ., Baton Rouge.

Kineke, G. C., K. J. Woolfe, S. A. Kuehl, J. D. Milliman, T. M. Dellapenna, and R. G. Purdon. 2000. Sediment export from the Sepik River, Papua New Guinea: evidence for a divergent sediment plume. Cont. Shelf Res. 20: 2239–2266.

Kottke, B., T. Schwenk, M. Breitzke, M. Wiedicke, H. R. Kudrass, and V. Spiess. 2003. Acoustic facies and depositional processes in the upper submarine canyon Swatch of No Ground (Bay of Bengal). Deep Sea Res. II. 50: 979–1001.

Kuehl, S.A., B. M. Levy, W. S. Moore, and M. A. Allison. 1997. Subaqueous delta of the Ganges–Brahmaputra river system. Mar. Geol. 144: 81–96.

Lehrter, J. C., M. C. Murrell, and J. C. Kurtz. 2009. Interactions between freshwater input, light, and phytoplankton dynamics on the Louisiana continental shelf. Cont. Shelf Res. 29: 1861–1872.

Lin, S., and J. Morse. 1991. Sulfate reduction and iron sulfide mineral formation in Gulf of Mexico anoxic sediments. Am. J. Sci. 291: 55–89.

Liu, H., and M. Dagg. 2003. Interactions between nutrients, phytoplankton growth, and grazing by micro- and meso-zooplankton in the plume of a large river. Mar. Ecol. Prog. Ser. 258: 31–42.

Liu, J. T., and H.-L. Lin. 2004. Sediment dynamics in a submarine canyon: a case of river–sea interaction. Mar. Geol. 207: 55–81.

Lohrenz, S. E., G. L. Fahnenstiel, D. G. Redalje, G. A. Lang, M. J. Dagg, T. E. Whitledge, and Q. Dortch. 1999. Nutrients, irradiance, and mixing as factors regulating primary production in coastal waters impacted by the Mississippi River plume. Cont. Shelf Res. 19: 1113–1141.

Lohrenz, S. E., G. L. Fahnenstiel, D. G. Redalje, G. A. Lang, X. Chen, and M. J. Dagg. 1997. Variations in primary production of northern Gulf of Mexico continental shelf waters linked to nutrient inputs from the Mississippi River. Mar. Ecol. Prog. Ser. 155: 45–54.

Lohrenz, S. E., M. J. Dagg, and T. E. Whitledge. 1990. Enhanced primary production at the plume/oceanic interface of the Mississippi River. Cont. Shelf Res. 10: 639–664.

Ludwig, W., and J. L. Probst. 1998. River sediment discharge to the oceans: present-day controls and global budgets. Am. J. Sci. 298: 265–295.

Mackenzie, F. T., and A. Lerman. 2006. Carbon in the Geobiosphere – Earth's Outer Shell. Springer, Dordrecht, the Netherlands.

Malcolm, R., and W. H. Durum. 1976. Organic carbon and nitrogen concentrations and organic load of six selected rivers of the United States. U.S. Geol. Survey Water Supply Paper, 1817F.

Mayer, L. M., L. L. Schick, M. Allison, K. Ruttenberg, and S. Bentley. 2007. Marine vs. terrigenous organic matter in Louisiana coastal sediments: The uses of Bromine: organic carbon ratios. Mar. Chem. 107: 244–254.

McKee, P. W., B. R. Clark, and J. B. Czarnecki. 2004. Conjunctive-use optimization model and sustainable yield estimation for the Sparta Aquifer of southeastern Arkansas and north-central Louisiana, p. 30. U.S. Geological Survey water resources investigation report 03-4231, Little Rock, AR.

McKee B., and R. Twilley. 1996. Ecosystem analysis of the Louisiana Bight and adjacent shelf environments. Vol. I The Fate of Organic Matter and Nutrients in the Sediments of the Louisiana Bight.

Mead, R., and M. A. Goñi. 2006. A lipid molecular marker assessment of sediments from the Northern Gulf of Mexico before and after the passage of Hurricane Lili. Org. Geochem. 37: 1115–1129.

Meade, R. H., and J. A. Moody. 2010. Causes for the decline in suspended-sediment discharge in the Mississippi River system, 1940–2007. Hydrolog. Proc. 24: 35–49.

Meade, R. H. 1995. Setting: Geology, hydrology, sediments, and engineering of the Mississippi River, pp. 13–28. *In* Meade, R. H. (eds.), Contaminants in the Mississippi River, 1987–92. U.S. Geological Survey Circular 1133, Denver, CO.

Meade, R. H., and R. Parker. 1985. Sediment in rivers of the United States, National Water Summary, U.S. Geological Survey Water Supply Paper. 2275: 49–60.

Mezic, I., S. Loire, V. A. Fonoberov, and P. Hogan. 2010. A new mixing diagnostic and Gulf Oil Spill movement. Science 330: 486–489.

Michels, K. H., A. Suckow, M. Breitzke, H. R. Kudrass, and B. Kottke. 2003. Sediment transport in the shelf canyon "Swatch of No Ground" (Bay of Bengal). Deep-Sea Res. II. 50: 1003–1022.

Milliman, J. D., and R. H. Meade. 1983. World-wide delivery of river sediment to the oceans. J. Geol. 91: 1–21.

Milliman, J. D., and J. P. M. Syvitski. 1992. Geomorphic/tectonic control of sediment discharge to the ocean: the importance of small mountainous rivers. J. Geol. 100: 525–544.

Morse, J. W., and P. M. Eldridge. 2007. A non-steady state diagenetic model for changes in sediment biogeochemistry in response to seasonally hypoxic/anoxic conditions in the "dead zone" of the Louisiana shelf. Mar. Chem. 106: 239–255.

Morse, J. W., and G. T. Rowe. 1999. Benthic biogeochemistry beneath the Mississippi River plume. Estuar. 22: 206–214.

Mossa, J. 1996. Sediment dynamics in the lowermost Mississippi River. Eng. Geol. 45: 457–479.

Mulder, T., and J. Alexander. 2001. The physical character of subaqueous sedimentary density flows and their deposits. Sedimentology. 48: 269–299.

Mullenbach, B. L., and C. A. Nittrouer. 2006. Decadal record of sediment export to the deep sea via Eel Canyon. Cont. Shelf Res. 26: 2157–2177.

Nelson, C. H., D. C. Twichell, W. C. Schwab, H. J. Lee, and H. H. Kenyon. 1992. Upper Pleistocene turbidite sand beds and chaotic silt beds in the channelized, distal, outer-fan lobes of the Mississippi fan. Geol. 20: 693–696.

Neill, C. F., and M. A. Allison. 2005. Subaqueous deltaic formation on the Atchafalaya Shelf, Louisiana. Mar. Geol. 214: 411–430.

Nelson, C. H., D. C. Twichell, W. C. Schwab, H. J. LEE, and H. H. Keny. 1992. Upper Pleistocene turbidite sand beds and chaotic silt beds in the channelized, distal, outer-fan lobes of the Mississippi Fan. Geology 20: 693–696.

Nittrouer, C. A., D. Mohrig, D., and M. A. Allison. 2011.Punctuated sand transport in the lowermost Mississippi River. J. Geophys. Res. – Earth Surface Processes. 116: 24, F04025.

Onstad, G. D., D. E. Canfield, P. D. Quay, and J. I. Hedges. 2000. Sources of particulate organic matter in rivers from the continental USA: lignin phenol and stable carbon isotope compositions. Geochim. Cosmochim. Acta. 64: 3539–3546.

Orton, G. J., and H. G. Reading. 1993. Variability of deltaic processes in terms of sediment supply, with particular emphasis on grain size. Sedimentology 40: 475–512.

Parsons, J. D., C. T. Friedrichs, P. A. Traykovski, D. Mohrig, J. Imran, J. P. M. Syvitski, G. Parker, P. Puig, J. L. Buttles, and M. H. Garcia. 2007. The mechanics of marine sediment gravity flows, pp. 275–337. *In* Continental Margin Sedimentation, IAS Special Publication.

Penland, S., R. Boyd, and J. R. Suter. 1988. Transgressive depositional systems of the Mississippi delta plain: a model for barrier shoreline and shelf sand development. J. Sediment. Res. 58: 932–949.

Postma, G. 1995. Sea-level-related architectural trends in coarse grained delta complexes. Sediment. Geol. 98: 3–12.

Presley, B. J., J. H. Trefry, and R. F. Shokes. 1980. Heavy metal inputs to Mississippi Delta sediments: a historical view. Water Air Soil Pollut. 13: 481–494.

Puig, P., and A. Palanques. 1998. Temporal variability and composition of settling particle fluxes on the Barcelona continental margin (Northwestern Mediterranean). J. Mar. Res. 56: 639–654.

Quigg, A., J. B. Sylvan, A. B. Gustafson, T. R. Fisher, R. L. Oliver, S. Tozzi, and J. W. Ammerman. 2011. Going West: Nutrient Limitation of Primary Production in the Northern Gulf of Mexico and the Importance of the Atchafalaya River. Aquat. Geochem. 17: 519–544.

Rabalais, N. N., R. E. Turner, Q. Dortch, D. Justic, V. J. Bierman, and W. J. Wiseman. 2002. Nutrient-enhanced productivity in the northern Gulf of Mexico: past, present and future. Hydrobiology 475: 39–63.

Rabalais, N. N., R. E. Turner, B. K. Sen. Gupta, D. F. Boesch, P. Chapman, and M. C. Murrell. 2007. Hypoxia in the Northern Gulf of Mexico: Does the Science Support the Plan to Reduce, Mitigate, and Control Hypoxia? Estuar. Coast. 30: 753–772.

Rabalais, N. N., W. J. Wiseman Jr., R. E. Turner, D. Justic, B. K. Sen. Gupta, Q. Dortch. 1996. Nutrient changes in the Mississippi River and system responses on the adjacent continental shelf. Estuaries 19: 386–407.

Redalje, D. G., S. E. Lohrenz, and G. L. Fahnensteil. 1994. The relationship between primary production and the vertical export of particulate organic matter in a river-impacted coastal system. Estuaries 17: 829–838.

Richey, J. E. 2004. Pathways of atmospheric CO_2 through fluvial systems, pp. 513–564. *In* C. B. Field, M. R. Raupach (eds.), The Global Carbon Cycle: Integrating Humans, Climate, and the Natural World. Island Press, Washington, DC.

Roberts, H. H. 1998. Delta switching: early responses to the Atchafalaya river diversion. J. Coast. Res. 14: 882–899.

Romankevich, E. A. 1984. Geochemistry of Organic Matter in the Ocean. New York-Heidelberg-Berlin-Tokyo, Springer-Verlag.

Rowe, G. T., M. E. C. Kaegi, J. W. Morse, G. S. Boland, and E. G. E. Briones. 2002. Sediment community metabolism associated with continental shelf hypoxia, Northern Gulf of Mexico. Estuaries 25: 1097–1106.

Rowe, G. T., and P. Chapman. 2002. Hypoxia in the northern Gulf of Mexico: some nagging questions. Gulf Mex. Sci. 20: 153–60.

Rowe, G. T., M. E. C. Kaegi, J. W. Morse, G. S. Boland, and E. G. E. Briones. 2002. Sediment community metabolism associated with continental shelf hypoxia, Northern Gulf of Mexico. Estuaries 25: 1097–1106.

Rowe, G. T. and R. W. Howarth. 1985. Early diagenesis of organic matter in sediments off the coast of Peru. Deep-Sea Res. 32: 43–55.

Safak, I., A. Sheremet, M. A. Allison, and T.-J. Hsui. 2010. Bottom turbulence on the muddy Atchafalaya Shelf, Louisiana, USA. J. Geophys. Res. 115: 15, C12019.

Sampere, T. P., T. S. Bianchi, M. A. Allison, and B. A. McKee. 2011. Burial and degradation of organic carbon in Louisiana shelf/slope sediments. Estuar. Coast. Shelf Sci. 95: 232–244.

Sampere, T. P., T. S. Bianchi, S. G. Wakeham, and M. A. Allison. 2008. Sources of organic matter in surface sediments of the Louisiana Continental margin: Effects of major depositional/transport pathways and Hurricane Ivan. Cont. Shelf Res. 28: 2472–2487.

Santschi, P. H., J. W. Murray, M. Baskaran, C. R. BenitezNelson, L. D. Guo, C. C. Hung, C. Lamborg, S. B. Moran, U. Passow, and M. Roy-Barman. 2006. Thorium speciation in seawater. Mar. Chem. 100: 250–268.

Sarmiento, J. L., and E. Sundquist, 1992. Oceanic uptake of anthropogenic CO_2: a new budget. Nature 356: 589–593.

Schlünz, B., and R. R. Schneider. 2000. Transport of riverine organic carbon to the oceans: Implications for the carbon cycle. Geologische Rundschau (International Journal of Earth Sciences). 88: 599–606.

Schwenk, T., V. Spieß, C. Hübscher, and M. Brietzke. 2003. Frequent channel avulsions within the active channel-levee system of the middle Bengal Fan – an exceptional channel-levee development derived from Parasound and Hydrosweep data. Deep Sea Res. II. 50: 1023–1045.

Shanmugan, G., R. J. Moiola, and J. E. Damuth. 1985. Eustatic control of submarine fan development, pp. 23–28. *In* A. H. Bouma, W. R. Normark, and M. E. Barnes (eds.), Submarine Fans and Related Turbidite Systems. New York, Springer-Verlag.

Sheremet, A., S. Jaramillo, S.-F. Su, M. A. Allison, and K. T. Holland. 2011. Wave-mud interaction over the muddy Atchafalaya subaqueous clinoform, Louisiana, United States: wave processes. J. Geophys. Res.-Oceans. 116: C06005.

Sheremet, A., A. J. Mehta, B. Liu, and G. W. Stone. 2005. Wave-sediment interaction on a muddy inner shelf during Hurricane Claudette. Estuar. Coast. Shelf Sci. 63: 225–233.

Shlemon, R. J. 1975. Subaqueous delta formation – Atchafalaya Bay, Louisiana, in Deltas, pp. 209–221. *In* M. L. Broussard (eds.), Houston Geological Society, Houston, TX.

Silverberg, N., B. Sundby, A. Mucci, S. Zhong, T. Arakaki, P. Hall, A. Landen, and A. Tengberg. 2000. Remineralization of organic carbon in eastern Canadian continental margin sediments. Deep-Sea Res. II. 47: 699–631.

Stelting, C. E., L. Droz, A. H. Bouma, J. M. Coleman, M. Cremer, A. W. Meyer, W. R. Normark, S. O'Connell, and D. A. V. Stow. 1986. Late Pleistocene seismic stratigraphy of the Mississippi Fan, pp. 437–456. *In* A. H. Bouma, J. M. Coleman, A. W. Meyer (eds.), Initial Reports of the Deep Sea Drilling Project 96.

Stow, D. A. V., D. G. Howell, and C. H. Nelson. 1985. Sedimentary, tectonic, and sea-level controls, pp. 15–22. *In* A. H. Bouma, W. R. Normark, and N. E. Barnes (eds.), Submarine Fans and Related Turbidite Systems. New York, Springer-Verlag.

Sylvan, J. B., Q. Dortch, D. M. Nelson, A. F. M. Brown, W. Morrison, and J. W. Ammerman. 2006. Phosphorus limits phytoplankton growth on the Louisiana shelf during the period of hypoxia formation. Environ. Sci. Technol. 40: 7548–7553.

Sylvan. J. B., A. Quigg, S. Tozzi, and J. W. Ammerman. 2007. Eutrophication induced phosphorus limitation in the Mississippi River plume: evidence from fast repetition rate fluorometry. Limnol. Oceanogr. 52: 2679–2685.

Tesi, T., P. Puig, A. Palanques, and M. A. Goñi. 2010. Lateral advection of organic matter in cascading – dominated submarine canyons. Prog. Oceanogr. 84: 185–203.

Tesi, T., L. Langone, M. A. Goñi, S. Miserocchi, and F. Bertasi. 2008. Changes in the composition of organic matter from prodeltaic sediments after a large flood event (Po River, Italy), Geochim. Cosmochim. Acta. 72: 2100–2114.

Törnqvist, T. E., T. R. Kidder, W. J. Autin, K. Van der Borg, A. F. M. De Jong, C. J. W. Klerks, E. M. A. Snijders, J. E. A. Storms, R. L. Van Dam, and M. C. Wiemann. 1996. A revised chronology for Mississippi River subdeltas. Science. 273: 1693–1696.

Trefry, J. H., D. B. Butterfield, S. Metz, G. J. Massoth, R. P. Trocine, and R. A. Feely. 1994. Trace metals in hydrothermal solutions from Cleft segment on the southern Juan de Fuca Ridge. J. Geophys. Res. 99: 4925–4935.

Turner, R. E., N. N. Rabalais, R. B. Alexander, G. McIsaac, and R. W. Howarth. 2007. Characterization of nutrient, organic carbon, and sediment loads and concentrations from the Mississippi River into the northern Gulf of Mexico. Estuar. Coasts. 30: 773–790.

Turner, R. E., N. N. Rabalais, E. M. Swenson, M. Kasprzak, and T. Romaire. 2005. Summer hypoxia in the northern Gulf of Mexico and its prediction from 1978 to 1995. Mar. Environ. Res. 59: 65–77.

Twichell, D. C., N. H. Kenyon, L. M. Parson, and B. A. McGregor. 1991. Depositional patterns of the Mississippi Fan surface: evidence from GLORIA II and high-resolution seismic pro¢les, pp. 349–363. *In* P. Weimer, M. H. Link (eds.), Seismic Facies and Sedimentary Processes of Submarine Fans and Turbidite Systems. Springer, New York.

Van Heerden, I. L., and H. H. Roberts. 1988. Facies development of the Atchafalaya Delta, Louisiana: A modern bayhead delta. Am. Assoc. Petrol. Geol. Bull. 72: 439–453.

Walker, N. A., and A. B. Hammack. 2000. Impacts of winter storms on circulation and sediment transport: Atchafalaya-Vermilion Bay Region, Louisiana, U.S.A. J. Coast. Res. 16: 996–1010.

Walsh, J., D. R. Corbett, D. Mallinson, M. A. Goni, and M. B. Dail. 2006. Mississippi Delta mudflow activity and 2005 Gulf Hurricanes. Eos Transact. 87: 477.

Wang, L., and D. Justic. 2009. A modeling study of the physical processes affecting the development of seasonal hypoxia over the inner Louisiana–Texas shelf: circulation and stratification. Cont. Shelf Res. 29: 1464–76.

Waterson, E. J., and E. A. Canuel. 2008. Sources of sedimentary organic matter in the Mississippi River and adjacent Gulf of Mexico as revealed by lipid biomarker and $\delta^{13}C$ TOC analyses. Org. Geochem. 39: 422–439.

Weaver, M. A., R. M. Zablotowicz, and M. A. Locke. 2004. Laboratory assessment of atrazine and fluometuron degradation in soils from a constructed wetland. Chemosph. 57: 853–862.

Webb. E. C., I. A. Mendelssohn, and B. J. Wiley. 1995. Causes for vegetation dieback in a Louisiana salt marsh: a bioassay approach. Aquat. Bot. 51: 281–289.

Weber, M. E., M. H. Wiedicke, H. R. Kudrass, C. Hübscher, and H. Erlenkeuser. 1997. Active growth of the Bengal Fan during sea-level rise and highstand. Geol. 25: 315–318.

Wei, C.-L., G. Rowe, G. F. Hubbard, A. H. Scheltema, G. D. F. Wilson, I. Petrescu, J. Foster, M. K. Wicksten, M. Chen, R. Davenport, Y. Soliman, and Y. Wang. 2010. Bathymetric zonation of deep-sea macrofauna in relation to export of surface phytoplankton production. Mar. Ecol. Prog. Ser. 399: 1–14.

Wells, J. T., and H. H. Roberts. 1980. Fluid mud dynamics and shoreline stabilization: Louisiana chenier plain, in Proceedings of the Seventeenth International Coastal Engineering Conference, American Society of Civil Engineering, Sydney, Australia. 17: 1382–1401.

Wilson. C. A., and M. A. Allison. 2008. Sediment dynamics and geomorphology of eroding marsh shorelines in southeastern Louisiana. Estuar. Coast. Shelf Sci. 80: 483–494.

Wiseman, W. J. Jr., N. N. Rabalais, R. E. Turner, S. P. Dinnel, and A. MacNaughton. 1997. Seasonal and interannual variability within the Louisiana coastal current: stratification and hypoxia. J. Mar. Sys. 12: 237–48.

Wright, L. D. 1985. River deltas, pp. 1–75. *In* R. A. Davis (eds.), Coastal Sedimentary Environments, 2nd ed. Springer-Verlag, New York.

Wright, L. D., and J. M. Coleman. 1974. Mississippi River mouth processes: effluent dynamics and morphologic development. J. Geol. 82: 751–778.

Wysocki, L. A., T. S. Bianchi, R. T. Powell, and N. Reuss. 2006. Spatial variability in the coupling of organic carbon, nutrients, and phytoplankton pigments in surface waters and sediments of the Mississippi river plume. Estuar. Coast. Shelf Sci. 69: 47–63.

20

Composition and fluxes of carbon and nutrient species from the Yukon River basin in a changing environment

L. Guo, R. G. Striegl, and R. Macdonald

1. Introduction

River export of carbon and nutrients and the interactions between river water and seawater at the land-ocean interface play important roles in regulating coastal environmental quality, material budgets, and biogeochemical processes in coastal marine ecosystems (Mayer et al. 1998; Lohrenz et al. 1999; Meybeck 2003; Bianchi et al. 2010). Terrestrial inputs are exceptionally important in the Arctic Ocean, which receives a disproportionately large amount of global runoff and associated nutrients and organic matter (Aagaard and Carmack 1989; Stein and Macdonald 2004; McGuire et al. 2009; Holmes et al. 2012).

The Yukon River contributes ~8% of the total freshwater input to the Arctic Ocean via Bering Strait (Aagaard and Carmack 1989), with an annual discharge of more than 2×10^{11} m^3 of freshwater and $\sim 60 \times 10^6$ tons of suspended sediments (Brabets et al. 2000; Holmes et al. 2002; Dornblaser and Striegl 2009). Recent studies demonstrate that northern ecosystems are sensitive to global and regional climate and environmental changes (Peterson et al. 2002; Hinzman et al. 2005; Post et al. 2009). Of special concern in the Yukon River Basin are permafrost degradation (e.g., Jorgenson et al. 2006; Guo et al. 2007; Douglas et al. 2011) and changes in hydrological cycling (Walvoord et al. 2012), vegetation, and landscape. Therefore, environmental and climate change in the Arctic terrestrial ecosystems, including increased total freshwater runoff and more frequent high-flow events in arctic rivers, promise to alter biogeochemical cycles within the drainage basin as well as the timing and magnitude of fluxes of carbon, nutrients, and other bioactive elements to coasts (Stein and Macdonald 2004; McGuire et al. 2009; Ping et al. 2011). For these reasons, the biogeochemical consequences of change in nutrient and carbon cycles within various river basins in northern high-latitude regions have become the focus of many recent studies.

Over the past decade, water chemistry, chemical and phase speciation of nutrients and carbon, and their export fluxes from the Yukon River have received increased attention. In 2001, the U.S. Geological Survey (USGS) launched a systematic study on the water quality of the Yukon River. In addition, sampling in the Yukon River basin also began under several federally funded research projects, including PARTNERS and the Arctic Great Rivers Observatory programs (e.g., McClelland et al. 2008; Holmes et al. 2012; Tank et al. 2012). Thus there is an increasing body of literature (Table 20.1) reporting recent changes in the Yukon River basin and the composition and fluxes of nutrient and carbon species (e.g., Brabets et al. 2000; Schuster 2003, 2005a, 2005b, 2006; Guo et al.

Table 20.1. *A compilation of recent studies on carbon, nutrients, and trace elements in the Yukon River basin*

References	Tributary (or location)	Subject(s)
Leenheer (1982)	Yukon at Pilot Station	DOC and POC
Holmes et al. (2002)	Yukon	Total suspend sediments
Carey (2003)	Wolf Creek/Granger Basin	Spring DOC
Guéguen et al. (2003)	Yukon at Stevens Village	C-DOM
Guo et al. (2003)	Chena River	DOC/COC/POC and radiocarbon
Guo et al. (2004a)	Yukon at Stevens Village	Dissolved inorganic and organic nutrients (N, P, and Si)
Striegl et al. (2005)	Yukon River	DOC and DIC
Belzile & Guo (2006)	Yukon at Pilot Station	DOC and fluorescence
Guéguen et al. (2006)	Yukon at Stevens Village	DOC, carbohydrates, CDOM
Guo & Macdonald (2006)	Yukon at Stevens Village	DIC, DOC, COC, POC, stable isotopes and radiocarbon
Zou et al. (2006)	Yukon at Stevens Village	Lipids in DOC/COC and POC, compound specific stable isotopes
Guo et al. (2007)	Yukon at Pilot Station	DOC and POC
Raymond et al. (2007)	Yukon at Pilot Station	DOC and its radiocarbon
Striegl et al. (2007)	Yukon, Tanana, and Porcupine Rivers	DOC, DIC, POC, and PIC
Walvoord & Striegl (2007)	Yukon River	Carbon and nitrogen
Cai et al. (2008a)	Chena River	DOC, DIC, and POC
Cai et al. (2008b)	Chena River	Nutrients (N, P, Si)
Cooper et al. (2008)	Yukon at PS	DOC and 18O
Spencer et al. (2008)	Yukon	CDOM, composition, and reactivity
Spencer et al. (2009)	Yukon	CDOM and export fluxes
O'Donnell et al. (2010)	Yukon River basin	DOM
Douglas et al. (2011)	Yukon/Chena	Permafrost
Schuster et al. (2011)	Yukon	Hg
Stedmon et al. (2011)	Yukon	CDOM
Douglas et al. (2012)	Chena River	DOC, TDN, major ions, and H, O and Sr isotopes
Guo et al. (2012)	Yukon at Pilot Station	DIC, DOC, POC, nutrients (N, P, and Si), and their fluxes; H and O isotopes
Holmes et al. (2012)	Yukon at Pilot Station	Nutrient and DOC fluxes
Stolpe et al. (2013a)	Six small rivers in Yukon basin	DOC, Fe, and other colloidal trace metals
Stolpe et al. (2013b)	Six small rivers in Yukon basin	Rare earth elements

2004a; Striegl et al. 2005; Guo and Macdonald 2006; Dornblaser and Striegl 2007; Striegl et al. 2007; Walvoord and Striegl 2007; Cai et al. 2008a, 2008b; Spencer et al. 2008, 2009; Guo et al. 2012).

In this chapter, we review and summarize recent studies conducted in the Yukon River basin and discuss recent advances in the biogeochemical cycling and export fluxes of nutrients, carbon, and other bioactive elements and their relationships with climate and environmental changes in the northern high-latitude regions.

2. The Yukon River Basin

The Yukon River Basin is one of the largest and most diverse ecosystems in the world. Perhaps just as important, the Yukon River is one of the few remaining large rivers that is not controlled by reservoirs or dams, thus allowing an unconfounded examination of the response of carbon and nutrient cycles to environmental changes in the basin. The Yukon River is about 3,340 km long and drains an area of roughly 855,000 km^2 in northwestern Canada and central Alaska in the United States (Fig. 20.1). It discharges into the Bering Sea, providing freshwater runoff, sediments, and nutrients for the eastern Bering Sea and Chukchi Sea ecosystems in the western Arctic Ocean (Brabets et al. 2000).

One of the most distinctive features for the Yukon River is that it is at high-latitude location (59–69°N) and flows from east to west, unlike other major arctic rivers that flow from south to north. Similar to other major arctic rivers, the Yukon River basin is seasonally snow covered and the river virtually frozen from October to May depending on the specific year, resulting in a stream flow pattern (Fig. 20.2) that is distinct from major world rivers in tropical, subtropical, and temperate regions.

Most flow occurs in the spring freshet and summer months through snowmelt, rainfall, and glacier melt, with very low discharge in winter months. Long-term averaged daily stream flow data collected at a downstream location near Pilot Station, Alaska, show a variation of freshwater discharge typical of arctic rivers (Fig. 20.2). The mean freshwater discharge during the winter season may be as low as 1,250 m^3/s, but the average freshwater discharge during spring freshet could be an order of magnitude higher, with a peak flow range between 13,000 and 35,000 m^3/s. In addition to the high variability in daily stream flow, carbon and nutrient species in river waters also show profoundly different concentrations and composition between seasons. Together, the heterogeneous flow pattern, seasonal ice cover, remoteness, and extreme weather conditions present considerable challenges in quantifying export fluxes of properties like sediments and nutrients owing to the nonlinear relationship between discharge and property in arctic rivers (e.g., Brabets et al. 2000; Finlay et al. 2006; Guo et al. 2012). Furthermore, the permafrost-dominated tributary basins and glacial-dominated flow regimes are likely to have separate responses in carbon and nutrient cycling to climate and environmental changes, complicating interpretation of overall changes in material fluxes and carbon and nutrient composition with time.

3. Variations in carbon species of river waters

The riverine export of carbon from drainage basins to the ocean represents a major component of the global carbon cycle (Cole et al. 2007). Carbon species in river waters can be classified into dissolved inorganic carbon (DIC), dissolved organic carbon (DOC), particulate inorganic carbon (PIC), and particulate organic carbon (POC). Through the use of ultrafiltration, the traditionally defined DOC can be further fractionated into two operationally defined DOC fractions: low-molecular-weight (LMW) DOC and high-molecular-weight (HMW) DOC or colloidal organic carbon (COC). Therefore, organic carbon in river waters may be partitioned into LMW-DOC, COC, and POC components (Guo and Macdonald 2006). These size classes of organic carbon species have been shown to contain different elemental and molecular compositions, to have different reactivities, to undergo different biogeochemical cycles, and to follow separate transport pathways in river systems, such as the Yukon River (Guo et al. 2003; Guo and Macdonald 2006; Zou et al. 2006; Spencer et al. 2009; O'Donnell et al. 2010). Despite these differences, measurements that distinguish all three OC phases remain scarce. Another operational classification of DOC is the fraction that biologically degrades within relevant river DOC transit times. This biodegradable DOC

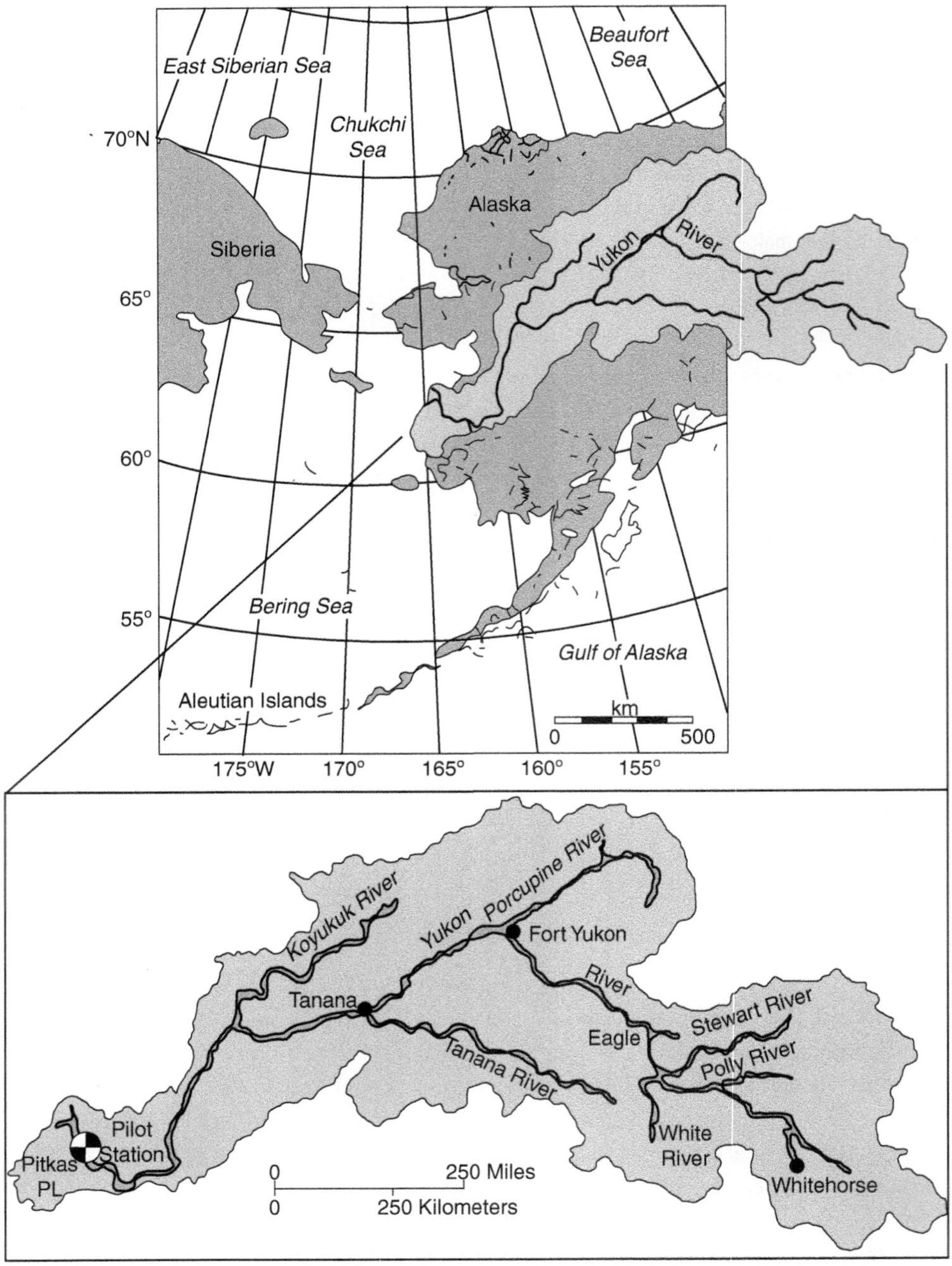

Figure 20.1. A map of the Yukon River basin showing its major tributaries and location in Canada and central Alaska.

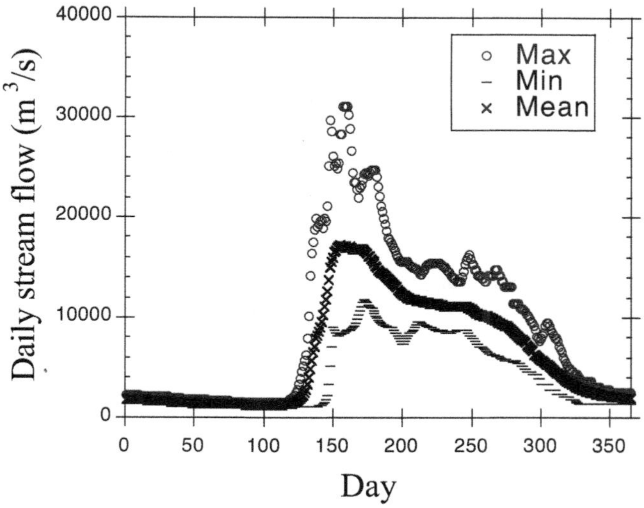

Figure 20.2. An example of variations in averaged daily stream flow (m^3/s) from the Yukon River at the downstream USGS hydrological station, Pilot Station, Alaska (data between 1975–2001 from USGS at http://waterdata.usgs.gov).

(BDOC) varies seasonally in amount and degradability in the Yukon River system (Wickland et al. 2012).

In addition to organic carbon, inorganic carbon components, variously expressed as DIC, alkalinity, pCO_2, and PIC, are important in elucidating carbon dynamics, transformation between organic and inorganic phases, and detailed biogeochemical cycling processes in fluvial systems (Cai 2003; Striegl et al. 2007, 2012; Dornblaser and Striegl 2009). Furthermore, chromophoric dissolved organic matter (CDOM), which absorbs visible and ultraviolet light, and fluorescent DOM (F-DOM), which emits fluorescence after absorption of light, have also been used recently to quantify the abundance, reactivity, and DOM molecular weight and to characterize sources and composition of DOM in Yukon River waters (Belzile and Guo 2006; Guéguen et al. 2006; Spencer et al. 2008, 2009; Stedmon et al. 2011; Dornblaser et al. 2012).

3.1. Seasonal and spatial variations in organic carbon species

Early studies have determined concentrations of DOC and POC in Yukon River waters that may be used together with flow to estimate organic carbon export fluxes (Table 20.1). Recent studies with better seasonal sampling resolution have shown an extraordinary seasonal range in DOC and POC concentrations in the Yukon River, with extremely high concentrations during spring freshet but very low concentrations during winter season under the ice (Guo and Macdonald 2006; Striegl et al. 2007; Cai et al. 2008a; Guo et al. 2012; and data available at http://arcticgreatrivers.org/). For example, concentrations of DOC in the Yukon River during 2004–2005 varied from ~180 μM during winter to as high as ~1,600 μM during ice melt, an almost 10-fold difference between seasons (Guo et al. 2012).

The concentration of COC followed the seasonal variation pattern of DOC. However, the percentage of COC in the bulk DOC pool remained fairly constant despite the high variability in concentration (Cai and Guo 2005; Guo and Macdonald 2006). On average, about 60–70% of the bulk DOC was measured in the >1-kDa colloidal fraction in the Yukon River (Guo et al. 2003; Belzile and Guo 2006; Guo and Macdonald 2006). Similar to DOC and COC, concentrations of POC also exhibited a large seasonal range, varying from 60–70 μM during the winter to 500–600 μM during spring freshet (Striegl et al. 2007; Guo et al. 2012). In addition to seasonal variations, both DOC and POC concentrations also showed a significant variability from year to year, as reported in Striegl et al. (2007) during 2001–2005 for the Yukon River and its major tributaries.

This distinct seasonal variation in river water organic carbon concentrations is unique to arctic rivers (Rember and Trefry 2004; Finlay et al. 2006; Guo et al. 2007; Holmes et al. 2011) and uncommon for subtropical or temperate rivers, especially engineered rivers with altered flows, such as the Mississippi River, where there is no obvious seasonality in DOC concentration (Duan et al. 2007; Guo 2010). We infer from these observations that it is natural rivers such as the Yukon that exhibit instantaneous responses in carbon dynamics when hydrological cycles in the basin are altered by climate or other environmental factors.

In addition to seasonal changes, DOC concentration also demonstrates an obvious spatial variability (e.g., Striegl et al. 2007). For example, annual average DOC concentration was the lowest in the Tanana River, highest in the Porcupine River and in between for the Yukon at Eagle/Stevens Village/Pilot Station. The maximum spring DOC concentration varied from as high as 4,617±1,775 μM in Wolf Creek/Granger Basin in the upper Yukon (Carey 2003) to 2,825 μM in the Yukon River at Stevens Village (Guo and Macdonald 2006), 1,187 μM in the Chena River (Cai et al. 2008a), and to 1,683 μM in the downstream Yukon River at Pilot Station (Guo et al. 2012). In addition, the spring DOC concentration also varied from year to year and between studies (Raymond et al. 2007; Striegl et al. 2007; Guo et al. 2012; PARTNERS, and data available at http://arcticgreatrivers.org/), likely resulting from the difference in timing of sampling relative to spring freshet and annual variability.

Although specific DOC concentration varied among rivers and tributaries (Striegl et al. 2007; O'Donnell et al. 2010; Frederick et al. 2011; Stolpe et al. 2013a), the DOC concentration during spring freshet seemed to decrease in general from the upper stream (e.g., the Granger Basin) to middle stream (e.g., Yukon at Stevens Village) to the lower stream (e.g., Yukon at Pilot Station) along the Yukon River. This decreasing trend was likely the result of mixing between high DOC rivers and lower DOC rivers or tributaries such as the Tanana River, which is dominated by glacial melt with relatively low DOC concentration throughout the year (Striegl et al. 2007). Therefore, DOC measured at downstream stations may not by itself provide sufficient information to decipher biogeochemical processes occurring upstream in the river basin even though this DOC represents the actual riverine export to the Bering Sea/Arctic Ocean.

The amount and chemical character of BDOC in the Yukon River and its tributaries also vary seasonally, with BDOC comprising about 12–18% of total annual DOC export (Wickland et al. 2012). The BDOC fraction of DOC is greatest under ice, but because DOC concentrations are lowest during winter, the actual amounts of BDOC in the river are greatest during spring flush. Although chemical composition of DOC was an important driver of BDOC, the overriding control on BDOC appears to be mineral nutrient availability (Wickland et al. 2012).

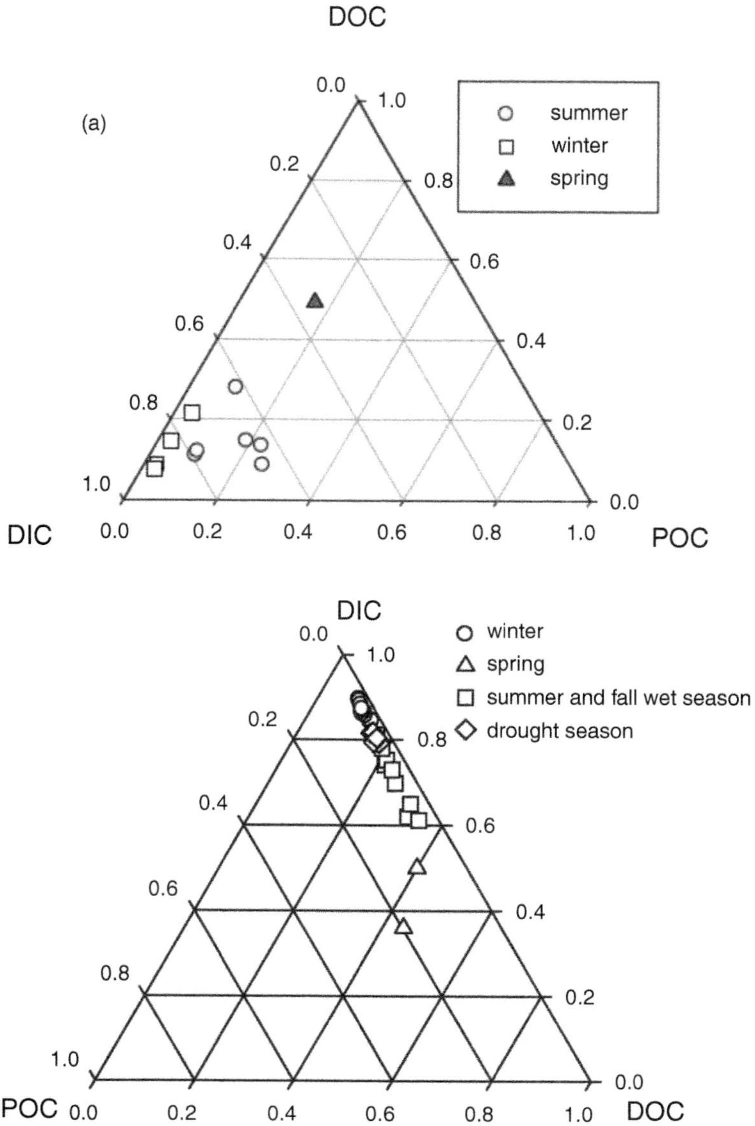

Figure 20.3. Examples of the partitioning of carbon between DIC, DOC, and POC pool in Yukon River waters at Pilot Station (upper panel, from Guo et al. 2012) and Chena River (lower panel, from Cai et al. 2008a).

3.2. Variations in inorganic carbon species

Inorganic carbon species in river waters include both DIC and PIC. Recent studies have shown that DIC is the predominant carbon species in Yukon River waters except during the spring freshet when DOC may become dominant (Fig. 20.3; Striegl et al. 2007; Cai et al. 2008a; Guo et al. 2012).

In contrast to organic carbon species, DIC concentration was highest during winter when the discharge is low and mostly comprised of groundwater (Fig. 20.2) and lowest during spring freshet in the Yukon River (Striegl et al. 2007; Cai et al. 2008a; Guo et al. 2012). For example, average winter DIC concentration in the Yukon River was up to 2,066±75 μM in the winter of 2004/2005 (Guo et al. 2012) and as high as 4,100±1.9 μM between 2001–2005 (Striegl et al. 2007), whereas DIC concentration during high flow in spring was in the range of 1,100–1,480 μM, depending on the year. Guo et al. (2012) reported an average DIC/DOC ratio of 5.8±3.2, ranging from 0.7 in spring freshet to 11.7 in late winter (April) before the onset of ice melt for the Yukon River during 2005. The averaged and open season DIC/DOC ratios appear similar for the Chena River (Cai et al. 2008a), a tributary of the Tanana/Yukon River. However, there appears to be a strong seasonality in that ratio and significant differences between tributaries. Based on loads during 2001–2005, DIC/DOC was 3.1 annually: 1.7 during spring and 6.0 during winter (Striegl et al. 2007). Thus changes in DIC/DOC ratio may provide a complementary indicator for climate and environmental change in the river basin (Striegl et al. 2005, 2007; Guo et al. 2012; Walvoord et al. 2012).

Interestingly, reported DIC concentrations during the winter season differ by up to a factor of two (2,066 vs. 4,100 μM), although numbers during spring high-flow season are somewhat similar. The extremely high winter DIC concentrations were also observed for different tributaries in the Porcupine River, Yukon River at Stevens Village, and Tanana River in the Yukon (Striegl et al. 2007), but not in the Yukon River at Eagle or the Chena River (Cai et al. 2008a). Unfortunately, riverine DIC concentrations were not directly measured by PARTNERS program for the Yukon River (http://arcticgreatrivers.org/), and very limited Yukon River DIC data are available in the literature. Striegl et al. (2007) reported the partial pressure of CO_2 (pCO_2) in Yukon River waters, with pressures varying from ~900 to 8,280 μ-atm depending on sampling season and tributary. The pCO_2 values in the river waters were generally higher than the atmospheric equilibrium value, suggesting that Yukon River waters are a source of CO_2 to the atmosphere, especially during the ice free seasons, likely a result of photochemical and biological degradation of riverine organic carbon. Predictors of DIC yield for the Yukon basin and across the circumboreal are consistent and include runoff, the extent of carbonate rocks, SO_4 yield, and permafrost extent (Tank et al. 2012). As permafrost thaws or becomes more discontinuous and groundwater contributions to river flow increase in these regions, DIC yield can also be expected to increase (Striegl et al. 2005; Walvoord and Striegl 2007; Stolpe et al. 2013b; Tank et al. 2012; Walvoord et al. 2012). Dissolved carbon dioxide ($CO_{2(aq)}$) is an important component of DIC that is receiving increased research attention because streams and rivers worldwide emit CO_2 to the atmosphere (e.g., Richey et al. 2002; Cole et al. 2007). CO_2 emission is quantitatively important in the Yukon River system C budget as well, with total CO_2 emissions from stream and river water surfaces equaling the total [IC + OC] lateral exports by the Yukon River at Pilot Station (Striegl et al. 2012). These emissions would be expected to increase with increased DIC yield, reinforcing the need for future studies to further investigate the abundance of DIC and its seasonal variations in this seasonally ice covered river basin.

Compared with DIC, POC, and DOC, concentrations of PIC are rarely reported for the Yukon River, with only one dataset available (Striegl et al. 2007). Based on their data, the average PIC concentration comprised on average about 10% of the DIC concentration, but could represent up to

64% of the POC in samples collected from the Yukon River during 2001–2005 (Striegl et al. 2007). PIC concentrations were also highly variable between rivers.

Stable carbon isotope composition of DIC in Yukon River waters during summer usually increases with increasing DIC concentration, although the opposite is true for some tributaries (Striegl et al. 2007). The slope value in the correlation between δ^{13}C-DIC and DIC concentration decreased from the upper Yukon to lower Yukon at Pilot Station, suggesting a change in DIC sources along the river. However, source apportionment for the riverine DIC pool will require quantitative measurements of end-member isotope signatures and multiple isotopes if there are more than two DIC sources (e.g., carbonate dissolution, organic respiration, and atmospheric CO_2).

3.3. *Chromophoric and fluorescent-DOM in Yukon River waters*

Optical properties, including UV absorbance, specific UV absorbance (SUVA), spectral slope values, and fluorescence excitation-emission matrix (EEMs), have been recently used to examine the source and composition of dissolved organic matter (DOM) in the Yukon River (e.g., Striegl et al. 2005; Belzile and Guo 2006; Guéguen et al. 2006; Spencer et al. 2008, 2009; O'Donnell et al. 2010; Stedmon et al. 2011). Striegl et al. (2005) showed a general increase in SUVA values with increasing DOC concentration or an increase in aromaticity from winter to summer to spring, consistent with source changes between seasons. Guéguen et al. (2006) reported consistently high $SUVA_{280}$ values with a weak correlation with DOC concentration. Spencer et al. (2009) showed a broader relation between freshwater discharge and $SUVA_{254}$ value or aromaticity of DOM. However, the relation between spectral slope value and discharge or DOC concentration was less clear. As also shown in Striegl et al. (2007), average $SUVA_{254}$ values in the Yukon River ranged from 1.9±4.2 L/mg-C/m during winter for the Yukon at Eagle to 3.5±2.3 L/mg-C/m during spring in the Porcupine River and Yukon River. Differences in spectral slope values (290–350 nm) and $SUVA_{254}$ values among sampling times and tributaries/rivers were also observed for the Yukon River (Spencer et al. 2008). O'Donnell et al. (2010) showed that $SUVA_{254}$ values in water from small rivers were relatively less variable during winter and spring, but had a high variability among rivers during the summer growing season. Stedmon et al. (2011) reported an average absorbance of 6.8 m^{-1} at 375 nm, spectral slope value of 17.5 μm^{-1} at 300–600 nm, and $SUVA_{375}$ of 2.37 m^2/g-C for water samples collected between April 2004 and August 2005 from the Yukon River. Note that different wavelengths or wavelength ranges were used for the calculation of DOM absorbance or absorption coefficient, spectral slope, and SUVA values by different researchers, making data comparisons difficult.

Using UV-visible spectroscopy and ultrafiltration methods, Belzile and Guo (2006) examined the distribution of aromatic DOM between the <1 kDa LMW and the >1 kDa HMW fractions. They found that aromatic DOM was predominantly partitioned in the >1 kDa HMW or colloidal fraction, with a very low $SUVA_{370}$ value measured for the <1 kDa LMW-DOM fraction (0.005 m^{-1} μM-C^{-1}) compared with bulk DOM and colloidal DOM (0.022–0.023 m^{-1} μM-C^{-1}). High aromaticity in the >1-kDa colloidal fraction is consistent with the dominance of COC over DOC fraction in Yukon River waters (Cai and Guo 2005; Belzile and Guo 2006; Guéguen et al. 2006). In addition, the spectral slope (derived from absorbance between 300–700 nm) was higher for the <1 kDa LMW-DOM fraction (~0.021 nm^{-1}) compared with bulk river water and the >1 kDa HMW colloidal fraction (<0.016 nm^{-1}), indicating that the spectral slope can be used as an indicator of DOM molecular weight (Belzile and Guo 2006; Zhou and Guo 2012).

Guéguen et al. (2006) showed that the HMW- and LMW-DOM fractions had distinctly different characteristic emission maximum wavelengths compared with the bulk DOM pool in addition to a marked difference in spectral slope values (Guéguen et al. 2003), indicating that riverine DOM is highly heterogeneous. The HMW-DOM fraction normally had a blue shift in emission maximum compared with the bulk DOM, whereas a red shift was found for the LMW-DOM fraction (Guéguen et al. 2006). This shift in emission maximum also varied with time of sampling, suggesting a change in DOM aromaticity and molecular weight composition from more HMW-DOM during ice melt to less HMW-DOM or increased LMW-DOM components during late summer/early fall in the Yukon River.

Based on measurements of $SUVA_{254}$ and spectral slope values, Spencer et al. (2009) further related these optical properties to the abundance of total lignin phenols and different phenol ratios including cinnamyl to vanillyl (C:V), syringyl to vanillyl (S:V), vanillic acid to vanillin $(Ad:Al)_V$, and syringic acid to syringaldehyde $(Ad:Al)_S$ ratios, in the XAD resin-extracted DOM fraction. They found a significant positive relationship between UV absorbance at 350 nm and the DOC concentration or total lignin phenol concentration in Yukon River waters. Although the specific phenol composition is highly variable between rivers and seasons, measurements of CDOM may provide a way to estimate the abundance of total lignin components in river waters and riverine DOM export (Spencer et al. 2009). Future deployments of optical sensors in various tributaries should provide high-resolution time series data for quantifying spatial and temporal variability of DOM in the Yukon River basin and for understanding the biogeochemical response in carbon cycles to climate and environmental changes.

4. Quantification of riverine export fluxes of carbon species

Unlike tropical and temperate rivers, the nonlinear relationship between freshwater discharge and the concentration of carbon species, especially DOC, observed in arctic rivers makes the quantification of riverine carbon export fluxes challenging (Finlay et al. 2006; Guo et al. 2012). As shown in many recent studies, neither DOC concentration nor discharge are proportionally distributed among seasons. Instead, spring freshet with high discharge and high DOC concentration may contribute the majority of the annual DOC flux (Raymond et al. 2007; Cai et al. 2008a; Striegl et al. 2012b). However, export fluxes during ice cover with low discharge and low DOC concentration may be overestimated if DOC is extrapolated from summer data, as pointed out by Guo et al. (2012).

In general, riverine export fluxes of carbon and other chemical species can be quantified using the following methods/models: (1) US Geological Survey's LOADEST model (http://water.usgs.gov/software/loadest/), which requires at least 12 flux measurements over the observed range of flow conditions, (2) regression between instantaneous carbon flux and discharge (e.g., Warnken and Santschi 2004), and (3) flow weighted mean concentration and discharge and other methods as discussed in Worrall and Burt (2007). However, the relationship between DOC concentration and the instantaneous freshwater flux is nonlinear for arctic rivers (e.g., Finlay et al. 2006), and the quantitative relation also varies with sampling frequency and timing. Therefore, the export fluxes of carbon species from northern high-latitude rivers are readily subject to overestimation or underestimation regardless of what method is used to calculate the flux (Guo et al. 2012). For example, Cai et al. (2008a) showed that, in the Chena River, spring freshet (16 days in 2005, or 4.4% of a year) contributed up to 22% of annual DOC export, 25% of annual POC export, and 8% of DIC, respectively. In addition, daily DOC flux during the open water period in the Yukon River varied widely and was up to two orders of magnitude higher than fluxes when the river was frozen. Thus, during spring freshet, considerable variation in daily DOC and POC fluxes requires more frequent sampling to reduce the uncertainty in carbon flux estimations.

For inorganic species such as DIC, seasonal variations are opposite to those of DOC and POC, with a dilution effect occurring during spring freshet and higher concentrations accumulating under the ice (Cai et al. 2008a; Guo et al. 2012). Therefore, export fluxes for DIC could be overestimated without spring freshet sampling but underestimated without winter sampling.

As shown in Table 20.2, carbon fluxes estimated by different studies for the same river may be significantly different, with errors or uncertainties similar to or higher than what might be expected from climate and environmental changes. Overall, uncertainties in riverine export fluxes of carbon must be reduced before the export fluxes can be confidently used to assess climate and environmental changes in the river basin.

5. Chemical and isotopic composition of organic matter

5.1. Elemental and molecular composition of riverine organic matter

A compilation of recent studies on elemental, molecular, and isotopic composition of organic matter from the Yukon River is given in Table 20.3. Concentrations of total fatty acids (FAs) have been reported for three size-fractionated organic matter pools, including LMW-DOM (<1 kDa), HMW or colloidal organic matter (COM, 1 kDa-0.45 μm), and particulate organic matter (POM), in the Yukon River (Zou et al. 2006). They found the content of total FAs (in μg/mg-OC) was significantly different among organic matter pools and among seasons (Fig. 20.4). In general, POM had the highest FAs, followed by the LMW-DOM and then the COM pool. During spring, the FAs-POM content was relatively low, but increased with time from spring to summer/fall, with a lower value observed during the rainy period in August. These results suggest that fatty acids are mostly produced by aquatic organisms. During periods of rain, river waters become turbid, depressing aquatic production.

Using pyrolysis-GC/MS techniques, Guo et al. (2003) also found a difference in molecular composition between dissolved, colloidal, and particulate phases in the Chena River. There was a progressive increase in polysaccharides (furfurals%) with increasing size fraction, indicating an increase in chemical reactivity from DOC to COC to POC in the Chena River.

The dissolved carbohydrate (CHO) concentration in the DOC pool generally decreased from spring to summer/fall in the Yukon River at Stevens Village and was significantly correlated to DOC concentration, indicating a terrestrial source (Guéguen et al. 2006). Guéguen et al. (2006) reported a DOC/CHO ratio, ranging from 3.7 to 4.6 in the Yukon River between spring freshet and late

Table 20.2. *Concentrations and fluxes of carbon species from the Yukon River*

Sampling location or tributary	Sampling date	DIC Conc. (mg/L)	DIC Flux (10^{12} g/y)	DOC Conc. (mg/L)	DOC Flux (10^{12} g/y)	POC Conc. (mg/L)	POC Flux (10^{12} g/y)	TOC Flux (10^{12} g/y)	References
Pilot Station	n/a	–	–	–	–	–	2.49	2.97	Leenheer (1982)
Pilot Station	2001–2003				–		–		Striegl et al. (2005)
Pilot Station	2005	–	–	6.2±4.1	1.70	–	–	–	Raymond et al. (2007)
Pilot Station	2001–2005		4.96		1.58		0.75		Striegl et al. (2007)
Chena River	2005	19.19±4.58	–	4.49±2.67	–	1.03±1.06	–	–	Cai et al. (2008)
Pilot Station	2004	–			1.64	–		–	Spencer et al. (2009)
	2005				1.86				Spencer et al. (2009)
Pilot Station	2004–2005	19.5±4.3	4.4±0.1	5.14±5.15	1.62±0.37	3.8±3.5	0.89±0.12	2.51	Guo et al. (2012)
Pilot Station	2001–2008	n/a			1.47		n/a		Holmes et al. (2012)

Table 20.3. *Chemical composition and isotopic signatures of dissolved (DOM), colloidal (COM), and particulate organic matter (POM) in the Yukon River*

Parameter	Location	DOM	COM	POM	Reference
C/N ratio	Chena River	39±4	48±0.9	16±1	Guo et al. (2003)
$\delta^{13}C$ (‰)	Chena River	−16.9±0.1	−27.4±0.1	−29.5±0.5	
$\delta^{15}N$ (‰)	Chena River	1.33±0.23	1.64±0.59	2.64±0.54	
^{14}C age	Chena River	600±45	Modern	n/a	
Total fatty acids (μg/mg-OC)	YR at Stevens Village	14.5±9.2[a]	5.6±2.9	112±51	Zou et al. (2006)
Carbohydrates (in CHO/OC ratio)	RR at Stevens Village	0.25	0.31	n/a	Gueguen et al. (2006)
C/N ratio	YR at Stevens Village	26.7±4.8[a]	45.8±3.2	15.0±1.4	Guo & Macdonald (2006)
$\delta^{13}C$ (‰)	"	−27.92±0.49[a]	−27.44±0.19	−26.22±0.70	"
$\delta^{15}N$ (‰)	"	−1.12±0.34[a]	−0.66±0.24	0.77±0.45	"
$\Delta^{14}C$ (‰)	"	−85±88[a]	73±41	−369±71	"
$\delta^{13}C$ (‰)	YR at Pilot Station	−26±1	n/a	n/a	Raymond et al. (2007)
$\Delta^{14}C$ (‰)	"	5±53	n/a	n/a	"
$\delta^{13}C$ (‰)	Chena River	n/a	n/a	−29.28±1.74	Cai et al. (2008a)
C/N ratio		41±13	n/a	32±12	Cai et al. (2008a)
Lignin phenol Σ8 (μg/L)	Yukon at PS	1.7–73	n/a		Spencer et al. (2008)
C/N ratio	Small rivers	1–50	n/a	n/a	O'Donnell et al. (2010)
C/N	Yukon at PS	41±10	n/a	27±8	Guo et al. (2012)

[a] Data from the <1-kDa low-molecular-weigh DOC.

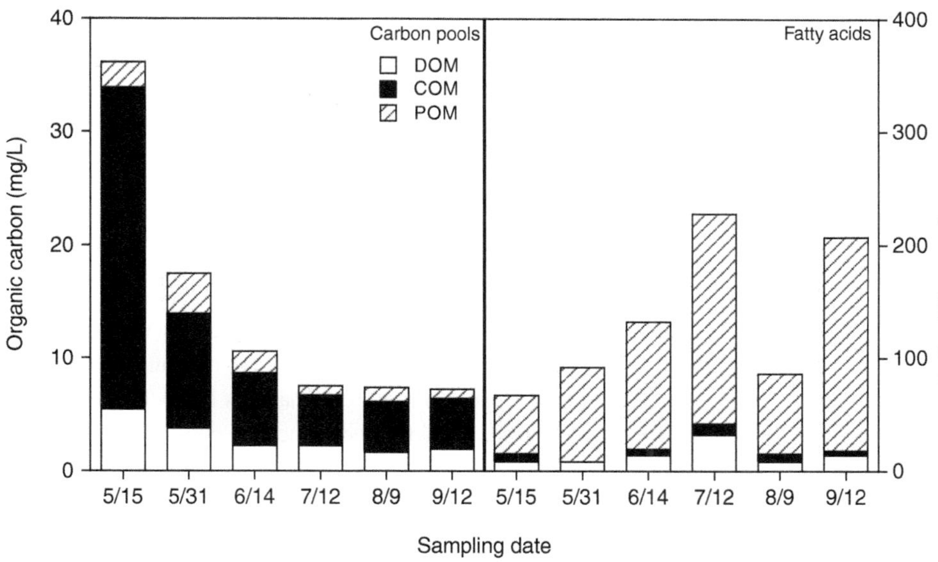

Figure 20.4. Temporal variations of total organic carbon and total fatty acids in dissolved organic matter (DOM, <1 kDa), colloidal organic matter (COM, 1 kDa-0.45 μm), and particulate organic matter (POM, >0.45 μm) pools (from Zou et al. 2006).

September. In other words, within the bulk DOC pool, 25±2% was carbohydrates. Interestingly, the CHO/OC ratio was lower in the LMW-DOC pool than the HMW-DOC pool (0.16 vs. 0.31). Higher CHO contents in the HMW-DOM pool in Yukon River waters are consistent with its contemporary ^{14}C age and likely higher chemical reactivity (Guo and Macdonald 2006). Within the dissolved CHO pool, on average, 75% was monosaccharides (MCHO) and 25% was polysaccharides (PCHO). However, the PCHO/CHO ratio decreased from 32% during spring freshet to 21% in late fall, whereas the MCHO/CHO ratio increased from 68% to 79%, indicating intensive degradation and transformation among organic pools/species during transport in the Yukon River. The concentration of different CHO species was also significantly correlated with DOC concentration within LMW- and HMW-DOC and the bulk DOC pools. Based on the overall low Chl-*a* concentrations and CHO abundance, 76±7% of the bulk DOC was estimated to be pedogenic DOM in the Yukon River (Guéguen et al. 2006).

The abundance and partitioning of CHO in the Chena River has been determined by Cai et al. (2008a) based on biweekly sampling during 2005–2006. Similar to the Yukon River, dissolved CHO co-varied with DOC in the Chena River, decreasing from spring to summer. On average, CHO comprised 15±4% of the DOC, with large variability between sampling seasons, whereas MCHO comprised 89±10% and PCHO made up 11±10% of the total dissolved CHO pool. In contrast to the CHO composition in fluvial systems with high percentage of monosaccharides, marine DOM contains mostly polysaccharides derived from freshly photosynthesized organic matter (Hung et al. 2003; Wang et al. 2010), indicating that terrestrial DOM in river waters is composed mostly of highly degraded organic matter derived from soil.

In addition to isotopic, elemental, and chemical composition, measurements of CDOM and fluorescence provide further information on organic molecular composition such as hydrophobicity, humic-like, and protein-like DOM components (Striegl et al. 2005, 2007; Spencer et al. 2008, 2009; O'Donnell et al. 2010). Based on the PARAFAC model and EEM data, O'Donnell et al. (2010) showed a generally higher protein% in fluorescent DOM collected during winter compared with samples collected in spring and summer. They also showed a large range of DOC/DON ratios, varying from 1 to about 50 for a suite of streams in the Yukon River basin. Extremely low DOC/DON ratio was ascribed to the production of bacterial DOM under the ice, which seems to be consistent with the high protein content implied by fluorescence signatures (O'Donnell et al. 2010).

5.2. Isotopic composition

Using filtration and ultrafiltration methods, bulk organic matter from the Chena River was size fractionated into dissolved LMW (<1 kDa), colloidal (1 kDa–0.4 μm), and particulate (>0.4 μm) phases (Guo et al. 2003). These authors found that riverine organic matter is highly heterogeneous, with distinctly different C/N ratios in dissolved, colloidal, and particulate pools (Table 20.3). In addition, stable isotopic composition, both $\delta^{13}C$ and $\delta^{15}N$, and radiocarbon signatures also varied between different size fractions of the organic pools, with a contemporary ^{14}C age for colloidal organic carbon and about 600 years BP for the LMW-DOC fraction. The $\delta^{13}C$ value for the <1 kDa DOC seemed to be significantly higher than for COC, POC, and C-3 plant values. Cai et al. (2008a) reported POC-$\delta^{13}C$ in the Chena River for a year's worth of biweekly samples, with relatively high

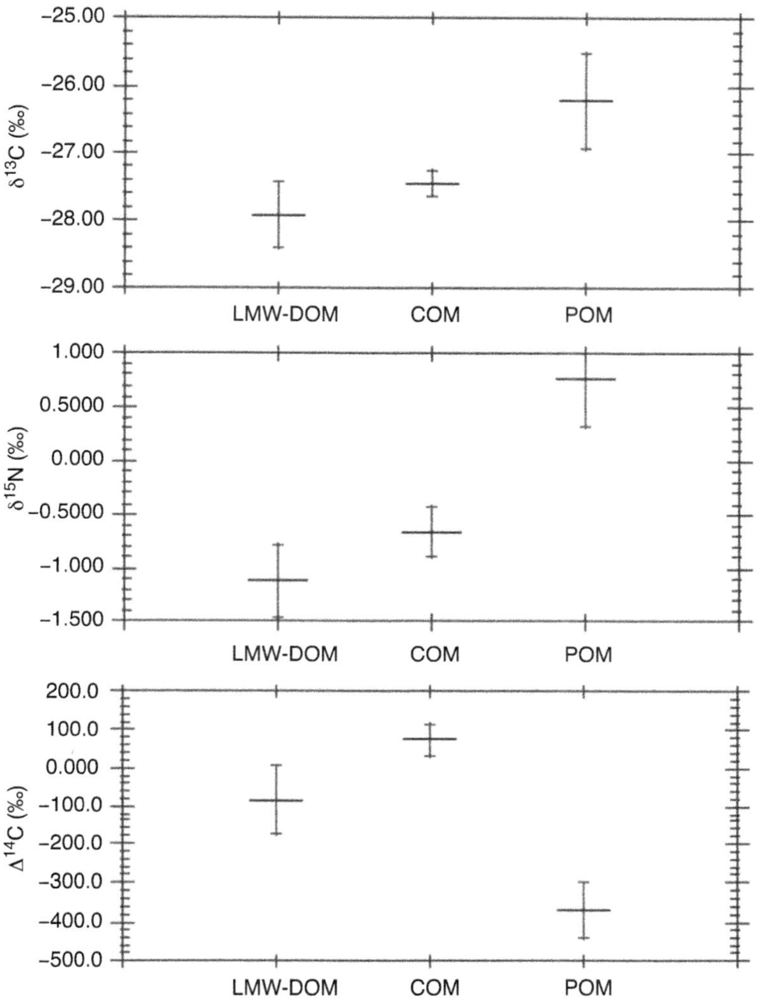

Figure 20.5. Examples of heterogeneity in isotopic composition ($\delta^{13}C$, $\delta^{15}N$, and $\Delta^{14}C$) of riverine organic matter between dissolved, colloidal, and particulate phases in Yukon River waters (from Guo and Macdonald 2006).

POC-$\delta^{13}C$ values (-25.87‰) during spring but lower POC-$\delta^{13}C$ values (-33.82‰) during summer/fall drought, indicating dynamic change in POC sources in the Chena River.

Similar to the heterogeneity observed for riverine organic matter from the Chena River, other examples, including $\delta^{13}C$, $\delta^{15}N$, and $\Delta^{14}C$, are given for the Yukon River in Figure 20.5. Both $\delta^{13}C$ and $\delta^{15}N$ increased progressively with increasing size fraction. Although $\Delta^{14}C$ values were also distinctly different between dissolved, colloidal, and particulate phases, the highest value or the youngest ^{14}C age was observed for the COM pool, which comprised 60–70% of the bulk DOC, followed by the LMW-DOM. The POM had the lowest $\Delta^{14}C$ value or the oldest ^{14}C age (Guo and Macdonald 2006).

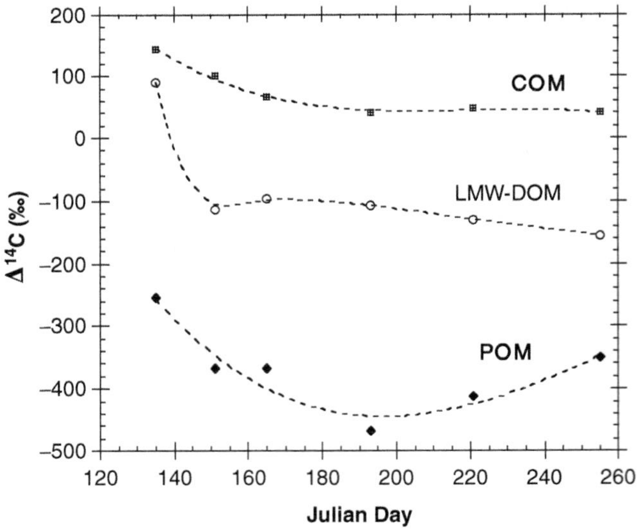

Figure 20.6. Variations of $\Delta^{14}C$ values (‰) of POM, COM, and LMW-DOM with sampling date. Notice the general decrease in $\Delta^{14}C$ values (increase in ^{14}C age) from COM to LMW-DOM to POM, and the decreasing trend from snowmelt to river ice-open season (from Guo and Macdonald 2006).

Distinctly different ^{14}C ages between OC size fractions indicates that dissolved and particulate organic matter transported in the Yukon River are derived from different sources with different turnover pathways and transport mechanisms that would respond differently to changing climate. This set of circumstances likely applies to many other rivers (e.g., Eglinton and Repeta 2006), but unfortunately, isotopic measurements for individual carbon phases in northern river waters remain scarce, with most studies reporting data for only a single carbon species or bulk organic/inorganic phases.

Contemporary and relatively young ^{14}C ages were also measured for the DOC pool in the Yukon River at Pilot Station (Guo et al. 2007; Raymond et al. 2007). The $\Delta^{14}C$ value seemed to increase with increasing discharge (Raymond et al. 2007), suggesting that DOC exported during spring freshet contained mostly younger carbon, whereas DOC exported during low or base flow contained older OC, as observed for the Yukon River at Steven Village (Guo and Macdonald 2006). Radiocarbon composition of different carbon pools also varied with sampling season (Fig. 20.6), with younger carbon exported during spring and relatively older carbon during summer/fall season. However, precipitation events during late summer/fall could flush surface POC, which exhibits slightly higher $\Delta^{14}C$ values (Fig. 20.6).

6. Variations in the abundance, speciation, and sources of nutrients (N, P, and Si)

Nitrogen (N), phosphorus (P), and silicon (Si) are essential nutrients for aquatic plants. Among these elements, N and P turn over much faster than Si in aquatic environments and potentially limit production and thus they have received greater attention. The abundance of dissolved silicate (Si) in river waters is normally higher, and therefore not limiting, but Si concentration may also be used to examine physicochemical and weathering processes in river basins (Millot et al. 2002; Guo et al. 2004a). Globally, riverine biogenic silica comprises about 10% of the dissolved silicate pool (Conley 1997; Hughes et al. 2011). To date, no biogenic silica data are available for the Yukon River.

In river waters, N species include inorganic and organic dissolved (DIN and DON) and particulate N (PIN and PON). Within the DIN pool, there are NO_3, NO_2, and NH_3 species, whereas PIN and PON are measured as total particulate N (PN) without differentiating the two components. Phosphorus species include dissolved inorganic P (DIP), dissolved organic P (DOP), particulate inorganic P (PIP), and particulate organic P (POP). In natural river waters, the concentration of DIP is usually low (Guo et al. 2004a; Lin et al. 2012), frequently below detection limits of traditional colorimetric methods. Thus many early studies reported only total dissolved P (TDP) without separating DIP from DOP, or even total phosphorus (TP) without separating particulate P from dissolved P. The abundance of Si is conventionally reported as dissolved silicate for the dissolved phase and biogenic silica for the particulate phase (Conley 2002; Guo et al. 2004a).

Guo et al. (2004a) measured nutrients (N, P, and Si) in dissolved, particulate, organic, and inorganic forms to examine temporal variations in nutrient concentrations and phase partitioning in the Yukon River at Stevens Village (SV) from May to September 2002. They found that both DIN and DIP concentrations in the Yukon River were much lower than those measured for world rivers, with average concentrations of 2.43 ± 0.63 μM-N and 0.053 ± 0.040 μM-P, respectively. Concentrations of dissolved silicate, however, were close to those of world rivers, with an average of 82 ± 21 μM-Si. Within the total N pool transported by the Yukon River, dissolved inorganic N comprised only $7\pm4\%$, and particulate N made up $25\pm10\%$, whereas dissolved organic N (DON) was the dominant N species, with an average of $67\pm10\%$. In contrast to N species, P was predominantly partitioned into the particulate phase with an average of $94\pm6\%$, leaving $4\pm5\%$ of the total P in the DOP phase and $\sim2\pm1\%$ in the DIP phase. It seems that the transformation between dissolved and particulate or inorganic and organic phases may play a critical role in controlling the abundance, bioavailability, and biogeochemical cycling of N and P in the Yukon River Basin.

In addition to measurements for the Yukon at Stevens Village (SV), nutrient speciation and partitioning have also been reported for the Chena River (Cai et al. 2008b) and the Yukon River at Pilot Station (Guo et al. 2012) using an approach similar to that used by Guo et al. (2004a). The partitioning of N and P between inorganic and organic, dissolved, and particulate phases differs distinctly between the Chena River and the Yukon River at SV (Fig. 20.7). As shown in Fig. 20.7, the predominant N species in the Yukon River at SV is DON followed by PON and DIN, whereas the predominant P species is particulate P followed by DOP and DIP, with <2% of total P residing in DIP. However, in the Chena River, where permafrost is mostly discontinuous and has experienced extensive thawing compared with the upper Yukon River (Jorgenson et al. 2006), DIN is the predominant N species, followed by DON and PN. For P species in the Chena River, %DIP is considerably higher than observed in the Yukon River at SV, although particulate P remains the predominant P species (Fig. 20.7). Carbon partitioning also shows distinctively different features between the Chena River (lower latitude) and the Yukon River at SV (higher latitude), with much higher %DIC observed in the Chena River (as with the Tanana), but higher DOC in the Yukon River at SV. Therefore, differences in the abundance and chemical speciation of nutrients in river waters are closely linked

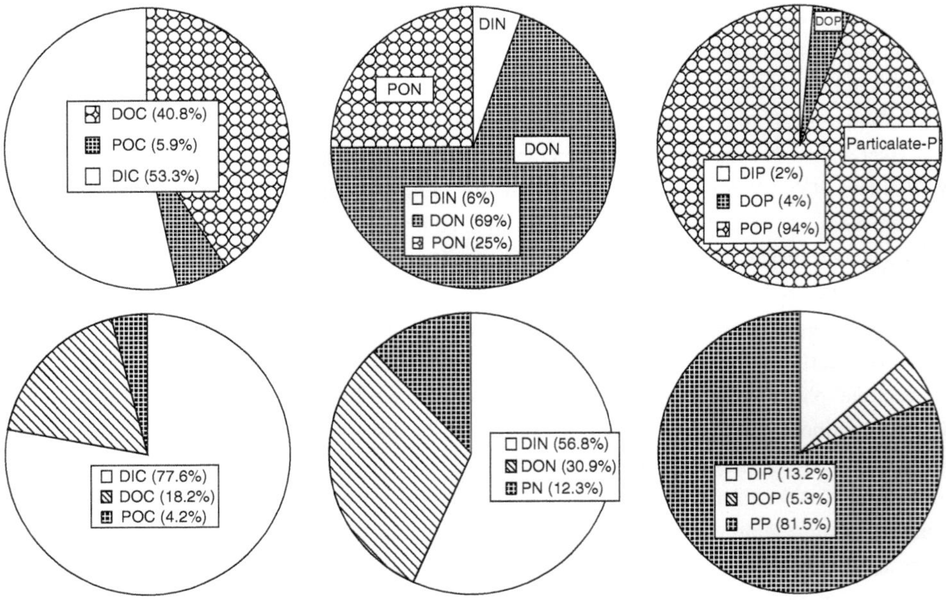

Figure 20.7. Variations in the partitioning and chemical speciation of C, N, and P between the Yukon River at Stevens Village (upper panel) and the Chena River (lower panel). The former basin has higher permafrost coverage and predominately organic C, N, and P species, whereas the latter has sporadic permafrost and thus mostly inorganic C and nutrient species.

to permafrost distribution, landscape and vegetation types, and climate and environmental changes in the river basin and groundwater inputs (Guo et al. 2004a; Walvoord and Striegl 2007; Guo et al. 2012).

Dornblaser and Striegl (2007) reported NH_4 = 2.3%, DIN = 16.4%, PN = 37.8%, and DON = 43.5% of the total N pool in the Yukon River at Pilot Station during 2001–2005. They also gave proportions for the Porcupine and Tanana Rivers and reported that particulate P was 98% of the total P pool in the Yukon River. Results from the Yukon River at Pilot Station reported by Guo et al. (2012) show that organic nutrient species had the highest concentration during spring freshet and the lowest during winter under the ice, indicating dominant sources from snowmelt and flushing of soils in the drainage basin. In contrast, inorganic species such as DIC and dissolved silicate had the highest concentrations in winter and the lowest during spring freshet, indicating dilution during snowmelt and sources from groundwater and leaching/weathering of the mineral layers. The contrasting relationship with discharge between organic and inorganic C and nutrient species points to hydrological control of nutrient and carbon concentrations but different sources and transport mechanisms for organic and inorganic species. In the Yukon River (at PS), particulate P is the predominant P species, whereas DIN, DON, and PN are of similar importance, with an average of 38±25%, 31±14% and 31±15%, respectively, although DIN in winter was the dominant N species transported in the Yukon River (Guo et al. 2012). As reported in Guo et al. (2012), the average DIN/DIP ratio is 110±124, with up to 442 observed in samples collected from under the ice, and 38–70 during the ice-open season. However, the TDN/TDP ratio is relatively low (about 49), as reported in Dornblaser and Striegl

(2007). Concentrations of DIP were not measured during the USGS and PARTNERS program, although total dissolved phosphorus (TDP=DIP+DOP) data are available (Dornblaser and Striegl 2007; McClelland et al. 2008).

Interestingly, nutrient concentrations for inorganic species, including DIN, DIP, and dissolved silicate in the Yukon River, seem to increase from upstream at Stevens Village to downstream at Pilot Station (Table 20.4), with the caution that sampling frequency and time intervals are not matched between the different studies. The Chena River had a higher dissolved silicate concentration than the Yukon River (Cai et al. 2008b), likely as a result of the geological setting, which includes hot springs along the Chena River. The concentrations of DIN and DIP also show a similar increasing trend along the Yukon River, although there are no winter data for the Yukon River at SV. The yearly average DIP concentration in the Chena River is similar to those of the Yukon River. However, the yearly average DIN concentration in the Chena River is much higher than those reported for the Yukon River (Table 20.4), although it is still considerably lower than found in temperate rivers. Higher DIN and dissolved silicate in the Chena River are consistent with measurements for the Tanana River (Dornblaser and Striegl 2007) and likely result from degradation of permafrost and the weathering of bedrock transmitted to the river through groundwater input (MacLean et al. 1999; Deluca et al. 2002; Jones et al. 2005; Petrone et al. 2006).

7. Export fluxes of nutrients from the Yukon River

In northern high-latitude rivers, concentrations of inorganic nutrients including N, P, and Si are very low, whereas their organic counterparts, such as DON and DOP, and particulate N and P could be extremely high (Fig. 20.7). As a result, nutrient export fluxes based on DIN and DIP by themselves may profoundly underestimate the effective flux because transformation of but a small fraction of organic and particulate nutrients into their dissolved inorganic counterparts could overwhelmingly surpass the fluxes of DIN and DIP (Mayer et al. 1998; Guo et al. 2004a; Lin et al. 2012). However, where a portion of the organic and particulate nutrients is not readily bioavailable in coastal marine environments, nutrient fluxes derived from total N and P would overestimate the biologically available fluxes from river basins to the ocean.

Table 20.4 lists recent studies of nutrients and their export fluxes from the Yukon River. The yields and export fluxes of total N and P from the Yukon River and its major tributaries, including Yukon River at Eagle, Porcupine River, Yukon River at Stevens Village, Tanana River, and Yukon River at Pilot Station, have been recently reported (Dornblaser and Striegl 2007). A comparison of the nutrient export fluxes reported by Dornblaser and Striegl (2007) for 2001–2005 with those of Guo et al. (2012) for 2004–2005, shows similar values for total N and P for the Yukon River, but there is a strong seasonal and annual variability. Based on the daily stream flow data of the USGS (see also Fig. 20.2), about 34–51% of the annual discharge occurs in spring. However, most inorganic chemical species have higher concentrations during winter when stream flow is at the lowest. Therefore, unlike the case of DOC and POC, nutrient fluxes are more evenly distributed among seasons. Continued warming would likely enhance the total export of nutrients from high-latitude rivers, whereas organic nutrient exports and their relative importance is more likely to decrease (Fig. 20.7).

Nutrient-specific fluxes normalized to drainage area in the Yukon River basin were 0.57 mmole-N/m^2/y for DIN, 0.012 mmole-P/m^2/y for DIP, and ~19 mmole-Si/m^2/y for dissolved silicate,

Table 20.4. *Concentrations and export fluxes (grams of N, P or Si per year) of nutrients species from the Yukon River*

Tributary or Sampling location	Sampling date	NO_3+NO_2 Conc. (μM)	NO_3+NO_2 Flux (10^{10} g/y)	DIP Conc. (μM)	DIP Flux (10^{10} g/y)	$Si(OH)_4$ Conc. (μM)	$Si(OH)_4$ Flux (10^{10} g/y)	References
Stevens Village	2002 (May–Sept)	2.4±0.6	–	0.05±0.04	–	82±21	–	Guo et al. (2004a)
Pilot Station (seasonal)	2001–2005	5.2–13.4	1.96	0.77–11.6	0.2[a]	n/a	–	Dornblaser and Striegl (2007)
Chena River (biweekly)	2004–2005	18.3±4.1[b]	n/a	0.15±0.07	n/a	157±40	n/a	Cai et al. (2008b)
Pilot Station (monthly)	2004–2005	11.1±1.5	2.85	0.145±0.04	0.15	134±43	81	Guo et al. (2012)
Pilot Station	2001–2008		2.6		0.02[a]		69	Holmes et al. (2012)

[a] From total dissolved phosphorus (TDP=DIP+DOP).
[b] NO3+NO2+NH3.

respectively. The relatively low specific fluxes of NO_3 and PO_4 in the Yukon River Basin reflect its pristine status with little anthropogenic influence, whereas the cold climate in the arctic/subarctic region may result in lower specific flux for dissolved silicate, in agreement with a general trend of increasing Si specific flux with decreasing latitude for global river systems (Guo et al. 2004a). Based on this trend, a warming climate, and thus deeper permafrost active layer in the Yukon River basin, would likely lead to enhanced export flux of dissolved silicate.

8. Role of organic matter in the transport of trace elements

It is clear that ongoing climate and environmental changes may have resulted in permafrost thawing and increasing terrestrial primary production, thus affecting mass remobilization of soil organic matter and the export of dissolved organic matter from land to the Arctic Ocean (Guo et al. 2004b; Frey and Smith 2005; Guo et al. 2007; Vonk et al. 2010; Ping et al. 2011). However, little is known about the transport and fluxes of trace elements accompanying the transport of DOM from the Yukon River and ice-covered rivers in general (Rember and Trefry 2004; Shiller 2010).

Concentrations of dissolved and colloidal trace metals in Yukon River waters have recently been measured for selected elements, including Li, V, Mn, Fe, Cu, Zn, Ni, Ru, Sr, Mo, Ba, and U (Schuster 2003; Stolpe et al. 2013a); strontium isotopes (Douglas et al. 2012); and rare earth elements (Stolpe et al. 2013b). Colloidal metal concentrations in the 0.02–0.4 μm fraction were also calculated from the difference between concentrations in the 0.4-μm filtrate and 0.02-μm filtrate (Schuster 2005a, 2005b). Using flow field-flow fractionation (FFF) techniques coupled with ICP-MS, Stolpe et al. (2013a, 2013b) have determined the abundance, size spectra, and binding of trace elements, including Fe and rare earth elements, for organic and iron-rich colloids in six small Alaskan rivers in the Yukon River basin. They found the presence of two major nanocolloids in river waters: "fulvic-rich colloids," which had a hydrodynamic diameter of 0.5–3 nm, and "Fe-rich colloids," in the <8-nm and 4–40-nm size range. Manganese, Co, Ni, Cu, and Zn were found to partition between the nanocolloid phases depending on the metal-organics stability constant, and Cr and Fe were found to bind to the Fe-rich colloids owing to their higher oxidation states (+III or +IV). Among the elements that co-varied with DOC, the nanocolloidal fractions were generally highest for Cu, Cr, Fe, and Ni; intermediate for Pb and Zn; and low for Co and Mn (Stolpe et al. 2013a). The consistently high concentrations of Fe measured during both spring flood and summer seasons suggest that Fe comes from two different sources: associated with organic matter leached from the top soil and associated with organic matter during spring melt, and soluble Fe (II) in sub-oxic groundwater entering the river during summer/fall season.

Based on their study on REEs, Stolpe et al (2013b) concluded that, as the export of organic matter increased during spring freshet, the abundance of fulvic-rich nanocolloids in river waters will likely also increase, resulting in a depletion in light-REEs and Ce. During summer, Fe in the form of large Fe-rich colloids and dissolved carbonate and LMW organic substances would dominate, which likely favors the transport of REEs by the Arctic rivers. The interactions between trace elements and organic matter and the fractionation of REEs are therefore affected by hydrological cycles, permafrost dynamics, changes in composition, and sources of organic matter in river waters.

For major ions, concentrations of Na, K, Mg, Ca, Sr, Ba, and Si were largely related to the weathering of the mineral soil, bedrock, and glaciers, with highest concentrations in late winter base flow immediately prior to spring melt, showing a covariation with alkalinity but not with DOC concentration (Douglas et al. 2013; Stolpe et al. 2013a). In the Chena River, $^{87}Sr/^{86}Sr$ ratios were lowest during winter base flow season and generally intermediate during fall recession (Douglas et al. 2013).

The transport and export of trace elements from arctic rivers are likely to be enhanced under warming conditions. However, the abundance, distribution, and fluxes of specific elements are highly dependent on the chemistry and complexation strength. Similar to the case of DOM, trace elements in river waters could provide new insights into dynamic cycling and its relationships with change in active layer thickness, hydrological flow patterns, source waters, vegetation types, and human activities in individual river basins.

9. Effects of hydrological regime and environmental change on nutrient and carbon biogeochemistry

Concentrations and chemistry of DOM and inorganic species vary among rivers and their source waters depending on features such as the extent of forest within the river basin (Spencer et al. 2008; O'Donnell et al. 2010). During spring freshet, riverine DOM is normally fresh, aromatic, and hydrophobic and contains more higher molecular weight organic materials (Striegl et al. 2005; Guo and Macdonald 2006; Spencer et al. 2008). In contrast, riverine DOM during summer and fall season is typically older, less aromatic, more hydrophilic, and of lower molecular weight (Guo and Macdonald 2006; Guo et al. 2007; Spencer et al. 2009). Therefore, active layer dynamics, hydrological flow patterns, landscape, and vegetation types all affect the quantity and quality of DOM transported in river waters. In other words, the composition and fluxes of DOM and other chemical species have the potential to be developed as proxies for climate and environmental changes and as tracers of biogeochemical processes and transport pathways.

With increased permafrost thawing, deepening of the summer active layer, and percolation of water into deeper layers of the thawed soil of high-latitude river systems, it can be expected that DOM, especially higher molecular weight components, will become more abundant during summer and more dominated by fulvic-rich and large Fe-rich nanocolloids from groundwater and less dominated by small organic/Fe-rich colloids from the upper organic-rich soil (Stolpe et al. 2013a). Under warming conditions, however, forest cover and land primary productivity will also increase (e.g., Hinzman et al. 2005), resulting in a better developed upper organic-rich soil and more DOM export, especially young DOM. In addition, accelerated melting of glaciers will dilute the high DOC waters, although this may also increase stream flow. Thus the overall effect on the export of carbon, nutrients, and trace elements will depend on the relative importance of all factors. The water chemistry in the lower Yukon River integrates the entire river basin containing complex and diverse ecosystems and tributaries, such as glacier-dominated, wetland-dominated, black water, forest, continuous and discontinuous permafrost, and different tree zones. In this complexity, drawing the linkage between river geochemistry and climate/environmental change may become a challenge. Perhaps a better way to understand change would be to target specific watersheds with consistent water sources, permafrost distribution, soil, and vegetation types that could serve as models for individual components of change and thus develop quantitative relationships between river water geochemistry and climate/environmental change in the basin (Guo et al. 2012; Douglas et al. 2013).

References

Aagaard K, and E.C. Carmack. 1989. The role of sea ice and other fresh water in the Arctic circulation. *J. Geophys. Res.* 94: 14485–14498.

Belzile, C. and L. Guo. 2006. Optical properties of low molecular weight and colloidal organic matter: Application of the ultrafiltration permeation model to DOM absorption and fluorescence. *Mar. Chem.* 98: 183–196.

Bianchi, T.S., S.F. DiMarco, J.H. Cowan Jr, R.D. Hetland, P. Chapman, J.W. Day, and M.A. Allison. 2010. The science of hypoxia in the Northern Gulf of Mexico: A review. *Sci. Total Environ.* 408: 1471–1484.

Brabets, T.P., B. Wang, and R.H. Meade. 2000. Environmental and hydrological overview of the Yukon River Basin, Alaska and Canada, USGS Water-resources investigations report 99-4204.

Cai, W. 2003. Riverine inorganic carbon flux and rate of biological uptake in the Mississippi River plume. *Geophys. Res. Lett.* 30: 1032.

Cai, Y. and L. Guo. 2005. Fluxes of colloidal organic carbon to the Arctic Ocean from North American Rivers. *Eos Trans. AGU* 86, Fall Meet. Suppl., Abstract #OS51B-0561.

Cai, Y., L.D. Guo, and T. Douglas. 2008a. Temporary variations in organic carbon species and fluxes from the Chena River, Alaska. *Limnol. Oceanogra.* 53: 1408–1419.

Cai, Y., L. Guo, T. Douglas, and T. Whitledge. 2008b. Seasonal variations in nutrient concentrations and speciation in the Chena River, Alaska. JGR-*Biogeosciences* 113: G030035.

Carey, S.K. 2003. Dissolved organic carbon fluxes in a discontinuous permafrost subarctic alpine catchment. Permafrost Periglac. *Process* 14: 161–171.

Cole, J.J., Y.T. Prairie, N.F. Caraco, W.H. McDowell, L.J. Tranvik, R.G. Striegl, C.M. Duarte, P. Kortelainen, J.A. Downing, J.J. Middleburg, and J. Melack. 2007. Plumbing the global carbon cycle: Integrating inland waters into the terrestrial carbon budget: *Ecosystems* 10: 172–185.

Conley, D.J. 2002. Terrestrial ecosystems and the global biogeochemical silica cycle. *Global Biogeochem. Cycles* 16: 1121.

Cooper, L.W., J.W. McClelland, R.M. Holmes, P.A. Raymond, J.J. Gibson, C.K. Guay, and B.J. Peterson. 2008. Flow-weighted Values of Runoff Tracers ($\delta^{18}O$, DOC, Ba, Alkalinity) from the Six Largest Arctic Rivers. *Geophys. Res. Lett* 35: L18606.

DeLuca, T. H., O. Zackrisson, M. C. Nilsson, and A. Sellstedt. 2002. Quantifying nitrogen-fixation in feather moss carpets of boreal forests. *Nature* 419: 917–920.

Dornblaser M.M. and R.G. Striegl. 2007. Nutrient (N, P) loads and yields at multiple scales and subbasin types in the Yukon River basin, Alaska. *J. Geophys. Res.* 112: G04S57.

Dornblaser, M.M. and R.G. Striegl. 2009. Suspended sediment and carbonate transport in the Yukon River Basin, Alaska: Fluxes and potential future responses to climate change. *Water Resources Res.* 45: W06411.

Dornblaser, M.M., R.G.M. Spencer, G.R. Aiken, and R.G. Striegl. 2012. Surmounting environmental influences to derive carbon dynamics in Arctic Rivers via in-situ chromophoric dissolved organic matter sensors. *Limnol. Oceanogra. Methods* (in review).

Douglas, T.A., D. Fortier, Y.L. Shur, M.Z. Kanevskiy, L. Guo, Y. Cai, and M.T. Bray. 2011. Geomorphic and geochemical characteristics of thermokarst cave ice in the CRREL Permafrost Tunnel, Alaska. *Permafrost and Periglacial Processes* 22:120–128.

Douglas, T.A., J.D. Blum, L. Guo, K. Keller, and J. Gleason. 2012. Hydrogeochemistry of seasonal flow regimes in the Chena River, a subarctic watershed draining discontinuous permafrost in interior Alaska. *Chem. Geol.* 335: 48–62.

Duan, S., T.S. Bianchi, A.M. Shiller, K. Dria, P.G. Hatcher, and K.R. Carman. 2007. Variability in the bulk composition and abundance of dissolved organic matter in the lower Mississippi and Pearl rivers, *J. Geophys. Res.* 112: G02024.

Eglinton, T.I. and D.J. Repeta, 2006. Organic matter in the contemporary ocean, pp. 145–180. In: *The Oceans and Marine Geochemistry* (ed. H. Elderfield), Vol 6 Treatise on Geochemistry (eds. H.D. Holland and K.K. Turekain), Elsevier-Pergamon, Oxford.

Finlay, J, J. Neff, S. Zimov, A. Davydova, and S. Davydov. 2006. Snowmelt dominance of DOC in high-latitude watersheds: implications for characterization and flux of river DOC. *Geophys. Res. Lett.* 33: 110401.

Frederick, Z.A., S.P. Anderson, and R.G. Striegl. 2011. Annual estimates of water and solute export from 42 tributaries to the Yukon River, Hydrol. *Process.* 26: 1949–1961.

Frey, K.E. and L.C. Smith. 2005. Amplified carbon release from vast West Siberian peatlands by 2100. *Geophys. Res. Lett.* 32: 109401.

Frey, K.E., and J.W. McClelland. 2009. Impacts of permafrost degradation on arctic river biogeochemistry. *Hydrol. Process.* 23: 169–182.

Guéguen, C., D. Wang, and L. Guo. 2003. Optical properties of natural organic matter in the Yukon River Basin. *Tohoku Geophys. J.* 36: 434–438.

Guéguen, C., L. Guo, D. Wang, N. Tanaka, and C.C. Hung. 2006. Chemical characteristics and origin of dissolved organic matter in the Yukon River. *Biogeochemistry* 77: 139–155.

Guo, L. 2010. Isotopic Composition and Export Fluxes of Organic Carbon Species from the Lower Mississippi River. 2010 AGU Fall Meeting, San Francisco, CA, December 13–17, 2010, abstract# OS22A-01.

Guo, L., J.K. Lehner., D.M. White, and D.S. Garland. 2003. Heterogeneity of natural organic matter from the Chena River, Alaska. *Water Res.* 37: 1015–1022.

Guo, L., J.Z. Zhang, and C. Guéguen. 2004a. Speciation and fluxes of nutrients (N, P, Si) from the upper Yukon River. *Global Biogeochem. Cycles* 18: GB1038.

Guo, L., I. Semiletov, O. Gustafsson, J. Ingri, P. Andersson, O. Dudarev, and D. White. 2004b. Characterization of Siberian Arctic estuarine sediments: Implications for terrestrial organic carbon export. *Global Biogeochem. Cycles* 18: GB1036.

Guo, L., and R.W. Macdonald. 2006. Source and transport of terrigenous organic matter in the upper Yukon River: evidence from isotope (^{13}C, ^{14}C and ^{15}N) composition of dissolved, colloidal and particulate phases. *Global Biogeochem. Cycles* 20: GB2011.

Guo, L., C.L. Ping, and R.W. Macdonald. 2007. Mobilization pathways of organic carbon from permafrost to arctic rivers in a changing climate. *Geophys. Res. Lett.* 34: L13603.

Guo, L., Y. Cai, C. Belzile, and R.W. Macdonald. 2012. Sources and export fluxes of inorganic and organic carbon and nutrient species from the seasonally ice-covered Yukon River. *Biogeochemistry* 107: 187–206.

Hinzman, L.D., N.D. Bettez, W.R. Bolton, et al. 2005. Evidence and implications of recent climate change in northern Alaska and arctic regions. *Climate Change* 72: 251–298.

Holmes, R.M., J.W. McClelland, B.J. Peterson, I.A. Shiklomanov, A.I. Shiklomanov, A.V. Zhulidov, V.V. Gordeev, and N.N. Bobrovitskaya. 2002. A circumpolar perspective on fluvial sediment flux to the Arctic Ocean. *Global Biogeochem. Cy.* 16: 1098.

Holmes, R.M., J.W. McClelland, B.J. Peterson, S.E. Tank, E. Bulygina, T.I. Eglinton, V.V. Gordeev, T.Y. Gurtovaya, P.A. Raymond, D.J. Repeta, R. Staples, R.G. Striegl, A.V. Zhulidov, and S.A. Zimov. 2012. Seasonal and annual fluxes of nutrients and organic matter from large rivers to the Arctic Ocean and surrounding seas. *Estuaries and Coasts* 35: 369–382.

Hope, D., M.F. Billett, and M.S. Cresser. 1994. A review of the export of carbon in river water: fluxes and processes. *Environ. Poll.* 84: 301–324.

Hughes, H. J., F. Sondag, C. Cocquyt, A. Laraque, A. Pandi, L. André, and D. Cardinal. 2011. Effect of seasonal biogenic silica variations on dissolved silicon fluxes and isotopic signatures in the Congo River. *Limnol. Oceanogr.* 56: 551–561.

Hung, C.-C., L. Guo, P.H. Santschi, N. Alvarado-Quiroz, and J. Haye. 2003. Distributions of carbohydrate species in the Gulf of Mexico. *Mar. Chem.* 8: 119–135.

Jones, J.B., K.C. Petrone, J.C. Finlay, L.D. Hinzman and W.R. Bolton. 2005. Nitrogen loss from watersheds of interior Alaska underlain with discontinuous permafrost. *Geophys. Res. Lett. 32*(2): L02401.

Jorgenson, M.T., Y.L. Shur, and E.R. Pullman. 2006. Abrupt increase in permafrost degradation in Arctic Alaska. *Geophys. Res. Lett.* 33: L02503.

Leenheer, L. 1982. United States Geological Survey Data Information Service. In: E.T. Degens (ed), Transport of carbon and minerals in major world rivers. *SCOPE/UNEP Sonderband* 52: 355–356.

Lin, P., M. Chen, and L. Guo. 2012. Speciation and transformation of phosphorus and its mixing behavior in the Bay of St. Louis estuary in the northern Gulf of Mexico. *Geochimca. Cosmochim. Acta.* 87: 283–298.

Lohrenz, S.E., G.L. Fahnenstiel, D.G. Redalje, G.A. Lang, M.J. Dagg, T.E. Whitledgee, and Q. Dortch. 1999. Nutrients, irradiance, and mixing as factors regulating primary production in coastal waters impacted by the Mississippi River plume. *Continent. Shelf Res.* 19: 1113–1141.

MacLean, R., M.W. Oswood, J.G. Irons III and W.H. McDowell. 1999. The effect of permafrost on stream biogeochemistry: a case study of two stream in the Alaskan (USA) taiga. *Biogeochemistry.* 47: 239–267.

Mayer, L.M., R.G. Keil, S.A. Macko, S.B. Joye, K.C. Ruttenberg, and R.C. Aller. 1998. Importance of suspended particulates in riverine delivery of bioavailable nitrogen to coastal zones. Global Biogeochem. *Cycles* 12: 573–579.

McClelland, J.W., R.M. Holmes, B.J. Peterson, et al. 2008. Development of a Pan-Arctic Database for River Chemistry, *Eos Trans. AGU* 89: 217.

McGuire, A.D., L. Anderson, T.R. Christensen, S. Dallimore, L. Guo, D. Hayes, M. Heimann, R.W. Macdonald, and N. Roulet. 2009. Sensitivity of the carbon cycle in the arctic to climate change (review). *Ecol. Monogr.* 79: 523–555.

Meybeck, M. 2003. Global analysis of river systems: from Earth system controls to Anthropocene syndromes. *Phil. Trans. R. Soc. Lond. B* 358: 1935–1955.

Millot, R., J. Gallardet, B. Dupre, and C.J. Allegre. 2002. The global control of silicate weathering rates and the coupling with physical erosion: New insights from rivers of the Canada Shield. *Earth Planet Sci. Lett.* 196: 83–98.

O'Donnell, J.A., G.R. Aiken, E.S. Kane, et al. 2010. Source water controls on the character and origin of dissolved organic matter in streams of the Yukon River basin, Alaska. *J. Geophys. Res.* 115: G03025.

Peterson, B.J., R.M. Holmes, J.W. McClelland, C.J. Vorosmarty, I.A. Shiklomanov, A.I. Shiklomanov, R.B. Lammers, and S. Rahmstorf. 2002. Increasing river discharge to the Arctic Ocean. *Science* 298: 2171–2173.

Petrone, K.C., J.B. Jones, L.D. Hinzman, and R.D. Boone. 2006. Seasonal export of carbon, nitrogen, and major solutes from Alaskan catchments with discontinuous permafrost. *JGR-Biogeosciences* 111 (G2): Art. No. G02020.

Ping, C.L., G.J. Michaelson, M.T. Jorgenson, J.M. Kimble, H. Epstein, V.E. Romanovsky, and D.A. Walker. 2008. High stocks of soil organic carbon in the North American Arctic region. *Nature Geoscience* 1: 615–619.

Ping, C.-L., G.J. Michaelson, L. Guo, M.T. Jorgenson, M. Kanevskiy, Y. Shur, F. Dou, and J. Liang. 2011. Soil carbon and material flux across the eroding coastline of the Beaufort Sea, Alaska. *JGR-Biogeosci.* 116: G02004.

Post, E., M.C. Forchhammer, M.S. Bret-Harte, et al. 2009. Ecological dynamics across the Arctic associated with recent climate change. *Science* 325:1355–1358.

Raymond, P.A., J.W. McClelland, R.M. Holmes, A.V. Zhulidov, K. Mull, B.J. Peterson, R.G. Striegl, G.R. Aiken, and T.Y. Gurtovaya. 2007. Flux and age of dissolved organic carbon exported to the Arctic Ocean: A carbon isotopic study of the five largest arctic rivers. *Global Biogeochem. Cycles* 21, GB4011.

Rember, R.D., and J.H. Trefry. 2004. Increased concentrations of dissolved trace metals and organic carbon during snowmelt in rivers of the Alaskan Arctic. *Geochim. Cosmochim. Acta.* 68: 477–489.

Richey, J.E., J.M. Melack, A.K. Aufdenkampe, V.M. Ballester, and L.L. Hess. 2002. Outgassing from Amazonian rivers and wetlands as a large tropical source of atmospheric CO_2. *Nature* 416: 617–620.

Schuster, P.F. 2003. Water and sediment quality in the Yukon River Basin, Alaska during water year 2001. U.S. Geol. Survey, Open-File Report, 03-427.

Schuster, P.F. 2005a. Water and sediment quality in the Yukon River Basin, Alaska, during water year 2002, *U.S. Geol. Surv. Open-File Rep., 2005–1199.*

Schuster, P.F. 2005b. Water and sediment quality in the Yukon River Basin, Alaska, during water year 2003, *U.S. Geol. Surv. Open-File Rep., 2005–1397.*

Schuster, P.F. 2006. Water and sediment quality in the Yukon River Basin, Alaska, during water year 2004, *U.S. Geol. Surv. Open-File Rep., 2006–1258.*

Schuster, P.F., R.G. Striegl, G.R. Aiken, et al. 2011. Mercury export from the Yukon River basin and potential response to a changing climate. *Environ. Sci. Technol.* 45: 9262–9267.

Shiller, A.M. 2010. Dissolved rare earth elements in a seasonally snow-covered, alpine/subalpine watershed, Loch Vale, Colorado. *Geochim. Cosmochim. Acta* 74: 2040–2052.

Spencer, R.G.M., G.R. Aiken, K.P. Wickland, R.G. Striegl, and P.J. Hernes. 2008. Seasonal and spatial variability in dissolved organic matter quantity and composition from the Yukon River basin, Alaska. *Global Biogeochem. Cycles* 22: GB4002.

Spencer, R.G.M., G.R. Aiken, K.D. Butler, M.M. Dornblaser, R.G. Striegl, and P.J. Hernes. 2009. Utilizing chromophoric dissolved organic matter measurements to derive export and reactivity of dissolved organic carbon exported to the Arctic Ocean: A case study of the Yukon River, Alaska. *Geophys. Res. Lett.* 36: L06401.

Stedmon, C.A., R.M.W. Amon, A.J. Rinehart and S.A. Walker. 2011. The supply and characteristics of colored dissolved organic matter (CDOM) in the Arctic Ocean: Pan Arctic trends and differences. *Mar. Chem.* 124: 1080118.

Stein, R., and R.W. Macdonald. 2004. *The Organic Carbon Cycle in the Arctic Ocean*. Springer, New York.

Stolpe, B., L. Guo, A.M. Shiller, and G.R. Aiken.2013a Abundance, size distributions and trace-element binding of organic and iron-rich nanocolloids in Alaskan rivers as revealed by flow field flow fractionation and ICP-MS. *Geochim. Cosmochim. Acta.* 105: 221–239.

Stolpe, B., L. Guo, and A.M. Shiller. 2013b. Binding and transport of rare earth elements by organic and iron-rich colloids in Alaskan rivers, as revealed by field-flow fractionation and ICP-MS. *Geochim. Cosmochim. Acta* 106, 446–462.

Striegl, R.G., G.R. Aiken, M.M. Dornblaser, P.A. Raymond, and K.P. Wickland. 2005. A decrease in discharge-normalized DOC export by the Yukon River during summer through autumn. *Geophys. Res. Lett.* 32: L21413.

Striegl, R.G., M.M. Dornblaser, G.R. Aiken, K.P. Wickland, and P.A. Raymond. 2007. Carbon export and cycling by the Yukon, Tanana, and Porcupine rivers, Alaska, 2001–2005. *Water Resour. Res.* 43: W02411.

Striegl, R.G., M.M. Dornblaser, C.P. McDonald, J.R. Rover, and E.G. Stets. 2012. Carbon dioxide and methane emissions from the Yukon River system. *Global Biogeochem. Cycles* 26: GB0E05.

Sturm, M., J. Schimel, G. Michaelson, et al. 2005. Winter biological processes could help convert arctic tundra to shrubland. *Bioscience* 55: 17–26.

Tank, S. E., K.E. Frey, R.G. Striegl, P.A. Raymond, R.M. Holmes, J.W. McClelland, and B.J. Peterson. 2012. Landscape-level controls on dissolved carbon flux from diverse catchments of the circumboreal. *Global Biogeochem. Cycles* 26: GB0E02.

Tarnocai, C., J.G. Canadell, E.A.G. Schuur, P. Kuhry, G. Mazhitova, and S. Zimov. 2009. Soil organic carbon pools in the northern circumpolar permafrost region. *Global Biogeochem. Cycl.* 23: GB2023.

Vonk, J.E., B.E. van Dongen, and Ö. Gustafsson. 2010. Selective preservation of old organic carbon fluvially released from sub-Arctic soils, *Geophys. Res. Lett.* 37: L11605.

Walvoord, M.A., and R.G. Striegl. 2007. Increased groundwater to stream discharge from permafrost thawing in the Yukon River basin: Potential impacts on lateral export of carbon and nitrogen. *Geophys. Res. Lett.* 34: L12402.

Walvoord, M.A., C.I. Voss, and T.P. Wellman. 2012. Influence of permafrost distribution on groundwater flow in the context of climate-driven permafrost thaw: Example from Yukon Flats Basin, Alaska, United States. *Water Resour. Res.* 48: W07524.

Wang, X., Y. Cai, and L. Guo. 2010. Preferential removal of dissolved carbohydrates during estuarine mixing in the Bay of Saint Louis in the northern Gulf of Mexico. *Mar. Chem.* 119: 130–138.

Warnken, K.W., and P.H. Santschi. 2004. Biogeochemical behavior of organic carbon in the Trinity River downstream of a large reservoir lake in Texas, USA. *Sci. Total Environ.* 329: 131–144.

Wickland, K.P., J.C. Neff, and G.R. Aiken. 2007. Dissolved organic carbon in Alaskan boreal forest: Sources, chemical characteristics, and biodegradability. *Ecosystems* 10: 1323–1340.

Wickland, K.P., G.R. Aiken, K. Butler, M.M. Dornblaser, R.G.M. Spencer, and R.G. Striegl. 2012. Biodegradability of dissolved organic carbon in the Yukon River and its tributaries: Seasonality and importance of inorganic nitrogen, *Global Biogeochem. Cycles* 26: GB0E03.

Worrall, F., and T.P. Burt. 2007. Flux of dissolved organic carbon from UK rivers. *Global Biogeochem. Cycles* 21: GB1013.

Zhou, Z. and L. Guo. 2012. Evolution of the optical properties of seawater influenced by the DWH oil spill in the Gulf of Mexico. *Environ. Res. Lett.* 7: 025301.

Zou, L., M.Y. Sun, and L. Guo. 2006. Temporal variations of organic carbon inputs into the upper Yukon River: Evidence from fatty acids and their stable carbon isotopic composition in dissolved, colloidal and particulate phases. *Org. Geochem.* 37: 944–956.

21

Fluxes, processing, and fate of riverine organic and inorganic carbon in the Arctic Ocean

P. J. Hernes, R. M. Holmes, P. A. Raymond, R. G. M. Spencer, and S. E. Tank

1. Introduction

Land and ocean are more tightly coupled in the Arctic than anywhere else on Earth. The strength of this linkage is such that estuarine conditions extend throughout the Arctic Ocean, and terrigenous materials are present in the Arctic Ocean at higher concentrations than in other ocean basins (Opsahl et al. 1999; Hernes and Benner 2006; McClelland et al. 2012). The Arctic is also the epicenter of global climate change, with warming approximately two-fold greater than the global average (Serreze and Barry 2011). This warming leads to permafrost thaw (Osterkamp 2007) that liberates ancient organic matter (OM) and likely accelerates biogeochemical fluxes from the pan-Arctic watershed to the Arctic Ocean. Thus climate change impacts on land-ocean linkage are particularly pronounced in the Arctic. However, despite intensified efforts to characterize and understand biogeochemical sources and cycling of terrestrial carbon in the Arctic, many gaps remain as a result of inadequate temporal and spatial coverage. This chapter addresses both our current state of knowledge as well as these gaps.

2. Hydrology and fluxes of Arctic rivers in comparison with global rivers

The watershed of the Arctic Ocean covers 16.8×10^6 km^2 (Fig. 21.1), the majority of which is in Russia (70%). The remainder is in Canada (17%), the United States (5%), Kazakhstan (4%), Mongolia (2%), and Norway, Finland, China, and Sweden (<1% combined). (Some researchers also include the Hudson Bay drainage in estimates of the pan-Arctic watershed, which increases the total area by a factor of 1.22 to 20.5×10^6 km^2.) Approximately 3,700 km^3 yr^{-1} discharge enters the ocean from the pan-Arctic watershed, roughly 10% of the global total (Holmes et al. 2012a). Two rivers have mean discharges exceeding 500 km^3 yr^{-1}, four have discharges exceeding 300 km^3 y^{-1}, nine have discharges exceeding 100 km^3 yr^{-1}, and 14 have discharges exceeding 25 km^3 y^{-1} (Holmes et al. 2013). As would be expected based on watershed area, the majority of river water enters the Arctic Ocean from Russia, with much lesser amounts from Canada, the United States, and Scandinavia.

A defining feature of Arctic rivers is their strong seasonality (McClelland et al. 2012). Winter discharge is low, but the transition from winter low flow to spring peak flow is extremely rapid during

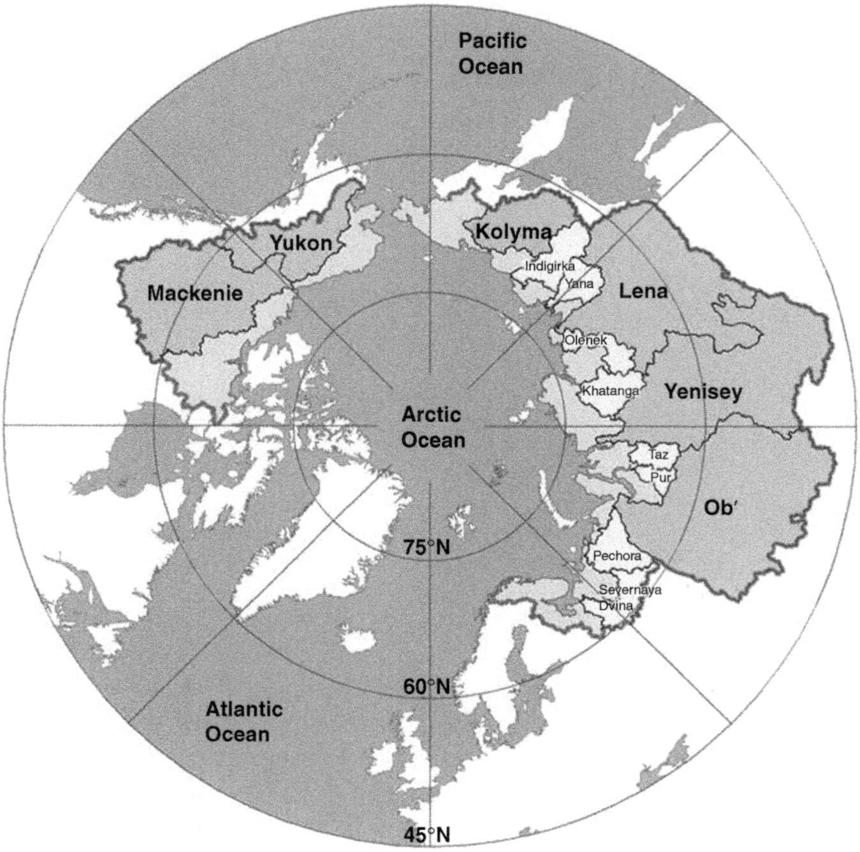

Figure 21.1. Map showing the catchments of the six biggest rivers (by discharge, darker shading) along with several smaller catchments (lighter shading) in the pan-Arctic system (16.8×10^6 km^2 in total, delineated by the bold gray line). The big six rivers account for 67% of the total area. (Some studies include the Hudson Bay drainage as part of the greater pan-Arctic system, which increases the area by a factor of 1.22, and the big six rivers account for 55% of this larger area.)

the snow-melt period (Fig. 21.2). Concomitant changes in the concentration and quality of riverine constituents make the annual freshet an extremely important period for investigating biogeochemical fluxes in Arctic rivers (Raymond et al. 2007; Holmes et al. 2008; Spencer et al. 2008; Mann et al. 2012).

Over recent decades, river discharge has been increasing throughout much of the pan-Arctic watershed (Peterson et al. 2002; McClelland et al. 2006; Déry et al. 2009; Rawlins et al. 2010; Holmes et al. 2013). Increases were first detected for the major Eurasian Arctic rivers, probably in part because of the availability of longer term datasets (Peterson et al. 2002). McClelland et al. (2006) investigated discharge trends for 30 rivers in the Eurasian and North American Arctic over the 1964–2000 timeframe. During that time, the net change in river discharge to the Arctic Ocean from these 30 rivers (5.6 km^3 per year) was the result of substantial increases from Eurasia tempered by

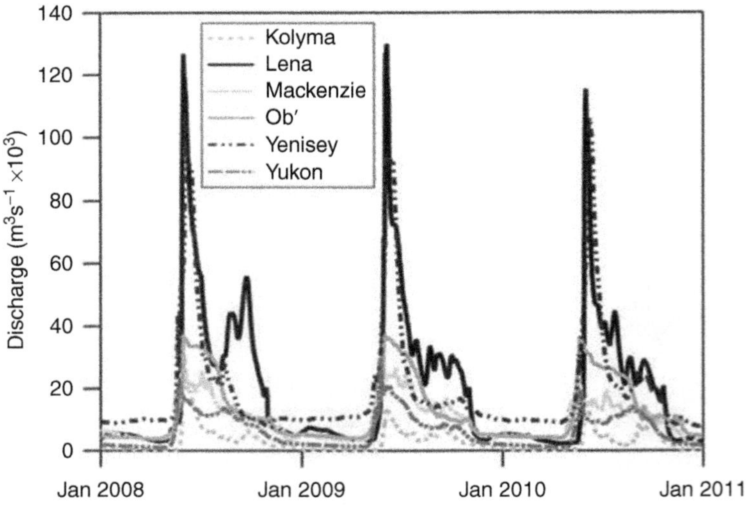

Figure 21.2. Freshwater discharge patterns (daily discharge measurements) for the six biggest Arctic rivers for the three-year period from January 2008 through January 2011.

smaller decreases from North America. More recent analyses, with data extending through 2007, have detected increases in North American Arctic river discharge as well (Déry et al. 2009). These increases are generally consistent with General Circulation Model predictions of increasing net precipitation with warming, although there is ongoing discussion about the roles that other factors such as fire, permafrost thaw, altered vegetation, and changing seasonality may play (Wu et al. 2005; Nohara et al. 2006).

Far fewer data are available to assess long-term trends in biogeochemical fluxes from Arctic rivers. Much excellent research has been done on individual rivers or collections of rivers within a single nation, but pan-Arctic syntheses have been hampered by analytical differences among studies or lack of seasonal coverage, which is essential for accurately estimating biogeochemical fluxes in Arctic rivers (Holmes et al. 2000; Zhulidov et al. 2000). The only international effort to collect a time series of biogeochemical samples from the downstream reaches of major Arctic rivers began in 2003 as the PARTNERS Project and in 2009 evolved into the Arctic Great Rivers Observatory (Holmes et al. 2013; www.arcticgreatrivers.org). In these projects, the Arctic's six largest rivers (Yenisey, Lena, Ob, Mackenzie, Yukon, Kolyma) are sampled multiple times per year, capturing the different phases of the hydrograph included during the ice-covered winter period. Although the resulting dataset is still short for detecting long-term change, it does allow determination of contemporary fluxes from the Arctic's six largest rivers against which to establish future changes. However, extrapolation to the remainder of the pan-Arctic watershed is complicated by the fact that the remainder of the basin differs in several important respects from the watersheds of the six largest rivers, including extent of permafrost and tundra as well as factors such as human population density (Holmes et al. 2013). Thus a true pan-Arctic synthesis will require additional sampling of smaller rivers as well as modeling to capture the unsampled portion of the basin.

3. Sources, composition, and fluxes of Arctic riverine organic matter exported to coastal systems

3.1. Arctic riverine tPOC and tDOC fluxes

As with riverine water discharge, fluxes of riverine organic carbon (OC) are dominated by the six major Arctic rivers spanning the Eurasian (Kolyma, Ob, Lena, Yenisey) and North American (Mackenzie, Yukon) continents (Fig. 21.1). The primary form of OC in Arctic rivers is dissolved organic carbon (DOC), also referred to as terrigenous DOC (tDOC). Arctic riverine flux estimates for tDOC have undergone various revisions over the past 10 years, with historic estimates of 18–26 Tg C yr^{-1} (Dittmar and Kattner 2003) that have more recently been increased to 25–36 Tg C yr^{-1}, with up to 18 Tg C yr^{-1} from the big six rivers (Raymond et al. 2007; Holmes et al. 2012a). The ranges at least partly derive from differences in how the pan-Arctic watershed is defined, i.e., the larger estimates include Hudson Bay drainage. Increasing estimates are driven largely by more accurately constrained tDOC concentration measurements over the dynamic spring flood event in which nearly half the tDOC (ranging from 36–63% for the big six) for the entire year is discharged in two months (Fig. 21.2; Holmes et al. 2012a). Flux estimates for the entire pan-Arctic region rely almost solely on measured areal tDOC yields from the six major rivers (67% of pan-Arctic drainage basin area without Hudson Bay, 55% when including Hudson Bay) scaled to the total area. A significant unknown is the accuracy of assuming that smaller, more northern watersheds (i.e., much greater percentage of tundra land use, per Lobbes et al. [2000]) yield water and tDOC with similar areal efficiencies compared with the large catchments of the big six rivers that extend much further south into forested regions. For example, average areal tDOC yields for nine minor Russian rivers with a total area of nearly 10^6 km^2 as reported in Lobbes et al. (2000) were significantly lower (~1 Mg tDOC km^{-2} yr^{-1}) compared with average yields from five of the big six (~1.5 Mg tDOC km^{-2} yr^{-1}, doesn't include the Yukon).

In contrast to tDOC, less is known about fluxes of terrestrial particulate OC (tPOC) from Arctic rivers. In part, this discrepancy between studies of Arctic riverine tDOC and tPOC derives from the relative importance of each toward total OC fluxes, with tDOC fluxes nearly an order of magnitude greater in the major Eurasian rivers. Fluxes of tPOC from 12 Russian rivers (including the Lena, Yenisey, and Kolyma Rivers, but not the Ob) have been estimated at 1.2 Tg C yr^{-1} (Lobbes et al. 2000), an order of magnitude lower than for tDOC. A five-year mean tPOC flux for the Yukon River (2001–2005) was measured at 0.75 Tg yr^{-1} (Striegl et al. 2007), whereas estimated averages for the Mackenzie River based on historical data are 2.1 Tg C yr^{-1} (Macdonald et al. 1998). Fluxes of tPOC from the Ob have been estimated at 0.27 Tg C yr^{-1} (Gebhardt et al. 2004); however, flux calculations for the Yenisey River from the same study are triple the estimates of Lobbes et al. (2000), 0.57 Tg C yr^{-1} versus 0.17 Tg C yr^{-1}. This highlights the level of uncertainty in tPOC fluxes as a whole, likely resulting from a combination of the location of measurements (tPOC measurements made upstream of the mouth assume 100% transfer efficiency to and through estuaries) and when they are made (most of these estimates do not include spring floods, with likely higher loads). Holmes et al. (2002) documented widely varying methodologies in tPOC flux estimates from Arctic rivers, as well as significant gaps that make it nearly impossible to assess any long-term changes in tPOC fluxes that could be attributed to either climate change or land-use change. Scaling tPOC fluxes from the big six rivers (4–4.5 Tg C yr^{-1}) to the entire pan-Arctic is even more problematic than for tDOC, as yields of sediments are tied to geology/topography (Chapter 8) and not simply area. Whereas the percentage of total OC derived from tPOC via the big six rivers is about 20%, the proportion varies widely from river to river, with the Eurasian rivers at about 10%, percentages from the Yukon River at roughly 30% (Striegl et al. 2007), and the Mackenzie River tPOC even greater at 60% of the total OC flux (Macdonald et al. 1998).

3.2. Biomarker compositions of tDOM and tPOM

Similar to OM fluxes, compositions of OM are distinctly seasonal, as demonstrated by changing lignin compositions across the hydrograph. The presence of lignin in any OM pool is an unambiguous indicator of a vascular plant source to that pool. In addition, the three primary components of lignin also contain useful source information, as syringyl phenols, S, are unique indicators of angiosperm tissues; cinnamyl phenols, C, are unique indicators of nonwoody tissues; and vanillyl phenols, V, are present in all lignin (Hedges and Mann 1979). Thus ratios of S:V and C:V are frequently used to discern sources of lignin, although the ratios are also sensitive to changes due to photooxidation (Opsahl and Benner 1998; Hernes and Benner 2003; Spencer et al. 2009b; Chapter 22) and partitioning between particulate and dissolved phases (Hernes et al. 2007). In the big six rivers, there is a distinct shift in lignin compositions of tDOM across the hydrograph, with decreases in both S:V and C:V on the order of 20–30% during the spring flush (Mann et al. Unpublished data). In an earlier study on just the Yukon River, Spencer et al. (2008) measured C:V and S:V ratios of 0.16 to 0.19 and 0.52 to 0.53, respectively, during monthly sampling in May and June at peak discharge, compared with values of 0.30 to 0.72 and 0.66 to 0.80, respectively, measured in the two preceding months and September. Although these changes could indicate a shift in sources toward gymnosperm woods, it is challenging to envision mechanisms for how this would occur, and thus it seems likely that other processes must contribute to the changing signature as well, such as fractionation from leaching and sorption (Hernes et al. 2007) or differential degradation. Seasonal variability in S:V and C:V within the Yukon River spans the entire range of S:V and C:V values measured in the snapshot sampling of Lobbes et al. (2000) in the Russian rivers, although they noted a positive correlation between S:V and the percentage of tundra within a catchment in both tDOM and tPOM, attributed to a higher proportion of angiosperms in the tundra. In contrast to tDOM, much less is known about temporal trends in tPOM, with only snapshot sampling of the Russian rivers (Lobbes et al. 2000) and June vs. July sampling of tPOM in the Mackenzie River (Goni et al. 2000). Samples in the latter, taken one month apart, revealed slight increases in S:V and slight decreases in C:V.

Within the syringyl and vanillyl phenol families, ratios of acids to aldehydes, (Ad:Al)v and (Ad:Al)s, are frequently used as indicators of diagenetic state, with higher ratios equating to more degradation. Little or no seasonal changes in (Ad:Al)v and (Ad:Al)s have been observed in Arctic rivers to date (Spencer et al. 2008; Mann et al. Unpublished data), although sampling resolution has been a limiting factor. Mackenzie River tPOM samples taken one month apart demonstrated significant increases in (Ad:Al)v and (Ad:Al)s (Goni et al. 2000), which together with S:V and C:V trends demonstrates decoupling between tDOM and tPOM. The latter is also indicated by ^{14}C ages, with tPOM much older than tDOM within Arctic rivers in general and the Mackenzie River in particular

(Guo et al. 2007). The general interpretation is that tPOM derives primarily from aged soils, whereas tDOM is more closely coupled to recent terrestrial primary production.

Beyond lignin, few other biomarker studies have been conducted on Arctic rivers. A snapshot of particulate amino acid and amino sugar measurements on the Ob and Yenisey rivers in August/September suggests limited in situ production of tPOM, as ratios of glucosamine to galactosamine are low and consistent with allochthonous sources (Unger et al. 2005). Light limitation was suggested as a major factor affecting in situ primary production (Unger et al. 2005).

Comparative carbon-normalized yields of biomarkers provide quantitative constraints on OM sources. Lignin carbon-normalized yields in tDOM vary seasonally, with values up to 0.8 mg 100 mg OC^{-1} during the spring freshet and values as low as 0.1 mg 100 mg OC^{-1} during low-flow winter conditions (Mann et al. Unpublished data). This indicates a higher contribution from vascular plant sources during the spring freshet and, in conjunction with other studies that demonstrate high bioavailability of tDOM in Arctic rivers in the spring (Holmes et al. 2008; Mann et al. 2012), runs counter to the prevailing paradigm that vascular plant-derived tDOM is more resistant to degradation than other sources. Because of the incomplete temporal coverage of carbon-normalized lignin yields in tPOM, source interpretations are ambiguous. Four measurements of Mackenzie River tPOM lignin in June/July ranged from 0.5 to 1.4 mg 100 mg OC^{-1} (Goni et al. 2000), indicating either a flashy system or strong local influence from compositionally distinct sources. Single tPOM measurements from the Kolyma, Lena, and Yenisey Rivers taken in summer were 0.1, 0.2, and 0.5 mg 100 mg OC^{-1}, respectively (Lobbes et al. 2000), either demonstrating widely different basin characteristics or reflecting similarly flashy systems, such as the Mackenzie River, that need adequate temporal sampling to properly characterize tPOM source dynamics. Carbon-normalized yields of amino acids in the Ob and Yenisey Rivers indicate that 20–30% of the carbon in tPOM is derived from amino acids (Unger et al. 2005).

3.3. Optical Properties of tDOM

Although biomarker measurements provide unambiguous compositional information about tDOM, extrapolating findings based on those measurements to the entire tDOM pool is not without risk, as biomarkers typically constitute at most only a few percent of the bulk pool. For this reason, it is advantageous to couple biomarker techniques with techniques that characterize bulk properties. Chromophoric dissolved organic matter (CDOM) represents the fraction of DOM that absorbs ultraviolet and visible light and is directly related to composition of unsaturated/aromatic compounds such as lignin and select amino acids. The component of CDOM that fluoresces, or FDOM, can yield even more specific information about chemical characteristics of DOM. In this regard, measurements of CDOM and FDOM have become increasingly popular because the measurements are fast, inexpensive, and contain a wealth of complementary information to biomarker techniques.

In Arctic rivers, CDOM and tDOC are strongly correlated (r^2 of 0.83; Fig. 21.3A) (Mann et al. Unpublished data), and hence the behavior of CDOM across the hydrograph in Arctic rivers tracks that of tDOC. The composition of CDOM also varies in Arctic rivers with changes in hydrology. The specific UV absorbance at 254 nm ($SUVA_{254}$) has been shown to be positively correlated to percent aromaticity (Weishaar et al. 2003) and typically increases with increasing discharge in Arctic

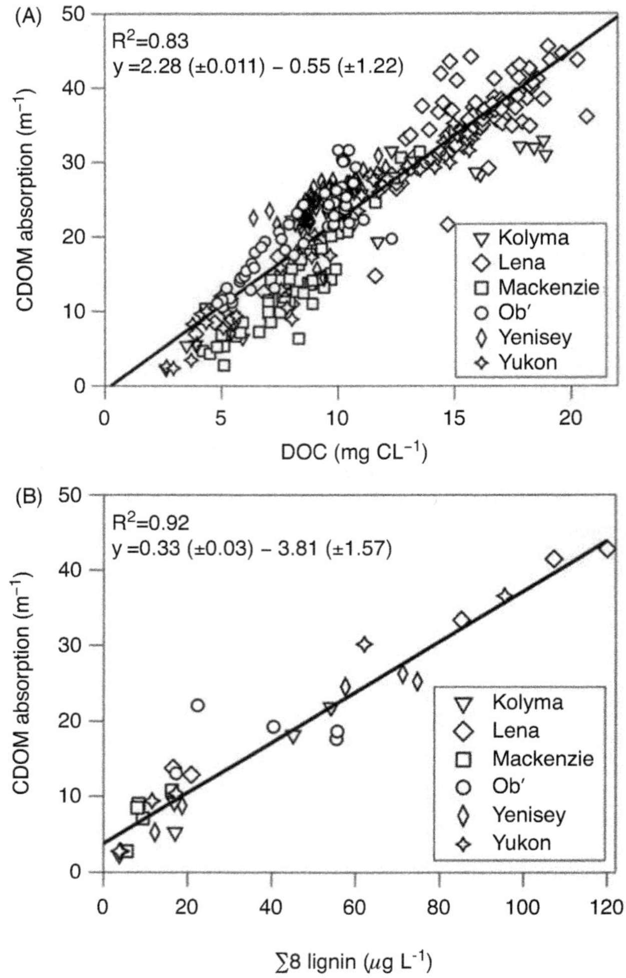

Figure 21.3. Relationships between CDOM absorption (a_{350}) for six largest Arctic rivers in 2009–2010 and (A) DOC concentration, (B) total lignin concentrations (from Mann et al. Unpublished data).

Rivers (Neff et al. 2006; Spencer et al. 2008; Spencer et al. 2009a; Stedmon et al. 2011; Mann et al. Unpublished data; Fig. 21.4A). The observed increase in $SUVA_{254}$ at high discharge indicates an increasing contribution of aromatic carbon exported at this time. Likewise, the spectral slope parameter (e.g., $S_{275-295}$) and the spectral slope ratio (S_R) have shown similar trends with increasing discharge in Arctic rivers, indicating export of tDOM that is relatively more aromatic and higher molecular weight in nature at elevated discharge (Spencer et al. 2008; Spencer et al. 2009a; Stedmon et al. 2011; Mann et al. Unpublished data; Fig. 21.4B), consistent with biomarker data indicating larger amounts of vascular plant-derived tDOM. Pan-Arctic CDOM compositional measurements all show similar trends across the hydrograph, but the Ob and Mackenzie consistently show export of less aromatic, higher molecular weight material (e.g., higher $S_{275-295}$ and lower $SUVA_{254}$)

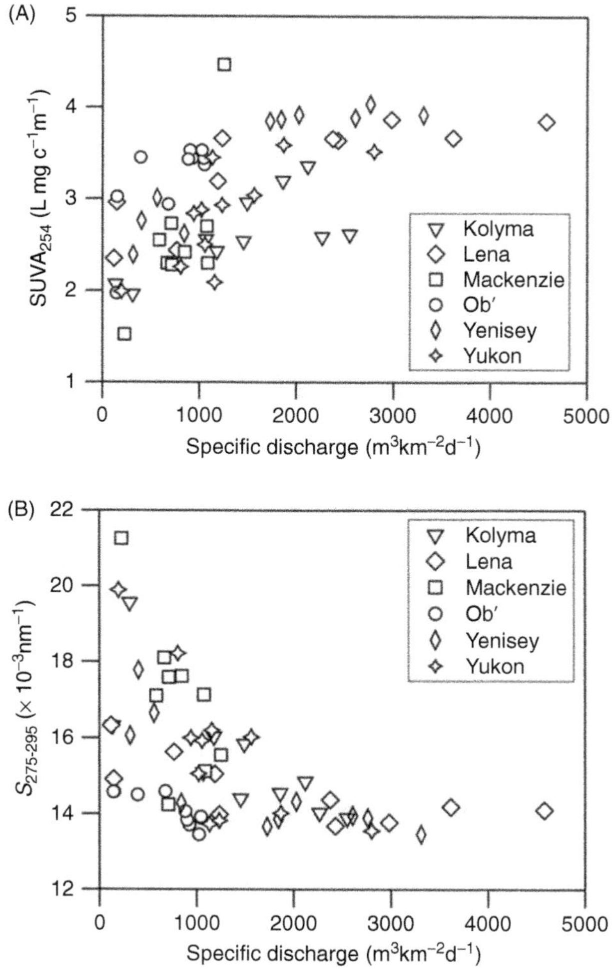

Figure 21.4. Optical measurements versus catchment area-normalized daily discharge for six largest Arctic rivers in 2009–2010. (A) $SUVA_{254}$, (B) spectral slope from 275–295 nm, $S_{275-295}$ (from Mann et al. Unpublished data).

at similar specific discharge compared to the other major Arctic rivers (Stedmon et al. 2011; Mann et al. Unpublished data; Fig. 21.4). This is likely a function of the extensive wetlands in the Ob as wetlands have been shown to produce CDOM with relatively high tDOC specific absorption (Helms et al. 2008; Fellman et al. 2009) and with respect to the Mackenzie may be due to sorption of the higher molecular weight, aromatic moieties to the elevated suspended sediment concentrations or elevated photodegradation in the extensive lakes of the Mackenzie watershed (Stedmon et al. 2011).

Similar to CDOM absorbance, comparable fluorescence parameters have been reported in a number of Arctic rivers and similar trends observed with shifts in hydrology (Cory et al. 2007; Balcarczyk et al. 2009; Walker et al. 2009; O'Donnell et al. 2010; Mann et al. 2012). Typically, the onset of the freshet leads to an increasing contribution to DOM fluorescence from humic-like fluorophores and a shift to higher emission wavelengths indicative of more conjugated, aromatic tDOM (O'Donnell et al. 2010; Mann et al. 2012). After the freshet, a blue shift (i.e., lowering in emission) has been observed in the fluorescence signature (Mann et al. 2012). Concurrent with the peak in discharge, protein-like fluorophores made up a smaller relative fraction of total fluorescence and then increased throughout the summer months. The fluorescence index parameter (FI: ratio of emission λ 470 to 520 nm at an excitation λ of 370 nm) (McKnight et al. 2001; Cory et al. 2010) has been linked to aromaticity and indicates the relative contribution of low-molecular-weight (LMW) nonaromatic DOM versus high-molecular-weight (HMW) aromatic material. In Arctic rivers, relatively high FI values have been observed under ice, indicating low aromaticity material with a microbial source or DOM derived from heavily degraded material (O'Donnell et al. 2010; Mann et al. 2012). Lowest FI values are observed at the peak of the freshet, highlighting tDOM compositions during this period that are more aromatic in nature and likely derived from terrestrial soils and vascular plant sources (Spencer et al. 2010; Mann et al. 2012). CDOM compositional measurements have also been linked to biolability and photoreactivity in Arctic rivers. For example, Mann et al. (2012) showed a positive correlation between humic-like fluorescence and biolability of tDOC in the Kolyma River across the hydrograph. This finding suggests that surface soils during the freshet provide biolabile tDOM, likely from vegetation inputs during the previous fall that have been frozen and then subsequently leached for the first time during the freshet. Similarly, Mann et al. (2012) linked $SUVA_{254}$ and the spectral slope parameter to photochemical degradation in Arctic rivers and therefore showed that optical measurements can be used to examine tDOM reactivity in these systems.

In addition to the various indices discussed previously, new emphasis is being placed on incorporating CDOM and FDOM measurements into quantitative predictive models for biomarkers (Hernes et al. 2009; Mann et al. Unpublished data). Development of rapid and inexpensive optical proxies for biomarkers will greatly facilitate future research capabilities in riverine research by providing a means for obtaining sufficient temporal and spatial coverage to capture variability.

4. Sources and fluxes of Arctic riverine DIC during transport

The riverine flux of dissolved inorganic C (DIC) has important consequences for C cycling at the landscape scale. Although the various forms of DIC (i.e., $CO_{2(aq)}/H_2CO_3$, HCO_3^-, and CO_3^{2-}) do speciate between one another, riverine DIC is largely derived from two primary sources. First, the decomposition of organic C as a result of photochemical or microbial processes results in its mineralization to CO_2, causing much tDOC mobilized from soils and transported to rivers to be released to the overlying atmosphere as CO_2 (e.g., Cole et al. 2007). In contrast, bicarbonate (HCO_3^-) and carbonate (CO_3^{2-}; which is minimal at pH <9) are predominantly derived from chemical weathering, which is one of the primary sinks for CO_2 on land (Berner et al. 1983). Weathering transforms carbonic acid (H_2CO_3) into HCO_3^-, typically as a result of the weathering of aluminosilicate (all HCO_3^- derived from CO_2; a relatively slow rate of reaction) or carbonate rocks (half of the resultant HCO_3^- derived from CO_2; a relatively rapid rate of reaction). As such, the flux and concentration of riverine DIC has important implications for C cycling across broad spatial scales.

An ongoing series of biogeochemical measurements from the mouths of the six largest Arctic rivers (the PARTNERS [2003–2007] and Arctic-GRO [2009-current] datasets) allows concentrations of CO_2 and total DIC to be calculated for the Ob, Yenisey, Lena, Kolyma, Yukon, and Mackenzie Rivers. Partial pressures of CO_2 (pCO_2) calculated using alkalinity, temperature, pH, and a correction for the contribution of organic ions to total alkalinity (Tank et al. 2012b; data available at www.aoncadis.org) indicate that like other large rivers worldwide (Cole and Caraco 2001), the big six Arctic rivers are typically saturated with CO_2 (Fig. 21.5a; Chapter 20). Values of pCO_2 at the mouths of these large rivers are highest under ice, when ice cover prevents air-water CO_2 exchange and thus allows the buildup of CO_2. Estimates of pCO_2 below atmospheric equilibrium occur sporadically on all rivers, typically toward the end of the summer, when within-river primary production is likely greatest (Fig. 21.5A). These trends are almost identical to those found for the downriver mainstem and large tributaries of the Yukon River, where pCO_2 is almost universally above atmospheric equilibrium, with the exception of select late-summer periods (Striegl et al. 2007; Chapter 20).

Although pCO_2 is typically above atmospheric equilibrium in the big six Arctic rivers, these rivers are also HCO_3^- rich (Fig. 21.5B), and thus the proportion of the total DIC pool that is present as dissolved $CO_2 + H_2CO_3$, or $H_2CO_3^*$ (represented as the ratio $H_2CO_3^*$:DIC) tends to be fairly low in these systems (Fig. 21.5C). Although $H_2CO_3^*$:DIC can reach high levels under ice, typically this ratio falls well below 0.2 during the open-water season. Rivers with the highest ratio of $H_2CO_3^*$:DIC also typically exhibit relatively low concentrations of HCO_3^-, compared with the other rivers in the dataset (Fig. 21.5C). Similar to trends in pCO_2, these rivers show clear seasonal variation in HCO_3^-, with the highest concentrations again occurring under ice, when high inputs of groundwater increase concentrations. In contrast, springtime (freshet) HCO_3^- concentrations are comparatively low, as a result of dilution at high flow (e.g., Raymond and Oh 2007; Tank et al. 2012b; Chapter 20). Values of HCO_3^- also show clear between-river variability, with the highest concentrations in the western North American rivers, and the lowest concentration measured in the Kolyma River of eastern Siberia (Tank et al. 2012b; Fig. 21.1B). These differences between rivers largely follow variations in lithology, which has been shown by numerous authors to be an important determinant of the flux of HCO_3^-, and other weathering ions, in rivers (e.g., Meybeck 1987; Gaillardet et al. 1999).

Values of pCO_2 in the downstream reaches of rivers are often not reflective of concentrations further upstream; pCO_2 is often markedly elevated in small headwater streams as a result of the breakdown of OM that occurs at the soil-water interface (e.g., Teodoru et al. 2009; Humborg et al. 2010). Synoptic surveys of sub-watershed concentrations of both total DIC and pCO_2 exist for several regions within the larger Arctic watershed. Although in some cases DIC and pCO_2 are calculated, rather than directly measured, these data still provide an overview of upstream DIC dynamics in Arctic systems. In Arctic landscapes that are peat rich, streams tend toward high pCO_2 in association with high DOC concentrations. For example, calculated pCO_2 for a series of summertime measurements from West Siberian streams are as high as 100-fold above atmospheric equilibrium (Frey et al. 2007). Similarly, stream reaches <10,000 m in the Kolyma region of East Siberia had summertime pCO_2 values ranging up to 25-fold atmospheric equilibrium in association with high concentrations of DOC (to 22 mg L^{-1}; Denfield et al. in press), and headwater streams from northern Sweden that eventually drain to the Baltic have also been shown to have high concentrations of CO_2-C

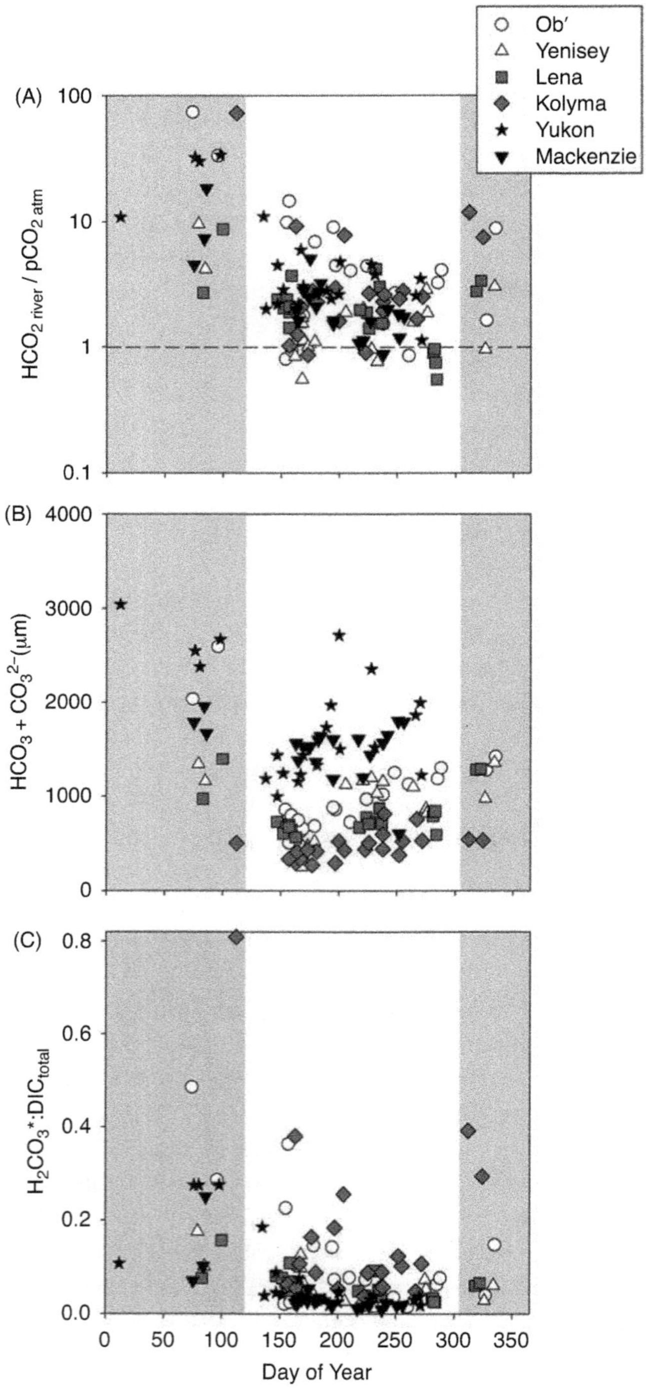

Figure 21.5. Seasonal variation the dissolved inorganic C system for the six largest Arctic rivers. Data for multiple years are shown by day of year; gray shading indicates the typical under-ice (November–April) period for these rivers. (A) the partial pressure of CO_2 relative to atmospheric pressures, here taken as 385 μatm. (B) the concentration of bicarbonate and carbonate ($H_2CO_3^- + CO_3^{2-}$). (C) the proportion of the total DIC pool that is CO_2 or H_2CO_3 (i.e., H_2CO_3*).

(to 6 mg L^{-1}) that increase in proportion to within-catchment peatland extent (Wallin et al. 2010). In less peat-rich regions, measurements of riverine pCO_2 tend to be lower, although still above atmospheric equilibrium. For example, summertime pCO_2 measurements from subwatersheds within the Yukon and Lena River basins range from slightly below saturation to 10-fold atmospheric equilibrium and again increase with increasing DOC (Huh and Edmond 1999; Huh et al. 1998; Dornblaser and Halm 2006; Halm and Dornblaser 2007).

Like pCO_2, concentrations of riverine HCO_3^- also vary significantly across the pan-Arctic region. In addition to the differences observed across the big six rivers, sub-watershed HCO_3^- concentrations also show wide variation, with summertime concentrations in tributary streams ranging from near-zero to greater than 40 mg C L^{-1} (Huh et al. 1998; Huh and Edmond 1999; Dornblaser and Halm 2006; Frey et al. 2007; Halm and Dornblaser 2007; Tank et al. 2012b). At both the full catchment and sub-watershed scale, variation in HCO_3^- flux in Arctic rivers has been shown to be broadly driven by three factors: lithology, runoff, and permafrost extent. At the sub-watershed scale, variation in these three determinants explains between 60% and 80% of between-catchment variation in HCO_3^- yield (riverine concentration normalized to watershed area) in pan-Arctic rivers (Tank et al. 2012b). Similarly, the relationship between these factors and the DIC flux from the six largest Arctic rivers has been used to estimate a 57 Tg C yr^{-1} flux of DIC to the Arctic Ocean, 30 Tg of which comes from the big six (Tank et al. in review).

Lithology is a well-known determinant of riverine HCO_3^- because of the control that rock type plays on weathering speed and the chemical composition of weathering products (e.g., Meybeck 1987; Gaillardet et al. 1999). Of the big six Arctic rivers, those with high HCO_3^- concentrations (the Yukon and Mackenzie) also display high proportions of carbonate-containing rock types, including pure carbonate, mixed sedimentary, and complex lithology classifications (Dürr et al. 2005). Clear lithology-driven variations in HCO_3^- flux are also evident at the subwatershed scale. For example, tributary rivers draining Precambrian basement (largely gneiss) terrains of the Lena watershed show low HCO_3^- concentrations when contrasted with tributaries draining regions of the watershed that are underlain by sedimentary rock (Huh et al. 1998; Huh and Edmond 1999). Similarly, rivers draining sub-watersheds of the Mackenzie system show marked variation between the high HCO_3^- concentration waters draining the carbonate rock-rich Mackenzie and Rocky Mountains, and the HCO_3^- poor waters draining the Precambrian basement terrain of the eastern shield (Millot et al. 2002; Millot et al. 2003).

Although HCO_3^- concentrations dilute with increasing runoff (e.g., Raymond and Oh 2007) the dilution effect is not strong enough to overcome the overall increase in weathering and HCO_3^- flux that occur with increasing discharge (Hartmann et al. 2009). Within the big six Arctic rivers, annual HCO_3^- flux is positively correlated with annual discharge (Tank et al. 2012b; data available at www.aoncadis.org). At the sub-watershed scale, increasing runoff has also been shown to be associated with an increase in the overall rate of weathering (Huh and Edmond 1999) and the specific weathering of carbonate rocks (Millot et al. 2003) in various Arctic regions. Recent climate-induced increases in overall runoff have been documented for both the Russian Arctic (Peterson et al. 2002) and, more recently, Arctic North America (Déry et al. 2009), suggesting a concomitant impact on weathering and HCO_3^- flux. Arctic riverine discharge is expected to continue to increase in coming years (Holland et al. 2007; Rawlins et al. 2010). However, because of the seasonal and discharge-specific variation in Arctic riverine HCO_3^- concentration (i.e., Fig. 21.5B), the effect of increasing Arctic discharge on HCO_3^- flux will largely be dependent on when, seasonally, changes in runoff occur.

There are several lines of evidence to suggest that permafrost is an additional, important regulator of HCO_3^- flux from northern catchments. Permafrost is known to inhibit surface to groundwater interactions (Ge et al. 2011) and thus the deeper flow paths that allow for interaction between water and mineral soils (reviewed by Frey and McClelland 2009). In the Yukon River basin, an increase in summertime HCO_3^- flux between the 1970s and 2000s (Striegl et al. 2005; Striegl et al. 2007) was attributed to decreases in permafrost extent and increasing groundwater flows during this same period (Walvoord and Striegl 2007; Chapter 20). Across broader spatial extents, permafrost extent has also been used to explain variation in Arctic riverine HCO_3^- flux at both the large river and sub-catchment scale (Tank et al. 2012b; Tank et al. 2012a). Finally, concentrations of non-HCO_3^- weathering ions (Ca, Mg, Na) and total inorganic solutes (Maclean et al. 1999; Petrone et al. 2006; Frey et al. 2007) have also been shown to be higher in low-permafrost catchments than in proximate high-permafrost catchments, suggesting that HCO_3^- should follow the same trend. Overall, this evidence suggests that climate-induced decreases in permafrost extent will act to increase the flux of HCO_3^- from northern watersheds, and that climate change, through its combined impact on runoff and permafrost extent, could cause measureable changes in HCO_3^- flux from northern watersheds.

In addition to its importance for processes occurring on land, the riverine flux of DIC has important implications for the biogeochemistry of the Arctic Ocean nearshore. In particular, river-influenced shelves can be disproportionately vulnerable to the effects of ocean acidification, the process by which the dissolution of CO_2 in ocean water decreases pH and thus the saturation state of aragonite (Ω_{arag}) and calcite (Ω_{calc}), the two main mineral forms of $CaCO_3$ (Doney et al. 2009). Because river water tends to be dilute in CO_3^{2-} relative to the ocean, depressions in $CaCO_3$ over river-influenced shelves act cumulatively with the global ocean acidification effect (Salisbury et al. 2008). The effects of ocean acidification are further enhanced in the Arctic, where (1) cold ocean waters increase CO_2 solubility and thus the concentration of $CaCO_3$ needed to reach saturation, (2) sea ice melt causes seasonal depressions in $CaCO_3$, and (3) the closed Arctic Ocean basin amplifies the importance of rivers in this system (Steinacher et al. 2009; Yamamoto-Kawai et al. 2009; Mathis et al. 2011; McClelland et al. 2012). As a result of this vulnerability, aragonite, the more soluble mineral form of $CaCO_3$, is predicted to reach widespread undersaturation (i.e., $\Omega_{arag} < 1$) in the Arctic Ocean within the next several decades (Steinacher et al. 2009).

Present-day instances of seasonal aragonite undersaturation have already been measured in several regions of the Arctic, in some cases where Ω_{arag} is documented to have been consistently above saturation in the recent, and preindustrial, past (Yamamoto-Kawai et al. 2009). Although much of the decrease in Arctic Ω_{arag} has been attributed to sea ice melt (Yamamoto-Kawai et al. 2009; Yamamoto-Kawai et al. 2011), clear regions of $\Omega_{arag}<1$ have been recorded near the outflows of both North American and Eurasian rivers (Bates et al. 2009; Chierici and Fransson 2009; Anderson et al. 2011). Although the effect of sea ice melt on Ω_{arag} can be expected to be greatest toward the late summer end of the melt season, fluxes of river water to the Arctic Ocean are greatest during the spring (May–June) freshet, when river water is also most dilute in DIC (Fig. 21.5). At the same time, concentrations of Arctic riverine tDOC are highest during the freshet period (Holmes et al. 2012), and the rapid mineralization of tDOC to CO_2 in the Arctic Ocean nearshore (Raymond et al. 2007; Alling et al. 2010) can be expected to further decrease Ω_{arag} as a result of CO_2-dependant decreases in pH. These seasonal patterns suggest that river water should have its greatest influence on Ω_{arag} during the spring to early summer (Mathis et al. 2011) and thus act to extend the period of Arctic Ω_{arag} depression caused by sea ice melt, in river-influenced shelves (Tank et al. 2012b). Siberian shelves, which receive relatively dilute riverine HCO_3^- fluxes but are strongly influenced by riverine tDOC and the coastal erosion of tPOC, should be particularly susceptible to the ongoing effects of ocean acidification when compared with other Arctic coastal regions (Anderson et al. 2011; Tank et al. 2012b). An increased focus on the collection of seasonally and geographically refined measurements of nearshore inorganic C chemistry will greatly improve our ability to predict future changes in the biogeochemistry of this region.

5. Evaluating sources and cycling of riverine OM utilizing natural abundance ^{14}C and ^{13}C

Streams and rivers of the Arctic drain some of the most organic rich soils in the world. These watersheds also have unique hydrology dominated by snowmelt and the spring freshet and draining permafrost soils that limit the interaction of tDOM with deeper soil mineral layers. The large stocks of old soil organic carbon and cold climate make understanding and monitoring arctic stream/river OM export particularly relevant with respect to climate change. The carbon isotopic signatures of riverine arctic OM also provides a potential tracer of riverine material in the Arctic Ocean (Karlsson et al. 2011; Griffith et al. 2012).

The first study of the age of tDOM in a large Arctic River demonstrated that it had a modern age (Neff et al. 2006). This was somewhat counterintuitive given the old age of soil organic carbon. The export of modern tDOM was affirmed at the pan-Arctic scale by Raymond et al. (2007). Raymond et al (2007) used a simple model to estimate that ~50% of the tDOM exported during the freshet is <5 years in age, with the remaining material only 5–20 years in age. tDOM in these large rivers is generally more ^{14}C-depleted, although still close to modern during low flow (Neff et al. 2006; Raymond et al. 2007).

There is variation in the age of tDOM exported from large rivers (Fig. 21.6). Flow-weighted tDOM age averages are generally more enriched compared with the atmosphere for the Lena, Yenisey, and Kolyma; close to the atmosphere for the Yukon and Ob; and depleted for the Mackenzie. The enriched values for the three Siberian rivers are consistent with flushing recent OM from upper soil layers (Tipping et al. 2010). This is also a dominant feature in other Arctic rivers, but these systems have secondary drivers that lead to the more depleted ages in the other rivers with land-use (Ob), glaciers (Yukon), groundwater, and large lakes (Mackenzie) as possibilities. Glaciers export ^{14}C-depleted tDOM directly (Hood et al. 2009; Stubbins et al. 2012) and glacial till, when dissolved, depletes ^{14}C-DIC and produces aged autochthonous OM (Striegl et al. 2005).

tPOC in large rivers is often highly ^{14}C-depleted, aged OM. The source of this old carbon is predominantly eroded peat (Guo et al. 2007). Reported ages for the Yukon and Mackenzie are ~6–7 kyr ^{14}C age (Guo et al. 2007; Striegl et al. 2007). There is also the potential for a significant contribution from ^{14}C-dead black carbon sources (Elmquist et al. 2008). These ages are confirmed in lake and river cores, which demonstrate highly aged tundra-derived plant material (Dickens et al.

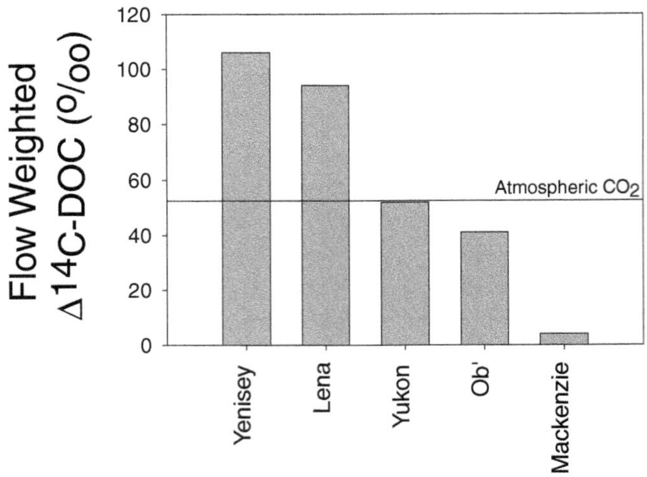

Figure 21.6. Flow-weighted DOC concentrations for the five largest Arctic rivers from Raymond et al. (2007). Samples are from 2004–2005. The line is the approximate Δ ^{14}C of atmospheric CO_2 during 2004–2005.

2011). Interestingly, despite the aged nature of Arctic River tPOC, it appears to be labile in estuarine environments (Sanchez-Garcia et al. 2011).

Moving away from the large watersheds of the Arctic provides additional information on the ages of OM. Studies in the Yukon have found younger tDOM in the summer for wetland-dominated rivers and predominantly old tPOC from meltwater mountainous/glacial, peat/permafrost, and groundwater-dominated watersheds (Striegl et al. 2007). Small stream-draining peatlands confirm that tDOM is enriched in ^{14}C compared with the atmosphere, but contain an aged tPOC component and demonstrate variation in both tPOC and tDOC with flow (Billett et al. 2007; Billett et al. 2012). In forested catchments, tDOM remains enriched, but can be aged during baseflow when groundwater contributions are high (Schiff et al. 1997).

It is possible that under future warming, stored organic matter could be mobilized and transported laterally as tDOM and tPOM. For tDOM, studies have argued that temperature increase may thaw permafrost and liberate tDOM (Freeman et al. 2001; Frey and Smith 2005), although other studies have argued that future warming may lead to increased utilization or land-altered flow paths that may lead to a decrease in tDOC concentration (Striegl et al. 2005). It is currently difficult to assess whether tDOM fluxes are changing at the pan-Arctic scale. For tPOM, destabilization of coastal and riverine banks could move large quantities of aged organic material to inland waters and coastal sediments (Goni et al. 2005; Guo et al. 2007). The fate of this material could be oxidation to CO_2.

6. Fate of riverine DOM in the arctic ocean: degradation, transport, and constraining flux models

Several studies over the past decade have focused on constraining turnover times for Arctic riverine tDOC in the Arctic Ocean, generally using a box model approach with calculations of the total tDOC (or proxies) in the Arctic Ocean divided by total annual fluxes of riverine tDOC (or proxies) derived from scaling the fluxes of one or more of the big six Arctic rivers. Using such an approach with lignin as a proxy for tDOC, Opsahl et al. (1999) calculated initial residence times of 1–6 years, indicating either rapid physical transport of tDOC out of the Arctic Ocean via surface currents or incorporation into deepwater, or transport in combination with photochemical and microbial degradation.

As research on Arctic rivers has intensified, a number of more recent studies have explored the potential of deriving improved dissolved lignin phenol and DOC flux estimates from Arctic rivers utilizing CDOM measurements (Spencer et al. 2009a; Mann et al. Unpublished data). These studies have resulted in pan-Arctic DOC and dissolved lignin loads that were higher than previous estimates because of high-resolution CDOM data, especially across the freshet period (Spencer et al. 2009a; Mann et al. Unpublished data). Using similar box model approaches to Opsahl et al. (1999), higher modeled flux to the Arctic Ocean leads to reduced residence time estimates for terrigenous DOC in the Polar Surface Waters, ranging from <1 to 4 years. These studies highlight the importance of accurately constraining fluxes and reservoirs and the utility of CDOM for that purpose in the Arctic (Spencer et al. 2009a; Mann et al. Unpublished data), but also the rapid cycling in general of riverine tDOC in Arctic systems.

Parsing out the fate of tDOC in the Arctic Ocean has numerous inherent uncertainties. Hansell et al. (2004) utilized a combination of isotopes, DIC production, and analysis of conservative mixing lines to determine a half-life for tDOC of 7.1± 3.0 years and an export term of 21–32% of total riverine input due to physical transport, suggesting that approximately three-quarters of tDOC is remineralized within the Arctic Ocean itself. Alling et al. (2010) employed a box model approach and conservative mixing lines in the Laptev and Eastern Siberian seas to estimate a 30–50% loss of tDOC during transport across the shelf. Benner et al. (2005) utilized lignin measurements in conjunction with stable isotopes to estimate a physical export term of 25–33% of tDOC to the surface North Atlantic and to North Atlantic Deep Water (NADW). Hernes and Benner (2006) measured elevated dissolved lignin concentrations in NADW in the western North Atlantic and determined that 10–16% of Arctic riverine tDOC entrainment in NADW could account for the elevated values. All these studies paint a consistent picture that two-thirds to three-quarters of tDOC must be remineralized within the Arctic Ocean. As ice coverage continues to shrink as a result of climate change, enhanced light availability has the potential to increase remineralization, either directly through increased photodegradation or indirectly through increased production of phytoplankton-derived DOC, which could in turn lead to greater microbial remineralization of tDOC through priming effects (Bianchi 2011).

Determining the relative amounts of degradation of tDOC due to photodegradation versus microbial degradation presents considerable challenges, in part because of the difficulties with upscaling bottle experiments, but also as a result of seasonal variability and rapid changes occurring in the pan-Arctic. CDOM plays a central role in the biogeochemistry of coastal waters as it determines the aquatic light field and is photoreactive, resulting in the production of a number of climatically active gases (Coble 2007; Stubbins et al. 2008; Del Vecchio et al. 2009). Therefore, any increase in CDOM load in Arctic rivers could reduce the photic depth, especially in the shelf regions and particularly with reductions in sea ice, could likely lead to an increasing role of photochemistry in Arctic biogeochemical cycles (Stedmon et al. 2011). The surface waters of the Arctic Ocean have been shown to absorb up to 30% more solar energy per unit area compared with other oceans as a result of the relatively higher CDOM amounts (Pegau 2002), and the potential for photodegradation during ice-free summer months appears high. Photodegradation experiments of Kolyma River tDOC indicated the potential losses of ~30% over 14 days (Mann et al. 2012). However, incubation studies in the Beaufort Sea suggested <1% tDOC removal due to photodegradation across the MacKenzie Shelf (Osburn et al. 2009), thus highlighting the complexities of upscaling bottle experiments to coastal-margin scales. Similarly, bioavailability incubations on Arctic riverine tDOC demonstrate high microbial degradation potentials (20–40%) over 1–3-month incubations (Holmes et al. 2008; Mann et al. 2012), but there is little direct confirmation of the extent of microbial degradation from shelf or open ocean measurements.

Clearly there is a need for better quantification of photodegradation and microbial degradation sinks of tDOC within the Arctic Ocean proper. Progress is also needed in modeling and scaling of fluxes, with a need for much better temporal and spatial coverage. Optical tools offer much potential for improved coverage as a result of strong linear relationships between CDOM and tDOC either in individual rivers or across a broad range of Arctic rivers (Spencer et al. 2008; Spencer et al. 2009a; Stedmon et al. 2011; Mann et al. Unpublished data; Fig. 21.3A). Furthermore, a number of studies have highlighted strong correlations in Arctic rivers between CDOM and biomarkers for terrigenous tDOC such as dissolved lignin phenols and dissolved black carbon, and thus CDOM may be employed to derive flux estimates of these analytically intensive and expensive measurements (Spencer et al. 2008; Spencer et al. 2009a; Mann et al. Unpublished data; Stubbins et al. Unpublished data; Fig. 21.3B). The ability to derive tDOC from CDOM in Arctic rivers opens up the potential to utilize in situ CDOM sensors that make high temporal resolution measurements to greatly improve tDOC flux estimates from these dynamic watersheds. The environmental influences (inner filter, temperature, and suspended sediment concentration effects) with respect to deploying in situ CDOM sensors in Arctic rivers are significant, but recent studies have shown potential to overcome these issues (Watras et al. 2011; Downing et al. 2012). The utilization of remote sensing techniques with high spatial resolution hyperspectral data, such as the Advanced Land Imager, have also shown potential for estimating CDOM and tDOC in Arctic rivers (Griffin et al. 2011). Therefore, a number of possibilities are available to future studies seeking to develop Arctic river CDOM fluxes or refine tDOC flux estimates into and export out of the Arctic Ocean via in situ or remote sensing technologies.

7. Conclusions

The pan-Arctic region is widely considered to be the "canary in the coal mine" in terms of the speed and magnitude with which it will be impacted by climate change. Therefore, it is imperative to continue characterizing baseline conditions for carbon budgets and cycling to more accurately capture changes in rates and fluxes of carbon in coming decades. Understanding the nature of these changes in the pan-Arctic may in turn facilitate more accurate predictions for climate change impacts in temperate and tropical river/coastal systems.

References

Alling, V., L. Sanchez-Garcia, D. Porcelli, S. Pugach, J. E. Vonk, B. van Dongen, C.-M. Mörth, L. G. Anderson, A. Sokolov, P. Andersson, C. Humborg, I. Semiletov, and Ö. Gustafsson. 2010.

Nonconservative behavior of dissolved organic carbon across the Laptev and East Siberian seas. Global Biogeochemical Cycles 24: GB4033.

Anderson, L. G., G. Bjork, S. Jutterstrom, I. Pipko, N. Shakhova, I. Semiletov, and I. Wahlstrom. 2011. East Siberian Sea, an Arctic region of very high biogeochemical activity. Biogeosciences 8: 1745–1754.

Balcarczyk, K., J. Jones, R. Jaffé, and N. Maie. 2009. Stream dissolved organic matter bioavailability and composition in watersheds underlain with discontinuous permafrost. Biogeochemistry 94: 255–270.

Bates, N. R., J. T. Mathis, and L. W. Cooper. 2009. Ocean acidification and biologically induced seasonality of carbonate mineral saturation states in the western Arctic Ocean. Journal of Geophysical Research-Oceans 114: C11007.

Benner, R., P. Louchouarn, and R. M. W. Amon. 2005. Terrigenous dissolved organic matter in the Arctic Ocean and its transport to surface and deep waters of the North Atlantic. Global Biogeochemical Cycles 19: GB2025.

Berner, R. A., A. C. Lasaga, and R. M. Garrels. 1983. The carbonate-silicate geochemical cycle and its effect on atmospheric carbon dioxide over the past 100 million years. American Journal of Science 283: 641–683.

Bianchi, T. S. 2011. The role of terrestrially derived organic carbon in the coastal ocean: A changing paradigm and the priming effect. Proceedings of the National Academy of Sciences of the United States of America 108: 19473–19481.

Billett, M. F., K. J. Dinsmore, R. P. Smart, M. H. Garnett, J. Holden, P. Chapman, A. J. Baird, R. Grayson, and A. W. Stott. 2012. Variable source and age of different forms of carbon released from natural peatland pipes. Journal of Geophysical Research-Biogeosciences 117: G02003.

Billett, M. F., M. H. Garnett, and F. Harvey. 2007. UK peatland streams release old carbon dioxide to the atmosphere and young dissolved organic carbon to rivers. Geophysical Research Letters 34: L23401.

Chierici, M., and A. Fransson. 2009. Calcium carbonate saturation in the surface water of the Arctic Ocean: undersaturation in freshwater influenced shelves. Biogeosciences 6: 2421–2431.

Coble, P. G. 2007. Marine Optical Biogeochemistry: The Chemistry of Ocean Color. Chemical Reviews 107: 402–418.

Cole, J. J., and N. F. Caraco. 2001. Carbon in catchments: Connecting terrestrial carbon losses with aquatic metabolism. Marine and Freshwater Research 52: 101–110.

Cole, J. J., Y. T. Prairie, N. F. Caraco, W. H. McDowell, L. J. Tranvik, R. G. Striegl, C. M. Duarte, P. Kortelainen, J. A. Downing, J. J. Middelburg, and J. Melack.2007. Plumbing the global carbon cycle: Integrating inland waters into the terrestrial carbon budget. Ecosystems 10: 171–184.

Cory, R. M., D. M. McKnight, Y.-P. Chin, P. Miller, and C. L. Jaros. 2007. Chemical characteristics of fulvic acids from Arctic surface waters: Microbial contributions and photochemical transformations. Journal of Geophysical Research-Biogeosciences 112: G04S51.

Cory, R. M., M. P. Miller, D. M. McKnight, J. J. Guerard, and P. L. Miller. 2010. Effect of instrument-specific response on the analysis of fulvic acid fluorescence spectra. Limnology and Oceanography: Methods 8: 67–78.

Del Vecchio, R., A. Subramaniam, S. Schollaert Uz, J. Ballabrera-Poy, C. W. Brown, and N. V. Blough. 2009. Decadal time-series of SeaWiFS retrieved CDOM absorption and estimated CO2 photoproduction on the continental shelf of the eastern United States. Geophysical Research Letters 36: L02602.

Denfield, B. A., K. E. Frey, W. Sobczak, and R. M. Holmes. In press. Summer CO^2 evasion from streams and rivers in the Kolyma River basin, Northeast Siberia. Polar Research.

Déry, S. J., M. A. Hernandez-Henriquez, J. E. Burford, and E. F. Wood. 2009. Observational evidence of an intensifying hydrological cycle in northern Canada. Geophysical Research Letters 36: L13402.

Dickens, A. F., J. Baldock, T. C. Kenna, and T. I. Eglinton. 2011. A depositional history of particulate organic carbon in a floodplain lake from the lower Ob' River, Siberia. Geochimica Et Cosmochimica Acta 75: 4796–4815.

Dittmar, T., and G. Kattner. 2003. The biogeochemistry of the river and shelf ecosystem of the Arctic Ocean: a review. Marine Chemistry 83: 103–120.

Doney, S. C., V. J. Fabry, R. A. Feely, and J. A. Kleypas. 2009. Ocean acidification: The other CO_2 problem. Annual Review of Marine Science 1: 169–192.

Dornblaser, M. M., and D. R. Halm (eds.). 2006. Water and sediment quality of the Yukon River and its tributaries, from Eagle to St. Marys, Alaska, 2002–2003. US Geological Survey Open File Report 2006–1228.

Downing, B. D., B. A. Pellerin, B. A. Bergamaschi, J. F. Saraceno, and T. E. C. Kraus. 2012. Seeing the light: The effects of particles, dissolved materials, and temperature on in situ measurements of DOM fluorescence in rivers and streams. Limnology and Oceanography: Methods 10: 767–775.

Dürr, H. H., M. Meybeck, and S. H. Durr. 2005. Lithologic composition of the Earth's continental surfaces derived from a new digital map emphasizing riverine material transfer. Global Biogeochemical Cycles 19: GB4S10.

Elmquist, M., I. Semiletov, L. D. Guo, and O. Gustafsson. 2008. Pan-Arctic patterns in black carbon sources and fluvial discharges deduced from radiocarbon and PAH source apportionment markers in estuarine surface sediments. Global Biogeochemical Cycles 22: GB2018.

Fellman, J., E. Hood, D. D'Amore, R. Edwards, and D. White. 2009. Seasonal changes in the chemical quality and biodegradability of dissolved organic matter exported from soils to streams in coastal temperate rainforest watersheds. Biogeochemistry 95: 277–293.

Freeman, C., C. D. Evans, D. T. Monteith, B. Reynolds, and N. Fenner. 2001. Export of organic carbon from peat soils. Nature 412: 785–785.

Frey, K. E., and J. W. McClelland. 2009. Impacts of permafrost degradation on arctic river biogeochemistry. Hydrological Processes 23: 169–182.

Frey, K. E., D. I. Siegel, and L. C. Smith. 2007. Geochemistry of west Siberian streams and their potential response to permafrost degradation. Water Resources Research 43: W03406.

Frey, K. E., and L. C. Smith. 2005. Amplified carbon release from vast West Siberian peatlands by 2100. Geophysical Research Letters 32: L09401.

Gaillardet, J., B. Dupre, P. Louvat, and C. J. Allegre. 1999. Global silicate weathering and CO2 consumption rates deduced from the chemistry of large rivers. Chemical Geology 159: 3–30.

Ge, S., J. Mckenzie, C. Voss, and Q. Wu. 2011. Exchange of groundwater and surface-water mediated by permafrost response to seasonal and long term air temperature variation. Geophysical Research Letters 38: L14402.

Gebhardt, A. C., B. Gaye-Haake, D. Unger, N. Lahajnar, and V. Ittekkot. 2004. Recent particulate organic carbon and total suspended matter fluxes from the Ob and Yenisei Rivers into the Kara Sea (Siberia). Marine Geology 207: 225–245.

Goni, M. A., M. B. Yunker, R. W. Macdonald, and T. I. Eglinton. 2000. Distribution and sources of organic biomarkers in arctic sediments from the Mackenzie River and Beaufort Shelf. Marine Chemistry 71: 23–51.

Goni, M. A., M. B. Yunker, R. W. Macdonald, and T. I. Eglinton. 2005. The supply and preservation of ancient and modem components of organic carbon in the Canadian Beaufort Shelf of the Arctic Ocean. Marine Chemistry 93: 53–73.

Griffin, C. G., K. E. Frey, J. Rogan, and R. M. Holmes. 2011. Spatial and interannual variability of dissolved organic matter in the Kolyma River, East Siberia, observed using satellite imagery. Journal of Geophysical Research-Biogeosciences 116: G03018.

Griffith, D. R., A. P. McNichol, L. Xu, F. A. McLaughlin, R. W. Macdonald, K. A. Brown, and T. I. Eglinton. 2012. Carbon dynamics in the western Arctic Ocean: insights from full-depth carbon isotope profiles of DIC, DOC, and POC. Biogeosciences 9: 1217–1224.

Guo, L. D., C. L. Ping, and R. W. Macdonald. 2007. Mobilization pathways of organic carbon from permafrost to arctic rivers in a changing climate. Geophysical Research Letters 34: L13603.

Halm, D. R., and M. M. Dornblaser (eds.). 2007. Water and sediment quality in the Yukon River and its tributaries between Atlin, British Columbia, Canada, and Eagle, Alaska, USA, 2004. US Geological Survey Open File Report 2007–1197.

Hansell, D. A., D. Kadko, and N. R. Bates. 2004. Degradation of terrigenous dissolved organic carbon in the western Arctic Ocean. Science 304: 858–861.

Hartmann, J., N. Jansen, H. H. Durr, S. Kempe, and P. Kohler. 2009. Global CO(2)-consumption by chemical weathering: What is the contribution of highly active weathering regions? Global and Planetary Change 69: 185–194.

Hedges, J. I., and D. C. Mann. 1979. The characterization of plant tissues by their lignin oxidation products. Geochimica Et Cosmochimica Acta 43: 1803–1807.

Helms, J. R., A. Stubbins, J. D. Ritchie, E. C. Minor, D. J. Kieber, and K. Mopper. 2008. Absorption spectral slopes and slope ratios as indicators of molecular weight, source, and photobleaching of chromophoric dissolved organic matter. Limnology and Oceanography 53: 955–969.

Hernes, P. J., and R. Benner. 2003. Photochemical and microbial degradation of dissolved lignin phenols: Implications for the fate of terrigenous dissolved organic matter in marine environments. Journal of Geophysical Research-Oceans 108: 3291.

Hernes, P. J., and R. Benner. 2006. Terrigenous organic matter sources and reactivity in the North Atlantic Ocean and a comparison to the Arctic and Pacific oceans. Marine Chemistry 100: 66–79.

Hernes, P. J., B. A. Bergamaschi, R. S. Eckard, and R. G. M. Spencer. 2009. Fluorescence-based proxies for lignin in freshwater dissolved organic matter. Journal of Geophysical Research-Biogeosciences 114: G00F03.

Hernes, P. J., A. C. Robinson, and A. K. Aufdenkampe. 2007. Fractionation of lignin during leaching and sorption and implications for organic matter freshness. Geophysical Research Letters 34: L17401.

Holland, M. M., J. Finnis, A. P. Barrett, and M. C. Serreze. 2007. Projected changes in arctic ocean freshwater budgets. Journal of Geophysical Research-Biogeosciences 112: G04S55.

Holmes, R. M. , J. W. McClelland, B. J. Peterson, S. E. Tank, E. Bulygina, T. I. Eglinton, V. V. Gordeev, T. Y. Gurtovaya, P. A. Raymond, D. J. Repeta, R. Staples, R. G. Striegl, A. V. Zhulidov, and S. A. Zimov. 2012. Seasonal and annual fluxes of nutrients and organic matter from large rivers to the Arctic Ocean and surrounding seas. Estuaries and Coasts 35: 369–382.

Holmes, R. M., M. T. Coe, G. J. Fiske, T. Gurtovaya, J. W. McClelland, A. I. Shiklomanov, R. G. M. Spencer, S. E. Tank, and A. V. Zhulidov. 2013. Climate change impacts on the hydrology and biogeochemistry of Arctic rivers. In: Goldman, C. R., Kumagai, M., and Robarts, R. D. (eds.), Climatic Change and Global Warming of Inland Waters: Impacts and Mitigation for Ecosystems and Societies. Wiley-Blackwell.

Holmes, R. M. , J. W. McClelland, B. J. Peterson, I. A. Shiklomanov, A. I. Shiklomanov, A. V. Zhulidov, V. V. Gordeev, and N. N. Bobrovitskaya. 2002. A circumpolar perspective on fluvial sediment flux to the Arctic ocean. Global Biogeochem. Cycles 16: 1098.

Holmes, R. M., J. W. McClelland, P. A. Raymond, B. B. Frazer, B. J. Peterson, and M. Stieglitz. 2008. Lability of DOC transported by Alaskan rivers to the arctic ocean. Geophysical Research Letters 35: L03402.

Holmes, R. M., B. J. Peterson, V. V. Gordeev, A. V. Zhulidov, M. Meybeck, R. B. Lammers, and C. J. Vörösmarty. 2000. Flux of nutrients from Russian rivers to the Arctic Ocean: Can we establish a baseline against which to judge future changes? Water Resources Research 36: 2309–2320.

Hood, E., J. Fellman, R. G. M. Spencer, P. J. Hernes, R. Edwards, D. D'Amore, and D. Scott. 2009. Glaciers as a source of ancient and labile organic matter to the marine environment. Nature 462: 1044–U1100.

Huh, Y., and J. M. Edmond. 1999. The fluvial geochemistry of the rivers of Eastern Siberia: III. Tributaries of the Lena and Anabar draining the basement terrain of the Siberian Craton and the Trans-Baikal Highlands. Geochimica Et Cosmochimica Acta 63: 967–987.

Huh, Y., M. Y. Tsoi, A. Zaitsev, and J. M. Edmond. 1998. The fluvial geochemistry of the rivers of eastern Siberia: I. Tributaries of the Lena River draining the sedimentary platform of the Siberian Craton. Geochimica Et Cosmochimica Acta 62: 1657–1676.

Humborg, C., C.-M. Mörth, M. Sundbom, H. Borg, T. Blenckner, R. Giesler, and V. Ittekkot. 2010. CO_2 supersaturation along the aquatic conduit in Swedish watersheds as a constrained by terrestrial respiration, aquatic respiration and weathering. Global Change Biology 16: 1966–1978.

Karlsson, E. S., A. Charkin, O. Dudarev, I. Semiletov, J. E. Vonk, L. Sanchez-Garcia, A. Andersson, and O. Gustafsson. 2011. Carbon isotopes and lipid biomarker investigation of sources, transport and degradation of terrestrial organic matter in the Buor-Khaya Bay, SE Laptev Sea. Biogeosciences 8: 1865–1879.

Lobbes, J. M., H. P. Fitznar, and G. Kattner. 2000. Biogeochemical characteristics of dissolved and particulate organic matter in Russian rivers entering the Arctic Ocean. Geochimica Et Cosmochimica Acta 64: 2973–2983.

Macdonald, R. W., S. M. Solomon, R. E. Cranston, H. E. Welch, M. B. Yunker, and C. Gobeil. 1998. A sediment and organic carbon budget for the Canadian Beaufort shelf. Marine Geology 144: 255–273.

Maclean, R., M. W. Oswood, J. G. Irons, and W. H. McDowell. 1999. The effect of permafrost on stream biogeochemistry: A case study of two streams in the Alaskan (USA) taiga. Biogeochemistry 47: 239–267.

Mann, P. J., A. Davydova, N. Zimov, R. G. M. Spencer, S. Davydov, E. B. Bulygina, S. A. Zimov, and R. M. Holmes. 2012. Controls on the composition and lability of dissolved organic matter in Siberia's Kolyma River Basin. Journal of Geophysical Research-Biogeosciences 117: G01028.

Mathis, J. T., J. N. Cross, and N. R. Bates. 2011. Coupling primary production and terrestrial runoff to ocean acidification and carbonate mineral suppression in the eastern Bering Sea. Journal of Geophysical Research-Oceans 116: C02030.

McClelland, J. W., S. J. Déry, B. J. Peterson, R. M. Holmes, and E. F. Wood. 2006. A pan-arctic evaluation of changes in river discharge during the latter half of the 20th century. Geophysical Research Letters 33: L06715.

McClelland, J. W., R. M. Holmes, K. H. Dunton, and R. W. Macdonald. 2012. The Arctic Ocean Estuary. Estuaries and Coasts 35: 353–368.

McClelland, J. W., R. M. Holmes, B. J. Peterson, R. Amon, T. Brabets, L. Cooper, J. Gibson, V. V. Gordeev, C. Guay, D. Milburn, R. Staples, P. A. Raymond, I. Shiklomanov, R. Striegl, A. Zhulidov, T. Gurtovaya, and S. Zimov. 2008. Development of a pan-Arctic database for river chemistry. EOS, Transactions American Geophysical Union 89.

McKnight, D. M., E. W. Boyer, P. K. Westerhoff, P. T. Doran, T. Kulbe, and D. T. Andersen. 2001. Spectrofluorometric characterization of dissolved organic matter for indication of precursor organic material and aromaticity. Limnology and Oceanography 46: 38–48.

Meybeck, M. 1987. Global chemical weathering of surficial rocks estimated from river dissolved loads. American Journal of Science 287: 401–428.

Millot, R., J. Gaillardet, B. Dupre, and C. J. Allegre. 2002. The global control of silicate weathering rates and the coupling with physical erosion: new insights from rivers of the Canadian Shield. Earth and Planetary Science Letters 196: 83–98.

Millot, R., J. Gaillardet, B. Dupre, and C. J. Allegre. 2003. Northern latitude chemical weathering rates: Clues from the Mackenzie River Basin, Canada. Geochimica Et Cosmochimica Acta 67: 1305–1329.

Neff, J. C., J. C. Finlay, S. A. Zimov, S. P. Davydov, J. J. Carrasco, E. A. G. Schuur, and A. I. Davydova. 2006. Seasonal changes in the age and structure of dissolved organic carbon in Siberian rivers and streams. Geophysical Research Letters 33: L23401.

Nohara, D., A. Kitoh, M. Hosaka, and T. Oki. 2006. Impact of climate change on river discharge projected by multimodel ensemble. Journal of Hydrometeorology 7: 1076–1089.

O'Donnell, J. A., G. R. Aiken, E. S. Kane, and J. B. Jones. 2010. Source water controls on the character and origin of dissolved organic matter in streams of the Yukon River basin, Alaska. Journal of Geophysical Research-Biogeosciences 115: G03025.

Opsahl, S., and R. Benner. 1998. Photochemical reactivity of dissolved lignin in river and ocean waters. Limnology and Oceanography 43: 1297–1304.

Opsahl, S., R. Benner, and R. M. W. Amon. 1999. Major flux of terrigenous dissolved organic matter through the Arctic Ocean. Limnology and Oceanography 44: 2017–2023.

Osburn, C. L., L. Retamal, and W. F. Vincent. 2009. Photoreactivity of chromophoric dissolved organic matter transported by the Mackenzie River to the Beaufort Sea. Marine Chemistry 115: 10–20.

Osterkamp, T. E. 2007. Characteristics of the recent warming of permafrost in Alaska. Journal of Geophysical Research-Earth Surface 112: F02S02.

Pegau, W. S. 2002. Inherent optical properties of the central Arctic surface waters. Journal of Geophysical Research-Oceans 107: 8035.

Peterson, B. J., R. M. Holmes, J. W. McClelland, C. J. Vorosmarty, R. B. Lammers, A. I. Shiklomanov, I. A. Shiklomanov, and S. Rahmstorf.2002. Increasing river discharge to the Arctic Ocean. Science 298: 2171–2173.

Petrone, K. C., J. B. Jones, L. D. Hinzman, and R. D. Boone. 2006. Seasonal export of carbon, nitrogen, and major solutes from Alaskan catchments with discontinuous permafrost Journal of Geophysical Research-Biogeosciences 111: G04003.

Rawlins, M. A., M. Steele, M. M. Holland, J. C. Adam, J. E. Cherry, J. A. Francis, P. Y. Groisman, L. D. Hinzman, T. G. Huntington, D. L. Kane, J. S. Kimball, R. Kwok, R. B. Lammers, C. M. Lee, D. P. Lettenmaier, K. C. McDonald, E. Podest, J. W. Pundsack, B. Rudels, M. C. Serreze, A. Shiklomanov, O. Skagseth, T. J. Troy, C. J. Vörösmarty, M. Wensnahan, E. F. Wood, R. Woodgate, D. Yang, K. Zhang, and T. Zhang. 2010. Analysis of the Arctic system for freshwater cycle intensification: Observations and expectations. Journal of Climate 23: 5715–5737.

Raymond, P. A., J. W. McClelland, R. M. Holmes, A. V. Zhulidov, K. Mull, B. J. Peterson, R. G. Striegl, G. R. Aiken, and T. Y. Gurtovaya.2007. Flux and age of dissolved organic carbon exported to the Arctic Ocean: A carbon isotopic study of the five largest arctic rivers. Global Biogeochemical Cycles 21: GB4011.

Raymond, P. A., and N. H. Oh. 2007. An empirical study of climatic controls on riverine C export from three major U.S. watersheds. Global Biogeochemical Cycles 21: GB2022.

Salisbury, J., M. Green, C. Hunt, and J. Campbell. 2008. Coastal acidification by rivers: A threat to shellfish? EOS, Transactions American Geophysical Union 89: 513–528.

Sanchez-Garcia, L., V. Alling, S. Pugach, J. Vonk, B. van Dongen, C. Humborg, O. Dudarev, I. Semiletov, and O. Gustafsson. 2011. Inventories and behavior of particulate organic carbon in the Laptev and East Siberian seas. Global Biogeochemical Cycles 25: GB2007.

Schiff, S. L., R. Aravena, S. E. Trumbore, M. J. Hinton, R. Elgood, and P. J. Dillon. 1997. Export of DOC from forested catchments on the Precambrian Shield of Central Ontario: Clues from ^{13}C and ^{14}C. Biogeochemistry 36: 43–65.

Serreze, M. C., and R. G. Barry. 2011. Processes and impacts of Arctic amplification: A research synthesis. Global and Planetary Change 77: 85–96.

Spencer, R. G. M., G. R. Aiken, K. D. Butler, M. M. Dornblaser, R. G. Striegl, and P. J. Hernes. 2009a. Utilizing chromophoric dissolved organic matter measurements to derive export and reactivity of dissolved organic carbon exported to the Arctic Ocean: A case study of the Yukon River, Alaska. Geophysical Research Letters 36: L06401.

Spencer, R. G. M., G. R. Aiken, R. Y. Dyda, K. D. Butler, B. A. Bergamaschi, and P. J. Hernes. 2010. Comparison of XAD with other dissolved lignin isolation techniques and a compilation of analytical improvements for the analysis of lignin in aquatic settings. Organic Geochemistry 41: 445–453.

Spencer, R. G. M., G. R. Aiken, K. P. Wickland, R. G. Striegl, and P. J. Hernes. 2008. Seasonal and spatial variability in dissolved organic matter quantity and composition from the Yukon River basin, Alaska. Global Biogeochemical Cycles 22: GB4002.

Spencer, R. G. M., A. Stubbins, P. J. Hernes, A. Baker, K. Mopper, A. K. Aufdenkampe, R. Y. Dyda, V. L. Mwamba, A. M. Mangangu, J. N. Wabakanghanzi, and J. Six. 2009b. Photochemical degradation of dissolved organic matter and dissolved lignin phenols from the Congo River. Journal of Geophysical Research-Biogeosciences 114: G03010.

Stedmon, C. A., R. M. W. Amon, A. J. Rinehart, and S. A. Walker. 2011. The supply and characteristics of colored dissolved organic matter (CDOM) in the Arctic Ocean: Pan Arctic trends and differences. Marine Chemistry 124: 108–118.

Steinacher, M., F. Joos, T. L. Frölicher, G.-K. Plattner, and S. C. Doney. 2009. Imminent ocean acidification in the Arctic projected with the NCAR global coupled carbon cycle-climate model. Biogeosciences 6: 515–533.

Striegl, R. G., G. R. Aiken, M. M. Dornblaser, P. A. Raymond, and K. P. Wickland. 2005. A decrease in discharge-normalized DOC export by the Yukon River during summer through autumn. Geophysical Research Letters 32: L21413.

Striegl, R. G., M. M. Dornblaser, G. R. Aiken, K. P. Wickland, and P. A. Raymond. 2007. Carbon export and cycling by the Yukon, Tanana, and Porcupine rivers, Alaska, 2001–2005. Water Resources Research 43: W02411.

Stubbins, A., E. Hood, P. A. Raymond, G. R. Aiken, R. L. Sleighter, P. J. Hernes, D. Butman, P. G. Hatcher, R. G. Striegl, P. Schuster, H. A. N. Abdulla, A. W. Vermilyea, D. T. Scott, and R. G. M. Spencer. 2012. Anthropogenic aerosols as a source of ancient dissolved organic matter in glaciers. Nature Geoscience 5: 198–201.

Stubbins, A., V. Hubbard, G. Uher, C. S. Law, R. C. Upstill-Goddard, G. R. Aiken, and K. Mopper. 2008. Relating carbon monoxide photoproduction to dissolved organic matter functionality. Environmental Science & Technology 42: 3271–3276.

Tank, S. E., K. E. Frey, R. G. Striegl, P. A. Raymond, R. M. Holmes, J. W. McClelland, and B. J. Peterson. 2012a. A land-to-ocean perspective on the magnitude, source and implication of DIC flux from major Arctic rivers to the Arctic Ocean. Global Biogeochemical Cycles 26: GB4018.

Tank, S. E., P. A. Raymond, R. G. Striegl, J. W. McClelland, R. M. Holmes, G. J. Fiske, and B. J. Peterson. 2012b. Landscape-level controls on dissolved carbon flux from diverse catchments of the circumboreal. Global Biogeochemical Cycles 26: GB0E02.

Teodoru, C. R., P. A. Del Giorgio, Y. T. Prairie, and M. Camire. 2009. Patterns in pCO_2 in boreal streams and rivers of northern Quebec, Canada. Global Biogeochemical Cycles 23: GB2012.

Tipping, E., M. F. Billett, C. L. Bryant, S. Buckingham, and S. A. Thacker. 2010. Sources and ages of dissolved organic matter in peatland streams: evidence from chemistry mixture modelling and radiocarbon data. Biogeochemistry 100: 121–137.

Unger, D., B. Gaye-Haake, K. Neumann, A. Catalina Gebhardt, and V. Ittekkot. 2005. Biogeochemistry of suspended and sedimentary material in the Ob and Yenisei rivers and Kara Sea: amino acids and amino sugars. Continental Shelf Research 25: 437–460.

Walker, S. A., R. M. W. Amon, C. Stedmon, S. Duan, and P. Louchouarn. 2009. The use of PARAFAC modeling to trace terrestrial dissolved organic matter and fingerprint water masses in coastal Canadian Arctic surface waters. Journal of Geophysical Research-Biogeosciences 114: G00F06.

Wallin, M., I. Buffam, M. Oquist, H. Laudon, and K. Bishop. 2010. Temporal and spatial variability of dissolved inorganic carbon in a boreal stream network: Concentrations and downstream fluxes. Journal of Geophysical Research-Biogeosciences 115: G02014.

Walvoord, M. A., and R. G. Striegl. 2007. Increased groundwater to stream discharge from permafrost thawing in the Yukon River basin: Potential impacts on lateral export of carbon and nitrogen. Geophysical Research Letters 34: L12402.

Watras, C. J., P. C. Hanson, T. L. Stacy, K. M. Morrison, J. Mather, Y.-H. Hu, and P. Milewski. 2011. A temperature compensation method for CDOM fluorescence sensors in freshwater. Limnology and Oceanography: Methods 9: 296–301.

Weishaar, J. L., G. R. Aiken, B. A. Bergamaschi, M. S. Fram, R. Fujii, and K. Mopper. 2003. Evaluation of specific ultraviolet absorbance as an indicator of the chemical composition and reactivity of dissolved organic carbon. Environmental Science & Technology 37: 4702–4708.

Wu, P., R. Wood, and P. Stott. 2005. Human influence on increasing Arctic river discharges. Geophysical Research Letters 32: L02703.

Yamamoto-Kawai, M., F. A. Mclaughlin, and E. C. Carmack. 2011. Effects of ocean acidification, warming and melting of sea ice on aragonite saturation of the Canada Basin surface water. Geophysical Research Letters 38: L03601.

Yamamoto-Kawai, M., F. A. Mclaughlin, E. C. Carmack, S. Nishino, and K. Shimada. 2009. Aragonite undersaturation in the Arctic Ocean: Effects of ocean acidification and sea ice melt. Science 326: 1098–1100.

Zhulidov, A. V., V. V. Khlobystov, R. D. Robarts, and D. F. Pavlov. 2000. Critical analysis of water quality monitoring in the Russian Federation and former Soviet Union. Canadian Journal of Fisheries and Aquatic Sciences 57: 1932–1939.

22

Geochemistry of the Congo River, estuary, and plume

R. G. M. Spencer, A. Stubbins, and J. Gaillardet

1. Introduction

The Congo is the largest river on the African continent and the second largest river in the World after the Amazon in terms of drainage basin size ($\sim 3.7 \times 10^6$ km^2) and water discharge ($\sim$45,000 m^3 s^{-1}; Runge 2007; Laraque et al. 2009; Chapter 3). The Congo watershed drains all of the Democratic Republic of Congo (formerly Zaire, Congo-Kinshasa) and regions of the Republic of Congo (Congo-Brazzaville), Angola, Central African Republic, Cameroon, Burundi, Rwanda, Tanzania, and Zambia. The Congo also drains the second largest area of tropical rainforest in the World ($\sim$18% of the total global tropical rainforest), and much of this rainforest, particularly within the borders of the Democratic Republic of Congo (DR Congo) remains in a pristine state (Laporte et al. 2007). Despite its obvious importance in regional and global hydrologic and biogeochemical cycles, the Congo has received limited study in comparison with other major global river systems.

The region has had a turbulent history, from initial contact with Europeans and proliferation of the slave trade, to exploration such as by Henry Morton Stanley, through to the Congo Free State run by King Leopold II that provides the backdrop to the haunting narrative of *Heart of Darkness* by Joseph Conrad. In modern times, the region has suffered from political instability, civil unrest, and failing states and has been the scene of brutal conflict resulting in the deaths of millions (Prunier 2008). However, the future is beginning to look hopeful as peace and stability return to the region, enabling trade and investment and also allowing for scientific research to be undertaken. Anthropogenic impacts such as industrial logging, conversion of pristine forest to agricultural lands, and settlement expansion are now growing threats within the Congo Basin (Laporte et al. 2007; Koenig 2008), and studies are urgently required to assess the effects of these increasing perturbations on hydrologic and biogeochemical cycles.

2. Hydrology of the Congo River

2.1. The Congo from source to sea

The complex sources of the Congo River consist of several small headwaters, swamps, and lakes that feed into the Lualaba River located in the southeastern part of the DR Congo, near the border with Zambia (Maurice 1946; Runge 2007). The Lualaba flows in a generally northern direction and

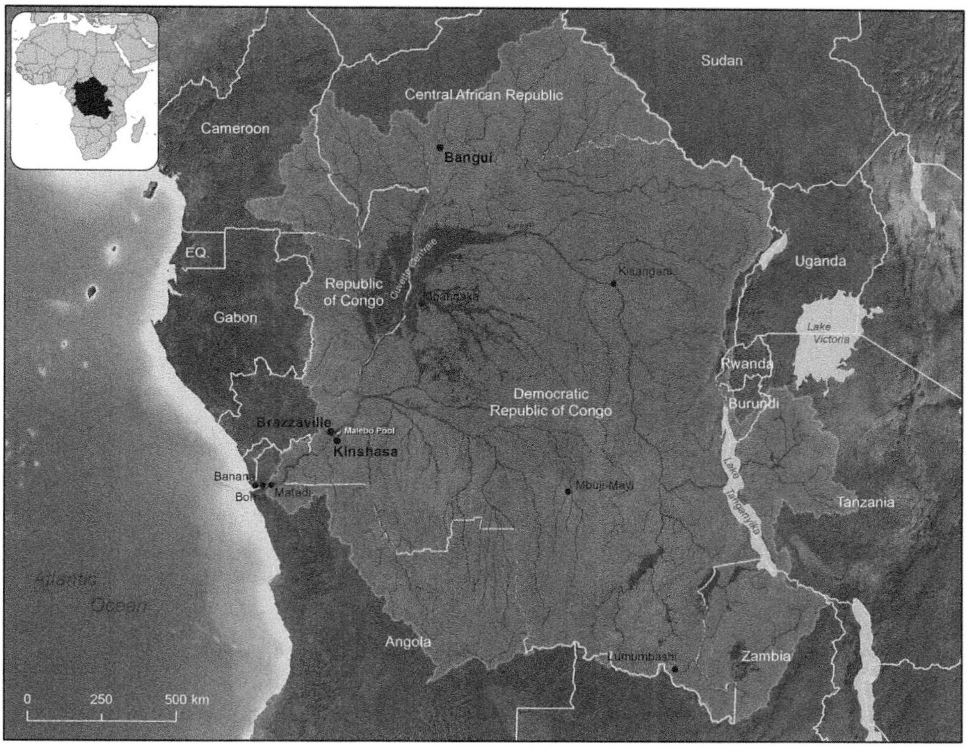

Figure 22.1. Map of the Congo watershed.

receives numerous tributaries that enter predominantly from the east, draining the western slopes of the Central African Rift valley. Near Kisangani where the Lualaba flows over the Wagenia Falls (Boyoma Falls, formerly Stanley Falls), it officially becomes known as the Congo. As the Congo flows on toward Mbandaka, it turns southwestward and passes through the World's largest swamp forest found in the central depression (Cuvette Congolaise or Cuvette Centrale), and the major tributaries of the Ubangi (Oubangui), Sangha, and the Kasai join prior to the Congo entering the Malebo Pool (Fig. 22.1). At this major widening of the Congo River, the cities of Kinshasa (population ~10 million, capital of DR Congo) and Brazzaville (population ~1.5 million, capital of Republic of Congo) lie across from one another at the western edge of the Malebo Pool. The Congo River then drops ~270 m through a series of falls and rapids (Livingstone Falls) over 350 km and narrows considerably down to the port of Matadi (Fig. 22.1). The lower Congo from the western edge of the Malebo Pool down to the Atlantic descends through 66 falls and rapids (Maurice 1946; Runge 2007), with a gradient typical of a mountain headwater stream (Jackson et al. 2009). The steep slope and large discharge has led to the development of some of the Earth's greatest cataracts, falls, plunge pools, and deepest river depths, with measurements potentially in excess of 200 m (Oberg et al. 2008; Jackson et al. 2009). In this reach of the Congo, high water velocity zones have also been recorded with depth-averaged flow velocities >4 m s^{-1} (Oberg et al. 2008; Jackson et al. 2009). Downstream, the

Congo widens again (up to 19 km), with abundant large sand bars, and at Boma (~100 km from the mouth; Fig. 22.1), the Congo becomes tidally influenced (Maurice 1946; Runge 2007).

2.2. Congo discharge overview

Hydrology data for the Congo watershed from the colonial era are quite abundant and detailed; however, after independence many gauging stations were abandoned, particularly those in rural areas (Runge 2007). A number of sites were maintained through government organizations (e.g., Regie de Voies Fluviales, D.R. Congo; Direction de la Meteorologie, Central African Republic) and through research institutions such as ORSTOM/IRD (http://www.ore-hybam.org/), so some long-term records exist (Olivry and Boulegue 1993; Laraque et al. 2001). Two flow regimes can be identified for the Congo Basin depending on the site location. First, tributaries that begin in the semi-humid regions that have distinct wet and dry seasons show a single-peak annual hydrograph. For example, tributaries north of the Equator such as the Ubangi River show a clear maximum in discharge from September to November (Bouillon et al. 2012). Southern tributaries below 5°S such as the Congo-Lualaba, Luvua, and Luapula Rivers also show a distinct peak in discharge but from March to May (Runge et al. 2007). Second, some rivers show a double discharge peak including the Congo near its mouth at Kinshasa-Brazzaville, which exhibits a bimodal hydrological cycle with maximum flows in November–December and May and minimum flows in August and March (Fig. 22.2). The greater discharge maximum in November–December relates to the period of increased discharge from the northern tributaries (e.g., the Ubangi) and is complemented by the southern tributaries, whose water discharges start to increase at about the same time. The smaller discharge maximum in May is due to an increase from the southern part of the Basin from savannah draining rivers (Coynel et al. 2005).

The Congo River discharge near its mouth at Kinshasa-Brazzaville exhibits an exceptionally stable intraannual regime (MaxQ/MinQ = 1.94, 1977–2006 inclusive; Spencer et al. 2012) as it straddles the Equator and thus receives rainfall in at least one part of its Basin throughout the year. As well as the discharge from northern and southern sub-watersheds, regular and high rainfall (1,800–2,400 mm yr^{-1}) falls on the central Basin, dampening extremes in flow (Maurice 1946; Runge et al. 2007). The Congo at Kinshasa-Brazzaville also has a very stable discharge regime from year to year, with an interannual ratio of 1.62 from 1977 to 2006 inclusive (Spencer et al. 2012). The monthly mean discharge at Kinshasa from 1977 to 2006 ranged from 30,060–54,538 $m^3\ s^{-1}$ (monthly median = 29,648–53,196 $m^3\ s^{-1}$; maximum = 35,488–63,516 $m^3\ s^{-1}$; minimum = 24,420–43,101 $m^3\ s^{-1}$; Fig. 22.2; Spencer et al. 2012). This is highly comparable to long-term hydrology data at the same location, highlighting the Congo as likely the most regular and uniform major river on Earth (Laraque et al. 2001; Coynel et al. 2005; Runge et al. 2007).

3. Inorganic and sediment geochemistry of the Congo River and tributaries

3.1. The inorganic poor waters of the Congo River system

The predominant feature with respect to the inorganic geochemistry of the Congo River system is that the dissolved load is extremely dilute. Amongst the main tributaries of the Congo River, the

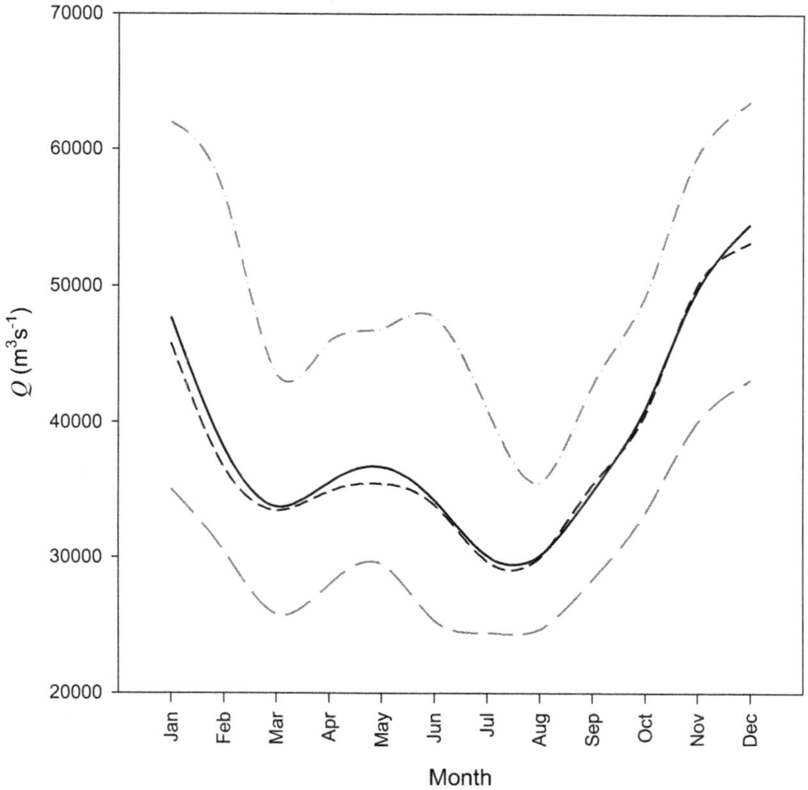

Figure 22.2 Monthly discharge data from the Congo River at Kinshasa-Brazzaville (30-yr average, 1977–2006); mean (solid black line); median (dashed black line); maximum (dash-dot gray line), and minimum (long dash gray line). Reproduced from Spencer et al. (2012).

total dissolved solutes (TDS) including dissolved silica (Si), major cations, and anions (Na^+, K^+, Ca^{2+}, Mg^{2+}, HCO_3^-, Cl^-, SO_4^{2-}) ranges from 11 mg L^{-1} for the Tékés to 44 mg L^{-1} for the Congo mainstem station before confluence with Ubangi. At Kinshasa-Brazzaville, TDS is 37 mg L^{-1}. These numbers were published for the year 1993 by Laraque et al. (2009), this year differing by less that 3% from interannual means for 1987–1994. Compared with the other large rivers of the world, these values are amongst the lowest, comparable to those found in the lowlands of the Amazon and Orinoco river systems and about four times less than the TDS median value deduced from the 60 largest rivers (150 mg L^{-1}; Gaillardet et al. 1999a). The low inorganic concentrations of the waters from the Congo Basin are due to high precipitation rates, the weakness of chemical weathering rates, and the dominant crystalline nature of the bedrock. Although dilute at a global scale, the rivers of the Congo Basin are relatively enriched in dissolved Si, trace elements, and dissolved organic carbon (DOC). The concentration of silicic acid (H_4SiO_4) for the Congo River at Kinshasa-Brazzaville is about 10 mg L^{-1}, a value higher than the global median value of 8 mg L^{-1} (Gaillardet et al. 1999a). The mass fraction of Si/TDS (not including DOC) ranges from 30% to 50%, highly comparable to the lowland rivers of the Amazon and Orinoco Basins (Edmond et al. 1995).

Congo tributaries have been separated historically into three main groups based on their solute load, pH, organic matter (OM) content, and dissolved Si content (Dupré et al. 1996; Négrel et al. 1993; Laraque et al. 1993). Aside from the special case of the Tékés Rivers (draining cation-depleted sandstone tablelands), the lowest TDS concentrations (<20 mg L^{-1}) in the Congo Basin are found in the Cuvette Centrale, a peculiar physiographic entity characterized by very flat relief, high rainfall, and dense, periodically flooded rainforest. This area has high OM export, very low sediment production, and a weathering regime dominated by OM complexation. River pH within the central zone of the Cuvette Centrale is <5, and more than half of the dissolved load is in the form of OM (e.g., Likouala aux Herbes River). In the outer part of the Cuvette Centrale, rivers have typically <50% of their dissolved load as OM, pH is >5, and they are less inorganic poor (e.g., Likouala Mossaka). In all OM-rich rivers of the Cuvette Centrale, dissolved Si contributes up to 50% of the total dissolved inorganic load. A final group of rivers is found outside of the Cuvette Centrale. They drain the vast majority of the Congo Basin, are the least inorganic poor in the Congo River system, and exhibit quasi-neutral pH values. The chemistry of the Congo headwaters is poorly known. In the east, influenced by the volcanic nature of the Tanganiyka-Kivu rift regions, local TDS values can be relatively elevated, in particular due to thermal spring inputs (Elsgaar and Prieur 2011).

3.2. Major dissolved solutes: chemical neutralization of soil CO_2

Riverine geochemistry reflects the neutralization of atmospheric and soil CO_2 by weathering reactions and the atmospheric input of rainwater to continental surfaces. In the Congo River and its main tributaries, the anions (Cl^-, SO_4^{2-}, and HCO_3^-) are clearly dominated by bicarbonate ions (HCO_3^-). In 1989, the HCO_3^- concentration (measured using a Gran titration) at Kinshasa-Brazzaville ranged from 9.1–21.4 mg L^{-1} (150–350 μmol L^{-1}) between the wet and dry water stages, respectively (Négrel et al. 1993; Chapter 7) and represented a fraction (in eq:eq) between 75% and 80% of the total anionic load. These anions are mostly compensated by Ca^{2+} Mg^{2+}, K^+, and Na^+ cations. The Ca/Na ratio ranged between 0.5 and 1, and the Ca/Mg ratio ranged (in 1989) between 0.7 and 1, typical of rivers influenced by the neutralization of silicate mineral by carbonic acid. These numbers are in agreement with those from temporal series data (1987–1989; Probst et al. 1992). The absence of a significant amount of nitrate detected during that period highlights little impact of anthropogenic activities in the Congo Basin.

Determining the source of dissolved cations in the Congo River and its major tributaries was undertaken utilizing an extensive database. Samples encompassed the largest tributaries, a series of rivers draining monolithological formations in the upper part of the Ubangi basin and between Kinshasa-Brazzaville and the Atlantic Ocean, as well as data from rainwaters collected in the central part of the Congo Basin (Négrel et al. 1993; Négrel and Dupré 1995a, 1995b). Major element concentration ratio and strontium (Sr) isotopes were used to show that the dissolved load of the Congo and its main tributaries results from the conservative mixing of waters having interacted with silicate and carbonate lithologies. The inverse mixing relationship between Sr isotopes and Sr/Na ratio (Fig. 22.3) shows that the Congo and its main tributaries fall on mixing curves between

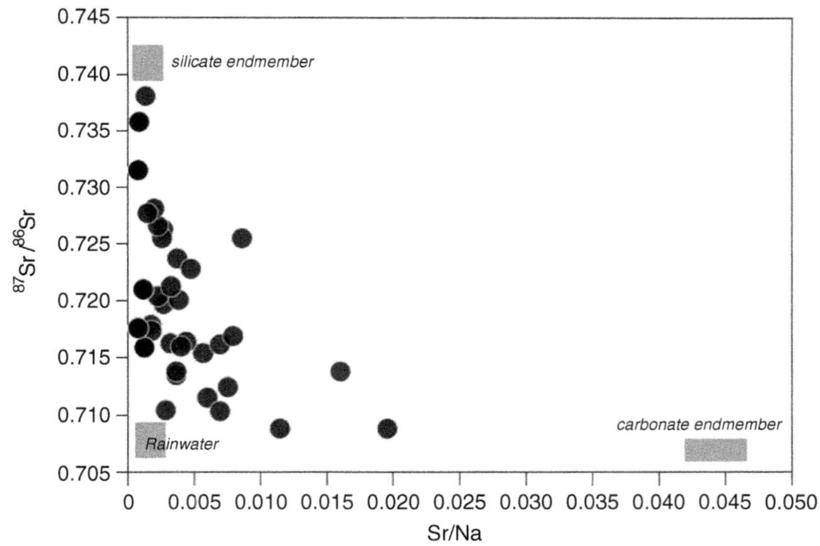

Figure 22.3. Mixing diagram using Sr isotopes to show that the inorganic solutes of the Congo River and its tributaries are derived from the mixing between waters interacting with carbonate and silicate minerals. The input of rainwater is also significant. Mixing curves are hyperbola in such a diagram. Reproduced from Negrel et al. (1993).

carbonate and silicate-draining water end-members. Chemical composition of the end-members was constrained from monolithological streams and rainwater, and the proportion of each end-member was determined using an elegant inverse modeling approach incorporating chemical (Ca/Na, Mg/Na, Sr/Na) and Sr isotopic ratios (Négrel et al. 1993). The results of the inversion of the Congo tributaries (Négrel and Dupré. 1995a), of the Ubangi at Bangui (Négrel and Dupré 1995b), and of the Congo at Kinshasa-Brazzaville (Négrel et al. 1993) show that rainwater input explains most of the Cl^- and SO_4^{2-} ions in solution, and 20% of the Na^+, Ca^{2+}, and Mg^{2+} ions. These proportions increase in the diluted rivers from the Cuvette Centrale, where they reach typical values of 40% for the same elements. It is important to note that the composition of rainwaters in the Congo Basin are significantly different from dilute seawater, and thus a significant amount of terrigenous particles contribute via dissolution to rainwater in addition to Atlantic sea salt. Whether this input is derived from the Congo Basin itself or from outside is still not clear, but should merit more attention given the major role of the atmospheric input to the waters of the Congo Basin (Dupré et al. 1994).

The inversion also shows that once corrected for atmospheric input, dissolved Na^+ is essentially derived from silicate weathering, whereas Ca^{2+} and Mg^{2+} are derived from both carbonate and silicate weathering. Proportions of Ca^{2+} and Mg^{2+} ions derived from carbonate at Kinshasa-Brazzaville were calculated by Négrel et al. (1995b) and vary between 30% and 56%, depending on water discharge. The presence of carbonate rocks in the Congo Basin, detected in the chemistry of the Congo River and its main tributaries, is supported by geological survey and remote sensing evidence that identified karst structures (hidden by the deep lateritic soils) (Alvarez 1997). More unexpected is the existence, deduced from the inversion procedure of a significant proportion of Na^+, Cl^-, and Mg^{2+} derived from saline rock formations. This is particularly true for the Congo River, where the dissolved Cl^- concentration is higher than the maximum riverine Cl^- concentration possible derived from atmospheric inputs and estimated based on watersheds exclusively covered by crystalline rocks (a value for the Congo Basin fixed at 0.88 mg L^{-1}, Négrel et al. 1993). The proportions of cations derived from the inversion can also be used to estimate the proportion of bicarbonate ions transported into the Congo River system that are derived from silicate weathering as opposed to carbonate weathering. Estimates suggest that 30% of HCO_3^- is derived from carbonate weathering and 66% from silicate weathering, giving a concentration of 5.5 mg L^{-1} (90 μmol L^{-1}) and 10.4 mg L^{-1} (170 μmol L^{-1}) for HCO_3^-, for carbonate and silicate weathering, respectively.

3.3. In-situ biogeochemical cycling of silica

Recent data on the abundance of Si isotopes (δ^{30}Si) in the river waters from the Congo at Kinshasa-Brazzaville (Hugues et al. 2010), across the hydrological cycle, showed that δ^{30}Si follows a seasonal pattern, with higher values at low water stages. These variations are consistent with the biological uptake of Si (that fractionates ^{30}Si and ^{28}Si isotopes). The good correlation between dissolved Si isotopic ratio and the biogenic concentration of Si in suspension (diatoms, phytoliths, and sponge spicule) provides additional support to biological uptake. The presence of diatoms in the total suspended solids (TSS) of the Congo River has been observed (Gasse et al. 1989). Hugues et al. (2010) estimated that 0–60% of the total Si flux of the Congo River is transported in the form of particulate biogenic Si depending on the season. The average value for 2006 to 2008 is 17%. Depending on the fate of diatoms (i.e., transported or deposited), the dissolved Si flux from the Congo to the Atlantic Ocean may be underestimated.

3.4. Dissolved trace elements: the organic footprint

Trace elements in the Congo River and its main tributaries were determined on the same sampling cruise in 1989 as described by Négrel et al. (1993) and were published by Dupré et al. (1996). Seyler and Elbaz-Poulichet (1993) also report trace element concentrations in the Ubangi River. Trace elements whose concentrations are correlated with major elements (Rb, Ba, Sr, U) have low concentrations compared with other regions of the world. Conversely, trace elements not correlated with major solutes (so-called insoluble elements) have concentrations that are amongst the highest globally (Gaillardet et al. 2007). For example, the concentration of neodymium (Nd) ranges from 0.3–0.8 mg L^{-1}, whereas the world average value in river water is 0.15 mg L^{-1} (Gaillardet et al. 2007). In the Congo River system, Dupré et al. (1996) reported an inverse correlation between rare Earth element's (REEs), Th, Al, Sc, Cr, Co, Hf, Cs, Fe, and Al concentrations with pH. The acidic blackwater rivers from the Cuvette Centrale clearly display the highest concentrations of insoluble elements. Dissolved organic carbon (DOC) concentrations were not determined on these samples, but based on studies by Viers et al. (1997) and Dupré et al. (1999) in a drainage basin in Cameroon having similar trace concentration and pH values, the negative correlation between trace element concentration and pH corresponds to a positive correlation between dissolved organic matter (DOM) and trace elements. In the Nyong Basin, Viers et al. (1997) and Dupré et al. (1999) performed ultrafiltration experiments (300 kD, 5 kD, 1 kD) that clearly demonstrated the close association of the most insoluble elements with DOM. Ultrafiltration experiments show that these trace elements are associated with organic colloids (defined as passing though 0.2 μm filters). In the Congo River, the apparent solubility of REEs, Al, and Fe is thus increased by the existence of organic molecules complexing them. By experimentally determining the partitioning of elements between the different size fractions, Dupré et al. (1999) showed that the association constant between metals and organic moieties is linearly correlated with the first hydrolysis constant of the element. The most complexed elements are Al, REEs, Th, and U. The metals (Co, Cr, Ni) have intermediate behavior, whereas dissolved Si is shown not to be associated with DOM. Seyler and Elbaz-Poulichet (1996) showed an increase in Ni, U, V, and Sb concentrations with water discharge, supporting their close association with OM. In the Congo River and its main tributaries, the insoluble element concentrations are thus good proxies for the presence of OM. The high affinity of DOM with Al shows the efficiency of the weathering environment of the Cuvette Centrale to dissolve primary or secondary alumino-silicate.

3.5. Inorganic geochemistry temporal variability

The temporal variability of Congo River inorganic geochemistry has been documented by Probst et al. (1992) and Laraque et al. (2009) based on times series from 1987 to 1989. The inorganic chemistry of the Congo River and its tributaries shows seasonal and interannual variations, mainly driven by variations in water discharge. The TDS concentration varied between 25–52 mg L^{-1} between the rainy and dry periods. The concentration-discharge relationships show that concentrations are not simply diluted when water discharge increases (Probst et al. 1992; Négrel and Dupré 1995a, 1995b). The HCO_3^- concentrations also vary by more than a factor of two in the Ubangi and Congo Rivers (Probst et al. 1994). With respect to Sr isotopes, important variations in Sr isotopic ratio ($^{87}Sr/^{86}Sr$) over the hydrological cycle at Bangui and Kinshasa-Brazzaville were reported by Négrel et al. (1995a, 1995b). A pure dilution would imply the constancy of Sr isotopic ratio with time. At Bangui, Sr isotopic variations indicate a mixing between at least two pools varying in contribution as a function of the hydrological cycle. Thus, at Kinshasa-Brazzaville, the Sr isotopic ratios reflect the hydrological contribution of the different tributaries across the hydrological cycle.

The interannual variability of the Congo River and its tributaries has received little attention to date. From 1987 to 1989, there was no significant change in the inorganic chemistry of the Congo (Probst et al. 1992). However, on a longer term, some variation may be expected given the hydrological variability of the Congo River (Laraque et al. 2001). The 1987–1989 period was during a "dry phase" of the Congo Basin and is therefore not representative of the 1900–2000 period, characterized by a stable phase (1900–1960), a wet phase (1960–1970), and a dry phase after 1980. The observed water discharge variations on the order of 15% around the mean average value imply that dissolved solute concentrations should have varied by about the same factor, being lower in the wet period and higher in the dry period.

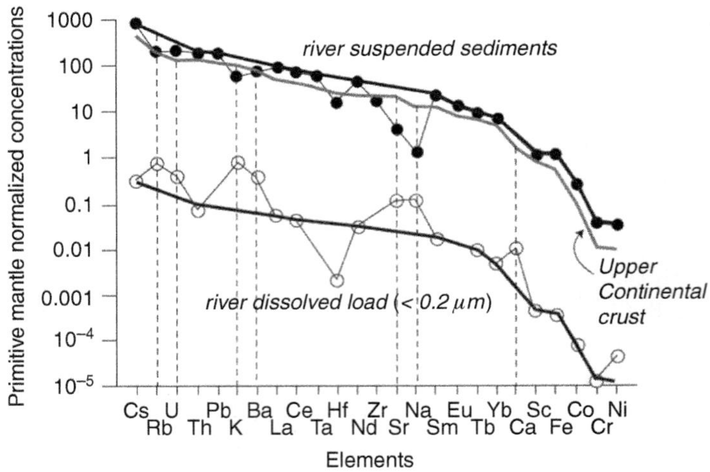

Figure 22.4. Diagram showing the normalized abundance of elements in the suspended material and dissolved load of the Congo River. Element concentrations are normalized to primitive mantle concentrations. Elements are ranked according to their decreasing abundance in the Upper Continental Crust from Taylor and MacLennan (1985). Reproduced from Gaillardet et al. (1995).

3.6. Inorganic geochemistry of the particulate load

The particulate load of the Congo and its tributaries consists of TSS and bedload. The amount of TSS in the Congo and its tributaries is low (10–30 mg L^{-1}, Coynel et al. 2005; Laraque et al. 2009) and constitutes 60–80% kaolinite, 10–15% illite, and 20% of oxyhydroxides and OM (Delaune et al. 1995). This mineral association reflects soil composition and the extremely severe weathering conditions prevailing in the Congo Basin.

The TSS inorganic geochemistry for the Congo and its main tributaries has been investigated by Dupré et al. (1996) and Allègre et al. (1996). Sr and Nd isotopic ratios have been used as provenance indicators due to the absence of fractionation during chemical weathering and mechanical transport. The Sr ($^{87}Sr/^{86}Sr = 0.723$–0.746), Nd ($\varepsilon Nd = -15‰$ to $-18‰$), and Pb ($^{206}Pb/^{204}Pb = 18.1$–19.7) isotopic ratios in TSS have typical crustal values consistent with the geology of the Congo shield and highlight that the sedimentary input of the eastern volcanic region is not apparent in the TSS load. The Nd model age, reflecting the mean extraction age of Congo bedrock material from the mantle, is typically 1.7–1.8 Ga (Allègre et al. 1996).

The major and trace element composition of Congo River TSS is shown in Fig. 22.4. Abundances have been compared to the upper continental crust from Taylor and MacLennan (1985) and normalized to the primitive mantle of the Earth. Three groups of elements can be depicted. The first corresponds to elements showing continental-like patterns. These elements are the most insoluble elements, mainly transported in solid form in the Congo Basin (except in the blackwaters of the Cuvette Centrale due to the absence of TSS and organic complexation). As shown for example by Dupré et al. (1996), 10–20% of total Cs, Th, REEs, and transition metals are transported as TSS in the largest rivers. Absolute concentrations for these elements, despite possible dilution by particulate organic matter (POM), are enriched in the TSS load compared with the Upper Continental Crust. The second group corresponds to Hf and Zr that show an important depletion in the TSS of the Congo rivers. Their enrichment in the bottom sands explains these negative depleted patterns, as Zr and Hf from the initial bedrocks are preferentially incorporated in zircons. The mineralogy of the bedload in the Ubangi River (Censier 1995), confirms the presence of quartz and heavy minerals such as zircon, disthene, epidote, and rutile in the bedload of the Congo. Finally, a group of elements is considerably depleted in the TSS load (Rb, U, K, Ba, Sr, Na, Ca). For example, Na in the TSS of the Congo River at Kinshasa-Brazzaville is over an order of magnitude lower than in the Upper Continental Crust. Elements of the third group are the soluble elements. Their strong depletion in TSS reflects the intensity of chemical weathering and is best explained by the severity of weathering conditions in the Congo Basin. If we compare the Congo with the other large river systems (Gaillardet et al. 1999b), it appears that the TSS of the Congo is the most depleted fluvial material on Earth, making the Congo Basin one of the most deeply weathering environments on the planet. The thickness of soils, optimum climatic conditions, and high biological productivity are probably the principal factors explaining why chemical denudation is so intense in the Congo Basin. In rivers from the Cuvette Centrale, little TSS is exported, in agreement with the Al complexation observed with OM.

3.7. Chemical and physical denudation rates of the Congo Basin

Denudation rates for the Congo Basin have been calculated based on river chemistry by a number of authors based on data collected in different time periods. The different estimates have been synthesized in Table 22.1. Chemical denudation rates are calculated based on solute concentrations corrected for atmospheric inputs. As this contribution is important in the Congo Basin, it is critical to estimate proper numbers of chemical denudation, and significant differences of weathering rates are found depending on whether sea-salt or rainwater corrections are applied. Based on the atmospheric correction (rainwater) applied by Négrel et al. (1993), Gaillardet et al. (1995) estimated a chemical denudation rate of 7.4 t km^2 yr^{-1}. Interannual data (1987–1989) from Probst et al. (1992), treated using the same correction for atmospheric inputs, gives 5.5 t km^2 yr^{-1}, an estimate lower than the oldest estimate by Nkounkou and Probst (1987) of 6.5 t km^2 yr^{-1}. Because the atmospheric correction performed by Négrel et al. (1993) uses the mean chemical composition of rainwater and the dissolved elements in rainwater could come from the Congo Basin itself, the weathering rates given previously are likely low estimates (by 20%). The CO_2 fluxes consumed by rock weathering are given in Table 22.1 and take into account that half of dissolved HCO_3^- derived from carbonate weathering reactions is derived from the bedrock and not from the atmosphere. The specific rate of CO_2 consumption in the Congo Basin is about 600 kg-C km^2 yr^{-1}, 60% of which is due to silicate weathering reactions according to Gaillardet et al. 1995.

The annual corresponding denudation and C consumption fluxes are given in Table 22.1. All estimates agree rather well and lead to weathering fluxes between 20 and 26 Mt yr^{-1} of solutes and 1.8–2.2 Tg-C yr^{-1}. These numbers are low at the global scale. According to the global estimates of Gaillardet et al. (1999a), the oceanic input of dissolved products of weathering due to the Congo River is only 2%. The consumption of atmospheric CO_2 by weathering reactions in the Congo River system accounts for 1% of the global flux. The relatively weak impact of the Congo River system

Table 22.1. *Erosion rates, chemical weathering and fluxes from the Congo Basin*

Study	TDS (mg L^{-1})	TDS flux (Tg yr^{-1})	DOC flux (Tg yr^{-1})	Chemical denudation rate (t km^2 yr^{-1})	CO_2 consumption flux (Tg yr^{-1})	CO_2 consumption flux – silicate weathering (Tg yr^{-1})	TSS (mg L^{-1})	TSS flux (Tg yr^{-1})	POC flux (Tg yr^{-1})	Physical erosion rate (t km^2 yr^{-1})	Chemical and physical erosion (mm kyr^{-1})
Probst et al. 1992	34*	42*	–	–	–	–	–	–	–	–	–
Probst et al. 1994	–	–	8.9	–	2.3	1.7	–	–	1.1	–	–
Gaillardet et al. 1995	21	25	–	7.4	2.1	1.3	21	25	–	8.1	5.7
Coynel et al. 2005	–	–	12.4	–	–	–	26.3	30.7	2.0	8.8	–
Laraque et al. 2009	37*	43*	–	12*	–	–	25	29	–	8.2	–

* Not corrected from atmospheric inputs; – no data.

on global CO_2 consumption by chemical weathering, despite optimum climatic conditions and the dense forest ecosystem, was pointed out by Edmond et al. (1994) as an argument to show the absence of temperature feedback on global weathering.

Mechanical erosion in the Congo Basin is quantified using a time series of TSS concentrations, and data from 1993 indicate average concentrations of TSS of 25 mg L^{-1} at Kinshasa-Brazzaville (Laraque et al. 2009). As a consequence, annual TSS yields and specific erosion fluxes are 8.2 t km^2 yr^{-1} and 29 Mt yr^{-1} respectively, and in good agreement with previous estimates (Laraque et al. 1995; Coynel et al. 2005). The transport of bottom sands is not included in these estimates, and it is usually thought that sands in large river systems do not exceed 10% of the total particulate flux. No estimate of the fraction of bedload transport has however been proposed for the Congo River. These estimates of physical erosion rates are in agreement with those determined using a geochemical mass budget model by Gaillardet et al. 1995. This model assumes that the TSS load and dissolved load (derived from silicate weathering) are geochemical complementary reservoirs with respect to the initial bedrock. Assuming that the representative bedrock in the Congo Basin is close to the UCC model of Taylor and Maclennan (1985), and a steady state of erosion (the soil keeps a constant thickness), the authors calculated that the TSS concentrations required to close the mass budget equation are in good agreement with the measured ones. Within error, this result shows that in the Congo Basin, the amount of soil formed each year by weathering reactions is equal to the amount of soil being eroded and transported to the Atlantic. Overall, the rates of mechanical erosion found in the Congo Basin are amongst the lowest found in the world. The flux of TSS delivered to the ocean by the Congo River is only 1% of the global input.

Finally, the chemical and physical denudation rates can be summed up to give the total denudation rate of the Congo River basin and translated in mm yr^{-1} using a mean density of 2.7. The Congo River system has typical denudation rates of 5–6 mm kyr^{-1}, amongst the lowest found at a worldwide scale (Berner and Berner 2012).

3.8. Weathering regimes

The inorganic chemistry of the Congo River and its tributaries is produced by a dominant weathering regime characterized by chemical reactions favoring intensive leaching of cations and Si. The residues of weathering, mostly kaolinite, gibbsite, and Fe oxides accumulate in deep soil profiles. This intense weathering regime results from optimum climatic conditions (warm and wet climate), the dense rainforest ecosystem, and low relief precluding the destruction of soils by erosive processes. With a mean denudation rate of 5–6 mm kyr^{-1} and a typical soil thickness of 15 m, the residence time of soil material is typically 3 Myr, sufficiently high to allow silicate minerals to weather completely to kaolinite. In the Cuvette Centrale, the abundance of DOM favors weathering conditions dominated by Al complexation. Recently reported data on Si isotopic ratio in the Congo River and main tributaries (Cardinal et al. 2010) is entirely consistent with this view. Isotopic signatures of water in equilibrium with kaolinite ($\delta^{30}Si$ = 1–1.2 ‰) are found in the major tributaries, whereas the blackwaters from the Centrale Cuvette have much lower values, consistent with dissolution of clay minerals or bedrock mineral ($\delta^{30}Si$ = –0.2 ‰). The lithospheric subsidence is probably responsible for the topographic depression and flooding of the Centrale Cuvette (Forte et al. 2010).

By contrast to this intense chemical weathering regime, the fluxes of dissolved material transported out and the consumption of CO_2 by weathering reactions in the Congo Basin are amongst the lowest at the global scale. Such conditions are encountered in the Congo and other parts of the world, where tectonic and climatic stability have led to the development of thick soils that ultimately protect the bedrock from chemical weathering (Godderis et al. 2008).

4. Carbon fluxes and organic matter geochemistry of the Congo River and tributaries

4.1. Flux of organic carbon via major tributaries

Of all the tributaries of the Congo, the Ubangi is by far the best studied in terms of organic carbon (OC) flux, predominantly at a site at Bangui (Central African Republic; Coynel et al. 2005; Laraque et al. 2009; Bouillon et al. 2012). The Ubangi is the second largest tributary of the Congo after the Kasai River, draining an area of 644,000 km^2, of which 489,000 km^2 (76%) is upstream of Bangui and drains predominantly dry and wooded savannah (Coynel et al. 2005; Bouillon et al. 2012). Between the well-studied site at Bangui and the Ubangi's confluence with the Congo mainstem, the tributary passes through the Cuvette Centrale, and its characteristics are known to change as TSS export rates decrease slightly and DOM concentrations increase dramatically (Laraque et al. 2009). Thus it seems likely that flux estimates for OC, especially DOC, derived from the study site at Bangui are a significant underestimate with respect to contribution to the Congo mainstem.

The two studies that have estimated particulate organic carbon (POC) flux from the Ubangi River at Bangui both derived a consistent flux of around 0.14–0.15 Tg-C yr^{-1} (Coynel et al. 2005; Bouillon et al. 2012). DOC flux from this site between the two studies showed more variability, with Coynel et al. (2005) estimating 0.54 Tg-C yr^{-1} and Bouillon et al. (2012) estimating 0.70 Tg-C yr^{-1}. Although the two studies took place at different times (1990–1996 vs. 2010–2011), the limited interannual variation in the Ubangi and the similarity in total discharge between the two periods (within $\sim$10%) means that hydrological differences alone are unlikely to explain the reported difference in DOC flux. Bouillon et al. (2012) suggested that their higher flux estimate is likely due to their higher temporal sampling resolution (28 samples during one year vs. 12 samples or less in one year), which results in a more representative capture of the distinct peak in discharge with elevated DOC concentrations. Aside from the Ubangi, few direct flux estimates exist for OC from Congo tributaries. Coynel et al. (2005) examined two relatively small tributaries draining distinct land cover types (the Mpoko River, 23,900 km^2, 86% savannah cover; and the Ngoko River, 67,000 km^2, 95% humid evergreen forest cover). The Mpoko was estimated to have a POC and DOC flux of 0.01 and 0.02 Tg-C yr^{-1}, respectively, and the Ngoko was estimated to have a POC and DOC flux of 0.06 and 0.26 Tg-C yr^{-1}, respectively (Coynel et al. 2005). The OC yield for the forested Ngoko watershed was therefore much higher than for the savannah Mpoko watershed, with a POC and DOC yield of 0.9 and 3.9 g-C m^2 yr^{-1} versus 0.3 and 0.9 g-C m^2 yr^{-1}, respectively (Coynel et al. 2005).

Laraque et al. (2009) examined DOM flux from a number of major tributaries utilizing a modeling approach and determined DOM as the difference between the weight of dry residue at 105°C and the weight of the TDS. Determination of DOM via this methodology is usually thought to result in a value approximately double that of measured DOC. Through this approach, Laraque et al. (2009)

calculated DOM fluxes at the outlet of a number of major tributaries, including the Ubangi (1.77 Tg-C yr^{-1}), Sangha (1.22 Tg-C yr^{-1}), the right bank (northern) tributaries (4.32 Tg-C yr^{-1}), and the Kasai (4.26 Tg-C yr^{-1}). These fluxes correspond to 11.4% for the Ubangi, 7.8% for the Sangha, 27.8% for the right bank tributaries, and 27.4% for the Kasai of the total DOM export from the Congo at Kinshasa-Brazzaville. Laraque et al. (2009) also provided TSS fluxes via the same approach and an estimate for the Ubangi of 2.72 Tg yr^{-1}. If we assume a %POC of 4–7, which is typical of the Congo and its tributaries (Mariotti et al. 1991; Coynel et al. 2005; Spencer et al. 2012), it results in a POC flux of 0.11–0.19 Tg-C yr^{-1} for the Ubangi, which is highly comparable to the estimate of 0.14–0.15 Tg-C yr^{-1} in Coynel et al. (2005) and Bouillon et al. (2012). Therefore, if we assume a %POC of 4–7 and apply it to the TSS flux data for the other tributaries in Laraque et al. (2009), this provides an estimated flux of 0.07–0.11, 0.20–0.34, and 0.32–0.56 Tg-C yr^{-1} for the Sangha, right bank tributaries, and the Kasai respectively.

4.2. Export of organic carbon to the Atlantic Ocean

The Congo has a very low TSS yield (8.2–8.8 t km^2 yr^{-1}), orders of magnitude below rivers such as the Brahmaputra, and despite ranking second in the world with respect to discharge, the Congo is only ranked twelfth in the world with respect to annual TSS load. However, Congo TSS has been reported to have a high POC content, and therefore the Congo ranks fifth in terms of annual POC flux to the oceans at approximately 2 Tg-C yr^{-1} (Mariotti et al. 1991; Coynel et al. 2005; Spencer et al. 2012). The POC yield of the Congo (0.5 g-C m^2 yr^{-1}) is low in comparison with other major tropical rivers (e.g., Amazon and Orinoco are 1 and 1.5 g-C m^2 yr^{-1}, respectively), but is higher than the major Eurasian Arctic Rivers (e.g., Lena, Ob, and Yenesei range from 0.1–0.2 g-C m^2 yr^{-1}). With respect to DOC, the Congo River is estimated to export 12.4 Tg-C yr^{-1}, which for comparison is equivalent to the DOC loads of the three largest Arctic Rivers (Yenisey, Lena, and Ob) combined, or greater than six times that of the Mississippi (Coynel et al. 2005; Raymond et al. 2007; Chapter 21). Near the mouth of the Congo River at Kinshasa-Brazzaville, POC and DOC concentrations are principally connected to discharge (Section 2.2), and a clear double hysteresis pattern can be seen for DOC in Fig. 22.5. The DOC yield of the Congo is high for a major river, ranking third globally behind the Amazon and the Orinoco with a yield of 3.3 g-C m^2 yr^{-1}, which for comparison is five times greater than the DOC yield of the Mississippi River (Bianchi et al. 2004; Coynel et al. 2005). Combining the estimated DOC and POC loads, the Congo River exports 14.4 Tg-C yr^{-1} of OC to the Atlantic Ocean and is the second major exporter of terrestrial OC to the ocean after the Amazon. The Congo therefore exports about seven times more OC than inorganic carbon (Section 3.7; Table 22.1).

4.3. Organic matter geochemistry of Congo tributaries

The Ubangi at Bangui is the most studied of the Congo tributaries with respect to organic matter quality. The Ubangi River OC load is dominated by DOC (typically 78% DOC/TOC), and the %POC of TSS is high in comparison with other rivers, ranging from 4.8–25.7% (Meybeck 1982; Coynel et al. 2005; Bouillon et al. 2012). Ratios of POC to particulate nitrogen (PN) are low at low flow conditions

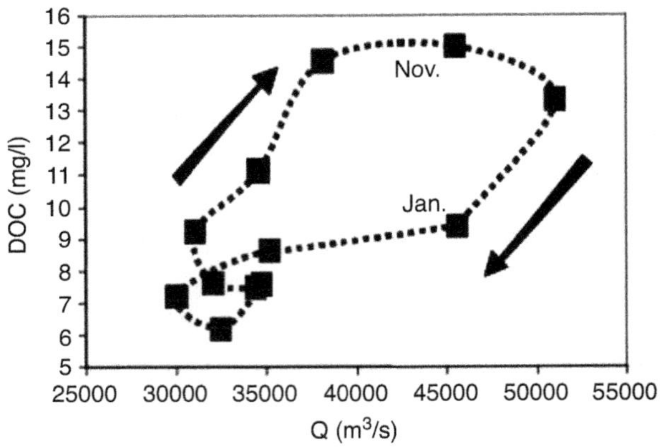

Figure 22.5. The relationship between monthly average DOC concentration and mean monthly water discharge for the Congo River at Kinshasa-Brazzaville. Reproduced from Coynel et al. (2005).

in the Ubangi (7.0±1.2), indicative of a phytoplankton source. At higher flow, POC:PN increases to 10.9±0.6, which in conjunction with a higher %POC at this time, suggests a shift toward vegetation inputs with increasing discharge (Bouillon et al. 2012). δ^{13}C-POC values for the Ubangi support this interpretation, as at low flow values of –28.6±1.3‰, show short-term variability that could reflect varying contributions of phytoplankton or changes in their δ^{13}C signatures due to variability in growth rate or δ^{13}C-DIC. Conversely, under high discharge conditions, δ^{13}C-POC values were –26.1±0.3‰, indicative of soil and litter-derived OM, and as the Ubangi drains a large area of savannah, it is unsurprising that the δ^{13}C-POC values are enriched compared with typical C_3 vegetation values (Bouillon et al. 2012; Chapter 8). δ^{13}C-DOC values for the Ubangi range between –31.8‰ and –27.1‰, with lowest values at peak flow and an apparent decoupling between POC and DOC pools related to discharge. In conjunction with the δ^{13}C-DOC data and utilizing dissolved lignin phenols as a tracer of vascular plant-derived OC, Bouillon et al. (2012) concluded that at least three different sources are required to explain the seasonal variability in DOM characteristics of the Ubangi River.

As well as providing a tracer of vascular plant-derived OC, lignin phenols also provide pertinent information that can be linked to source and biogeochemical processing (Hedges and Mann 1979; Spencer et al. 2009, 2010a; Chapter 21). Syringyl phenols (S) are solely found in angiosperms, and cinnamyl phenols (C) are exclusively found in non-woody tissue, so ratios of these phenols to ubiquitous vanillyl phenols (V) can differentiate OM sources from angiosperm and gymnosperm plant types (S:V) and non-woody and woody tissues (C:V; Hedges and Mann 1979). A number of studies have shown that increased biogeochemical processing of OM results in increased quantities of oxidized lignin phenols (Opsahl and Benner 1995; Spencer et al. 2009). This increase leads to a higher yield of acidic phenols during CuO oxidation, so, with certain caveats (Hernes et al. 2007), acid:aldehyde ratios (Ad:Al) of lignin phenols can be used to investigate the relative degree of degradation of OM. For the Ubangi River, lignin phenol composition was observed to change across the hydrograph with relatively low lignin carbon-normalized yields (Λ_8) at low discharge (0.25 ± 0.05 [mg{100 mg OC}$^{-1}$]) to higher Λ_8 at elevated flow (0.64 ± 0.11 [mg{100 mg OC}$^{-1}$]), reflecting a greater relative contribution of vascular plant-derived carbon at high flow (Bouillon et al. 2012). With respect to lignin compositional parameters, (Ad:Al)v and C:V values were observed to increase and decrease with increasing flow respectively, which is in stark contrast to data from a rainforest dominated watershed (Epulu River; north-east DR Congo), which showed very minimal seasonal variability in C:V and (Ad:Al)v (Spencer et al. 2010b; Bouillon et al. 2012). Epulu River Λ_8 values were also higher than Ubangi values, ranging from 0.78 to 1.47 (mg[100 mg OC]$^{-1}$), highlighting the greater vascular plant inputs in the Epulu watershed. The lignin compositional data for the rainforest dominated Epulu River clearly suggest a single source such as surface litter and organic-rich soil layers, whereas in the savannah-dominated Ubangi, the lignin signatures are consistent with seasonal drainage through deeper and more well-developed soils that contain gradients in OM concentration and composition (Spencer et al. 2010b; Bouillon et al. 2012). The lignin compositional measurements reported in the Epulu River and the Ubangi River at high discharge appear to be comparable to those observed on the lower Congo River mainstem (Spencer et al. 2010b, 2012; Bouillon et al. 2012), representing a typical rainforest-dominated signature for the Congo Basin. Ubangi River low-discharge lignin compositional measurements appear distinct and are therefore proposed by Bouillon et al. (2012) to represent baseflow signatures for a savannah-dominated watershed. This is further supported by the high C:V values in the Ubangi at low flow, as grasses in the Amazon have been shown to have a high C:V near 1.0, and increased C:V ratios have been reported for the savannah-draining Luilu River in comparison with the Congo River mainstem (Hedges et al. 1986; Spencer et al. 2012).

Aside from the Ubangi River, there is currently only a smattering of OM geochemistry data from a number of tributaries throughout the Congo Basin, including the Epulu, Mpoko, Ngoko, Luilu, Djoue, Djili, and Loua. Comparing the savannah-draining Mpoko River to the forested Ngoko River, POC content ranged from 3.9% to 4.2% and 6.8% to 7.8%, respectively, and DOC/TOC was around 74% and 81%, respectively, highlighting differences with respect to OC exported by forested and savannah tributaries (Coynel et al. 2005). With respect to the Luilu, Djoue, Djili, and Loua tributaries, %OC, C/N, and $\delta^{13}C$ data exist with limited sample number ($n = 1$–4). As a brief summary for these four rivers, typically %OC ranged from 7.3 to 19.0, C/N from 5.1 to 20.0, and $\delta^{13}C$-POC from −26.6‰ to −28.3‰ (Mariotti et al. 1991; Spencer et al. 2012). Generally, there is a paucity of data for OM biogeochemistry in tropical ecosystems, with the striking exception of the Amazon, and although tropical rainforest and savannah-dominated systems do not undergo the temporal variability with respect to seasonal climatic extremes observed in temperate and northern high-latitude rivers, seasonal variability in both OM quantity and quality has been shown in the Congo Basin (Coynel et al. 2005; Spencer et al. 2010b; Bouillon et al. 2012). In the Epulu River, temporal variability in DOM quantity and quality was attributed to greater surface runoff and leaching of organic rich horizons during the wet period and increased flow path, residence time, and thus greater microbial mineralization in the intermediary period in conjunction with an already well-leached source material in the dry period/post flush (Striegl et al. 2005; Spencer et al. 2010b; Chapter 20). This results in DOM exported during the flushing period of the Epulu River having not only a higher DOC concentration, but also a higher molecular weight and more aromatic character, as evidenced by Λ_8 and optical measurements such as the specific UV absorbance at 254 nm ($SUVA_{254}$), which has been shown to increase with increasing aromaticity (Fig. 22.6; Weishaar et al. 2003; Spencer et al. 2010b; Chapter 21).

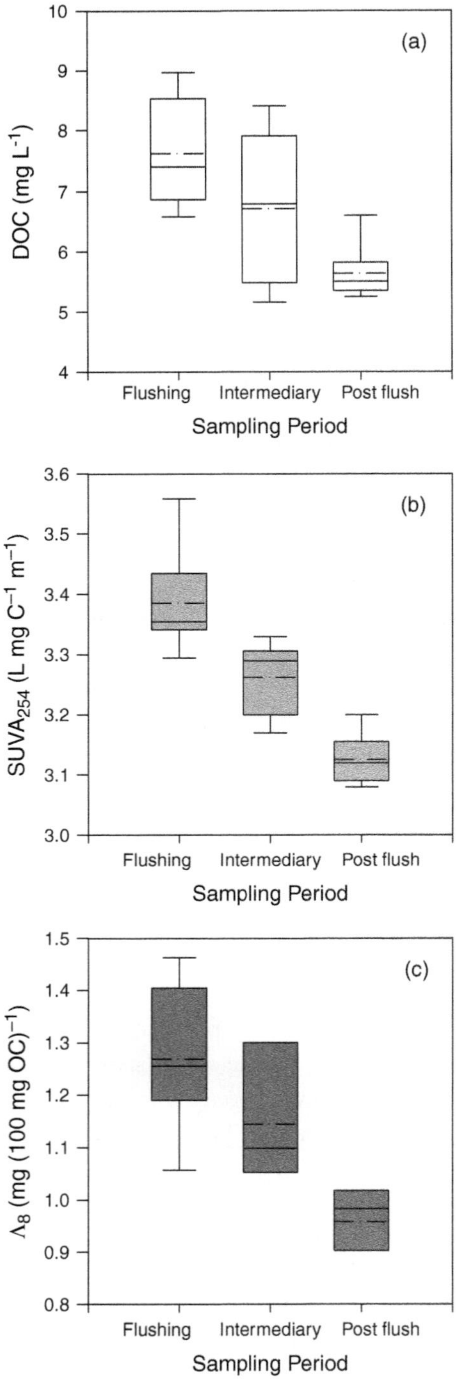

Figure 22.6. Box plots of **(a)** DOC; **(b)** $SUVA_{254}$, and **(c)** Λ_8 during the flushing, intermediary, and post-flush sampling periods on the Epulu River. The black dash-dotted line and the solid black line in the box represent the mean and median, respectively. The horizontal edges of the box denote the 25th and 75th percentiles, and the error bars denote the 10th and 90th percentiles (reproduced from Spencer et al. [2010b]).

In high-latitude northern rivers, DOM exported during the spring freshet has also been shown to have elevated Λ_8 and $SUVA_{254}$ values and to be simultaneously younger and more biolabile and photoreactive (Raymond et al. 2007; Holmes et al. 2008; Mann et al. 2012). Therefore, it seems likely that temporal dynamics in tropical rivers also impact on the fate of the exported DOM.

4.4. Organic matter geochemistry of the Congo River mainstem

At the Kinshasa-Brazzaville station, the %POC of TSS is typically 6.1–7.4%, and the percent DOC of TOC ranges from 82–89% (Mariotti et al. 1991; Coynel et al. 2005; Spencer et al. 2012). If the POM is divided into the coarse (>63 μm; CPOM) and fine (0.7–63 μm; FPOM) fractions, mainstem CPOM and FPOM %OC values (5.7–11.8% and 7.0–8.8%, respectively) are elevated in comparison with similar samples from the Amazon Basin, where typically CPOM and FPOM %OC range from 0.3–4.7% and 0.4–2.0%, and Amazon mainstem sites at Obidos are typically around 1.0% and 1.2%, respectively (Hedges et al. 1986, 2000; Aufdenkampe et al. 2007; Spencer et al. 2012). With respect to DOM and POM, the Congo appears to be a major exporter of terrigenous derived OM, as evidenced by high Λ_8 values, and in this respect is similar to other major tropical rivers (Hedges et al. 1986, 2000; Alin et al. 2008). Congo mainstem CPOM, FPOM, and DOM all appear to predominantly originate from vegetation and soil inputs, as evidenced by elemental, isotopic, and lignin phenol data; however, FPOM is derived from much older carbon pools ($\Delta^{14}C = -61.7\pm-15.2$‰) compared with CPOM and DOM (mean $\Delta^{14}C = 42.4\pm18.4$‰ and 79.2 ± 11.1‰, respectively) (Spencer et al. 2012). Congo mainstem CPOM appears to be well-preserved higher vascular plant material due its modern radiocarbon age, elevated C/N (17.2–20.4), and Λ_8 values (4.56–5.01 [mg{100 mg OC}$^{-1}$]). In relation to CPOM, the aged FPOM fraction (320–580 yBP ^{14}C ages) is comparatively degraded, as demonstrated by its nitrogen enrichment (C/N 11.4–12.5), lower Λ_8 (2.80–2.97 [mg{100 mg OC}$^{-1}$]) and elevated lignin Ad:Al values, similar to soil-derived OM (Spencer et al. 2012). Little modification in OM concentrations or signatures has been observed from the Kinshasa-Brazzaville station to the head of the estuary (~350 km), allowing future studies to assess seasonal and long-term OM dynamics with respect to OM export from this logistical feasible location.

5. Congo estuarine and plume processes

The Congo estuary begins just upstream of Malela (0 km in Fig. 22.7). Depth then increases rapidly to ~100 m at the head of the canyon that transects the estuary and extends out to the abyssal ocean (Section 6). At the coast the estuarine canyon reaches a depth of ~425 m, where it is flanked by the two giant spits of Ponta do Padrao to the south and Banana to the north (Eisma and van Bennekom 1978). The estuarine physical system was first described by Commander Purey-Cust (1900) aboard *HMS Rambler* (1899) and is summarized by Eisma and van Bennekom (1978). Tides are semi-diurnal and of modest strength, with springs reaching 1.9 m and neaps of 0.3 m at Banana. The tidal wave is retarded and flattened by the great flow of water from the river. By Malela tidal range is ~90% of that experienced at the coast, although tides are synchronous with those at Banana. Tidal range then diminishes up river to ~10% at Boma, where high and low tides occur ~4 hours later than at Banana.

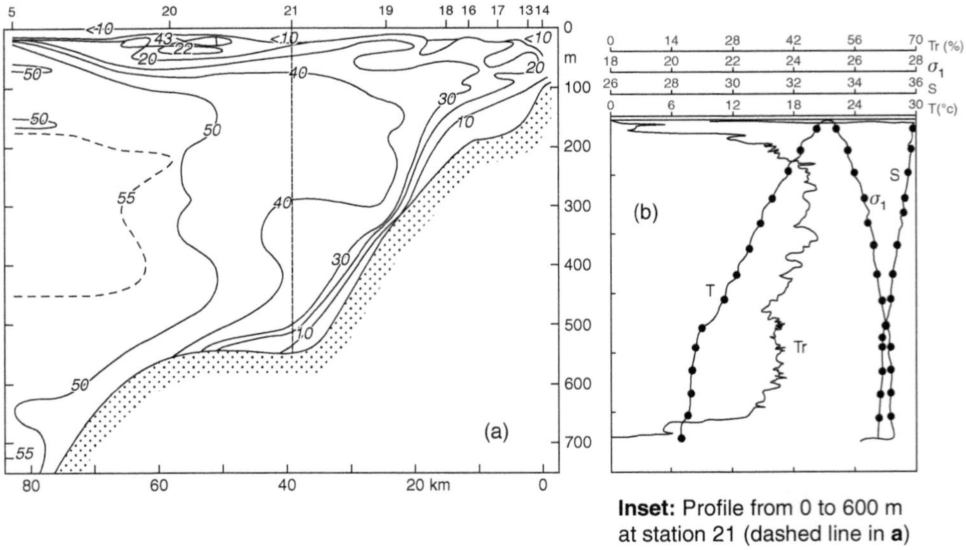

Figure 22.7. (a) Distribution of turbidity (% light transmission) in the Congo estuary and across the shelf, axial to the canyon. **(b)** Depth profile from 0–600 m at station 21 (black dash vertical line in **a**) of salinity (S), temperature (T), density (σ_t) and turbidity (Tr). Both adapted from Pak et al. (1984).

The incision of the canyon head enables cool, near-full salinity water (>35) to penetrate deep into the estuary to just below Malela. A surface layer (~10 m) of river water overflows the saline incursion. The surface layer travels centrally within the plume maintained by a strong pyconocline and deepens to ~20 m within the mid-estuary. By the mouth at Banana, salinity in the 2–3 m surface plume is about 0.6 (Eisma and Kalf 1984). The core of saline water moves upstream in the estuary and spills over into the adjacent mangroves, presumably entraining OC from these habitats (Eisma and van Bennekom 1978). The salt balance of the estuary is maintained by export in a layer of low-salinity water that flows seaward at the interface between the saline core and freshwater skin of the estuary.

Beyond the mouth of the estuary, the plume (salinity ≤35) is maintained as a surface lens that extends 800 km from shore west to ~8° E (Cadee 1984; Pak et al. 1984) and south to 8°15′ S (Fuglister 1960). The plume starts as a jet stream of low-salinity, high-velocity waters, invariably heading WNW nearshore (Meulenbergh 1968). Offshore the plume is diverted by seasonally shifting winds. In February and March, the plume heads NW, keeping close to the coast and colliding with the low-salinity waters of the Bight of Biafra (Koleshnikov 1973). Through May, June, and August, the plume heads due west, then WNW in November (Eisma and van Bennekom 1978).

5.1. Organic matter in the surface plume

The incision of the canyon produces a near-unique three-tiered vertical structure of sediment and OM distributions within the Congo estuary consisting of a freshwater surface lens rich in fine sediment and DOC, a mid-layer of clear near-full salinity seawater, and a base layer enriched in TSS

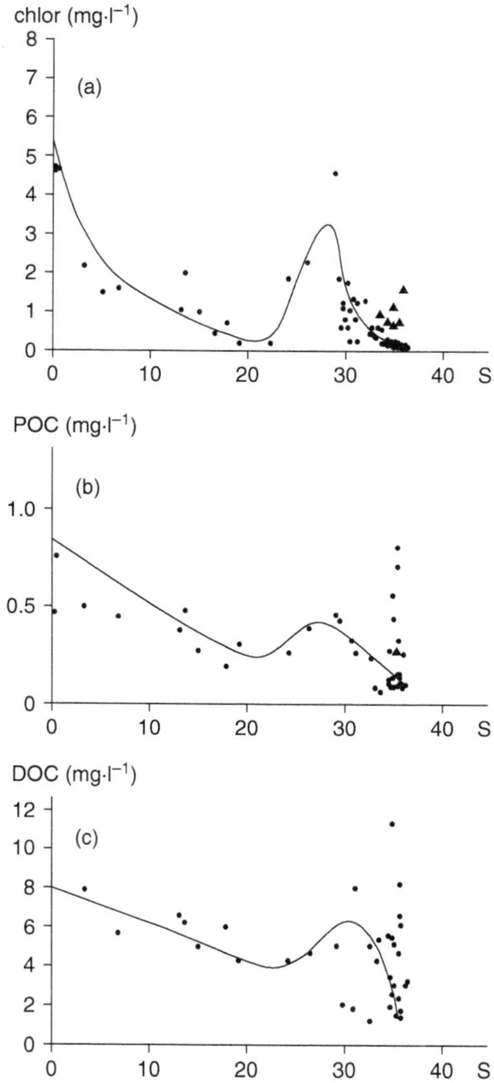

Figure 22.8. Chlorophyll *a* (**a**), POC (**b**), and DOC (**c**) in the Congo river, estuary, and plume plotted against salinity, separately for surface (•), bottom (×), and upwelling influenced surface samples offshore (▲). Adapted from Cadee et al. (1984) where the lines were fitted through the surface values by eye.

(Fig. 22.7, Cadee 1984; Pak et al. 1984). The surface film of low-salinity water carries TSS, POC, and DOC seaward, although the highest levels of near-surface particulates occur just below the halocline (Fig. 22.7; Pak et al. 1984). Within the physical confines of the estuary, surface concentrations of chlorophyll, TSS, POC, and DOC all decrease with increasing salinity, with POC and DOC showing near-conservative mixing between salinities of ~0 and 20 (Fig. 22.8; Cadee 1982, 1984; Pak et al.

1984). However, approximately 25 km beyond the mouth of the estuary between salinities of 20 to 30, a peak in chlorophyll coincides with increases in TSS (Fig. 22.7), POC, and DOC (Fig. 22.8), suggesting production of new POC and DOC fed by nutrient inputs from river-driven upwelling (Cadee 1984). Heading further offshore, chlorophyll, TSS load, POC, and DOC all decrease gradually until the plume is no longer distinguishable at ~800 km from coast. Throughout the plume's expanse, it deposits POM to the seafloor, although this makes only a modest contribution to seafloor sediment accumulation and OC burial rates (Section 6).

5.2. A surface photo-reactor

The confinement of DOC within the surface waters of the estuary and plume produces conditions favoring the extensive photochemical degradation of exported Congo River DOM. The proportion of DOC exported from the Congo that is photo-mineralized in the estuary and plume is unknown. However, ~45% of the DOC exported from the Congo was found to be photo-labile on irradiation with simulated sunlight (Spencer et al. 2009), indicating that photo-chemistry has the capacity to remove a considerable fraction of Congo River DOC within the estuary and plume prior to its export to the wider Atlantic Ocean. Photo-irradiation also altered the quality of Congo DOM, preferentially degrading CDOM and lignin, enriching δ^{13}C-DOC, and shifting the molecular fingerprint of DOM toward a less condensed, less aromatic composition (Spencer et al. 2009; Stubbins et al. 2010). For instance, during Congo River water irradiation, Λ_8 decreased >20-fold from 0.72 to 0.03 (mg[100 mg OC^{-1}]); Spencer et al. 2009), and the DOC-specific absorption coefficient at 325 nm was reduced from from 1.90 m^2g^{-1} to 0.19 m^2g^{-1}. This latter value is within the range reported for open ocean waters of the Atlantic Ocean (0.1–0.3 m^2 g^{-1}; Nelson et al. 2007). Carbon-normalized light absorbance and lignin are often used as indicators of the terrestrial contribution to the total DOC pool in marine waters (Hernes and Benner 2006). The photo-bleaching of lignin and CDOM absorption indicate that solar irradiation can shift the identifying signatures of Congo River DOM away from their original, highly aromatic terrestrial character to a signal consistent with the low aromatic signal of open-ocean DOM. The surviving photo-resistant pool, stripped of its terrestrial signatures, constitutes ~55% of the total DOC exported from the Congo River, suggesting that a considerable fraction of photo-altered terrestrial DOC could go untraced in the oceans.

5.3. Particle-rich estuarine bottom waters

As discussed previously, intruding seawater extends throughout the estuary, producing a clear water layer with high salinity, low TSS, and low POC at mid-depths, below which lies the TSS and POC-rich bottom layer (Fig. 22.7; Cadee 1984; Pak et al. 1984). The TSS and POC in the bottom layer are derived from the settling of riverine TSS (Cadee 1984; Pak et al. 1984). The net upstream flow of seawater at depth concentrates suspended material, including POC, right at the head of the estuary (Eisma and Kalf 1984). The resultant accumulation of POC drives high mineralization rates, which in turn drawdown oxygen concentrations (<1% saturation) and produce nitrate and ammonia (van Bennekom et al. 1978). In addition, the bottom waters at the head of the estuarine canyon have high DOC concentrations, apparently due to the dissolution of DOC from the accumulated POC

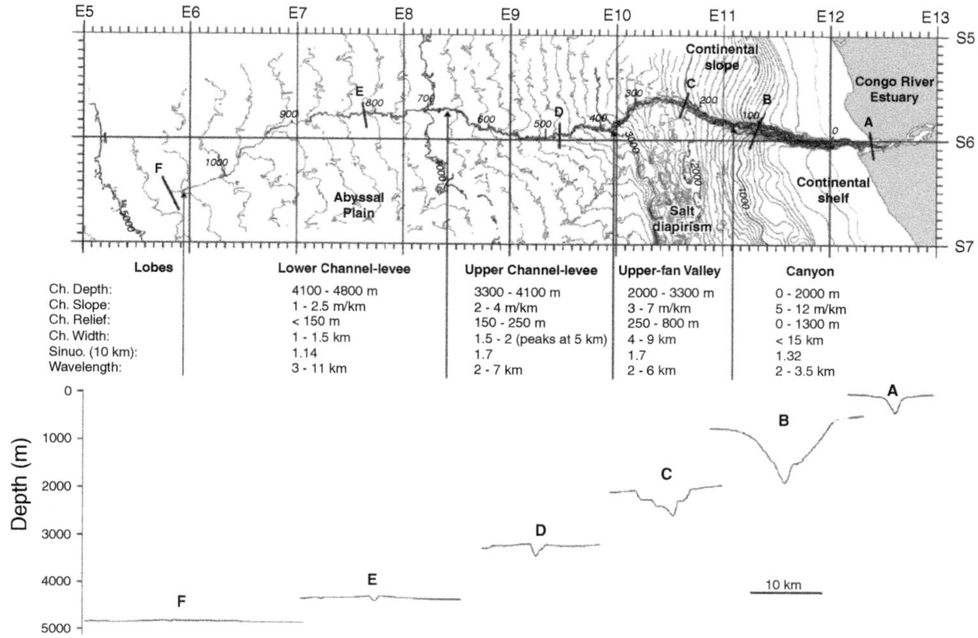

Figure 22.9. Channel morphology of the Congo Canyon and associated channel-levee system. Reproduced from Savoye et al. (2009).

(Cadee 1984). Downstream of this area of intense biogeochemical activity, TSS remains at elevated concentrations and flows out to sea at depth, with the majority of turbidite flows following the well-worn path of the canyon to the abyssal plain (Section 6).

6. The Congo Canyon: from land to the abyssal plain

The Congo Canyon and associated channel-levee system extends approximately due west from within the Congo estuary (Section 5) out ~800 km to the Congo Fan in the deep ocean. The canyon is well developed, having been in its current position since the Miocene-Pliocene boundary ~5 million years ago (Ferry et al. 2004). Over this time, tubidite flows have carved a massive incision transecting the continental shelf and continuing out across the continental slope to the abyssal plain (Fig. 22.9). Starting 30 km inland near Kisanga and reaching a depth of ~100 m by Malela (Heezen et al. 1964), the canyon then deepens rapidly (slope 5–12 m km^{-1}) into a V-shaped feature reaching a depth of 400–500 m by the coast (Shepard and Emery 1973). It then deepens further, reaching a width of ~12 km and a maximum relief of ~1,300 m on the continental shelf and slope (Fig. 22.9), where the canyon thalweg is approximately 2 km wide and 350 m deep (Savoye et al. 2009). The continuation of the canyon once the seafloor drops below 1,000 m has been termed the "Upper-Fan Valley" to denote a decrease in its relief (250–800 m) and width (4–9 km). The Upper-Fan Valley continues westward out beyond the foot of the continental slope break (~3,000 m). From there the feature is reclassified first as the upper, then the lower channel-levee (Babonneau 2002) as it gradually decreases in relief and width until petering out into the Congo deep-sea fan, a massive depositional feature on the abyssal floor.

Uniquely for major world rivers, most of which migrated from their historic canyons during the last increase in sea-level, the alignment of the contemporary Congo River with its ancient deep-sea canyon provides an incision, transecting the shelf (~65–80 km wide, average depth ~200 m) and enabling the efficient injection of riverine sediments directly to the abyssal plain. Riverine sediment inputs to the canyon are generally fine-grained and organic-rich. In addition, the canyon receives inputs of sediment from longshore drift of sands northward into the canyon. Reworking of deposited sediments and the slumping of the canyon's muddy walls also provides inputs of organic-rich sediments (Eisma and Kalf 1984). Despite these inputs, there appears to be no deposition within the canyon on geological timescales, as high-energy turbidite flows sweep the canyon clean (Savoye et al. 2009). Consequently, the canyon floor is comprised of coarse sandy sediments with minimal OM content (Savoye et al. 2009). Downstream of the true canyon, the morphology of the levee-channel system is shaped by a mixture of depositional and erosional processes. The depositional levees can reach relief heights of 350 m and widths of 30 km (Babonneau 2002), decreasing in height and width downslope until they end in the deep-sea fan (~5,100 m depth), the terminal site of submarine, gravity-driven turbidite deposition for the Congo River canyon system.

The Congo Fan is a major morphological feature of the Atlantic Ocean, the main depositional feature in the SE Atlantic, and one of the largest active depocenters in the World. As such it is of major interest for the study of ocean floor biogeochemical and physical processes (Savoye et al. 2009). The Congo deep-sea fan has three main lobes, the northern and southern fans, which are no longer active sites for deposition, and the active Axial Fan. The northern fan was the main site for deposition between 780 and 540 ka and the southern between 540 and 210 ka (Wefer et al. 1998). For the last 210 thousand years, the contemporary Axial Fan has been active. As its name suggests, the Axial Fan extends axially in a westward direction from the end of the canyon/levee-channel system between the relict northern and southern fans (Savoye et al. 2009). The active fan covers some 300,000 km^2 of the abyssal ocean off the Congo-Angola Shelf, an area approximately equivalent to that of the Amazon Fan, which is disconnected from its contemporary river (Savoye et al. 2009).

6.1. Organic matter deposition

Globally, river-dominated margins are the site of significant OM sedimentation and burial (Hedges and Keil 1995; Burdige 2005; Bianchi and Allison 2009). The efficiency with which riverine POC is delivered to the seafloor and subsequently buried varies with the quality of riverine particulates exported and the physical nature of the river-coastal margin system (Blair and Aller 2012; Fig. 22.10). In a study of TSS fluxes to the Congo continental shelf and abyssal plain, Rabouille et al. (2009) estimated sedimentation rates of 0.95–2.0 mm yr^{-1} between approximately 1,000 to 4,000 m deep, at sites off axis from the canyon. At these sites, POC deposition rates were 0.9–2.2 mol-C m^{-2} yr^{-1}, which is five to nine times greater than the downward transport of POC from the water column (Rabouille et al. 2009). Deposition rates of Si also exceed inputs from pelagic rain by a factor of six to eight (Ragueneau et al. 2009), indicating that cross-slope processes are important for the transfer of POC and Si to depth. For the two sites on the continental shelf at ~4,000 m, POC burial rates were 0.3 and 0.56 mol-C m^{-2} yr^{-1}, translating to preservation efficiencies of 33% and

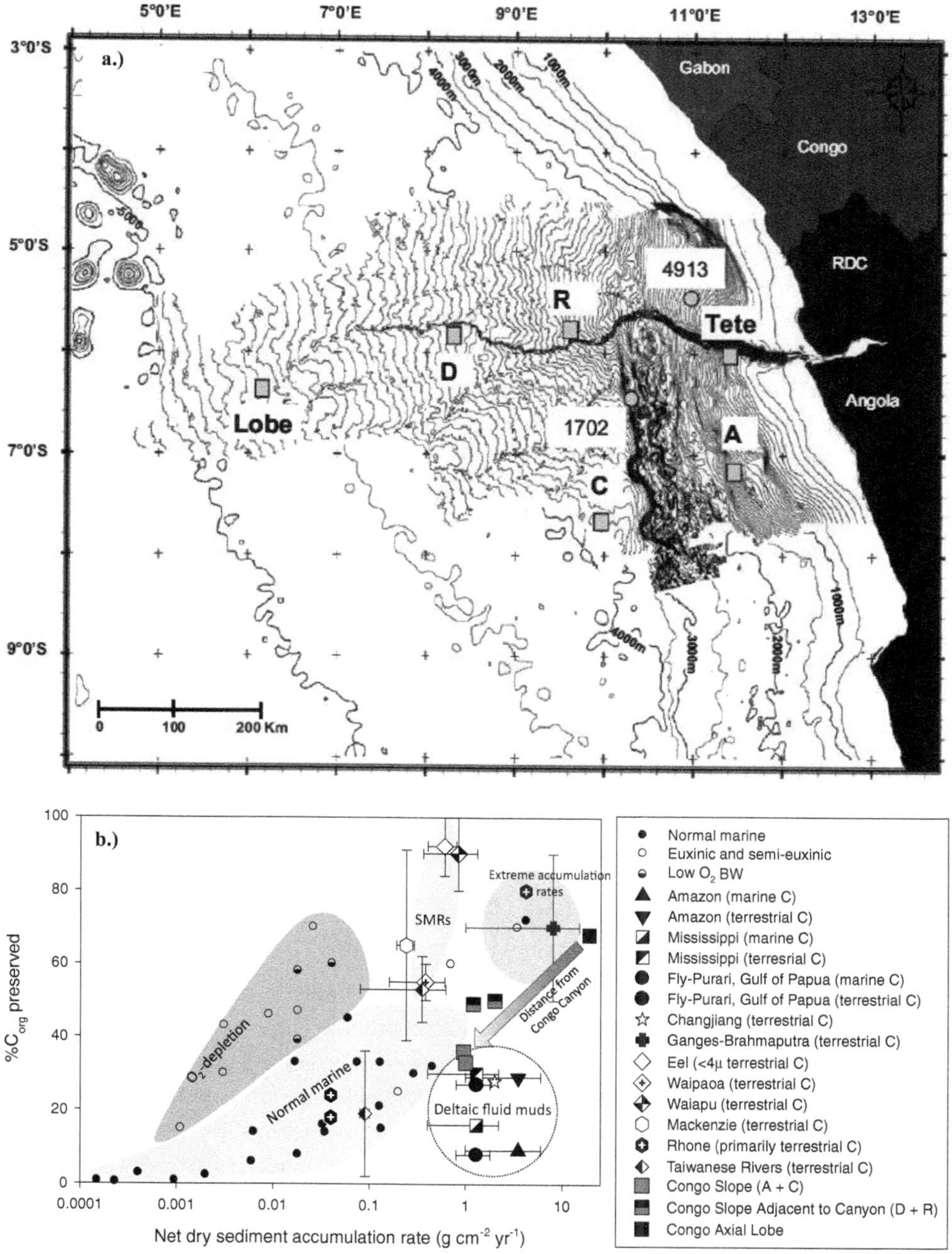

Figure 22.10. **a**) Bathymétric map of the Congo Canyon and Channel showing the stations examined by Rabouille et al. (2009). **b**) The percentage of organic carbon (C_{org}) deposited on the seafloor that is buried versus net sediment accumulation rate reproduced from Blair and Aller (2012) with addition of Congo data from Rabouille et al. (2009). Congo stations indicated in the legend refer to panel **a**).

49% (Rabouille et al. 2009). These efficiencies are far greater than reported for regions of the ocean floor at depths exceeding 4,000 m that typically bury between 1% and 5% of deposited POC (Berger et al. 1989; Jahnke 1996). The ratios of Si:C accumulation rates (0.55–0.57) are extremely close to mean Si:C ratios reported for continental shelf deposits globally (0.6, DeMaster 2002), suggesting that shelf particulates undergo minimal alteration as they travel down to deeper points of deposition (Ragueneau et al. 2009).

In addition to downslope transport, the Congo Canyon/levee-channel system provides for rapid export and deposition of particulates beyond the coastal shelf to the deep sea fan. The Lobe site studied by Rabouille et al. (2009) and Ragueneau et al. (2009) lies at ~4,800 m within the lower reaches of the levee-channel system, and deposition here is taken to be broadly representative of the fan. At the Lobe site, rates of sedimentation (19 mm yr^{-1}) and POC deposition (~19 mol-C m^{-2} yr^{-1}) are about 10–20 times greater than the rates observed further up the slope, but peripheral to the channel (Rabouille et al. 2009), demonstrating the efficiency with which the canyon-channel system exports TSS and POC to the abyssal ocean. Furthermore, due to efficient export to the Lobe site, total POC deposition exceeds particulate rain by a factor of 9, and two-thirds of the POC deposited ends up buried in the sediments (~13 mol-C m^{-2} yr^{-1}; Rabouille et al. 2009). Scaling rates to the surface area of the active depocenter (~3,000 km^2; Sanvoye et al. 2009) yields a fan-wide deposition rate of ~0.7 Tg-C yr^{-1} and a burial rate of ~0.4 Tg-C yr^{-1} (Rabouille et al. 2009). This suggests that about a third of the Congo River POC export (~2 Tg-C yr^{-1} Coynel et al. 2005) reaches the depocenter and one-fifth is buried. Comparison of the burial term to the rates for the South Atlantic (~1.8 Tg-C yr^{-1}) and global deep-sea burial (10–20 Tg-C yr^{-1}; Berner 1989; Hedges and Keil 1995; Jahnke 1996) reveal that the unique confluence of high riverine export from the Congo River, the efficient injection of POC to the deep ocean through the canyon, and the subsequent burial of ~66% of deposited OM, combine to make the Congo Deep Sea Fan a globally significant site for POC burial. The high absolute sedimentation rates and %POC burial rates in the Congo Fan (Axial Lobe) are evident when contrasted with deposition in other global systems (Fig. 22.10; Blair and Aller 2012).

Acknowledgments

We thank numerous colleagues for their assistance in Congo scientific research programs, including those from 1987–1994 (PIRAT and PEGI, coordinated by Institut de recherche pour le developpement and Centre national de la recherche scientifique) and more recently the group involving scientists from University of California Davis, Woods Hole Research Center, and researchers involved with the Tervuren Congo expeditions. R.G.M.S. is very grateful for support from NSF-ETBC #0851101 and for Johan Six for sharing his infectious love of the Congo and its people. Finally, we would like to thank the people of the Congo Basin for their kindness, hospitality, and joie de vivre.

References

Alin, S.R., R. Aalto, M.A. Goni, J.R. Richey, and W.E. Dietrich. 2008. Biogeochemical characterization of carbon sources in the Strickland and Fly rivers, Papua New Guinea. J. Geophys. Res. Earth Surface. 113, F01S05, doi: 10.1029/2006JF000625.

Allègre C.J., B. Dupré, P. Négrel, and J. Gaillardet. 1996. Sr-Nd-Pb isotopic systematics of the Congo and Amazon River systems: constraints on erosion processes. Chem. Geol. 131, 93–112.

Alvarez, P. 1997. Morphologies karstiques et implications minières en République Centrafricaine. J. African Earth Sci. 25, 293–305.

Aufdenkampe, A.K., E. Mayorga, J.I. Hedges, C. Llerena, P.D. Quay, J. Gudeman, A.V. Krusche, and J.E. Richey. 2007. Organic matter in the Peruvian headwaters of the Amazon: Compositional evolution from the Andes to the lowland Amazon mainstem. Org. Geochem. 38, 337–364.

Babonneau, N., B. Savoye, M. Cremer, and B. Klein. 2002. Morphology and architecture of the present canyon and channel system of the Zaire. Deep-Sea Fan. Mar. Petrol. Geol. 19, 445–467.

Berger, W.H., V.S. Smetacek, and G. Wefer. 1989. Ocean productivity and paleoproductivity-an overview. In: Berger, W.H., Smetacek, V.S., Wefer, G. (eds.), Productivity of the Oceans: Present and Past, Dahlem Konferenzen. Wiley, Berlin, pp. 1–34.

Berner, R.A. 1989. Biogeochemical cycles of carbon and sulfur and their effect on atmospheric oxygen over phanearozoic time. Pal. Pal. Pal. 73, 97–122.

Bianchi, T.S., T. Filley, K. Dria, and P. Hatcher. 2004. Temporal variability in sources of dissolved organic carbon in the lower Mississippi River. Geochim. Cosmochim. Acta 68, 959–967.

Bianchi, T.S., and M.A. Allison. 2009. Large-river delta-front estuaries as natural "recorders" of global environmental change. Proc. Nat. Acad. Sci. 106, 8085–8092.

Blair, N.E. and R.C. Aller. 2012. The fate of terrestrial organic carbon in the marine environment. Ann. Rev. Mar. Sci. 4, 401–423.

Bouillon, S., A. Yambele, R.G.M. Spencer, D.P. Gillikin, P.J. Hernes, J. Six, R. Merckx, and A.V. Borges. 2012. Organic matter sources, fluxes and greenhouse gases in the Oubangui River (Congo River basin). Biogeosciences 9, 2045–2062.

Burdige, D.J. 2005. Burial of terrestrial organic matter in marine sediments: a re-assessment. Glob. Biogeochem. Cycles 19, GB4011.

Cadee, G.C. 1982. Organic carbon and phytoplankton in Zaire river, estuary and plume. Mitt. geol. palaont. Inst. Univ. Hamburg. 52, 429–431.

Cadee, G.C. 1984. Particulate and dissolved organic carbon and chlorophyll A in the Zaire river, estuary and plume. Neth. J. Sea Res. 17, 426–440.

Cardinal D., J. Gaillardet, H.J. Hughes, S. Opfergeld, and L. André. 2010. Contrasting silicon isotope signatures in rivers from the Congo Basin and the specific behaviour of organic-rich waters. Geophys. Res. Lett. 37. doi:10.1029/2010GL043413.

Censier C. 1993. Dynamique sédimentaire de la charge de fond sableuse du cours moyen et inférieur de la Sangha (République du Congo). Grands Bassins fluviaux périatlantiques: Congo, Niger, Amazone. Actes du colloque Pegi INSU-CNRS-ORSTOM, pp. 229–238.

Coynel, A., P. Seyler, H. Etcheber, M. Meybeck, and D. Orange. 2005. Spatial and seasonal dynamics of total suspended sediment and organic carbon species in the Congo River. Glob. Biogeochem. Cyc. 19, GB4019, doi:10.1029/2004GB002335.

DeMaster, D.J. 2002. The accumulation and cycling of biogenic silica in the Southern Ocean: revisiting the marine silica budget. Deep-Sea Res. Part II 49, 3155–3167.

Dupré, B., P. Négrel, F. Seimbille, and C. Allègre. 1994. $^{87}Sr/^{86}Sr$ ratio variation during a rain event. Atmos. Env. 28, 617–620.

Dupré, B., J. Gaillardet, D. Rouseau, and C.J. Allègre. 1996. Major and trace elements of riverborne material: the Congo Basin. Geochim. Cosmochim. Acta 60, 1301–1321.

Dupré, B., J. Viers, J.L. Dandurand, M. Polvé, P. Bénézeth, P. Vervier, and J.J. Braun. 1999. Major and trace elements associated with colloids in organic-rich river waters: ultrafiltration of natural and spiked solutions. Chem. Geol. 160, 63–80.

Edmond J.M., M.R. Palmer, C.I. Measures, B. Grant, and R.F. Stallard. 1995. Fluvial geochemistry and denudation rate of the Guyana Shield. Geochim. Cosmochim. Acta 59, 3301–3325.
Eisma, D. and A.J. Van Bennekom. 1978. The Zaire river and estuary and the Zaire outflow in the Atlantic ocean. Neth. J. Sea Res. 12, 255–272.
Eisma, D. and J. Kalf. 1984. Dispersal of Zaire river suspended matter in the estuary and the Angola Basin. Neth. J. Sea Res. 17, 385–411.
Elsgaard, L. and D. Prieur. 2011. Hydrothermal vents in Lake Tanganyika harbor spore-forming thermophiles with extremely rapid growth. J. Great Lakes Res. 37, 203–206.
Ferry, J.-N., N. Babonneau, T. Mulder, O. Parize, and S. Raillard. 2004. Morphogenesis of Congo submarine canyon and valley: implications about the theories of the canyons formation. Geodynamica Acta 17, 241–251.
Forte, A.M., S. Quéré, R. Moucha, N.A. Simmons, S.P. Grand, J.X. Mitrovica, and D.B. Rowley. 2010. Joint seismic–geodynamic-mineral physical modelling of African geodynamics: a reconciliation of deep-mantle convection with surface geophysical constraints. Earth Planet. Sci. Lett. 295, 329–341.
Fuglister, F. 1960. Atlantic Ocean atlas of temperature and salinity. Woods Hole Oceanographic Institute.
Gaillardet J., B. Dupré and C.J. Allègre. 1995. A global mass budget applied to the Congo Basin Rivers: Erosion rates and continental composition. Geochim. Cosmochim. Acta 59, 3469–3485.
Gaillardet J., B. Dupré, C.J. Allègre, and P. Négrel. 1997. Chemical and Physical weathering rates in the Amazon River Basin. Chem. Geol. 142, 141–173.
Gaillardet J., B. Dupré, P. Louvat and C.J. Allègre. 1999a. Global silicate weathering of silicates estimated from large river geochemistry. Chem. Geol. 159, 3–30.
Gaillardet J., B. Dupré, and C.J. Allègre 1999b. Geochemistry of large river suspended sediments: silicate weathering or crustal recycling? Geochim. Cosmochim. Acta. 63, 4037–4051.
Gaillardet J., J. Viers, and B. Dupré. 2007. Trace elements in river saters. In K. Turekian and D. Holland (eds.), Treatise of Geochemistry 5, pp. 225–273.
Gasse, F., B. Stabell, E. Fourtanier, and Y. Van Iperen. 1989. Freshwater diatom influx in intertropical Atlantic: relationships with continental records from Africa. Quat. Res. 32, 229–243.
Godderis, Y., Y. Donnadieu, M. Tombozafyc, and C. Dessert. 2008. Shield effect on continental weathering: Implication for climatic evolution of the Earth at the geological timescale. Geoderma 145, 439–448.
Hedges, J.I., and D.C. Mann. 1979. Characterization of plant tissues by their lignin oxidation products. Geochim. Cosmochim. Acta 43, 1803–1807.
Hedges, J.I., W.A. Clark, P.D. Quay, J.E. Richey, A.H. Devol, and U.D.M. Santos. 1986. Compositions and fluxes of particulate organic material in the Amazon River. Limnol. Oceanogr. 31, 717–738.
Hedges, J.I., and R.G. Keil. 1995. Sedimentary organic-matter preservation – an assessement and speculative synthesis. Mar. Chem. 49, 81–115.
Hedges, J.I., E. Mayorga, E. Tsamakis, M.E. McClain, A.K. Aufdenkampe, P. Quay, J.E. Richey, R. Benner, S. Opsahl, B. Black, T. Pimental, J. Quintanilla, and L. Maurice. 2000. Organic matter in Bolivian tributaries of the Amazon River: A comparison to the lower main stream. Limnol. Oceanogr. 45, 1449–1466.
Heezen, B.C., R.J. Menzies, E.D. Schneider, W.M. Ewing and N.C.C. Granelli. 1964. Congo submarine canyon. Bull. Am. Ass. Petrol. Geol. 48, 1126–1149.
Hernes, P.J. and R. Benner. 2006. Terrigenous organic matter sources and reactivity in the North Atlantic Ocean and a comparison to the Arctic and Pacific Oceans. Mar. Chem. 100, 66–79.

Hernes, P.J., A.C. Robinson, and A.K. Aufdenkampe. 2007. Fractionation of lignin during leaching and sorption and implications for organic matter "freshness". Geophys. Res. Lett. 34, L17401, doi:10.1029/2007GL031017.

Holmes, R.M., J.W. McClelland, P.A. Raymond, B.B. Frazer, B.J. Peterson, and M. Stieglitz. 2008. Lability of DOC transported by Alaskan rivers to the Arctic Ocean. Geophys. Res. Lett. 35, L03402, doi:10.1029/2007GL032837.

Hugues, H.J., F. Sondag, C. Cocquyt, A. Laraque, A. Pandi, L. André, and D. Cardinal. 2011. Effect of seasonal biogenic silica variations on dissolved silicon fluxes and isotopic signatures in the Congo River. Limnol. Oceanogr. 56, 551–561.

Jackson, P.R., K.A. Oberg, N. Gardiner, and J.M. Shelton. 2009. Velocity Mapping in the Lower Congo River: A First Look at the Unique Bathymetry and Hydrodynamics of Bulu Reach: Proc. IAHR, pp.1007–1014.

Jahnke, R.A. 1996. The global ocean flux of particulate organic carbon: areal distribution and magnitude. Glob. Biogeochem. Cycles 10, 71–88.

Koenig, R. 2008. Critical time for African rainforests. Science 320, 1439–1441.

Koleshnikov, A.G. 1973. Equalant I and Equalant II, Physical Oceanography. Unesco, Paris. 1, 1–289.

Labat, D., J. Ronchail, and J.L. Guyot. 2005. Recent advances in wavelet analyses: Part 2-Amazon, Parana, Orinoco and Congo discharge time scale variability. J. Hyd. 314, 289–311.

Laporte, N., J.A. Stabach, R. Grosch, T.S. Lin, and S.J. Goetz. 2007. Expansion of industrial logging in Central Africa. Science 316, 1451.

Laraque A., J.P. Bricquet, J.C. Olivry, and M. Berthelot. 1995. Transports solides et dissous du fleuve Congo (Bilan de six annees d'observations). Grands Bassins fluviaux périatlantiques: Congo, Niger, Amazone. Actes du colloque Pegi INSU-CNRS-ORSTOM, pp. 133–145.

Laraque A., G. Mahe, D. Orange, and B. Marieu. 2001. Spatiotemporal variations in hydrological regimes within Central Africa during the XXth century. J. Hyd. 245, 104–117.

Laraque A., J.P. Bricquet, A. Pandi, and J.C. Olivry. 2009. A review of material transport by the Congo River and its tributaries. Hyd. Proc. 23, 3216–3224.

Mann, P.J., A. Davydova, N. Zimov, R.G.M. Spencer, S. Davydov, E. Bulygina, S. Zimov, and R.M. Holmes. 2012. Controls on the composition and lability and dissolved organic matter in Siberia's Kolyma River basin. J. Geophys Res. Biogeo. 117, G01028, doi: 10.1029/2011JG001798.

Maurice, R. 1946. Le Congo Physique. Impr. de l'Academie.

Mariotti, A., G. Francois, P. Giresse, and K. Mouzeo. 1991. Carbon isotope composition and geochemistry of particulate organic matter in the Congo River (Central Africa): Application to the study of Quaternary sediments off the mouth of the river. Chem. Geo. 86, 345–357.

Meulenbergh, J. 1968. Diffusion des eaux du fleuve Congo dans les eaux de l'Atlantique sud. Cl. Sci. Tech. 16, 1–149.

Meybeck, M. 1982. Carbon, nitrogen, and phosphorus transport by world rivers. Am. J. Sci. 282, 401–450.

Négrel, P., C.J. Allègre, B. Dupré, and E. Lewin. 1993. Erosion sources determined by inversion of major and trace element ratios in river water: The Congo Basin case. Earth Planet. Sci. Lett. 120: 59–76.

Négrel, P. and B. Dupré. 1995a. Seasonal fluctuations of major, trace element and Sr isotopic composition of the Congo River, implications for the hydrological functioning and input to the ocean. Grands Bassins fluviaux périatlantiques: Congo, Niger, Amazone. Actes du colloque Pegi INSU-CNRS-ORSTOM, pp. 39–50.

Négrel, P. and B. Dupré. 1995b. Temporal variations of Sr isotopic ratios, major and trace elements composition of the Ubangi river basin: implication for the source of material. Grands Bassins fluviaux périatlantiques: Congo, Niger, Amazone. Actes du colloque Pegi INSU-CNRS-ORSTOM, pp. 181–198.

Nelson, N.B., D.A. Siegel, C.A. Carlson, C. Swan, W.M. Smethie, Jr., and S. Khatiwala. 2007. Hydrography of chromophoric dissolved organic matter in the North Atlantic. Deep-Sea Res. I 54, 710–731.

Nkounkou, R.R., and J.L. Probst. 1987. Hydrology and geochemistry of the Congo River system. Scope/UNEP Sonderband Heft. 64, 483–508.

Oberg, K.A., J.M. Shelton, N. Gardiner, and P.R. Jackson. 2008. Discharge and Other Hydraulic Measurements for Characterizing the Hydraulics of Lower Congo River: Proc. IAHR.

Olivry, J.C. and J. Boulegue. 1993. Grands Bassins Fluviaux Peri Atlantiques: Congo, Niger, Amazone, ORSTOM, Paris.

Opsahl, S., and R. Benner. 1995. Early diagenesis of vascular plant tissues, lignin and cutin decomposition and biogeochemical implications. Geochim. Cosmochim. Acta 59, 4889–4904.

Pak, H., J.R.V. Zaneveld and R.W. Spinrad. 1984. Vertical distribution of suspended particulate matter in the Zaire river, estuary and plume. Neth. J. Sea Res. 17, 412–425.

Probst, J.L., R. Nkounkou, G. Krempp, J.P. Briquet, J.P. Thiebaux, J. P. and J.C. Olivry. 1992. Dissolved major elements exported by the Congo and Ubangi rivers during the period 1987–89. J. Hyd. 135, 237–257.

Probst, J.L., J. Mortatti, and Y. Tardy. 1994. Carbon river fluxes and weathering CO2 consumption in the Congo and Amazon River basins. Appl. Geochem. 9, 1–13.

Prunier, G. 2008. Africa's World War: Congo, the Rwandan genocide, and the Making of a Continental Catastrophe. Oxford University Press.

Purey-Cust, H.E. 1900. Report on the undercurrents in the river Congo, West Africa. Hydrographic Department, London.

Rabouille, C., J.-C. Caprais, B. Lansard, P. Crassous, K. Dedieu, J.-L. Reyss, and A. Khripounoff. 2009. In situ measurements of sediment oxygen consumption and organic matter budget in the Southeast Atlantic continental margin close to the Congo Canyon. Deep-Sea Res. II 56, 2223–2238.

Ragueneau, O., A. Regaudie-de-Gioux, B. Moriceau, M. Gallinari, A. Vangriesheim, F. Baurand, and A. Khripounoff. 2009. A benthic Si mass balance on the Congo Margin: Origin of the 4,000m DSi anomaly and implications for the transfer of Si from land to ocean. Deep-Sea Res. II 56, 2196–2206.

Raymond, P.A., J.W. McClelland, R.M. Holmes, A.V. Zhulidov, K. Mull, B.J. Peterson, R.G. Striegl, G.R. Aiken, and T.Y. Gurtovaya, T.Y. 2007. Flux and age of dissolved organic carbon exported to the Arctic Ocean: A carbon isotopic study of the five largest rivers. Glob. Biogeochem. Cyc. 21, GB4011, doi:10.1029/2007GB002934.

Runge, J. 2007. The Congo River, central Africa. In: Gupta, A. (ed.), Large Rivers: Geomorphology and Management. Wiley, pp. 293–309.

Savoye, B., N. Babonneau, B. Dennielou and M. Bez. 2009. Geological overview of the Angola–Congo margin, the Congo deep-sea fan and its submarine valleys. Deep Sea Res. II 56, 2169–2182.

Seyler, P., and F. Elbaz-Poulichet. 1996. Biogeochemical control on the temporal variability of trace element conentrations in the Oubangui River, Central African Republic. J. Hyd. 180, 319–332.

Shepard, F.P. and K.O. Emery. 1973. Congo submarine canyon and fan valley. Bull. Am. Ass. Petrol. Geol. 57, 1679–1691.

Spencer, R.G.M., A. Stubbins, P.J. Hernes, A. Baker, K. Mopper, A.K. Aufdenkampe, R.Y. Dyda, V.L. Mwamba, A.M. Mangangu, J.N. Wabakanghanzi, and J. Six. 2009. Photochemical degradation of dissolved organic matter and dissolved lignin phenols from the Congo River. J. Geophys. Res. Biogeosci. 114, G03010, doi: 10.1029/2009JG000968.

Spencer, R.G.M., G.R. Aiken, R.Y. Dyda, K.D. Butler, B.A. Bergamaschi, and P.J. Hernes. 2010a. Comparison of XAD with other dissolved lignin isolation techniques and a compilation of analytical improvements for the analysis of lignin in aquatic settings. Org. Geochem. 41, 445–453.

Spencer, R.G.M., P.J. Hernes, R. Ruf, A. Baker, R.Y. Dyda, A. Stubbins, and J. Six. 2010b. Temporal controls on dissolved organic matter and lignin biogeochemistry in a pristine tropical river, Democratic Republic of Congo. J. Geophys. Res. Biogeosci. 115, G03013, doi: 10.1029/2009JG001180.

Spencer, R.G.M., P.J. Hernes, A.K. Aufdenkampe, A. Baker, P. Gulliver, A. Stubbins, G.R. Aiken, R.Y. Dyda, K.D. Butler, V. L. Mwamba, A.M. Mangangu, J.N. Wabakanghanzi, and J. Six. 2012. An initial investigation into the organic matter biogeochemistry of the Congo River. Geochim. Cosmochim. Acta 84, 614–627.

Striegl, R.G., G.R. Aiken, M.M. Dornblaser, P.A. Raymond, and K.P. Wickland. 2005. A decrease in discharge-normalized DOC export by the Yukon River during summer through autumn. Geophys. Res. Lett. 32, L21413, doi:10.1029/2005GL024413.

Stubbins, A., R.G.M. Spencer, H. Chen, P.G. Hatcher, K. Mopper, P.J. Hernes, V.L. Mwamba, A.M. Mangangu, J.N. Wabakanghanzi, and J. Six. 2010. Illuminated darkness: Molecular signatures of Congo River dissolved organic matter and its photochemical alteration as revealed by ultrahigh precision mass spectrometry. Limnol. Oceanogr. 55, 1467–1477.

Taylor S.R. and S.M. McLennan. 1985. The Continental Crust: Its Composition and Evolution. Blackwell. Oxford.

van Bennekom, A.J., G.W. Beroer, W. Helder and R.T.P. de Vries. 1978. Nutrient distribution in the Zaire estuary and river plume. Neth. J. Sea Res. 12, 296–323.

Viers J., B. Dupré, M. Polvé, J. Schott, J.L. Dandurand, and J.J. Braun.1997. Chemical weathering in the drainage basin of a tropical watershed (Nsimi-Zoetele site, Cameroon): comparison between organic-poor and organic-rich waters. Chem. Geol. 140: 181–206.

Wefer, G., W.H. Berger and C. Richter. 1998. Proc. ODP Initial Reports, 175. Ocean Drilling Program, College Station, TX.

Weishaar, J.L., G.R. Aiken, B.A. Bergamaschi, M.S. Fram, R. Fuji, and K. Mopper. 2003. Evaluation of specific ultraviolet absorbance as an indicator of the chemical composition and reactivity of dissolved organic carbon. Environ. Sci. Technol. 37, 4702–4708.

23

The Nile delta in the anthropocene: drivers of coastal change and impacts on land-ocean material transfer

W. Moufaddal

1. Introduction

The Nile delta is one of the earliest recognized deltaic systems in the world, and its fertile land has been a focal point for emergence and flourishing of the ancient pharaonic civilization, as well as modern life in Egypt (Fig. 23.1). Despite its historical importance, it represents only about 2.3% of Egypt's surface area. This delta accommodates about 45% of human population and supports at least 45% of total agricultural land, 60% of the industry, and most of the large cities, trade, and fishing activities of the country (Sestini 1992).

The origin of the Nile delta can be dated back to the late Miocene, when it was no more than a marine embayment and part of the Tethys Sea in the southeastern Mediterranean (Goudie 2005). It began to take its modern form in the late Pliocene, when the Nile began to advance northward across this embayment. During the low sea-level stands of the Pleistocene, large quantities of sand and mud were transported via the paleo-Nile and dispersed far into the embayment, forming a large submarine fan (Sestini 1992; Hamza 2009). Continuation of this process for several thousands of years has eventually resulted in emergence of a new arcuate shape land and typical prograding delta during the Holocene about 7,500 years ago (Stanley and Warne 1994).

The Nile River (~6,650 km long) has been the major source of all sediments and fertile soil of the delta since its inception in late Miocene until recently. About 93–98% of this sediment was derived from the Ethiopian highlands, which is drained by the Blue Nile and Atbara Rivers (Shukri 1950; Hassan 1981; Stanley et al. 1992). Growth of the delta and deposition of new sediments (mainly silt and clay) on the top soil was very active, particularly at the time of the seasonal flooding during August–October (Stanley 1996). This resulted in seaward progradation of the delta over the past 5,000 years by as much as 50 km north of the present coastline (Stanley 1988, 1990). The annual flood also enriched the delta shelf and the receiving basin in the southeastern Mediterranean with inorganic nutrients and organic matter that supported a classic massive phytoplankton bloom and, consequently, productive wild fisheries off the delta coast (Halim 1976; Sharaf El Din 1977; Halim et al. 1995).

Historical documents show that there have been at least seven major distributaries of the Nile River crossing the delta during middle to late Holocene (pre-Dynastic to Roman time). These are from east to west: the Pelusiac, the Tanitic, the Mendesian, the Phatnitic (or Phatmetic), the Sebennytic, the Bolbitine, and the Canopic (Toussoun 1934; United Nations Development Programme/United Nation Educational, Science and Cultural Organization [UNDP/UNESCO] 1978; Sestini 1992). Currently,

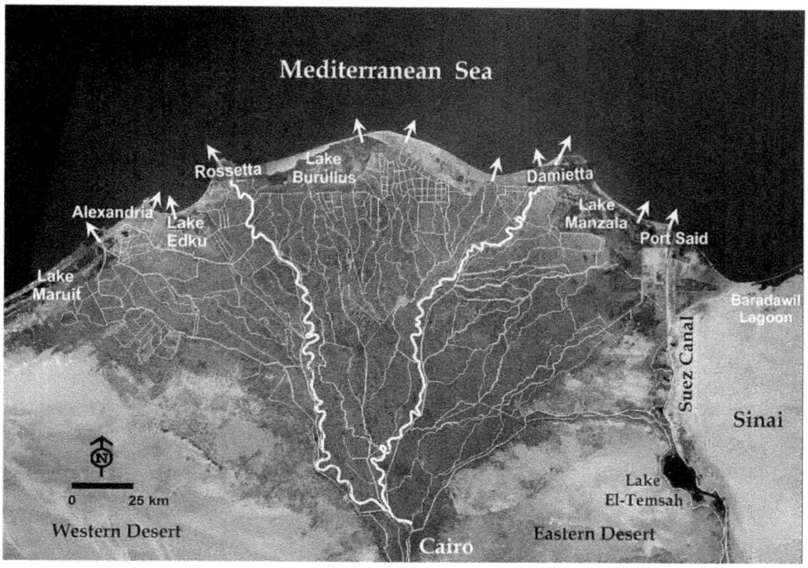

Figure 23.1. Satellite image for the Nile delta showing the coastal lakes, the two main branches (Rosetta and Damietta) of the Nile, and the intensive irrigation canals and drains, which channelize the whole delta and form extensive waterway system. White arrows indicate locations of major urban outfalls, drains, and lakes outlets along the delta coast.

the Nile delta is drained by only two distributary branches (Rosetta and Damietta), with minimal input of sediment and natural fertilizers primarily because of damming of the river and channelization of the delta (Stanley 1996). The Rosetta and Damietta branches (Fig. 23.1) are believed to be the relict of the Bolbitine and the Phatnitic branches (Stanley et al. 1992). The eastern branches (Tanitic, Pelusiac) and perhaps the Sebennitic branch have silted up and disappeared about 1,000 years ago, whereas the other western branches are thought to be disappeared much earlier, in the second to fifth centuries (Toussoun 1934).

Geographically, the Nile delta encompasses about 22,000 km^2 of subaerial delta plain and comparable area offshore. The latter (Nile delta shelf) is part of the Levantine basin, and hence its hydrodynamic regime is controlled by the overall regime and surface circulation of the Levant and SE Mediterranean, which consists of sub-basin scale cyclonic and anti-cyclonic gyres in addition to the so-called Mid-Mediterranean Jet. The latter is a major stream flowing eastward (counterclockwise) cross-basin and transferring the Atlantic water eastward within the Levantine (Millot 2005). Unlike the other parts of the Levant, the continental shelf off the Nile delta is very wide, particularly north of Rosetta and Port Said, where it reaches the 200 m depth contour at about 60 km offshore. West of Alexandria as well as east of Port Said, it is narrower, only 10 km wide. The delta shelf consists of a series of terraces separated by low slopes that are bisected by drowned channels and a major submarine canyon fronting Rosetta promontory (Summerhayes et al. 1978).

Tidal range along the delta coasts is low (25–30 cm), and winds and waves are blown predominantly from the northwest direction oblique to the shore, generating in this way moderate to strong

(20–100 cm s^{-1} in velocity) eastward longshore currents dominating the entire delta coast and the inner shelf (Inman and Jenkins 1984). These currents and the other prevailing coastal processes are responsible for transport of eroded sand as well as anthropogenic effluents and agricultural drainage, which discharged from lakes outlets and urban outfalls towards the east along the north Sinai and Levant margins as far as Lebanese coast and Turkey (Sestini 1992). Notable similarities in physical processes and form of shelf and slope sedimentation are observed in some other large river deltas systems, particularly, the Amazon (see Chapter 4).

The subaerial plain hosts four shallow brackish water lakes (Fig. 23.1), in addition to one relatively open lagoon at its extreme eastern side. From the west to east, these are the Maruit, Idku, Burullus, Manzala, and Bardawil. These lakes used to represent ~25% of the total Mediterranean wetlands (Sestini 1992). Land reclamation, dumping, and drying up of lakes water bodies and other improper human practices led to substantial shrink of their total area and, consequently, reduction of this percentage (Moufaddal et al. 2008). With the exception of Maruit, all the other lakes are connected to the sea through outlets or narrow openings dissecting the sandy barriers and maintaining open exchange with the sea. These lakes are highly productive and periodically experience eutrophication problems, where they receive increasing amounts of agricultural runoff and sewage effluent through agricultural drains and urban outfalls (Nixon 2003). Lake Manzala is the most productive, with productivity reaching more than 1 g C m^{-2} hr^{-1} (Abdel-Moati and El-Sammak 1997). The subaerial section is also intersected by more than 10,000 km of irrigation canals and drains (Fig. 23.1) that mostly flow northward and forming the extensive drainage waterway system throughout the delta (Stanley 1996). These canals eventually discharge to one of the Nile branches or to the northern delta lakes, some directly to the Mediterranean Sea off the delta (Hamza 2009).

Population density is very high in the delta, particularly near its apex (Greater Cairo) and the northern and central parts. Roughly 40% of Egypt's population lives in major cities such as Alexandria and Cairo, where urbanization and migration from rural areas to large urban centers has been intense, especially into these two cities. Major cities of the delta include the Greater Cairo and Alexandria, Egypt's second largest city and largest industrial center of the country. Domestic effluents and waste waters of these cities are discharged into agricultural drains or directly to the coastal lakes. The main land-use activities in the delta include agriculture, industry, tourism, and trade. Aquaculture is also very active and widespread in the northern delta, particularly around the coastal lakes, and fish farming is now considered one of the most widespread land-use activities in the northern delta (Moufaddal et al. 2008).

Continuous input of anthropogenic nutrient-rich effluents from urban outfalls, fish farming, and other anthropogenic sources is believed to be responsible for enriching the coastal waters and the delta shelf with nutrients and, therefore, stimulating growth of a modern massive and continuous phytoplankton bloom off the delta (El-Sayed and van Dijken 1995; Halim et al. 1995; Nixon 2003). This bloom is evident in recent satellite images most of the year, where it can be seen as a massive bloom of dark green to brownish green color in front of the Nile delta and Sinai coasts. Chlorophyll-*a* (Chl-*a*) concentrations can be as high as 80 μg L^{-1}, in particular within coastal areas that lie in front of the Nile mouths, lakes outlets, harbors, agricultural drains, and urban outfalls (Coastal Water Monitoring Program of the Environmental Information and Monitoring Program [EIMP/CWMP] 2007).

Like most other deltas of the world, the Nile delta is a very dynamic coastal system, and therefore, its stability, sedimentology, and other aspects have been subject to long- and short-term changes across its history through natural and man-made processes. These changes have been intense and accelerated substantially through the last 100 years owing to human intervention to the drainage system and impacts of sea-level rise, land subsidence, and other factors. The construction of the Aswan High Dam (AHD) in the mid-1960s and other subsequent anthropogenic activities have resulted in multiple effects and long-term changes in the northern delta, particularly in coastal stability, marine biogeochemistry, and fisheries.

The geo-indicators of change of the coastal stability include erosion and accretion of the delta shoreline at alarming rates and reshaping of the delta promontories, subsidence and tectonic tilting of the northern parts of the delta at accelerated rates, and salt-water intrusion to underground water to distance of ~130 km from the delta coastline (Kashef 1983). The bio-indicators for change of biogeochemical cycle include disappearance of the historical Nile bloom after the final closure of the AHD, then its recovery in ~early 1980s primarily through increase of nutrient loading from the ever-increasing anthropogenic activities (Nixon 2003). There has been also a contemporaneous major fall and then rise of the coastal and marine fisheries off the delta after closing of the AHD in the mid-1960s.

2. Threats and major environmental problems

Today, the Nile delta, and particularly its coastal zone, is in state of disequilibrium because of the impacts of accelerating land subsidence and sea-level rise (SLR), which are considered as the most alarming natural risks, particularly at the low-lying areas of the coastal plain and the northern delta. Estimated trends for relative SLR at Alexandria, Burullus, and Port Said are 1.6, 2.3, and 5.3 mm yr^{-1}, respectively. These values combine the effect of both SLR and land subsidence (UNEP 2009). The potential impacts of SLR on low-lying coastal areas and coastal systems in the delta have been evaluated in recent years. For example, Inman and Jenkins (1984) estimated that rise in sea level over the past 100 years accounts for a shoreline retreat of 10–15 meters along the shores of the delta. Estimates of Milliman et al. (1989) indicate that under a worst-case scenario (rise of sea level 100 cm by 2100), habitable land loss in Egypt between 1989 and 2100 could be 24% and 36%, respectively. According to Ericson et al. (2006), about 2.1% of the total delta area and 1.3 million people will potentially be impacted by year 2050 due to SLR of ~6 mm yr^{-1}.

More recently, Ghoniem (2009) reported that that an SLR of only 1 m would flood much of the Nile delta, inundating about one-third (~34%) of its land, placing important coastal cities such as Alexandria, Idku, Damietta, and Port Said at a great risk and possibly causing displacement of about 8.5% of the nation's population to other locations. In the extreme case of 5 m SLR, more than half (~58%) of the Nile delta will be facing destructive impacts, which would threaten at least 10 major cities (among them Alexandria, Damanhur, Damietta, Mansura, and Port Said), flooding productive agricultural lands and threatening in this way the livelihood of about 14% of the country's population (Ghoniem 2009). Other expected impacts of SLR include displacement of more than 2 million people in large coastal cities (El-Raey 1997), loss of wetlands and agriculture land, aggravation of the problem of saline seawater intrusion to groundwater aquifers in the northern parts of the delta and its advancement farther landward, and salinization of the farmland as a result of increase the rate in saline seepage to the topsoil of the delta (UNEP 2009). Investigations have already proved that salt-water intrusion has been extended to a distance of about 130 km from the Mediterranean coast (Kashef 1983).

Land subsidence is a natural phenomenon in most of fluvio-marine deltas and is caused by sediment compaction, isostatic loading, and the underlying tectonic movements (Ericson et al. 2006). It has been an active phenomenon in the modern Nile delta since its formation in the Holocene (Stanley 1988, 1990). It is also believed to be responsible for some major changes in the paleography of Nile delta. According to Stanley (1988), subsidence of the northwestern part of the delta during the late Pleistocene and Holocene was rapid and likely to be responsible for the diversion and concentration of former major distributary channels of the Nile into this region. Age and thickness of accumulated deltaic sediments play an important role in controlling rate of subsidence. In this context, Becker and Sultan (2009) reported that sediment age rather than thickness alone might be the dominant factor controlling ongoing subsidence in the Nile delta. The problem of land subsidence has been aggravated in the Nile delta through intensive local groundwater withdrawal and hydrocarbon extraction. Results of investigations carried out by Stanley and coworkers (e.g., Stanley 1988, 1990; Warne and Stanley 1993; Stanley 2001; Stanley 2005) indicate that average subsidence rates for the past 7,500 years in the northern part of the Nile delta vary between 0.5 and 5 mm yr^{-1} (~2 mm near Alexandria at the west up to 5 mm near Port Said at the east). The northeastern part of the delta is subsiding at more accelerated rates (2–8 mm yr^{-1}) because of hosting comparatively higher Holocene sediment thickness than the other parts of the delta (Becker and Sultan 2009). Modern rates of subsidence are generally high compared with average Holocene rates. Highest modern rates are observed over the youngest (<3,500 year old) sediments under the terminus of the Damietta branch (up to 8 mm yr^{-1}) and around Lake Manzala (4–6 mm yr^{-1}), which is assumed to be underlain by the highest sediment thickness in the northern part of the delta (Becker and Sultan 2009). Rate of sea-level change is an important factor to consider in subsidence calculations. According to Stanley (1988), combined effects of subsidence, eustatic SLR, and the reduced sediment supply could cause a relative SLR of 1 m or more by the year 2100, a level that would submerge much of the delta region within 30 km of the coast.

The Nile delta coast is also considered to be one of the most anthropogenically altered coastal deltaic systems in the world (Bianchi and Allison 2009), whereby it has been subjected to drastic changes through the past 100 years as a result of river basin alterations and land-use transformation. Anthropogenic drivers to these changes include (1) diversion of the natural Nile flow system and flood patterns through emplacement of series of dams along the River (Fig. 23.2) since 1902, (2) intensive channelization of the delta, (3) discharge of untreated effluents and other waste waters to the coastal waters in the northern lakes and delta shelf, (4) drying up and reclamation of parts of the northern lakes and wetland areas for fish farming and other uses, and (5) rapid urbanization and unplanned coastal development. A combination of these factors has turned the delta into a destructional phase and changed it from a river- or fluvial-dominated delta into wave-dominated one (Stanley 1996). It has also led to the appearance of a series of environmental problems in the delta plain. The most alarming problems include erosion of delta coasts, collapse of coastal fisheries, marine pollution, degradation of the northern lakes, salinization and waterlogging of farmland, and intrusion of saltwater to the underground freshwater (Moufaddal 2005).

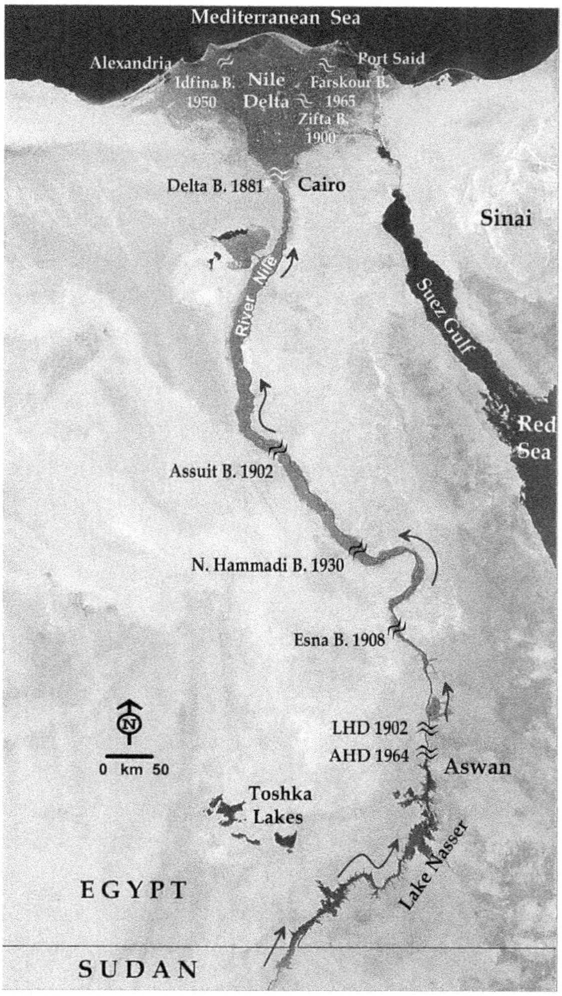

Figure 23.2. Locations and construction dates of the dams and barrages (abbreviated B.) constructed on the Nile River in Egypt since the late nineteenth century (adapted from Frihy and Lawrence 2004).

Since the final closure of the AHD, the sediment discharge and flow from the Nile became negligible. In effect to this, most of the delta coasts are currently eroding and retreating rapidly and at very alarming rates. Coastal erosion is more pronounced along the delta promontories and headlands (Rosetta, Damietta, and Burullus) and can be as much as 100 m yr^{-1} or even more (Frihy et al. 2009; Hereher 2011). A portion of the eroded material is transported away to the embayments and saddles between delta promontories forming accretional features, as well as to the lake inlets and river mouths, causing sedimentation problems and, consequently, navigation hazards within these areas (Frihy and Lawrence 2004). Emplacement of the AHD on the Nile upstream and other subsequent changes (cut-off riverine sediments and natural fertilizers, freshwater diversion, and channelization of the delta, etc.) also negatively impacted the nutrient supply and disrupted the biogeochemical cycle in the receiving basin off the delta coast and the Levantine basin. Biological productivity has been severely affected by a more than 95% decrease in phosphates and nitrates (Halim et al. 1967), and the historical Nile phytoplankton bloom disappeared for a while from the delta front and shelf areas (Aleem 1972; Halim 1976; Halim et al. 1995; Nixon 2003). Also, fish catches and the coastal and marine fisheries yielded from the Egyptian Mediterranean and the Nile delta have been collapsed (Aleem 1972) during the 1960s until the early 1980s (Nixon 2003).

However, the recent anthropogenic N and P inputs to the river and increased domestic sewage and untreated agricultural drainage, which eventually discharged to the coastal waters off the delta coast, seem to have exerted, over time, a counter positive effect and replaced the effect of the pre-dam annual flood, contributing in this way to the reappearance of the Nile bloom (Nixon 2003; Halim 2004). Previous studies (e.g., Dowidar 1984; Halim et al. 1995; El-Sayed and van Dijken 1995; Nixon 2003) indicated that the post-dam phytoplankton bloom (>1965) is different from the pre-dam one in many aspects (see the next section on the difference between the classic and modern Nile blooms) and that there were several shifts in the Chl-*a* concentrations and phytoplankton levels in the Nile delta shelf during the post-dam years (Lavender et al. 2009).

Coastal pollution and degradation state of water quality along the delta coast and inner shelf is currently one of the most alarming environmental problems. Very large amounts of raw sewage, agricultural runoff, and industrial effluents are being discharged on daily basis, mostly untreated (Halim and Abou Shouk 2000), to the coastal waters, particularly from land-based sources in the Nile delta and adjacent coasts, to deteriorate in this way the coastal lakes and the inshore waters with a variety of chemical and organic pollutants. The absence of the dilution effect of the Nile freshwater through complete damming of the River (Moufaddal 2005), together with increased pressure from population growth and rapid urbanization, has accentuated this problem (Sestini 1992). Change of irrigation system into the perennial one after construction of the AHD in 1964 has also exacerbated the problem of waterlogging and salinitization of the delta farmland. Soil salinity and waterlogging conditions represent a major threat to the sustainability of irrigated agriculture lands in Egypt. About 40% of the cultivated area in the Nile delta is affected to some extent by waterlogging and soil salinity as sole elements or together (Awaad and Ahmed 2010).

3. Anthropogenic drivers of coastal change and impacts on land-ocean material transfer

Apart from the effect of the natural factors such as climatic variations, sea level changes, and land subsidence, which have played an important role in fluctuation of the sediment and nutrient budgets of the Nile, the modern configuration and ocean biogeochemistry of the Nile delta has been also largely affected by human intervention to the Nile flow system, land-use transformation, and other large-scale anthropogenic influences.

The year 1881 marks the first event in the history of human intervention to the modern delta and the Nile River discharge, when the delta barrages were constructed near Cairo (Fig. 23.2) in an attempt to impound the floodwater and improve the irrigation in the lower delta (Halim 1990). In 1902, the Nile was dammed for the first time through the Low Aswan Dam (LAD), which was built

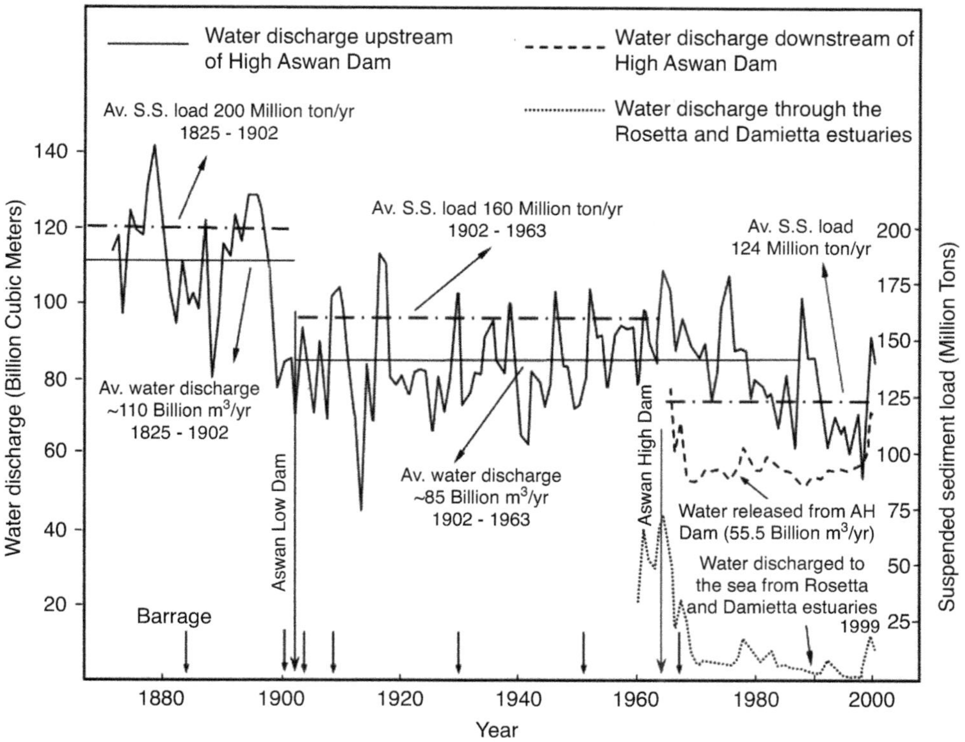

Figure 23.3. Long-term variations in water discharge and suspended sediment load of the Nile as recorded upstream of the AHD between 1871 and 2000 (after Fanos et al. 2001; Frihy and Lawrence 2004).

at the former First Cataract of the Nile and raised twice, between 1907–1912 and 1929–1933, for reducing risk of flooding and increasing economic and agriculture production in the Lower Egypt. The LAD was provided with sluices to let the floodwaters flow with the sediment load (Halim 1990). However, its design wasn't efficient enough to provide a complete and adequate protection against the risk of floods. At this point, the average Nile discharge was estimated to be about 84×10^9 m^3 yr^{-1} (Fig. 23.3), mostly in the flood season (White 1988; El-Sayed 1990; Halim et al. 1995). The annual flood was fairly active and strong enough to make the Nile stream and freshwater flow sometimes extend as much as 80 km seaward of the delta and to a depth of about 100 m (Sharaf El Din 1977; Summerhayes et al. 1978).

A total of 90×10^9 kg of nutrient-rich sediments on average were also discharged into the sea each year (Fig. 23.3), mostly during the flood season (Sestini 1992), with up to 98% of this load derived from the Ethiopian highlands (Shukri 1950; Hassan 1981). The Nile sediment load was variable (Fig. 23.3) and during floods was composed of 25% sand, 45% silt, and 30% clay (Sharaf El Din 1977). The effect of this was pronounced in the form of a surface plume of suspended silt and clay (1–4 g L^{-1}) carried eastward along Levant margins by the predominant longshore currents (Summerhayes et al. 1978), as far as the Lebanese coast, and associated with it a massive phytoplankton bloom in front of the Nile delta (Sestini 1992; Halim et al. 1995). The effect of the sediment influx on delta coasts has long been difficult to quantify (White 1988; El-Sayed 1990) as a result of lack of time series and reference data, but analysis of historical maps and some available information indicate that the delta coast, particularly along its promontories, has witnessed an overall constructional phase before damming of the river and progressively prograded seaward for about 3–4 km throughout the eighteenth and most of the nineteenth century (Blodget et al. 1991). However, this trend has been reversed at the beginning of the twentieth century. The Rosetta promontory, for example, retreated about 1.5 km between 1909 and 1965 (Inman and Jenkins 1984).

The reverse pattern of erosion along the delta coast is likely attributed to the effect of impounding of the annual flood and laden sediments by the LAD. Change of the precipitation pattern in east Africa is another possible reason whereby it might have led to decrease of the Nile water hydrology and the Nile discharge. Rossignol-Strick (1983) reported that the Nile discharge was subjected to an overall decrease of about 25% around year 1910 owing to reduction of the summer monsoon and equatorial rains over Ethiopia and east Africa (Fig. 23.3).

The impact of the LAD on the marine productivity is also not clear, but the available information indicates that it is not likely to have induced any significant effect. The reason is that the LAD has been designed with movable sluices to allow the floodwater (and less effectively, the sediment load) to pass through the dam to the main course of the river, where it is carried away to the lower delta and continental shelf (El-Sayed and van Dijken 1995).

The second stage of damming of the river began in 1960 through construction of the AHD, another higher dam near Aswan at about 6.5 km upriver from the older low dam (Fig. 23.2). The main purpose for emplacement of the AHD was to further protect the country from the risk of periodic flood and draughts through controlling of the annual flood and Nile drainage, but its basic functions also included increasing farmland and electricity production (Smith 2003).

The AHD was designed to store the Nile flow and the annual floodwater in a separate and extensive artificial reservoir (Lake Nasser) ahead, so that there would be no excess beyond the actual needs of 55.5 billion cubic meters, as permitted under the Nile Waters Agreements of 1929 and 1959 (White 1988). Storage of the Nile flow in Lake Nasser permitted the year-round irrigation of all cultivated lands in the delta and changed the irrigation system from basin to perennial (Sestini 1992). With closure of the AHD in 1964, the Nile flood has been completely impounded, and the majority of the fluvial material is blocked behind the dam in Lake Nasser (Stanley 1988). As a result, physical parameters and biological productivity of the coastal and shelf waters off the delta and the Levant have been strongly affected. Earlier studies (e.g., Halim et al. 1967; Sharaf El Din 1977) reported increase of transparency and decrease of the salinity and dissolved oxygen levels after closure of the high dam. Post-dam reduction in water discharge and sediment supply (Fig. 23.3) has also inevitably triggered an enormous change in sedimentary processes and has therefore led to further reshaping and reorientation of extensive sections of the Nile delta and Sinai coasts, where the strong predominantly eastward-directed coastal currents have accelerated erosion and increased sediment transport along these sections (UNDP/UNESCO 1978; Stanley 1988). It has been estimated (Hereher 2011) that the Rosetta promontory had lost 12.29 km^2 of land between 1973 and 2008, with most land loss and shoreline retreat occurring between 1973 and 1978 (0.55 km^2 yr^{-1} and 132 m yr^{-1}, respectively). Currently, the Rosetta and Damietta promontories are rapidly eroding at rates of about 100 m yr^{-1}

and 14 m yr^{-1}, respectively (Frihy et al. 2009). However, acceleration of coastal erosion along the wave-dominated delta during recent times is not solely attributed to damming of the river and the AHD impacts, but also to some additional anthropogenic factors such as channelization of the delta and water diversion projects. Natural forces such as coastal processes, eustatic SLR, and subsidence have also contributed to this phenomenon.

Marine biological productivity has also been temporarily affected owing to dramatic decrease of the net sediment and nutrient loads of the Nile (Halim et al. 1967). As a result, the historical Nile phytoplankton has temporarily disappeared from the coastal and shelf waters off the delta and the eastern Egyptian Mediterranean coast. The disappearance of the bloom impacted negatively on the marine fisheries, with the total Egyptian Mediterranean coastal fish catch decreasing dramatically in the late 1960s and 1970s to less than 25% of its pre-dam total yield during the same period (Halim et al. 1995; El-Sayed and van Dijken 1995; Nixon 2003; Oczkowski et al. 2009).

Impounding of the Nile by the AHD has also deleteriously impacted the coastal water quality and therefore accentuated the coastal water pollution at the front of the delta through disruption of most of freshwater discharge and hence the dilution effect to the domestic effluents and other land-based pollutants, which are discharged in very large quantities on a daily basis from the nearby urban centers to the coastal lakes and inshore waters off the Nile delta. This problem has been more aggravated by three additional factors (Moufaddal 2005). The first is the improper design and placement of most of the urban outfalls relative to the coastal waters. For example, of 18 outfalls dumping waste into the Alexandrian coastal water, only one exceeded 730 m in length, whereas the length of any of the other outfalls does not exceed 40 m at maximum (El-Sharkawi 1990). Moreover, most of these outfalls are placed at shallow depths, and therefore, waste plumes are more likely to rise to the surface in a very short time (Roberts et al. 1989). The second factor is dumping of immense volumes of domestic effluents and agricultural drainage wastes predominantly into the shallow bays (as in the case of El-Tabia and El-Max pumping stations, which discharge their wastes into Abu Qir and El-Max bays, respectively) and near estuaries that have a residence time longer than that of the open-coastal waters. The consequences of long residence times are that effluent may accumulate over time, entailing bays to have high concentrations of pollutants compared with the more open areas (Parnell 2003). The third is the nature of the coastal current system and hydrographic regime prevailing along the delta front and the Levantine margin, which does not aid pollutants to be dispersed away to the open sea. On the contrary, the east-trending longshore current, which predominant along the delta and the Egyptian Mediterranean coast, promotes transport of industrial effluents, domestic sewage, and all other material discharged from the city of Alexandria, for example, toward the east and close to the delta and Sinai coasts. Surprisingly, role of these factors and disruption of the AHD dilution effect in aggravation of the coastal pollution didn't receive much attention from the marine scientific community and has largely escaped detailed analysis from recent research.

Channelization of the delta and construction of artificial waterways for irrigation (Fig. 23.1) represents another important factor in controlling of the sediment distribution and nutrient budget of the Nile. Change of the irrigation system into perennial after construction of the AHD has permitted reclamation of new agricultural land areas and allied drainage canals and artificial waterways in the delta and the nearby desert areas. Stanley (1996) reported that more than 10,000 km of irrigation and drainage canals, most only 2–3 km apart, have been excavated in the delta during the period from 1967 to the mid-1990s. According to the same study, some canals, originally dug in Dynastic time (Butzer 1976), have also been expanded and enlarged over the years. Initiation of this dense network of closely spaced artificial waterways and the subsequent major new irrigation projects (e.g., El-Sallam Canal in the northeastern delta and Toshka in the southwestern desert) after impounding of the Nile flow have trapped much of the remaining fertile sediment of the Nile and prevented it from reaching the coastline (Fanos 1995; Stanley and Warne 1998) and have therefore contributed to aggravation of the problem of coastal erosion and land loss in the northern Nile delta (Stanley 1996).

Complete entrapment of the nutrient-rich material in Lake Nasser and drainage canals has also led to decrease of fertility of the delta farmland (Orlova and Zenkovitch 1974). This imposed farmers in the delta to rely on artificial fertilizers, particularly superphosphates and nitrates (Sestini 1992), in an attempt to increase fertility of the farmland and compensate for the deficit in the riverine organic matter and natural fertilizers after closure of the AHD. Land-use change and, particularly, transformation of farmland and parts of the coastal lakes into fish farms is a widespread activity in the northern Nile delta and represents a new important source for input of nutrients to the coastal inshore waters. The Nile delta hosts highly developed agricultural areas representing about 45% of the total cultivated land in the country. Despite urban encroachment and transformation of farmland into other uses, the total agriculture area has increased about 4×10^6 feddan ($\sim17\times10^3$ km^2) between 1965 and 2000 through reclamation of parts of the coastal lakes and wetlands in the northern delta and desert fringes west and east of the delta.

The intensive agricultural farming and widespread use of the soil in the delta contribute to nutrient loading through excess use of synthetic fertilizers and production of high amounts of solid and liquid manure, with discharge of drainage water into the coastal lakes or directly to the inshore area. It is of note that the total annual fertilizer consumption in Egypt has increased almost four-fold in about three decades, when it climbed to about 1.3×10^9 kg from 1965 to 2002 (Ozckowski et al. 2009). Blockage of the Nile sediments and attempts of the delta famers to improve fertility of their farmland is largely responsible for this climb. Fish farming and aquaculture activities, in general, are also growing in importance and have been expanding in the delta since the early 1980s, particularly near Alexandria and along the south and southeast margins of the coastal lakes, as a result of the poor return from the salty soils of the newly reclaimed lands and the higher economic return compared with agriculture (Sestini 1992; Nixon 2003). It has been estimated (Moufaddal et al. 2008) that the area exploited for fish farming around lakes Edku and Burullus has increased from 62.9 km^2 to 65 km^2 and from 330 km^2 to 364 km^2; i.e., a 2.1-km^2 and 16-km^2 net increase, between July 2002 and May 2007, respectively. Most of this increase was at the expense of lake water areas through dumping and reclamation their southern and eastern shores. More recently, El Banna and Frihy (2009) showed that a total area of about 48 km^2 of fish farms has been newly established between 1982 and 2002 in the Manzala lake area at the expense of the adjacent sand dune fields, sabkhas, marshes, and salt pans occurring south of the lake.

Most of the aquaculture in the northern delta is semi-intensive and known for its excess use of fertilizers and organic wastes (commercial feed, poultry waste, agricultural byproducts) (Oczkowski and Nixon 2008), with drainage of their nutrient-laden waters to the coastal lakes and then to the inshore waters. Effluents and drainage water discharged from this intensive fish farming, represent, therefore, an additional source of nutrient input and contribute in this way to enhancement of the biological productivity and stimulation growth of local algal blooms and eutrophication problems in the northern delta lakes as well as in the recipient inshore waters, particularly in the front of the drainage canals and lakes outlets.

There is a great lack of quantitative data regarding the total nutrient load of northern delta fish farming to the Mediterranean Sea or coastal lakes. To the author's knowledge, research works and attempts aiming to assess the total load or the amount of nitrogen and phosphorus inputs from fish farming or other anthropogenic sources in the Nile delta, or its impact on the coastal lakes and marine environment, are generally very scarce or almost absent. However, the available information indicate that biological productivity in the coastal lakes has at least doubled in the last three decades due to the progressive increase in discharge of agricultural drainage and domestic waste water. It has also been estimated that the percentage of phosphorous increased 15 times between 1963 and 1978 in Lake Maruit (Sestini 1992), which is has been long known as one of the oldest and largest fish farming areas in the northern delta. Fish farming is likely to be responsible for part of this change through supply of more nutrient-rich drainage and organic matter to water bodies of these lakes. The progressive increase of the total aquaculture production in Egypt from $\sim 62 \times 10^6$ kg in year 1995 to 226×10^6 kg in year 1999 at an annual growth of about 25–43% (Basurco 2001) gives additional supporting evidence, as this rise should be accompanied by a similar increase in the nutrients released from fish farms that are mostly concentrated in the northern Nile delta. Also, increase in nitrogen and phosphorus inputs from intensive mariculture in the whole Mediterranean region (including the Nile delta and the Levant) by about 25 and 22 times, respectively, during the period from 1990 to 2000 (UNEP/MAP/MED POL 2004) should have promoted high primary productivity in the shelf and coastal water off the delta.

Population growth and rapid urbanization in the Nile delta represent additional factors in controlling land-ocean material transfer, as they can impact ocean biogeochemistry and marine productivity both positively, through discharge of increased quantities of domestic sewage to the coastal waters, and negatively, through suppressing transfer of nutrients and organic matter to the inshore area by conversion of farmland and wetlands into built-up areas. Human population in the large cities and urban areas of the Nile delta has doubled over the last three decades. Associated with this, expansion of public water and sewer systems and a significant and effective increase of per capita calorie and mean consumption eventually translate to more input of nitrogen (Oczkowski et al. 2009). The total quantity of the drainage and domestic waste waters that are discharged into the sea and coastal lakes each year equals 12×10^9 m^3 (Oosterbaan 1999). Only minor amounts of these waste waters are subjected to primary treatment (without removal of nutrients) before dumping into the coast (Halim and Abou Shouk 2000; Oczkowski et al. 2009). UNEP/FAO/WHO (1996) introduced a simple Population Shoreline Density index to show the potential regional and local consequences of population densities of all Mediterranean countries relative to their shoreline length on nutrient loads and eutrophication. Accordingly, Egypt has been given the highest index, with generation of about 73×10^6 kg yr^{-1} nitrogen and $\sim 9 \times 10^6$ kg yr^{-1} phosphorus (assuming 4.4 kg N and 0.54 kg P excreted per capita yr^{-1}). This shows that progressive increase of population could contribute to the total nutrient budget and therefore cause enhancement of marine productivity of coastal waters near major coastal cities and urban centers. Population growth and urbanization can also cause an opposite effect through encroachment of agricultural land, coastal lakes, and wetlands that are considered as prime sources for nutrient input. For example, the Manzala lake, one of the largest and most productive wetland areas in the northern delta, has been reduced by about 50% between 1955 and 2002 as a result of expansion of the nearby urban centers, land reclamation, and other human factors (El Banna and Frihy 2009).

Rapid urbanization contributed also to shoreline instability through uncontrolled coastal development, implementation of seawater structure of improper design, contour alteration of the shore area, and removal of coastal sand dunes. For example, the northeastern coast of the Nile delta at Damietta harbor and Ras el-Bar has been subject to high rates of erosion (-14 m yr^{-1}) and accretion (15 m yr^{-1}), respectively, primarily because of the impact of man-made coastal protection structures and partly because of dominated natural processes (El Banna and Frihy 2009). Earlier studies (e.g., Fanos et al. 1995; Frihy and Lawrence 2004; El-Sayed et al. 2005) reported comparable changes along the Nile delta coast owing to implementation of other coastal projects.

4. The classic versus modern phytoplankton Nile bloom

Based on the previous discussion, the modern history of the Nile delta can be broadly subdivided into two distinct stages; the pre- and post-high dam. Biogeochemically, there have been large-scale changes and major shifts in the levels of the phytoplankton chlorophyll and nature of the Nile bloom (Fig. 23.4) after construction of the AHD and complete impounding of the River by the other subsequent anthropogenic activities. The coastal fisheries seem to have also been impacted by these changes where there were comparative shifts in sardine and shrimp catches. The following section aims to show the main biogeochemical aspects and spatial characteristics of the Nile phytoplankton bloom during the pre- and post-AHD conditions.

4.1. The classic Nile bloom and pre-dam conditions

Prior to construction of the AHD, most of the annual Nile freshwater and sediments were discharged along with their dissolved and adsorbed nutrients to the receiving basin off the Nile delta coast (Halim 1976; Stanley 1996), between 134 and 56×10^9 kg of sediment (Shukri 1950) and 84 to 34×10^9 m^3 of water each year (Halim 1960) (Fig. 23.3). Most of this discharge was from the two mouths (Rosetta and Damietta) of the Nile during the flood season, with average flow rates of 6×10^3 m^3 s^{-1} or even more (Vörösmarty et al. 1999). The outflow then progressed rapidly eastward along the Egyptian Mediterranean coast as a turbid coastal plume as much as 80 km seaward of the delta (Sharaf El Din 1977), showing in this way sharp boundaries with a sudden discontinuity in salinity, density, turbidity, and nutrients (Halim 1960) and pronounced effects on water quality (for a detailed explanation of characteristics of river plume systems and their effects on coastal biological and physical processes, please refer to Chapter 3). At this stage, the river carried in suspension 1–4 g L^{-1} of sediment, most of which was silt (45%) and clay (35%) fractions (Emelyanov and Shimkus 1972; Summerhayes et al. 1978). The outflow also spread considerable amounts of nutrients along the neritic and oceanic waters off the delta coasts, which were of positive effect on marine productivity and the whole food chain of the area (Halim et al. 1995). During an average flood, the total discharge of nutrients was

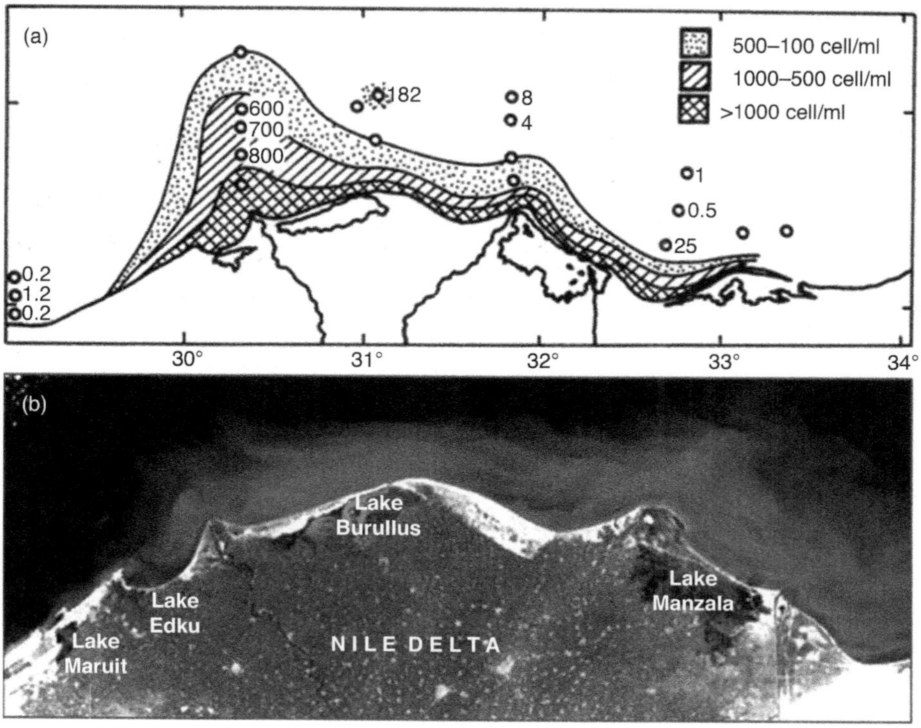

Figure 23.4. Comparison between the classic (a) and the modern (b) Nile phytoplankton blooms as appeared on product of mapping of pre-dam in situ measurements (after Halim 1976) and recent satellite (MODIS) image taken on February 5, 2003 (image courtesy of the MODIS Rapid Response Team at NASA/GSFC).

estimated to be 5.5×10^6 kg of phosphate and 280×10^6 kg of silicate per year (Halim 1976; El-Sayed and van Dijken 1995).

It has been estimated (Halim 1991, 2004; Halim et al. 1967, 1995) that the 1964 flood season, which is considered to be the last high flood before closing of the AHD, released about 8.2×10^6 kg of dissolved phosphate and 410×10^6 kg silicates into the delta shelf. The amount of the adsorbed fraction of phosphate is believed to be at least five times higher than the dissolved fraction (Halim et al. 1995). Estimates of Nixon (2003) indicate that the annual Nile flood was delivering from 7 to 11×10^6 kg of phosphorous, 7×10^6 kg of nitrogen, and 110×10^6 kg of silica. In any case, inflow of these nutrients with the annual floodwater was enough to trigger an exceptionally large and dense (5–10 m depth of the upper surface water layer) phytoplankton bloom off the Nile delta (Halim 1976). The standing crop of this bloom increased by 200–fold, from $50–150 \times 10^3$ cell L^{-1} to $2–20 \times 10^6$ cell L^{-1} (Fig. 23.4a) in the circumlittoral belt within a few days, with persistence of the bloom for several months during the flood season (Halim et al. 1967, 1995). This bloom was initiated within a few hours of the flood discharge along the continental shelf waters off the delta (Halim 1991) and used to have a characteristic brownish green color because of mixing of the reddish brown color of the suspended sediments and the greenish color of the Nile bloom. The phytoplankton assemblage of the seasonal bloom was exclusively composed of diatoms (Halim 1960; Aleem and Dowidar 1967). The bloom magnitude was variable from year to year and from one location to another along the delta coast, where it has been largely controlled by volume of the incoming flow and laden sediments, prevailing hydrodynamics, and concentration of natural fertilizers in the flood water. It was generally much larger and massive in Rosetta estuary than Damietta (Fig. 23.4a), with a variation of standing crop in the order of 1 million individual cells up to 10 million or more per liter (Halim et al. 1967).

A continuous supply of nutrients during the flood sustained the bloom for several months in spite of the heavy grazing by zooplankton and planktivorous fish (Halim et al. 1995). The Nile bloom also supported a productive marine fishery, especially of sardines and prawns. Between 1962 and 1965, an average of 37×10^6 kg of sardines were caught per year (Wadie 1982; Nixon 2003). Estimates of Aleem (1972) show that the marine fisheries landed from the Egyptian Mediterranean coasts in 1964 account for 23% of the total fish catch of Egypt.

4.2. The modern Nile bloom and post-dam conditions

Damming of the Nile at Aswan in 1965 resulted in a control of the river drainage and cut off most of the sediments and the adsorbed and dissolved nutrients that were input to the lower and subaqueous delta plains during the annual flood. The volume of Nile water outflow to the Egyptian coast has dropped from about 53×10^9 m^3 in 1964 to only 18×10^9 m^3 and 12×10^9 m^3 in 1965 and 1966, respectively (Halim 1976). This amount has further dropped to its present day level of 2.5–4×10^9 m^3 (Halim et al. 1995) or less. Blockage of this material has suppressed the fertilization of the shelf waters and eventually has led to a disappearance of the classic Nile bloom. Studies of Aleem (1972), Halim (1976), and Dowidar (1984) reported that the post-dam phytoplankton standing crop in the Nile delta shelf rarely exceeded a few thousand cells per liter, compared with ~10 million cells per liter during any pre-dam flood. In addition, the composition and relative abundance of phytoplankton populations has been altered significantly. Diatoms are no longer the dominant component of the coastal waters, with the dinoflagellates becoming equally important (Halim 1976; Dowidar 1984; Halim 2004).

Reduction of the natural fertilizers and the subsequent suppression of the Nile bloom also led to a collapse of fisheries. The decline in the catch of sardines was especially dramatic, from an annual average of almost 37×10^6 kg between 1962 and 1965 to 6.5×10^6 kg from 1966 through to 1970 (Wadie 1982). El Sayed and van Dijken (1995) also showed that the sardine fishery total catch decreased from 18×10^6 kg in 1962 to about only 55×10^4 kg in 1968 and 1969. Fisheries bottom trawl surveys off the western part of the Nile delta showed an almost 80% decline in average fish and shrimp abundance between 1960–1961 and 1969–1970 (Wadie 1982). The decline in marine fisheries off the Nile delta persisted for about 15 years; it began to recover again after the early 1980s. By the late 1980s, total fish catch from the Egyptian Mediterranean coast exceeded the levels prior to the closing of the High Dam (Halim et al. 1995; El Sayed and van Dijken 1995; Nixon 2003; Halim 2004). Recovery of the marine fisheries could suggest a contemporaneous enhancement in rates of primary production and a return of the Nile phytoplankton bloom (Nixon 2003; Lavender et al. 2009). However, it might also be possible that increasing of fishing grounds and fishing techniques effort after resolution of the military conflict and signing a peace treaty with Israel in 1979 contributed to this. According to Halim (2004), both decline and recovery of fisheries yields during the 1967–1980 period cannot be taken as outcome of a single event. Taking this into consideration, the recovery can be interpreted as the result of a combination of factors; the restoration of normal conditions of fishing activities accompanied by improvement of fishing techniques (Halim 2004) and progressive increase of agricultural effluents and other nutrient-rich anthropogenic runoff (Nixon 2003; Halim 2004; Lavender et al. 2009).

Lack of historical in situ measurements and time-series data, particularly on phytoplankton variability in the delta shelf, makes it very difficult to determine the exact timing or duration of recovery of the Nile bloom. However, a large number of studies have pointed to reappearance of the Nile bloom and predominance of higher phytoplankton concentrations along the inner delta shelf and adjacent coastal areas (e.g., Dowidar 1984; Sestini 1992; El-Sayed and van Dijken 1995; Halim et al. 1995; Nixon 2003). For example, the in situ measurements presented by Dowidar (1984) on the seasonal distribution of Chl-*a* and primary productivity in 1982 revealed that the seasonal cycle of chlorophyll biomass was unimodal, with a conspicuous peak in the winter and another less pronounced one in the spring. The seasonal Chl-*a* concentration taken in that year varied between 0.6 and 6.0 $\mu g\ L^{-1}$ in winter, between 0.2 and 2.5 $\mu g\ L^{-1}$ in spring, between 0.02 and 0.5 $\mu g\ L^{-1}$ in summer, and between 0.06 and 0.6 $\mu g\ L^{-1}$ in autumn. Other previous local studies (e.g., Fahmy et al. 1996; Labib 2000; Shams El-Din and Dorgham 2007), have reported very high Chl-a concentrations along the various parts of the delta shelf and inshore areas. More recent studies (e.g. Shams El-Din and Dorgham 2007; Moufaddal and Lavender 2009; Dorgham 2011) and satellite observations also document predominance of higher phytoplankton concentrations (Fig. 23.4b) during most of the year along the inner delta shelf and adjacent coastal areas.

Evaluation of the recent general trend and pattern of distribution of the monthly satellite-derived Chl-*a* in the Nile delta shelf during the period 1997 to 2006 and other results of the study of Lavender et al. (2009), reveal a stimulation of localized blooms with very strong color signatures in front of the Nile delta and also confirm the temporal and spatial shift in the Nile phytoplankton bloom compared to the pre-Dam conditions. Results also indicate that there is a general upward trend in the mean surface Chl-*a* concentration during that period. This upward trend seems to manifest the combined downstream effects of explosive growth in population, industry, agriculture and other sources of surface nutrients draining from the Nile delta into the Mediterranean (Lavender et al. 2009). Continuous surface runoff from the two branches of the Nile, and the other land-based sources (urban outfalls, lakes outlets, pumping stations), supply the coastal waters with organic and inorganic nutrients and so is the likely explanation for the persistence of the modern bloom spreading across the coastal and inshore areas of the delta most of the year. The presence of these blooms very close to urban outfalls and Nile estuaries, together with their strong signatures, support the idea that the surface runoff and discharge of anthropogenic nutrients from land-based sources is a key factor for their growth (Lavender et al. 2009).

Available information on the pelagic to demersal (P/D) ratio for the period 1962–1986 (Halim et al. 1995) tends to also support the theory of nutrient loading and its role in recovery of the Nile bloom and possibly the coastal fisheries. As shown in Fig. 23.5, the P/D ratio mainly remained above average until the mid-1960s; afterward, in 1966, the ratio dropped abruptly below the average

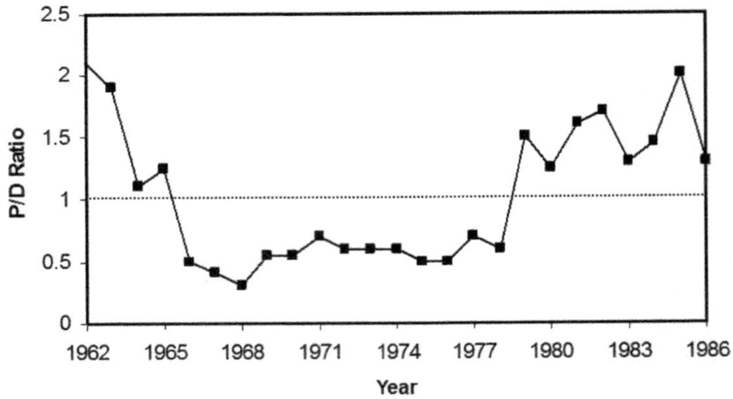

Figure 23.5. The pelagic to demersal (P/D) ratio for fish landings off the Nile delta during 1962–1986 (after Halim et al. 1995).

and remained at around 0.5 until 1979 before it returned to the pre-dam values of the early 1980s. The timing of the decline of the P/D ratio below the long-term average and its recovery coincides with construction of the AHD and subsequent disappearance of the historical Nile phytoplankton bloom in the mid-1960s and then with the return of this bloom (in a new modern form) in the early 1980s (Halim et al. 1995).

5. Conclusions, future directions, and research priorities

The Nile delta has subjected to long- and short-term historical changes from human-made and natural drivers. Most drastic changes occurred in the post-dam period, during which the final closure of the AHD and the other subsequent anthropogenic activities greatly affected the coastal stability and marine biogeochemistry in the delta plain. Changes in the sediment and nutrient budgets are already well recognized and established. However, to date, there is no accurate or detailed estimation to volume of this change or to contribution of the different anthropogenic activities or other drivers to the nutrient budget. Also, the role of some new anthropogenic and natural drivers and their impacts on the nutrient input has been ignored or not studied well. For example, the northwestern Egyptian Mediterranean coast from Al-Agamy (west of Alexandria) to Marsa Matruh was almost untouched and least developed in the early 1970s. However, this ~250-km long coast has completely changed since the early 1980s through coastal tourism development and now hosts thousands of tourist hotels and summer homes that become active and very busy in the summer, increasing in this way the nutrient runoff to the local coastal waters and also to the adjacent Nile delta shelf (through transportation of the discharged domestic wastes by the east-trending longshore current).

Similarly, the role and contribution of airborne nitrogen deposition to marine biological productivity in the delta shelf is not well studied or documented. It is already well recognized that anthropogenic nitrogen can enter the seawater from the atmosphere and is accounting for about one-third of the ocean's external (nonrecycled) nitrogen supply and up to 3% of the annual new marine biological production (Duce et al. 2008). Atmospheric emission of N species has also reportedly increased

250% and deposition to marine and terrestrial ecosystems increased over 200% during the period from 1600 to 1990 (Galloway and Cowling 2002). This shows that airborne nitrogen deposition runoff might be of value for marine ecosystems and biological productivity in the delta shelf. In any case, changes in the nutrient budget of the delta estuaries and coastal and marine ecosystems are expected to continue through the next few decades owing to some new anthropogenic drivers. These include future treatment and reuse of irrigation water and agricultural drainage and additional diversion of canal waters for new irrigation projects, for example, El-Sallam in the northeastern delta and Toshka (South Valley) in the southwestern desert.

In this context, the GEO (Group on Earth Observation) and World Bank are currently cofinancing major integrated projects for management of water irrigation and improvement of the sewerage and sanitation that aim to treat rural domestic sewage in the delta. Also, the government of Egypt is planning to divert additional canal water for new irrigation developments: some 30×10^8 m^3 yr^{-1} for the Toshka project in the Upper Egypt and about 10×10^8 m^3 yr^{-1} for the El-Sallam Canal project in Sinai. These two factors will likely cause decrease of the nutrient budget into half of the present supply through reduction of the drainage nutrient-rich water that are eventually discharged into the sea and lakes (Oosterbaan 1999). Counter effects are also expected through further population growth, increased development, additional use of synthetic fertilizers, and animal manure in the new cultivated land areas in the northeastern delta and north Sinai (El-Sallam Canal project area). This shows that there is an urgent need to adopt long-term studies and monitoring programs for assessment of current sediment and nutrient budgets. Further work is required to provide a precise quantitative assessment of impact of the current and future human activities on land-ocean material fluxes and the net effect of increasing and decreasing these fluxes on marine biological productivity and coastal stability in the delta. Particular attention must be paid to evaluation of total nutrient budget. The case is different for the sediment budget, which hasn't been subjected to any further change since it has been lowered to its minimal level after construction of the AHD and the subsequent delta channelization. Use and application of nitrogen and carbon fluxes modeling is also highly desired and would be very useful for assessment and simulation of the present situation and future changes in the biogeochemical fluxes to the Nile delta estuaries and coastal waters.

References

Abdel-Moati, M. A. R., and A. A. El-Sammak. 1997. Man-made impact on the geochemistry of the Nile Delta Lakes. A study of metals concentrations in sediments. Water, Air & Soil Pollut. 97: 413–429.

Aleem, A. A. 1972. Effect of river outflow management on marine life. Mar. Biol. 15: 200–208.

Aleem, A. A., and N. Dowidar. 1967. Phytoplankton production in relation with nutrients along the Egyptian Mediterranean coast. Proc. Int. Conf. Trop. Oceanography, pp. 305–327.

Awaad, M. S., and A. R. Ahmed. 2010. The negative role of soil salinity and waterlogging on crop productivity in the north-eastern region of the Nile Delta, Egypt. Res. J. Agric. Biol. Sci. 6: 378–385.

Basurco, B. 2001. Mediterranean Aquaculture: Marine Fish Farming Development. Option Mediterraneennes – International Centre for Advanced Mediterranean Agronomic Studies 19.

Becker, R. H., and M. Sultan. 2009. Land subsidence in the Nile Delta: Inferences from radar interferometry. Holocene 19: 949–954.

Bianchi, T. S, and M. A. Allison. 2009. Large-river delta-front estuaries as natural "recorders" of global environmental change. Proc. Natl. Acad. Sci. 106: 8085–8092.

Blodget H., P. T. Taylor, and J. H. Roak. 1991. Shoreline changes along the Rosetta-Nile Promontory: monitoring with satellite observations. Mar. Geol. 99: 67–77.

Butzer, K. W. 1976. Early Hydraulic Civilization in Egypt. University of Chicago Press, Chicago.

Dorgham, M. M. 2011. Eutrophication Problem in Egypt. *In* A. A. Ansari, S. Singh Gill, G. R. Lanza, W. Rast (eds.), Eutrophication: Causes, Consequences and Control. Springer, New York.

Dowidar, N. M. 1984. Phytoplankton biomass and primary productivity of the southeastern Mediterranean. Deep Sea Res. 31: 983–1000.

Duce R. A., J. LaRoche, K. Altieri, K. R. Arrigo, A. R. Baker, et al. 2008. Impacts of atmospheric anthropogenic nitrogen on the open ocean. Science 320:893–897.

EIMP/CWMP. 2007. Annual report on status of the water quality of the Egyptian Mediterranean coastal waters during 2006. The Egyptian Environment Affair Agency, Cairo. Available at: http://www.eeaa.gov.eg/eimp/reports/ArrAnualMed2006.pdf (accessed March 30, 2012).

El Banna, M. M., and O. E. Frihy. 2009. Human-induced changes in the geomorphology of the northeastern coast of the Nile delta, Egypt. Geomorphology 107: 72–78.

El-Raey, M. 1997. Vulnerability assessment of the coastal zone of the Nile delta of Egypt, to the impacts of sea level rise. J. Ocean Coast. Manag. 37: 29–40.

El-Sharkawi, F. 1990. Aquatic pollution (Alexandria region), pp. 55–71. *In* Regional Symposium for Environmental Studies. UNARC, Alexandria.

El-Sayed, M. Kh. 1990. Effect of river damming and deficit of sediment flux on the continental shelf: a case study of the Nile delta. *In* IOC-ICSEM Workshop on Ocean Sciences in Non-Living Resources (OSNLR), Intergovernmental Oceanographic Commission Workshop Report No. 89, Perpignan, France.

El-Sayed, M. Kh, N. A. Younan, A. M. Fanos and Kh. H. Baghdady. 1995. Accretion and Erosion Patterns along Rosetta Promontory, Nile Delta Coast. J. Coast. Res. 21: 412–420.

El-Sayed, S., and G. van Dijken. 1995. The southeastern Mediterranean ecosystem revisited: Thirty years after the construction of the Aswan High Dam. Quarterdeck 3: 4–7.

Emelyanov, E. M. and K. M. Shimkus. 1972. Suspended matter in the Mediterranean Sea, pp. 417–439. *In* D. J. Stanley (ed.), The Mediterranean Sea – A Natural Sedimentation Laboratory. Dowden, Hutchinson & Ross, Stroudsburg, PA.

Ericson, J., C. Vörösmarty, S. Lawrence, L. Ward, and M. Meybeck. 2006. Effective sea-level rise and deltas: Causes of change and human dimension implications. J. Glob. Planet. Change 50: 63–82.

Fahmy M. A, M. M. Abbas, and A. I. Beltagy. 1996. Distribution of nutrient salts in the coastal Egyptian Mediterranean waters after 30 years of the High Dam erection. Bull. Natl. Inst. Oceanogr. Fish., Egypt 22: 267–291.

Fanos, A. M. 1995. The impact of human activities on the erosion and accretion of the Nile Delta coast. J. Coast. Res. 11: 821–833.

Fanos, A. M., A. A. Khafagy, and R. G. Dean. 1995. Protective works on the Nile Delta coast. J. Coast. Res. 11: 516–528.

Fanos A. M., A. A. Khafagy, and M. M. El Kady. 2001. Variations of the Nile River discharge and sediment regime during the period from 1800 to 2000 and its effects on the Nile Delta coast. 8th Int. Symp. on River Sedimentation, Cairo.

Frihy, O. E., and D. Lawrence. 2004. Evolution of the modern Nile delta promontories: development of accretional features during shoreline retreat. Environ. Geo. 46: 914–993.

Frihy, O. E., W. M. Moufaddal, A. A. El-Gindy, and E. Deabes. 2009. Comparison of annual rates of beach changes determined from land survey and remote sensing techniques along the Nile delta. Int. J. Eco. & Develop. 12: 42–58.

Galloway, J. N., and E. B. Cowling. 2002. Reactive nitrogen and the world: 200 years of change. Ambio 31: 64–71.

Ghoniem, M. 2009. A remote sensing study of some impacts of global warming on the Arab region. *In* M. Tolba and N. Saab (eds.), Arab Environment Climate Change: Impact of Climate Change on Arab Countries. Arab Forum for Environment & Development (AFED), Beirut, Lebanon.

Goudie, A. 2005. The drainage of Africa since the Cretaceous. Geomorphology 67: 437–456.

Halim, Y. 1960. Observations on the Nile bloom of phytoplankton in the Mediterranean. J. Cons. Int. Explor. Mer. 26: 57–67.

Halim, Y. 1976. The Nile and the East Levantine Sea, Past and Present, pp. 76–84. *In* R. Natarajan (ed.), Recent Researches in Estuarine Biology. Hindustan Publishing Corporation, Delhi, India.

Halim, Y. 1990. Manipulations of hydrological cycles. *In* Technical Annexes to the Report on the State of Marine Environment. UNEP Regional Seas Reports and Studies No 114/1. Annex VI: 231 B 319.

Halim, Y. 1991. The impact of human alterations of the hydrological cycle on ocean margins, pp. 301–325. *In* R. F. C. Mantoura, J.-M. Martin, and R. Wollast (eds.), Ocean Margin Processes in Global Change, Dahlem Workshop Reports: Physical, Chemical, and Earth Sci. Res., Rept. 9. John Wiley & Sons, New York.

Halim, Y. 2004. The Nile and the Levantine Sea. ATOMS-Med Workshop, Monaco.

Halim, Y. and F. Abou Shouk. 2000. Human impact on Alexandria's marine environment. *In* H. M. Mostafa, D. Grimal and D. Nakashima (eds.), Underwater Archeology and Coastal Management, Focus on Alexandria. UNESCO-Cairo, Paris.

Halim, Y., S. A. Morcos, S. Rizkalla and M. Kh. El-Sayed. 1995. The impact of the Nile and the Suez Canal on the living marine resources of the Egyptian Mediterranean waters (1958–1986). *In* Effects of Riverine Inputs on Coastal Ecosystems and Fisheries Resources. FAO Fisheries Technical paper 349.

Halim, Y., S. K. Guergues, and H. H. Saleh. 1967. Hydrographic conditions and plankton in the South East Mediterranean during the last normal Nile flood (1964). Int. Rev. Ges. Hydrobiol. 52: 401–25.

Hamza, W. 2009. The Nile Delta, pp. 75–94. *In* H. J. Dumont (ed.), The Nile: Origin, Environment, Limnology and Human Use, Monographiae Biologicae, Vol. 89.

Hassan, F. A. 1981. Historical Nile floods and their implications for climatic change. Science 212: 1142–1145.

Hereher, M. E. 2011. Mapping coastal erosion at the Nile Delta western promontory using Landsat imagery. Enviro. Earth Sci. 64: 1117–1125.

Inman, D. L. and S. A. Jenkins. 1984. The Nile littoral cell and man's impact on the coastal zone of the southeastern Mediterranean. Scripps Institution of Oceanography. Reference Series, vol. 84–31. University of California, La Jolla.

Kashef, A. I. 1983. Salt water intrusion in the Nile delta. Ground Water 21: 160–167.

Labib, W. 2000. Short-term variations in phytoplankton and ambient conditions in a highly dynamic Egyptian marine basin. Vie Et Milieu 50: 29–39.

Lavender, S., W. M. Moufaddal, and Y. Pradhan. 2009. Assessment of temporal shifts of chlorophyll levels in the Egyptian Mediterranean shelf and satellite detection of the Nile bloom. Egypt. J. Aquat. Res. 35: 121–135.

Milliman, J. D., J. M. Broadus, and F. Gable. 1989. Environmental and economic implications of rising sea-level and subsiding deltas: the Nile and Bengal examples. Ambio 18: 340–345.

Millot, C. and I. Taupier-Letage. 2005. Circulation in the Mediterranean Sea, pp. 29–66. *In* A. Saliot (ed.), Handbook of Environmental Chemistry, Vol. 1, Springer-Verlag, Berlin Heidelberg.

Moufaddal, W. M. 2005. State of coastal water along the Nile Delta front, Egypt, as revealed by satellite imagery. Int. J. Rem. Sens. 26: 4877–4853.

Moufaddal W. M., E. El-Sayed, and E. Deghedy. 2008. Updating morphometric and idaphic information of lakes Edku and Burullus (NW Nile delta, Egypt) with the aid of satellite remote sensing. Egypt. J. Aquat. Res. 34: 291–310.

Moufaddal, W. M. and S. Lavender. 2009. Biogeochemical response to mesoscale circulation features and the high levels of the Nile flood in the SE Levantine basin, as revealed by ocean-color remote sensing. Egypt. J. Aquat. Res. 35: 431–443.

Nixon, S. W. 2003. Replacing the Nile: Are Anthropogenic Nutrients Providing the fertility once brought to the Mediterranean by a Great River?. Ambio 32: 30–39.

Oczkowski, A. and S. W. Nixon. 2008. Increasing nutrient concentrations and the rise and fall of a coastal fishery; a review of data from the Nile Delta, Egypt. Estuar. Coast. Shelf Sci. 77: 309–319.

Oczkowski, A. J., S. W. Nixon, S. L. Granger, A. M. El-Sayed, and R. A. McKinney. 2009. Anthropogenic enhancement of Egypt's Mediterranean fishery. PNAS 106: 1364–1367.

Oosterbaan, R. J. 1999. Impacts of the irrigation improvement project, Egypt, on drainage requirements and water savings. Technical Report. International Institute for Land Reclamation and Improvement (ILRI), Wageningen, The Netherlands.

Orlova, G. and V. Zenkovitch. 1974. Erosion of the shores of the Nile delta. Geoforum 18: 68–72.

Parnell, E. P. 2003. The effects of sewage discharge on water quality and phytoplankton of Hawaiian coastal waters. Mar. Enviro. Res. 55: 293–311.

Roberts, P., W. Snyder, and D. Baumgartner. 1989. Ocean outfalls. I. Submerged waste field formation. II. Spatial evolution of submerged waste field. J. Hydraul. Eng. 115: 1–48.

Rossignol-Strick, M. 1983. African monsoon: an intermediate climate response to orbital insulation. Nature 303: 46–49.

Sestini, G. 1992. Implications of climatic changes for the Nile Delta, pp. 535–601. *In* L. Jefric, J. D. Milliman and G. Sestini (eds.), Climatic Changes and the Mediterranean. E. Arnold, London.

Shams El-Din N. G. and M. M. Dorgham. 2007. Phytoplankton community in Abu Qir Bay as a hot spot on the Southeastern Mediterranean Coast. Egypt J. Aquat. Res. 33: 163–182.

Sharaf El Din, S. H. 1977. Effect of the Aswan High Dam on the Nile flood and on the estuarine and coastal circulation pattern along the Mediterranean Egyptian coast. Limnol. Oceanogr. 22: 194–207.

Shukri, N. M. 1950. The mineralogy of some Nile sediments. Quart. J. Geol. Soc. London 105: 511–534.

Smith, E. S. 2003. The Aswan High Dam at Thirty: An Environmental Impact Assessment. *In* T. L. Crisman, L. J. Chapman, C. A. Chapman, and L. S. Kaufman (eds.), Conservation, Ecology, and Management of African Fresh Waters, University Press of Florida.

Stanley, D. J. 1988. Subsiding in the northeastern Nile delta, rapid rates, possible causes and consequences. Science 240: 497–500.

Stanley, D. J. 1990. Recent subsidence and northeast tilting of the Nile delta, Egypt. Mar. Geol. 94: 147–154.

Stanley, D. J. 1996. Nile delta: extreme case of sediment entrapment on a delta plain and consequent coastal land loss. Mar Geol. 129: 189–195.

Stanley, D. J. 2001. Dating modern deltas: progress, problems, and prognostics. Ann. Rev. Earth Planet. Sci. 29: 257–94.

Stanley, D. J. 2005. Submergence and burial of ancient coastal sites on the subsiding Nile delta margin, Egypt. Méditerranée 1–2: 65–74.

Stanley, D. J., and A. G. Warne. 1994. Worldwide Initiation of Holocene Marine Deltas by Deceleration of Sea-Level Rise. Science 265: 228–231.

Stanley, D. J., A. G. Warne, H. R. Davis, M. P. Bernasconi, and Z. Chen. 1992. Late Quaternary evolution of the north-central Nile Delta between Manzala and Burullus lagoons, Egypt. Nat. Geogr. Res. Explor. 8: 22–51.

Stanley, D. J. and A. G. Warne. 1998. Nile delta in its destruction phase. J. Coast. Res. 14: 794–825.

Summerhayes, C. P., G. Sestini, R. Misdorp, and N. Marks. 1978. Nile delta: nature and evolution of continental shelf sediments. Mar. Geol. 27: 43–65.

Toussoun, O. 1934. Les ruines sous-marines de la Baie d'Abuquir. Bull. Soc. Archeol. Alex. 29: 342–352.

UNDP/UNESCO. 1978. Coastal protection studies. Project Findings and recommendations. Paris, UNDP/ EGY/73/063.

UNEP. 2009. Adaptation to Climate Change in the Nile Delta through Integrated Coastal Zone Management. UNEP Project document # 3748.

UNEP/FAO/WHO. 1996. Assessment of the state of eutrophication in the Mediterranean sea. MAP Technical Reports Series No. 106. UNEP, Athens,

UNEP/MAP/MED POL. 2004. Mariculture in the Mediterranean. MAP Technical Reports Series No. 140, UNEP, Athens.

Vörösmarty, C. J., B. M. Fekete, M. Routhier, and T. Baker. 1999. Global River Discharge Database, Version 1.1 (Riv. Dis. v1.0 supplement). Available at: http://www.rivdis.sr.unh.edu (accessed March 30, 2012).

Wadie, W. F. 1982. Observations on catch of the most important fishes along the Egyptian continental shelf in the South-Eastern part of the Mediterranean Sea. Bull. Nat. Inst. Oceanogr. Fish. Egypt 8: 213–227.

Warne, A. G., and J. D. Stanley. 1993. Archaeology to refine Holocene subsidence rates along the Nile Delta margin. Egypt. Geol. 21: 715–18.

White, G. F. 1988. The environmental effects of the High Dam at Aswan. Environment 30: 5–40.

24

Fate of nutrients in the aquatic continuum of the Seine River and its estuary: modeling the impacts of human activity changes in the watershed

J. Garnier, P. Passy, V. Thieu, J. Callens, M. Silvestre, and G. Billen

1. Introduction

Coastal marine eutrophication is presently considered a major threat at the global scale (Diaz and Rosenberg 2008), and the link between the very diverse manifestations of this phenomenon and the excessive discharge of nutrients by rivers draining highly populated continental areas has often been stressed. However, it is not simply the absolute value of nutrient discharge, but rather the balance between riverine inputs of nitrogen (N), phosphorus (P), and silica (Si), that is the primary cause of marine eutrophication, and silica must be taken into account as a nutrient for diatom growth (Officer and Ryther 1980; Conley et al. 1993; Turner and Rabalais, 1994; Justic et al. 1995a, 1995b; Garnier et al. 1995; Turner et al. 1998; Conley 1999; Billen and Garnier 1997; Humborg et al. 2000, 2008). It is therefore of crucial importance to fully understand the processes that determine the fluxes of C, N, P, and Si delivered by rivers (Cugier 1999; Cugier et al. 2005; Billen et al. 2007; Lancelot et al. 2007; Garnier et al. 2010a; Lancelot et al. 2011). Yet these processes are complex because between their land-based sources, mostly from agriculture and urban wastewater management, and their ultimate delivery into marine waters, nutrients have to travel downward through a whole continuum of aquatic systems, where they may be transformed, retained, or eliminated, thus changing their concentrations and ratios (Thieu et al. 2009).

In this respect, the Seine River system offers a particularly striking and well-documented example. The Seine basin, with 17 million inhabitants over an area of 76,215 km^2, is subject to considerable urban and agricultural pressures (Billen and Garnier 1999; Billen et al. 2001). The population density of more than 200 inh. km^{-2} has increased by 7% during the last 20 years, and two-thirds of the population is concentrated within the Paris conurbation (Table 24.1). The Seine basin is occupied by intensive agriculture, with arable land dominating in the centre of the basin, whereas some grassland and forests are distributed at the periphery (Table 24.1).

This has been intensively studied for more than 20 years in the scope of two long-term French research projects, namely the PIREN-Seine program, centered on the freshwater river system (Billen et al. 2007), and the Seine-Aval program, devoted to the estuary (Even et al. 2007; Garnier et al. 2008, 2010b).

In this chapter we describe the major processes controlling the transfer and biogeochemical transformation of C, N, P, and Si within the Seine land-to-sea continuum, along with the modeling approaches that have been developed to represent the complex interactions within this system and

Table 24.1. *General characteristics of the watershed of Seine River at the limit of the Seine Estuary (at Poses) and of the watershed of the estuary itself (fluvial estuary from Poses to Caudebec, and saline estuary downstream from Caudebec, i.e., a total length of 150 km)*

	Seine (at Poses)	Estuary
Mean discharge (m^3 s^{-1}) 1984-2007	480	
Area (km^2)	71,530	4,685
Drainage density ($km.km^{-2}$)	0.29	0.19
Maximum altitude (m)	910	338
Mean altitude (m)	97	134
Arable land, %	52.7	42.9
Grassland, %	10.2	26.4
Forest, %	25.8	20.5
Population 1990	15,382,830	1,066,209
Population 2006	16,440,175	1,090 229

predict their responses to changes in the pressures exerted by human activity within the watershed.

Similar to the structure of the models themselves, our analysis starts from the processes occurring in watershed soils and determining the diffuse sources of biogenic elements to headwaters, continues with the processes occurring within the water bodies during their downstream travel through the drainage network, then examines the estuarine processes, before using the suite of modeling tools to predict the river's nutrient delivery to the coastal zone resulting from future scenarios of wastewater treatment and agricultural practices.

2. From agricultural soils and aquifers to surface waters

The Seine basin covers one of the most intensive agricultural areas of the world, specialized in synthetic fertilizer-based cereal production. Losses from cropland thus account for by far the greatest diffuse source of nutrients to water resources (Fig. 24.1). We briefly examine the mechanisms involved in N, P, and Si transfer from agricultural soil to ground and surface water.

Nitrogen, because of the high solubility of nitrate, its dominant inorganic form, is highly mobile and subject to leaching from agricultural soil, particularly during periods when they are kept without plant cover after harvest.

Looking into the long-term relationship between crop production (in terms of N content of harvested crops) and the total nitrogen fertilization (total N inputs to the soil through manure and synthetic fertilizer application, atmospheric deposition, and crop symbiotic N_2 fixation) (Fig. 24.2) reveals that the increased yield over the last few decades was obtained through a continuous increase of total N fertilization, but in the end results in a loss of N use efficiency and an increase of the N surplus, which can be considered the potential source of N leaching. Soil leaching of nitrate from excess fertilization of agricultural soils is the primary cause of groundwater nitrate contamination. The large, unconfined aquifers that characterize the central region of the Seine basin are not the site

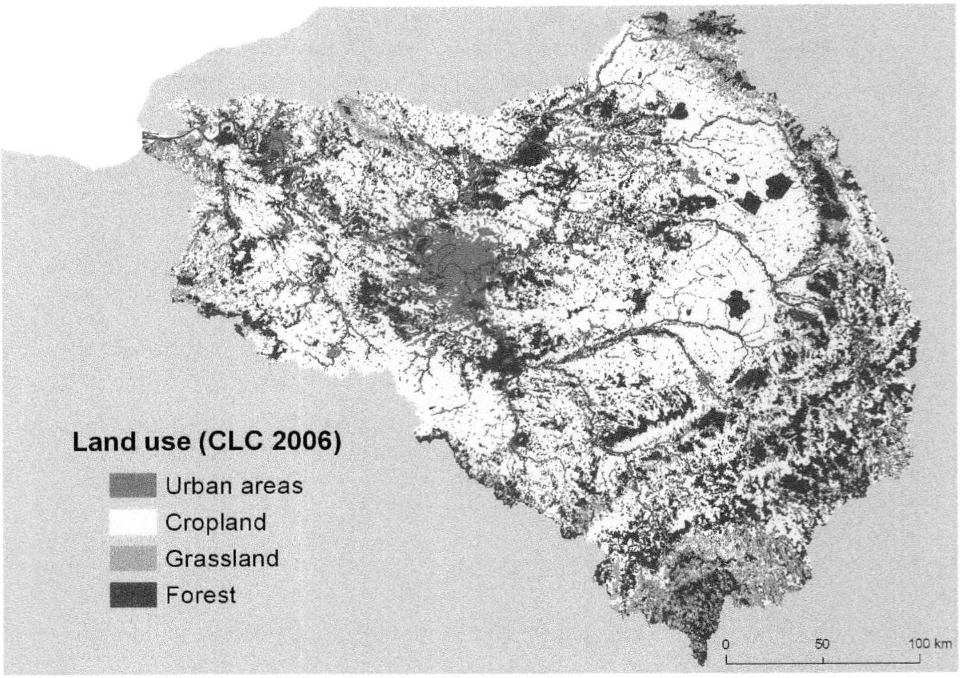

Figure 24.1. The Seine basin and its estuary. Drainage network and land use.

of significant denitrification (Vilain et al. 2012). However, because of the often very long residence time in these aquifers, their nitrogen concentration may not yet be in equilibrium with the nitrate concentration in the recharge water, so that, until equilibrium is reached, the aquifer stores nitrogen and releases less nitrogen than it receives.

Stream waters are fed from both base flow coming from aquifers and rapid subsurface flow, with a concentration close to the sub-root water composition. The seasonal variations in the contribution of these two components of discharge and their difference in concentration are responsible for the seasonality of nitrogen concentration in head waters (Fig. 24.3). Before they reach surface water, however, both base flow and subsurface flow have to cross the biogeochemically active riparian

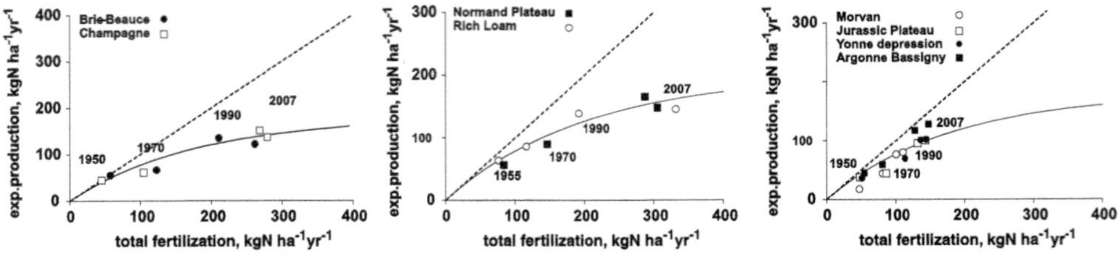

Figure 24.2. Long-term variation of the N content of crop harvest and total N fertilization of arable land (fertilizer and manure input, crop symbiotic N fixation, and atmospheric deposition) in different agricultural regions of the Seine watershed.

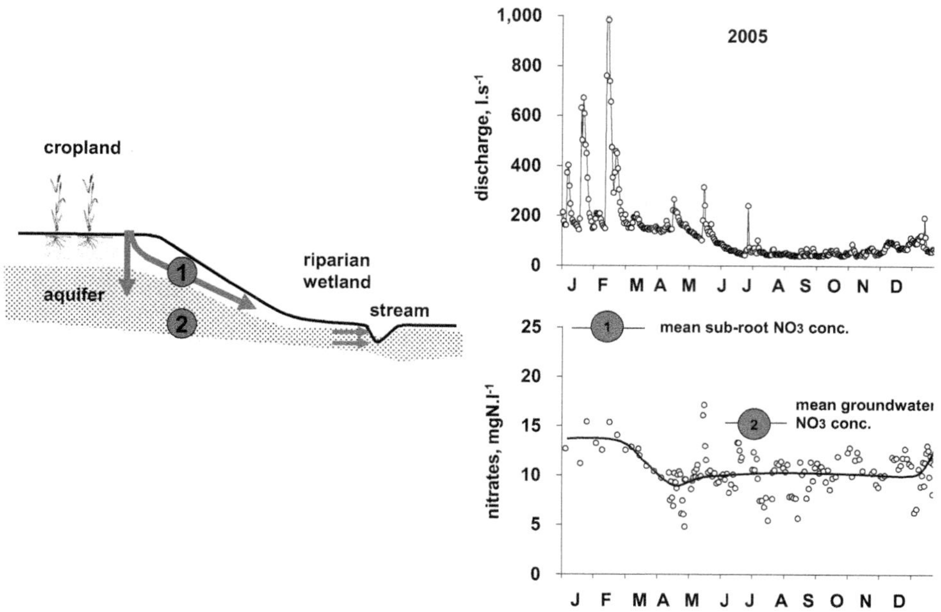

Figure 24.3. Seasonal variations of nitrate concentration in a rural headwater stream in the Brie region (east of Paris). The nitrate concentrations in the sub-root water (collected in tile-drains) and in the underlying aquifer are indicated for comparison: the difference between nitrate concentration in the stream and those in the sub-root and groundwater reflects riparian retention (data from Billy et al. 2011).

zone, which is often the site of significant denitrification. For this reason, the nitrate concentration in headwaters is significantly lower than in the sub-root and phreatic waters feeding them (Fig. 24.3).

Unlike nitrogen, phosphorus is tightly retained by soils, in an adsorbed or insoluble form. Over-fertilization of agricultural land should progressively decrease the capacity of soils to absorb phosphorus, but until then, erosion of particles is the main transfer process of phosphorus to rural headwaters. This delayed response of the saturated P-binding soils and the progress made in controlling point source pollution from household and industrial wastewater during the last two decades have resulted in a recent shift in the apportionment of point sources and diffuse sources, which now play a significant role (Némery and Garnier 2007a; Grizzetti et al. 2012, Fig. 24.4).

In-stream processes of phosphorus retention mostly comprise the uptake by planktonic algae followed by their immobilization within the sediment as well as the deposition of particulate phosphorus associated with suspended solids having adsorbed dissolved phosphorus released by point sources. In the Seine, outside periods of high aquatic vegetation development where biological uptake occurs, the dissolved orthophosphate concentrations in these waters closely correspond to the adsorption-desorption equilibrium with arable soil particles (Garnier et al. 2005; Billen et al. 2007).

Although silica was long considered to be mainly released from geochemical rock-weathering processes only dependent on lithology and climate (Meybeck 1986), several authors (Cochran and

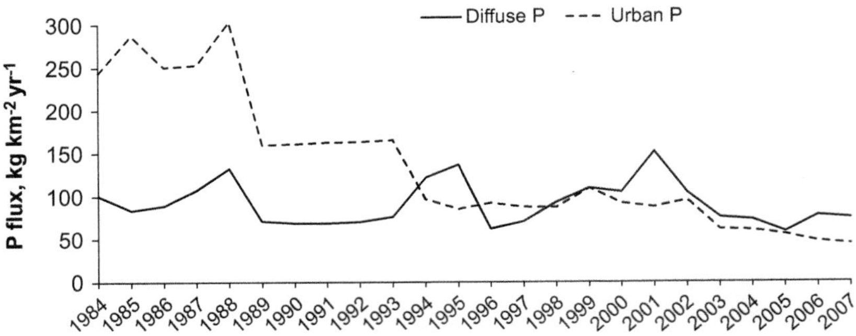

Figure 24.4. Interannual variations, in yearly averages, of phosphorus diffuse and point source inputs to the watershed.

Berner 1996; Conley 2002; Humborg et al. 2004; Struyf et al. 2010) have shown the influence of vegetation on silica export from soils to the water system. According to their view, very ancient agricultural land could well release less dissolved silica than a forested system, whereas recently cleared forested soils would present maximum silica release for several decades after having been put into cultivation. Garnier et al. (2002a) and Sferratore et al. (2006) have estimated the amount of particulate biogenic silica brought by soil erosion in the form of phytoliths in the Seine basin. This represents a maximum of 10% of the diffuse inputs of dissolved silica.

3. Changes in urban sources and wastewater treatment

Point sources are industrial and urban wastewaters (including households and indirect industrial emissions) discharged into the river network via wastewater treatment plants (WWTP); these data are currently provided by the French Seine Normandie Water Agency. Each WWTP is characterized by a number of inhabitant equivalents treated and the type of treatment applied (namely primary, secondary, or tertiary treatment, including P removal, nitrification, and denitrification). On average, 1,700 effluent outlets have been listed over the Seine basin, all geo-referenced for the needs of the modeling approach.

Whereas recent data have been fully documented since the 2000s with the application of the European Directive concerning urban wastewater treatment (UWWTD, 91/271/CEE), the data were reconstructed for the 1980s and 1990s, based on the population changes and the rate of connexion to sewer systems within the Seine watershed, the year of each WWTP implementation, and various assumptions (see later) on wastewater treatment progression.

The amount of P discharged per inhabitant equivalent (inh. equiv.) was at a maximum of 4 gP.day^{-1}.inh. equiv.$^{-1}$ owing to the use of polyphosphates in washing powders in the 1980s. In the late 1980s, phosphorus decreased significantly with the improvement of the organic matter abatement in wastewater treatment plants, and in the mid-1990s, phosphorus decreased again when polyphosphates were banned from washing powders in France, although this occurred later than the other surrounding countries because of the resistance of the French industrial lobbies (Billen et al. 1999) (Fig. 24.5). In addition, a gradual P treatment in WWTP was implemented after the release

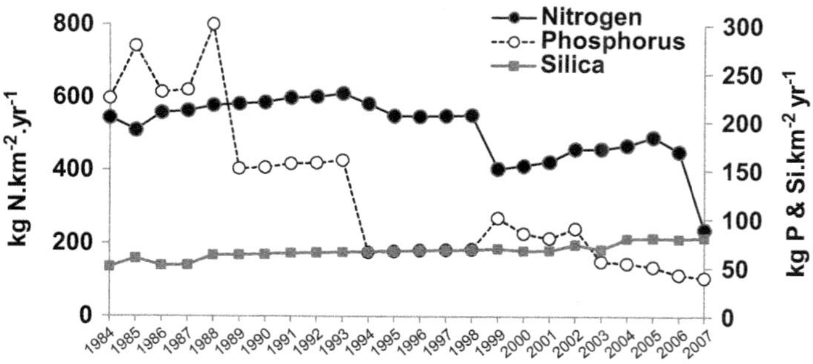

Figure 24.5. Changes in total nitrogen, phosphorus, and silica (dissolved + biogenic) urban inputs to the surface waters of the Seine during the period from 1984 to 2007.

of the European Directive 91/271/CEE concerning urban wastewater treatment, which imposed a specific treatment of P in the largest WWTPs. The specific P discharge decreased to 2 gP.day^{-1}. inh. equiv.$^{-1}$ (Garnier et al. 2006) and is even lower today; the value of specific P discharge is indeed close to the physiological release, i.e., 0.8 gP .day^{-1}. capita^{-1} (Verbanck et al. 1994). The recommendation of a decrease by about 85% of the P discharged into the river has now been nearly attained in the Seine and in other European rivers, e.g., the Rhine (see Chapter 25).

Regarding nitrogen, recommendations were given to cities located within nitrate-vulnerable zones (European Water Framework Directive, EU WFD 2000/60/EC) to implement a tertiary treatment so that the N discharge would be reduced by 70%. In the Seine River basin, most of Paris's wastewaters are treated at the Achères plant, one of the largest in Europe, with a treated volume of 1,700,000 m^3 of wastewater per day. In 2007, the Achères WWTP was upgraded to nitrify all NH_4 and to denitrify about 50% of NO_3. Whereas many of the small WWTPs in the basin already denitrify 70%, the latter objective will be reached in 2015 for the largest Achères WWTP.

For silica, based on the study by Sferratore et al. (2006), where silica was measured in the treated effluents of the Achères WWTP, but also compared with estimates from human diet and detergent consumption leading to a silica inhabitant equivalent (0.8 gSi. day^{-1}.inh. equiv.$^{-1}$), we calculated the amount of silica input to the drainage network of the Seine, using the total inhabitant equivalent in the Seine and its evolution (Fig. 24.5). The variations in silica load throughout the period are rather small compared with those of P and N. Indeed, despite a 40% increase of the number of inhabitant equivalents in the last 25 years, land-based emissions of silica that relatively stabilized over this period remain the main silica source of the Seine River system.

4. Transfer in the drainage network: the modeling approach

The RIVERSTRAHLER model (Billen et al. 1994; Garnier et al. 1995; Garnier et al. 2002b) describes the drainage network of any river system as a combination of basins, represented as a regular scheme of the confluence of tributaries of increasing stream order, each characterized by mean morphologic properties, connected to branches, represented with a higher spatial resolution. The advantage of this representation of the drainage network is that it takes into account, with a reasonable calculation time, both the processes occurring in small first-order streams (i.e., headwater streams) and those occurring in large tributaries. The water flows in the hydrographical network are calculated from the specific discharges generated within the watershed of the different sub-basins and branches considered. The RIVERSTRAHLER model is coupled to a GIS-interface (SENEQUE), designed to calculate the seasonal and spatial variations of water quality within drainage networks ranging in scale from 10 to more than 100,000 km^2 (Ruelland et al. 2007; Billen et al. 2007; Thieu et al. 2009; Le et al. 2010).

The essence of the SENEQUE/RIVERSTRAHLER model is that it couples these water flows that are routed through the defined structure of basins and branches with a model describing biological, microbiological, and physicochemical processes that occur within the water bodies. The module representing the kinetics of the processes is known as the RIVE Model. The state variables comprise nutrients, oxygen, suspended matter, dissolved and particulate nonliving organic carbon, and algal, bacterial, and zooplankton biomasses. Most of the processes that are important in the transformations, elimination, and/or storage of nutrients during their transfer within the network of rivers and streams are explicitly calculated, including algal primary production, aerobic and anaerobic organic matter degradation by planktonic, as well as benthic bacteria with coupled oxidant consumption and nutrient remineralization, nitrification and denitrification, and phosphate reversible adsorption onto suspended matter and subsequent sedimentation. Garnier et al. (2002b) provided a detailed description of the RIVE Model and of the physiological parameters used; the module of the benthic processes was revisited by Thouvenot-Korppoo et al. (2009) (Fig. 24.6).

Besides morphological and climatic constraints, the SENEQUE/RIVERSTRAHLER model takes into account diffuse and point sources of nutrients from land-based anthropogenic sources as described previously.

Because the kinetics and the parameters of the model, as well as their values, were determined independently from the observations, the comparison of the simulations with the observations, i.e., the available data of the variables taken into account in the model, is a true validation of the model. This validation exercise is done at Poses, the outlet of the Seine River, where most of the observations are available, although the model is run in the fluvial freshwater estuary up to Caudebec, the entrance of the saline zone (110 km farther). This validation is fully described in Passy et al. (2013), and only nitrate, ammonium, and total phosphorus are shown here (Fig. 24.7).

The model then calculates N, P, and Si deliveries to the estuary as well as the retention and/or elimination within the river continuum for the whole period (see also Thieu et al. 2009, for a wet and dry year and three drainage networks).

The interannual average in N and Si fluxes at the outlet of the river amounted to 1,890 kgN km^{-2} yr^{-1}, and 990 kgSi km^{-2} yr^{-1}, on average, with year-to-year variations related to the discharge variations for nitrogen and silica. This strong parallel between nutrient fluxes and water discharges does not show up as clearly for phosphorus (average flux, 145 kgP km^{-2} yr^{-1}), because point sources contributed greater quantities. The recent improvement of urban wastewater treatment thus led to a significant decrease of the phosphorus exported by the Seine River (Fig. 24.8).

The calculated difference between the flux delivered at the river and the sum of nutrient (point and diffuse) inputs from the watershed to the hydrosphere represents the overall retention in the drainage network; roughly, an average of 40% is found for all three elements, N, P, and Si (Fig. 24.8). The model explicitly calculates the contribution of various processes to this retention. Sediment storage,

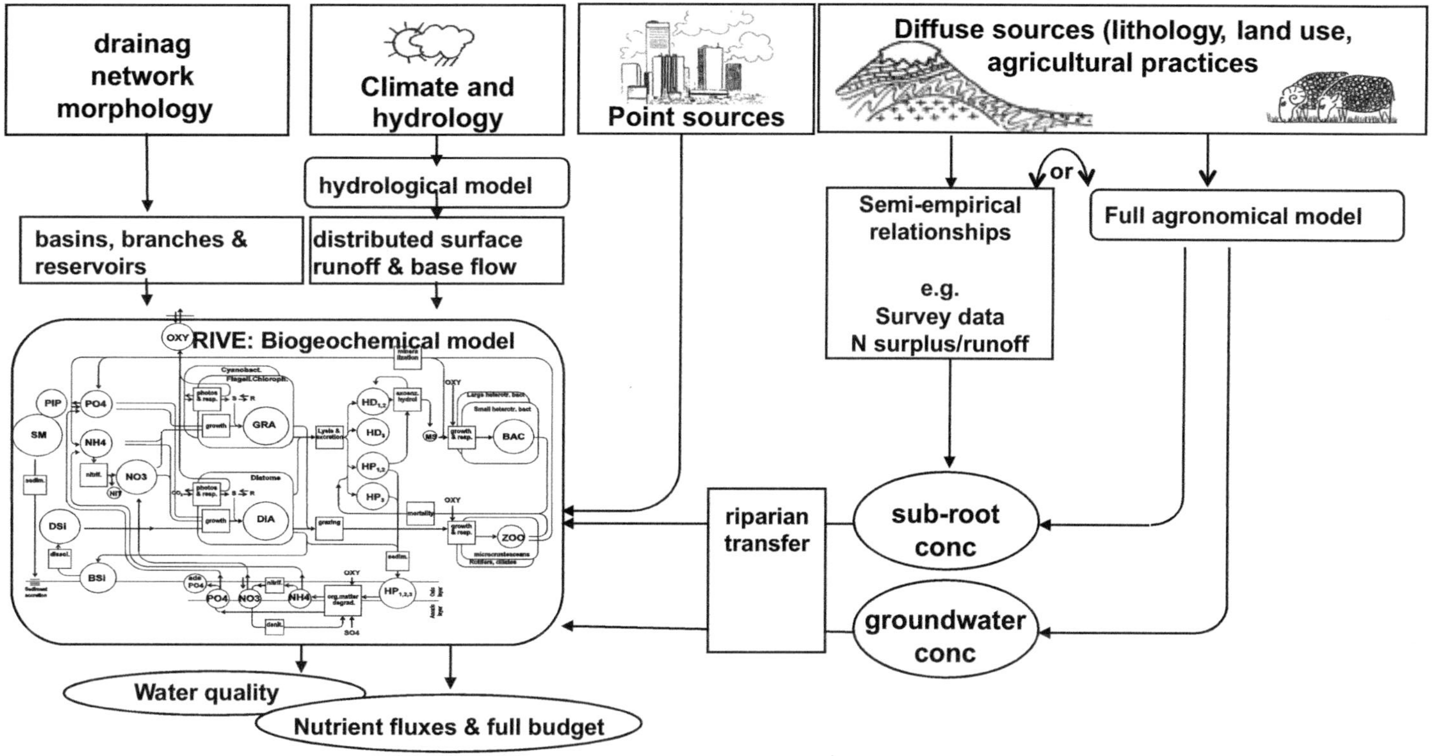

Figure 24.6. Description of the SENEQUE/RIVERSTRAHLER model. Drainage network morphology, climate, and hydrology, as well as diffuse and point sources, are the main constraints to be documented for implementation of the model. The discharge is broken down into two components and routed in the sub-basins classified into order rivers to the main branches and through reservoirs when they exist. Diffuse sources as given by the agronomical model or semi-empirical relationships for sub-root concentration and ground water concentration are calibrated for a riparian transfer filter. The variables (31 in all, e.g., nutrients and carbon in their various forms, etc.) feed the RIVE model of the ecological processes.

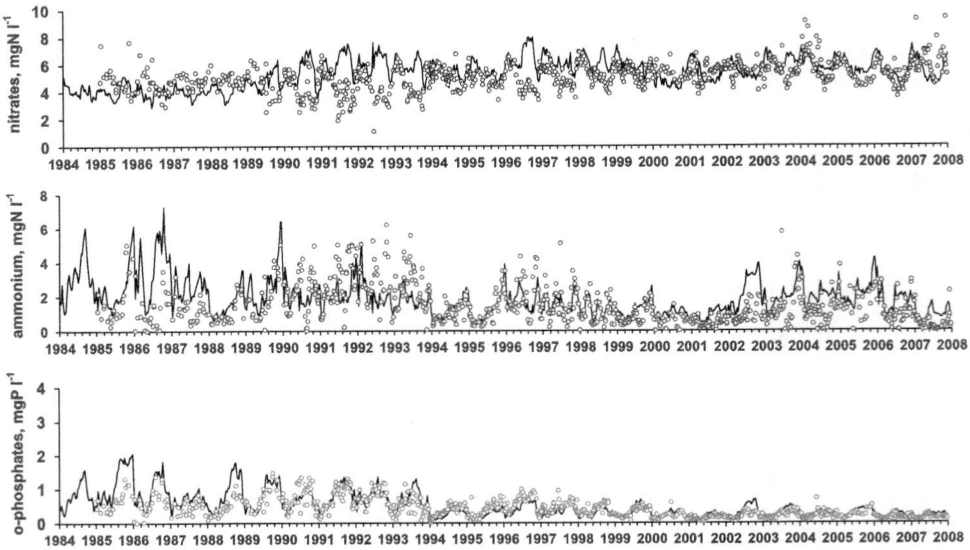

Figure 24.7. Validation of the SENEQUE/RIVERSTRAHLER model at the outlet of the Seine basin at Poses, the limit between the Seine River and the fluvial estuary: comparison of the calculations by the model with the observations (from top to bottom: nitrate and ammonium, mgN L^{-1}, total phosphorus, mgP L^{-1}).

although playing a minor role in nitrogen retention, is the major retention process for both silica and phosphorus (Fig. 24.8). Nitrogen retention/elimination is the highest in the riparian zones (26%), followed by the storage of nitrogen in groundwater (12%), with benthic denitrification accounting for 6% (twice the sediment storage). Nitrogen fluxes in the Seine basin are dominated by dissolved inorganic nitrogen, mostly nitrate. Because the sediment of the Seine River is not anoxic, except for a short distance downstream from the outlet of wastewater treatment plants, benthic denitrification remains relatively low. However, in the riparian zone, at the saturated interfaces with groundwaters and sub-root flows, denitrification is a significant microbial process for nitrogen elimination (Burt et al. 2002; Hoffmann et al. 2006; Jahangir et al. 2012). Differently from nitrogen, phosphorus may be retained in sediments at the river bottom when originating from particles in terrestrial organic matter/vegetation (Pärn et al. 2012), and phosphates essentially brought by domestic effluent have a strong propensity to be adsorbed onto particles that also settle on the bottom sediment (Owens and Walling 2002; Némery and Garnier 2007a, 2007b). Regarding silica, although essentially under a dissolved form from rock weathering, it reaches sediment via the diatom-weighted silica wall (frustules) (Garnier et al. 1995).

5. Transfer through the estuarine zone

As seen in the previous sections, the drainage network as a whole can be seen as a very active reactor, transferring, transforming, retaining, or eliminating a large fraction of the elements brought in from the watershed before they reach the river outlet. The estuarine zone similarly represents a second

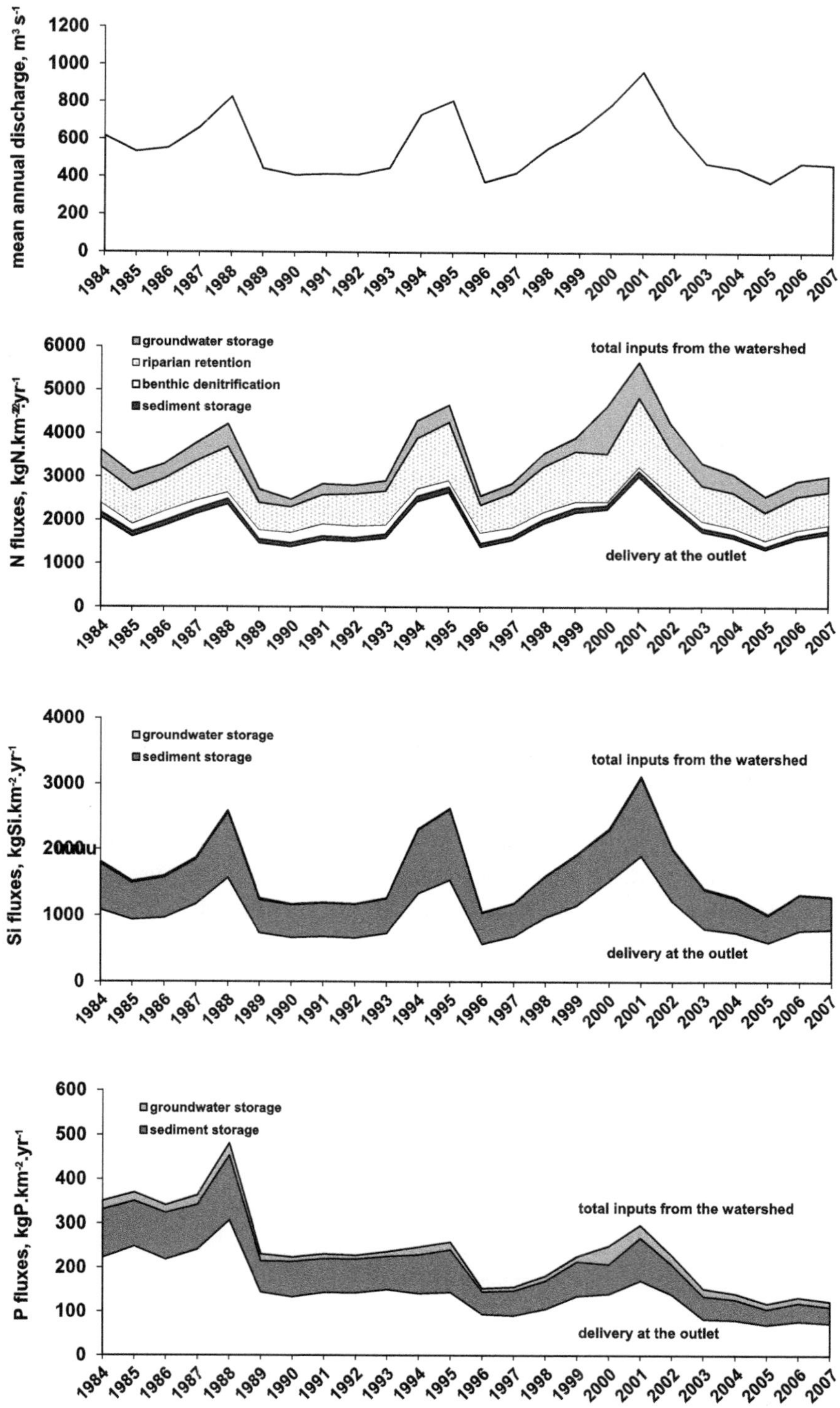

Figure 24.8. Interannual variations, in yearly average, of the discharge and specific nitrogen, silica, phosphorus fluxes entering the Seine saline estuary (Caudebec, bottom curve). The modeling approach allows comparing these water fluxes to the total input from the watershed (top curve). The areas between these two curves represent the retention/elimination and decomposition into sediment storage, benthic denitrification, groundwater storage, and riparian retention/elimination.

reactor through which the river loading has to transit before reaching the sea. The Seine Estuary is a typical macrotidal estuary, characterized by the presence, outside winter periods of high discharge, of a distinct turbidity maximum zone (TMZ). Such estuaries are known to be biogeochemically very active areas (Deborde et al. 2007; Etcheber et al. 2007; Garnier et al. 2008, 2010b).

A detailed modeling approach of the processes occurring in the Seine Estuary has been developed by coupling the RIVE module, briefly described previously, to a detailed 1D-hydro-sedimentary model (Le Hir et al. 2001): the SiAM-1D/RIVE model (Even et al. 2007). A 3D version is under development. In parallel, a more simplified and generic approach, called LIFT (Lumped Idealisation of the ecological Functioning in estuarine Turbidity maximum, Garnier et al. 2008, 2010b), has been developed to assess the nutrient behavior in the TMZ and the salinity gradient of the Seine Estuary. LIFT is a 0D model describing the turbidity maximum zone as a reactor with homogeneously distributed particulate material where freshwater flowing through it progressively mixes with seawater, whereas dissolved species react with particulate material.

The limit conditions of the LIFT model are given by the calculated value of the water quality variables provided by the SENEQUE/RIVERSTRAHLER model at the head of the salinity gradient (here Caudebec, 310 km downstream from Paris). The total volume of the TMZ is estimated at $5.3.10^8$ m^{-3} and the surface area at $7.3.10^7$ m^2, corresponding to the maximum domain of the TMZ extension throughout the seasons (see Garnier et al. 2008).

The suspended solid stock, i.e., the total amount of particles within the TMZ, was calculated from experimentally determined average concentrations, multiplied by the volume (Garnier et al. 2008, 2010b). When the upstream discharge (Q) is below a critical flow rate (Q*, calibrated to 400 m^3 s^{-1}), particulate matter from the upstream watershed accumulates in the TMZ, whereas above the critical flow rate, particles are washed out and evacuated to the sea (Garnier et al. 2008).

The results of the LIFT calculations, when averaged at the annual scale over the last 25 years (Fig. 24.9), do not show significant retention in the TMZ, neither for total nitrogen and silica (about 1%) nor for phosphorus (5%). A 10% rate of retention is found for total organic carbon (Fig. 24.9). This is consistent with previous observations that the mixing diagrams of the Seine Estuary indicated more conservative mixing than major transformations (Garnier et al. 2010b). Therefore, at an annual scale, the role of the Seine Estuary in possibly eliminating a fraction of nutrients that could reduce the eutrophication problem at the marine coastal zone is limited.

Estuaries are receptacles of all materials transferred from upstream rivers, and estuarine retention was shown to be inversely correlated with the mean residence time of water (Nixon et al. 1996).

Biogeochemical exchanges are found in estuaries (including their wetlands), but also on the continental shelves where intense transformations (e.g., at the inner and outer estuaries) may be important for the N, P, and Si fluxes reaching the open ocean (Nedwell et al. 2002; McKee et al. 2004). However, many large estuaries such as the Seine have been considerably modified for navigation (Cuvilliez et al. 2009). Decreasing the connectivity with saline and freshwater marshes has reduced the residence time of the water and particles, so that the transformation and retention/elimination of nutrients can be lowered due to the decreasing opportunity for exchanges across the interfaces (e.g., water/particle, particle/organisms) (Van Damme et al. 2009; Statham 2011). Such a behavior is also valid for the Rhine delta (Chapter 25).

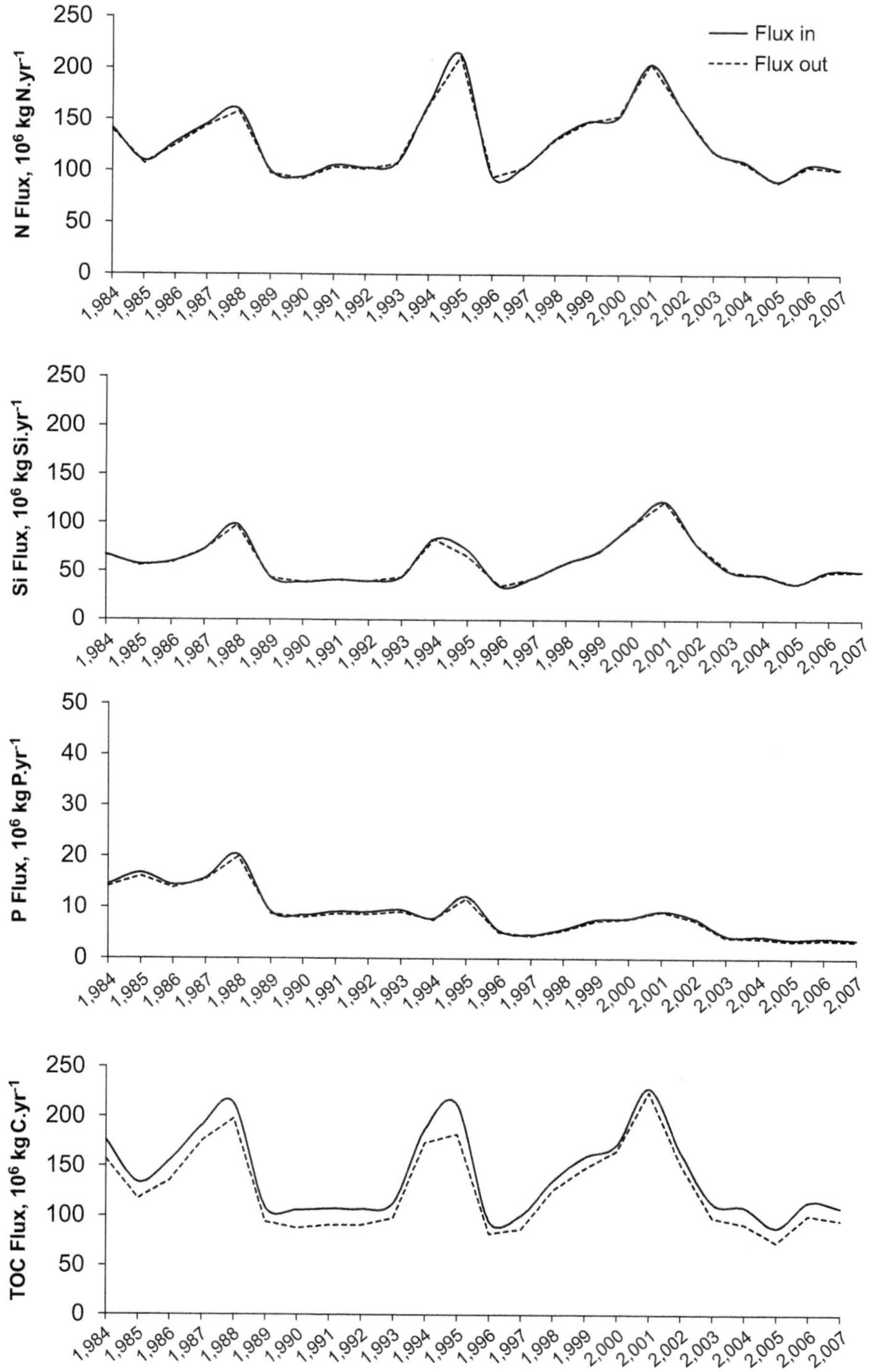

Figure 24.9. Interannual variations, expressed as the yearly average of total nitrogen, silica, phosphorus, and organic carbon fluxes entering the Seine saline estuary (Caudebec, in) and leaving the estuary.

Calculated values by the models (RIVERSTRAHLER-in and LIFT-out). The difference indicates the retention by the turbidity maximum zone of the Seine estuary.

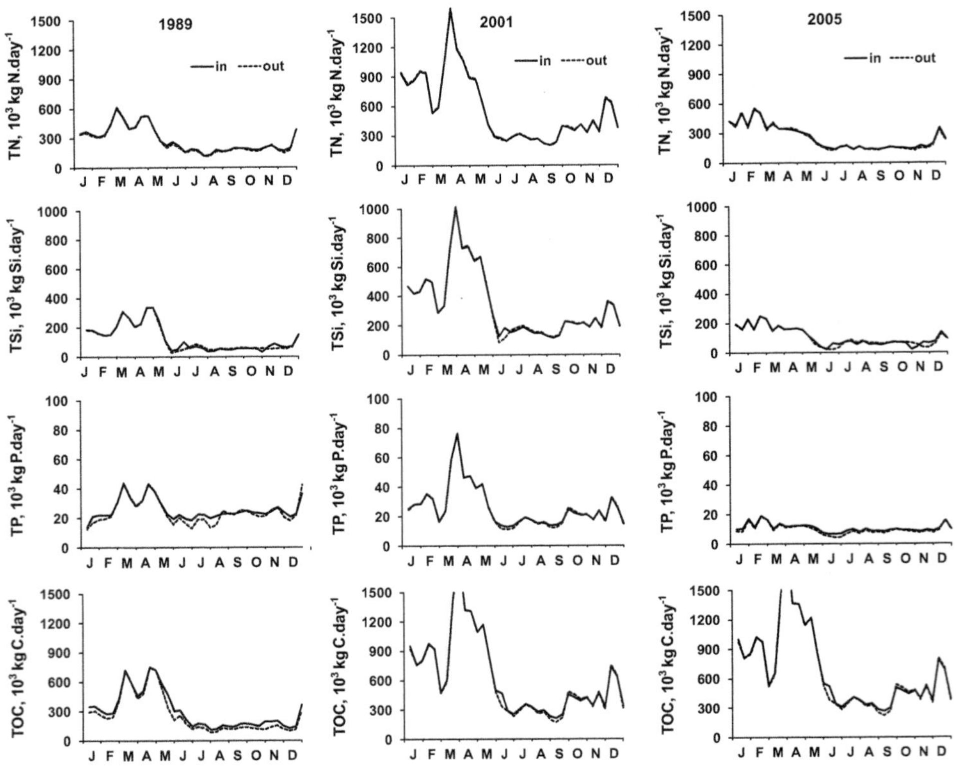

Figure 24.10. Seasonal variations for two dry years (1989 and 2005) and one wet year (2001) of daily total nitrogen (TN), silica (TSi), phosphorus (TP), and organic carbon (TOC) fluxes entering the Seine saline estuary (Caudebec, in) and leaving the estuary. Calculations by the models (RIVERSTRAHLER-in and LIFT-out). The difference indicates the retention by the turbidity maximum zone of the Seine Estuary.

However, when looking at the seasonal scale, the LIFT calculations sometimes reveal significant nutrient retention of some nutrients for short periods of time, which may affect the nutrient balance (N:P:Si) and temporarily modify the species succession and biomass amplitude at the coast (Cugier et al. 2005; Garnier et al. 2010b). Figure 24.10 illustrates this with the case of two dry years (1989 and 2005) at the beginning and the end of the 24-year period studied, when the phosphorus load in the river was respectively at its maximum and minimum, and one wet year, 2001, the wettest of the period studied.

At the seasonal scale, although still low, a retention in the TMZ is evidenced, as a whole more in summer (May–September) than in winter (October–April); on average over the 24 years, total organic carbon shows the highest seasonal retention (22% in summer vs. 11% in winter), followed by phosphorus (10% in summer vs. 4% in winter) and nitrogen (4% in summer vs. 1% in winter), with no retention obtained for silica (see Fig. 24.10).

Such slight transformations were also found in the summer period for a very short salinity gradient (3 km) of a small river entering the French Arcachon Bay (Canton et al. 2012), whereas the

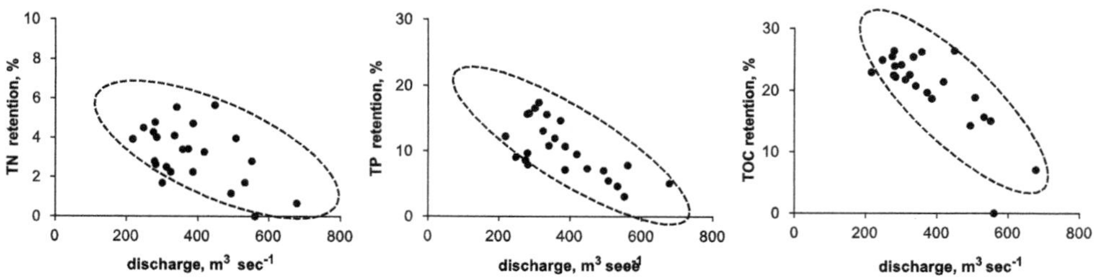

Figure 24.11. Relationships between the percentages of total nitrogen, phosphorus and carbon retention and the associated discharges. Average summer values (May–September) for the 24 years.

Humber United Kingdom estuary with a tidal length of 150 km, similar to the Seine, appeared to be a strong buffer for phosphates, removing 85% of the load (Sanders et al. 1997).

Too few studies are available that quantify the buffer effect of estuaries, but their residence time is a good indicator for inferring retention (Nixon et al. 1996). Further combinations of residence time with the rivers' geomorphology, based on the world typology by Meybeck et al. (2007) and Dürr et al. (2011), would help refine the calculation of the deliveries to coastal waters.

When plotting the average of the summer N, P, and C retention of the Seine estuary for each of the 24 years against the corresponding average discharge, a clearly inverse trend is shown, illustrating the role of residence time in favoring biogeochemical processes in the TMZ and its possible buffer role (Fig. 24.11). The wettest year in 2001 led to the lowest retention for all elements. Figure 24.11 also evidences the range of summer retention, from 7% to 30%, 0% to 6%, and 3% to 18% for total carbon, nitrogen, and phosphorus, respectively.

The seasonal trend for silica retention (not shown) is not so clear, as silica concentration is related to diatom dynamics (uptake of dissolved silica, grazing, sedimentation, etc.), with maximum growth in spring, from March to June, depending on the discharge, but also in autumn, when the temperature decreases.

6. Management in the watershed and nutrient deliveries to the coastal zone

The suite of models described previously for simulating nutrient fluxes across the river network and the estuarine zone also allows exploring prospective scenarios simulating possible future changes in human activity in the watershed.

As an example, three scenarios are considered here, based on the hydrological conditions of 2006 as a reference year, because it represents the median discharge over the last 24-year period (480 m^3 sec^{-1}, annual average).

The incontrovertible scenario simulating the implementation of the European Directive dealing with the improvement of wastewater treatment plants (91/271/CEE) was first tested by considering a 70% decrease in nitrogen and a 90% decrease in phosphorus within each WWTP treating more than 2000 inhabitant equivalents.

The "Grenelle de l'Environnement," a corpus of French laws legislating key environmental issues, strongly recommended organic farming within catchments dedicated as drinking water protection areas (DWPAs). Organic farming, besides completely banishing the use of pesticides, avoids the use of synthetic fertilizers, which are replaced by green organic fertilizers and new agriculture practices, such as long crop rotations including legumes. A decrease in nitrogen leaching is therefore expected, reducing both nitrate concentrations and fluxes in surface water, but also over the long term in aquifers (Drinkwater and Snapp 2007; Hansen et al. 2000; Korsaeth and Eltun 2000). A second scenario therefore combines the first one with a scenario implementing organic farming on DWPAs. The surface areas of these catchments account for about 30% of the total area of the whole Seine basin.

To quantify the sensitivity of the model to agricultural measures, the third scenario consisted of implementing organic agriculture at the scale of the entire agricultural land of the basin, a theoretical scenario.

Compared with the reference situation (Ref.), the implementation of wastewater measures (WWT) should reduce the nitrogen flux by 25%, a further reduction of 10% being expected with the addition of the agricultural measure (Org. F DWPA + WWT), although without changes in the estuarine retention, maintained at a few percent. N retention is proportionally greater (10%) with the more drastic scenario generalizing organic agriculture, which cuts the fluxes in half (Org. F. + WWT) (Fig. 24.12).

Regarding silica and carbon, their fluxes are not significantly modified by the tested scenarios, and their retention in the estuary reaches 2% and 20%, respectively (Fig. 24.12).

Although phosphorus has already decreased considerably in wastewater treatment plants since the 2000s, an additional 35% reduction of the flux would be expected with the full application of the regulation, with no further significant decrease with the agricultural measures. These measures would be accompanied by a 3% increase in phosphorus retention in the estuary.

Algae take up nutrients according to N:P, Si:P ratios, which represent their physiological needs (Redfield ratios: Redfield et al. 1963). Comparing these ratios with those of the fluxes brought into the sea (Fig. 24.13) thus allow us to determine which element controls phytoplankton production and the kind of algal community that will dominate (diatoms vs. non siliceaous algae) (Officer and Ryther 1980, Billen and Garnier 1997; Garnier et al. 2010a).

In the 1980s, the TN:TP ratio of the nutrient delivered to the sea was close to the Redfield value (7, in $g.g^{-1}$) but rapidly increased to more than 15 in the mid-1990s and further in the 2000s. This is clearly the result of both an increase of N issued from the watershed (mainly from agricultural emission) and of a decrease of the sources of P (mainly from urban sources). Interestingly, the effect of watershed and in-stream processes (retaining N more efficiently than P) tends to reduce the N:P ratio at the entrance of the estuary, whereas estuarine retention (more efficient for P than for N) accentuates the increasing trend of the N:P ratio.

The TSi:TP ratio of the fluxes delivered to the sea were much lower than the Redfield value (14, in $g.g^{-1}$) in the 1980–1995 period, indicating silica limitation of the diatom growth. Since then, it has regularly increased, and nowadays reaches a value close to the Redfield equilibrium (Fig. 24.13), mainly because of the reduction of P sources from the watershed. In-stream retention processes in the drainage network do not appear to bear a significant contribution to the increase of the Si:P ratio, as they similarly affect P and Si. The estuarine TMZ, however, significantly affects the Si:P ratio and contributes to bring it back closer to the Redfield value. As a result of TN:TP ratio increase over the period studied and a TSi:TP ratio that tended to Redfield equilibrum, TN fluxes delivered are in excess

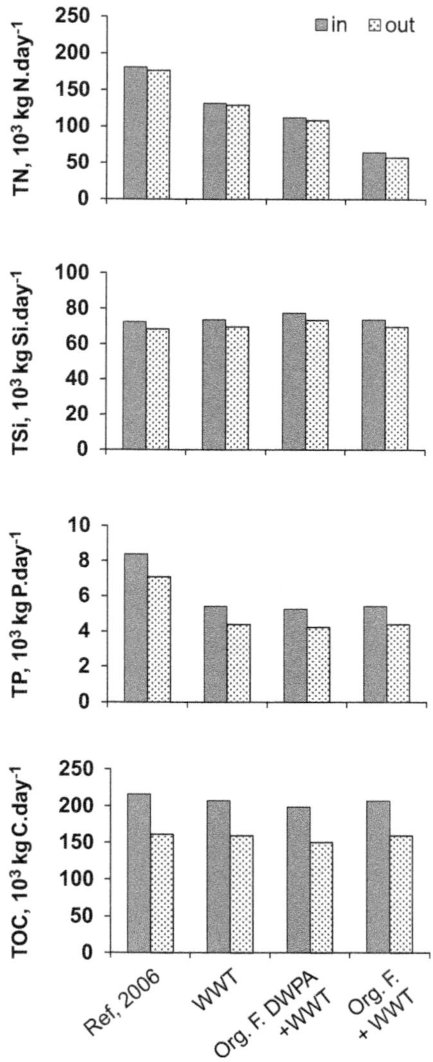

Figure 24.12. Comparison of the average nitrogen, silica, phosphorus, and organic carbon daily summer fluxes (May–September) entering and leaving the Seine estuarine zone for the year 2006 (Ref) and for various scenarios explored by the model (WWT: treatment of 70% nitrogen and of 90% phosphorus urban loading; Org. F. DWPA + WWT: organic farming in the catchments dedicated as drinking water protection areas + WWT; Org. F. + WWT: organic farming at the scale of the Seine watershed + WWT).

relatively to Si ones. This result differs from the Mississippi and Atchafalaya rivers, where an excess of both N and P relatively to Si is mentioned (see Chapter 18). However, although annual average values of the Redfield ratios are well informative about the potential eutrophication, the knowledge of their seasonal variations is required to better help contribute to the control of environmental nuisances,

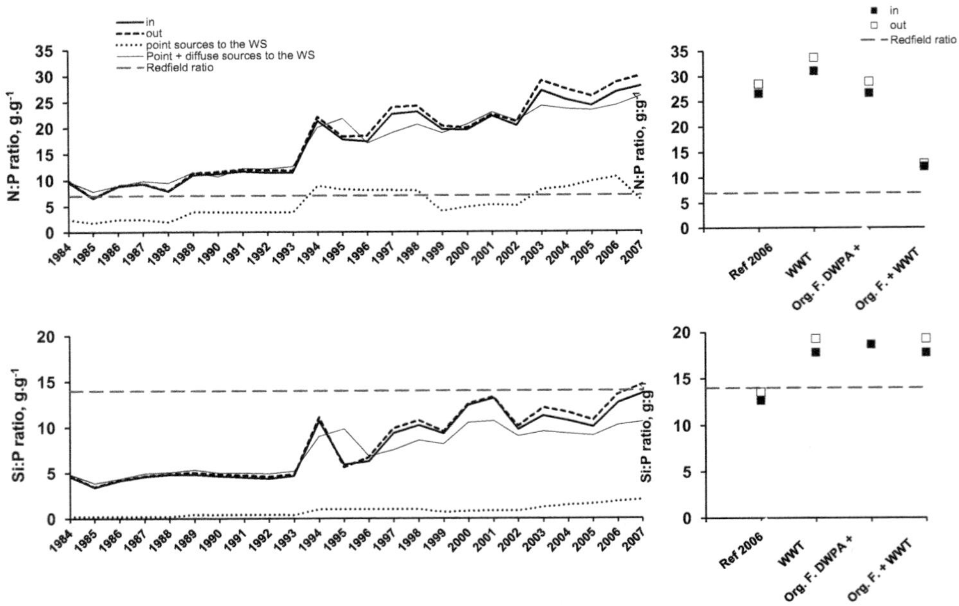

Figure 24.13. **Left.** Interannual variations, expressed as yearly averages of N:P and Si:P Redfield ratios entering the Seine saline estuary (Caudebec, in) and leaving the estuary(out). Calculated values by the models (RIVERSTRAHLER-in and LIFT-out). Input to the watershed as point sources and as point + diffuse sources. **Right.** Ratios for the year 2006 (Ref) and for the various scenarios explored by the model (WWT: treatment of 70% nitrogen and of 90% phosphorus urban loading; Org. F. DWPA + WWT: organic farming in the catchments dedicated as drinking water protection areas + WWT; Org. F. + WWT: organic farming at the scale of the Seine watershed + WWT).

namely eutrophication, but also hypoxia, etc. (Garnier et al. 2010b; Howarth et al. 2010; see also Chapter 16).

Examining the Redfield ratios from the results of the scenarios, it appears that further treatments in WWTPs would still increase the N:P divergence from the equilibrium, because the abatement is greater for P than for N, whereas Si:P would confirm the shift from a Si limitation to a P limitation (Fig. 24.13). Agricultural scenarios that would reduce N inputs without affecting P and Si (and Si:P) would lead to a return of the N:P ratio closer to the Redfield equilibrium, especially for the unlikely scenario of organic agriculture at the scale of the whole watershed.

Although this is far from being the case at the global scale, point sources of urban nutrient pollution are now about to come under control in most developed countries. Agriculture, however, is still a growing source of nutrient contamination of hydrosystems. In particular, intensive agriculture and its use of industrial fertilizers in Europe and North America have led to the massive introduction of reactive nitrogen into the biosphere, which subsequently cascades through a number of environmental compartments until it is finally denitrified back to the atmosphere (Galloway et al. 2003; Sutton et al. 2011).

7. Conclusion

The Seine represents up to ca. 80–85% of the freshwater inflow in the Seine Bight, and algal blooms are generally observed in the plume of the Seine River just downstream from the TMZ. The TMZ role of the Seine Estuary is rather low in its present geomorphology.

A remarkable decrease of P fluxes has been observed in the Seine, as in most Atlantic rivers (Romero et al. 2012), up to 60%, which is higher than the OSPAR P objective that aims at a 50% reduction. Silica fluxes, from diffuse natural sources, are not modified, and their interannual variations are essentially governed by local and regional hydrology.

Similar to the trends observed at the Atlantic coast (Romero et al. 2012), N deliveries to the Seine Bight should remain high, despite the ongoing decrease via further nitrogen treatment in WWTPs, especially the Achères WWTP downstream from Paris, which treats 40% of the Seine basin population's wastewaters. Unless a strong European political commitment moves agriculture practices toward less synthetic N fertilization and less intensive modes of agricultural production and includes organic agriculture and/or new agricultural practices, nitrogen will remain largely in excess with regard to phosphorus and silica at the outlet of the Seine continuum.

The respective overall retentions in the Seine TMZ (on the order of 10% P, nearly 0% Si, 3% N) in fact accentuate the observed trends: (1) the excess of nitrogen with regard to phosphorus increases along the studied period and more downstream the estuary; (2) silica regularly increases with regard to phosphorus up to the Redfield ratio, thanks in part to the estuarine phosphorus retention.

The mechanistic SENEQUE/RIVERSTRAHLER model, combined with the LIFT model along the Seine continuum, does not describe the terrestrial processes controlling diffuse nutrient sources from agriculture. These are instead imposed as a forcing function or as boundary conditions, obtained from other model results, observation surveys, or literature data (Thieu et al. 2011; Passy et al. 2013). A challenge will be in the next few years to combine the SENEQUE/RIVERSTRAHLER drainage network model with a comprehensive, deterministic model of nutrient cycling in agricultural and semi-natural soils, such as the public-domain SWAT Model (Soil and Water Assessment Tool; Srinivasan et al. 1998; Neitsch et al. 2005). This will make it possible to refine the agriculture scenarios necessary to establish functional links with the adjoining terrestrial, freshwater, and marine systems (Elliott and Whitfield 2011). This is a major. challenge for the next decade to combat eutrophication, to protect our freshwater resources, and to promote sustainable agriculture. Our modeling tool could be a candidate at multiregional, continental to global scales for better understanding of in-stream nutrient transformation, retention, and elimination within the drainage network resulting from biological and physical-chemical processes (Kroeze et al. 2012).

References

Billen G., Garnier J., Hanset Ph. (1994). Modelling phytoplankton development in whole drainage networks: the RIVERSTRAHLER model applied to the Seine river system. Hydrobiologia, 289:119–137.

Billen G., Garnier J. (1997). The Phison River Plume: coastal eutrophication in response to changes in land use and water management in the watershed. Aquat. Microb. Ecol., 13: 3–17.

Billen G., Garnier J., Deligne. C, Billen C. (1999). Estimates of early-industrial inputs of nutrients to river systems: implication for coastal eutrophication. Sci. Total Environ, 243/244: 43–52.

Billen G., Garnier J. (1999). Nitrogen transfers through the Seine drainage network: a budget based on the application of the Riverstrahler model. Hydrobiologia, 410: 139–150.

Billen G., Garnier J., Ficht A., Cun C. (2001). Modelling the response of water quality in the Seine Estuary to human activity in its watershed over the last 50 years. Estuaries, 24: 977–993.

Billen G., Garnier J., Nemery J., Sebilo M., Sferratore A., Benoit P., Barles S, Benoit M. (2007). A long term view of nutrient transfers through the Seine river continuum. Sci. Total Environ., 275: 80–97.

Billen, G., Garnier, J. (2007). River basin nutrient delivery to the coastal sea: assessing its potential to sustain new production of non siliceous algae. Mar. Chem., 106: 148–160.

Billy C., Birgand F., Sebilo M., Billen G., Tournebize J., Kao C. (2011). Nitrate dynamics in artificially drained nested watersheds. Phys. Chem. Earth, 36: 506–514.

Burt T.P., Pinay G., Matheson F.E., Haycock N.E., Butturini A., Clement J.C. (2002). Water table fluctuations in the riparian zone: comparative results from a pan-European experiment. J. Hydrol., 265: 129–148.

Canton M., Anschutz P., Poirier D., Chassagne R., Deborde J., Savoye N. (2012). The buffering capacity of a small estuary on nutrient fluxes originating from its catchment (Leyre estuary, SW France). Estuar. Coast. Shelf Sci., 99: 171–181.

Cochran M.F., Berner R.A. (1996). Promotion of chemical weathering by higher plants: field observations on Hawaiian basalts. Chem. Geol., 132: 71–77.

Conley D.J., Schelske C.L., Stoermer E.F. (1993). Modification of the biogeochemical cycle of silica with eutrophication. Mar. Ecol. Prog. Ser., 101: 179–192.

Conley D.J. (1999). Biogeochemical nutrient cycles and nutrient management strategies. Hydrobiologia, 410: 87–96.

Conley D.J. (2002). Terrestrial ecosystems and the global biogeochemical silica cycle. Global Biogeochem. Cycles, 16: 1121–1129.

Cugier Ph. (1999). Modélisation du devenir à moyen terme dans l'eau et le sédiment des éléments majeurs (N, P, Si, O) rejetés par la Seine en baie de Seine. Thèse de doctorat, Univ. de Caen.

Cugier Ph., Billen G., Guillaud, J.F., Garnier, J., Ménesguen A. (2005). Modelling the eutrophication of the Seine Bight (France) under historical, present and future riverine nutrient loading. J. Hydrol., 304: 381–396.

Cuvilliez A., Deloffre J., Lafite R., Bessineton C. (2009). Morphological responses of an estuarine intertidal mudflat to constructions since 1978 to 2005: The Seine estuary (France). Geomorphology, 104: 165–174.

Deborde J., Anschutz P., Chaillou G., Etcheber H., Commarieu M.V., Lecroart P., Abril G. (2007). The dynamics of phosphorus in turbid estuarine systems: example of the Gironde estuary (France). Limnol. Oceanogr., 52: 862–872.

Diaz R.J., Rosenberg R. (2008). Spreading dead zones and consequences for marine ecosystems. Science, 321 (5891): 926–929.

Drinkwater L.E., Snapp S.S. (2007). Nutrients in Agroecosystems: Rethinking the Management Paradigm. Adv. Agron., 92:163–186.

Dürr H.H., Laruelle G.G., Van Kepen C.M., Slomp C.P., Meybeck M., Middelkoop H. (2011). Worldwide typology of nearshore coastal systems: defining the estuarine filter of river inputs to the oceans. Estuar. Coasts, 34: 441–458.

Elliott M., Whitfield A.K. (2011). Challenging paradigms in estuarine ecology and management. Estuar. Coast. Shelf Sci., 94: 306–314.

Etcheber H., Taillez A., Abril G., Garnier J., Servais P., Moatar F., Commarieu M.V. (2007). Particulate organic carbon in the Estuarine Turbidity Maxima of the Gironde, Loire and Seine Estuaries: origin and lability. Hydrobiologia, 588: 245–259.

EU-WFD (Water Framework Directive). OJ L 327/1, 22.12. 2000, pp. 1–72.

Even S. Billen G., Bacq N., Théry S., Ruelland D., Garnier J., Cugier Ph., Poulin M., Blanc S., Lamy F., Paffoni C. (2007). New modelling tools for water quality management of regional river systems: simulating a prospective scenario in the scope of the European Water Framework Directive in the Seine River Basin. Sci. Total Environ., 375: 274–291.

Galloway J.N., Aber J.D., Erisman J.W., Seitzinger S.P., Howarth R.W., Cowling E.B., Cosby B.J. (2003). The nitrogen cascade. BioScience, 53: 341–356.

Garnier J., Billen G., Coste M. (1995). Seasonal succession of diatoms and Chlorophyceae in the drainage network of the river Seine: Observations and modelling. Limnol. Oceanogr., 40: 750–765.

Garnier J., d'Ayguesvives A., Billen G., Conley D., Sferratore A. (2002a). Silica dynamics in the hydrographic network of the Seine River. Oceanis, 28: 487–508.

Garnier J., Billen G., Hannon E., Fonbonne S., Videnina Y., Soulie M. (2002b). Modeling transfer and retention of nutrients in the drainage network of the Danube River. Estuar. Coast. Shelf Sci., 54: 285–308.

Garnier J., Némery J., Billen G., Théry S. (2005). Nutrient dynamics and control of eutrophication in the Marne River system: modelling the role of exchangeable phosphorus. J. Hydrol., 304: 397–412.

Garnier J., Laroche L., Pinault S. (2006). Determining the domestic specific loads of two wastewater plants of the Paris conurbation (France) with contrasted treatments: a step for exploring the effects of the application of the European Directive. Water Res., 40: 3257–3266.

Garnier J., Billen G., Even S., Etcheber H., Servais P. (2008). Organic matter dynamics and budgets in the maximum turbidity zone of the Seine Estuary (France). Estuar., Coast. Schelf Sci., 77: 150–162.

Garnier J., Beusen A., Thieu V., Billen G., Bouwman L. (2010a). N:P:Si nutrient export ratios and ecological consequences in coastal seas evaluated by the ICEP approach. Special issue "Past and Future Trends in Nutrient Export from Global Watersheds and Impacts on Water Quality and Eutrophication." Global Biogeochem. Cycles, 24: GB0A05.

Garnier J., Billen G., Némery J., Sebilo M. (2010b). Transformations of nutrients (N, P, Si) in the turbidity maximum zone of the Seine estuary and export to the sea. Estuar. Coast. Shelf Sci., 90: 129–141.

Grizzetti B., Bouraoui F., Aloe A. (2012). Changes of nitrogen and phosphorus loads to European seas. Global Change Biol., 18: 769–782.

Hansen B., Kristensen E.S., Grant R., Høgh-Jensen H., Simmelsgaard S.E., Olesen J.E. (2000). Nitrogen leaching from conventional versus organic farming systems – a systems modelling approach. Eur. J. Agron., 13: 65–82.

Hoffmann C.C., Berg P., Dahl M., Larsen S.E., Andersen H.E., Andersen B. (2006). Groundwater flow and transport of nutrients through a riparian meadow – Field data and modeling. J. Hydrol., 331: 315–335.

Howarth R., Chan F., Conley D.J Garnier J., Doney S.C., Marino R., Billen G. (2010). Coupled biogeochemical cycles: Eutrophication and hypoxia in temperate estuaries and coastal marine ecosystems. Frontiers Ecol. Environ., 9: 18–26.

Humborg C., Conley D. J., Rahm L., Wulff F., Cociasu A., Ittekot V. (2000). Silicon retention in river basins: far-reaching effects on biogeochemistry and aquatic food webs in coastal marine environments. Ambio, 29: 45–51.

Humborg C., Smedberg E., Blomqvist S., Mörth C. M., Brink J., Rahm L., Danielsson Å., Sahlberg J. (2004). Nutrient variations in boreal and subarctic Swedish rivers: Landsdcape controle on land-sea fluxes. Limnol. Oceanogr., 49: 1871–1883.

Humborg C., Smedberg E., Rodriguez-Medina M., Mörth C.-M. (2008). Changes in dissolved silicate loads to the Baltic Sea – The effects of lakes and reservoirs. J. Mar. Syst., 73: 223–235.

Jahangir M.M.R., Khalil M.I., Johnston P., Cardenas L.M., Hatch D.J., Butler M., Barrette M., O'flaherty V., Richards K.G. (2012). Denitrification potential in subsoils: A mechanism to reduce nitrate leaching to groundwater. Agric. Ecosys. Environ., 147: 13–23.

Justic D., Rabalais N.N., Turner R.E., Dortch Q. (1995a). Changes in nutrient structure of river-dominated coastal waters: stoichiometric nutrient balance and its consequences. Estuar. Coast. Shelf Sci., 40: 339–356.

Justic D., Rabalais N.N., Turner R.E. (1995b). Stoichiometric nutrient balance and origin of coastal eutrophication. Mar. Pollut. Bull., 30: 41–46.

Korsaeth A., Eltun R. (2000). Nitrogen mass balances in conventional, integrated and ecological cropping systems and the relationship between balance calculations, and nitrogen runoff in an 8-year field experiment. Agric. Ecosys. Environ., 79: 199–214.

Kroeze C., Bouwman L., Seitzinger S. (2012). Modeling global nutrient export from watersheds. Curr. Opin. Environ., Sustain., 4: 195–202.

Lancelot C., Gypens N., Billen G., Garnier J., Roubeix V. (2007). Linking marine eutrophication to land use: an integrated river-ocean mathematical tool: The Southern Bight of the North Sea over the past 50 years. J. Mar. Syst., 64: 216–228.

Lancelot C., Thieu V., Polard A., Garnier J., Billen G., Hecq W., Gypens N. (2011). Ecological and economic effectiveness of nutrient reduction policies on coastal Phaeocystis colony blooms in the Southern North Sea: an integrated modeling approach. Sci. Total Environ., 409: 2179–2191.

Le Hir P., Ficht A., Silva-Jacinto R., Lesueur P., Dupont J.-P., Lafite R., Brenon I., Thouvenin B., Cugier Ph. (2001). Fine sediment transport and accumulations at the mouth of the Seine Estuary (France). Estuaries, 24: 950–963.

Le T.P.Q., Billen G., Garnier J., Théry S., Ruelland D., Nguyem X.A., Chau V.M. (2010). Modelling nutrient transfer in the sub-tropical Red River system (China and Vietnam): implementation of the Seneque/Riverstrahler model. J. Asian Earth Sci., 37: 259–274.

Meybeck M. (1986). Composition chimique des ruisseaux non pollués de France. Sci. Geol. Bull. (Strasbourg), 39: 3–77.

Meybeck M., Dürr H.H., Roussennac S., Ludwig W. (2007). Regional seas and their interception of riverine fluxes to oceans. Mar. Chem., 106: 301–325.

McKee B.A., Aller R.C., Allison M.A., Bianchi T.S., Kineke G.C. (2004). Transport and transformation of dissolved and particulate materials on continental margins influenced by major rivers:benthic boundary layer and seabed processes. Cont. Shelf Res., 24: 899–926.

Nedwell D.B., Dong L.F., Sage A., Underwood G.J.C. (2002). Variations of the Nutrients Loads to the Mainland U.K. Estuaries: Correlation with Catchment Areas, Urbanization and Coastal Eutrophication. Estuar. Coast. Shelf Sci., 54: 951–970.

Neitsch S.L., Arnold J.G., Kiniry J.R., Williams J.R. (2005). Soil and Water Assessment Tool (SWAT), Theoretical Documentation. Blackland Research Center, Grassland. Soil and Water Research Laboratory, Agricultural Research Service. Temple, TX.

Némery J., Garnier, J. (2007a). Dynamics of particulate phosphorus in the Seine estuary (France). Hydrobiologia, 588: 271–290.

Némery J., Garnier J. (2007b). Origin and fate of phosphorus in the Seine watershed (France): Agricultural and hydrographic P budgets. J. Geophys. Res., 112: G03012.

Nixon S.W., Ammerman J., Atkinson L., Berounsky V., Billen G., Boicourt W., Boyton W., Church T., DiToro D., Elmgren R., Garber J., Giblin A., Jahnke R., Owens N., Pilson M.E.Q.,

Seitzinger S. (1996). The fate of nitrogenandphosphorus at the landesea margin of the North Atlantic Ocean. Biogeochemistry, 35: 141–180.

Officer C.B., Ryther J.H. (1980). The possible importance of silicon in marine eutrophication. Mar. Ecol. Prog. Ser., 3: 83–91.

Owens P. N., Walling D.E. (2002). The phosphorus content of fluvial sediment in rural and industrialized river basins. Water Res., 36: 685–701.

Pärn J., Pinay G., Mander U. (2012). Indicators of nutrients transport from agricultural catchments under temperate climate: A review. Ecol. Indic., 22: 4–15.

Passy P., Gypens N., Billen G., Garnier J., Thieu V., Rousseau V., Callens J., Parent Y., Lancelot C. (2013). A Model reconstruction of riverine nutrient fluxes and eutrophication in the Belgian Coastal Zone since 1984. J. Mar. Syst., doi: 10.1016/j.jmarsys.2013.05.005.

Romero E., Garnier J., Lassaletta L., Billen G., Le Gendre R., Riou P., Cugier Ph. (2012). Large-scale patterns of river inputs in SW Europe: seasonal and interannual variations and potential eutrophication effects at the coastal zone. Biogeochemistry, 113: 481–505.

Redfield, A. C., Ketchum B. H., Richards F.A. (1963). The influence of organisms on the composition of sea-water. In M. N. Hill (ed.), The Sea, John Wiley & Sons, New York, pp. 12–37.

Ruelland D., Billen G., Brunstein D., Garnier J. (2007). SENEQUE: a multi-scaled GIS interface to the RIVERSTRAHLER model of the biogeochemical functioning of river systems. Sci. Total Environ., 375: 257–273.

Sanders R. J., Jickells T., Malcolm S., Brown J., Kirkwood D., Reeve A., Taylor J., Horrobin T., Ashcroft C. (1997). Nutrient fluxes through the Humber estuary. J. Sea Res., 37: 3–23.

Servais P., Garnier J. (2006). Organic carbon and bacterial heterotrophic activity in the maximum turbidity zone of the Seine estuary (France). Aquatic Sci., 68: 78–85.

Sferratore A., Garnier J., Billen G., Conley D., Pinault S. (2006). Silica diffuse and point sources in the Seine watershed. Environ. Sci. Technol., 40: 6630–6635.

Srinivasan R., Ramanarayanan T.S., Arnold J.G., Bednarz S.T. (1998). Large area hydrologic modelling and assessment part II: model application. J. Am. Water Resour. Assoc., 34: 91–101.

Statham P.J. (2011). Nutrients in estuaries – An overview and the potential impacts of climate change. Sci. Total Environ., 434: 213–227.

Struyf E., Smis A., Van Damme S., Garnier J., Govers G., Van Wesemael B., Conley D.J., Batelaan O., Frot E., Clymans W., Vandevenne F., Lancelot C., Goos P., Meire P. (2010). Historical land use change has lowered base-line silica mobilization from landscapes. Nat. commun., 1–7.

Sutton M.A., Howarth C.M., Erisman J.W., Billen G., Bleeker A., Grennfelt P., van Grinsven H., Grizzetti B. (2011). The European Nitrogen Assessment: Sources, Effect and Policy perspectives. Cambridge University Press.

Thieu V., Billen G., Garnier J. (2009). Nutrient transfer in three contrasting NW European watersheds: the Seine, Somme, and Scheldt Rivers. A comparative application of the Seneque/Riverstrahler model. Water Res., 43: 1740–1754.

Thieu V., Billen G., Garnier J., Benoit M. (2011). Nitrogen cycling in a hypothetical scenario of generalised organic agriculture in the Seine, Somme and Scheldt watersheds. Reg. Environ. Change, 11: 359–370.

Turner R.E., Rabalais N.N. (1994). Evidence for coastal eutrophication near the Mississippi River Delta. Nature, 368: 619–621.

Turner R.E., Quresh N.A., Rabalais N.N., Dortch Q. Justic D., Shaw R., Cope J. (1998). Fluctuating silicate: nitrate ratios and coastal plankton food webs. Proc. Natl. Acad. Sci., 95: 13048–13051.

Thouvenot-Korppoo M., Billen G., Garnier J. (2009). Modelling benthic denitrification processes over a whole drainage network. J. Hydrol., 379: 239–250.

UWWTD (Urban wastewater treatment Directive (1991). 91/271/CEE du Conseil, du 21 mai 1991, relative au traitement des eaux urbaines résiduaires, JO L 135 du 30.5.1991, pp. 40–52.

Van Damme S., Dehairs F., Tackx M., Beauchard O., Struyf E., Gribsholt B., Van Cleemput O., Meire P. (2009). Tidal exchange between a freshwater tidal marsh and an impacted estuary: the Scheldt estuary, Belgium. Estuar. Coast. Shelf Sci., 85: 197–207.

Verbanck M., Vanderborght J.-P., Wollast R. (1994). Major ion content of urban wastewater: assessment of per capita loading. J. Water Pollut. Cont. Fed., 61: 1722–1728.

Vilain G., Garnier J., Tallec G., Tournebize J. (2012). Indirect N_2O emissions from shallow groundwater in an agricultural catchment (Seine Basin, France). Biogeochemistry, 111: 253–271.

25

Anthropogenic changes in sediment and nutrient retention in the Rhine delta

H. Middelkoop, M. van der Perk, and G. Erkens

1. Introduction

The Rhine delta in the Netherlands is one of the largest and most complete sediment traps in the Rhine river basin (Hoffmann et al. 2007) and presently covers an area of about 3,100 km^2. Located at the mouth of a large densely populated river basin and draining an area of 160,800 km^2 (Fig. 25.1; CHR/KHR, 1978), the delta also receives large amounts of contaminants and nutrients. At present, the Rhine river discharges on average about 2,200 m^3 s^{-1}, with an average annual suspended sediment load of 3.1×10^9 kg yr^{-1} (Middelkoop and Van Haselen 1999), and annual nutrient loads of about 190–225 Gg N per year and 7.5–14 Gg P per year (data: live.waterbase.nl). Since the early Holocene, the delta has formed a major trap of Rhine sediment as a result of rapid aggradation governed by eustatic sea level rise and land subsidence. This unique relatively complete geological record has been extensively studied, and an extremely detailed database of the Holocene delta architecture has been established (Berendsen and Stouthamer 2000, 2001; Gouw and Erkens 2007; Gouw 2008). To date, the lower Rhine distributaries in the delta still function as a filter in the continent-to-coast flux of sediments and nutrients, before these reach the estuary. This chapter first addresses sediment and nutrient fluxes to the delta across a range of time scales from the Rhine basin, where land use, agriculture, population, and technological development have dramatically changed and have been relatively well documented by geologic, historic, and observational data. The chapter focuses then on the retention of sediment and nutrients within fluvial part of the delta, upstream of the estuary. Sediment deposition is described for the natural "pristine" delta and later periods of increasing human influence until contemporary sediment budgets. Estimates of nutrient retention are given for present-day conditions, followed by an outlook to the future for the functioning of the lower Rhine floodplains as filters of sediment and nutrients.

2. Historic changes in the Rhine basin: land use, population, industry, and engineering as controls of sediment and nutrient fluxes

Sediment and nutrient fluxes to the delta have been increasingly determined by humans since the earliest impacts on the basin by deforestation and subsequent agricultural development, population growth, industrialization, and river engineering.

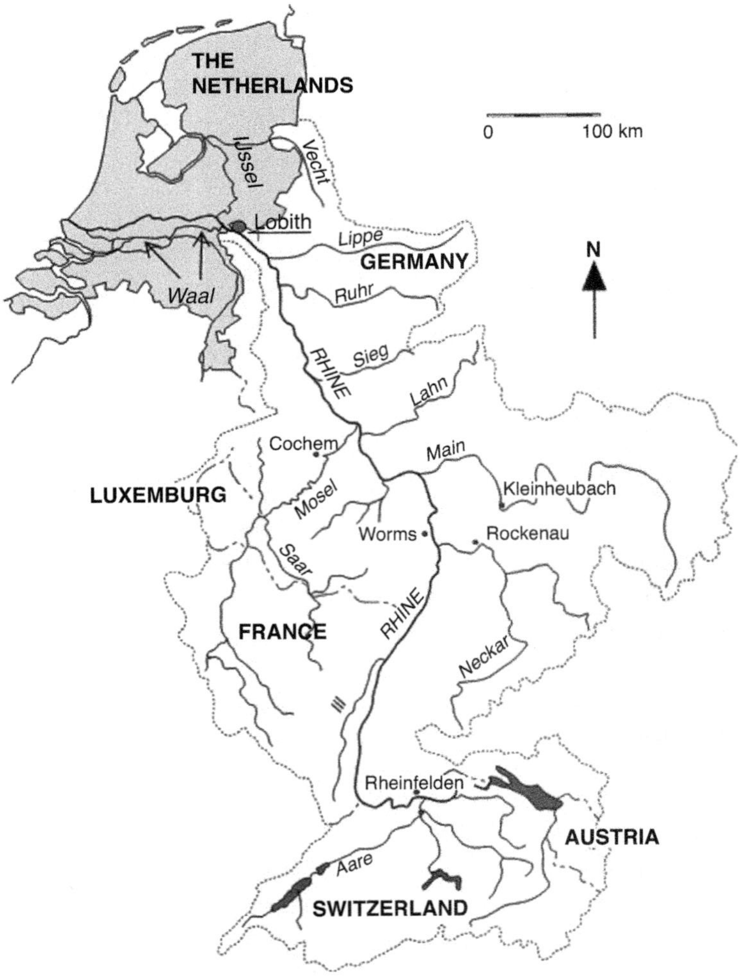

Figure 25.1. The Rhine basin.

2.1. Land-use change

The earliest indications of human impact on the pristine vegetation in the Rhine basin go back to Neolithic (ca. 5500–4700 B.P.) to Bronze age (ca. 4000–2500 B.P.) (Dreibrodt et al. 2009). The earliest deforestation may have been too small to cause increased erosion beyond the scale of the agricultural fields; during the Bronze age, erosion mainly caused the progressive filling of colluvia on the lower hillslopes (Erkens 2009). From the Iron age onward, however, the increase in floodplain deposits indicates that the eroded sediment indeed reached the stream system of the Rhine river (Garbe-Schönberg et al. 1998; Zolitschka 1998; Erkens 2009). The sedimentary records show a major period with increased sediment input during the Roman Empire Period (ca 2000–1600 B.P.), followed by low input during the subsequent Dark Ages, until high inputs occurred from the Late

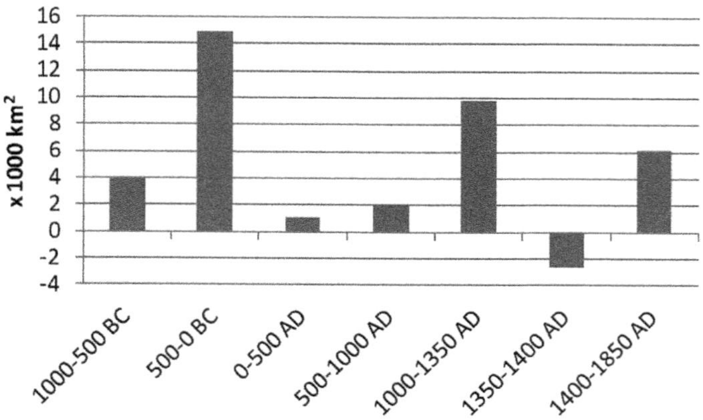

Figure 25.2. Deforestation in the Rhine basin after 1000 B.P. – negative values = increase in forested area (data from: Kaplan et al. 2009).

Medieval times (ca A.D. 1000) onward. Considering the Rhine basin as major part of central Europe, the reconstructions made by Kaplan et al. (2009) provide an excellent quantitative overview of the prehistoric deforestation history of the basin from 1000 B.C. onward until the era of industrialization began around 1850 A.D. Kaplan et al. (2009) estimated that by 1000 B.C. about 60–70% of Central and Western Europe were still forested and showed major periods of deforestation in Germany between 500–0 B.C. and 1000–1350 A.D. (Figure 25.2).

2.2. Economic development and population

In Germany, large-scale industrial development and industrialization started in the 1870s, with the development of coal mining and iron and steel production, followed by chemical and electric industries, which made the country a major industrial player by the onset of World War I (Solsten 1995). Within the Rhine basin, the Ruhr area, the upper Rhine valley, and the Neckar valley were major industrial regions. Population increased from 49.7 million in 1871 to 65.3 million in 1911. After World War I and the period of great economic depression, industry entered a booming growth period in the 1930–1940s until the collapse of the country by the end of World War II. Post-war economy developed rapidly in the 1950s as an "economic miracle," reaching consolidation in the 1960. After the 1970s, oil shocks, and recession period, further growth started in the 1980s. This economic development has dramatically affected the sediment quality in the fluvial system. This is well reflected in the heavy metal concentrations present in sediments from floodplain lakes in the Netherlands that show the two periods of industrial prosperity and environmental contamination in the 1930s and 1960s as distinct peaks in metal concentrations (Middelkoop 2000).

In the late 1960 and 1970s, the awareness arose that the Rhine was nearly "biologically dead," and German legislation started on waste disposal and emissions, and in 1974 the Federal Environmental Agency was established. An International Commission for the Protection of the Rhine (IKSR) was already established in 1950, as a platform of riparian states for protection of the environment of the

Rhine river (www.iksr.org; IKSR 1992). Subsequent years showed major reductions in emissions of contaminants and nutrients.

2.3. River Engineering

Along with industrial development, growing urban areas, and the need for improved water ways, major engineering works were carried out on the main Rhine course in the nineteenth century. Most extensive was the so-called Tulla-rectification carried out between 1817 and 1876 in the upper Rhine between the Swiss border and the city of Worms, aiming at improving navigation and reducing flood risk. Meanders were cut off, new straight channels were dug, and the main channel of the Rhine was deepened, whereas small islands were removed. As a result, the length of this stretch reduced from 355 km to 275 km (Kalweit 1993). A further development occurred in the 1970s, when the entire bed was impounded at two large barrages. This resulted in a 60% loss of the adjacent floodplain between 1955 and 1977 and a dramatic decrease of the natural flood retention area of this reach. Regulation of the Lower and Middle Rhine started in the 1880s: meanders were cut off and the river was channelized between embankments. Also, the main tributaries (Neckar, Main, Mosel) were regulated to improve navigability.

The lower Rhine distributaries of the delta in the Netherlands have a long history of human impact. Between A.D. 1000 and 1350, the lower Rhine distributaries were embanked (Middelkoop 1997; Hesselink 2002). As a result, the floodplains were reduced to 500–1,000-m wide areas between the river channel and the river dikes. Within the confinement of the river dikes, the river deposited and eroded sand and clay. The straightening of the meander bands, fixing their banks and narrowing the river by groynes (so-called normalization works) after the 1850s, lateral channel migration, and associated bank erosion halted. Overbank deposition during flood events continued, gradually raising the floodplain surface, whereas transport and sedimentation of sand became almost exclusively limited to the channel, except during extreme flood events.

At present, the total surface area of the embanked floodplains in the Netherlands is 286 km^2. A typical profile in the embanked floodplains consists of a clay layer of up to about 2 m in thickness, on top of sandy channel/point-bar deposits, with occasional residual channel fills of several meters (Middelkoop 1997; Hesselink et al. 2003). The principal land use in the embanked floodplains is pasture land; in some areas nature has been created. The main industrial activity is structural-ceramics manufacture, including clay extraction.

3. The Rhine delta as record of holocene variations in sediment trapping

The Rhine delta (Fig. 25.3) has been functioning as a near complete sediment trap for Rhine sediments since the onset of deltaic deposition about 9,000 years ago (Erkens 2009). Most sediment was deposited in a back-barrier area in the central part of The Netherlands, where an up to 20-m thick stack of Holocene fluvial deposits aggradated under a rising relative sea level (Berendsen and Stouthamer 2000, 2001; Erkens et al. 2006). Earliest aggradation occurred in the downstream part of the delta and progressively shifted upstream to reach around 5000 B.P. the upstream part of the delta (Gouw and

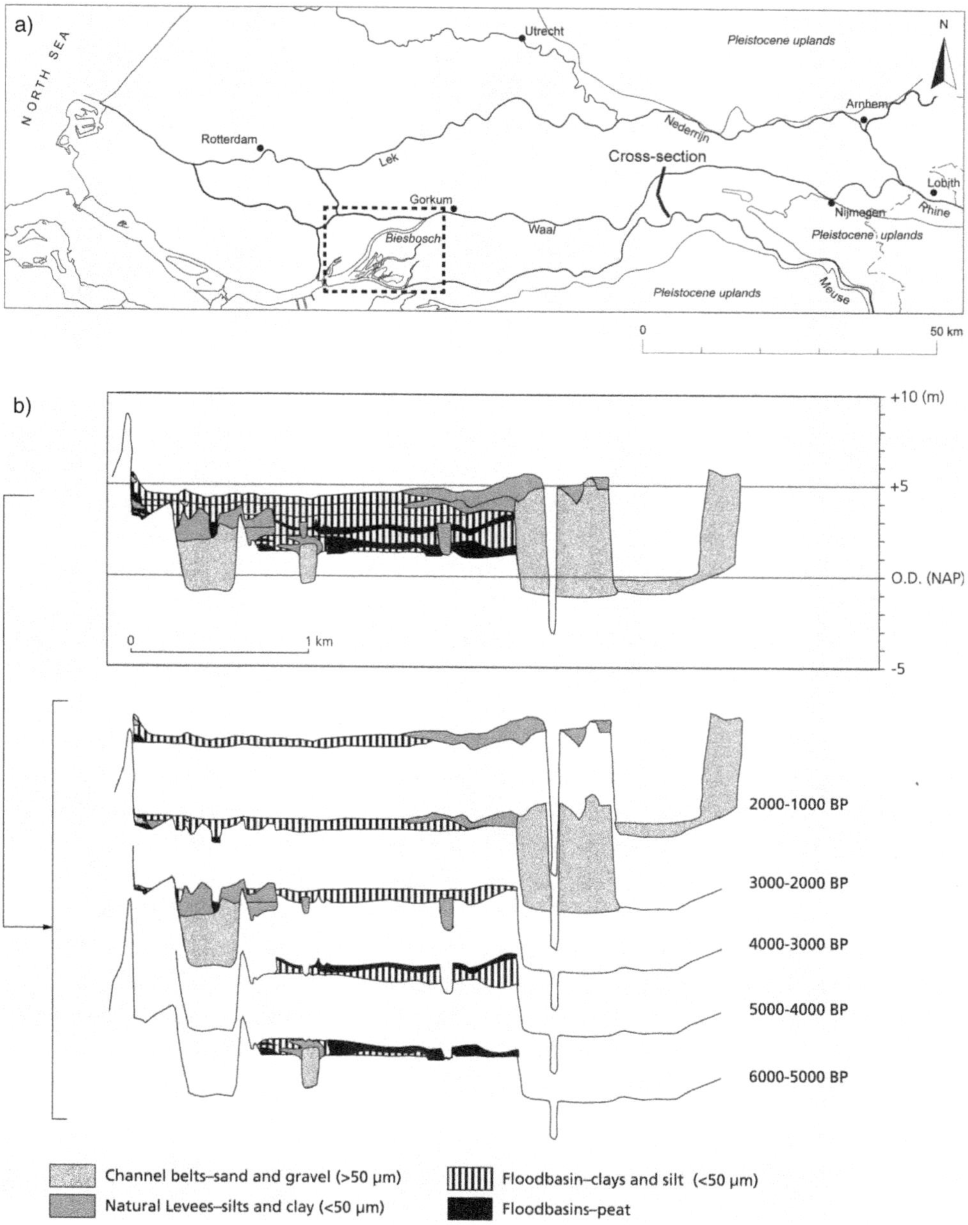

Figure 25.3. (a) The Rhine-Meuse delta. The dashed box indicates the inland delta of the Waal formed after A.D. 1421. (b) Cross-section with reconstructed ages in 1,000-year time slices (Erkens 2009).

Erkens 2007). The delta comprises flood basin deposits of which consist near the delta apex of 1 m mostly clastic deposits, increasing to about 10 m in the near-coastal area, where flood basins organic beds of increasing thickness occur, intercalated by clastic overbank deposits. These flood basins are intersected by the sand bodies of fossil river channel belts that are meandering in the upstream part of the delta and become narrower and larger in number toward the west.

3.1. Holocene trapping amounts

Erkens (2009) estimated the amounts of sediment trapped in the Holocene Rhine delta for 1,000-year time slices by using an extensive database of 200,000 borehole descriptions of the delta (Berendsen and Stouthamer 2000, 2001). The total mass of Holocene overbank fines deposited since 9000 B.P. in the entire Rhine delta amounts to about 15.6×10^{12} kg, thereby accounting for eroded amounts of sediment in subsequent time slices. Between 6000 and 3000 B.P., when the rate of eustatic sea level rise had strongly decreased, about 4.0×10^{12} kg overbank fines from the Rhine were deposited in the delta, which corresponds to an average sedimentation rate of 1.3×10^{9} kg year^{-1} for this period. A moderate relative sea level rise continued to occur, caused by land subsidence, mainly owing to compaction of previously deposited deltaic sediments (Erkens 2009). Under these conditions, the delta gradually expanded in upstream direction, and considerable peat formation occurred, indicating that sediment supply was not sufficient to fill the available accommodation space (Erkens 2009). It is thus likely that the trapping efficiency of the delta was relatively high, such that the bulk of the overbank sediment delivered to the delta was trapped on its way to the North Sea. After 3000 B.P., overbank sedimentation increased across the entire delta, peat formation in the flood basins almost ceased, and two northern distributaries (IJssel, Vecht) became active (Berendsen and Stouthamer 2000; Erkens 2009). Between 3000 and 1000 B.P., a total mass of 4.2×10^{12} kg sediment was deposited, which is an average rate of 2.1×10^{9} kg year^{-1}. This enhanced deposition must be attributed to increased sediment delivery from the Rhine drainage basin, as it matches the documented increase in erosion and sediment delivery in upstream valleys, resulting from progressive deforestation from the Bronze Age onward and deforestation widely expanding during the Iron, Roman, and Middle Ages (Kalis et al. 2003; Mäckel et al. 2003; Hoffmann et al. 2007, 2009; Erkens 2009; Kaplan et al. 2009). Although the climate has varied as well in this period, with climate optima during Roman and Middle Ages, the changes in climate were too subtle to explain the major increases in hillslope erosion reported from upstream valleys and sediment deposition in the Rhine delta.

3.2. Deposition after embankment a.d. 1350–1850

Numerous old river maps of the Rhine branches from the sixteenth through twentieth centuries often show in great detail the channel planform and floodplain development after embankment (Middelkoop 1997; Wolfert 2001; Hesselink 2002). Coring data in combination with these maps (see Hesselink 2002) demonstrates that the present-day total amount of overbank fines contained in the embanked floodplains is 0.64×10^{12} kg.

In addition to the embanked floodplains, an inland fluvial delta developed by the Waal in the western part of the delta (Fig. 25.3) after a large part of land near the city of Dordrecht was eroded by

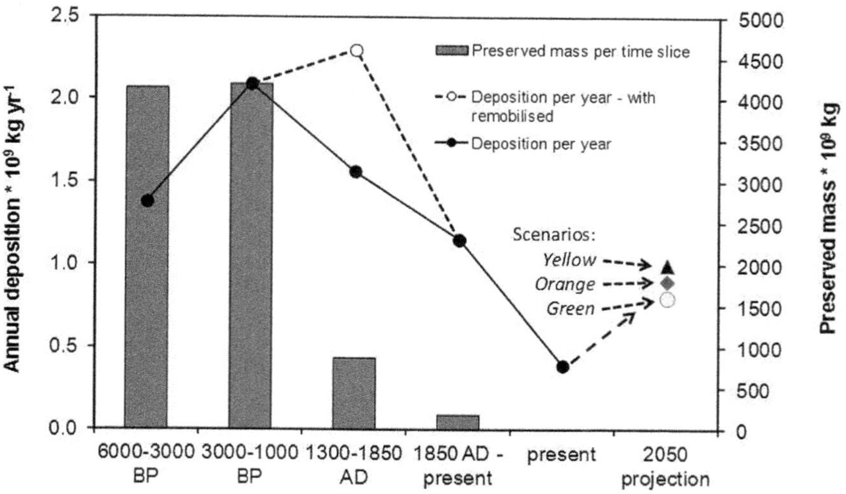

Figure 25.4. Amounts of overbank fines deposited in the Rhine delta over different time slices. Bars indicate preserved amounts; dashed line is amount per year for each time slice. The higher value for the period 1300–1850 represents annual average deposition rate without accounting for subsequent erosion during the same time slice.

the St-Elisabeth Flood of 1421. Detailed reconstructions of the sedimentary structure of the deltaic deposits built-up here by the Waal River (Kleinhans et al. 2010) showed that until the newly formed delta was embanked in the nineteenth century, about 0.17 km^3 overbank fines (0.22×10^{12} kg) were deposited here.

In the past centuries, increasing amounts of clay were excavated from the floodplain for brick production. Van der Meulen et al. (2009) estimated of the post A.D. 1850 clay extraction for brick production at 0.13 km^3, which corresponds to a clay extraction rate of 1.1 million m^3 yr^{-1}. Although this clay was excavated after A.D. 1850, most of it was deposited before A.D. 1850.

When adding the total amount of pre-1850 embanked floodplain clays, the floodplain clays in the Biesbosch splay, and the amount of excavated clay, the resulting total amount of floodplain deposition between 1300 and 1850 is 0.86×10^{12} kg (Fig. 25.4), which is equivalent to a volume of 0.66 km^3. This corresponds to a net accumulation rate of 2.55×10^9 kg yr^{-1}. However, the *actual* deposition rate will have been considerably larger. Old river maps and the current floodplain morphology indicate that about 90% of the Waal embanked floodplain, about 80% of the embanked Nederrijn-Lek floodplain, and 60% of the IJssel floodplain was reworked between 1300 and 1850 because of lateral bank erosion of the river channel. Using these reconstructed erosion values, the derived accumulation rates were converted to the actual deposition rates for the 1300–1850 period. This results in a total reconstructed deposition of 1.6×10^{12} kg and an actual deposition rate of 2.9×10^9 kg yr^{-1}. The latter amount is similar to the present-day annual flux of sediment to the delta. Assuming that the trapping efficiency of the embanked floodplains was smaller than the natural delta, this means that the sediment delivery during the 1300–1850 period was considerably larger than to date. This fits well with the land use reconstructions and soil erosion estimates from this period (e.g., Lang et al. 2003; Kaplan et al. 2009) and with high suspended load values measured in the Meuse river during the late nineteenth century (Ward 2008).

3.3. Post-normalization deposition rates: a.d. 1850–present

The normalization works around 1850 coincided with the onset of the Industrial Revolution and inherent increase in sediment contamination in the Rhine River (Middelkoop 2000). This contamination allows identifying the overbank sediment deposited after the 1850 normalizations from the older uncontaminated sediment. Post-1850 sediments of the lower Rhine floodplains were identified by increased concentrations of Pb, Zn, Cu, and Cd in the soil profiles along transects perpendicular to the main channel (Middelkoop 1997, 2000, 2002; Hobo et al. 2010). Using these heavy metal profiles in floodplain sediments, Middelkoop (2002) found average overbank deposition rates to vary between 0.2 and 10 mm per year over the past century. Sedimentation rates on the Waal floodplain generally have been higher than in the Nederrijn-Lek and IJssel floodplains, with lower rates in distal and higher floodplain parts, and behind minor embankments. Hobo et al. (2010) found local deposition rates varying between 3 and 10 mm per year for the past decades, using OSL dating and heavy metal profile characteristics. Based on the depth of metal contamination, the total volume of floodplain sediments deposited since 1850 is about 0.66 km^3, equivalent to a mass of 0.14×10^{12} kg. The reconstructed sedimentation rates represent annual amounts of clay deposition in the order of $0.4–3.7\times10^9$ kg yr^{-1} when accumulated over the entire embanked floodplain of the lower Rhine.

Accounting for clay extraction, the estimated total deposition is 0.17×10^{12} kg. On average this is about 1.39×10^9 kg per year since A.D. 1850. The trapped amounts of sediment in the floodplains after channel normalization were thus considerably lower than in the preceding period. There are different causes for this. First, the main channel no longer could migrate laterally and erode its floodplain, which prevented the "resetting" of the overbank deposition process. Floodplain sedimentation has resulted in increasingly higher floodplain surfaces with decreasing flooding frequencies and inherent lower deposition rates (cf. Middelkoop 1997). Second, the construction of minor embankments further prevented flooding and sediment deposition on the embanked floodplains. Finally, sediment yield from the Rhine basin might have decreased. In the course of the twentieth century, large areas in the Rhine basin became reforested, which must have reduced soil erosion and sediment yield. The large-scale river works that were undertaken in the Rhine basin (Kalweit et al. 1993) comprised protection of channel banks from erosion, preventing them to function as sediment source, whereas many stretches of the upstream Rhine tributaries were regulated by weirs that trap sediment.

4. Contemporary sediment deposition in the lower Rhine delta

At present, about 3.1×10^9 kg – with year-to-year variations between about 2.8×10^9 and 3.4×10^9 kg – suspended sediment reaches the Rhine delta at the German-Dutch border each year. This sediment is distributed over the lower Rhine branches along with the discharge distribution; the Waal river carries about 2×10^9 kg per year. Sediment trap measurements show that present-day sediment deposition on the embanked floodplains during individual flood events is in the order of 0.5–7.5 mm per year (Asselman and Middelkoop 1998; Middelkoop and Asselman 1998; Thonon et al. 2007). Model

Table 25.1. *Present-day suspended sediment trapping in the Rhine delta*

	Compartment	Amount per year [10^9 kg yr^{-1}]	Sediment trapping efficiency From Total	Per branch
Inflow	Rhine at Lobith	3.10	–	–
Deposition	Rhine branches	0.39	0.13	–
	IJssel	0.09	0.03	0.26
	Nederrijn-Lek	0.12	0.04	0.17
	Waal	0.19	0.06	0.09
	Clay extraction	0.68	–	–
	Rhine Estuary	2.50	0.81	–
Outflow	North Sea	0.20	0.06	–

simulations show that overbank sedimentation rates for different locations and reaches of the Rhine floodplains vary between 0.1 and 9 mm per year (Middelkoop & Van der Perk 1998; Straatsma et al. 2009). The average annual accumulation for the embanked floodplains of all Rhine distributaries together is 0.39×10^9 kg yr^{-1}, of which 0.09×10^9 kg yr^{-1} is deposited on the IJssel floodplain, 0.12×10^9 kg yr^{-1} on the Nederrijn-Lek floodplain, and 0.19×10^9 kg yr^{-1} on the Waal floodplain (Table 25.1). Sediment trapping efficiencies (defined as the trapped amount divided by the total amount entering each branch) are 0.26 for the IJssel, 0.17 for the Nederrijn-Lek, and 0.09 for the Waal. This larger trapping for the IJssel is due to its large floodplain area compared with the smaller main channel, whereas the large main channel and relatively narrow floodplain area along the Waal result in a low trapping efficiency of this branch. Particularly during high peak flows, overbank flow velocities on the Waal floodplain are too high to result in settling of all sediment before the flow returns to the main channel. The total annual conveyance loss owing to overbank sedimentation is 13% of the total annual suspended sediment load entering the delta at Lobith.

When compared with the preceding periods (Fig. 25.4), the embanked floodplains along the lower Rhine distributaries have become very inefficient traps of sediment. The largest trapping of suspended sediment presently occurs in the Rhine estuary, in the deep Rotterdam harbor area and Haringvliet (the presently closed estuary mouth south from Rotterdam), where each year 2.5×10^9 kg of fluvial fines settle, which is 81% of the incoming fluvial suspended sediment (Ten Brinke 2005). About 0.2×10^9 kg yr^{-1}, which is 6% of the incoming flux at Lobith, is discharged into the North Sea (Table 25.1).

5. Nutrient tluxes through the Rhine: decadal trends and variability

Deforestation and the increase of agricultural land use has likely affected the nutrient concentrations in the Rhine river basin, but present-day reconstruction techniques do not allow to distinguish nutrient concentrations that occurred just before the introduction and large-scale application of artificial fertilizers (~A.D. 1950) from concentrations under pristine conditions. Based on model calculations, Behrendt et al. (2002) estimated the natural background nutrient concentrations in the Rhine River at the German-Dutch border to be approximately 0.20±0.04 mg l^{-1} for total nitrogen and 0.008±0.003 mg l^{-1} for total phosphorus (TP).

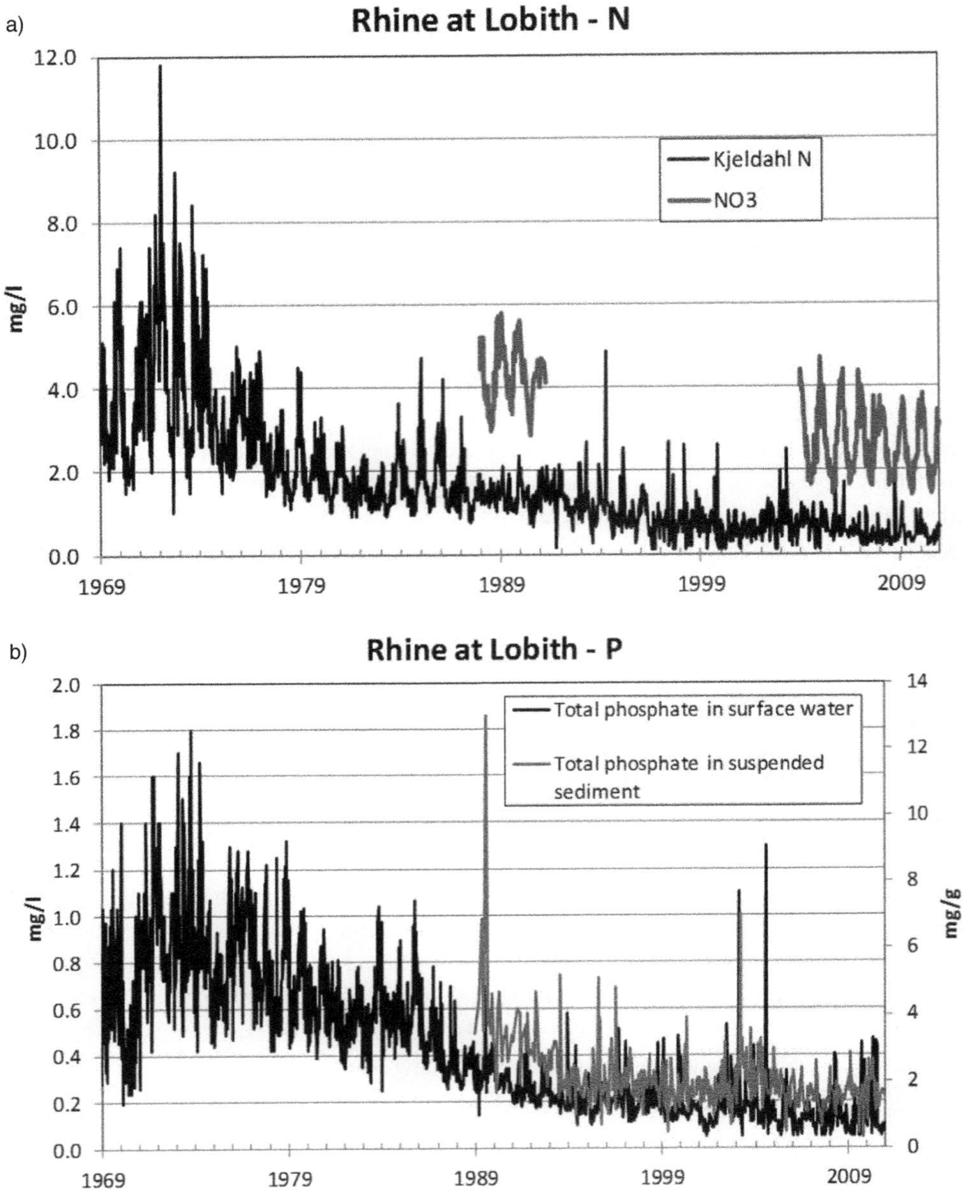

Figure 25.5. Nutrient concentrations in the Rhine at Lobith. (a) Nitrogen; (b) phosphorus (data: live.waterbase.nl).

Over the past 5 years, nutrient loads of the Rhine at Lobith were in the order 190–225 Gg N yr^{-1} per year and 7.5–14 Gg P yr^{-1} (data: live.waterbase.nl). These values are considerably lower than a few decades ago. Figure 25.5 shows the concentrations of N, total phosphorus (TP), and particulate phosphorus (PP) measured at 10-daily to biweekly intervals in the Rhine at the Lobith monitoring station. Both N and P graphs show highest concentrations (>6 mg N L^{-1} and > 1.2 mg P L^{-1}) in the late 1960s to early 1970, followed by a rapid decrease into the 1980s and 1990s, which has leveled off in the past decennium to values around 0.5 mg N L^{-1} and 0.08 mg P L^{-1}. Furthermore, with exception of a few events, seasonal variability in measured concentrations has greatly decreased. However, it must be reminded that the biweekly sampling interval may show considerable variation between successive measurements and that these still provide incomplete estimates of total loads. This is particularly the case for periods with peak flows that might be not well sampled, while in these periods (that occur in winter) concentrations can be high. Nevertheless, the record shows a clear annual cycle in concentrations, with high winter values and lower summer values. Low summer nutrient concentrations arise from denitrification and nutrient cycling by biota, which is intensified during low-flow conditions with low suspended sediment concentrations.

The decreasing trend in nutrient loads from the Rhine basin toward the delta and the underlying changes in nutrient emissions in the basin and Germany have been extensively investigated by Behrendt and De Wit (Behrendt 1988, 1993, 1996; Behrendt et al. 2002; De Wit and Behrendt 1999; De Wit 1999, 2001). Between the 1970s and 1990s, major reductions in nutrient emissions from point and diffuse sources in the Rhine basin were achieved (Figs. 25.6 and 25.7). Emissions of N decreased by about 30% as a result of a reduction by 47% in point sources, i.e., sewerage systems. The Ruhr area, with high population density, remains a hotspot of N emission. A decrease in N emission from agriculture has been achieved as well (about 16% until the 1990s), but because of long residence time of ground water (>20 years), the effect of reduced emissions is not yet visible in the river loads. P emissions were reduced by about 68% for the Rhine basin, mainly because of an 80% input reduction from point sources. Diffuse P inputs have been only reduced by about 17%. These trends are similar to those observed in, for example, the Seine River in France, shown in Chapter 24. To date, point sources contribute now about 40% of P inputs in Rhine basin. The wastewater treatment plant pathway of P is about 30% of the total P inputs, erosion 22%, groundwater 15%, and the rest comes from surface runoff and paved areas. Inputs from France, Switzerland, and Austria contribute about 30% of total N in the Rhine and 40% of total P emissions into the Rhine (Behrendt et al. 2002).

6. Nutrient retention along the lower Rhine distributaries

Nutrient retention in the fluvial part of the Rhine delta occurs primarily through nutrient uptake by aquatic organisms (e.g., algae and macrophytes) and their associations (e.g., periphyton), denitrification in water, bed sediments, and inundated floodplain soils, and due to sedimentation. For N and particularly for P, sedimentation contributes most to the total retention. Van der Lee et al. (2004) concluded that the contribution of denitrification is insignificant in the total N (TN) retention along the lower Rhine distributaries. Annual average retention of N in the lower Rhine branches is less than 3% of the total N load, whereas TP retention is 5–18% (Van der Lee et al. 2004). However, because considerable – but unknown – amounts of nutrients (particularly N) are emitted within the Netherlands to the rivers through groundwater seepage and by tributaries (Oenema et al. 2005), estimates of denitrification rates based on concentration changes along the river are highly unreliable. Retention occurs at different rates for different distributaries, largely depending on the efficiency of the

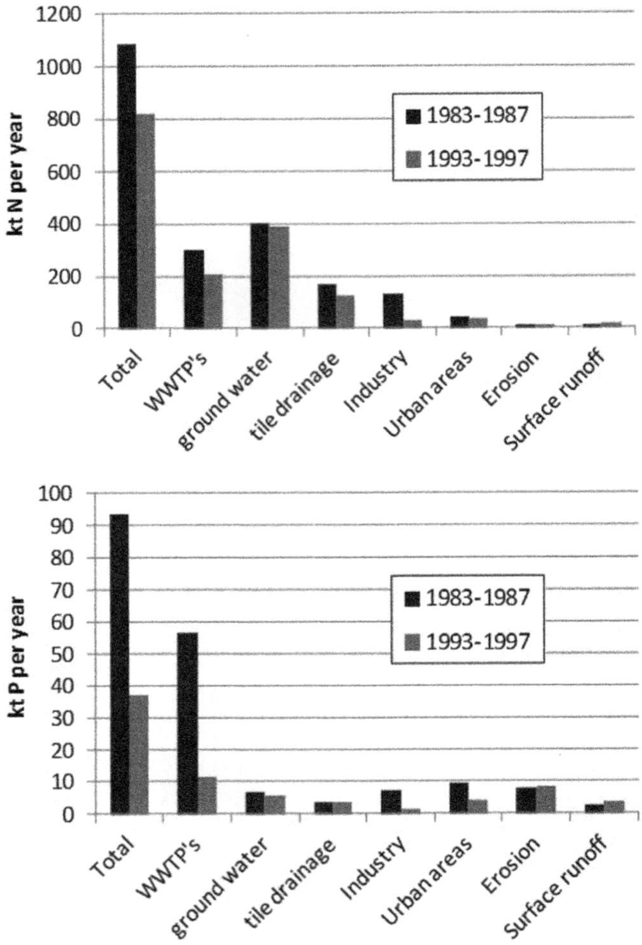

Figure 25.6. Changes in (a) nitrogen and (b) phosphorus emissions in Germany between 1980s and 1990s (data: Behrendt et al. 2002).

floodplains in trapping sediment (Olde Venterink et al. 2003; Van der Lee et al. 2004). In the IJssel branch, which has relatively large floodplain areas, annual retention of phosphorus is in the order of 20% of the annual load, which is much larger than nitrogen retention (2–3%) (Van der Lee et al. 2004), owing to the larger proportion of particulate-bound P in the TP load (50–70%) when compared with nitrogen (about 2%). P retention along the Waal branch (<5%) is considerably smaller than along the IJssel, owing to the smaller sediment trapping efficiency of the Waal floodplain. Furthermore, considerable P retention may occur in the IJssel river during summer low-flow periods (up to 0.6% of P per river km), resulting in the highest relative retention because TP concentrations are lower during summer (Olde Venterink et al. 2003). Still, when comparing observation records of TP concentrations in the Rhine at Lobith with those measured in the Lower IJssel at Kampen, major fluctuations in P

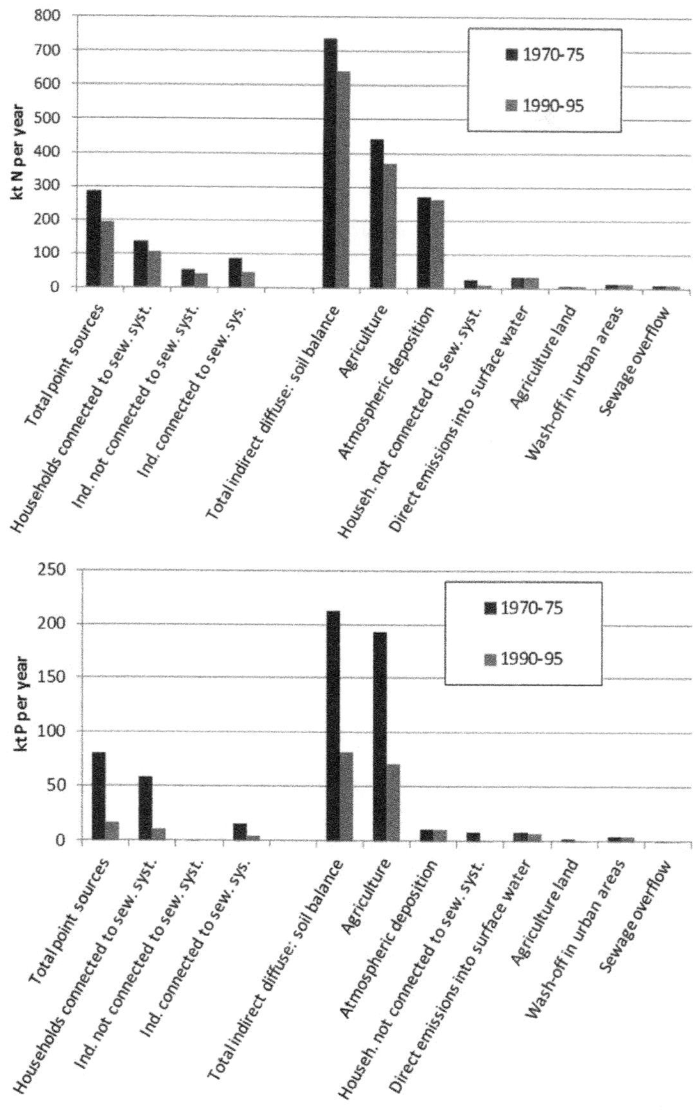

Figure 25.7. Changes in (a) nitrogen and (b) phosphorus emissions in the Rhine basin the 1970s and 1990s (data: De Wit 1999).

retention between these stations become apparent, demonstrating a considerable uncertainty in the estimates of annual retention.

Nitrogen retention in the lower Rhine distributaries appears negligible: although winter N concentrations are clearly higher (1–2.5 mg N L^{-1}; Fig. 25.5) than those in summer (0.2–0.5 mg N L^{-1}), due to lower summer concentrations of NO_3, no downstream decrease in TN load occurs (cf. Olde Venterink et al. 2003).

In conclusion, N retention in the lower Rhine distributaries is low, owing to the small proportion of particulate N deposited during floods, whereas the relatively deep rivers (due to the normalizations) and turbid water prevent high N uptake rates by biota in the main channel. P retention is strongly determined by its larger fraction PP and occurs because of overbank sedimentation and, during low-flow periods, even within the main channel.

Overall, as shown in previous chapters for other rivers around the world, human alterations have had a major impact both on the amounts and locale of sediment and nutrient sequestration within the Rhine delta.

7. Future retention in the Rhine delta

Future changes in the retention of suspended sediment and nutrients by the lower Rhine distributaries depends on a combination of factors, including natural and human control, both in the upstream basin and in the delta. As demonstrated in previous chapters for, e.g., the Mississippi (Chapter 16) and Seine (Chapter 24) Rivers, future scenarios should consider both types of forcings. First, climate change may considerably affect river discharge. Previous studies have demonstrated that climate change may lead to increased peak flows of the Rhine, causing more frequent overbank flooding (Middelkoop et al. 2001; Shabalova et al. 2003). Depending on the scenario considered, peak flows may increase in magnitude by about 5% by the year 2050. Furthermore, increased precipitation amounts may enhance soil erosion, potentially increasing sediment yield to the river (Van Dijk and Kwaad 1998; Asselman et al. 2003). Although basin-wide erosion may increase by about 10%, sediment transport to the delta is expected to be insignificant, owing to the inefficient sediment delivery from the hillslopes to the streams. Future sediment delivery also depends on land use change. Increased crop production and European Union regulations may lead to a reduction of the area of arable land in the basin. The combined result of increased precipitation and land use changes may lead to a decrease in the suspended sediment load of the Rhine. Model simulations by Asselman et al. (2003) suggest a reduction between 13% and 19% for the year 2100. Nevertheless, due to increased flooding frequency, overbank deposition may decrease to a smaller extent, or still increase.

The landscape of the embanked floodplains along the lower Rhine distributaries may drastically change in the future, depending on the measures that may be undertaken for flood reduction, nature development, recreation, infrastructure, and other use of the floodplain area. Measures include the removal of minor embankments, lowering the floodplain surface, and digging secondary channels, all enhancing water flow over the floodplain and potentially increasing deposition of overbank fines and nutrient retention. Depending on the management policy that may be adopted in the Netherlands, different strategies might be implemented. Straatsma et al. (2009) developed three fundamentally different scenarios for the layout of the floodplains of the River Waal in the Netherlands for 2050 using value-based methodology:

1. A relativistic scenario ("green"); measures comprise side channels, floodplain lowering projects, three dike section relocations, minor embankment removal, and natural management of the floodplain.

2. A materialistic scenario ("orange"); characterized by a productive-efficient layout. Measures comprise the construction of deep side channels, 51 floodplain-lowering projects, minor embankment removal, and emphasized agricultural production.
3. A systemic scenario ("yellow"); a diverse pattern of city expansion, nature development, agricultural production, and innovative groyne lowering. It comprises floodplain lowering, minor embankment removal outside the production regions, and seven dike repositioning projects.

Because each of these results in a very different layout of the future floodplains, this may considerably change sediment trapping and nutrient retention.

7.1. Suspended sediment trapping

The combined effects of these changing boundary conditions on suspended sediment trapping of the embanked floodplains was determined in a modeling study for the Waal distributary, using climate and land-use impacts on the upstream boundary conditions of discharge and suspended sediment loads and the floodplain layouts according to the three different floodplain management strategies (Straatsma et al. 2009). Increased floodplain inundation due to climate change will increase annual overbank deposition by 58%, in spite of the reduced sediment load. However, landscaping measures will more than double the floodplain sedimentation rate, which will increase by 112% (green scenario), 143% (yellow scenario), and 148% (orange scenario) when compared with the present-day situation. Spatial differentiation between the scenarios is considerable; large increases in deposition will occur where minor embankments are removed, floodplain surface is lowered, and side channels are dug.

7.2. Nutrient retention

Emission rates of N and P in the Rhine basin have shown little decrease in recent years, and diffuse sources have become an increasingly large part of the total emissions. Seitzinger et al. (2010) and Bouwman et al. (2009) provided global and regional estimates of the future Millennium Ecosystem Assessment Scenarios. Because these do not give nutrient emission values at the scale or the Rhine basin, some best guess estimates of nutrient loading of the Rhine River for the year 2050 were made on the basis of these scenarios. Particulate-bound N and P loads are generally estimated to decrease owing to a reduction of the area of agriculture land and inherent soil erosion (Seitzinger et al. 2010). However, this process is already considered in our estimates of future sediment delivery. Emissions will furthermore decrease along with agricultural development and further reduction of point source emissions. For the Rhine basin, we estimate that a further efficiency in N use for agriculture and improved wastewater sanitation may lead to another reduction by 10–20% in N and P loads exported by the Rhine into the delta by the year 2050.

Adopting the conclusion of Van der Lee et al. (2004) that overbank deposition forms the largest contribution to nutrient retention along the lower Rhine distributaries, future nutrient retention will be largely determined by changes in sediment trapping. Nevertheless, higher summer temperatures, along with lower summer flow in the Rhine (Middelkoop et al. 2001), may lead to enhanced denitrification and fine sediment deposition in the main channel, particularly in the IJssel distributary. However, as the IJssel branch carries about one-ninth of total Rhine discharge, and this occurs during a few months of low flow annually, this contribution to the total retention remains small. Also, inundation periods of the floodplain will change, enhancing denitrification in the water column and floodplains. Still, even when the amounts would double (from 0.06% to 1.2% for N; cf. Van der Lee et al. 2005), the absolute values remain small when compared with overbank sedimentation. We therefore consider here estimates of changes in nutrient retention as a result of changes in overbank sediment trapping. We assumed total nutrient loads, as well as particulate-bound concentrations to decrease by 10% to 20%. By combining the changes in sediment trapping depending on the floodplain management scenarios with these reductions, we calculated N and P retention estimates for the Waal distributary projected to the year 2050 (Table 25.2).

The results show that changes in TN loads reaching the estuary will be almost entirely determined by the emission reduction from the upstream basin, i.e., a decrease by 10% and 20% for the two emission scenarios respectively. These reductions are underestimates, because they do not account for enhanced denitrification. Because trapping of PP by floodplain deposition may double under the floodplain management scenarios, the P loads reaching the estuary show a considerably larger decrease: by about 15% and 25% for both emission scenarios, respectively. The difference in PP retention among the floodplain management scenarios is relatively small.

8. What is the future role of the Rhine delta?

Sediment trapping and nutrient retention by the lower Rhine distributaries has been greatly affected by human activities. The amounts of sediment transported toward the delta have varied in response to a long history of land-use change and river engineering. Present-day suspended sediment loads are probably higher than those under pristine (pre Bronze Age) conditions because deforestation enhanced sediment production by surface erosion. However, present-day values are lower than early nineteenth century values, when bank erosion provided an additional sediment source, and less sediment was trapped in regulated reaches behind weirs. However, it is hard to provide quantitative estimates. With ongoing river engineering along the lower Rhine distributaries, the embanked floodplains have become inefficient sediment traps. The present-day high floodplains only trap about 13% of the annual incoming amount of suspended sediment.

Nutrient flows toward the delta have known a very different development and varied dramatically during the past century. Highest loads occurred during the late 1960, in a period of intensive industrialization and agriculture, dense population, increasing connectivity to wastewater systems providing a direct connection to the river, and still inefficient wastewater treatment. The major reduction in emissions led to a great decrease in nutrient loads in subsequent decades, reaching near-stable values in recent years. Nutrient retention along the lower Rhine distributaries mainly occurs through deposition of particle-bound nutrients (mostly P) with over bank fines. The low trapping efficiency of the floodplains, combined with the relatively deep and fast-flowing river channels, render nutrient retention in the lower Rhine branches inefficient: the Waal distributary that carries about two-thirds of all Rhine discharge retains less than 1% of the total N load and about 5% of that total P load.

Table 25.2. *Retention of suspended sediment and nutrients in the Rhine delta: present situation and for future scenarios projected to 2050*

			Present scenario, load reduction: −10%				Scenario, load reduction: −20%				
			Floodplain management				Floodplain management				Source
	N	(2004–2010)	Present	Green	Orange	Yellow	Present	Green	Orange	Yellow	*italic numbers are calculated*
1	Total Rhine load (kt yr^{-1})	175	158	158	158	158	140	140	140	140	Data from: live.waterbase.nl
2	Total Waal load (kt yr^{-1})	*117*	*105*	*105*	*105*	*105*	*93*	*93*	*93*	*93*	*3/2 of total Rhine load*
3	PN conc. (mg g^{-1})	*2.5*	*2.3*	*2.3*	*2.3*	*2.3*	*2.0*	*2.0*	*2.0*	*2.0*	*Present: calculated from 4 and 5; scen.: 10% and 20% reduction*
4	Flpl. sed. (Mton yr^{-1})	0.20	0.31	0.36	0.42	0.41	0.31	0.36	0.42	0.41	Straatsma et al. (2010)
5	N flpl. sed. (kt yr^{-1})	*0.51*	*0.70*	*0.82*	*0.97*	*0.93*	*0.62*	*0.73*	*0.86*	*0.83*	*calculated from 2 and 6*
6	N retention (%)	0.62	0.7	0.8	0.9	0.9	0.7	0.8	0.9	0.9	present: from Van der Lee et al. (2005); *Scen.: calc. from 2 and 5*
7	N to estuary (kt yr^{-1})	116	*104*	*104*	*104*	*104*	*93*	*93*	*92*	*93*	*calculated from 2 and 5*
8	d N to estuary (factor)	*1*	*0.90*	*0.90*	*0.90*	*0.90*	*0.80*	*0.80*	*0.80*	*0.80*	*calculated from 7 (present and scenario)*
	P										
1	Total Rhine load (kt yr^{-1})	9.5	8.6	8.6	8.6	8.6	7.6	7.6	7.6	7.6	Data from: live.waterbase.nl
2	Total Waal load (kt yr^{-1})	*6.3*	*5.7*	*5.7*	*5.7*	*5.7*	*5.1*	*5.1*	*5.1*	*5.1*	*3/2 of total Rhine load*
3	PP conc. (mg g^{-1})	1.6	*1.4*	*1.4*	*1.4*	*1.4*	*1.3*	*1.3*	*1.3*	*1.3*	Present: data from waterbase.nl; *scen.: 10% and 20% reduction*
4	Flpl. sed. (Mton yr^{-1})	0.20	0.31	0.36	0.42	0.41	0.31	0.36	0.42	0.41	Straatsma et al. (2010)
5	P flpl. sed. (kton yr^{-1})	*0.32*	*0.44*	*0.52*	*0.61*	*0.58*	*0.39*	*0.46*	*0.54*	*0.52*	*calculated from 3 and 4*
6	P retention (%)	*5.0*	*7.7*	*9.0*	*10.7*	*10.3*	*7.7*	*9.0*	*10.7*	*10.3*	*calculated from 2 and 5*
7	P to estuary (kton yr^{-1})	*6.0*	*5.3*	*5.2*	*5.1*	*5.1*	*4.7*	*4.6*	*4.5*	*4.5*	*calculated from 2 and 5*
8	d P to estuary (factor)	*1*	*0.87*	*0.86*	*0.85*	*0.85*	*0.78*	*0.77*	*0.75*	*0.76*	*calculated from 7 (present and scenario)*

Future developments may considerably change this situation. Considering developments in the upstream basin climate change seems the most important control, as this causes an increase in flooding frequency of the embanked floodplains, enhancing overbank sedimentation. Land-use changes and a further reduction in nutrient emission may in contrast lead to a reduction in sediment and nutrient loads reaching the delta.

Considerable impacts are to be expected from landscaping measures along the lower Rhine distributaries. Depending on the policy adopted, landscaping measures will lead to a substantial increase in residence time of water in the embanked floodplain sections and concurrent floodplain deposition and nutrient retention, including denitrification. If the Dutch society decides to carry out large-scale measures within the floodplain areas in view of flood protection and ecological restoration, the floodplains may regain a significant role in sediment and nutrient retention. Sediment trapping and P retention rates may more than double. The amounts of suspended sediment reaching the estuary downstream of the Waal may decrease by more than 50%, and P loads may decrease by 15–25%. Nitrogen loads reduce mainly due to reduced emissions; together with increased denitrification and deposition, this may result in a 10–20% decreased annual flux into the estuary.

Obtaining more reliable estimates of future nutrient fluxes through the Rhine and other large rivers to the coastal zone under changing climate and human activities will require spatially explicit and mechanistic modeling instruments for nutrient cycling within soils, transfer to the drainage system, retention by riparian zones, and within-stream processes (cf. Loos et al. 2009, see also Chapter 24). The challenge here lies in the balance between model complexity and the availability of data as model input and for model testing.

The present first estimates indicate that future developments within the upstream basin may reduce the amounts of sediment and nutrients carried by the Rhine into the delta. Within the delta, the floodplains may become more efficient in trapping sediment and nutrients, in particular of P. Overall, the estuary will experience small but noticeably decreasing inputs of suspended sediment and nutrients.

References

Asselman, N.E.M. and H. Middelkoop. 1998. Temporal variability of contemporary floodplain sedimentation in the Rhine-Meuse delta, The Netherlands. Earth Surface Processes and Landforms 23: 595–609.

Asselman, N.E.M. and M. Van Wijngaarden. 2002. Development and application of a 1D floodplain sedimentation model for the River Rhine in the Netherlands. Journal of Hydrology 268: 127–142.

Asselman, N.E.M., H. Middelkoop and P.M. Van Dijk. 2003. The impact of changes in climate and land use on soil erosion, transport and deposition of suspended sediment in the River Rhine. Hydrological Processes 17: 3225–3244.

Berendsen, H.J.A. and E. Stouthamer. 2000. Late-Weichselian and Holocene palaeogeography of the Rhine-Meuse delta, The Netherlands. Palaeo3 161: 311–335.

Berendsen, H.J.A. and E. Stouthamer. 2001. Palaeogeographic Development of the Rhine-Meuse Delta, The Netherlands. Van Gorcum, Assen.

Behrendt, H. 1988. Changes in nonpoint nutrient Loading into European freshwaters: Trends and consequences since 1950 and not-impossible changes until 2080. Working Paper WP-88-026, International Institute For Applied Systems Analysis, Laxenburg, Austria.

Behrendt, H. 1993. Point and diffuse loads of selected pollutants in the River Rhine and its main tributaries. Report RR-93-1. International Institute For Applied Systems Analysis, Laxenburg, Austria.

Behrendt, H. 1996. Inventories of point and diffuse sources and estimated nutrient loads: a comparison for different river basins in central Europe. Water, Science and Technology 33: 99–107.

Behrendt, H., M. Kornmilch, D. Opitz, O. Schmoll and G. Scholz. 2002. Estimation of the nutrient inputs into river systems – experiences from German rivers. Regional Environmental Change 3: 107–117.

Bouwman, A.F., A.H.W. Beusen and G. Billen. 2009. Human alteration of the global nitrogen and phosphorus soil balances for the period 1970–2050. Global Biogeochemical Cycles 23, GB0A04.

CHR/KHR. 1978. Das Rheingebiet, Hydrologische Monographie / Le bassin du Rhin. Monographie Hydrologique. CHR/KHR Koblenz.

De Wit, M.J.M. 1999. Nutrient fluxes in the Rhine and Elbe basins. PhD thesis, Utrecht University, the Netherlands. Netherlands Geographical Studies 259.

De Wit, M.J.M. 2001. Nutrient fluxes at the river basin scale. I: The PolFlow model. Hydrological Processes 15: 743–759.

De Wit, M.J.M. and H. Behrendt. 1999. Nitrogen and phosphorus emissions from soil to surface water in the Rhine and Elbe basins. Water Science and Technology 39: 109–116.

Dreibrodt, S., C. Lubos, B. Terhorst, B. Damm and H-R. Bork. 2009. Historical soil erosion by water in Germany: Scales and archives, chronology, research perspectives. Quaternary International 222: 80–95.

Erkens, G., K.M. Cohen, M. Gouw, H. Middelkoop and W.Z. Hoek. 2006. Holocene sediment budgets of the Rhine Delta (The Netherlands): a record of changing sediment delivery, pp. 406–415. *In* J.S. Rowan, R.W. Duck and A. Werritty (eds.), Sediment Dynamics and the Hydromorphology of Fluvial Systems. IAHS Publication 306.

Erkens, G. 2009. Sediment dynamics in the Rhine catchment: Quantification of fluvial response to climate change and human impact. PhD thesis, Utrecht University, the Netherlands. Netherlands Geographical Studies 388.

Garbe-Schönberg, C., J. Wiethold, D. Butenhoff, C. Utech and P. Stoffers. 1998. Geochemical and palynological record in annually laminated lake sediments from Lake Belau (Schleswig-Holstein) reflecting palaeoecology and human impact over 9000 a. Meyninia 50: 47–70.

Gouw, M.J.P. 2008. Alluvial architecture of the Holocene Rhine-Meuse delta (the Netherlands). Sedimentology 55: 1487–1516.

Gouw, M.J.P. and G Erkens. 2007. Architecture of the Holocene Rhine-Meuse delta (the Netherlands) A result of changing external controls. Netherlands Journal of Geosciences 86: 23–54.

Hesselink, A.W. 2002. History makes a river – morphological changes and human interference in the river Rhine, The Netherlands. PhD thesis, Utrecht University, the Netherlands. Netherlands Geographical Studies 292.

Hesselink, A.W., H.J.T. Weerts and H.J.A. Berendsen. 2003. Alluvial architecture of the human-influenced river Rhine, The Netherlands. Sedimentary Geology 161: 229–248.

Hobo, N., B. Makaske, H. Middelkoop and J. Wallinga. 2010. Reconstruction of floodplain sedimentation rates: a combination of methods to optimise estimates. Earth Surface Processes and Landforms 35: 1499–1515.

Hoffmann, T., G. Erkens, K.M. Cohen, P. Houben, J. Seidel and R. Dikau. 2007. Holocene floodplain sediment storage and hillslope erosion within the Rhine catchment. The Holocene 17: 105–118.

Hoffmann, T., G. Erkens, R. Gerlach, J. Klostermann and A. Lang. 2009. Trends and controls of Holocene floodplain sedimentation in the Rhine catchment. Catena 77: 96–106.

IKSR (1992). Aktionsprogramm Rhein. Bestandsaufnahme der Einleitungen prioritärer Stoffe 1992. International Commission for the Protection of the Rhine. Koblenz.

Kalis, A.J., J. Merkt and J. Wunderlich. 2003. Environmental changes during the Holocene climatic optimum in central Europe – human impact and natural causes. Quaternary Science Reviews 22: 33–79.

Kalweit, H. (ed.) 1993. Der Rhein unter der Einwirkung des Menschen – Ausbau, Schifffahrt, Wasserwirtschaft / Le Rhin sous l'influende de l'homme – Aménagement, navigation, gestion des eaux. CHR/KHR report I-11. CHR/KHR Koblenz.

Kaplan, J.O., K.M. Krumhardt and N. Zimmermann. 2009. The prehistoric and preindustrial deforestation of Europe. Quaternary Science Reviews 28: 3016–3034.

Kleinhans, M.G., H.J.T. Weerts and K.M. Cohen. 2010. Avulsion in action: reconstruction and modelling sedimentation pace and upstream flood water levels following a Medieval tidal-river diversion catastrophe (Biesbosch, The Netherlands, 1421–1750 ad). Geomorphology 118: 65–79.

Lang, A., H.R. Bork, R. Mäckel, N. Preston, J. Wunderlich and R. Dikau. 2003. Changes in sediment flux and storage within a fluvial system: some examples from the Rhine catchment. Hydrological Processes 17: 3321–3334.

Loos, S., H. Middelkoop, M. Van der Perk and L.P.H. Van Beek. 2009. Large scale nutrient modelling using globally available datasets: A test for the Rhine basin. Journal of Hydrology 369: 403–415.

Mäckel, R., R. Schneider and J. Seidel. 2003. Anthropogenic impact on the Landscape of Southern Badenia (Germany) during the Holocene – documented by colluvial and alluvial sediments. Archaeometry 45: 487–501.

Middelkoop, H. 1997. Embanked floodplains in the Netherlands PhD thesis, Utrecht University, the Netherlands. Netherlands Geographical Studies 224.

Middelkoop, H. 2000. Heavy-metal pollution of the river Rhine and Meuse floodplains in the Netherlands. Netherlands Journal of Geosciences 79: 411–428.

Middelkoop, H., K. Daamen, D. Gellens, W. Grabs, J.C.J. Kwadijk, H. Lang, B.W.A.H. Parmet, B. Schädler, J. Schulla and K. Wilke. 2001. Impact of climate change on hydrological regimes and water resources management in the Rhine basin. Climatic Change 49: 105–128.

Middelkoop, H. 2002. Reconstructing floodplain sedimentation rates from heavy metal profiles by inverse modelling. Hydrological Processes 16: 47–64.

Middelkoop, H. and N.E.M. Asselman. 1998. Spatial variability of floodplain sedimentation at the event scale in the Rhine-Meuse delta, The Netherlands. Earth Surface Processes and Landforms 23: 561–573.

Middelkoop, H. and M. Van der Perk. 1998. Modelling spatial patterns of overbank sedimentation on embanked floodplains. Geografiska Annaler 80A: 95–109.

Middelkoop, H. and C.O.G. Van Haselen. 1999. Twice a river. Rhine and Meuse in the Netherlands. RIZA Report No. 99.033. RIZA, Arnhem.

Oenema, O., L. Van Liere and O. Schoumans. 2005. Effects of lowering nitrogen and phosphorus surpluses in agriculture on the quality of groundwater and surface water in the Netherlands. Journal of Hydrology 304: 298–301.

Olde Venterink, H., F. Wiegman, G.E.M. Van der Lee and J.E. Vermaat. 2003. Role of active floodplains for nutrient retention in the river Rhine. Journal of Environmental Quality 32: 1430–1435.

Seitzinger, S.P., E. Mayorga, A.F. Bouwman, C. Kroeze, A.W. Beusen, G. Billen, G. Van Drecht, E. Dumont, B.M. Fekete, J. Garnier and J.A. Harrison. 2010. Global river nutrient export: A scenario analysis of past and future trends, Global Biogeochemical Cycles 24: GB0A08.

Shabalova, M.V., W.P.A. Van Deursen and T.A. Buishand. 2003. Assessing future discharge of the river Rhine using regional climate model integrations and a hydrological model. Climate Research 23: 233–246.

Solsten, E. (ed.) 1995. Germany: A Country Study. Washington: GPO for the Library of Congress.

Straatsma, M.W., A. Schipper, M. Van der Perk, N.G.M. Van den Brink, H. Middelkoop and R.S.E.W. Leuven. 2009. Impact of value-driven scenarios on the geomorphology and ecology of lower Rhine floodplains under a changing climate. Landscape and Urban Planning 92: 160–174.

Thonon, I., K. De Jong, M. Van der Perk and H. Middelkoop. 2007. Modelling floodplain sedimentation using particle tracking. Hydrological Process 21: 1402–1412.

Van der Lee, G.E.M., H. Olde Venterink and N.E.M. Asselman. 2004. Nutrient retention in floodplains of the Rhine distributaries in the Netherlands. River Research and Applications 20: 315–325.

Van der Meulen, M.J., A.P. Wiersma, M. Van der Perk, H. Middelkoop and N. Hobo. 2009. Sediment management and the renewability of floodplain clay for structural ceramics. Journal of Soils and Sediments 9: 627–639.

Van Dijk, P.M. and F.J.P.M. Kwaad. 1998. Modelling suspended sediment supply to the river Rhine drainage network: a methodological study. International Association of Hydrological Sciences Publications 249: 165–176.

Ward, P.J. 2008. River Meuse suspended sediment yield: a new estimate and past estimates revisited. Netherlands Journal of Geosciences 87: 189–193.

Wolfert, H.P. 2001. Geomorphological Change and River Rehabilitation: Case Studies on Lowland Fluvial Systems in the Netherlands. PhD Thesis, Scientific Publications 6, Alterra Green World Research, Wageningen.

Zolitschka, B. 1998. A 14.000 year sediment yield record from western Germany based on annually laminated lake sediments. Geomorphology 22: 1–17.

Index